INTRODUCTION TO CODE BREAKER

The Code Breaker book has been produced as a reference and explanation for over 2700 EOBD / ISO fault codes. Although it is not intended to be a "Diagnostic" book, it does provide a wealth of guidance for many diagnostic procedures and checks that can be carried out when a fault code is retrieved from a vehicle. The reality is that the EOBD / ISO fault codes are applicable to a wide range of makes and models, therefore the applicable diagnostics and checks will vary with each make and model. However, there are many generic procedures that are applicable to some components, irrespective of which vehicle or system the components are fitted to.

Making use of Chapters 1, 2 and 3

Within the first 3 chapters, we have provided educational information relating to how and why fault codes are activated, along with many notes as to how different vehicle manufacturers apply codes in different ways. I would strongly recommend that you read the first 3 chapters, due to the additional understanding that they provide of how fault codes are applied.

Whilst carrying out the research for this book, we have been able (for most of the fault codes) to locate at least one example of where a fault code is applied by a vehicle manufacturer; in many cases however, some fault codes appear to be only applicable to vehicles used in specific markets. There is of course a trend where technologies and components used in one market place often migrate to other markets, so we cannot assume that some of the more obscure components and systems will not find their way into other markets.

Keeping Up to Date

You will find enclosed, within the protective packaging, an "Update Card"; because this book is unlikely to require a full update for some considerable time, we have decided to provide an "occasional" update service (free of charge); by following the instructions on the card and filling in the appropriate details on the web-site, we will then be able to provide you with any changes or amendments that we produce.

I hope that you find that the Code Breaker book is a useful workshop reference tool and also a means of providing additional informati   tion to help with the repair and maintenance of motor vehicles.

Peter Coombes

FOREWORD

By Mike Owen
Head of Aftermarket Independent Garage Association at the RMIF

Welcome to the new Code Breaker book from Peter Coombes and his team at Mototek Solutions – the recognized experts in the matter of technical information and diagnostics.

The metamorphosis of cars from mechanical to electro-mechanical has radically changed the landscape on which the vehicle repairer has to work. Most of us, who have either been in the industry for many years or have recently joined, are equally excited and bemused by this change and need to adapt in order to survive.

Information is king! I can remember looking at one of the first solid state ignition units on a 1976 Chrysler Alpine with a sense of foreboding and apprehension; this was quickly allayed once fear was replaced with understanding. For the current technologies, the Code Breaker book is therefore intended to help with spreading understanding of the latest error codes and the associated components, to all of those who wish to continue in the motor industry of the future.

New power sources, being planned for the future to meet stringent environmental requirements, will accelerate the requirement for new technology; your requirement will be to keep abreast of these changes - and how to work with the new technologies. Any source of information that assists you in this process of technology change will become an essential workshop tool, and a journal such as this that brings information in an understandable manner isn't just worth having it's a must. Don't put it on the shelf as a reference document, put it in your tool box, read it, use it; its content will become more and more important every day.

Mike Owen

MAIN CONTENTS

Chapter 1:
     UNDERSTANDING FAULT CODES AND FAULT CODE DEFINITIONS…………..………………………Page 4

Chapter 2:
     SENSOR, ACTUATOR AND COMPONENT CHECKS………………………………………………...Page 19

Chapter 3:
     COMPONENT LOCATIONS……………………………………………………………………….……Page 53

Chapter 4:
     ISO / EOBD FAULT CODES………………………………………………………………………....…Page 63

Chapter 1

UNDERSTANDING FAULT CODES
AND
FAULT CODE DEFINITIONS

1. THE DO'S AND DON'TS OF USING FAULT CODES..5
   1.1   Terminology...5
   1.2   The Essential Things To Remember 5
   1.3   THE DO'S...5
   1.4   The DON'TS..6

2. HOW FAULTS ARE DETECTED 7
   2.1   On Board Diagnosis (OBD)..7
   2.2   Circuit / Electrical faults 7
   2.3   System Performance and Range / Performance faults 7
   2.4   Using "Logic" to detect a fault 8

3. MONITORING THE SIGNALS USED IN SENSORS AND ACTUATORS............................8
   3.1   Monitoring signal voltage, frequency, pulse width, etc 8

4. IDENTIFYING FAULTY SIGNALS 10
   4.1   Electrical fault or system fault 10
   4.2   Signal "Operating Range"..10

5. UNDERSTANDING FREQUENTLY USED FAULT CODE DEFINITIONS.........................13
   5.1   Defining the "Circuit" faults 13
   5.2   Frequently used fault code definitions for Sensors 14
   5.3   Frequently used fault code definitions for actuators 16

1. THE DO'S AND DON'TS OF USING FAULT CODES

1.1 TERMINOLOGY

1.1.1 The ISO / EOBD code descriptions / definitions are often provided by vehicle manufacturers and will therefore refer to a specific component or system used by a manufacturer. In addition, some codes are applicable to vehicles produced / sold in other markets (other than Europe) and might therefore use terminology that is not commonly used in Europe. Wherever possible therefore, the information contained within this publication makes use of standard terms. An important example is for those electronic control units that are used to control different vehicle systems. The term "Control Unit" or "Main Control Unit" is frequently used to indicate the main engine control unit (rather than use PCM, ECM or ECU); where another control unit communicates with the main control unit, it is often referred to as a "Control Module" e.g. Air Conditioning Control module.

1.1.2 The terms "Sensor" and "Actuator" are used throughout the publication; these are standard terms used by virtually all manufacturers. A sensor is a component that detects or senses an activity or value e.g. movement or temperature. An actuator is a device that often creates motion e.g. a motor or solenoid, but there is often a mechanical device connected to an electric actuator and these mechanical devices can often also be referred to as actuators. In virtually all cases within this publication reference to an actuator will usually relate to the electric actuator rather than the mechanical actuator?

1.2 THE ESSENTIAL THINGS TO REMEMBER

1.2.1 A fault code can indicate the problem area; it doesn't necessarily define the fault.

Some fault code definitions or descriptions can be very specific about the nature and location of a fault; however, many fault codes are less specific and can be activated because the signal in the circuit (sensor signal or actuator control signal) is incorrect. The fault could therefore be related to a wiring or connection fault, a sensor or actuator fault or, another type of fault in the applicable system.

In addition, many fault codes are activated because the response of a sensor or actuator is not correct due to a fault with another system. The most common example (and one mentioned a few times within this publication) is with Oxygen / Lambda sensors. Lambda sensor signals are dependent on many factors including engine condition, exhaust air leaks etc. For example, a small air leak in the exhaust system can cause a significant change to the Lambda sensor signal and this can be interpreted as a Lambda sensor fault.

The rule therefore is to try and "prove" what the fault is or, what is causing the fault before condemning a sensor or actuator (or any other component / system).

1.2.2 In researching the information for this publication, it has been possible to identify application of codes (by a vehicle manufacturer) for virtually every P0, P2 and P3 fault code; however, vehicle manufacturers use different terminologies for components and also, different systems may share the same terminologies or names.

For example, some fault codes refer to a "power steering control circuit" and, because there are different types of power steering systems, the fault code could therefore refer to different control systems. For a mechanically driven hydraulic power steering pump, the fault code could refer to the control of a "pressure control valve"; however, if the hydraulic pump is driven by an electric motor, the fault could refer to the control of the electric motor. In addition, if the vehicle has full electronic control such as where a motor is attached to the steering column, the fault code could refer to the control of the steering column motor.

In many cases therefore, it might be necessary to identify what components are fitted to the applicable system as well as identifying the wiring circuit as well as obtaining any other applicable information.

1.2.3 Within this publication, there are indications as to the possible cause of a fault and, recommendations for checks that can be carried out. Because each manufacturer can apply slightly different methods or values to identify whether or not there is a fault, the indications and recommendations provided must therefore cover a wide range of vehicles and systems. As mentioned many times within this publication, it will often be beneficial to refer to vehicle specific information, which can help identity why a fault code has been activated and, what checks are therefore applicable.

1.2.4 Although faults are often detected by the control unit because the sensor or actuator signal is incorrect (which is detected when a system is operating), many faults can be detected when the ignition is initially switched on. The control unit can pass a small current or inject a signal into a circuit, and then monitor whether the current / signal returns back to the control unit. It can therefore be possible for a control unit to activate a code before the engine is running or the vehicle is being driven.

1.3 THE DO'S

1.3.1 Always check for other fault codes; they may seem unconnected but, it is often the case that one code refers to a symptom but a second code can indicate the cause of the fault.

1.3.2 Always check whether a signal is passing through a CAN-Bus system. Sensor signals are often shared between control units e.g. the signal from a wheel speed sensor could pass to the ABS / Traction control system, also to the transmission control system and to the engine control system; the speed sensor signal could also be used for the simple task of indicating vehicle speed at the instrument system (dashboard display). With CAN-Bus systems, the speed signal can pass to one control module (for this example the braking module), which then passes a signal out to the other control units. A fault that is defined as a circuit problem could therefore be caused by a CAN-Bus fault.

1.3.3 Always check for the more basic causes of a fault e.g. mechanical faults.

A common example is where the fault code refers to a problem with the "Camshaft Position Sensor" or, sometimes the code refers to the "Camshaft / Crankshaft Position Sensor Correlation" (this can be where the signals from the two sensors are not synchronised e.g.

when the Crankshaft sensor indicates TDC for Cylinders 1 & 4, the Camshaft sensor is indicating that the Camshaft is out of phase or incorrectly timed). The fault can often be caused by a slack timing belt / chain, which is causing a fluctuation in the Camshaft rotation (thus causing the Camshaft position sensor signal to be erratic or slightly out of phase with the Crankshaft position sensor signal).

1.3.4 Wherever possible, try to establish if the information provided by a sensor signal is correct or incorrect e.g. if the fault relates to a temperature value such as coolant temperature, try to measure the temperature with a separate gauge / thermometer to prove if the sensor information is correct / incorrect.

1.3.5 Wherever possible, try to obtain a "Live Data" reading for the affected system or component. The Live Data values, which can be displayed via many types of diagnostic equipment, provides a clearer picture of the measurements being made by sensors and a clearer picture of sensor signal values and actuator control signals.

However, it is also advantageous if the signals can be checked using Oscilloscopes or other test equipment (multi-meters etc).

1.4 THE DON'TS

1.4.1 Don't assume that a code definition is providing the final diagnosis. The fault code might refer to a symptom but not the cause.

Probably the most common example of misinterpreting fault code definitions relates to Lambda / Oxygen sensors. The Lambda sensors monitor the oxygen content in the exhaust gas thus providing an indication of fuel mixture (air / fuel ratio); if the oxygen content is too high or low, the Lambda sensor provides the applicable signal to the engine system control unit (ECU or PCM), which will respond by altering the amount of petrol being injected. However, if the control unit is unable to provide sufficient correction to the fuelling or, if the incorrect mixture problem remains, this could be identified as a Lambda sensor fault. There are however many factors that affect the oxygen content in the exhaust gas, which include: air leaks in the exhaust and intake systems, incorrect fuel control, high or low fuel pressure etc. Therefore, as an example, if the Lambda sensor signal indicated a lack of oxygen (an excess of petrol / rich mixture) it could be a sensor related fault but, it is just as likely that the fault is being caused by: high fuel pressure, a leaking fuel injector or a fault with another sensor that is causing an excess of petrol to be delivered by the injectors.

Another example could be a fault code that is referring to a "coolant temperature sensor" (or sensor circuit) problem but, this doesn't necessarily mean that the temperature sensor is faulty. It could of course be a fault with the sensor circuit wiring, the sensor or the control unit; however, (depending on the code used), it could actually be a fault with the cooling system (e.g. low coolant level, thermostat problem or cooling fans not working) that is causing the coolant temperature to be incorrect and therefore the sensor signal is incorrect.

1.4.2 Don't assume that all vehicle manufacturers use the codes in the same way, and don't assume that all manufacturers use the same terminology.

The EOBD codes were originally designed to allow some flexibility in their application and therefore, manufacturers are able to interpret a code in different ways. The result is that different manufacturers could apply different codes for the same fault, but it also allows for different faults to be allocated to a single code.

Example 1. Fault Codes P0001 – P0004; these codes refer to a fuel volume regulator, which can be used on Common Rail Diesel high pressure pumps. The regulator controls the volume of fuel delivered to match the demand e.g. idle speed or high load conditions. However, the volume regulator does not control the pressure in the system, which is controlled by a separate pressure regulating device but, manufacturers have used these codes to refer to a Fuel pressure Regulator on some vehicles.

Example 2. Fault Code P0715: Input / Turbine speed sensor circuit. The code refers to the transmission input / turbine speed (turbine refers to the torque converter). In most cases, there is unlikely to be a speed sensor for the transmission input speed because this will be the same speed as the crankshaft. Therefore, the Transmission Control Unit can use a speed signal from the Engine Control Unit (possibly through the CAN-Bus network) or, obtain a signal from the engine speed sensor. There are however instances where this transmission code has been used to refer to the "Diesel pump speed sensor" (rotary type pump); this is because the pump speed is used to indicate engine speed on some vehicles (and therefore used for the transmission input speed). Also note that some manufacturers refer to the Torque converter Turbine as being the output side of the torque converter whilst other manufacturers refer to the turbine being the input side (the torque converter output speed is not always the same as the engine speed due to torque converter slip); it can therefore be confusing to rely on the terminology used by one manufacturer.

Example 3. An oxygen sensor can be referred to by many different terms such as an O2 sensor, an HO2 sensor or a Lambda or Oxygen sensor. Whilst most of these terms are commonly used and generally understood, it does provide a simple example of the differences in terminology across different manufacturers.

2. HOW FAULTS ARE DETECTED

2.1 ON BOARD DIAGNOSIS (OBD)

2.1.1 The EOBD fault codes, which are defined by ISO (International Standards Organisation) in conjunction with the American based SAE (Society of Automobile Engineers), cover a wide range of system and component failures.

2.1.2 Faults are detected by the vehicle's Electronic Control Units, which monitor the sensor and actuator signals; these signals should conform to specified values and criteria e.g. signal voltage, signal frequency or other values. If the signals do not conform to the expected values, or the actuator response to a control signal is not as expected, then it will be assumed that there is a fault. A significant amount of the control units processing power is devoted to monitoring signals and systems and performing diagnostic processes.

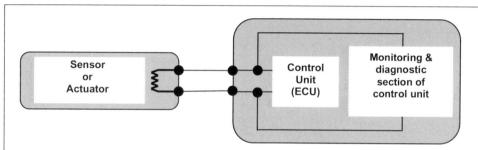

For sensors, there are usually 2 or 3 connections to the control unit (providing earth path, signal and reference voltage as applicable). For actuators, there can be one or more connections to the control unit, depending on the type of actuator and how the actuator is controlled.

The control unit can monitor the electrical values in the signal wire and the earth path for sensors and actuators; note that where independent circuits provide supply voltage to actuators (e.g. from a relay), a "sense" wire or connection can be provided back to the control unit to enable the control unit to monitor the supply voltage.

Fig 1.1   Sensors and actuators connected to the Monitoring Diagnostic section of the control unit

2.1.3 When a vehicle system is controlled by an Electronic Control Unit or Module (ECU / ECM), the control unit is usually receiving information from sensors e.g. temperature, speed etc (information can also be received from other control units). A control unit will also be providing control signals to actuators e.g. motors, solenoids etc (signals can also be passed to other control units). The sensors provide the information via electrical signals that will change with the changes in conditions e.g. if the temperature changes, the electrical signal from the temperature sensor changes. The control signals (that are provided by the control unit to the actuators) are also electrical signals, and when the control unit changes the signal value, this causes the actuator to operate e.g. an electric motor will rotate.

Control units have an on board monitoring and diagnostic process, which effectively enables the control unit to measure or detect the electrical values in the circuits and, to also monitor responses of actuators when a control signal is changed; the control unit then compares the electrical values and responses against data stored in its memory.

2.2 CIRCUIT / ELECTRICAL FAULTS

2.2.1 The control unit is obviously connected to the sensor and actuator circuits; therefore, the control unit can detect if there is a circuit / electrical fault. In effect, the control unit has its own built-in measurement system that can measure voltage, current, frequency etc, and this enables the control unit to detect if the electrical values in the circuits are correct or incorrect. However, to establish if an electrical value is correct or incorrect, the control unit also has the correct values stored in its memory; it therefore monitors the electrical values in the circuit and then looks up (in the memory) what the correct values should be. In reality, this is the same process that a technician performs when a multimeter is used to measure the electrical value in a circuit and then compares the measured value against the correct specification (which is stored in a book or a technical information system). Part of the On Board Diagnostic system therefore simply compares actual electrical values in a circuit and compares them with what they should be.

2.3 SYSTEM PERFORMANCE AND RANGE / PERFORMANCE FAULTS

2.3.1 There are many instances where the electrical circuits and the signals are good (signal values are within normal operating range) but, the sensor or actuator is not operating correctly; these faults are often referred to as "Circuit Range / Performance" faults or "System Performance" faults.

2.3.2 A simple example for a sensor, is where the control unit is expecting the coolant temperature to increase by an expected amount after a cold start (temperature should increase after defined period of time) but, the actual temperature increase is not as large as expected (expected increase is stored in the memory). The fault could in fact be a sensor or wiring fault but, it could also be because the cooling system has a fault (e.g. thermostat stuck open) and the temperature is therefore not increasing as quickly or as high as expected. The fault code could indicate that the coolant temperature is "Too Low", which is a cooling system "Performance" problem; however, another code could indicate a "Circuit Range / Performance" fault for the coolant temperature sensor circuit.

2.3.3 A simple example for an actuator is where a control signal is passed to a solenoid that operates an idle air valve; this should result in additional air passing through the intake system to increase the idle speed. If the idle speed does not increase (even when the control signal continues to request that the air valve opens more), this could in fact be a solenoid or wiring fault but, it could also be because the idle air system has a fault (e.g. blocked air passage) and the idle speed is therefore not increasing as expected (expected increase for a particular control signal value is stored in the memory). The fault code could indicate that the idle speed is "Too Low", which is an idle system "Performance" problem;

however, another fault code could indicate a "Circuit Range / Performance" fault for the solenoid circuit.

2.3.4 For both the sensor and the actuator examples, note that there is a difference between a "Circuit Range / Performance" fault and the "System Performance" problem but, both can be indicators that the" response" to changes in conditions or changes in control signals are not as expected.

As with the circuit / electrical faults (section 2.2), the Range / Performance and Range problems can be detected because the memory in the control unit stores certain values that should be matched when changes occur i.e. the responses should match expected values. The control unit is therefore monitoring the changes / responses in sensor signals, which should match changes in conditions; and the control unit also monitors system performance or response, which should again match expected values.

2.3.5 Note that in many cases, the control unit can monitor system performance because sensors can indicate the responses. For example, if a control signal is passed to a throttle motor (to open the throttle butterfly in the intake system); a throttle position sensor will pass a signal back to the control unit to indicate whether the throttle motor has opened the throttle butterfly to the correct angle. In this case therefore, the control unit monitors the system performance by using the information provided by the throttle position sensor. If the control unit is "requesting" that the throttle motor should change the throttle angle e.g. open the throttle, but the throttle position sensor does not detect the throttle angle change then this can be identified as a throttle motor fault; the fault code indicate a throttle motor "Circuit Range / Performance" fault or throttle motor "Performance" fault.

2.4   USING "LOGIC" TO DETECT A FAULT

2.4.1 In some cases, when an incorrect sensor signal or actuator response does not match the expected value, this can be regarded as being implausible or improbable; in effect, the signal or response shouldn't happen under the existing operating conditions.

Using the example, where the throttle movement / angle is monitored by a throttle position sensor, if the control unit requests that the throttle should open fully (full load) and the throttle sensor indicates that the throttle motor has responded correctly, this would then indicate correct operation. However, the control unit would then expect an increase in airflow through the intake system, which could be indicated by an airflow sensor. If the throttle is fully open, and the engine speed is increasing but, the airflow sensor signal <u>does not</u> indicate an increase in airflow, this will be regarded as implausible. A fault code could therefore indicate that there is an airflow sensor fault (likely to be a "Circuit Range / Performance" fault or possibly a "Performance" fault) or a fault code could indicate that there is an "Airflow Sensor / Throttle Position Sensor Correlation" fault i.e. the signals from the two sensors are indicating different operating conditions.

2.4.2 The control unit is therefore using "Logic" (often referred to as evaluation logic) to establish that a fault exists in the airflow sensor. Note that the airflow sensor signal could

be within the expected operating range, but the signal value is not indicating the expected increase in airflow.

3.   MONITORING THE SIGNALS USED IN SENSORS AND ACTUATORS

3.1   MONITORING SIGNAL VOLTAGE, FREQUENCY, PULSE WIDTH, ETC

3.1.1 Sensor Signals
Probably the simplest signal is a voltage output that is either "on" or "off"; an example of this could be a switch (Fig 1.2). However, sensors also provide signals where the voltage rises and falls e.g. a temperature sensor, where the voltage progressively changes with temperature (Fig 1.3).

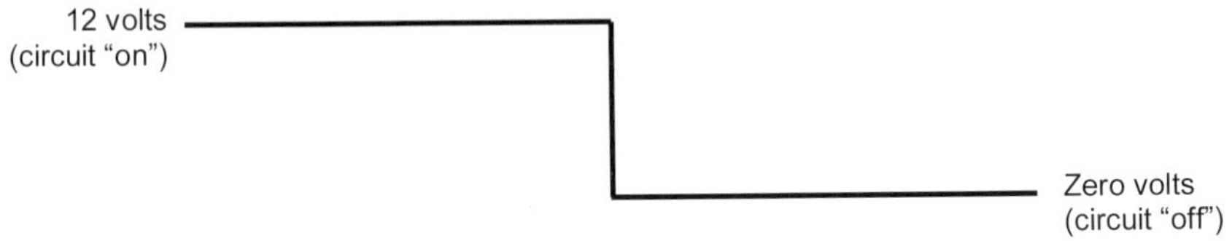

12 volts
(circuit "on")

Zero volts
(circuit "off")

Fig 1.2   A simple on / off signal that could be produced by a simple sensor e.g. a switch

Other sensors provide digital signals, where the number of pulses changes (frequency) with changes in conditions (Fig 1.4); an example is for some manifold pressure sensors, where the pulse frequency increases and decreases with changes in manifold pressure.

Another type of sensor signal is the analogue type speed / position signal, which has been used on crankshaft position / speed sensors, on wheel speed sensors, on transmission speed sensors etc. Fig 1.5a shows the typical analogue output signal for a speed sensor and 1.5b shows the same signal but with a reference point missing; this can be used to indicate a specific reference point e.g. on a crankshaft position sensor, it could indicate TDC. The control unit can convert the signal into a digital signal but, as with the true digital signal, the control unit will count the pulses (frequency) to identify speed information.

A control unit can therefore monitor different aspects of sensor signals, these include:
* the voltage for a simple "on" or "off" sensor signal e.g. a switch.
* the voltage or voltage change of a signal e.g. temperature sensor voltage.
* the frequency of a signal e.g. a digital signal from a manifold sensor or, an analogue or digital signal from a speed sensor.

Note that sensors can create a signal or voltage, or they can modify an existing signal or voltage that is passed to the sensor from a control unit.

The control unit monitors a signal to ensure that the signal values remain within a defined operating range for each type of signal e.g. a voltage range or a frequency range; if the signal value is outside of the range, this would then be regarded as a fault. Fig 1.3 shows the voltage range produced by a temperature sensor, which is between 4.7 and 0.3 volts. If the sensor was fitted to a cooling system, these extreme limits of the voltage range could exceed any minimum or maximum temperature for normal engine operation. If the voltage exceeded 4.7 volts or fell below 0.3 volts this would be regarded as a circuit or sensor fault.

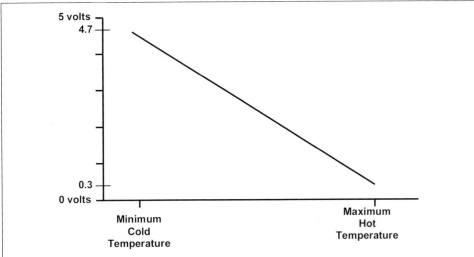

Fig 1.3   Sensor signal where the voltage changes with changes in conditions. In this example the voltage reduces as the temperature increases

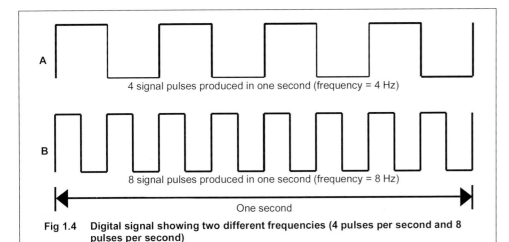

Fig 1.4   Digital signal showing two different frequencies (4 pulses per second and 8 pulses per second)

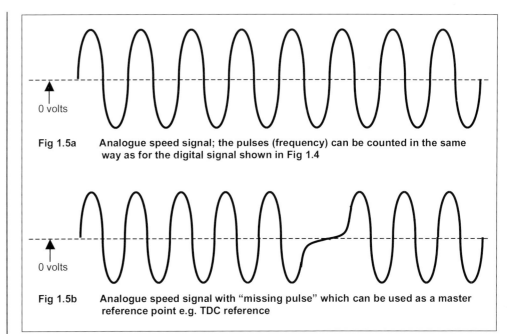

Fig 1.5a   Analogue speed signal; the pulses (frequency) can be counted in the same way as for the digital signal shown in Fig 1.4

Fig 1.5b   Analogue speed signal with "missing pulse" which can be used as a master reference point e.g. TDC reference

3.1.2 Actuator Control Signals

Actuators e.g. motors, solenoids etc, can be operated or controlled by switching on and off the power supply or, making and breaking the earth circuit; this will provide a simple on or off operation. A light bulb is a simple example of where the power supply is switched on (or the earth circuit is completed), for as long as it is required for the light to be "on".

However, actuators are used for many tasks that involve movement of a component to an accurate position (e.g. throttle motor) or for controlling pressure or fluid flow e.g. fuel injector. It is therefore necessary to control the actuator with a varying control signal; the value of the signal will be controlled by the system control unit, which will also monitor the signal to ensure that it is correct.

The process of switching on and off a circuit (as for the light bulb) can also be used to control different actuators e.g. switching on and off a solenoid that could be opening / closing a vacuum valve. The control signal will be a very simple on and off pulse, but it can happen in just a few thousandths of a second. In many cases, the control signal consists of many thousands of very short pulses.

For solenoids and other actuators, it is quite common for the control unit to switch the earth path of the actuator (with the power supply being provided all the time that the engine is running e.g. supplied via a relay); therefore the control signal will exist on the solenoid earth circuit (which will pass through the control unit to earth). A simple actuator signal can therefore be a rapid on / off pulse, with the length or duration of the "on" pulse dictating how long the solenoid is switched on (refer to Fig 1.6a and 1.6b). Note that when a circuit is switched via the earth path, voltage will exist in the earth circuit (at supply voltage level) when the earth path is "open" (not completed); when the earth circuit is "closed" (completed); the voltage will then reduce to zero (or almost zero) in the earth circuit.

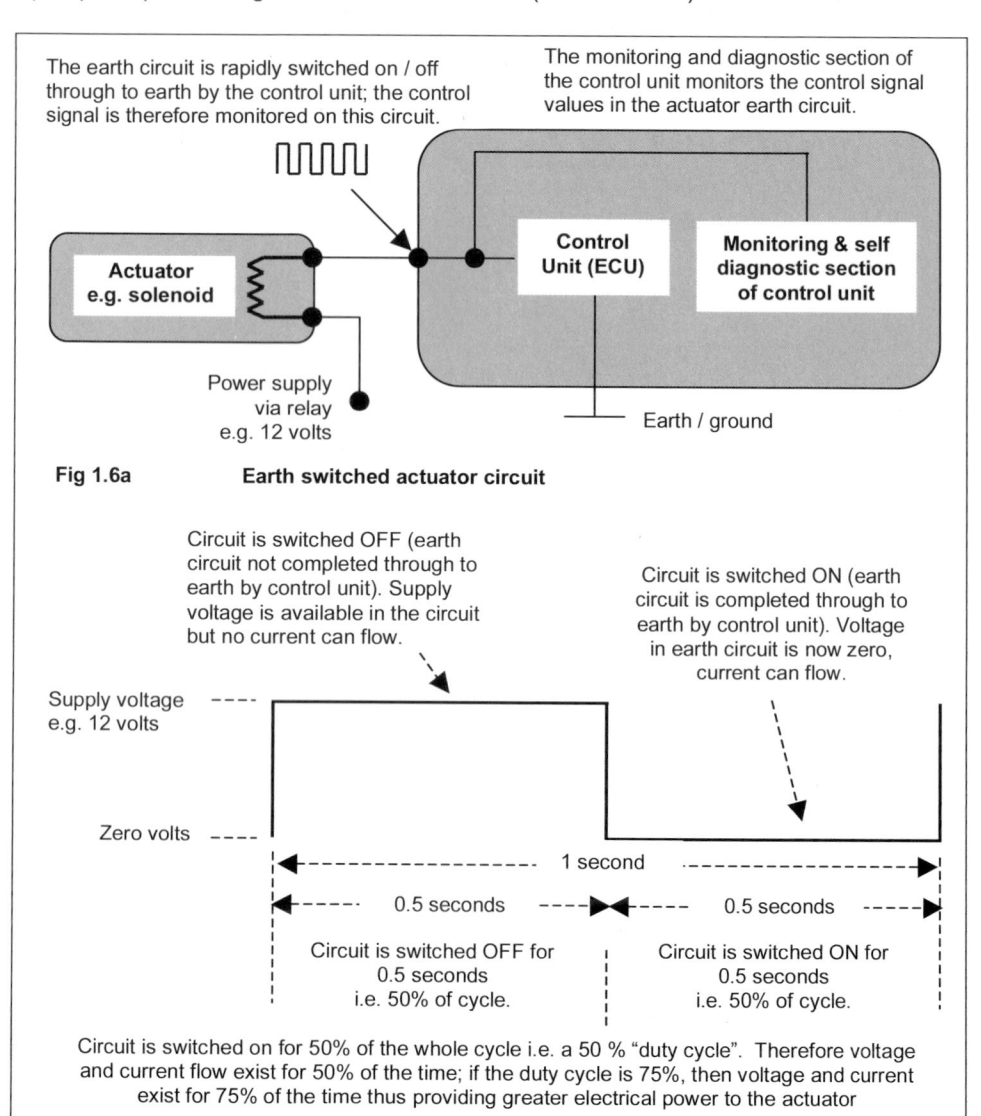

The earth circuit is rapidly switched on / off through to earth by the control unit; the control signal is therefore monitored on this circuit.

The monitoring and diagnostic section of the control unit monitors the control signal values in the actuator earth circuit.

Actuator e.g. solenoid

Control Unit (ECU)

Monitoring & self diagnostic section of control unit

Power supply via relay e.g. 12 volts

Earth / ground

Fig 1.6a   **Earth switched actuator circuit**

Circuit is switched OFF (earth circuit not completed through to earth by control unit). Supply voltage is available in the circuit but no current can flow.

Circuit is switched ON (earth circuit is completed through to earth by control unit). Voltage in earth circuit is now zero, current can flow.

Supply voltage e.g. 12 volts

Zero volts

1 second

0.5 seconds

0.5 seconds

Circuit is switched OFF for 0.5 seconds i.e. 50% of cycle.

Circuit is switched ON for 0.5 seconds i.e. 50% of cycle.

Circuit is switched on for 50% of the whole cycle i.e. a 50 % "duty cycle". Therefore voltage and current flow exist for 50% of the time; if the duty cycle is 75%, then voltage and current exist for 75% of the time thus providing greater electrical power to the actuator

Fig 1.6b   **Digital control signal for an earth switched actuator circuit**

The duration of the "on" pulse can be quoted as a percentage of the on / off cycle e.g. 50%, which is referred to as the "duty cycle". Increasing the duty cycle percentage ("on" time) will allow current to flow through to the actuator for a greater percentage of the time thus providing increased electrical power to the actuator; if the actuator is a solenoid that opens a valve, the increase in duty cycle will open the valve more. By varying the duty cycle, the control unit can alter the movement of an actuator (solenoid or motor) to a greater or lesser extent; it can also increase or decrease the speed of a motor.

Note that digital control signals can consist of a series of on / off pulses; the number of pulses occurring in one second (frequency) is often measured in Hertz. For example, 1 KHz (1 Kilo Hertz) equals one thousand pulses per second.

Also note that the process of altering the pulse width on a control signal (duty cycle) is often referred to as: Pulse Width Modulation (PWM).

4.   IDENTIFYING FAULTY SIGNALS

4.1   ELECTRICAL FAULT OR SYSTEM FAULT

4.1.1   As explained in section 3, the control unit can monitor voltages, frequencies or pulse widths (duty cycles) for sensor and actuator signals, but the control unit then needs to be able to identify faulty signals. Note that a high percentage of electrical faults that are detected are caused by faults in the wiring / connections or, an electrical related fault in the sensor or actuator. Where electrical faults occur, it generally results in the signal being non-existent, constant value or corrupt; but the signals will often be out of the normal operating range.

However, other faults can exist that will cause the signal to be incorrect but, the signal can still be within operating range i.e. within the expected values for normal operation. The fault can therefore be related to the operation of the system rather than just an electrical fault in the circuit. An example could be the signal from a coolant temperature sensor that indicates that the coolant temperature is "too hot". The fault could be caused by insufficient coolant, or the cooling fans not working; the sensor signal is within normal operating range but not correct for the operating conditions. The fault is therefore a system fault.

4.2   SIGNAL "OPERATING RANGE"

4.2.1   Operating range for circuits that are "on" or "off"
For circuits that are either switched on or switched off (without any variation in the signal voltage), the control unit can check for the voltage in the circuit, which for example, could be either battery voltage or zero volts; the normal operating voltage range will therefore be between zero and battery volts. The control unit will therefore need to identify if the correct voltage exists at the correct time.

A simple example is for the starter motor circuit, where the power supply to the starter solenoid should be at battery voltage during cranking, and zero volts when the ignition switch is in the normal "run" position. If however there is a fault in the circuit or relay, that causes voltage to be applied from the relay to the starter solenoid after the engine has started (ignition switch now in the "run" position), this will be regarded as a fault. In this example, the normal operating range in the circuit when the engine is running should be zero volts (probably with a small tolerance). Therefore, if voltage exists in the circuit when it should be at zero volts, the fault code can indicate "Starter Relay Circuit High".

The control unit applies 5 volts to the sensor circuit at terminal A. When the circuit is connected to the sensor, the voltage at terminal A will be dependent on the resistance of the sensor resistor (which is dependent on temperature). The voltage will however be within the normal operating range. Terminal B is on the earth / return circuit and will be zero volts.

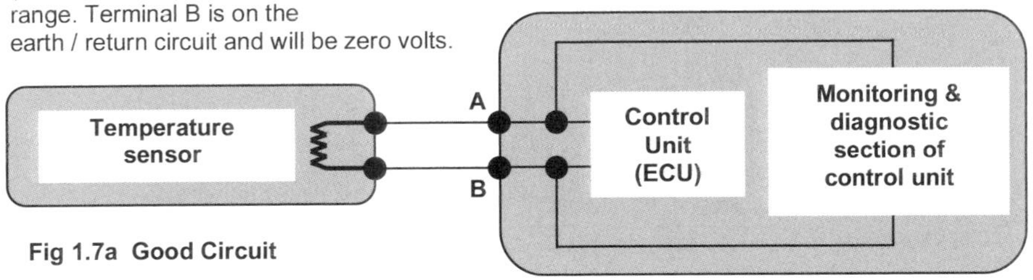

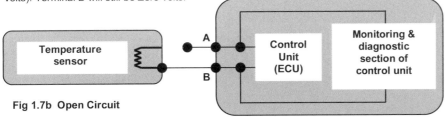

Fig 1.7a Good Circuit

If there is a break in the circuit (open circuit) at any point from terminal A through the sensor and back to terminal B, the voltage at terminal A will be 5 volts (control unit is still providing the 5 volts). Terminal B will still be Zero volts.

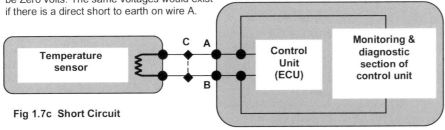

Fig 1.7b Open Circuit

If there is a "short circuit " at point C or at any point between the A and B wires, this is effectively a short to earth. The voltage at terminal A will be Zero volts. The voltage at Terminal B will still be Zero volts. The same voltages would exist if there is a direct short to earth on wire A.

Fig 1.7c Short Circuit

Fig 1.7 (a, b, c)   Voltages in a sensor circuit (with a good circuit and with faults)

4.2.2 Operating range for circuits where the voltage changes

Temperature sensors are an example of where the voltage in the circuit will change, with the changes in temperature. In the example (Fig 1.7a) the control unit provides a regulated voltage to the sensor circuit of 5 volts (applied by the control unit at terminal A); the possible minimum and maximum voltages in the circuit would therefore be zero volts or 5 volts. However, when the circuit is connected to the sensor, the resistance in the sensor (which forms part of a series resistance circuit) causes the voltage at terminal A to reduce to below 5 volts (but it will be above zero volts); in fact, for this example it should be within the normal operating range of 0.3 to 4.7 volts (Fig 1.8). The exact voltage will depend on the sensor resistor value. Note that the sensor is part of a series resistance circuit (with other resistances contained within the control unit); if the value of one resistor changes, it changes the voltage in the circuit between the resistors. Therefore, when the sensor resistor changes with temperature changes, it changes the voltage in the sensor circuit.

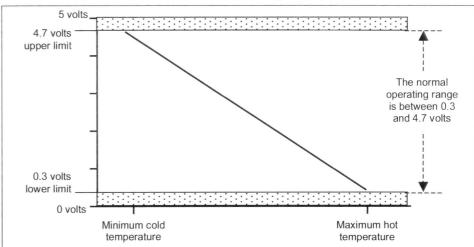

In this example, the normal operating range for the temperature sensor is between 0.3 and 4.7 volts. If the voltage is outside of this range (above or below the range limits), it will be regarded as a fault. Note that different systems will have different voltage ranges.

Fig 1.8   Operating range for a sensor signal where the voltage changes

If there is an "open circuit" (as shown in Fig 1.7b), the voltage at terminal A will be at the maximum i.e. 5 volts (there is no resistance in the circuit to reduce the voltage). The control unit detects that the voltage is at the maximum value (above the normal operating range) and this will be regarded as a fault. The fault code could indicate "Circuit High".

If there is a short circuit between the two wires (A and B), or a direct short from wire A to earth, the voltage at terminal A will be zero (Fig 1.7c). The control unit detects that the voltage is at the minimum value (below the normal operating range) and this will be regarded as a fault. The fault code could indicate "Circuit Low"

Therefore, if a fault occurs in the circuit, such as a short or open circuit; the voltage at terminal A will be either at 5 volts or zero volts. It would therefore not be practicable to allow the normal sensor signal to reach 5 volts or zero volts during normal operation because the control unit could not detect the difference between a fault and an acceptable signal value. A normal operating limit is therefore established, which is slightly above the minimum of zero volts, and slightly below the maximum of 5 volts (refer to Fig 1.8). Our example uses the values of 0.3 to 4.7 volts, which is the "normal operating limit" for the sensor in this example (always refer to vehicle specific information for max and min limits).

Note that the extreme limits of the normal operating range will represent temperatures that are beyond the normal operating limits of the vehicle system e.g. if the temperature sensor is used to check coolant temperature, the limits of 0.3 volts and 4.7 volts would represent temperatures much higher and lower than even extreme coolant temperatures.

4.2.3 Operating range for circuits where the frequency changes

Frequency signals can be used to provide speed information. If, for example, the sensor indicates vehicle speed then the signal range could in fact start at zero pulses (vehicle not moving); this would be within the normal operating range. However, if the control unit is receiving other information that indicates that the vehicle is moving (e.g. transmission speed signal) but there is still no vehicle speed signal, then the control unit would register this as a fault. For this type of fault, the signal is within normal operating range; the fault would be detected using "logic" or a calculation based on information from other sensors.

For most frequency based signals (refer to Fig 1.9 and 1.10), if the signal is out of normal operating range, the signal will most likely be at zero volts (with no pulses) or at maximum system voltage (also no pulses). The causes of the fault are likely to be short or open circuits in the wiring or sensor (the same as those wiring faults shown in Fig 1.7b and 1.7c). For those sensors being provided with a power supply or reference voltage (e.g. Hall Effect sensors), the supply or reference voltage might not exist due to another fault.

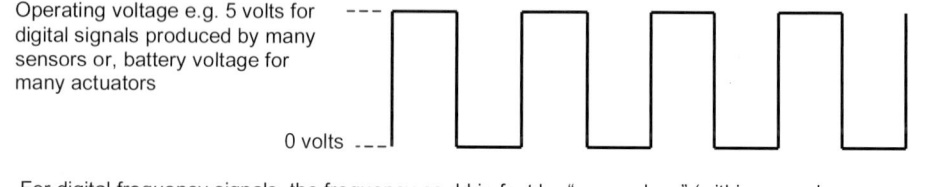

For digital frequency signals, the frequency could in fact be "zero pulses" (within normal operating range) if the speed is zero e.g. vehicle speed when the vehicle is not moving. Therefore, a signal that is out of normal range is likely to be a signal with no pulses detected when the control unit assesses that a frequency should exist. If there is a short or open circuit in the wiring or sensor, the signal is most likely to be a constant value of zero volts or, at supply voltage e.g. 5 volts for a sensor.

Fig 1.9   Operating range for a digital frequency / speed signal

---

4.2.4 Note that the control unit can monitor the operating voltage of a digital signal to ensure that it is at the correct level e.g. 5 volts for a sensor or possibly battery voltage for an actuator signal. For analogue signals, the control unit can monitor the peak voltage to check for a weak signal. If the signal does not contain any pulses i.e. it is a constant voltage at either full voltage or at zero voltage, and the voltage value is outside of normal operating range, the fault code could indicate "Circuit High" or "Circuit Low".

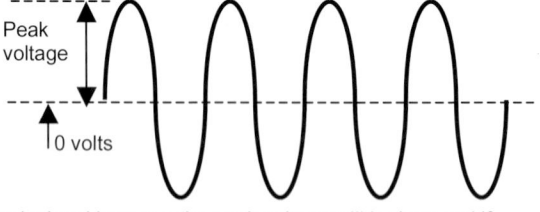

The control unit will however detect a weak signal because the peak voltage will be low; and if the control unit does not see the expected increase in peak voltage, this can be regarded as a fault

For analogue frequency signals, the frequency could in fact be zero pulses (within normal operating range) if the speed is zero e.g. vehicle speed sensor when the vehicle is not moving. Therefore, a signal that is out of normal range is likely to be a signal with no pulses detected when the control unit assesses that a frequency should exist. If there is a short or open circuit in the wiring or sensor, the signal is most likely to be a constant value of zero volts.

Fig 1.10   Operating range for an analogue frequency / speed signal

4.2.5 Operating range for circuits where the pulse width changes

Pulse width changes (duty cycle change) are associated with control signals passing from the control unit to an actuator e.g. solenoid, motor etc (refer to Fig 1.11). Therefore, if the control signal exists but the actual pulse width / duty cycle is an incorrect value, this is likely to be a control unit related fault.

A normal operating range for pulse width / duty cycle values could exist, but the maximum and minimum values will depend on which actuator or system is being controlled. Using the illustration in Fig 1.11, the smaller pulse width could represent the minimum value (minimum duty / cycle value e.g. 33%); the larger pulse width could represent the maximum value (maximum duty cycle e.g. 67%). If the duty cycle values are within normal operating range, but are too high or too low for the operating conditions, this will normally be referred to a "Range / Performance" fault or, the fault code might indicate that a fault exists in the system being controlled by the actuator.

If a circuit fault exists e.g. short or open circuit, this can affect the voltage level that exists in the circuit (it will also affect the current flow in the circuit which can also be monitored on some systems). Note that a control unit can perform a basic continuity check on a circuit by briefly passing a small current through the circuit e.g. when ignition is initially switched on. If there is a short or open circuit, this would be detected during the basic continuity check, which can be monitored by the control unit.

Note that actuators can be controlled by switching the earth path; however, some actuators are controlled by switching the "positive" circuit (power supply). Depending on whether the circuit is "earth circuit switched" or "power circuit switched" will dictate what will happen in the circuit when a short or open circuit exists (fault code could indicate "High" or "Low" depending on which method of switching is used); the result will also depend on where in the circuit the short or open circuit problem occurs.

However, a short or open circuit for an actuator circuit will often result in the fault code indicating an undefined "Circuit" problem as well as "Circuit High", "Circuit Low" or "Circuit / Open". Although the code is indicating the signal is out of the normal operating range, this will be due to the voltage or current being incorrect (or no circuit continuity).

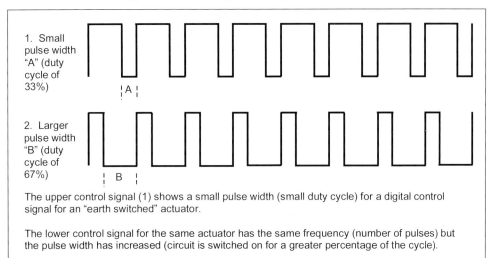

1. Small pulse width "A" (duty cycle of 33%)

2. Larger pulse width "B" (duty cycle of 67%)

The upper control signal (1) shows a small pulse width (small duty cycle) for a digital control signal for an "earth switched" actuator.

The lower control signal for the same actuator has the same frequency (number of pulses) but the pulse width has increased (circuit is switched on for a greater percentage of the cycle).

Fig 1.11   Control signal where the pulse width / duty cycle changes

5.   UNDERSTANDING FREQUENTLY USED FAULT CODE DEFINITIONS

5.1   DEFINING THE "CIRCUIT" FAULTS

5.1.1   Many faults can be classified as relatively simple "electrical failures" of a sensor, an actuator or a wiring circuit; these "electrical failures" usually revolve around a wiring or sensor / actuator fault and can be as simple as a short circuit or open circuit (which would cause the signal to be outside of the normal operating range i.e. out of tolerance). It is possible to group different types of electrical failures into a small number of categories, and the ISO / EOBD fault code definitions do in fact make use of standard phrases for many fault codes; some examples of these phrases are shown in the following list:

* Circuit (or Circuit Malfunction)
* Circuit / Open
* Circuit High
* Circuit Low
* Circuit Range / Performance (this definition can refer to an electrical fault, it can also refer to a system fault as explained in 5.1.3 and 5.2.5).
* Circuit Intermittent

Many fault definitions are preceded with the word "Circuit", which indicates that a fault is likely to be related to a signal problem caused by a fault in the sensor or actuator circuits.

5.1.2   One example of a simple fault definition for a sensor or an actuator is when there is no voltage or signal present in the circuit (the control unit would normally be expecting some level of voltage or signal activity); the voltage in the circuit will probably be zero volts, which is obviously lower than the control unit is expecting. Although the control unit cannot establish the exact cause of the fault, it recognises that there is no voltage or signal in the circuit and, the control unit will therefore activate a fault code for the sensor or actuator. The fault code definition could include the phrase: "Circuit Low" which indicates that the voltage or signal value is lower than the normal operating value i.e. the signal is out of tolerance. Therefore, the phrase "Circuit Low" appears many times within the range of fault code definitions, simply because electrical circuits across all of the sensors and actuators can fail in a similar way thus allowing this standard phrase to be used.
Therefore, for many different fault codes, there are some consistent or similar code definitions e.g. a coolant temperature sensor can have exactly the same electrical failures as an air temperature sensor; the fault code descriptions for the two sensors will therefore be virtually identical. These commonly used definitions (and others) are explained within the following pages for sensors and for actuators.

5.1.3   Other types of fault can occur because a sensor or actuator does not respond as expected; these types of faults often have the terms "Range / Performance" or "Performance" in the EOBD code definition. A simple example is when the engine coolant temperature is too high; the temperature sensor signal would therefore be indicating a high temperature i.e. a temperature value that is not normally expected by the control unit (under the existing operating conditions). Note that the signal voltage will be within the normal operating range even though the indicated temperature is incorrect.

The fault could be caused by a faulty cooling system .e.g. thermostat fault or low coolant level; in this example, a fault code description could therefore refer to the "coolant" temperature being too high or "Over Temperature". Note that the code indicates that the "Temperature" is high, whereas circuit faults are defined as "Circuit High". However, a manufacturer could take the option of using another fault code, which could have the following definition: "Temperature Sensor Circuit Range / Performance". The definition indicates that the signal voltage is within normal operating range but the signal voltage is not as expected, or not responding as expected; this could be caused by an electrical fault with the sensor or wiring, but it could also be caused by a cooling system fault. In reality, both fault code definitions could be used and, in some cases, there might be more than one code relating to the same fault.

5.1.4 When a fault code definition indicates that the sensor or actuator is not responding as expected (as discussed in the previous paragraph), it is not necessarily indicating that the sensor or actuator is at fault. If for example, the coolant temperature sensor signal indicates that the coolant temperature has not increased very much after the engine has been running for some time (following a cold start), this can be regarded as a Range / Performance problem because the control unit was expecting a more substantial increase in temperature i.e. the signal change is not as expected (which could be a sensor / wiring fault or a coolant temperature fault).

5.2   FREQUENTLY USED FAULT CODE DEFINITIONS FOR SENSORS

5.2.1 "Circuit" or "Circuit Malfunction"
Using a coolant temperature sensor as the example, the information in this book will appear similar to the example shown in Fig 1.12.

The EOBD Fault Code description does not specify what the fault is in the circuit, and is therefore referred to (within this book) as being and "undefined fault".

The signal from the sensor could therefore be:

* Out of normal operating range (above or below the maximum / minimum limits).
* A constant value i.e. the voltage could be in or out of the operating range but remaining at a fixed voltage value.
* Non-existent i.e. no voltage / zero voltage
* Corrupt e.g. the signal could be affected by another electrical circuit (electrical interference); this can prevent the control unit from being able to "read" the signal.

When the fault is undefined, it will be necessary to carry out a full range of checks, which will include checking for short / open circuits, high resistances in the wiring and connections and also checking the sensor operation. It is also possible that the control unit is faulty. Refer to Figs 1.17 and 1.18, for examples of how the "Circuit" could fail.

5.2.2 Circuit High
Using a coolant temperature sensor as the example, the information in this book will appear similar to the example shown in Fig 1.13.

The EOBD Fault Code description indicates that value of the signal in the circuit is "High". The signal voltage at the control unit terminal (where the sensor signal is normally received) is therefore above the normal operating range (above the maximum limit).

The control unit will be monitoring the signal circuit at the terminal where the signal is normally received (it can also monitor the earth circuit). A typical voltage range could be in the region of 0.3 to 4.7 volts (the range will vary across different systems and manufacturers) therefore, in this example, the control unit will have detected a voltage value that is above 4.7 volts (most likely to be approximately 5 volts, which is a typical reference voltage value). Note that the reference voltage is provided by the control unit to the sensor circuit; when the circuit is connected across the temperature sensor resistance,

the voltage will reduce because the sensor resistance is now part of a series resistance circuit (exact voltage will depend on the resistance value, which in turn is dependent on temperature).

Fault Code	EOBD / ISO Definition	Component / System Description	Meaningful Description and quick check
P0115	Engine coolant Temperature Sensor 1 Circuit	The coolant temperature sensor signal is used for control of fuel, ignition and emissions control. The sensor information can be used for cooling fan control as well as for other vehicle systems. If there is more than one sensor, it can provide independent measurement for banks on a "V" engine or, it could provide information to different systems e.g. engine management, cooling fan control and driver instrumentation.	The voltage, frequency or value of the temperature sensor signal is incorrect (undefined fault). Sensor voltage / signal could be: out of normal operating range, constant value, non-existent, corrupt. Possible fault in the temperature sensor or wiring e.g. short / open circuits, circuit resistance, incorrect sensor resistance, or fault with the reference / operating voltage; interference from other circuits can also affect sensor signals. Refer to list of sensor / actuator checks on page 19.

Fig 1.12   Sensor fault code, indicating a "Circuit" fault

Fault Code	EOBD / ISO Definition	Component / System Description	Meaningful Description and quick check
P0118	Engine Coolant Temperature Sensor 1 Circuit High	The coolant temperature sensor signal is used for control of fuel, ignition and emissions control. The sensor information can be used for cooling fan control as well as for other vehicle systems. If there is more than one sensor, it can provide independent measurement for banks on a "V" engine or, it could provide information to different systems e.g. engine management, cooling fan control and driver instrumentation.	The voltage, frequency or value of the temperature sensor signal is either at full value (e.g. battery voltage with no signal or frequency) or, the signal exists but it is above the normal operating range (higher than the maximum operating limit). Possible fault in the temperature sensor or wiring e.g. short / open circuits, circuit resistance, incorrect sensor resistance or, fault with the reference / operating voltage; interference from other circuits can also affect sensor signals. Refer to list of sensor / actuator checks on page 19.

Fig 1.13   Sensor fault code, indicating a "Circuit High" fault

A likely fault is an open circuit; the open circuit could exist anywhere in the wiring circuit including within the sensor or a connector plug (refer to Fig 1.17).

5.2.3 Circuit Low

Using a coolant temperature sensor as the example, the information in this book will appear similar to the example shown in Fig 1.14.

Fault Code	EOBD / ISO Definition	Component / System Description	Meaningful Description and quick check
P0117	Engine Coolant Temperature Sensor 1 Circuit Low	The coolant temperature sensor signal is used for control of fuel, ignition and emissions control. The sensor information can be used for cooling fan control as well as for other vehicle systems. If there is more than one sensor, it can provide independent measurement for banks on a "V" engine or, it could provide information to different systems e.g. engine management, cooling fan control and driver instrumentation.	The voltage, frequency or value of the temperature sensor signal is either "zero" (no signal) or, the signal exists but it is below the normal operating range (lower than the minimum operating limit). Possible fault in the temperature sensor or wiring e.g. short / open circuits, circuit resistance, incorrect sensor resistance or, fault with the reference / operating voltage; interference from other circuits can also affect sensor signals. Refer to list of sensor / actuator checks on page 19.

Fig 1.14   Sensor fault code, indicating a "Circuit Low" fault

The EOBD Fault Code description indicates that value of the signal in the circuit is "Low". The signal voltage at the control unit terminal (where the sensor signal is normally received) is therefore below the normal operating range (below the minimum limit).

The control unit will be monitoring the signal circuit (it can also monitor the earth circuit). A typical voltage range could be in the region of 0.3 to 4.7 volts (the range will vary across different systems and manufacturers) therefore, in this example, the control unit will have detected a voltage value below 0.3 volts (most likely to be zero volts).

A likely fault is a short circuit from the signal wire to earth or, to the sensor earth path; the short could exist anywhere in the wiring circuit or within the sensor). It is possible on some systems that an open circuit could cause the same fault but, for most temperature sensors an open circuit will result in a high voltage as explained in 5.2.2.

5.2.4 Circuit Range / Performance

Using a coolant temperature sensor as the example, the information in this book will appear similar to the example shown in Fig 1.15. Note that this definition can refer to an electrical fault in the sensor or circuit but, it can also refer to a system fault.

The EOBD Fault Code description indicates that value of the signal in the circuit is "within the normal operating range" (within the maximum and minimum limits). The control unit will be monitoring the signal circuit (it can also monitor the earth circuit). In this example therefore, the voltage detected by the control unit will be within the typical voltage range of

0.3 to 4.7 volts (the range will vary across different systems and manufacturers); the control unit cannot detect / identify an electrical circuit or sensor fault because the voltage is within normal range but, this does not necessarily mean that there isn't an electrical fault - it is simply an indication that an electrical fault is not identifiable.

Fault Code	EOBD / ISO Definition	Component / System Description	Meaningful Description and quick check
P0116	Engine Coolant Temperature Sensor 1 Circuit Range /Performance	The coolant temperature sensor signal is used for control of fuel, ignition and emissions control. The sensor information can be used for cooling fan control as well as for other vehicle systems. If there is more than one sensor, it can provide independent measurement for banks on a "V" engine or, it could provide information to different systems e.g. engine management, cooling fan control and driver instrumentation.	The voltage, frequency or value of the temperature sensor signal is within the normal operating range / tolerance but the signal is not plausible or is incorrect due to an undefined fault. The sensor signal is not changing / responding as expected to the changes in conditions. Possible fault in the temperature sensor or wiring e.g. short / open circuits, circuit resistance, incorrect sensor resistance or, fault with the reference / operating voltage; interference from other circuits can also affect sensor signals. Refer to list of sensor / actuator checks on page 19. It is also possible that the temperature sensor is operating correctly but the temperature being measured by the sensor is unacceptable or incorrect for the operating conditions e.g. cooling system fault causing unacceptable coolant temperatures; if possible, use alternative method of checking temperature to establish if sensor system is operating correctly and whether temperature is correct / incorrect.

Fig 1.15   Sensor fault code, indicating a "Circuit Range / Performance" fault

However, the voltage detected by the control unit is not matching the expected values e.g. the temperature indicated is "cold" but the control unit is expecting a "hot" value (possibly because the engine has been running for a considerable amount of time). It is therefore possible that the temperature sensor is operating correctly but the coolant temperature being indicated by the sensor is incorrect for the operating conditions (this could indicate a cooling system fault). As another example, the control unit can switch on the cooling fans, which should then reduce the coolant temperature but, the sensor signal does not change to indicate the temperature reduction; although other fault codes could be used to indicate a cooling system fault, the sensor "Range / Performance" code could also be used because the signal does not match the expected change in value.

Unfortunately, it is also possible that there is an electrical fault, which could be causing the signal value to be incorrect (although it is within operating range). Possible faults include high circuit resistance, shorts to other circuits (live and earth circuits), sensor failure or a control unit fault. In this example, it will be advisable to try and check the temperature using a separate gauge; this will indicate whether the coolant temperature is incorrect or whether the temperature is correct but the sensor signal is incorrect.

Wherever possible, try to obtain a "Live Data" reading for the affected system or component. The Live Data values, which can be displayed via many types of diagnostic equipment, provides a clearer picture of the measurements being made by sensors and a clearer picture of sensor signal values and actuator control signals. However, it is also advantageous if the signals can be checked using Oscilloscopes or other test equipment (multi-meters etc).

5.2.5 Circuit Intermittent
Using a coolant temperature sensor as the example, the information in this book will appear similar to the example shown in Fig 1.16.

Fault Code	EOBD / ISO Definition	Component / System Description	Meaningful Description and quick check
P0119	Engine Coolant Temperature Sensor 1 Circuit Intermittent	The coolant temperature sensor signal is used for control of fuel, ignition and emissions control. The sensor information can be used for cooling fan control as well as for other vehicle systems. If there is more than one sensor, it can provide independent measurement for banks on a "V" engine or, it could provide information to different systems e.g. engine management, cooling fan control and driver instrumentation.	The voltage, frequency or value of the temperature sensor signal is intermittent (likely to be out of normal operating range / tolerance when intermittent fault is detected) or, the signal / voltage is erratic (e.g. signal changes are irregular / unpredictable). Possible fault with wiring or connections (including internal sensor connections); also interference from other circuits can affect the signal. Check the sensor signal, using "live data" or other test equipment and wiggle wiring / connections to try and recreate the fault. Refer to list of sensor / actuator checks on page 19.

Fig 1.16   Sensor fault code, indicating a "Circuit Intermittent" fault

The EOBD Fault Code description indicates that the signal in the circuit is "Intermittent". The signal could be within normal operating range (in this example, between 0.3 and 4.7 volts) but it is possible that the signal will drop out of the operating range when the intermittent fault occurs; it is also possible that the signal is erratic e.g. the signal could remain within the operating range but the value is changing erratically (signal is not changing smoothly or occasionally the value is jumping unexpectedly).

The faults are most likely to revolve around poor connection or wiring faults (or intermittent breaks in the wiring); it is also possible that there is an internal fault with the sensor. The signal should be checked using either "Live Data" or a multi-meter / oscilloscope and, at the same time the wring / connections should be moved / wiggled to try and recreate the fault.

An intermittent fault code could be activated if the measurement or value being detected by the sensor is changing erratically. An example could be a pressure sensor in a hydraulic circuit e.g. a transmission system, which is detecting a momentary / occasional pressure change (possibly caused by a pressure regulator occasionally failing); this could cause an occasional (but very brief) increase or decrease in pressure. The pressure signal will therefore momentarily change, which could be regarded as an erratic signal.
Another fault that can activate an intermittent / erratic fault code is interference from another electrical circuit. A genuine problem occurs on some vehicles where the wiring from the coolant temperature sensor passes close to the high secondary voltage of the ignition system (ignition coil or Plug leads on older systems). When the high voltage is produced for the spark plug, it induces a voltage "spike" into the temperature sensor circuit; this can then cause the control unit to activate the intermittent / erratic fault code.

5.3   FREQUENTLY USED FAULT CODE DEFINITIONS FOR ACTUATORS

5.3.1   Note. Whereas sensor signals are generally created by the sensor and therefore detected by the control unit, an actuator control signal is created by the control unit and passed to the actuator. Therefore, if an actuator signal has an incorrect value (within normal operating range), it is possible that the control unit is providing an incorrect signal due to a fault within the control unit or, the control unit is providing the incorrect signal because the actuator is not responding as expected and the control unit is attempting to overcome the actuator response problem.

Also note that a control unit could be receiving incorrect or abnormal information from a sensor, which is then causing the control unit to provide an incorrect actuator signal. The control unit could however recognise that the control signal is in fact incorrect for the operating conditions (indicated by other sensors) and it could therefore activate a fault code indicating a "performance" related problem with the actuator signal but, it might not be able to define why the fault exists.

5.3.2   When the control unit activates a "Circuit" related fault (low, high etc), it will be necessary to identify whether the control unit switches the actuator earth path or the power supply; this will affect the interpretation of the fault description.

Also note that the interpretation of the fault code can also depend on the type of control signal being provided to the actuator e.g. a simple voltage on / off signal or, a pulse width modulated signal (duty cycle control).

5.3.3   Most of the frequently used fault code definitions for actuators use the same phrases as those definitions used for sensors, and in fact most of the faults associated with the actuator code definitions are the same as for sensors. The information contained in section

5.2 (relating to sensor code definitions) will therefore provide an understanding of the definitions used for actuators. Where applicable in this section, reference is therefore made to the appropriate paragraphs of section 5.2.

5.3.4 Many actuator faults can be classified as relatively simple "electrical failures" of an actuator or wiring circuit; these "electrical failures" usually revolve around a wiring or actuator fault and can be as simple as a short circuit or open circuit (which would cause the signal to be outside of the normal operating range i.e. out of tolerance). It is possible to group different types of electrical failures into a small number of categories, and the ISO / EOBD fault code definitions do in fact make use of standard phrases for many fault codes; some examples of these phrases are shown below:

* Circuit
* Circuit / Open
* Circuit High
* Circuit Low
* Circuit Range / Performance (this definition can refer to an electrical fault, it can also refer to a system fault as explained in 5.3.9).
* Circuit Intermittent

Note that many fault definitions are preceded with the word "Circuit", which indicates that the fault is likely to be related to a control signal problem caused by an electrical fault in the actuator or actuator circuit.

5.3.5 Circuit
When an actuator fault code is defined with the phrase "Circuit", the meaning is much the same as for a when a sensor has a circuit fault (section 5.2.1). The EOBD Fault Code description does not specify what the fault is in the circuit, and is therefore referred to (within this book) as being an "undefined fault".

The control signal could therefore be:
*   Out of normal operating range (above or below the maximum / minimum limits).
*   A constant value i.e. the voltage frequency or other signal value could be in or out of the operating range but remaining at a fixed value.
*   Non-existent i.e. no voltage or signal i.e. zero voltage
*   Corrupt e.g. the signal could be corrupted or affected by another electrical circuit (electrical interference); this can prevent the control unit from being able to "read" the signal.

When the fault is undefined, it will be necessary to carry out a full range of checks, which will include checking for short / open circuits, high resistances in the wiring and connections and also checking the actuator operation. It is also possible that the control unit is faulty. Refer to Figs 1.17 and 1.18, for examples of how the "Circuit" could fail.

5.3.6 Circuit / Open
Although the "Circuit / Open" fault code description could indicate an undefined fault i.e. the same as for when a "Circuit" description is provided (section 5.3.4), the addition of the word

"Open" is likely to indicate that the control unit has detected an open circuit. The control unit could detect the fault in a number of ways such as: not being able to detect any power supply voltage in the circuit (power supply could pass from a relay, through the actuator to the actuator earth circuit, which is then switched to earth by the control unit); if the control unit is unable to detect the voltage, this could indicate an open circuit fault but, it could also indicate that there is no power supply to the actuator.

The control unit could also be providing a "Test" signal or pass a current through the circuit when the ignition is initially switched on; if there is no current flow in the circuit or the test signal is not detected returning back through the circuit, this could be regarded as an open circuit fault.

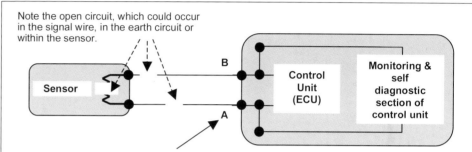

The normal operating voltage range (at terminal A) could be between 0.3 and 4.7 volts (depending on temperature which will change the resistance value in the sensor and the voltage in the circuit). An open circuit will cause the voltage in the signal wire to be at the "Reference Voltage" value (typically around 5 volts).

Fig 1.17   Simple example of a sensor circuit where the Fault code will indicate "Circuit High" due to an open circuit fault

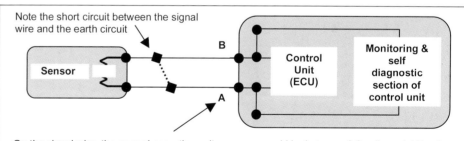

On the signal wire, the normal operating voltage range could be between 0.3 volts and 4.7 volts depending on temperature (which will change the resistance value in the sensor). The short circuit from the signal wire A, to the earth path B, will cause the signal voltage to drop to zero

Fig 1.18   Simple example of a sensor circuit where the Fault code will indicate "Circuit Low" due to a short circuit fault

5.3.7 Circuit High

The interpretation of the fault will depend on whether the control unit is switching the earth path or the power supply circuit.

The control unit could be detecting a high voltage in the circuit, which could be caused by a short circuit to another power supply or there is an open circuit, which is preventing the circuit passing through to earth.

It is also possible that the control signal is above the normal operating range, but this is likely to be a control unit related fault.

5.3.8 Circuit low

It will depend on whether the control unit is switching the earth path or the power supply circuit as to the interpretation of the fault.

The control unit could be detecting a low voltage in the circuit, which could be caused by a short to earth (on the earth path of an earth switched actuator). The fault could also be caused by an open circuit in or no power supply (again on an earth switched circuit).

It is also possible that the control signal is above the normal operating range, but this is likely to be a control unit related fault.

5.3.9 Circuit intermittent

As with a sensor circuit (section 5.2.5), the fault is likely to be caused by a poor / loose connection in the wiring, in the sensor or in a connector.

5.3.10 Circuit Range / Performance

When a "Range / Performance" fault code is activated, it is likely that the control signal is within the normal operating range but, it is incorrect for the operating conditions.

However, it is also possible that the control unit is passing a specific control signal to the actuator but, the actuator response is not as expected. An example is where a control signal should be causing the idle air solenoid valve to open, thus increasing the idle speed to a desired value e.g. 1000 rpm. If the idle speed does not increase sufficiently (due to an idle valve problem or a possible blockage in an air passage), the control unit could in fact provide a larger duty cycle in an attempt to raise the idle speed. The control signal required to achieve the desired idle speed would therefore be incorrect (compared to the value stored in the memory), and this could be regarded as a "Range / Performance" fault.

Range performance faults may therefore be caused by system faults such as the faulty idle speed valve but the fault can still be caused by electrical problems such as short / open circuit, high circuit resistances etc.

Chapter 2

SENSOR, ACTUATOR AND COMPONENT CHECKS

1. SENSORS: IDENTIFICATION AND GENERAL SENSOR CHECKS 20
2. TEMPERATURE SENSORS 21
3. POSITION SENSORS USING POTENTIOMETERS E.G. THROTTLE 22
4. POSITION SWITCHES E.G. THROTTLE, BRAKE OR CLUTCH 24
5. OTHER TYPES OF SWITCH (E.G. PRESSURE OR LEVEL) 26
6. SPEED / POSITION SENSORS (ANALOGUE SIGNAL) 26
7. SPEED / POSITION SENSORS (DIGITAL SIGNAL) 28
8. PRESSURE / VACUUM SENSORS (ANALOGUE SIGNAL) 31
9. VACUUM / PRESSURE SENSORS (DIGITAL SIGNAL) 33
10. AIR FLOW SENSORS 34
11. PROXIMITY SENSORS (MAGNETIC) 36
12. O2 / LAMBDA SENSORS 36
13. ACTUATORS: IDENTIFICATION AND GENERAL ACTUATOR CHECKS 41
14. SOLENOIDS / SOLENOID VALVES 42
15. INJECTORS (SOLENOID AND PIEZO) 44
16. MOTORS AND STEPPER MOTORS 44
17. IDLE SPEED CONTROL SOLENOIDS AND MOTORS 47
18. ELECTRIC HEATERS / HEATING ELEMENTS 47
19. DIESEL GLOW PLUGS 49
20. RELAYS 49
21. CAN-BUS 51

NOTE. Refer to the contents list on the previous page to select the applicable sensor type; the list identifies sensors either by a specific type (e.g. Position Switches) or a sensor task (e.g. Temperature Sensor).

If it is not initially possible to identify the exact type of sensor e.g. how the sensor operates and what type of signal is produced by the sensor, use the information contained in Section 1 below for the initial checks.

The following information provides basic guidelines to tests / checks for components and systems; due to the variations across different manufacturers and systems, it is always advisable to refer to vehicle or component specific information to identify specifications, signal type / values and dedicated test procedure

1. SENSORS: IDENTIFICATION AND GENERAL SENSOR CHECKS

1.1 NECESSITY TO IDENTIFY THE SENSOR TYPE

1.1.1 Vehicle and system manufacturers can use different types of sensors for specific tasks e.g. a crankshaft position sensor could be an "Inductive" type sensor with an analogue signal or it could be a "Hall Effect" sensor with a digital signal. When testing the sensors it is therefore essential to know the type of sensor, the wiring for the sensor and the signal values for the sensor (as well as the type of signal being provided e.g. analogue or digital).

1.1.2 In many cases, it can be initially difficult to identify a particular sensor type fitted to a vehicle; this can be due to difficult location and / or lack of applicable information. In some cases, prior experience and knowledge may be sufficient to identify a sensor type and its operation e.g. most temperature sensors operate in the same way and therefore checks will be similar across most vehicle models; however, it is often necessary to know how the sensor is wired to the system as well as knowing what the signal values should be.

Wherever possible therefore, it is advisable to refer to any available vehicle specific information to obtain the necessary knowledge and understanding of the sensor and the applicable wiring circuit. There are however some basic checks that can be carried out, even if it is not initially possible to identify the sensor and wiring; these checks are briefly highlighted in section 1.2.

1.1.3 It is important to note that applying the incorrect tests to a sensor or a sensor circuit can result in incorrect diagnosis but, more importantly, it is also possible to damage some types of sensor if the incorrect tests are applied.

1.1.4 Note. The use of incompatible equipment when carrying out resistance checks (and some other checks) on some electronic components can cause component damage.

1.2 BASIC CHECKS FOR ALL SENSORS

1.2.1 If the Fault Code definition includes the phrases: "Circuit ", "Circuit High", "Circuit Low" or "Circuit Intermittent", it is quite possible that an electrical fault exists in the circuit. Initial checks should therefore include testing for short or open circuits and poor connections on all applicable wiring between the sensor and the control unit. Note that a short circuit can exist between the sensor wiring and another circuit as well as just between sensor wires or shorts to earth.

Note that the problem could also be caused by a fault in the sensor but testing the sensor will most likely require more specific sensor or vehicle information.

1.2.2 If the Fault Code definition includes the phrase "Circuit Range / Performance", this could still also be caused by an electrical or sensor fault (1.2.1); however, it is also possible that the sensor is operating correctly but the system or measurement being monitored by the sensor is incorrect e.g. coolant temperature is incorrect for the operating conditions. It will therefore be necessary to also check whether the system being monitored by the sensor is incorrect or correct e.g. use an alternative method of checking the coolant temperature (or whatever system is being monitored by the sensor).

1.2.3 If possible, use the Live Data facility (available on some diagnostic equipment) to obtain an indication of the values being provided by the sensor to the control unit. If for example the fault relates to a coolant temperature, check the Live Data values to identify what temperature value is being passed to the control unit; this may help identify whether the fault is sensor / wiring related or whether the fault is a system fault e.g. a cooling system fault.

1.2.4 Although it is obviously preferable to refer to the applicable specifications and compare signal values against the specified values, it is however possible for many sensors and systems to detect faults because there is no change in the signal value or the signal is corrupt.

Check the sensor signal, either using the Live Data facility or with an oscilloscope / multimeter. Although there are different sensor signal types, it should be possible to check for a good signal e.g. no interference. Check for a change in the signal with changes in operating conditions e.g. for a pressure sensor, check for changes in the signal voltage or frequency when the pressure changes; if necessary simulate the system operation e.g. use a pressure / vacuum pump to provide different pressure / vacuum levels to the sensor to enable sensor operation to be monitored.

1.2.5 Also refer to Chapter 1 for an indication of how fault code definitions can identify certain types of fault.

2. TEMPERATURE SENSORS

2.1 OPERATION

2.1.1 Note. If the temperature sensor is combined within another sensor e.g. Intake Air temperature sensor combined into an airflow sensor, it might be necessary to refer to vehicle or component specific information to identify the wiring and specific tests etc.

2.1.2 The sensor normally has two wires, both of which are usually connected to the control unit. A regulated voltage of typically around 5 volts (some applications may use a different voltage value) originates in the control unit and is passed to the sensor on one of the wires (this is the signal wire); the circuit then passes through the sensor resistor back to the control unit to earth (Fig 2.1a). Most temperature sensors form part of a series resistance circuit, with the sensor being one of the resistors, and other resistances being contained within the control unit. The resistor within the sensor changes its resistance value when the temperature changes, this then causes the voltage in the signal wire to change; the exact voltage value in the circuit will depend on the resistance value (ohms), which is dependant on the temperature.

2.2 CIRCUIT OR CIRCUIT / OPEN

2.2.1 When a fault code definition states "Circuit", without any other information (e.g. low, high), it is likely that control unit has detected an incorrect sensor signal value. However, the control unit might not be programmed to specify the exact nature of the incorrect signal. Although it is likely that an electrical fault exists (refer to sections 2.3, 2.4 and 2.5), it is also possible the sensor is operating correctly but the temperature being indicated by the sensor is incorrect (refer to section 2.6). If the code definitions states "Circuit / Open", it is likely that the control unit is not detecting the sensor signal and assumes that there is an open circuit (refer to sections 2.3 and 2.4).

2.3 CIRCUIT LOW

2.3.1 When a fault code definition states "Circuit Low", this will usually indicate that the voltage at the signal terminal (terminal A in Fig 2.1c) is below the normal operating range (refer to Chapter 1 - section 4.2). The fact that the voltage is outside the normal operating range indicates a probable circuit or sensor fault, rather than an incorrect temperature.

2.3.2 A "Circuit Low" fault can be caused by a short between the signal wire and earth (or to the sensor earth wire - terminal B in Fig 2.1c); if the circuit is shorted to earth, the voltage at terminal A will be zero. Check all applicable wiring between the sensor and the control unit for a short circuit to earth or to another wire. Note that a short to another circuit could cause an incorrect circuit resistance which could result in the circuit voltage being too low.

2.3.3 Check across the sensor terminals for the correct resistance throughout the temperature range (refer to vehicle specific information); ensure that there is not a short

circuit within the sensor and that the sensor resistance changes correctly with changes in temperature.

2.3.4 If the sensor resistance is correct and there are no wiring faults, it is possible that there is a control unit fault; also check for regulated voltage output from control unit.

2.4 CIRCUIT HIGH

2.4.1 When a fault code definition states "Circuit High", this will usually indicate that the voltage at the signal terminal (terminal A in Fig 2.1b) is above the normal operating range (refer to Chapter 1 - section 4.2). The fact that the voltage is outside the normal operating range indicates a probable circuit or sensor fault, rather than an incorrect temperature.

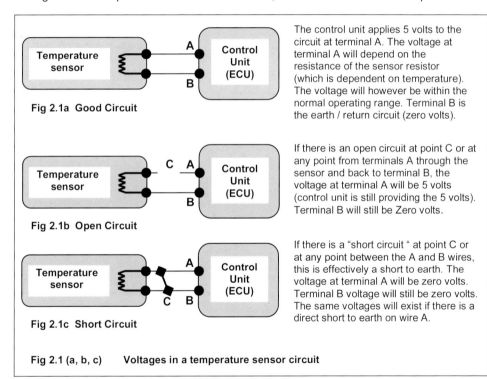

Fig 2.1a Good Circuit

The control unit applies 5 volts to the circuit at terminal A. The voltage at terminal A will depend on the resistance of the sensor resistor (which is dependent on temperature). The voltage will however be within the normal operating range. Terminal B is the earth / return circuit (zero volts).

Fig 2.1b Open Circuit

If there is an open circuit at point C or at any point from terminals A through the sensor and back to terminal B, the voltage at terminal A will be 5 volts (control unit is still providing the 5 volts). Terminal B will still be Zero volts.

Fig 2.1c Short Circuit

If there is a "short circuit" at point C or at any point between the A and B wires, this is effectively a short to earth. The voltage at terminal A will be zero volts. Terminal B voltage will still be zero volts. The same voltages will exist if there is a direct short to earth on wire A.

Fig 2.1 (a, b, c)   Voltages in a temperature sensor circuit

2.4.2 A "Circuit High" fault can be caused by an open circuit in the wiring or sensor (or sensor / harness plug disconnected). When an open circuit exists (Fig 2.1b), the voltage at the control unit "signal" terminal will be approximately 5 volts (or whatever value is used for the regulated voltage); this is the available voltage in the circuit (provided by the control unit) and, due to the open circuit condition, there is no current flow and therefore the resistances do not affect the voltage. Check all applicable wiring between the sensor and the control unit for an open circuit, including the earth circuit.

2.4.3 Check across the sensor terminals for correct resistance throughout the temperature range (refer to vehicle specific information); ensure that there is not an open circuit within the sensor and that the sensor resistance changes correctly with temperature changes.

2.4.4 If the sensor resistance is correct and there are no wiring faults, it is possible that there is a control unit fault; also check for regulated voltage output from control unit.

2.5   CIRCUIT INTERMITTENT

2.5.1 When a fault code definition states "Circuit Intermittent", this will usually indicate that the voltage at the signal terminal (terminal A in Fig 2.1a) is unstable or erratic; the voltage is likely to be outside of the normal operating range when the fault occurs (refer to Chapter 1 - section 4.2). Intermittent faults can be caused by loose wiring or connections causing an intermittent short or open circuit; refer to sections 2.3 and 2.4 for additional information. Note that the connection fault could be inside the sensor or possibly inside the control unit / connector plug. Wherever possible, check the signal and at the same time wiggle the wiring and connections to check for intermittent breaks or short circuits.

2.6   CIRCUIT RANGE / PERFORMANCE

2.6.1 When a fault code definition states "Circuit Range / Performance", it indicates that the voltage at the signal terminal (terminal A in Fig 2.1a) is within the normal operating range (refer to Chapter 1 - section 4.2); it is therefore possible that the sensor is operating correctly but the temperature value indicated by the sensor signal is not close to the value expected by the control unit. It is however also possible that a faulty sensor or, a short circuit to another wire could result in the signal still being within normal operating range.

2.6.2 If possible, check the signal value from the sensor using a Live Data facility on the diagnostic equipment (the live data might display the temperature or it might display a signal voltage value); alternatively, use a multi-meter to check the voltage at the control unit signal terminal (it might be necessary to refer to vehicle specific information to identify correct live data values or voltage values).

Check the temperature of the system being measured e.g. water, oil etc, and compare with the live data temperature value (or calculate temperature using the signal voltage value). Note that it is advisable to check the sensor values over a wide operating range. If the live data value (or calculated temperature) is significantly different from the actual temperature, this indicates a possible sensor or wiring related fault (refer to sections 2.3 and 2.4).

2.6.3 If the live data value (or calculated temperature) is the same as the actual temperature, this indicates that the sensor and circuit are probably good; therefore the problem is possibly due to the temperature being incorrect for the operating conditions. It will therefore be necessary to look for causes of incorrect temperatures e.g. on a cooling system the coolant level could be too low causing a high temperature value.

3.   POSITION SENSORS USING POTENTIOMETERS E.G. THROTTLE

3.1   OPERATION

3.1.1 Potentiometers are used to convert mechanical movement into an electrical signal (a progressive voltage change). Potentiometers usually have three wires: an earth connection, a reference / supply voltage and a signal wire (all three wires are usually connected to the control unit). The reference voltage (regulated voltage of typically 5 volts provided by control unit) is applied to one end of a resistive track, with the earth path being connected to the other end. The signal wire is attached to a "wiper" or contact that moves across the resistive track; at one end of the track, the resistance between the reference voltage connection and the wiper will be high, and at the other end the resistance will be low. As the wiper moves across the resistive track, the resistance will change and therefore the voltage will also change. The control unit will monitor the voltage on the signal wire and fault codes will generally be activated because the signal voltage is incorrect.

3.2   CIRCUIT OR CIRCUIT / OPEN

3.2.1 When a fault code definition states "Circuit" without any other information (e.g. low, high), it is likely that control unit has detected an incorrect sensor signal value. However, the control unit might not be programmed to specify the exact nature of the incorrect signal. Although it is likely that an electrical fault exists (refer to sections 3.3, 3.4 and 3.5), it is also possible the sensor is operating correctly but the information being indicated by the sensor is incorrect e.g. throttle position is incorrect compared with information from other sensors (also refer to 3.6). If the code definitions states "Circuit / Open", it is likely that the control unit is not detecting the sensor signal and assumes that there is an open circuit (refer to 3.3 and 3.4).

3.3   CIRCUIT LOW

3.3.1 When a fault code definition states "Circuit Low", this will usually indicate that the voltage at the signal wire terminal (terminal C in Fig 2.2a) is below the normal operating range (refer to Chapter 1 - section 4.2). The fact that the voltage is outside the normal operating range indicates a probable circuit or sensor fault, rather than an incorrect position or movement being indicated.

3.3.2 A "Circuit Low" fault can be caused by an open circuit in the signal wire or the reference / supply voltage wire (Fig 2.2b); in both cases voltage is unable to pass through the circuit back to the control unit.

A "Circuit Low" fault can also be caused by a short circuit (Fig 2.2d), either between the signal or reference voltage wires to earth (or to the sensor earth / return wire). Note that a short to another circuit could cause an incorrect circuit resistance which could result in the circuit voltage being too low. If there is a fault with the resistive track in the potentiometer e.g. a break / open circuit, this could also cause a "Circuit Low" code to be activated.

3.3.3 Check for the correct resistance between the sensor resistive track terminals (refer to vehicle specific information); check the resistance between the sensor reference / supply voltage terminal and the sensor signal terminal and at the same time operate the potentiometer to check the resistance throughout the sensor operating range e.g. on a throttle position sensor, check from closed to open throttle positions. Check that the sensor resistance changes smoothly / progressively when the potentiometer is operated.

3.3.4 If the sensor resistance is correct and there are no wiring faults, it is possible that there is a control unit fault; also check for correct regulated voltage output from control unit.

Potentiometer / sensor

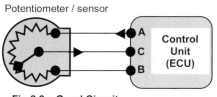

Fig 2.2a Good Circuit

The control unit applies a reference / supply voltage (typically 5 volts but can be different on some systems) to the circuit at terminal A, which passes across the resistive track in the sensor. Terminal B is the earth or return path. Terminal C is connected to the signal wire; the voltage at the terminal will therefore depend on the position of the wiper, which in turn will depend on the position of the component being monitored e.g. throttle position.

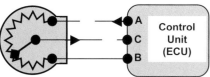

Fig 2.2b Open Circuit in signal or supply circuit

If there is an open circuit fault in the supply circuit A or in the signal circuit C (including an open connection at the wiper) the voltage at terminal C will be zero; this is likely to activate a "Circuit Low" fault code.

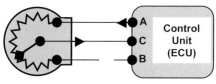

Fig 2.2c Open Circuit in earth / return circuit

If there is an open circuit fault in the earth / return circuit B (including an open circuit in the control unit) the voltage at terminal C will be probably be 5 volts (or whatever supply / reference voltage is used); this is likely to activate a "Circuit High" fault code.

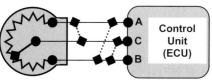

Fig 2.2d Short Circuit

If there is a short circuit between the A and C wires, the voltage at C will be 5 volts (or whatever supply / reference voltage is used); this is likely to activate a "Circuit High" fault code.
If there is a short between A and B or C and B, this is effectively a short to earth, the voltage at C will be zero volts; this is likely to activate a "Circuit Low" fault code.

Fig 2.2 (a, b, c, d)   Faults in Potentiometer type sensor circuit

3.4   CIRCUIT HIGH

3.4.1 When a fault code definition states "Circuit High", this will usually indicate that the voltage at the signal wire terminal (terminal C in Fig 2.2a) is above the normal operating range (refer to Chapter 1 - section 4.2). The fact that the voltage is outside the normal operating range indicates a probable circuit or sensor fault, rather than an incorrect position or movement being indicated.

3.4.2 A "Circuit High" fault can be caused by a short circuit between the signal wire and the reference / supply voltage wire (Fig 2.5d). If there is a fault with the resistive track in the potentiometer e.g. internal short, this could also cause a "Circuit High" code to be activated.

3.4.3 Check for the correct resistance between the sensor resistive track terminals (refer to vehicle specific information); check the resistance between the sensor reference / supply voltage terminal and the sensor signal terminal and at the same time operate the potentiometer to check the resistance throughout the sensor operating range e.g. on a throttle position sensor, check from closed to open throttle positions. Check that the sensor resistance changes smoothly and progressively when the potentiometer is operated.

3.4.4 If the sensor resistance is correct and there are no wiring faults, it is possible that there is a control unit fault; also check for correct regulated voltage output from control unit.

3.5   CIRCUIT INTERMITTENT

3.5.1 When a fault code definition states "Circuit Intermittent", this will usually indicate that the voltage at the signal terminal (terminal C in Fig 2.2a) is unstable or erratic; the voltage is likely to be outside of the normal operating range when the fault occurs (refer to Chapter 1 - section 4.2). Intermittent faults can be caused by loose wiring or connections causing an intermittent short or open circuit; refer to sections 3.3 and 3.4 for additional information. Note that the connection fault could be inside the sensor or possibly inside the control unit / connector plug. Wherever possible, check the signal and at the same time wiggle the wiring and connections to check for intermittent breaks or short circuits.

3.6   CIRCUIT RANGE / PERFORMANCE

3.6.1 When a fault code definition states "Circuit Range / Performance", it indicates that the voltage at the signal terminal (terminal C in Fig 2.2a) is within the normal operating range (refer to Chapter 1 - section 4.2); it is therefore possible that the sensor is operating correctly but the information indicated by the sensor signal e.g. throttle position is not close to the value expected by the control unit e.g. sensor information indicates throttle fully open, but other sensors indicate engine at low rpm with no load. It is however also possible that a faulty sensor or, a short circuit to another wire could result in the signal voltage still being within normal operating range.

3.6.2 If possible, check the signal value from the sensor using a Live Data facility on the diagnostic equipment (the live data might display a signal voltage value or it might indicate the angular position of the sensor or component being monitored e.g. throttle open 45%);

alternatively, use a multi-meter or oscilloscope to check the voltage at the control unit signal terminal (it might be necessary to refer to vehicle specific information to identify correct live data values or voltage values).

3.6.3 Check the operation of the system or component being monitored by the sensor e.g. check throttle movement and position; compare the actual position or movement with the live data values or signal. Note that it is advisable to check the sensor values over a wide operating range. If the live data value or signal voltage does not correspond with the position or movement, this indicates a possible sensor or wiring related fault (refer to sections 3.3 and 3.4). Check that the sensor signal changes smoothly and progressively when the potentiometer is operated.

3.6.4 If the live data value or signal voltage appears to be correct for the position and movement of the component, this indicates that the sensor and circuit are probably good; therefore the problem is possibly caused by faults in other systems e.g. if the throttle position is incorrect, it could be caused by a throttle motor or control unit fault.

4. POSITION SWITCHES E.G. THROTTLE, BRAKE OR CLUTCH

4.1 OPERATION

4.1.1 There are different wiring circuits that can be used to connect a switch to the control unit; Figs 2.3 and 2.4 illustrate the variations. Most of the checks identified in this section refer to the commonly used wiring circuits as illustrated in Fig 2.3 but, Fig 2.4 provides an indication of faults that can occur in the alternative wiring. When checking switches and circuits, it is advisable to identify which type of wiring circuit is used.

4.1.2 Switches normally have two wires, both of which can be connected to the control unit. On some systems, a regulated voltage (typically around 5 volts but some applications may use a different voltage value) originates in the control unit and passes to the switch on one of the wires; the voltage then passes across the switch contacts back to the control unit (Fig 2.3a). Note that on some systems, battery voltage (via a relay or common supply) could be provided to the switch; the control unit will therefore detect full battery voltage when the switch contacts are closed.

It is also possible to use the switch as part of an earth circuit (Fig 2.4a); in these instances, with the switch contacts open, available voltage would be detected at the input terminal of the switch (typically a 5 volt regulated voltage provided by the control unit). When the contacts close, the earth circuit will be completed and the voltage at the switch input terminal (and control unit terminal) will now reduce to zero. It may be necessary to check which type of wiring and voltage level is used in the switch circuit. The following checks relate to the more commonly used wiring circuits where the voltage passes across the contacts and back to the control unit.

Switches normally provide a simple "ON" or "OFF" signal to the control unit i.e. the voltage will either exist in the signal circuit or, the voltage will be zero.

4.2 CIRCUIT

4.2.1 When a fault code definition states "Circuit" or "Circuit Malfunction", without any other information (e.g. low, high), it is likely that control unit has detected an incorrect switch signal value e.g. switch stuck ON or stuck OFF;. However, the control unit might not be programmed to specify the exact nature of the incorrect signal. Although it is possible that an electrical fault exists (sections 4.3, 4.4 and 4.5), it is also possible the switch is operating correctly but the information indicated by the switch is incorrect; this could be caused by the switch / voltage being permanently ON or permanently OFF, but it could also be caused by the switch / voltage being ON or OFF at the incorrect time (section 4.6).

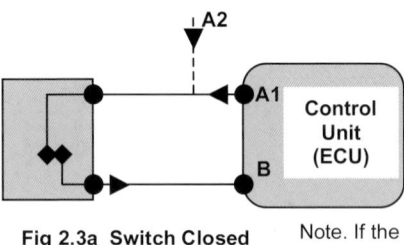

Fig 2.3a Switch Closed

The control unit applies a reference / supply voltage from terminal A1 to the switch circuit (reference voltage is typically 5 volts but can be different on some systems). On some systems, a separate supply (possibly battery voltage) can be applied at A2, instead of the reference voltage. When the switch contacts are closed, the voltage will be detected at Terminal B of the control unit.

Note. If the contacts are closed when they should be open, voltage will exist at terminal B at the incorrect time; this can be regarded as a "Circuit High" Fault

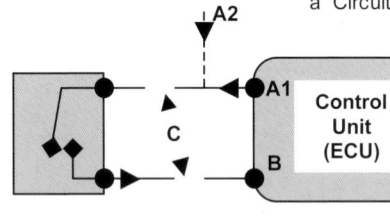

Fig 2.3b Switch Open or an open circuit

If the switch contacts are open, there will be zero voltage at Terminal B; however, a break in the wiring (open circuit) in any part of the circuit (point C) will have the same effect as an open switch. If there is an open circuit or, the switch does not close as expected (therefore no voltage will be detected at Terminal B), this can be regarded as a "Circuit Low" fault.

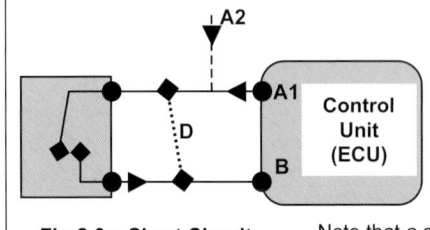

Fig 2.3c Short Circuit

If there is a short circuit between the A and B wires (point D), this is effectively the same as when the contacts are closed i.e. voltage will exist at Terminal B. Therefore a short circuit between A and B or, if the contacts remain closed (not opening as expected), this can be regarded as a as a "Circuit High" fault.

Note that a short to earth at any point in the circuit will result in zero volts at Terminal B (Circuit Low); if the voltage source is a separate battery voltage supply (A2) it is likely to blow a fuse, and if the voltage source is the reference voltage from terminal A1 it will earth the circuit or disable the reference voltage supply.

Fig 2.3 (a, b, c)   Switch circuit where switch output passes to control unit

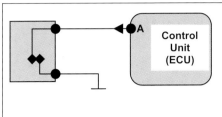

Fig 2.4a Switch Closed or short circuit

The control unit applies a reference voltage from terminal A to the switch circuit (reference voltage is typically 5 volts but can be different on some systems). When the switch contacts are closed, the circuit will be directly connected to earth and the voltage at Terminal A will be zero volts.

A short circuit between the switch input and output wires will have the same effect as if the switch contacts remain permanently closed i.e. voltage at terminal A will be permanently at zero; this could be regarded as a "Circuit Low" Fault.

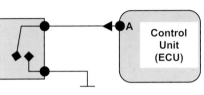

Fig 2.4b Switch Open or an open circuit

If the switch contacts are open, the voltage at terminal A will be 5 volts (or whatever reference voltage is used).

An open circuit in any part of the wiring or switch, will have the same effect as the switch contacts remaining permanently open i.e. the voltage at Terminal A will be permanently at 5 volts (or whatever reference voltage is used); this could be regarded as a "Circuit High" fault.

Fig 2.4 (a, b)   Switch circuit where the switch output passes to earth

4.3   CIRCUIT LOW

4.3.1   When a fault code definition states "Circuit Low", this can indicate that the voltage at the switch output terminal and at the control unit signal terminal (terminal B in Fig 2.3a) is below the normal operating range (refer to Chapter 1 - section 4.2).

4.3.2   A "Circuit Low" fault can be caused by an open circuit in the wiring on the switch input or output circuit (Fig 2.3b), but it could also be caused by the switch contacts not closing / opening correctly. Note that on those systems where the switch completes an earth circuit from the control unit, a short to earth could also result in zero volts on the signal wire.

4.3.3   A "Circuit Low" fault can also be caused if the switch is not operating correctly e.g. the switch is not adjusted correctly. The control unit could be expecting voltage to exist in the switch signal / output circuit when the switch contacts are closed e.g. when the throttle is closed; however, if the switch is not adjusted correctly and the contacts do not close, the voltage will be zero at all times (voltage is therefore Low when it should be High).

4.3.4   Check for any adjustments or settings for the switch and check that the switch contacts close and open when the mechanism operating the switch is in the applicable position e.g. when the throttle or brake pedals are depressed and released.

4.3.5   Check all wiring connected to the switch for open or short circuits. Check for correct voltage being applied to the switch input terminal. Check continuity (zero or almost zero ohms) across the switch contacts when the contacts are closed

4.4   CIRCUIT HIGH

4.4.1   When a fault code definition states "Circuit High", this can indicate that the voltage at the switch output terminal and at the control unit signal terminal (terminal B in Fig 2.3a) is above the normal operating range (refer to Chapter 1 - section 4.2).

4.4.2   A "Circuit High" fault can be caused by a short circuit in the wiring between the switch input and output circuits (Fig 2.3c) or a short to another power supply, but it could also be caused by the switch contacts not closing / opening correctly. Note that on those systems where the switch completes an earth circuit from the control unit, an open circuit in the wiring or switch contacts not closing could also result in a "Circuit High" fault code.

4.4.3   A "Circuit High" fault can also be caused if the switch is not operating correctly e.g. the switch is not adjusted correctly. The control unit could be expecting "zero voltage" to exist in the switch signal / output circuit when the switch contacts are open e.g. when the brake pedal is released (brake off); however, if the switch is not adjusted correctly and the contacts do not open, voltage will exist in the signal circuit at all times (voltage is therefore High when it should be Low).

4.4.4   Check for any adjustments or settings for the switch and check that the switch contacts close and open when the mechanism operating the switch is in the applicable position e.g. when the throttle or brake pedals are depressed and released.

4.4.5   Check all wiring connected to the switch for open or short circuits. Check for correct voltage being applied to the switch input terminal. Check continuity (zero or almost zero ohms) across the switch contacts when the contacts are closed.

4.5   CIRCUIT INTERMITTENT

4.5.1   When a fault code definition states "Circuit Intermittent", this will usually indicate that the voltage at the switch out put terminal or control unit signal terminal (terminal A in Fig 2.3a) is unstable or erratic. Intermittent faults can be caused by loose wiring or connections causing an intermittent short or open circuit; refer to sections 2.3 and 2.4 for additional information. Note that the connection fault could be inside the switch or possibly inside the control unit / connector plug. Wherever possible, check the signal and wiggle the wiring and connections to check for intermittent breaks or short circuits (also refer to 4.6.2).

4.6   CIRCUIT RANGE / PERFORMANCE

4.6.1   When a fault code definition states "Circuit Range / Performance", it indicates that the voltage at the signal terminal (terminal B in Fig 2.3a) is within the normal operating range (refer to Chapter 1 - section 4.2); it is therefore possible that the switch is operating correctly but the information indicated by the switch signal e.g. throttle position is not occurring at the expected time e.g. switch signal indicates throttle fully closed, but other sensors indicate engine is operating at high rpm and high load. It is however also possible that a faulty switch or a short circuit to another wire could result in the signal voltage still being within normal operating range.

4.6.2 If possible, check the switch signal value using a Live Data facility on the diagnostic equipment (the live data might display a signal voltage value or it might indicate the operation of the switch e.g. Brake Pedal On); alternatively, use a multi-meter to check the voltage at the control unit / switch signal terminal (it might be necessary to refer to vehicle specific information to identify correct live data values or voltage values).

Check the switch operation with the live data or switch signal voltage value, check that the switch signal or live data is correct for the conditions e.g. if voltage exists in the circuit at the correct time (when the switch is intended to be open or closed).

If the live data value or signal voltage indicates that the switch is not operating correctly, this could be a possible switch or wiring related fault (refer to sections 4.3 and 4.4) but it can also indicate incorrect switch adjustment or incorrect operation of the mechanism that activates the switch. If the live data value or switch signal voltage indicates that the switch is operating correctly i.e. switch is opening and closing at the correct time, it is possible that a fault exists in the control unit.

5. OTHER TYPES OF SWITCH (E.G. PRESSURE OR LEVEL)

5.1.1 Refer to Section 4 for information on checking switches, but note that a switch related fault code could be activated because the system being monitored by the switch is incorrect. As an example, if a switch is closed because the pressure in a hydraulic system is too low or too high, the control unit could activate a "Circuit Range / Performance" fault code or other code that indicates a possible circuit fault. It is therefore advisable to check the measurement being detected by the switch e.g. hydraulic pressure or fluid level, as well as checking the switch operation and switch circuit.

6. SPEED / POSITION SENSORS (ANALOGUE SIGNAL)

6.1 OPERATION

6.1.1 Analogue type speed / position sensors are commonly used to indicate the engine speed or vehicle speed. Sensors can be detecting speed of rotation of crankshafts, camshafts, and Diesel pumps as well as wheel speed and transmission shaft speed. In addition, by providing specific reference points on a trigger disc, it is possible to identify exact positions of rotation e.g. a reference point on a crankshaft trigger disc could indicate Top Dead Centre (TDC) for a piston in a cylinder (e.g. cylinders 1 & 4 on a 4 cylinder engine). Note that earlier speed / position sensors were located in a distributor body and could be analogue or digital sensors, refer to Speed / Position sensors (digital signal) in section 7.

6.1.2 Analogue speed / position sensors are generally referred to as being passive sensors i.e. they do not contain any electronics within the sensor. The sensors usually operate by inducing an electrical signal into a circuit; this is achieved by passing reference points through a magnetic field. The reference points are located on a trigger disc (toothed wheel or similar device), and the sensor contains a coil of wire, which is coiled around or adjacent to a permanent magnet. When a reference point passes through the sensor, it causes a change in the magnetic field which in turn, induces a small voltage in the coil of wire. Note that when the reference point approaches and then leaves the sensor, it creates a positive and then a negative voltage i.e. an alternating current (AC). Also note that the voltage generally increases with the speed of rotation.

6.1.3 The speed / position sensor provides a series of pulses (one for each reference point). The control unit can therefore monitor the frequency of the pulses to indicate speed of rotation, as well as monitoring the voltage of the signal. If there is a special reference point e.g. a missing tooth on crankshaft trigger discs to indicate a TDC position, this will also be monitored by the control unit.

6.1.4 For some sensors e.g. some camshaft position sensors, the sensor signal can function as a secondary engine speed sensor (crankshaft sensor being the normally primary signal). However, the camshaft sensor will be indicating the relative position between the crankshaft and camshaft (camshaft / valve timing on variable valve timing systems). It is therefore possible that a camshaft sensor fault code is activated because the control unit detects incorrect camshaft position (cam / valve timing incorrect). For this type of position sensor, the following checks still apply but the sensor signal might only provide a position reference, and not speed information.

6.1.5 For some speed sensors (typically wheel speed sensors that can be used to indicate vehicle speed as well as providing speed information for ABS etc), the control unit checks the continuity of the wiring and sensor by briefly providing a small electrical current through the circuit (usually when the ignition switch is initially switched on); this check is in addition to monitoring the sensor signal during normal operation.

6.1.6 Note that analogue speed sensors normally have two wires connecting to the control unit; a signal wire and earth / return path. Sensors can be fitted with a "screen" or shield wire (to prevent interference); the screen wire can in fact be the return circuit for the sensor and is usually wrapped around the signal wire.

6.2 CIRCUIT OR CIRCUIT / OPEN

6.2.1 When a fault code definition states "Circuit", without any other information (e.g. low, high), it is likely that control unit has detected an incorrect sensor signal value. However, the control unit might not be programmed to specify the exact nature of the incorrect signal. Although it is likely that an electrical fault exists (refer to 6.3, 6.4 and 6.5), it is also possible the sensor is operating correctly but the measurement being indicated by sensor is incorrect (refer to section 6.6). If the code definitions states "Circuit / Open", it is likely that the control unit is not detecting the sensor signal and assumes that there is an open circuit (Fig 2.5b) or, the control unit has performed a continuity check and detected an open circuit (refer to sections 6.3 and 6.4).

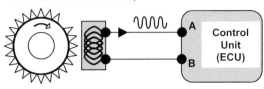

Trigger disc (note missing tooth for master reference)

Fig 2.5a Good Circuit

When the trigger disc rotates, it causes an AC voltage to be induced into the coil of wire (located in the sensor). The AC signal passes to the signal Terminal A. The 2nd wire provides a return / earth path to Terminal B, but note that the 2nd wire can function as a screen or shield to electrical interference and can be wrapped around the signal wire.

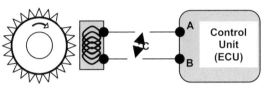

Fig 2.5b Open Circuit in signal or earth / return circuit

If there is an open circuit fault (C) in the signal wire to Terminal A or in the earth / return wire to Terminal B (or an open circuit sensor coil), there will be no signal or voltage at terminal A; this will be regarded as "Circuit Low" fault or a "No Signal" fault.

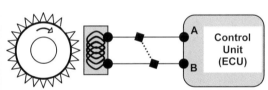

Fig 2.5c Short Circuit

If there is a short circuit between the A and B wires (D), or the A wire is shorted to earth, there will be no signal or voltage at terminal A, this will be regarded as "Circuit Low" fault or a "No Signal" fault. Note that a complete short in the sensor coil could activate the same fault code.

Note. A "Circuit High" fault code is only likely to be activated if there is a short to another wire / circuit that is carrying a voltage.

Fig 2.5 (a, b, c)   Faults in speed / position sensor circuit (inductive sensor with analogue signal)

6.3   CIRCUIT LOW

6.3.1   When a fault code definition states "Circuit Low", this will usually indicate that either the signal frequency or the signal voltage is below the normal operating range (refer to Chapter 1 - section 4.2). If possible, check the signal value using the Live Data facility or with an oscilloscope to identify which aspect of the signal is low e.g. voltage or frequency.

6.3.2   If the signal voltage is low but a signal still exists, it can be caused due to slow rotation speed e.g. slow cranking engine speed. Low signal voltage (Fig 2.6) can also be

caused by a faulty sensor e.g. weak magnet or internally shorted coil (check sensor resistance value), or it can also be caused if the air gap between the sensor and the trigger disc is incorrect (check for any air gap adjustment and check if sensor is fitted correctly). Also check for contamination around the sensor and the trigger disc (e.g. metallic particles).

6.3.3   A "Circuit Low" fault code can also be activated if there is a short or open circuit in the sensor wiring (including the screen wire) or, a short to another circuit or earth (Fig 2.5b and 2.5c).

6.3.4   If the control unit detects a low signal frequency, this can possibly be detected due to information from other sensors providing conflicting information e.g. crankshaft sensor and one camshaft sensor indicate engine speed is at 2000 rpm, but the faulty signal from the 2nd camshaft sensor is indicating a lower speed. If possible, use an alternative method of checking the speed e.g. engine speed or vehicle speed, and compare with the live data values to establish if the sensor is providing an incorrect speed signal.

6.3.5   Check for rotation of the trigger disc; if the trigger disc is not turning with the shaft, the sensor will not be able to produce a signal.

6.3.6   If there are no obvious faults, it is possible that there is a control unit fault.

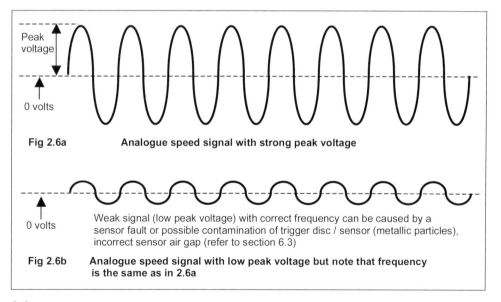

Fig 2.6a   Analogue speed signal with strong peak voltage

Weak signal (low peak voltage) with correct frequency can be caused by a sensor fault or possible contamination of trigger disc / sensor (metallic particles), incorrect sensor air gap (refer to section 6.3)

Fig 2.6b   Analogue speed signal with low peak voltage but note that frequency is the same as in 2.6a

6.4   CIRCUIT HIGH

6.4.1   When a fault code definition states "Circuit High", this will usually indicate that either the signal frequency or signal voltage is above the normal operating range (refer to

Chapter 1 - section 4.2). If possible, check the signal using the Live Data facility (if available) or an oscilloscope to identify which aspect of the signal is high e.g. voltage or frequency.

6.4.2   If the signal voltage is high, it can be caused by a short circuit to another wire / circuit that is carrying a voltage, check for short circuits on the signal wire and on the screen wire. Note that it is unusual for inductive sensors to produce excessive voltage unless the rotation speed is excessive.

6.4.3   A "Circuit High" fault code can possibly be detected due to information from other sensors providing conflicting information e.g. crankshaft sensor and one camshaft sensor indicate engine speed is at 2000 rpm, but the faulty signal from the 2nd camshaft sensor is indicating a higher speed. If possible, use an alternative method of checking the speed of the system being monitored e.g. engine speed or vehicle speed, and compare with the live data values to establish if the sensor is providing an incorrect speed signal.

6.4.4   If there are no obvious faults, it is possible that there is a control unit fault.

6.5   CIRCUIT INTERMITTENT OR INTERMITTENT / ERRATIC

6.5.1   When a fault code definition states "Circuit Intermittent" or "Intermittent / Erratic", this will usually indicate that the signal is unstable or erratic; the signal could be outside of the normal operating range when the fault occurs (refer to Chapter 1 - section 4.2). Intermittent faults can be caused by loose wiring or connections causing an intermittent short or open circuit; refer to sections 6.3 and 6.4 for additional information. Note that the connection fault could be inside the sensor or possibly inside the control unit / connector plug. Wherever possible, check the signal and at the same time wiggle the wiring and connections to check for intermittent breaks or short circuits.

6.5.2   An erratic signal can be caused by fluctuations in the trigger disc due to a mechanical fault e.g. a loose timing belt can cause the camshaft rotation to fluctuate which can then cause an erratic signal form the camshaft sensor. Check any drive mechanism to the system being measured e.g. camshaft drive etc and check that the trigger disc is not loose. Also note that clutch judder on a manual transmission has been known to cause an Intermittent / Erratic speed sensor fault code to be activated, and for automatic transmissions it is possible that poor operation of the gear change clutches can cause judder and erratic signals form the gearbox shaft speed sensors.

6.6   CIRCUIT RANGE / PERFORMANCE

6.6.1   When a fault code definition states "Circuit Range / Performance", it indicates that the signal is within the normal operating range (refer to Chapter 1 - section 4.2); it is therefore possible that the sensor is operating correctly but the value indicated by the sensor signal is not close to the value expected by the control unit, or it conflicts with information from other sensors. It is however also possible that a faulty sensor e.g. weak magnet or internally shorted coil (check sensor resistance value) could cause a Range / Performance code to be activated; also check for a short circuit to another wire that is carrying a voltage.

6.6.2   A Range / Performance fault code can be activated if the sensor signal is indicating a value that is unexpected or does not match the operating conditions also, the signal may not have changed as expected. For example, a crankshaft speed sensor could indicate a constant low engine speed but a camshaft sensor is indicating that the speed has changed.

If available, check the signal value from the sensor using a Live Data facility on the diagnostic equipment (the live data might display frequency or it might display a speed value); alternatively, use an oscilloscope / multi-meter to check the signal. Check that the speed indicated is correct. If the indicated speed appears to be incorrect, check trigger disc for contamination (metallic particles), check the air gap between the sensor and the trigger (check for any air gap adjustment and check if sensor is fitted correctly).

Check for a good signal and check that there is no interference from another circuit e.g. ignition high voltage (check that the sensor screen wire is not damaged or and does not have a short to another wire). Also check for signal fluctuation (refer to section 6.5.2).

6.6.3   Note that it is advisable to check the sensor signals over a wide speed range.

6.7   NO SIGNAL

6.7.1   If a fault code definition includes the phrase "No Signal", there are a number of possible faults preventing the signal from being detected. Although it is likely that a sensor or wiring fault is causing the problem, carry out checks detailed in sections 6.2 to 6.6.

7.   SPEED / POSITION SENSORS (DIGITAL SIGNAL)

7.1   OPERATION

7.1.1   Digital type speed / position sensors are commonly used to indicate the engine speed or vehicle speed (often via a wheel speed sensor, which will also be used for ABS, traction control etc). Sensors can be detecting speed of rotation of crankshafts, camshafts, and Diesel pumps as well as wheel speed and transmission shaft speed. In addition, by providing specific reference points on a rotor or trigger disc, it is possible to identify exact positions of rotation e.g. a reference point on a crankshaft trigger disc could indicate Top Dead Centre (TDC) for a piston in a cylinder (e.g. cylinders 1 & 4 on a 4 cylinder engine).

7.1.2   Digital speed / position sensors often contain electronics within the sensor, and can be referred to as "Active" sensors. There are a number of methods of producing a digital signal but there are two commonly used types i.e. Hall Effect and Magneto Resistive. One main advantage of the digital speed / position systems, is that the signal voltage doesn't increase / decrease with changes in speed; therefore the signal does not become weak at slow speeds.

7.1.3   **The Hall Effect systems** (Fig 2.7) use a change in a magnetic field, which acts on an electronic integrated circuit (Hall chip). The magnetic field (from a fixed magnet) can be blocked thus preventing it from influencing the electronic chip and, by blocking and unblocking the magnetic field, the Hall chip then effectively functions as a switch for the sensor circuit. A rotor or trigger disc, which has a specific number of cut-outs, passes between the magnet and the Hall chip; when the cut outs are aligned with the Hall chip, the magnetic field can then influence the chip causing the switching action.

A supply voltage (typically 5 volts but can vary with different systems) is provided to the Hall chip, along with an earth circuit; when the Hall chip is switching on / off, voltage from the supply passes out from the chip to a third connection in a series of on / off pulses (digital signal as shown in Fig 2.9).

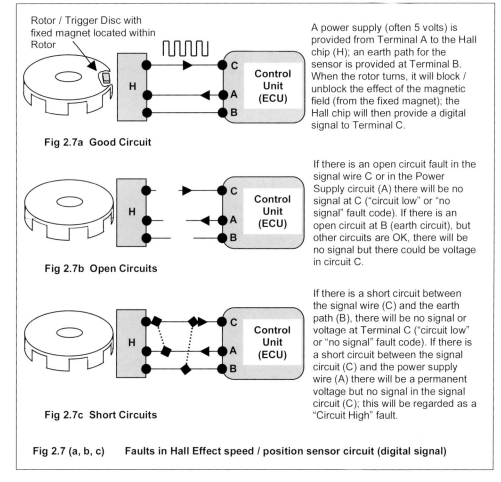

Fig 2.7a Good Circuit

A power supply (often 5 volts) is provided from Terminal A to the Hall chip (H); an earth path for the sensor is provided at Terminal B. When the rotor turns, it will block / unblock the effect of the magnetic field (from the fixed magnet); the Hall chip will then provide a digital signal to Terminal C.

Fig 2.7b Open Circuits

If there is an open circuit fault in the signal wire C or in the Power Supply circuit (A) there will be no signal at C ("circuit low" or "no signal" fault code). If there is an open circuit at B (earth circuit), but other circuits are OK, there will be no signal but there could be voltage in circuit C.

Fig 2.7c Short Circuits

If there is a short circuit between the signal wire (C) and the earth path (B), there will be no signal or voltage at Terminal C ("circuit low" or "no signal" fault code). If there is a short circuit between the signal circuit (C) and the power supply wire (A) there will be a permanent voltage but no signal in the signal circuit (C); this will be regarded as a "Circuit High" fault.

Fig 2.7 (a, b, c)   Faults in Hall Effect speed / position sensor circuit (digital signal)

7.1.4 The Magneto Resistive sensors (Fig 2.8), which are now often used for wheel / vehicle speed sensors, also make use of changes in a magnetic field to produce a signal.

A permanent magnet within the sensor is influenced by the movement of a trigger or tone wheel; the trigger wheel has a specific number of teeth or reference points which influence the magnetic field when they pass the magnet (when the trigger wheel is rotating).

The sensor contains an electronic integrated circuit (micro chip) which is provided with a supply voltage (often in the region of 12 volts); the chip is located close to the magnet. When the trigger wheel rotates and changes the magnetic field, this affects the resistance of part of the integrated circuit. The result is that voltage passes out of the integrated circuit on a second wire but, the voltage output consists of a series of pulses (digital signal as shown in Fig 2.9); the voltage typically oscillates between just under 1 volt to just under 2 volts, therefore it is not strictly an on / off pulse but a pulse that operates between two voltage ranges.

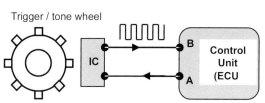

Fig 2.8a Good Circuit

A power supply is provided from terminal A to the integrated circuit / microchip (IC). Note that a magnet is contained within the same assembly as the integrated circuit. When the tone wheel turns, the teeth / reference points cause a change in the magnetic filed which in turn, causes the integrated circuit to produce a pulsed output (digital signal) in circuit B.

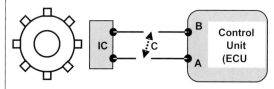

Fig 2.8b Open Circuit in signal or earth / return circuit

If there is an open circuit fault (C) in the signal wire to Terminal B or in the power supply wire from Terminal A (or an open circuit within the integrated circuit), there will be no signal or voltage at terminal B; this will be regarded as "Circuit Low" fault or a "No Signal" fault.

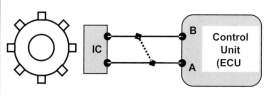

Fig 2.8c Short Circuit

If there is a short circuit between the power supply wire A and the signal wire B, there will be no signal at B and there will be a constant voltage (same as the supply voltage). This could activate a "Circuit High" or "No Signal" fault code. If however wires A or B were shorted to earth, this will cause the voltage in wire B to fall to zero volts, which will probably activate a "Circuit Low" or "No Signal" fault code.

Fig 2.8 (a, b, c)   Faults in Magneto Resistive speed / position sensor circuit (digital signal)

7.1.5 The digital speed / position sensors provide a series of pulses (one for each reference point). The control unit can therefore monitor the frequency of the pulses to indicate speed of rotation, as well as monitoring the voltage of the signal and the power input voltage to the sensor. If there is a special reference point e.g. a missing tooth on crankshaft trigger disc to indicate a TDC position, this will also be monitored by the control unit.

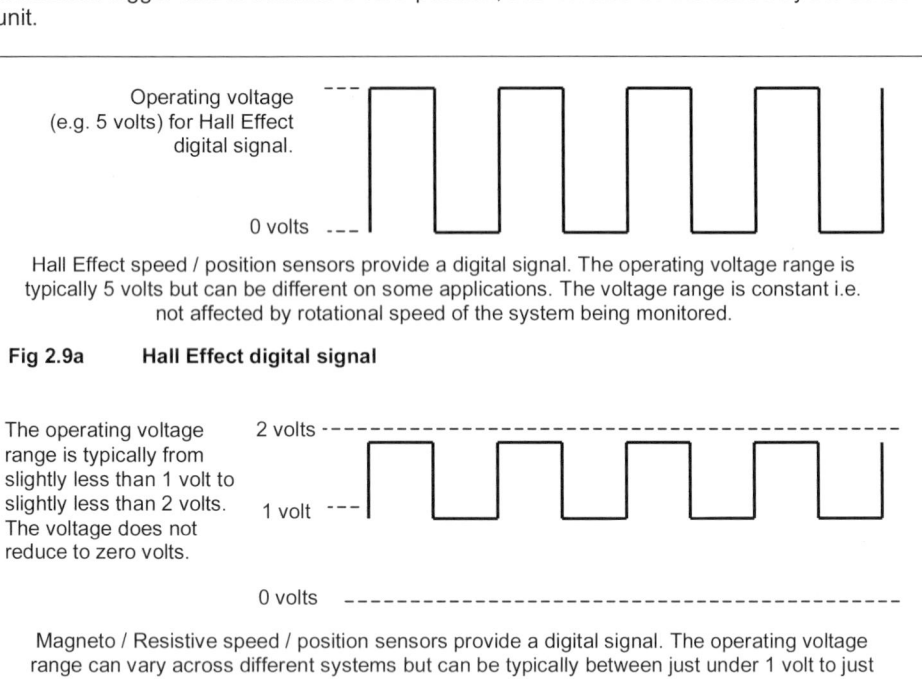

Hall Effect speed / position sensors provide a digital signal. The operating voltage range is typically 5 volts but can be different on some applications. The voltage range is constant i.e. not affected by rotational speed of the system being monitored.

Fig 2.9a   Hall Effect digital signal

The operating voltage range is typically from slightly less than 1 volt to slightly less than 2 volts. The voltage does not reduce to zero volts.

Magneto / Resistive speed / position sensors provide a digital signal. The operating voltage range can vary across different systems but can be typically between just under 1 volt to just under 2 volts. The signal voltage is constant i.e. not affected by rotation speed of the system being monitored.

Note that for both types of sensor, a missing trigger / reference point on the rotor or tone wheel can provide a "Master Position" reference e.g. TDC for Number 1 cylinder, The missing reference will create a "Missing Pulse" on the signal.

Fig 2.9b   Magneto / Resistive digital signal

Fig 2.9   Digital Speed / Position signals (Hall Effect & Magneto / Resistive)

7.1.6 For some sensors e.g. some camshaft position sensors, the sensor signal can function as a secondary engine speed sensor (crankshaft sensor being the normally primary signal). However, the camshaft sensor will be indicating the relative position between the crankshaft and camshaft (camshaft / valve timing on variable valve timing systems). It is therefore possible that a camshaft sensor fault code is activated because the control unit detects incorrect camshaft position (cam / valve timing incorrect). For this

type of position sensor, the following checks still apply but the sensor signal might only provide a position reference, and not speed information.

7.2   CIRCUIT OR CIRCUIT / OPEN

7.2.1 When a fault code definition states "Circuit", without any other information (e.g. low, high), it is likely that control unit has detected an incorrect sensor signal value. However, the control unit might not be programmed to specify the exact nature of the incorrect signal. Although it is likely that an electrical fault exists (refer to 7.3, 7.4 and 7.5), it is also possible the sensor is operating correctly but the measurement being indicated by sensor is incorrect (refer to section 7.6). If the code definitions states "Circuit / Open", it is likely that the control unit is not detecting the sensor signal and assumes that there is an open circuit (Figs 2.7 and 2.8).

7.3   CIRCUIT LOW

7.3.1 When a fault code definition states "Circuit Low", this will usually indicate that either the signal frequency or the signal voltage is below the normal operating range (refer to Chapter 1 - section 4.2). If available, check the signal value using the Live Data facility, or use an oscilloscope / multi-meter to identify which aspect of the signal is low e.g. voltage or frequency. It may be necessary to refer to vehicle specific information to establish the correct voltage for the signal.

7.3.2 Low signal voltage for most digital sensor signals is likely to be caused by a low voltage power supply, poor connections (high resistance connections and wiring) or faulty sensor e.g. weak magnet or internal fault. Also check for contamination around the sensor and the trigger disc (e.g. metallic particles).

A "Circuit Low" fault code can be activated if there is a short or open circuit in the sensor wiring (on the signal wire or supply voltage wire) or, a short to another circuit or to earth (Figs 2.7 and 2.8).

7.3.3 If the control unit detects a low signal frequency, this can possibly be detected due to information from other sensors providing conflicting information e.g. crankshaft sensor and one camshaft sensor indicate engine speed is at 2000 rpm, but the faulty signal from the 2nd camshaft sensor is indicating a lower speed. If possible, use an alternative method of checking the speed e.g. engine speed or vehicle speed, and compare with the live data values to establish if the sensor is providing an incorrect speed signal.

7.3.4 Check for rotation of the trigger disc; if the trigger disc is not turning with the shaft, the sensor will not be able to produce a signal.

7.3.5 If there are no obvious faults, it is possible that there is a control unit fault.

7.4   CIRCUIT HIGH

7.4.1 When a fault code definition states "Circuit High", this will usually indicate that either the signal frequency or the signal voltage is above the normal operating range (refer to Chapter 1 - section 4.2) but, this would also indicate excessive speed of the system being monitored. If possible, check the signal value using the Live Data facility or with an oscilloscope / multi-meter to identify which aspect of the signal is high e.g. voltage or frequency. It may be necessary to refer to vehicle specific information to establish the correct voltage for the signal.

7.4.2 If the signal voltage is high, it can be caused by a short circuit to another wire / circuit that is carrying a voltage, check for short circuits on the signal wire and for the correct supply voltage to the sensor (Figs 2.7 and 2.8).

7.4.3 If the control unit detects a high signal frequency, this can possibly be detected due to information from other sensors providing conflicting information e.g. crankshaft sensor and one camshaft sensor indicate engine speed is at 2000 rpm, but the faulty signal from the 2nd camshaft sensor is indicating a higher speed. If possible, use an alternative method of checking the speed of the system being monitored e.g. engine speed or vehicle speed, and compare with the live data values to establish if the sensor is providing an incorrect speed signal.

7.4.4 If there are no obvious faults, it is possible that there is a control unit fault.

7.5 CIRCUIT INTERMITTENT OR INTERMITTENT / ERRATIC

7.5.1 When a fault code definition states "Circuit Intermittent" or "Intermittent / Erratic", this will usually indicate that the signal is unstable or erratic; the signal could be outside of the normal operating range when the fault occurs (refer to Chapter 1 - section 4.2). Intermittent faults can be caused by loose wiring or connections causing an intermittent short or open circuit; refer to sections 6.3 and 6.4 for additional information. Note that the connection fault could be inside the sensor or possibly inside the control unit / connector plug. Wherever possible, check the signal and at the same time wiggle the wiring and connections to check for intermittent breaks or short circuits.

7.5.2 An erratic signal can be caused by fluctuations in the trigger disc due to a mechanical fault e.g. a loose timing belt can cause the camshaft rotation to fluctuate which can then cause an erratic signal form the camshaft sensor. Check any drive mechanism to the system being measured e.g. camshaft drive etc and check that the trigger disc is not loose. Also note that clutch judder on a manual transmission has been known to cause an Intermittent / Erratic speed sensor fault code to be activated, and for automatic transmissions it is possible that poor operation of the gear change clutches can cause judder and erratic signals from the gearbox shaft speed sensors.

7.6 CIRCUIT RANGE / PERFORMANCE

7.6.1 When a fault code definition states "Circuit Range / Performance", it indicates that the signal is within the normal operating range (refer to Chapter 1 - section 4.2); it is therefore possible that the sensor is operating correctly but the value indicated by the sensor signal is not close to the value expected by the control unit, or it conflicts with information from other sensors. It is however also possible that a faulty sensor e.g. weak magnet or internal fault could cause a Range / Performance code to be activated.

7.6.2 A Range / Performance fault code can be activated if the sensor signal is indicating a value that is unexpected or does not match the operating conditions also, the signal may not have changed as expected. For example, a crankshaft speed sensor could indicate a constant low engine speed but a camshaft sensor is indicating that the speed has changed.

If available, check the signal value from the sensor using a Live Data facility on the diagnostic equipment (the live data might display frequency or it might display a speed value); alternatively, use an oscilloscope / multi-meter to check the signal. Check that the speed indicated is correct. If the indicated speed appears to be incorrect, check trigger disc for contamination (metallic particles).

Check for a good signal and check that there is no interference from another circuit e.g. ignition high voltage (check that the sensor screen wire is not damaged or and does not have a short to another wire). Also check for signal fluctuation (refer to section 7.5.2).

7.7 NO SIGNAL

7.7.1 If a fault code definition includes the phrase "No Signal", there are a number of possible faults preventing the signal from being detected. Although it is likely that a sensor or wiring fault is causing the problem, carry out checks detailed in sections 7.2 to 7.6.

8. PRESSURE / VACUUM SENSORS (ANALOGUE SIGNAL)

8.1 OPERATION

8.1.1 Pressure / Vacuum sensors (Fig 2.10) are used to convert pressure and vacuum values into an electrical signal; for analogue sensors, the electrical signal is a progressive voltage change. Analogue vacuum / pressure sensors usually have three wires: an earth connection, a reference / supply voltage and a signal wire (all three wires are usually connected to the control unit). The reference voltage (regulated voltage of typically 5 volts provided by control unit) is applied to the sensor (which contains electronic components); the earth path is also connected to the sensor and to the control unit. The output voltage / signal is provided from the sensor electronics to the control unit on the third wire. The electronics produce an output voltage signal where the voltage increases and decreases with the changes in vacuum / pressure (Fig 2.11). The control unit will monitor the voltage on the signal wire and fault codes will generally be activated because the signal voltage is incorrect. For most systems, the normal operating range for the signal will reach a maximum of just under 5 volts and a minimum of just above zero volts.

8.2 CIRCUIT OR CIRCUIT / OPEN

8.2.1 When a fault code definition states "Circuit" without any other information (e.g. low, high), it is likely that control unit has detected an incorrect sensor signal value. However, the control unit might not be programmed to specify the exact nature of the incorrect signal. Although it is likely that an electrical fault exists (refer to sections 8.3, 8.4 and 8.5), it is also possible that the sensor is operating correctly but the information being indicated by sensor is incorrect e.g. the vacuum / pressure value is incorrect compared with information from other sensors (also refer to 8.6). If the code definitions states "Circuit / Open", it is likely that the control unit is not detecting the sensor signal and assumes that there is an open circuit (Fig 2.10).

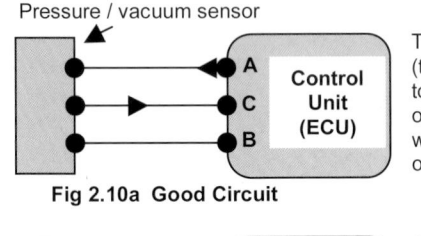

Pressure / vacuum sensor

Fig 2.10a Good Circuit

The control unit applies a reference / supply voltage (typically 5 volts but can be different on some systems) to the sensor from terminal A. Terminal B is the earth or return path. Terminal C is connected to the signal wire; the voltage at the terminal will therefore depend on the pressure / vacuum level.

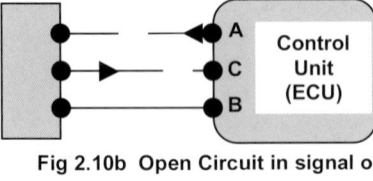

Fig 2.10b Open Circuit in signal or supply circuit

If there is an open circuit fault in the supply circuit A or in the signal circuit C (including an open connection within the sensor) the voltage at terminal C will be zero; this is likely to activate a "Circuit Low" fault code.

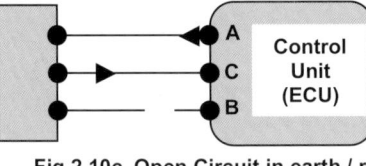

Fig 2.10c Open Circuit in earth / return circuit

If there is an open circuit fault in the earth / return circuit B (including an open circuit in the control unit) the voltage at terminal C will be probably be 5 volts (or whatever supply / reference voltage is used); this is likely to activate a "Circuit High" fault code.

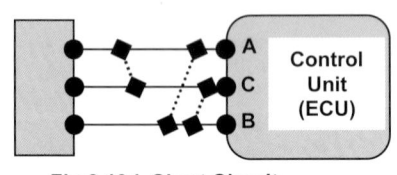

Fig 2.10d Short Circuit

If there is a short circuit between A and C, the voltage at C will be 5 volts (or whatever supply / reference voltage is used); this is likely to activate a "Circuit High" fault code.
If there is a short between A and B or C and B, this is effectively a short to earth, voltage at C will be zero volts; this is likely to activate a "Circuit Low" fault code.

Fig 2.10 (a, b, c, d)   Faults in Pressure / Vacuum sensor circuit

8.3   CIRCUIT LOW

8.3.1   When a fault code definition states "Circuit Low", this will usually indicate that the voltage at the signal wire terminal (Fig 2.10) is below the normal operating range (refer to Chapter 1 - section 4.2). The fact that the voltage is outside the normal operating range indicates a probable circuit or sensor fault, rather than an incorrect vacuum / pressure being indicated.

8.3.2   A "Circuit Low" fault can be caused by an open circuit in the signal wire or the reference / supply voltage wire (Fig 2.10); in both cases voltage is unable to pass through the circuit back to the control unit.

A "Circuit Low" fault can also be caused by a short circuit, either between the signal or reference voltage wires to earth (or to the sensor earth / return wire). Note that a short to another circuit could cause an incorrect circuit resistance which could result in the circuit voltage being too low. If there is an internal fault with the sensor electronics, this could also cause a "Circuit Low" code to be activated.

8.3.3   Also check for correct regulated reference voltage / power supply from control unit, and ensure that the earth path does not have a short or open circuit.

8.4   CIRCUIT HIGH

8.4.1   When a fault code definition states "Circuit High", this will usually indicate that the voltage at the signal wire terminal (Fig 2.10) is above the normal operating range (refer to Chapter 1 - section 4.2). The fact that the voltage is outside the normal operating range indicates a probable circuit or sensor fault, rather than an incorrect vacuum / pressure being indicated.

8.4.2   A "Circuit High" fault can be caused by a short circuit between the signal wire and the reference / supply voltage wire (Fig 2.10). If there is an internal fault with the sensor, this could also cause a "Circuit High" code to be activated

8.4.3   Also check for correct regulated reference voltage / power supply from control unit, and ensure that the earth path does not have a short or open circuit.

8.5   CIRCUIT INTERMITTENT

8.5.1   When a fault code definition states "Circuit Intermittent", this will usually indicate that the voltage at the signal terminal (Fig 2.10) is unstable or erratic; the voltage is likely to be outside of the normal operating range when the fault occurs (refer to Chapter 1 - section 4.2). Intermittent faults can be caused by loose wiring or connections causing an intermittent short or open circuit; refer to sections 8.3 and 8.4 for additional information. Note that the connection fault could be inside the sensor or possibly inside the control unit / connector plug. Wherever possible, check the signal and at the same time wiggle the wiring and connections to check for intermittent breaks or short circuits.

It is also possible that an erratic vacuum / pressure value can cause an erratic sensor signal; this could in some circumstances activate the "Circuit Intermittent" fault code.

8.6   CIRCUIT RANGE / PERFORMANCE

8.6.1   When a fault code definition states "Circuit Range / Performance", it indicates that the voltage at the signal terminal (Fig 2.10) is within the normal operating range (refer to Chapter 1 - section 4.2); it is therefore possible that the sensor is operating correctly but the information indicated by the sensor signal is not close to the value expected by the control unit e.g. sensor information is not plausible or does not match the operating conditions indicated by other sensors. It is however also possible that a faulty sensor or, a short circuit to another wire could result in the signal voltage still being within normal operating range.

8.6.2 If available, check the signal value from the sensor using a Live Data facility on the diagnostic equipment (the live data might display a signal voltage value or it might indicate a vacuum / pressure value; alternatively, use a multi-meter or oscilloscope to check the voltage at the control unit signal terminal (it might be necessary to refer to vehicle specific information to identify correct live data values or voltage values).

8.6.3 Check the pressure of the system being monitored by the sensor using a separate vacuum / pressure gauge, and compare the actual pressure with the live data values or the value indicated by the sensor signal. Note that it is advisable to check the sensor values over a wide operating range. If the live data value or signal voltage does not correspond with the actual pressure, this indicates a possible sensor or wiring related fault (refer to sections 8.3 and 8.4). Check that the sensor signal changes smoothly and progressively when the vacuum / pressure changes.

8.6.4 If possible, apply vacuum / pressure to the sensor to check whether the sensor signal changes correctly with the changes in vacuum / pressure.

8.6.5 If the live data value or signal voltage appears to be correct when compared with actual vacuum pressure values, this indicates that the sensor and circuit are probably good; therefore the problem is possibly caused by faults in system being monitored by the sensor; it will be necessary to compare the actual vacuum / pressure with the expected values to establish if the system vacuum / pressures are incorrect.

9. VACUUM / PRESSURE SENSORS (DIGITAL SIGNAL)

9.1   OPERATION

9.1.1 Vacuum / pressure sensors (Fig 2.10) are used to convert pressure and vacuum values into an electrical signal; for digital sensors, the electrical signal is a typical digital on / off pulse where the frequency of the signal changes with changes in vacuum / pressure. Digital vacuum / pressure sensors usually have three wires: an earth connection, a reference / supply voltage and a signal wire (all three wires are usually connected to the control unit). The reference voltage (regulated voltage of typically 5 volts provided by control unit) is applied to the sensor (which contains electronic components); the earth path is also connected to the sensor and to the control unit. The output signal is provided from the sensor electronics to the control unit on the third wire. The electronics produce an output signal where the frequency increases and decreases with the changes in vacuum / pressure (Fig 2.11). The control unit will monitor the frequency on the signal wire and fault codes will generally be activated because the signal frequency is incorrect, or the voltage range of the signal is incorrect (the signal will typically operate within the range of 0 to 5 volts, depending on the reference / power supply voltage).

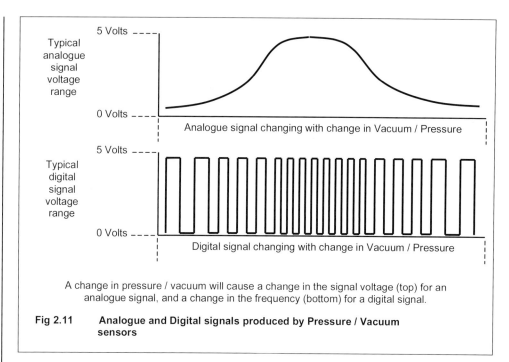

A change in pressure / vacuum will cause a change in the signal voltage (top) for an analogue signal, and a change in the frequency (bottom) for a digital signal.

Fig 2.11   **Analogue and Digital signals produced by Pressure / Vacuum sensors**

9.2   CIRCUIT OR CIRCUIT / OPEN

9.2.1 When a fault code definition states "Circuit" without any other information (e.g. low, high), it is likely that control unit has detected an incorrect sensor signal value. However, the control unit might not be programmed to specify the exact nature of the incorrect signal. Although it is likely that an electrical fault exists (refer to sections 9.3, 9.4 and 9.5), it is also possible that the sensor is operating correctly but the information being indicated by sensor is incorrect e.g. the vacuum / pressure value is incorrect compared with information from other sensors (also refer to 9.6). If the definitions states "Circuit / Open", it is likely that the control unit is not detecting the sensor signal and assumes that there is an open circuit.

9.3   CIRCUIT LOW

9.3.1 When a fault code definition states "Circuit Low", this will usually indicate that the frequency or voltage at the signal wire terminal (Fig 2.10) is below the normal operating range (refer to Chapter 1 - section 4.2). The fact that the signal is outside the normal operating range indicates a probable circuit or sensor fault, rather than an incorrect vacuum / pressure being indicated.

9.3.2 A "Circuit Low" fault can be caused by an open circuit in the signal wire or the reference / supply voltage wire (Fig 2.10); in both cases the signal or voltage is unable to pass through the circuit back to the control unit.

A "Circuit Low" fault can also be caused by a short circuit (Fig 2.10), either between the signal or reference voltage wires to earth (or to the sensor earth / return wire). Note that a short to another circuit could cause an incorrect circuit resistance which could result in the no signal but with a fixed voltage level. If there is an internal fault with the sensor electronics, this could also cause a "Circuit Low" code to be activated.

9.3.3 Also check for correct regulated reference voltage / power supply from control unit, and ensure that the earth path does not have a short or open circuit.

9.4   CIRCUIT HIGH

9.4.1 When a fault code definition states "Circuit High", this will usually indicate that the frequency or voltage at the signal wire terminal (Fig 2.10) is above the normal operating range (refer to Chapter 1 - section 4.2). The fact that the signal is outside the normal operating range indicates a probable circuit or sensor fault, rather than an incorrect vacuum / pressure being indicated.

9.4.2 A "Circuit High" fault can be caused by a short circuit between the signal wire and the reference / supply voltage wire (Fig 2.10). If there is an internal fault with the sensor electronics, this could also cause a "Circuit High" code to be activated

9.4.3 Also check for correct regulated reference voltage / power supply from control unit, and ensure that the earth path does not have a short or open circuit.

9.5   CIRCUIT INTERMITTENT

9.5.1 When a fault code definition states "Circuit Intermittent", this will usually indicate that the signal frequency or voltage at the signal terminal (Fig 2.10) is unstable or erratic; the frequency or voltage is likely to be outside of the normal operating range when the fault occurs (refer to Chapter 1 - section 4.2). Intermittent faults can be caused by loose wiring or connections causing an intermittent short or open circuit; refer to sections 9.3 and 9.4 for additional information. Note that the connection fault could be inside the sensor or possibly inside the control unit / connector plug. Wherever possible, check the signal and at the same time wiggle the wiring and connections to check for intermittent breaks or short circuits.

It is also possible that an erratic vacuum / pressure value can cause an erratic sensor signal; this could in some circumstances activate the "Circuit Intermittent" fault code.

9.6   CIRCUIT RANGE / PERFORMANCE

9.6.1 When a fault code definition states "Circuit Range / Performance", it indicates that the frequency or voltage at the signal terminal (Fig 2.10) is within the normal operating range (refer to Chapter 1 - section 4.2); it is therefore possible that the sensor is operating correctly but the information indicated by the sensor signal is not close to the value expected by the control unit e.g. sensor information is not plausible or does not match the operating conditions indicated by other sensors.

9.6.2 If available, check the signal value from the sensor using a Live Data facility on the diagnostic equipment (the live data might display a signal value or it might indicate a vacuum / pressure value); alternatively, use a multi-meter or oscilloscope to check the voltage at the control unit signal terminal (it might be necessary to refer to vehicle specific information to identify correct live data values or voltage values).

9.6.3 Check the vacuum / pressure of the system being monitored by the sensor using a separate vacuum / pressure gauge, and compare the actual pressure with the live data values or the value indicated by the sensor signal. Note that it is advisable to check the sensor values over a wide operating range. If the live data value or signal voltage does not correspond with the actual pressure, this indicates a possible sensor or wiring related fault (refer to sections 9.3 and 9.4). Check that the sensor signal changes smoothly and progressively when the vacuum / pressure changes.

9.6.4 If possible, apply vacuum / pressure to the sensor to check whether the sensor signal changes correctly with the changes in vacuum / pressure.

9.6.5 If the live data value or signal appears to be correct when compared with actual vacuum pressure values, this indicates that the sensor and circuit are probably good; therefore the problem is possibly caused by faults in system being monitored by the sensor; it will be necessary to compare the actual vacuum / pressure with the expected values to establish if the system vacuum / pressures are incorrect.

10.   AIR FLOW SENSORS

10.1   OPERATION

10.1.1 Air Flow sensors (Fig 2.12) are used to convert the flow of air into an electrical signal; the most common application of air flow sensors is for monitoring the "Mass" of air passing through the engine intake system. Note that the air flow sensor can be combined with an "intake air" temperature sensor, which can provide a separate signal to the control unit (refer to temperature sensor checks on page 21 for information / checks). The signal provided by the air flow sensor can be a progressive voltage change or, a digital signal; the changes in air flow will therefore result in a change in signal voltage or signal frequency.

Air flow sensors usually have at least three wires: an earth connection, a reference / supply voltage and a signal wire (all three wires are usually connected to the control unit). The reference voltage (regulated voltage of typically 5 volts or, battery voltage) is provided by the control unit to the sensor (which contains electronic components); the earth path is also connected to the sensor and to the control unit. The output signal is provided from the sensor electronics to the control unit on the third wire. The electronics produce an output signal where the voltage or frequency increases and decreases with the changes in air flow (Fig 2.13). The control unit will monitor the signal and fault codes will generally be activated because the signal voltage or frequency is incorrect.

If a temperature sensor is combined with the air flow sensor, the sensor could have additional wires (reference voltage / signal wire and earth path); refer to temperature sensor checks in section 2. Note that the earth could be shared with the air flow sensor and in some cases, the temperature sensor signal is fed to the air flow sensor electronics and the value is then integrated in to the air flow sensor signal.

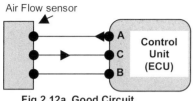

Fig 2.12a Good Circuit

The control unit applies a reference / supply voltage (typically 5 volts but can be different on some systems) to the sensor from terminal A. Terminal B is the earth or return path. Terminal C is connected to the signal wire; the voltage at the terminal will therefore depend on the Air Flow level (mass or volume depending on design).

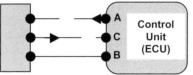

Fig 2.12b Open Circuit in signal or supply circuit

If there is an open circuit fault in the supply circuit A or in the signal circuit C (including an open connection within the sensor) the voltage at terminal C will be zero; this is likely to activate a "Circuit Low" fault code.

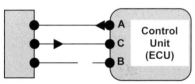

Fig 2.12c Open Circuit in earth / return circuit

If there is an open circuit fault in the earth / return circuit B (including an open circuit in the control unit) the voltage at terminal C will be probably be 5 volts (or whatever supply / reference voltage is used); this is likely to activate a "Circuit High" fault code.

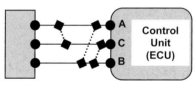

Fig 2.12d Short Circuit

If there is a short circuit between A and C, the voltage at C will be 5 volts (or whatever supply / reference voltage is used); this is likely to activate a "Circuit High" fault code.
If there is a short between A and B or C and B, this is effectively a short to earth, voltage at C will be zero volts; this is likely to activate a "Circuit Low" fault code.

Fig 2.12 (a, b, c, d)   Faults in Air Flow sensor circuit

10.2   CIRCUIT OR CIRCUIT / OPEN

10.2.1 When a fault code definition states "Circuit" without any other information (e.g. low, high), it is likely that control unit has detected an incorrect sensor signal value. However, the control unit might not be programmed to specify the exact nature of the incorrect signal. Although it is likely that an electrical fault exists (refer to sections 10.3, 10.4 and 10.5), it is also possible that the sensor is operating correctly but the information being indicated by the sensor is incorrect e.g. the air flow value is incorrect compared with information from

other sensors (also refer to 10.6). If the code definitions states "Circuit / Open", it is likely that the control unit is not detecting the sensor signal and assumes that there is an open circuit (refer to 10.3 and 10.4).

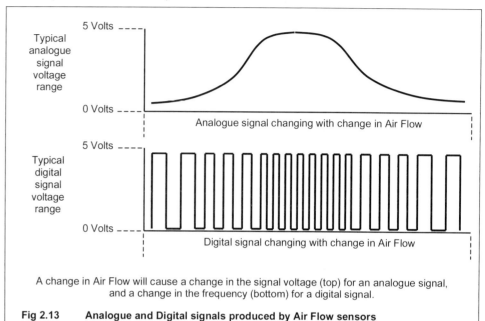

A change in Air Flow will cause a change in the signal voltage (top) for an analogue signal, and a change in the frequency (bottom) for a digital signal.

Fig 2.13   Analogue and Digital signals produced by Air Flow sensors

10.3   CIRCUIT LOW

10.3.1 When a fault code definition states "Circuit Low", this will usually indicate that the frequency or voltage at the signal wire terminal (Fig 2.12) is below the normal operating range (refer to Chapter 1 - section 4.2). The fact that the signal is outside the normal operating range indicates a probable circuit or sensor fault, rather than an incorrect air flow being indicated.

10.3.2 A "Circuit Low" fault can be caused by an open circuit in the signal wire or the reference / supply voltage wire (Fig 2.12); in both cases the signal or voltage is unable to pass through the circuit back to the control unit.

A "Circuit Low" fault can also be caused by a short circuit (Fig 2.12), either between the signal or reference voltage wires to earth (or to the sensor earth / return wire). Note that a short to another circuit could cause an incorrect circuit resistance which could result in "no signal" but with a fixed voltage level. If there is an internal fault with the sensor electronics, this could also cause a "Circuit Low" code to be activated.

10.3.3 Also check for correct regulated reference voltage / power supply from control unit, and ensure that the earth path does not have a short or open circuit.

10.4 CIRCUIT HIGH

10.4.1 When a fault code definition states "Circuit High", this will usually indicate that the frequency or voltage at the signal wire terminal (Fig 2.12) is above the normal operating range (refer to Chapter 1 - section 4.2). The fact that the signal is outside the normal operating range indicates a probable circuit or sensor fault, rather than an incorrect air flow being indicated.

10.4.2 A "Circuit High" fault can be caused by a short circuit between the signal wire and the reference / supply voltage wire (Fig 2.12). If there is an internal fault with the sensor electronics, this could also cause a "Circuit High" code to be activated

10.4.3 Also check for correct regulated reference voltage / power supply from control unit, and ensure that the earth path does not have a short or open circuit.

10.5 CIRCUIT INTERMITTENT

10.5.1 When a fault code definition states "Circuit Intermittent", this will usually indicate that the signal frequency or voltage at the signal terminal (Fig 2.12) is unstable or erratic; the frequency or voltage is likely to be outside of the normal operating range when the fault occurs (refer to Chapter 1 - section 4.2). Intermittent faults can be caused by loose wiring or connections causing an intermittent short or open circuit; refer to sections 10.3 and 10.4 for additional information. Note that the connection fault could be inside the sensor or possibly inside the control unit / connector plug. Wherever possible, check the signal and at the same time wiggle the wiring and connections to check for intermittent breaks or short circuits.

10.6 CIRCUIT RANGE / PERFORMANCE

10.6.1 When a fault code definition states "Circuit Range / Performance", it indicates that the frequency or voltage at the signal terminal (Fig 2.12) is within the normal operating range (refer to Chapter 1 - section 4.2); it is therefore possible that the sensor is operating correctly but the information indicated by the sensor signal is not close to the value expected by the control unit e.g. sensor information is not plausible or does not match the operating conditions indicated by other sensors.

10.6.2 If available, check the signal value from the sensor using a Live Data facility on the diagnostic equipment (the live data might display a signal value or it might indicate the air flow value); alternatively, use a multi-meter or oscilloscope to check the voltage at the control unit signal terminal (it might be necessary to refer to vehicle specific information to identify correct live data values or voltage values).

10.6.3 Check the operation of the sensor by increasing / decreasing the air flow through the system, and check the signal and / or live data values when the air flow changes. If the live data value or signal voltage indicates a lack of response form the sensor, this indicates a possible sensor or wiring related fault (refer to sections 10.3 and 10.4). Check that the sensor signal changes smoothly and progressively when the air flow changes.

11. PROXIMITY SENSORS (MAGNETIC)

11.1 OPERATION

11.1.1 Proximity or movement sensors are used to convert movement into an electrical signal; the sensors can be typically used to monitor minor changes in movement for a mechanism or to detect when a component or mechanism reaches a particular position. The signal from the sensors is usually a voltage output that is either on or off.

There are different types of magnetic proximity sensors. One type, known as a read switch, can use the proximity of a metallic object (which could form part of a linkage or mechanism) to cause a magnetic element of a switch assembly to move; the movement can open or close a switch contact. Other types of proximity sensors can use a similar principle of movement of metallic objects to affect a magnetic field but, the change in magnetic field acts on a electronic circuit which can then function as a switch in a circuit or, produce a small electric pulse / signal.

11.1.2 Proximity sensors can have two connections; a live and return path (or the sensor could form part of an earth circuit). Some sensors could have three connections i.e. a power supply, an earth and a signal wire. Whichever type of sensor is used, the signal is likely to be "on / off" or, a small electrical pulse that can be monitored by a control unit.

11.1.3 For the two wire proximity sensors, the checks for the wiring and circuits will be similar to that for conventional switches; refer to section 4 for additional information. If the sensor has three connections, check for short / open circuits in the wiring, check for a good earth or return path connection and check for a power supply to the sensor (likely to be provided by the control unit and could be a regulated voltage e.g. 5 volts).

12. O2 / LAMBDA SENSORS

12.1 IMPORTANT NOTES

12.1.1 Many Lambda sensor related fault codes are activated because the sensor signal value is incorrect. However, it is important to note that in many cases, the fault is not caused by a faulty sensor but the code is activated because the signal is incorrect due to incorrect oxygen content in the exhaust gas or, the oxygen content is not changing as expected. Therefore, many Lambda sensor fault codes can be activated due to other faults e.g. fuelling problems, engine problems, mechanical problems and exhaust gas leaks etc. Always check for other fault codes and check for other problems.

12.1.2 Sensors can have as many as five connections through to the control unit, it will be necessary to identify the purpose of each connection to ensure that the checks are carried out on the correct wires / connections.

12.2 OPERATION

12.2.1 O2 / Lambda sensors (also referred to as Oxygen sensors) monitor the oxygen contact in the exhaust gas. Although there are a few variations to the basic Lambda sensor, the sensors can use three basic principles to measure the oxygen as briefly outlined below. There are however different applications of the sensors which are also outlined in the following paragraphs.

Note that in general, the signals provided by the sensors oscillate within their defined operating range (ranges are detailed in the descriptions below). The oscillation, which can be typically around 2 cycles a second, is a function of the sensor and the minor changes that occur in the fuelling control

Zirconia type Lambda sensors. In simple terms, many Lambda sensors produce a small voltage / current when oxygen ions pass through a porous ceramic type material (these are referred to as Zirconia type sensors). The ceramic acts as a partition within the sensor; on one side of the partition is the atmosphere (containing approximately 20.8% oxygen) whilst on the other side of the partition is the exhaust gas (typically less than 1% oxygen).

When the ceramic material is heated (using exhaust gas heat and a heating element on many sensors) , because there is a higher amount of oxygen in the atmosphere, oxygen ions will pass through the ceramic to the other side of the partition (in effect trying to achieve the same percentage of oxygen on both sides of the partition). The flow of oxygen ions through the porous ceramic causes a small voltage / current to be produced across the ceramic; by attaching terminals to the ceramic, the voltage / current can be used as a reference to the oxygen passing through the ceramic.

The lower the amount of oxygen in the exhaust gas (richer mixture), the greater is the flow of oxygen ions from the atmosphere; and the greater the oxygen flow, the higher the voltage produced i.e. a rich mixture (low oxygen) produces a higher voltage and a weak mixture (high oxygen) produces a lower voltage. For this type of Lambda sensor, the output voltage is typically between 0.2 and 0.8 volts when the air / fuel ratio (exhaust gas oxygen content) is within the "Lambda window" (normal variations of slightly weak to slightly rich mixture); the optimum value is around 0.45 to 0.5 volts when the Lambda value = 1 (ideal air / fuel ratio).

Titania type Lambda sensors. The Titania type sensors use a slightly different process to produce an electrical signal; the flow of oxygen ions is passed across a different type of material which, again in simple terms, has a resistance value that is affected by the amount of oxygen passing through it (the amount of oxygen passing through depends on how much or little oxygen exists in the exhaust gas). A voltage / current is applied to one side of the material and, when its resistance changes due to different levels of oxygen flowing through it, this affects the voltage passing out on the other side of the material; the output voltage from the material is then used as a reference to the oxygen content in the exhaust gas. For the Titania type sensors, the voltage applied to the sensor is typically either approximately 1 volt or 5 volts; the output signal is therefore typically in the range of 0 to 1 volt or 0 to 5 volts.

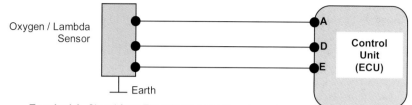

Terminal A. Signal from Zirconia type sensing element.
The sensor is attached to the exhaust system which provides an earth path.
Terminal D. Sensor heater power supply
Terminal E. Sensor heater earth path
The voltage / current in the heater circuit can be regulated by the control unit using a duty cycle / pulse width modulated signal.

Fig 2.14a   Zirconia type Oxygen Lambda sensor wiring (with heater)

Note. Titania type sensors may have 2 wires for the signal circuit (instead of the single wire used on Zirconia types (refer to section 12.2).

Terminal A. Signal from reference cell.
The sensor is attached to the exhaust system which provides an earth path.
Terminals B and C. Positive and Negative current supply for reference cell (refer to section 12.2 – Broadband Sensors)
Terminal D. Sensor heater power supply
Terminal E. Sensor heater earth path
The voltage / current in the heater circuit can be regulated by the control unit using a duty cycle / pulse width modulated signal. Note however, other systems can use a simple on / off control of the heater (possibly switched via a relay).

Fig 2.14b   Broadband type Oxygen Lambda sensor wiring (with heater)

Fig 2.14   Examples of Oxygen / Lambda sensor wiring (refer to applicable paragraphs contained within section12.2 for wiring checks)

Note that Titania sensors, as with Zirconia types, can produce a higher voltage with a richer mixture but, with some Titania sensors the voltage change can be the reverse i.e. higher with a weak mixture and lower with a rich mixture.

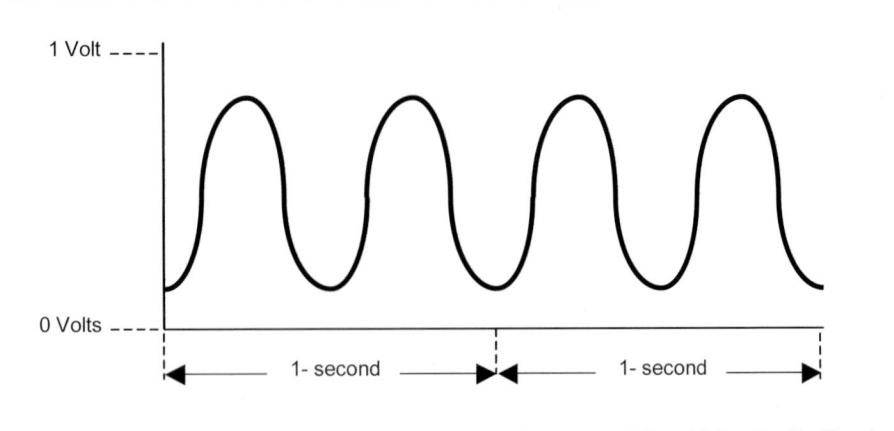

1 Volt - - - -

0 Volts - - - -

|← 1- second →|← 1- second →|

For a Zirconia type sensor, the voltage range is typically between 0.2 and 0.8 volts. For Titania type sensors, the voltage can be between 0 and 1 volt or, on some versions between 0 and 5 volts. Note that for Post-Cat sensors, the voltage range is usually much narrower because the oxygen variation is less.

For broadband sensors, the voltage range will be typically the same as for a standard Zirconia type due to the fact that the control unit provides a "Pumping Current" to maintain the oxygen content (in the reference cell) within the correct limits; the pumping current required to maintain the oxygen content in the reference cell is then used as an indicator of oxygen content in the exhaust gas (air / fuel ratio).

The example shows 2 oscillations / cycles per second, which is a typical frequency for a Zirconia sensor signal under stable operating conditions. Different operating conditions and different sensor types may provide different frequencies.

Fig 2.15   Oxygen / Lambda Sensor signal

Broadband sensors. The normal Zirconia and Titania sensors operate within a small air / fuel ratio range (the Lambda window); this is acceptable for engines that operate within this mixture window. However, for Diesel engines (which operate on a very wide range of air / fuel ratios) and for most Direct Injection petrol engines (which operate with a wider range of air / fuel ratios than non Direct injection engines), the Lambda sensor must be able to monitor a much wider range of air / fuel ratios i.e. oxygen levels have a wider range.

Broadband sensors can use a Zirconia type sensing element as a reference cell but, because the reference cell will be operating outside of its normal range (high oxygen levels due to very weak air / fuel ratios), it is necessary to remove the excess oxygen from the reference cell so that the cell is operating at normal levels. A current is applied to the reference cell which causes the oxygen ions to move out of or into the cell; the amount of current required will depend on the amount of excess oxygen in the reference cell. In summary, a small positive or negative current (milliamps) is applied to the reference cell to maintain the oxygen content at the desired value (Lambda 1); the amount of current required is dependent on the excess oxygen content in the cell (which is dependent on the amount of oxygen in the exhaust gas). The current is referred to as the "Pumping Current" and the amount or value of the pumping current required is an indicator of exhaust gas oxygen content (air / fuel ratio).

Heated and unheated Lambda sensors. For the Lambda sensor to function, it must be at a fairly high temperature. The heat is provided by the exhaust gas but, to help raise the temperature more rapidly (after cold starts), many Lambda sensors have an electrical heater (the heater can also help to stabilise the temperature).

Pre and Post-Cat sensors. A Lambda sensor fitted in front of the main catalytic converter (Pre-Cat) monitors the oxygen content (air / fuel ratio) from the engine; this allows the control unit to make any mixture corrections so that the oxygen content is as close as possible to Lambda 1, which ensures that the catalytic converter is as efficient as possible.

A Post-Cat sensor checks the efficiency of the catalytic converter by monitoring the oxygen content coming out of the converter; this detects whether the catalytic converter has used the oxygen in the process of burning / oxidising excess HC and CO. The oxygen content should be zero or very close to zero and if it is too high, this can indicate an inefficient catalytic converter.

A third Lambda sensor can be fitted behind another catalytic converter (usually the last converter in the exhaust system). The third sensor again monitors the exhaust gas oxygen content coming out of the converter but, it may also be combined with a sensor that monitors Oxides of Nitrogen (NOx); this sensor is effectively a "selective" oxygen sensor (it monitors the oxygen that has been combined with Nitrogen i.e. NOx). The converter operates in the same way as a traditional 3-way catalytic converter but, it stores or accumulates Oxides of Nitrogen; however, at specific intervals, the air / fuel ratio is set to be rich (for a few seconds) and the rich mixture (which is lacking in oxygen) combines with the oxygen in the NOX thus leaving Nitrogen which is effectively harmless. The sensor therefore detects the changes in oxygen during normal running but also during the recovery phase (to ensure that the oxygen is recovered from the NOx).

12.2.2 Depending on which type of Lambda sensor is fitted, it can have just one connection or as many as five connections to the control unit. Note that for many Lambda sensors, the earth path can be through the sensor body to the exhaust pipe and engine. One wire will provide the signal; two wires can be used for the heater power and earth connections, and two wires can be used for the circuit carrying the pumping current on Broadband sensors.

12.3   CIRCUIT (SENSOR SIGNAL FAULT)

12.3.1 When a fault code definition states "Circuit" without any other information (e.g. low, high), it is likely that control unit has detected an incorrect sensor signal value. However, the control unit might not be programmed to specify the exact nature of the incorrect signal. Although it is possible that an electrical fault exists (refer to sections 12.4, 12.5 and 12.6), it is also possible that the sensor is operating correctly but the information being indicated by sensor is incorrect e.g. the oxygen content is incorrect compared with information from other sensors or not as expected (also refer to 12.7).

12.4 CIRCUIT LOW (SENSOR SIGNAL FAULT)

12.4.1 When a fault code definition states "Circuit Low", this will usually indicate that the frequency or voltage at the signal wire terminal is below the normal operating range (refer to Chapter 1 - section 4.2). The fact that the voltage is outside the normal operating range indicates a probable circuit / sensor fault, rather than an incorrect oxygen level.

12.4.2 A "Circuit Low" fault can be caused by an open circuit in the signal wire or, for Titania sensors, an open circuit in the reference / supply voltage wire.

A "Circuit Low" fault can also be caused by a short circuit, either between the signal or reference voltage wires to earth. Note that a short to another circuit could cause an incorrect circuit resistance, which could result in the no signal but with a fixed voltage level. If there is an internal fault with the sensor electronics, this could also cause a "Circuit Low" code to be activated.

On Titania sensors, also check for correct regulated reference voltage / power supply from control unit.

12.4.3 For Broadband sensors, the fault code can refer to the positive or negative current being low; this can also be caused by a short or open circuit in the wires carrying the pumping current or, a sensor fault.

12.5 CIRCUIT HIGH (SENSOR SIGNAL FAULT)

12.5.1 When a fault code definition states "Circuit High", this will usually indicate that the frequency or voltage at the signal wire terminal is above the normal operating range (refer to Chapter 1 - section 4.2). The fact that the voltage is outside the normal operating range indicates a probable circuit or sensor fault, rather than an incorrect oxygen level.

12.5.2 A "Circuit High" fault can be caused by a short circuit between the signal wire and the reference / supply voltage wire.

On Titania sensors, also check for correct regulated reference voltage / power supply from control unit.

12.5.3 For Broadband sensors, the fault code can refer to the positive or negative current being high; this can also be caused by a short or open circuit in the wires carrying the pumping current or, a sensor fault.

12.6 CIRCUIT INTERMITTENT (SENSOR SIGNAL FAULT)

12.6.1 When a fault code definition states "Circuit Intermittent", this will usually indicate that the signal frequency or voltage at the signal terminal is unstable or erratic; the frequency or voltage is likely to be outside of the normal operating range when the fault occurs (refer to

Chapter 1 - section 4.2). Intermittent faults can be caused by loose wiring or connections causing an intermittent short or open circuit; refer to sections 12.4 and 12.5 for additional information. Note that the connection fault could be inside the sensor or possibly inside the control unit / connector plug. Wherever possible, check the signal and at the same time wiggle the wiring and connections to check for intermittent breaks or short circuits.

It is also possible that an erratic oxygen value can cause an erratic sensor signal; this could in some circumstances activate the "Circuit Intermittent" fault code.

12.7 CIRCUIT RANGE / PERFORMANCE (SENSOR SIGNAL FAULT)

12.7.1 When a fault code definition states "Circuit Range / Performance", it indicates that the frequency or voltage at the signal terminal is within the normal operating range (refer to Chapter 1 - section 4.2); it is therefore possible that the sensor is operating correctly but the information indicated by the sensor signal is not close to the value expected by the control unit e.g. sensor information is not plausible or does not match the operating conditions indicated by other sensors.

12.7.2 If available, check the signal value from the sensor using a Live Data facility on the diagnostic equipment; alternatively, use a multi-meter or oscilloscope to check the voltage at the control unit signal terminal (it might be necessary to refer to vehicle specific information to identify correct live data values or voltage values).

12.7.3 Check the signal value at different operating conditions e.g. simulated load, light throttle opening etc. If the live data value or signal voltage does not correspond with the operating conditions or does not respond as expected, this indicates a possible sensor or wiring related fault (refer to sections 9.3 and 9.4).

However, it is also possible that an engine / fuelling related fault or an exhaust leak will cause an incorrect oxygen level and therefore an incorrect sensor signal.

12.8 NO ACTIVITY (SENSOR SIGNAL FAULT)

12.8.1 Where the fault code indicates "No Activity", this can be caused by a sensor or wiring fault (refer to 12.4, 12.5 and 12.6) but, it can also be caused by a constant value of oxygen in the exhaust gas (engine or fuelling related fault). Check for other fault codes and faults that could affect the oxygen content.

12.9 LEAN / STUCK LEAN

12.9.1 The fault code indicates an excess of oxygen, which can be caused by a lean / weak mixture or an exhaust leak. Check for other fault codes or faults that could cause a fuelling fault. Also note that a small misfire will increase the oxygen content but, the misfire should be detected by other sensors, and an appropriate fault code should be activated.

It is also possible that a sensor or wiring fault (short / open circuit) can cause the signal to be biased towards a "Lean" value (refer to 12.4, 12.5 and 12.6).

12.9.2 If the fault is identified as occurring at a particular operating condition e.g. at idle, check systems and components that might only be influencing the mixture under the specified conditions e.g. a small intake leak might only be significant at idle speed, but an air flow sensor might fail under load conditions.

12.10   RICH / STUCK RICH

12.10.1   The fault code indicates a deficiency of oxygen, which can be caused by a rich mixture. Check for other fault codes or faults that could cause a fuelling fault.

It is also possible that a sensor or wiring fault (short / open circuit) can cause the signal to be biased towards a "Rich" value (refer to 12.4, 12.5 and 12.6).

12.10.2   If the fault is identified as occurring at a particular operating condition e.g. at idle, check systems and components that might only be influencing the mixture under the specified conditions e.g. a throttle position sensor might be adjusted incorrectly at closed throttle position (idle speed).

12.11   FUEL TRIM

12.11.1   There are a number of codes referring to fuel trim related faults. The fuel trim is the process where the control unit alters the fuelling within defined limits of adjustment. If the trim process has to be constantly maintained at an extreme end of the tolerance or outside of normal limits to compensate for an incorrect oxygen content / mixture strength, this can be regarded as a trim fault.

The fault code can identify Rich or Lean trim faults, in which case refer to sections 12.9 and 12.10; if the fault code does not specify Rich or Lean, refer to section 12. 7. It is also possible that a trim problem can be caused by a control unit fault.

12.11.2   For Post-Cat sensors (assuming that there is no Pre-Cat sensor or trim fault), the fault could be caused by an air leak between the Pre-Cat and Post-Cat sensors (possibly in the catalytic converter) or the catalytic converter is not operating efficiently (oxygen content Post-Cat is incorrect but Pre-Cat values are good).

12.12   SLOW RESPONSE: SENSOR SIGNAL FAULT

12.12.1   A slow response can refer to the response of the sensor signal when conditions change or, the control unit alters the mixture and the sensor signal response to the change in oxygen is slow. However, the code can refer to the frequency of the sensor signal, which oscillates either side of the ideal value i.e. oscillate from slightly rich to slightly weak; the frequency will vary on different sensors and system but can be in the region of 2 cycles a second.

12.12.2   A slow response can be caused by a faulty sensor but also check that the mixture is changing quickly enough with changes in conditions; simulate a load condition, light load etc and observe the changes by checking the live data or using an oscilloscope / multimeter. If the mixture is slow to respond, checking the fuelling system and sensors; also check operation of catalytic converters and check for exhaust system blockages. Also note that a failed sensor heater can cause a slow response.

12.12.3   Check the frequency of the sensor signal; if it is slow and does not change with changes in fuelling or operating conditions, it could indicate a sensor fault.

12.13   REFERENCE VOLTAGE FAULTS

12.13.1   The term reference voltage is likely to refer to the signal being provided by the reference cell of a broadband sensor (refer to 12.2.1). The code can refer to the "performance" of the reference cell signal, which is identifying that the reference voltage value is incorrect due to an undefined fault. The fault can be caused a faulty sensor or, by short open circuits in the wiring. It also possible that the oxygen content in the exhaust gas is incorrect for the operating conditions and therefore, the reference cell signal will be incorrect; check for any other fault codes or faults that could cause incorrect oxygen content.

12.14   PUMPING CURRENT – NEGATIVE / POSITIVE CURRENT FAULTS

12.14.1   Pumping current faults relate to the current that is required to correct the oxygen content in the Broadband sensor reference cell (refer to section 12.2.1).

12.14.2   A High or Low reference in the fault code definition can indicate that the current level is out of normal operating range and this can be caused by a short / open circuit in the pump current circuits; it is also possible that there is a sensor fault.

12.14.3   It also possible that pumping current faults can be caused due to the oxygen content in the exhaust gas being incorrect for the operating conditions and therefore, the current requirement is out of normal range; check for any other fault codes or faults that could cause incorrect oxygen content.

12.15   O2 / LAMBDA SENSOR HEATER

12.15.1   Lambda sensors can be fitted with an integrated heater to help raise and stabilise the sensor temperature thus allowing the sensor to quickly provide an accurate signal. The heater can be provided with a simple on / off power supply or, the current or voltage can be varied to allow some control over the heater operation.

It may be necessary to identify the wiring and control of the heater, which could be switched on / off using a relay (controlled by the engine control unit) or, the heater could be directly connected to the control unit.

12.15.2   The fault code could indicate a "Circuit", "Circuit Low" or "Circuit High" fault, which is likely to be caused by a short or open circuit in the wiring or a power supply fault. Also check for continuity / correct resistance across the sensor heater (refer to appropriate specifications).

12.15.3   Some fault codes refer to the "Heater Resistance"; this could be detected due to the way in which the control unit monitors the circuit, and the monitoring process has identified an incorrect current or voltage in the circuit which can be caused by an incorrect heater resistance. Check for continuity / correct resistance across the sensor heater (refer to appropriate specifications). Also check for a high resistance or short / open circuit in the wiring.

NOTE. Refer to the list below (section 13) or the contents list on page 19 to select the applicable actuator type; the list identifies actuators either by a specific type (e.g. solenoid) or a task (e.g. injector).

If it is not initially possible to identify the exact type of actuator e.g. how the sensor operates and what type of signal is produced by the sensor, use the information contained in Section 1 below for the initial checks.

The following information provides basic guidelines to tests / checks for components and systems; due to the variations across different manufacturers and systems, it is always advisable to refer to vehicle or component specific information to identify specifications, signal type / values and dedicated test procedure

13. ACTUATORS: IDENTIFICATION AND GENERAL ACTUATOR CHECKS

13.1 ACTUATORS TYPES COVERED IN THIS CHAPTER

13.1.1 This chapter covers the more commonly used actuators, which are listed below; note that most actuators are adaptations of either solenoids or motors and therefore the information provided can be used as a guideline for most actuator checks. However, it is always advisable to refer to vehicle or component specific information.

Solenoids / Solenoid Valves	Page 43
Injectors (Solenoid and Piezo)	Page 44
Motors and Stepper Motors	Page 44
Idle speed control solenoids and motors	Page 47
Heaters	Page 47

13.2 IDENTIFYING THE ACTUATOR TYPE

13.2.1 Vehicle and system manufacturers can use different types of actuators for specific tasks but the many actuators are based on either solenoids or motors. The major differences are often related to the control signals and the wring for the actuators. Wherever possible therefore, it is preferable to refer to circuit diagrams and any available specifications for control signal values.

In many cases, it is however initially difficult to identify a particular actuator type fitted to a vehicle, this can be due to difficult location and / or lack of applicable information. In some cases, prior experience and knowledge may be sufficient to identify an actuator type and its operation; however, it is often necessary to know how the actuator is wired to the system as well as knowing what the control signal values should be. Because it is not always easy to access the vehicle or component specific information, the information in this book does provide a list of basic checks that can be useful, even if it is not initially possible to identify the actuator and wiring.

13.2.2 It is important to note that applying the incorrect tests to an actuator or an actuator circuit can result in incorrect diagnosis but, more importantly, there is always a possibility that damage could be caused to the actuator system or control unit if the incorrect tests are applied.

13.3 BASIC CHECKS FOR ALL ACTUATORS

13.3.1 If the Fault Code definition includes the phrases: "Circuit ", "Circuit / Open", "Circuit High", "Circuit Low" or Circuit Intermittent", it is quite possible that an electrical fault exists in the circuit. Initial checks should therefore include testing for short or open circuits and poor connections on all applicable wiring between the actuator and the control unit or to relays and power supplies / earths. Note that a short circuit can exist between the actuator wiring and another circuit as well as just between actuator wires or shorts to earth. Note that problems can also be caused by a fault in the actuator (short / open circuits etc).

If the Fault Code definition includes the phrase "Circuit Range / Performance", this could still also be caused by a wiring or actuator fault; however, it is also possible that the actuator is operating correctly but the system or component being operated by the actuator is faulty e.g. a solenoid could be operating an idle speed air valve; if there is a blockage in the port passing the idle air, this could affect the idle speed performance or response to the idle valve positioning. Although the fault code refers to the actuator (solenoid) as having a range / performance problem, the fault code could be activated because the idle speed is not responding as expected. It can therefore be necessary to also check the operation of the system being controlled by the actuator.

13.3.2 If available, use the Live Data facility (available on some diagnostic equipment) to obtain an indication of the control signal values being provided by the control unit. Wiring faults e.g. short / open circuits will often cause the control signal to have a zero value or no signal; this is especially true where the control unit is providing a switched earth path for the actuator e.g. injectors. If the earth path / control circuit for the injector is shorted to earth, the voltage in the control circuit will remain at zero volts.

13.3.3 Wherever possible, check for changes in control signal values with changes in operating conditions e.g. by switching on electrical loads or air / conditioning at idle speed, this should cause the control signal to change for the idle valve actuator. Although it is obviously preferable to refer to the applicable specifications and compare signal values against the specified values, it is however possible to detect faults because there is no change in the signal value or the signal is corrupt.

Also refer to Chapter 1 for an indication of how fault code definitions can identify certain types of fault.

14. SOLENOIDS / SOLENOID VALVES

14.1 OPERATION

14.1.1 A solenoid is used to produce linear motion in response to a control signal. Many solenoids produce a very small movement, which can be used to open a small valve e.g. until recently, virtually all fuel injectors were solenoids that opened a small valve / nozzle to allow very small quantities of fuel to pass through; the solenoid and valve movement can be as little as 0.15 mm (0.006 inch). In other applications, a solenoid can be used to operate a linkage or mechanism and the amount of movement could be 1 cm or more.

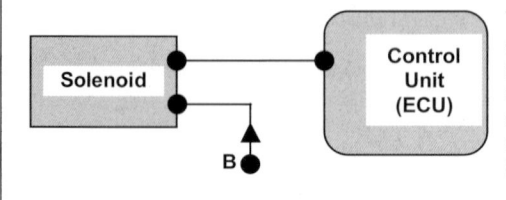

The solenoid receives a power supply from connection B; the power supply is often provided via a relay. The solenoid can be switched on / off (on the earth circuit) by the control unit at Terminal A. Many solenoids are operated using a digital control signal (duty cycle / pulse width modulated).

Fig 2.16a Solenoid circuit with Earth Switching by ECU

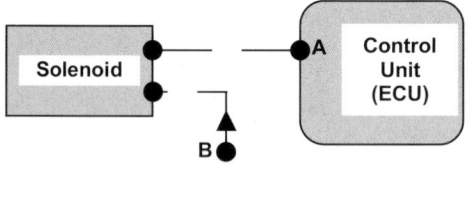

If there is an open circuit fault in the control circuit wire to Terminal A, or in the power supply circuit (B), there will be no signal or voltage at terminal on the earth path to Terminal A, which is likely to be regarded as "Circuit Low" fault. A "Circuit Low" fault code could also be activated if there is no power supply or an open circuit in the solenoid winding.

Fig 2.16b Open Circuit in control circuit or power supply circuit

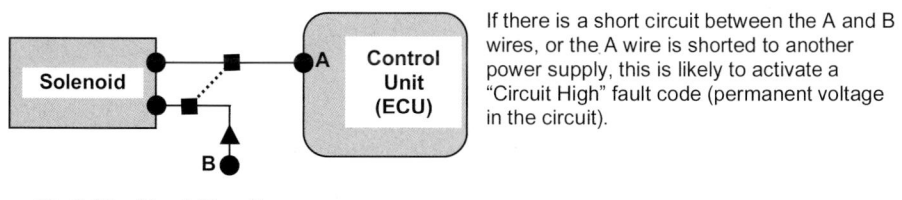

If there is a short circuit between the A and B wires, or the A wire is shorted to another power supply, this is likely to activate a "Circuit High" fault code (permanent voltage in the circuit).

Fig 2.16c Short Circuit

Fig 2.16 (a, b, c)   Faults in Solenoid circuit (earth switched solenoid with separate power supply)

The solenoid operates by placing an armature (e.g. an iron shaft) inside a wire coil. When a current is passed through the coil, it creates a magnetic field which moves the armature (a spring can be used to return the armature to the "rest" position. Depending on which way current flows through the coil, will dictate which direction the armature moves. The greater the current, the more powerful the solenoid or, the more rapidly it will move.

A solenoid can simply be switched on or off; in these cases, the solenoid will therefore either be at the "rest" position (off) or at full movement / stroke of the armature (on). However, in many cases, a solenoid can be controlled using a varying current / voltage, which can then control the position of the armature to any position between "rest" and "full stroke". The varying control signal is often created by controlling the duty cycle (or pulse width) of the control signal (refer to chapter 1, section 3.1.2). This type of solenoid control, allows solenoids to be used to accurately control vacuum valves, which then regulate the vacuum being applied to a vacuum operated actuator e.g. an EGR valve; by varying the vacuum, the amount that the EGR valve opens can therefore be controlled.

14.1.2 Solenoids will usually have two connections: a power supply and an earth connection. Depending on the application of the solenoid and how it is controlled, the power supply could be provided via a relay with a direct earth connection; the control unit could therefore switch on the relay, which in turn will provide the power supply to the solenoid (solenoid will be either on or off). For other applications, the earth path for the solenoid could be switched by the control unit e.g. an injector. The power supply could still be provided by a relay (which would be providing power whilst the engine is running) but, the control unit will be able to switch the injector solenoid on and off via the injector earth circuit. The control unit is able to switch the injector solenoid so that the injector can open for less than two thousandths of a second and, the injector can be opened and closed once and sometimes twice for each complete cycle of one cylinder.

14.2 CIRCUIT OR CIRCUIT / OPEN

14.2.1 When a fault code definition states "Circuit", without any other information (e.g. low, high), it is likely that control unit has detected an incorrect control signal value or a "self test" has identified a circuit fault. However, the control unit might not be programmed to specify the exact nature of the fault. It is likely that an electrical fault exists; refer to sections 14.3, 14.4 and 14.5. If the code definitions states "Circuit / Open", it is likely that the control unit is not detecting any current / voltage in the circuit and therefore assumes that there is an open circuit (refer to sections 14.3 and 14.4).

Note: If a control unit is switching the earth path for the solenoid, when the solenoid is "off" i.e. the control unit is not completing the earth circuit, the power supply voltage will exist in the earth circuit (until the earth circuit is switched to earth); if the voltage is not detected in the circuit then this can be regarded as a "Circuit / Open" fault.

14.3 CIRCUIT LOW

14.3.1 When a fault code definition states "Circuit Low", this will usually indicate that the voltage or value of the control signal (refer to the Note in section 14.2 and refer to Fig 2.18) is below the normal operating range (refer to Chapter 1 - section 4.2). If the voltage in the circuit is constantly low (outside the normal operating range) this could indicate a possible circuit / solenoid fault or, a possible control unit fault.

14.3.2 Depending on whether the control unit switches the earth path or the power supply to the solenoid will dictate how the control unit interprets a "Circuit Low" fault. A "Circuit Low" fault can be caused by an open circuit in the wiring or solenoid; for systems where the control unit switches the power supply, the fault could be caused by a short circuit. The fault could also be caused if there is no power supply to the solenoid.

14.3.3 Check across the solenoid terminals for continuity and the correct resistance; ensure that there is not a short circuit within the solenoid.

14.3.4 If the solenoid resistance is correct and there are no wiring faults, it is possible that there is a control unit fault.

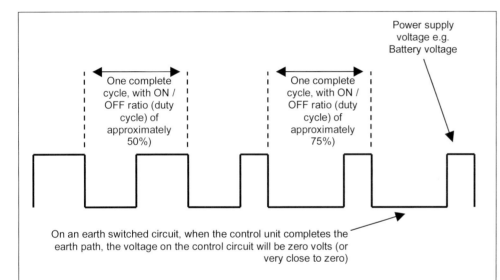

Typical digital control signal for a solenoid. Note that the control signal is Pulse Width Modulated (Duty Cycle control). To alter the average current / voltage in the solenoid circuit, the pulse width is lengthened or shortened (increase or decrease of ON time pulse).

Note that on some applications, small voltage spikes can be produced due to solenoid action; these spikes can often be viewed on the oscilloscope waveform.

For those solenoids that are either switched on or off (without duty cycle control), the voltage in the control circuit (usually the earth circuit) will be either zero volts or supply voltage.

Fig 2.17   Solenoid control signal (duty cycle / pulse width modulated digital signal)

14.4   CIRCUIT HIGH

14.4.1 When a fault code definition states "Circuit High", this will usually indicate that the voltage or value of the control signal is above the normal operating range (refer to Chapter 1 - section 4.2). If the voltage in the circuit is constantly high (outside the normal operating range) e.g. not reducing when the control unit attempts to provide a control signal, this would indicate a possible circuit / solenoid fault or, a possible control unit fault. However, if the control unit is switching the power supply to the solenoid (as opposed to the earth circuit), the fault could be caused by an open circuit in the wiring to the solenoid, within the solenoid or on the solenoid earth circuit.

14.4.2 Depending on whether the control unit switches the solenoid earth path or the power supply, this will dictate how the control unit interprets a "Circuit High" fault. A "Circuit High" fault can be caused by a short circuit in the wiring or solenoid (short to positive) or, for systems where the control unit switches the power supply, the fault could be caused by an open circuit; a short to another positive circuit can also cause a "Circuit High" fault code.

14.4.3 Check across the solenoid terminals for continuity and the correct resistance; ensure that there is not a short circuit within the solenoid.

14.4.4 If the solenoid resistance is correct and there are no wiring faults, it is possible that there is a control unit fault.

14.5   CIRCUIT INTERMITTENT

14.5.1 When a fault code definition states "Circuit Intermittent", this will usually indicate that the voltage or signal value is unstable or erratic; the voltage / signal is likely to be outside of the normal operating range when the fault occurs (refer to Chapter 1 - section 4.2). Intermittent faults can be caused by loose wiring or connections causing an intermittent short or open circuit; refer to sections 14.3 and 14.4 for additional information. Note that the connection fault could be inside the solenoid or possibly inside the control unit / connector plug. Wherever possible, check the signal and at the same time wiggle the wiring and connections to check for intermittent breaks or short circuits.

14.6   CIRCUIT RANGE / PERFORMANCE OR STUCK

14.6.1 When a fault code definition states "Circuit Range / Performance", it indicates that the voltage or signal value is within the normal operating range (refer to Chapter 1 - section 4.2); it is therefore possible that the solenoid is operating correctly but the solenoid response to the control signal is not as expected. An example could be where the control signal should cause the solenoid to operate and open a hydraulic circuit in an automatic transmission system (to change / shift to another gear); if the gear change / shift does not occur or does not occur at the correct speed, it is possible that the control unit would activate the Range / Performance fault code (note that other codes could also be activated that indicate a transmission problem).

It is however also possible that a faulty solenoid or, a short circuit to another circuit (carrying a voltage) could result in the signal still being within normal operating range.

14.6.2 If available, check the control signal value using a Live Data facility on the diagnostic equipment; alternatively, use a multi-meter to check the signal (it might be necessary to refer to vehicle specific information to identify correct live data values or voltage values).

Check the operation of the system being controlled by the solenoid e.g. check operation of vacuum / pressure valves and check that linkages / mechanisms being controlled are free to move and are not affecting solenoid operation or movement.

15. INJECTORS (SOLENOID AND PIEZO)

15.1 OPERATION

15.1.1 Injectors are generally either solenoids or Piezo type. For Solenoid type injectors, refer to section 14 for information relating to checks. Section 14 is also applicable for Piezo type injectors (wiring is effectively the same) but note that on Piezo systems, there is no energising coil. The Piezo type injectors are constructed using many thin layers of Piezo ceramic material; when a voltage is applied to the Piezo layers, it causes a small expansion of the layers, which is used to lift the nozzle of its seat.

15.1.2 For solenoid and Piezo injectors, a control signal is passed to the injector circuit which causes a small movement of a valve or nozzle; the valve will lift off its seat and allow fuel to pass through (typical nozzle movement is as little as 0.15 mm / 0.006 inch).

On some systems (Common Rail Diesel), a valve within the injector body lifts off its seat due to the solenoid action; however, when this valve lifts, it allows High Pressure Fuel in the injector to lift the fuel injector spray Nozzle off its seat (thus allowing fuel to flow through the nozzle). However, for most injectors, the result is the same i.e. a control signal causes the opening of a valve, which either directly or indirectly causes the injector nozzle to open.

Note. On some Diesel systems (commonly used on American Diesels), the solenoid controlled valve is used to control engine oil that has been pumped to a very high pressure; when the solenoid valve opens, the high pressure engine oil is allowed to act on an injector causing it to open. The solenoid valve therefore receives the control signal.

15.1.3 Solenoid Injectors can be switched on the earth circuit with a power supply being provided via a relay (refer to Fig 2.16); the engine control unit will then complete the earth circuit to switch on the injector. However, many injectors have the earth and power supply connected direct to the control unit (refer to Fig 2.18).

Note that some Diesel systems have a separate Diesel Injector control module which acts as the power stage for the injectors.

15.1.4 Refer to section 14 for a list of checks applicable to the different fault codes; note that resistance checks are not normally applicable to Piezo injectors.

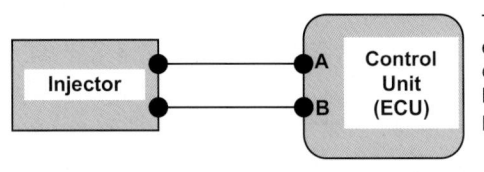

Fig 2.18a Injector circuit with Earth and Power Supply connected via control module

The injector receives a power supply and earth path at terminals A and B. The injector can be switched on / off on the earth circuit by the control unit but on some systems, the power supply circuit can be switched

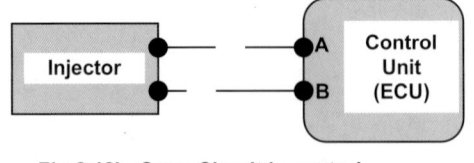

Fig 2.18b Open Circuit in control circuit or power supply circuit

If there is an open circuit fault in the control circuit wire, or in the power supply circuit, there will be no signal or voltage at terminal on the earth path which is likely to be regarded as "Circuit Low" fault. A "Circuit Low" fault code could also be activated if there is no power supply or an open circuit in the solenoid winding.

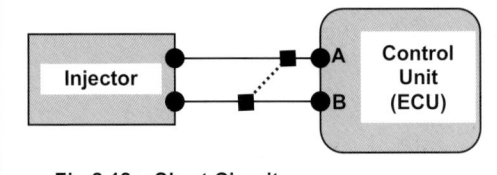

Fig 2.18c Short Circuit

If there is a short circuit between the A and B wires, or the control circuit is shorted to another power supply, this is likely to activate a "Circuit High" fault code (permanent in the circuit).

Fig 2.18 (a, b, c) Faults in Injector circuit (where the earth and power supply are connected to the control module)

16. MOTORS AND STEPPER MOTORS

16.1 OPERATION

16.1.1 Refer to the wiring diagrams shown in this section for examples of different methods for wiring motors; refer to vehicle specific wiring if necessary. If the Control Unit switches a relay, which in turn controls the Motor (refer to Fig 2.19), checks will have to be made on the relay control circuit.

Note. For an example of a motor control signal (often a digital signal with duty cycle / pulse width control), refer to Fig 2.17.

16.1.2 Motors can be used to produce a continuous rotation e.g. a fuel pump or, can be used to position component to a specific position e.g. a throttle position or stepper motor.

Motors that produce continuous rotation generally have two connections i.e. a power supply and earth. Motors that are used for positioning of components can have a number of independent circuits, each of which can be used to enable the motor to rotate through a small defined angle or, to rotate the motor in the opposite direction.

A motor (or one of the motor circuits) can simply be switched on or off; in these cases, the motor will therefore either be at the "rest" position (off) or rotating. However, in many cases, a motor can be controlled using a varying current / voltage, which can then control the speed of the motor. The varying control signal is often created by controlling the duty cycle / pulse width of the control signal (refer to chapter 1, section 3.1.2), and in some applications the polarity in the circuit can be reversed to enable the motor to rotate in the opposite direction.

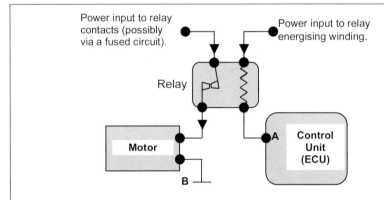

The motor receives a power supply from the relay contacts and the motor is earthed at connection B. The relay energising winding is switched by the control unit at terminal A. If the fault code description refers to the relay, a relay is used to switch the power to the motor, it will be necessary to check the relay operation and wiring as applicable.

Fig 2.19   Motor control circuit using a relay (which is controlled by the ECU)

16.1.3 For conventional motors e.g. motors providing continuous rotation, the power or earth circuit could be via a relay with the other circuit being controlled by a control module or the main control unit. For stepper motors and motors used for positioning of components, both connections to the motor (or to each of the motor circuits) will often be directly connected to the control module or main control unit.

16.2   CIRCUIT OR CIRCUIT / OPEN

16.2.1 When a fault code definition states "Circuit", without any other information (e.g. low, high), it is likely that the control unit has detected an incorrect control signal value or, a "self test" has identified a circuit fault. However, the control unit might not be programmed

to specify the exact nature of the fault. It is likely that an electrical fault exists; refer to sections 16.3, 16.4 and 16.5. If the code definitions states "Circuit / Open", it is likely that the control unit is not detecting any current / voltage in the circuit and therefore assumes that there is an open circuit (refer to sections 16.3 and 16.4).

Note: If a control unit is switching the earth path for the motor, when the motor is "off" i.e. the control unit is not completing the earth circuit, the power supply voltage will exist in the earth circuit (until the earth circuit is switched to earth); if the voltage is not detected in the circuit then this can be regarded as a "Circuit / Open" fault.

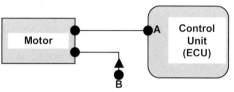

The motor receives a power supply from connection B; the power supply will normally be provided via a relay. The motor is switched on / off on the motor earth circuit by the control unit at terminal A. Many motors are operated using a digital control signal (duty cycle / pulse width modulated).

Fig 2.20a Motor Circuit with Earth Switching by ECU

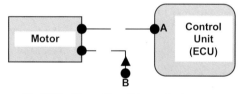

If there is an open circuit fault in the control circuit wire to Terminal A, or in the power supply circuit (B), there will be no signal or voltage at terminal A which is likely to be regarded as "Circuit Low" fault. A "Circuit Low" fault code could also be activated if there is no power supply (relay fault etc) or an open circuit in the solenoid winding.

Fig 2.20b Open Circuit in control circuit or power supply circuit

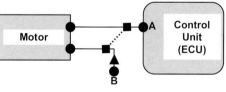

If there is a short circuit between the A and B wires, or the A wire is shorted to another power supply, this is likely to activate a "Circuit High" fault code (permanent voltage being applied at terminal A).

Fig 2.20c Short Circuit

Note. A "Circuit High" fault code is only likely to be activated if there is a short to another wire / circuit that is carrying a voltage.

Fig 2.20 (a, b, c)   Faults in a Motor circuit (earth switched motor with separate power supply via relay or other source)

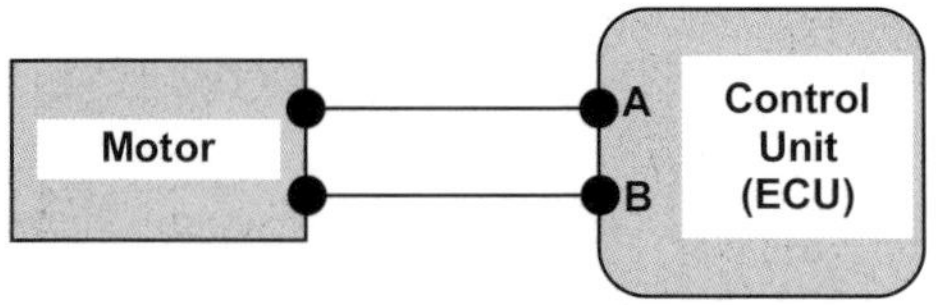

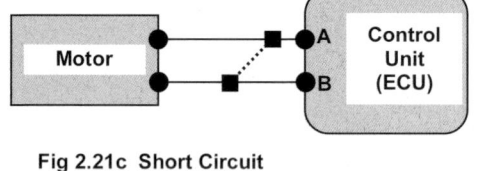

The motor receives a power supply and earth path at terminals A and B. The motor can be switched on / off on the earth circuit by the control unit but on some systems, the power supply circuit can be switched

Fig 2.21a Motor circuit with Earth and Power Supply connected via control module

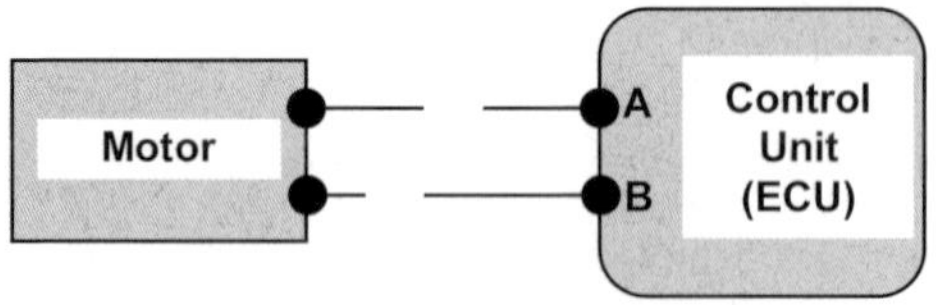

If there is an open circuit fault in the control circuit wire, or in the power supply circuit, there will be no signal or voltage at terminal on the earth path which is likely to be regarded as "Circuit Low" fault. A "Circuit Low" fault code could also be activated if there is no power supply or an open circuit in the solenoid winding.

Fig 2.21b Open Circuit in control circuit or power supply circuit

If there is a short circuit between the A and B wires, or the control circuit is shorted to another power supply, this is likely to activate a "Circuit High" fault code (permanent in the circuit).

Fig 2.21c Short Circuit

Fig 2.21 (a, b, c)   Faults in motor circuit (where the earth and power supply are connected to the control module)

16.3   CIRCUIT LOW

16.3.1 When a fault code definition states "Circuit Low", this will usually indicate that the voltage or value of the control signal (refer to the Note in section 16.2 and refer to Fig 2.20) is below the normal operating range (refer to Chapter 1 - section 4.2). If the voltage in the circuit is constantly low (outside the normal operating range) this could indicate a possible circuit / motor fault or, a possible control unit fault.

16.3.2 Depending on whether the control unit switches the earth path or the power supply (or both) to the motor will dictate how the control unit interprets a circuit low fault. A "Circuit Low" fault can be caused by an open circuit in the wiring or motor or, for systems where the control unit switches the power supply; the fault could be caused by a short circuit .The fault could also be caused if there is no power supply.

16.3.3 Check across the motor terminals for continuity and the correct resistance; ensure that there is not a short circuit within the motor.

16.3.4 If the motor resistance is correct and there are no wiring faults, it is possible that there is a control unit fault.

16.4   CIRCUIT HIGH

16.4.1 When a fault code definition states "Circuit High", this will usually indicate that the voltage or value of the control signal is above the normal operating range (refer to Chapter 1 - section 4.2). If the voltage in the circuit is constantly high (outside the normal operating range) e.g. not reducing when the control unit attempts to provide a control signal, this would indicate a possible circuit / motor fault or, a possible control unit fault. However, if the control unit is switching the power supply to the motor (as opposed to the earth circuit), the fault could be caused by an open circuit in the wiring to the motor, within the motor or on the motor earth circuit.

16.4.2 Depending on whether the control unit switches the earth path or the power supply (or both) to the motor will dictate how the control unit interprets a "Circuit Low" fault. A "Circuit High" fault can be caused by a short circuit in the wiring or motor (short to positive). For systems where the control unit switches the power supply, the fault could be caused by an open circuit; a short to another positive circuit can also cause a "Circuit High" fault code.

16.4.3 Check across the motor terminals for continuity and the correct resistance; ensure that there is not a short circuit within the motor.

16.4.4 If the motor resistance is correct and there are no wiring faults, it is possible that there is a control unit fault.

16.5   CIRCUIT INTERMITTENT

16.5.1 When a fault code definition states "Circuit Intermittent", this will usually indicate that the voltage or signal value is unstable or erratic; the voltage / signal is likely to be outside of the normal operating range when the fault occurs (refer to Chapter 1 - section 4.2). Intermittent faults can be caused by loose wiring or connections causing an intermittent short or open circuit; refer to sections 16.3 and 16.4 for additional information. Note that the connection fault could be inside the motor or possibly inside the control unit / connector plug. Wherever possible, check the signal and at the same time wiggle the wiring and connections to check for intermittent breaks or short circuits.

16.6   CIRCUIT RANGE / PERFORMANCE OR STUCK

16.6.1 When a fault code definition states "Circuit Range / Performance", it indicates that the voltage or signal value is within the normal operating range (refer to Chapter 1 - section 4.2); it is therefore possible that the motor is operating correctly but the motor response to the control signal is not as expected. An example could be where the control signal should cause the motor to operate and open a hydraulic circuit in an automatic transmission system (to change / shift to another gear); if the gear change / shift does not occur or does not occur at the correct speed, it is possible that the control unit would activate the "Range

/ Performance" fault code (note that other codes could also be activated that indicate a transmission problem).

It is however also possible that a faulty solenoid or, a short circuit to another circuit (carrying a voltage) could result in the signal still being within normal operating range.

16.6.2 If available, check the control signal value using a Live Data facility on the diagnostic equipment; alternatively, use a multi-meter to check the signal (it might be necessary to refer to vehicle specific information to identify correct live data values or voltage values).

Check the operation of the system being controlled by the motor e.g. check operation of vacuum / pressure valves and check that linkages / mechanisms being controlled are free to move and are not affecting motor operation or movement.

17. IDLE SPEED CONTROL SOLENOIDS AND MOTORS

17.1 OPERATION

17.1.1 Solenoid based idle speed control valves can be used as a simple devices to allow additional air through a port; the valve could therefore be either open or closed but many idle speed control valves will be controlled so that they can accurately be positioned and therefore regulate the air flow at idle speed. The solenoid will be controlled either with an On / Off signal or, more likely it will be controlled using a digital control signal with duty cycle / pulse width modulation as illustrated in Fig 2.17.

Whichever type of solenoid control is used, the solenoid is likely to have two wiring connections. Refer to section 14 for checks and additional information.

17.1.2 There are generally two types of idle speed control motors. The older types used a motor to control an air valve; the valve then controls the air passing through a throttle by-pass port. Depending on the design, the motor could rotate through a limited angle e.g. 60 degrees of rotation, to alter the valve position, whilst other types used a motor that rotated a number of turns and connected to a linkage e.g. a threaded shaft that opened / closed the air valve (note that these motors were often "Stepper Motors" which allowed them to be stopped in a wide range of positions.

It is now normal on modern vehicles for motors to be used to adjust the position of the throttle butterfly / plate. The idle speed is therefore controlled using angular position of the throttle butterfly.

For all idle speed control valves that are actuated by motors, refer to section 16 for checks and additional information and refer to Fig 2.17 for an example of the control signal.

18. ELECTRIC HEATERS / HEATING ELEMENTS

18.1 OPERATION

18.1.1 Heaters / Heating Elements can be used for many applications including: heaters for Oxygen / Lambda sensors, Diesel Glow Plugs, Cooling System thermostats, Catalytic Converters, Intake Air etc.

18.1.2 Heaters / heater elements generally operate by applying a current through a specific material (usually a metallic based material); the current flow causes the material to increase in temperature / heat. Depending on the application of the heater element, the current / voltage supply could be simply on or off. However, for many applications, the voltage / current supply to the heater is regulated so that the heat produced by the element is also regulated. Voltage / current control can be achieved using a digital control signal with duty cycle / pulse width control (refer to Chapter 1, section 4.2.5).

It is often the case that a high current is allowed to flow through the heating element initially (to create a rapid increase in temperature), and the current is then reduced sufficiently to maintain a desired or stable temperature.

Heating elements are often constructed using a material that increases its resistance with the increase in temperature (Positive Temperature Co-efficient or PTC). With the PTC type elements, a fixed voltage / current can be applied and when the element is cold (low resistance) a high current will flow thus causing rapid heating; due to the increase in temperature, the element resistance increases thus causing a reduction in current flow which can prevent the element from overheating and being damaged, and it also helps maintain a stable or desired temperature. Some types of Diesel Glow Plug make use of PTC type current control.

18.1.3 Heating elements will usually have two connections: a power supply and an earth connection; it is possible to control either the power supply or the earth connection for switching on / off the circuit or for controlling the current flow. Note that heaters can use the heater body as an earth path; this is normal practice for Diesel Glow Plugs.

18.1.4 Heater elements can be switched on/ off by a relay (in the earth or the power supply circuit). However, where current control is used to regulate the current passing to the heater element, the applicable control unit or control module can be used to produce the control signal e.g. a Diesel Glow Plug Control Module.

18.2 CIRCUIT OR CIRCUIT / OPEN

18.2.1 When a fault code definition states "Circuit", without any other information (e.g. low, high), it is likely that the control unit has detected an incorrect control signal value or a "self test" has identified a circuit fault. However, the control unit might not be programmed to specify the exact nature of the fault. It is likely that an electrical fault exists; refer to sections 18.3, 18.4 and 18.5. If the code definitions states "Circuit / Open", it is likely that the control unit is not detecting any current / voltage in the circuit and therefore assumes that there is an open circuit (refer to sections 18.3 and 18.4).

Note: If a control unit is switching the earth path for the heater element, when the heater element is "off" i.e. the control unit is not completing the earth circuit, the power supply voltage will exist in the earth circuit (until the earth circuit is switched to earth); if the voltage is not detected in the circuit then this can be regarded as a "Circuit / Open" fault.

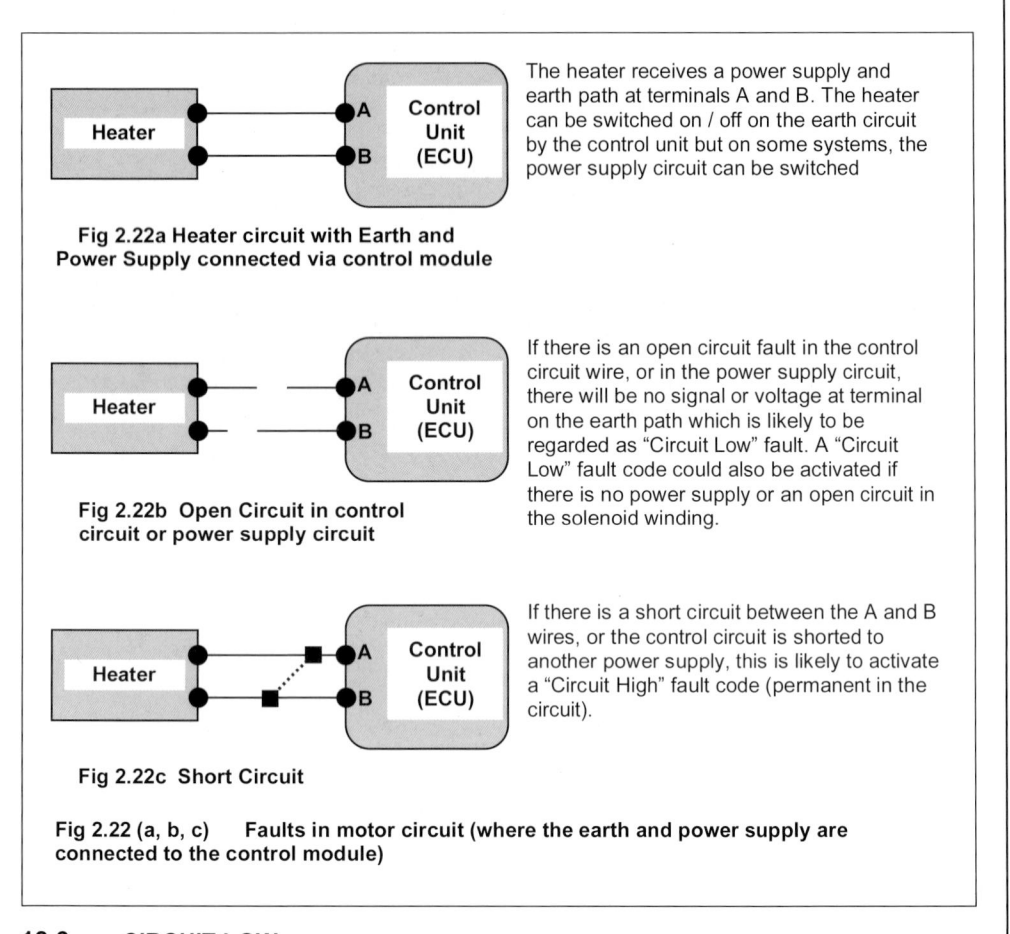

The heater receives a power supply and earth path at terminals A and B. The heater can be switched on / off on the earth circuit by the control unit but on some systems, the power supply circuit can be switched

Fig 2.22a Heater circuit with Earth and Power Supply connected via control module

If there is an open circuit fault in the control circuit wire, or in the power supply circuit, there will be no signal or voltage at terminal on the earth path which is likely to be regarded as "Circuit Low" fault. A "Circuit Low" fault code could also be activated if there is no power supply or an open circuit in the solenoid winding.

Fig 2.22b Open Circuit in control circuit or power supply circuit

If there is a short circuit between the A and B wires, or the control circuit is shorted to another power supply, this is likely to activate a "Circuit High" fault code (permanent in the circuit).

Fig 2.22c Short Circuit

Fig 2.22 (a, b, c)   Faults in motor circuit (where the earth and power supply are connected to the control module)

18.3   CIRCUIT LOW

18.3.1 When a fault code definition states "Circuit Low", this will usually indicate that the voltage or value of the control signal (refer to the Note in section 18.2 and refer to Fig 2.22) is below the normal operating range (refer to Chapter 1 - section 4.2). If the voltage in the circuit is constantly low (outside the normal operating range) this could indicate a possible circuit / heater element fault or, a possible control unit fault.

18.3.2 Depending on whether the control unit switches the earth path or the power supply to the solenoid, this will dictate how the control unit interprets a "Circuit High" fault. A "Circuit Low" fault can be caused by an open circuit in the wiring or heater element; for systems where the control unit switches the power supply, the fault could be caused by a short circuit (e.g. short to earth). The fault could also be caused if there is no power supply to the heater element.

18.3.3 Check across the heater element terminals for continuity and the correct resistance; ensure that there is not a short circuit within the heater element. If the element resistance is correct and there are no wiring faults, it is possible that there is a control unit fault.

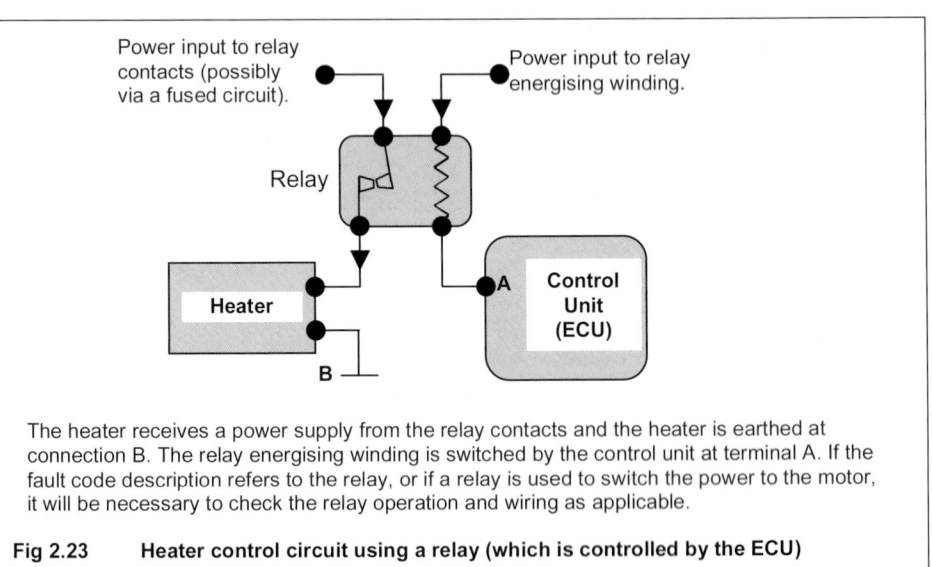

The heater receives a power supply from the relay contacts and the heater is earthed at connection B. The relay energising winding is switched by the control unit at terminal A. If the fault code description refers to the relay, or if a relay is used to switch the power to the motor, it will be necessary to check the relay operation and wiring as applicable.

Fig 2.23   Heater control circuit using a relay (which is controlled by the ECU)

18.4   CIRCUIT HIGH

18.4.1 When a fault code definition states "Circuit High", this will usually indicate that the voltage or value of the control signal is above the normal operating range (refer to Chapter 1 - section 4.2). If the voltage in the circuit is constantly high (outside the normal operating range) e.g. not reducing when the control unit attempts to provide a control signal, this would indicate a possible circuit / heater element fault or, a possible control unit fault. However, if the control unit is switching the power supply to the heater element (as opposed to the earth circuit), the fault could be caused by an open circuit in the wiring to the heater element, within the heater element or on the heater element earth circuit; a short to another positive circuit can also cause a "Circuit High" fault code.

18.4.2 Depending on whether the control unit switches the earth path or the power supply to the heater element will dictate how the control unit interprets a "Circuit High" fault. A "Circuit High" fault can be caused by a short circuit in the wiring or heater element (short to positive) or, for systems where the control unit switches the power supply, the fault could be caused by an open circuit.

18.4.3 Check across the heater element terminals for continuity and the correct resistance; ensure that there is not a short circuit within the heater element. If the heater resistance is correct and there are no wiring faults, it is possible that there is a control unit fault.

18.5   CIRCUIT INTERMITTENT

18.5.1 When a fault code definition states "Circuit Intermittent", this will usually indicate that the voltage or signal value is unstable or erratic; the voltage / signal is likely to be outside of the normal operating range when the fault occurs (refer to Chapter 1 - section 4.2). Intermittent faults can be caused by loose wiring or connections causing an intermittent short or open circuit; refer to sections 18.3 and 18.4 for additional information. Note that the connection fault could be inside the heater element or possibly inside the control unit / connector plug. Wherever possible, check the signal and at the same time wiggle the wiring and connections to check for intermittent breaks or short circuits.

18.6   CIRCUIT RANGE / PERFORMANCE OR STUCK

18.6.1 When a fault code definition states "Circuit Range / Performance", it indicates that the voltage or signal value is within the normal operating range (refer to Chapter 1 - section 4.2); it is therefore possible that the heater element is operating correctly but the heater element response to the control signal is not as expected or the control signal is not altering as expected. There are very few Range / Performance fault codes relating to heaters due to the difficulty of monitoring the heater response; it is therefore likely that the fault will relate to an electrical problem, refer to sections 18.3 and 18.4 and note that it is possible that a faulty heater element or, a short circuit to another wire could result in the signal still being within normal operating range.

18.6.2 If available, check the control signal value using a Live Data facility on the diagnostic equipment; alternatively, use a multi-meter to check the signal (it might be necessary to refer to vehicle specific information to identify correct live data values or voltage values).

Check the operation of the heater element and check that there is no other fault that could influence heater element operation e.g. a build up of deposits on the heater element.

19.   DIESEL GLOW PLUGS

19.1.1 Note. Diesel Glow plugs are heaters and therefore covered in section 18. Glow plugs are primarily used to aid starting on cold engines by providing rapid and supplementary heating within the combustion chamber; note however that some modern systems also use the glow plugs under post-start conditions as well as pre-start.
Note that there are many different types and versions of Diesel Glow Plugs, and it is therefore essential to ensure that replacement units are correct for the vehicle. If the engine suffers with starting or operating faults (primarily on cold engines) it is also advisable to check that the Glow Plugs previously fitted are not incorrect.

20.   RELAYS

20.1   OPERATION

20.1.1 Relays are generally a relative simple device that enables a low current or low voltage circuit to switch a high current / high voltage circuit e.g. a relay can be switched using a low current circuit controlled by an engine control unit, and the relay then switches on the high current to an electric fuel pump. Relays are often remotely mounted but, relays can be integrated into a control unit or module.

20.1.2 In general, relays make use of an "energising winding" or coil through which a low current is passed; when the low current passes through the winding, this creates a magnetic field that causes a set of contacts to close. The contacts then carry the high current to the actuator e.g. an actuator such as a fuel pump or a set of fuel injectors.

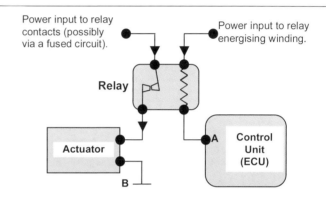

A relay is generally used to provide power in a high current circuit, this allows the control unit (or a simple switch) to control the low current circuit (the energising winding circuit). The above example shows a simple relay and wiring but note that there are many different types of relays and different wiring arrangements; it may therefore be necessary to refer to vehicle specific information to identify the relay and wiring.
In the above example, the actuator receives a power supply from the relay contacts and the actuator is earthed at connection B. The relay energising winding is switched by the control unit at terminal A. If the fault code description refers to the relay or, a relay is used to switch the power to the actuator, it will be necessary to check the relay operation and wiring as applicable.

Fig 2.24   A simple relay and wiring

20.1.3 There are many variations of relay and they can contain more than one set of contacts and more than one energising winding. Also note that a set of contacts could in fact be "double contacts" i.e. a central contact is fed with the high current power supply and when the energising winding is off (central contact in the rest position), the high current could pass from the central contact to one output terminal; when the energising winding is then switched on, the central contact can move away from the first output terminal (breaking the circuit) and the central contact will then connect to a second output terminal (completing a second circuit). It may therefore be necessary to refer to vehicle specific information to identify the type of relay and the wiring connected to the relay.

20.1.4 A fault code could refer to a "Relay Control Circuit" which is likely to refer to the circuit for the energising winding (often controlled by the applicable control unit or module which can switch the earth path or the power circuit for the energising winding). However, a fault code could refer to the circuit being switched by the relay (the high current circuit), in which case the fault code description may refer to the component or circuit receiving the high current e.g. a starter motor circuit.

If however a fault code description does not specify the exact relay circuit, a fault code could refer to the high current circuit (power supply input / output circuits and relay contacts) or to the low current circuit (energising winding circuit).

20.1.5 Note that there are references to a sense circuit for some fault codes. The control unit could have a connection to the output of the relay (high current output from the relay to the component); this will provide the control unit with an indication of whether the relay has switched on the high current circuit.

20.2   CIRCUIT OR CIRCUIT / OPEN

20.2.1 When a fault code definition states "Circuit", without any other information (e.g. low, high), it is likely that the control unit has detected an incorrect voltage / value in the applicable relay circuit. However, the control unit might not be programmed to specify the exact nature of the fault. It is likely that an electrical fault exists; refer to sections 20.3, 20.4 and 20.5. If the code definitions states "Circuit / Open", it is likely that the control unit is not detecting any current / voltage in the applicable circuit and therefore assumes that there is an open circuit (refer to sections 20.3 and 20.4).

Note: If a control unit is switching the earth path for the relay energising winding, when the control unit has switched the circuit off i.e. not completing the earth circuit, the power supply voltage will still exist in the earth circuit (until the earth circuit is switched to earth); if the voltage is not detected in the circuit, this can be regarded as a "Circuit / Open" fault.

20.3   CIRCUIT LOW

20.3.1 When a fault code definition states "Circuit Low", this will usually indicate that the voltage in the circuit (refer to the Note in section 20.2 and refer to Fig 2.24) is below the normal operating range (refer to Chapter 1 - section 4.2). If the voltage in the circuit is constantly low (i.e. does not increase as expected when the relay operates or when the control unit switches on/ off the energising winding circuit), this could indicate a possible circuit / energising winding fault or, a possible fault with the operation of the relay contacts and the high current circuit; if the fault code refers to the high current circuit, the fault could be detected due to a lack of output voltage / current from the relay which could be caused by a relay or wiring fault.

20.3.2 Depending on whether the control unit switches the earth path or the power supply to the relay energising winding, this will dictate how the control unit interprets a "Circuit Low" fault. A "Circuit Low" fault can be caused by an open circuit in the wiring or energising winding. However, if the control unit switches the power supply for the winding, the fault could be caused by a short circuit (e.g. short to earth). The fault could also be caused if there is no power supply to the energising winding (earth switched circuits).

20.3.3 Check across the energising winding terminals for continuity and the correct resistance; ensure that there is not a short circuit within the relay.

20.4   CIRCUIT HIGH

20.4.1 When a fault code definition states "Circuit High", this will usually indicate that the voltage in the circuit is above the normal operating range (refer to Chapter 1 - section 4.2). If the voltage in the circuit is constantly high (i.e. does not decrease as expected when the relay operates or when the control unit switches on / off the energising winding circuit), this could indicate a possible circuit / energising winding fault or, a possible fault with the operation of the relay contacts and the high current circuit.

20.4.2 Depending on whether the control unit switches the earth path or the power supply to the energising winding, this will dictate how the control unit interprets a "Circuit High" fault. A "Circuit High" fault can be caused by a short circuit in the wiring (short to positive) or, for systems where the control unit switches the power supply, the fault could be caused by an open circuit.

20.4.3 Check across the energising winding terminals for continuity and the correct resistance; ensure that there is not a short circuit.

20.5   CIRCUIT INTERMITTENT

20.5.1 When a fault code definition states "Circuit Intermittent", this will usually indicate that the voltage in the circuit is unstable or erratic. Intermittent faults can be caused by loose wiring or connections causing an intermittent short or open circuit; refer to sections 20.3 and 20.4 for additional information. Note that the connection fault could be inside the relay or possibly inside the control unit / connector plug. Wherever possible, check the voltage and at the same time wiggle the wiring and connections to check for intermittent breaks or short circuits.

20.6   CIRCUIT RANGE / PERFORMANCE OR STUCK

20.6.1 When a fault code definition states "Circuit Range / Performance", it indicates that the voltage is within the normal operating range (refer to Chapter 1 - section 4.2) but the relay response to the control signal from the control unit is not as expected or the voltage not altering as expected. The fault is likely to relate to either the wiring or to a relay problem. Refer to sections 20.3 and 20.4 for indications of wiring checks. Also check that the relay contacts open / close when the relay energising winding is switched. Check energising winding resistance and check for high resistance or non closure of relay contacts.

21. CAN-BUS

21.1 OPERATION

21.1.1 IMPORTANT NOTE. The following information is provided as a basic guide to CAN-Bus system operation. There are many variations of CAN-Bus systems and it may therefore be necessary to refer to vehicle specific information for relevant details of operation and testing.

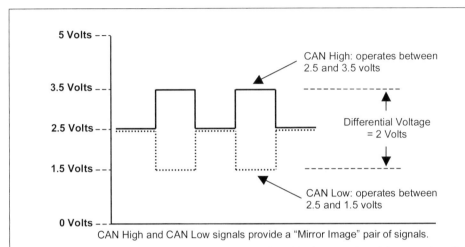

CAN High: operates between 2.5 and 3.5 volts

Differential Voltage = 2 Volts

CAN Low: operates between 2.5 and 1.5 volts

CAN High and CAN Low signals provide a "Mirror Image" pair of signals.

The two Mirror Image signals (in a twisted pair circuit) enable the control modules to analyse the data, by monitoring the voltage difference between CAN High & CAN Low (2 volts in this example). CAN High & CAN Low can both be affected by interference e.g. a positive voltage spike; therefore, interference on CAN High will be replicated on the CAN Low signal. Interference in CAN High would be counteracted by interference in CAN Low and the voltage difference would remain at 2 volts i.e. a usable signal will be transmitted even with interference.
Note that the voltages used in the above illustration provide an example of the voltage levels that can be used. However, different systems can use different voltage levels and, different voltage levels can be used for transmission of High, Medium and Low speed signals.

Fig 2.25   CAN High and CAN Low signals

21.1.2 CAN-Bus systems (CAN = Communication or Controller Area Network, BUS = a means of transporting data / information). A CAN-Bus system provides a means of communication between different control units / modules; in effect, this allows information and data to be passed between different vehicle systems. CAN-Bus for the automotive industry is in effect a "Set of Rules" or a Protocol that has been widely adopted for automotive applications.

There are a number of advantages to using a Can-Bus system, but essentially, it enables information to be shared e.g. a single sensor can provide information that is shared across

many control units. Additionally, one control unit can communicate with another control unit to allow the two different controllers to perform related but different tasks e.g. when a transmission control unit intends to select another gear (gear shift), it can request the engine control unit to reduce the engine torque / power to help improve the smoothness of the gear change and to reduce the potential shock or excessive loading of components (such actions can also help improve emissions).

Because of the ability of modern control units to operate and calculate at high speed, the CAN-Bus system must therefore also be able to transmit data / information at high speed; also, a CAN-Bus system can transmit many different signals within the system at the same time. Additionally, the transmission of data / information must be performed reliably / accurately so that there are no faults caused by electrical interference or other problems.

21.1.3 Automotive CAN-Bus systems often make use of a pair of wires twisted together (referred to as a twisted pair) to transmit the data / information; one wire carries a signal, with the other wire carrying a "Mirror Image" of the same signal (refer to Fig 2.25); these two signals are referred to as "CAN High" and "CAN Low". By using CAN High and CAN Low signals, it is possible to reduce the problems caused by interference. Note that some systems also use a Fibre Optic to carry the data / information and on some circuits, a single wire can be used.

21.1.4 There are a number of arrangements for connecting the various control units together (refer to Fig 2.26); the three main arrangements are:

*   Ring   * Line (linear)   * Star (possibly using a central or main control unit).

21.1.5 Signals can be transmitted at different speeds, which are usually referred to as High Speed CAN, Medium speed CAN and Low Speed CAN. The different speeds are utilised for transmission of signals from different systems e.g. Engine system control signals or Entertainment signals (Radio etc). The different speeds can also be utilised for prioritising the important signals e.g. High Speed for engine control signals, which would be more important than door mirror adjustment signals. Note that High, Medium and Low speed signals can all be transmitted on the same circuit, but different voltage levels can be used for the signals (refer to Fig 2.25).

High Speed:	Generally used for Drivetrain / Powertrain system communication
Medium Speed:	Generally used for Body system communication
Low Speed:	Some in car Infotainment systems (Sat-Nav, telephone etc).

21.1.6 Other variations of CAN systems include the MOST system (Media Oriented Systems Transport); the MOST system is capable of transmitting higher volumes of data at high speed which is required for Multi-media, Personal Computer and TV signals. Also, new developments include Flex-Ray which is again a system for providing faster transmission speeds.

21.1.7 CAN systems make use of "Gateways". Different elements of the CAN system may not be able to communicate due to different Speeds (High, Medium Low etc), and the CAN system also has to communicate with other systems e.g. a diagnostic tool. A Gateway functions as a "Translator or Converter", effectively converting one format of signal into a different format that can then be "Read" by another network or system.

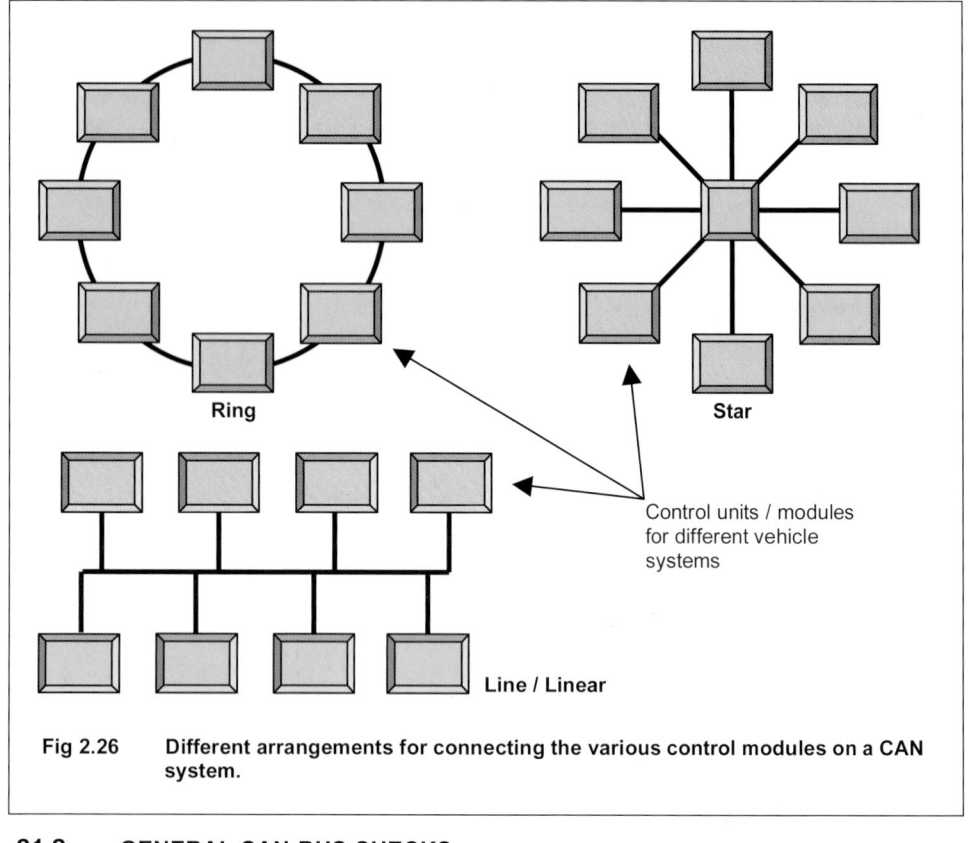

Ring

Star

Line / Linear

Control units / modules for different vehicle systems

Fig 2.26   **Different arrangements for connecting the various control modules on a CAN system.**

21.2   GENERAL CAN-BUS CHECKS

21.2.1 Due to the variations in design and operation of CAN-Bus systems, it may be necessary to refer to vehicle specific information.

21.2.2 As with other electronic systems, CAN-Bus systems rely on signals passing between two components i.e. for CAN-Bus systems the signals pass between electronic modules. Although some systems do make use of Fibre Optics for transfer of information, many systems still rely on a conventional wire (usually a twisted pair(refer to 21.1.3). Checks for wiring and connection faults are therefore still relevant to most CAN-Bus systems.

21.2.3 Where a Fault Code description refers to "Circuit / Open", "Circuit Low" or "Circuit High", it indicates a circuit related fault which is most likely to be caused by short or open circuits in the wiring or connections.

21.2.4 Note. The two wires of the twisted pair are usually connected with resistors (often one at each end of the circuit), the resistors are typically around 120 ohms but could vary across different systems), it is therefore possible to check for the correct resistance between the two wires (with the wires disconnected from the system at both ends); because there may be two resistors (which are effectively connected in parallel to the twisted wires), the resistance measurement across the wires will be half of the value of the resistors (this is assuming both resistors are the same value).

21.3   CAN HIGH AND CAN LOW CHECKS

21.3.1 The Fault codes can refer to faults in the CAN High and CAN Low circuits; this refers to each of the two "Mirrored Signals" (Fig 2.25) carried by the twisted pair wires. Note that one wire in the pair carries the CAN High signal with the other wire carrying the CAN Low signal therefore it will be necessary to identify which wire is High or Low. It is possible to check that the two signals exist (using appropriate Oscilloscope or alternative equipment). Check for clean signals without interference and check that the signals mirror each other.

21.3.2 A continuity check can be carried out across each of the wires in the twisted pair to ensure that there are no open circuits. Also check the connection between the two wires (refer to 21.2.4).

21.4   HIGH SPEED, MEDIUM SPEED, LOW SPEED

21.4.1 Where a fault code refers to a High Medium or Low speed signal, it will be necessary to identify which modules and system use the specified speed; refer to paragraph 21.1.5 for examples, but note that system specific information may be required for accurate identification. By establishing which modules or systems are applicable, this can assist in narrowing down which parts of the system to check.

21.5   LOST COMMUNICATION

21.5.1 Where a fault code refers to Lost Communication, the fault can be caused by wiring / connection faults (refer to 21.3) or possible control unit faults.

21.6   INVALID DATA

21.6.1 Invalid data can be caused by wiring or connections faults (refer to 21.3) or due to a module fault. Note that for most Invalid Data fault codes, the description identifies the module or part of the CAN system that is affected.

Chapter 3

COMPONENT LOCATIONS

1.   ISO AND VEHICLE MANUFACTURER COMPONENT LOCATION..54

2.   EXAMPLES OF LOCATIONS FOR OXYGEN / LAMBDA SENSORS AND CATALYTIC CONVERTERS 56

3.   EXAMPLES OF COMPONENT / SENSOR LOCATIONS ON TURBO / SUPERCHARGED ENGINES 60

1. ISO AND VEHICLE MANUFACTURER COMPONENT LOCATION

1.1 ISO STANDARD FOR COMPONENT LOCATION

1.1.1 Within the Standards for ISO / EOBD fault Codes, there is a standard for the identification of some components. The standard primarily relates to emission related devices and includes "Cylinder Bank" identification along with associated sensor for the intake and exhaust system e.g. Oxygen / Lambda sensors and pressure sensors.

1.1.2 There is also some identification for component locations on Turbo / Supercharged engines where the different layouts can affect the position of some sensors etc.

1.1.3 The following pages provide illustrations that provide a guide to the standard component locations.

1.2 GENERAL GUIDELINES FOR ISO COMPONENT LOCATION

1.2.1 The following guidelines are based on the ISO standard for component location. Refer to the illustrations on pages 54 – 57 for additional information.

1.2.2 Cylinder Banks (Bank 1 / Bank 2)

* Where an engine has a single bank of cylinders i.e. in-line engines, if there is more than one sensor (e.g. more than one Oxygen / Lambda Sensor) a fault code can still refer to "Bank 1"; therefore any sensor identified as being associated with Bank 1 on an in-line engine, will be associated with that single bank of cylinders.

Note however that fault codes do not always refer to Bank 1 for in-line engines; the use of "Bank 1" is not consistent and generally applies if there is more than one sensor of the same type.

* Bank 1 on a V type engine will contain Cylinder 1 (refer to vehicle specific information to identify Cylinder 1). Therefore, an Oxygen / Lambda sensor that is identified as being "Bank 1 - Sensor 1" will be fitted to the exhaust system for the bank of cylinders that contains Cylinder 1.

If the Oxygen / Lambda sensor is identified as being "Bank 2 - Sensor 1", it will be fitted to the exhaust system for the cylinder bank that DOES NOT contain Cylinder 1.

1.2.3 Oxygen / Lambda sensors (Sensor 1, 2 or 3)

* There can be 3 Oxygen / Lambda sensors fitted to an exhaust system (with 2 catalytic converters); this could therefore mean that there are 6 Oxygen / Lambda sensors on a V type engine that has two independent exhaust systems.

* Sensor 1 is the pre-cat sensor

* Sensor 2 is the post cat sensor (fitted after the front or first catalytic converter)

* Sensor 3 is the post-cat sensor (fitted after the rear or second catalytic converter).

* As an example, if a fault code description indicates a fault with an Oxygen / Lambda Sensor (often referred to as O2 sensors in the fault codes), the description could identify "Bank 2 - Sensor 3"; this would therefore refer to the Post-Cat sensor (after the second catalytic converter) on Bank 2 (which is the Bank of cylinders that DOES NOT contain Cylinder 1).

Note: The ISO fault codes can include the letter "a" or "b" for some Oxygen Lambda sensor fault codes e.g. P2236^a. The suffix letter "a" can refer to a heated Oxygen / Lambda sensor and, the suffix letter "b" can refer to a Wide band / broad band sensor. The Fault code definition associated with the fault code does however provide sufficient indication of the nature of the fault, and therefore the suffix letters "a" and "b" are not used within this book for oxygen / Lambda sensor codes.

1.2.4 Intake / exhaust system and fuel Sensor locations

* Where more than one sensor of the same type is fitted to a system e.g. Intake temperature sensors, the ISO standard indicates that the sensors should be numbered according to their position in the system. For the intake / exhaust system, the location numbering starts at the fresh air intake and progresses through the system to the engine, and then to the exhaust tailpipe. Therefore, Intake air temperature sensor 1 will be nearest to the fresh air intake (start of the intake system).

* For the fuel system, the numbering starts at the fuel tank and passes through to the engine e.g. Fuel pressure regulator 1 should be the first pressure regulator in the system (assuming that there is more than one regulator).

* Although the ISO numbering standard should be followed by vehicle manufacturers, it is possible that in some cases e.g. on a V type engine, manufacturers may have used the numbering to identify Sensor 1 or Sensor 2 as being related to one of the banks of cylinders.

1.2.5 Front / Rear and Left / Right

* A reference to Front / rear or Left / Right should be relative to sitting in the driver's seat.

1.2.6 Letters A, B, C etc used within the fault code number

* Where a capital letter is used as part of the fault code number e.g. P242A (which is an exhaust gas temperature sensor code), this usually indicates that the code has been inserted in the fault code numbering sequence as an addition; this enables newer or additional fault codes to be inserted in an appropriate position i.e. to keep related codes together (in this example, it keeps Exhaust Gas Temperature Sensor codes in the same group).

1.2.7   Letters A or B (as used by vehicle manufacturers)

*   A capital letter can also be used as part of the fault code description; the letters are used to identify a component e.g. "Fuel Level Sensor B" or "Shift Solenoid F". The use of the letter "B" or "F" in these examples is defined by the vehicle manufacturer and it might therefore be necessary to refer to manufacturer specific information.

1.2.8   A or B Camshaft Location

*   A fault code description relating to an engine camshaft can include the letters "A" or "B" at the beginning of the description e.g. "A" Camshaft Position Actuator Circuit / Open.

The Letter A refers to either: the Front, the Left or the Intake camshaft (as viewed from the drivers seat); if applicable, a Cylinder Bank number can also be included (1 or 2) to identify the Cylinder Bank (refer to 1.2.2).

The Letter B refers to either: the Rear, the Right or the Exhaust camshaft (as viewed from the drivers seat); with the Bank number (1 or 2) identifying the Cylinder Bank (refer to 1.2.2).

Note: If the fault code description includes the Letters "A" or "B" after the description instead of at the beginning e.g. Camshaft Position Sensor "A" this could be a sensor location defined by the vehicle manufacturer (refer to 1.2.7)

1.2.9   A or B Rocker Arm Actuator Location

*   A fault code description relating to an engine Rocker Arm Actuator (possibly used for Rocker Arm actuation of variable valve timing systems) can include the letters "A" or "B" e.g. "A" Rocker Arm Actuator Position Sensor Circuit High.

The Letter "A" refers to either: the Front, the Left or the Intake Rocker Arm Actuator (as viewed from the drivers seat); with the Bank number (1 or 2) identifying the Cylinder Bank (refer to 1.2.2).

The Letter "B" refers to either: the Rear, the Right or the Rocker Arm Actuator (as viewed from the drivers seat); with the Bank number (1 or 2) identifying the Cylinder Bank (refer to 1.2.2).

2. EXAMPLES OF LOCATIONS FOR OXYGEN / LAMBDA SENSORS AND CATALYTIC CONVERTERS

* **Sensor 1.** Oxygen / Lambda sensor 1 monitors the oxygen content in the exhaust gas passing from the engine; effectively an indicator of Air / Fuel Ratio.
* **Sensor 2.** Oxygen / Lambda sensor 2 monitors the oxygen content in the exhaust gas passing from the front Catalytic Converter; this indicates efficiency of the Converter.

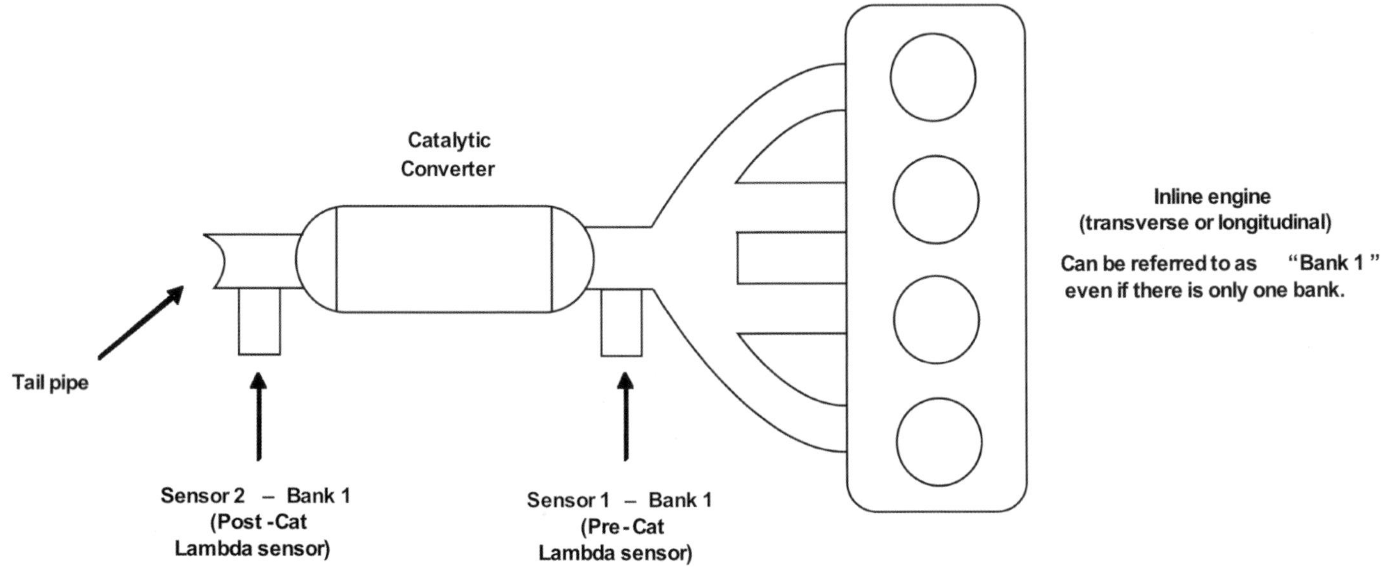

Catalytic
Converter

Inline engine
(transverse or longitudinal)

Can be referred to as "Bank 1"
even if there is only one bank.

Tail pipe

Sensor 2 – Bank 1
(Post-Cat
Lambda sensor)

Sensor 1 – Bank 1
(Pre-Cat
Lambda sensor)

Fig 3.1   Engine with single Catalytic Converter; fitted with Pre-Cat and Post-Cat Oxygen / Lambda sensors.

*   **Sensor 1.** Oxygen / Lambda sensor 1 monitors the oxygen content in the exhaust gas passing from the engine; effectively an indicator of Air / Fuel Ratio.
*   **Sensor 2.** Oxygen / Lambda sensor 2 monitors the oxygen content in the exhaust gas passing from the front Catalytic Converter; this indicates efficiency of the Converter.
*   **Sensor 3.** Oxygen / Lambda sensor 3 is used to monitor oxygen content in exhaust gas passing from the rear Catalytic Converter but, the sensor can be selective to enable it to monitor the NOx reduction process in the rear converter (the rear converter can be a NOx accumulator as well as a 3-way Catalytic Converter); the sensor information can be used to regulate the recovery NOx process

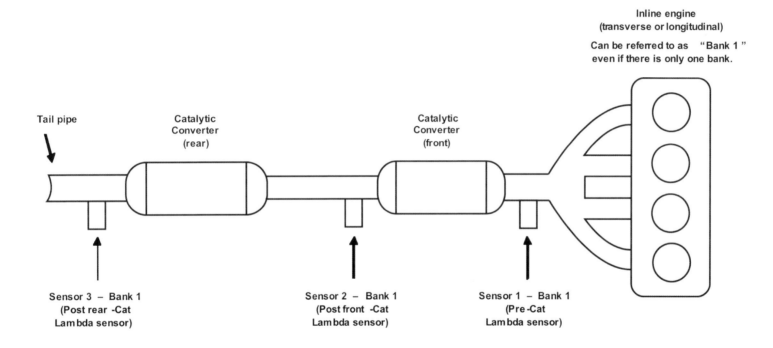

Fig 3.2   **Engine with two Catalytic Converters; fitted with a Pre-Cat and two Post-Cat Oxygen / Lambda sensors.**

* **Sensor 1.** Oxygen / Lambda sensor 1 monitors the oxygen content in the exhaust gas passing from the engine; effectively an indicator of Air / Fuel Ratio.
* **Sensor 2.** Oxygen / Lambda sensor 2 monitors the oxygen content in the exhaust gas passing from the front Catalytic Converter; this indicates efficiency of the Converter.
* **Sensor 3.** Oxygen / Lambda sensor 3 is used to monitor oxygen content in exhaust gas passing from the rear Catalytic Converter but, the sensor can be selective to enable it to monitor the NOx reduction process in the rear converter (the rear converter can be a NOx accumulator as well as a 3-way Catalytic Converter); the sensor information can be used to regulate the recovery NOx process

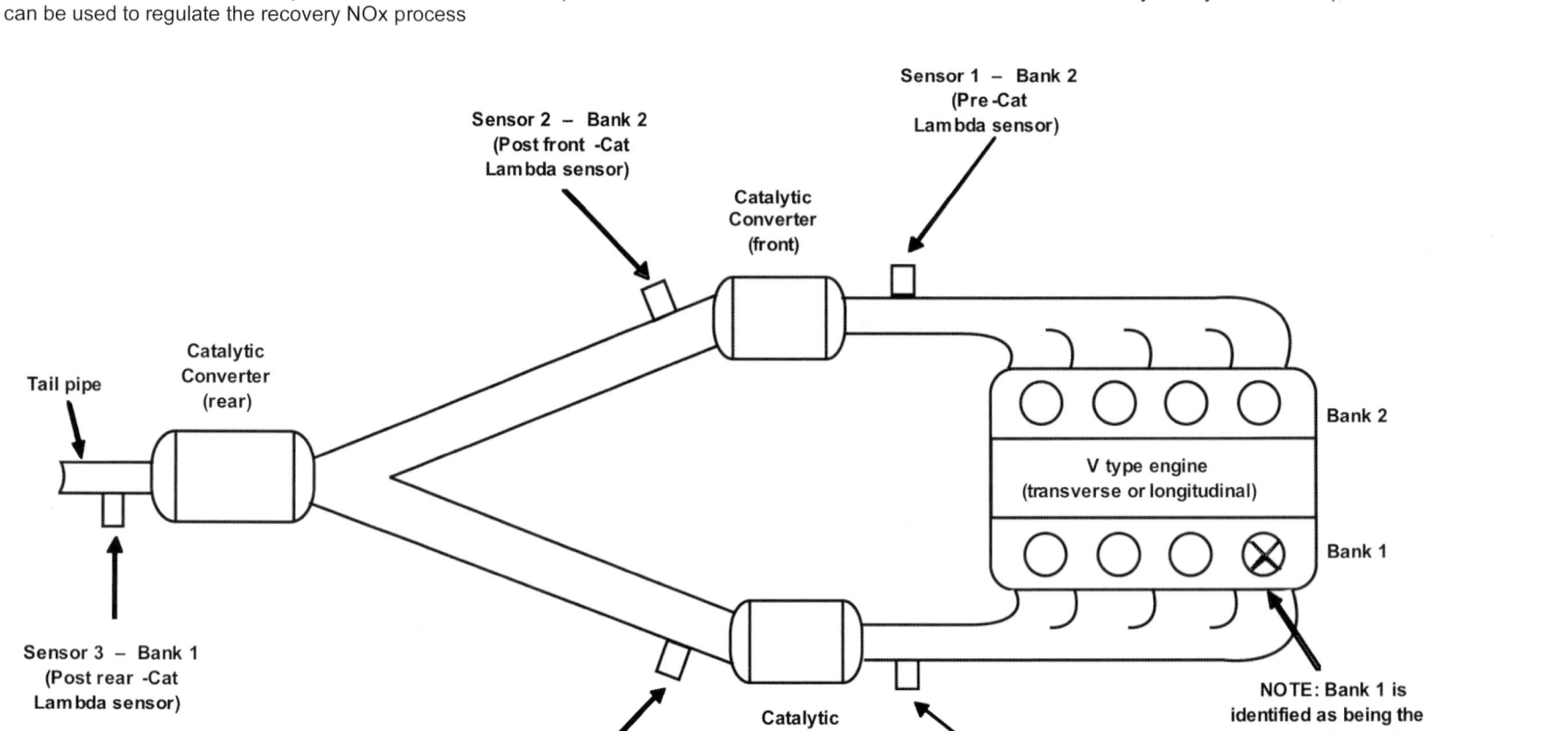

Fig 3.3 V-Engine with two "Front" Catalytic Converters and a single / shared "Rear" catalytic Converter; each cylinder bank has a Pre-Cat and a Post-Cat Oxygen / Lambda sensor (for the front Catalytic Converter), and a single / shared post cat sensor for the rear Catalytic Converter.

- **Sensor 1.** Oxygen / Lambda sensor 1 monitors the oxygen content in the exhaust gas passing from the engine; effectively an indicator of Air / Fuel Ratio.
- **Sensor 2.** Oxygen / Lambda sensor 2 monitors the oxygen content in the exhaust gas passing from the front Catalytic Converter; this indicates efficiency of the Converter.
- **Sensor 3.** Oxygen / Lambda sensor 3 is used to monitor oxygen content in exhaust gas passing from the rear Catalytic Converter but, the sensor can be selective to enable it to monitor the NOx reduction process in the rear converter (the rear converter can be a NOx accumulator as well as a 3-way Catalytic Converter); the sensor information can be used to regulate the recovery NOx process

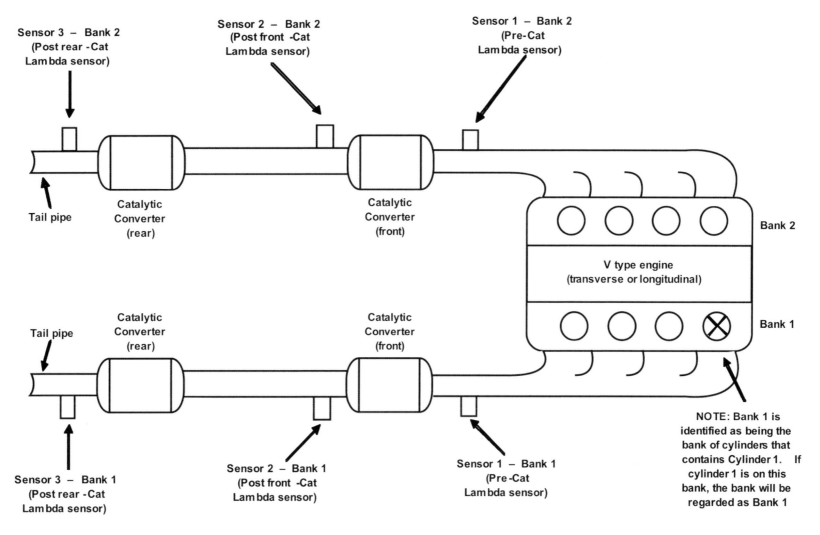

Fig 3.4   V-Engine with a "Front" Catalytic Converter and a "Rear" Catalytic Converter for each bank; each bank has its own pre-cat oxygen / Lambda / sensor and two Post-Cat Oxygen / Lambda sensors.

3. EXAMPLES OF COMPONENT / SENSOR LOCATIONS ON TURBO / SUPERCHARGED ENGINES

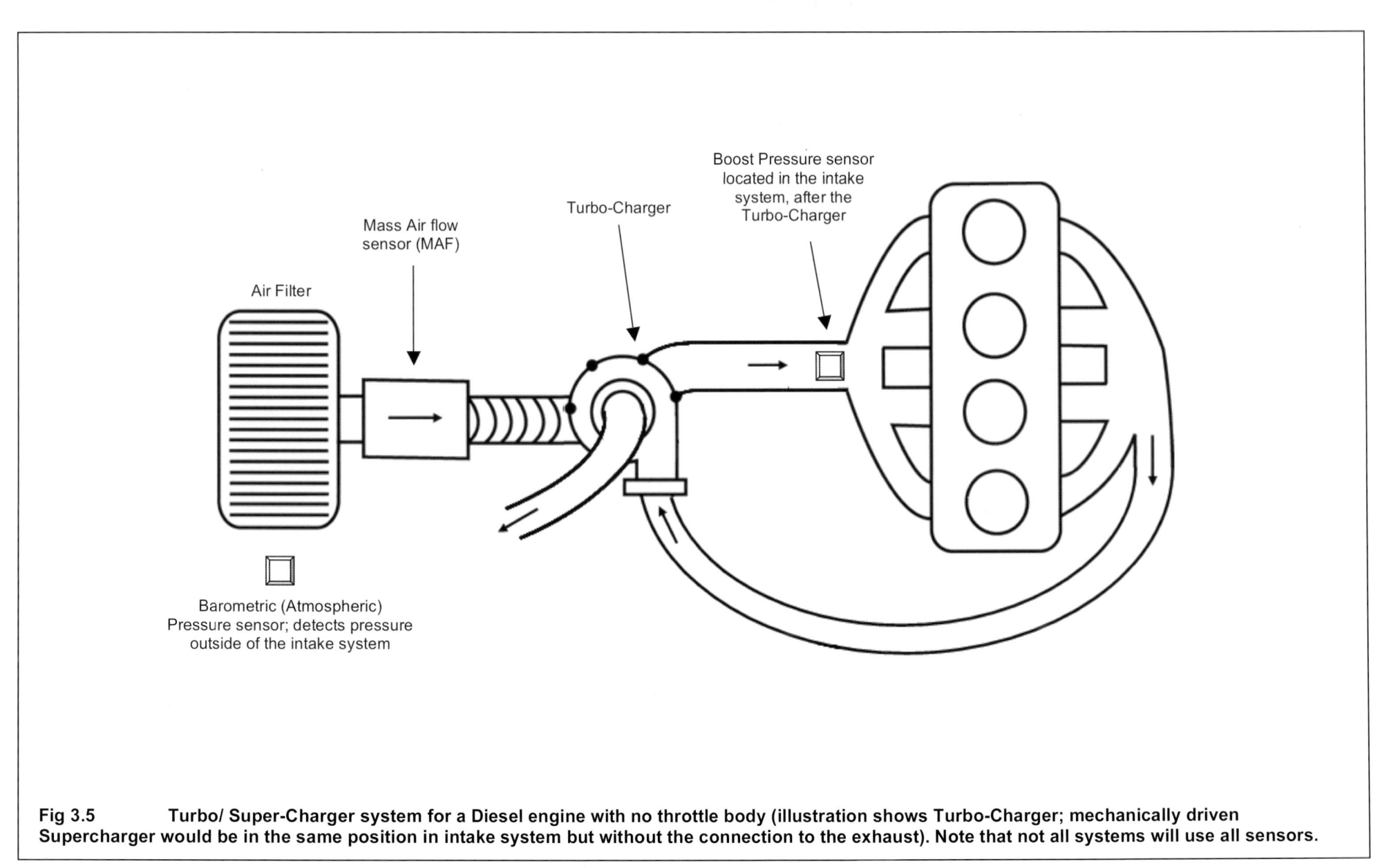

Fig 3.5 **Turbo/ Super-Charger system for a Diesel engine with no throttle body (illustration shows Turbo-Charger; mechanically driven Supercharger would be in the same position in intake system but without the connection to the exhaust). Note that not all systems will use all sensors.**

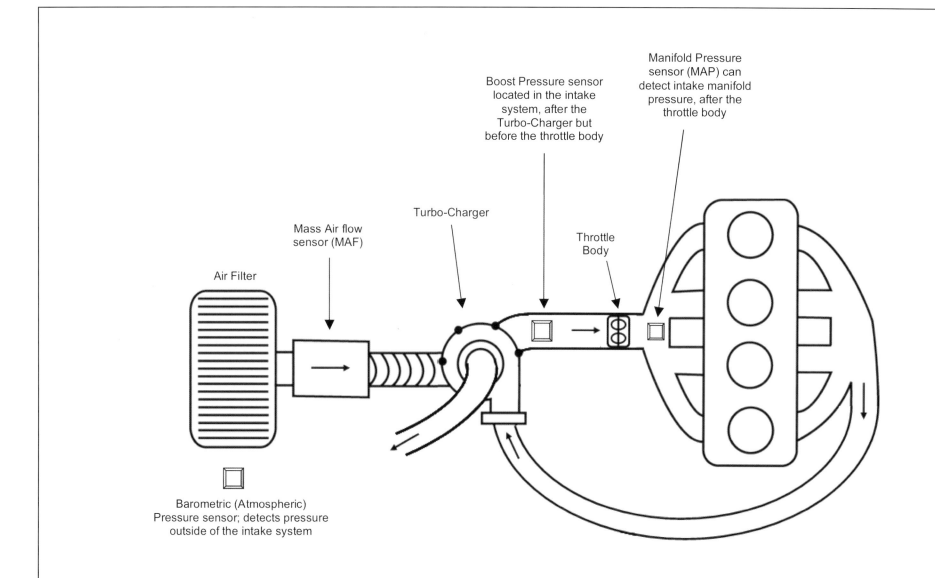

Manifold Pressure sensor (MAP) can detect intake manifold pressure, after the throttle body

Boost Pressure sensor located in the intake system, after the Turbo-Charger but before the throttle body

Turbo-Charger

Mass Air flow sensor (MAF)

Throttle Body

Air Filter

Barometric (Atmospheric) Pressure sensor; detects pressure outside of the intake system

Fig 3.6   Turbo/ Super-Charger system with throttle body after the Turbo-Charger (illustration shows Turbo-Charger but mechanically driven Supercharger would be located in the same position). Note that not all systems will use all identified sensors.

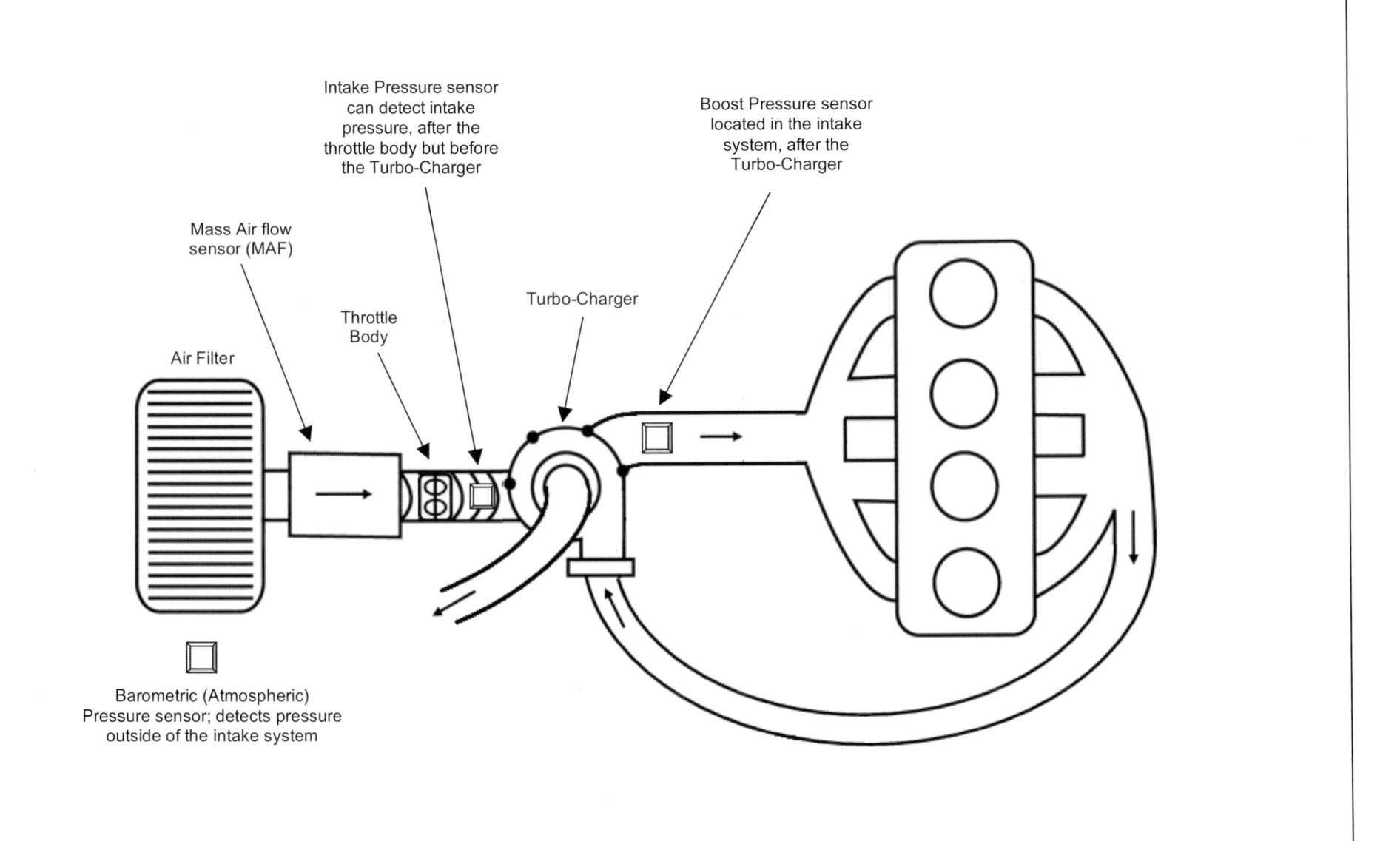

Intake Pressure sensor can detect intake pressure, after the throttle body but before the Turbo-Charger

Boost Pressure sensor located in the intake system, after the Turbo-Charger

Mass Air flow sensor (MAF)

Throttle Body

Turbo-Charger

Air Filter

Barometric (Atmospheric) Pressure sensor; detects pressure outside of the intake system

Fig 3.7   Turbo/ Super-Charger system with throttle body after the Turbo-Charger (illustration shows Turbo-Charger but mechanically driven Supercharger would be located in the same position). Note that not all systems will use all identified sensors.

Chapter 4

ISO / EOBD FAULT CODES

P0xxx   ISO / EOBD FAULT CODES..64

P2xxx   ISO / EOBD FAULT CODES...213

P3xxx   ISO / EOBD FAULT CODES 324

U0xxx   ISO / EOBD FAULT CODES 334

NOTE 1. Check for other fault codes that could provide additional information.   **NOTE 2.** Communication between control units can pass via a CAN-Bus system; refer to Page 51 for CAN-Bus checks.

NOTE 3. If a fault cannot be located, it is also possible that the control unit is at fault.   **NOTE 4.** Refer to Page 53 for list of pages that show the ISO standard for component locations e.g. Sensor A - Bank 1.

Fault code	EOBD / ISO Description	Component / System Description	Meaningful Description and Quick Check
P0000			There are no faults allocated to this code. However, the code may be displayed on some diagnostic tools and a message may indicate "No fault found" or "No fault detected by On-Board Diagnostics".
P0001	Fuel Volume Regulator Control Circuit / Open	The fuel volume delivered by a pump can be controlled by a volume regulator; the regulator will be operated by an actuator e.g. solenoid. On many "common rail" systems, when fuel demand is low (e.g. idle / light load), one of the three pumping elements in the high pressure pump is disabled (by the action of a solenoid). For distributor pumps, volume regulators can be referred to as a "Fuel Metering Solenoid". Note, some manufacturers refer to the "Fuel Pressure Regulator" for this fault code, but volume and pressure regulation can be separate functions.	The voltage, frequency or value of the control signal in the regulator circuit is incorrect. Signal could be out of normal operating range, constant value, non-existent, corrupt. The fault code indicates a possible "open circuit" but other undefined faults in the regulator or wiring could activate the same code e.g. regulator power supply, short circuit, high circuit resistance or incorrect regulator resistance. Note: the control unit could be providing the correct signal but it is being affected by the circuit fault. Refer to list of sensor / actuator checks on page 19.
P0002	Fuel Volume Regulator Control Circuit Range / Performance	The fuel volume delivered by a pump can be controlled by a volume regulator; the regulator will be operated by an actuator e.g. solenoid. On many "common rail" systems, when fuel demand is low (e.g. idle / light load), one of the three pumping elements in the high pressure pump is disabled (by the action of a solenoid). For distributor pumps, volume regulators can be referred to as a "Fuel Metering Solenoid". Note, some manufacturers refer to the "Fuel Pressure Regulator" for this fault code, but volume and pressure regulation can be separate functions.	The voltage, frequency or value of the control signal in the regulator circuit is within the normal operating range / tolerance, but the signal might be incorrect for the operating conditions. It is also possible that the regulator response (or response of the system controlled by the regulator) differs from the expected / desired response (incorrect response for the control signal provided by the control unit). Possible fault in regulator or wiring, but also possible that the regulator (or mechanism / system controlled by the regulator) is not operating or moving correctly. Note: the control unit could be providing the correct signal but it is being affected by the circuit fault. Refer to list of sensor / actuator checks on page 19. Also check that when the regulator is provided with a control signal, the system being controlled by the regulator (e.g. a valve) operates as required. If a valve leaks or does not open / close properly this will cause incorrect response to the normal control signal; the control unit could then provide a different control signal in an attempt to obtain the correct response (which could be regarded as an incorrect signal).
P0003	Fuel Volume Regulator Control Circuit Low	The fuel volume delivered by a pump can be controlled by a volume regulator; the regulator will be operated by an actuator e.g. solenoid. On many "common rail" systems, when fuel demand is low (e.g. idle / light load), one of the three pumping elements in the high pressure pump is disabled (by the action of a solenoid). For distributor pumps, volume regulators can be referred to as a "Fuel Metering Solenoid". Note, some manufacturers refer to the "Fuel Pressure Regulator" for this fault code, but volume and pressure regulation can be separate functions.	The voltage, frequency or value of the control signal in the regulator circuit is either "zero" (no signal) or, the signal exists but is below the normal operating range. Possible fault with regulator or wiring (e.g. regulator power supply, short / open circuit or resistance) that is causing a signal value to be lower than the minimum operating limit. Note: the control unit could be providing the correct signal but it is being affected by the circuit fault. Refer to list of sensor / actuator checks on page 19.
P0004	Fuel Volume Regulator Control Circuit High	The fuel volume delivered by a pump can be controlled by a volume regulator; the regulator will be operated by an actuator e.g. solenoid. On many "common rail" systems, when fuel demand is low (e.g. idle / light load), one of the three pumping elements in the high pressure pump is disabled (by the action of a solenoid). For distributor pumps, volume regulators can be referred to as a "Fuel Metering Solenoid". Note, some manufacturers refer to the "Fuel Pressure Regulator" for this fault code, but volume and pressure regulation can be separate functions.	The voltage, frequency or value of the control signal in the regulator circuit is either at full value (e.g. battery voltage with no on / off signal) or, the signal exists but it is above the normal operating range. Possible fault with regulator or wiring (e.g. regulator power supply, short / open circuit or resistance) that is causing a signal value to be higher than the maximum operating limit. Note: the control unit could be providing the correct signal but it is being affected by the circuit fault. Refer to list of sensor / actuator checks on page 19.
P0005	Fuel Shutoff Valve "A" Control Circuit / Open	A fuel shut off valve is usually located in fuel injection pump. The valve is operated by an electrical actuator (usually a solenoid) and is used to switch off the fuel supply when engine is stopped. Note: some systems move the fuel quantity control (in the diesel pump) to the "zero" fuel delivery position. Note that different fuel systems may use a different type of shut off valve.	Identify the type of shutoff valve; note the signal from the control unit could be a simple on / off on the valve earth circuit or, it could be frequency / duty cycle signal (on fuel quantity control systems). The voltage, frequency or value of the control signal in the actuator circuit is incorrect. Signal / voltage could be out of normal operating range, constant value, non-existent, corrupt. The fault code indicates a possible "open circuit" but other undefined faults in the actuator or wiring could activate the same code e.g. actuator power supply, short circuit, high circuit resistance or incorrect actuator resistance. Note: the control unit could be providing the correct signal but it is being affected by a circuit fault. Refer to list of sensor / actuator checks on page 19.
P0006	Fuel Shutoff Valve "A" Control Circuit Low	A fuel shut off valve is usually located in fuel injection pump. The valve is operated by an electrical actuator (usually a solenoid) and is used to switch off the fuel supply when engine is stopped. Note: some systems move the fuel quantity control (in the diesel pump) to the "zero" fuel delivery position. Note that different fuel systems may use a different type of shut off valve.	Identify the type of shutoff valve; note the signal from the control unit could be a simple on / off on the valve earth or power circuit or, it could be frequency / duty cycle signal (on fuel quantity control systems). The voltage, frequency or value of the control signal in the actuator circuit is either "zero" (no signal) or, the signal / voltage exists but is below the normal operating range. Possible fault with actuator or wiring (e.g. actuator power supply, short / open circuit or resistance) that is causing a signal value to be lower than the minimum operating limit. Note: the control unit could be providing the correct signal but it is being affected by a circuit fault. Refer to list of sensor / actuator checks on page 19.
P0007	Fuel Shutoff Valve "A" Control Circuit High	A fuel shut off valve is usually located in fuel injection pump. The valve is operated by an electrical actuator (usually a solenoid) and is used to switch off the fuel supply when engine is stopped. Note: some systems move the fuel quantity control (in the diesel pump) to the "zero" fuel delivery position. Note that different fuel systems may use a different type of shut off valve.	Identify the type of shutoff valve; note the signal from the control unit could be a simple on / off on the valve earth circuit or, it could be frequency / duty cycle signal (on fuel quantity control systems). The voltage, frequency or value of the control signal in the actuator circuit is either at full value (e.g. battery voltage with no on / off signal) or, the signal exists but it is above the normal operating range. Possible fault with actuator or wiring (e.g. actuator power supply, short / open circuit or resistance) that is causing a signal value to be higher than the maximum operating limit. Note: the control unit could be providing the correct signal but it is being affected by a circuit fault. Refer to list of sensor / actuator checks on page 19.

NOTE 1. Check for other fault codes that could provide additional information.   NOTE 2. Communication between control units can pass via a CAN-Bus system; refer to Page 51 for CAN-Bus checks.

NOTE 3. If a fault cannot be located, it is also possible that the control unit is at fault.   NOTE 4. Refer to Page 53 for list of pages that show the ISO standard for component locations e.g. Sensor A - Bank 1.

code	Description		
P0008	Engine Position System Performance Bank 1	Engine position systems are used to indicate angular position of crankshaft and / or camshaft to enable accurate control of engine functions e.g. fuel, ignition, emissions system, valve timing etc; sensors can include crankshaft and camshaft position or diesel pump speed / position sensor (rotary pumps). The position system is normally used to provide the engine speed signal.	The fault code indicates that there is an undefined fault with the engine position system. Although there could be a sensor or electrical fault (which would result in a sensor signal that is either out of normal operating range, constant value, non-existent or corrupt), it is also possible that the signal from the position sensor(s) is within normal operating range but not correct for the operating conditions (as indicated by other sensors) or not as expected by the control unit i.e. sensor / sensor signal performance is incorrect due to unidentified fault. Identify the type of position sensor fitted, and Refer to list of sensor / actuator checks on page 19. Check (as applicable) the position reference system for damaged reference points e.g. reference teeth on rotor / disc etc.
P0009	Engine Position System Performance Bank 2	Engine position systems are used to indicate angular position of crankshaft and / or camshaft to enable accurate control of engine functions e.g. fuel, ignition, emissions system, valve timing etc; sensors can include crankshaft and camshaft position or diesel pump speed / position sensor (rotary pumps). The position system is normally used to provide the engine speed signal.	The fault code indicates that there is an undefined fault with the engine position system. Although there could be a sensor or electrical fault (which would result in a sensor signal that is either out of normal operating range, constant value, non-existent or corrupt), it is also possible that the signal from the position sensor(s) is within normal operating range but not correct for the operating conditions (as indicated by other sensors) or not as expected by the control unit i.e. sensor / sensor signal performance is incorrect due to unidentified fault. Identify the type of position sensor fitted, and Refer to list of sensor / actuator checks on page 19. Check (as applicable) the position reference system for damaged reference points e.g. reference teeth on rotor / disc etc.
P000A	"A" Camshaft Position Slow Response Bank 1	Refers to operation of variable valve timing system, which is used to alter camshaft position / valve timing; the control unit will control the actuator (e.g. solenoid) which then regulates oil pressure or other mechanism to alter camshaft position.	The control unit is providing a control signal to the actuator but the control unit has detected a slow or implausible response for the change in camshaft / actuator position. The fault could be related to an actuator / wiring fault e.g. power supply, short / open circuit, incorrect actuator resistance. Refer to list of sensor / actuator checks on page 19. It is also possible that the oil pressure system, or other mechanisms used to move the camshaft, could be faulty. A position sensor (if fitted) would be indicating the "slow response / position change"; also check position sensor operation.
P000B	"B" Camshaft Position Slow Response Bank 1	Refers to operation of variable valve timing system, which is used to alter camshaft position / valve timing; the control unit will control the actuator (e.g. solenoid) which then regulates oil pressure or other mechanism to alter camshaft position.	The control unit is providing a control signal to the actuator but the control unit has detected a slow or implausible response for the change in camshaft / actuator position. The fault could be related to an actuator / wiring fault e.g. power supply, short / open circuit, incorrect actuator resistance. Refer to list of sensor / actuator checks on page 19. It is also possible that the oil pressure system, or other mechanism used to move the camshaft, could be faulty. A position sensor (if fitted) would be indicating the "slow response / position change"; also check position sensor operation.
P000C	"A" Camshaft Position Slow Response Bank 2	Refers to operation of variable valve timing system, which is used to alter camshaft position / valve timing; the control unit will control the actuator (e.g. solenoid) which then regulates oil pressure or other mechanism to alter camshaft position.	The control unit is providing a control signal to the actuator but the control unit has detected a slow or implausible response for the change in camshaft / actuator position. The fault could be related to an actuator / wiring fault e.g. power supply, short / open circuit, incorrect actuator resistance. Refer to list of sensor / actuator checks on page 19. It is also possible that the oil pressure system, or other mechanisms used to move the camshaft, could be faulty. A position sensor (if fitted) would be indicating the "slow response / position change"; also check position sensor operation.
P000D	"B" Camshaft Position Slow Response Bank 2	Refers to operation of variable valve timing system, which is used to alter camshaft position / valve timing; the control unit will control the actuator (e.g. solenoid) which then regulates oil pressure or other mechanism to alter camshaft position.	The control unit is providing a control signal to the actuator but the control unit has detected a slow or implausible response for the change in camshaft / actuator position. The fault could be related to an actuator / wiring fault e.g. power supply, short / open circuit, incorrect actuator resistance. Refer to list of sensor / actuator checks on page 19. It is also possible that the oil pressure system, or other mechanisms used to move the camshaft, could be faulty. A position sensor (if fitted) would be indicating the "slow response / position change"; also check position sensor operation.
P000E	ISO / SAE reserved		Not used, reserved for future allocation.
P000F	ISO / SAE reserved		Not used, reserved for future allocation.
P0010	"A" Camshaft Position Actuator Circuit / Open Bank 1	Refers to operation of variable valve timing system, which is used to alter camshaft position / valve timing; the control unit will control the actuator (e.g. solenoid) which then regulates oil pressure or other mechanism to alter camshaft position.	The voltage, frequency or value of the control signal in the actuator circuit is incorrect. Signal could be out of normal operating range, constant value, non-existent, corrupt. The fault code indicates a possible "open circuit" but other undefined faults in the actuator or wiring could activate the same code e.g. actuator power supply, short circuit, high circuit resistance or incorrect actuator resistance. Note: the control unit could be providing the correct signal but it is being affected by the circuit fault. Refer to list of sensor / actuator checks on page 19.
P0011	"A" Camshaft Position -Timing Over-Advanced or System Performance Bank 1	Refers to operation of variable valve timing system, which is used to alter camshaft position / valve timing; the control unit will control the actuator (e.g. solenoid) which then regulates oil pressure or other mechanism to alter camshaft position.	The signal from the control unit to the timing actuator is within normal operating range but when the actuator has been provided with a specific control signal, the control unit has detected an incorrect / implausible response within the system (over advanced valve timing). The fault could be detected due to the relationship of the signals from the camshaft and crankshaft position sensors. Check operation of variable valve timing actuator. Refer to list of sensor / actuator checks on page 19. It is also possible that the oil pressure system, or other mechanisms used to move the camshaft, could be faulty; also check camshaft / crankshaft position sensors.
P0012	"A" Camshaft Position -Timing Over-Retarded Bank 1	Refers to operation of variable valve timing system, which is used to alter camshaft position / valve timing; the control unit will control the actuator (e.g. solenoid) which then regulates oil pressure or other mechanism to alter camshaft position.	The signal from the control unit to the timing actuator is within normal operating range but when the actuator has been provided with a specific control signal, the control unit has detected an incorrect / implausible response within the system (over retarded valve timing). The fault could be detected due to the relationship of the signals from the camshaft and crankshaft position sensors. Check operation of variable valve timing actuator. Refer to list of sensor / actuator checks on page 19. It is also possible that the oil pressure system, or other mechanisms used to move the camshaft, could be faulty; also check camshaft / crankshaft position sensors.
P0013	"B" Camshaft Position -Actuator Circuit / Open Bank 1	Refers to operation of variable valve timing system, which is used to alter camshaft position / valve timing; the control unit will control the actuator (e.g. solenoid) which then regulates oil pressure or other mechanism to alter camshaft position.	The voltage, frequency or value of the control signal in the actuator circuit is incorrect. Signal could be out of normal operating range, constant value, non-existent, corrupt. The fault code indicates a possible "open circuit" but other undefined faults in the actuator or wiring could activate the same code e.g. actuator power supply, short circuit, high circuit resistance or incorrect actuator resistance. Note: the control unit could be providing the correct signal but it is being affected by the circuit fault. Refer to list of sensor / actuator checks on page 19.

NOTE 1. Check for other fault codes that could provide additional information. **NOTE 2.** Communication between control units can pass via a CAN-Bus system; refer to Page 51 for CAN-Bus checks.

NOTE 3. If a fault cannot be located, it is also possible that the control unit is at fault. **NOTE 4.** Refer to Page 53 for list of pages that show the ISO standard for component locations e.g. Sensor A - Bank 1.

Fault code	EOBD / ISO Description	Component / System Description	Meaningful Description and Quick Check
P0014	"B" Camshaft Position -Timing Over-Advanced or System Performance Bank 1	Refers to operation of variable valve timing system, which is used to alter camshaft position / valve timing; the control unit will control the actuator (e.g. solenoid) which then regulates oil pressure or other mechanism to alter camshaft position.	The signal from the control unit to the timing actuator is within normal operating range but when the actuator has been provided with a specific control signal, the control unit has detected an incorrect / implausible response within the system (over advanced valve timing). The fault could be detected due to the relationship of the signals from the camshaft and crankshaft position sensors. Check operation of variable valve timing actuator. Refer to list of sensor / actuator checks on page 19. It is also possible that the oil pressure system, or other mechanisms used to move the camshaft, could be faulty; also check camshaft / crankshaft position sensors.
P0015	"B" Camshaft Position -Timing Over-Retarded Bank 1	Refers to operation of variable valve timing system, which is used to alter camshaft position / valve timing; the control unit will control the actuator (e.g. solenoid) which then regulates oil pressure or other mechanism to alter camshaft position.	The signal from the control unit to the timing actuator is within normal operating range but when the actuator has been provided with a specific control signal, the control unit has detected an incorrect / implausible response within the system (over retarded valve timing). The fault could be detected due to the relationship of the signals from the camshaft and crankshaft position sensors. Check operation of variable valve timing actuator. Refer to list of sensor / actuator checks on page 19. It is also possible that the oil pressure system, or other mechanisms used to move the camshaft, could be faulty; also check camshaft / crankshaft position sensors.
P0016	Crankshaft Position - Camshaft Position Correlation Bank 1 Sensor A	Sensors indicate angular position of crankshaft and camshaft to enable accurate control of engine functions e.g. fuel, ignition, emissions system, valve timing etc.	The signals from the two sensors are providing conflicting or implausible information. Signals from Crankshaft speed / position sensor and Camshaft position sensor are indicating incorrect or fluctuating relationship between crankshaft and camshaft position. It is possible that one of the sensors is operating incorrectly therefore check each sensor system for a fault with the sensor / wiring e.g. short / open circuits, circuit resistance, incorrect sensor resistance or reference / operating voltage fault. Refer to list of sensor / actuator checks on page 19. Check camshaft drive belt / chain for correct tension and condition, check static camshaft / valve timing. If applicable, check operation of variable valve timing mechanism.
P0017	Crankshaft Position - Camshaft Position Correlation Bank 1 Sensor B	Sensors indicate angular position of crankshaft and camshaft to enable accurate control of engine functions e.g. fuel, ignition, emissions system, valve timing etc.	The signals from the two sensors are providing conflicting or implausible information. Signals from Crankshaft speed / position sensor and Camshaft position sensor are indicating incorrect or fluctuating relationship between crankshaft and camshaft position. It is possible that one of the sensors is operating incorrectly therefore check each sensor system for a fault with the sensor / wiring e.g. short / open circuits, circuit resistance, incorrect sensor resistance or reference / operating voltage fault. Refer to list of sensor / actuator checks on page 19. Check camshaft drive belt / chain for correct tension and condition, check static camshaft / valve timing. If applicable, check operation of variable valve timing mechanism.
P0018	Crankshaft Position - Camshaft Position Correlation Bank 2 Sensor A	Sensors indicate angular position of crankshaft and camshaft to enable accurate control of engine functions e.g. fuel, ignition, emissions system, valve timing etc.	The signals from the two sensors are providing conflicting or implausible information. Signals from Crankshaft speed / position sensor and Camshaft position sensor are indicating incorrect or fluctuating relationship between crankshaft and camshaft position. It is possible that one of the sensors is operating incorrectly therefore check each sensor system for a fault with the sensor / wiring e.g. short / open circuits, circuit resistance, incorrect sensor resistance or reference / operating voltage fault. Refer to list of sensor / actuator checks on page 19. Check camshaft drive belt / chain for correct tension and condition, check static camshaft / valve timing. If applicable, check operation of variable valve timing mechanism.
P0019	Crankshaft Position - Camshaft Position Correlation Bank 2 Sensor B	Sensors indicate angular position of crankshaft and camshaft to enable accurate control of engine functions e.g. fuel, ignition, emissions system, valve timing etc.	The signals from the two sensors are providing conflicting or implausible information. Signals from Crankshaft speed / position sensor and Camshaft position sensor are indicating incorrect or fluctuating relationship between crankshaft and camshaft position. It is possible that one of the sensors is operating incorrectly therefore check each sensor system for a fault with the sensor / wiring e.g. short / open circuits, circuit resistance, incorrect sensor resistance or reference / operating voltage fault. Refer to list of sensor / actuator checks on page 19. Check camshaft drive belt / chain for correct tension and condition, check static camshaft / valve timing. If applicable, check operation of variable valve timing mechanism.
P0020	"A" Camshaft Position Actuator Circuit / Open Bank 2	Refers to operation of variable valve timing system, which is used to alter camshaft position / valve timing; the control unit will control the actuator (e.g. solenoid) which then regulates oil pressure or other mechanism to alter camshaft position.	The voltage, frequency or value of the control signal in the actuator circuit is incorrect. Signal could be out of normal operating range, constant value, non-existent, corrupt. The fault code indicates a possible "open circuit" but other undefined faults in the actuator or wiring could activate the same code e.g. actuator power supply, short circuit, high circuit resistance or incorrect actuator resistance. Note: the control unit could be providing the correct signal but it is being affected by the circuit fault. Refer to list of sensor / actuator checks on page 19.
P0021	"A" Camshaft Position -Timing Over-Advanced or System Performance Bank 2	Refers to operation of variable valve timing system, which is used to alter camshaft position / valve timing; the control unit will control the actuator (e.g. solenoid) which then regulates oil pressure or other mechanism to alter camshaft position.	The signal from the control unit to the timing actuator is within normal operating range but when the actuator has been provided with a specific control signal, the control unit has detected an incorrect / implausible response within the system (over advanced valve timing). The fault could be detected due to the relationship of the signals from the camshaft and crankshaft position sensors. Check operation of variable valve timing actuator. Refer to list of sensor / actuator checks on page 19. It is also possible that the oil pressure system, or other mechanisms used to move the camshaft, could be faulty; also check camshaft / crankshaft position sensors.
P0022	"A" Camshaft Position -Timing Over-Retarded Bank 2	Refers to operation of variable valve timing system, which is used to alter camshaft position / valve timing; the control unit will control the actuator (e.g. solenoid) which then regulates oil pressure or other mechanism to alter camshaft position.	The signal from the control unit to the timing actuator is within normal operating range but when the actuator has been provided with a specific control signal, the control unit has detected an incorrect / implausible response within the system (over retarded valve timing). The fault could be detected due to the relationship of the signals from the camshaft and crankshaft position sensors. Check operation of variable valve timing actuator. Refer to list of sensor / actuator checks on page 19. It is also possible that the oil pressure system, or other mechanisms used to move the camshaft, could be faulty; also check camshaft / crankshaft position sensors.
P0023	"B" Camshaft Position -Actuator Circuit / Open Bank 2	Refers to operation of variable valve timing system, which is used to alter camshaft position / valve timing; the control unit will control the actuator (e.g. solenoid) which then regulates oil pressure or other mechanism to alter camshaft position.	The voltage, frequency or value of the control signal in the actuator circuit is incorrect. Signal could be out of normal operating range, constant value, non-existent, corrupt. The fault code indicates a possible "open circuit" but other undefined faults in the actuator or wiring could activate the same code e.g. actuator power supply, short circuit, high circuit resistance or incorrect actuator resistance. Note: the control unit could be providing the correct signal but it is being affected by the circuit fault. Refer to list of sensor / actuator checks on page 19.

NOTE 1. Check for other fault codes that could provide additional information.   **NOTE 2.** Communication between control units can pass via a CAN-Bus system; refer to Page 51 for CAN-Bus checks.

NOTE 3. If a fault cannot be located, it is also possible that the control unit is at fault.   **NOTE 4.** Refer to Page 53 for list of pages that show the ISO standard for component locations e.g. Sensor A - Bank 1.

code	Description		
P0024	"B" Camshaft Position -Timing Over- Advanced or System Performance Bank 2	Refers to operation of variable valve timing system, which is used to alter camshaft position / valve timing; the control unit will control the actuator (e.g. solenoid) which then regulates oil pressure or other mechanism to alter camshaft position.	The signal from the control unit to the timing actuator is within normal operating range but when the actuator has been provided with a specific control signal, the control unit has detected an incorrect / implausible response within the system (over advanced valve timing). The fault could be detected due to the relationship of the signals from the camshaft and crankshaft position sensors. Check operation of variable valve timing actuator. Refer to list of sensor / actuator checks on page 19. It is also possible that the oil pressure system, or other mechanisms used to move the camshaft, could be faulty; also check camshaft / crankshaft position sensors.
P0025	"B" Camshaft Position -Timing Over- Retarded Bank 2	Refers to operation of variable valve timing system, which is used to alter camshaft position / valve timing; the control unit will control the actuator (e.g. solenoid) which then regulates oil pressure or other mechanism to alter camshaft position.	The signal from the control unit to the timing actuator is within normal operating range but when the actuator has been provided with a specific control signal, the control unit has detected an incorrect / implausible response within the system (over retarded valve timing). The fault could be detected due to the relationship of the signals from the camshaft and crankshaft position sensors. Check operation of variable valve timing actuator. Refer to list of sensor / actuator checks on page 19. It is also possible that the oil pressure system, or other mechanisms used to move the camshaft, could be faulty; also check camshaft / crankshaft position sensors.
P0026	Intake Valve Control Solenoid Circuit Range / Performance Bank 1	The valve control solenoid is used to alter camshaft position / valve timing / valve lift (depending on system); the control solenoid regulates oil pressure (or other mechanism) which alters the position as required.	The voltage, frequency or value of the control signal in the regulator circuit is within the normal operating range / tolerance, but the signal might be incorrect for the operating conditions. It is also possible that the regulator response (or response of system controlled by the regulator) differs from the expected / desired response (incorrect response to the control signal provided by the control unit). Possible fault in regulator or wiring, but also possible that the regulator (or mechanism / system controlled by the regulator) is not operating or moving correctly. Note: the control unit could be providing the correct signal but it is being affected by a circuit fault. Refer to list of sensor / actuator checks on page 19. Also check for possible fault with the camshaft position sensor or sensor used to detect position change of the operating mechanism (check for other related codes). Ensure engine static valve timing is correct. For systems with hydraulic actuation, check engine oil pressure, oil viscosity and quality. For other mechanisms carry out checks applicable to the system.
P0027	Exhaust Valve Control Solenoid Circuit Range / Performance Bank 1	The valve control solenoid is used to alter camshaft position / valve timing / valve lift (depending on system); the control solenoid regulates oil pressure (or other mechanism) which alters the position as required.	The voltage, frequency or value of the control signal in the regulator circuit is within the normal operating range / tolerance, but the signal might be incorrect for the operating conditions. It is also possible that the regulator response (or response of system controlled by the regulator) differs from the expected / desired response (incorrect response to the control signal provided by the control unit). Possible fault in regulator or wiring, but also possible that the regulator (or mechanism / system controlled by the regulator) is not operating or moving correctly. Note: the control unit could be providing the correct signal but it is being affected by a circuit fault. Refer to list of sensor / actuator checks on page 19. Also check for possible fault with the camshaft position sensor or sensor used to detect position change of the operating mechanism (check for other related codes). Ensure engine static valve timing is correct. For systems with hydraulic actuation, check engine oil pressure, oil viscosity and quality. For other mechanisms carry out checks applicable to the system.
P0028	Intake Valve Control Solenoid Circuit Range / Performance Bank 2	The valve control solenoid is used to alter camshaft position / valve timing / valve lift (depending on system); the control solenoid regulates oil pressure (or other mechanism) which alters the position as required.	The voltage, frequency or value of the control signal in the regulator circuit is within the normal operating range / tolerance, but the signal might be incorrect for the operating conditions. It is also possible that the regulator response (or response of system controlled by the regulator) differs from the expected / desired response (incorrect response to the control signal provided by the control unit). Possible fault in regulator or wiring, but also possible that the regulator (or mechanism / system controlled by the regulator) is not operating or moving correctly. Note: the control unit could be providing the correct signal but it is being affected by a circuit fault. Refer to list of sensor / actuator checks on page 19. Also check for possible fault with the camshaft position sensor or sensor used to detect position change of the operating mechanism (check for other related codes). Ensure engine static valve timing is correct. For systems with hydraulic actuation, check engine oil pressure, oil viscosity and quality. For other mechanisms carry out checks applicable to the system.
P0029	Exhaust Valve Control Solenoid Circuit Range / Performance Bank 2	The valve control solenoid is used to alter camshaft position / valve timing / valve lift (depending on system); the control solenoid regulates oil pressure (or other mechanism) which alters the position as required.	The voltage, frequency or value of the control signal in the regulator circuit is within the normal operating range / tolerance, but the signal might be incorrect for the operating conditions. It is also possible that the regulator response (or response of system controlled by the regulator) differs from the expected / desired response (incorrect response to the control signal provided by the control unit). Possible fault in regulator or wiring, but also possible that the regulator (or mechanism / system controlled by the regulator) is not operating or moving correctly. Note: the control unit could be providing the correct signal but it is being affected by a circuit fault. Refer to list of sensor / actuator checks on page 19. Also check for possible fault with the camshaft position sensor or sensor used to detect position change of the operating mechanism (check for other related codes). Ensure engine static valve timing is correct. For systems with hydraulic actuation, check engine oil pressure, oil viscosity and quality. For other mechanisms carry out checks applicable to the system.
P0030	HO2S Heater Control Circuit Bank 1 Sensor 1	The heater control for the HO2S (oxygen / Lambda) sensor is used to heat up and provide stable temperature for the sensor (primarily after cold starts). The heating element can receive a power supply via a relay (often the main system relay) and the heater earth path can be switched to earth via the control unit (this is the heater control circuit), the control unit is therefore able to control the operation of the heating element by controlling the earth circuit. Note that some systems may also provide the power supply to the heating element via the control unit. It may be necessary to refer to vehicle specific information to identify the type of Oxygen / Lambda sensor and the wiring.	The fault code can relate to the earth path from the heater element to the control unit (which can be the control circuit for the heater), but note that in some cases (depending on the wiring for the heater and the manufacturers interpretation of the code), the code could relate to the heater power supply circuit. The fault code indicates that there is an undefined fault in the sensor heater control circuit, which is causing an incorrect voltage / current in the circuit; this is likely to be: non-existent, constant value (but incorrect for the operating conditions), corrupt or, is out of normal operating range (out of tolerance). The fault could be caused by a faulty sensor heater element (open circuit or high resistance) or a wiring fault (short / open circuit). Also check heater supply voltage (could be provided via a relay or from a control unit). Refer to list of sensor / actuator checks on page 19.

NOTE 1. Check for other fault codes that could provide additional information.   NOTE 2. Communication between control units can pass via a CAN-Bus system; refer to Page 51 for CAN-Bus checks.   67

NOTE 3. If a fault cannot be located, it is also possible that the control unit is at fault.   NOTE 4. Refer to Page 53 for list of pages that show the ISO standard for component locations e.g. Sensor A - Bank 1.

Fault code	EOBD / ISO Description	Component / System Description	Meaningful Description and Quick Check
P0031	HO2S Heater Control Circuit Low Bank 1 Sensor 1	The heater control for the HO2S (oxygen / Lambda) sensor is used to heat up and provide stable temperature for the sensor (primarily after cold starts). The heating element can receive a power supply via a relay (often the main system relay) and the heater earth path can be switched to earth via the control unit (this is the heater control circuit), the control unit is therefore able to control the operation of the heating element by controlling the earth circuit. Note that some systems may also provide the power supply to the heating element via the control unit. It may be necessary to refer to vehicle specific information to identify the type of Oxygen / Lambda sensor and the wiring.	The fault code can relate to the earth path from the heater element to the control unit (which can be the control circuit for the heater) but, depending on the wiring for the heater and the manufacturer's interpretation of the code, the code could relate to the heater power supply circuit. The fault code indicates that the voltage / current in the sensor heater control circuit is low e.g. zero volts, when the control unit is expecting to detect a high voltage e.g. battery volts (when the control unit is not completing the earth circuit for the heater, battery voltage would normally be detected in this circuit). The fault is likely to be caused by an open circuit in the wiring or heater element (or short to earth); it also possible that there is no power supply (either from the relay or control unit, as applicable). Check power supply and wiring to heater, check continuity / resistance of heater element (as applicable), and check wiring through to the control unit for short / open circuit to heater. Refer to list of sensor / actuator checks on page 19.
P0032	HO2S Heater Control Circuit High Bank 1 Sensor 1	The heater control for the HO2S (oxygen / Lambda) sensor is used to heat up and provide stable temperature for the sensor (primarily after cold starts). The heating element can receive a power supply via a relay (often the main system relay) and the heater earth path can be switched to earth via the control unit (this is the heater control circuit), the control unit is therefore able to control the operation of the heating element by controlling the earth circuit. Note that some systems may also provide the power supply to the heating element via the control unit. It may be necessary to refer to vehicle specific information to identify the type of Oxygen / Lambda sensor and the wiring.	The fault code can relate to the earth path from the heater element to the control unit (which can be the control circuit for the heater) but, depending on the wiring for the heater and the manufacturer's interpretation of the code, the code could relate to the heater power supply circuit. The fault code indicates that the voltage / current in the sensor heater control circuit is high e.g. battery voltage when it should be low e.g. zero volts (it would normally be zero / close to zero, when the control unit completes the earth circuit). The fault could also be caused by a short from the heater earth circuit to another circuit (e.g. power supply). Check all applicable wiring for short / open circuits, and check for good connections between the heater and the control unit. Refer to list of sensor / actuator checks on page 19.
P0033	Turbocharger / Supercharger Bypass Valve Control Circuit	Turbo / supercharger systems have different methods of regulating boost pressure e.g. wastegate or variable geometry control. Some systems (blow through systems where the turbo / supercharger is located before / upstream of the throttle body), have a bypass valve (blow off / dump valve) which opens to relieve pressure build up and surging in the intake system during closed throttle deceleration. The bypass valve actuator is often a solenoid valve which controls the vacuum used to open / close the bypass valve.	The voltage, frequency or value of the control signal in the actuator circuit is incorrect (undefined fault). Signal could be out of normal operating range, constant value, non-existent, corrupt. Possible fault with actuator or wiring e.g. actuator power supply, short / open circuit, high circuit resistance or incorrect actuator resistance. Note: the control unit could be providing the correct signal but it is being affected by a circuit fault. Refer to list of sensor / actuator checks on page 19.
P0034	Turbocharger / Supercharger Bypass Valve Control Circuit Low	Turbo / supercharger systems have different methods of regulating boost pressure e.g. wastegate or variable geometry control. Some systems (blow through systems where the turbo / supercharger is located before / upstream of the throttle body), have a bypass valve (blow off / dump valve) which opens to relieve pressure build up and surging in the intake system during closed throttle deceleration. The bypass valve actuator is often a solenoid valve which controls the vacuum used to open / close the bypass valve.	The voltage, frequency or value of the control signal in the actuator circuit is either "zero" (no signal) or, the signal exists but is below the normal operating range. Possible fault with actuator or wiring (e.g. actuator power supply, short / open circuit or resistance) that is causing a signal value to be lower than the minimum operating limit. Note: the control unit could be providing the correct signal but it is being affected by a circuit fault. Refer to list of sensor / actuator checks on page 19.
P0035	Turbocharger / Supercharger Bypass Valve Control Circuit High	Turbo / supercharger systems have different methods of regulating boost pressure e.g. wastegate or variable geometry control. Some systems (blow through systems where the turbo / supercharger is located before / upstream of the throttle body), have a bypass valve (blow off / dump valve) which opens to relieve pressure build up and surging in the intake system during closed throttle deceleration. The bypass valve actuator is often a solenoid valve which controls the vacuum used to open / close the bypass valve.	The voltage, frequency or value of the control signal in the actuator circuit is either at full value (e.g. battery voltage with no on / off signal) or, the signal exists but it is above the normal operating range. Possible fault with actuator or wiring (e.g. actuator power supply, short / open circuit or resistance) that is causing a signal value to be higher than the maximum operating limit. Note: the control unit could be providing the correct signal but it is being affected by a circuit fault. Refer to list of sensor / actuator checks on page 19.
P0036	HO2S Heater Control Circuit Bank 1 Sensor 2	The heater control for the HO2S (oxygen / Lambda) sensor is used to heat up and provide stable temperature for the sensor (primarily after cold starts). The heating element can receive a power supply via a relay (often the main system relay) and the heater earth path can be switched to earth via the control unit (this is the heater control circuit), the control unit is therefore able to control the operation of the heating element by controlling the earth circuit. Note that some systems may also provide the power supply to the heating element via the control unit. It may be necessary to refer to vehicle specific information to identify the type of Oxygen / Lambda sensor and the wiring.	The fault code can relate to the earth path from the heater element to the control unit (which can be the control circuit for the heater), but note that in some cases (depending on the wiring for the heater and the manufacturers interpretation of the code), the code could relate to the heater power supply circuit. The fault code indicates that there is an undefined fault in the sensor heater control circuit, which is causing an incorrect voltage / current in the circuit; this is likely to be: non-existent, constant value (but incorrect for the operating conditions), corrupt or, is out of normal operating range (out of tolerance). The fault could be caused by a faulty sensor heater element (open circuit or high resistance) or a wiring fault (short / open circuit). Also check heater supply voltage (could be provided via a relay or from a control unit). Refer to list of sensor / actuator checks on page 19.
P0037	HO2S Heater Control Circuit Low Bank 1 Sensor 2	The heater control for the HO2S (oxygen / Lambda) sensor is used to heat up and provide stable temperature for the sensor (primarily after cold starts). The heating element can receive a power supply via a relay (often the main system relay) and the heater earth path can be switched to earth via the control unit (this is the heater control circuit), the control unit is therefore able to control the operation of the heating element by controlling the earth circuit. Note that some systems may also provide the power supply to the heating element via the control unit. It may be necessary to refer to vehicle specific information to identify the type of Oxygen / Lambda sensor and the wiring.	The fault code can relate to the earth path from the heater element to the control unit (which can be the control circuit for the heater) but, depending on the wiring for the heater and the manufacturer's interpretation of the code, the code could relate to the heater power supply circuit. The fault code indicates that the voltage / current in the sensor heater control circuit is low e.g. zero volts, when the control unit is expecting to detect a high voltage e.g. battery volts (when the control unit is not completing the earth circuit for the heater, battery voltage would normally be detected in this circuit). The fault is likely to be caused by an open circuit in the wiring or heater element (or short to earth); it also possible that there is no power supply (either from the relay or control unit, as applicable). Check power supply and wiring to heater, check continuity / resistance of heater element (as applicable), and check wiring through to the control unit for short / open circuit to heater. Refer to list of sensor / actuator checks on page 19.

NOTE 1. Check for other fault codes that could provide additional information.   NOTE 2. Communication between control units can pass via a CAN-Bus system; refer to Page 51 for CAN-Bus checks.

NOTE 3. If a fault cannot be located, it is also possible that the control unit is at fault.   NOTE 4. Refer to Page 53 for list of pages that show the ISO standard for component locations e.g. Sensor A - Bank 1.

Code	Description		
P0038	HO2S Heater Control Circuit High Bank 1 Sensor 2	The heater control for the HO2S (oxygen / Lambda) sensor is used to heat up and provide stable temperature for the sensor (primarily after cold starts). The heating element can receive a power supply via a relay (often the main system relay) and the heater earth path can be switched to earth via the control unit (this is the heater control circuit), the control unit is therefore able to control the operation of the heating element by controlling the earth circuit. Note that some systems may also provide the power supply to the heating element via the control unit. It may be necessary to refer to vehicle specific information to identify the type of Oxygen / Lambda sensor and the wiring.	The fault code can relate to the earth path from the heater element to the control unit (which can be the control circuit for the heater) but, depending on the wiring for the heater and the manufacturer's interpretation of the code, the code could relate to the heater power supply circuit. The fault code indicates that the voltage / current in the sensor heater control circuit is high e.g. battery voltage when it should be low e.g. zero volts (it would normally be zero / close to zero, when the control unit completes the earth circuit). The fault could also be caused by a short from the heater earth circuit to another circuit (e.g. power supply). Check all applicable wiring for short / open circuits, and check for good connections between the heater and the control unit. Refer to list of sensor / actuator checks on page 19.
P0039	Turbocharger / Supercharger Bypass Valve Control Circuit Range / Performance	Turbo / supercharger systems have different methods of regulating boost pressure e.g. wastegate or variable geometry control. Some systems (blow through systems where the turbo / supercharger is located before / upstream of the throttle body), have a bypass valve (blow off / dump valve) which opens to relieve pressure build up and surging in the intake system during closed throttle deceleration. The bypass valve actuator is often a solenoid valve which controls the vacuum used to open / close the bypass valve.	The voltage, frequency or value of the control signal in the actuator circuit is within the normal operating range / tolerance, but the signal might be incorrect for the operating conditions. It is also possible that the actuator response (or response of system controlled by the actuator) differs from the expected / desired response (incorrect response to the control signal provided by the control unit). Possible fault in actuator or wiring, but also possible that the actuator (or mechanism / system controlled by the actuator) is not operating or moving correctly. Note: the control unit could be providing the correct signal but it is being affected by a circuit fault. Refer to list of sensor / actuator checks on page 19. Also check bypass valve operation to ensure the valve is not stuck etc. Note that the control unit provides an opening signal to the actuator / bypass valve dependent on information from other sensors e.g. boost pressure, throttle position etc. If a sensor is faulty e.g. boost pressure, when the bypass valve is opened to reduce the pressure, the information from the sensor could still be indicating high pressure; the control unit could assess this as a bypass valve fault. Check for other fault codes and check sensor operation.
P0040	O2 Sensor Signals Swapped Bank 1 Sensor 1 / Bank 2 Sensor 1	Note: Many O2 / Lambda sensor related fault codes are activated due to faults in other systems (engine, fuelling, misfires etc). O2 (Oxygen / Lambda) sensors detect the exhaust gas oxygen content, which is also an indicator of air / fuel ratio (rich / lean). Oxygen content is indicated by a "Lambda" value. The air / fuel ratio (and therefore oxygen content) is controlled by the engine control unit to maximise efficiency of catalytic converters and reduce emissions of harmful gases.	The control unit has been able to identify that the signals from the two identified O2 / Lambda sensors are swapped i.e. signal from one sensor is swapped with the signal from the other sensor. It is likely that the harness connector plugs for the two identified sensors have been connected incorrectly (the connector plugs are most likely identical). Locate the wiring harness for the sensors and reconnect correctly. Also check Lambda sensor wiring and connections in case any wiring repairs have been carried out that could have resulted in the signal wires for the two sensors being swapped. For information on location and identification of the sensors, Refer to list of sensor / actuator checks on page 19.
P0041	O2 Sensor Signals Swapped Bank 1 Sensor 2 / Bank 2 Sensor 2	Note: Many O2 / Lambda sensor related fault codes are activated due to faults in other systems (engine, fuelling, misfires etc). O2 (Oxygen / Lambda) sensors detect the exhaust gas oxygen content, which is also an indicator of air / fuel ratio (rich / lean). Oxygen content is indicated by a "Lambda" value. The air / fuel ratio (and therefore oxygen content) is controlled by the engine control unit to maximise efficiency of catalytic converters and reduce emissions of harmful gases.	The control unit has been able to identify that the signals from the two identified O2 / Lambda sensors are swapped i.e. signal from one sensor is swapped with the signal from the other sensor. It is likely that the harness connector plugs for the two identified sensors have been connected incorrectly (the connector plugs are most likely identical). Locate the wiring harness for the sensors and reconnect correctly. Also check Lambda sensor wiring and connections in case any wiring repairs have been carried out that could have resulted in the signal wires for the two sensors being swapped. For information on location and identification of the sensors, Refer to list of sensor / actuator checks on page 19.
P0042	HO2S Heater Control Circuit Bank 1 Sensor 3	The heater control for the HO2S (oxygen / Lambda) sensor is used to heat up and provide stable temperature for the sensor (primarily after cold starts). The heating element can receive a power supply via a relay (often the main system relay) and the heater earth path can be switched to earth via the control unit (this is the heater control circuit), the control unit is therefore able to control the operation of the heating element by controlling the earth circuit. Note that some systems may also provide the power supply to the heating element via the control unit. It may be necessary to refer to vehicle specific information to identify the type of Oxygen / Lambda sensor and the wiring.	The fault code can relate to the earth path from the heater element to the control unit (which can be the control circuit for the heater), but note that in some cases (depending on the wiring for the heater and the manufacturers interpretation of the code), the code could relate to the heater power supply circuit. The fault code indicates that there is an undefined fault in the sensor heater control circuit, which is causing an incorrect voltage / current in the circuit; this is likely to be: non-existent, constant value (but incorrect for the operating conditions), corrupt or, is out of normal operating range (out of tolerance). The fault could be caused by a faulty sensor heater element (open circuit or high resistance) or a wiring fault (short / open circuit). Also check heater supply voltage (could be provided via a relay or from a control unit). Refer to list of sensor / actuator checks on page 19.
P0043	HO2S Heater Control Circuit Low Bank 1 Sensor 3	The heater control for the HO2S (oxygen / Lambda) sensor is used to heat up and provide stable temperature for the sensor (primarily after cold starts). The heating element can receive a power supply via a relay (often the main system relay) and the heater earth path can be switched to earth via the control unit (this is the heater control circuit), the control unit is therefore able to control the operation of the heating element by controlling the earth circuit. Note that some systems may also provide the power supply to the heating element via the control unit. It may be necessary to refer to vehicle specific information to identify the type of Oxygen / Lambda sensor and the wiring.	The fault code can relate to the earth path from the heater element to the control unit (which can be the control circuit for the heater) but, depending on the wiring for the heater and the manufacturer's interpretation of the code, the code could relate to the heater power supply circuit. The fault code indicates that the voltage / current in the sensor heater control circuit is low e.g. zero volts, when the control unit is expecting to detect a high voltage e.g. battery volts (when the control unit is not completing the earth circuit for the heater, battery voltage would normally be detected in this circuit). The fault is likely to be caused by an open circuit in the wiring or heater element (or short to earth); it also possible that there is no power supply (either from the relay or control unit, as applicable). Check power supply and wiring to heater, check continuity / resistance of heater element (as applicable), and check wiring through to the control unit for short / open circuit to heater. Refer to list of sensor / actuator checks on page 19.

NOTE 1. Check for other fault codes that could provide additional information.
NOTE 2. Communication between control units can pass via a CAN-Bus system; refer to Page 51 for CAN-Bus checks.
NOTE 3. If a fault cannot be located, it is also possible that the control unit is at fault.
NOTE 4. Refer to Page 53 for list of pages that show the ISO standard for component locations e.g. Sensor A - Bank 1.

Fault code	EOBD / ISO Description	Component / System Description	Meaningful Description and Quick Check
P0044	HO2S Heater Control Circuit High Bank 1 Sensor 3	The heater control for the HO2S (oxygen / Lambda) sensor is used to heat up and provide stable temperature for the sensor (primarily after cold starts). The heating element can receive a power supply via a relay (often the main system relay) and the heater earth path can be switched to earth via the control unit (this is the heater control circuit), the control unit is therefore able to control the operation of the heating element by controlling the earth circuit. Note that some systems may also provide the power supply to the heating element via the control unit. It may be necessary to refer to vehicle specific information to identify the type of Oxygen / Lambda sensor and the wiring.	The fault code can relate to the earth path from the heater element to the control unit (which can be the control circuit for the heater) but, depending on the wiring for the heater and the manufacturer's interpretation of the code, the code could relate to the heater power supply circuit. The fault code indicates that the voltage / current in the sensor heater control circuit is high e.g. battery voltage when it should be low e.g. zero volts (it would normally be zero / close to zero, when the control unit completes the earth circuit). The fault could also be caused by a short from the heater earth circuit to another circuit (e.g. power supply). Check all applicable wiring for short / open circuits, and check for good connections between the heater and the control unit. Refer to list of sensor / actuator checks on page 19.
P0045	Turbocharger / Supercharger Boost Control Solenoid "A" Circuit / Open	The boost control solenoid is generally used on turbochargers that have: Variable geometry, Variable vane /nozzle. The geometry change of the turbine blades or nozzle blades regulates boost pressure instead of a boost pressure wastegate. The solenoid controls the vacuum passing to the geometry control mechanism. If the turbo / supercharger does use a "wastegate" type of control mechanism, similar checks will apply.	The voltage, frequency or value of the control signal in the solenoid circuit is incorrect. Signal could be out of normal operating range, constant value, non-existent, corrupt. The fault code indicates a possible "open circuit" but other undefined faults in the solenoid or wiring could activate the same code e.g. solenoid power supply, short circuit, high circuit resistance or incorrect solenoid resistance. Note: the control unit could be providing the correct signal but it is being affected by the circuit fault. Refer to list of sensor / actuator checks on page 19.
P0046	Turbocharger / Supercharger Boost Control Solenoid "A" Circuit Range / Performance	The boost control solenoid is generally used on turbochargers that have: Variable geometry, Variable vane /nozzle. The geometry change of the turbine blades or nozzle blades regulates boost pressure instead of a boost pressure wastegate. The solenoid controls the vacuum passing to the geometry control mechanism. If the turbo / supercharger does use a "wastegate" type of control mechanism, similar checks will apply.	The voltage, frequency or value of the control signal in the solenoid circuit is within the normal operating range / tolerance, but the signal might be incorrect for the operating conditions. It is also possible that the solenoid response (or response of system controlled by the solenoid) differs from the expected / desired response (incorrect response to the control signal provided by the control unit). Possible fault in solenoid or wiring, but also possible that the solenoid (or mechanism / system controlled by the solenoid) is not operating or moving correctly. Note: the control unit could be providing the correct signal but it is being affected by a circuit fault. Refer to list of sensor / actuator checks on page 19. Note that the solenoid could be operating correctly but the vacuum valve being operated by the solenoid could be faulty (leaking etc). Also check for leaks or blockages in vacuum pipes Check that when vacuum is applied to the vacuum actuator, the boost control linkage / mechanism is moving correctly.
P0047	Turbocharger / Supercharger Boost Control Solenoid "A" Circuit Low	The boost control solenoid is generally used on turbochargers that have: Variable geometry, Variable vane /nozzle. The geometry change of the turbine blades or nozzle blades regulates boost pressure instead of a boost pressure wastegate. The solenoid controls the vacuum passing to the geometry control mechanism. If the turbo / supercharger does use a "wastegate" type of control mechanism, similar checks will apply.	The voltage, frequency or value of the control signal in the solenoid circuit is either "zero" (no signal) or, the signal exists but is below the normal operating range. Possible fault with solenoid or wiring (e.g. solenoid power supply, short / open circuit or resistance) that is causing a signal value to be lower than the minimum operating limit. Note: the control unit could be providing the correct signal but it is being affected by a circuit fault. Refer to list of sensor / actuator checks on page 19.
P0048	Turbocharger / Supercharger Boost Control Solenoid "A" Circuit High	The boost control solenoid is generally used on turbochargers that have: Variable geometry, Variable vane /nozzle. The geometry change of the turbine blades or nozzle blades regulates boost pressure instead of a boost pressure wastegate. The solenoid controls the vacuum passing to the geometry control mechanism. If the turbo / supercharger does use a "wastegate" type of control mechanism, similar checks will apply.	The voltage, frequency or value of the control signal in the solenoid circuit is either at full value (e.g. battery voltage with no on / off signal) or, the signal exists but it is above the normal operating range. Possible fault with solenoid or wiring (e.g. solenoid power supply, short / open circuit or resistance) that is causing a signal value to be higher than the maximum operating limit. Note: the control unit could be providing the correct signal but it is being affected by a circuit fault. Refer to list of sensor / actuator checks on page 19.
P0049	Turbocharger / Supercharger Turbine Overspeed	Speed sensors can be used to detect the turbine speed on turbo / supercharger systems.	The control unit has detected excessive turbine speed (over speed) in the Turbo / Super charger. The fault could be a genuine overspeed problem or, it could possibly be a speed sensor fault. Refer to list of sensor / actuator checks on page 19.
P004A	Turbocharger / Supercharger Boost Control Solenoid "B" Circuit / Open	The boost control solenoid is generally used on turbochargers that have: Variable geometry, Variable vane /nozzle. The geometry change of the turbine blades or nozzle blades regulates boost pressure instead of a boost pressure wastegate. The solenoid controls the vacuum passing to the geometry control mechanism. If the turbo / supercharger does use a "wastegate" type of control mechanism, similar checks will apply.	The voltage, frequency or value of the control signal in the solenoid circuit is incorrect. Signal could be out of normal operating range, constant value, non-existent, corrupt. The fault code indicates a possible "open circuit" but other undefined faults in the solenoid or wiring could activate the same code e.g. solenoid power supply, short circuit, high circuit resistance or incorrect solenoid resistance. Note: the control unit could be providing the correct signal but it is being affected by the circuit fault. Refer to list of sensor / actuator checks on page 19.
P004B	Turbocharger / Supercharger Boost Control Solenoid "B" Circuit Range / Performance	The boost control solenoid is generally used on turbochargers that have: Variable geometry, Variable vane /nozzle. The geometry change of the turbine blades or nozzle blades regulates boost pressure instead of a boost pressure wastegate. The solenoid controls the vacuum passing to the geometry control mechanism. If the turbo / supercharger does use a "wastegate" type of control mechanism, similar checks will apply.	The voltage, frequency or value of the control signal in the solenoid circuit is within the normal operating range / tolerance, but the signal might be incorrect for the operating conditions. It is also possible that the solenoid response (or response of system controlled by the solenoid) differs from the expected / desired response (incorrect response to the control signal provided by the control unit). Possible fault in solenoid or wiring, but also possible that the solenoid (or mechanism / system controlled by the solenoid) is not operating or moving correctly. Note: the control unit could be providing the correct signal but it is being affected by a circuit fault. Refer to list of sensor / actuator checks on page 19. Note that the solenoid could be operating correctly but the vacuum valve being operated by the solenoid could be faulty (leaking etc). Also check for leaks or blockages in vacuum pipes Check that when vacuum is applied to the vacuum actuator, the boost control linkage / mechanism is moving correctly.
P004C	Turbocharger / Supercharger Boost Control Solenoid "B" Circuit Low	The boost control solenoid is generally used on turbochargers that have: Variable geometry, Variable vane /nozzle. The geometry change of the turbine blades or nozzle blades regulates boost pressure instead of a boost pressure wastegate. The solenoid controls the vacuum passing to the geometry control mechanism. If the turbo / supercharger does use a "wastegate" type of control mechanism, similar checks will apply.	The voltage, frequency or value of the control signal in the solenoid circuit is either "zero" (no signal) or, the signal exists but is below the normal operating range. Possible fault with solenoid or wiring (e.g. solenoid power supply, short / open circuit or resistance) that is causing a signal value to be lower than the minimum operating limit. Note: the control unit could be providing the correct signal but it is being affected by a circuit fault. Refer to list of sensor / actuator checks on page 19.

NOTE 1. Check for other fault codes that could provide additional information. **NOTE 2.** Communication between control units can pass via a CAN-Bus system; refer to Page 51 for CAN-Bus checks.

NOTE 3. If a fault cannot be located, it is also possible that the control unit is at fault NOTE 4. Refer to Page 53 for list of pages that show the ISO standard for component locations e.g. Sensor A - Bank 1.

code	Description		
P004D	Turbocharger / Supercharger Boost Control Solenoid "B" Circuit High	The boost control solenoid is generally used on turbochargers that have: Variable geometry, Variable vane /nozzle. The geometry change of the turbine blades or nozzle blades regulates boost pressure instead of a boost pressure wastegate. The solenoid controls the vacuum passing to the geometry control mechanism. If the turbo / supercharger does use a "wastegate" type of control mechanism, similar checks will apply.	The voltage, frequency or value of the control signal in the solenoid circuit is either at full value (e.g. battery voltage with no on / off signal) or, the signal exists but it is above the normal operating range. Possible fault with solenoid or wiring (e.g. solenoid power supply, short / open circuit or resistance) that is causing a signal value to be higher than the maximum operating limit. Note: the control unit could be providing the correct signal but it is being affected by a circuit fault. Refer to list of sensor / actuator checks on page 19.
P004E	Turbocharger / Supercharger Boost Control Solenoid "A" Circuit Intermittent / Erratic	The boost control solenoid is generally used on turbochargers that have: Variable geometry, Variable vane /nozzle. The geometry change of the turbine blades or nozzle blades regulates boost pressure instead of a boost pressure wastegate. The solenoid controls the vacuum passing to the geometry control mechanism. If the turbo / supercharger does use a "wastegate" type of control mechanism, similar checks will apply.	The voltage, frequency or value of the control signal in the solenoid circuit is intermittent (likely to be out of normal operating range / tolerance when intermittent fault is detected) or, the signal is erratic e.g. signal changes are irregular / unpredictable. Possible fault with solenoid or wiring (e.g. internal / external connections, broken wiring) that is causing intermittent or erratic signal to exist. Note: the control unit could be providing the correct signal but it is being affected by a circuit fault. Refer to list of sensor / actuator checks on page 19.
P004F	Turbocharger / Supercharger Boost Control Solenoid "B" Circuit Intermittent / Erratic	The boost control solenoid is generally used on turbochargers that have: Variable geometry, Variable vane /nozzle. The geometry change of the turbine blades or nozzle blades regulates boost pressure instead of a boost pressure wastegate. The solenoid controls the vacuum passing to the geometry control mechanism. If the turbo / supercharger does use a "wastegate" type of control mechanism, similar checks will apply.	The voltage, frequency or value of the control signal in the solenoid circuit is intermittent (likely to be out of normal operating range / tolerance when intermittent fault is detected) or, the signal is erratic e.g. signal changes are irregular / unpredictable. Possible fault with solenoid or wiring (e.g. internal / external connections, broken wiring) that is causing intermittent or erratic signal to exist. Note: the control unit could be providing the correct signal but it is being affected by a circuit fault. Refer to list of sensor / actuator checks on page 19.
P0050	HO2S Heater Control Circuit Bank 2 Sensor 1	The heater control for the HO2S (oxygen / Lambda) sensor is used to heat up and provide stable temperature for the sensor (primarily after cold starts). The heating element can receive a power supply via a relay (often the main system relay) and the heater earth path can be switched to earth via the control unit (this is the heater control circuit), the control unit is therefore able to control the operation of the heating element by controlling the earth circuit. Note that some systems may also provide the power supply to the heating element via the control unit. It may be necessary to refer to vehicle specific information to identify the type of Oxygen / Lambda sensor and the wiring.	The fault code can relate to the earth path from the heater element to the control unit (which can be the control circuit for the heater), but note that in some cases (depending on the wiring for the heater and the manufacturers interpretation of the code), the code could relate to the heater power supply circuit. The fault code indicates that there is an undefined fault in the sensor heater control circuit, which is causing an incorrect voltage / current in the circuit; this is likely to be: non-existent, constant value (but incorrect for the operating conditions), corrupt or, is out of normal operating range (out of tolerance). The fault could be caused by a faulty sensor heater element (open circuit or high resistance) or a wiring fault (short / open circuit). Also check heater supply voltage (could be provided via a relay or from a control unit). Refer to list of sensor / actuator checks on page 19.
P0051	HO2S Heater Control Circuit Low Bank 2 Sensor 1	The heater control for the HO2S (oxygen / Lambda) sensor is used to heat up and provide stable temperature for the sensor (primarily after cold starts). The heating element can receive a power supply via a relay (often the main system relay) and the heater earth path can be switched to earth via the control unit (this is the heater control circuit), the control unit is therefore able to control the operation of the heating element by controlling the earth circuit. Note that some systems may also provide the power supply to the heating element via the control unit. It may be necessary to refer to vehicle specific information to identify the type of Oxygen / Lambda sensor and the wiring.	The fault code can relate to the earth path from the heater element to the control unit (which can be the control circuit for the heater) but, depending on the wiring for the heater and the manufacturer's interpretation of the code, the code could relate to the heater power supply circuit. The fault code indicates that the voltage / current in the sensor heater control circuit is low e.g. zero volts, when the control unit is expecting to detect a high voltage e.g. battery volts (when the control unit is not completing the earth circuit for the heater, battery voltage would normally be detected in this circuit). The fault is likely to be caused by an open circuit in the wiring or heater element (or short to earth); it also possible that there is no power supply (either from the relay or control unit, as applicable). Check power supply and wiring to heater, check continuity / resistance of heater element (as applicable), and check wiring through to the control unit for short / open circuit to heater. Refer to list of sensor / actuator checks on page 19.
P0052	HO2S Heater Control Circuit High Bank 2 Sensor 1	The heater control for the HO2S (oxygen / Lambda) sensor is used to heat up and provide stable temperature for the sensor (primarily after cold starts). The heating element can receive a power supply via a relay (often the main system relay) and the heater earth path can be switched to earth via the control unit (this is the heater control circuit), the control unit is therefore able to control the operation of the heating element by controlling the earth circuit. Note that some systems may also provide the power supply to the heating element via the control unit. It may be necessary to refer to vehicle specific information to identify the type of Oxygen / Lambda sensor and the wiring.	The fault code can relate to the earth path from the heater element to the control unit (which can be the control circuit for the heater) but, depending on the wiring for the heater and the manufacturer's interpretation of the code, the code could relate to the heater power supply circuit. The fault code indicates that the voltage / current in the sensor heater control circuit is high e.g. battery voltage when it should be low e.g. zero volts (it would normally be zero / close to zero, when the control unit completes the earth circuit). The fault could also be caused by a short from the heater earth circuit to another circuit (e.g. power supply). Check all applicable wiring for short / open circuits, and check for good connections between the heater and the control unit. Refer to list of sensor / actuator checks on page 19.
P0053	HO2S Heater Resistance Bank 1 Sensor 1	The heater control for the HO2S (oxygen / Lambda) sensor is used to heat up and provide stable temperature for the sensor (primarily after cold starts). The heating element can receive a power supply via a relay (often the main system relay) or from the control unit on some systems; the heater earth path can be switched to earth via the control unit (this is the heater control circuit). It may be necessary to refer to vehicle specific information to identify the type of Oxygen / Lambda sensor and the wiring, and also the resistance value for the heater element.	The control unit monitors the voltage / current in the Oxygen / Lambda sensor heater circuit (the control unit is connected to the circuit to enable switching the heater on / off). The fault code indicates that the resistance of the heater element is incorrect; this is likely to be detected due to incorrect voltage / current in the circuit. The likely cause of the fault is an incorrect heater element resistance but it could also be caused by high resistance in the wiring and connections. Check the heater element resistance, check all wiring (including power supply circuit) for high resistance, short / open circuits; also check for correct power supply voltage. Refer to list of sensor / actuator checks on page 19.

NOTE 1. Check for other fault codes that could provide additional information. NOTE 2. Communication between control units can pass via a CAN-Bus system; refer to Page 51 for CAN-Bus checks.

NOTE 3. If a fault cannot be located, it is also possible that the control unit is at fault. NOTE 4. Refer to Page 53 for list of pages that show the ISO standard for component locations e.g. Sensor A - Bank 1.

Fault code	EOBD / ISO Description	Component / System Description	Meaningful Description and Quick Check
P0054	HO2S Heater Resistance Bank 1 Sensor 2	The heater control for the HO2S (oxygen / Lambda) sensor is used to heat up and provide stable temperature for the sensor (primarily after cold starts). The heating element can receive a power supply via a relay (often the main system relay) or from the control unit on some systems; the heater earth path can be switched to earth via the control unit (this is the heater control circuit). It may be necessary to refer to vehicle specific information to identify the type of Oxygen / Lambda sensor and the wiring, and also the resistance value for the heater element.	The control unit monitors the voltage / current in the Oxygen / Lambda sensor heater circuit (the control unit is connected to the circuit to enable switching the heater on / off). The fault code indicates that the resistance of the heater element is incorrect; this is likely to be detected due to incorrect voltage / current in the circuit. The likely cause of the fault is an incorrect heater element resistance but it could also be caused by high resistance in the wiring and connections. Check the heater element resistance, check all wiring (including power supply circuit) for high resistance, short / open circuits; also check for correct power supply voltage. Refer to list of sensor / actuator checks on page 19.
P0055	HO2S Heater Resistance Bank 1 Sensor 3	The heater control for the HO2S (oxygen / Lambda) sensor is used to heat up and provide stable temperature for the sensor (primarily after cold starts). The heating element can receive a power supply via a relay (often the main system relay) or from the control unit on some systems; the heater earth path can be switched to earth via the control unit (this is the heater control circuit). It may be necessary to refer to vehicle specific information to identify the type of Oxygen / Lambda sensor and the wiring, and also the resistance value for the heater element.	The control unit monitors the voltage / current in the Oxygen / Lambda sensor heater circuit (the control unit is connected to the circuit to enable switching the heater on / off). The fault code indicates that the resistance of the heater element is incorrect; this is likely to be detected due to incorrect voltage / current in the circuit. The likely cause of the fault is an incorrect heater element resistance but it could also be caused by high resistance in the wiring and connections. Check the heater element resistance, check all wiring (including power supply circuit) for high resistance, short / open circuits; also check for correct power supply voltage. Refer to list of sensor / actuator checks on page 19.
P0056	HO2S Heater Control Circuit Bank 2 Sensor 2	The heater control for the HO2S (oxygen / Lambda) sensor is used to heat up and provide stable temperature for the sensor (primarily after cold starts). The heating element can receive a power supply via a relay (often the main system relay) and the heater earth path can be switched to earth via the control unit (this is the heater control circuit), the control unit is therefore able to control the operation of the heating element by controlling the earth circuit. Note that some systems may also provide the power supply to the heating element via the control unit. It may be necessary to refer to vehicle specific information to identify the type of Oxygen / Lambda sensor and the wiring.	The fault code can relate to the earth path from the heater element to the control unit (which can be the control circuit for the heater), but note that in some cases (depending on the wiring for the heater and the manufacturers interpretation of the code), the code could relate to the heater power supply circuit. The fault code indicates that there is an undefined fault in the sensor heater control circuit, which is causing an incorrect voltage / current in the circuit; this is likely to be: non-existent, constant value (but incorrect for the operating conditions), corrupt or, is out of normal operating range (out of tolerance). The fault could be caused by a faulty sensor heater element (open circuit or high resistance) or a wiring fault (short / open circuit). Also check heater supply voltage (could be provided via a relay or from a control unit). Refer to list of sensor / actuator checks on page 19.
P0057	HO2S Heater Control Circuit Low Bank 2 Sensor 2	The heater control for the HO2S (oxygen / Lambda) sensor is used to heat up and provide stable temperature for the sensor (primarily after cold starts). The heating element can receive a power supply via a relay (often the main system relay) and the heater earth path can be switched to earth via the control unit (this is the heater control circuit), the control unit is therefore able to control the operation of the heating element by controlling the earth circuit. Note that some systems may also provide the power supply to the heating element via the control unit. It may be necessary to refer to vehicle specific information to identify the type of Oxygen / Lambda sensor and the wiring.	The fault code can relate to the earth path from the heater element to the control unit (which can be the control circuit for the heater) but, depending on the wiring for the heater and the manufacturer's interpretation of the code, the code could relate to the heater power supply circuit. The fault code indicates that the voltage / current in the sensor heater control circuit is low e.g. zero volts, when the control unit is expecting to detect a high voltage e.g. battery volts (when the control unit is not completing the earth circuit for the heater, battery voltage would normally be detected in this circuit). The fault is likely to be caused by an open circuit in the wiring or heater element (or short to earth); it also possible that there is no power supply (either from the relay or control unit, as applicable). Check power supply and wiring to heater, check continuity / resistance of heater element (as applicable), and check wiring through to the control unit for short / open circuit to heater. Refer to list of sensor / actuator checks on page 19.
P0058	HO2S Heater Control Circuit High Bank 2 Sensor 2	The heater control for the HO2S (oxygen / Lambda) sensor is used to heat up and provide stable temperature for the sensor (primarily after cold starts). The heating element can receive a power supply via a relay (often the main system relay) and the heater earth path can be switched to earth via the control unit (this is the heater control circuit), the control unit is therefore able to control the operation of the heating element by controlling the earth circuit. Note that some systems may also provide the power supply to the heating element via the control unit. It may be necessary to refer to vehicle specific information to identify the type of Oxygen / Lambda sensor and the wiring.	The fault code can relate to the earth path from the heater element to the control unit (which can be the control circuit for the heater) but, depending on the wiring for the heater and the manufacturer's interpretation of the code, the code could relate to the heater power supply circuit. The fault code indicates that the voltage / current in the sensor heater control circuit is high e.g. battery voltage when it should be low e.g. zero volts (it would normally be zero / close to zero, when the control unit completes the earth circuit). The fault could also be caused by a short from the heater earth circuit to another circuit (e.g. power supply). Check all applicable wiring for short / open circuits, and check for good connections between the heater and the control unit. Refer to list of sensor / actuator checks on page 19.
P0059	HO2S Heater Resistance Bank 2 Sensor 1	The heater control for the HO2S (oxygen / Lambda) sensor is used to heat up and provide stable temperature for the sensor (primarily after cold starts). The heating element can receive a power supply via a relay (often the main system relay) or from the control unit on some systems; the heater earth path can be switched to earth via the control unit (this is the heater control circuit). It may be necessary to refer to vehicle specific information to identify the type of Oxygen / Lambda sensor and the wiring, and also the resistance value for the heater element.	The control unit monitors the voltage / current in the Oxygen / Lambda sensor heater circuit (the control unit is connected to the circuit to enable switching the heater on / off). The fault code indicates that the resistance of the heater element is incorrect; this is likely to be detected due to incorrect voltage / current in the circuit. The likely cause of the fault is an incorrect heater element resistance but it could also be caused by high resistance in the wiring and connections. Check the heater element resistance, check all wiring (including power supply circuit) for high resistance, short / open circuits; also check for correct power supply voltage. Refer to list of sensor / actuator checks on page 19.
P0060	HO2S Heater Resistance Bank 2 Sensor 2	The heater control for the HO2S (oxygen / Lambda) sensor is used to heat up and provide stable temperature for the sensor (primarily after cold starts). The heating element can receive a power supply via a relay (often the main system relay) or from the control unit on some systems; the heater earth path can be switched to earth via the control unit (this is the heater control circuit). It may be necessary to refer to vehicle specific information to identify the type of Oxygen / Lambda sensor and the wiring, and also the resistance value for the heater element.	The control unit monitors the voltage / current in the Oxygen / Lambda sensor heater circuit (the control unit is connected to the circuit to enable switching the heater on / off). The fault code indicates that the resistance of the heater element is incorrect; this is likely to be detected due to incorrect voltage / current in the circuit. The likely cause of the fault is an incorrect heater element resistance but it could also be caused by high resistance in the wiring and connections. Check the heater element resistance, check all wiring (including power supply circuit) for high resistance, short / open circuits; also check for correct power supply voltage. Refer to list of sensor / actuator checks on page 19.

NOTE 1. Check for other fault codes that could provide additional information. NOTE 2. Communication between control units can pass via a CAN-Bus system; refer to Page 51 for CAN-Bus checks.

NOTE 3. If a fault cannot be located, it is also possible that the control unit is at fault. NOTE 4. Refer to Page 53 for list of pages that show the ISO standard for component locations e.g. Sensor A - Bank 1.

code	Description		
P0061	HO2S Heater Resistance Bank 2 Sensor 3	The heater control for the HO2S (oxygen / Lambda) sensor is used to heat up and provide stable temperature for the sensor (primarily after cold starts). The heating element can receive a power supply via a relay (often the main system relay) or from the control unit on some systems; the heater earth path can be switched to earth via the control unit (this is the heater control circuit). It may be necessary to refer to vehicle specific information to identify the type of Oxygen / Lambda sensor and the wiring, and also the resistance value for the heater element.	The control unit monitors the voltage / current in the Oxygen / Lambda sensor heater circuit (the control unit is connected to the circuit to enable switching the heater on / off) The fault code indicates that the resistance of the heater element is incorrect; this is likely to be detected due to incorrect voltage / current in the circuit. The likely cause of the fault is an incorrect heater element resistance but it could also be caused by high resistance in the wiring and connections. Check the heater element resistance, check all wiring (including power supply circuit) for high resistance, short / open circuits; also check for correct power supply voltage. Refer to list of sensor / actuator checks on page 19.
P0062	HO2S Heater Control Circuit Bank 2 Sensor 3	The heater control for the HO2S (oxygen / Lambda) sensor is used to heat up and provide stable temperature for the sensor (primarily after cold starts). The heating element can receive a power supply via a relay (often the main system relay) and the heater earth path can be switched to earth via the control unit (this is the heater control circuit), the control unit is therefore able to control the operation of the heating element by controlling the earth circuit. Note that some systems may also provide the power supply to the heating element via the control unit. It may be necessary to refer to vehicle specific information to identify the type of Oxygen / Lambda sensor and the wiring.	The fault code can relate to the earth path from the heater element to the control unit (which can be the control circuit for the heater), but note that in some cases (depending on the wiring for the heater and the manufacturers interpretation of the code), the code could relate to the heater power supply circuit. The fault code indicates that there is an undefined fault in the sensor heater control circuit, which is causing an incorrect voltage / current in the circuit; this is likely to be: non-existent, constant value (but incorrect for the operating conditions), corrupt or, is out of normal operating range (out of tolerance). The fault could be caused by a faulty sensor heater element (open circuit or high resistance) or a wiring fault (short / open circuit). Also check heater supply voltage (could be provided via a relay or from a control unit). Refer to list of sensor / actuator checks on page 19.
P0063	HO2S Heater Control Circuit Low Bank 2 Sensor 3	The heater control for the HO2S (oxygen / Lambda) sensor is used to heat up and provide stable temperature for the sensor (primarily after cold starts). The heating element can receive a power supply via a relay (often the main system relay) and the heater earth path can be switched to earth via the control unit (this is the heater control circuit), the control unit is therefore able to control the operation of the heating element by controlling the earth circuit. Note that some systems may also provide the power supply to the heating element via the control unit. It may be necessary to refer to vehicle specific information to identify the type of Oxygen / Lambda sensor and the wiring.	The fault code can relate to the earth path from the heater element to the control unit (which can be the control circuit for the heater) but, depending on the wiring for the heater and the manufacturer's interpretation of the code, the code could relate to the heater power supply circuit. The fault code indicates that the voltage / current in the sensor heater control circuit is low e.g. zero volts, when the control unit is expecting to detect a high voltage e.g. battery volts (when the control unit is not completing the earth circuit for the heater, battery voltage would normally be detected in this circuit). The fault is likely to be caused by an open circuit in the wiring or heater element (or short to earth); it also possible that there is no power supply (either from the relay or control unit, as applicable). Check power supply and wiring to heater, check continuity / resistance of heater element (as applicable), and check wiring through to the control unit for short / open circuit to heater. Refer to list of sensor / actuator checks on page 19.
P0064	HO2S Heater Control Circuit High Bank 2 Sensor 3	The heater control for the HO2S (oxygen / Lambda) sensor is used to heat up and provide stable temperature for the sensor (primarily after cold starts). The heating element can receive a power supply via a relay (often the main system relay) and the heater earth path can be switched to earth via the control unit (this is the heater control circuit), the control unit is therefore able to control the operation of the heating element by controlling the earth circuit. Note that some systems may also provide the power supply to the heating element via the control unit. It may be necessary to refer to vehicle specific information to identify the type of Oxygen / Lambda sensor and the wiring.	The fault code can relate to the earth path from the heater element to the control unit (which can be the control circuit for the heater) but, depending on the wiring for the heater and the manufacturer's interpretation of the code, the code could relate to the heater power supply circuit. The fault code indicates that the voltage / current in the sensor heater control circuit is high e.g. battery voltage when it should be low e.g. zero volts (it would normally be zero / close to zero, when the control unit completes the earth circuit). The fault could also be caused by a short from the heater earth circuit to another circuit (e.g. power supply). Check all applicable wiring for short / open circuits, and check for good connections between the heater and the control unit. Refer to list of sensor / actuator checks on page 19.
P0065	Air Assisted Injector Control Range / Performance	Air assisted injectors are an adaptation of normal fuel injectors but they have the addition of an air injection facility which helps fuel atomisation. Also refer to fault code P0261 for additional information on injectors.	Identify the type of injector and the wiring / control modules used on the system. The voltage, frequency or value of the control signal in the injector circuit is within the normal operating range / tolerance, but the signal might be incorrect for the operating conditions. It is also possible that the injector response (or response of system controlled by the injector) differs from the expected / desired response (incorrect response to the control signal provided by the control unit). Possible fault in injector or wiring, but also possible that the injector (or mechanism / system controlled by the injector) is not operating or moving correctly. Note: the control unit could be providing the correct signal but it is being affected by a circuit fault. Refer to list of sensor / actuator checks on page 19.
P0066	Air Assisted Injector Control Circuit or Circuit Low	Air assisted injectors are an adaptation of normal fuel injectors but they have the addition of an air injection facility which helps fuel atomisation. Also refer to fault code P0261 for additional information on injectors.	Identify the type of injector and the wiring / control modules used on the system. The voltage, frequency or value of the control signal in the injector circuit is either "zero" (no signal) or, the signal exists but is below the normal operating range. Possible fault with injector or wiring (e.g. injector power supply, short / open circuit or resistance) that is causing a signal value to be lower than the minimum operating limit. Note: the control unit could be providing the correct signal but it is being affected by a circuit fault. Refer to list of sensor / actuator checks on page 19. It is possible that the control unit is detecting "no voltage" in the circuit, which could be caused by faults with the power supply, connections or open circuit in wiring or injector.
P0067	Air Assisted Injector Control Circuit High	Air assisted injectors are an adaptation of normal fuel injectors but they have the addition of an air injection facility which helps fuel atomisation. Also refer to fault code P0262 for additional information on injectors.	Identify the type of injector and the wiring / control modules used on the system. The voltage, frequency or value of the control signal in the injector circuit is either at full value (e.g. battery voltage with no on / off signal) or, the signal exists but it is above the normal operating range. Possible fault with injector or wiring (e.g. injector power supply, short / open circuit or resistance) that is causing a signal value to be higher than the maximum operating limit. Note: the control unit could be providing the correct signal but it is being affected by a circuit fault. Refer to list of sensor / actuator checks on page 19. It is possible that the when the control unit completes the earth path for the injector, the voltage on the earth circuit remains at supply voltage level e.g. battery voltage instead of dropping to zero (or close to zero); this could be caused by a open circuit, poor earth for the control unit or control unit fault.

NOTE 1. Check for other fault codes that could provide additional information.   **NOTE 2.** Communication between control units can pass via a CAN-Bus system; refer to Page 51 for CAN-Bus checks.   73

NOTE 3. If a fault cannot be located, it is also possible that the control unit is at fault.   **NOTE 4.** Refer to Page 53 for list of pages that show the ISO standard for component locations e.g. Sensor A - Bank 1.

Fault code	EOBD / ISO Description	Component / System Description	Meaningful Description and Quick Check
P0068	MAP / MAF -Throttle Position Correlation	The MAP (Manifold Absolute Pressure) sensor or MAF (Mass Airflow) sensor (depending on which sensor is fitted) is used in conjunction with the Throttle position sensor to indicate load and airflow to the control unit.	The signals from the different sensors are providing conflicting or implausible information e.g. MAP or MAF sensor or indicates "full load" but Throttle Position sensor indicates "closed throttle". If possible, use alternative method of checking the measurements to establish if one of the sensors / wiring is faulty or whether the measured values are incorrect. It is possible that one of the sensors is operating incorrectly, therefore check each sensor for a fault with the sensor / wiring e.g. short / open circuits, circuit resistance, incorrect sensor resistance or, reference / operating voltage fault. Refer to list of sensor / actuator checks on page 19. Also check for intake airflow restrictions, intake air leaks and blockage or leak in MAP sensor air pipe (if applicable).
P0069	Manifold Absolute Pressure -Barometric Pressure Correlation	The MAP (Manifold Absolute Pressure) sensor is used to detect the vacuum / pressure in the intake manifold and the Barometric sensor provides a reference to the ambient air pressure.	The signals from the different sensors are providing conflicting or implausible information e.g. the MAP sensor and Barometric pressure sensor signal values are different when the engine is not running (ignition on); however, both signal values should be the same under these conditions (the control unit could also carry out checks under engine running conditions). It is possible that one of the sensors is operating incorrectly, therefore check each sensor for a fault with the sensor / wiring e.g. short / open circuits, circuit resistance, incorrect sensor resistance or, reference / operating voltage fault. Refer to list of sensor / actuator checks on page 19. Also check for intake airflow restrictions, intake air leaks and blockage or leak in sensor air pipes.
P006A	MAP - Mass or Volume Air Flow Correlation	The MAP (Manifold Absolute Pressure) sensor is used to detect the vacuum / pressure in the intake manifold. The Mass / volume air flow is used to detect air flow in the intake system.	The signals from the different sensors are providing conflicting or implausible information e.g. the MAP sensor could be indicating a "high engine load" condition but the air flow sensor indicates a "low engine load condition". It is possible that one of the sensors is operating incorrectly, therefore check each sensor for a fault with the sensor / wiring e.g. short / open circuits, circuit resistance, incorrect sensor resistance or, reference / operating voltage fault. Refer to list of sensor / actuator checks on page 19. Also check for intake airflow restrictions, intake air leaks and blockage or leak in MAP sensor air pipe (if applicable).
P006B	MAP - Exhaust Pressure Correlation	The MAP (Manifold Absolute Pressure) sensor is used to detect the vacuum / pressure in the intake manifold and the Exhaust pressure sensor indicates back pressure in the exhaust system.	The signals from the different sensors are providing conflicting or implausible information e.g. the MAP sensor could be indicating "normal operation" but the exhaust pressure sensor indicates a high exhaust pressure, which could indicate a blocked exhaust / catalytic converter. It is possible that one of the sensors is operating incorrectly, therefore check each sensor for a fault with the sensor / wiring e.g. short / open circuits, circuit resistance, incorrect sensor resistance or, reference / operating voltage fault. Refer to list of sensor / actuator checks on page 19. Check for intake airflow restrictions, intake air leaks, exhaust system blockage or leaks and check for blockage or leak in sensor air pipes (if applicable).
P006C	MAP - Turbocharger / Supercharger Inlet Pressure Correlation	The turbo / supercharger boost control (usually performed by the main control unit) can make use of information from different sensors to calculate the required boost pressure under different conditions. For this fault code, the sensors identified are the MAP (Manifold Absolute Pressure) sensor and inlet / intake pressure sensor (usually measuring intake pressure ahead of the turbo / supercharger).	The signals from the two sensors are providing conflicting or implausible information e.g. when the engine is not running (ignition on), the MAP sensor and intake pressure sensor signal values are different; however, both signal values should be the same under these conditions (the control unit could also carry out checks under engine running conditions). If possible, use alternative method of checking the measurements to establish if one of the sensors / wiring is faulty or whether the measured values are incorrect. It is possible that one of the sensors is operating incorrectly therefore check each sensor for a fault with the sensor / wiring e.g. short / open circuits, circuit resistance, incorrect sensor resistance or reference / operating voltage fault. Refer to list of sensor / actuator checks on page 19. Check for intake airflow restrictions, intake air leaks and blockage or leak in sensor air pipes.
P006D	Barometric Pressure - Turbocharger / Supercharger Inlet Pressure Correlation	The turbo / supercharger boost control (usually performed by the main control unit) can make use of information from different sensors to calculate the required boost pressure under different conditions. For this fault code, the sensors identified are the Barometric Pressure sensor (ambient air pressure) and inlet / intake pressure sensor (usually measuring intake pressure after the throttle body but ahead of the turbo / supercharger).	The signals from the different sensors are providing conflicting or implausible information e.g. when the engine is not running (ignition on), the Barometric sensor and intake pressure sensor signal values are different; however, both signal values should be the same under these conditions (the control unit could also carry out checks under engine running conditions). It is possible that one of the sensors is operating incorrectly, therefore check each sensor for a fault with the sensor / wiring e.g. short / open circuits, circuit resistance, incorrect sensor resistance or, reference / operating voltage fault. Refer to list of sensor / actuator checks on page 19. Also check for intake airflow restrictions, intake air leaks and blockage or leak in sensor air pipes.
P006E		ISO / SAE reserved	Not used, reserved for future allocation.
P006F		ISO / SAE reserved	Not used, reserved for future allocation.
P0070	Ambient Air Temperature Sensor Circuit	The ambient temperature sensor can be used for fine tuning of fuel, ignition, emissions control and boost pressure control, but can also be used for other vehicle systems e.g. air conditioning, driver display etc.	The voltage, frequency or value of the temperature sensor signal is incorrect (undefined fault). Sensor voltage / signal could be: out of normal operating range, constant value, non-existent, corrupt. Possible fault in the temperature sensor or wiring e.g. short / open circuits, circuit resistance, incorrect sensor resistance or, fault with the reference / operating voltage; interference from other circuits can also affect sensor signals. Refer to list of sensor / actuator checks on page 19.
P0071	Ambient Air Temperature Sensor Range / Performance	The ambient temperature sensor can be used for fine tuning of fuel, ignition, emissions control and boost pressure control, but can also be used for other vehicle systems e.g. air conditioning, driver display etc.	The voltage, frequency or value of the temperature sensor signal is within the normal operating range / tolerance but the signal is not plausible or is incorrect due to an undefined fault. The sensor signal might not match the operating conditions indicated by other sensors or, the sensor signal is not changing / responding as expected to the changes in conditions. Possible fault in the temperature sensor or wiring e.g. short / open circuits, circuit resistance, incorrect sensor resistance or, fault with the reference / operating voltage; interference from other circuits can also affect sensor signals. Refer to list of sensor / actuator checks on page 19. It is also possible that the temperature sensor is operating correctly but the temperature being measured by the sensor is unacceptable or incorrect for the operating conditions; if possible, use alternative method of checking temperature to establish if sensor system is operating correctly and whether temperature is correct / incorrect.
P0072	Ambient Air Temperature Sensor Circuit Low	The ambient temperature sensor can be used for fine tuning of fuel, ignition, emissions control and boost pressure control, but can also be used for other vehicle systems e.g. air conditioning, driver display etc.	The voltage, frequency or value of the temperature sensor signal is either "zero" (no signal) or, the signal exists but it is below the normal operating range (lower than the minimum operating limit). Possible fault in the temperature sensor or wiring e.g. short / open circuits, circuit resistance, incorrect sensor resistance or, fault with the reference / operating voltage; interference from other circuits can also affect sensor signals. Refer to list of sensor / actuator checks on page 19.

Fault code	Description		
P0073	Ambient Air Temperature Sensor Circuit High	The ambient temperature sensor can be used for fine tuning of fuel, ignition, emissions control and boost pressure control, but can also be used for other vehicle systems e.g. air conditioning, driver display etc.	The voltage, frequency or value of the temperature sensor signal is either at full value (e.g. battery voltage with no signal or frequency) or, the signal exists but it is above the normal operating range (higher than the maximum operating limit). Possible fault in the temperature sensor or wiring e.g. short / open circuits, circuit resistance, incorrect sensor resistance or, fault with the reference / operating voltage; interference from other circuits can also affect sensor signals. Refer to list of sensor / actuator checks on page 19.
P0074	Ambient Air Temperature Sensor Circuit Intermittent	The ambient temperature sensor can be used for fine tuning of fuel, ignition, emissions control and boost pressure control, but can also be used for other vehicle systems e.g. air conditioning, driver display etc.	The voltage, frequency or value of the temperature sensor signal is intermittent (likely to be out of normal operating range / tolerance when intermittent fault is detected) or, the signal / voltage is erratic (e.g. signal changes are irregular / unpredictable). Possible fault with wiring or connections (including internal sensor connections); also interference from other circuits can affect the signal. Refer to list of sensor / actuator checks on page 19.
P0075	Intake Valve Control Solenoid Circuit Bank 1	The valve control solenoid is used to alter camshaft position / valve timing / valve lift (depending on system); the control solenoid regulates oil pressure (or other mechanism) which alters the position as required.	The voltage, frequency or value of the control signal in the solenoid circuit is incorrect (undefined fault). Signal could be:- out of normal operating range, constant value, non-existent, corrupt. Possible fault with solenoid or wiring e.g. solenoid power supply, short / open circuit, high circuit resistance or incorrect solenoid resistance. Note: the control unit could be providing the correct signal but it is being affected by the circuit fault. Refer to list of sensor / actuator checks on page 19.
P0076	Intake Valve Control Solenoid Circuit Low Bank 1	The valve control solenoid is used to alter camshaft position / valve timing / valve lift (depending on system); the control solenoid regulates oil pressure (or other mechanism) which alters the position as required.	The voltage, frequency or value of the control signal in the solenoid circuit is either "zero" (no signal) or, the signal exists but is below the normal operating range. Possible fault with solenoid or wiring (e.g. solenoid power supply, short / open circuit or resistance) that is causing a signal value to be lower than the minimum operating limit. Note: the control unit could be providing the correct signal but it is being affected by a circuit fault. Refer to list of sensor / actuator checks on page 19.
P0077	Intake Valve Control Solenoid Circuit High Bank 1	The valve control solenoid is used to alter camshaft position / valve timing / valve lift (depending on system); the control solenoid regulates oil pressure (or other mechanism) which alters the position as required.	The voltage, frequency or value of the control signal in the solenoid circuit is either at full value (e.g. battery voltage with no on / off signal) or, the signal exists but it is above the normal operating range. Possible fault with solenoid or wiring (e.g. solenoid power supply, short / open circuit or resistance) that is causing a signal value to be higher than the maximum operating limit. Note: the control unit could be providing the correct signal but it is being affected by a circuit fault. Refer to list of sensor / actuator checks on page 19.
P0078	Exhaust Valve Control Solenoid Circuit Bank 1	The valve control solenoid is used to alter camshaft position / valve timing / valve lift (depending on system); the control solenoid regulates oil pressure (or other mechanism) which alters the position as required.	The voltage, frequency or value of the control signal in the solenoid circuit is incorrect (undefined fault). Signal could be:- out of normal operating range, constant value, non-existent, corrupt. Possible fault with solenoid or wiring e.g. solenoid power supply, short / open circuit, high circuit resistance or incorrect solenoid resistance. Note: the control unit could be providing the correct signal but it is being affected by the circuit fault. Refer to list of sensor / actuator checks on page 19.
P0079	Exhaust Valve Control Solenoid Circuit Low Bank 1	The valve control solenoid is used to alter camshaft position / valve timing / valve lift (depending on system); the control solenoid regulates oil pressure (or other mechanism) which alters the position as required.	The voltage, frequency or value of the control signal in the solenoid circuit is either "zero" (no signal) or, the signal exists but is below the normal operating range. Possible fault with solenoid or wiring (e.g. solenoid power supply, short / open circuit or resistance) that is causing a signal value to be lower than the minimum operating limit. Note: the control unit could be providing the correct signal but it is being affected by a circuit fault. Refer to list of sensor / actuator checks on page 19.
P0080	Exhaust Valve Control Solenoid Circuit High Bank 1	The valve control solenoid is used to alter camshaft position / valve timing / valve lift (depending on system); the control solenoid regulates oil pressure (or other mechanism) which alters the position as required.	The voltage, frequency or value of the control signal in the solenoid circuit is either at full value (e.g. battery voltage with no on / off signal) or, the signal exists but it is above the normal operating range. Possible fault with solenoid or wiring (e.g. solenoid power supply, short / open circuit or resistance) that is causing a signal value to be higher than the maximum operating limit. Note: the control unit could be providing the correct signal but it is being affected by a circuit fault. Refer to list of sensor / actuator checks on page 19.
P0081	Intake Valve Control Solenoid Circuit Bank 2	The valve control solenoid is used to alter camshaft position / valve timing / valve lift (depending on system); the control solenoid regulates oil pressure (or other mechanism) which alters the position as required.	The voltage, frequency or value of the control signal in the solenoid circuit is incorrect (undefined fault). Signal could be:- out of normal operating range, constant value, non-existent, corrupt. Possible fault with solenoid or wiring e.g. solenoid power supply, short / open circuit, high circuit resistance or incorrect solenoid resistance. Note: the control unit could be providing the correct signal but it is being affected by the circuit fault. Refer to list of sensor / actuator checks on page 19.
P0082	Intake Valve Control Solenoid Circuit Low Bank 2	The valve control solenoid is used to alter camshaft position / valve timing / valve lift (depending on system); the control solenoid regulates oil pressure (or other mechanism) which alters the position as required.	The voltage, frequency or value of the control signal in the solenoid circuit is either "zero" (no signal) or, the signal exists but is below the normal operating range. Possible fault with solenoid or wiring (e.g. solenoid power supply, short / open circuit or resistance) that is causing a signal value to be lower than the minimum operating limit. Note: the control unit could be providing the correct signal but it is being affected by a circuit fault. Refer to list of sensor / actuator checks on page 19.
P0083	Intake Valve Control Solenoid Circuit High Bank 2	The valve control solenoid is used to alter camshaft position / valve timing / valve lift (depending on system); the control solenoid regulates oil pressure (or other mechanism) which alters the position as required.	The voltage, frequency or value of the control signal in the solenoid circuit is either at full value (e.g. battery voltage with no on / off signal) or, the signal exists but it is above the normal operating range. Possible fault with solenoid or wiring (e.g. solenoid power supply, short / open circuit or resistance) that is causing a signal value to be higher than the maximum operating limit. Note: the control unit could be providing the correct signal but it is being affected by a circuit fault. Refer to list of sensor / actuator checks on page 19.
P0084	Exhaust Valve Control Solenoid Circuit Bank 2	The valve control solenoid is used to alter camshaft position / valve timing / valve lift (depending on system); the control solenoid regulates oil pressure (or other mechanism) which alters the position as required.	The voltage, frequency or value of the control signal in the solenoid circuit is incorrect (undefined fault). Signal could be:- out of normal operating range, constant value, non-existent, corrupt. Possible fault with solenoid or wiring e.g. solenoid power supply, short / open circuit, high circuit resistance or incorrect solenoid resistance. Note: the control unit could be providing the correct signal but it is being affected by the circuit fault. Refer to list of sensor / actuator checks on page 19.
P0085	Exhaust Valve Control Solenoid Circuit Low Bank 2	The valve control solenoid is used to alter camshaft position / valve timing / valve lift (depending on system); the control solenoid regulates oil pressure (or other mechanism) which alters the position as required.	The voltage, frequency or value of the control signal in the solenoid circuit is either "zero" (no signal) or, the signal exists but is below the normal operating range. Possible fault with solenoid or wiring (e.g. solenoid power supply, short / open circuit or resistance) that is causing a signal value to be lower than the minimum operating limit. Note: the control unit could be providing the correct signal but it is being affected by a circuit fault. Refer to list of sensor / actuator checks on page 19.

NOTE 1. Check for other fault codes that could provide additional information. **NOTE 2.** Communication between control units can pass via a CAN-Bus system; refer to Page 51 for CAN-Bus checks.

NOTE 3. If a fault cannot be located, it is also possible that the control unit is at fault. **NOTE 4.** Refer to Page 53 for list of pages that show the ISO standard for component locations e.g. Sensor A - Bank 1.

Fault code	EOBD / ISO Description	Component / System Description	Meaningful Description and Quick Check
P0086	Exhaust Valve Control Solenoid Circuit High Bank 2	The valve control solenoid is used to alter camshaft position / valve timing / valve lift (depending on system); the control solenoid regulates oil pressure (or other mechanism) which alters the position as required.	The voltage, frequency or value of the control signal in the solenoid circuit is either at full value (e.g. battery voltage with no on / off signal) or, the signal exists but it is above the normal operating range. Possible fault with solenoid or wiring (e.g. solenoid power supply, short / open circuit or resistance) that is causing a signal value to be higher than the maximum operating limit. Note: the control unit could be providing the correct signal but it is being affected by a circuit fault. Refer to list of sensor / actuator checks on page 19.
P0087	Fuel Rail / System Pressure -Too Low	A sensor indicates the fuel pressure to the control unit, which can then control a number of different functions (depending on the system). On many systems, the control unit will control fuel pressure via a pressure regulator. Where more than one sensor is fitted, this can be due to the monitoring of low and high pressure circuits or for engines using separate fuel delivery systems e.g. "V" engines. It may be necessary to refer to vehicle specific information to identify location of the sensor.	The control unit has detected a low fuel rail pressure under all or some operating conditions (low pressure likely to be detected by pressure sensor signal). Check fuel pressure using separate pressure test equipment and compare against specifications. If possible also check live data readings. If the actual pressure is correct but sensor signal / live data indicates low pressure, check for a possible pressure sensor / wiring fault (refer to fault codes P0190-P0194). If the actual fuel pressure is low, check pressure regulator system (refer to fault codes P0089-P0092). Also check condition of fuel system (blockages, leaks etc) and check fuel pumps (primary and main pumps as applicable).
P0088	Fuel Rail / System Pressure -Too High	A sensor indicates the fuel pressure to the control unit, which can then control a number of different functions (depending on the system). On many systems, the control unit will control fuel pressure via a pressure regulator. Where more than one sensor is fitted, this can be due to the monitoring of low and high pressure circuits or for engines using separate fuel delivery systems e.g. "V" engines. It may be necessary to refer to vehicle specific information to indentify location of the sensor.	The control unit has detected a high fuel rail pressure under all or some operating conditions (high pressure likely to be detected by pressure sensor signal). Check fuel pressure using separate pressure test equipment and compare against specifications. If possible also check live data readings. If the actual pressure is correct but sensor signal / live data indicates high pressure, check for a possible pressure sensor / wiring fault (refer to fault codes P0190-P0194). If the actual fuel pressure is high, check pressure regulator system (refer to fault codes P0089-P0092). Also check condition of fuel system (blockages on return lines etc) and check fuel pumps (primary and main pumps as applicable).
P0089	Fuel Pressure Regulator 1 Performance	The fuel pressure regulator is used to control / maintain the fuel system pressure at the specified value. Regulators are normally controlled by an electrical actuator (usually a solenoid) and pressure sensors are normally used to monitor fuel pressure. Refer to vehicle specific information to identify regulator 1 and 2 (if more than one regulator is fitted).	The control unit has detected an undefined fault / performance problem with the fuel pressure regulator / fuel pressure system. Note that the fault does not indicate an electrical fault with the regulator / wiring but checks should be made to ensure that the wiring and regulator are not at fault. Refer to list of sensor / actuator checks on page 19. The fault could be detected because the fuel pressure sensor is detecting incorrect pressure under some or all operating conditions. Check fuel pressure if possible with a separate gauge and if necessary check operation of pressure sensor and fuel system.
P0090	Fuel Pressure Regulator 1 Control Circuit	The fuel pressure regulator is used to control / maintain the fuel system pressure at the specified value. Regulators are normally controlled by an electrical actuator (usually a solenoid). Refer to vehicle specific information to identify regulator 1 and 2 (if more than one regulator is fitted).	The voltage, frequency or value of the control signal in the regulator circuit is incorrect (undefined fault). Signal could be:- out of normal operating range, constant value, non-existent, corrupt. Possible fault with regulator or wiring e.g. regulator power supply, short / open circuit, high circuit resistance or incorrect regulator resistance. Note: the control unit could be providing the correct signal but it is being affected by the circuit fault. Refer to list of sensor / actuator checks on page 19.
P0091	Fuel Pressure Regulator 1 Control Circuit Low	The fuel pressure regulator is used to control / maintain the fuel system pressure at the specified value. Regulators are normally controlled by an electrical actuator (usually a solenoid). Refer to vehicle specific information to identify regulator 1 and 2 (if more than one regulator is fitted).	The voltage, frequency or value of the control signal in the regulator circuit is either "zero" (no signal) or, the signal exists but is below the normal operating range. Possible fault with regulator or wiring (e.g. regulator power supply, short / open circuit or resistance) that is causing a signal value to be lower than the minimum operating limit. Note: the control unit could be providing the correct signal but it is being affected by the circuit fault. Refer to list of sensor / actuator checks on page 19.
P0092	Fuel Pressure Regulator 1 Control Circuit High	The fuel pressure regulator is used to control / maintain the fuel system pressure at the specified value. Regulators are normally controlled by an electrical actuator (usually a solenoid). Refer to vehicle specific information to identify regulator 1 and 2 (if more than one regulator is fitted).	The voltage, frequency or value of the control signal in the regulator circuit is either at full value (e.g. battery voltage with no on / off signal) or, the signal exists but it is above the normal operating range. Possible fault with regulator or wiring (e.g. regulator power supply, short / open circuit or resistance) that is causing a signal value to be higher than the maximum operating limit. Note: the control unit could be providing the correct signal but it is being affected by the circuit fault. Refer to list of sensor / actuator checks on page 19.
P0093	Fuel System Leak Detected -Large Leak	The fuel system will normally operate at a regulated pressure. If the control unit detects a loss of pressure but there is no fault detected on the pressure regulator or pressure sensor, this will be regarded as a leak. Note that some vehicles are fitted with EVAP (evaporative emission system) leak detection systems; on some of these systems, the EVAP leak test process also embraces part of the main fuel system. If necessary refer to vehicle specific information to identify the process for leak detection.	Control unit detects a large leak; the leak is likely to be detected due to a loss of pressure as indicated by the by the fuel pressure sensor. If the system uses an electronically controlled pressure regulator, the control unit will be therefore be providing a signal to the regulator that should be causing the pressure to increase; however, if the pressure is not increasing as expected, this could also be used as information that indicates a pressure loss. If there are no detectable leaks it is possible that there is a fault with the pressure regulator or pressure sensor. Refer to fault codes P0190 - P0194 for indication of fuel pressure regulator checks, and codes P0090-P0092 for fuel pressure sensor checks.
P0094	Fuel System Leak Detected -Small Leak	The fuel system will normally operate at a regulated pressure. If the control unit detects a loss of pressure but there is no fault detected on the pressure regulator or pressure sensor, this will be regarded as a leak. Note that some vehicles are fitted with EVAP (evaporative emission system) leak detection systems; on some of these systems, the EVAP leak test process also embraces part of the main fuel system. If necessary refer to vehicle specific information to identify the process for leak detection.	Control unit detects a small leak; the leak is likely to be detected due to a loss of pressure as indicated by the by the fuel pressure sensor. If the system uses an electronically controlled pressure regulator, the control unit will be therefore be providing a signal to the regulator that should be causing the pressure to increase; however, if the pressure is not increasing as expected, this could also be used as information that indicates a pressure loss. If there are no detectable leaks it is possible that there is a fault with the pressure regulator or pressure sensor. Refer to fault codes P0190 - P0194 for indication of fuel pressure regulator checks, and codes P0090-P0092 for fuel pressure sensor checks.
P0095	Intake Air Temperature Sensor 2 Circuit	The intake air temperature sensor can be used for fine tuning of fuel, ignition and emissions control. On turbo / supercharged engines, the intake air temperature can form part of the calculation for controlling boost pressure or controlling airflow through the intercooler.	The voltage, frequency or value of the temperature sensor signal is incorrect (undefined fault). Sensor voltage / signal could be: out of normal operating range, constant value, non-existent, corrupt. Possible fault in the temperature sensor or wiring e.g. short / open circuits, circuit resistance, incorrect sensor resistance or, fault with the reference / operating voltage; interference from other circuits can also affect sensor signals. Refer to list of sensor / actuator checks on page 19.

NOTE 1. Check for other fault codes that could provide additional information.   **NOTE 2.** Communication between control units can pass via a CAN-Bus system; refer to Page 51 for CAN-Bus checks.

NOTE 3. If a fault cannot be located, it is also possible that the control unit is at fault.   **NOTE 4.** Refer to Page 53 for list of pages that show the ISO standard for component locations e.g. Sensor A - Bank 1.

Code	Fault Location	Description	Probable Cause
P0096	Intake Air Temperature Sensor 2 Circuit Range / Performance	The intake air temperature sensor can be used for fine tuning of fuel, ignition and emissions control. On turbo / supercharged engines, the intake air temperature can form part of the calculation for controlling boost pressure or controlling airflow through the intercooler.	The voltage, frequency or value of the temperature sensor signal is within the normal operating range / tolerance but the signal is not plausible or is incorrect due to an undefined fault. The sensor signal might not match the operating conditions indicated by other sensors or, the sensor signal is not changing / responding as expected to the changes in conditions. Possible fault in the temperature sensor or wiring e.g. short / open circuits, circuit resistance, incorrect sensor resistance or, fault with the reference / operating voltage;; interference from other circuits can also affect sensor signals. Refer to list of sensor / actuator checks on page 19. It is also possible that the temperature sensor is operating correctly but the temperature being measured by the sensor is unacceptable or incorrect for the operating conditions; if possible, use alternative method of checking temperature to establish if sensor system is operating correctly and whether temperature is correct / incorrect.
P0097	Intake Air Temperature Sensor 2 Circuit Low	The intake air temperature sensor can be used for fine tuning of fuel, ignition and emissions control. On turbo / supercharged engines, the intake air temperature can form part of the calculation for controlling boost pressure or controlling airflow through the intercooler.	The voltage, frequency or value of the temperature sensor signal is either "zero" (no signal) or, the signal exists but it is below the normal operating range (lower than the minimum operating limit). Possible fault in the temperature sensor or wiring e.g. short / open circuits, circuit resistance, incorrect sensor resistance or, fault with the reference / operating voltage; interference from other circuits can also affect sensor signals. Refer to list of sensor / actuator checks on page 19.
P0098	Intake Air Temperature Sensor 2 Circuit High	The intake air temperature sensor can be used for fine tuning of fuel, ignition and emissions control. On turbo / supercharged engines, the intake air temperature can form part of the calculation for controlling boost pressure or controlling airflow through the intercooler.	The voltage, frequency or value of the temperature sensor signal is either at full value (e.g. battery voltage with no signal or frequency) or, the signal exists but it is above the normal operating range (higher than the maximum operating limit). Possible fault in the temperature sensor or wiring e.g. short / open circuits, circuit resistance, incorrect sensor resistance or, fault with the reference / operating voltage; interference from other circuits can also affect sensor signals. Refer to list of sensor / actuator checks on page 19.
P0099	Intake Air Temperature Sensor 2 Circuit Intermittent / Erratic	The intake air temperature sensor can be used for fine tuning of fuel, ignition and emissions control. On turbo / supercharged engines, the intake air temperature can form part of the calculation for controlling boost pressure or controlling airflow through the intercooler.	The voltage, frequency or value of the temperature sensor signal is intermittent (likely to be out of normal operating range / tolerance when intermittent fault is detected) or, the signal / voltage is erratic (e.g. signal changes are irregular / unpredictable). Possible fault with wiring or connections (including internal sensor connections); also interference from other circuits can affect the signal. Refer to list of sensor / actuator checks on page 19.
P009A	Intake Air Temperature / Ambient Air Temperature Correlation	The intake air temperature sensor can be used for fine tuning of fuel, ignition and emissions control. The ambient temperature sensor can also be used for fine tuning of engine control but can also be used for other vehicle systems e.g. air conditioning, driver display etc.	The signals from the different sensors are providing conflicting or implausible information e.g. Ambient air temperature sensor indicates "cold" ambient air but the intake air temperature sensor indicates "hot" intake air temperature. It is possible that one of the sensor signals is incorrect; if possible check the operation of each sensor and note that with the engine off (ignition on) the sensors should indicate the same temperature (check air temperature with independent thermometer to identify which sensor is correct). Check the sensors and wiring for faults e.g. short / open circuits, circuit resistance, incorrect sensor resistance or reference / operating voltage fault; interference from other circuits can affect sensor signals. Refer to list of sensor / actuator checks on page 19.
P0100	Mass or Volume Air Flow "A" Circuit	Mass and volume air flow sensors are used to detect intake air flow thus providing an indication of fuelling and ignition requirements (and other engine control functions). Mass airflow sensors provide a signal that indicates the actual air "Mass" but the volume airflow sensor signal forms part of a calculation, made by the control unit, to determine the air mass.	The voltage, frequency or value of the airflow sensor signal is incorrect (undefined fault). Sensor voltage / signal could be: out of normal operating range, constant value, non-existent, corrupt. Possible fault in the airflow sensor or wiring e.g. short / open circuits, circuit resistance, internal sensor circuitry fault or, fault with the reference / operating voltage; interference from other circuits can also affect sensor signals. Refer to list of sensor / actuator checks on page 19.
P0101	Mass or Volume Air Flow "A" Circuit Range / Performance	Mass and volume air flow sensors are used to detect intake air flow thus providing an indication of fuelling and ignition requirements (and other engine control functions). Mass airflow sensors provide a signal that indicates the actual air "Mass" but the volume airflow sensor signal forms part of a calculation, made by the control unit, to determine the air mass.	The voltage, frequency or value of the airflow sensor signal is within the normal operating range / tolerance but the signal is not plausible or is incorrect due to an undefined fault. The sensor signal might not match the operating conditions indicated by other sensors or, the sensor signal is not changing / responding as expected to the changes in conditions. Possible fault in the airflow sensor or wiring e.g. short / open circuits, circuit resistance, internal sensor circuitry fault or, fault with the reference / operating voltage; interference from other circuits can also affect sensor signals. Refer to list of sensor / actuator checks on page 19. It is also possible that the airflow sensor is operating correctly but the airflow being measured by the sensor is unacceptable or incorrect for the operating conditions. Check for intake system air leaks or restrictions (and exhaust restrictions / blockage) that could be affecting the volume or mass of air passing through the sensor.
P0102	Mass or Volume Air Flow "A" Circuit Low	Mass and volume air flow sensors are used to detect intake air flow thus providing an indication of fuelling and ignition requirements (and other engine control functions). Mass airflow sensors provide a signal that indicates the actual air "Mass" but the volume airflow sensor signal forms part of a calculation, made by the control unit, to determine the air mass.	The voltage, frequency or value of the airflow sensor signal is either "zero" (no signal) or, the signal exists but it is below the normal operating range (lower than the minimum operating limit). Possible fault in the airflow sensor or wiring e.g. short / open circuits, circuit resistance, internal sensor circuitry fault or, fault with the reference / operating voltage; interference from other circuits can also affect sensor signals. Refer to list of sensor / actuator checks on page 19.
P0103	Mass or Volume Air Flow "A" Circuit High	Mass and volume air flow sensors are used to detect intake air flow thus providing an indication of fuelling and ignition requirements (and other engine control functions). Mass airflow sensors provide a signal that indicates the actual air "Mass" but the volume airflow sensor signal forms part of a calculation, made by the control unit, to determine the air mass.	The voltage, frequency or value of the sensor signal is either at full value (e.g. battery voltage with no signal or frequency) or, the signal exists but it is above the normal operating range (higher than the maximum operating limit). Possible fault in the sensor or wiring e.g. short / open circuits, circuit resistance, incorrect sensor resistance or, fault with the reference / operating voltage; interference from other circuits can also affect sensor signals. Refer to list of sensor / actuator checks on page 19.
P0104	Mass or Volume Air Flow "A" Circuit Intermittent	Mass and volume air flow sensors are used to detect intake air flow thus providing an indication of fuelling and ignition requirements (and other engine control functions). Mass airflow sensors provide a signal that indicates the actual air "Mass" but the volume airflow sensor signal forms part of a calculation, made by the control unit, to determine the air mass.	The voltage, frequency or value of the airflow sensor signal is intermittent (likely to be out of normal operating range / tolerance when intermittent fault is detected) or, the signal / voltage is erratic (e.g. signal changes are irregular / unpredictable). Possible fault with wiring or connections (including internal sensor connections); also interference from other circuits can affect the signal. Refer to list of sensor / actuator checks on page 19.

NOTE 1. Check for other fault codes that could provide additional information. **NOTE 2.** Communication between control units can pass via a CAN-Bus system; refer to Page 51 for CAN-Bus checks.

NOTE 3. If a fault cannot be located, it is also possible that the control unit is at fault. **NOTE 4.** Refer to Page 53 for list of pages that show the ISO standard for component locations e.g. Sensor A - Bank 1.

Fault code	EOBD / ISO Description	Component / System Description	Meaningful Description and Quick Check
P0105	Manifold Absolute Pressure / Barometric Pressure Circuit	Manifold Absolute Pressure (MAP) sensor detects manifold vacuum / pressure (after the throttle butterfly valve); the sensor is able to detect barometric (ambient) air pressure with the engine off (ignition on) but intake manifold pressure varies with barometric pressure and the combined sensor accounts for the two values. The sensor can also be used to indicate boost pressure sensor on turbocharged engines.	The voltage, frequency or value of the pressure sensor signal is incorrect (undefined fault). Sensor voltage / signal could be: out of normal operating range, constant value, non-existent, corrupt. Possible fault in the pressure sensor or wiring e.g. short / open circuits, circuit resistance, incorrect sensor resistance (if applicable) or, fault with the reference / operating voltage; interference from other circuits can also affect sensor signals. Refer to list of sensor / actuator checks on page 19.
P0106	Manifold Absolute Pressure / Barometric Pressure Circuit Range / Performance	Manifold Absolute Pressure (MAP) sensor detects manifold vacuum / pressure (after the throttle butterfly valve); the sensor is able to detect barometric (ambient) air pressure with the engine off (ignition on) but intake manifold pressure varies with barometric pressure and the combined sensor accounts for the two values. The sensor can also be used to indicate boost pressure sensor on turbocharged engines.	The voltage, frequency or value of the pressure sensor signal is within the normal operating range / tolerance but the signal is not plausible or is incorrect due to an undefined fault. The sensor signal might not match the operating conditions indicated by other sensors or, the sensor signal is not changing / responding as expected to the changes in conditions. Possible fault in the pressure sensor or wiring e.g. short / open circuits, circuit resistance, incorrect sensor resistance (if applicable) or, fault with the reference / operating voltage; interference from other circuits can also affect sensor signals. Refer to list of sensor / actuator checks on page 19. Also check any air passages / pipes to the sensor for leaks / blockages etc. It is also possible that the pressure sensor is operating correctly but the pressure being measured by the sensor is unacceptable or incorrect for the operating conditions; if possible, use alternative method of checking pressure to establish if sensor system is operating correctly and whether the pressure indicated by the sensor is correct / incorrect.
P0107	Manifold Absolute Pressure / Barometric Pressure Circuit Low	Manifold Absolute Pressure (MAP) sensor detects manifold vacuum / pressure (after the throttle butterfly valve); the sensor is able to detect barometric (ambient) air pressure with the engine off (ignition on) but intake manifold pressure varies with barometric pressure and the combined sensor accounts for the two values. The sensor can also be used to indicate boost pressure sensor on turbocharged engines.	The voltage, frequency or value of the pressure sensor signal is either "zero" (no signal) or, the signal exists but it is below the normal operating range (lower than the minimum operating limit). Possible fault in the pressure sensor or wiring e.g. short / open circuits, circuit resistance, incorrect sensor resistance (if applicable) or, fault with the reference / operating voltage; interference from other circuits can also affect sensor signals. Refer to list of sensor / actuator checks on page 19.
P0108	Manifold Absolute Pressure / Barometric Pressure Circuit High	Manifold Absolute Pressure (MAP) sensor detects manifold vacuum / pressure (after the throttle butterfly valve); the sensor is able to detect barometric (ambient) air pressure with the engine off (ignition on) but intake manifold pressure varies with barometric pressure and the combined sensor accounts for the two values. The sensor can also be used to indicate boost pressure sensor on turbocharged engines.	The voltage, frequency or value of the pressure sensor signal is either at full value (e.g. battery voltage with no signal or frequency) or, the signal exists but it is above the normal operating range (higher than the maximum operating limit). Possible fault in the pressure sensor or wiring e.g. short / open circuits, circuit resistance, incorrect sensor resistance (if applicable) or, fault with the reference / operating voltage; interference from other circuits can also affect sensor signals. Refer to list of sensor / actuator checks on page 19.
P0109	Manifold Absolute Pressure / Barometric Pressure Circuit Intermittent	Manifold Absolute Pressure (MAP) sensor detects manifold vacuum / pressure (after the throttle butterfly valve); the sensor is able to detect barometric (ambient) air pressure with the engine off (ignition on) but intake manifold pressure varies with barometric pressure and the combined sensor accounts for the two values. The sensor can also be used to indicate boost pressure sensor on turbocharged engines.	The voltage, frequency or value of the pressure sensor signal is intermittent (likely to be out of normal operating range / tolerance when intermittent fault is detected) or, the signal / voltage is erratic (e.g. signal changes are irregular / unpredictable). Possible fault with wiring or connections (including internal sensor connections); also interference from other circuits can affect the signal. Refer to list of sensor / actuator checks on page 19. It is also possible that the pressure could be changing erratically, which could cause the fault code to be activated; if necessary, check pressure with a separate gauge to check for erratic pressure changes.
P010A	Mass or Volume Air Flow "B" Circuit	Mass and volume air flow sensors are used to detect intake air flow thus providing an indication of fuelling and ignition requirements (and other engine control functions). Mass airflow sensors provide a signal that indicates the actual air "Mass" but the volume airflow sensor signal forms part of a calculation, made by the control unit, to determine the air mass.	The voltage, frequency or value of the airflow sensor signal is incorrect (undefined fault). Sensor voltage / signal could be: out of normal operating range, constant value, non-existent, corrupt. Possible fault in the airflow sensor or wiring e.g. short / open circuits, circuit resistance, internal sensor circuitry fault or, fault with the reference / operating voltage; interference from other circuits can also affect sensor signals. Refer to list of sensor / actuator checks on page 19.
P010B	Mass or Volume Air Flow "B" Circuit Range / Performance	Mass and volume air flow sensors are used to detect intake air flow thus providing an indication of fuelling and ignition requirements (and other engine control functions). Mass airflow sensors provide a signal that indicates the actual air "Mass" but the volume airflow sensor signal forms part of a calculation, made by the control unit, to determine the air mass.	The voltage, frequency or value of the airflow sensor signal is within the normal operating range / tolerance but the signal is not plausible or is incorrect due to an undefined fault. The sensor signal might not match the operating conditions indicated by other sensors or, the sensor signal is not changing / responding as expected to the changes in conditions. Possible fault in the airflow sensor or wiring e.g. short / open circuits, circuit resistance, internal sensor circuitry fault or, fault with the reference / operating voltage; interference from other circuits can also affect sensor signals. Refer to list of sensor / actuator checks on page 19. It is also possible that the airflow sensor is operating correctly but the airflow being measured by the sensor is unacceptable or incorrect for the operating conditions. Check for intake system air leaks or restrictions that could be affecting the volume or mass of air passing through the sensor.
P010C	Mass or Volume Air Flow "B" Circuit Low	Mass and volume air flow sensors are used to detect intake air flow thus providing an indication of fuelling and ignition requirements (and other engine control functions). Mass airflow sensors provide a signal that indicates the actual air "Mass" but the volume airflow sensor signal forms part of a calculation, made by the control unit, to determine the air mass.	The voltage, frequency or value of the airflow sensor signal is either "zero" (no signal) or, the signal exists but it is below the normal operating range (lower than the minimum operating limit). Possible fault in the airflow sensor or wiring e.g. short / open circuits, circuit resistance, internal sensor circuitry fault or, fault with the reference / operating voltage; interference from other circuits can also affect sensor signals. Refer to list of sensor / actuator checks on page 19.
P010D	Mass or Volume Air Flow "B" Circuit High	Mass and volume air flow sensors are used to detect intake air flow thus providing an indication of fuelling and ignition requirements (and other engine control functions). Mass airflow sensors provide a signal that indicates the actual air "Mass" but the volume airflow sensor signal forms part of a calculation, made by the control unit, to determine the air mass.	The voltage, frequency or value of the sensor signal is either at full value (e.g. battery voltage with no signal or frequency) or, the signal exists but it is above the normal operating range (higher than the maximum operating limit). Possible fault in the sensor or wiring e.g. short / open circuits, circuit resistance, incorrect sensor resistance or, fault with the reference / operating voltage; interference from other circuits can also affect sensor signals. Refer to list of sensor / actuator checks on page 19.

NOTE 1. Check for other fault codes that could provide additional information.

NOTE 2. Communication between control units can pass via a CAN-Bus system; refer to Page 51 for CAN-Bus checks.

NOTE 3. If a fault cannot be located, it is also possible that the control unit is at fault.

NOTE 4. Refer to Page 53 for list of pages that show the ISO standard for component locations e.g. Sensor A - Bank 1.

Code	Fault Location	Description	Possible Cause
P010E	Mass or Volume Air Flow "B" Circuit Intermittent / Erratic	Mass and volume air flow sensors are used to detect intake air flow thus providing an indication of fuelling and ignition requirements (and other engine control functions). Mass airflow sensors provide a signal that indicates the actual air "Mass" but the volume airflow sensor signal forms part of a calculation, made by the control unit, to determine the air mass.	The voltage, frequency or value of the airflow sensor signal is intermittent (likely to be out of normal operating range / tolerance when intermittent fault is detected) or, the signal / voltage is erratic (e.g. signal changes are irregular / unpredictable). Possible fault with wiring or connections (including internal sensor connections); also interference from other circuits can affect the signal. Refer to list of sensor / actuator checks on page 19.
P010F	Mass or Volume Air Flow Sensor A / B Correlation	The mass or volume airflow sensors provide an indication of airflow to the control unit which enables calculations for engine control functions e.g. fuel, ignition emission control systems etc. When more than one airflow sensor is used, each of the airflow sensors will normally be sensing airflow for a bank of cylinders on a V engine.	The signals from the two sensors (A and B) are providing conflicting or implausible information e.g. Airflow sensor A indicates high airflow or "full load" but Airflow sensor B indicates a lower airflow; it is therefore possible that one bank of cylinders is inducing less air than the other bank (possibly due to mechanical or other faults). It is possible that one of the sensors is operating incorrectly, therefore check each sensor for a fault with the sensor / wiring e.g. short / open circuits, circuit resistance, incorrect sensor resistance or reference / operating voltage fault. Refer to list of sensor / actuator checks on page 19. Check for intake system leaks and blockages or other faults that could cause a difference in the airflow being induced through the two different banks of cylinders.
P0110	Intake Air Temperature Sensor 1 Circuit	The intake air temperature sensor can be used for fine tuning of fuel, ignition and emissions control. On turbo / supercharged engines, the intake air temperature can form part of the calculation for controlling boost pressure or controlling airflow through the intercooler.	The voltage, frequency or value of the temperature sensor signal is incorrect (undefined fault). Sensor voltage / signal could be: out of normal operating range, constant value, non-existent, corrupt. Possible fault in the temperature sensor or wiring e.g. short / open circuits, circuit resistance, incorrect sensor resistance or, fault with the reference / operating voltage; interference from other circuits can also affect sensor signals. Refer to list of sensor / actuator checks on page 19.
P0111	Intake Air Temperature Sensor 1 Circuit Range / Performance	The intake air temperature sensor can be used for fine tuning of fuel, ignition and emissions control. On turbo / supercharged engines, the intake air temperature can form part of the calculation for controlling boost pressure or controlling airflow through the intercooler.	The voltage, frequency or value of the temperature sensor signal is within the normal operating range / tolerance but the signal is not plausible or is incorrect due to an undefined fault. The sensor signal might not match the operating conditions indicated by other sensors or, the sensor signal is not changing / responding as expected to the changes in conditions. Possible fault in the temperature sensor or wiring e.g. short / open circuits, circuit resistance, incorrect sensor resistance or, fault with the reference / operating voltage; interference from other circuits can also affect sensor signals. Refer to list of sensor / actuator checks on page 19. It is also possible that the temperature sensor is operating correctly but the temperature being measured by the sensor is unacceptable or incorrect for the operating conditions; if possible, use alternative method of checking temperature to establish if sensor system is operating correctly and whether temperature is correct / incorrect.
P0112	Intake Air Temperature Sensor 1 Circuit Low	The intake air temperature sensor can be used for fine tuning of fuel, ignition and emissions control. On turbo / supercharged engines, the intake air temperature can form part of the calculation for controlling boost pressure or controlling airflow through the intercooler.	The voltage, frequency or value of the temperature sensor signal is either "zero" (no signal) or, the signal exists but it is below the normal operating range (lower than the minimum operating limit). Possible fault in the temperature sensor or wiring e.g. short / open circuits, circuit resistance, incorrect sensor resistance or, fault with the reference / operating voltage; interference from other circuits can also affect sensor signals. Refer to list of sensor / actuator checks on page 19.
P0113	Intake Air Temperature Sensor 1 Circuit High	The intake air temperature sensor can be used for fine tuning of fuel, ignition and emissions control. On turbo / supercharged engines, the intake air temperature can form part of the calculation for controlling boost pressure or controlling airflow through the intercooler.	The voltage, frequency or value of the temperature sensor signal is either at full value (e.g. battery voltage with no signal or frequency) or, the signal exists but it is above the normal operating range (higher than the maximum operating limit). Possible fault in the temperature sensor or wiring e.g. short / open circuits, circuit resistance, incorrect sensor resistance or, fault with the reference / operating voltage; interference from other circuits can also affect sensor signals. Refer to list of sensor / actuator checks on page 19.
P0114	Intake Air Temperature Sensor 1 Circuit Intermittent	The intake air temperature sensor can be used for fine tuning of fuel, ignition and emissions control. On turbo / supercharged engines, the intake air temperature can form part of the calculation for controlling boost pressure or controlling airflow through the intercooler.	The voltage, frequency or value of the temperature sensor signal is intermittent (likely to be out of normal operating range / tolerance when intermittent fault is detected) or, the signal / voltage is erratic (e.g. signal changes are irregular / unpredictable). Possible fault with wiring or connections (including internal sensor connections); also interference from other circuits can affect the signal. Refer to list of sensor / actuator checks on page 19.
P0115	Engine Coolant Temperature Sensor 1 Circuit	The coolant temperature sensor signal is used for control of fuel, ignition and emissions control. The sensor information can be used for cooling fan control as well as for other vehicle systems. If there is more than one sensor, it can provide independent measurement for banks on a "V" engine or, it could provide information to different systems e.g. engine management, cooling fan control and driver instrumentation.	The voltage, frequency or value of the temperature sensor signal is incorrect (undefined fault). Sensor voltage / signal could be: out of normal operating range, constant value, non-existent, corrupt. Possible fault in the temperature sensor or wiring e.g. short / open circuits, circuit resistance, incorrect sensor resistance or, fault with the reference / operating voltage; interference from other circuits can also affect sensor signals. Refer to list of sensor / actuator checks on page 19.
P0116	Engine Coolant Temperature Sensor 1 Circuit Range / Performance	The coolant temperature sensor signal is used for control of fuel, ignition and emissions control. The sensor information can be used for cooling fan control as well as for other vehicle systems. If there is more than one sensor, it can provide independent measurement for banks on a "V" engine or, it could provide information to different systems e.g. engine management, cooling fan control and driver instrumentation.	The voltage, frequency or value of the temperature sensor signal is within the normal operating range / tolerance but the signal is not plausible or is incorrect due to an undefined fault. The sensor signal is not changing / responding as expected to the changes in conditions. Possible fault in the temperature sensor or wiring e.g. short / open circuits, circuit resistance, incorrect sensor resistance or, fault with the reference / operating voltage; interference from other circuits can also affect sensor signals. Refer to list of sensor / actuator checks on page 19. It is also possible that the temperature sensor is operating correctly but the temperature being measured by the sensor is unacceptable or incorrect for the operating conditions e.g. cooling system fault causing unacceptable coolant temperatures; if possible, use alternative method of checking temperature to establish if sensor system is operating correctly and whether temperature is correct / incorrect
P0117	Engine Coolant Temperature Sensor 1 Circuit Low	The coolant temperature sensor signal is used for control of fuel, ignition and emissions control. The sensor information can be used for cooling fan control as well as for other vehicle systems. If there is more than one sensor, it can provide independent measurement for banks on a "V" engine or, it could provide information to different systems e.g. engine management, cooling fan control and driver instrumentation.	The voltage, frequency or value of the temperature sensor signal is either "zero" (no signal) or, the signal exists but it is below the normal operating range (lower than the minimum operating limit). Possible fault in the temperature sensor or wiring e.g. short / open circuits, circuit resistance, incorrect sensor resistance or, fault with the reference / operating voltage; interference from other circuits can also affect sensor signals. Refer to list of sensor / actuator checks on page 19.

Fault code	EOBD / ISO Description	Component / System Description	Meaningful Description and Quick Check
P0118	Engine Coolant Temperature Sensor 1 Circuit High	The coolant temperature sensor signal is used for control of fuel, ignition and emissions control. The sensor information can be used for cooling fan control as well as for other vehicle systems. If there is more than one sensor, it can provide independent measurement for banks on a "V" engine or, it could provide information to different systems e.g. engine management, cooling fan control and driver instrumentation.	The voltage, frequency or value of the temperature sensor signal is either at full value (e.g. battery voltage with no signal or frequency) or, the signal exists but it is above the normal operating range (higher than the maximum operating limit). Possible fault in the temperature sensor or wiring e.g. short / open circuits, circuit resistance, incorrect sensor resistance or, fault with the reference / operating voltage; interference from other circuits can also affect sensor signals. Refer to list of sensor / actuator checks on page 19.
P0119	Engine Coolant Temperature Sensor 1 Circuit Intermittent	The coolant temperature sensor signal is used for control of fuel, ignition and emissions control. The sensor information can be used for cooling fan control as well as for other vehicle systems. If there is more than one sensor, it can provide independent measurement for banks on a "V" engine or, it could provide information to different systems e.g. engine management, cooling fan control and driver instrumentation.	The voltage, frequency or value of the temperature sensor signal is intermittent (likely to be out of normal operating range / tolerance when intermittent fault is detected) or, the signal / voltage is erratic (e.g. signal changes are irregular / unpredictable). Possible fault with wiring or connections (including internal sensor connections); also interference from other circuits can affect the signal. Check the sensor signal, using "live data" or other test equipment and wiggle wiring / connections to try and recreate the fault. Refer to list of sensor / actuator checks on page 19.
P011A	Engine Coolant Temperature Sensor 1 / 2 Correlation	The coolant temperature sensor signal is used for control of fuel, ignition and emissions control. The sensor information can be used for cooling fan control as well as for other vehicle systems. If there is more than one sensor, it can provide independent measurement for banks on a "V" engine or, it could provide information to different systems e.g. engine management, cooling fan control and driver instrumentation.	The signals from the two sensors are providing conflicting or implausible information e.g. Coolant sensor 1 indicates a significantly different temperature to Coolant sensor 2. Identify the functions of the two sensors e.g. independent temperature measurement for banks on a V engine or, for different engine systems. If possible, use alternative method of checking temperature to establish if each sensor system is operating correctly and whether temperature is correct / incorrect. It is possible that one of the sensors is operating incorrectly therefore check each sensor for a fault with the sensor / wiring e.g. short / open circuits, circuit resistance, incorrect sensor resistance or reference / operating voltage fault. Refer to list of sensor / actuator checks on page 19.
P0120	Throttle / Pedal Position Sensor / Switch "A" Circuit	The throttle / pedal position sensor / switch could be located on the throttle body, throttle cable / linkage or on the throttle pedal. If a switch is fitted, it could be used to indicate "Throttle Closed", "Throttle Open" or both positions. If a position sensor is fitted, it can indicate all angles of throttle opening as well as the speed of throttle opening (rate of change). Systems can be fitted with combinations of sensors and switches, which can be separate or combined into a single unit.	The sensor assembly may contain more than one switch or sensor, refer to vehicle specific information to establish which sensor or switch is affected. The voltage, frequency or value of the pedal position sensor / switch signal is incorrect (undefined fault). Sensor voltage / signal could be: out of normal operating range, constant value, non-existent, corrupt. Possible fault in the pedal position sensor / switch or wiring e.g. short / open circuits, circuit resistance, incorrect sensor resistance or, fault with the reference / operating voltage; the switch contacts (if applicable) could also be faulty e.g. not opening / closing correctly or have a high contact resistance. Refer to list of sensor / actuator checks on page 19.
P0121	Throttle / Pedal Position Sensor / Switch "A" Circuit Range / Performance	The throttle / pedal position sensor / switch could be located on the throttle body, throttle cable / linkage or on the throttle pedal. If a switch is fitted, it could be used to indicate "Throttle Closed", "Throttle Open" or both positions. If a position sensor is fitted, it can indicate all angles of throttle opening as well as the speed of throttle opening (rate of change). Systems can be fitted with combinations of sensors and switches, which can be separate or combined into a single unit.	The sensor assembly may contain more than one switch or sensor, refer to vehicle specific information to establish which sensor or switch is affected. The voltage, frequency or value of the pedal position sensor / switch signal is within the normal operating range / tolerance but the signal is not plausible or is incorrect due to an undefined fault. The sensor / switch signal might not match the operating conditions indicated by other sensors or, the sensor / switch signal is not changing / responding as expected. Possible fault in the pedal position sensor / switch or wiring e.g. short / open circuits, circuit resistance, incorrect sensor resistance or, fault with the reference / operating voltage; also check switch contacts (if applicable) to ensure that they open / close and check continuity of switch contacts (ensure that there is not a high resistance). Refer to list of sensor / actuator checks on page 19. It is also possible that the pedal position sensor / switch is operating correctly but the position being indicated by the sensor is unacceptable or incorrect for the operating conditions indicated by other sensors or, is not as expected by the control unit; check mechanism / linkage used to operate switch / sensor, and check base settings for throttle mechanism and for the sensor / switch.
P0122	Throttle / Pedal Position Sensor / Switch "A" Circuit Low	The throttle / pedal position sensor / switch could be located on the throttle body, throttle cable / linkage or on the throttle pedal. If a switch is fitted, it could be used to indicate "Throttle Closed", "Throttle Open" or both positions. If a position sensor is fitted, it can indicate all angles of throttle opening as well as the speed of throttle opening (rate of change). Systems can be fitted with combinations of sensors and switches, which can be separate or combined into a single unit.	Note. The sensor assembly may contain more than one switch or sensor, refer to vehicle specific information to establish which sensor or switch is affected. The voltage, frequency or value of the pedal position sensor / switch signal is either "zero" (no signal) or, the signal exists but it is below the normal operating range (lower than the minimum operating limit). Possible fault in the pedal position sensor / switch or wiring e.g. short / open circuits, circuit resistance, incorrect sensor resistance or, fault with the reference / operating voltage; the switch contacts (if applicable) could also be faulty e.g. not opening / closing correctly or have a high contact resistance. Refer to list of sensor / actuator checks on page 19. Note that the fault code can be activated for systems with position switches, if the switch output remains at the low value e.g. zero volts and does not rise to the high value e.g. battery voltage (the control unit will detect no switch signal change but it has detected change in engine operating conditions).
P0123	Throttle / Pedal Position Sensor / Switch "A" Circuit High	The throttle / pedal position sensor / switch could be located on the throttle body, throttle cable / linkage or on the throttle pedal. If a switch is fitted, it could be used to indicate "Throttle Closed", "Throttle Open" or both positions. If a position sensor is fitted, it can indicate all angles of throttle opening as well as the speed of throttle opening (rate of change). Systems can be fitted with combinations of sensors and switches, which can be separate or combined into a single unit.	The sensor assembly may contain more than one switch or sensor, refer to vehicle specific information to establish which sensor or switch is affected. The voltage, frequency or value of the pedal position sensor / switch signal is either at full value (e.g. battery voltage with no signal or frequency) or, the signal exists but it is above the normal operating range (higher than the maximum operating limit). Possible fault in the pedal position sensor / switch or wiring e.g. short / open circuits, circuit resistance, incorrect sensor resistance or, fault with the reference / operating voltage; the switch contacts (if applicable) could also be faulty e.g. not opening / closing correctly or have a high contact resistance. Refer to list of sensor / actuator checks on page 19. Note that the fault code can be activated for systems with position switches, if the switch output remains at the high value e.g. battery voltage and does not fall to the low value e.g. zero volts (the control unit will detect no switch signal change but it has detected change in engine operating conditions).
P0124	Throttle / Pedal Position Sensor / Switch "A" Circuit Intermittent	The throttle / pedal position sensor / switch could be located on the throttle body, throttle cable / linkage or on the throttle pedal. If a switch is fitted, it could be used to indicate "Throttle Closed", "Throttle Open" or both positions. If a position sensor is fitted, it can indicate all angles of throttle opening as well as the speed of throttle opening (rate of change). Systems can be fitted with combinations of sensors and switches, which can be separate or combined into a single unit.	The sensor assembly may contain more than one switch or sensor, refer to vehicle specific information to establish which sensor or switch is affected. The voltage, frequency or value of the pedal position sensor / switch signal is intermittent (likely to be out of normal operating range / tolerance when intermittent fault is detected) or, the signal / voltage is erratic (e.g. signal changes are irregular / unpredictable). Possible fault with wiring or connections (including internal sensor / switch connections); the switch contacts (if applicable) could also be faulty e.g. not opening / closing correctly or have a high contact resistance. Refer to list of sensor / actuator checks on page 19.

NOTE 1. Check for other fault codes that could provide additional information.

NOTE 2. Communication between control units can pass via a CAN-Bus system; refer to Page 51 for CAN-Bus checks.

NOTE 3. If a fault cannot be located, it is also possible that the control unit is at fault

NOTE 4. Refer to Page 53 for list of pages that show the ISO standard for component locations e.g. Sensor A - Bank 1.

Code	Description		
P0125	Insufficient Coolant Temperature for Closed Loop Fuel Control	The coolant temperature sensor signal is used for control of fuel, ignition and emissions control. The sensor information can be used for cooling fan control as well as for other vehicle systems. Closed loop control refers to the process of monitoring the air / fuel ratio (mixture) via the Lambda sensor signal (detecting oxygen content in exhaust gas) and correcting the mixture as required by altering injected fuel quantity.	The signal value from engine coolant temperature sensor is indicating that the engine coolant is not at a sufficient operating temperature for effective "closed loop" emissions control operation (closed loop Lambda control) e.g. the coolant temperature has not reached the required value within the expected time. Note that closed loop control is normally inoperative during the very early stages after cold starts but, after a short period the coolant temperature should be high enough to allow closed loop control to function. If possible, use alternative method of checking temperature to establish if the sensor system is operating correctly and whether temperature is correct / incorrect. Check cooling system operation e.g. thermostat operation etc. to ensure that the coolant temperature is increasing normally. If it appears that the temperature sensor is at fault, check for a fault with the sensor / wiring e.g. short / open circuits, circuit resistance, incorrect sensor resistance or reference / operating voltage fault. Refer to list of sensor / actuator checks on page 19.
P0126	Insufficient Coolant Temperature for Stable Operation	The coolant temperature sensor signal is used for control of fuel, ignition and emissions control. The sensor information can be used for cooling fan control as well as for other vehicle systems.	The signal value from engine coolant temperature sensor is indicating that the engine coolant is not at a sufficient operating temperature for normal operation or, the temperature is too low for other functions or processes to take place e.g. operation of the Exhaust Gas Recirculation (EGR) system. If possible, use alternative method of checking temperature to establish if the sensor system is operating correctly and whether temperature is correct / incorrect. Check cooling system operation e.g. thermostat operation etc. to ensure that the coolant temperature is increasing normally. If it appears that the temperature sensor is at fault, check for a fault with the sensor / wiring e.g. short / open circuits, circuit resistance, incorrect sensor resistance or reference / operating voltage fault. Refer to list of sensor / actuator checks on page 19.
P0127	Intake Air Temperature Too High	The intake air temperature sensor can be used for fine tuning of fuel, ignition and emissions control. On turbo / supercharged engines, the intake air temperature can form part of the calculation for controlling boost pressure or controlling airflow through the intercooler.	The voltage, frequency or value of the temperature sensor signal is within the normal operating range / tolerance but is indicating a high temperature value. The control unit has not detected an electrical fault, therefore the assessment is that the temperature is too high for the operating conditions (indicated by other sensors) or, the temperature does not match values expected by the control unit. If possible, check the temperature using a separate thermometer and compare with the value indicated by the sensor. If the sensor does appear to be faulty, check temperature sensor / wiring e.g. short / open circuits, circuit resistance, incorrect sensor resistance or reference / operating voltage fault; interference from other circuits can also affect sensor signals. Refer to list of sensor / actuator checks on page 19.
P0128	Coolant Thermostat (Coolant Temperature Below Thermostat Regulating Temperature)	The thermostat can operate in traditional way i.e. opening to increase coolant flow when coolant has reached specified temperature, some thermostats have the addition of an electrical heater to influence thermostat operation under different operating conditions.	The signal value from the temperature sensor, combined with information from other sensors, is indicating that the engine coolant temperature has not reached the expected temperature (with regard to the operating conditions). Possible causes include faulty thermostat (stuck open); check operation of electrically heated thermostat (if applicable). Refer to Fault code P0597 for additional information on electrically heated thermostats.
P0129	Barometric Pressure Too Low	The barometric / atmospheric air pressure sensor signal can be used for fine tuning of the fuel, ignition and emissions control systems. Barometric pressure can also be used in the calculations for boost pressure control on turbo / supercharger systems. A low Barometric pressure (high altitudes or extreme weather conditions) can occur, which could result in certain engine management / emission control functions not being effective or possible.	The fault code identifies that the sensor signal is indicating a low Barometric / atmospheric pressure. Although this could be a genuine low pressure problem, it is also possible that a Barometric pressure sensor fault could exist (the sensor signal e.g. frequency or voltage, could still be within normal operating limits but indicating a low pressure). It is possible that the pressure indicated by the barometric pressure sensor is being compared with other sensor signals e.g. Manifold Absolute Pressure (MAP) sensor, and that the indicated pressure values from the different sensors provide conflicting information. Check the Barometric pressure with a separate gauge to establish if the pressure value indicated by the sensor is correct or incorrect. If the sensor signal is incorrect Refer to list of sensor / actuator checks on page 19. Also check any air passages / pipes to the sensor for leaks / blockages etc. If the sensor is indicating the correct pressure, it is likely that the control unit has activated the code because it is unable to perform certain tasks at low pressures.
P012A	Turbocharger / Supercharger Inlet Pressure Sensor Circuit	The inlet pressure sensor is normally used to detect the pressure in the intake system prior to / upstream of the turbo / supercharger. The location and function of the inlet pressure sensor can depend on the type of turbo / super charger system, refer to Chapter 3 for examples of possible locations.	The voltage, frequency or value of the pressure sensor signal is incorrect (undefined fault). Sensor voltage / signal could be: out of normal operating range, constant value, non-existent, corrupt. Possible fault in the pressure sensor or wiring e.g. short / open circuits, circuit resistance, incorrect sensor resistance (if applicable) or, fault with the reference / operating voltage; interference from other circuits can also affect sensor signals. Refer to list of sensor / actuator checks on page 19.
P012B	Turbocharger / Supercharger Inlet Pressure Sensor Circuit Range / Performance	The inlet pressure sensor is normally used to detect the pressure in the intake system prior to / upstream of the turbo / supercharger. The location and function of the inlet pressure sensor can depend on the type of turbo / super charger system, refer to Chapter 3 for examples of possible locations.	The voltage, frequency or value of the pressure sensor signal is within the normal operating range / tolerance but the signal is not plausible or is incorrect due to an undefined fault. The sensor signal might not match the operating conditions indicated by other sensors or, the sensor signal is not changing / responding as expected to the changes in conditions. Possible fault in the pressure sensor or wiring e.g. short / open circuits, circuit resistance, incorrect sensor resistance (if applicable) or, fault with the reference / operating voltage; interference from other circuits can also affect sensor signals. Refer to list of sensor / actuator checks on page 19. Also check any air passages / pipes to the sensor for leaks / blockages etc. It is also possible that the pressure sensor is operating correctly but the boost pressure being measured by the sensor is unacceptable or incorrect for the operating conditions; if possible, use alternative method of checking pressure to establish if sensor system is operating correctly and whether the pressure indicated by the sensor is correct / incorrect.
P012C	Turbocharger / Supercharger Inlet Pressure Sensor Circuit Low	The inlet pressure sensor is normally used to detect the pressure in the intake system prior to / upstream of the turbo / supercharger. The location and function of the inlet pressure sensor can depend on the type of turbo / super charger system, refer to Chapter 3 for examples of possible locations.	The voltage, frequency or value of the pressure sensor signal is either "zero" (no signal) or, the signal exists but it is below the normal operating range (lower than the minimum operating limit). Possible fault in the pressure sensor or wiring e.g. short / open circuits, circuit resistance, incorrect sensor resistance (if applicable) or, fault with the reference / operating voltage; interference from other circuits can also affect sensor signals. Refer to list of sensor / actuator checks on page 19.
P012D	Turbocharger / Supercharger Inlet Pressure Sensor Circuit High	The inlet pressure sensor is normally used to detect the pressure in the intake system prior to / upstream of the turbo / supercharger. The location and function of the inlet pressure sensor can depend on the type of turbo / super charger system, refer to Chapter XX for examples of possible locations.	The voltage, frequency or value of the pressure sensor signal is either at full value (e.g. battery voltage with no signal or frequency) or, the signal exists but it is above the normal operating range (higher than the maximum operating limit). Possible fault in the pressure sensor or wiring e.g. short / open circuits, circuit resistance, incorrect sensor resistance (if applicable) or, fault with the reference / operating voltage; interference from other circuits can also affect sensor signals. Refer to list of sensor / actuator checks on page 19.

NOTE 1. Check for other fault codes that could provide additional information.

NOTE 2. Communication between control units can pass via a CAN-Bus system; refer to Page 51 for CAN-Bus checks.

NOTE 3. If a fault cannot be located, it is also possible that the control unit is at fault.

NOTE 4. Refer to Page 53 for list of pages that show the ISO standard for component locations e.g. Sensor A - Bank 1.

Fault code	EOBD / ISO Description	Component / System Description	Meaningful Description and Quick Check
P012E	Turbocharger / Supercharger Inlet Pressure Sensor Circuit Intermittent / Erratic	The inlet pressure sensor is normally used to detect the pressure in the intake system prior to / upstream of the turbo / supercharger. The location and function of the inlet pressure sensor can depend on the type of turbo / super charger system, refer to Chapter XX for examples of possible locations.	The voltage, frequency or value of the pressure sensor signal is intermittent (likely to be out of normal operating range / tolerance when intermittent fault is detected) or, the signal / voltage is erratic (e.g. signal changes are irregular / unpredictable). Possible fault with wiring or connections (including internal sensor connections); also interference from other circuits can affect the signal. Refer to list of sensor / actuator checks on page 19.
P012F		ISO / SAE reserved	Not used, reserved for future allocation.
P0130	O2 Sensor Circuit Bank 1 Sensor 1	Note: Many O2 / Lambda sensor related fault codes are activated due to faults in other systems (engine, fuelling, misfires etc). O2 (Oxygen / Lambda) sensors detect the exhaust gas oxygen content, which is also an indicator of air / fuel ratio (rich / lean). Oxygen content is indicated by a "Lambda" value. The air / fuel ratio (and therefore oxygen content) is controlled by the engine control unit to maximise efficiency of catalytic converters and reduce emissions of harmful gases.	For a detailed list of checks and identification of the sensor location, Refer to list of sensor / actuator checks on page 19. Different types of O2 / Lambda sensors (e.g. Zirconia or Titania / wideband etc) provide different signals and values, it will be necessary to identify the type of sensor fitted to the vehicle. The fault code indicates that the voltage, frequency or value of the O2 / Lambda sensor signal is incorrect (undefined fault). The signal value could be: out of normal operating range, constant value, non-existent, corrupt. The problem could be related to a sensor or wiring fault; check wiring between sensor and control unit for short / open circuits and high resistance. It is also possible that the oxygen content in the exhaust gas is incorrect for the operating conditions (causing sensor signal to be out of normal operating range). Check sensor signal and signal response under different operating conditions (load, constant rpm etc). If the sensor appears to be operating correctly but the signal value is incorrect, this could indicate an engine / fuelling fault, including misfires and exhaust air leaks.
P0131	O2 Sensor Circuit Low Voltage Bank 1 Sensor 1	Note: Many O2 / Lambda sensor related fault codes are activated due to faults in other systems (engine, fuelling, misfires etc). O2 (Oxygen / Lambda) sensors detect the exhaust gas oxygen content, which is also an indicator of air / fuel ratio (rich / lean). Oxygen content is indicated by a "Lambda" value. The air / fuel ratio (and therefore oxygen content) is controlled by the engine control unit to maximise efficiency of catalytic converters and reduce emissions of harmful gases.	For a detailed list of checks and identification of the sensor location, Refer to list of sensor / actuator checks on page 19. Different types of O2 / Lambda sensors (e.g. Zirconia or Titania / wideband etc) provide different signals and values, it will be necessary to identify the type of sensor fitted to the vehicle. The fault code indicates that the voltage of the O2 / Lambda sensor signal is low (possibly out of normal operating range). The problem could be related to a sensor or wiring fault; check wiring between sensor and control unit for short / open circuits and high resistance. It is also possible that the oxygen content in the exhaust gas is incorrect for the operating conditions (causing sensor signal voltage to be low). Check sensor signal and signal response under different operating conditions (load, constant rpm etc). If the sensor appears to be operating correctly but the signal value is low, this could indicate an engine / fuelling fault, including misfires and exhaust air leaks. Note that depending on sensor design, a high voltage can indicate low or high oxygen content (rich or lean mixtures).
P0132	O2 Sensor Circuit High Voltage Bank 1 Sensor 1	Note: Many O2 / Lambda sensor related fault codes are activated due to faults in other systems (engine, fuelling, misfires etc). O2 (Oxygen / Lambda) sensors detect the exhaust gas oxygen content, which is also an indicator of air / fuel ratio (rich / lean). Oxygen content is indicated by a "Lambda" value. The air / fuel ratio (and therefore oxygen content) is controlled by the engine control unit to maximise efficiency of catalytic converters and reduce emissions of harmful gases.	For a detailed list of checks and identification of the sensor location, Refer to list of sensor / actuator checks on page 19. Different types of O2 / Lambda sensors (e.g. Zirconia or Titania / wideband etc) provide different signals and values, it will be necessary to identify the type of sensor fitted to the vehicle. The fault code indicates that the voltage of the O2 / Lambda sensor signal is high (possibly out of normal operating range). The problem could be related to a sensor or wiring fault; check wiring between sensor and control unit for short / open circuits and high resistance. It is also possible that the oxygen content in the exhaust gas is incorrect for the operating conditions (causing sensor signal voltage to be high). Check sensor signal and signal response under different operating conditions (load, constant rpm etc). If the sensor appears to be operating correctly but the signal value is high, this could indicate an engine / fuelling fault, including misfires and exhaust air leaks. Note that depending on sensor design, a high voltage can indicate low or high oxygen content (rich or lean mixtures).
P0133	O2 Sensor Circuit Slow Response Bank 1 Sensor 1	Note: Many O2 / Lambda sensor related fault codes are activated due to faults in other systems (engine, fuelling, misfires etc). O2 (Oxygen / Lambda) sensors detect the exhaust gas oxygen content, which is also an indicator of air / fuel ratio (rich / lean). Oxygen content is indicated by a "Lambda" value. The air / fuel ratio (and therefore oxygen content) is controlled by the engine control unit to maximise efficiency of catalytic converters and reduce emissions of harmful gases.	For a detailed list of checks and identification of the sensor location, Refer to list of sensor / actuator checks on page 19. Different types of O2 / Lambda sensors (e.g. Zirconia or Titania / wideband etc) provide different signals and values, it will be necessary to identify the type of sensor fitted to the vehicle. The fault code indicates that the response (change in value) of the O2 / Lambda sensor signal is slow when there is a change in operating conditions. It is also possible that the control unit has detected a slow or reduced signal frequency (the signal produced by many sensors oscillates at an expected frequency). The problem could be related to a sensor or wiring fault; check wiring between sensor and control unit for short / open circuits and high resistance. It is also possible that other faults can cause slow response of the sensor, especially air leaks close to the sensor. Check sensor signal response under different operating conditions (load, constant rpm etc) to establish if the sensor response is slow or whether other faults are causing a slow change in the exhaust gas oxygen content.
P0134	O2 Sensor Circuit No Activity Detected Bank 1 Sensor 1	Note: Many O2 / Lambda sensor related fault codes are activated due to faults in other systems (engine, fuelling, misfires etc). O2 (Oxygen / Lambda) sensors detect the exhaust gas oxygen content, which is also an indicator of air / fuel ratio (rich / lean). The air / fuel ratio is controlled by the engine control unit so that the oxygen content is as close as possible to a value of "Lambda 1" (improves emissions reduction efficiency of 3-way catalytic converters). For some operating conditions and for certain phases of operation of direct injection engines, the air / fuel ratio can be different to Lambda 1; Refer to list of sensor / actuator checks on page 19 for additional information.	For a detailed list of checks and identification of the sensor location, Refer to list of sensor / actuator checks on page 19. Different types of O2 / Lambda sensors (e.g. Zirconia or Titania / wideband etc) provide different signals and values, it will be necessary to identify the type of sensor fitted to the vehicle. The fault code indicates that there is no signal activity (voltage change, frequency etc) from the O2 / Lambda sensor (undefined fault). The signal could be a constant value with no change or, non-existent. The problem could be caused by a sensor or wiring fault; check wiring between sensor and control unit for short / open circuits and high resistance. It is also possible that the oxygen content in the exhaust gas is incorrect, which is causing the sensor signal to remain stuck outside of normal operating range (stuck high or low). Check sensor signal and signal response under different operating conditions (load, constant rpm etc). If the sensor appears to be operating correctly but the signal value is incorrect, this could indicate an engine / fuelling fault, including misfires and exhaust air leaks.

NOTE 1. Check for other fault codes that could provide additional information. NOTE 2. Communication between control units can pass via a CAN-Bus system; refer to Page 51 for CAN-Bus checks.
NOTE 3. If a fault cannot be located, it is also possible that the control unit is at fault. NOTE 4. Refer to Page 53 for list of pages that show the ISO standard for component locations e.g. Sensor A - Bank 1.

code	Description		
P0135	O2 Sensor Heater Circuit Bank 1 Sensor 1	The heater control for the HO2S (oxygen / Lambda) sensor is used to heat up and provide stable temperature for the sensor (primarily after cold starts). The heating element can receive a power supply via a relay (often the main system relay) and the heater earth path can be switched to earth via the control unit (this is the heater control circuit), the control unit is therefore able to control the operation of the heating element by controlling the earth circuit. Note that some systems may also provide the power supply to the heating element via the control unit. It may be necessary to refer to vehicle specific information to identify the type of Oxygen / Lambda sensor and the wiring.	The fault code not does identify which part of the heater circuit i.e. the power supply to the heater or the earth path (usually via the control unit) but monitoring of the circuit is carried out on the earth path for most systems. However, checks should embrace all applicable heater wiring, the power supply and the heater element. The fault code indicates that there is an undefined fault in the Oxygen / Lambda sensor heater circuit; the fault is causing an incorrect voltage / current in the circuit; this is likely to be: non-existent, constant value (but incorrect for the operating conditions), corrupt or, is out of normal operating range (out of tolerance). The fault could be caused by a faulty sensor heater element (open circuit or high resistance) or a wiring fault (short / open circuit). Also check heater supply voltage (could be reduced due to a high resistance or wiring fault). Refer to list of sensor / actuator checks on page 19.
P0136	O2 Sensor Circuit Bank 1 Sensor 2	Note: Many O2 / Lambda sensor related fault codes are activated due to faults in other systems (engine, fuelling, misfires etc). O2 (Oxygen / Lambda) sensors detect the exhaust gas oxygen content, which is also an indicator of air / fuel ratio (rich / lean). Oxygen content is indicated by a "Lambda" value. The air / fuel ratio (and therefore oxygen content) is controlled by the engine control unit to maximise efficiency of catalytic converters and reduce emissions of harmful gases.	For a detailed list of checks and identification of the sensor location, Refer to list of sensor / actuator checks on page 19. Different types of O2 / Lambda sensors (e.g. Zirconia or Titania / wideband etc) provide different signals and values, it will be necessary to identify the type of sensor fitted to the vehicle. The fault code indicates that the voltage, frequency or value of the O2 / Lambda sensor signal is incorrect (undefined fault). The signal value could be: out of normal operating range, constant value, non-existent, corrupt. The problem could be related to a sensor or wiring fault; check wiring between sensor and control unit for short / open circuits and high resistance. It is also possible that the oxygen content in the exhaust gas is incorrect for the operating conditions (causing sensor signal to be out of normal operating range). Check sensor signal and signal response under different operating conditions (load, constant rpm etc). If the sensor appears to be operating correctly but the signal value is incorrect, this could indicate an engine / fuelling fault, including misfires and exhaust air leaks.
P0137	O2 Sensor Circuit Low Voltage Bank 1 Sensor 2	Note: Many O2 / Lambda sensor related fault codes are activated due to faults in other systems (engine, fuelling, misfires etc). O2 (Oxygen / Lambda) sensors detect the exhaust gas oxygen content, which is also an indicator of air / fuel ratio (rich / lean). Oxygen content is indicated by a "Lambda" value. The air / fuel ratio (and therefore oxygen content) is controlled by the engine control unit to maximise efficiency of catalytic converters and reduce emissions of harmful gases.	For a detailed list of checks and identification of the sensor location, Refer to list of sensor / actuator checks on page 19. Different types of O2 / Lambda sensors (e.g. Zirconia or Titania / wideband etc) provide different signals and values, it will be necessary to identify the type of sensor fitted to the vehicle. The fault code indicates that the voltage of the O2 / Lambda sensor signal is low (possibly out of normal operating range). The problem could be related to a sensor or wiring fault; check wiring between sensor and control unit for short / open circuits and high resistance. It is also possible that the oxygen content in the exhaust gas is incorrect for the operating conditions (causing sensor signal voltage to be low). Check sensor signal and signal response under different operating conditions (load, constant rpm etc). If the sensor appears to be operating correctly but the signal value is low, this could indicate an engine / fuelling fault, including misfires and exhaust air leaks. Note that depending on sensor design, a high voltage can indicate low or high oxygen content (rich or lean mixtures).
P0138	O2 Sensor Circuit High Voltage Bank 1 Sensor 2	Note: Many O2 / Lambda sensor related fault codes are activated due to faults in other systems (engine, fuelling, misfires etc). O2 (Oxygen / Lambda) sensors detect the exhaust gas oxygen content, which is also an indicator of air / fuel ratio (rich / lean). Oxygen content is indicated by a "Lambda" value. The air / fuel ratio (and therefore oxygen content) is controlled by the engine control unit to maximise efficiency of catalytic converters and reduce emissions of harmful gases.	For a detailed list of checks and identification of the sensor location, Refer to list of sensor / actuator checks on page 19. Different types of O2 / Lambda sensors (e.g. Zirconia or Titania / wideband etc) provide different signals and values, it will be necessary to identify the type of sensor fitted to the vehicle. The fault code indicates that the voltage of the O2 / Lambda sensor signal is high (possibly out of normal operating range). The problem could be related to a sensor or wiring fault; check wiring between sensor and control unit for short / open circuits and high resistance. It is also possible that the oxygen content in the exhaust gas is incorrect for the operating conditions (causing sensor signal voltage to be high). Check sensor signal and signal response under different operating conditions (load, constant rpm etc). If the sensor appears to be operating correctly but the signal value is high, this could indicate an engine / fuelling fault, including misfires and exhaust air leaks. Note that depending on sensor design, a high voltage can indicate low or high oxygen content (rich or lean mixtures).
P0139	O2 Sensor Circuit Slow Response Bank 1 Sensor 2	Note: Many O2 / Lambda sensor related fault codes are activated due to faults in other systems (engine, fuelling, misfires etc). O2 (Oxygen / Lambda) sensors detect the exhaust gas oxygen content, which is also an indicator of air / fuel ratio (rich / lean). Oxygen content is indicated by a "Lambda" value. The air / fuel ratio (and therefore oxygen content) is controlled by the engine control unit to maximise efficiency of catalytic converters and reduce emissions of harmful gases.	For a detailed list of checks and identification of the sensor location, Refer to list of sensor / actuator checks on page 19. Different types of O2 / Lambda sensors (e.g. Zirconia or Titania / wideband etc) provide different signals and values, it will be necessary to identify the type of sensor fitted to the vehicle. The fault code indicates that the response (change in value) of the O2 / Lambda sensor signal is slow when there is a change in operating conditions. It is also possible that the control unit has detected a slow or reduced signal frequency (the signal produced by many sensors oscillates at an expected frequency). The problem could be related to a sensor or wiring fault; check wiring between sensor and control unit for short / open circuits and high resistance. It is also possible that other faults can cause slow response of the sensor, especially air leaks close to the sensor. Check sensor signal response under different operating conditions (load, constant rpm etc) to establish if the sensor response is slow or whether other faults are causing a slow change in the exhaust gas oxygen content.
P0140	O2 Sensor Circuit No Activity Detected Bank 1 Sensor 2	Note: Many O2 / Lambda sensor related fault codes are activated due to faults in other systems (engine, fuelling, misfires etc). O2 (Oxygen / Lambda) sensors detect the exhaust gas oxygen content, which is also an indicator of air / fuel ratio (rich / lean). The air / fuel ratio is controlled by the engine control unit so that the oxygen content is as close as possible to a value of "Lambda 1" (improves emissions reduction efficiency of 3-way catalytic converters). For some operating conditions and for certain phases of operation of direct injection engines, the air / fuel ratio can be different to Lambda 1; Refer to list of sensor / actuator checks on page 19 for additional information.	For a detailed list of checks and identification of the sensor location, Refer to list of sensor / actuator checks on page 19. Different types of O2 / Lambda sensors (e.g. Zirconia or Titania / wideband etc) provide different signals and values, it will be necessary to identify the type of sensor fitted to the vehicle. The fault code indicates that there is no signal activity (voltage change, frequency etc) from the O2 / Lambda sensor (undefined fault). The signal could be a constant value with no change or, non-existent. The problem could be caused by a sensor or wiring fault; check wiring between sensor and control unit for short / open circuits and high resistance. It is also possible that the oxygen content in the exhaust gas is incorrect, which is causing the sensor signal to remain stuck outside of normal operating range (stuck high or low). Check sensor signal and signal response under different operating conditions (load, constant rpm etc). If the sensor appears to be operating correctly but the signal value is incorrect, this could indicate an engine / fuelling fault, including misfires and exhaust air leaks.

Fault code	EOBD / ISO Description	Component / System Description	Meaningful Description and Quick Check
P0141	O2 Sensor Heater Circuit Bank 1 Sensor 2	The heater control for the HO2S (oxygen / Lambda) sensor is used to heat up and provide stable temperature for the sensor (primarily after cold starts). The heating element can receive a power supply via a relay (often the main system relay) and the heater earth path can be switched to earth via the control unit (this is the heater control circuit), the control unit is therefore able to control the operation of the heating element by controlling the earth circuit. Note that some systems may also provide the power supply to the heating element via the control unit. It may be necessary to refer to vehicle specific information to identify the type of Oxygen / Lambda sensor and the wiring.	The fault code not does identify which part of the heater circuit i.e. the power supply to the heater or the earth path (usually via the control unit) but monitoring of the circuit is carried out on the earth path for most systems. However, checks should embrace all applicable heater wiring, the power supply and the heater element. The fault code indicates that there is an undefined fault in the Oxygen / Lambda sensor heater circuit; the fault is causing an incorrect voltage / current in the circuit; this is likely to be: non-existent, constant value (but incorrect for the operating conditions), corrupt or, is out of normal operating range (out of tolerance). The fault could be caused by a faulty sensor heater element (open circuit or high resistance) or a wiring fault (short / open circuit). Also check heater supply voltage (could be reduced due to a high resistance or wiring fault). Refer to list of sensor / actuator checks on page 19.
P0142	O2 Sensor Circuit Bank 1 Sensor 3	Note: Many O2 / Lambda sensor related fault codes are activated due to faults in other systems (engine, fuelling, misfires etc). O2 (Oxygen / Lambda) sensors detect the exhaust gas oxygen content, which is also an indicator of air / fuel ratio (rich / lean). The air / fuel ratio is controlled by the engine control unit so that the oxygen content is as close as possible to a value of "Lambda 1" (improves emissions reduction efficiency of 3-way catalytic converters). For some operating conditions and for certain phases of operation of direct injection engines, the air / fuel ratio can be different to Lambda 1; Refer to list of sensor / actuator checks on page 19 for additional information.	For a detailed list of checks and identification of the sensor location, Refer to list of sensor / actuator checks on page 19. Different types of O2 / Lambda sensors (e.g. Zirconia or Titania / wideband etc) provide different signals and values, it will be necessary to identify the type of sensor fitted to the vehicle. The fault code indicates that the voltage, frequency or value of the O2 / Lambda sensor signal is incorrect (undefined fault). The signal value could be: out of normal operating range, constant value, non-existent, corrupt. The problem could be related to a sensor or wiring fault; check wiring between sensor and control unit for short / open circuits and high resistance. It is also possible that the oxygen content in the exhaust gas is incorrect for the operating conditions (causing sensor signal to be out of normal operating range). Check sensor signal and signal response under different operating conditions (load, constant rpm etc). If the sensor appears to be operating correctly but the signal value is incorrect, this could indicate an engine / fuelling fault, including misfires and exhaust air leaks.
P0143	O2 Sensor Circuit Low Voltage Bank 1 Sensor 3	Note: Many O2 / Lambda sensor related fault codes are activated due to faults in other systems (engine, fuelling, misfires etc). O2 (Oxygen / Lambda) sensors detect the exhaust gas oxygen content, which is also an indicator of air / fuel ratio (rich / lean). Oxygen content is indicated by a "Lambda" value. The air / fuel ratio (and therefore oxygen content) is controlled by the engine control unit to maximise efficiency of catalytic converters and reduce emissions of harmful gases.	For a detailed list of checks and identification of the sensor location, Refer to list of sensor / actuator checks on page 19. Different types of O2 / Lambda sensors (e.g. Zirconia or Titania / wideband etc) provide different signals and values, it will be necessary to identify the type of sensor fitted to the vehicle. The fault code indicates that the voltage of the O2 / Lambda sensor signal is low (possibly out of normal operating range). The problem could be related to a sensor or wiring fault; check wiring between sensor and control unit for short / open circuits and high resistance. It is also possible that the oxygen content in the exhaust gas is incorrect for the operating conditions (causing sensor signal voltage to be low). Check sensor signal and signal response under different operating conditions (load, constant rpm etc). If the sensor appears to be operating correctly but the signal value is high, this could indicate an engine / fuelling fault, including misfires and exhaust air leaks. Note that depending on sensor design, a high voltage can indicate low or high oxygen content (rich or lean mixtures).
P0144	O2 Sensor Circuit High Voltage Bank 1 Sensor 3	Note: Many O2 / Lambda sensor related fault codes are activated due to faults in other systems (engine, fuelling, misfires etc). O2 (Oxygen / Lambda) sensors detect the exhaust gas oxygen content, which is also an indicator of air / fuel ratio (rich / lean). Oxygen content is indicated by a "Lambda" value. The air / fuel ratio (and therefore oxygen content) is controlled by the engine control unit to maximise efficiency of catalytic converters and reduce emissions of harmful gases.	For a detailed list of checks and identification of the sensor location, Refer to list of sensor / actuator checks on page 19. Different types of O2 / Lambda sensors (e.g. Zirconia or Titania / wideband etc) provide different signals and values, it will be necessary to identify the type of sensor fitted to the vehicle. The fault code indicates that the voltage of the O2 / Lambda sensor signal is high (possibly out of normal operating range). The problem could be related to a sensor or wiring fault; check wiring between sensor and control unit for short / open circuits and high resistance. It is also possible that the oxygen content in the exhaust gas is incorrect for the operating conditions (causing sensor signal voltage to be high). Check sensor signal and signal response under different operating conditions (load, constant rpm etc). If the sensor appears to be operating correctly but the signal value is high, this could indicate an engine / fuelling fault, including misfires and exhaust air leaks. Note that depending on sensor design, a high voltage can indicate low or high oxygen content (rich or lean mixtures).
P0145	O2 Sensor Circuit Slow Response Bank 1 Sensor 3	Note: Many O2 / Lambda sensor related fault codes are activated due to faults in other systems (engine, fuelling, misfires etc). O2 (Oxygen / Lambda) sensors detect the exhaust gas oxygen content, which is also an indicator of air / fuel ratio (rich / lean). Oxygen content is indicated by a "Lambda" value. The air / fuel ratio (and therefore oxygen content) is controlled by the engine control unit to maximise efficiency of catalytic converters and reduce emissions of harmful gases.	For a detailed list of checks and identification of the sensor location, Refer to list of sensor / actuator checks on page 19. Different types of O2 / Lambda sensors (e.g. Zirconia or Titania / wideband etc) provide different signals and values, it will be necessary to identify the type of sensor fitted to the vehicle. The fault code indicates that the response (change in value) of the O2 / Lambda sensor signal is slow when there is a change in operating conditions. It is also possible that the control unit has detected a slow / reduced signal frequency (the signal produced by many sensors oscillates at an expected frequency). The problem could be related to a sensor or wiring fault; check wiring between sensor and control unit for short / open circuits and high resistance. It is also possible that other faults can cause slow response of the sensor, especially air leaks close to the sensor. Check sensor signal response under different operating conditions (load, constant rpm etc) to establish if the sensor response is slow or whether other faults are causing a slow change in the exhaust gas oxygen content.
P0146	O2 Sensor Circuit No Activity Detected Bank 1 Sensor 3	Note: Many O2 / Lambda sensor related fault codes are activated due to faults in other systems (engine, fuelling, misfires etc). O2 (Oxygen / Lambda) sensors detect the exhaust gas oxygen content, which is also an indicator of air / fuel ratio (rich / lean). The air / fuel ratio is controlled by the engine control unit so that the oxygen content is as close as possible to a value of "Lambda 1" (improves emissions reduction efficiency of 3-way catalytic converters). For some operating conditions and for certain phases of operation of direct injection engines, the air / fuel ratio can be different to Lambda 1; Refer to list of sensor / actuator checks on page 19 for additional information.	For a detailed list of checks and identification of the sensor location, Refer to list of sensor / actuator checks on page 19. Different types of O2 / Lambda sensors (e.g. Zirconia or Titania / wideband etc) provide different signals and values, it will be necessary to identify the type of sensor fitted to the vehicle. The fault code indicates that there is no signal activity (voltage change, frequency etc) from the O2 / Lambda sensor (undefined fault). The signal could be a constant value with no change or, non-existent. The problem could be caused by a sensor or wiring fault; check wiring between sensor and control unit for short / open circuits and high resistance. It is also possible that the oxygen content in the exhaust gas is incorrect, which is causing the sensor signal to remain stuck outside of normal operating range (stuck high or low). Check sensor signal and signal response under different operating conditions (load, constant rpm etc). If the sensor appears to be operating correctly but the signal value is incorrect, this could indicate an engine / fuelling fault, including misfires and exhaust air leaks.

NOTE 1. Check for other fault codes that could provide additional information.
NOTE 3. If a fault cannot be located, it is also possible that the control unit is at fault.
NOTE 2. Communication between control units can pass via a CAN-Bus system; refer to Page 51 for CAN-Bus checks.
NOTE 4. Refer to Page 53 for list of pages that show the ISO standard for component locations e.g. Sensor A - Bank 1.

Code	Description		
P0147	O2 Sensor Heater Circuit Bank 1 Sensor 3	The heater control for the HO2S (oxygen / Lambda) sensor is used to heat up and provide stable temperature for the sensor (primarily after cold starts). The heating element can receive a power supply via a relay (often the main system relay) and the heater earth path can be switched to earth via the control unit (this is the heater control circuit), the control unit is therefore able to control the operation of the heating element by controlling the earth circuit. Note that some systems may also provide the power supply to the heating element via the control unit. It may be necessary to refer to vehicle specific information to identify the type of Oxygen / Lambda sensor and the wiring.	The fault code not does identify which part of the heater circuit i.e. the power supply to the heater or the earth path (usually via the control unit) but monitoring of the circuit is carried out on the earth path for most systems. However, checks should embrace all applicable heater wiring, the power supply and the heater element. The fault code indicates that there is an undefined fault in the Oxygen / Lambda sensor heater circuit; the fault is causing an incorrect voltage / current in the circuit; this is likely to be: non-existent, constant value (but incorrect for the operating conditions), corrupt or, is out of normal operating range (out of tolerance). The fault could be caused by a faulty sensor heater element (open circuit or high resistance) or a wiring fault (short / open circuit). Also check heater supply voltage (could be reduced due to a high resistance or wiring fault). Refer to list of sensor / actuator checks on page 19.
P0148	Fuel Delivery Error	Fuel delivery errors can possibly refer to a number of problems with the fuel system. However, some applications of the code refer to injection occurring when the control unit is requesting "fuel shut off" e.g. during "overrun" conditions (throttle pedal in the idle position, with the vehicle coasting or going downhill). For Diesel engines with rotary type pumps (non-common rail systems); the control unit can activate the fuel metering control system to provide Zero fuel (a delivery sensor located in the pumping chamber will still detect fuel delivery even though the control unit has requested fuel cut off). On common rail systems, the code has been used to indicate fuel metering control problems i.e. the fuel metering control valve is not responding to the control unit commands (fuel quantity / pressure is not altering as expected).	For this fault code, there are a number of possible interpretations of the code definition. It will be necessary to identify the type of fuel system and the control functions used on the system. For rotary pump systems (non-common rail) checks should be carried out on operation of the fuel metering control valve to establish if the fuel is being cut off at appropriate times; check metering valve control solenoid and mechanism. On common rail systems, check operation of the fuel metering control system (volume and pressure control valves as applicable) and check system pressures / fuel delivery.
P0149	Fuel Timing Error	Depending on the fuel system, pump timing may be dependent on signals from Crankshaft position / speed sensor, Camshaft position sensor and / or Pump position sensor; therefore check for other fault codes that might indicate applicable sensor faults. Also note that an injector needle lift sensor can be used that will indicate actual fuel delivery timing, this sensor should therefore also be checked (if fitted). Additionally, for systems fitted with Diesel injector pumps (non-common rail systems), incorrect fuel pressure / quantity, or air in the fuel system will influence injection timing.	Control Unit has detected a fuel delivery / timing error. Undefined fault in Fuel pump or associated system. Carry out routine checks on fuel system, including checks for leaks, blockages etc. Also check whether system may require bleeding. Check operation of injection timing control system. Check operation of crankshaft, camshaft and pump speed sensors. Also check static pump timing and camshaft timing. Check for wear or slackness in timing belt / chain / pump drive mechanism.
P0150	O2 Sensor Circuit Bank 2 Sensor 1	Note: Many O2 / Lambda sensor related fault codes are activated due to faults in other systems (engine, fuelling, misfires etc). O2 (Oxygen / Lambda) sensors detect the exhaust gas oxygen content, which is also an indicator of air / fuel ratio (rich / lean). Oxygen content is indicated by a "Lambda" value. The air / fuel ratio (and therefore oxygen content) is controlled by the engine control unit to maximise efficiency of catalytic converters and reduce emissions of harmful gases.	For a detailed list of checks and identification of the sensor location, Refer to list of sensor / actuator checks on page 19. Different types of O2 / Lambda sensors (e.g. Zirconia or Titania / wideband etc) provide different signals and values, it will be necessary to identify the type of sensor fitted to the vehicle. The fault code indicates that the voltage, frequency or value of the O2 / Lambda sensor signal is incorrect (undefined fault). The signal value could be: out of normal operating range, constant value, non-existent, corrupt. The problem could be related to a sensor or wiring fault; check wiring between sensor and control unit for short / open circuits and high resistance. It is also possible that the oxygen content in the exhaust gas is incorrect for the operating conditions (causing sensor signal to be out of normal operating range). Check sensor signal and signal response under different operating conditions (load, constant rpm etc). If the sensor appears to be operating correctly but the signal value is incorrect, this could indicate an engine / fuelling fault, including misfires and exhaust air leaks.
P0151	O2 Sensor Circuit Low Voltage Bank 2 Sensor 1	Note: Many O2 / Lambda sensor related fault codes are activated due to faults in other systems (engine, fuelling, misfires etc). O2 (Oxygen / Lambda) sensors detect the exhaust gas oxygen content, which is also an indicator of air / fuel ratio (rich / lean). Oxygen content is indicated by a "Lambda" value. The air / fuel ratio (and therefore oxygen content) is controlled by the engine control unit to maximise efficiency of catalytic converters and reduce emissions of harmful gases.	For a detailed list of checks and identification of the sensor location, Refer to list of sensor / actuator checks on page 19. Different types of O2 / Lambda sensors (e.g. Zirconia or Titania / wideband etc) provide different signals and values, it will be necessary to identify the type of sensor fitted to the vehicle. The fault code indicates that the voltage of the O2 / Lambda sensor signal is low (possibly out of normal operating range). The problem could be related to a sensor or wiring fault; check wiring between sensor and control unit for short / open circuits and high resistance. It is also possible that the oxygen content in the exhaust gas is incorrect for the operating conditions (causing sensor signal voltage to be low). Check sensor signal and signal response under different operating conditions (load, constant rpm etc). If the sensor appears to be operating correctly but the signal value is high, this could indicate an engine / fuelling fault, including misfires and exhaust air leaks. Note that depending on sensor design, a high voltage can indicate low or high oxygen content (rich or lean mixtures).
P0152	O2 Sensor Circuit High Voltage Bank 2 Sensor 1	Note: Many O2 / Lambda sensor related fault codes are activated due to faults in other systems (engine, fuelling, misfires etc). O2 (Oxygen / Lambda) sensors detect the exhaust gas oxygen content, which is also an indicator of air / fuel ratio (rich / lean). Oxygen content is indicated by a "Lambda" value. The air / fuel ratio (and therefore oxygen content) is controlled by the engine control unit to maximise efficiency of catalytic converters and reduce emissions of harmful gases.	For a detailed list of checks and identification of the sensor location, Refer to list of sensor / actuator checks on page 19. Different types of O2 / Lambda sensors (e.g. Zirconia or Titania / wideband etc) provide different signals and values, it will be necessary to identify the type of sensor fitted to the vehicle. The fault code indicates that the voltage of the O2 / Lambda sensor signal is high (possibly out of normal operating range). The problem could be related to a sensor or wiring fault; check wiring between sensor and control unit for short / open circuits and high resistance. It is also possible that the oxygen content in the exhaust gas is incorrect for the operating conditions (causing sensor signal voltage to be high). Check sensor signal and signal response under different operating conditions (load, constant rpm etc). If the sensor appears to be operating correctly but the signal value is high, this could indicate an engine / fuelling fault, including misfires and exhaust air leaks. Note that depending on sensor design, a high voltage can indicate low or high oxygen content (rich or lean mixtures).

NOTE 1. Check for other fault codes that could provide additional information. **NOTE 2.** Communication between control units can pass via a CAN-Bus system; refer to Page 51 for CAN-Bus checks.

NOTE 3. If a fault cannot be located, it is also possible that the control unit is at fault. **NOTE 4.** Refer to Page 53 for list of pages that show the ISO standard for component locations e.g. Sensor A - Bank 1.

Fault code	EOBD / ISO Description	Component / System Description	Meaningful Description and Quick Check
P0153	O2 Sensor Circuit Slow Response Bank 2 Sensor 1	Note: Many O2 / Lambda sensor related fault codes are activated due to faults in other systems (engine, fuelling, misfires etc). O2 (Oxygen / Lambda) sensors detect the exhaust gas oxygen content, which is also an indicator of air / fuel ratio (rich / lean). Oxygen content is indicated by a "Lambda" value. The air / fuel ratio (and therefore oxygen content) is controlled by the engine control unit to maximise efficiency of catalytic converters and reduce emissions of harmful gases.	For a detailed list of checks and identification of the sensor location, Refer to list of sensor / actuator checks on page 19. Different types of O2 / Lambda sensors (e.g. Zirconia or Titania / wideband etc) provide different signals and values, it will be necessary to identify the type of sensor fitted to the vehicle. The fault code indicates that the response (change in value) of the O2 / Lambda sensor signal is slow when there is a change in operating conditions. It is also possible that the control unit has detected a slow or reduced signal frequency (the signal produced by many sensors oscillates at an expected frequency). The problem could be related to a sensor or wiring fault; check wiring between sensor and control unit for short / open circuits and high resistance. It is also possible that other faults can cause slow response of the sensor, especially air leaks close to the sensor. Check sensor signal response under different operating conditions (load, constant rpm etc) to establish if the sensor response is slow or whether other faults are causing a slow change in the exhaust gas oxygen content.
P0154	O2 Sensor Circuit No Activity Detected Bank 2 Sensor 1	Note: Many O2 / Lambda sensor related fault codes are activated due to faults in other systems (engine, fuelling, misfires etc). O2 (Oxygen / Lambda) sensors detect the exhaust gas oxygen content, which is also an indicator of air / fuel ratio (rich / lean). The air / fuel ratio is controlled by the engine control unit so that the oxygen content is as close as possible to a value of "Lambda 1" (improves emissions reduction efficiency of 3-way catalytic converters). For some operating conditions and for certain phases of operation of direct injection engines, the air / fuel ratio can be different to Lambda 1; Refer to list of sensor / actuator checks on page 19 for additional information.	For a detailed list of checks and identification of the sensor location, Refer to list of sensor / actuator checks on page 19. Different types of O2 / Lambda sensors (e.g. Zirconia or Titania / wideband etc) provide different signals and values, it will be necessary to identify the type of sensor fitted to the vehicle. The fault code indicates that there is no signal activity (voltage change, frequency etc) from the O2 / Lambda sensor (undefined fault). The signal could be a constant value with no change or, non-existent. The problem could be caused by a sensor or wiring fault; check wiring between sensor and control unit for short / open circuits and high resistance. It is also possible that the oxygen content in the exhaust gas is incorrect, which is causing the sensor signal to remain stuck outside of normal operating range (stuck high or low). Check sensor signal and signal response under different operating conditions (load, constant rpm etc). If the sensor appears to be operating correctly but the signal value is incorrect, this could indicate an engine / fuelling fault, including misfires and exhaust air leaks.
P0155	O2 Sensor Heater Circuit Bank 2 Sensor 1	The heater control for the HO2S (oxygen / Lambda) sensor is used to heat up and provide stable temperature for the sensor (primarily after cold starts). The heating element can receive a power supply via a relay (often the main system relay) and the heater earth path can be switched to earth via the control unit (this is the heater control circuit), the control unit is therefore able to control the operation of the heating element by controlling the earth circuit. Note that some systems may also provide the power supply to the heating element via the control unit. It may be necessary to refer to vehicle specific information to identify the type of Oxygen / Lambda sensor and the wiring.	The fault code not does identify which part of the heater circuit i.e. the power supply to the heater or the earth path (usually via the control unit) but monitoring of the circuit is carried out on the earth path for most systems. However, checks should embrace all applicable heater wiring, the power supply and the heater element. The fault code indicates that there is an undefined fault in the Oxygen / Lambda sensor heater circuit; the fault is causing an incorrect voltage / current in the circuit; this is likely to be: non-existent, constant value (but incorrect for the operating conditions), corrupt or, is out of normal operating range (out of tolerance). The fault could be caused by a faulty sensor heater element (open circuit or high resistance) or a wiring fault (short / open circuit). Also check heater supply voltage (could be reduced due to a high resistance or wiring fault). Refer to list of sensor / actuator checks on page 19.
P0156	O2 Sensor Circuit Bank 2 Sensor 2	Note: Many O2 / Lambda sensor related fault codes are activated due to faults in other systems (engine, fuelling, misfires etc). O2 (Oxygen / Lambda) sensors detect the exhaust gas oxygen content, which is also an indicator of air / fuel ratio (rich / lean). Oxygen content is indicated by a "Lambda" value. The air / fuel ratio (and therefore oxygen content) is controlled by the engine control unit to maximise efficiency of catalytic converters and reduce emissions of harmful gases.	For a detailed list of checks and identification of the sensor location, Refer to list of sensor / actuator checks on page 19. Different types of O2 / Lambda sensors (e.g. Zirconia or Titania / wideband etc) provide different signals and values, it will be necessary to identify the type of sensor fitted to the vehicle. The fault code indicates that the voltage, frequency or value of the O2 / Lambda sensor signal is incorrect (undefined fault). The signal value could be: out of normal operating range, constant value, non-existent, corrupt. The problem could be related to a sensor or wiring fault; check wiring between sensor and control unit for short / open circuits and high resistance. It is also possible that the oxygen content in the exhaust gas is incorrect for the operating conditions (causing sensor signal to be out of normal operating range). Check sensor signal and signal response under different operating conditions (load, constant rpm etc). If the sensor appears to be operating correctly but the signal value is incorrect, this could indicate an engine / fuelling fault, including misfires and exhaust air leaks.
P0157	O2 Sensor Circuit Low Voltage Bank 2 Sensor 2	Note: Many O2 / Lambda sensor related fault codes are activated due to faults in other systems (engine, fuelling, misfires etc). O2 (Oxygen / Lambda) sensors detect the exhaust gas oxygen content, which is also an indicator of air / fuel ratio (rich / lean). Oxygen content is indicated by a "Lambda" value. The air / fuel ratio (and therefore oxygen content) is controlled by the engine control unit to maximise efficiency of catalytic converters and reduce emissions of harmful gases.	For a detailed list of checks and identification of the sensor location, Refer to list of sensor / actuator checks on page 19. Different types of O2 / Lambda sensors (e.g. Zirconia or Titania / wideband etc) provide different signals and values, it will be necessary to identify the type of sensor fitted to the vehicle. The fault code indicates that the voltage of the O2 / Lambda sensor signal is low (possibly out of normal operating range). The problem could be related to a sensor or wiring fault; check wiring between sensor and control unit for short / open circuits and high resistance. It is also possible that the oxygen content in the exhaust gas is incorrect for the operating conditions (causing sensor signal voltage to be low). Check sensor signal and signal response under different operating conditions (load, constant rpm etc). If the sensor appears to be operating correctly but the signal value is high, this could indicate an engine / fuelling fault, including misfires and exhaust air leaks. Note that depending on sensor design, a high voltage can indicate low or high oxygen content (rich or lean mixtures).
P0158	O2 Sensor Circuit High Voltage Bank 2 Sensor 2	Note: Many O2 / Lambda sensor related fault codes are activated due to faults in other systems (engine, fuelling, misfires etc). O2 (Oxygen / Lambda) sensors detect the exhaust gas oxygen content, which is also an indicator of air / fuel ratio (rich / lean). Oxygen content is indicated by a "Lambda" value. The air / fuel ratio (and therefore oxygen content) is controlled by the engine control unit to maximise efficiency of catalytic converters and reduce emissions of harmful gases.	For a detailed list of checks and identification of the sensor location, Refer to list of sensor / actuator checks on page 19. Different types of O2 / Lambda sensors (e.g. Zirconia or Titania / wideband etc) provide different signals and values, it will be necessary to identify the type of sensor fitted to the vehicle. The fault code indicates that the voltage of the O2 / Lambda sensor signal is high (possibly out of normal operating range). The problem could be related to a sensor or wiring fault; check wiring between sensor and control unit for short / open circuits and high resistance. It is also possible that the oxygen content in the exhaust gas is incorrect for the operating conditions (causing sensor signal voltage to be high). Check sensor signal and signal response under different operating conditions (load, constant rpm etc). If the sensor appears to be operating correctly but the signal value is high, this could indicate an engine / fuelling fault, including misfires and exhaust air leaks. Note that depending on sensor design, a high voltage can indicate low or high oxygen content (rich or lean mixtures).

NOTE 1. Check for other fault codes that could provide additional information.   NOTE 2. Communication between control units can pass via a CAN-Bus system; refer to Page 51 for CAN-Bus checks.   86

NOTE 3. If a fault cannot be located, it is also possible that the control unit is at fault   NOTE 4. Refer to Page 53 for list of pages that show the ISO standard for component locations e.g. Sensor A - Bank 1.

code	Description		
P0159	O2 Sensor Circuit Slow Response Bank 2 Sensor 2	Note: Many O2 / Lambda sensor related fault codes are activated due to faults in other systems (engine, fuelling, misfires etc). O2 (Oxygen / Lambda) sensors detect the exhaust gas oxygen content, which is also an indicator of air / fuel ratio (rich / lean). Oxygen content is indicated by a "Lambda" value. The air / fuel ratio (and therefore oxygen content) is controlled by the engine control unit to maximise efficiency of catalytic converters and reduce emissions of harmful gases.	For a detailed list of checks and identification of the sensor location, Refer to list of sensor / actuator checks on page 19. Different types of O2 / Lambda sensors (e.g. Zirconia or Titania / wideband etc) provide different signals and values, it will be necessary to identify the type of sensor fitted to the vehicle. The fault code indicates that the response (change in value) of the O2 / Lambda sensor signal is slow when there is a change in operating conditions. It is also possible that the control unit has detected a slow or reduced signal frequency (the signal produced by many sensors oscillates at an expected frequency). The problem could be related to a sensor or wiring fault; check wiring between sensor and control unit for short / open circuits and high resistance. It is also possible that other faults can cause slow response of the sensor, especially air leaks close to the sensor. Check sensor signal response under different operating conditions (load, constant rpm etc) to establish if the sensor response is slow or whether other faults are causing a slow change in the exhaust gas oxygen content.
P0160	O2 Sensor Circuit No Activity Detected Bank 2 Sensor 2	Note: Many O2 / Lambda sensor related fault codes are activated due to faults in other systems (engine, fuelling, misfires etc). O2 (Oxygen / Lambda) sensors detect the exhaust gas oxygen content, which is also an indicator of air / fuel ratio (rich / lean). The air / fuel ratio is controlled by the engine control unit so that the oxygen content is as close as possible to a value of "Lambda 1" (improves emissions reduction efficiency of 3-way catalytic converters). For some operating conditions and for certain phases of operation of direct injection engines, the air / fuel ratio can be different to Lambda 1; Refer to list of sensor / actuator checks on page 19 for additional information.	For a detailed list of checks and identification of the sensor location, Refer to list of sensor / actuator checks on page 19. Different types of O2 / Lambda sensors (e.g. Zirconia or Titania / wideband etc) provide different signals and values, it will be necessary to identify the type of sensor fitted to the vehicle. The fault code indicates that there is no signal activity (voltage change, frequency etc) from the O2 / Lambda sensor (undefined fault). The signal could be a constant value with no change or, non-existent. The problem could be caused by a sensor or wiring fault; check wiring between sensor and control unit for short / open circuits and high resistance. It is also possible that the oxygen content in the exhaust gas is incorrect, which is causing the sensor signal to remain stuck outside of normal operating range (stuck high or low). Check sensor signal and signal response under different operating conditions (load, constant rpm etc). If the sensor appears to be operating correctly but the signal value is incorrect, this could indicate an engine / fuelling fault, including misfires and exhaust air leaks.
P0161	O2 Sensor Heater Circuit Bank 2 Sensor 2	The heater control for the HO2S (oxygen / Lambda) sensor is used to heat up and provide stable temperature for the sensor (primarily after cold starts). The heating element can receive a power supply via a relay (often the main system relay) and the heater earth path can be switched to earth via the control unit (this is the heater control circuit), the control unit is therefore able to control the operation of the heating element by controlling the earth circuit. Note that some systems may also provide the power supply to the heating element via the control unit. It may be necessary to refer to vehicle specific information to identify the type of Oxygen / Lambda sensor and the wiring.	The fault code not does identify which part of the heater circuit i.e. the power supply to the heater or the earth path (usually via the control unit) but monitoring of the circuit is carried out on the earth path for most systems. However, checks should embrace all applicable heater wiring, the power supply and the heater element. The fault code indicates that there is an undefined fault in the Oxygen / Lambda sensor heater circuit; the fault is causing an incorrect voltage / current in the circuit; this is likely to be: non-existent, constant value (but incorrect for the operating conditions), corrupt or, is out of normal operating range (out of tolerance). The fault could be caused by a faulty sensor heater element (open circuit or high resistance) or a wiring fault (short / open circuit). Also check heater supply voltage (could be reduced due to a high resistance or wiring fault). Refer to list of sensor / actuator checks on page 19.
P0162	O2 Sensor Circuit Bank 2 Sensor 3	Note: Many O2 / Lambda sensor related fault codes are activated due to faults in other systems (engine, fuelling, misfires etc). O2 (Oxygen / Lambda) sensors detect the exhaust gas oxygen content, which is also an indicator of air / fuel ratio (rich / lean). Oxygen content is indicated by a "Lambda" value. The air / fuel ratio (and therefore oxygen content) is controlled by the engine control unit to maximise efficiency of catalytic converters and reduce emissions of harmful gases.	For a detailed list of checks and identification of the sensor location, Refer to list of sensor / actuator checks on page 19. Different types of O2 / Lambda sensors (e.g. Zirconia or Titania / wideband etc) provide different signals and values, it will be necessary to identify the type of sensor fitted to the vehicle. The fault code indicates that the voltage, frequency or value of the O2 / Lambda sensor signal is incorrect (undefined fault). The signal value could be: out of normal operating range, constant value, non-existent, corrupt. The problem could be related to a sensor or wiring fault; check wiring between sensor and control unit for short / open circuits and high resistance. It is also possible that the oxygen content in the exhaust gas is incorrect for the operating conditions (causing sensor signal to be out of normal operating range). Check sensor signal and signal response under different operating conditions (load, constant rpm etc). If the sensor appears to be operating correctly but the signal value is incorrect, this could indicate an engine / fuelling fault, including misfires and exhaust air leaks.
P0163	O2 Sensor Circuit Low Voltage Bank 2 Sensor 3	Note: Many O2 / Lambda sensor related fault codes are activated due to faults in other systems (engine, fuelling, misfires etc). O2 (Oxygen / Lambda) sensors detect the exhaust gas oxygen content, which is also an indicator of air / fuel ratio (rich / lean). Oxygen content is indicated by a "Lambda" value. The air / fuel ratio (and therefore oxygen content) is controlled by the engine control unit to maximise efficiency of catalytic converters and reduce emissions of harmful gases.	For a detailed list of checks and identification of the sensor location, Refer to list of sensor / actuator checks on page 19. Different types of O2 / Lambda sensors (e.g. Zirconia or Titania / wideband etc) provide different signals and values, it will be necessary to identify the type of sensor fitted to the vehicle. The fault code indicates that the voltage of the O2 / Lambda sensor signal is low (possibly out of normal operating range). The problem could be related to a sensor or wiring fault; check wiring between sensor and control unit for short / open circuits and high resistance. It is also possible that the oxygen content in the exhaust gas is incorrect for the operating conditions (causing sensor signal voltage to be low). Check sensor signal and signal response under different operating conditions (load, constant rpm etc). If the sensor appears to be operating correctly but the signal value is high, this could indicate an engine / fuelling fault, including misfires and exhaust air leaks. Note that depending on sensor design, a high voltage can indicate low or high oxygen content (rich or lean mixtures).
P0164	O2 Sensor Circuit High Voltage Bank 2 Sensor 3	Note: Many O2 / Lambda sensor related fault codes are activated due to faults in other systems (engine, fuelling, misfires etc). O2 (Oxygen / Lambda) sensors detect the exhaust gas oxygen content, which is also an indicator of air / fuel ratio (rich / lean). Oxygen content is indicated by a "Lambda" value. The air / fuel ratio (and therefore oxygen content) is controlled by the engine control unit to maximise efficiency of catalytic converters and reduce emissions of harmful gases.	For a detailed list of checks and identification of the sensor location, Refer to list of sensor / actuator checks on page 19. Different types of O2 / Lambda sensors (e.g. Zirconia or Titania / wideband etc) provide different signals and values, it will be necessary to identify the type of sensor fitted to the vehicle. The fault code indicates that the voltage of the O2 / Lambda sensor signal is high (possibly out of normal operating range). The problem could be related to a sensor or wiring fault; check wiring between sensor and control unit for short / open circuits and high resistance. It is also possible that the oxygen content in the exhaust gas is incorrect for the operating conditions (causing sensor signal voltage to be high). Check sensor signal and signal response under different operating conditions (load, constant rpm etc). If the sensor appears to be operating correctly but the signal value is high, this could indicate an engine / fuelling fault, including misfires and exhaust air leaks. Note that depending on sensor design, a high voltage can indicate low or high oxygen content (rich or lean mixtures).

Fault code	EOBD / ISO Description	Component / System Description	Meaningful Description and Quick Check
P0165	O2 Sensor Circuit Slow Response Bank 2 Sensor 3	Note: Many O2 / Lambda sensor related fault codes are activated due to faults in other systems (engine, fuelling, misfires etc). O2 (Oxygen / Lambda) sensors detect the exhaust gas oxygen content, which is also an indicator of air / fuel ratio (rich / lean). Oxygen content is indicated by a "Lambda" value. The air / fuel ratio (and therefore oxygen content) is controlled by the engine control unit to maximise efficiency of catalytic converters and reduce emissions of harmful gases.	For a detailed list of checks and identification of the sensor location, Refer to list of sensor / actuator checks on page 19. Different types of O2 / Lambda sensors (e.g. Zirconia or Titania / wideband etc) provide different signals and values, it will be necessary to identify the type of sensor fitted to the vehicle. The fault code indicates that the response (change in value) of the O2 / Lambda sensor signal is slow when there is a change in operating conditions. It is also possible that the control unit has detected a slow or reduced signal frequency (the signal produced by many sensors oscillates at an expected frequency). The problem could be related to a sensor or wiring fault; check wiring between sensor and control unit for short / open circuits and high resistance. It is also possible that other faults can cause slow response of the sensor, especially air leaks close to the sensor. Check sensor signal response under different operating conditions (load, constant rpm etc) to establish if the sensor response is slow or whether other faults are causing a slow change in the exhaust gas oxygen content.
P0166	O2 Sensor Circuit No Activity Detected Bank 2 Sensor 3	Note: Many O2 / Lambda sensor related fault codes are activated due to faults in other systems (engine, fuelling, misfires etc). O2 (Oxygen / Lambda) sensors detect the exhaust gas oxygen content, which is also an indicator of air / fuel ratio (rich / lean). The air / fuel ratio is controlled by the engine control unit so that the oxygen content is as close as possible to a value of "Lambda 1" (improves emissions reduction efficiency of 3-way catalytic converters). For some operating conditions and for certain phases of operation of direct injection engines, the air / fuel ratio can be different to Lambda 1; Refer to list of sensor / actuator checks on page 19 for additional information.	For a detailed list of checks and identification of the sensor location, Refer to list of sensor / actuator checks on page 19. Different types of O2 / Lambda sensors (e.g. Zirconia or Titania / wideband etc) provide different signals and values, it will be necessary to identify the type of sensor fitted to the vehicle. The fault code indicates that there is no signal activity (voltage change, frequency etc) from the O2 / Lambda sensor (undefined fault). The signal could be a constant value with no change or, non-existent. The problem could be caused by a sensor or wiring fault; check wiring between sensor and control unit for short / open circuits and high resistance. It is also possible that the oxygen content in the exhaust gas is incorrect, which is causing the sensor signal to remain stuck outside of normal operating range (stuck high or low). Check sensor signal and signal response under different operating conditions (load, constant rpm etc). If the sensor appears to be operating correctly but the signal value is incorrect, this could indicate an engine / fuelling fault, including misfires and exhaust air leaks.
P0167	O2 Sensor Heater Circuit Bank 2 Sensor 3	The heater control for the HO2S (oxygen / Lambda) sensor is used to heat up and provide stable temperature for the sensor (primarily after cold starts). The heating element can receive a power supply via a relay (often the main system relay) and the heater earth path can be switched to earth via the control unit (this is the heater control circuit), the control unit is therefore able to control the operation of the heating element by controlling the earth circuit. Note that some systems may also provide the power supply to the heating element via the control unit. It may be necessary to refer to vehicle specific information to identify the type of Oxygen / Lambda sensor and the wiring.	The fault code not does identify which part of the heater circuit i.e. the power supply to the heater or the earth path (usually via the control unit) but monitoring of the circuit is carried out on the earth path for most systems. However, checks should embrace all applicable heater wiring, the power supply and the heater element. The fault code indicates that there is an undefined fault in the Oxygen / Lambda sensor heater circuit; the fault is causing an incorrect voltage / current in the circuit; this is likely to be: non-existent, constant value (but incorrect for the operating conditions), corrupt or, is out of normal operating range (out of tolerance). The fault could be caused by a faulty sensor heater element (open circuit or high resistance) or a wiring fault (short / open circuit). Also check heater supply voltage (could be reduced due to a high resistance or wiring fault). Refer to list of sensor / actuator checks on page 19.
P0168	Fuel Temperature Too High	The fuel temperature can be used to fine tune the engine control (fuelling etc). On Diesel engines, a temperature sensor can be used to check for excessively low fuel temperatures which can cause fuel waxing. A sensor can also be used on Common Rail Diesel systems, to monitor the fuel returning to the tank; if the temperature is high, the fuel can be diverted through a cooler.	The fault code indicates that the fuel temperature is above acceptable or expected temperature. Identify the fuel system and whether a temperature cooler is fitted. The temperature could be high due to unusual operating conditions e.g. continuous low speed operation (lack of cooling airflow) or due to very low fuel level (can sometimes cause high fuel temperatures). For Diesel fuel systems that use a heater (usually in the filter assembly), check that the heater is not operating at incorrect times. Clear the fault code and drive vehicle under normal conditions; if the fault code is re-activated, further checks will be required on the fuel system.
P0169	Incorrect Fuel Composition	The fuel composition sensor is used when mixed fuels are used e.g. ethanol with petrol; the sensor monitors the mix of the two fuels. If necessary, refer to vehicle specific information to identify the sensor operation, location etc.	The trouble code indicates that the mix or composition of the fuel is incorrect. The engine will be set to operate (timing, fuelling etc) using a certain percentage mix of the fuels and it is likely that the mix is incorrect for the engine settings. It will be necessary to identify if and why the fuel is incorrect e.g. incorrect fuel put into tank etc. It is also possible that there is a fault with the composition sensor (refer to fault code P0176 for additional information).
P0170	Fuel Trim Bank 1	Note: Many O2 / Lambda sensor related fault codes are activated due to faults in other systems (engine, fuelling, misfires etc). Fuel trim refers to the monitoring and control of the air / fuel ratio. The control unit monitors the Oxygen / Lambda sensor signal (indicates exhaust gas oxygen content). The control unit adjusts the fuelling to maintain the oxygen content within defined operating limits (helps efficient catalytic converter operation). If the control unit applies a fuelling adjustment, but this adjustment does not alter the oxygen content as expected, this can be regarded as a "fuel trim" fault. Fuel trim can apply to Pre-cat oxygen measurements (adjusting to a pre-defined air / fuel ratio), but it can also apply to Post-cat measurements (measuring after the catalyst indicates if there is a need for any fine tuning of the air / fuel ratio to maximise catalyst operation).	The fault code does not define whether the problem is due to Lean or Rich mixture (excess or lack of oxygen). The fault code indicates that there is a fuel trim problem (likely to be a pre-cat measurement). The fault could be caused by a sensor / wiring fault; check wiring between sensor and control unit for short / open circuits and high resistance. Note that the sensor could also be contaminated or faulty. It is also possible that the exhaust gas oxygen content is incorrect for the operating conditions (the sensor might therefore be operating correctly). Check sensor signal and signal response under different operating conditions (load, constant rpm etc). If the sensor does appear to be operating correctly, the fault could be caused by an engine / fuelling fault, or other fault causing incorrect oxygen content in the exhaust gas. For a detailed list of checks, Refer to list of sensor / actuator checks on page 19. Different types of O2 / Lambda sensors (e.g. Zirconia or Titania / wideband etc) provide different signals and values, it will be necessary to identify the type of sensor fitted to the vehicle.
P0171	System Too Lean Bank 1	Note: Many O2 / Lambda sensor related fault codes are activated due to faults in other systems (engine, fuelling, misfires etc). The control unit is able to detect the ratio of the air / fuel mixture by monitoring the information from different sensors. The O2 (Oxygen / Lambda) sensors (primarily the pre-cat sensor) will indicate the oxygen content in the exhaust gas, which is an indicator of air / fuel ratio. The control unit will also monitor other sensors to identify the operating conditions (e.g. heavy or light load etc) and (using the O2 sensor signal), assess whether the air / fuel ratio is correct. Note that faulty engine operation e.g. misfires and exhaust air leaks can cause increased oxygen in the exhaust gas which will affect the oxygen sensor signal.	For a detailed list of checks and identification of the sensor location, Refer to list of sensor / actuator checks on page 19. Different types of O2 / Lambda sensors (e.g. Zirconia or Titania / wideband etc) provide different signals and values, it will be necessary to identify the type of sensor fitted to the vehicle. The fault code indicates a lean mixture exists, possibly during all operating conditions. The fault could be caused by a genuine lean mixture or, the fault code could be activated due to an Oxygen / Lambda sensor fault; also note that exhaust air leaks can cause a high oxygen content to exist in the exhaust gas, which will cause a "lean mixture" sensor signal. Check sensor signal and signal response under different operating conditions (load, constant rpm etc). If the sensor appears to be operating correctly but the signal value is indicating a genuine lean mixture, this could indicate an engine related fault e.g. incorrect mixtures, fuel pressure, misfires, air leaks, EVAP system faults or other engine / engine system problems.

NOTE 1. Check for other fault codes that could provide additional information.   **NOTE 2.** Communication between control units can pass via a CAN-Bus system; refer to Page 51 for CAN-Bus checks.

NOTE 3. If a fault cannot be located, it is also possible that the control unit is at fault.   **NOTE 4.** Refer to Page 53 for list of pages that show the ISO standard for component locations e.g. Sensor A - Bank 1.

code	Description		
P0172	System Too Rich Bank 1	Note: Many O2 / Lambda sensor related fault codes are activated due to faults in other systems (engine, fuelling, misfires etc). The control unit is able to detect the ratio of the air / fuel mixture by monitoring the information from different sensors. The O2 (Oxygen / Lambda) sensors (primarily the pre-cat sensor) will indicate the oxygen content in the exhaust gas, which is an indicator of air / fuel ratio. The control unit will also monitor other sensors to identify the operating conditions (e.g. heavy or light load etc) and (using the O2 sensor signal), assess whether the air / fuel ratio is correct.	For a detailed list of checks and identification of the sensor location, Refer to list of sensor / actuator checks on page 19. Different types of O2 / Lambda sensors (e.g. Zirconia or Titania / wideband etc) provide different signals and values, it will be necessary to identify the type of sensor fitted to the vehicle. The fault code indicates a rich mixture exists, possibly during all operating conditions. The fault could be caused by a genuine rich mixture or, the fault code could be activated due to an Oxygen / Lambda sensor fault. Check sensor signal and signal response under different operating conditions (load, constant rpm etc). If the sensor appears to be operating correctly but the signal value is indicating a genuine rich mixture, this could indicate an engine related fault e.g. incorrect mixtures, fuel pressure, leaking injector or other engine / engine system problems.
P0173	Fuel Trim Bank 2	Note: Many O2 / Lambda sensor related fault codes are activated due to faults in other systems (engine, fuelling, misfires etc). Fuel trim refers to the monitoring and control of the air / fuel ratio. The control unit monitors the Oxygen / Lambda sensor signal (indicates exhaust gas oxygen content). The control unit adjusts the fuelling to maintain the oxygen content within defined operating limits (helps efficient catalytic converter operation). If the control unit applies a fuelling adjustment, but this adjustment does not alter the oxygen content as expected, this can be regarded as a "fuel trim" fault. Fuel trim can apply to Pre-cat oxygen measurements (adjusting to a pre-defined air / fuel ratio), but it can also apply to Post-cat measurements (measuring after the catalyst indicates if there is a need for any fine tuning of the air / fuel ratio to maximise catalyst operation).	The fault code does not define whether the problem is due to Lean or Rich mixture (excess or lack of oxygen). The fault code indicates that there is a fuel trim problem (likely to be a pre-cat measurement). The fault could be caused by a sensor / wiring fault; check wiring between sensor and control unit for short / open circuits and high resistance. Note that the sensor could also be contaminated or faulty. It is also possible that the exhaust gas oxygen content is incorrect for the operating conditions (the sensor might therefore be operating correctly). Check sensor signal and signal response under different operating conditions (load, constant rpm etc). If the sensor does appear to be operating correctly, the fault could be caused by an engine / fuelling fault, or other fault causing incorrect oxygen content in the exhaust gas. For a detailed list of checks, Refer to list of sensor / actuator checks on page 19. Different types of O2 / Lambda sensors (e.g. Zirconia or Titania / wideband etc) provide different signals and values, it will be necessary to identify the type of sensor fitted to the vehicle.
P0174	System Too Lean Bank 2	Note: Many O2 / Lambda sensor related fault codes are activated due to faults in other systems (engine, fuelling, misfires etc). The control unit is able to detect the ratio of the air / fuel mixture by monitoring the information from different sensors. The O2 (Oxygen / Lambda) sensors (primarily the pre-cat sensor) will indicate the oxygen content in the exhaust gas, which is an indicator of air / fuel ratio. The control unit will also monitor other sensors to identify the operating conditions (e.g. heavy or light load etc) and (using the O2 sensor signal), assess whether the air / fuel ratio is correct. Note that faulty engine operation e.g. misfires and exhaust air leaks can cause increased oxygen in the exhaust gas which will affect the oxygen sensor signal.	For a detailed list of checks and identification of the sensor location, Refer to list of sensor / actuator checks on page 19. Different types of O2 / Lambda sensors (e.g. Zirconia or Titania / wideband etc) provide different signals and values, it will be necessary to identify the type of sensor fitted to the vehicle. The fault code indicates a lean mixture exists, possibly during all operating conditions. The fault could be caused by a genuine lean mixture or, the fault code could be activated due to an Oxygen / Lambda sensor fault; also note that exhaust air leaks can cause a high oxygen content to exist in the exhaust gas, which will cause a "lean mixture" sensor signal. Check sensor signal and signal response under different operating conditions (load, constant rpm etc). If the sensor appears to be operating correctly but the signal value is indicating a genuine lean mixture, this could indicate an engine related fault e.g. incorrect mixtures, fuel pressure, misfires, air leaks, EVAP system faults or other engine / engine system problems.
P0175	System Too Rich Bank 2	Note: Many O2 / Lambda sensor related fault codes are activated due to faults in other systems (engine, fuelling, misfires etc). The control unit is able to detect the ratio of the air / fuel mixture by monitoring the information from different sensors. The O2 (Oxygen / Lambda) sensors (primarily the pre-cat sensor) will indicate the oxygen content in the exhaust gas, which is an indicator of air / fuel ratio. The control unit will also monitor other sensors to identify the operating conditions (e.g. heavy or light load etc) and (using the O2 sensor signal), assess whether the air / fuel ratio is correct. Note that faulty engine operation e.g. misfires and exhaust air leaks can cause increased oxygen in the exhaust gas which will affect the oxygen sensor signal.	For a detailed list of checks and identification of the sensor location, Refer to list of sensor / actuator checks on page 19. Different types of O2 / Lambda sensors (e.g. Zirconia or Titania / wideband etc) provide different signals and values, it will be necessary to identify the type of sensor fitted to the vehicle. The fault code indicates a rich mixture exists, possibly during all operating conditions. The fault could be caused by a genuine rich mixture or, the fault code could be activated due to an Oxygen / Lambda sensor fault. Check sensor signal and signal response under different operating conditions (load, constant rpm etc). If the sensor appears to be operating correctly but the signal value is indicating a genuine rich mixture, this could indicate an engine related fault e.g. incorrect mixtures, fuel pressure, leaking injector or other engine / engine system problems.
P0176	Fuel Composition Sensor Circuit	The fuel composition sensor is used when mixed fuels are used e.g. ethanol with petrol; the sensor monitors the mix of the two fuels. If necessary, refer to vehicle specific information to identify the sensor operation, location etc.	The voltage, frequency or value of the sensor signal is incorrect (undefined fault). Sensor voltage / signal could be: out of normal operating range, constant value, non-existent, corrupt. Possible fault in the sensor or wiring e.g. short / open circuits, circuit resistance, incorrect sensor resistance or, fault with the reference / operating voltage; interference from other circuits can also affect sensor signals. Refer to list of sensor / actuator checks on page 19.
P0177	Fuel Composition Sensor Circuit Range / Performance	The fuel composition sensor is used when mixed fuels are used e.g. ethanol with petrol; the sensor monitors the mix of the two fuels. If necessary, refer to vehicle specific information to identify the sensor operation, location etc.	The voltage, frequency or value of the sensor signal is within the normal operating range / tolerance but the signal is not plausible or is incorrect due to an undefined fault. The sensor signal might not match the operating conditions indicated by other sensors or, the sensor signal is not changing / responding as expected to the changes in conditions. Possible fault in the sensor or wiring e.g. short / open circuits, circuit resistance, incorrect sensor resistance or, fault with the reference / operating voltage; interference from other circuits can also affect sensor signals. Refer to list of sensor / actuator checks on page 19. It is also possible that the sensor is operating correctly but the measurement being made by the sensor (fuel composition) is unacceptable or incorrect for the operating conditions.
P0178	Fuel Composition Sensor Circuit Low	The fuel composition sensor is used when mixed fuels are used e.g. ethanol with petrol; the sensor monitors the mix of the two fuels. If necessary, refer to vehicle specific information to identify the sensor operation, location etc.	The voltage, frequency or value of the sensor signal is either "zero" (no signal) or, the signal exists but it is below the normal operating range (lower than the minimum operating limit). Possible fault in the sensor or wiring e.g. short / open circuits, circuit resistance, incorrect sensor resistance or, fault with the reference / operating voltage; interference from other circuits can also affect sensor signals. Refer to list of sensor / actuator checks on page 19.
P0179	Fuel Composition Sensor Circuit High	The fuel composition sensor is used when mixed fuels are used e.g. ethanol with petrol; the sensor monitors the mix of the two fuels. If necessary, refer to vehicle specific information to identify the sensor operation, location etc.	The voltage, frequency or value of the sensor signal is either at full value (e.g. battery voltage with no signal or frequency) or, the signal exists but it is above the normal operating range (higher than the maximum operating limit). Possible fault in the sensor or wiring e.g. short / open circuits, circuit resistance, incorrect sensor resistance or, fault with the reference / operating voltage; interference from other circuits can also affect sensor signals. Refer to list of sensor / actuator checks on page 19.

Fault code	EOBD / ISO Description	Component / System Description	Meaningful Description and Quick Check
P0180	Fuel Temperature Sensor "A" Circuit	The fuel temperature can be used to fine tune the engine control (fuelling etc). On Diesel engines, a temperature sensor can be used to check for excessively low fuel temperatures which can cause fuel waxing. A sensor can also be used on Common Rail Diesel systems, to monitor the fuel returning to the tank; if the temperature is high, the fuel can be diverted through a cooler.	The voltage, frequency or value of the temperature sensor signal is incorrect (undefined fault). Sensor voltage / signal could be: out of normal operating range, constant value, non-existent, corrupt. Possible fault in the temperature sensor or wiring e.g. short / open circuits, circuit resistance, incorrect sensor resistance or, fault with the reference / operating voltage; interference from other circuits can also affect sensor signals. Refer to list of sensor / actuator checks on page 19.
P0181	Fuel Temperature Sensor "A" Circuit Range / Performance	The fuel temperature can be used to fine tune the engine control (fuelling etc). On Diesel engines, a temperature sensor can be used to check for excessively low fuel temperatures which can cause fuel waxing. A sensor can also be used on Common Rail Diesel systems, to monitor the fuel returning to the tank; if the temperature is high, the fuel can be diverted through a cooler.	The voltage, frequency or value of the temperature sensor signal is within the normal operating range / tolerance but the signal is not plausible or is incorrect due to an undefined fault. The sensor signal might not match the operating conditions indicated by other sensors or, the sensor signal is not changing / responding as expected to the changes in conditions. Possible fault in the temperature sensor or wiring e.g. short / open circuits, circuit resistance, incorrect sensor resistance or, fault with the reference / operating voltage; interference from other circuits can also affect sensor signals. Refer to list of sensor / actuator checks on page 19. It is also possible that the temperature sensor is operating correctly but the temperature being measured by the sensor is unacceptable or incorrect for the operating conditions; if possible, use alternative method of checking temperature to establish if sensor system is operating correctly and whether temperature is correct / incorrect.
P0182	Fuel Temperature Sensor "A" Circuit Low	The fuel temperature can be used to fine tune the engine control (fuelling etc). On Diesel engines, a temperature sensor can be used to check for excessively low fuel temperatures which can cause fuel waxing. A sensor can also be used on Common Rail Diesel systems, to monitor the fuel returning to the tank; if the temperature is high, the fuel can be diverted through a cooler.	The voltage, frequency or value of the temperature sensor signal is either "zero" (no signal) or, the signal exists but it is below the normal operating range (lower than the minimum operating limit). Possible fault in the temperature sensor or wiring e.g. short / open circuits, incorrect sensor resistance or, fault with the reference / operating voltage; interference from other circuits can also affect sensor signals. Refer to list of sensor / actuator checks on page 19.
P0183	Fuel Temperature Sensor "A" Circuit High	The fuel temperature can be used to fine tune the engine control (fuelling etc). On Diesel engines, a temperature sensor can be used to check for excessively low fuel temperatures which can cause fuel waxing. A sensor can also be used on Common Rail Diesel systems, to monitor the fuel returning to the tank; if the temperature is high, the fuel can be diverted through a cooler.	The voltage, frequency or value of the temperature sensor signal is either at full value (e.g. battery voltage with no signal or frequency) or, the signal exists but it is above the normal operating range (higher than the maximum operating limit). Possible fault in the temperature sensor or wiring e.g. short / open circuits, circuit resistance, incorrect sensor resistance or, fault with the reference / operating voltage; interference from other circuits can also affect sensor signals. Refer to list of sensor / actuator checks on page 19.
P0184	Fuel Temperature Sensor "A" Circuit Intermittent	The fuel temperature can be used to fine tune the engine control (fuelling etc). On Diesel engines, a temperature sensor can be used to check for excessively low fuel temperatures which can cause fuel waxing. A sensor can also be used on Common Rail Diesel systems, to monitor the fuel returning to the tank; if the temperature is high, the fuel can be diverted through a cooler.	The voltage, frequency or value of the temperature sensor signal is intermittent (likely to be out of normal operating range / tolerance when intermittent fault is detected) or, the signal / voltage is erratic (e.g. signal changes are irregular / unpredictable). Possible fault with wiring or connections (including internal sensor connections); also interference from other circuits can affect the signal. Refer to list of sensor / actuator checks on page 19.
P0185	Fuel Temperature Sensor "B" Circuit	The fuel temperature can be used to fine tune the engine control (fuelling etc). On Diesel engines, a temperature sensor can be used to check for excessively low fuel temperatures which can cause fuel waxing. A sensor can also be used on Common Rail Diesel systems, to monitor the fuel returning to the tank; if the temperature is high, the fuel can be diverted through a cooler.	The voltage, frequency or value of the temperature sensor signal is incorrect (undefined fault). Sensor voltage / signal could be: out of normal operating range, constant value, non-existent, corrupt. Possible fault in the temperature sensor or wiring e.g. short / open circuits, circuit resistance, incorrect sensor resistance or, fault with the reference / operating voltage; interference from other circuits can also affect sensor signals. Refer to list of sensor / actuator checks on page 19.
P0186	Fuel Temperature Sensor "B" Circuit Range / Performance	The fuel temperature can be used to fine tune the engine control (fuelling etc). On Diesel engines, a temperature sensor can be used to check for excessively low fuel temperatures which can cause fuel waxing. A sensor can also be used on Common Rail Diesel systems, to monitor the fuel returning to the tank; if the temperature is high, the fuel can be diverted through a cooler.	The voltage, frequency or value of the temperature sensor signal is within the normal operating range / tolerance but the signal is not plausible or is incorrect due to an undefined fault. The sensor signal might not match the operating conditions indicated by other sensors or, the sensor signal is not changing / responding as expected to the changes in conditions. Possible fault in the temperature sensor or wiring e.g. short / open circuits, circuit resistance, incorrect sensor resistance or, fault with the reference / operating voltage; interference from other circuits can also affect sensor signals. Refer to list of sensor / actuator checks on page 19. It is also possible that the temperature sensor is operating correctly but the temperature being measured by the sensor is unacceptable or incorrect for the operating conditions; if possible, use alternative method of checking temperature to establish if sensor system is operating correctly and whether temperature is correct / incorrect.
P0187	Fuel Temperature Sensor "B" Circuit Low	The fuel temperature can be used to fine tune the engine control (fuelling etc). On Diesel engines, a temperature sensor can be used to check for excessively low fuel temperatures which can cause fuel waxing. A sensor can also be used on Common Rail Diesel systems, to monitor the fuel returning to the tank; if the temperature is high, the fuel can be diverted through a cooler.	The voltage, frequency or value of the temperature sensor signal is either "zero" (no signal) or, the signal exists but it is below the normal operating range (lower than the minimum operating limit). Possible fault in the temperature sensor or wiring e.g. short / open circuits, circuit resistance, incorrect sensor resistance or, fault with the reference / operating voltage; interference from other circuits can also affect sensor signals. Refer to list of sensor / actuator checks on page 19.
P0188	Fuel Temperature Sensor "B" Circuit High	The fuel temperature can be used to fine tune the engine control (fuelling etc). On Diesel engines, a temperature sensor can be used to check for excessively low fuel temperatures which can cause fuel waxing. A sensor can also be used on Common Rail Diesel systems, to monitor the fuel returning to the tank; if the temperature is high, the fuel can be diverted through a cooler.	The voltage, frequency or value of the temperature sensor signal is either at full value (e.g. battery voltage with no signal or frequency) or, the signal exists but it is above the normal operating range (higher than the maximum operating limit). Possible fault in the temperature sensor or wiring e.g. short / open circuits, circuit resistance, incorrect sensor resistance or, fault with the reference / operating voltage; interference from other circuits can also affect sensor signals. Refer to list of sensor / actuator checks on page 19.
P0189	Fuel Temperature Sensor "B" Circuit Intermittent	The fuel temperature can be used to fine tune the engine control (fuelling etc). On Diesel engines, a temperature sensor can be used to check for excessively low fuel temperatures which can cause fuel waxing. A sensor can also be used on Common Rail Diesel systems, to monitor the fuel returning to the tank; if the temperature is high, the fuel can be diverted through a cooler.	The voltage, frequency or value of the temperature sensor signal is intermittent (likely to be out of normal operating range / tolerance when intermittent fault is detected) or, the signal / voltage is erratic (e.g. signal changes are irregular / unpredictable). Possible fault with wiring or connections (including internal sensor connections); also interference from other circuits can affect the signal. Refer to list of sensor / actuator checks on page 19.

NOTE 1. Check for other fault codes that could provide additional information. **NOTE 2.** Communication between control units can pass via a CAN-Bus system; refer to Page 51 for CAN-Bus checks.
NOTE 3. If a fault cannot be located, it is also possible that the control unit is at fault. **NOTE 4.** Refer to Page 53 for list of pages that show the ISO standard for component locations e.g. Sensor A - Bank 1.

90

code	Description		
P018A	Fuel Pressure Sensor "B" Circuit	A sensor indicates the fuel pressure to the control unit, which can then control a number of different functions (depending on the system). On many systems, the control unit will control fuel pressure via a pressure regulator. Where more than one sensor is fitted, this can be due to the monitoring of low and high pressure circuits or for engines using separate fuel delivery systems e.g. "V" engines. It may be necessary to refer to vehicle specific information to identify location of the sensor.	The voltage, frequency or value of the pressure sensor signal is incorrect (undefined fault). Sensor voltage / signal could be: out of normal operating range, constant value, non-existent, corrupt. Possible fault in the pressure sensor or wiring e.g. short / open circuits, circuit resistance, incorrect sensor resistance or, fault with the reference / operating voltage; interference from other circuits can also affect sensor signals. Refer to list of sensor / actuator checks on page 19.
P018B	Fuel Pressure Sensor "B" Circuit Range / Performance	A sensor indicates the fuel pressure to the control unit, which can then control a number of different functions (depending on the system). On many systems, the control unit will control fuel pressure via a pressure regulator. Where more than one sensor is fitted, this can be due to the monitoring of low and high pressure circuits or for engines using separate fuel delivery systems e.g. "V" engines. It may be necessary to refer to vehicle specific information to identify location of the sensor.	The voltage, frequency or value of the sensor signal is within the normal operating range / tolerance but the signal is not plausible or is incorrect due to an undefined fault. The sensor signal might not match the operating conditions indicated by other sensors or, the sensor signal is not changing / responding as expected to the changes in conditions. Possible fault in the sensor or wiring e.g. short / open circuits, circuit resistance, incorrect sensor resistance or, fault with the reference / operating voltage; interference from other circuits can also affect sensor signals. Refer to list of sensor / actuator checks on page 19. It is also possible that the sensor is operating correctly but the pressure being measured by the sensor is unacceptable or incorrect for the operating conditions. Check the fuel pressure using a separate gauge; if pressure is incorrect check pressure regulator (refer to fault codes P0089-P0092) and check fuel system for faults (blockages, leaks, fuel pump etc).
P018C	Fuel Pressure Sensor "B" Circuit Low	A sensor indicates the fuel pressure to the control unit, which can then control a number of different functions (depending on the system). On many systems, the control unit will control fuel pressure via a pressure regulator. Where more than one sensor is fitted, this can be due to the monitoring of low and high pressure circuits or for engines using separate fuel delivery systems e.g. "V" engines. It may be necessary to refer to vehicle specific information to identify location of the sensor.	The voltage, frequency or value of the pressure sensor signal is either "zero" (no signal) or, the signal exists but it is below the normal operating range (lower than the minimum operating limit). Possible fault in the pressure sensor or wiring e.g. short / open circuits, circuit resistance, incorrect sensor resistance or, fault with the reference / operating voltage; interference from other circuits can also affect sensor signals. Refer to list of sensor / actuator checks on page 19.
P018D	Fuel Pressure Sensor "B" Circuit High	A sensor indicates the fuel pressure to the control unit, which can then control a number of different functions (depending on the system). On many systems, the control unit will control fuel pressure via a pressure regulator. Where more than one sensor is fitted, this can be due to the monitoring of low and high pressure circuits or for engines using separate fuel delivery systems e.g. "V" engines. It may be necessary to refer to vehicle specific information to identify location of the sensor.	The voltage, frequency or value of the pressure sensor signal is either at full value (e.g. battery voltage with no signal or frequency) or, the signal exists but it is above the normal operating range (higher than the maximum operating limit). Possible fault in the pressure sensor or wiring e.g. short / open circuits, circuit resistance, incorrect sensor resistance or, fault with the reference / operating voltage; interference from other circuits can also affect sensor signals. Refer to list of sensor / actuator checks on page 19.
P018E	Fuel Pressure Sensor "B" Circuit Intermittent / Erratic	A sensor indicates the fuel pressure to the control unit, which can then control a number of different functions (depending on the system). On many systems, the control unit will control fuel pressure via a pressure regulator. Where more than one sensor is fitted, this can be due to the monitoring of low and high pressure circuits or for engines using separate fuel delivery systems e.g. "V" engines. It may be necessary to refer to vehicle specific information to identify location of the sensor.	The voltage, frequency or value of the pressure sensor signal is intermittent (likely to be out of normal operating range / tolerance when intermittent fault is detected) or, the signal / voltage is erratic (e.g. signal changes are irregular / unpredictable). Possible fault with wiring or connections (including internal sensor connections); also interference from other circuits can affect the signal. Refer to list of sensor / actuator checks on page 19. It is also possible that the pressure could be changing erratically, which could cause the fault code to be activated; if necessary, check pressure with a separate gauge to check for erratic pressure changes.
P018F	ISO / SAE reserved		Not used, reserved for future allocation.
P0190	Fuel Rail Pressure Sensor "A" Circuit	A sensor indicates the fuel pressure to the control unit, which can then control a number of different functions (depending on the system). On many systems, the control unit will control fuel pressure via a pressure regulator. Where more than one sensor is fitted, this can be due to the monitoring of low and high pressure circuits or for engines using separate fuel delivery systems e.g. "V" engines. It may be necessary to refer to vehicle specific information to identify location of the sensor.	The voltage, frequency or value of the pressure sensor signal is incorrect (undefined fault). Sensor voltage / signal could be: out of normal operating range, constant value, non-existent, corrupt. Possible fault in the pressure sensor or wiring e.g. short / open circuits, circuit resistance, incorrect sensor resistance or, fault with the reference / operating voltage; interference from other circuits can also affect sensor signals. Refer to list of sensor / actuator checks on page 19.
P0191	Fuel Rail Pressure Sensor "A" Circuit Range / Performance	A sensor indicates the fuel pressure to the control unit, which can then control a number of different functions (depending on the system). On many systems, the control unit will control fuel pressure via a pressure regulator. Where more than one sensor is fitted, this can be due to the monitoring of low and high pressure circuits or for engines using separate fuel delivery systems e.g. "V" engines. It may be necessary to refer to vehicle specific information to identify location of the sensor.	The voltage, frequency or value of the sensor signal is within the normal operating range / tolerance but the signal is not plausible or is incorrect due to an undefined fault. The sensor signal might not match the operating conditions indicated by other sensors or, the sensor signal is not changing / responding as expected to the changes in conditions. Possible fault in the sensor or wiring e.g. short / open circuits, circuit resistance, incorrect sensor resistance or, fault with the reference / operating voltage; interference from other circuits can also affect sensor signals. Refer to list of sensor / actuator checks on page 19. It is also possible that the sensor is operating correctly but the pressure being measured by the sensor is unacceptable or incorrect for the operating conditions. Check the fuel pressure using a separate gauge; if pressure is incorrect check pressure regulator (refer to fault codes P0089-P0092) and check fuel system for faults (blockages, leaks, fuel pump etc).
P0192	Fuel Rail Pressure Sensor "A" Circuit Low	A sensor indicates the fuel pressure to the control unit, which can then control a number of different functions (depending on the system). On many systems, the control unit will control fuel pressure via a pressure regulator. Where more than one sensor is fitted, this can be due to the monitoring of low and high pressure circuits or for engines using separate fuel delivery systems e.g. "V" engines. It may be necessary to refer to vehicle specific information to identify location of the sensor.	The voltage, frequency or value of the pressure sensor signal is either "zero" (no signal) or, the signal exists but it is below the normal operating range (lower than the minimum operating limit). Possible fault in the pressure sensor or wiring e.g. short / open circuits, circuit resistance, incorrect sensor resistance or, fault with the reference / operating voltage; interference from other circuits can also affect sensor signals. Refer to list of sensor / actuator checks on page 19.

NOTE 1. Check for other fault codes that could provide additional information. **NOTE 2.** Communication between control units can pass via a CAN-Bus system; refer to Page 51 for CAN-Bus checks.
NOTE 3. If a fault cannot be located, it is also possible that the control unit is at fault. **NOTE 4.** Refer to Page 53 for list of pages that show the ISO standard for component locations e.g. Sensor A - Bank 1.

91

Fault code	EOBD / ISO Description	Component / System Description	Meaningful Description and Quick Check
P0193	Fuel Rail Pressure Sensor "A" Circuit High	A sensor indicates the fuel pressure to the control unit, which can then control a number of different functions (depending on the system). On many systems, the control unit will control fuel pressure via a pressure regulator. Where more than one sensor is fitted, this can be due to the monitoring of low and high pressure circuits or for engines using separate fuel delivery systems e.g. "V" engines. It may be necessary to refer to vehicle specific information to identify location of the sensor.	The voltage, frequency or value of the pressure sensor signal is either at full value (e.g. battery voltage with no signal or frequency) or, the signal exists but it is above the normal operating range (higher than the maximum operating limit). Possible fault in the pressure sensor or wiring e.g. short / open circuits, circuit resistance, incorrect sensor resistance or, fault with the reference / operating voltage; interference from other circuits can also affect sensor signals. Refer to list of sensor / actuator checks on page 19.
P0194	Fuel Rail Pressure Sensor "A" Circuit Intermittent / Erratic	A sensor indicates the fuel pressure to the control unit, which can then control a number of different functions (depending on the system). On many systems, the control unit will control fuel pressure via a pressure regulator. Where more than one sensor is fitted, this can be due to the monitoring of low and high pressure circuits or for engines using separate fuel delivery systems e.g. "V" engines. It may be necessary to refer to vehicle specific information to identify location of the sensor.	The voltage, frequency or value of the pressure sensor signal is intermittent (likely to be out of normal operating range / tolerance when intermittent fault is detected) or, the signal / voltage is erratic (e.g. signal changes are irregular / unpredictable). Possible fault with wiring or connections (including internal sensor connections); also interference from other circuits can affect the signal. Refer to list of sensor / actuator checks on page 19. It is also possible that the pressure could be changing erratically, which could cause the fault code to be activated; if necessary, check pressure with a separate gauge to check for erratic pressure changes.
P0195	Engine Oil Temperature Sensor	The engine oil temperature is critical to engine operation / performance / durability; it is also possible that operation of some functions e.g. emission devices, or engine system control could be dependent on the oil temperature being within certain values. Many vehicles make use of oil cooler systems to regulate the oil temperature.	The code indicates a sensor related fault rather than a temperature problem. The voltage, frequency or value of the temperature sensor signal is incorrect (undefined fault). The fault code does not identify the exact type of fault therefore a full range of checks should be carried out on the sensor and wiring. Possible fault in the temperature sensor or wiring e.g. short / open circuits, circuit resistance, incorrect sensor resistance or, fault with the reference / operating voltage; interference from other circuits can also affect sensor signals. Refer to list of sensor / actuator checks on page 19.
P0196	Engine Oil Temperature Sensor Range / Performance	The engine oil temperature is critical to engine operation / performance / durability; it is also possible that operation of some functions e.g. emission devices, or engine system control could be dependent on the oil temperature being within certain values. Many vehicles make use of oil cooler systems to regulate the oil temperature.	The voltage, frequency or value of the temperature sensor signal is within the normal operating range / tolerance but the signal is not plausible or is incorrect due to an undefined fault. The sensor signal might not match the operating conditions indicated by other sensors or, the sensor signal is not changing / responding as expected to the changes in conditions. Possible fault in the temperature sensor or wiring e.g. short / open circuits, circuit resistance, incorrect sensor resistance or, fault with the reference / operating voltage; interference from other circuits can also affect sensor signals. Refer to list of sensor / actuator checks on page 19. Note that the sensor signal could pass via a CAN-Bus system before passing to a control unit. It is also possible that the temperature sensor is operating correctly but the temperature being measured by the sensor is unacceptable or incorrect for the operating conditions; check for correct oil grade, oil quantity and quality. If applicable, check oil cooler operation and check for restriction in cooling air etc.
P0197	Engine Oil Temperature Sensor Low	The engine oil temperature is critical to engine operation / performance / durability; it is also possible that operation of some functions e.g. emission devices, or engine system control could be dependent on the oil temperature being within certain values. Many vehicles make use of oil cooler systems to regulate the oil temperature.	The voltage, frequency or value of the temperature sensor signal is either "zero" (no signal) or, the signal exists but it is below the normal operating range (lower than the minimum operating limit). Possible fault in the temperature sensor or wiring e.g. short / open circuits, circuit resistance, incorrect sensor resistance or, fault with the reference / operating voltage; interference from other circuits can also affect sensor signals. Refer to list of sensor / actuator checks on page 19.
P0198	Engine Oil Temperature Sensor High	The engine oil temperature is critical to engine operation / performance / durability; it is also possible that operation of some functions e.g. emission devices, or engine system control could be dependent on the oil temperature being within certain values. Many vehicles make use of oil cooler systems to regulate the oil temperature.	The voltage, frequency or value of the temperature sensor signal is either at full value (e.g. battery voltage with no signal or frequency) or, the signal exists but it is above the normal operating range (higher than the maximum operating limit). Possible fault in the temperature sensor or wiring e.g. short / open circuits, circuit resistance, incorrect sensor resistance or, fault with the reference / operating voltage; interference from other circuits can also affect sensor signals. Refer to list of sensor / actuator checks on page 19.
P0199	Engine Oil Temperature Sensor Intermittent	The engine oil temperature is critical to engine operation / performance / durability; it is also possible that operation of some functions e.g. emission devices, or engine system control could be dependent on the oil temperature being within certain values. Many vehicles make use of oil cooler systems to regulate the oil temperature.	The voltage, frequency or value of the temperature sensor signal is intermittent (likely to be out of normal operating range / tolerance when intermittent fault is detected) or, the signal / voltage is erratic (e.g. signal changes are irregular / unpredictable). Possible fault with wiring or connections (including internal sensor connections); also interference from other circuits can affect the signal. Refer to list of sensor / actuator checks on page 19. It is also possible that the fault could be caused by rapidly fluctuating oil temperature.
P0200	Injector Circuit / Open	Most injectors have traditionally used a solenoid (contained within the injector body) to open the injector. Later injector designs make use of Piezo actuators instead of solenoids. Note that injector operating voltages on some applications are considerably higher than "battery voltage" traditionally used on solenoid injectors; these systems can use a separate injector power module to deliver the power and control signal to the injectors.	Identify the type of injector and the wiring / control modules used on the system. The voltage, frequency or value of the control signal in the injector circuit is incorrect. Signal could be:- out of normal operating range, constant value, non-existent, corrupt. The fault code indicates a possible "open circuit" but other undefined faults in the injector or wiring could activate the same code e.g. injector power supply, short circuit, high circuit resistance or incorrect injector resistance (solenoid type injectors). Note: the control unit could be providing the correct signal but it is being affected by a circuit fault. Identify the type of injector and the wiring / control modules used on the system. Refer to list of sensor / actuator checks on page 19. Note that for all systems, initial checks should include checks for short / open circuits, poor connections. Note that the wiring should be checked through to the control unit, or to the power module (if fitted) and from the power module to the control unit. If other injectors are operating correctly, it indicates a likely wiring / connection fault but it is also possible that the injector is faulty.

NOTE 1. Check for other fault codes that could provide additional information.   NOTE 2. Communication between control units can pass via a CAN-Bus system; refer to Page 51 for CAN-Bus checks.

NOTE 3. If a fault cannot be located, it is also possible that the control unit is at fault.   NOTE 4. Refer to Page 53 for list of pages that show the ISO standard for component locations e.g. Sensor A - Bank 1.

P0201	Injector Circuit / Open -Cylinder 1	Most injectors have traditionally used a solenoid (contained within the injector body) to open the injector. Later injector designs make use of Piezo actuators instead of solenoids. Note that injector operating voltages on some applications are considerably higher than "battery voltage" traditionally used on solenoid injectors; these systems can use a separate injector power module to deliver the power and control signal to the injectors.	Identify the type of injector and the wiring / control modules used on the system. The voltage, frequency or value of the control signal in the injector circuit is incorrect. Signal could be:- out of normal operating range, constant value, non-existent, corrupt. The fault code indicates a possible "open circuit" but other undefined faults in the injector or wiring could activate the same code e.g. injector power supply, short circuit, high circuit resistance or incorrect injector resistance (solenoid type injectors). Note: the control unit could be providing the correct signal but it is being affected by a circuit fault. Identify the type of injector and the wiring / control modules used on the system. Refer to list of sensor / actuator checks on page 19. Note that for all systems, initial checks should include checks for short / open circuits, poor connections. Note that the wiring should be checked through to the control unit, or to the power module (if fitted) and from the power module to the control unit. If other injectors are operating correctly, it indicates a likely wiring / connection fault but it is also possible that the injector is faulty.
P0202	Injector Circuit / Open -Cylinder 2	Most injectors have traditionally used a solenoid (contained within the injector body) to open the injector. Later injector designs make use of Piezo actuators instead of solenoids. Note that injector operating voltages on some applications are considerably higher than "battery voltage" traditionally used on solenoid injectors; these systems can use a separate injector power module to deliver the power and control signal to the injectors.	Identify the type of injector and the wiring / control modules used on the system. The voltage, frequency or value of the control signal in the injector circuit is incorrect. Signal could be:- out of normal operating range, constant value, non-existent, corrupt. The fault code indicates a possible "open circuit" but other undefined faults in the injector or wiring could activate the same code e.g. injector power supply, short circuit, high circuit resistance or incorrect injector resistance (solenoid type injectors). Note: the control unit could be providing the correct signal but it is being affected by a circuit fault. Identify the type of injector and the wiring / control modules used on the system. Refer to list of sensor / actuator checks on page 19. Note that for all systems, initial checks should include checks for short / open circuits, poor connections. Note that the wiring should be checked through to the control unit, or to the power module (if fitted) and from the power module to the control unit. If other injectors are operating correctly, it indicates a likely wiring / connection fault but it is also possible that the injector is faulty.
P0203	Injector Circuit / Open -Cylinder 3	Most injectors have traditionally used a solenoid (contained within the injector body) to open the injector. Later injector designs make use of Piezo actuators instead of solenoids. Note that injector operating voltages on some applications are considerably higher than "battery voltage" traditionally used on solenoid injectors; these systems can use a separate injector power module to deliver the power and control signal to the injectors.	Identify the type of injector and the wiring / control modules used on the system. The voltage, frequency or value of the control signal in the injector circuit is incorrect. Signal could be:- out of normal operating range, constant value, non-existent, corrupt. The fault code indicates a possible "open circuit" but other undefined faults in the injector or wiring could activate the same code e.g. injector power supply, short circuit, high circuit resistance or incorrect injector resistance (solenoid type injectors). Note: the control unit could be providing the correct signal but it is being affected by a circuit fault. Identify the type of injector and the wiring / control modules used on the system. Refer to list of sensor / actuator checks on page 19. Note that for all systems, initial checks should include checks for short / open circuits, poor connections. Note that the wiring should be checked through to the control unit, or to the power module (if fitted) and from the power module to the control unit. If other injectors are operating correctly, it indicates a likely wiring / connection fault but it is also possible that the injector is faulty.
P0204	Injector Circuit / Open -Cylinder 4	Most injectors have traditionally used a solenoid (contained within the injector body) to open the injector. Later injector designs make use of Piezo actuators instead of solenoids. Note that injector operating voltages on some applications are considerably higher than "battery voltage" traditionally used on solenoid injectors; these systems can use a separate injector power module to deliver the power and control signal to the injectors.	Identify the type of injector and the wiring / control modules used on the system. The voltage, frequency or value of the control signal in the injector circuit is incorrect. Signal could be:- out of normal operating range, constant value, non-existent, corrupt. The fault code indicates a possible "open circuit" but other undefined faults in the injector or wiring could activate the same code e.g. injector power supply, short circuit, high circuit resistance or incorrect injector resistance (solenoid type injectors). Note: the control unit could be providing the correct signal but it is being affected by a circuit fault. Identify the type of injector and the wiring / control modules used on the system. Refer to list of sensor / actuator checks on page 19. Note that for all systems, initial checks should include checks for short / open circuits, poor connections. Note that the wiring should be checked through to the control unit, or to the power module (if fitted) and from the power module to the control unit. If other injectors are operating correctly, it indicates a likely wiring / connection fault but it is also possible that the injector is faulty.
P0205	Injector Circuit / Open -Cylinder 5	Most injectors have traditionally used a solenoid (contained within the injector body) to open the injector. Later injector designs make use of Piezo actuators instead of solenoids. Note that injector operating voltages on some applications are considerably higher than "battery voltage" traditionally used on solenoid injectors; these systems can use a separate injector power module to deliver the power and control signal to the injectors.	Identify the type of injector and the wiring / control modules used on the system. The voltage, frequency or value of the control signal in the injector circuit is incorrect. Signal could be:- out of normal operating range, constant value, non-existent, corrupt. The fault code indicates a possible "open circuit" but other undefined faults in the injector or wiring could activate the same code e.g. injector power supply, short circuit, high circuit resistance or incorrect injector resistance (solenoid type injectors). Note: the control unit could be providing the correct signal but it is being affected by a circuit fault. Identify the type of injector and the wiring / control modules used on the system. Refer to list of sensor / actuator checks on page 19. Note that for all systems, initial checks should include checks for short / open circuits, poor connections. Note that the wiring should be checked through to the control unit, or to the power module (if fitted) and from the power module to the control unit. If other injectors are operating correctly, it indicates a likely wiring / connection fault but it is also possible that the injector is faulty.
P0206	Injector Circuit / Open -Cylinder 6	Most injectors have traditionally used a solenoid (contained within the injector body) to open the injector. Later injector designs make use of Piezo actuators instead of solenoids. Note that injector operating voltages on some applications are considerably higher than "battery voltage" traditionally used on solenoid injectors; these systems can use a separate injector power module to deliver the power and control signal to the injectors.	Identify the type of injector and the wiring / control modules used on the system. The voltage, frequency or value of the control signal in the injector circuit is incorrect. Signal could be:- out of normal operating range, constant value, non-existent, corrupt. The fault code indicates a possible "open circuit" but other undefined faults in the injector or wiring could activate the same code e.g. injector power supply, short circuit, high circuit resistance or incorrect injector resistance (solenoid type injectors). Note: the control unit could be providing the correct signal but it is being affected by a circuit fault. Identify the type of injector and the wiring / control modules used on the system. Refer to list of sensor / actuator checks on page 19. Note that for all systems, initial checks should include checks for short / open circuits, poor connections. Note that the wiring should be checked through to the control unit, or to the power module (if fitted) and from the power module to the control unit. If other injectors are operating correctly, it indicates a likely wiring / connection fault but it is also possible that the injector is faulty.

NOTE 1. Check for other fault codes that could provide additional information. **NOTE 2.** Communication between control units can pass via a CAN-Bus system; refer to Page 51 for CAN-Bus checks.

NOTE 3. If a fault cannot be located, it is also possible that the control unit is at fault. **NOTE 4.** Refer to Page 53 for list of pages that show the ISO standard for component locations e.g. Sensor A - Bank 1.

Fault code	EOBD / ISO Description	Component / System Description	Meaningful Description and Quick Check
P0207	Injector Circuit / Open -Cylinder 7	Most injectors have traditionally used a solenoid (contained within the injector body) to open the injector. Later injector designs make use of Piezo actuators instead of solenoids. Note that injector operating voltages on some applications are considerably higher than "battery voltage" traditionally used on solenoid injectors; these systems can use a separate injector power module to deliver the power and control signal to the injectors.	Identify the type of injector and the wiring / control modules used on the system. The voltage, frequency or value of the control signal in the injector circuit is incorrect. Signal could be:- out of normal operating range, constant value, non-existent, corrupt. The fault code indicates a possible "open circuit" but other undefined faults in the injector or wiring could activate the same code e.g. injector power supply, short circuit, high circuit resistance or incorrect injector resistance (solenoid type injectors). Note: the control unit could be providing the correct signal but it is being affected by a circuit fault. Identify the type of injector and the wiring / control modules used on the system. Refer to list of sensor / actuator checks on page 19. Note that for all systems, initial checks should include checks for short / open circuits, poor connections. Note that the wiring should be checked through to the control unit, or to the power module (if fitted) and from the power module to the control unit. If other injectors are operating correctly, it indicates a likely wiring / connection fault but it is also possible that the injector is faulty.
P0208	Injector Circuit / Open -Cylinder 8	Most injectors have traditionally used a solenoid (contained within the injector body) to open the injector. Later injector designs make use of Piezo actuators instead of solenoids. Note that injector operating voltages on some applications are considerably higher than "battery voltage" traditionally used on solenoid injectors; these systems can use a separate injector power module to deliver the power and control signal to the injectors.	Identify the type of injector and the wiring / control modules used on the system. The voltage, frequency or value of the control signal in the injector circuit is incorrect. Signal could be:- out of normal operating range, constant value, non-existent, corrupt. The fault code indicates a possible "open circuit" but other undefined faults in the injector or wiring could activate the same code e.g. injector power supply, short circuit, high circuit resistance or incorrect injector resistance (solenoid type injectors). Note: the control unit could be providing the correct signal but it is being affected by a circuit fault. Identify the type of injector and the wiring / control modules used on the system. Refer to list of sensor / actuator checks on page 19. Note that for all systems, initial checks should include checks for short / open circuits, poor connections. Note that the wiring should be checked through to the control unit, or to the power module (if fitted) and from the power module to the control unit. If other injectors are operating correctly, it indicates a likely wiring / connection fault but it is also possible that the injector is faulty.
P0209	Injector Circuit / Open -Cylinder 9	Most injectors have traditionally used a solenoid (contained within the injector body) to open the injector. Later injector designs make use of Piezo actuators instead of solenoids. Note that injector operating voltages on some applications are considerably higher than "battery voltage" traditionally used on solenoid injectors; these systems can use a separate injector power module to deliver the power and control signal to the injectors.	Identify the type of injector and the wiring / control modules used on the system. The voltage, frequency or value of the control signal in the injector circuit is incorrect. Signal could be:- out of normal operating range, constant value, non-existent, corrupt. The fault code indicates a possible "open circuit" but other undefined faults in the injector or wiring could activate the same code e.g. injector power supply, short circuit, high circuit resistance or incorrect injector resistance (solenoid type injectors). Note: the control unit could be providing the correct signal but it is being affected by a circuit fault. Identify the type of injector and the wiring / control modules used on the system. Refer to list of sensor / actuator checks on page 19. Note that for all systems, initial checks should include checks for short / open circuits, poor connections. Note that the wiring should be checked through to the control unit, or to the power module (if fitted) and from the power module to the control unit. If other injectors are operating correctly, it indicates a likely wiring / connection fault but it is also possible that the injector is faulty.
P020A	Cylinder 1 Injection Timing	There are various methods of controlling and identifying injection timing (depending on the fuel system fitted to the vehicle). It will therefore be necessary to identify the fuel system and the method used to control / check injection timing.	Refer to vehicle specific information to identify the timing monitoring process and timing control (individual injector timing can be altered by the control unit by altering the injector control signal). The timing can be monitored in a number of ways e.g. injector lift / movement sensor but, some systems use a pressure sensor in the combustion chamber (sensor located within in the glow plug) to monitor start of combustion, which can be influenced by injection timing. The fault code indicates that there is an undefined fault with the Fuel Injection Timing (on the specified cylinder) but the timing will be outside of expected values for the operating conditions. The fault could be caused by an injector / fuel delivery related fault, but for those Diesel systems using combustion pressure monitoring, if any faults delay or advance the start of combustion, the injection timing can be altered to compensate. If the injection timing adjustments do not compensate for the combustion timing error, this can be regarded as an injection timing fault.
P020B	Cylinder 2 Injection Timing	There are various methods of controlling and identifying injection timing (depending on the fuel system fitted to the vehicle). It will therefore be necessary to identify the fuel system and the method used to control / check injection timing.	Refer to vehicle specific information to identify the timing monitoring process and timing control (individual injector timing can be altered by the control unit by altering the injector control signal). The timing can be monitored in a number of ways e.g. injector lift / movement sensor but, some systems use a pressure sensor in the combustion chamber (sensor located within in the glow plug) to monitor start of combustion, which can be influenced by injection timing. The fault code indicates that there is an undefined fault with the Fuel Injection Timing (on the specified cylinder) but the timing will be outside of expected values for the operating conditions. The fault could be caused by an injector / fuel delivery related fault, but for those Diesel systems using combustion pressure monitoring, if any faults delay or advance the start of combustion, the injection timing can be altered to compensate. If the injection timing adjustments do not compensate for the combustion timing error, this can be regarded as an injection timing fault.
P020C	Cylinder 3 Injection Timing	There are various methods of controlling and identifying injection timing (depending on the fuel system fitted to the vehicle). It will therefore be necessary to identify the fuel system and the method used to control / check injection timing.	Refer to vehicle specific information to identify the timing monitoring process and timing control (individual injector timing can be altered by the control unit by altering the injector control signal). The timing can be monitored in a number of ways e.g. injector lift / movement sensor but, some systems use a pressure sensor in the combustion chamber (sensor located within in the glow plug) to monitor start of combustion, which can be influenced by injection timing. The fault code indicates that there is an undefined fault with the Fuel Injection Timing (on the specified cylinder) but the timing will be outside of expected values for the operating conditions. The fault could be caused by an injector / fuel delivery related fault, but for those Diesel systems using combustion pressure monitoring, if any faults delay or advance the start of combustion, the injection timing can be altered to compensate. If the injection timing adjustments do not compensate for the combustion timing error, this can be regarded as an injection timing fault.
P020D	Cylinder 4 Injection Timing	There are various methods of controlling and identifying injection timing (depending on the fuel system fitted to the vehicle). It will therefore be necessary to identify the fuel system and the method used to control / check injection timing.	Refer to vehicle specific information to identify the timing monitoring process and timing control (individual injector timing can be altered by the control unit by altering the injector control signal). The timing can be monitored in a number of ways e.g. injector lift / movement sensor but, some systems use a pressure sensor in the combustion chamber (sensor located within in the glow plug) to monitor start of combustion, which can be influenced by injection timing. The fault code indicates that there is an undefined fault with the Fuel Injection Timing (on the specified cylinder) but the timing will be outside of expected values for the operating conditions. The fault could be caused by an injector / fuel delivery related fault, but for those Diesel systems using combustion pressure monitoring, if any faults delay or advance the start of combustion, the injection timing can be altered to compensate. If the injection timing adjustments do not compensate for the combustion timing error, this can be regarded as an injection timing fault.

NOTE 1. Check for other fault codes that could provide additional information.
NOTE 2. Communication between control units can pass via a CAN-Bus system; refer to Page 51 for CAN-Bus checks.
NOTE 3. If a fault cannot be located, it is also possible that the control unit is at fault.
NOTE 4. Refer to Page 53 for list of pages that show the ISO standard for component locations e.g. Sensor A - Bank 1.

code	Description		
P020E	Cylinder 5 Injection Timing	There are various methods of controlling and identifying injection timing (depending on the fuel system fitted to the vehicle). It will therefore be necessary to identify the fuel system and the method used to control / check injection timing.	Refer to vehicle specific information to identify the timing monitoring process and timing control (individual injector timing can be altered by the control unit by altering the injector control signal). The timing can be monitored in a number of ways e.g. injector lift / movement sensor but, some systems use a pressure sensor in the combustion chamber (sensor located within in the glow plug) to monitor start of combustion, which can be influenced by injection timing. The fault code indicates that there is an undefined fault with the Fuel Injection Timing (on the specified cylinder) but the timing will be outside of expected values for the operating conditions. The fault could be caused by an injector / fuel delivery related fault, but for those Diesel systems using combustion pressure monitoring, if any faults delay or advance the start of combustion, the injection timing can be altered to compensate. If the injection timing adjustments do not compensate for the combustion timing error, this can be regarded as an injection timing fault.
P020F	Cylinder 6 Injection Timing	There are various methods of controlling and identifying injection timing (depending on the fuel system fitted to the vehicle). It will therefore be necessary to identify the fuel system and the method used to control / check injection timing.	Refer to vehicle specific information to identify the timing monitoring process and timing control (individual injector timing can be altered by the control unit by altering the injector control signal). The timing can be monitored in a number of ways e.g. injector lift / movement sensor but, some systems use a pressure sensor in the combustion chamber (sensor located within in the glow plug) to monitor start of combustion, which can be influenced by injection timing. The fault code indicates that there is an undefined fault with the Fuel Injection Timing (on the specified cylinder) but the timing will be outside of expected values for the operating conditions. The fault could be caused by an injector / fuel delivery related fault, but for those Diesel systems using combustion pressure monitoring, if any faults delay or advance the start of combustion, the injection timing can be altered to compensate. If the injection timing adjustments do not compensate for the combustion timing error, this can be regarded as an injection timing fault.
P0210	Injector Circuit / Open -Cylinder 10	Most injectors have traditionally used a solenoid (contained within the injector body) to open the injector. Later injector designs make use of Piezo actuators instead of solenoids. Note that injector operating voltages on some applications are considerably higher than "battery voltage" traditionally used on solenoid injectors; these systems can use a separate injector power module to deliver the power and control signal to the injectors.	Identify the type of injector and the wiring / control modules used on the system. The voltage, frequency or value of the control signal in the injector circuit is incorrect. Signal could be:- out of normal operating range, constant value, non-existent, corrupt. The fault code indicates a possible "open circuit" but other undefined faults in the injector or wiring could activate the same code e.g. injector power supply, short circuit, high circuit resistance or incorrect injector resistance (solenoid type injectors). Note: the control unit could be providing the correct signal but it is being affected by a circuit fault. Identify the type of injector and the wiring / control modules used on the system. Refer to list of sensor / actuator checks on page 19. Note that for all systems, initial checks should include checks for short / open circuits, poor connections. Note that the wiring should be checked through to the control unit, or to the power module (if fitted) and from the power module to the control unit. If other injectors are operating correctly, it indicates a likely wiring / connection fault but it is also possible that the injector is faulty.
P0211	Injector Circuit / Open -Cylinder 11	Most injectors have traditionally used a solenoid (contained within the injector body) to open the injector. Later injector designs make use of Piezo actuators instead of solenoids. Note that injector operating voltages on some applications are considerably higher than "battery voltage" traditionally used on solenoid injectors; these systems can use a separate injector power module to deliver the power and control signal to the injectors.	Identify the type of injector and the wiring / control modules used on the system. The voltage, frequency or value of the control signal in the injector circuit is incorrect. Signal could be:- out of normal operating range, constant value, non-existent, corrupt. The fault code indicates a possible "open circuit" but other undefined faults in the injector or wiring could activate the same code e.g. injector power supply, short circuit, high circuit resistance or incorrect injector resistance (solenoid type injectors). Note: the control unit could be providing the correct signal but it is being affected by a circuit fault. Identify the type of injector and the wiring / control modules used on the system. Refer to list of sensor / actuator checks on page 19. Note that for all systems, initial checks should include checks for short / open circuits, poor connections. Note that the wiring should be checked through to the control unit, or to the power module (if fitted) and from the power module to the control unit. If other injectors are operating correctly, it indicates a likely wiring / connection fault but it is also possible that the injector is faulty.
P0212	Injector Circuit / Open -Cylinder 12	Most injectors have traditionally used a solenoid (contained within the injector body) to open the injector. Later injector designs make use of Piezo actuators instead of solenoids. Note that injector operating voltages on some applications are considerably higher than "battery voltage" traditionally used on solenoid injectors; these systems can use a separate injector power module to deliver the power and control signal to the injectors.	Identify the type of injector and the wiring / control modules used on the system. The voltage, frequency or value of the control signal in the injector circuit is incorrect. Signal could be:- out of normal operating range, constant value, non-existent, corrupt. The fault code indicates a possible "open circuit" but other undefined faults in the injector or wiring could activate the same code e.g. injector power supply, short circuit, high circuit resistance or incorrect injector resistance (solenoid type injectors). Note: the control unit could be providing the correct signal but it is being affected by a circuit fault. Identify the type of injector and the wiring / control modules used on the system. Refer to list of sensor / actuator checks on page 19. Note that for all systems, initial checks should include checks for short / open circuits, poor connections. Note that the wiring should be checked through to the control unit, or to the power module (if fitted) and from the power module to the control unit. If other injectors are operating correctly, it indicates a likely wiring / connection fault but it is also possible that the injector is faulty.
P0213	Cold Start Injector 1	Cold start injectors have been used on some systems to provide additional fuel under extreme cold start conditions. Refer to vehicle specific information to identify the wiring and control for the injector, which could make use of a relay (switched via the control unit) or be directly switched by the control unit or a separate module.	The fault code does not identify whether the fault is related to an electrical problem or to a fuel delivery fault. However, it is likely that the fault has been detected due to incorrect voltages or values in the circuit. Check for short / open circuits in the wiring and connections, also check injector resistance and check for any control devices for the injector (e.g. temperature dependent switches etc). Check fuel delivery to the injector.
P0214	Cold Start Injector 2	Cold start injectors have been used on some systems to provide additional fuel under extreme cold start conditions. Refer to vehicle specific information to identify the wiring and control for the injector, which could make use of a relay (switched via the control unit) or be directly switched by the control unit or a separate module.	The fault code does not identify whether the fault is related to an electrical problem or to a fuel delivery fault. However, it is likely that the fault has been detected due to incorrect voltages or values in the circuit. Check for short / open circuits in the wiring and connections, also check injector resistance and check for any control devices for the injector (e.g. temperature dependent switches etc). Check fuel delivery to the injector.
P0215	Engine Shutoff Solenoid	Some Diesel engines (usually common rail) use an air flap / butterfly in the intake or in the inlet ports. The flap can be vacuum operated, with the vacuum regulated by a solenoid valve; the flap system provides a smoother engine shut off compared to systems that shut off the fuel. Other systems can use fuel shut off solenoid valves (refer to fault codes P0005-6-7 for additional information).	The control unit has detected an undefined fault with the shutoff solenoid system; the fault could be detected due to an electrical problem with the solenoid / wiring (e.g. solenoid power supply, short / open circuit or resistance). Refer to list of sensor / actuator checks on page 19. The fault could also be detected because sensors are detecting incorrect position of the shutoff mechanism; check operation of any position sensors (if applicable) and check mechanism for freedom of movement.

NOTE 1. Check for other fault codes that could provide additional information.   **NOTE 2.** Communication between control units can pass via a CAN-Bus system; refer to Page 51 for CAN-Bus checks.

NOTE 3. If a fault cannot be located, it is also possible that the control unit is at fault.   **NOTE 4.** Refer to Page 53 for list of pages that show the ISO standard for component locations e.g. Sensor A - Bank 1.

Fault code	EOBD / ISO Description	Component / System Description	Meaningful Description and Quick Check
P0216	Injector / Injection Timing Control Circuit	There are various methods of controlling and identifying injection timing (depending on the fuel system fitted to the vehicle). It will therefore be necessary to identify the fuel system and the method used to control / check injection timing.	The control signal for the Fuel Injection Timing Control is assessed by the Control Unit as being incorrect. Undefined fault in Fuel Injection Timing Control circuit. The exact fault will depend on the type of fuel system being used. The fault code relates to a timing control signal fault which could be applicable to the signal passing from the main Diesel Control Unit to the Diesel Timing Control unit (on those systems where two control units are used). It could also relate to the signal being provided to the timing device fitted to the Diesel injection pump. Also refer to fault code P020A for information on other timing control and timing monitoring systems.
P0217	Engine Coolant Over Temperature Condition	The coolant temperature sensor signal is used for control of fuel, ignition and emissions control. The sensor information can be used for cooling fan control as well as for other vehicle systems.	The fault code indicates that the engine coolant temperature sensor is indicating excessive temperature for the operating conditions or does not match values expected by the control unit. The fault could be related to a genuine over temperature condition but could also be caused by a faulty temperature sensor. If possible, use alternative method of checking temperature to establish if the sensor system is operating correctly and whether temperature is correct / incorrect. If the coolant temperature is too high, check cooling system e.g. coolant level, thermostat operation, cooling fan operation etc. If it appears that the temperature sensor is at fault, check for a fault with the sensor / wiring e.g. short / open circuits, circuit resistance, incorrect sensor resistance or reference / operating voltage fault. Refer to list of sensor / actuator checks on page 19.
P0218	Transmission Fluid Over Temperature Condition	The information from the transmission fluid temperature sensor can be used to alter transmission operation during cold running / warm up and also influence the gear change strategy; under certain conditions engine control could be influenced by transmission oil temperature. The fault code relates to a transmission fluid temperature problem. Note the transmission fluid can be cooled by an intercooler / radiator.	The fault could be related to a genuine over temperature condition but could also be caused by a faulty temperature sensor. The voltage, frequency or value of the temperature sensor signal is within the normal operating range / tolerance but is indicating a high temperature value. The control unit has not detected an electrical fault, therefore the assessment is that the temperature is too high for the operating conditions (indicated by other sensors) or, the temperature does not match values expected by the control unit. If possible, check the temperature using a separate thermometer and compare with the value indicated by the sensor. If the sensor does appear to be faulty, Refer to list of sensor / actuator checks on page 19. If temperature is too high, check operation of intercooler (where fitted), check oil level, oil condition and grade; also check for other faults that could cause high oil temperature.
P0219	Engine Overspeed Condition	Note that although the fault could relate to a problem with excessive engine speed, it is also possible that the overspeed condition could have occurred during a process that could be engine speed dependent (e.g. emissions control function).	The control unit has detected an engine overspeed condition during normal vehicle operation; if no obvious fault exists but fault code re-occurs (but engine speed has remained within rpm limits) it might be necessary to refer to vehicle specific information to check for any processes that are engine speed dependent e.g. an emissions control function. It is also possible that a fault exists with the speed sensor. Refer to list of sensor / actuator checks on page 19.
P021A	Cylinder 7 Injection Timing	There are various methods of controlling and identifying injection timing (depending on the fuel system fitted to the vehicle). It will therefore be necessary to identify the fuel system and the method used to control / check injection timing.	Refer to vehicle specific information to identify the timing monitoring process and timing control (individual injector timing can be altered by the control unit by altering the injector control signal). The timing can be monitored in a number of ways e.g. injector lift / movement sensor but, some systems use a pressure sensor in the combustion chamber (sensor located within in the glow plug) to monitor start of combustion, which can be influenced by injection timing. The fault code indicates that there is an undefined fault with the Fuel Injection Timing (on the specified cylinder) but the timing will be outside of expected values for the operating conditions. The fault could be caused by an injector / fuel delivery related fault, but for those Diesel systems using combustion pressure monitoring, if any faults delay or advance the start of combustion, the injection timing can be altered to compensate. If the injection timing adjustments do not compensate for the combustion timing error, this can be regarded as an injection timing fault.
P021B	Cylinder 8 Injection Timing	There are various methods of controlling and identifying injection timing (depending on the fuel system fitted to the vehicle). It will therefore be necessary to identify the fuel system and the method used to control / check injection timing.	Refer to vehicle specific information to identify the timing monitoring process and timing control (individual injector timing can be altered by the control unit by altering the injector control signal). The timing can be monitored in a number of ways e.g. injector lift / movement sensor but, some systems use a pressure sensor in the combustion chamber (sensor located within in the glow plug) to monitor start of combustion, which can be influenced by injection timing. The fault code indicates that there is an undefined fault with the Fuel Injection Timing (on the specified cylinder) but the timing will be outside of expected values for the operating conditions. The fault could be caused by an injector / fuel delivery related fault, but for those Diesel systems using combustion pressure monitoring, if any faults delay or advance the start of combustion, the injection timing can be altered to compensate. If the injection timing adjustments do not compensate for the combustion timing error, this can be regarded as an injection timing fault.
P021C	Cylinder 9 Injection Timing	There are various methods of controlling and identifying injection timing (depending on the fuel system fitted to the vehicle). It will therefore be necessary to identify the fuel system and the method used to control / check injection timing.	Refer to vehicle specific information to identify the timing monitoring process and timing control (individual injector timing can be altered by the control unit by altering the injector control signal). The timing can be monitored in a number of ways e.g. injector lift / movement sensor but, some systems use a pressure sensor in the combustion chamber (sensor located within in the glow plug) to monitor start of combustion, which can be influenced by injection timing. The fault code indicates that there is an undefined fault with the Fuel Injection Timing (on the specified cylinder) but the timing will be outside of expected values for the operating conditions. The fault could be caused by an injector / fuel delivery related fault, but for those Diesel systems using combustion pressure monitoring, if any faults delay or advance the start of combustion, the injection timing can be altered to compensate. If the injection timing adjustments do not compensate for the combustion timing error, this can be regarded as an injection timing fault.
P021D	Cylinder 10 Injection Timing	There are various methods of controlling and identifying injection timing (depending on the fuel system fitted to the vehicle). It will therefore be necessary to identify the fuel system and the method used to control / check injection timing.	Refer to vehicle specific information to identify the timing monitoring process and timing control (individual injector timing can be altered by the control unit by altering the injector control signal). The timing can be monitored in a number of ways e.g. injector lift / movement sensor but, some systems use a pressure sensor in the combustion chamber (sensor located within in the glow plug) to monitor start of combustion, which can be influenced by injection timing. The fault code indicates that there is an undefined fault with the Fuel Injection Timing (on the specified cylinder) but the timing will be outside of expected values for the operating conditions. The fault could be caused by an injector / fuel delivery related fault, but for those Diesel systems using combustion pressure monitoring, if any faults delay or advance the start of combustion, the injection timing can be altered to compensate. If the injection timing adjustments do not compensate for the combustion timing error, this can be regarded as an injection timing fault.

NOTE 1. Check for other fault codes that could provide additional information.　　**NOTE 2.** Communication between control units can pass via a CAN-Bus system; refer to Page 51 for CAN-Bus checks.
NOTE 3. If a fault cannot be located, it is also possible that the control unit is at fault.　　**NOTE 4.** Refer to Page 53 for list of pages that show the ISO standard for component locations e.g. Sensor A - Bank 1.

Code	Description		
P021E	Cylinder 11 Injection Timing	There are various methods of controlling and identifying injection timing (depending on the fuel system fitted to the vehicle). It will therefore be necessary to identify the fuel system and the method used to control / check injection timing.	Refer to vehicle specific information to identify the timing monitoring process and timing control (individual injector timing can be altered by the control unit by altering the injector control signal). The timing can be monitored in a number of ways e.g. injector lift / movement sensor but, some systems use a pressure sensor in the combustion chamber (sensor located within in the glow plug) to monitor start of combustion, which can be influenced by injection timing. The fault code indicates that there is an undefined fault with the Fuel Injection Timing (on the specified cylinder) but the timing will be outside of expected values for the operating conditions. The fault could be caused by an injector / fuel delivery related fault, but for those Diesel systems using combustion pressure monitoring, if any faults delay or advance the start of combustion, the injection timing can be altered to compensate. If the injection timing adjustments do not compensate for the combustion timing error, this can be regarded as an injection timing fault.
P021F	Cylinder 12 Injection Timing	There are various methods of controlling and identifying injection timing (depending on the fuel system fitted to the vehicle). It will therefore be necessary to identify the fuel system and the method used to control / check injection timing.	Refer to vehicle specific information to identify the timing monitoring process and timing control (individual injector timing can be altered by the control unit by altering the injector control signal). The timing can be monitored in a number of ways e.g. injector lift / movement sensor but, some systems use a pressure sensor in the combustion chamber (sensor located within in the glow plug) to monitor start of combustion, which can be influenced by injection timing. The fault code indicates that there is an undefined fault with the Fuel Injection Timing (on the specified cylinder) but the timing will be outside of expected values for the operating conditions. The fault could be caused by an injector / fuel delivery related fault, but for those Diesel systems using combustion pressure monitoring, if any faults delay or advance the start of combustion, the injection timing can be altered to compensate. If the injection timing adjustments do not compensate for the combustion timing error, this can be regarded as an injection timing fault.
P0220	Throttle / Pedal Position Sensor / Switch "B" Circuit	The throttle / pedal position sensor / switch could be located on the throttle body, throttle cable / linkage or on the throttle pedal. If a switch is fitted, it could be used to indicate "Throttle Closed", "Throttle Open" or both positions. If a position sensor is fitted, it can indicate all angles of throttle opening as well as the speed of throttle opening (rate of change). Systems can be fitted with combinations of sensors and switches, which can be separate or combined into a single unit.	The sensor assembly may contain more than one switch or sensor, refer to vehicle specific information to establish which sensor or switch is affected. The voltage, frequency or value of the pedal position sensor / switch signal is incorrect (undefined fault). Sensor voltage / signal could be: out of normal operating range, constant value, non-existent, corrupt. Possible fault in the pedal position sensor / switch or wiring e.g. short / open circuits, circuit resistance, incorrect sensor resistance or, fault with the reference / operating voltage; the switch contacts (if applicable) could also be faulty e.g. not opening / closing correctly or have a high contact resistance. Refer to list of sensor / actuator checks on page 19.
P0221	Throttle / Pedal Position Sensor / Switch "B" Circuit Range / Performance	The throttle / pedal position sensor / switch could be located on the throttle body, throttle cable / linkage or on the throttle pedal. If a switch is fitted, it could be used to indicate "Throttle Closed", "Throttle Open" or both positions. If a position sensor is fitted, it can indicate all angles of throttle opening as well as the speed of throttle opening (rate of change). Systems can be fitted with combinations of sensors and switches, which can be separate or combined into a single unit.	The sensor assembly may contain more than one switch or sensor, refer to vehicle specific information to establish which sensor or switch is affected. The voltage, frequency or value of the pedal position sensor / switch signal is within the normal operating range / tolerance but the signal is not plausible or is incorrect due to an undefined fault. The sensor / switch signal might not match the operating conditions indicated by other sensors or, the sensor / switch signal is not changing / responding as expected. Possible fault in the pedal position sensor / switch or wiring e.g. short / open circuits, circuit resistance, incorrect sensor resistance or, fault with the reference / operating voltage; also check switch contacts (if applicable) to ensure that they open / close and check continuity of switch contacts (ensure that there is not a high resistance). Refer to list of sensor / actuator checks on page 19. It is also possible that the pedal position sensor / switch is operating correctly but the position being indicated by the sensor is unacceptable or incorrect for the operating conditions indicated by other sensors or, is not as expected by the control unit; check mechanism / linkage used to operate switch / sensor, and check base settings for throttle mechanism and for the sensor / switch.
P0222	Throttle / Pedal Position Sensor / Switch "B" Circuit Low	The throttle / pedal position sensor / switch could be located on the throttle body, throttle cable / linkage or on the throttle pedal. If a switch is fitted, it could be used to indicate "Throttle Closed", "Throttle Open" or both positions. If a position sensor is fitted, it can indicate all angles of throttle opening as well as the speed of throttle opening (rate of change). Systems can be fitted with combinations of sensors and switches, which can be separate or combined into a single unit.	Note. The sensor assembly may contain more than one switch or sensor, refer to vehicle specific information to establish which sensor or switch is affected. The voltage, frequency or value of the pedal position sensor / switch signal is either "zero" (no signal) or, the signal exists but it is below the normal operating range (lower than the minimum operating limit). Possible fault in the pedal position sensor / switch or wiring e.g. short / open circuits, circuit resistance, incorrect sensor resistance or, fault with the reference / operating voltage; the switch contacts (if applicable) could also be faulty e.g. not opening / closing correctly or have a high contact resistance. Refer to list of sensor / actuator checks on page 19. Note that the fault code can be activated for systems with position switches, if the switch output remains at the low value e.g. zero volts and does not rise to the high value e.g. battery voltage (the control unit will detect no switch signal change but it has detected change in engine operating conditions.
P0223	Throttle / Pedal Position Sensor / Switch "B" Circuit High	The throttle / pedal position sensor / switch could be located on the throttle body, throttle cable / linkage or on the throttle pedal. If a switch is fitted, it could be used to indicate "Throttle Closed", "Throttle Open" or both positions. If a position sensor is fitted, it can indicate all angles of throttle opening as well as the speed of throttle opening (rate of change). Systems can be fitted with combinations of sensors and switches, which can be separate or combined into a single unit.	The sensor assembly may contain more than one switch or sensor, refer to vehicle specific information to establish which sensor or switch is affected. The voltage, frequency or value of the pedal position sensor / switch signal is either at full value (e.g. battery voltage with no signal or frequency) or, the signal exists but it is above the normal operating range (higher than the maximum operating limit). Possible fault in the pedal position sensor / switch or wiring e.g. short / open circuits, circuit resistance, incorrect sensor resistance or, fault with the reference / operating voltage; the switch contacts (if applicable) could also be faulty e.g. not opening / closing correctly or have a high contact resistance. Refer to list of sensor / actuator checks on page 19. Note that the fault code can be activated for systems with position switches, if the switch output remains at the high value e.g. battery voltage and does not fall to the low value e.g. zero volts (the control unit will detect no switch signal change but it has detected change in engine operating conditions).
P0224	Throttle / Pedal Position Sensor / Switch "B" Circuit Intermittent	The throttle / pedal position sensor / switch could be located on the throttle body, throttle cable / linkage or on the throttle pedal. If a switch is fitted, it could be used to indicate "Throttle Closed", "Throttle Open" or both positions. If a position sensor is fitted, it can indicate all angles of throttle opening as well as the speed of throttle opening (rate of change). Systems can be fitted with combinations of sensors and switches, which can be separate or combined into a single unit.	The sensor assembly may contain more than one switch or sensor, refer to vehicle specific information to establish which sensor or switch is affected. The voltage, frequency or value of the pedal position sensor / switch signal is intermittent (likely to be out of normal operating range / tolerance when intermittent fault is detected) or, the signal / voltage is erratic (e.g. signal changes are irregular / unpredictable). Possible fault with wiring or connections (including internal sensor / switch connections); the switch contacts (if applicable) could also be faulty e.g. not opening / closing correctly or have a high contact resistance. Refer to list of sensor / actuator checks on page 19.

Fault code	EOBD / ISO Description	Component / System Description	Meaningful Description and Quick Check
P0225	Throttle / Pedal Position Sensor / Switch "C" Circuit	The throttle / pedal position sensor / switch could be located on the throttle body, throttle cable / linkage or on the throttle pedal. If a switch is fitted, it could be used to indicate "Throttle Closed", "Throttle Open" or both positions. If a position sensor is fitted, it can indicate all angles of throttle opening as well as the speed of throttle opening (rate of change). Systems can be fitted with combinations of sensors and switches, which can be separate or combined into a single unit.	The sensor assembly may contain more than one switch or sensor, refer to vehicle specific information to establish which sensor or switch is affected. The voltage, frequency or value of the pedal position sensor / switch signal is incorrect (undefined fault). Sensor voltage / signal could be: out of normal operating range, constant value, non-existent, corrupt. Possible fault in the pedal position sensor / switch or wiring e.g. short / open circuits, circuit resistance, incorrect sensor resistance or, fault with the reference / operating voltage; the switch contacts (if applicable) could also be faulty e.g. not opening / closing correctly or have a high contact resistance. Refer to list of sensor / actuator checks on page 19.
P0226	Throttle / Pedal Position Sensor / Switch "C" Circuit Range / Performance	The throttle / pedal position sensor / switch could be located on the throttle body, throttle cable / linkage or on the throttle pedal. If a switch is fitted, it could be used to indicate "Throttle Closed", "Throttle Open" or both positions. If a position sensor is fitted, it can indicate all angles of throttle opening as well as the speed of throttle opening (rate of change). Systems can be fitted with combinations of sensors and switches, which can be separate or combined into a single unit.	The sensor assembly may contain more than one switch or sensor, refer to vehicle specific information to establish which sensor or switch is affected. The voltage, frequency or value of the pedal position sensor / switch signal is within the normal operating range / tolerance but the signal is not plausible or is incorrect due to an undefined fault. The sensor / switch signal might not match the operating conditions indicated by other sensors or, the sensor / switch signal is not changing / responding as expected. Possible fault in the pedal position sensor / switch or wiring e.g. short / open circuits, circuit resistance, incorrect sensor resistance or, fault with the reference / operating voltage; also check switch contacts (if applicable) to ensure that they open / close and check continuity of switch contacts (ensure that there is not a high resistance). Refer to list of sensor / actuator checks on page 19. It is also possible that the pedal position sensor / switch is operating correctly but the position being indicated by the sensor is unacceptable or incorrect for the operating conditions indicated by other sensors or, is not as expected by the control unit; check mechanism / linkage used to operate switch / sensor, and check base settings for throttle mechanism and for the sensor / switch.
P0227	Throttle / Pedal Position Sensor / Switch "C" Circuit Low	The throttle / pedal position sensor / switch could be located on the throttle body, throttle cable / linkage or on the throttle pedal. If a switch is fitted, it could be used to indicate "Throttle Closed", "Throttle Open" or both positions. If a position sensor is fitted, it can indicate all angles of throttle opening as well as the speed of throttle opening (rate of change). Systems can be fitted with combinations of sensors and switches, which can be separate or combined into a single unit.	Note. The sensor assembly may contain more than one switch or sensor, refer to vehicle specific information to establish which sensor or switch is affected. The voltage, frequency or value of the pedal position sensor / switch signal is either "zero" (no signal) or, the signal exists but it is below the normal operating range (lower than the minimum operating limit). Possible fault in the pedal position sensor / switch or wiring e.g. short / open circuits, circuit resistance, incorrect sensor resistance or, fault with the reference / operating voltage; the switch contacts (if applicable) could also be faulty e.g. not opening / closing correctly or have a high contact resistance. Refer to list of sensor / actuator checks on page 19. Note that the fault code can be activated for systems with position switches, if the switch output remains at the low value e.g. zero volts and does not rise to the high value e.g. battery voltage (the control unit will detect no switch signal change but it has detected change in engine operating conditions).
P0228	Throttle / Pedal Position Sensor / Switch "C" Circuit High	The throttle / pedal position sensor / switch could be located on the throttle body, throttle cable / linkage or on the throttle pedal. If a switch is fitted, it could be used to indicate "Throttle Closed", "Throttle Open" or both positions. If a position sensor is fitted, it can indicate all angles of throttle opening as well as the speed of throttle opening (rate of change). Systems can be fitted with combinations of sensors and switches, which can be separate or combined into a single unit.	The sensor assembly may contain more than one switch or sensor, refer to vehicle specific information to establish which sensor or switch is affected. The voltage, frequency or value of the pedal position sensor / switch signal is either at full value (e.g. battery voltage with no signal or frequency) or, the signal exists but it is above the normal operating range (higher than the maximum operating limit). Possible fault in the pedal position sensor / switch or wiring e.g. short / open circuits, circuit resistance, incorrect sensor resistance or, fault with the reference / operating voltage; the switch contacts (if applicable) could also be faulty e.g. not opening / closing correctly or have a high contact resistance. Refer to list of sensor / actuator checks on page 19. Note that the fault code can be activated for systems with position switches, if the switch output remains at the high value e.g. battery voltage and does not fall to the low value e.g. zero volts (the control unit will detect no switch signal change but it has detected change in engine operating conditions).
P0229	Throttle / Pedal Position Sensor / Switch "C" Circuit Intermittent	The throttle / pedal position sensor / switch could be located on the throttle body, throttle cable / linkage or on the throttle pedal. If a switch is fitted, it could be used to indicate "Throttle Closed", "Throttle Open" or both positions. If a position sensor is fitted, it can indicate all angles of throttle opening as well as the speed of throttle opening (rate of change). Systems can be fitted with combinations of sensors and switches, which can be separate or combined into a single unit.	The sensor assembly may contain more than one switch or sensor, refer to vehicle specific information to establish which sensor or switch is affected. The voltage, frequency or value of the pedal position sensor / switch signal is intermittent (likely to be out of normal operating range / tolerance when intermittent fault is detected) or, the signal / voltage is erratic (e.g. signal changes are irregular / unpredictable). Possible fault with wiring or connections (including internal sensor / switch connections); the switch contacts (if applicable) could also be faulty e.g. not opening / closing correctly or have a high contact resistance. Refer to list of sensor / actuator checks on page 19.
P022A	Charge Air Cooler Bypass Control "A" Circuit / Open	The bypass control system for the charge air cooler (intercooler) diverts the intake air around or through the intercooler to assist in regulating the intake air temperature. The bypass control valve could be controlled using vacuum (regulated by a solenoid valve) or could be controlled by a motor. If necessary, refer to vehicle specific information to identify the type of actuator used in the system e.g. solenoid, motor etc.	The voltage, frequency or value of the control signal in the actuator circuit is incorrect. Signal could be:- out of normal operating range, constant value, non-existent, corrupt. The fault code indicates a possible "open circuit" but other undefined faults in the actuator or wiring could activate the same code e.g. actuator power supply, short circuit, high circuit resistance or incorrect actuator resistance. Note: the control unit could be providing the correct signal but it is being affected by the circuit fault. Refer to list of sensor / actuator checks on page 19.
P022B	Charge Air Cooler Bypass Control "A" Circuit Low	The bypass control system for the charge air cooler (intercooler) diverts the intake air around or through the intercooler to assist in regulating the intake air temperature. The bypass control valve could be controlled using vacuum (regulated by a solenoid valve) or could be controlled by a motor. If necessary, refer to vehicle specific information to identify the type of actuator used in the system e.g. solenoid, motor etc.	The voltage, frequency or value of the control signal in the actuator circuit is either "zero" (no signal) or, the signal exists but is below the normal operating range. Possible fault with actuator or wiring (e.g. actuator power supply, short / open circuit or resistance) that is causing a signal value to be lower than the minimum operating limit. Note: the control unit could be providing the correct signal but it is being affected by the circuit fault. Refer to list of sensor / actuator checks on page 19.
P022C	Charge Air Cooler Bypass Control "A" Circuit High	The bypass control system for the charge air cooler (intercooler) diverts the intake air around or through the intercooler to assist in regulating the intake air temperature. The bypass control valve could be controlled using vacuum (regulated by a solenoid valve) or could be controlled by a motor. If necessary, refer to vehicle specific information to identify the type of actuator used in the system e.g. solenoid, motor etc.	The voltage, frequency or value of the control signal in the actuator circuit is either at full value (e.g. battery voltage with no on / off signal) or, the signal exists but it is above the normal operating range. Possible fault with actuator or wiring (e.g. actuator power supply, short / open circuit or resistance) that is causing a signal value to be higher than the maximum operating limit. Note: the control unit could be providing the correct signal but it is being affected by the circuit fault. Refer to list of sensor / actuator checks on page 19.

NOTE 1. Check for other fault codes that could provide additional information.
NOTE 3. If a fault cannot be located, it is also possible that the control unit is at fault

NOTE 2. Communication between control units can pass via a CAN-Bus system; refer to Page 51 for CAN-Bus checks.
NOTE 4. Refer to Page 53 for list of pages that show the ISO standard for component locations e.g. Sensor A - Bank 1.

98

code	Description		
P022D	Charge Air Cooler Bypass Control "B" Circuit / Open	The bypass control system for the charge air cooler (intercooler) diverts the intake air around or through the intercooler to assist in regulating the intake air temperature. The bypass control valve could be controlled using vacuum (regulated by a solenoid valve) or could be controlled by a motor. If necessary, refer to vehicle specific information to identify the type of actuator used in the system e.g. solenoid, motor etc.	The voltage, frequency or value of the control signal in the actuator circuit is incorrect. Signal could be:- out of normal operating range, constant value, non-existent, corrupt. The fault code indicates a possible "open circuit" but other undefined faults in the actuator or wiring could activate the same code e.g. actuator power supply, short circuit, high circuit resistance or incorrect actuator resistance. Note: the control unit could be providing the correct signal but it is being affected by the circuit fault. Refer to list of sensor / actuator checks on page 19.
P022E	Charge Air Cooler Bypass Control "B" Circuit Low	The bypass control system for the charge air cooler (intercooler) diverts the intake air around or through the intercooler to assist in regulating the intake air temperature. The bypass control valve could be controlled using vacuum (regulated by a solenoid valve) or could be controlled by a motor. If necessary, refer to vehicle specific information to identify the type of actuator used in the system e.g. solenoid, motor etc.	The voltage, frequency or value of the control signal in the actuator circuit is either "zero" (no signal) or, the signal exists but is below the normal operating range. Possible fault with actuator or wiring (e.g. actuator power supply, short / open circuit or resistance) that is causing a signal value to be lower than the minimum operating limit. Note: the control unit could be providing the correct signal but it is being affected by the circuit fault. Refer to list of sensor / actuator checks on page 19.
P022F	Charge Air Cooler Bypass Control "B" Circuit High	The bypass control system for the charge air cooler (intercooler) diverts the intake air around or through the intercooler to assist in regulating the intake air temperature. The bypass control valve could be controlled using vacuum (regulated by a solenoid valve) or could be controlled by a motor. If necessary, refer to vehicle specific information to identify the type of actuator used in the system e.g. solenoid, motor etc.	The voltage, frequency or value of the control signal in the actuator circuit is either at full value (e.g. battery voltage with no on / off signal) or, the signal exists but it is above the normal operating range. Possible fault with actuator or wiring (e.g. actuator power supply, short / open circuit or resistance) that is causing a signal value to be higher than the maximum operating limit. Note: the control unit could be providing the correct signal but it is being affected by the circuit fault. Refer to list of sensor / actuator checks on page 19.
P0230	Fuel Pump Primary Circuit	Electric fuel pumps are normally provided with a power supply from the pump relay. Some systems can use two power supply circuits from the relay; one circuit provides battery voltage (pump runs at high speed for high fuel demand / starting). The second circuit contains a resistor to reduce the voltage / current (pump runs at lower speeds for low fuel demand); the selection of the two circuits is made by the control unit, which controls the relay operation.	Although it is likely that the primary circuit refers to the "full battery voltage" circuit, refer to vehicle specific information for confirmation of wiring; note that any wiring checks should include wiring between the pump motor and the relay, and between relay and control unit. The voltage in the pump motor circuit is incorrect (undefined fault). Voltage could be:- out of normal operating range, constant value, non-existent, corrupt. Possible fault with motor or wiring e.g. motor power supply from relay, short / open circuit, high circuit resistance or incorrect motor resistance. Refer to list of sensor / actuator checks on page 19. Check relay operation to ensure that it is switching between the primary and secondary circuits.
P0231	Fuel Pump Secondary Circuit Low	Electric fuel pumps are normally provided with a power supply from the pump relay. Some systems can use two power supply circuits from the relay; one circuit provides battery voltage (pump runs at high speed for high fuel demand / starting). The second circuit contains a resistor to reduce the voltage / current (pump runs at lower speeds for low fuel demand); the selection of the two circuits is made by the control unit, which controls the relay operation.	Although it is likely that the secondary circuit refers to the circuit containing the resistor (low speed circuit), refer to vehicle specific information for confirmation of wiring; note that any wiring checks should include wiring between the pump motor and the relay (including resistor), and between the relay and control unit. The voltage, frequency or value of the control signal in the motor circuit is either "zero" (no voltage) or, the control signal exists but voltage is below the normal operating range. Possible fault with motor, wiring or resistor (e.g. motor power supply, short / open circuit or resistance) that is causing the voltage to be lower than the minimum operating limit. Refer to list of sensor / actuator checks on page 19. If the circuit contains a resistor, check the resistance and compare against specification.
P0232	Fuel Pump Secondary Circuit High	Fuel pumps are normally provided with a power supply from the relay; systems can use two power supply circuits from the relay. One circuit provides battery voltage (pump runs at high speed for high fuel demand / starting). The other circuit contains a resistor to reduce the voltage / current (pump runs at lower speeds for low fuel demand); the selection of the two circuits is made by the control unit which controls the relay operation.	The voltage in the motor circuit is either at full value all the time (e.g. battery voltage with no reduced voltage via the resistor circuit) or, voltage exists but it is above the normal operating range. Possible fault with motor, wiring or resistor (e.g. motor power supply, short / open circuit or resistance) that is causing a signal value to be higher than the normal maximum operating limit. Refer to list of sensor / actuator checks on page 19. If the circuit contains a resistor, check the resistance and compare against specification. Check relay operation to ensure that it is switching between the primary and secondary circuits.
P0233	Fuel Pump Secondary Circuit Intermittent	Fuel pumps are normally provided with a power supply from the pump relay; systems can use two power supply circuits from the relay. One circuit provides battery voltage (pump runs at high speed for high fuel demand / starting). The other circuit contains a resistor to reduce the voltage / current (pump runs at lower speeds for low fuel demand); the selection of the two circuits is made by the control unit which controls the relay operation.	Although it is likely that the secondary circuit refers to the circuit containing the resistor (low speed circuit), refer to vehicle specific information for confirmation of wiring; note that any wiring checks should include wiring between the pump motor and the relay (including resistor), and between the relay and control unit. The voltage in the motor circuit is intermittent (likely to be out of normal operating range / tolerance when intermittent fault is detected) or, the voltage is erratic. Possible fault with motor or wiring (e.g. internal / external connections, broken wiring). Refer to list of sensor / actuator checks on page 19.
P0234	Turbocharger / Supercharger Overboost Condition	Different types of turbo / supercharger systems can be fitted to produce additional engine performance. For this fault code, it may be necessary to identify the system and components used.	Control unit has detected that the turbo / super charger boost limit has been exceeded during normal vehicle operation. The fault could be related to a boost pressure control system fault or be caused by a fault in the boost pressure sensor causing an incorrect boost pressure signal to be passed to the control unit. Refer to list of sensor / actuator checks on page 19.
P0235	Turbocharger / Supercharger Boost Sensor "A" Circuit	The boost sensor is used to detect the turbo / supercharger boost pressure under different operating conditions; the boost pressure sensor signal is used by the control unit to ensure that the correct level of boost is provided but not exceeded. Note that the boost pressure and Manifold Absolute Pressure (MAP) sensors can be a combined unit.	The voltage, frequency or value of the pressure sensor signal is incorrect (undefined fault). Sensor voltage / signal could be: out of normal operating range, constant value, non-existent, corrupt. Possible fault in the pressure sensor or wiring e.g. short / open circuits, circuit resistance, incorrect sensor resistance (if applicable) or, fault with the reference / operating voltage; interference from other circuits can also affect sensor signals. Refer to list of sensor / actuator checks on page 19.

NOTE 1. Check for other fault codes that could provide additional information.   **NOTE 2.** Communication between control units can pass via a CAN-Bus system; refer to Page 51 for CAN-Bus checks.   99
NOTE 3. If a fault cannot be located, it is also possible that the control unit is at fault.   **NOTE 4.** Refer to Page 53 for list of pages that show the ISO standard for component locations e.g. Sensor A - Bank 1.

Fault code	EOBD / ISO Description	Component / System Description	Meaningful Description and Quick Check
P0236	Turbocharger / Supercharger Boost Sensor "A" Circuit Range / Performance	The boost sensor is used to detect the turbo / supercharger boost pressure under different operating conditions; the boost pressure sensor signal is used by the control unit to ensure that the correct level of boost is provided but not exceeded. Note that the boost pressure and Manifold Absolute Pressure (MAP) sensors can be a combined unit.	The voltage, frequency or value of the pressure sensor signal is within the normal operating range / tolerance but the signal is not plausible or is incorrect due to an undefined fault. The sensor signal might not match the operating conditions indicated by other sensors or, the sensor signal is not changing / responding as expected to the changes in conditions. Possible fault in the pressure sensor or wiring e.g. short / open circuits, circuit resistance, incorrect sensor resistance (if applicable) or, fault with the reference / operating voltage; interference from other circuits can also affect sensor signals. Refer to list of sensor / actuator checks on page 19. Also check any air passages / pipes to the sensor for leaks / blockages etc. It is also possible that the pressure sensor is operating correctly but the boost pressure being measured by the sensor is unacceptable or incorrect for the operating conditions; if possible, use alternative method of checking pressure to establish if sensor system is operating correctly and whether the pressure indicated by the sensor is correct / incorrect.
P0237	Turbocharger / Supercharger Boost Sensor "A" Circuit Low	The boost sensor is used to detect the turbo / supercharger boost pressure under different operating conditions; the boost pressure sensor signal is used by the control unit to ensure that the correct level of boost is provided but not exceeded. Note that the boost pressure and Manifold Absolute Pressure (MAP) sensors can be a combined unit.	The voltage, frequency or value of the pressure sensor signal is either "zero" (no signal) or, the signal exists but it is below the normal operating range (lower than the minimum operating limit). Possible fault in the pressure sensor or wiring e.g. short / open circuits, circuit resistance, incorrect sensor resistance (if applicable) or, fault with the reference / operating voltage; interference from other circuits can also affect sensor signals. Refer to list of sensor / actuator checks on page 19.
P0238	Turbocharger / Supercharger Boost Sensor "A" Circuit High	The boost sensor is used to detect the turbo / supercharger boost pressure under different operating conditions; the boost pressure sensor signal is used by the control unit to ensure that the correct level of boost is provided but not exceeded. Note that the boost pressure and Manifold Absolute Pressure (MAP) sensors can be a combined unit.	The voltage, frequency or value of the pressure sensor signal is either at full value (e.g. battery voltage with no signal or frequency) or, the signal exists but it is above the normal operating range (higher than the maximum operating limit). Possible fault in the pressure sensor or wiring e.g. short / open circuits, circuit resistance, incorrect sensor resistance (if applicable) or, fault with the reference / operating voltage; interference from other circuits can also affect sensor signals. Refer to list of sensor / actuator checks on page 19.
P0239	Turbocharger / Supercharger Boost Sensor "B" Circuit	The boost sensor is used to detect the turbo / supercharger boost pressure under different operating conditions; the boost pressure sensor signal is used by the control unit to ensure that the correct level of boost is provided but not exceeded. Note that the boost pressure and Manifold Absolute Pressure (MAP) sensors can be a combined unit.	The voltage, frequency or value of the pressure sensor signal is incorrect (undefined fault). Sensor voltage / signal could be: out of normal operating range, constant value, non-existent, corrupt. Possible fault in the pressure sensor or wiring e.g. short / open circuits, circuit resistance, incorrect sensor resistance (if applicable) or, fault with the reference / operating voltage; interference from other circuits can also affect sensor signals. Refer to list of sensor / actuator checks on page 19.
P023A	Charge Air Cooler Coolant Pump Control Circuit / Open	The charge air coolant pump is used to regulate the coolant flow through the intercooler (turbocharger / supercharger system).	Note that the speed / power for some electric motors can be regulated using a frequency / pulse width control signal (provided by the control unit on the earth or power circuit). The voltage, frequency or value of the control signal in the pump motor circuit is incorrect. Signal could be:- out of normal operating range, constant value, non-existent, corrupt. The fault code indicates a possible "open circuit" but other undefined faults in the motor or wiring could activate the same code e.g. motor power supply, short circuit, high circuit resistance or incorrect motor resistance. Note: the control unit could be providing the correct signal but it is being affected by a circuit fault. Identify the type of motor control (if applicable) and the wiring used on the system. Refer to list of sensor / actuator checks on page 19.
P023B	Charge Air Cooler Coolant Pump Control Circuit Low	The charge air coolant pump is used to regulate the coolant flow through the intercooler (turbocharger / supercharger system).	Note that the speed / power for some electric motors can be regulated using a frequency / pulse width control signal (provided by the control unit on the earth or power circuit). The voltage, frequency or value of the control signal in the motor circuit is either "zero" (no voltage or signal) or, the voltage / signal exists but is below the normal operating range. Possible fault with motor or wiring e.g. motor power supply (relay if fitted), short / open circuit or resistance that is causing a signal value to be lower than the minimum operating limit. Note: the control unit or relay could be providing the correct voltage / signal but it is being affected by a circuit fault. Refer to list of sensor / actuator checks on page 19.
P023C	Charge Air Cooler Coolant Pump Control Circuit High	The charge air coolant pump is used to regulate the coolant flow through the intercooler (turbocharger / supercharger system).	Note that the speed / power for some electric motors can be regulated using a frequency / pulse width control signal (provided by the control unit on the earth or power circuit). The voltage, frequency or value of the voltage / control signal in the motor circuit is either at full value (e.g. battery voltage with no on / off signal) or, the voltage / signal exists but it is above the normal operating range. Possible fault with motor or wiring e.g. motor power supply (relay if fitted), short / open circuit or resistance that is causing a voltage / signal value to be higher than the maximum operating limit. Note: the control unit could be providing the correct voltage / signal but it is being affected by a circuit fault. Refer to list of sensor / actuator checks on page 19.
P023D	Manifold Absolute Pressure - Turbocharger / Supercharger Boost Sensor "A" Correlation	The turbo / supercharger boost control (usually performed by the main control unit) can make use of information from different sensors to calculate the required boost pressure under different conditions. For this fault code, the sensors identified are the MAP (Manifold Absolute Pressure) sensor and a boost pressure sensor.	The signals from the two sensors are providing conflicting or implausible information e.g. when the engine is not running (ignition on), the MAP sensor and boost pressure sensor signal values are different; however, both signal values should be the same under these conditions (the control unit could also carry out checks under engine running conditions). If possible, use alternative method of checking the measurements to establish if one of the sensors / wiring is faulty or whether the measured values are incorrect. It is possible that one of the sensors is operating incorrectly therefore check each sensor for a fault with the sensor / wiring e.g. short / open circuits, circuit resistance, incorrect sensor resistance or reference / operating voltage fault. Refer to list of sensor / actuator checks on page 19. It is also possible that there is a fault with the boost pressure control systems or turbo / supercharger (boost pressure does not correlate with MAP sensor value, or is not as expected by the control unit). Check boost pressure under operating conditions, also for intake airflow restrictions, intake air leaks and blockage or leak in sensor air pipes.

NOTE 1. Check for other fault codes that could provide additional information.   NOTE 2. Communication between control units can pass via a CAN-Bus system; refer to Page 51 for CAN-Bus checks.   100

NOTE 3. If a fault cannot be located, it is also possible that the control unit is at fault.   NOTE 4. Refer to Page 53 for list of pages that show the ISO standard for component locations e.g. Sensor A - Bank 1.

code	Description		
P023E	Manifold Absolute Pressure - Turbocharger / Supercharger Boost Sensor "B" Correlation	The turbo / supercharger boost control (usually performed by the main control unit) can make use of information from different sensors to calculate the required boost pressure under different conditions. For this fault code, the sensors identified are the MAP (Manifold Absolute Pressure) sensor and a boost pressure sensor.	The signals from the two sensors are providing conflicting or implausible information e.g. when the engine is not running (ignition on), the MAP sensor and boost pressure sensor signal values are different; however, both signal values should be the same under these conditions (the control unit could also carry out checks under engine running conditions). If possible, use alternative method of checking the measurements to establish if one of the sensors / wiring is faulty or whether the measured values are incorrect. It is possible that one of the sensors is operating incorrectly therefore check each sensor for a fault with the sensor / wiring e.g. short / open circuits, circuit resistance, incorrect sensor resistance or reference / operating voltage fault. Refer to list of sensor / actuator checks on page 19. It is also possible that there is a fault with the boost pressure control systems or turbo / supercharger (boost pressure does not correlate with MAP sensor value, or is not as expected by the control unit). Check boost pressure under operating conditions, also for intake airflow restrictions, intake air leaks and blockage or leak in sensor air pipes.
P023F		ISO / SAE reserved	Not used, reserved for future allocation.
P0240	Turbocharger / Supercharger Boost Sensor "B" Circuit Range / Performance	The boost sensor is used to detect the turbo / supercharger boost pressure under different operating conditions; the boost pressure sensor signal is used by the control unit to ensure that the correct level of boost is provided but not exceeded. Note that the boost pressure and Manifold Absolute Pressure (MAP) sensors can be a combined unit.	The voltage, frequency or value of the pressure sensor signal is within the normal operating range / tolerance but the signal is not plausible or is incorrect due to an undefined fault. The sensor signal might not match the operating conditions indicated by other sensors or, the sensor signal is not changing / responding as expected to the changes in conditions. Possible fault in the pressure sensor or wiring e.g. short / open circuits, circuit resistance, incorrect sensor resistance (if applicable) or, fault with the reference / operating voltage; interference from other circuits can also affect sensor signals. Refer to list of sensor / actuator checks on page 19. Also check any air passages / pipes to the sensor for leaks / blockages etc. It is also possible that the pressure sensor is operating correctly but the boost pressure being measured by the sensor is unacceptable or incorrect for the operating conditions; if possible, use alternative method of checking pressure to establish if sensor system is operating correctly and whether the pressure indicated by the sensor is correct / incorrect.
P0241	Turbocharger / Supercharger Boost Sensor "B" Circuit Low	The boost sensor is used to detect the turbo / supercharger boost pressure under different operating conditions; the boost pressure sensor signal is used by the control unit to ensure that the correct level of boost is provided but not exceeded. Note that the boost pressure and Manifold Absolute Pressure (MAP) sensors can be a combined unit.	The voltage, frequency or value of the pressure sensor signal is either "zero" (no signal) or, the signal exists but it is below the normal operating range (lower than the minimum operating limit). Possible fault in the pressure sensor or wiring e.g. short / open circuits, circuit resistance, incorrect sensor resistance (if applicable) or, fault with the reference / operating voltage; interference from other circuits can also affect sensor signals. Refer to list of sensor / actuator checks on page 19.
P0242	Turbocharger / Supercharger Boost Sensor "B" Circuit High	The boost sensor is used to detect the turbo / supercharger boost pressure under different operating conditions; the boost pressure sensor signal is used by the control unit to ensure that the correct level of boost is provided but not exceeded. Note that the boost pressure and Manifold Absolute Pressure (MAP) sensors can be a combined unit.	The voltage, frequency or value of the pressure sensor signal is either at full value (e.g. battery voltage with no signal or frequency) or, the signal exists but it is above the normal operating range (higher than the maximum operating limit). Possible fault in the pressure sensor or wiring e.g. short / open circuits, circuit resistance, incorrect sensor resistance (if applicable) or, fault with the reference / operating voltage; interference from other circuits can also affect sensor signals. Refer to list of sensor / actuator checks on page 19.
P0243	Turbocharger / Supercharger Wastegate Solenoid "A"	The wastegate solenoid is used to regulate the vacuum that operates the wastegate valve actuator mechanism.	The control unit has detected an undefined fault with the wastegate solenoid system; the fault could be detected due to an electrical problem with the solenoid / wiring (e.g. solenoid power supply, short / open circuit or resistance). Refer to list of sensor / actuator checks on page 19. Although the fault is likely to be related to a solenoid / wiring fault, the control unit could be activating the fault code because of incorrect boost pressure or wastegate operation; if necessary, also check wastegate operation and boost pressure
P0244	Turbocharger / Supercharger Wastegate Solenoid "A" Range / Performance	The wastegate solenoid is used to regulate the vacuum that operates the wastegate valve actuator mechanism.	The voltage, frequency or value of the control signal in the solenoid circuit is within the normal operating range / tolerance, but the signal might be incorrect for the operating conditions. It is also possible that the solenoid response (or response of system controlled by the solenoid) differs from the expected / desired response (incorrect response to the control signal provided by the control unit). Possible fault in solenoid or wiring, but also possible that the solenoid (or mechanism / system controlled by the solenoid) is not operating or moving correctly. Note: the control unit could be providing the correct signal but it is being affected by a circuit fault. Refer to list of sensor / actuator checks on page 19. Note that the solenoid could be operating correctly but the vacuum valve being operated by the solenoid could be faulty (leaking etc). Also check for leaks or blockages in vacuum pipes Check for that when vacuum is applied to the vacuum actuator, the wastegate control linkage / mechanism is moving correctly.
P0245	Turbocharger / Supercharger Wastegate Solenoid "A" Low	The wastegate solenoid is used to regulate the vacuum that operates the wastegate valve actuator mechanism.	The voltage, frequency or value of the control signal in the solenoid circuit is either "zero" (no signal) or, the signal exists but is below the normal operating range. Possible fault with solenoid or wiring (e.g. solenoid power supply, short / open circuit or resistance) that is causing a signal value to be lower than the minimum operating limit. Note: the control unit could be providing the correct signal but it is being affected by a circuit fault. Refer to list of sensor / actuator checks on page 19.
P0246	Turbocharger / Supercharger Wastegate Solenoid "A" High	The wastegate solenoid is used to regulate the vacuum that operates the wastegate valve actuator mechanism.	The voltage, frequency or value of the control signal in the solenoid circuit is either at full value (e.g. battery voltage with no on / off signal) or, the signal exists but it is above the normal operating range. Possible fault with solenoid or wiring (e.g. solenoid power supply, short / open circuit or resistance) that is causing a signal value to be higher than the maximum operating limit. Note: the control unit could be providing the correct signal but it is being affected by a circuit fault. Refer to list of sensor / actuator checks on page 19.
P0247	Turbocharger / Supercharger Wastegate Solenoid "B"	The wastegate solenoid is used to regulate the vacuum that operates the wastegate valve actuator mechanism.	The control unit has detected an undefined fault with the wastegate solenoid system; the fault could be detected due to an electrical problem with the solenoid / wiring (e.g. solenoid power supply, short / open circuit or resistance). Refer to list of sensor / actuator checks on page 19. Although the fault is likely to be related to a solenoid / wiring fault, the control unit could be activating the fault code because of incorrect boost pressure or wastegate operation; if necessary, also check wastegate operation and boost pressure

NOTE 1. Check for other fault codes that could provide additional information. NOTE 2. Communication between control units can pass via a CAN-Bus system; refer to Page 51 for CAN-Bus checks. 101
NOTE 3. If a fault cannot be located, it is also possible that the control unit is at fault. NOTE 4. Refer to Page 53 for list of pages that show the ISO standard for component locations e.g. Sensor A - Bank 1.

Fault code	EOBD / ISO Description	Component / System Description	Meaningful Description and Quick Check
P0248	Turbocharger / Supercharger Wastegate Solenoid "B" Range / Performance	The wastegate solenoid is used to regulate the vacuum that operates the wastegate valve actuator mechanism.	The voltage, frequency or value of the control signal in the solenoid circuit is within the normal operating range / tolerance, but the signal might be incorrect for the operating conditions. It is also possible that the solenoid response (or response of system controlled by the solenoid) differs from the expected / desired response (incorrect response to the control signal provided by the control unit). Possible fault in solenoid or wiring, but also possible that the solenoid (or mechanism / system controlled by the solenoid) is not operating or moving correctly. Note: the control unit could be providing the correct signal but it is being affected by a circuit fault. Refer to list of sensor / actuator checks on page 19. Note that the solenoid could be operating correctly but the vacuum valve being operated by the solenoid could be faulty (leaking etc). Also check for leaks or blockages in vacuum pipes. Check that when vacuum is applied to the wastegate actuator, the wastegate control linkage / mechanism is moving correctly.
P0249	Turbocharger / Supercharger Wastegate Solenoid "B" Low	The wastegate solenoid is used to regulate the vacuum that operates the wastegate valve actuator mechanism.	The voltage, frequency or value of the control signal in the solenoid circuit is either "zero" (no signal) or, the signal exists but is below the normal operating range. Possible fault with solenoid or wiring (e.g. solenoid power supply, short / open circuit or resistance) that is causing a signal value to be lower than the minimum operating limit. Note: the control unit could be providing the correct signal but it is being affected by a circuit fault. Refer to list of sensor / actuator checks on page 19.
P024A	Charge Air Cooler Bypass Control "A" Range / Performance	The bypass control system for the charge air cooler (intercooler) diverts the intake air around or through the intercooler to assist in regulating the intake air temperature. If necessary, refer to vehicle specific information to identify the type of actuator used in the system e.g. solenoid, motor etc. If the actuator appears to be responding to the control signal but the response of the charge air cooler bypass system is incorrect e.g. not operating or controlling the intake air temperature correctly, this could indicate a fault with other components in the bypass control system or a fault with the intake temperature sensor.	The voltage, frequency or value of the control signal in the actuator circuit is within the normal operating range / tolerance, but the signal might be incorrect for the operating conditions. It is also possible that the actuator response (or response of system controlled by the actuator) differs from the expected / desired response (incorrect response for the control signal provided by the control unit). Possible fault in actuator or wiring, but also possible that the actuator (or mechanism / system controlled by the actuator) is not operating or moving correctly. Note: the control unit could be providing the correct signal but it is being affected by the circuit fault. Refer to list of sensor / actuator checks on page 19.
P024B	Charge Air Cooler Bypass Control "A" Stuck	The bypass control system for the charge air cooler (intercooler) diverts the intake air around or through the intercooler to assist in regulating the intake air temperature. If necessary, refer to vehicle specific information to identify the type of actuator used in the system e.g. solenoid, motor etc.	The control unit is providing a control signal to the actuator but the control unit has detected that the actuator is stuck in a fixed position or is not responding to the control signal. The fault could be related to an actuator / wiring fault e.g. power supply, short / open circuit, incorrect actuator resistance. Refer to list of sensor / actuator checks on page 19. Note that the control unit could be detecting "no response" from the system or mechanism being controlled by the actuator; check system (as well as the actuator) for correct operation, movement and response. A position sensor (if fitted) would be indicating the "stuck position"; also check position sensor operation.
P024C	Charge Air Cooler Bypass Position Sensor "A" Circuit	The bypass control system for the charge air cooler (intercooler) diverts the intake air around or through the intercooler to assist in regulating the intake air temperature. A sensor can be used to detect the position of the bypass air control valve.	The voltage, frequency or value of the position sensor signal is incorrect (undefined fault). Sensor voltage / signal value could be: out of normal operating range, constant value, non-existent, corrupt. Possible fault in the position sensor or wiring e.g. short / open circuits, circuit resistance, incorrect sensor resistance (if applicable) or, fault with the reference / operating voltage; interference from other circuits can also affect sensor signals. Also check earth connections and cable screening. Refer to list of sensor / actuator checks on page 19.
P024D	Charge Air Cooler Bypass Position Sensor "A" Circuit Range / Performance	The bypass control system for the charge air cooler (intercooler) diverts the intake air around or through the intercooler to assist in regulating the intake air temperature. A sensor can be used to detect the position of the bypass air control valve.	The voltage, frequency or value of the position sensor signal is within the normal operating range / tolerance but the signal is not plausible or is incorrect due to an undefined fault. The sensor signal might not match the operating conditions indicated by other sensors or, the sensor signal is not changing / responding as expected to the requested [change in position of the bypass valve. Possible fault in the position sensor or wiring e.g. short / open circuits, circuit resistance, incorrect sensor resistance or, fault with the reference / operating voltage; interference from other circuits can also affect sensor signals. Refer to list of sensor / actuator checks on page 19. It is also possible that the sensor is operating correctly but the position being detected by the sensor is unacceptable or incorrect for the operating conditions.
P024E	Charge Air Cooler Bypass Position Sensor "A" Circuit Low	The bypass control system for the charge air cooler (intercooler) diverts the intake air around or through the intercooler to assist in regulating the intake air temperature. A sensor can be used to detect the position of the bypass air control valve.	The voltage, frequency or value of the position sensor signal is either "zero" (no signal) or, the signal exists but it is below the normal operating range (lower than the minimum operating limit). Possible fault in the position sensor or wiring e.g. short / open circuits, circuit resistance, incorrect sensor resistance or, fault with the reference / operating voltage; interference from other circuits can also affect sensor signals. Refer to list of sensor / actuator checks on page 19.
P024F	Charge Air Cooler Bypass Position Sensor "A" Circuit High	The bypass control system for the charge air cooler (intercooler) diverts the intake air around or through the intercooler to assist in regulating the intake air temperature. A sensor can be used to detect the position of the bypass air control valve.	The voltage, frequency or value of the position sensor signal is either at full value (e.g. battery voltage with no signal or frequency) or, the signal exists but it is above the normal operating range (higher than the maximum operating limit). Possible fault in the position sensor or wiring e.g. short / open circuits, circuit resistance, incorrect sensor resistance or, fault with the reference / operating voltage; interference from other circuits can also affect sensor signals. Refer to list of sensor / actuator checks on page 19.
P0250	Turbocharger / Supercharger Wastegate Solenoid "B" High	The wastegate solenoid is used to regulate the vacuum that operates the wastegate valve actuator mechanism.	The voltage, frequency or value of the control signal in the solenoid circuit is either at full value (e.g. battery voltage with no on / off signal) or, the signal exists but it is above the normal operating range. Possible fault with solenoid or wiring (e.g. solenoid power supply, short / open circuit or resistance) that is causing a signal value to be higher than the maximum operating limit. Note: the control unit could be providing the correct signal but it is being affected by a circuit fault. Refer to list of sensor / actuator checks on page 19.
P0251	Injection Pump Fuel Metering Control "A" (Cam / Rotor / Injector)	Depending on fuel system design, different mechanisms are used to control the fuel metering, (timing and quantity); the mechanisms can use an electrical actuator (e.g. solenoid) to control the metering mechanism. Refer to system specific information to identify the metering control system	The control unit has detected an undefined fault with the fuel metering control system; the fault could be detected due to an electrical problem with the metering control actuator / wiring. Refer to list of sensor / actuator checks on page 19. The fault could also be detected because position sensors are detecting incorrect position of the metering control system; also check operation of position sensors.

NOTE 1. Check for other fault codes that could provide additional information.   NOTE 2. Communication between control units can pass via a CAN-Bus system; refer to Page 51 for CAN-Bus checks.   102

NOTE 3. If a fault cannot be located, it is also possible that the control unit is at fault.   NOTE 4. Refer to Page 53 for list of pages that show the ISO standard for component locations e.g. Sensor A - Bank 1.

code	Description		
P0252	Injection Pump Fuel Metering Control "A" Range / Performance (Cam / Rotor / Injector)	Depending on fuel system design, different mechanisms are used to control the fuel metering, (timing and quantity); the mechanisms can use an electrical actuator (e.g. solenoid) to control the metering mechanism. Refer to system specific information to identify the metering control system	The voltage, frequency or value of the control signal in the actuator circuit is within the normal operating range / tolerance, but the signal might be incorrect for the operating conditions. It is also possible that the actuator response (or response of system controlled by the actuator) differs from the expected / desired response (incorrect response for the control signal provided by the control unit). Possible fault in actuator or wiring, but also possible that the actuator (or mechanism / system controlled by the actuator) is not operating or moving correctly. Note: the control unit could be providing the correct signal but it is being affected by the circuit fault. Refer to list of sensor / actuator checks on page 19. If the actuator appears to be responding to the control signal but the response of the fuel metering system is incorrect, this could indicate a fault with the fuel metering mechanism.
P0253	Injection Pump Fuel Metering Control "A" Low (Cam / Rotor / Injector)	Depending on fuel system design, different mechanisms are used to control the fuel metering, (timing and quantity); the mechanisms can use an electrical actuator (e.g. solenoid) to control the metering mechanism. Refer to system specific information to identify the metering control system	The voltage, frequency or value of the control signal in the actuator circuit is either "zero" (no signal) or, the signal exists but is below the normal operating range. Possible fault with actuator or wiring (e.g. actuator power supply, short / open circuit or resistance) that is causing a signal value to be lower than the minimum operating limit. Note: the control unit could be providing the correct signal but it is being affected by the circuit fault. Refer to list of sensor / actuator checks on page 19.
P0254	Injection Pump Fuel Metering Control "A" High (Cam / Rotor / Injector)	Depending on fuel system design, different mechanisms are used to control the fuel metering, (timing and quantity); the mechanisms can use an electrical actuator (e.g. solenoid) to control the metering mechanism. Refer to system specific information to identify the metering control system	The voltage, frequency or value of the control signal in the actuator circuit is either at full value (e.g. battery voltage with no on / off signal) or, the signal exists but it is above the normal operating range. Possible fault with actuator or wiring (e.g. actuator power supply, short / open circuit or resistance) that is causing a signal value to be higher than the maximum operating limit. Note: the control unit could be providing the correct signal but it is being affected by the circuit fault. Refer to list of sensor / actuator checks on page 19.
P0255	Injection Pump Fuel Metering Control "A" Intermittent (Cam / Rotor / Injector)	Depending on fuel system design, different mechanisms are used to control the fuel metering, (timing and quantity); the mechanisms can use an electrical actuator (e.g. solenoid) to control the metering mechanism. Refer to system specific information to identify the metering control system	The voltage, frequency or value of the control signal in the actuator circuit is intermittent (likely to be out of normal operating range / tolerance when intermittent fault is detected) or, the signal is erratic e.g. signal changes are irregular / unpredictable. Possible fault with actuator or wiring (e.g. internal / external connections, broken wiring) that is causing intermittent or erratic signal to exist. Note: the control unit could be providing the correct signal but it is being affected by the circuit fault. Refer to list of sensor / actuator checks on page 19.
P0256	Injection Pump Fuel Metering Control "B" (Cam / Rotor / Injector)	Depending on fuel system design, different mechanisms are used to control the fuel metering, (timing and quantity); the mechanisms can use an electrical actuator (e.g. solenoid) to control the metering mechanism. Refer to system specific information to identify the metering control system	The control unit has detected an undefined fault with the fuel metering control system; the fault could be detected due to an electrical problem with the metering control actuator / wiring. Refer to list of sensor / actuator checks on page 19. The fault could also be detected because position sensors are detecting incorrect position of the metering control system; also check operation of position sensors.
P0257	Injection Pump Fuel Metering Control "B" Range / Performance (Cam / Rotor / Injector)	Depending on fuel system design, different mechanisms are used to control the fuel metering, (timing and quantity); the mechanisms can use an electrical actuator (e.g. solenoid) to control the metering mechanism. Refer to system specific information to identify the metering control system	The voltage, frequency or value of the control signal in the actuator circuit is within the normal operating range / tolerance, but the signal might be incorrect for the operating conditions. It is also possible that the actuator response (or response of system controlled by the actuator) differs from the expected / desired response (incorrect response for the control signal provided by the control unit). Possible fault in actuator or wiring, but also possible that the actuator (or mechanism / system controlled by the actuator) is not operating or moving correctly. Note: the control unit could be providing the correct signal but it is being affected by the circuit fault. Refer to list of sensor / actuator checks on page 19. If the actuator appears to be responding to the control signal but the response of the fuel metering system is incorrect, this could indicate a fault with the fuel metering mechanism.
P0258	Injection Pump Fuel Metering Control "B" Low (Cam / Rotor / Injector)	Depending on fuel system design, different mechanisms are used to control the fuel metering, (timing and quantity); the mechanisms can use an electrical actuator (e.g. solenoid) to control the metering mechanism. Refer to system specific information to identify the metering control system	The voltage, frequency or value of the control signal in the actuator circuit is either "zero" (no signal) or, the signal exists but is below the normal operating range. Possible fault with actuator or wiring (e.g. actuator power supply, short / open circuit or resistance) that is causing a signal value to be lower than the minimum operating limit. Note: the control unit could be providing the correct signal but it is being affected by the circuit fault. Refer to list of sensor / actuator checks on page 19.
P0259	Injection Pump Fuel Metering Control "B" High (Cam / Rotor / Injector)	Depending on fuel system design, different mechanisms are used to control the fuel metering, (timing and quantity); the mechanisms can use an electrical actuator (e.g. solenoid) to control the metering mechanism. Refer to system specific information to identify the metering control system	The voltage, frequency or value of the control signal in the actuator circuit is either at full value (e.g. battery voltage with no on / off signal) or, the signal exists but it is above the normal operating range. Possible fault with actuator or wiring (e.g. actuator power supply, short / open circuit or resistance) that is causing a signal value to be higher than the maximum operating limit. Note: the control unit could be providing the correct signal but it is being affected by the circuit fault. Refer to list of sensor / actuator checks on page 19.
P025A	Fuel Pump Module Control Circuit / Open	A fuel pump module can be used to regulate the speed of the fuel pump to match fuel demand. The module, which will regulate the current being supplied to the fuel pump, is usually controlled by the main engine control unit (depending on fuel demand). The control circuit is likely to be the circuit connecting the fuel pump control module to the main control unit.	The fault code indicates that there is an incorrect signal value in the applicable control circuit. The code refers to a possible "open" circuit, but the incorrect signal is likely to be: constant value, non-existent, corrupt or signal is out of normal operating range (out of tolerance). The fault could be caused by faulty wiring between the fuel pump module and the main control unit; check for short / open circuits and high resistance in the wiring. It is also possible that a fuel pump module could have an internal fault.
P025B	Fuel Pump Module Control Circuit Range / Performance	A fuel pump module can be used to regulate the speed of the fuel pump to match fuel demand. The module, which will regulate the current being supplied to the fuel pump, is usually controlled by the main engine control unit (depending on fuel demand). The control circuit is likely to be the circuit connecting the fuel pump control module to the main control unit.	The fault code indicates that the control signal to the fuel pump module is within normal operating range / tolerance but not correct for the operating conditions i.e. the pump module and therefore the pump response is not as expected when a specific control signal is provided by the main control unit. The fault could be caused by faulty wiring between the fuel pump module and the main control unit; check for short / open circuits and high resistance in the wiring. It is also possible that the fuel pump is unable to respond correctly due to fuel pump or fuel system faults (e.g. blocked fuel filter etc). Also check that the fuel pump module is providing a changing control signal to the pump e.g. voltage or current change.
P025C	Fuel Pump Module Control Circuit Low	A fuel pump module can be used to regulate the speed of the fuel pump to match fuel demand. The module, which will regulate the current being supplied to the fuel pump, is usually controlled by the main engine control unit (depending on fuel demand). The control circuit is likely to be the circuit connecting the fuel pump control module to the main control unit.	The fault code indicates that the signal value in the control circuit is low. The fault could be caused by faulty wiring between the fuel pump module and the main control unit; check for short / open circuits and high resistance in the wiring. It is also possible that a fuel pump module could have an internal fault.

Fault code	EOBD / ISO Description	Component / System Description	Meaningful Description and Quick Check
P025D	Fuel Pump Module Control Circuit High	A fuel pump module can be used to regulate the speed of the fuel pump to match fuel demand. The module, which will regulate the current being supplied to the fuel pump, is usually controlled by the main engine control unit (depending on fuel demand). The control circuit is likely to be the circuit connecting the fuel pump control module to the main control unit.	The fault code indicates that the signal value in the control circuit is high. The fault could be caused by faulty wiring between the fuel pump module and the main control unit; check for short / open circuits and high resistance in the wiring. It is also possible that a fuel pump module could have an internal fault.
P0260	Injection Pump Fuel Metering Control "B" Intermittent (Cam / Rotor / Injector)	Depending on fuel system design, different mechanisms are used to control the fuel metering, (timing and quantity); the mechanisms can use an electrical actuator (e.g. solenoid) to control the metering mechanism. Refer to system specific information to identify the metering control system	The voltage, frequency or value of the control signal in the actuator circuit is intermittent (likely to be out of normal operating range / tolerance when intermittent fault is detected) or, the signal is erratic e.g. signal changes are irregular / unpredictable. Possible fault with actuator or wiring (e.g. internal / external connections, broken wiring) that is causing intermittent or erratic signal to exist. Note: the control unit could be providing the correct signal but it is being affected by the circuit fault. Refer to list of sensor / actuator checks on page 19.
P0261	Cylinder 1 Injector Circuit Low	Most injectors have traditionally used a solenoid (contained within the injector body) to open the injector. Later injector designs make use of Piezo actuators instead of solenoids. Note that injector operating voltages on some applications are considerably higher than "battery voltage" traditionally used on solenoid injectors; these systems can use a separate injector power module to deliver the power and control signal to the injectors.	Identify the type of injector and the wiring / control modules used on the system. The voltage, frequency or value of the control signal in the injector circuit is either "zero" (no signal) or, the signal exists but is below the normal operating range. Possible fault with injector or wiring (e.g. injector power supply, short / open circuit or resistance) that is causing a signal value to be lower than the minimum operating limit. Note: the control unit could be providing the correct signal but it is being affected by a circuit fault. Refer to list of sensor / actuator checks on page 19. It is possible that the control unit is detecting "no voltage" in the circuit, which could be caused by faults with the power supply, connections or open circuit in wiring or injector.
P0262	Cylinder 1 Injector Circuit High	Most injectors have traditionally used a solenoid (contained within the injector body) to open the injector. Later injector designs make use of Piezo actuators instead of solenoids. Note that injector operating voltages on some applications are considerably higher than "battery voltage" traditionally used on solenoid injectors; these systems can use a separate injector power module to deliver the power and control signal to the injectors.	Identify the type of injector and the wiring / control modules used on the system. The voltage, frequency or value of the control signal in the injector circuit is either at full value (e.g. battery voltage with no on / off signal) or, the signal exists but it is above the normal operating range. Possible fault with injector or wiring (e.g. injector power supply, short / open circuit or resistance) that is causing a signal value to be higher than the maximum operating limit. Note: the control unit could be providing the correct signal but it is being affected by a circuit fault. Refer to list of sensor / actuator checks on page 19. It is possible that the when the control unit completes the earth path for the injector, the voltage on the earth circuit remains at supply voltage level e.g. battery voltage instead of dropping to zero (or close to zero); this could be caused by a open circuit, poor earth for the control unit or control unit fault.
P0263	Cylinder 1 Contribution / Balance	The control unit checks the signal from engine speed sensor to monitor crankshaft acceleration / speed during each power stroke; this provides an indication of the power / torque contribution provided by the cylinder (compared to the other cylinders). Any fault with the cylinder, including mechanical faults, fuelling, ignition (petrol engines) or other faults affecting combustion, will affect the power / torque produced by the cylinder.	The fault code indicates that the contribution from the identified cylinder is different when compared to other cylinders (likely to be a lower contribution but, in some circumstances the contribution could be higher). Check ignition (petrol engines), check fuelling and injection timing (as applicable) and check condition of injectors (leaking / dribbling injectors, injector, spray pattern etc). Also check mechanical condition of the cylinder (compression etc), and check for any other aspect of the engine operation that could affect just one cylinder (air leaks etc). It is also possible that there could be an engine position / speed sensor related fault (if necessary, Refer to list of sensor / actuator checks on page 19).
P0264	Cylinder 2 Injector Circuit Low	Most injectors have traditionally used a solenoid (contained within the injector body) to open the injector. Later injector designs make use of Piezo actuators instead of solenoids. Note that injector operating voltages on some applications are considerably higher than "battery voltage" traditionally used on solenoid injectors; these systems can use a separate injector power module to deliver the power and control signal to the injectors.	Identify the type of injector and the wiring / control modules used on the system. The voltage, frequency or value of the control signal in the injector circuit is either "zero" (no signal) or, the signal exists but is below the normal operating range. Possible fault with injector or wiring (e.g. injector power supply, short / open circuit or resistance) that is causing a signal value to be lower than the minimum operating limit. Note: the control unit could be providing the correct signal but it is being affected by a circuit fault. Refer to list of sensor / actuator checks on page 19. It is possible that the control unit is detecting "no voltage" in the circuit, which could be caused by faults with the power supply, connections or open circuit in wiring or injector.
P0265	Cylinder 2 Injector Circuit High	Most injectors have traditionally used a solenoid (contained within the injector body) to open the injector. Later injector designs make use of Piezo actuators instead of solenoids. Note that injector operating voltages on some applications are considerably higher than "battery voltage" traditionally used on solenoid injectors; these systems can use a separate injector power module to deliver the power and control signal to the injectors.	Identify the type of injector and the wiring / control modules used on the system. The voltage, frequency or value of the control signal in the injector circuit is either at full value (e.g. battery voltage with no on / off signal) or, the signal exists but it is above the normal operating range. Possible fault with injector or wiring (e.g. injector power supply, short / open circuit or resistance) that is causing a signal value to be higher than the maximum operating limit. Note: the control unit could be providing the correct signal but it is being affected by a circuit fault. Refer to list of sensor / actuator checks on page 19. It is possible that the when the control unit completes the earth path for the injector, the voltage on the earth circuit remains at supply voltage level e.g. battery voltage instead of dropping to zero (or close to zero); this could be caused by a open circuit, poor earth for the control unit or control unit fault.
P0266	Cylinder 2 Contribution / Balance	The control unit checks the signal from engine speed sensor to monitor crankshaft acceleration / speed during each power stroke; this provides an indication of the power / torque contribution provided by the cylinder (compared to the other cylinders). Any fault with the cylinder, including mechanical faults, fuelling, ignition (petrol engines) or other faults affecting combustion, will affect the power / torque produced by the cylinder.	The fault code indicates that the contribution from the identified cylinder is different when compared to other cylinders (likely to be a lower contribution but, in some circumstances the contribution could be higher). Check ignition (petrol engines), check fuelling and injection timing (as applicable) and check condition of injectors (leaking / dribbling injectors, injector, spray pattern etc). Also check mechanical condition of the cylinder (compression etc), and check for any other aspect of the engine operation that could affect just one cylinder (air leaks etc). It is also possible that there could be an engine position / speed sensor related fault (if necessary, Refer to list of sensor / actuator checks on page 19).
P0267	Cylinder 3 Injector Circuit Low	Most injectors have traditionally used a solenoid (contained within the injector body) to open the injector. Later injector designs make use of Piezo actuators instead of solenoids. Note that injector operating voltages on some applications are considerably higher than "battery voltage" traditionally used on solenoid injectors; these systems can use a separate injector power module to deliver the power and control signal to the injectors.	Identify the type of injector and the wiring / control modules used on the system. The voltage, frequency or value of the control signal in the injector circuit is either "zero" (no signal) or, the signal exists but is below the normal operating range. Possible fault with injector or wiring (e.g. injector power supply, short / open circuit or resistance) that is causing a signal value to be lower than the minimum operating limit. Note: the control unit could be providing the correct signal but it is being affected by a circuit fault. Refer to list of sensor / actuator checks on page 19. It is possible that the control unit is detecting "no voltage" in the circuit, which could be caused by faults with the power supply, connections or open circuit in wiring or injector.

code	Description		
P0268	Cylinder 3 Injector Circuit High	Most injectors have traditionally used a solenoid (contained within the injector body) to open the injector. Later injector designs make use of Piezo actuators instead of solenoids. Note that injector operating voltages on some applications are considerably higher than "battery voltage" traditionally used on solenoid injectors; these systems can use a separate injector power module to deliver the power and control signal to the injectors.	Identify the type of injector and the wiring / control modules used on the system. The voltage, frequency or value of the control signal in the injector circuit is either at full value (e.g. battery voltage with no on / off signal) or, the signal exists but it is above the normal operating range. Possible fault with injector or wiring (e.g. injector power supply, short / open circuit or resistance) that is causing a signal value to be higher than the maximum operating limit. Note: the control unit could be providing the correct signal but it is being affected by a circuit fault. Refer to list of sensor / actuator checks on page 19. It is possible that the when the control unit completes the earth path for the injector, the voltage on the earth circuit remains at supply voltage level e.g. battery voltage instead of dropping to zero (or close to zero); this could be caused by a open circuit, poor earth for the control unit or control unit fault.
P0269	Cylinder 3 Contribution / Balance	The control unit checks the signal from engine speed sensor to monitor crankshaft acceleration / speed during each power stroke; this provides an indication of the power / torque contribution provided by the cylinder (compared to the other cylinders). Any fault with the cylinder, including mechanical faults, fuelling, ignition (petrol engines) or other faults affecting combustion, will affect the power / torque produced by the cylinder.	The fault code indicates that the contribution from the identified cylinder is different when compared to other cylinders (likely to be a lower contribution but, in some circumstances the contribution could be higher). Check ignition (petrol engines), check fuelling and injection timing (as applicable) and check condition of injectors (leaking / dribbling injectors, injector, spray pattern etc). Also check mechanical condition of the cylinder (compression etc), and check for any other aspect of the engine operation that could affect just one cylinder (air leaks etc). It is also possible that there could be an engine position / speed sensor related fault (if necessary, Refer to list of sensor / actuator checks on page 19).
P0270	Cylinder 4 Injector Circuit Low	Most injectors have traditionally used a solenoid (contained within the injector body) to open the injector. Later injector designs make use of Piezo actuators instead of solenoids. Note that injector operating voltages on some applications are considerably higher than "battery voltage" traditionally used on solenoid injectors; these systems can use a separate injector power module to deliver the power and control signal to the injectors.	Identify the type of injector and the wiring / control modules used on the system. The voltage, frequency or value of the control signal in the injector circuit is either "zero" (no signal) or, the signal exists but is below the normal operating range. Possible fault with injector or wiring (e.g. injector power supply, short / open circuit or resistance) that is causing a signal value to be lower than the minimum operating limit. Note: the control unit could be providing the correct signal but it is being affected by a circuit fault. Refer to list of sensor / actuator checks on page 19. It is possible that the control unit is detecting "no voltage" in the circuit, which could be caused by faults with the power supply, connections or open circuit in wiring or injector.
P0271	Cylinder 4 Injector Circuit High	Most injectors have traditionally used a solenoid (contained within the injector body) to open the injector. Later injector designs make use of Piezo actuators instead of solenoids. Note that injector operating voltages on some applications are considerably higher than "battery voltage" traditionally used on solenoid injectors; these systems can use a separate injector power module to deliver the power and control signal to the injectors.	Identify the type of injector and the wiring / control modules used on the system. The voltage, frequency or value of the control signal in the injector circuit is either at full value (e.g. battery voltage with no on / off signal) or, the signal exists but it is above the normal operating range. Possible fault with injector or wiring (e.g. injector power supply, short / open circuit or resistance) that is causing a signal value to be higher than the maximum operating limit. Note: the control unit could be providing the correct signal but it is being affected by a circuit fault. Refer to list of sensor / actuator checks on page 19. It is possible that the when the control unit completes the earth path for the injector, the voltage on the earth circuit remains at supply voltage level e.g. battery voltage instead of dropping to zero (or close to zero); this could be caused by a open circuit, poor earth for the control unit or control unit fault.
P0272	Cylinder 4 Contribution / Balance	The control unit checks the signal from engine speed sensor to monitor crankshaft acceleration / speed during each power stroke; this provides an indication of the power / torque contribution provided by the cylinder (compared to the other cylinders). Any fault with the cylinder, including mechanical faults, fuelling, ignition (petrol engines) or other faults affecting combustion, will affect the power / torque produced by the cylinder.	The fault code indicates that the contribution from the identified cylinder is different when compared to other cylinders (likely to be a lower contribution but, in some circumstances the contribution could be higher). Check ignition (petrol engines), check fuelling and injection timing (as applicable) and check condition of injectors (leaking / dribbling injectors, injector, spray pattern etc). Also check mechanical condition of the cylinder (compression etc), and check for any other aspect of the engine operation that could affect just one cylinder (air leaks etc). It is also possible that there could be an engine position / speed sensor related fault (if necessary, Refer to list of sensor / actuator checks on page 19).
P0273	Cylinder 5 Injector Circuit Low	Most injectors have traditionally used a solenoid (contained within the injector body) to open the injector. Later injector designs make use of Piezo actuators instead of solenoids. Note that injector operating voltages on some applications are considerably higher than "battery voltage" traditionally used on solenoid injectors; these systems can use a separate injector power module to deliver the power and control signal to the injectors.	Identify the type of injector and the wiring / control modules used on the system. The voltage, frequency or value of the control signal in the injector circuit is either "zero" (no signal) or, the signal exists but is below the normal operating range. Possible fault with injector or wiring (e.g. injector power supply, short / open circuit or resistance) that is causing a signal value to be lower than the minimum operating limit. Note: the control unit could be providing the correct signal but it is being affected by a circuit fault. Refer to list of sensor / actuator checks on page 19. It is possible that the control unit is detecting "no voltage" in the circuit, which could be caused by faults with the power supply, connections or open circuit in wiring or injector.
P0274	Cylinder 5 Injector Circuit High	Most injectors have traditionally used a solenoid (contained within the injector body) to open the injector. Later injector designs make use of Piezo actuators instead of solenoids. Note that injector operating voltages on some applications are considerably higher than "battery voltage" traditionally used on solenoid injectors; these systems can use a separate injector power module to deliver the power and control signal to the injectors.	Identify the type of injector and the wiring / control modules used on the system. The voltage, frequency or value of the control signal in the injector circuit is either at full value (e.g. battery voltage with no on / off signal) or, the signal exists but it is above the normal operating range. Possible fault with injector or wiring (e.g. injector power supply, short / open circuit or resistance) that is causing a signal value to be higher than the maximum operating limit. Note: the control unit could be providing the correct signal but it is being affected by a circuit fault. Refer to list of sensor / actuator checks on page 19. It is possible that the when the control unit completes the earth path for the injector, the voltage on the earth circuit remains at supply voltage level e.g. battery voltage instead of dropping to zero (or close to zero); this could be caused by a open circuit, poor earth for the control unit or control unit fault.
P0275	Cylinder 5 Contribution / Balance	The control unit checks the signal from engine speed sensor to monitor crankshaft acceleration / speed during each power stroke; this provides an indication of the power / torque contribution provided by the cylinder (compared to the other cylinders). Any fault with the cylinder, including mechanical faults, fuelling, ignition (petrol engines) or other faults affecting combustion, will affect the power / torque produced by the cylinder.	The fault code indicates that the contribution from the identified cylinder is different when compared to other cylinders (likely to be a lower contribution but, in some circumstances the contribution could be higher). Check ignition (petrol engines), check fuelling and injection timing (as applicable) and check condition of injectors (leaking / dribbling injectors, injector, spray pattern etc). Also check mechanical condition of the cylinder (compression etc), and check for any other aspect of the engine operation that could affect just one cylinder (air leaks etc). It is also possible that there could be an engine position / speed sensor related fault (if necessary, Refer to list of sensor / actuator checks on page 19).
P0276	Cylinder 6 Injector Circuit Low	Most injectors have traditionally used a solenoid (contained within the injector body) to open the injector. Later injector designs make use of Piezo actuators instead of solenoids. Note that injector operating voltages on some applications are considerably higher than "battery voltage" traditionally used on solenoid injectors; these systems can use a separate injector power module to deliver the power and control signal to the injectors.	Identify the type of injector and the wiring / control modules used on the system. The voltage, frequency or value of the control signal in the injector circuit is either "zero" (no signal) or, the signal exists but is below the normal operating range. Possible fault with injector or wiring (e.g. injector power supply, short / open circuit or resistance) that is causing a signal value to be lower than the minimum operating limit. Note: the control unit could be providing the correct signal but it is being affected by a circuit fault. Refer to list of sensor / actuator checks on page 19. It is possible that the control unit is detecting "no voltage" in the circuit, which could be caused by faults with the power supply, connections or open circuit in wiring or injector.

NOTE 1. Check for other fault codes that could provide additional information. **NOTE 2.** Communication between control units can pass via a CAN-Bus system; refer to Page 51 for CAN-Bus checks.

NOTE 3. If a fault cannot be located, it is also possible that the control unit is at fault. **NOTE 4.** Refer to Page 53 for list of pages that show the ISO standard for component locations e.g. Sensor A - Bank 1.

Fault code	EOBD / ISO Description	Component / System Description	Meaningful Description and Quick Check
P0277	Cylinder 6 Injector Circuit High	Most injectors have traditionally used a solenoid (contained within the injector body) to open the injector. Later injector designs make use of Piezo actuators instead of solenoids. Note that injector operating voltages on some applications are considerably higher than "battery voltage" traditionally used on solenoid injectors; these systems can use a separate injector power module to deliver the power and control signal to the injectors.	Identify the type of injector and the wiring / control modules used on the system. The voltage, frequency or value of the control signal in the injector circuit is either at full value (e.g. battery voltage with no on / off signal) or, the signal exists but it is above the normal operating range. Possible fault with injector or wiring (e.g. injector power supply, short / open circuit or resistance) that is causing a signal value to be higher than the maximum operating limit. Note: the control unit could be providing the correct signal but it is being affected by a circuit fault. Refer to list of sensor / actuator checks on page 19. It is possible that the when the control unit completes the earth path for the injector, the voltage on the earth circuit remains at supply voltage level e.g. battery voltage instead of dropping to zero (or close to zero); this could be caused by a open circuit, poor earth for the control unit or control unit fault.
P0278	Cylinder 6 Contribution / Balance	The control unit checks the signal from engine speed sensor to monitor crankshaft acceleration / speed during each power stroke; this provides an indication of the power / torque contribution provided by the cylinder (compared to the other cylinders). Any fault with the cylinder, including mechanical faults, fuelling, ignition (petrol engines) or other faults affecting combustion, will affect the power / torque produced by the cylinder.	The fault code indicates that the contribution from the identified cylinder is different when compared to other cylinders (likely to be a lower contribution but, in some circumstances the contribution could be higher). Check ignition (petrol engines), check fuelling and injection timing (as applicable) and check condition of injectors (leaking / dribbling injectors, injector, spray pattern etc). Also check mechanical condition of the cylinder (compression etc), and check for any other aspect of the engine operation that could affect just one cylinder (air leaks etc). It is also possible that there could be an engine position / speed sensor related fault (if necessary, Refer to list of sensor / actuator checks on page 19).
P0279	Cylinder 7 Injector Circuit Low	Most injectors have traditionally used a solenoid (contained within the injector body) to open the injector. Later injector designs make use of Piezo actuators instead of solenoids. Note that injector operating voltages on some applications are considerably higher than "battery voltage" traditionally used on solenoid injectors; these systems can use a separate injector power module to deliver the power and control signal to the injectors.	Identify the type of injector and the wiring / control modules used on the system. The voltage, frequency or value of the control signal in the injector circuit is either "zero" (no signal) or, the signal exists but is below the normal operating range. Possible fault with injector or wiring (e.g. injector power supply, short / open circuit or resistance) that is causing a signal value to be lower than the minimum operating limit. Note: the control unit could be providing the correct signal but it is being affected by a circuit fault. Refer to list of sensor / actuator checks on page 19. It is possible that the control unit is detecting "no voltage" in the circuit, which could be caused by faults with the power supply, connections or open circuit in wiring or injector.
P0280	Cylinder 7 Injector Circuit High	Most injectors have traditionally used a solenoid (contained within the injector body) to open the injector. Later injector designs make use of Piezo actuators instead of solenoids. Note that injector operating voltages on some applications are considerably higher than "battery voltage" traditionally used on solenoid injectors; these systems can use a separate injector power module to deliver the power and control signal to the injectors.	Identify the type of injector and the wiring / control modules used on the system. The voltage, frequency or value of the control signal in the injector circuit is either at full value (e.g. battery voltage with no on / off signal) or, the signal exists but it is above the normal operating range. Possible fault with injector or wiring (e.g. injector power supply, short / open circuit or resistance) that is causing a signal value to be higher than the maximum operating limit. Note: the control unit could be providing the correct signal but it is being affected by a circuit fault. Refer to list of sensor / actuator checks on page 19. It is possible that the when the control unit completes the earth path for the injector, the voltage on the earth circuit remains at supply voltage level e.g. battery voltage instead of dropping to zero (or close to zero); this could be caused by a open circuit, poor earth for the control unit or control unit fault.
P0281	Cylinder 7 Contribution / Balance	The control unit checks the signal from engine speed sensor to monitor crankshaft acceleration / speed during each power stroke; this provides an indication of the power / torque contribution provided by the cylinder (compared to the other cylinders). Any fault with the cylinder, including mechanical faults, fuelling, ignition (petrol engines) or other faults affecting combustion, will affect the power / torque produced by the cylinder.	The fault code indicates that the contribution from the identified cylinder is different when compared to other cylinders (likely to be a lower contribution but, in some circumstances the contribution could be higher). Check ignition (petrol engines), check fuelling and injection timing (as applicable) and check condition of injectors (leaking / dribbling injectors, injector, spray pattern etc). Also check mechanical condition of the cylinder (compression etc), and check for any other aspect of the engine operation that could affect just one cylinder (air leaks etc). It is also possible that there could be an engine position / speed sensor related fault (if necessary, Refer to list of sensor / actuator checks on page 19).
P0282	Cylinder 8 Injector Circuit Low	Most injectors have traditionally used a solenoid (contained within the injector body) to open the injector. Later injector designs make use of Piezo actuators instead of solenoids. Note that injector operating voltages on some applications are considerably higher than "battery voltage" traditionally used on solenoid injectors; these systems can use a separate injector power module to deliver the power and control signal to the injectors.	Identify the type of injector and the wiring / control modules used on the system. The voltage, frequency or value of the control signal in the injector circuit is either "zero" (no signal) or, the signal exists but is below the normal operating range. Possible fault with injector or wiring (e.g. injector power supply, short / open circuit or resistance) that is causing a signal value to be lower than the minimum operating limit. Note: the control unit could be providing the correct signal but it is being affected by a circuit fault. Refer to list of sensor / actuator checks on page 19. It is possible that the control unit is detecting "no voltage" in the circuit, which could be caused by faults with the power supply, connections or open circuit in wiring or injector.
P0283	Cylinder 8 Injector Circuit High	Most injectors have traditionally used a solenoid (contained within the injector body) to open the injector. Later injector designs make use of Piezo actuators instead of solenoids. Note that injector operating voltages on some applications are considerably higher than "battery voltage" traditionally used on solenoid injectors; these systems can use a separate injector power module to deliver the power and control signal to the injectors.	Identify the type of injector and the wiring / control modules used on the system. The voltage, frequency or value of the control signal in the injector circuit is either at full value (e.g. battery voltage with no on / off signal) or, the signal exists but it is above the normal operating range. Possible fault with injector or wiring (e.g. injector power supply, short / open circuit or resistance) that is causing a signal value to be higher than the maximum operating limit. Note: the control unit could be providing the correct signal but it is being affected by a circuit fault. Refer to list of sensor / actuator checks on page 19. It is possible that the when the control unit completes the earth path for the injector, the voltage on the earth circuit remains at supply voltage level e.g. battery voltage instead of dropping to zero (or close to zero); this could be caused by a open circuit, poor earth for the control unit or control unit fault.
P0284	Cylinder 8 Contribution / Balance	The control unit checks the signal from engine speed sensor to monitor crankshaft acceleration / speed during each power stroke; this provides an indication of the power / torque contribution provided by the cylinder (compared to the other cylinders). Any fault with the cylinder, including mechanical faults, fuelling, ignition (petrol engines) or other faults affecting combustion, will affect the power / torque produced by the cylinder.	The fault code indicates that the contribution from the identified cylinder is different when compared to other cylinders (likely to be a lower contribution but, in some circumstances the contribution could be higher). Check ignition (petrol engines), check fuelling and injection timing (as applicable) and check condition of injectors (leaking / dribbling injectors, injector, spray pattern etc). Also check mechanical condition of the cylinder (compression etc), and check for any other aspect of the engine operation that could affect just one cylinder (air leaks etc). It is also possible that there could be an engine position / speed sensor related fault (if necessary, Refer to list of sensor / actuator checks on page 19).
P0285	Cylinder 9 Injector Circuit Low	Most injectors have traditionally used a solenoid (contained within the injector body) to open the injector. Later injector designs make use of Piezo actuators instead of solenoids. Note that injector operating voltages on some applications are considerably higher than "battery voltage" traditionally used on solenoid injectors; these systems can use a separate injector power module to deliver the power and control signal to the injectors.	Identify the type of injector and the wiring / control modules used on the system. The voltage, frequency or value of the control signal in the injector circuit is either "zero" (no signal) or, the signal exists but is below the normal operating range. Possible fault with injector or wiring (e.g. injector power supply, short / open circuit or resistance) that is causing a signal value to be lower than the minimum operating limit. Note: the control unit could be providing the correct signal but it is being affected by a circuit fault. Refer to list of sensor / actuator checks on page 19. It is possible that the control unit is detecting "no voltage" in the circuit, which could be caused by faults with the power supply, connections or open circuit in wiring or injector.

NOTE 1. Check for other fault codes that could provide additional information. **NOTE 2.** Communication between control units can pass via a CAN-Bus system; refer to Page 51 for CAN-Bus checks.
NOTE 3. If a fault cannot be located, it is also possible that the control unit is at fault. **NOTE 4.** Refer to Page 53 for list of pages that show the ISO standard for component locations e.g. Sensor A - Bank 1.

code	Description		
P0286	Cylinder 9 Injector Circuit High	Most injectors have traditionally used a solenoid (contained within the injector body) to open the injector. Later injector designs make use of Piezo actuators instead of solenoids. Note that injector operating voltages on some applications are considerably higher than "battery voltage" traditionally used on solenoid injectors; these systems can use a separate injector power module to deliver the power and control signal to the injectors.	Identify the type of injector and the wiring / control modules used on the system. The voltage, frequency or value of the control signal in the injector circuit is either at full value (e.g. battery voltage with no on / off signal) or, the signal exists but it is above the normal operating range. Possible fault with injector or wiring (e.g. injector power supply, short / open circuit or resistance) that is causing a signal value to be higher than the maximum operating limit. Note: the control unit could be providing the correct signal but it is being affected by a circuit fault. Refer to list of sensor / actuator checks on page 19. It is possible that the when the control unit completes the earth path for the injector, the voltage on the earth circuit remains at supply voltage level e.g. battery voltage instead of dropping to zero (or close to zero); this could be caused by a open circuit, poor earth for the control unit or control unit fault.
P0287	Cylinder 9 Contribution / Balance	The control unit checks the signal from engine speed sensor to monitor crankshaft acceleration / speed during each power stroke; this provides an indication of the power / torque contribution provided by the cylinder (compared to the other cylinders). Any fault with the cylinder, including mechanical faults, fuelling, ignition (petrol engines) or other faults affecting combustion, will affect the power / torque produced by the cylinder.	The fault code indicates that the contribution from the identified cylinder is different when compared to other cylinders (likely to be a lower contribution but, in some circumstances the contribution could be higher). Check ignition (petrol engines), check fuelling and injection timing (as applicable) and check condition of injectors (leaking / dribbling injectors, injector, spray pattern etc). Also check mechanical condition of the cylinder (compression etc), and check for any other aspect of the engine operation that could affect just one cylinder (air leaks etc). It is also possible that there could be an engine position / speed sensor related fault (if necessary, Refer to list of sensor / actuator checks on page 19).
P0288	Cylinder 10 Injector Circuit Low	Most injectors have traditionally used a solenoid (contained within the injector body) to open the injector. Later injector designs make use of Piezo actuators instead of solenoids. Note that injector operating voltages on some applications are considerably higher than "battery voltage" traditionally used on solenoid injectors; these systems can use a separate injector power module to deliver the power and control signal to the injectors.	Identify the type of injector and the wiring / control modules used on the system. The voltage, frequency or value of the control signal in the injector circuit is either "zero" (no signal) or, the signal exists but is below the normal operating range. Possible fault with injector or wiring (e.g. injector power supply, short / open circuit or resistance) that is causing a signal value to be lower than the minimum operating limit. Note: the control unit could be providing the correct signal but it is being affected by a circuit fault. Refer to list of sensor / actuator checks on page 19. It is possible that the control unit is detecting "no voltage" in the circuit, which could be caused by faults with the power supply, connections or open circuit in wiring or injector.
P0289	Cylinder 10 Injector Circuit High	Most injectors have traditionally used a solenoid (contained within the injector body) to open the injector. Later injector designs make use of Piezo actuators instead of solenoids. Note that injector operating voltages on some applications are considerably higher than "battery voltage" traditionally used on solenoid injectors; these systems can use a separate injector power module to deliver the power and control signal to the injectors.	Identify the type of injector and the wiring / control modules used on the system. The voltage, frequency or value of the control signal in the injector circuit is either at full value (e.g. battery voltage with no on / off signal) or, the signal exists but it is above the normal operating range. Possible fault with injector or wiring (e.g. injector power supply, short / open circuit or resistance) that is causing a signal value to be higher than the maximum operating limit. Note: the control unit could be providing the correct signal but it is being affected by a circuit fault. Refer to list of sensor / actuator checks on page 19. It is possible that the when the control unit completes the earth path for the injector, the voltage on the earth circuit remains at supply voltage level e.g. battery voltage instead of dropping to zero (or close to zero); this could be caused by a open circuit, poor earth for the control unit or control unit fault.
P0290	Cylinder 10 Contribution / Balance	The control unit checks the signal from engine speed sensor to monitor crankshaft acceleration / speed during each power stroke; this provides an indication of the power / torque contribution provided by the cylinder (compared to the other cylinders). Any fault with the cylinder, including mechanical faults, fuelling, ignition (petrol engines) or other faults affecting combustion, will affect the power / torque produced by the cylinder.	The fault code indicates that the contribution from the identified cylinder is different when compared to other cylinders (likely to be a lower contribution but, in some circumstances the contribution could be higher). Check ignition (petrol engines), check fuelling and injection timing (as applicable) and check condition of injectors (leaking / dribbling injectors, injector, spray pattern etc). Also check mechanical condition of the cylinder (compression etc), and check for any other aspect of the engine operation that could affect just one cylinder (air leaks etc). It is also possible that there could be an engine position / speed sensor related fault (if necessary, Refer to list of sensor / actuator checks on page 19).
P0291	Cylinder 11 Injector Circuit Low	Most injectors have traditionally used a solenoid (contained within the injector body) to open the injector. Later injector designs make use of Piezo actuators instead of solenoids. Note that injector operating voltages on some applications are considerably higher than "battery voltage" traditionally used on solenoid injectors; these systems can use a separate injector power module to deliver the power and control signal to the injectors.	Identify the type of injector and the wiring / control modules used on the system. The voltage, frequency or value of the control signal in the injector circuit is either "zero" (no signal) or, the signal exists but is below the normal operating range. Possible fault with injector or wiring (e.g. injector power supply, short / open circuit or resistance) that is causing a signal value to be lower than the minimum operating limit. Note: the control unit could be providing the correct signal but it is being affected by a circuit fault. Refer to list of sensor / actuator checks on page 19. It is possible that the control unit is detecting "no voltage" in the circuit, which could be caused by faults with the power supply, connections or open circuit in wiring or injector.
P0292	Cylinder 11 Injector Circuit High	Most injectors have traditionally used a solenoid (contained within the injector body) to open the injector. Later injector designs make use of Piezo actuators instead of solenoids. Note that injector operating voltages on some applications are considerably higher than "battery voltage" traditionally used on solenoid injectors; these systems can use a separate injector power module to deliver the power and control signal to the injectors.	Identify the type of injector and the wiring / control modules used on the system. The voltage, frequency or value of the control signal in the injector circuit is either at full value (e.g. battery voltage with no on / off signal) or, the signal exists but it is above the normal operating range. Possible fault with injector or wiring (e.g. injector power supply, short / open circuit or resistance) that is causing a signal value to be higher than the maximum operating limit. Note: the control unit could be providing the correct signal but it is being affected by a circuit fault. Refer to list of sensor / actuator checks on page 19. It is possible that the when the control unit completes the earth path for the injector, the voltage on the earth circuit remains at supply voltage level e.g. battery voltage instead of dropping to zero (or close to zero); this could be caused by a open circuit, poor earth for the control unit or control unit fault.
P0293	Cylinder 11 Contribution / Balance	The control unit checks the signal from engine speed sensor to monitor crankshaft acceleration / speed during each power stroke; this provides an indication of the power / torque contribution provided by the cylinder (compared to the other cylinders). Any fault with the cylinder, including mechanical faults, fuelling, ignition (petrol engines) or other faults affecting combustion, will affect the power / torque produced by the cylinder.	The fault code indicates that the contribution from the identified cylinder is different when compared to other cylinders (likely to be a lower contribution but, in some circumstances the contribution could be higher). Check ignition (petrol engines), check fuelling and injection timing (as applicable) and check condition of injectors (leaking / dribbling injectors, injector, spray pattern etc). Also check mechanical condition of the cylinder (compression etc), and check for any other aspect of the engine operation that could affect just one cylinder (air leaks etc). It is also possible that there could be an engine position / speed sensor related fault (if necessary, Refer to list of sensor / actuator checks on page 19).
P0294	Cylinder 12 Injector Circuit Low	Most injectors have traditionally used a solenoid (contained within the injector body) to open the injector. Later injector designs make use of Piezo actuators instead of solenoids. Note that injector operating voltages on some applications are considerably higher than "battery voltage" traditionally used on solenoid injectors; these systems can use a separate injector power module to deliver the power and control signal to the injectors.	Identify the type of injector and the wiring / control modules used on the system. The voltage, frequency or value of the control signal in the injector circuit is either "zero" (no signal) or, the signal exists but is below the normal operating range. Possible fault with injector or wiring (e.g. injector power supply, short / open circuit or resistance) that is causing a signal value to be lower than the minimum operating limit. Note: the control unit could be providing the correct signal but it is being affected by a circuit fault. Refer to list of sensor / actuator checks on page 19. It is possible that the control unit is detecting "no voltage" in the circuit, which could be caused by faults with the power supply, connections or open circuit in wiring or injector.

NOTE 1. Check for other fault codes that could provide additional information.

NOTE 3. If a fault cannot be located, it is also possible that the control unit is at fault.

NOTE 2. Communication between control units can pass via a CAN-Bus system; refer to Page 51 for CAN-Bus checks.

NOTE 4. Refer to Page 53 for list of pages that show the ISO standard for component locations e.g. Sensor A - Bank 1.

Fault code	EOBD / ISO Description	Component / System Description	Meaningful Description and Quick Check
P0295	Cylinder 12 Injector Circuit High	Most injectors have traditionally used a solenoid (contained within the injector body) to open the injector. Later injector designs make use of Piezo actuators instead of solenoids. Note that injector operating voltages on some applications are considerably higher than "battery voltage" traditionally used on solenoid injectors; these systems can use a separate injector power module to deliver the power and control signal to the injectors.	Identify the type of injector and the wiring / control modules used on the system. The voltage, frequency or value of the control signal in the injector circuit is either at full value (e.g. battery voltage with no on / off signal) or, the signal exists but it is above the normal operating range. Possible fault with injector or wiring (e.g. injector power supply, short / open circuit or resistance) that is causing a signal value to be higher than the maximum operating limit. Note: the control unit could be providing the correct signal but it is being affected by a circuit fault. Refer to list of sensor / actuator checks on page 19. It is possible that the when the control unit completes the earth path for the injector, the voltage on the earth circuit remains at supply voltage level e.g. battery voltage instead of dropping to zero (or close to zero); this could be caused by a open circuit, poor earth for the control unit or control unit fault.
P0296	Cylinder 12 Contribution / Balance	The control unit checks the signal from engine speed sensor to monitor crankshaft acceleration / speed during each power stroke; this provides an indication of the power / torque contribution provided by the cylinder (compared to the other cylinders). Any fault with the cylinder, including mechanical faults, fuelling, ignition (petrol engines) or other faults affecting combustion, will affect the power / torque produced by the cylinder.	The fault code indicates that the contribution from the identified cylinder is different when compared to other cylinders (likely to be a lower contribution but, in some circumstances the contribution could be higher). Check ignition (petrol engines), check fuelling and injection timing (as applicable) and check condition of injectors (leaking / dribbling injectors, injector, spray pattern etc). Also check mechanical condition of the cylinder (compression etc), and check for any other aspect of the engine operation that could affect just one cylinder (air leaks etc). It is also possible that there could be an engine position / speed sensor related fault (if necessary, Refer to list of sensor / actuator checks on page 19).
P0297	Vehicle Overspeed Condition	Note that the vehicle speed will be indicated by a vehicle speed sensor which could be an independent device (transmission or drive train located) or the vehicle speed could be indicated by a wheel speed sensor (e.g. ABS wheel speed sensor); refer to vehicle specific information if necessary. Identify the type of sensor e.g. Inductive, Hall effect etc; this will dictate the checks that are applicable.	The vehicle overspeed condition does not necessarily indicate that the vehicle has exceeded its maximum limit. The overspeed fault condition could have occurred during a process that is vehicle speed dependent e.g. on some vehicles the Diesel Particulate regeneration process will only occur within certain defined vehicle speeds. It will be necessary to refer to vehicle specific information to indentify which process or system is being influenced or affected by the overspeed condition (the process or system being affected will be vehicle / manufacturer specific). It is also possible that a fault exists with the speed sensor. Refer to list of sensor / actuator checks on page 19.
P0298	Engine Oil Over Temperature	The engine oil temperature is critical to engine operation / performance / durability; it is also possible that operation of some functions e.g. emission devices, or engine system control could be dependent on the oil temperature being within certain values. Many vehicles make use of oil cooler systems to regulate the oil temperature.	The voltage, frequency or value of the temperature sensor signal is within the normal operating range / tolerance but is indicating a high temperature value. The control unit has not detected an electrical fault, therefore the assessment is that the temperature is too high for the operating conditions (indicated by other sensors) or, the temperature does not match values expected by the control unit. It is also possible that the temperature is too high for operation of other systems e.g. emission control device. If possible, check the temperature using a separate thermometer and compare with the value indicated by the sensor. Check oil quantity and condition. If the sensor does appear to be faulty, check temperature sensor / wiring e.g. short / open circuits, circuit resistance, incorrect sensor resistance or, fault with the reference / operating voltage; interference from other circuits can also affect sensor signals. Refer to list of sensor / actuator checks on page 19.
P0299	Turbocharger / Supercharger Underboost	Different types of turbo / supercharger systems can be fitted to produce additional engine performance. For this fault code, it may be necessary to identify the system and components used.	Control unit has detected that the expected turbo / super charger boost level has not been achieved during normal vehicle operation. The fault could be related to a boost pressure control system fault or be caused by a fault in the boost pressure sensor causing an incorrect boost pressure signal to be passed to the control unit. Refer to list of sensor / actuator checks on page 19. Underboost conditions also can be caused by intake and exhaust system blockages / leaks.
P0300	Random / Multiple Cylinder Misfire Detected	A common method of detecting misfires is to monitor the signal from engine speed sensor to check crankshaft acceleration / speed during each power stroke; this provides an indication of the power / torque contribution provided by the cylinder (compared to the other cylinders). Any fault with the cylinder, including mechanical faults, fuelling, ignition (petrol engines) or other faults affecting combustion, will affect the power / torque produced by the cylinder. Misfires can also be detected by the oxygen / lambda sensor, due to high oxygen content (unburned oxygen) in the exhaust gas. Also note that there is reference to manufacturers monitoring the strength of exhaust pressure pulses (which will be weak if a misfire exists); a weak pressure pulse is then linked to an individual cylinder by using cylinder identification sensors (cam shaft sensor etc).	The fault code indicates that there is a random misfire but it is not identifiable to specific cylinders. It is likely that the fault is causing erratic fluctuations of the crankshaft speed (or exhaust pressure pulses) which are not consistent to any one cylinder or group of cylinders. The fault could be related to any aspect of the engine operation including: fuelling, ignition (petrol engines), air leaks, mechanical condition etc. Also check (as applicable) for faults with variable valve timing, EGR and other emission control systems. It is also possible that there could be an engine position / speed sensor related fault (if necessary, Refer to list of sensor / actuator checks on page 19).
P0301	Cylinder 1 Misfire Detected	A common method of detecting misfires is to monitor the signal from engine speed sensor to check crankshaft acceleration / speed during each power stroke; this provides an indication of the power / torque contribution provided by the cylinder (compared to the other cylinders). Any fault with the cylinder, including mechanical faults, fuelling, ignition (petrol engines) or other faults affecting combustion, will affect the power / torque produced by the cylinder. Misfires can also be detected by the oxygen / lambda sensor, due to high oxygen content (unburned oxygen) in the exhaust gas. Also note that there is reference to manufacturers monitoring the strength of exhaust pressure pulses (which will be weak if a misfire exists); a weak pressure pulse is then linked to an individual cylinder by using cylinder identification sensors (cam shaft sensor etc).	The fault code indicates that the contribution from the identified cylinder is different when compared to other cylinders (likely to be a lower contribution but, in some circumstances the contribution could be higher). If the system monitors exhaust pressure pulses, it could have detected a weaker pulse for a specific cylinder). Check ignition (petrol engines), check fuelling and injection timing (as applicable) and check condition of injectors (leaking / dribbling injectors, injector, spray pattern etc). Also check mechanical condition of the cylinder (compression etc), and check for any other aspect of the engine operation that could affect just one cylinder (air leaks etc). It is also possible that there could be an engine position / speed sensor related fault (if necessary, Refer to list of sensor / actuator checks on page 19).

code	Description		
P0302	Cylinder 2 Misfire Detected	The control unit checks the signal from engine speed sensor to monitor crankshaft acceleration / speed during each power stroke; this provides an indication of the power / torque contribution provided by the cylinder (compared to the other cylinders). Any fault with the cylinder, including mechanical faults, fuelling, ignition (petrol engines) or other faults affecting combustion, will affect the power / torque produced by the cylinder. Misfires can also be detected by the oxygen / lambda sensor, due to high oxygen content (unburned oxygen) in the exhaust gas; however this detection process will not identify the cylinder.	The fault code indicates that the contribution from the identified cylinder is different when compared to other cylinders (likely to be a lower contribution but, in some circumstances the contribution could be higher). If the system monitors exhaust pressure pulses, it could have detected a weaker pulse for a specific cylinder). Check ignition (petrol engines), check fuelling and injection timing (as applicable) and check condition of injectors (leaking / dribbling injectors, injector, spray pattern etc). Also check mechanical condition of the cylinder (compression etc), and check for any other aspect of the engine operation that could affect just one cylinder (air leaks etc). It is also possible that there could be an engine position / speed sensor related fault (if necessary, Refer to list of sensor / actuator checks on page 19).
P0303	Cylinder 3 Misfire Detected	A common method of detecting misfires is to monitor the signal from engine speed sensor to check crankshaft acceleration / speed during each power stroke; this provides an indication of the power / torque contribution provided by the cylinder (compared to the other cylinders). Any fault with the cylinder, including mechanical faults, fuelling, ignition (petrol engines) or other faults affecting combustion, will affect the power / torque produced by the cylinder. Misfires can also be detected by the oxygen / lambda sensor, due to high oxygen content (unburned oxygen) in the exhaust gas. Also note that there is reference to manufacturers monitoring the strength of exhaust pressure pulses (which will be weak if a misfire exists); a weak pressure pulse is then linked to an individual cylinder by using cylinder identification sensors (cam shaft sensor etc).	The fault code indicates that the contribution from the identified cylinder is different when compared to other cylinders (likely to be a lower contribution but, in some circumstances the contribution could be higher). If the system monitors exhaust pressure pulses, it could have detected a weaker pulse for a specific cylinder). Check ignition (petrol engines), check fuelling and injection timing (as applicable) and check condition of injectors (leaking / dribbling injectors, injector, spray pattern etc). Also check mechanical condition of the cylinder (compression etc), and check for any other aspect of the engine operation that could affect just one cylinder (air leaks etc). It is also possible that there could be an engine position / speed sensor related fault (if necessary, Refer to list of sensor / actuator checks on page 19).
P0304	Cylinder 4 Misfire Detected	A common method of detecting misfires is to monitor the signal from engine speed sensor to check crankshaft acceleration / speed during each power stroke; this provides an indication of the power / torque contribution provided by the cylinder (compared to the other cylinders). Any fault with the cylinder, including mechanical faults, fuelling, ignition (petrol engines) or other faults affecting combustion, will affect the power / torque produced by the cylinder. Misfires can also be detected by the oxygen / lambda sensor, due to high oxygen content (unburned oxygen) in the exhaust gas. Also note that there is reference to manufacturers monitoring the strength of exhaust pressure pulses (which will be weak if a misfire exists); a weak pressure pulse is then linked to an individual cylinder by using cylinder identification sensors (cam shaft sensor etc).	The fault code indicates that the contribution from the identified cylinder is different when compared to other cylinders (likely to be a lower contribution but, in some circumstances the contribution could be higher). If the system monitors exhaust pressure pulses, it could have detected a weaker pulse for a specific cylinder). Check ignition (petrol engines), check fuelling and injection timing (as applicable) and check condition of injectors (leaking / dribbling injectors, injector, spray pattern etc). Also check mechanical condition of the cylinder (compression etc), and check for any other aspect of the engine operation that could affect just one cylinder (air leaks etc). It is also possible that there could be an engine position / speed sensor related fault (if necessary, Refer to list of sensor / actuator checks on page 19).
P0305	Cylinder 5 Misfire Detected	A common method of detecting misfires is to monitor the signal from engine speed sensor to check crankshaft acceleration / speed during each power stroke; this provides an indication of the power / torque contribution provided by the cylinder (compared to the other cylinders). Any fault with the cylinder, including mechanical faults, fuelling, ignition (petrol engines) or other faults affecting combustion, will affect the power / torque produced by the cylinder. Misfires can also be detected by the oxygen / lambda sensor, due to high oxygen content (unburned oxygen) in the exhaust gas. Also note that there is reference to manufacturers monitoring the strength of exhaust pressure pulses (which will be weak if a misfire exists); a weak pressure pulse is then linked to an individual cylinder by using cylinder identification sensors (cam shaft sensor etc).	The fault code indicates that the contribution from the identified cylinder is different when compared to other cylinders (likely to be a lower contribution but, in some circumstances the contribution could be higher). If the system monitors exhaust pressure pulses, it could have detected a weaker pulse for a specific cylinder). Check ignition (petrol engines), check fuelling and injection timing (as applicable) and check condition of injectors (leaking / dribbling injectors, injector, spray pattern etc). Also check mechanical condition of the cylinder (compression etc), and check for any other aspect of the engine operation that could affect just one cylinder (air leaks etc). It is also possible that there could be an engine position / speed sensor related fault (if necessary, Refer to list of sensor / actuator checks on page 19).
P0306	Cylinder 6 Misfire Detected	A common method of detecting misfires is to monitor the signal from engine speed sensor to check crankshaft acceleration / speed during each power stroke; this provides an indication of the power / torque contribution provided by the cylinder (compared to the other cylinders). Any fault with the cylinder, including mechanical faults, fuelling, ignition (petrol engines) or other faults affecting combustion, will affect the power / torque produced by the cylinder. Misfires can also be detected by the oxygen / lambda sensor, due to high oxygen content (unburned oxygen) in the exhaust gas. Also note that there is reference to manufacturers monitoring the strength of exhaust pressure pulses (which will be weak if a misfire exists); a weak pressure pulse is then linked to an individual cylinder by using cylinder identification sensors (cam shaft sensor etc).	The fault code indicates that the contribution from the identified cylinder is different when compared to other cylinders (likely to be a lower contribution but, in some circumstances the contribution could be higher). If the system monitors exhaust pressure pulses, it could have detected a weaker pulse for a specific cylinder). Check ignition (petrol engines), check fuelling and injection timing (as applicable) and check condition of injectors (leaking / dribbling injectors, injector, spray pattern etc). Also check mechanical condition of the cylinder (compression etc), and check for any other aspect of the engine operation that could affect just one cylinder (air leaks etc). It is also possible that there could be an engine position / speed sensor related fault (if necessary, Refer to list of sensor / actuator checks on page 19).

NOTE 1. Check for other fault codes that could provide additional information.   **NOTE 2.** Communication between control units can pass via a CAN-Bus system; refer to Page 51 for CAN-Bus checks.   **109**

NOTE 3. If a fault cannot be located, it is also possible that the control unit is at fault.   **NOTE 4.** Refer to Page 53 for list of pages that show the ISO standard for component locations e.g. Sensor A - Bank 1.

Fault code	EOBD / ISO Description	Component / System Description	Meaningful Description and Quick Check
P0307	Cylinder 7 Misfire Detected	A common method of detecting misfires is to monitor the signal from engine speed sensor to check crankshaft acceleration / speed during each power stroke; this provides an indication of the power / torque contribution provided by the cylinder (compared to the other cylinders). Any fault with the cylinder, including mechanical faults, fuelling, ignition (petrol engines) or other faults affecting combustion, will affect the power / torque produced by the cylinder. Misfires can also be detected by the oxygen / lambda sensor, due to high oxygen content (unburned oxygen) in the exhaust gas. Also note that there is reference to manufacturers monitoring the strength of exhaust pressure pulses (which will be weak if a misfire exists); a weak pressure pulse is then linked to an individual cylinder by using cylinder identification sensors (cam shaft sensor etc).	The fault code indicates that the contribution from the identified cylinder is different when compared to other cylinders (likely to be a lower contribution but, in some circumstances the contribution could be higher). If the system monitors exhaust pressure pulses, it could have detected a weaker pulse for a specific cylinder). Check ignition (petrol engines), check fuelling and injection timing (as applicable) and check condition of injectors (leaking / dribbling injectors, injector, spray pattern etc). Also check mechanical condition of the cylinder (compression etc), and check for any other aspect of the engine operation that could affect just one cylinder (air leaks etc). It is also possible that there could be an engine position / speed sensor related fault (if necessary, Refer to list of sensor / actuator checks on page 19).
P0308	Cylinder 8 Misfire Detected	A common method of detecting misfires is to monitor the signal from engine speed sensor to check crankshaft acceleration / speed during each power stroke; this provides an indication of the power / torque contribution provided by the cylinder (compared to the other cylinders). Any fault with the cylinder, including mechanical faults, fuelling, ignition (petrol engines) or other faults affecting combustion, will affect the power / torque produced by the cylinder. Misfires can also be detected by the oxygen / lambda sensor, due to high oxygen content (unburned oxygen) in the exhaust gas. Also note that there is reference to manufacturers monitoring the strength of exhaust pressure pulses (which will be weak if a misfire exists); a weak pressure pulse is then linked to an individual cylinder by using cylinder identification sensors (cam shaft sensor etc).	The fault code indicates that the contribution from the identified cylinder is different when compared to other cylinders (likely to be a lower contribution but, in some circumstances the contribution could be higher). If the system monitors exhaust pressure pulses, it could have detected a weaker pulse for a specific cylinder). Check ignition (petrol engines), check fuelling and injection timing (as applicable) and check condition of injectors (leaking / dribbling injectors, injector, spray pattern etc). Also check mechanical condition of the cylinder (compression etc), and check for any other aspect of the engine operation that could affect just one cylinder (air leaks etc). It is also possible that there could be an engine position / speed sensor related fault (if necessary, Refer to list of sensor / actuator checks on page 19).
P0309	Cylinder 9 Misfire Detected	A common method of detecting misfires is to monitor the signal from engine speed sensor to check crankshaft acceleration / speed during each power stroke; this provides an indication of the power / torque contribution provided by the cylinder (compared to the other cylinders). Any fault with the cylinder, including mechanical faults, fuelling, ignition (petrol engines) or other faults affecting combustion, will affect the power / torque produced by the cylinder. Misfires can also be detected by the oxygen / lambda sensor, due to high oxygen content (unburned oxygen) in the exhaust gas. Also note that there is reference to manufacturers monitoring the strength of exhaust pressure pulses (which will be weak if a misfire exists); a weak pressure pulse is then linked to an individual cylinder by using cylinder identification sensors (cam shaft sensor etc).	The fault code indicates that the contribution from the identified cylinder is different when compared to other cylinders (likely to be a lower contribution but, in some circumstances the contribution could be higher). If the system monitors exhaust pressure pulses, it could have detected a weaker pulse for a specific cylinder). Check ignition (petrol engines), check fuelling and injection timing (as applicable) and check condition of injectors (leaking / dribbling injectors, injector, spray pattern etc). Also check mechanical condition of the cylinder (compression etc), and check for any other aspect of the engine operation that could affect just one cylinder (air leaks etc). It is also possible that there could be an engine position / speed sensor related fault (if necessary, Refer to list of sensor / actuator checks on page 19).
P0310	Cylinder 10 Misfire Detected	A common method of detecting misfires is to monitor the signal from engine speed sensor to check crankshaft acceleration / speed during each power stroke; this provides an indication of the power / torque contribution provided by the cylinder (compared to the other cylinders). Any fault with the cylinder, including mechanical faults, fuelling, ignition (petrol engines) or other faults affecting combustion, will affect the power / torque produced by the cylinder. Misfires can also be detected by the oxygen / lambda sensor, due to high oxygen content (unburned oxygen) in the exhaust gas. Also note that there is reference to manufacturers monitoring the strength of exhaust pressure pulses (which will be weak if a misfire exists); a weak pressure pulse is then linked to an individual cylinder by using cylinder identification sensors (cam shaft sensor etc).	The fault code indicates that the contribution from the identified cylinder is different when compared to other cylinders (likely to be a lower contribution but, in some circumstances the contribution could be higher). If the system monitors exhaust pressure pulses, it could have detected a weaker pulse for a specific cylinder). Check ignition (petrol engines), check fuelling and injection timing (as applicable) and check condition of injectors (leaking / dribbling injectors, injector, spray pattern etc). Also check mechanical condition of the cylinder (compression etc), and check for any other aspect of the engine operation that could affect just one cylinder (air leaks etc). It is also possible that there could be an engine position / speed sensor related fault (if necessary, Refer to list of sensor / actuator checks on page 19).
P0311	Cylinder 11 Misfire Detected	A common method of detecting misfires is to monitor the signal from engine speed sensor to check crankshaft acceleration / speed during each power stroke; this provides an indication of the power / torque contribution provided by the cylinder (compared to the other cylinders). Any fault with the cylinder, including mechanical faults, fuelling, ignition (petrol engines) or other faults affecting combustion, will affect the power / torque produced by the cylinder. Misfires can also be detected by the oxygen / lambda sensor, due to high oxygen content (unburned oxygen) in the exhaust gas. Also note that there is reference to manufacturers monitoring the strength of exhaust pressure pulses (which will be weak if a misfire exists); a weak pressure pulse is then linked to an individual cylinder by using cylinder identification sensors (cam shaft sensor etc).	The fault code indicates that the contribution from the identified cylinder is different when compared to other cylinders (likely to be a lower contribution but, in some circumstances the contribution could be higher). If the system monitors exhaust pressure pulses, it could have detected a weaker pulse for a specific cylinder). Check ignition (petrol engines), check fuelling and injection timing (as applicable) and check condition of injectors (leaking / dribbling injectors, injector, spray pattern etc). Also check mechanical condition of the cylinder (compression etc), and check for any other aspect of the engine operation that could affect just one cylinder (air leaks etc). It is also possible that there could be an engine position / speed sensor related fault (if necessary, Refer to list of sensor / actuator checks on page 19).

NOTE 1. Check for other fault codes that could provide additional information.   NOTE 2. Communication between control units can pass via a CAN-Bus system; refer to Page 51 for CAN-Bus checks.

NOTE 3. If a fault cannot be located, it is also possible that the control unit is at fault.   NOTE 4. Refer to Page 53 for list of pages that show the ISO standard for component locations e.g. Sensor A - Bank 1.

code	Description		
P0312	Cylinder 12 Misfire Detected	A common method of detecting misfires is to monitor the signal from engine speed sensor to check crankshaft acceleration / speed during each power stroke; this provides an indication of the power / torque contribution provided by the cylinder (compared to the other cylinders). Any fault with the cylinder, including mechanical faults, fuelling, ignition (petrol engines) or other faults affecting combustion, will affect the power / torque produced by the cylinder. Misfires can also be detected by the oxygen / lambda sensor, due to high oxygen content (unburned oxygen) in the exhaust gas. Also note that there is reference to manufacturers monitoring the strength of exhaust pressure pulses (which will be weak if a misfire exists); a weak pressure pulse is then linked to an individual cylinder by using cylinder identification sensors (cam shaft sensor etc).	The fault code indicates that the contribution from the identified cylinder is different when compared to other cylinders (likely to be a lower contribution but, in some circumstances the contribution could be higher). If the system monitors exhaust pressure pulses, it could have detected a weaker pulse for a specific cylinder). Check ignition (petrol engines), check fuelling and injection timing (as applicable) and check condition of injectors (leaking / dribbling injectors, injector, spray pattern etc). Also check mechanical condition of the cylinder (compression etc), and check for any other aspect of the engine operation that could affect just one cylinder (air leaks etc). It is also possible that there could be an engine position / speed sensor related fault (if necessary, Refer to list of sensor / actuator checks on page 19).
P0313	Misfire Detected With Low Fuel	A common method of detecting misfires is to monitor the signal from engine speed sensor to check crankshaft acceleration / speed during each power stroke; this provides an indication of the power / torque contribution provided by the cylinder (compared to the other cylinders). Also refer to Fault code P0300 and P0312 for additional information relating to misfire detection.	The fault code indicates that there is a random misfire but it is not identifiable to specific cylinders. It is likely that the fault is causing erratic fluctuations of the crankshaft speed (or affecting other monitoring processes), which are not consistent to any one cylinder or group of cylinders. The fault could be related to any aspect of the engine operation including: fuelling, ignition (petrol engines), air leaks, mechanical condition etc, but the control unit has also identified that the fuel level is low. Check fuel level and add fuel if required. Clear fault code. If the same fault is re-activated, check fuel level sensor system (refer to fault code P0461 for additional information).
P0314	Single Cylinder Misfire (Cylinder not Specified)	A common method of detecting misfires is to monitor the signal from engine speed sensor to check crankshaft acceleration / speed during each power stroke; this provides an indication of the power / torque contribution provided by the cylinder (compared to the other cylinders). Also refer to Fault code P0300 and P0312 for additional information relating to misfire detection.	The fault code indicates that the contribution from the identified cylinder is different when compared to other cylinders (likely to be a lower contribution but, in some circumstances the contribution could be higher). If the system monitors exhaust pressure pulses, it could have detected a weaker pulse for a specific cylinder). Check ignition (petrol engines), check fuelling and injection timing (as applicable) and check condition of injectors (leaking / dribbling injectors, injector, spray pattern etc). Also check mechanical condition of the cylinder (compression etc), and check for any other aspect of the engine operation that could affect just one cylinder (air leaks etc). It is also possible that there could be an engine position / speed sensor related fault (if necessary, Refer to list of sensor / actuator checks on page 19).
P0315	Crankshaft Position System Variation Not Learned	The "crankshaft position system" refers to the system used to detect the position / speed of the crankshaft. Note that some systems (usually Diesel engines) could use the signal from the camshaft sensor or the fuel pump position / speed sensor.	The control unit is not fully recognising the crankshaft positions sensor / signal; this is likely to be caused by the fitting of a new sensor or control unit (control unit programme / software could also have been corrupted or erased). It may be necessary to perform a re-initialisation of the control unit (especially if new components have been fitted) refer to vehicle specific information to identify the process. If it is suspected that the sensor is faulty, Refer to list of sensor / actuator checks on page 19.
P0316	Engine Misfire Detected on Start-up (First 1000 Revolutions)	A common method of detecting misfires is to monitor the signal from engine speed sensor to check crankshaft acceleration / speed during each power stroke; this provides an indication of the power / torque contribution provided by the cylinder (compared to the other cylinders). Also refer to Fault code P0300 and P0312 for additional information relating to misfire detection.	The fault code indicates that there is a misfire (on unidentified cylinder or cylinders) immediately after engine start. It is likely that the fault is causing erratic fluctuations of the crankshaft speed, which are not consistent to any one cylinder or group of cylinders. The fault could be related to any aspect of the engine operation but it is likely to be specific to a function or controls on immediate post start conditions e.g. post start glow plug control, lack of fuel enrichment, ignition systems problem (spark plugs etc). Also check checks on all engine systems including: fuelling, ignition (petrol engines), air leaks, mechanical condition etc. Also check (as applicable) for faults with variable valve timing, EGR and other emission control systems. It is also possible that there could be an engine position / speed sensor related fault (if necessary, Refer to list of sensor / actuator checks on page 19).
P0317	Rough Road Hardware Not Present	Road sensors (usually accelerometers) indicate rough terrain conditions; the sensor information can be used for transmission control and for engine control. The engine speed can fluctuate rapidly when a vehicle passes over rough terrain, which can be regarded as an engine related fault e.g. misfire or reduced cylinder performance because of sudden change in engine speed. If the rough road sensor indicates rough conditions, the engine speed fluctuations will be ignored. Note that some vehicles use the signal from a wheel speed sensor (ABS sensor) to indicate rough conditions (wheel speed will rapidly change when a bump or dip is encountered).	The fault code indicates that a rough road sensor is not fitted or cannot be detected by the control unit; this can occur if the sensor or control unit has been changed and the control unit is not recognising the rough road sensor; it may be necessary to re-initialise the control unit (refer to vehicle specific information). It s possible that a sensor has been changed and not connected correctly to the system / control unit or, it is also possible that a replacement control unit cannot detect the sensor, because the vehicle is not fitted with a sensor and the control unit is configured incorrectly for the vehicle.
P0318	Rough Road Sensor "A" Signal Circuit	Road sensors (usually accelerometers) indicate rough terrain conditions; the sensor information can be used for transmission control and for engine control. The engine speed can fluctuate rapidly when a vehicle passes over rough terrain, which can be regarded as an engine related fault e.g. misfire or reduced cylinder performance because of sudden change in engine speed. If the rough road sensor indicates rough conditions, the engine speed fluctuations will be ignored. Note that some vehicles use the signal from a wheel speed sensor (ABS sensor) to indicate rough conditions (wheel speed will rapidly change when a bump or dip is encountered).	Identify the type of sensor system used on the vehicle e.g. a separate sensor such as an accelerometer or, whether a wheel speed sensor is used. The voltage, frequency or value of the sensor signal is incorrect (undefined fault). Sensor voltage / signal could be: out of normal operating range, constant value, non-existent, corrupt. Possible fault in the sensor or wiring e.g. short / open circuits, circuit resistance, incorrect sensor resistance or, fault with the reference / operating voltage; interference from other circuits can also affect sensor signals. If a wheel speed sensor is used, Refer to list of sensor / actuator checks on page 19. If a separate sensor is used, it may be necessary to refer to vehicle specific information to identify specific test procedures.

NOTE 1.   Check for other fault codes that could provide additional information.   NOTE 2.   Communication between control units can pass via a CAN-Bus system; refer to Page 51 for CAN-Bus checks.

NOTE 3.   If a fault cannot be located, it is also possible that the control unit is at fault.   NOTE 4.   Refer to Page 53 for list of pages that show the ISO standard for component locations e.g. Sensor A - Bank 1.

Fault code	EOBD / ISO Description	Component / System Description	Meaningful Description and Quick Check
P0319	Rough Road Sensor "B" Signal Circuit	Road sensors (usually accelerometers) indicate rough terrain conditions; the sensor information can be used for transmission control and for engine control. The engine speed can fluctuate rapidly when a vehicle passes over rough terrain, which can be regarded as an engine related fault e.g. misfire or reduced cylinder performance because of sudden change in engine speed. If the rough road sensor indicates rough conditions, the engine speed fluctuations will be ignored. Note that some vehicles use the signal from a wheel speed sensor (ABS sensor) to indicate rough conditions (wheel speed will rapidly change when a bump or dip is encountered).	Identify the type of sensor system used on the vehicle e.g. a separate sensor such as an accelerometer or, whether a wheel speed sensor is used. The voltage, frequency or value of the sensor signal is incorrect (undefined fault). Sensor voltage / signal could be: out of normal operating range, constant value, non-existent, corrupt. Possible fault in the sensor or wiring e.g. short / open circuits, circuit resistance, incorrect sensor resistance or, fault with the reference / operating voltage; interference from other circuits can also affect sensor signals. If a wheel speed sensor is used, Refer to list of sensor / actuator checks on page 19. If a separate sensor is used, it may be necessary to refer to vehicle specific information to identify specific test procedures.
P0320	Ignition / Distributor Engine Speed Input Circuit	For engines fitted with spark distributor systems, an engine speed sensor is often located within the distributor body. The speed sensor signal can be passed to the main engine control unit or passed to a separate amplifier, which then passes a speed signal to the control unit. The fault code could refer to the signal from the speed sensor or from the amplifier. If necessary, refer to vehicle specific information to identify the type of system.	Identify the type of sensor e.g. Inductive, Hall effect etc; this will dictate the checks that are applicable. The voltage, frequency or value of the speed sensor signal (from the sensor or from the amplifier) is incorrect (undefined fault). Signal value could be: out of normal operating range, constant value, non-existent, corrupt. As applicable, check wiring from the speed sensor to the control unit and / or wiring between the amplifier and the control unit. Possible fault in the speed sensor or wiring e.g. short / open circuits, circuit resistance, incorrect sensor resistance (if applicable) or, fault with the reference / operating voltage; interference from other circuits can also affect sensor signals. Also check earth connections and cable screening. Refer to list of sensor / actuator checks on page 19.
P0321	Ignition / Distributor Engine Speed Input Circuit Range / Performance	For engines fitted with spark distributor systems, an engine speed sensor is often located within the distributor body. The speed sensor signal can be passed to the main engine control unit or passed to a separate amplifier, which then passes a speed signal to the control unit. The fault code could refer to the signal from the speed sensor or from the amplifier. If necessary, refer to vehicle specific information to identify the type of system.	Identify the type of sensor e.g. Inductive, Hall effect etc; this will dictate the checks that are applicable. The voltage, frequency or value of the speed sensor signal (from the sensor or from the amplifier) is within the normal operating range / tolerance but the signal is not plausible or is incorrect due to an undefined fault. The speed signal might not match the operating conditions indicated by other sensors, or the speed signal is not changing / responding as expected. As applicable, check wiring from the speed sensor to the control unit and / or wiring between the amplifier and the control unit. Possible fault in the speed sensor or wiring e.g. short / open circuits, circuit resistance, incorrect sensor resistance or, fault with the reference / operating voltage; interference from other circuits can also affect sensor signals. Refer to list of sensor / actuator checks on page 19. If possible, use alternative method of checking speed to establish if sensor system is operating correctly and whether speed is correct / incorrect. Also check condition of reference teeth on trigger disc / rotor and ensure rotor is turning with shaft.
P0322	Ignition / Distributor Engine Speed Input Circuit No Signal	For engines fitted with spark distributor systems, an engine speed sensor is often located within the distributor body. The speed sensor signal can be passed to the main engine control unit or passed to a separate amplifier, which then passes a speed signal to the control unit. The fault code could refer to the signal from the speed sensor or from the amplifier. If necessary, refer to vehicle specific information to identify the type of system.	Identify the type of sensor e.g. Inductive, Hall effect etc; this will dictate the checks that are applicable. The voltage, frequency or value of the speed signal is incorrect (undefined fault resulting in no signal being detected). Sensor voltage / signal value could be: constant value or non-existent. As applicable, check wiring from the speed sensor to the control unit and / or wiring between the amplifier and the control unit. Possible fault in the speed sensor or wiring e.g. short / open circuits, circuit resistance, incorrect sensor resistance (if applicable) or, fault with the reference / operating voltage; interference from other circuits can also affect sensor signals Also check earth connections and cable screening. Refer to list of sensor / actuator checks on page 19.
P0323	Ignition / Distributor Engine Speed Input Circuit Intermittent	For engines fitted with spark distributor systems, an engine speed sensor is often located within the distributor body. The speed sensor signal can be passed to the main engine control unit or passed to a separate amplifier, which then passes a speed signal to the control unit. The fault code could refer to the signal from the speed sensor or from the amplifier. If necessary, refer to vehicle specific information to identify the type of system.	Identify the type of sensor e.g. Inductive, Hall effect etc; this will dictate the checks that are applicable. The voltage, frequency or value of the speed sensor signal is intermittent (likely to be out of normal operating range / tolerance when intermittent fault is detected) or, the signal / voltage is erratic (e.g. signal changes are irregular / unpredictable). As applicable, check wiring from the speed sensor to the control unit and / or wiring between the amplifier and the control unit. Possible fault with wiring or connections (including internal sensor connections); also interference from other circuits can affect the signal. Refer to list of sensor / actuator checks on page 19. Also check condition of reference teeth on trigger disc / rotor and ensure rotor is turning with shaft and that there is no slack in the rotor shaft or drive mechanism that could cause erratic shaft movement.
P0324	Knock Control System Error	Knock sensors (usually fitted to the cylinder block or cylinder head) are used to detect combustion knock. Depending the engine and systems, the knock system (usually integrated into the main control unit) will then alter engine operation e.g. alter ignition timing, fuelling or reduce boost pressure, to reduce / eliminate the knock.	The fault code is indicating that there is an undefined fault with the knock sensor system, this could be related to a sensor fault, a circuit fault or a fault with the control unit. It is also possible that there is knock being detected that cannot be corrected by the normal process e.g. retarding the ignition timing or altering the fuelling. If the knock sensor appears to be operating correctly and the wiring is good the fault could be caused by engine condition or other engine control systems that are causing knock to exist; it is also possible that the control unit is faulty.
P0325	Knock Sensor 1 Circuit Bank 1 or Single Sensor	Knock sensors (usually fitted to the cylinder block or cylinder head) are used to detect combustion knock. Depending the engine and systems, the knock system (usually integrated into the main control unit) will then alter engine operation e.g. alter ignition timing, fuelling or reduce boost pressure, to reduce / eliminate the knock.	The voltage, frequency or value of the knock sensor signal is incorrect (undefined fault). Knock sensor voltage / signal could be: out of normal operating range, constant value, non-existent, corrupt. Possible fault in the knock sensor or wiring e.g. short / open circuits, circuit resistance; interference from other circuits can also affect knock sensor signals.
P0326	Knock Sensor 1 Circuit Range / Performance Bank 1 or Single Sensor	Knock sensors (usually fitted to the cylinder block or cylinder head) are used to detect combustion knock. Depending the engine and systems, the knock system (usually integrated into the main control unit) will then alter engine operation e.g. alter ignition timing, fuelling or reduce boost pressure, to reduce / eliminate the knock.	The voltage, frequency or value of the knock sensor signal is within the normal operating range / tolerance but the signal is not plausible or is incorrect due to an undefined fault. The knock sensor signal might not match the operating conditions indicated by other sensors or, the knock sensor signal is not changing / responding as expected (when adjustments have been made to reduce knock). Possible fault in the knock sensor or wiring e.g. short / open circuits, circuit resistance; interference from other circuits can also affect knock sensor signals. It is also possible that the knock sensor is operating correctly but the measurement being made by the knock sensor is unacceptable or incorrect for the operating conditions ("real combustion knock" detected). Check for any indication of combustion and check for possible causes (combustion chamber condition, faulty injectors, ignition / fuelling problems, incorrect boost pressure, high air temperature etc).

NOTE 1. Check for other fault codes that could provide additional information. NOTE 2. Communication between control units can pass via a CAN-Bus system; refer to Page 51 for CAN-Bus checks.
NOTE 3. If a fault cannot be located, it is also possible that the control unit is at fault. NOTE 4. Refer to Page 53 for list of pages that show the ISO standard for component locations e.g. Sensor A - Bank 1.

Code	Description		
P0327	Knock Sensor 1 Circuit Low Bank 1 or Single Sensor	Knock sensors (usually fitted to the cylinder block or cylinder head) are used to detect combustion knock. Depending the engine and systems, the knock system (usually integrated into the main control unit) will then alter engine operation e.g. alter ignition timing, fuelling or reduce boost pressure, to reduce / eliminate the knock.	The voltage, frequency or value of the knock sensor signal is either "zero" (no signal) or, the signal exists but it is below the normal operating range (lower than the minimum operating limit). Possible fault in the knock sensor or wiring e.g. short / open circuits, circuit resistance; interference from other circuits can also affect knock sensor signals.
P0328	Knock Sensor 1 Circuit High Bank 1 or Single Sensor	Knock sensors (usually fitted to the cylinder block or cylinder head) are used to detect combustion knock. Depending the engine and systems, the knock system (usually integrated into the main control unit) will then alter engine operation e.g. alter ignition timing, fuelling or reduce boost pressure, to reduce / eliminate the knock.	The voltage, frequency or value of the knock sensor signal is either at full value (e.g. battery voltage with no signal or frequency) or, the signal exists but it is above the normal operating range (higher than the maximum operating limit). Possible fault in the knock sensor or wiring e.g. short / open circuits, circuit resistance; interference from other circuits can also affect knock sensor signals.
P0329	Knock Sensor 1 Circuit Intermittent Bank 1 or Single Sensor	Knock sensors (usually fitted to the cylinder block or cylinder head) are used to detect combustion knock. Depending the engine and systems, the knock system (usually integrated into the main control unit) will then alter engine operation e.g. alter ignition timing, fuelling or reduce boost pressure, to reduce / eliminate the knock.	The voltage, frequency or value of the knock sensor signal is intermittent (likely to be out of normal operating range / tolerance when intermittent fault is detected) or, the signal / voltage is erratic (e.g. signal changes are irregular / unpredictable). Possible fault with wiring or connections; also interference from other circuits can affect the signal. Although the fault code relates to a sensor circuit / signal fault it is also possible that the control unit is detecting an erratic knock sensor signal that is being caused by combustion knock; Check for any indication of combustion and check for possible causes (combustion chamber condition, faulty injectors etc).
P0330	Knock Sensor 2 Circuit Bank 2	Knock sensors (usually fitted to the cylinder block or cylinder head) are used to detect combustion knock. Depending the engine and systems, the knock system (usually integrated into the main control unit) will then alter engine operation e.g. alter ignition timing, fuelling or reduce boost pressure, to reduce / eliminate the knock.	The voltage, frequency or value of the knock sensor signal is incorrect (undefined fault). Knock sensor voltage / signal could be: out of normal operating range, constant value, non-existent, corrupt. Possible fault in the knock sensor or wiring e.g. short / open circuits, circuit resistance; interference from other circuits can also affect knock sensor signals.
P0331	Knock Sensor 2 Circuit Range / Performance Bank 2	Knock sensors (usually fitted to the cylinder block or cylinder head) are used to detect combustion knock. Depending the engine and systems, the knock system (usually integrated into the main control unit) will then alter engine operation e.g. alter ignition timing, fuelling or reduce boost pressure, to reduce / eliminate the knock.	The voltage, frequency or value of the knock sensor signal is within the normal operating range / tolerance but the signal is not plausible or is incorrect due to an undefined fault. The knock sensor signal might not match the operating conditions indicated by other sensors or, the knock sensor signal is not changing / responding as expected (when adjustments have been made to reduce knock). Possible fault in the knock sensor or wiring e.g. short / open circuits, circuit resistance; interference from other circuits can also affect knock sensor signals. It is also possible that the knock sensor is operating correctly but the measurement being made by the knock sensor is unacceptable or incorrect for the operating conditions ("real combustion knock" detected). Check for any indication of combustion knock and check for possible causes (combustion chamber condition, faulty injectors, ignition / fuelling problems, incorrect boost pressure, high air temperature etc).
P0332	Knock Sensor 2 Circuit Low Bank 2	Knock sensors (usually fitted to the cylinder block or cylinder head) are used to detect combustion knock. Depending the engine and systems, the knock system (usually integrated into the main control unit) will then alter engine operation e.g. alter ignition timing, fuelling or reduce boost pressure, to reduce / eliminate the knock.	The voltage, frequency or value of the knock sensor signal is either "zero" (no signal) or, the signal exists but it is below the normal operating range (lower than the minimum operating limit). Possible fault in the knock sensor or wiring e.g. short / open circuits, circuit resistance; interference from other circuits can also affect knock sensor signals.
P0333	Knock Sensor 2 Circuit High Bank 2	Knock sensors (usually fitted to the cylinder block or cylinder head) are used to detect combustion knock. Depending the engine and systems, the knock system (usually integrated into the main control unit) will then alter engine operation e.g. alter ignition timing, fuelling or reduce boost pressure, to reduce / eliminate the knock.	The voltage, frequency or value of the knock sensor signal is either at full value (e.g. battery voltage with no signal or frequency) or, the signal exists but it is above the normal operating range (higher than the maximum operating limit). Possible fault in the knock sensor or wiring e.g. short / open circuits, circuit resistance; interference from other circuits can also affect knock sensor signals.
P0334	Knock Sensor 2 Circuit Intermittent Bank 2	Knock sensors (usually fitted to the cylinder block or cylinder head) are used to detect combustion knock. Depending the engine and systems, the knock system (usually integrated into the main control unit) will then alter engine operation e.g. alter ignition timing, fuelling or reduce boost pressure, to reduce / eliminate the knock.	The voltage, frequency or value of the knock sensor signal is intermittent (likely to be out of normal operating range / tolerance when intermittent fault is detected) or, the signal / voltage is erratic (e.g. signal changes are irregular / unpredictable). Possible fault with wiring or connections; also interference from other circuits can affect the signal. Although the fault code relates to a sensor circuit / signal fault it is also possible that the control unit is detecting an erratic knock sensor signal that is being caused by combustion knock; Check for any indication of combustion and check for possible causes (combustion chamber condition, faulty injectors etc).
P0335	Crankshaft Position Sensor "A" Circuit	The crankshaft position sensor provides position (and usually speed information) to the control unit. The information is used in calculations for fuelling, ignition timing, variable valve timing and emissions control functions; the information is also used for other vehicle systems such as transmission, stability control systems etc.	Identify the type of sensor e.g. Inductive, Hall effect etc; this will dictate the checks that are applicable. The voltage, frequency or value of the position / speed sensor signal is incorrect (undefined fault). Sensor voltage / signal value could be: out of normal operating range, constant value, non-existent, corrupt. Possible fault in the position / speed sensor or wiring e.g. short / open circuits, circuit resistance, incorrect sensor resistance (if applicable) or, fault with the reference / operating voltage; interference from other circuits can also affect sensor signals. Also check earth connections and cable screening. Refer to list of sensor / actuator checks on page 19.
P0336	Crankshaft Position Sensor "A" Circuit Range / Performance	The crankshaft position sensor provides position (and usually speed information) to the control unit. The information is used in calculations for fuelling, ignition timing, variable valve timing and emissions control functions; the information is also used for other vehicle systems such as transmission, stability control systems etc.	Identify the type of sensor e.g. Inductive, Hall effect etc; this will dictate the checks that are applicable. The voltage, frequency or value of the speed sensor signal is within the normal operating range / tolerance but the signal is not plausible or is incorrect due to an undefined fault. The sensor signal might not match the operating conditions indicated by other sensors or, the sensor signal is not changing / responding as expected to the changes in conditions. Possible fault in the speed sensor or wiring e.g. short / open circuits, circuit resistance, incorrect sensor resistance or, fault with the reference / operating voltage; interference from other circuits can also affect sensor signals. Refer to list of sensor / actuator checks on page 19. It is also possible that the sensor is operating correctly but the position / speed being measured by the sensor is unacceptable or incorrect for the operating conditions. Also check condition of reference teeth on trigger disc / rotor and ensure rotor is turning with shaft.

NOTE 1. Check for other fault codes that could provide additional information. NOTE 2. Communication between control units can pass via a CAN-Bus system; refer to Page 51 for CAN-Bus checks.
NOTE 3. If a fault cannot be located, it is also possible that the control unit is at fault. NOTE 4. Refer to Page 53 for list of pages that show the ISO standard for component locations e.g. Sensor A - Bank 1.

Fault code	EOBD / ISO Description	Component / System Description	Meaningful Description and Quick Check
P0337	Crankshaft Position Sensor "A" Circuit Low	The crankshaft position sensor provides position (and usually speed) information) to the control unit. The information is used in calculations for fuelling, ignition timing, variable valve timing and emissions control functions; the information is also used for other vehicle systems such as transmission, stability control systems etc.	Identify the type of sensor e.g. Inductive, Hall effect etc; this will dictate the checks that are applicable. The voltage, frequency or value of the position / speed sensor signal is either "zero" (no signal) or, the signal exists but it is below the normal operating range (lower than the minimum operating limit). Possible fault in the position / speed sensor or wiring e.g. short / open circuits, circuit resistance, incorrect sensor resistance or, fault with the reference / operating voltage; interference from other circuits can also affect sensor signals. Refer to list of sensor / actuator checks on page 19.
P0338	Crankshaft Position Sensor "A" Circuit High	The crankshaft position sensor provides position (and usually speed) information) to the control unit. The information is used in calculations for fuelling, ignition timing, variable valve timing and emissions control functions; the information is also used for other vehicle systems such as transmission, stability control systems etc.	Identify the type of sensor e.g. Inductive, Hall effect etc; this will dictate the checks that are applicable. The voltage, frequency or value of the position / speed sensor signal is either at full value (e.g. battery voltage with no signal or frequency) or, the signal exists but it is above the normal operating range (higher than the maximum operating limit). Possible fault in the position / speed sensor or wiring e.g. short / open circuits, circuit resistance, incorrect sensor resistance or, fault with the reference / operating voltage; interference from other circuits can also affect sensor signals. Refer to list of sensor / actuator checks on page 19.
P0339	Crankshaft Position Sensor "A" Circuit Intermittent	The crankshaft position sensor provides position (and usually speed) information) to the control unit. The information is used in calculations for fuelling, ignition timing, variable valve timing and emissions control functions; the information is also used for other vehicle systems such as transmission, stability control systems etc.	Identify the type of sensor e.g. Inductive, Hall effect etc; this will dictate the checks that are applicable. The voltage, frequency or value of the position / speed sensor signal is intermittent (likely to be out of normal operating range / tolerance when intermittent fault is detected) or, the signal / voltage is erratic (e.g. signal changes are irregular / unpredictable). Possible fault with wiring or connections (including internal sensor connections); also interference from other circuits can affect the signal. Refer to list of sensor / actuator checks on page 19. Also check condition of reference teeth on trigger disc / rotor and ensure rotor is turning with shaft.
P0340	Camshaft Position Sensor "A" Circuit Bank 1 or Single Sensor	The sensor is used to detect camshaft position (angular position relative to crankshaft position). sensor information can be used to enable accurate control of ignition timing, fuel injection timing and variable valve timing. Identify the type of sensor e.g. Inductive, Hall effect etc; this will dictate the checks that are applicable.	Identify the type of sensor e.g. Inductive, Hall effect etc; this will dictate the checks that are applicable. The voltage, frequency or value of the position / speed sensor signal is incorrect (undefined fault). Sensor voltage / signal value could be: out of normal operating range, constant value, non-existent, corrupt. Possible fault in the position / speed sensor or wiring e.g. short / open circuits, circuit resistance, incorrect sensor resistance (if applicable) or, fault with the reference / operating voltage; interference from other circuits can also affect sensor signals. Also check earth connections and cable screening. Refer to list of sensor / actuator checks on page 19.
P0341	Camshaft Position Sensor "A" Circuit Range / Performance Bank 1 or Single Sensor	The sensor is used to detect camshaft position (angular position relative to crankshaft position); sensor information can be used to enable accurate control of ignition timing, fuel injection timing and variable valve timing. Identify the type of sensor e.g. Inductive, Hall effect etc; this will dictate the checks that are applicable.	Identify the type of sensor e.g. Inductive, Hall effect etc; this will dictate the checks that are applicable. The voltage, frequency or value of the position / speed sensor signal is within the normal operating range / tolerance but the signal is not plausible or is incorrect due to an undefined fault. The sensor signal might not match the operating conditions indicated by other sensors or, the sensor signal is not changing / responding as expected to the changes in conditions. Possible fault in the speed sensor or wiring e.g. short / open circuits, circuit resistance, incorrect sensor resistance or, fault with the reference / operating voltage; interference from other circuits can also affect sensor signals. Refer to list of sensor / actuator checks on page 19. Check camshaft drive belt / chain for wear or slack adjustment (can cause erratic camshaft rotation). It is also possible that the sensor is operating correctly but the position / speed being measured by the sensor is unacceptable or incorrect for the operating conditions. Also check condition of reference teeth on trigger disc / rotor.
P0342	Camshaft Position Sensor "A" Circuit Low Bank 1 or Single Sensor	The sensor is used to detect camshaft position (angular position relative to crankshaft position); sensor information can be used to enable accurate control of ignition timing, fuel injection timing and variable valve timing. Identify the type of sensor e.g. Inductive, Hall effect etc; this will dictate the checks that are applicable.	Identify the type of sensor e.g. Inductive, Hall effect etc; this will dictate the checks that are applicable. The voltage, frequency or value of the position / speed sensor signal is either "zero" (no signal) or, the signal exists but it is below the normal operating range (lower than the minimum operating limit). Possible fault in the position / speed sensor or wiring e.g. short / open circuits, circuit resistance, incorrect sensor resistance or, fault with the reference / operating voltage; interference from other circuits can also affect sensor signals. Refer to list of sensor / actuator checks on page 19.
P0343	Camshaft Position Sensor "A" Circuit High Bank 1 or SingleSensor	The sensor is used to detect camshaft position (angular position relative to crankshaft position); sensor information can be used to enable accurate control of ignition timing, fuel injection timing and variable valve timing. Identify the type of sensor e.g. Inductive, Hall effect etc; this will dictate the checks that are applicable.	Identify the type of sensor e.g. Inductive, Hall effect etc; this will dictate the checks that are applicable. The voltage, frequency or value of the position / speed sensor signal is either at full value (e.g. battery voltage with no signal or frequency) or, the signal exists but it is above the normal operating range (higher than the maximum operating limit). Possible fault in the position / speed sensor or wiring e.g. short / open circuits, circuit resistance, incorrect sensor resistance or, fault with the reference / operating voltage; interference from other circuits can also affect sensor signals. Refer to list of sensor / actuator checks on page 19.
P0344	Camshaft Position Sensor "A" Circuit Intermittent Bank 1 or Single Sensor	The sensor is used to detect camshaft position (angular position relative to crankshaft position); sensor information can be used to enable accurate control of ignition timing, fuel injection timing and variable valve timing. Identify the type of sensor e.g. Inductive, Hall effect etc; this will dictate the checks that are applicable.	Identify the type of sensor e.g. Inductive, Hall effect etc; this will dictate the checks that are applicable. The voltage, frequency or value of the position / speed sensor signal is intermittent (likely to be out of normal operating range / tolerance when intermittent fault is detected) or, the signal / voltage is erratic (e.g. signal changes are irregular / unpredictable). Possible fault with wiring or connections (including internal sensor connections); also interference from other circuits can affect the signal. Refer to list of sensor / actuator checks on page 19. Check camshaft drive belt / chain for wear or slack adjustment (can cause erratic camshaft rotation). Also check condition of reference teeth on trigger disc / rotor and ensure rotor is turning with shaft.
P0345	Camshaft Position Sensor "A" Circuit Bank 2	The sensor is used to detect camshaft position (angular position relative to crankshaft position); sensor information can be used to enable accurate control of ignition timing, fuel injection timing and variable valve timing. Identify the type of sensor e.g. Inductive, Hall effect etc; this will dictate the checks that are applicable.	Identify the type of sensor e.g. Inductive, Hall effect etc; this will dictate the checks that are applicable. The voltage, frequency or value of the position / speed sensor signal is incorrect (undefined fault). Sensor voltage / signal value could be: out of normal operating range, constant value, non-existent, corrupt. Possible fault in the position / speed sensor or wiring e.g. short / open circuits, circuit resistance, incorrect sensor resistance (if applicable) or, fault with the reference / operating voltage; interference from other circuits can also affect sensor signals. Also check earth connections and cable screening. Refer to list of sensor / actuator checks on page 19.

NOTE 1. Check for other fault codes that could provide additional information. **NOTE 2.** Communication between control units can pass via a CAN-Bus system; refer to Page 51 for CAN-Bus checks.

NOTE 3. If a fault cannot be located, it is also possible that the control unit is at fault. **NOTE 4.** Refer to Page 53 for list of pages that show the ISO standard for component locations e.g. Sensor A - Bank 1.

114

Code	Description		
P0346	Camshaft Position Sensor "A" Circuit Range / Performance Bank 2	The sensor is used to detect camshaft position (angular position relative to crankshaft position); sensor information can be used to enable accurate control of ignition timing, fuel injection timing and variable valve timing. Identify the type of sensor e.g. Inductive, Hall effect etc; this will dictate the checks that are applicable.	Identify the type of sensor e.g. Inductive, Hall effect etc; this will dictate the checks that are applicable. The voltage, frequency or value of the position / speed sensor signal is within the normal operating range / tolerance but the signal is not plausible or is incorrect due to an undefined fault. The sensor signal might not match the operating conditions indicated by other sensors or, the sensor signal is not changing / responding as expected to the changes in conditions. Possible fault in the speed sensor or wiring e.g. short / open circuits, circuit resistance, incorrect sensor resistance or, fault with the reference / operating voltage; interference from other circuits can also affect sensor signals. Refer to list of sensor / actuator checks on page 19. Check camshaft drive belt / chain for wear or slack adjustment (can cause erratic camshaft rotation). It is also possible that the sensor is operating correctly but the position / speed being measured by the sensor is unacceptable or incorrect for the operating conditions. Also check condition of reference teeth on trigger disc / rotor.
P0347	Camshaft Position Sensor "A" Circuit Low Bank 2	The sensor is used to detect camshaft position (angular position relative to crankshaft position); sensor information can be used to enable accurate control of ignition timing, fuel injection timing and variable valve timing. Identify the type of sensor e.g. Inductive, Hall effect etc; this will dictate the checks that are applicable.	Identify the type of sensor e.g. Inductive, Hall effect etc; this will dictate the checks that are applicable. The voltage, frequency or value of the position / speed sensor signal is either "zero" (no signal) or, the signal exists but it is below the normal operating range (lower than the minimum operating limit). Possible fault in the position / speed sensor or wiring e.g. short / open circuits, circuit resistance, incorrect sensor resistance or, fault with the reference / operating voltage; interference from other circuits can also affect sensor signals. Refer to list of sensor / actuator checks on page 19.
P0348	Camshaft Position Sensor "A" Circuit High Bank 2	The sensor is used to detect camshaft position (angular position relative to crankshaft position); sensor information can be used to enable accurate control of ignition timing, fuel injection timing and variable valve timing. Identify the type of sensor e.g. Inductive, Hall effect etc; this will dictate the checks that are applicable.	Identify the type of sensor e.g. Inductive, Hall effect etc; this will dictate the checks that are applicable. The voltage, frequency or value of the position / speed sensor signal is either at full value (e.g. battery voltage with no signal or frequency) or, the signal exists but it is above the normal operating range (higher than the maximum operating limit). Possible fault in the position / speed sensor or wiring e.g. short / open circuits, circuit resistance, incorrect sensor resistance or, fault with the reference / operating voltage; interference from other circuits can also affect sensor signals. Refer to list of sensor / actuator checks on page 19.
P0349	Camshaft Position Sensor "A" Circuit Intermittent Bank 2	The sensor is used to detect camshaft position (angular position relative to crankshaft position); sensor information can be used to enable accurate control of ignition timing, fuel injection timing and variable valve timing. Identify the type of sensor e.g. Inductive, Hall effect etc; this will dictate the checks that are applicable.	Identify the type of sensor e.g. Inductive, Hall effect etc; this will dictate the checks that are applicable. The voltage, frequency or value of the position / speed sensor signal is intermittent (likely to be out of normal operating range / tolerance when intermittent fault is detected) or, the signal / voltage is erratic (e.g. signal changes are irregular / unpredictable). Possible fault with wiring or connections (including internal sensor connections); also interference from other circuits can affect the signal. Refer to list of sensor / actuator checks on page 19. Check camshaft drive belt / chain for wear or slack adjustment (can cause erratic camshaft rotation). Also check condition of reference teeth on trigger disc / rotor and ensure rotor is turning with shaft.
P0350	Ignition Coil Primary / Secondary Circuit	The control unit (or separate ignition module if fitted) normally "switches on / off" the current to the ignition coil(s) on the coil earth circuit; this circuit combined with the coil power supply circuit are referred to as the "primary circuit". The control unit monitors the voltage, current and / or dwell time (coil charge time) in the primary circuit; if the control unit detects a measurement that is outside of expected values, this will be assessed as a coil circuit fault. Refer to vehicle specific information to indentify the specified coil.	Fault detection is on the primary circuit but secondary circuit faults can influence primary circuit performance and voltage. The fault code does not define the exact fault, but the control unit will have detected an ignition related problem that is likely to be caused by a coil primary or secondary fault; note that the spark plug forms part of the secondary circuit. Due to the different types of coils and coil switching systems, checks will be applicable to the system. Initial checks should include all primary circuit connections, correct power supply voltage (ignition on / cranking) and resistance check on coil primary and secondary windings (refer to vehicle specific information). Also check secondary circuit connections as applicable (e.g. coil leads where fitted, coil to plug connections etc). Check for poor insulation on secondary circuit (arcing or tracking of high voltage).
P0351	Ignition Coil "A" Primary / Secondary Circuit	The control unit (or separate ignition module if fitted) normally "switches on / off" the current to the ignition coil(s) on the coil earth circuit; this circuit combined with the coil power supply circuit are referred to as the "primary circuit". The control unit monitors the voltage, current and / or dwell time (coil charge time) in the primary circuit; if the control unit detects a measurement that is outside of expected values, this will be assessed as a coil circuit fault. Refer to vehicle specific information to indentify the specified coil.	Fault detection is on the primary circuit but secondary circuit faults can influence primary circuit performance and voltage. The fault code does not define the exact fault, but the control unit will have detected an ignition related problem that is likely to be caused by a coil primary or secondary fault; note that the spark plug forms part of the secondary circuit. Due to the different types of coils and coil switching systems, checks will be applicable to the system. Initial checks should include all primary circuit connections, correct power supply voltage (ignition on / cranking) and resistance check on coil primary and secondary windings (refer to vehicle specific information). Also check secondary circuit connections as applicable (e.g. coil leads where fitted, coil to plug connections etc). Check for poor insulation on secondary circuit (arcing or tracking of high voltage).
P0352	Ignition Coil "B" Primary / Secondary Circuit	The control unit (or separate ignition module if fitted) normally "switches on / off" the current to the ignition coil(s) on the coil earth circuit; this circuit combined with the coil power supply circuit are referred to as the "primary circuit". The control unit monitors the voltage, current and / or dwell time (coil charge time) in the primary circuit; if the control unit detects a measurement that is outside of expected values, this will be assessed as a coil circuit fault. Refer to vehicle specific information to indentify the specified coil.	Fault detection is on the primary circuit but secondary circuit faults can influence primary circuit performance and voltage. The fault code does not define the exact fault, but the control unit will have detected an ignition related problem that is likely to be caused by a coil primary or secondary fault; note that the spark plug forms part of the secondary circuit. Due to the different types of coils and coil switching systems, checks will be applicable to the system. Initial checks should include all primary circuit connections, correct power supply voltage (ignition on / cranking) and resistance check on coil primary and secondary windings (refer to vehicle specific information). Also check secondary circuit connections as applicable (e.g. coil leads where fitted, coil to plug connections etc). Check for poor insulation on secondary circuit (arcing or tracking of high voltage).
P0353	Ignition Coil "C" Primary / Secondary Circuit	The control unit (or separate ignition module if fitted) normally "switches on / off" the current to the ignition coil(s) on the coil earth circuit; this circuit combined with the coil power supply circuit are referred to as the "primary circuit". The control unit monitors the voltage, current and / or dwell time (coil charge time) in the primary circuit; if the control unit detects a measurement that is outside of expected values, this will be assessed as a coil circuit fault. Refer to vehicle specific information to indentify the specified coil.	Fault detection is on the primary circuit but secondary circuit faults can influence primary circuit performance and voltage. The fault code does not define the exact fault, but the control unit will have detected an ignition related problem that is likely to be caused by a coil primary or secondary fault; note that the spark plug forms part of the secondary circuit. Due to the different types of coils and coil switching systems, checks will be applicable to the system. Initial checks should include all primary circuit connections, correct power supply voltage (ignition on / cranking) and resistance check on coil primary and secondary windings (refer to vehicle specific information). Also check secondary circuit connections as applicable (e.g. coil leads where fitted, coil to plug connections etc). Check for poor insulation on secondary circuit (arcing or tracking of high voltage).

NOTE 1. Check for other fault codes that could provide additional information. NOTE 2. Communication between control units can pass via a CAN-Bus system; refer to Page 51 for CAN-Bus checks.

NOTE 3. If a fault cannot be located, it is also possible that the control unit is at fault. NOTE 4. Refer to Page 53 for list of pages that show the ISO standard for component locations e.g. Sensor A - Bank 1.

Fault code	EOBD / ISO Description	Component / System Description	Meaningful Description and Quick Check
P0354	Ignition Coil "D" Primary / Secondary Circuit	The control unit (or separate ignition module if fitted) normally "switches on / off" the current to the ignition coil(s) on the coil earth circuit; this circuit combined with the coil power supply circuit are referred to as the "primary circuit". The control unit monitors the voltage, current and / or dwell time (coil charge time) in the primary circuit; if the control unit detects a measurement that is outside of expected values, this will be assessed as a coil circuit fault. Refer to vehicle specific information to indentify the specified coil.	Fault detection is on the primary circuit but secondary circuit faults can influence primary circuit performance and voltage. The fault code does not define the exact fault, but the control unit will have detected an ignition related problem that is likely to be caused by a coil primary or secondary fault; note that the spark plug forms part of the secondary circuit. Due to the different types of coils and coil switching systems, checks will be applicable to the system. Initial checks should include all primary circuit connections, correct power supply voltage (ignition on / cranking) and resistance check on coil primary and secondary windings (refer to vehicle specific information). Also check secondary circuit connections as applicable (e.g. coil leads where fitted, coil to plug connections etc). Check for poor insulation on secondary circuit (arcing or tracking of high voltage).
P0355	Ignition Coil "E" Primary / Secondary Circuit	The control unit (or separate ignition module if fitted) normally "switches on / off" the current to the ignition coil(s) on the coil earth circuit; this circuit combined with the coil power supply circuit are referred to as the "primary circuit". The control unit monitors the voltage, current and / or dwell time (coil charge time) in the primary circuit; if the control unit detects a measurement that is outside of expected values, this will be assessed as a coil circuit fault. Refer to vehicle specific information to indentify the specified coil.	Fault detection is on the primary circuit but secondary circuit faults can influence primary circuit performance and voltage. The fault code does not define the exact fault, but the control unit will have detected an ignition related problem that is likely to be caused by a coil primary or secondary fault; note that the spark plug forms part of the secondary circuit. Due to the different types of coils and coil switching systems, checks will be applicable to the system. Initial checks should include all primary circuit connections, correct power supply voltage (ignition on / cranking) and resistance check on coil primary and secondary windings (refer to vehicle specific information). Also check secondary circuit connections as applicable (e.g. coil leads where fitted, coil to plug connections etc). Check for poor insulation on secondary circuit (arcing or tracking of high voltage).
P0356	Ignition Coil "F" Primary / Secondary Circuit	The control unit (or separate ignition module if fitted) normally "switches on / off" the current to the ignition coil(s) on the coil earth circuit; this circuit combined with the coil power supply circuit are referred to as the "primary circuit". The control unit monitors the voltage, current and / or dwell time (coil charge time) in the primary circuit; if the control unit detects a measurement that is outside of expected values, this will be assessed as a coil circuit fault. Refer to vehicle specific information to indentify the specified coil.	Fault detection is on the primary circuit but secondary circuit faults can influence primary circuit performance and voltage. The fault code does not define the exact fault, but the control unit will have detected an ignition related problem that is likely to be caused by a coil primary or secondary fault; note that the spark plug forms part of the secondary circuit. Due to the different types of coils and coil switching systems, checks will be applicable to the system. Initial checks should include all primary circuit connections, correct power supply voltage (ignition on / cranking) and resistance check on coil primary and secondary windings (refer to vehicle specific information). Also check secondary circuit connections as applicable (e.g. coil leads where fitted, coil to plug connections etc). Check for poor insulation on secondary circuit (arcing or tracking of high voltage).
P0357	Ignition Coil "G" Primary / Secondary Circuit	The control unit (or separate ignition module if fitted) normally "switches on / off" the current to the ignition coil(s) on the coil earth circuit; this circuit combined with the coil power supply circuit are referred to as the "primary circuit". The control unit monitors the voltage, current and / or dwell time (coil charge time) in the primary circuit; if the control unit detects a measurement that is outside of expected values, this will be assessed as a coil circuit fault. Refer to vehicle specific information to indentify the specified coil.	Fault detection is on the primary circuit but secondary circuit faults can influence primary circuit performance and voltage. The fault code does not define the exact fault, but the control unit will have detected an ignition related problem that is likely to be caused by a coil primary or secondary fault; note that the spark plug forms part of the secondary circuit. Due to the different types of coils and coil switching systems, checks will be applicable to the system. Initial checks should include all primary circuit connections, correct power supply voltage (ignition on / cranking) and resistance check on coil primary and secondary windings (refer to vehicle specific information). Also check secondary circuit connections as applicable (e.g. coil leads where fitted, coil to plug connections etc). Check for poor insulation on secondary circuit (arcing or tracking of high voltage).
P0358	Ignition Coil "H" Primary / Secondary Circuit	The control unit (or separate ignition module if fitted) normally "switches on / off" the current to the ignition coil(s) on the coil earth circuit; this circuit combined with the coil power supply circuit are referred to as the "primary circuit". The control unit monitors the voltage, current and / or dwell time (coil charge time) in the primary circuit; if the control unit detects a measurement that is outside of expected values, this will be assessed as a coil circuit fault. Refer to vehicle specific information to indentify the specified coil.	Fault detection is on the primary circuit but secondary circuit faults can influence primary circuit performance and voltage. The fault code does not define the exact fault, but the control unit will have detected an ignition related problem that is likely to be caused by a coil primary or secondary fault; note that the spark plug forms part of the secondary circuit. Due to the different types of coils and coil switching systems, checks will be applicable to the system. Initial checks should include all primary circuit connections, correct power supply voltage (ignition on / cranking) and resistance check on coil primary and secondary windings (refer to vehicle specific information). Also check secondary circuit connections as applicable (e.g. coil leads where fitted, coil to plug connections etc). Check for poor insulation on secondary circuit (arcing or tracking of high voltage).
P0359	Ignition Coil "I" Primary / Secondary Circuit	The control unit (or separate ignition module if fitted) normally "switches on / off" the current to the ignition coil(s) on the coil earth circuit; this circuit combined with the coil power supply circuit are referred to as the "primary circuit". The control unit monitors the voltage, current and / or dwell time (coil charge time) in the primary circuit; if the control unit detects a measurement that is outside of expected values, this will be assessed as a coil circuit fault. Refer to vehicle specific information to indentify the specified coil.	Fault detection is on the primary circuit but secondary circuit faults can influence primary circuit performance and voltage. The fault code does not define the exact fault, but the control unit will have detected an ignition related problem that is likely to be caused by a coil primary or secondary fault; note that the spark plug forms part of the secondary circuit. Due to the different types of coils and coil switching systems, checks will be applicable to the system. Initial checks should include all primary circuit connections, correct power supply voltage (ignition on / cranking) and resistance check on coil primary and secondary windings (refer to vehicle specific information). Also check secondary circuit connections as applicable (e.g. coil leads where fitted, coil to plug connections etc). Check for poor insulation on secondary circuit (arcing or tracking of high voltage).
P0360	Ignition Coil "J" Primary / Secondary Circuit	The control unit (or separate ignition module if fitted) normally "switches on / off" the current to the ignition coil(s) on the coil earth circuit; this circuit combined with the coil power supply circuit are referred to as the "primary circuit". The control unit monitors the voltage, current and / or dwell time (coil charge time) in the primary circuit; if the control unit detects a measurement that is outside of expected values, this will be assessed as a coil circuit fault. Refer to vehicle specific information to indentify the specified coil.	Fault detection is on the primary circuit but secondary circuit faults can influence primary circuit performance and voltage. The fault code does not define the exact fault, but the control unit will have detected an ignition related problem that is likely to be caused by a coil primary or secondary fault; note that the spark plug forms part of the secondary circuit. Due to the different types of coils and coil switching systems, checks will be applicable to the system. Initial checks should include all primary circuit connections, correct power supply voltage (ignition on / cranking) and resistance check on coil primary and secondary windings (refer to vehicle specific information). Also check secondary circuit connections as applicable (e.g. coil leads where fitted, coil to plug connections etc). Check for poor insulation on secondary circuit (arcing or tracking of high voltage).
P0361	Ignition Coil "K" Primary / Secondary Circuit	The control unit (or separate ignition module if fitted) normally "switches on / off" the current to the ignition coil(s) on the coil earth circuit; this circuit combined with the coil power supply circuit are referred to as the "primary circuit". The control unit monitors the voltage, current and / or dwell time (coil charge time) in the primary circuit; if the control unit detects a measurement that is outside of expected values, this will be assessed as a coil circuit fault. Refer to vehicle specific information to indentify the specified coil.	Fault detection is on the primary circuit but secondary circuit faults can influence primary circuit performance and voltage. The fault code does not define the exact fault, but the control unit will have detected an ignition related problem that is likely to be caused by a coil primary or secondary fault; note that the spark plug forms part of the secondary circuit. Due to the different types of coils and coil switching systems, checks will be applicable to the system. Initial checks should include all primary circuit connections, correct power supply voltage (ignition on / cranking) and resistance check on coil primary and secondary windings (refer to vehicle specific information). Also check secondary circuit connections as applicable (e.g. coil leads where fitted, coil to plug connections etc). Check for poor insulation on secondary circuit (arcing or tracking of high voltage).

NOTE 1. Check for other fault codes that could provide additional information. **NOTE 2.** Communication between control units can pass via a CAN-Bus system; refer to Page 51 for CAN-Bus checks.
NOTE 3. If a fault cannot be located, it is also possible that the control unit is at fault. **NOTE 4.** Refer to Page 53 for list of pages that show the ISO standard for component locations e.g. Sensor A - Bank 1.
116

Code	Description		
P0362	Ignition Coil "L" Primary / Secondary Circuit	The control unit (or separate ignition module if fitted) normally "switches on / off" the current to the ignition coil(s) on the coil earth circuit; this circuit combined with the coil power supply circuit are referred to as the "primary circuit". The control unit monitors the voltage, current and / or dwell time (coil charge time) in the primary circuit; if the control unit detects a measurement that is outside of expected values, this will be assessed as a coil circuit fault. Refer to vehicle specific information to indentify the specified coil.	Fault detection is on the primary circuit but secondary circuit faults can influence primary circuit performance and voltage. The fault code does not define the exact fault, but the control unit will have detected an ignition related problem that is likely to be caused by a coil primary or secondary fault; note that the spark plug forms part of the secondary circuit. Due to the different types of coils and coil switching systems, checks will be applicable to the system. Initial checks should include all primary circuit connections, correct power supply voltage (ignition on / cranking) and resistance check on coil primary and secondary windings (refer to vehicle specific information). Also check secondary circuit connections as applicable (e.g. coil leads where fitted, coil to plug connections etc). Check for poor insulation on secondary circuit (arcing or tracking of high voltage).
P0363	Misfire Detected - Fuelling Disabled	A common method of detecting misfires is to monitor the signal from engine speed sensor to check crankshaft acceleration / speed during each power stroke; this provides an indication of the power / torque contribution provided by the cylinder (compared to the other cylinders). Also refer to Fault code P0300 and P0312 for additional information relating to misfire detection.	The fault code indicates that there is a misfire one or multiple cylinders. The misfire is regarded as sufficiently bad to justify switching off the fuel; this could be due to the potential to damage components e.g. catalytic converter or, it could cause excessive emissions. It is likely that the fault is causing erratic fluctuations of the crankshaft speed (or exhaust pressure pulses), which are not consistent to any one cylinder or group of cylinders. The fault could be related to any aspect of the engine operation including: fuelling, ignition (petrol engines), air leaks, mechanical condition etc. Also check (as applicable) for faults with variable valve timing, EGR and other emission control systems. It is also possible that there could be an engine position / speed sensor related fault (if necessary, Refer to list of sensor / actuator checks on page 19).
P0364		ISO / SAE reserved	
P0365	Camshaft Position Sensor "B" Circuit Bank 1	The sensor is used to detect camshaft position (angular position relative to crankshaft position); sensor information can be used to enable accurate control of ignition timing, fuel injection timing and variable valve timing. Identify the type of sensor e.g. Inductive, Hall effect etc; this will dictate the checks that are applicable.	Identify the type of sensor e.g. Inductive, Hall effect etc; this will dictate the checks that are applicable. The voltage, frequency or value of the position / speed sensor signal is incorrect (undefined fault). Sensor voltage / signal value could be: out of normal operating range, constant value, non-existent, corrupt. Possible fault in the position / speed sensor or wiring e.g. short / open circuits, circuit resistance, incorrect sensor resistance (if applicable) or, fault with the reference / operating voltage; interference from other circuits can also affect sensor signals. Also check earth connections and cable screening. Refer to list of sensor / actuator checks on page 19.
P0366	Camshaft Position Sensor "B" Circuit Range / Performance Bank 1	The sensor is used to detect camshaft position (angular position relative to crankshaft position); sensor information can be used to enable accurate control of ignition timing, fuel injection timing and variable valve timing. Identify the type of sensor e.g. Inductive, Hall effect etc; this will dictate the checks that are applicable.	Identify the type of sensor e.g. Inductive, Hall effect etc; this will dictate the checks that are applicable. The voltage, frequency or value of the position / speed sensor signal is within the normal operating range / tolerance but the signal is not plausible or is incorrect due to an undefined fault. The sensor signal might not match the operating conditions indicated by other sensors or, the sensor signal is not changing / responding as expected to the changes in conditions. Possible fault in the speed sensor or wiring e.g. short / open circuits, circuit resistance, incorrect sensor resistance or, fault with the reference / operating voltage; interference from other circuits can also affect sensor signals. Refer to list of sensor / actuator checks on page 19. Check camshaft drive belt / chain for wear or slack adjustment (can cause erratic camshaft rotation). It is also possible that the sensor is operating correctly but the position / speed being measured by the sensor is unacceptable or incorrect for the operating conditions. Also check condition of reference teeth on trigger disc / rotor.
P0367	Camshaft Position Sensor "B" Circuit Low Bank 1	The sensor is used to detect camshaft position (angular position relative to crankshaft position); sensor information can be used to enable accurate control of ignition timing, fuel injection timing and variable valve timing. Identify the type of sensor e.g. Inductive, Hall effect etc; this will dictate the checks that are applicable.	Identify the type of sensor e.g. Inductive, Hall effect etc; this will dictate the checks that are applicable. The voltage, frequency or value of the position / speed sensor signal is either "zero" (no signal) or, the signal exists but it is below the normal operating range (lower than the minimum operating limit). Possible fault in the position / speed sensor or wiring e.g. short / open circuits, circuit resistance, incorrect sensor resistance or, fault with the reference / operating voltage; interference from other circuits can also affect sensor signals. Refer to list of sensor / actuator checks on page 19.
P0368	Camshaft Position Sensor "B" Circuit High Bank 1	The sensor is used to detect camshaft position (angular position relative to crankshaft position); sensor information can be used to enable accurate control of ignition timing, fuel injection timing and variable valve timing. Identify the type of sensor e.g. Inductive, Hall effect etc; this will dictate the checks that are applicable.	Identify the type of sensor e.g. Inductive, Hall effect etc; this will dictate the checks that are applicable. The voltage, frequency or value of the position / speed sensor signal is either at full value (e.g. battery voltage with no signal or frequency) or, the signal exists but it is above the normal operating range (higher than the maximum operating limit). Possible fault in the position / speed sensor or wiring e.g. short / open circuits, circuit resistance, incorrect sensor resistance or, fault with the reference / operating voltage; interference from other circuits can also affect sensor signals. Refer to list of sensor / actuator checks on page 19.
P0369	Camshaft Position Sensor "B" Circuit Intermittent Bank 1	The sensor is used to detect camshaft position (angular position relative to crankshaft position); sensor information can be used to enable accurate control of ignition timing, fuel injection timing and variable valve timing. Identify the type of sensor e.g. Inductive, Hall effect etc; this will dictate the checks that are applicable.	Identify the type of sensor e.g. Inductive, Hall effect etc; this will dictate the checks that are applicable. The voltage, frequency or value of the position / speed sensor signal is intermittent (likely to be out of normal operating range / tolerance when intermittent fault is detected) or, the signal / voltage is erratic (e.g. signal changes are irregular / unpredictable). Possible fault with wiring or connections (including internal sensor connections); also interference from other circuits can affect the signal. Refer to list of sensor / actuator checks on page 19. Check camshaft drive belt / chain for wear or slack adjustment (can cause erratic camshaft rotation). Also check condition of reference teeth on trigger disc / rotor and ensure rotor is turning with shaft.
P0370	Timing Reference High Resolution Signal "A"	A "timing reference" is required to enable accurate Diesel injection timing (or ignition on petrol engines). For Distributor type Diesel pumps, a timing reference sensor indicates the position of the pump timing control device, which is then compared to the TDC sensor signal (crank or cam sensor signal); the control unit then identifies if the delivery / injection timing is correct. Some systems may use a pressure sensor to detect pressure in the pumping element chamber (provides a delivery timing reference). Other Diesel systems can use an injector needle lift sensor to detect injection timing. If the fault code is used on a petrol engine, this could refer to a crankshaft or camshaft sensor.	Identify the type of timing reference sensor(s) fitted to the system and Refer to list of sensor / actuator checks on page 19. The fault code indicates an undefined signal fault with the sensor or sensor signal. The fault could be related to a sensor / wiring problem but it could also be caused by a fault in the Diesel pump timing control device.

NOTE 1. Check for other fault codes that could provide additional information.
NOTE 2. Communication between control units can pass via a CAN-Bus system; refer to Page 51 for CAN-Bus checks.
NOTE 3. If a fault cannot be located, it is also possible that the control unit is at fault.
NOTE 4. Refer to Page 53 for list of pages that show the ISO standard for component locations e.g. Sensor A - Bank 1.

Fault code	EOBD / ISO Description	Component / System Description	Meaningful Description and Quick Check
P0371	Timing Reference High Resolution Signal "A" Too Many Pulses	A "timing reference" is required to enable accurate Diesel injection timing (or ignition on petrol engines). For Distributor type Diesel pumps, a timing reference sensor indicates the position of the pump timing control device, which is then compared to the TDC sensor signal (crank or cam sensor signal); the control unit then identifies if the delivery / injection timing is correct. Some systems may use a pressure sensor to detect pressure in the pumping element chamber (provides a delivery timing reference). Other Diesel systems can use an injector needle lift sensor to detect injection timing. If the fault code is used on a petrol engine, this could refer to a crankshaft or camshaft sensor.	Identify the type of timing reference sensor(s) fitted to the system and Refer to list of sensor / actuator checks on page 19. The fault code indicates that the sensor signal contains too many reference pulses. The fault could be related to a sensor / problem but it could also be caused by a fault with the trigger wheel / disc (if applicable) or interference for other electrical circuits.
P0372	Timing Reference High Resolution Signal "A" Too Few Pulses	A "timing reference" is required to enable accurate Diesel injection timing (or ignition on petrol engines). For Distributor type Diesel pumps, a timing reference sensor indicates the position of the pump timing control device, which is then compared to the TDC sensor signal (crank or cam sensor signal); the control unit then identifies if the delivery / injection timing is correct. Some systems may use a pressure sensor to detect pressure in the pumping element chamber (provides a delivery timing reference). Other Diesel systems can use an injector needle lift sensor to detect injection timing. If the fault code is used on a petrol engine, this could refer to a crankshaft or camshaft sensor.	Identify the type of timing reference sensor(s) fitted to the system and Refer to list of sensor / actuator checks on page 19. The fault code indicates that the sensor signal contains too few reference pulses. The fault could be related to a sensor / problem but it could also be caused by a fault with the trigger wheel / disc (if applicable) or interference for other electrical circuits.
P0373	Timing Reference High Resolution Signal "A" Intermittent / Erratic Pulses	A "timing reference" is required to enable accurate Diesel injection timing (or ignition on petrol engines). For Distributor type Diesel pumps, a timing reference sensor indicates the position of the pump timing control device, which is then compared to the TDC sensor signal (crank or cam sensor signal); the control unit then identifies if the delivery / injection timing is correct. Some systems may use a pressure sensor to detect pressure in the pumping element chamber (provides a delivery timing reference). Other Diesel systems can use an injector needle lift sensor to detect injection timing. If the fault code is used on a petrol engine, this could refer to a crankshaft or camshaft sensor.	Identify the type of timing reference sensor(s) fitted to the system and Refer to list of sensor / actuator checks on page 19. The fault code indicates that the signal value from sensor is intermittent or occasionally incorrect (likely to be out of normal operating range / tolerance when intermittent fault is detected); in some cases, the signal could be erratic e.g. the change in signal value is too rapid or too slow compared with the normal / expected change. Likely faults are poor connections / broken wiring but it is possible that the sensor has an internal connection fault. It is also possible that the sensor is operating correctly but another fault is causing the measurement to change erratically.
P0374	Timing Reference High Resolution Signal "A" No Pulse	A "timing reference" is required to enable accurate Diesel injection timing (or ignition on petrol engines). For Distributor type Diesel pumps, a timing reference sensor indicates the position of the pump timing control device, which is then compared to the TDC sensor signal (crank or cam sensor signal); the control unit then identifies if the delivery / injection timing is correct. Some system may use a pressure sensor to detect pressure in the pumping element chamber (provides a delivery timing reference). Other Diesel systems can use an injector needle lift sensor to detect injection timing. If the fault code is used on a petrol engine, this could refer to a crankshaft or camshaft sensor.	Identify the type of timing reference sensor(s) fitted to the system and Refer to list of sensor / actuator checks on page 19. The fault code indicates that there are no signal pulses being detected by the control unit. The fault is likely to be caused by a wiring or sensor failure but it could also be caused by a fault with the trigger wheel / disc (if applicable) or interference for other electrical circuits.
P0375	Timing Reference High Resolution Signal "B"	A "timing reference" is required to enable accurate Diesel injection timing (or ignition on petrol engines). For Distributor type Diesel pumps, a timing reference sensor indicates the position of the pump timing control device, which is then compared to the TDC sensor signal (crank or cam sensor signal); the control unit then identifies if the delivery / injection timing is correct. Some systems may use a pressure sensor to detect pressure in the pumping element chamber (provides a delivery timing reference). Other Diesel systems can use an injector needle lift sensor to detect injection timing. If the fault code is used on a petrol engine, this could refer to a crankshaft or camshaft sensor.	Identify the type of timing reference sensor(s) fitted to the system and Refer to list of sensor / actuator checks on page 19. The fault code indicates an undefined signal fault with the sensor or sensor signal. The fault could be related to a sensor / wiring problem but it could also be caused by a fault in the Diesel pump timing control device.
P0376	Timing Reference High Resolution Signal "B" Too Many Pulses	A "timing reference" is required to enable accurate Diesel injection timing (or ignition on petrol engines). For Distributor type Diesel pumps, a timing reference sensor indicates the position of the pump timing control device, which is then compared to the TDC sensor signal (crank or cam sensor signal); the control unit then identifies if the delivery / injection timing is correct. Some systems may use a pressure sensor to detect pressure in the pumping element chamber (provides a delivery timing reference). Other Diesel systems can use an injector needle lift sensor to detect injection timing. If the fault code is used on a petrol engine, this could refer to a crankshaft or camshaft sensor.	Identify the type of timing reference sensor(s) fitted to the system and Refer to list of sensor / actuator checks on page 19. The fault code indicates that the sensor signal contains too many reference pulses. The fault could be related to a sensor / problem but it could also be caused by a fault with the trigger wheel / disc (if applicable) or interference for other electrical circuits.

NOTE 1. Check for other fault codes that could provide additional information.

NOTE 2. Communication between control units can pass via a CAN-Bus system; refer to Page 51 for CAN-Bus checks.

NOTE 4. Refer to Page 53 for list of pages that show the ISO standard for component locations e.g. Sensor A - Bank 1.

Code	Name	Description	Diagnosis
P0377	Timing Reference High Resolution Signal "B" Too Few Pulses	A "timing reference" is required to enable accurate Diesel injection timing (or ignition on petrol engines). For Distributor type Diesel pumps, a timing reference sensor indicates the position of the pump timing control device, which is then compared to the TDC sensor signal (crank or cam sensor signal); the control unit then identifies if the delivery / injection timing is correct. Some systems may use a pressure sensor to detect pressure in the pumping element chamber (provides a delivery timing reference). Other Diesel systems can use an injector needle lift sensor to detect injection timing. If the fault code is used on a petrol engine, this could refer to a crankshaft or camshaft sensor.	Identify the type of timing reference sensor(s) fitted to the system and Refer to list of sensor / actuator checks on page 19. The fault code indicates that the sensor signal contains too few reference pulses. The fault could be related to a sensor / problem but it could also be caused by a fault with the trigger wheel / disc (if applicable) or interference for other electrical circuits.
P0378	Timing Reference High Resolution Signal "B" Intermittent / Erratic Pulses	A "timing reference" is required to enable accurate Diesel injection timing (or ignition on petrol engines). For Distributor type Diesel pumps, a timing reference sensor indicates the position of the pump timing control device, which is then compared to the TDC sensor signal (crank or cam sensor signal); the control unit then identifies if the delivery / injection timing is correct. Some systems may use a pressure sensor to detect pressure in the pumping element chamber (provides a delivery timing reference). Other Diesel systems can use an injector needle lift sensor to detect injection timing. If the fault code is used on a petrol engine, this could refer to a crankshaft or camshaft sensor.	Identify the type of timing reference sensor(s) fitted to the system and Refer to list of sensor / actuator checks on page 19. The fault code indicates that the signal value from sensor is intermittent or occasionally incorrect (likely to be out of normal operating range / tolerance when intermittent fault is detected); in some cases, the signal could be erratic e.g. the change in signal value is too rapid or too slow compared with the normal / expected change. Likely faults are poor connections / broken wiring but it is possible that the sensor has an internal connection fault. It is also possible that the sensor is operating correctly but another fault is causing the measurement to change erratically.
P0379	Timing Reference High Resolution Signal "B" No Pulses	A "timing reference" is required to enable accurate Diesel injection timing (or ignition on petrol engines). For Distributor type Diesel pumps, a timing reference sensor indicates the position of the pump timing control device, which is then compared to the TDC sensor signal (crank or cam sensor signal); the control unit then identifies if the delivery / injection timing is correct. Some system may use a pressure sensor to detect pressure in the pumping element chamber (provides a delivery timing reference). Other Diesel systems can use an injector needle lift sensor to detect injection timing. If the fault code is used on a petrol engine, this could refer to a crankshaft or camshaft sensor.	Identify the type of timing reference sensor(s) fitted to the system and Refer to list of sensor / actuator checks on page 19. The fault code indicates that there are no signal pulses being detected by the control unit. The fault is likely to be caused by a wiring or sensor failure but it could also be caused by a fault with the trigger wheel / disc (if applicable) or interference for other electrical circuits.
P0380	Glow Plug / Heater Circuit "A"	It will be necessary to identify the type of glow plug control system. The glow plugs are usually switched / controlled by a "Glow Plug Control Module" (which can be integrated into the main Diesel control unit or, can be a separate unit which receives a control signal from the Diesel control unit). Depending on system design, the voltage applied to the glow plugs can be full battery voltage or, on many systems the circuit voltage / current is regulated by using a digital control signal (voltage / current is dependent on signal pulse width / duty cycle). Note that different types of glow plug can be specific to different types of glow plug control systems, and also note that the voltage / current level in the circuit can alter during the period of operation (including pre and post start conditions).	Refer to vehicle specific information for information on the glow plug control system and design of glow plug; the exact interpretation of the fault code will depend on the system design, but note that the system could be controlling glow plugs individually or in groups. It is likely that the fault code relates to the power supply circuit to the glow plugs (from the applicable control module). The fault code indicates that there is an undefined fault in the glow plug circuit. It is likely that the fault is detected due to incorrect voltage or signal value in the circuit during glow plug operation (either on one or a group of glow plugs). Check the circuit between each glow plug and the control module for short / open circuit and high resistance. Check the resistance of the glow plugs (refer to vehicle specific information especially for ceramic element glow plugs). Refer to list of sensor / actuator checks on page 19.
P0381	Glow Plug / Heater Indicator Circuit	The glow plugs are usually switched on / off by a "Glow Plug Control Module", which can be controlled by the main Diesel Engine control unit. The control module is usually a separate item but it is possible that it could form part of the main control unit. The Glow Plug Module will receive a control signal from the Diesel control unit, and the module will switch on / off the power supply to the glow plugs. Note that glow plug heater indicator light could be switched on by the main diesel control unit or by the Glow plug control module.	Refer to vehicle specific information to identify the wiring circuit for the indicator light; it is likely that the light will receive a power supply from a common power supply within the instrument system and, the earth path for the light will be completed by the applicable control unit. The voltage or value in the indicator circuit is incorrect (undefined fault). The fault is likely to be detected by the control unit, which has detected that there is no voltage available in the circuit (or there is no continuity in the circuit) either due to a power supply fault or there is a short / open circuit in the wiring. Check wiring and connections between the indicator light and the control unit. Also check the condition of the light.
P0382	Glow Plug / Heater Circuit "B"	The glow plugs are usually switched on / off by a "Glow Plug Control Module", which can be controlled by the main Diesel Engine control unit. The control module can be a separate item but it can form part of the main control unit. The Glow Plug Module will receive a control signal from the Diesel control unit, and the module will switch on / off the power supply to the glow plugs (the module can contain an internal relay / power stage, although it is possible to use an external relay). Note that some systems will switch groups of glow plugs e.g. 2, 3 or 4 glow plugs together.	Refer to vehicle specific information for information on the glow plug control system and design of glow plug; the exact interpretation of the fault code will depend on the system design, but note that the system could be controlling glow plugs individually or in groups. It is likely that the fault code relates to the power supply circuit to the glow plugs (from the applicable control module). The fault code indicates that there is an undefined fault in the glow plug circuit. It is likely that the fault is detected due to incorrect voltage or signal value in the circuit during glow plug operation (either on one or a group of glow plugs). Check the circuit between each glow plug and the control module for short / open circuit and high resistance. Check the resistance of the glow plugs (refer to vehicle specific information especially for ceramic element glow plugs). Refer to list of sensor / actuator checks on page 19.

NOTE 1. Check for other fault codes that could provide additional information.
NOTE 3. If a fault cannot be located, it is also possible that the control unit is at fault.
NOTE 2. Communication between control units can pass via a CAN-Bus system; refer to Page 51 for CAN-Bus checks.
NOTE 4. Refer to Page 53 for list of pages that show the ISO standard for component locations e.g. Sensor A - Bank 1.

Fault code	EOBD / ISO Description	Component / System Description	Meaningful Description and Quick Check
P0383	Glow Plug Control Module Control Circuit Low	The glow plugs are usually switched on / off by a "Glow Plug Control Module", which can be controlled by the main Diesel Engine control unit. The control module can be a separate item but it can form part of the main control unit. The Glow Plug Module will receive a control signal from the Diesel control unit, and the module will switch on / off the power supply to the glow plugs (the module can contain an internal relay / power stage, although it is possible to use an external relay). Note that some systems will switch groups of glow plugs e.g. 2, 3 or 4 glow plugs together.	Refer to vehicle specific information for the glow plug control system and design of glow plug. The exact interpretation of the fault code will depend on the system monitoring / diagnostic process and system design; it is however likely that the fault code relates to circuit between the main Diesel control unit and the glow plug control module (the circuit carries the signal to the glow plug control module to instruct the module to switch on the glow plugs). The voltage, frequency or value of the control signal in the control circuit is either "zero" (no signal / voltage) or, the signal / voltage exists but it is below the normal operating range. Note: The Diesel control unit could be providing the correct signal but it is being affected by a circuit fault. Check the circuit between the Diesel control unit and the glow plug control module for short / open circuit and high resistance. Refer to list of sensor / actuator checks on page 19. Note that it is possible that a manufacturer could apply the code to the circuit between the glow plug control module and the glow plugs; if necessary refer to Fault Code P0380 for additional information.
P0384	Glow Plug Control Module Control Circuit High	The glow plugs are usually switched on / off by a "Glow Plug Control Module", which can be controlled by the main Diesel Engine control unit. The control module can be a separate item but it can form part of the main control unit. The Glow Plug Module will receive a control signal from the Diesel control unit, and the module will switch on / off the power supply to the glow plugs (the module can contain an internal relay / power stage, although it is possible to use an external relay). Note that some systems will switch groups of glow plugs e.g. 2, 3 or 4 glow plugs together.	Refer to vehicle specific information for the glow plug control system and design of glow plug. The exact interpretation of the fault code will depend on the system monitoring / diagnostic process and system design; it is however likely that the fault code relates to circuit between the main Diesel control unit and the glow plug control module (the circuit carries the signal to the glow plug control module to switch on the glow plugs). The voltage, frequency or value of the control signal in the actuator circuit is either at full value (e.g. battery voltage with no on / off signal) or, the signal exists but it is above the normal operating range. Note: The Diesel control unit could be providing the correct signal but it is being affected by a circuit fault. Check the circuit between the Diesel control unit and the glow plug control module for short / open circuit and high resistance. Refer to list of sensor / actuator checks on page 19. Note that it is possible that a manufacturer could apply the code to the circuit between the glow plug control module and the glow plugs; if necessary refer to Fault Code P0380 for additional information.
P0385	Crankshaft Position Sensor "B" Circuit	The crankshaft position sensor provides position (and usually speed information) to the control unit. The information is used in calculations for fuelling, ignition timing, variable valve timing and emissions control functions; the information is also used for other vehicle systems such as transmission, stability control systems etc.	Identify the type of sensor e.g. Inductive, Hall effect etc; this will dictate the checks that are applicable. The voltage, frequency or value of the position / speed sensor signal is incorrect (undefined fault). Sensor voltage / signal value could be: out of normal operating range, constant value, non-existent, corrupt. Possible fault in the position / speed sensor or wiring e.g. short / open circuits, circuit resistance, incorrect sensor resistance (if applicable) or, fault with the reference / operating voltage; interference from other circuits can also affect sensor signals. Also check earth connections and cable screening. Refer to list of sensor / actuator checks on page 19.
P0386	Crankshaft Position Sensor "B" Circuit Range / Performance	The crankshaft position sensor provides position (and usually speed information) to the control unit. The information is used in calculations for fuelling, ignition timing, variable valve timing and emissions control functions; the information is also used for other vehicle systems such as transmission, stability control systems etc.	Identify the type of sensor e.g. Inductive, Hall effect etc; this will dictate the checks that are applicable. The voltage, frequency or value of the speed sensor signal is within the normal operating range / tolerance but the signal is not plausible or is incorrect due to an undefined fault. The sensor signal might not match the operating conditions indicated by other sensors or, the sensor signal is not changing / responding as expected to the changes in conditions. Possible fault in the speed sensor or wiring e.g. short / open circuits, circuit resistance, incorrect sensor resistance or, fault with the reference / operating voltage; interference from other circuits can also affect sensor signals. Refer to list of sensor / actuator checks on page 19. It is also possible that the sensor is operating correctly but the position / speed being measured by the sensor is unacceptable or incorrect for the operating conditions. Also check condition of reference teeth on trigger disc / rotor and ensure rotor is turning with shaft.
P0387	Crankshaft Position Sensor "B" Circuit Low	The crankshaft position sensor provides position (and usually speed information) to the control unit. The information is used in calculations for fuelling, ignition timing, variable valve timing and emissions control functions; the information is also used for other vehicle systems such as transmission, stability control systems etc.	Identify the type of sensor e.g. Inductive, Hall effect etc; this will dictate the checks that are applicable. The voltage, frequency or value of the position / speed sensor signal is either "zero" (no signal) or, the signal exists but it is below the normal operating range (lower than the minimum operating limit). Possible fault in the position / speed sensor or wiring e.g. short / open circuits, circuit resistance, incorrect sensor resistance or, fault with the reference / operating voltage; interference from other circuits can also affect sensor signals. Refer to list of sensor / actuator checks on page 19.
P0388	Crankshaft Position Sensor "B" Circuit High	The crankshaft position sensor provides position (and usually speed information) to the control unit. The information is used in calculations for fuelling, ignition timing, variable valve timing and emissions control functions; the information is also used for other vehicle systems such as transmission, stability control systems etc.	Identify the type of sensor e.g. Inductive, Hall effect etc; this will dictate the checks that are applicable. The voltage, frequency or value of the position / speed sensor signal is either at full value (e.g. battery voltage with no signal or frequency) or, the signal exists but it is above the normal operating range (higher than the maximum operating limit). Possible fault in the position / speed sensor or wiring e.g. short / open circuits, circuit resistance, incorrect sensor resistance or, fault with the reference / operating voltage; interference from other circuits can also affect sensor signals. Refer to list of sensor / actuator checks on page 19.
P0389	Crankshaft Position Sensor "B" Circuit Intermittent	The crankshaft position sensor provides position (and usually speed information) to the control unit. The information is used in calculations for fuelling, ignition timing, variable valve timing and emissions control functions; the information is also used for other vehicle systems such as transmission, stability control systems etc.	Identify the type of sensor e.g. Inductive, Hall effect etc; this will dictate the checks that are applicable. The voltage, frequency or value of the position / speed sensor signal is intermittent (likely to be out of normal operating range / tolerance when intermittent fault is detected) or, the signal / voltage is erratic (e.g. signal changes are irregular / unpredictable). Possible fault with wiring or connections (including internal sensor connections); also interference from other circuits can affect the signal. Refer to list of sensor / actuator checks on page 19. Also check condition of reference teeth on trigger disc / rotor and ensure rotor is turning with shaft.
P0390	Camshaft Position Sensor "B" Circuit Bank 2	The sensor is used to detect camshaft position (angular position relative to crankshaft position); sensor information can be used to enable accurate control of ignition timing, fuel injection timing and variable valve timing.	Identify the type of sensor e.g. Inductive, Hall effect etc; this will dictate the checks that are applicable. The voltage, frequency or value of the position / speed sensor signal is incorrect (undefined fault). Sensor voltage / signal value could be: out of normal operating range, constant value, non-existent, corrupt. Possible fault in the position / speed sensor or wiring e.g. short / open circuits, circuit resistance, incorrect sensor resistance (if applicable) or, fault with the reference / operating voltage; interference from other circuits can also affect sensor signals. Also check earth connections and cable screening. Refer to list of sensor / actuator checks on page 19.

NOTE 1. Check for other fault codes that could provide additional information. NOTE 2. Communication between control units can pass via a CAN-Bus system; refer to Page 51 for CAN-Bus checks.

NOTE 3. If a fault cannot be located, it is also possible that the control unit is at fault NOTE 4. Refer to Page 53 for list of pages that show the ISO standard for component locations e.g. Sensor A - Bank 1.

Fault code	Description		
P0391	Camshaft Position Sensor "B" Circuit Range / Performance Bank 2	The sensor is used to detect camshaft position (angular position relative to crankshaft position); sensor information can be used to enable accurate control of ignition timing, fuel injection timing and variable valve timing.	Identify the type of sensor e.g. Inductive, Hall effect etc; this will dictate the checks that are applicable. The voltage, frequency or value of the position / speed sensor signal is within the normal operating range / tolerance but the signal is not plausible or is incorrect due to an undefined fault. The sensor signal might not match the operating conditions indicated by other sensors or, the sensor signal is not changing / responding as expected to the changes in conditions. Possible fault in the speed sensor or wiring e.g. short / open circuits, circuit resistance, incorrect sensor resistance or, fault with the reference / operating voltage; interference from other circuits can also affect sensor signals. Refer to list of sensor / actuator checks on page 19. Check camshaft drive belt / chain for wear or slack adjustment (can cause erratic camshaft rotation). It is also possible that the sensor is operating correctly but the position / speed being measured by the sensor is unacceptable or incorrect for the operating conditions. Also check condition of reference teeth on trigger disc / rotor.
P0392	Camshaft Position Sensor "B" Circuit Low Bank 2	The sensor is used to detect camshaft position (angular position relative to crankshaft position); sensor information can be used to enable accurate control of ignition timing, fuel injection timing and variable valve timing.	Identify the type of sensor e.g. Inductive, Hall effect etc; this will dictate the checks that are applicable. The voltage, frequency or value of the position / speed sensor signal is either "zero" (no signal) or, the signal exists but it is below the normal operating range (lower than the minimum operating limit). Possible fault in the position / speed sensor or wiring e.g. short / open circuits, circuit resistance, incorrect sensor resistance or, fault with the reference / operating voltage; interference from other circuits can also affect sensor signals. Refer to list of sensor / actuator checks on page 19.
P0393	Camshaft Position Sensor "B" Circuit High Bank 2	The sensor is used to detect camshaft position (angular position relative to crankshaft position); sensor information can be used to enable accurate control of ignition timing, fuel injection timing and variable valve timing.	Identify the type of sensor e.g. Inductive, Hall effect etc; this will dictate the checks that are applicable. The voltage, frequency or value of the position / speed sensor signal is either at full value (e.g. battery voltage with no signal or frequency) or, the signal exists but it is above the normal operating range (higher than the maximum operating limit). Possible fault in the position / speed sensor or wiring e.g. short / open circuits, circuit resistance, incorrect sensor resistance, fault with the reference / operating voltage; interference from other circuits can also affect sensor signals. Refer to list of sensor / actuator checks on page 19.
P0394	Camshaft Position Sensor "B" Circuit Intermittent Bank 2	The sensor is used to detect camshaft position (angular position relative to crankshaft position); sensor information can be used to enable accurate control of ignition timing, fuel injection timing and variable valve timing.	Identify the type of sensor e.g. Inductive, Hall effect etc; this will dictate the checks that are applicable. The voltage, frequency or value of the position / speed sensor signal is intermittent (likely to be out of normal operating range / tolerance when intermittent fault is detected) or, the signal / voltage is erratic (e.g. signal changes are irregular / unpredictable). Possible fault with wiring or connections (including internal sensor connections); also interference from other circuits can affect the signal. Refer to list of sensor / actuator checks on page 19. Check camshaft drive belt / chain for wear or slack adjustment (can cause erratic camshaft rotation). Also check condition of reference teeth on trigger disc / rotor and ensure rotor is turning with shaft.
P0395		ISO / SAE reserved	Not used, reserved for future allocation.
P0396		ISO / SAE reserved	Not used, reserved for future allocation.
P0397		ISO / SAE reserved	Not used, reserved for future allocation.
P0398		ISO / SAE reserved	Not used, reserved for future allocation.
P0399		ISO / SAE reserved	Not used, reserved for future allocation.
P0400	Exhaust Gas Recirculation Flow	The measurement of gas flow within an Exhaust Gas Recirculation (EGR) system is often a calculated value; the calculation will be based on information provided by other sensors (primarily intake airflow sensor but, throttle position sensor, temperature sensor can also form part of the calculation). It is also possible to make use of an EGR valve lift / movement sensor to enable the control unit to assess the movement and position of the EGR valve. Also refer to additional information attached to fault code P0404.	The fault code indicates that the EGR gas flow is calculated or measured as being incorrect or not as expected by the control unit. The checks required will be depend on the system design; refer to vehicle specific information to identify the type of system. Checks should however be made on the operation of the EGR valve; check that the valve opens / closes correctly (use vacuum pump to test operation if applicable), check for contamination of valve and ports. Check EGR valve wiring and check vacuum hoses for leaks and blockages; also check vacuum level available from the vacuum source (e.g. vacuum reservoir / pump). Also check for leaks or blockages on the pipes carrying the exhaust gases. Check operation of other applicable sensors; other related fault codes (if available) may provide a more definitive fault description or might indicate a fault in another related component. Note: With calculated flow systems, problems are often encountered where an airflow sensor fault causes the EGR flow system code to be activated; another common fault is where the vacuum level is low (leak in vacuum system etc) can also cause the EGR flow code to be activated.
P0401	Exhaust Gas Recirculation Flow Insufficient Detected	The measurement of gas flow within an Exhaust Gas Recirculation (EGR) system is often a calculated value; the calculation will be based on information provided by other sensors (primarily intake airflow sensor but, throttle position sensor, temperature sensor can also form part of the calculation). It is also possible to make use of an EGR valve lift / movement sensor to enable the control unit to assess the movement and position of the EGR valve. Also refer to additional information attached to fault code P0404.	The fault code indicates that the EGR gas flow is calculated or measured as being insufficient or too low. The checks required will be depend on the system design; refer to vehicle specific information to identify the type of system. Checks should however be made on the operation of the EGR valve; check that the valve opens / closes correctly (use vacuum pump to test operation if applicable), check for contamination of valve and ports. Check EGR valve wiring and check vacuum hoses for leaks and blockages; also check vacuum level available from the vacuum source (e.g. vacuum reservoir / pump). Also check for leaks or blockages on the pipes carrying the exhaust gases. Check operation of other applicable sensors; other related fault codes (if available) may provide a more definitive fault description or might indicate a fault in another related component. Note: With calculated flow systems, problems are often encountered where an airflow sensor fault causes the EGR flow system code to be activated; another common fault is where the vacuum level is low (leak in vacuum system etc) can also cause the EGR flow code to be activated.
P0402	Exhaust Gas Recirculation Flow Excessive Detected	The measurement of gas flow within an Exhaust Gas Recirculation (EGR) system is often a calculated value; the calculation will be based on information provided by other sensors (primarily intake airflow sensor but, throttle position sensor, temperature sensor can also form part of the calculation). It is also possible to make use of an EGR valve lift / movement sensor to enable the control unit to assess the movement and position of the EGR valve. Also refer to additional information attached to fault code P0404.	The fault code indicates that the EGR gas flow is calculated or measured as being excessive or too high. The checks required will be depend on the system design; refer to vehicle specific information to identify the type of system. Checks should however be made on the operation of the EGR valve; check that the valve opens / closes correctly (use vacuum pump to test operation if applicable), check for contamination of valve and ports. Check EGR valve wiring and check vacuum hoses for leaks and blockages; also check vacuum level available from the vacuum source (e.g. vacuum reservoir / pump). Where applicable, check operation of other applicable sensors (airflow etc); other related fault codes (if available) may provide a more definitive fault description or might indicate a fault in another related component. Note: With calculated flow systems, problems are often encountered where an airflow sensor fault causes the EGR flow system code to be activated; another common fault is where the vacuum level is low (leak in vacuum system etc) can also cause the EGR flow code to be activated.

NOTE 1. Check for other fault codes that could provide additional information.
NOTE 2. Communication between control units can pass via a CAN-Bus system; refer to Page 51 for CAN-Bus checks.
NOTE 3. If a fault cannot be located, it is also possible that the control unit is at fault.
NOTE 4. Refer to Page 53 for list of pages that show the ISO standard for component locations e.g. Sensor A - Bank 1.

121

Fault code	EOBD / ISO Description	Component / System Description	Meaningful Description and Quick Check
P0403	Exhaust Gas Recirculation Control Circuit	Refer to vehicle specific information to identify the components used on the vehicle's EGR system. EGR systems make use of various methods to control the main EGR valve, with intake vacuum being used on many systems to open the valve or, vacuum can be applied from a pump; the vacuum being regulated by an electric actuator e.g. solenoid. Other systems use an electric actuator acting directly on the EGR valve (these valves may also include cooling as well as temperature monitoring).	The voltage, frequency or value of the control signal in the actuator circuit is incorrect (undefined fault). Signal could be:- out of normal operating range, constant value, non-existent, corrupt. Possible fault with actuator or wiring e.g. actuator power supply, short / open circuit, high circuit resistance or incorrect actuator resistance. Note: the control unit could be providing the correct signal but it is being affected by the circuit fault. Refer to list of sensor / actuator checks on page 19.
P0404	Exhaust Gas Recirculation Control Circuit Range / Performance	Refer to vehicle specific information to identify the components used on the vehicle's EGR system. EGR systems make use of various methods to control the main EGR valve, with intake vacuum being used on many systems to open the valve; the vacuum being regulated by an electric actuator e.g. solenoid. Other systems use an electric actuator acting directly on the EGR valve (these valves may also include cooling as well as temperature monitoring).	The voltage, frequency or value of the control signal in the actuator circuit is within the normal operating range / tolerance, but the signal might be incorrect for the operating conditions. It is also possible that the actuator response (or response of system controlled by the actuator) differs from the expected / desired response (incorrect response for the control signal provided by the control unit). Possible fault in actuator or wiring, but also possible that the actuator (or mechanism / system controlled by the actuator) is not operating or moving correctly. Note: the control unit could be providing the correct signal but it is being affected by the circuit fault. Refer to list of sensor / actuator checks on page 19.
P0405	Exhaust Gas Recirculation Sensor "A" Circuit Low	Different EGR systems can use different types of sensor to monitor EGR valve operation. Systems can use a "Lift" or position sensor (often a potentiometer connected to the valve) to monitor EGR valve movement; other systems can use temperature sensors or pressure sensors to monitor the changes in the exhaust gas temperature or pressure in the EGR system. Also note that the effects of the EGR valve operation can be monitored by the intake airflow sensor; when the EGR valve opens, less air will be drawn through the intake system and this change is detected by the airflow sensor.	It will be necessary to identify the type of sensor used for monitoring EGR operation (e.g. lift / position, temperature or pressure). Note that for many systems, the intake airflow sensor can be the primary sensor for EGR operation or, it may provide additional information alongside other EGR sensors; it might therefore be necessary to check airflow sensor operation as well as checking EGR sensors. The voltage, frequency or value of the EGR sensor signal is either "zero" (no signal) or, the signal exists but it is below the normal operating range (lower than the minimum operating limit). Possible fault in the sensor or wiring e.g. short / open circuits, circuit resistance, incorrect sensor resistance or, fault with the reference / operating voltage; interference from other circuits can also affect sensor signals. Refer to list of sensor / actuator checks on page 19. Also refer to fault code P0472 for pressure sensor checks.
P0406	Exhaust Gas Recirculation Sensor "A" Circuit High	Different EGR systems can use different types of sensor to monitor EGR valve operation. Systems can use a "Lift" or position sensor (often a potentiometer connected to the valve) to monitor EGR valve movement; other systems can use temperature sensors or pressure sensors to monitor the changes in the exhaust gas temperature or pressure in the EGR system. Also note that the effects of the EGR valve operation can be monitored by the intake airflow sensor; when the EGR valve opens, less air will be drawn through the intake system and this change is detected by the airflow sensor.	It will be necessary to identify the type of sensor used for monitoring EGR operation (e.g. lift / position, temperature or pressure). Note that for many systems, the intake airflow sensor can be the primary sensor for EGR operation or, it may provide additional information alongside other EGR sensors; it might therefore be necessary to check airflow sensor operation as well as checking EGR sensors. The voltage, frequency or value of the EGR sensor signal is either at full value (e.g. battery voltage with no signal or frequency) or, the signal exists but it is above the normal operating range (higher than the maximum operating limit). Possible fault in the sensor or wiring e.g. short / open circuits, circuit resistance, incorrect sensor resistance or, fault with the reference / operating voltage; interference from other circuits can also affect sensor signals. Refer to list of sensor / actuator checks on page 19. Also refer to fault code P0473 for pressure sensor checks.
P0407	Exhaust Gas Recirculation Sensor "B" Circuit Low	Different EGR systems can use different types of sensor to monitor EGR valve operation. Systems can use a "Lift" or position sensor (often a potentiometer connected to the valve) to monitor EGR valve movement; other systems can use temperature sensors or pressure sensors to monitor the changes in the exhaust gas temperature or pressure in the EGR system. Also note that the effects of the EGR valve operation can be monitored by the intake airflow sensor; when the EGR valve opens, less air will be drawn through the intake system and this change is detected by the airflow sensor.	It will be necessary to identify the type of sensor used for monitoring EGR operation (e.g. lift / position, temperature or pressure). Note that for many systems, the intake airflow sensor can be the primary sensor for EGR operation or, it may provide additional information alongside other EGR sensors; it might therefore be necessary to check airflow sensor operation as well as checking EGR sensors. The voltage, frequency or value of the EGR sensor signal is either "zero" (no signal) or, the signal exists but it is below the normal operating range (lower than the minimum operating limit). Possible fault in the sensor or wiring e.g. short / open circuits, circuit resistance, incorrect sensor resistance or, fault with the reference / operating voltage; interference from other circuits can also affect sensor signals. Refer to list of sensor / actuator checks on page 19. Also refer to fault code P0472 for pressure sensor checks.
P0408	Exhaust Gas Recirculation Sensor "B" Circuit High	Different EGR systems can use different types of sensor to monitor EGR valve operation. Systems can use a "Lift" or position sensor (often a potentiometer connected to the valve) to monitor EGR valve movement; other systems can use temperature sensors or pressure sensors to monitor the changes in the exhaust gas temperature or pressure in the EGR system. Also note that the effects of the EGR valve operation can be monitored by the intake airflow sensor; when the EGR valve opens, less air will be drawn through the intake system and this change is detected by the airflow sensor.	It will be necessary to identify the type of sensor used for monitoring EGR operation (e.g. lift / position, temperature or pressure). Note that for many systems, the intake airflow sensor can be the primary sensor for EGR operation or, it may provide additional information alongside other EGR sensors; it might therefore be necessary to check airflow sensor operation as well as checking EGR sensors. The voltage, frequency or value of the EGR sensor signal is either at full value (e.g. battery voltage with no signal or frequency) or, the signal exists but it is above the normal operating range (higher than the maximum operating limit). Possible fault in the sensor or wiring e.g. short / open circuits, circuit resistance, incorrect sensor resistance or, fault with the reference / operating voltage; interference from other circuits can also affect sensor signals. Refer to list of sensor / actuator checks on page 19. Also refer to fault code P0473 for pressure sensor checks.
P0409	Exhaust Gas Recirculation Sensor "A" Circuit	Different EGR systems can use different types of sensor to monitor EGR valve operation. Systems can use a "Lift" or position sensor (often a potentiometer connected to the valve) to monitor EGR valve movement; other systems can use temperature sensors or pressure sensors to monitor the changes in the exhaust gas temperature or pressure in the EGR system. Also note that the effects of the EGR valve operation can be monitored by the intake airflow sensor; when the EGR valve opens, less air will be drawn through the intake system and this change is detected by the airflow sensor.	It will be necessary to identify the type of sensor used for monitoring EGR operation (e.g. lift / position, temperature or pressure). Note that for many systems, the intake airflow sensor can be the primary sensor for EGR operation or, it may provide additional information alongside other EGR sensors; it might therefore be necessary to check airflow sensor operation as well as checking EGR sensors. The voltage, frequency or value of the EGR sensor signal is incorrect (undefined fault). Sensor voltage, frequency or signal value could be: out of normal operating range, constant value, non-existent, corrupt. Possible fault in the sensor or wiring e.g. short / open circuits, circuit resistance, incorrect sensor resistance or, fault with the reference / operating voltage; interference from other circuits can also affect sensor signals. Refer to list of sensor / actuator checks on page 19. Also refer to fault code P0470 for pressure sensor checks.
P040A	Exhaust Gas Recirculation Temperature Sensor "A" Circuit	Temperature sensors can be fitted to the Exhaust Gas Recirculation valve (usually the electrically operated valves) to monitor the temperature of the gas passing out of the valve to the intake system. Note, on systems with an EGR cooler (to cool exhaust gas being returned to the intake system), the temperature sensor monitors the temperature as it leaves the cooler.	The voltage, frequency or value of the temperature sensor signal is incorrect (undefined fault). Sensor voltage / signal could be: out of normal operating range, constant value, non-existent, corrupt. Possible fault in the temperature sensor or wiring e.g. short / open circuits, circuit resistance, incorrect sensor resistance or, fault with the reference / operating voltage; interference from other circuits can also affect sensor signals. Refer to list of sensor / actuator checks on page 19.

NOTE 1. Check for other fault codes that could provide additional information.

NOTE 2. Communication between control units can pass via a CAN-Bus system; refer to Page 51 for CAN-Bus checks.

NOTE 3. If a fault cannot be located, it is also possible that the control unit is at fault.

NOTE 4. Refer to Page 53 for list of pages that show the ISO standard for component locations e.g. Sensor A - Bank 1.

Fault code	OBD / SC Description		
P040B	Exhaust Gas Recirculation Temperature Sensor "A" Circuit Range / Performance	Temperature sensors can be fitted to the Exhaust Gas Recirculation valve (usually the electrically operated valves) to monitor the temperature of the gas passing out of the valve to the intake system. Note, on systems with an EGR cooler (to cool exhaust gas being returned to the intake system), the temperature sensor monitors the temperature as it leaves the cooler.	The voltage, frequency or value of the temperature sensor signal is within the normal operating range / tolerance but the signal is not plausible or is incorrect due to an undefined fault. The sensor signal might not match the operating conditions indicated by other sensors or, the sensor signal is not changing / responding as expected to the changes in conditions. Possible fault in the temperature sensor or wiring e.g. short / open circuits, circuit resistance, incorrect sensor resistance or, fault with the reference / operating voltage; interference from other circuits can also affect sensor signals. Refer to list of sensor / actuator checks on page 19. It is also possible that the temperature sensor is operating correctly but the temperature being measured by the sensor is unacceptable or incorrect for the operating conditions. Check for other faults that could cause high temperatures in the EGR systems or high exhaust gas temperatures.
P040C	Exhaust Gas Recirculation Temperature Sensor "A" Circuit Low	Temperature sensors can be fitted to the Exhaust Gas Recirculation valve (usually the electrically operated valves) to monitor the temperature of the gas passing out of the valve to the intake system. Note, on systems with an EGR cooler (to cool exhaust gas being returned to the intake system), the temperature sensor monitors the temperature as it leaves the cooler.	The voltage, frequency or value of the temperature sensor signal is either "zero" (no signal) or, the signal exists but it is below the normal operating range (lower than the minimum operating limit). Possible fault in the temperature sensor or wiring e.g. short / open circuits, circuit resistance, incorrect sensor resistance or, fault with the reference / operating voltage; interference from other circuits can also affect sensor signals. Refer to list of sensor / actuator checks on page 19.
P040D	Exhaust Gas Recirculation Temperature Sensor "A" Circuit High	Temperature sensors can be fitted to the Exhaust Gas Recirculation valve (usually the electrically operated valves) to monitor the temperature of the gas passing out of the valve to the intake system. Note, on systems with an EGR cooler (to cool exhaust gas being returned to the intake system), the temperature sensor monitors the temperature as it leaves the cooler.	The voltage, frequency or value of the temperature sensor signal is either at full value (e.g. battery voltage with no signal or frequency) or, the signal exists but it is above the normal operating range (higher than the maximum operating limit). Possible fault in the temperature sensor or wiring e.g. short / open circuits, circuit resistance, incorrect sensor resistance or, fault with the reference / operating voltage; interference from other circuits can also affect sensor signals. Refer to list of sensor / actuator checks on page 19.
P040E	Exhaust Gas Recirculation Temperature Sensor "A" Circuit Intermittent / Erratic	Temperature sensors can be fitted to the Exhaust Gas Recirculation valve (usually the electrically operated valves) to monitor the temperature of the gas passing out of the valve to the intake system. Note, on systems with an EGR cooler (to cool exhaust gas being returned to the intake system), the temperature sensor monitors the temperature as it leaves the cooler.	The voltage, frequency or value of the temperature sensor signal is intermittent (likely to be out of normal operating range / tolerance when intermittent fault is detected) or, the signal / voltage is erratic (e.g. signal changes are irregular / unpredictable). Possible fault with wiring or connections (including internal sensor connections); also interference from other circuits can affect the signal. Refer to list of sensor / actuator checks on page 19.
P040F	Exhaust Gas Recirculation Temperature Sensor "A" / "B" Correlation	Temperature sensors can be fitted to the Exhaust Gas Recirculation valve (usually the electrically operated valves) to monitor the temperature of the gas passing out of the valve to the intake system. Note, on systems with an EGR cooler (to cool exhaust gas being returned to the intake system), the temperature sensor monitors the temperature as it leaves the cooler.	There can be two exhaust gas recirculation (EGR) systems (and two temperature sensors) fitted to an engine, usually one for each bank of a V engine. The fault code indicates that the two EGR temperatures sensors are providing conflicting information, which could be caused a sensor failure or, there could be a problem related to the temperature of the exhaust gas on one cylinder bank. Refer to list of sensor / actuator checks on page 19.
P0410	Secondary Air Injection System	After cold starts, the secondary air injection system pumps air (oxygen) into the exhaust gas (which is rich with HC and CO), this helps the HC / CO to combust (combustion heat in exhaust gas helps to heat the catalyst); the additional air can be detected in a number of ways including using a Lambda (oxygen) sensor (detecting oxygen level) or a flow / pressure sensor. Some systems direct air from the main intake system and then use the main airflow sensor to detect increase in air flow. Note that some systems (pulse air systems) use negative exhaust pressure pulses to draw air into the exhaust system instead of an air pump. If additional air is not detected or quantity is incorrect, this is regarded as a secondary air system fault. It may be necessary to refer to vehicle specific information to identify the system and components used.	The fault code indicates that the secondary air system is not functioning; undefined fault. The fault could be detected due to incorrect secondary air flow, or because the control unit has detected an electrical fault in the circuits. Check for other fault codes that could provide additional information. Refer to fault code P0411 for additional information.
P0411	Secondary Air Injection System Incorrect Flow Detected	After cold starts, the secondary air injection system pumps air (oxygen) into the exhaust gas (which is rich with HC and CO), this helps the HC / CO to combust (combustion heat in exhaust gas helps to heat the catalyst); the additional air can be detected in a number of ways including using a Lambda (oxygen) sensor (detecting oxygen level) or a flow / pressure sensor. Some systems direct air from the main intake system and then use the main airflow sensor to detect increase in air flow. Note that some systems (pulse air systems) use negative exhaust pressure pulses to draw air into the exhaust system instead of an air pump. If additional air is not detected or quantity is incorrect, this is regarded as a secondary air system fault. It may be necessary to refer to vehicle specific information to identify the system and components used.	The fault code indicates that the air flow provided by the secondary air system (into the exhaust system) is incorrect / not as expected by the control unit; the flow could be a calculated value based on the Lambda sensor signal or could be detected by the secondary air "flow / pressure" sensor. Check operation of the secondary air system; depending on the system design, this could involve checks on the secondary air control valves and switching valve control systems (vacuum control, vacuum pipes etc) and air pump. Check for blockages in the pipes and air filter. Check all applicable wiring and connections for short / open circuits and high resistance. Also check for correct supply voltage to applicable secondary air system components. Check that the non-return valve (in the air pipe or combined with switching valve) is operating correctly. Check for other fault codes e.g. flow / pressure sensor (if fitted) and oxygen / Lambda sensor fault codes that might indicate if the oxygen / Lambda sensor is at fault.
P0412	Secondary Air Injection System Switching Valve "A" Circuit	After cold starts, the secondary air injection system pumps air (oxygen) into the exhaust gas (which is rich with HC and CO), the additional air helps the HC / CO to combust (combustion heat in exhaust gas helps to heat the catalyst). Note that some systems (pulse air systems) use negative exhaust pressure pulses to draw air into the exhaust system instead of a pump. The switching valve is operated by an electric actuator (usually a solenoid), and is used to regulate the vacuum that opens the secondary air valve (which allows the air into the exhaust gas).	The voltage, frequency or value of the control signal in the actuator circuit is incorrect (undefined fault). Signal could be:- out of normal operating range, constant value, non-existent, corrupt. Possible fault with actuator or wiring e.g. actuator power supply, short / open circuit, high circuit resistance or incorrect actuator resistance. Note: the control unit could be providing the correct signal but it is being affected by a circuit fault. Refer to list of sensor / actuator checks on page 19.

NOTE 1. Check for other fault codes that could provide additional information.   **NOTE 2.** Communication between control units can pass via a CAN-Bus system; refer to Page 51 for CAN-Bus checks.   **123**

NOTE 3. If a fault cannot be located, it is also possible that the control unit is at fault.   **NOTE 4.** Refer to Page 53 for list of pages that show the ISO standard for component locations e.g. Sensor A - Bank 1.

Fault code	EOBD / ISO Description	Component / System Description	Meaningful Description and Quick Check
P0413	Secondary Air Injection System Switching Valve "A" Circuit Open	After cold starts, the secondary air injection system pumps air (oxygen) into the exhaust gas (which is rich with HC and CO), the additional air helps the HC / CO to combust (combustion heat in exhaust gas helps to heat the catalyst). Note that some systems (pulse air systems) use negative exhaust pressure pulses to draw air into the exhaust system instead of a pump. The switching valve is operated by an electric actuator (usually a solenoid), and is used to regulate the vacuum that opens the secondary air valve (which allows the air into the exhaust gas).	The voltage, frequency or value of the control signal in the actuator circuit is incorrect. Signal could be:- out of normal operating range, constant value, non-existent, corrupt. The fault code indicates a possible "open circuit" but other undefined faults in the actuator or wiring could activate the same code e.g. actuator power supply, short circuit, high circuit resistance or incorrect actuator resistance. Note: the control unit could be providing the correct signal but it is being affected by a circuit fault. Refer to list of sensor / actuator checks on page 19.
P0414	Secondary Air Injection System Switching Valve "A" Circuit Shorted	After cold starts, the secondary air injection system pumps air (oxygen) into the exhaust gas (which is rich with HC and CO), the additional air helps the HC / CO to combust (combustion heat in exhaust gas helps to heat the catalyst). Note that some systems (pulse air systems) use negative exhaust pressure pulses to draw air into the exhaust system instead of a pump. The switching valve is operated by an electric actuator (usually a solenoid), and is used to regulate the vacuum that opens the secondary air valve (which allows the air into the exhaust gas).	The voltage, frequency or value of the control signal in the actuator circuit is incorrect. The control unit detects the fault by assessing electrical values in the circuit e.g. voltage, current etc. Signal / voltage is likely to be "zero" (no signal or voltage), which could indicate that the circuit is shorted to earth; it is also possible that the fault could relate to a short to "positive" (power supply), which would result in a permanent voltage existing in the circuit. Check actuator and wiring for a short circuit but note that other faults could activate the same fault code; check actuator and wiring for open circuits, high resistance, also check power supply to actuator. Refer to list of sensor / actuator checks on page 19.
P0415	Secondary Air Injection System Switching Valve "B" Circuit	After cold starts, the secondary air injection system pumps air (oxygen) into the exhaust gas (which is rich with HC and CO), the additional air helps the HC / CO to combust (combustion heat in exhaust gas helps to heat the catalyst). Note that some systems (pulse air systems) use negative exhaust pressure pulses to draw air into the exhaust system instead of a pump. The switching valve is operated by an electric actuator (usually a solenoid), and is used to regulate the vacuum that opens the secondary air valve (which allows the air into the exhaust gas).	The voltage, frequency or value of the control signal in the actuator circuit is incorrect (undefined fault). Signal could be:- out of normal operating range, constant value, non-existent, corrupt. Possible fault with actuator or wiring e.g. actuator power supply, short / open circuit, high circuit resistance or incorrect actuator resistance. Note: the control unit could be providing the correct signal but it is being affected by a circuit fault. Refer to list of sensor / actuator checks on page 19.
P0416	Secondary Air Injection System Switching Valve "B" Circuit Open	After cold starts, the secondary air injection system pumps air (oxygen) into the exhaust gas (which is rich with HC and CO), the additional air helps the HC / CO to combust (combustion heat in exhaust gas helps to heat the catalyst). Note that some systems (pulse air systems) use negative exhaust pressure pulses to draw air into the exhaust system instead of a pump. The switching valve is operated by an electric actuator (usually a solenoid), and is used to regulate the vacuum that opens the secondary air valve (which allows the air into the exhaust gas).	The voltage, frequency or value of the control signal in the actuator circuit is incorrect. Signal could be:- out of normal operating range, constant value, non-existent, corrupt. The fault code indicates a possible "open circuit" but other undefined faults in the actuator or wiring could activate the same code e.g. actuator power supply, short circuit, high circuit resistance or incorrect actuator resistance. Note: the control unit could be providing the correct signal but it is being affected by a circuit fault. Refer to list of sensor / actuator checks on page 19.
P0417	Secondary Air Injection System Switching Valve "B" Circuit Shorted	After cold starts, the secondary air injection system pumps air (oxygen) into the exhaust gas (which is rich with HC and CO), the additional air helps the HC / CO to combust (combustion heat in exhaust gas helps to heat the catalyst). Note that some systems (pulse air systems) use negative exhaust pressure pulses to draw air into the exhaust system instead of a pump. The switching valve is operated by an electric actuator (usually a solenoid), and is used to regulate the vacuum that opens the secondary air valve (which allows the air into the exhaust gas).	The voltage, frequency or value of the control signal in the actuator circuit is incorrect. The control unit detects the fault by assessing electrical values in the circuit e.g. voltage, current etc. Signal / voltage is likely to be "zero" (no signal or voltage), which could indicate that the circuit is shorted to earth; it is also possible that the fault could relate to a short to "positive" (power supply), which would result in a permanent voltage existing in the circuit. Check actuator and wiring for a short circuit but note that other faults could activate the same fault code; check actuator and wiring for open circuits, high resistance, also check power supply to actuator. Refer to list of sensor / actuator checks on page 19.
P0418	Secondary Air Injection System Control "A" Circuit	After cold starts, the secondary air injection system pumps air (oxygen) into the exhaust gas (which is rich with HC and CO), this helps the HC / CO to combust (combustion heat in exhaust gas helps to heat the catalyst). Note that some systems (pulse air systems) use negative exhaust pressure pulses to draw air into the exhaust system instead of a pump. The term "control circuit" could refer to the circuit connecting the control unit to the switching valves or the air pump, depending on the system. Refer to fault codes P0412 and P2444 for additional information.	The voltage, frequency or value of the control signal in the actuator circuit is incorrect (undefined fault). Signal could be:- out of normal operating range, constant value, non-existent, corrupt. Possible fault with actuator or wiring e.g. actuator power supply, short / open circuit, high circuit resistance or incorrect actuator resistance. Note: the control unit could be providing the correct signal but it is being affected by the circuit fault. Refer to list of sensor / actuator checks on page 19.
P0419	Secondary Air Injection System Control "B" Circuit	After cold starts, the secondary air injection system pumps air (oxygen) into the exhaust gas (which is rich with HC and CO), this helps the HC / CO to combust (combustion heat in exhaust gas helps to heat the catalyst). Note that some systems (pulse air systems) use negative exhaust pressure pulses to draw air into the exhaust system instead of a pump. The term "control circuit" could refer to the circuit connecting the control unit to the switching valves or the air pump, depending on the system. Refer to fault codes P0412 and P2444 for additional information.	The voltage, frequency or value of the control signal in the actuator circuit is incorrect (undefined fault). Signal could be:- out of normal operating range, constant value, non-existent, corrupt. Possible fault with actuator or wiring e.g. actuator power supply, short / open circuit, high circuit resistance or incorrect actuator resistance. Note: the control unit could be providing the correct signal but it is being affected by the circuit fault. Refer to list of sensor / actuator checks on page 19.

code	Description		
P041A	Exhaust Gas Recirculation Temperature Sensor "B" Circuit	Temperature sensors can be fitted to the Exhaust Gas Recirculation valve (usually the electrically operated valves) to monitor the temperature of the gas passing out of the valve to the intake system. Note, on systems with an EGR cooler (to cool exhaust gas being returned to the intake system), the temperature sensor monitors the temperature as it leaves the cooler.	The voltage, frequency or value of the temperature sensor signal is incorrect (undefined fault). Sensor voltage / signal could be: out of normal operating range, constant value, non-existent, corrupt. Possible fault in the temperature sensor or wiring e.g. short / open circuits, circuit resistance, incorrect sensor resistance or, fault with the reference / operating voltage; interference from other circuits can also affect sensor signals. Refer to list of sensor / actuator checks on page 19.
P041B	Exhaust Gas Recirculation Temperature Sensor "B" Circuit Range / Performance	Temperature sensors can be fitted to the Exhaust Gas Recirculation valve (usually the electrically operated valves) to monitor the temperature of the gas passing out of the valve to the intake system. Note, on systems with an EGR cooler (to cool exhaust gas being returned to the intake system), the temperature sensor monitors the temperature as it leaves the cooler.	The voltage, frequency or value of the temperature sensor signal is within the normal operating range / tolerance but the signal is not plausible or is incorrect due to an undefined fault. The sensor signal might not match the operating conditions indicated by other sensors or, the sensor signal is not changing / responding as expected to the changes in conditions. Possible fault in the temperature sensor or wiring e.g. short / open circuits, circuit resistance, incorrect sensor resistance or, fault with the reference / operating voltage; interference from other circuits can also affect sensor signals. Refer to list of sensor / actuator checks on page 19. It is also possible that the temperature sensor is operating correctly but the temperature being measured by the sensor is unacceptable or incorrect for the operating conditions. Check for other faults that could cause high temperatures in the EGR systems or high exhaust gas temperatures.
P041C	Exhaust Gas Recirculation Temperature Sensor "B" Circuit Low	Temperature sensors can be fitted to the Exhaust Gas Recirculation valve (usually the electrically operated valves) to monitor the temperature of the gas passing out of the valve to the intake system. Note, on systems with an EGR cooler (to cool exhaust gas being returned to the intake system), the temperature sensor monitors the temperature as it leaves the cooler.	The voltage, frequency or value of the temperature sensor signal is either "zero" (no signal) or, the signal exists but it is below the normal operating range (lower than the minimum operating limit). Possible fault in the temperature sensor or wiring e.g. short / open circuits, circuit resistance, incorrect sensor resistance or, fault with the reference / operating voltage; interference from other circuits can also affect sensor signals. Refer to list of sensor / actuator checks on page 19.
P041D	Exhaust Gas Recirculation Temperature Sensor "B" Circuit High	Temperature sensors can be fitted to the Exhaust Gas Recirculation valve (usually the electrically operated valves) to monitor the temperature of the gas passing out of the valve to the intake system. Note, on systems with an EGR cooler (to cool exhaust gas being returned to the intake system), the temperature sensor monitors the temperature as it leaves the cooler.	The voltage, frequency or value of the temperature sensor signal is either at full value (e.g. battery voltage with no signal or frequency) or, the signal exists but it is above the normal operating range (higher than the maximum operating limit). Possible fault in the temperature sensor or wiring e.g. short / open circuits, circuit resistance, incorrect sensor resistance or, fault with the reference / operating voltage; interference from other circuits can also affect sensor signals. Refer to list of sensor / actuator checks on page 19.
P041E	Exhaust Gas Recirculation Temperature Sensor "B" Circuit Intermittent / Erratic	Temperature sensors can be fitted to the Exhaust Gas Recirculation valve (usually the electrically operated valves) to monitor the temperature of the gas passing out of the valve to the intake system. Note, on systems with an EGR cooler (to cool exhaust gas being returned to the intake system), the temperature sensor monitors the temperature as it leaves the cooler.	The voltage, frequency or value of the temperature sensor signal is intermittent (likely to be out of normal operating range / tolerance when intermittent fault is detected) or, the signal / voltage is erratic (e.g. signal changes are irregular / unpredictable). Possible fault with wiring or connections (including internal sensor connections); also interference from other circuits can affect the signal. Refer to list of sensor / actuator checks on page 19.
P041F	ISO / SAE reserved		Not used, reserved for future allocation.
P0420	Catalyst System Efficiency Below Threshold Bank 1	The control unit monitors the Lambda (oxygen) sensor signals for the pre-cat and post cat Lambda sensors (the intake air flow value can also be used in the calculation for checking the main converter efficiency); if the pre-cat sensor reading is good but the post cat sensor reading is indicating that the oxygen is not being burnt within the catalytic converter, this can indicate low catalytic converter efficiency.	The fault code indicates that the catalytic converter efficiency is below an acceptable level; the problem could occur during all driving conditions or only at specific times e.g. warm up period. Due to the different arrangements for catalytic converters and Lambda sensors it may be advisable to refer to vehicle specific information to identify the types of catalytic converters and Lambda sensors fitted to the vehicle. If possible, check exhaust gas readings to check whether the catalyst is operating efficiently or inefficiently. Check the pre-cat and post-cat Lambda sensor signals to ensure that the sensors are operating correctly (Refer to list of sensor / actuator checks on page 19). If it appears that the catalytic converter is faulty, check for other faults that could influence the catalytic converter operation e.g. fuelling, misfires, air leaks (exhaust and intake) and secondary air injection system. Check for other fault codes that could indicate related faults.
P0421	Warm Up Catalyst Efficiency Below Threshold Bank 1	The warm up catalyst is usually a small catalyst which can heat up quickly after engine starts; the warm up catalyst will be closest to the engine. The control unit monitors the Lambda (oxygen) sensor signals for the pre-cat (if fitted) and post cat Lambda sensors; if the pre-cat sensor reading is good but the post cat sensor reading is indicating that the oxygen is not being burnt within the catalytic converter, this can indicate low catalytic converter efficiency.	The fault code indicates that the catalytic converter efficiency is below an acceptable level; the problem could occur during all driving conditions but it is likely that the fault is relating to the warm up catalyst operation during the immediate post start / warm up period. Due to the different arrangements for catalytic converters and Lambda sensors it may be advisable to refer to vehicle specific information to identify the types of catalytic converters and Lambda sensors fitted to the vehicle. If possible, check exhaust gas readings to check whether the catalyst is operating efficiently or inefficiently. Check the Lambda sensor signals to ensure that the sensors are operating correctly (Refer to list of sensor / actuator checks on page 19). If it appears that the catalytic converter is faulty, check for other faults that could influence the catalytic converter operation e.g. fuelling, misfires, air leaks (exhaust and intake) and secondary air injection system. Check for other fault codes that could indicate related faults.
P0422	Main Catalyst Efficiency Below Threshold Bank 1	The control unit monitors the Lambda (oxygen) sensor signals for the pre-cat and post cat Lambda sensors (the intake air flow value can also be used in the calculation for checking the main converter efficiency); if the pre-cat sensor reading is good but the post cat sensor reading is indicating that the oxygen is not being burnt within the catalytic converter, this can indicate low catalytic converter efficiency.	The fault code indicates that the catalytic converter efficiency is below an acceptable level. Due to the different arrangements for catalytic converters and Lambda sensors it may be advisable to refer to vehicle specific information to identify the types of catalytic converters and Lambda sensors fitted to the vehicle. If possible, check exhaust gas readings to check whether the catalyst is operating efficiently or inefficiently. Check the pre-cat and post-cat Lambda sensor signals to ensure that the sensors are operating correctly (Refer to list of sensor / actuator checks on page 19). If it appears that the catalytic converter is faulty, check for other faults that could influence the catalytic converter operation e.g. fuelling, misfires, air leaks (exhaust and intake) and secondary air injection system. Check for other fault codes that could indicate related faults.

Fault code	EOBD / ISO Description	Component / System Description	Meaningful Description and Quick Check
P0423	Heated Catalyst Efficiency Below Threshold Bank 1	Heated catalysts are used to rapidly increase catalyst temperature / efficiency after engine starting. Some systems do have an electrical heater, but the manufacturer could use the code for catalysts that have other forms of heating e.g. retarding the ignition timing or injecting additional fuel after combustion; this causes late burning of the fuel in the exhaust system thus heating the catalyst. Secondary air injection is also used to achieve the same result. The control unit monitors the Lambda (oxygen) sensor signals for the pre-cat and post cat Lambda sensors (the intake air flow value can also be used in the calculation for checking the main converter efficiency); if the pre-cat sensor reading is good but the post cat sensor reading is indicating that the oxygen is not being burnt within the catalytic converter, this can indicate low catalytic converter efficiency.	The fault code indicates that the catalytic converter efficiency is below an acceptable level. Due to the different arrangements for catalytic converters and Lambda sensors it may be advisable to refer to vehicle specific information to identify the types of catalytic converters and Lambda sensors fitted to the vehicle. If possible, check exhaust gas readings to check whether the catalyst is operating efficiently or inefficiently. Check the pre-cat and post-cat Lambda sensor signals to ensure that the sensors are operating correctly (Refer to list of sensor / actuator checks on page 19). If it appears that the catalytic converter is faulty, check for other faults that could influence the catalytic converter operation e.g. fuelling, misfires, air leaks (exhaust and intake) and secondary air injection system. Check for other fault codes that could indicate related faults.
P0424	Heated Catalyst Temperature Below Threshold Bank 1	Heated catalysts are used to rapidly increase catalyst temperature / efficiency after engine starting. Some systems do have an electrical heater, but the manufacturer could use the code for catalysts that have other forms of heating e.g. retarding the ignition timing or injecting additional fuel after combustion; this causes late burning of the fuel in the exhaust system thus heating the catalyst. Secondary air injection is also used to achieve the same result.	The fault code relates to the signal from the temperature sensor (fitted in or close to the heated catalytic converter), which is indicating that the catalyst temperature is too low for efficient operation; this could occur after cold starts if the catalyst takes too long to heat up. Identify if the catalysts has an electrical heater and check heater operation (refer to vehicle specific information to identify wiring etc). It is likely that the heater will be provided with a power supply via a relay (usually switched by the control unit); check relay operation and power supply, check wiring between heater and relay and between relay and control unit. If the system makes use of other forms of catalyst heating, check the operation as applicable.
P0425	Catalyst Temperature Sensor Circuit Bank 1 Sensor 1	The sensor is used to ensure that the catalyst does not overheat, which will rapidly cause deterioration of the catalyst.	The voltage, frequency or value of the temperature sensor signal is incorrect (undefined fault). Sensor voltage / signal could be: out of normal operating range, constant value, non-existent, corrupt. Possible fault in the temperature sensor or wiring e.g. short / open circuits, circuit resistance, incorrect sensor resistance or, fault with the reference / operating voltage; interference from other circuits can also affect sensor signals. Refer to list of sensor / actuator checks on page 19. The code refers to a sensor / circuit fault but if the sensor does appear to be operating correctly, check for other faults that could affect catalyst temperatures e.g. high exhaust gas temperatures due to engine faults.
P0426	Catalyst Temperature Sensor Circuit Range / Performance Bank 1 Sensor 1	The sensor is used to ensure that the catalyst does not overheat, which will rapidly cause deterioration of the catalyst.	The voltage, frequency or value of the temperature sensor signal is within the normal operating range / tolerance but the signal is not plausible or is incorrect due to an undefined fault. The sensor signal might not match the operating conditions indicated by other sensors or, the sensor signal is not changing / responding as expected to the changes in conditions. Possible fault in the temperature sensor or wiring e.g. short / open circuits, circuit resistance, incorrect sensor resistance or, fault with the reference / operating voltage; interference from other circuits can also affect sensor signals. Refer to list of sensor / actuator checks on page 19. It is also possible that the temperature sensor is operating correctly but the catalyst temperature is unacceptable or incorrect for the operating conditions. Check for faults that could affect catalyst temperatures e.g. high exhaust gas temperatures caused by engine faults.
P0427	Catalyst Temperature Sensor Circuit Low Bank 1 Sensor 1	The sensor is used to ensure that the catalyst does not overheat, which will rapidly cause deterioration of the catalyst.	The voltage, frequency or value of the temperature sensor signal is either "zero" (no signal) or, the signal exists but it is below the normal operating range (lower than the minimum operating limit). Possible fault in the temperature sensor or wiring e.g. short / open circuits, circuit resistance, incorrect sensor resistance or, fault with the reference / operating voltage; interference from other circuits can also affect sensor signals. Refer to list of sensor / actuator checks on page 19.
P0428	Catalyst Temperature Sensor Circuit High Bank 1 Sensor 1	The sensor is used to ensure that the catalyst does not overheat, which will rapidly cause deterioration of the catalyst.	The voltage, frequency or value of the temperature sensor signal is either at full value (e.g. battery voltage with no signal or frequency) or, the signal exists but it is above the normal operating range (higher than the maximum operating limit). Possible fault in the temperature sensor or wiring e.g. short / open circuits, circuit resistance, incorrect sensor resistance or, fault with the reference / operating voltage; interference from other circuits can also affect sensor signals. Refer to list of sensor / actuator checks on page 19.
P0429	Catalyst Heater Control Circuit Bank 1	Heated catalysts are used to rapidly increase catalyst temperature / efficiency after engine starting.	The voltage, frequency or value of the control signal in the heater circuit is incorrect (undefined fault). Signal could be:- out of normal operating range, constant value, non-existent, corrupt. Possible fault with heater or wiring e.g. heater power supply, short / open circuit, high circuit resistance or incorrect heater resistance. Note: the control unit could be providing the correct signal but it is being affected by a circuit fault. Refer to list of sensor / actuator checks on page 19.
P042A	Catalyst Temperature Sensor Circuit Bank 1 Sensor 2	The sensor is used to ensure that the catalyst does not overheat, which will rapidly cause deterioration of the catalyst.	The voltage, frequency or value of the temperature sensor signal is incorrect (undefined fault). Sensor voltage / signal could be: out of normal operating range, constant value, non-existent, corrupt. Possible fault in the temperature sensor or wiring e.g. short / open circuits, circuit resistance, incorrect sensor resistance or, fault with the reference / operating voltage; interference from other circuits can also affect sensor signals. Refer to list of sensor / actuator checks on page 19. The code refers to a sensor / circuit fault but if the sensor does appear to be operating correctly, check for other faults that could affect catalyst temperatures e.g. high exhaust gas temperatures due to engine faults.
P042B	Catalyst Temperature Sensor Circuit Range / Performance Bank 1 Sensor 2	The sensor is used to ensure that the catalyst does not overheat, which will rapidly cause deterioration of the catalyst.	The voltage, frequency or value of the temperature sensor signal is within the normal operating range / tolerance but the signal is not plausible or is incorrect due to an undefined fault. The sensor signal might not match the operating conditions indicated by other sensors or, the sensor signal is not changing / responding as expected to the changes in conditions. Possible fault in the temperature sensor or wiring e.g. short / open circuits, circuit resistance, incorrect sensor resistance or, fault with the reference / operating voltage; interference from other circuits can also affect sensor signals. Refer to list of sensor / actuator checks on page 19. It is also possible that the temperature sensor is operating correctly but the catalyst temperature is unacceptable or incorrect for the operating conditions. Check for faults that could affect catalyst temperatures e.g. high exhaust gas temperatures caused by engine faults.

NOTE 1. Check for other fault codes that could provide additional information.
NOTE 3. If a fault cannot be located, it is also possible that the control unit is at fault.
NOTE 2. Communication between control units can pass via a CAN-Bus system; refer to Page 51 for CAN-Bus checks.
NOTE 4. Refer to Page 53 for list of pages that show the ISO standard for component locations e.g. Sensor A - Bank 1.

code	Description		
P042C	Catalyst Temperature Sensor Circuit Low Bank 1 Sensor 2	The sensor is used to ensure that the catalyst does not overheat, which will rapidly cause deterioration of the catalyst.	The voltage, frequency or value of the temperature sensor signal is either "zero" (no signal) or, the signal exists but it is below the normal operating range (lower than the minimum operating limit). Possible fault in the temperature sensor or wiring e.g. short / open circuits, circuit resistance, incorrect sensor resistance or, fault with the reference / operating voltage; interference from other circuits can also affect sensor signals. Refer to list of sensor / actuator checks on page 19.
P042D	Catalyst Temperature Sensor Circuit High Bank 1 Sensor 2	The sensor is used to ensure that the catalyst does not overheat, which will rapidly cause deterioration of the catalyst.	The voltage, frequency or value of the temperature sensor signal is either at full value (e.g. battery voltage with no signal or frequency) or, the signal exists but it is above the normal operating range (higher than the maximum operating limit). Possible fault in the temperature sensor or wiring e.g. short / open circuits, circuit resistance, incorrect sensor resistance or, fault with the reference / operating voltage; interference from other circuits can also affect sensor signals. Refer to list of sensor / actuator checks on page 19.
P042E		ISO / SAE reserved	Not used, reserved for future allocation.
P042F		ISO / SAE reserved	Not used, reserved for future allocation.
P0430	Catalyst System Efficiency Below Threshold Bank 2	The control unit monitors the Lambda (oxygen) sensor signals for the pre-cat and post cat Lambda sensors (the intake air flow value can also be used in the calculation for checking the main converter efficiency); if the pre-cat sensor reading is good but the post cat sensor reading is indicating that the oxygen is not being burnt within the catalytic converter, this can indicate low catalytic converter efficiency.	The fault code indicates that the catalytic converter efficiency is below an acceptable level; the problem could occur during all driving conditions or only at specific times e.g. warm up period. Due to the different arrangements for catalytic converters and Lambda sensors it may be advisable to refer to vehicle specific information to identify the types of catalytic converters and Lambda sensors fitted to the vehicle. If possible, check exhaust gas readings to check whether the catalyst is operating efficiently or inefficiently. Check the pre-cat and post-cat Lambda sensor signals to ensure that the sensors are operating correctly (Refer to list of sensor / actuator checks on page 19). If it appears that the catalytic converter is faulty, check for other faults that could influence the catalytic converter operation e.g. fuelling, misfires, air leaks (exhaust and intake) and secondary air injection system. Check for other fault codes that could indicate related faults.
P0431	Warm Up Catalyst Efficiency Below Threshold Bank 2	The warm up catalyst is usually a small catalyst which can heat up quickly after engine starts; the warm up catalyst will be closest to the engine. The control unit monitors the Lambda (oxygen) sensor signals for the pre-cat (if fitted) and post cat Lambda sensors; if the pre-cat sensor reading is good but the post cat sensor reading is indicating that the oxygen is not being burnt within the catalytic converter, this can indicate low catalytic converter efficiency.	The fault code indicates that the catalytic converter efficiency is below an acceptable level; the problem could occur during all driving conditions but it is likely that the fault is relating to the warm up catalyst operation during the immediate post start / warm up period. Due to the different arrangements for catalytic converters and Lambda sensors it may be advisable to refer to vehicle specific information to identify the types of catalytic converters and Lambda sensors fitted to the vehicle. If possible, check exhaust gas readings to check whether the catalyst is operating efficiently or inefficiently. Check the Lambda sensor signals to ensure that the sensors are operating correctly (Refer to list of sensor / actuator checks on page 19). If it appears that the catalytic converter is faulty, check for other faults that could influence the catalytic converter operation e.g. fuelling, misfires, air leaks (exhaust and intake) and secondary air injection system. Check for other fault codes that could indicate related faults.
P0432	Main Catalyst Efficiency Below Threshold Bank 2	The control unit monitors the Lambda (oxygen) sensor signals for the pre-cat and post cat Lambda sensors (the intake air flow value can also be used in the calculation for checking the main converter efficiency); if the pre-cat sensor reading is good but the post cat sensor reading is indicating that the oxygen is not being burnt within the catalytic converter, this can indicate low catalytic converter efficiency.	The fault code indicates that the catalytic converter efficiency is below an acceptable level; the problem could occur during all driving conditions or only at specific times. Due to the different arrangements for catalytic converters and Lambda sensors it may be advisable to refer to vehicle specific information to identify the types of catalytic converters and Lambda sensors fitted to the vehicle. If possible, check exhaust gas readings to check whether the catalyst is operating efficiently or inefficiently. Check the pre-cat and post-cat Lambda sensor signals to ensure that the sensors are operating correctly (Refer to list of sensor / actuator checks on page 19). If it appears that the catalytic converter is faulty, check for other faults that could influence the catalytic converter operation e.g. fuelling, misfires, air leaks (exhaust and intake) and secondary air injection system. Check for other fault codes that could indicate related faults.
P0433	Heated Catalyst Efficiency Below Threshold Bank 2	Heated catalysts are used to rapidly increase catalyst temperature / efficiency after engine starting. Some systems do have an electrical heater, but the manufacturer could use the code for catalysts that have other forms of heating e.g. retarding the ignition timing or injecting additional fuel after combustion; this causes late burning of the fuel in the exhaust system thus heating the catalyst. Secondary air injection is also used to achieve the same result. The control unit monitors the Lambda (oxygen) sensor signals for the pre-cat and post cat Lambda sensors (the intake air flow value can also be used in the calculation for checking the main converter efficiency); if the pre-cat sensor reading is good but the post cat sensor reading is indicating that the oxygen is not being burnt within the catalytic converter, this can indicate low catalytic converter efficiency.	The fault code indicates that the catalytic converter efficiency is below an acceptable level; the problem could occur during all driving conditions or only at specific times. Due to the different arrangements for catalytic converters and Lambda sensors it may be advisable to refer to vehicle specific information to identify the types of catalytic converters and Lambda sensors fitted to the vehicle. If possible, check exhaust gas readings to check whether the catalyst is operating efficiently or inefficiently. Check the pre-cat and post-cat Lambda sensor signals to ensure that the sensors are operating correctly (Refer to list of sensor / actuator checks on page 19). If the catalyst fitted to the vehicle is the electrically heated type, check for heater operation and values (resistances etc), check heater power supply and wiring. Check for other faults that could influence the catalytic converter operation e.g. fuelling, misfires, air leaks (exhaust and intake) and secondary air injection system. Check for other fault codes that could indicate related faults.
P0434	Heated Catalyst Temperature Below Threshold Bank 2	Heated catalysts are used to rapidly increase catalyst temperature / efficiency after engine starting. Some systems do have an electrical heater, but the manufacturer could use the code for catalysts that have other forms of heating e.g. retarding the ignition timing or injecting additional fuel after combustion; this causes late burning of the fuel in the exhaust system thus heating the catalyst. Secondary air injection is also used to achieve the same result.	The fault code relates to the signal from the temperature sensor (fitted in or close to the heated catalytic converter), which is indicating that the catalyst temperature is too low for efficient operation; this could occur after cold starts if the catalyst takes too long to heat up. Identify if the catalysts has an electrical heater and check heater operation (refer to vehicle specific information to identify wiring etc). It is likely that the heater will be provided with a power supply via a relay (usually switched by the control unit); check relay operation and power supply, check wiring between heater and relay and between relay and control unit. If the system makes use of other forms of catalyst heating, check the operation as applicable.
P0435	Catalyst Temperature Sensor Circuit Bank 2 Sensor 1	The sensor is used to ensure that the catalyst does not overheat, which will rapidly cause deterioration of the catalyst.	The voltage, frequency or value of the temperature sensor signal is incorrect (undefined fault). Sensor voltage / signal could be: out of normal operating range, constant value, non-existent, corrupt. Possible fault in the temperature sensor or wiring e.g. short / open circuits, circuit resistance, incorrect sensor resistance or, fault with the reference / operating voltage; interference from other circuits can also affect sensor signals. Refer to list of sensor / actuator checks on page 19. The code refers to a sensor / circuit fault but if the sensor does appear to be operating correctly, check for other faults that could affect catalyst temperatures e.g. high exhaust gas temperatures due to engine faults.

NOTE 1. Check for other fault codes that could provide additional information. **NOTE 2.** Communication between control units can pass via a CAN-Bus system; refer to Page 51 for CAN-Bus checks.

NOTE 3. If a fault cannot be located, it is also possible that the control unit is at fault. **NOTE 4.** Refer to Page 53 for list of pages that show the ISO standard for component locations e.g. Sensor A - Bank 1.

Fault code	EOBD / ISO Description	Component / System Description	Meaningful Description and Quick Check
P0436	Catalyst Temperature Sensor Circuit Range / Performance Bank 2 Sensor 1	The sensor is used to ensure that the catalyst does not overheat, which will rapidly cause deterioration of the catalyst.	The voltage, frequency or value of the temperature sensor signal is within the normal operating range / tolerance but the signal is not plausible or is incorrect due to an undefined fault. The sensor signal might not match the operating conditions indicated by other sensors or, the sensor signal is not changing / responding as expected to the changes in conditions. Possible fault in the temperature sensor or wiring e.g. short / open circuits, circuit resistance, incorrect sensor resistance or, fault with the reference / operating voltage; interference from other circuits can also affect sensor signals. Refer to list of sensor / actuator checks on page 19. It is also possible that the temperature sensor is operating correctly but the catalyst temperature is unacceptable or incorrect for the operating conditions. Check for faults that could affect catalyst temperatures e.g. high exhaust gas temperatures caused by engine faults.
P0437	Catalyst Temperature Sensor Circuit Low Bank 2 Sensor 1	The sensor is used to ensure that the catalyst does not overheat, which will rapidly cause deterioration of the catalyst.	The voltage, frequency or value of the temperature sensor signal is either "zero" (no signal) or, the signal exists but it is below the normal operating range (lower than the minimum operating limit). Possible fault in the temperature sensor or wiring e.g. short / open circuits, circuit resistance, incorrect sensor resistance or, fault with the reference / operating voltage; interference from other circuits can also affect sensor signals. Refer to list of sensor / actuator checks on page 19.
P0438	Catalyst Temperature Sensor Circuit High Bank 2 Sensor 1	The sensor is used to ensure that the catalyst does not overheat, which will rapidly cause deterioration of the catalyst.	The voltage, frequency or value of the temperature sensor signal is either at full value (e.g. battery voltage with no signal or frequency) or, the signal exists but it is above the normal operating range (higher than the maximum operating limit). Possible fault in the temperature sensor or wiring e.g. short / open circuits, circuit resistance, incorrect sensor resistance or, fault with the reference / operating voltage; interference from other circuits can also affect sensor signals. Refer to list of sensor / actuator checks on page 19.
P0439	Catalyst Heater Control Circuit Bank 2	Heated catalysts are used to rapidly increase catalyst temperature / efficiency after engine starting.	The voltage, frequency or value of the control signal in the heater circuit is incorrect (undefined fault). Signal could be:- out of normal operating range, constant value, non-existent, corrupt. Possible fault with heater or wiring e.g. heater power supply, short / open circuit, high circuit resistance or incorrect heater resistance. Note: the control unit could be providing the correct signal but it is being affected by a circuit fault. Refer to list of sensor / actuator checks on page 19.
P043A	Catalyst Temperature Sensor Circuit Bank 2 Sensor 2	The sensor is used to ensure that the catalyst does not overheat, which will rapidly cause deterioration of the catalyst.	The voltage, frequency or value of the temperature sensor signal is incorrect (undefined fault). Sensor voltage / signal could be: out of normal operating range, constant value, non-existent, corrupt. Possible fault in the temperature sensor or wiring e.g. short / open circuits, circuit resistance, incorrect sensor resistance or, fault with the reference / operating voltage; interference from other circuits can also affect sensor signals. Refer to list of sensor / actuator checks on page 19. The code refers to a sensor / circuit fault but if the sensor does appear to be operating correctly, check for other faults that could affect catalyst temperatures e.g. high exhaust gas temperatures due to engine faults.
P043B	Catalyst Temperature Sensor Circuit Range / Performance Bank 2 Sensor 2	The sensor is used to ensure that the catalyst does not overheat, which will rapidly cause deterioration of the catalyst.	The voltage, frequency or value of the temperature sensor signal is within the normal operating range / tolerance but the signal is not plausible or is incorrect due to an undefined fault. The sensor signal might not match the operating conditions indicated by other sensors or, the sensor signal is not changing / responding as expected to the changes in conditions. Possible fault in the temperature sensor or wiring e.g. short / open circuits, circuit resistance, incorrect sensor resistance or, fault with the reference / operating voltage; interference from other circuits can also affect sensor signals. Refer to list of sensor / actuator checks on page 19. It is also possible that the temperature sensor is operating correctly but the catalyst temperature is unacceptable or incorrect for the operating conditions. Check for faults that could affect catalyst temperatures e.g. high exhaust gas temperatures caused by engine faults.
P043C	Catalyst Temperature Sensor Circuit Low Bank 2 Sensor 2	The sensor is used to ensure that the catalyst does not overheat, which will rapidly cause deterioration of the catalyst.	The voltage, frequency or value of the temperature sensor signal is either "zero" (no signal) or, the signal exists but it is below the normal operating range (lower than the minimum operating limit). Possible fault in the temperature sensor or wiring e.g. short / open circuits, circuit resistance, incorrect sensor resistance or, fault with the reference / operating voltage; interference from other circuits can also affect sensor signals. Refer to list of sensor / actuator checks on page 19.
P043D	Catalyst Temperature Sensor Circuit High Bank 2 Sensor 2	The sensor is used to ensure that the catalyst does not overheat, which will rapidly cause deterioration of the catalyst.	The voltage, frequency or value of the temperature sensor signal is either at full value (e.g. battery voltage with no signal or frequency) or, the signal exists but it is above the normal operating range (higher than the maximum operating limit). Possible fault in the temperature sensor or wiring e.g. short / open circuits, circuit resistance, incorrect sensor resistance or, fault with the reference / operating voltage; interference from other circuits can also affect sensor signals. Refer to list of sensor / actuator checks on page 19.
P043E	Evaporative Emission System Leak Detection Reference Orifice Low Flow	On some EVAP systems, a pump is used to provide pressure / vacuum to the fuel tank and EVAP system for leak detection (engine off); the pump applies the pressure / vacuum via a calibrated orifice. The control unit "learns" and then monitors the pump motor current required to achieve or maintain the pressure / vacuum level during the leak test cycle, and the pressure / vacuum level is also monitored.	The fault code indicates a low pressure / vacuum flow problem through the calibrated orifice. The fault could be related to a blockage in the calibrated orifice or, caused by a monitoring system fault (control unit, pressure sensor, pressure / vacuum pump etc). Refer to vehicle specific information to identify location and operation of the monitoring system.
P043F	Evaporative Emission System Leak Detection Reference Orifice High Flow	On some EVAP systems, a pump is used to provide pressure / vacuum to the fuel tank and EVAP system for leak detection (engine off); the pump applies the pressure / vacuum via a calibrated orifice. The control unit "learns" and then monitors the pump motor current required to achieve or maintain the pressure / vacuum level. The control unit will check the current draw to the pump motor at each leak test; if the current value is different from the "learnt" value, this will indicate a possible leak.	The fault code indicates a high pressure / vacuum flow problem through the calibrated orifice. The fault could be related to a problem with the calibrated orifice but, it is also possible that a leak exists in the EVAP system which is causing high flow. The fault could also be caused by a monitoring system fault (control unit, pressure sensor, pressure / vacuum pump etc). Refer to vehicle specific information to identify location and operation of the monitoring system.

NOTE 1. Check for other fault codes that could provide additional information.   **NOTE 2.** Communication between control units can pass via a CAN-Bus system; refer to Page 51 for CAN-Bus checks.   128

NOTE 3. If a fault cannot be located, it is also possible that the control unit is at fault.   **NOTE 4.** Refer to Page 53 for list of pages that show the ISO standard for component locations e.g. Sensor A - Bank 1.

code	Description		
P0440	Evaporative Emission System	The EVAP system removes fuel vapours from the canister and passes them to the intake manifold (under specific engine operating conditions); this provides a momentary enrichment of the mixture which is detected by the Lambda sensor. If the enrichment is not as expected, the engine control unit will regard this as a fault.	The fault code indicates an undefined fault with the EVAP system (probably assessed by the control unit due to incorrect or no enrichment). Refer to vehicle specific information to identify the components used on the EVAP system. Check condition of EVAP canister (canister can become flooded with petrol, also canister granules can deteriorate and will not store fuel vapours). Check for correct operation of purge valve, vent valve and other system valves / sensors. Check all hoses for connections and leaks. Check all applicable wiring for short / open circuits.
P0441	Evaporative Emission System Incorrect Purge Flow	The EVAP system removes fuel vapours from the canister and passes them to the intake manifold (under specific engine operating conditions); this provides a momentary enrichment of the mixture which is detected by the Lambda sensor. If the enrichment is not as expected, the engine control unit will regard this as a fault. Some systems use various sensors to monitor purge operation including pressure / vacuum changes within the fuel tank / EVAP system. Refer to vehicle specific information to identify the method of assessing purge flow.	The fault code indicates that the purge flow is not as expected during the purge cycle. Refer to vehicle specific information to identify the components used on the EVAP system. Check condition of EVAP canister (canister can become flooded with petrol, also canister granules can deteriorate and will not store fuel vapours). Check for correct operation of purge valve, vent valve and other system valves / sensors. Check all hoses for connections and leaks. Check all applicable wiring for short / open circuits.
P0442	Evaporative Emission System Leak Detected (small leak)	EVAP systems can use two types of leak detection. One type applies intake manifold vacuum via the purge valve at idle (the vent valve is closed thus sealing the EVAP system); if there is a leak, it will affect manifold vacuum / air flow. The second type has a pressure / vacuum pump in the tank, which applies pressure / vacuum when all EVAP system valves are closed; a pressure / vacuum sensor monitors the pressure / vacuum build up.	The fault code indicates that small leak has been detected (during the leak test) the fault is detected either by a change in manifold vacuum or, because of pressure / vacuum loss. Identify the components and design of the EVAP system and identify the type of leak test system. Depending on system design and the type of leak test, the problem could include fuel tank system and fuel filler cap. Check for other fault codes that could identify related faults. Check all applicable pipes, hoses and connections for leaks. Check for a leak in the vent valve and vent valve connections, also check purge valve connections for a leak. Note that the code refers to a "small" leak which could indicate minor problems with the hoses, connections etc.
P0443	Evaporative Emission System Purge Control Valve Circuit	EVAP systems make use of a canister to collect fuel vapours and then release them (under controlled conditions) to the engine intake system. The EVAP purge control valve is usually operated by an electrical actuator (usually a solenoid), which is provided by a control signal from the control unit. When the valve is opened, it allows fuel vapour to pass from the canister into the vehicle intake system (the valve is also opened during some leak detection checks).	The voltage, frequency or value of the control signal in the actuator circuit is incorrect (undefined fault). Signal could be:- out of normal operating range, constant value, non-existent, corrupt. Possible fault with actuator or wiring e.g. actuator power supply, short / open circuit, high circuit resistance or incorrect actuator resistance. Note: the control unit could be providing the correct signal but it is being affected by a circuit fault. Refer to list of sensor / actuator checks on page 19.
P0444	Evaporative Emission System Purge Control Valve Circuit Open	EVAP systems make use of a canister to collect fuel vapours and then release them (under controlled conditions) to the engine intake system. The EVAP purge control valve is usually operated by an electrical actuator (usually a solenoid), which is provided by a control signal from the control unit. When the valve is opened, it allows fuel vapour to pass from the canister into the vehicle intake system (the valve is also opened during some leak detection checks).	The voltage, frequency or value of the control signal in the actuator circuit is incorrect. Signal could be:- out of normal operating range, constant value, non-existent, corrupt. The fault code indicates a possible "open circuit" but other undefined faults in the actuator or wiring could activate the same code e.g. actuator power supply, short circuit, high circuit resistance or incorrect actuator resistance. Note: the control unit could be providing the correct signal but it is being affected by a circuit fault. Refer to list of sensor / actuator checks on page 19.
P0445	Evaporative Emission System Purge Control Valve Circuit Shorted	EVAP systems make use of a canister to collect fuel vapours and then release them (under controlled conditions) to the engine intake system. The EVAP purge control valve is usually operated by an electrical actuator (usually a solenoid), which is provided by a control signal from the control unit. When the valve is opened, it allows fuel vapour to pass from the canister into the vehicle intake system (the valve is also opened during some leak detection checks).	The voltage, frequency or value of the control signal in the actuator circuit is incorrect. The communication signal value / voltage is likely to be "zero" (no signal or voltage), which could indicate that the circuit is shorted to earth; it is also possible that the fault could relate to a short to "positive" (power supply), which would result in a permanent voltage existing in the circuit. Check actuator and wiring for a short circuit but note that other faults could activate the same fault code; check actuator and wiring for open circuits, high resistance, also check power supply to actuator. Refer to list of sensor / actuator checks on page 19.
P0446	Evaporative Emission System Vent Control Circuit	EVAP systems make use of a canister to collect fuel vapours and then release them (under controlled conditions) to the engine intake system. The vent control valve is operated by an electric actuator (usually a solenoid), which is provided by a control signal from the control unit. The valve allows air to enter the canister (when vapours are being drawn out of the canister) but, the valve is closed during the leak test cycle. Note that "Purge Valves" have also been referred to as "Vent Valves" on some vehicles and documentation (refer to P0443 for information on purge valves).	The voltage, frequency or value of the control signal in the actuator circuit is incorrect (undefined fault). Signal could be:- out of normal operating range, constant value, non-existent, corrupt. Possible fault with actuator or wiring e.g. actuator power supply, short / open circuit, high circuit resistance or incorrect actuator resistance. Note: the control unit could be providing the correct signal but it is being affected by a circuit fault. Refer to list of sensor / actuator checks on page 19.
P0447	Evaporative Emission System Vent Control Circuit Open	EVAP systems make use of a canister to collect fuel vapours and then release them (under controlled conditions) to the engine intake system. The vent control valve is operated by an electric actuator (usually a solenoid), which is provided by a control signal from the control unit. The valve allows air to enter the canister (when vapours are being drawn out of the canister) but, the valve is closed during the leak test cycle. Note that "Purge Valves" have also been referred to as "Vent Valves" on some vehicles and documentation (refer to P0443 for information on purge valves).	The voltage, frequency or value of the control signal in the actuator circuit is incorrect. Signal could be:- out of normal operating range, constant value, non-existent, corrupt. The fault code indicates a possible "open circuit" but other undefined faults in the actuator or wiring could activate the same code e.g. actuator power supply, short circuit, high circuit resistance or incorrect actuator resistance. Note: the control unit could be providing the correct signal but it is being affected by a circuit fault. Refer to list of sensor / actuator checks on page 19.

NOTE 1. Check for other fault codes that could provide additional information.   **NOTE 2.** Communication between control units can pass via a CAN-Bus system; refer to Page 51 for CAN-Bus checks.

NOTE 3. If a fault cannot be located, it is also possible that the control unit is at fault.   **NOTE 4.** Refer to Page 53 for list of pages that show the ISO standard for component locations e.g. Sensor A - Bank 1.

Fault code	EOBD / ISO Description	Component / System Description	Meaningful Description and Quick Check
P0448	Evaporative Emission System Vent Control Circuit Shorted	EVAP systems make use of a canister to collect fuel vapours and then release them (under controlled conditions) to the engine intake system. The vent control valve is operated by an electric actuator (usually a solenoid), which is provided by a control signal from the control unit. The valve allows air to enter the canister (when vapours are being drawn out of the canister) but, the valve is closed during the leak test cycle. Note that "Purge Valves" have also been referred to as "Vent Valves" on some vehicles and documentation (refer to P0443 for information on purge valves).	The voltage, frequency or value of the control signal in the actuator circuit is incorrect. The communication signal value / voltage is likely to be "zero" (no signal or voltage), which could indicate that the circuit is shorted to earth; it is also possible that the fault could relate to a short to "positive" (power supply), which would result in a permanent voltage existing in the circuit. Check actuator and wiring for a short circuit but note that other faults could activate the same fault code; check actuator and wiring for open circuits, high resistance, also check power supply to actuator. Refer to list of sensor / actuator checks on page 19.
P0449	Evaporative Emission System Vent Valve / Solenoid Circuit	EVAP systems make use of a canister to collect fuel vapours and then release them (under controlled conditions) to the engine intake system. The vent control valve is operated by an electric actuator (usually a solenoid), which is provided by a control signal from the control unit. The valve allows air to enter the canister (when vapours are being drawn out of the canister) but, the valve is closed during the leak test cycle. Note that "Purge Valves" have also been referred to as "Vent Valves" on some vehicles and documentation (refer to P0443 for information on purge valves).	The voltage, frequency or value of the control signal in the valve actuator / solenoid circuit is incorrect (undefined fault). Signal could be:- out of normal operating range, constant value, non-existent, corrupt. Possible fault with actuator / solenoid or wiring e.g. actuator / solenoid power supply, short / open circuit, high circuit resistance or incorrect actuator / solenoid resistance. Note: the control unit could be providing the correct signal but it is being affected by a circuit fault. Refer to list of sensor / actuator checks on page 19.
P0450	Evaporative Emission System Pressure Sensor / Switch	EVAP systems make use of a canister to collect fuel vapours and then release them (under controlled conditions) to the engine intake system. The EVAP pressure (or vacuum) sensor is used to check the EVAP system pressure / vacuum during a leak test cycle. On some systems, the sensor could also be used to measure fuel system pressure / vacuum during normal operation. Systems can apply the engine vacuum during leak test cycles or, systems can use a pressure / vacuum pump. The pressure / vacuum sensor or switch is used to indicate pressure / vacuum loss (to identify any leaks).	The voltage, frequency or value of the pressure sensor / switch signal is incorrect (undefined fault). Sensor / switch voltage / signal could be: out of normal operating range, constant value, non-existent, corrupt. Possible fault in the pressure sensor / switch or wiring e.g. short / open circuits, circuit resistance, incorrect sensor resistance (if applicable) or, fault with the reference / operating voltage; interference from other circuits can also affect sensor signals. Where a pressure switch is fitted, it is possible that the switch contacts are not opening / closing correctly or the switch contacts have a high resistance. Refer to list of sensor / actuator checks on page 19.
P0451	Evaporative Emission System Pressure Sensor / Switch Range / Performance	EVAP systems make use of a canister to collect fuel vapours and then release them (under controlled conditions) to the engine intake system. The EVAP pressure (or vacuum) sensor is used to check the EVAP system pressure / vacuum during a leak test cycle. On some systems, the sensor could also be used to measure fuel system pressure / vacuum during normal operation. Systems can apply the engine vacuum during leak test cycles or, systems can use a pressure / vacuum pump. The pressure / vacuum sensor or switch is used to indicate pressure / vacuum loss (to identify any leaks).	The voltage, frequency or value of the pressure sensor / switch signal is within the normal operating range / tolerance but the signal is not plausible or is incorrect due to an undefined fault. The sensor / switch signal might not match the operating conditions or, the sensor signal is not changing / responding as expected to the changes in conditions. Possible fault in the pressure sensor / switch or wiring e.g. short / open circuits, circuit resistance, incorrect sensor resistance (if applicable) or, fault with the reference / operating voltage; interference from other circuits can also affect sensor signals. Where a pressure switch is fitted, it is possible that the switch contacts are not opening / closing correctly or the switch contacts have a high resistance. Refer to list of sensor / actuator checks on page 19. It is also possible that the pressure sensor is operating correctly but the pressure / vacuum being measured by the sensor is unacceptable or incorrect for the operating conditions; if possible, use alternative method of checking pressure / vacuum to establish whether the pressure indicated by the sensor is correct / incorrect.
P0452	Evaporative Emission System Pressure Sensor / Switch Low	EVAP systems make use of a canister to collect fuel vapours and then release them (under controlled conditions) to the engine intake system. The EVAP pressure (or vacuum) sensor is used to check the EVAP system pressure / vacuum during a leak test cycle. On some systems, the sensor could also be used to measure fuel system pressure / vacuum during normal operation. Systems can apply the engine vacuum during leak test cycles or, systems can use a pressure / vacuum pump. The pressure / vacuum sensor or switch is used to indicate pressure / vacuum loss (to identify any leaks).	The voltage, frequency or value of the pressure sensor / switch signal is either "zero" (no signal) or, the signal exists but it is below the normal operating range (lower than the minimum operating limit). Possible fault in the pressure sensor / switch or wiring e.g. short / open circuits, circuit resistance, incorrect sensor resistance (if applicable) or, fault with the reference / operating voltage; interference from other circuits can also affect sensor / switch signals. Where a pressure switch is fitted, it is possible that the switch contacts are not opening / closing correctly or, the contacts are sticking (signal is remaining low when it should have risen to the high value); on some circuits, a high resistance on the contacts could also cause a low signal value to exist. Refer to list of sensor / actuator checks on page 19.
P0453	Evaporative Emission System Pressure Sensor / Switch High	EVAP systems make use of a canister to collect fuel vapours and then release them (under controlled conditions) to the engine intake system. The EVAP pressure (or vacuum) sensor is used to check the EVAP system pressure / vacuum during a leak test cycle. On some systems, the sensor could also be used to measure fuel system pressure / vacuum during normal operation. Systems can apply the engine vacuum during leak test cycles or, systems can use a pressure / vacuum pump. The pressure / vacuum sensor or switch is used to indicate pressure / vacuum loss (to identify any leaks).	The voltage, frequency or value of the pressure sensor / switch signal is either at full value (e.g. battery voltage with no signal or frequency) or, the signal exists but it is above the normal operating range (higher than the maximum operating limit). Possible fault in the pressure sensor / switch or wiring e.g. short / open circuits, circuit resistance, incorrect sensor resistance (if applicable) or, fault with the reference / operating voltage; interference from other circuits can also affect sensor / switch signals. Where a pressure switch is fitted, it is possible that the switch contacts are not opening / closing correctly or, the contacts are sticking (signal is remaining high when it should have dropped to low value); on some circuits, a high resistance on the contacts could also cause a high signal value to exist. Refer to list of sensor / actuator checks on page 19.
P0454	Evaporative Emission System Pressure Sensor / Switch Intermittent	EVAP systems make use of a canister to collect fuel vapours and then release them (under controlled conditions) to the engine intake system. The EVAP pressure (or vacuum) sensor is used to check the EVAP system pressure / vacuum during a leak test cycle. On some systems, the sensor could also be used to measure fuel system pressure / vacuum during normal operation. Systems can apply the engine vacuum during leak test cycles or, systems can use a pressure / vacuum pump. The pressure / vacuum sensor or switch is used to indicate pressure / vacuum loss (to identify any leaks).	The voltage, frequency or value of the pressure sensor / switch signal is intermittent (likely to be out of normal operating range / tolerance when intermittent fault is detected) or, the signal / voltage is erratic (e.g. signal changes are irregular / unpredictable). Possible fault with wiring or connections (including internal sensor / switch connections); also interference from other circuits can affect the signal. Where a pressure switch is fitted, it is possible that the switch contacts are not opening / closing correctly or the switch contacts have a high resistance. Refer to list of sensor / actuator checks on page 19. It is also possible that the pressure / vacuum could be changing erratically, which could cause the fault code to be activated; if necessary, check pressure with a separate gauge to check for erratic pressure changes.

code	Description		
P0455	Evaporative Emission System Leak Detected (large leak)	EVAP systems can use two types of leak detection. One type applies intake manifold vacuum via the purge valve at idle (the vent valve is closed thus sealing the EVAP system); if there is a leak, it will affect manifold vacuum / air flow. The second type has a pressure / vacuum pump in the tank, which applies pressure / vacuum when all EVAP system valves are closed; a pressure / vacuum sensor monitors the pressure / vacuum build up.	The fault code indicates that a large leak has been detected (during the leak test); the fault is detected either by a change in manifold vacuum or, because of pressure / vacuum loss. Identify the components and design of the EVAP system and identify the type of leak test system. Depending on system design and the type of leak test, the problem could include fuel tank system and fuel filler cap. Check for other fault codes that could identify related faults. Check all applicable pipes, hoses and connections for leaks. Check for a leak in the vent valve and vent valve connections, also check purge valve connections for a leak.
P0456	Evaporative Emission System Leak Detected (very small leak)	EVAP systems can use two types of leak detection. One type applies intake manifold vacuum via the purge valve at idle (the vent valve is closed thus sealing the EVAP system); if there is a leak, it will affect manifold vacuum / air flow. The second type has a pressure / vacuum pump in the tank, which applies pressure / vacuum when all EVAP system valves are closed; a pressure / vacuum sensor monitors the pressure / vacuum build up.	The fault code indicates that a very small leak has been detected (during the leak test) the fault is detected either by a change in manifold vacuum or, because of pressure / vacuum loss. Identify the components and design of the EVAP system and identify the type of leak test system. Depending on system design and the type of leak test, the problem could include fuel tank system and fuel filler cap. Check for other fault codes that could identify related faults. Check all applicable pipes, hoses and connections for leaks. Check for a leak in the vent valve and vent valve connections, also check purge valve connections for a leak. Note that the code refers to a "very small" leak which could indicate very minor problems with the hoses, connections etc.
P0457	Evaporative Emission System Leak Detected (fuel cap loose / off)	EVAP systems can use two types of leak detection. One type applies intake manifold vacuum via the purge valve at idle (the vent valve is closed thus sealing the EVAP system); if there is a leak, it will affect manifold vacuum / air flow. The second type has a pressure / vacuum pump in the tank, which applies pressure / vacuum when all EVAP system valves are closed; a pressure / vacuum sensor monitors the pressure / vacuum build up.	The fault code indicates that a leak has been detected which is likely to be caused by a fuel filler cap that is loose or not attached correctly; the fault is detected either by a change in manifold vacuum or, because of pressure / vacuum loss. Note that some systems have additional sensors that may allow detection of a loose or disconnected fuel filler cap. Check all applicable pipes, hoses and connections for leaks. Check for a leak in the vent valve and vent valve connections, also check purge valve connections for a leak.
P0458	Evaporative Emission System Purge Control Valve Circuit Low	EVAP systems make use of a canister to collect fuel vapours and then release them (under controlled conditions) to the engine intake system. The EVAP purge control valve is usually operated by an electrical actuator (usually a solenoid), which is provided by a control signal from the control unit. When the valve is opened, it allows fuel vapour to pass from the canister into the vehicle intake system (the valve is also opened during some leak detection checks).	The voltage, frequency or value of the control signal in the actuator circuit is either "zero" (no signal) or, the signal exists but is below the normal operating range. Possible fault with actuator or wiring (e.g. actuator power supply, short / open circuit or resistance) that is causing a signal value to be lower than the minimum operating limit. Note: the control unit could be providing the correct signal but it is being affected by a circuit fault. Refer to list of sensor / actuator checks on page 19.
P0459	Evaporative Emission System Purge Control Valve Circuit High	EVAP systems make use of a canister to collect fuel vapours and then release them (under controlled conditions) to the engine intake system. The EVAP purge control valve is usually operated by an electrical actuator (usually a solenoid), which is provided by a control signal from the control unit. When the valve is opened, it allows fuel vapour to pass from the canister into the vehicle intake system (the valve is also opened during some leak detection checks).	The voltage, frequency or value of the control signal in the actuator circuit is either at full value (e.g. battery voltage with no on / off signal) or, the signal exists but it is above the normal operating range. Possible fault with actuator or wiring (e.g. actuator power supply, short / open circuit or resistance) that is causing a signal value to be higher than the maximum operating limit. Note: the control unit could be providing the correct signal but it is being affected by a circuit fault. Refer to list of sensor / actuator checks on page 19.
P0460	Fuel Level Sensor "A" Circuit	Note that there could be more than one fuel level sensor; one sensor being used for the fuel gauge and the second providing a signal to the control unit for other purposes such as assisting Evaporative Emission control calculations. It is therefore possible that a fault has been detected due to the difference in the signals between the two sensors (the control unit may be able to assess which sensor is at fault due to fuel used compared with distance travelled).	The voltage, frequency or value of the level sensor signal is incorrect (undefined fault). Sensor voltage / signal could be: out of normal operating range, constant value, non-existent, corrupt. Possible fault in the level sensor or wiring e.g. short / open circuits, circuit resistance, incorrect sensor resistance or, fault with the reference / operating voltage; interference from other circuits can also affect sensor signals. Refer to list of sensor / actuator checks on page 19.
P0461	Fuel Level Sensor "A" Circuit Range / Performance	Note that there could be more than one fuel level sensor; one sensor being used for the fuel gauge and the second providing a signal to the control unit for other purposes such as assisting Evaporative Emission control calculations. It is therefore possible that a fault has been detected due to the difference in the signals between the two sensors (the control unit may be able to assess which sensor is at fault due to fuel used compared with distance travelled).	The voltage, frequency or value of the level sensor signal is within the normal operating range / tolerance but the signal is not plausible or is incorrect due to an undefined fault. The sensor signal might not match the operating conditions indicated by other sensors or, the sensor signal is not changing / responding as expected to the changes in conditions. Possible fault in the level sensor or wiring e.g. short / open circuits, circuit resistance, incorrect sensor resistance or, fault with the reference / operating voltage; interference from other circuits can also affect sensor signals. Refer to list of sensor / actuator checks on page 19. It is also possible that the sensor is operating correctly but the measurement being made by the sensor is unacceptable or incorrect (does not match the values expected / calculated by the control unit).
P0462	Fuel Level Sensor "A" Circuit Low	Note that there could be more than one fuel level sensor; one sensor being used for the fuel gauge and the second providing a signal to the control unit for other purposes such as assisting Evaporative Emission control calculations. It is therefore possible that a fault has been detected due to the difference in the signals between the two sensors (the control unit may be able to assess which sensor is at fault due to fuel used compared with distance travelled).	The voltage, frequency or value of the level sensor signal is either "zero" (no signal) or, the signal exists but it is below the normal operating range (lower than the minimum operating limit). Possible fault in the level sensor or wiring e.g. short / open circuits, circuit resistance, incorrect sensor resistance or, fault with the reference / operating voltage; interference from other circuits can also affect sensor signals. Refer to list of sensor / actuator checks on page 19.
P0463	Fuel Level Sensor "A" Circuit High	Note that there could be more than one fuel level sensor; one sensor being used for the fuel gauge and the second providing a signal to the control unit for other purposes such as assisting Evaporative Emission control calculations. It is therefore possible that a fault has been detected due to the difference in the signals between the two sensors (the control unit may be able to assess which sensor is at fault due to fuel used compared with distance travelled).	The voltage, frequency or value of the level sensor signal is either at full value (e.g. battery voltage with no signal or frequency) or, the signal exists but it is above the normal operating range (higher than the maximum operating limit). Possible fault in the level sensor or wiring e.g. short / open circuits, circuit resistance, incorrect sensor resistance or, fault with the reference / operating voltage; interference from other circuits can also affect sensor signals. Refer to list of sensor / actuator checks on page 19.

NOTE 1. Check for other fault codes that could provide additional information.

NOTE 2. Communication between control units can pass via a CAN-Bus system; refer to Page 51 for CAN-Bus checks.

NOTE 3. If a fault cannot be located, it is also possible that the control unit is at fault.

NOTE 4. Refer to Page 53 for list of pages that show the ISO standard for component locations e.g. Sensor A - Bank 1.

Fault code	EOBD / ISO Description	Component / System Description	Meaningful Description and Quick Check
P0464	Fuel Level Sensor "A" Circuit Intermittent	Note that there could be more than one fuel level sensor; one sensor being used for the fuel gauge and the second providing a signal to the control unit for other purposes such as assisting Evaporative Emission control calculations. It is therefore possible that a fault has been detected due to the difference in the signals between the two sensors (the control unit may be able to assess which sensor is at fault due to fuel used compared with distance travelled).	The voltage, frequency or value of the level sensor signal is intermittent (likely to be out of normal operating range / tolerance when intermittent fault is detected) or, the signal / voltage is erratic (e.g. signal changes are irregular / unpredictable). Possible fault with wiring or connections (including internal sensor connections); also interference from other circuits can affect the signal. Refer to list of sensor / actuator checks on page 19.
P0465	EVAP Purge Flow Sensor Circuit	The EVAP system removes fuel vapours from the canister and passes them to the intake manifold (under specific engine operating conditions). Some systems use a sensor to monitor the purge flow; the sensor is usually located in the vacuum pipe between the canister purge valve and the engine. Flow sensors can operate using a Thermistor (resistor that changes resistance with changes in temperature); when the purge flow valve opens, the flow through the pipe will cause a temperature change. Refer to vehicle specific information to identify the type of purge flow sensor fitted.	The voltage, frequency or value of the sensor signal is incorrect (undefined fault). Sensor voltage / signal could be: out of normal operating range, constant value, non-existent, corrupt. Possible fault in the sensor or wiring e.g. short / open circuits, circuit resistance, incorrect sensor resistance or, fault with the reference / operating voltage; interference from other circuits can also affect sensor signals. Refer to list of sensor / actuator checks on page 19.
P0466	EVAP Purge Flow Sensor Circuit Range / Performance	The EVAP system removes fuel vapours from the canister and passes them to the intake manifold (under specific engine operating conditions). Some systems use a sensor to monitor the purge flow; the sensor is usually located in the vacuum pipe between the canister purge valve and the engine. Flow sensors can operate using a Thermistor (resistor that changes resistance with changes in temperature); when the purge flow valve opens, the flow through the pipe will cause a temperature change. Refer to vehicle specific information to identify the type of purge flow sensor fitted.	The voltage, frequency or value of the sensor signal is within the normal operating range / tolerance but the signal is not plausible or is incorrect due to an undefined fault. The sensor signal may not be changing / responding as expected i.e. when the purge valve is open or closed. Possible fault in the sensor or wiring e.g. short / open circuits, circuit resistance, incorrect sensor resistance or, fault with the reference / operating voltage; interference from other circuits can also affect sensor signals. Refer to list of sensor / actuator checks on page 19. It is also possible that the sensor is operating correctly but the measurement being made by the sensor is unacceptable due to a fault in the EVAP purge system; this could be caused by reduced flow through the purge pipe (e.g. blockages in pipe or purge valve not opening correctly) or alternatively, there could be a flow through the purge pipe at incorrect times (e.g. purge valve remaining open or leak in pipe etc)..
P0467	EVAP Purge Flow Sensor Circuit Low	The EVAP system removes fuel vapours from the canister and passes them to the intake manifold (under specific engine operating conditions). Some systems use a sensor to monitor the purge flow; the sensor is usually located in the vacuum pipe between the canister purge valve and the engine. Flow sensors can operate using a Thermistor (resistor that changes resistance with changes in temperature); when the purge flow valve opens, the flow through the pipe will cause a temperature change. Refer to vehicle specific information to identify the type of purge flow sensor fitted.	The voltage, frequency or value of the sensor signal is either "zero" (no signal) or, the signal exists but it is below the normal operating range (lower than the minimum operating limit). Possible fault in the sensor or wiring e.g. short / open circuits, circuit resistance, incorrect sensor resistance or, fault with the reference / operating voltage; interference from other circuits can also affect sensor signals. Refer to list of sensor / actuator checks on page 19.
P0468	EVAP Purge Flow Sensor Circuit High	The EVAP system removes fuel vapours from the canister and passes them to the intake manifold (under specific engine operating conditions). Some systems use a sensor to monitor the purge flow; the sensor is usually located in the vacuum pipe between the canister purge valve and the engine. Flow sensors can operate using a Thermistor (resistor that changes resistance with changes in temperature); when the purge flow valve opens, the flow through the pipe will cause a temperature change. Refer to vehicle specific information to identify the type of purge flow sensor fitted.	The voltage, frequency or value of the sensor signal is either at full value (e.g. battery voltage with no signal or frequency) or, the signal exists but it is above the normal operating range (higher than the maximum operating limit). Possible fault in the sensor or wiring e.g. short / open circuits, circuit resistance, incorrect sensor resistance or, fault with the reference / operating voltage; interference from other circuits can also affect sensor signals. Refer to list of sensor / actuator checks on page 19.
P0469	EVAP Purge Flow Sensor Circuit Intermittent	The EVAP system removes fuel vapours from the canister and passes them to the intake manifold (under specific engine operating conditions); this provides a momentary enrichment of the mixture which is detected by the Lambda sensor. If the enrichment is not as expected, the engine control unit will regard this as a fault. Some systems use various sensors to monitor purge operation including pressure / vacuum changes within the fuel tank / EVAP system. Refer to vehicle specific information to identify the method of assessing purge flow.	The voltage, frequency or value of the sensor signal is intermittent (likely to be out of normal operating range / tolerance when intermittent fault is detected) or, the signal / voltage is erratic (e.g. signal changes are irregular / unpredictable). Possible fault with wiring or connections (including internal sensor connections); also interference from other circuits can affect the signal. Refer to list of sensor / actuator checks on page 19.
P0470	Exhaust Pressure Sensor "A" Circuit	Exhaust pressure sensors can be used for a number of different tasks e.g. the exhaust pressure signal is used for control of emissions related devices such as EGR systems, but the pressure sensors have also been used to monitor the exhaust pressure for systems where the exhaust gas is used to provide rapid heating for the cooling system (refer to fault code P0475). Exhaust pressure sensors have also been used to detect misfire conditions (refer to fault code P0300). If necessary, refer to vehicle specific information to identify the location and function of the pressure sensor.	The voltage, frequency or value of the pressure sensor signal is incorrect (undefined fault). Sensor voltage / signal could be: out of normal operating range, constant value, non-existent, corrupt. Possible fault in the pressure sensor or wiring e.g. short / open circuits, circuit resistance, incorrect sensor resistance (if applicable) or, fault with the reference / operating voltage; interference from other circuits can also affect sensor signals. Refer to list of sensor / actuator checks on page 19.

NOTE 1. Check for other fault codes that could provide additional information. NOTE 2. Communication between control units can pass via a CAN-Bus system; refer to Page 51 for CAN-Bus checks.
NOTE 3. If a fault cannot be located, it is also possible that the control unit is at fault. NOTE 4. Refer to Page 53 for list of pages that show the ISO standard for component locations e.g. Sensor A - Bank 1.

132

code	Description		
P0471	Exhaust Pressure Sensor "A" Circuit Range / Performance	Exhaust pressure sensors can be used for a number of different tasks e.g. the exhaust pressure signal is used for control of emissions related devices such as EGR systems, but the pressure sensors have also been used to monitor the exhaust pressure for systems where the exhaust gas is used to provide rapid heating for the cooling system (refer to fault code P0475). Exhaust pressure sensors have also been used to detect misfire conditions (refer to fault code P0300). If necessary, refer to vehicle specific information to identify the location and function of the pressure sensor.	The voltage, frequency or value of the pressure sensor signal is within the normal operating range / tolerance but the signal is not plausible or is incorrect due to an undefined fault. The sensor signal might not match the operating conditions indicated by other sensors or, the sensor signal is not changing / responding as expected to the changes in conditions. Possible fault in the pressure sensor or wiring e.g. short / open circuits, circuit resistance, incorrect sensor resistance (if applicable) or, fault with the reference / operating voltage; interference from other circuits can also affect sensor signals. Refer to list of sensor / actuator checks on page 19. Also check any air passages / pipes to the sensor for leaks / blockages etc. It is also possible that the pressure sensor is operating correctly but the pressure being measured by the sensor is unacceptable or incorrect for the operating conditions; if possible, use alternative method of checking pressure to establish if sensor system is operating correctly and whether the pressure indicated by the sensor is correct / incorrect.
P0472	Exhaust Pressure Sensor "A" Circuit Low	Exhaust pressure sensors can be used for a number of different tasks e.g. the exhaust pressure signal is used for control of emissions related devices such as EGR systems, but the pressure sensors have also been used to monitor the exhaust pressure for systems where the exhaust gas is used to provide rapid heating for the cooling system (refer to fault code P0475). Exhaust pressure sensors have also been used to detect misfire conditions (refer to fault code P0300). If necessary, refer to vehicle specific information to identify the location and function of the pressure sensor.	The voltage, frequency or value of the pressure sensor signal is either "zero" (no signal) or, the signal exists but it is below the normal operating range (lower than the minimum operating limit). Possible fault in the pressure sensor or wiring e.g. short / open circuits, circuit resistance, incorrect sensor resistance (if applicable) or, fault with the reference / operating voltage; interference from other circuits can also affect sensor signals. Refer to list of sensor / actuator checks on page 19.
P0473	Exhaust Pressure Sensor "A" Circuit High	Exhaust pressure sensors can be used for a number of different tasks e.g. the exhaust pressure signal is used for control of emissions related devices such as EGR systems, but the pressure sensors have also been used to monitor the exhaust pressure for systems where the exhaust gas is used to provide rapid heating for the cooling system (refer to fault code P0475). Exhaust pressure sensors have also been used to detect misfire conditions (refer to fault code P0300). If necessary, refer to vehicle specific information to identify the location and function of the pressure sensor.	The voltage, frequency or value of the pressure sensor signal is either at full value (e.g. battery voltage with no signal or frequency) or, the signal exists but it is above the normal operating range (higher than the maximum operating limit). Possible fault in the pressure sensor or wiring e.g. short / open circuits, circuit resistance, incorrect sensor resistance (if applicable) or, fault with the reference / operating voltage; interference from other circuits can also affect sensor signals. Refer to list of sensor / actuator checks on page 19.
P0474	Exhaust Pressure Sensor "A" Circuit Intermittent / Erratic	Exhaust pressure sensors can be used for a number of different tasks e.g. the exhaust pressure signal is used for control of emissions related devices such as EGR systems, but the pressure sensors have also been used to monitor the exhaust pressure for systems where the exhaust gas is used to provide rapid heating for the cooling system (refer to fault code P0475). Exhaust pressure sensors have also been used to detect misfire conditions (refer to fault code P0300). If necessary, refer to vehicle specific information to identify the location and function of the pressure sensor.	The voltage, frequency or value of the pressure sensor signal is intermittent (likely to be out of normal operating range / tolerance when intermittent fault is detected) or, the signal / voltage is erratic (e.g. signal changes are irregular / unpredictable). Possible fault with wiring or connections (including internal sensor connections); also interference from other circuits can affect the signal. Refer to list of sensor / actuator checks on page 19. It is also possible that the pressure could be changing erratically, which could cause the fault code to be activated; if necessary, check pressure with a separate gauge to check for erratic pressure changes.
P0475	Exhaust Pressure Control Valve	Exhaust pressure controls valves have been used (primarily on commercial vehicles) to regulate exhaust pressure after cold starts; by regulating / creating a back pressure in the exhaust, hot gasses are forced through a heat exchanger (to heat the engine coolant) thus achieving a more rapid warm up of the engine and cab heating system. Note that manufacturers could apply this code for other applications where exhaust gas control is used e.g. on an EGR system.	Identify the type of pressure control valve actuator (e.g. solenoid). The fault code indicates an unidentified malfunction with the control valve / actuator (probably detected due to an incorrect control signal value). The voltage, frequency or value of the control signal in the actuator circuit is incorrect (undefined fault). Signal could be:- out of normal operating range, constant value, non-existent, corrupt. Possible fault with actuator or wiring e.g. actuator power supply, short / open circuit, high circuit resistance or incorrect actuator resistance. Note: the control unit could be providing the correct signal but it is being affected by a circuit fault. Refer to list of sensor / actuator checks on page 19.
P0476	Exhaust Pressure Control Valve Range / Performance	Exhaust pressure controls valves have been used (primarily on commercial vehicles) to regulate exhaust pressure after cold starts; by regulating / creating a back pressure in the exhaust, hot gasses are forced through a heat exchanger (to heat the engine coolant) thus achieving a more rapid warm up of the engine and cab heating system. Note that manufacturers could apply this code for other applications where exhaust gas control is used e.g. on an EGR system.	Identify the type of pressure control valve actuator (e.g. solenoid). The voltage, frequency or value of the control signal in the actuator circuit is within the normal operating range / tolerance, but the signal might be incorrect for the operating conditions indicated by sensors. It is also possible that the actuator response (or response of the exhaust pressure control valve system) differs from the expected / desired response (incorrect response to the control signal provided by the control unit). Possible fault in actuator or wiring, but also possible that the actuator (or pressure control valve system) is not operating or moving correctly. Note: the control unit could be providing the correct signal but it is being affected by a circuit fault. Refer to list of sensor / actuator checks on page 19.
P0477	Exhaust Pressure Control Valve Low	Exhaust pressure controls valves have been used (primarily on commercial vehicles) to regulate exhaust pressure after cold starts; by regulating / creating a back pressure in the exhaust, hot gasses are forced through a heat exchanger (to heat the engine coolant) thus achieving a more rapid warm up of the engine and cab heating system. Note that manufacturers could apply this code for other applications where exhaust gas control is used e.g. on an EGR system.	Identify the type of pressure control valve actuator (e.g. solenoid). The voltage, frequency or value of the control signal in the actuator circuit is either "zero" (no signal) or, the signal exists but is below the normal operating range. Possible fault with actuator or wiring (e.g. actuator power supply, short / open circuit or resistance) that is causing a signal value to be lower than the minimum operating limit. Note: the control unit could be providing the correct signal but it is being affected by a circuit fault. Refer to list of sensor / actuator checks on page 19.

NOTE 1. Check for other fault codes that could provide additional information. **NOTE 2.** Communication between control units can pass via a CAN-Bus system; refer to Page 51 for CAN-Bus checks.

NOTE 3. If a fault cannot be located, it is also possible that the control unit is at fault. **NOTE 4.** Refer to Page 53 for list of pages that show the ISO standard for component locations e.g. Sensor A - Bank 1.

Fault code	EOBD / ISO Description	Component / System Description	Meaningful Description and Quick Check
P0478	Exhaust Pressure Control Valve High	Exhaust pressure controls valves have been used (primarily on commercial vehicles) to regulate exhaust pressure after cold starts; by regulating / creating a back pressure in the exhaust, hot gasses are forced through a heat exchanger (to heat the engine coolant) thus achieving a more rapid warm up of the engine and cab heating system. Note that manufacturers could apply this code for other applications where exhaust gas control is used e.g. on an EGR system.	Identify the type of pressure control valve actuator (e.g. solenoid). The voltage, frequency or value of the control signal in the actuator circuit is either at full value (e.g. battery voltage with no on / off signal) or, the signal exists but it is above the normal operating range. Possible fault with actuator or wiring (e.g. actuator power supply, short / open circuit or resistance) that is causing a signal value to be higher than the maximum operating limit. Note: the control unit could be providing the correct signal but it is being affected by a circuit fault. Refer to list of sensor / actuator checks on page 19.
P0479	Exhaust Pressure Control Valve Intermittent	Exhaust pressure controls valves have been used (primarily on commercial vehicles) to regulate exhaust pressure after cold starts; by regulating / creating a back pressure in the exhaust, hot gasses are forced through a heat exchanger (to heat the engine coolant) thus achieving a more rapid warm up of the engine and cab heating system. Note that manufacturers could apply this code for other applications where exhaust gas control is used e.g. on an EGR system.	Identify the type of pressure control valve actuator (e.g. solenoid). The voltage, frequency or value of the control signal in the actuator circuit is intermittent (likely to be out of normal operating range / tolerance when intermittent fault is detected) or, the signal is erratic e.g. signal changes are irregular / unpredictable. Possible fault with actuator or wiring (e.g. internal / external connections, broken wiring) that is causing intermittent or erratic signal to exist. Note: the control unit could be providing the correct signal but it is being affected by a circuit fault. Refer to list of sensor / actuator checks on page 19.
P047A	Exhaust Pressure Sensor "B" Circuit	Exhaust pressure sensors can be used for a number of different tasks e.g. the exhaust pressure signal is used for control of emissions related devices such as EGR systems, but the pressure sensors have also been used to monitor the exhaust pressure for systems where the exhaust gas is used to provide rapid heating for the cooling system (refer to fault code P0475). Exhaust pressure sensors have also been used to detect misfire conditions (refer to fault code P0300). If necessary, refer to vehicle specific information to identify the location and function of the pressure sensor.	The voltage, frequency or value of the pressure sensor signal is incorrect (undefined fault). Sensor voltage / signal could be: out of normal operating range, constant value, non-existent, corrupt. Possible fault in the pressure sensor or wiring e.g. short / open circuits, circuit resistance, incorrect sensor resistance (if applicable) or, fault with the reference / operating voltage; interference from other circuits can also affect sensor signals. Refer to list of sensor / actuator checks on page 19.
P047B	Exhaust Pressure Sensor "B" Circuit Range / Performance	Exhaust pressure sensors can be used for a number of different tasks e.g. the exhaust pressure signal is used for control of emissions related devices such as EGR systems, but the pressure sensors have also been used to monitor the exhaust pressure for systems where the exhaust gas is used to provide rapid heating for the cooling system (refer to fault code P0475). Exhaust pressure sensors have also been used to detect misfire conditions (refer to fault code P0300). If necessary, refer to vehicle specific information to identify the location and function of the pressure sensor.	The voltage, frequency or value of the pressure sensor signal is within the normal operating range / tolerance but the signal is not plausible or is incorrect due to an undefined fault. The sensor signal might not match the operating conditions indicated by other sensors or, the sensor signal is not changing / responding as expected to the changes in conditions. Possible fault in the pressure sensor or wiring e.g. short / open circuits, circuit resistance, incorrect sensor resistance (if applicable) or, fault with the reference / operating voltage; interference from other circuits can also affect sensor signals. Refer to list of sensor / actuator checks on page 19. Also check any air passages / pipes to the sensor for leaks / blockages etc. It is also possible that the pressure sensor is operating correctly but the pressure being measured by the sensor is unacceptable or incorrect for the operating conditions; if possible, use alternative method of checking pressure to establish if sensor system is operating correctly and whether the pressure indicated by the sensor is correct / incorrect.
P047C	Exhaust Pressure Sensor "B" Circuit Low	Exhaust pressure sensors can be used for a number of different tasks e.g. the exhaust pressure signal is used for control of emissions related devices such as EGR systems, but the pressure sensors have also been used to monitor the exhaust pressure for systems where the exhaust gas is used to provide rapid heating for the cooling system (refer to fault code P0475). Exhaust pressure sensors have also been used to detect misfire conditions (refer to fault code P0300). If necessary, refer to vehicle specific information to identify the location and function of the pressure sensor.	The voltage, frequency or value of the pressure sensor signal is either "zero" (no signal) or, the signal exists but it is below the normal operating range (lower than the minimum operating limit). Possible fault in the pressure sensor or wiring e.g. short / open circuits, circuit resistance, incorrect sensor resistance (if applicable) or, fault with the reference / operating voltage; interference from other circuits can also affect sensor signals. Refer to list of sensor / actuator checks on page 19.
P047D	Exhaust Pressure Sensor "B" Circuit High	Exhaust pressure sensors can be used for a number of different tasks e.g. the exhaust pressure signal is used for control of emissions related devices such as EGR systems, but the pressure sensors have also been used to monitor the exhaust pressure for systems where the exhaust gas is used to provide rapid heating for the cooling system (refer to fault code P0475). Exhaust pressure sensors have also been used to detect misfire conditions (refer to fault code P0300). If necessary, refer to vehicle specific information to identify the location and function of the pressure sensor.	The voltage, frequency or value of the pressure sensor signal is either at full value (e.g. battery voltage with no signal or frequency) or, the signal exists but it is above the normal operating range (higher than the maximum operating limit). Possible fault in the pressure sensor or wiring e.g. short / open circuits, circuit resistance, incorrect sensor resistance (if applicable) or, fault with the reference / operating voltage; interference from other circuits can also affect sensor signals. Refer to list of sensor / actuator checks on page 19.
P047E	Exhaust Pressure Sensor "B" Circuit Intermittent / Erratic	Exhaust pressure sensors can be used for a number of different tasks e.g. the exhaust pressure signal is used for control of emissions related devices such as EGR systems, but the pressure sensors have also been used to monitor the exhaust pressure for systems where the exhaust gas is used to provide rapid heating for the cooling system (refer to fault code P0475). Exhaust pressure sensors have also been used to detect misfire conditions (refer to fault code P0300). If necessary, refer to vehicle specific information to identify the location and function of the pressure sensor.	The voltage, frequency or value of the pressure sensor signal is intermittent (likely to be out of normal operating range / tolerance when intermittent fault is detected) or, the signal / voltage is erratic (e.g. signal changes are irregular / unpredictable). Possible fault with wiring or connections (including internal sensor connections); also interference from other circuits can affect the signal. Refer to list of sensor / actuator checks on page 19. It is also possible that the pressure could be changing erratically, which could cause the fault code to be activated; if necessary, check pressure with a separate gauge to check for erratic pressure changes.
P047F	ISO / SAE reserved		Not used, reserved for future allocation.

NOTE 1. Check for other fault codes that could provide additional information. NOTE 2. Communication between control units can pass via a CAN-Bus system; refer to Page 51 for CAN-Bus checks.

NOTE 3. If a fault cannot be located, it is also possible that the control unit is at fault. NOTE 4. Refer to Page 53 for list of pages that show the ISO standard for component locations e.g. Sensor A - Bank 1.

Code	Description		
P0480	Fan 1 Control Circuit	The fan control system helps regulate coolant temperature to suit the operating conditions; a fan control module can be used, which is controlled by the main engine control unit (note that air con control units can also control fan operation). A fan speed sensor can be used to indicate fan speed / performance. The control unit may also perform a test under specific conditions e.g. engine at idle, by applying full speed control signal and then monitoring the fan speed.	Note that the speed / power for some electric motors can be regulated using a frequency / pulse width control signal (provided by the control module on the earth or power circuit). Identify the type of speed control and wiring. The voltage, frequency or value of the control signal in the motor circuit is incorrect (undefined fault). Signal could be:- out of normal operating range, constant value, non-existent, corrupt. Possible fault with motor or wiring e.g. motor power supply, short / open circuit, high circuit resistance or incorrect motor resistance. Note: the control unit or relay (if fitted) could be providing the correct voltage / signal but it is being affected by a circuit fault. Refer to list of sensor / actuator checks on page 19.
P0481	Fan 2 Control Circuit	The fan control system helps regulate coolant temperature to suit the operating conditions; a fan control module can be used, which is controlled by the main engine control unit (note that air con control units can also control fan operation). A fan speed sensor can be used to indicate fan speed / performance. The control unit may also perform a test under specific conditions e.g. engine at idle, by applying full speed control signal and then monitoring the fan speed.	Note that the speed / power for some electric motors can be regulated using a frequency / pulse width control signal (provided by the control module on the earth or power circuit). Identify the type of speed control and wiring. The voltage, frequency or value of the control signal in the motor circuit is incorrect (undefined fault). Signal could be:- out of normal operating range, constant value, non-existent, corrupt. Possible fault with motor or wiring e.g. motor power supply, short / open circuit, high circuit resistance or incorrect motor resistance. Note: the control unit or relay (if fitted) could be providing the correct voltage / signal but it is being affected by a circuit fault. Refer to list of sensor / actuator checks on page 19.
P0482	Fan 3 Control Circuit	The fan control system helps regulate coolant temperature to suit the operating conditions; a fan control module can be used, which is controlled by the main engine control unit (note that air con control units can also control fan operation). A fan speed sensor can be used to indicate fan speed / performance. The control unit may also perform a test under specific conditions e.g. engine at idle, by applying full speed control signal and then monitoring the fan speed.	Note that the speed / power for some electric motors can be regulated using a frequency / pulse width control signal (provided by the control module on the earth or power circuit). Identify the type of speed control and wiring. The voltage, frequency or value of the control signal in the motor circuit is incorrect (undefined fault). Signal could be:- out of normal operating range, constant value, non-existent, corrupt. Possible fault with motor or wiring e.g. motor power supply, short / open circuit, high circuit resistance or incorrect motor resistance. Note: the control unit or relay (if fitted) could be providing the correct voltage / signal but it is being affected by a circuit fault. Refer to list of sensor / actuator checks on page 19.
P0483	Fan Rationality Check	The fan control system helps regulate coolant temperature to suit the operating conditions; a fan control module can be used, which is controlled by the main engine control unit (note that air con control units can also control fan operation). A fan speed sensor can be used to indicate fan speed / performance. The control unit may also perform a test under specific conditions e.g. engine at idle, by applying full speed control signal and then monitoring the fan speed.	The control unit has identified a fan control system problem or, the cooling system is not responding as expected when the fan system is operated e.g. coolant temperature remains high when fan control system switches on the cooling fans. The fault could be caused by a fan system fault or a cooling system fault e.g. low coolant level or restricted airflow through radiator. Possible fault with fan motor or wiring e.g. motor power supply, short / open circuit, high circuit resistance or incorrect motor resistance. Note: the control unit / relay could be providing the correct voltage / signal but it is being affected by a circuit fault. Refer to list of sensor / actuator checks on page 19. Also refer to fault code P0480 for additional information.
P0484	Fan Circuit Over Current	The fan control system helps regulate coolant temperature to suit the operating conditions; a fan control module can be used, which is controlled by the main engine control unit (note that air con control units can also control fan operation). A fan speed sensor can be used to indicate fan speed / performance. The control unit may also perform a test under specific conditions e.g. engine at idle, by applying full speed control signal and then monitoring the fan speed.	The control unit or fan control module has identified that the fan motor is drawing excessive current. It might be necessary to refer to vehicle specific information to identify the circuit and components. The fault could be caused by an electrical fault or a partial or full seizure of the motor / fan assembly. Possible fault with motor or wiring e.g. short / open circuit or incorrect motor resistance. Refer to list of sensor / actuator checks on page 19. Also refer to fault code P0480 for additional information.
P0485	Fan Power / Ground Circuit	The fan control system helps regulate coolant temperature to suit the operating conditions; a fan control module can be used, which is controlled by the main engine control unit (note that air con control units can also control fan operation). A fan speed sensor can be used to indicate fan speed / performance. The control unit may also perform a test under specific conditions e.g. engine at idle, by applying full speed control signal and then monitoring the fan speed.	Note that the speed / power for some electric motors can be regulated using a frequency / pulse width control signal (provided by the control module on the earth or power circuit). Identify the type of speed control and wiring. The voltage, frequency or value of the control signal in the motor circuit is incorrect (undefined fault). Signal could be:- out of normal operating range, constant value, non-existent, corrupt. Possible fault with motor or wiring e.g. motor power supply, short / open circuit, high circuit resistance or incorrect motor resistance. Note: the control unit or relay (if fitted) could be providing the correct voltage / signal but it is being affected by a circuit fault. Refer to list of sensor / actuator checks on page 19.
P0486	Exhaust Gas Recirculation Sensor "B" Circuit	Different EGR systems can use different types of sensor to monitor EGR valve operation. Systems can use a "Lift" or position sensor (often a potentiometer connected to the valve) to monitor EGR valve movement; other systems can use temperature sensors or pressure sensors to monitor the changes in the exhaust gas temperature or pressure in the EGR system. Also note that the effects of the EGR valve operation can be monitored by the intake airflow sensor; when the EGR valve opens, less air will be drawn through the intake system and this change is detected by the airflow sensor.	It will be necessary to identify the type of sensor used for monitoring EGR operation (e.g. lift / position, temperature or pressure). Note that for many systems, the intake airflow sensor can be the primary sensor for EGR operation or, it may provide additional information alongside other EGR sensors; it might therefore be necessary to check airflow sensor operation as well as checking EGR sensors. The voltage, frequency or value of the EGR sensor signal is incorrect (undefined fault). Sensor voltage, frequency or signal value could be: out of normal operating range, constant value, non-existent, corrupt. Possible fault in the sensor or wiring e.g. short / open circuits, circuit resistance, incorrect sensor resistance or, fault with the reference / operating voltage; interference from other circuits can also affect sensor signals. Refer to list of sensor / actuator checks on page 19. Also refer to fault code P0470 for pressure sensor checks.
P0487	Exhaust Gas Recirculation Throttle Control Circuit "A" / Open	The EGR throttle can be fitted to Diesel engines and to some petrol engines (usually Direct injection engines where under certain driving conditions, the normal throttle is held fully open). The EGR throttle can be located in the intake system and, will be closed off when the EGR valve is opened; partially closing the EGR throttle reduces intake air volume (air volume is replaced by EGR gases) and it also creates a negative pressure / vacuum in the intake to assist in drawing in the EGR gases. The EGR throttle can be controlled by an electric actuator (usually a motor).	The voltage, frequency or value of the control signal in the actuator circuit is incorrect. Signal could be:- out of normal operating range, constant value, non-existent, corrupt. The fault code indicates a possible "open circuit" but other undefined faults in the actuator or wiring could activate the same code e.g. actuator power supply, short circuit, high circuit resistance or incorrect actuator resistance. Note: the control unit could be providing the correct signal but it is being affected by the circuit fault. Refer to list of sensor / actuator checks on page 19.

NOTE 1. Check for other fault codes that could provide additional information.　　**NOTE 2.** Communication between control units can pass via a CAN-Bus system; refer to Page 51 for CAN-Bus checks.

NOTE 3. If a fault cannot be located, it is also possible that the control unit is at fault.　　**NOTE 4.** Refer to Page 53 for list of pages that show the ISO standard for component locations e.g. Sensor A - Bank 1.

Fault code	EOBD / ISO Description	Component / System Description	Meaningful Description and Quick Check
P0488	Exhaust Gas Recirculation Throttle Control Circuit "A" Range / Performance	The EGR throttle can be fitted to Diesel engines and to some petrol engines (usually Direct injection engines where under certain driving conditions, the normal throttle is held fully open). The EGR throttle can be located in the intake system and, will be closed off when the EGR valve is opened; partially closing the EGR throttle reduces intake air volume (air volume is replaced by EGR gases) and it also creates a negative pressure / vacuum in the intake to assist in drawing in the EGR gases. The EGR throttle can be controlled by an electric actuator (usually a motor).	The voltage, frequency or value of the control signal in the actuator circuit is within the normal operating range / tolerance, but the signal might be incorrect for the operating conditions. It is also possible that the actuator response (or response of system controlled by the actuator) differs from the expected / desired response (incorrect response for the control signal provided by the control unit). Possible fault in actuator or wiring, but also possible that the actuator (or mechanism / system controlled by the actuator) is not operating or moving correctly. Note: the control unit could be providing the correct signal but it is being affected by the circuit fault. Refer to list of sensor / actuator checks on page 19.
P0489	Exhaust Gas Recirculation Control Circuit "A" Low	The EGR throttle can be fitted to Diesel engines and to some petrol engines (usually Direct injection engines where under certain driving conditions, the normal throttle is held fully open). The EGR throttle can be located in the intake system and, will be closed off when the EGR valve is opened; partially closing the EGR throttle reduces intake air volume (air volume is replaced by EGR gases) and it also creates a negative pressure / vacuum in the intake to assist in drawing in the EGR gases. The EGR throttle can be controlled by an electric actuator (usually a motor).	The voltage, frequency or value of the control signal in the actuator circuit is either "zero" (no signal) or, the signal exists but is below the normal operating range. Possible fault with actuator or wiring (e.g. actuator power supply, short / open circuit or resistance) that is causing a signal value to be lower than the minimum operating limit. Note: the control unit could be providing the correct signal but it is being affected by the circuit fault. Refer to list of sensor / actuator checks on page 19.
P0490	Exhaust Gas Recirculation Control Circuit "A" High	The EGR throttle can be fitted to Diesel engines and to some petrol engines (usually Direct injection engines where under certain driving conditions, the normal throttle is held fully open). The EGR throttle can be located in the intake system and, will be closed off when the EGR valve is opened; partially closing the EGR throttle reduces intake air volume (air volume is replaced by EGR gases) and it also creates a negative pressure / vacuum in the intake to assist in drawing in the EGR gases. The EGR throttle can be controlled by an electric actuator (usually a motor).	The voltage, frequency or value of the control signal in the actuator circuit is either at full value (e.g. battery voltage with no on / off signal) or, the signal exists but it is above the normal operating range. Possible fault with actuator or wiring (e.g. actuator power supply, short / open circuit or resistance) that is causing a signal value to be higher than the maximum operating limit. Note: the control unit could be providing the correct signal but it is being affected by the circuit fault. Refer to list of sensor / actuator checks on page 19.
P0491	Secondary Air Injection System Insufficient Flow Bank 1	After cold starts, the secondary air injection system pumps air (oxygen) into the exhaust gas (which is rich with HC and CO), this helps the HC / CO to combust (combustion heat in exhaust gas helps to heat the catalyst); the additional air can be detected in a number of ways including using a Lambda (oxygen) sensor (detecting oxygen level) or a flow / pressure sensor. Some systems direct air from the main intake system and then use the main airflow sensor to detect increase in air flow. Note that some systems (pulse air systems) use negative exhaust pressure pulses to draw air into the exhaust system instead of an air pump. If additional air is not detected or quantity is incorrect, this is regarded as a secondary air system fault. It may be necessary to refer to vehicle specific information to identify the system and components used.	The fault code indicates that the air flow provided by the secondary air system (into the exhaust system) is incorrect / insufficient; the flow could be a calculated value based on the Lambda sensor signal or could be detected by the secondary air "flow / pressure" sensor. The fault could be caused by an electrical or a mechanical fault in the secondary air system. Check operation of the secondary air system; depending on the system design, this could involve checks on the secondary air control valves, valve control systems (vacuum control, vacuum pipes etc) and air pump. Check for blockages in the pipes and air filter. Check all applicable wiring and connections for short / open circuits and high resistance. Also check for correct supply voltage to applicable secondary air system components. Check that the non-return valve (in the air pipe or combined with switching valve) is operating correctly. Check for other fault codes e.g. flow / pressure sensor (if fitted) and oxygen / Lambda sensor fault codes that might indicate if the oxygen / Lambda sensor is at fault.
P0492	Secondary Air Injection System Insufficient Flow Bank 2	After cold starts, the secondary air injection system pumps air (oxygen) into the exhaust gas (which is rich with HC and CO), this helps the HC / CO to combust (combustion heat in exhaust gas helps to heat the catalyst); the additional air can be detected in a number of ways including using a Lambda (oxygen) sensor (detecting oxygen level) or a flow / pressure sensor. Some systems direct air from the main intake system and then use the main airflow sensor to detect increase in air flow. Note that some systems (pulse air systems) use negative exhaust pressure pulses to draw air into the exhaust system instead of an air pump. If additional air is not detected or quantity is incorrect, this is regarded as a secondary air system fault. It may be necessary to refer to vehicle specific information to identify the system and components used.	The fault code indicates that the air flow provided by the secondary air system (into the exhaust system) is incorrect / insufficient; the flow could be a calculated value based on the Lambda sensor signal or could be detected by the secondary air "flow / pressure" sensor. The fault could be caused by an electrical or a mechanical fault in the secondary air system. Check operation of the secondary air system; depending on the system design, this could involve checks on the secondary air control valves, valve control systems (vacuum control, vacuum pipes etc) and air pump. Check for blockages in the pipes and air filter. Check all applicable wiring and connections for short / open circuits and high resistance. Also check for correct supply voltage to applicable secondary air system components. Check that the non-return valve (in the air pipe or combined with switching valve) is operating correctly. Check for other fault codes e.g. flow / pressure sensor (if fitted) and oxygen / Lambda sensor fault codes that might indicate if the oxygen / Lambda sensor is at fault.
P0493	Fan Overspeed	The fan control system helps regulate coolant temperature to suit the operating conditions; a fan control module can be used, which is controlled by the main engine control unit (note that air / con control units can also control fan operation). A fan speed sensor can be used to indicate fan speed / performance. The control unit may also perform a test under specific conditions e.g. engine at idle, by applying full speed control signal and then monitoring the fan speed.	The control unit detects the fan motor is operating at a higher speed than expected (either during normal operation or during a test) or the fan speed has briefly exceeded the maximum specified value. The fault could be caused by a control system problem (control unit or fan control module selecting the incorrect speed for the conditions). Note that some systems may use a resistor pack (different resistances used for different speed requirements). Also check motor and wiring e.g. short / open circuit or incorrect motor resistance, and check control resistance values (if applicable). Refer to list of sensor / actuator checks on page 19. Also refer to fault code P0480 for additional information.

NOTE 1. Check for other fault codes that could provide additional information.
NOTE 3. If a fault cannot be located, it is also possible that the control unit is at fault

NOTE 2. Communication between control units can pass via a CAN-Bus system; refer to Page 51 for CAN-Bus checks.
NOTE 4. Refer to Page 53 for list of pages that show the ISO standard for component locations e.g. Sensor A - Bank 1.

code	Description		
P0494	Fan Speed Low	The fan control system helps regulate coolant temperature to suit the operating conditions; a fan control module can be used, which is controlled by the main engine control unit (note that air / con control units can also control fan operation). A fan speed sensor can be used to indicate fan speed / performance. The control unit may also perform a test under specific conditions e.g. engine at idle, by applying full speed control signal and then monitoring the fan speed.	The control unit detects the fan motor is operating at a lower speed than expected (either during normal operation or during a test). The fault could be caused by a control system problem (control unit or fan control module selecting the incorrect speed for the conditions). Note that some systems may use a resistor pack (different resistances used for different speed requirements). Also check motor and wiring e.g. short / open circuit, high circuit resistance or incorrect motor resistance, and check control resistance values (if applicable). Refer to list of sensor / actuator checks on page 19. Also refer to fault code P0480 for additional information.
P0495	Fan Speed High	The fan control system helps regulate coolant temperature to suit the operating conditions; a fan control module can be used, which is controlled by the main engine control unit (note that air con control units can also control fan operation). A fan speed sensor can be used to indicate fan speed / performance. The control unit may also perform a test under specific conditions e.g. engine at idle, by applying full speed control signal and then monitoring the fan speed.	The control unit detects the fan motor is operating at a higher speed than expected (either during normal operation or during a test). The fault could be caused by a control system problem (control unit or fan control module selecting the incorrect speed for the conditions). Note that some systems may use a resistor pack (different resistances used for different speed requirements). Also check motor and wiring e.g. short / open circuit or incorrect motor resistance, and check control resistance values (if applicable). Refer to list of sensor / actuator checks on page 19. Also refer to fault code P0480 for additional information.
P0496	Evaporative Emission System High Purge Flow	The EVAP system removes fuel vapours from the canister and passes them to the intake manifold (under specific engine operating conditions); this provides a momentary enrichment of the mixture which is detected by the Lambda sensor. If the enrichment is not as expected, the engine control unit will regard this as a fault. Some systems use various sensors to monitor purge operation including pressure / vacuum changes within the fuel tank / EVAP system. Refer to vehicle specific information to identify the method of assessing purge flow.	The fault code indicates that the purge flow is higher than expected during the purge cycle. Check condition of EVAP canister (canister can become flooded with petrol, also canister granules can deteriorate and will not store fuel vapours). Check for correct operation of purge valve, vent valve and other system valves / sensors. Check all hoses for connections and leaks. Check all applicable wiring for short / open circuits.
P0497	Evaporative Emission System Low Purge Flow	The EVAP system removes fuel vapours from the canister and passes them to the intake manifold (under specific engine operating conditions); this provides a momentary enrichment of the mixture which is detected by the Lambda sensor. If the enrichment is not as expected, the engine control unit will regard this as a fault. Some systems use various sensors to monitor purge operation including pressure / vacuum changes within the fuel tank / EVAP system. Refer to vehicle specific information to identify the method of assessing purge flow.	The fault code indicates that the purge flow is lower than expected during the purge cycle. Check condition of EVAP canister, check for correct operation of purge valve, vent valve and other system valves / sensors. Check all hoses for connections and leaks, also check for blockages in the pipes and valves. Check all applicable wiring for short / open circuits.
P0498	Evaporative Emission System Vent Valve Control Circuit Low	EVAP systems make use of a canister to collect fuel vapours and then release them (under controlled conditions) to the engine intake system. The vent control valve is operated by an electric actuator (usually a solenoid), which is provided by a control signal from the control unit. The valve allows air to enter the canister (when vapours are being drawn out of the canister) but, the valve is closed during the leak test cycle. Note that "Purge Valves" have also been referred to as "Vent Valves" on some vehicles and documentation (refer to P0443 for information on purge valves).	The voltage, frequency or value of the control signal in the actuator circuit is either "zero" (no signal) or, the signal exists but is below the normal operating range. Possible fault with actuator or wiring (e.g. actuator power supply, short / open circuit or resistance) that is causing a signal value to be lower than the minimum operating limit. Note: the control unit could be providing the correct signal but it is being affected by a circuit fault. Refer to list of sensor / actuator checks on page 19.
P0499	Evaporative Emission System Vent Valve Control Circuit High	EVAP systems make use of a canister to collect fuel vapours and then release them (under controlled conditions) to the engine intake system. The vent control valve is operated by an electric actuator (usually a solenoid), which is provided by a control signal from the control unit. The valve allows air to enter the canister (when vapours are being drawn out of the canister) but, the valve is closed during the leak test cycle. Note that "Purge Valves" have also been referred to as "Vent Valves" on some vehicles and documentation (refer to P0443 for information on purge valves).	The voltage, frequency or value of the control signal in the actuator circuit is either at full value (e.g. battery voltage with no on / off signal) or, the signal exists but it is above the normal operating range. Possible fault with actuator or wiring (e.g. actuator power supply, short / open circuit or resistance) that is causing a signal value to be higher than the maximum operating limit. Note: the control unit could be providing the correct signal but it is being affected by a circuit fault. Refer to list of sensor / actuator checks on page 19.
P0500	Vehicle Speed Sensor "A"	Note that the vehicle speed sensor could be an independent device (transmission or drive train located) or the vehicle speed could be indicated by a wheel speed sensor (ABS); refer to vehicle specific information if necessary. Identify the type of sensor e.g. Inductive, Hall effect etc; this will dictate the checks that are applicable.	The voltage, frequency or value of the speed sensor signal is incorrect (undefined fault). Sensor voltage / signal value could be: out of normal operating range, constant value, non-existent, corrupt. Possible fault in the speed sensor or wiring e.g. short / open circuits, circuit resistance, incorrect sensor resistance (if applicable) or, fault with the reference / operating voltage; interference from other circuits can also affect sensor signals Also check earth connections and cable screening. Refer to list of sensor / actuator checks on page 19. Note, because the fault is undefined, it is also possible that the speed sensor is operating correctly but the speed being measured by the sensor is unacceptable or incorrect for the operating conditions; if possible, use alternative method of checking speed to establish if sensor system is operating correctly and whether speed is correct / incorrect. Also check condition of reference teeth on trigger disc / rotor and ensure rotor is turning with shaft.

NOTE 1. Check for other fault codes that could provide additional information.   **NOTE 2.** Communication between control units can pass via a CAN-Bus system; refer to Page 51 for CAN-Bus checks.   **137**

NOTE 3. If a fault cannot be located, it is also possible that the control unit is at fault.   **NOTE 4.** Refer to Page 53 for list of pages that show the ISO standard for component locations e.g. Sensor A - Bank 1.

Fault code	EOBD / ISO Description	Component / System Description	Meaningful Description and Quick Check
P0501	Vehicle Speed Sensor "A" Range / Performance	Note that the vehicle speed sensor could be an independent device (transmission or drive train located) or the vehicle speed could be indicated by a wheel speed sensor (ABS); refer to vehicle specific information if necessary. Identify the type of sensor e.g. Inductive, Hall effect etc; this will dictate the checks that are applicable.	The voltage, frequency or value of the speed sensor signal is within the normal operating range / tolerance but the signal is not plausible or is incorrect due to an undefined fault. The sensor signal might not match the operating conditions indicated by other sensors or, the sensor signal is not changing / responding as expected to the changes in conditions. Possible fault in the speed sensor or wiring e.g. short / open circuits, circuit resistance, incorrect sensor resistance or, fault with the reference / operating voltage; interference from other circuits can also affect sensor signals. Refer to list of sensor / actuator checks on page 19. It is also possible that the speed sensor is operating correctly but the speed being measured by the sensor is unacceptable or incorrect for the operating conditions; if possible, use alternative method of checking speed to establish if sensor system is operating correctly and whether speed is correct / incorrect. Also check condition of reference teeth on trigger disc / rotor and ensure rotor is turning with shaft.
P0502	Vehicle Speed Sensor "A" Circuit Low	Note that the vehicle speed sensor could be an independent device (transmission or drive train located) or the vehicle speed could be indicated by a wheel speed sensor (ABS); refer to vehicle specific information if necessary. Identify the type of sensor e.g. Inductive, Hall effect etc; this will dictate the checks that are applicable.	The voltage, frequency or value of the speed sensor signal is either "zero" (no signal) or, the signal exists but it is below the normal operating range (lower than the minimum operating limit). Possible fault in the speed sensor or wiring e.g. short / open circuits, circuit resistance, incorrect sensor resistance or, fault with the reference / operating voltage; interference from other circuits can also affect sensor signals. Refer to list of sensor / actuator checks on page 19.
P0503	Vehicle Speed Sensor "A" Intermittent / Erratic / High	Note that the vehicle speed sensor could be an independent device (transmission or drive train located) or the vehicle speed could be indicated by a wheel speed sensor (ABS); refer to vehicle specific information if necessary. Identify the type of sensor e.g. Inductive, Hall effect etc; this will dictate the checks that are applicable.	Identify the type of sensor e.g. Inductive, Hall effect etc; this will dictate the checks that are applicable. The fault code indicates 3 possible types of failure. For the "intermittent / erratic" failure, the voltage, frequency or value of the speed sensor signal is intermittent (likely to be out of normal operating range / tolerance when intermittent fault is detected) or, the signal / voltage is erratic (e.g. signal changes are irregular / unpredictable). Possible fault with wiring or connections; also interference from other circuits can affect the signal. Check condition of reference teeth on trigger disc / rotor and ensure rotor is turning. For the "high" failure, the voltage, frequency or value of the speed sensor signal is either at full value (e.g. battery voltage with no signal or frequency) or, the signal exists but it is above the normal operating range (higher than the maximum operating limit). Carry out the same sensor and wiring checks identified for the intermittent / erratic fault. Refer to list of sensor / actuator checks on page 19.
P0504	Brake Switch "A" / "B" Correlation	In addition to normal brake light operation, the brake switch signal can be used in the control of engine systems, vehicle stability systems, transmission and other vehicle systems. Where more than one brake switch is used, the signals can be passed to the different vehicle systems e.g. one for stability control / ABS (and other systems), and one for brake light operation. Note that some brake switches can be conventional type switches operated by brake pedal movement, but some types are "Pressure Switches" which operate due to brake line pressure (can be fitted into ABS modulator assembly).	The signals from the two Brake switches are providing conflicting or implausible information e.g. one switch is "ON" (indicating brake operation or brake pedal depressed) but the other switch is indicating "OFF" (brake system not operating). Check the operation of the switches to establish if one of the switches is faulty or if there is a wiring fault; check for short or open circuits in the wiring, check that switch contacts are closing / opening correctly and check for high resistance at switch contacts. Check for reference voltage / power supply to the switches. Also check the mechanism operating the switches e.g. pedal and switch adjustment or, check that there is sufficient brake line pressure to operate switch (where applicable). Refer to list of sensor / actuator checks on page 19.
P0505	Idle Air Control System	The idle air control system is used to maintain the idle speed at the desired value and, to alter the idle speed with different temperature and load conditions (e.g. for cold running or when air conditioning is operating). Many different idle air control valves are used; it will be necessary to identify the types of air valves and actuators. Note that many vehicles are fitted with electronic throttle control which also controls the idle speed.	The fault code is indicating that there is an unidentified problem with the idle air control system. Identify the idle system used and components fitted to the system e.g. types of actuator such as throttle control or solenoid valve). Refer to list of sensor / actuator checks on page 19. In all cases, check connections between the idle control actuator (solenoid or motor etc) and the control unit; check for short / open circuits, high resistances, poor connections and power supply. Check for air leaks in the intake system that could allow air to bypass the throttle butterfly, also check for blockages or contamination in the idle air passages and around the throttle butterfly (especially on electronic controlled throttle systems). Check that the idle air valve / actuator is not seized or sticking (including electronic controlled throttles). Note that other engine or engine management system faults could affect the idle speed; check for other fault codes.
P0506	Idle Air Control System RPM Lower Than Expected	The idle air control system is used to maintain the idle speed at the desired value and, to alter the idle speed with different temperature and load conditions (e.g. for cold running or when air conditioning is operating). Many different idle air control valves are used; it will be necessary to identify the types of air valves and actuators. Note that many vehicles are fitted with electronic throttle control which also controls the idle speed.	The fault code is indicating that the idle speed is lower than expected under some or all operating conditions e.g. different temperature and load conditions. Identify the idle system used and components fitted to the system e.g. types of actuator such as throttle control or solenoid valve). Refer to list of sensor / actuator checks on page 19. In all cases, check connections between the idle control actuator (solenoid or motor etc) and the control unit; check for short / open circuits, high resistances, poor connections and power supply. Check for blockages or contamination in the idle air passages and around the throttle butterfly (especially on electronic controlled throttle systems), also check for intake system restrictions (air filter etc.). Check that the idle air valve / actuator is not seized or sticking (including electronic controlled throttles). Note that other engine or engine management system faults could affect the idle speed; check for other fault codes.
P0507	Idle Air Control System RPM Higher Than Expected	The idle air control system is used to maintain the idle speed at the desired value and, to alter the idle speed with different temperature and load conditions (e.g. for cold running or when air conditioning is operating). Many different idle air control valves are used; it will be necessary to identify the types of air valves and actuators. Note that many vehicles are fitted with electronic throttle control which also controls the idle speed.	The fault code is indicating that the idle speed is higher than expected under some or all operating conditions e.g. different temperature and load conditions. Identify the idle system used and components fitted to the system e.g. types of actuator such as throttle control or solenoid valve). Refer to list of sensor / actuator checks on page 19. In all cases, check connections between the idle control actuator (solenoid or motor etc) and the control unit; check for short / open circuits, high resistances, poor connections and power supply. Check for air leaks in the intake system that could allow air to bypass the throttle butterfly, also check that the idle air valve / actuator is not seized or sticking (including electronic controlled throttles). Note that other engine or engine management system faults could affect the idle speed; check for other fault codes.
P0508	Idle Air Control System Circuit Low	The idle air control system is used to maintain the idle speed at the desired value and, to alter the idle speed with different temperature and load conditions (e.g. for cold running or when air conditioning is operating). Many different idle air control valves are used; it will be necessary to identify the types of air valves and actuators. Note that many vehicles are fitted with electronic throttle control which also controls the idle speed.	Identify the type of electric actuator used to operate the idle air valve e.g. solenoid, motor etc. The voltage, frequency or value of the control signal in the actuator circuit is either "zero" (no signal) or, the signal exists but is below the normal operating range. Possible fault with actuator or wiring (e.g. actuator power supply, short / open circuit or resistance) that is causing a signal value to be lower than the minimum operating limit. Note: the control unit could be providing the correct signal but it is being affected by a circuit fault. Refer to list of sensor / actuator checks on page 19.

NOTE 1. Check for other fault codes that could provide additional information. **NOTE 2.** Communication between control units can pass via a CAN-Bus system; refer to Page 51 for CAN-Bus checks.
NOTE 3. If a fault cannot be located, it is also possible that the control unit is at fault. **NOTE 4.** Refer to Page 53 for list of pages that show the ISO standard for component locations e.g. Sensor A - Bank 1.

code	Description		
P0509	Idle Air Control System Circuit High	The idle air control system is used to maintain the idle speed at the desired value and, to alter the idle speed with different temperature and load conditions (e.g. for cold running or when air conditioning is operating). Many different idle air control valves are used; it will be necessary to identify the types of air valves and actuators. Note that many vehicles are fitted with electronic throttle control which also controls the idle speed.	Identify the type of electric actuator used to operate the idle air valve e.g. solenoid, motor etc. The voltage, frequency or value of the control signal in the actuator circuit is either at full value (e.g. battery voltage with no on / off signal) or, the signal exists but it is above the normal operating range. Possible fault with actuator or wiring (e.g. actuator power supply, short / open circuit or resistance) that is causing a signal value to be higher than the maximum operating limit. Note: the control unit could be providing the correct signal but it is being affected by a circuit fault. Refer to list of sensor / actuator checks on page 19.
P050A	Cold Start Idle Air Control System Performance	The idle air control system is used to maintain the idle speed at the desired value and, to alter the idle speed with different temperature and load conditions (e.g. for cold running or when air conditioning is operating). Some engines are fitted with more than one idle air valve; the main valve (or electronic throttle control on some engines controls idle air for most operating conditions but a separate valve can be used to add additional idle air after cold starts.	The fault code is indicating that there is an unidentified problem with the cold start idle air control system. Refer to list of sensor / actuator checks on page 19. In all cases, check connections between the cold start valve actuator and the control unit; check for short / open circuits, high resistances, poor connections and power supply. Check that the cold start idle air valve / actuator is not seized or sticking. Note that other engine or engine management system faults could affect the idle speed; check for other fault codes.
P050B	Cold Start Ignition Timing Performance	The ignition system "spark" timing will normally be altered to suit cold start and immediate post start conditions. The control unit is detecting a difference between the actual timing and the timing that should occur under cold start conditions.	It is possible that the fault is a control unit / software related fault, if possible check for any available software upgrades and refer to vehicle specific information if possible to identify timing specifications under cold start conditions. It is possible that the control unit could rely on information from ambient and cooling temperature to establish cold start conditions. Therefore, check for other fault codes and temperature sensor operation; if possible, refer to ""live data" for temperature sensor values and also, check "live data" for actual timing value and compare with specified timing command from control unit.
P050C	Cold Start Engine Coolant Temperature Performance	The engine control unit will monitor the coolant temperature (via the temperature sensor) after a cold start; the temperature increase (as the engine warms up) will be monitored over a time period and compared against expected values.	The fault code indicates that the coolant temperature is not correct or not changing / responding as expected after a cold start. Check cooling system operation and coolant level. Check thermostat operation and (if applicable) check operation of electrically heated thermostat. Check operation of coolant temperature sensor, which could be providing incorrect signal to control unit. Refer to list of sensor / actuator checks on page 19.
P050D	Cold Start Rough Idle	The control unit can use information from various sensors to identify engine performance e.g. crankshaft position sensor which will indicate the power / torque contribution provided by each cylinder (also refer to fault codes P0263 and P0300 for additional information).	The fault code indicates that the engine is not operating correctly after a cold start; check for other fault codes that might indicate additional information relating to the fault. Depending on the engine type, checks should be carried out on fuelling, ignition, and all mechanical aspects. Check for air leaks and injector faults. Check for correct operation of coolant temperature sensor.
P050E		ISO / SAE reserved	Not used, reserved for future allocation.
P050F		ISO / SAE reserved	Not used, reserved for future allocation.
P0510	Closed Throttle Position Switch	The throttle / pedal position sensor / switch could be located on the throttle body, throttle cable / linkage or on the throttle pedal. The "closed throttle" position signal can be used for engine control (fuelling, ignition emissions etc) but it can be used for control of transmission, air conditioning and other vehicle systems.	Note. The throttle switch /sensor assembly may contain more than one switch or sensor, refer to vehicle specific information to establish which sensor or switch is affected. The switch signal is incorrect (undefined fault), the signal could be: out of normal operating range, constant value, non-existent, corrupt. Possible fault in the position switch or wiring e.g. short / open circuits, circuit resistance, incorrect sensor resistance or, fault with the reference / operating voltage; the switch contacts could also be faulty e.g. not opening / closing correctly, stuck closed or open, or have a high contact resistance. Refer to list of sensor / actuator checks on page 19.
P0511	Idle Air Control Circuit	The idle air control system is used to maintain the idle speed at the desired value and, to alter the idle speed with different temperature and load conditions (e.g. for cold running or when air conditioning is operating). Many different idle air control valves are used; it will be necessary to identify the types of air valves and actuators. Note that many vehicles are fitted with electronic throttle control which also controls the idle speed.	Identify the type of electric actuator used to operate the idle air valve e.g. solenoid, motor etc. The voltage, frequency or value of the control signal in the actuator circuit is incorrect (undefined fault). Signal could be:- out of normal operating range, constant value, non-existent, corrupt. Possible fault with actuator or wiring e.g. actuator power supply, short / open circuit, high circuit resistance or incorrect actuator resistance. Note: the control unit could be providing the correct signal but it is being affected by a circuit fault. Refer to list of sensor / actuator checks on page 19.
P0512	Starter Request Circuit	Depending on the system, the starter relay is usually connected to the control unit (control unit completes earth circuit for relay energising winding, therefore the control unit is effectively requesting the relay to operate the starter motor); because the control unit is part of the energising winding circuit it is therefore able to monitor voltage in the circuit. Note that on other systems, the power feed to the relay energising winding (from the ignition switch start position) could also be regarded as part of the "starter request" circuit; some systems provide a separate monitoring wire from this circuit to the control unit.	Refer to list of sensor / actuator checks on page 19 for notes on relay checks. The voltage in the relay energising winding circuit is incorrect (undefined fault). The fault could be related to: the power supply to the energising winding (usually from ignition switch "start" position), the relay switched earth circuit (through to the control unit) or, a fault in relay energising winding. Check all wiring connected to the relay for short / open circuits, check energising winding for an open circuit. Check for power supply to energising winding (might only be available with ignition switch in start position). Also check "start inhibit" switch (if applicable), which can form part of the energising winding circuit on some vehicles. Note that some systems may be linked to an immobilzer which could influence the starter request signal.
P0513	Incorrect Immobilzer Key	It may be necessary to refer to vehicle specific information to identify the operation and components involved for the immobilizer system.	The fault code indicates that the control unit / immobilizer system does not recognise the key (or other device) being used to enter / start the vehicle. The fault could be related to a control unit or key programming error (new key or control unit fitted). If possible, carry out reprogramming / learning process to enable the system to recognise the key.

NOTE 1. Check for other fault codes that could provide additional information.
NOTE 3. If a fault cannot be located, it is also possible that the control unit is at fault.

NOTE 2. Communication between control units can pass via a CAN-Bus system; refer to Page 51 for CAN-Bus checks.
NOTE 4. Refer to Page 53 for list of pages that show the ISO standard for component locations e.g. Sensor A - Bank 1.

Fault code	EOBD / ISO Description	Component / System Description	Meaningful Description and Quick Check
P0514	Battery Temperature Sensor Circuit Range / Performance	The temperature sensor can be used to monitor the battery temperature, which can increase due to the heat caused by excessive charging / discharging conditions as well due to high ambient heat. Note that on some vehicles, the charging rate is regulated depending on battery temperature.	The voltage, frequency or value of the temperature sensor signal is within the normal operating range / tolerance but the signal is not plausible or is incorrect due to an undefined fault. The sensor signal might not match the operating conditions indicated by other sensors or, the sensor signal is not changing / responding as expected to the changes in conditions. Possible fault in the temperature sensor or wiring e.g. short / open circuits, circuit resistance, incorrect sensor resistance or, fault with the reference / operating voltage; interference from other circuits can also affect sensor signals. Refer to list of sensor / actuator checks on page 19. It is also possible that the temperature sensor is operating correctly but the temperature is unacceptable or incorrect for the operating conditions; if possible, establish whether temperature is correct / incorrect. If battery temperature is high, this can be caused by faults with the battery compartment cooling fan (if fitted) and excessive charge / discharge conditions; check charging voltage current, also check battery condition.
P0515	Battery Temperature Sensor Circuit	The temperature sensor can be used to monitor the battery temperature, which can increase due to the heat caused by excessive charging / discharging conditions as well due to high ambient heat. Note that on some vehicles, the charging rate is regulated depending on battery temperature.	The voltage, frequency or value of the temperature sensor signal is incorrect (undefined fault). Sensor voltage / signal could be: out of normal operating range, constant value, non-existent, corrupt. Possible fault in the temperature sensor or wiring e.g. short / open circuits, circuit resistance, incorrect sensor resistance or, fault with the reference / operating voltage; interference from other circuits can also affect sensor signals. Refer to list of sensor / actuator checks on page 19.
P0516	Battery Temperature Sensor Circuit Low	The temperature sensor can be used to monitor the battery temperature, which can increase due to the heat caused by excessive charging / discharging conditions as well due to high ambient heat. Note that on some vehicles, the charging rate is regulated depending on battery temperature.	The voltage, frequency or value of the temperature sensor signal is either "zero" (no signal) or, the signal exists but it is below the normal operating range (lower than the minimum operating limit). Possible fault in the temperature sensor or wiring e.g. short / open circuits, circuit resistance, incorrect sensor resistance or, fault with the reference / operating voltage; interference from other circuits can also affect sensor signals. Refer to list of sensor / actuator checks on page 19.
P0517	Battery Temperature Sensor Circuit High	The temperature sensor can be used to monitor the battery temperature, which can increase due to the heat caused by excessive charging / discharging conditions as well due to high ambient heat. Note that on some vehicles, the charging rate is regulated depending on battery temperature.	The voltage, frequency or value of the temperature sensor signal is either at full value (e.g. battery voltage with no signal or frequency) or, the signal exists but it is above the normal operating range (higher than the maximum operating limit). Possible fault in the temperature sensor or wiring e.g. short / open circuits, circuit resistance, incorrect sensor resistance or, fault with the reference / operating voltage; interference from other circuits can also affect sensor signals. Refer to list of sensor / actuator checks on page 19.
P0518	Idle Air Control Circuit Intermittent	The idle air control system is used to maintain the idle speed at the desired value and, to alter the idle speed with different temperature and load conditions (e.g. for cold running or when air conditioning is operating). Many different idle air control valves are used; it will be necessary to identify the types of air valves and actuators. Note that many vehicles are fitted with electronic throttle control which also controls the idle speed.	The voltage, frequency or value the control signal in the actuator circuit is intermittent (likely to be out of normal operating range / tolerance when intermittent fault is detected) or, the signal is erratic e.g. signal changes are irregular / unpredictable. Possible fault with actuator or wiring (e.g. internal / external connections, broken wiring) that is causing intermittent or erratic signal to exist. Note: the control unit could be providing the correct signal but it is being affected by a circuit fault. Refer to list of sensor / actuator checks on page 19.
P0519	Idle Air Control System Performance	The idle air control system is used to maintain the idle speed at the desired value and, to alter the idle speed with different temperature and load conditions (e.g. for cold running or when air conditioning is operating). Many different idle air control valves are used; it will be necessary to identify the types of air valves and actuators. Note that many vehicles are fitted with electronic throttle control which also controls the idle speed.	Identify the type of electric actuator used to operate the idle air valve e.g. solenoid, motor etc. The voltage, frequency or value of the control signal in the actuator circuit is within the normal operating range / tolerance, but the signal might be incorrect for the operating conditions. It is also possible that the actuator response (or response of system controlled by the actuator) differs from the expected / desired response (incorrect response to the control signal provided by the control unit). Possible fault in actuator or wiring, but also possible that the actuator (or mechanism / system controlled by the actuator) is not operating or moving correctly (check idle vale assembly for blockages, leaks or contamination). Note: the control unit could be providing the correct signal but it is being affected by a circuit fault. Refer to list of sensor / actuator checks on page 19, refer to fault code P0505 for additional information.
P0520	Engine Oil Pressure Sensor / Switch Circuit	An oil pressure sensor can provide a signal that changes in direct proportion to the changes in oil pressure; this can therefore allow a pressure gauge (driver display) to indicate the full range of oil pressure under all conditions. However some sensors and switches may only indicate a good or bad pressure value; this would be used to illuminate the warning light for low oil pressure.	The voltage, frequency or value of the pressure sensor signal is incorrect (undefined fault). Sensor voltage / signal could be: out of normal operating range, constant value, non-existent, corrupt. Possible fault in the pressure sensor or wiring e.g. short / open circuits, circuit resistance, incorrect sensor resistance (if applicable) or, fault with the reference / operating voltage; interference from other circuits can also affect sensor signals. Refer to list of sensor / actuator checks on page 19. In all cases, check oil pressure with a separate pressure gauge to establish whether the sensor / switch is at fault or if the oil pressure is incorrect. If a switch is used to indicate low oil pressure, the signal from the switch will effectively be either on or off i.e. acceptable oil pressure or unacceptable oil pressure; it is possible that the switch signal is indicating "acceptable" oil pressure with the engine not running (ignition on) which could activate the fault code.
P0521	Engine Oil Pressure Sensor / Switch Range / Performance	An oil pressure sensor can provide a signal that changes in direct proportion to the changes in oil pressure; this can therefore allow a pressure gauge (driver display) to indicate the full range of oil pressure under all conditions. However some sensors and switches may only indicate a good or bad pressure value; this would be used to illuminate the warning light for low oil pressure.	The voltage, frequency or value of the pressure sensor signal is within the normal operating range / tolerance but the signal is not plausible or is incorrect due to an undefined fault. The sensor signal might not match the operating conditions indicated by other sensors or, the sensor signal is not changing / responding as expected. Possible fault in the pressure sensor or wiring e.g. short / open circuits, circuit resistance, incorrect sensor resistance (if applicable) or, fault with the reference / operating voltage; interference from other circuits can also affect sensor signals. Refer to list of sensor / actuator checks on page 19. In all cases, check oil pressure with a separate pressure gauge to establish whether the sensor / switch is at fault. If a switch is used to indicate low oil pressure, the signal from the switch will effectively be either on or off i.e. acceptable oil pressure or unacceptable oil pressure; it is possible that the pressure switch signal is indicating "acceptable" oil pressure with the engine not running (ignition on) which could activate the fault code. Also check oil quantity / condition.

Code	Description		
P0522	Engine Oil Pressure Sensor / Switch Low	An oil pressure sensor can provide a signal that changes in direct proportion to the changes in oil pressure; this can therefore allow a pressure gauge (driver display) to indicate the full range of oil pressure under all conditions. However some sensors and switches may only indicate a good or bad pressure value; this would be used to illuminate the warning light for low oil pressure.	The voltage, frequency or value of the pressure sensor signal is either "zero" (no signal) or, the signal exists but it is below the normal operating range (lower than the minimum operating limit). Possible fault in the pressure sensor or wiring e.g. short / open circuits, circuit resistance, incorrect sensor resistance (if applicable) or, fault with the reference / operating voltage; interference from other circuits can also affect sensor signals. Refer to list of sensor / actuator checks on page 19. In all cases, check oil pressure with a separate pressure gauge to establish whether the sensor / switch is at fault or if the oil pressure is incorrect. If a switch is used to indicate low oil pressure, the signal from the switch will effectively be either on or off i.e. acceptable oil pressure or unacceptable oil pressure; it is possible that the pressure switch signal is stuck in the low or off value position at all times, which could activate the fault code.
P0523	Engine Oil Pressure Sensor / Switch High	An oil pressure sensor can provide a signal that changes in direct proportion to the changes in oil pressure; this can therefore allow a pressure gauge (driver display) to indicate the full range of oil pressure under all conditions. However some sensors and switches may only indicate a good or bad pressure value; this would be used to illuminate the warning light for low oil pressure.	The voltage, frequency or value of the pressure sensor signal is either at full value (e.g. battery voltage with no signal or frequency) or, the signal exists but it is above the normal operating range (higher than the maximum operating limit). Possible fault in the pressure sensor or wiring e.g. short / open circuits, circuit resistance, incorrect sensor resistance (if applicable) or, fault with the reference / operating voltage; interference from other circuits can also affect sensor signals. Refer to list of sensor / actuator checks on page 19. In all cases, check oil pressure with a separate pressure gauge to establish whether the sensor / switch is at fault or if the oil pressure is incorrect. If a switch is used to indicate low oil pressure, the signal from the switch will effectively be either on or off i.e. acceptable oil pressure or unacceptable oil pressure; it is possible that the pressure switch signal is indicating "acceptable" oil pressure with the engine not running (ignition on) which could activate the fault code.
P0524	Engine Oil Pressure Too Low	An oil pressure sensor can provide a signal that changes in direct proportion to the changes in oil pressure; this can therefore allow a pressure gauge (driver display) to indicate the full range of oil pressure under all conditions. However some sensors and switches may only indicate a good or bad pressure value; this would be used to illuminate the warning light for low oil pressure.	The fault code indicates that the oil pressure sensor signal is indicating that oil pressure is too low. The sensor signal is within normal operating range / tolerance but is below expected values or not acceptable for the operating conditions indicated by other sensors e.g. engine RPM. Check oil pressure using an independent gauge. Although the fault code refers to low oil pressure, the fault could also be activated due to a sensor related fault. Refer to list of sensor / actuator checks on page 19.
P0525	Cruise Control Servo Control Circuit Range / Performance	Due to the different designs of cruise control operation, it may be necessary to refer to vehicle specific information to identify components and system operation. For cruise control systems that make use of vacuum (applied to a vacuum servo unit) to move the throttle system (to hold / adjust speed), vacuum control valves can be used to regulate the vacuum being applied to the servo and a vent valve can also be used to release the vacuum from the servo. The valves will be opened using an electric actuator e.g. a solenoid. Note that on some systems, the vacuum control valve and the vent valve can be a combined unit (e.g. double action solenoid valve).	The voltage, frequency or value of the control signal in the actuator circuit is within the normal operating range / tolerance, but the signal might be incorrect for the operating conditions. It is also possible that the actuator response (or response of system controlled by the actuator) differs from the expected / desired response (incorrect response for the control signal provided by the control unit). Possible fault in actuator or wiring, but also possible that the actuator (or mechanism / system controlled by the actuator) is not operating or moving correctly. Note: the control unit could be providing the correct signal but it is being affected by the circuit fault. Refer to list of sensor / actuator checks on page 19.
P0526	Fan Speed Sensor Circuit	The fan control system helps regulate coolant temperature to suit the operating conditions; a fan control module can be used, which is controlled by the main engine control unit (note that air / con control units can also control fan operation). A fan speed sensor can be used to indicate fan speed / performance. A fan speed sensor can be used to indicate fan speed / performance. Identify the type of sensor e.g. Inductive, Hall effect etc; this will dictate the checks that are applicable.	The voltage, frequency or value of the speed sensor signal is incorrect (undefined fault). Sensor voltage / signal value could be: out of normal operating range, constant value, non-existent, corrupt. Possible fault in the speed sensor or wiring e.g. short / open circuits, circuit resistance, incorrect sensor resistance (if applicable) or, fault with the reference / operating voltage; interference from other circuits can also affect sensor signals Also check earth connections and cable screening. Refer to list of sensor / actuator checks on page 19.
P0527	Fan Speed Sensor Circuit Range / Performance	The fan control system helps regulate coolant temperature to suit the operating conditions; a fan control module can be used, which is controlled by the main engine control unit (note that air / con control units can also control fan operation). A fan speed sensor can be used to indicate fan speed / performance. A fan speed sensor can be used to indicate fan speed / performance. Identify the type of sensor e.g. Inductive, Hall effect etc; this will dictate the checks that are applicable.	The voltage, frequency or value of the speed sensor signal is within the normal operating range / tolerance but the signal is not plausible or is incorrect due to an undefined fault. The sensor signal might not match the operating conditions indicated by other sensors or, the sensor signal is not changing / responding as expected to the changes in conditions. Possible fault in the speed sensor or wiring e.g. short / open circuits, circuit resistance, incorrect sensor resistance, or fault with the reference / operating voltage; interference from other circuits can also affect sensor signals. Refer to list of sensor / actuator checks on page 19. It is also possible that the speed sensor is operating correctly but the speed being measured by the sensor is unacceptable or incorrect for the operating conditions. Check for free rotation of the fan motor and if applicable to the sensor design, check condition of reference teeth on trigger disc / rotor and ensure rotor is turning with shaft.
P0528	Fan Speed Sensor Circuit No Signal	The fan control system helps regulate coolant temperature to suit the operating conditions; a fan control module can be used, which is controlled by the main engine control unit (note that air / con control units can also control fan operation). A fan speed sensor can be used to indicate fan speed / performance. A fan speed sensor can be used to indicate fan speed / performance. Identify the type of sensor e.g. Inductive, Hall effect etc; this will dictate the checks that are applicable.	The voltage, frequency or value of the speed sensor signal is incorrect (undefined fault resulting in no signal being detected). Sensor voltage / signal value could be: constant value or non-existent. Possible fault in the speed sensor or wiring e.g. short / open circuits, circuit resistance, incorrect sensor resistance (if applicable) or, fault with the reference / operating voltage; interference from other circuits can also affect sensor signals Also check earth connections and cable screening. Refer to list of sensor / actuator checks on page 19. It is also possible that the speed sensor is operating correctly but the motor is not turning or the sensor trigger / reference mechanism is faulty e.g. check teeth on trigger disc / rotor and ensure rotor is turning with shaft.
P0529	Fan Speed Sensor Circuit Intermittent	The fan control system helps regulate coolant temperature to suit the operating conditions; a fan control module can be used, which is controlled by the main engine control unit (note that air / con control units can also control fan operation). A fan speed sensor can be used to indicate fan speed / performance. A fan speed sensor can be used to indicate fan speed / performance. Identify the type of sensor e.g. Inductive, Hall effect etc; this will dictate the checks that are applicable.	The voltage, frequency or value of the speed sensor signal is intermittent (likely to be out of normal operating range / tolerance when intermittent fault is detected) or, the signal / voltage is erratic (e.g. signal changes are irregular / unpredictable). Possible fault with wiring or connections (including internal sensor connections); also interference from other circuits can affect the signal. Refer to list of sensor / actuator checks on page 19. If applicable to the sensor design, check condition of reference teeth on trigger disc / rotor and ensure rotor is turning with shaft.

NOTE 1. Check for other fault codes that could provide additional information. **NOTE 2.** Communication between control units can pass via a CAN-Bus system; refer to Page 51 for CAN-Bus checks.

NOTE 3. If a fault cannot be located, it is also possible that the control unit is at fault. **NOTE 4.** Refer to Page 53 for list of pages that show the ISO standard for component locations e.g. Sensor A - Bank 1.

Fault code	EOBD / ISO Description	Component / System Description	Meaningful Description and Quick Check
P0530	A / C Refrigerant Pressure Sensor "A" Circuit	The information from the pressure sensor can assist in control of the air conditioning system. The information can be also used by the main control unit to indicate the load demand of the air conditioning system, which allows the control unit to regulate the idle speed, as well as for safety and other control purposes. It is normal practice to fit pressure sensors on the "High Pressure Side" of the system but it may be necessary to refer to vehicle specific information to identify location / function of the sensor.	The voltage, frequency or value of the pressure sensor signal is incorrect (undefined fault). Sensor voltage / signal could be: out of normal operating range, constant value, non-existent, corrupt. Possible fault in the pressure sensor or wiring e.g. short / open circuits, circuit resistance, incorrect sensor resistance (if applicable) or, fault with the reference / operating voltage; interference from other circuits can also affect sensor signals. Refer to list of sensor / actuator checks on page 19.
P0531	A / C Refrigerant Pressure Sensor "A" Circuit Range / Performance	The information from the pressure sensor can assist in control of the air conditioning system. The information can be also used by the main control unit to indicate the load demand of the air conditioning system, which allows the control unit to regulate the idle speed, as well as for safety and other control purposes. It is normal practice to fit pressure sensors on the "High Pressure Side" of the system but it may be necessary to refer to vehicle specific information to identify location / function of the sensor.	The voltage, frequency or value of the pressure sensor signal is within the normal operating range / tolerance but the signal is not plausible or is incorrect due to an undefined fault. The sensor signal might not match the operating conditions indicated by other sensors or, the sensor signal is not changing / responding as expected to the changes in conditions. Possible fault in the pressure sensor or wiring e.g. short / open circuits, circuit resistance, incorrect sensor resistance (if applicable) or, fault with the reference / operating voltage; interference from other circuits can also affect sensor signals. Refer to list of sensor / actuator checks on page 19. It is also possible that the pressure sensor is operating correctly but the refrigerant pressure is unacceptable or incorrect for the operating conditions; if possible, use alternative method of checking pressure to establish if sensor system is operating correctly and whether the pressure indicated by the sensor is correct / incorrect
P0532	A / C Refrigerant Pressure Sensor "A" Circuit Low	The information from the pressure sensor can assist in control of the air conditioning system. The information can be also used by the main control unit to indicate the load demand of the air conditioning system, which allows the control unit to regulate the idle speed, as well as for safety and other control purposes. It is normal practice to fit pressure sensors on the "High Pressure Side" of the system but it may be necessary to refer to vehicle specific information to identify location / function of the sensor.	The voltage, frequency or value of the pressure sensor signal is either "zero" (no signal) or, the signal exists but it is below the normal operating range (lower than the minimum operating limit). Possible fault in the pressure sensor or wiring e.g. short / open circuits, circuit resistance, incorrect sensor resistance (if applicable) or, fault with the reference / operating voltage; interference from other circuits can also affect sensor signals. Refer to list of sensor / actuator checks on page 19.
P0533	A / C Refrigerant Pressure Sensor "A" Circuit High	The information from the pressure sensor can assist in control of the air conditioning system. The information can be also used by the main control unit to indicate the load demand of the air conditioning system, which allows the control unit to regulate the idle speed, as well as for safety and other control purposes. It is normal practice to fit pressure sensors on the "High Pressure Side" of the system but it may be necessary to refer to vehicle specific information to identify location / function of the sensor.	The voltage, frequency or value of the pressure sensor signal is either at full value (e.g. battery voltage with no signal or frequency) or, the signal exists but it is above the normal operating range (higher than the maximum operating limit). Possible fault in the pressure sensor or wiring e.g. short / open circuits, circuit resistance, incorrect sensor resistance (if applicable) or, fault with the reference / operating voltage; interference from other circuits can also affect sensor signals. Refer to list of sensor / actuator checks on page 19.
P0534	A / C Refrigerant Charge Loss	The pressure sensor is used to detect the refrigerant pressure (usually on the high pressure side of the air conditioning system). The pressure signal value assists in the control of the A / C system, but the pressure signal can also be used by the main control unit to assess the load demand of the air conditioning system thus allowing the main control unit to regulate the idle speed as necessary (as well as performing safety and other control functions).	The fault code indicates a loss of refrigerant, which will usually be detected by a loss of pressure in the A / C system. It will be necessary to perform A / C system checks (pressures and refrigerant quantities etc) to establish whether the refrigerant is low or whether there is another type of A / C system fault. Note that it is possible that the pressure sensor or sensor wiring is faulty; check for short or open circuits and check for reference voltage / power supply to sensor. If A / C system pressure is correct, check sensor signal for correct value (refer to vehicle specific information). Refer to list of sensor / actuator checks on page 19.
P0535	A / C Evaporator Temperature Sensor Circuit	The sensor is used to measure the temperature around the air conditioning evaporator unit (to check for evaporator and system operation); the sensor is usually located on the outside of the evaporator, typically at the coldest point. If necessary, refer to vehicle specific information to identify exact location of sensor.	The voltage, frequency or value of the temperature sensor signal is incorrect (undefined fault). Sensor voltage / signal could be: out of normal operating range, constant value, non-existent, corrupt. Possible fault in the temperature sensor or wiring e.g. short / open circuits, circuit resistance, incorrect sensor resistance or, fault with the reference / operating voltage; interference from other circuits can also affect sensor signals. Refer to list of sensor / actuator checks on page 19. The fault code relates to a sensor / sensor circuit fault but if the sensor does appear to be operating correctly, check operation of air conditioning system.
P0536	A / C Evaporator Temperature Sensor Circuit Range / Performance	The sensor is used to measure the temperature around the air conditioning evaporator unit (to check for evaporator and system operation); the sensor is usually located on the outside of the evaporator, typically at the coldest point. If necessary, refer to vehicle specific information to identify exact location of sensor.	The voltage, frequency or value of the temperature sensor signal is within the normal operating range / tolerance but the signal is not plausible or is incorrect due to an undefined fault. The sensor signal might not match the operating conditions indicated by other sensors or, the sensor signal is not changing / responding as expected to the changes in conditions. Possible fault in the temperature sensor or wiring e.g. short / open circuits, circuit resistance, incorrect sensor resistance or, fault with the reference / operating voltage; interference from other circuits can also affect sensor signals. Refer to list of sensor / actuator checks on page 19. It is also possible that the temperature sensor is operating correctly but the temperature being measured by the sensor is unacceptable or incorrect for the operating conditions; if possible, use alternative method of checking temperature to establish if sensor system is operating correctly and whether temperature is correct / incorrect. If temperature is incorrect, check system pressures and refrigerant quantity.
P0537	A / C Evaporator Temperature Sensor Circuit Low	The sensor is used to measure the temperature around the air conditioning evaporator unit (to check for evaporator and system operation); the sensor is usually located on the outside of the evaporator, typically at the coldest point. If necessary, refer to vehicle specific information to identify exact location of sensor.	The voltage, frequency or value of the temperature sensor signal is either "zero" (no signal) or, the signal exists but it is below the normal operating range (lower than the minimum operating limit). Possible fault in the temperature sensor or wiring e.g. short / open circuits, circuit resistance, incorrect sensor resistance or, fault with the reference / operating voltage; interference from other circuits can also affect sensor signals. Refer to list of sensor / actuator checks on page 19.
P0538	A / C Evaporator Temperature Sensor Circuit High	The sensor is used to measure the temperature around the air conditioning evaporator unit (to check for evaporator and system operation); the sensor is usually located on the outside of the evaporator, typically at the coldest point. If necessary, refer to vehicle specific information to identify exact location of sensor.	The voltage, frequency or value of the temperature sensor signal is either at full value (e.g. battery voltage with no signal or frequency) or, the signal exists but it is above the normal operating range (higher than the maximum operating limit). Possible fault in the temperature sensor or wiring e.g. short / open circuits, circuit resistance, incorrect sensor resistance or, fault with the reference / operating voltage; interference from other circuits can also affect sensor signals. Refer to list of sensor / actuator checks on page 19.

NOTE 1. Check for other fault codes that could provide additional information.

NOTE 2. Communication between control units can pass via a CAN-Bus system; refer to Page 51 for CAN-Bus checks.

NOTE 3. If a fault cannot be located, it is also possible that the control unit is at fault

NOTE 4. Refer to Page 53 for list of pages that show the ISO standard for component locations e.g. Sensor A - Bank 1.

code	Description		
P0539	A / C Evaporator Temperature Sensor Circuit Intermittent	The sensor is used to measure the temperature around the air conditioning evaporator unit (to check for evaporator and system operation); the sensor is usually located on the outside of the evaporator, typically at the coldest point. If necessary, refer to vehicle specific information to identify exact location of sensor.	The voltage, frequency or value of the temperature sensor signal is intermittent (likely to be out of normal operating range / tolerance when intermittent fault is detected) or, the signal / voltage is erratic (e.g. signal changes are irregular / unpredictable). Possible fault with wiring or connections (including internal sensor connections); also interference from other circuits can affect the signal. Refer to list of sensor / actuator checks on page 19.
P053A	Positive Crankcase Ventilation Heater Control Circuit / Open	A heater can be fitted to the Positive Crankcase Ventilation system (PCV) to maintain the temperature of the gasses passing from the crankcase to the induction system and aid combustion of the vented crankcase fumes.	The voltage, frequency or value of the control signal in the heater circuit is incorrect. Signal could be:- out of normal operating range, constant value, non-existent, corrupt. The fault code indicates a possible "open circuit" but other undefined faults in the heater or wiring could activate the same code e.g. heater power supply, short / circuit, high circuit resistance or incorrect heater resistance. Note: the control unit could be providing the correct signal but it is being affected by a circuit fault. Refer to list of sensor / actuator checks on page 19.
P053B	Positive Crankcase Ventilation Heater Control Circuit Low	A heater can be fitted to the Positive Crankcase Ventilation system (PCV) to maintain the temperature of the gasses passing from the crankcase to the induction system and aid combustion of the vented crankcase fumes.	The voltage, frequency or value of the control signal in the heater circuit is either "zero" (no signal) or, the signal exists but is below the normal operating range. Possible fault with heater or wiring (e.g. heater power supply, short / open circuit or resistance) that is causing a signal value to be lower than the minimum operating limit. Note: the control unit could be providing the correct signal but it is being affected by a circuit fault. Refer to list of sensor / actuator checks on page 19.
P053C	Positive Crankcase Ventilation Heater Control Circuit High	A heater can be fitted to the Positive Crankcase Ventilation system (PCV) to maintain the temperature of the gasses passing from the crankcase to the induction system and aid combustion of the vented crankcase fumes.	The voltage, frequency or value of the control signal in the heater circuit is either at full value (e.g. battery voltage with no on / off signal) or, the signal exists but it is above the normal operating range. Possible fault with heater or wiring (e.g. heater power supply, short / open circuit or resistance) that is causing a signal value to be higher than the maximum operating limit. Note: the control unit could be providing the correct signal but it is being affected by a circuit fault. Refer to list of sensor / actuator checks on page 19.
P0540	Intake Air Heater "A" Circuit	Note that some heaters are simple electrical heating elements but some may use the heater element to ignite a dedicated fuel supply. The heater could be provided with a power supply via a relay, with the control unit switching the relay on / off; alternatively, a control unit / module could be providing a direct switching of the heater. Refer to vehicle specific information to identify the type of heater used and the wiring / control for the heater.	The heater circuit is likely to refer to the circuit connecting the heater element to a power supply and earth path. The voltage, frequency or value of the control signal in the heater circuit is incorrect (undefined fault). Signal could be:- out of normal operating range, constant value, non-existent, corrupt. Possible fault with heater or wiring e.g. heater power supply, short / open circuit, high circuit resistance or incorrect heater resistance. Note: If the heater is directly controlled via a control unit / module, the control unit could be providing the correct control signal but it is being affected by a circuit fault. Refer to list of sensor / actuator checks on page 19.
P0541	Intake Air Heater "A" Circuit Low	Note that some heaters are simple electrical heating elements but some may use the heater element to ignite a dedicated fuel supply. The heater could be provided with a power supply via a relay, with the control unit switching the relay on / off; alternatively, a control unit / module could be providing a direct switching of the heater. Refer to vehicle specific information to identify the type of heater used and the wiring / control for the heater.	The heater circuit is likely to refer to the circuit connecting the heater element to a power supply and earth path. The voltage, frequency or value of the signal in the heater circuit is either "zero" (no signal) or, the signal exists but is below the normal operating range. Possible fault with heater or wiring (e.g. heater power supply, short / open circuit or resistance) that is causing a signal value to be lower than the minimum operating limit. Note: If the heater is directly controlled via a control unit / module, the control unit could be providing the correct control signal but it is being affected by a circuit fault. Refer to list of sensor / actuator checks on page 19.
P0542	Intake Air Heater "A" Circuit High	Note that some heaters are simple electrical heating elements but some may use the heater element to ignite a dedicated fuel supply. The heater could be provided with a power supply via a relay, with the control unit switching the relay on / off; alternatively, a control unit / module could be providing a direct switching of the heater. Refer to vehicle specific information to identify the type of heater used and the wiring / control for the heater.	The heater circuit is likely to refer to the circuit connecting the heater element to a power supply and earth path. The voltage, frequency or value of the control in the heater circuit is either at full value (e.g. battery voltage with no on / off signal) or, the signal exists but it is above the normal operating range. Possible fault with heater or wiring (e.g. heater power supply, short / open circuit or resistance) that is causing a signal value to be higher than the maximum operating limit. Note: If the heater is directly controlled via a control unit / module, the control unit could be providing the correct control signal but it is being affected by a circuit fault. Refer to list of sensor / actuator checks on page 19.
P0543	Intake Air Heater "A" Circuit Open	Note that some heaters are simple electrical heating elements but some may use the heater element to ignite a dedicated fuel supply. The heater could be provided with a power supply via a relay, with the control unit switching the relay on / off; alternatively, a control unit / module could be providing a direct switching of the heater. Refer to vehicle specific information to identify the type of heater used and the wiring / control for the heater.	The heater circuit is likely to refer to the circuit connecting the heater element to a power supply and earth path. The voltage, frequency or value of the signal in the heater circuit is non-existent, out of normal operating range, constant value, corrupt, but the fault likely to be caused by an open circuit . Possible fault with heater or wiring e.g. heater power supply, short / open circuit, high circuit resistance or incorrect heater resistance. Note: If the heater is directly controlled via a control unit / module, the control unit could be providing the correct control signal but it is being affected by a circuit fault. Refer to list of sensor / actuator checks on page 19.
P0544	Exhaust Gas Temperature Sensor Circuit Bank 1 Sensor 1	The exhaust gas temperature sensor signal is used to indicate if the temperature is outside of acceptable limits (some emissions control functions are more efficient when the gas temperature is within specified limits). Excessive temperatures can rapidly cause deterioration of the catalyst and other emissions control components. High gas temperatures are often caused by faults in the engine operation e.g. misfires, air leaks, fuelling etc.	The voltage, frequency or value of the temperature sensor signal is incorrect (undefined fault). Sensor voltage / signal could be: out of normal operating range, constant value, non-existent, corrupt. Possible fault in the temperature sensor or wiring e.g. short / open circuits, circuit resistance, incorrect sensor resistance or, fault with the reference / operating voltage; interference from other circuits can also affect sensor signals. Refer to list of sensor / actuator checks on page 19.
P0545	Exhaust Gas Temperature Sensor Circuit Low Bank 1 Sensor 1	The exhaust gas temperature sensor signal is used to indicate if the temperature is outside of acceptable limits (some emissions control functions are more efficient when the gas temperature is within specified limits). Excessive temperatures can rapidly cause deterioration of the catalyst and other emissions control components. High gas temperatures are often caused by faults in the engine operation e.g. misfires, air leaks, fuelling etc.	The voltage, frequency or value of the temperature sensor signal is either "zero" (no signal) or, the signal exists but it is below the normal operating range (lower than the minimum operating limit). Possible fault in the temperature sensor or wiring e.g. short / open circuits, circuit resistance, incorrect sensor resistance or, fault with the reference / operating voltage; interference from other circuits can also affect sensor signals. Refer to list of sensor / actuator checks on page 19.

NOTE 1. Check for other fault codes that could provide additional information.　　NOTE 2. Communication between control units can pass via a CAN-Bus system; refer to Page 51 for CAN-Bus checks.

NOTE 3. If a fault cannot be located, it is also possible that the control unit is at fault.　　NOTE 4. Refer to Page 53 for list of pages that show the ISO standard for component locations e.g. Sensor A - Bank 1.

Fault code	EOBD / ISO Description	Component / System Description	Meaningful Description and Quick Check
P0546	Exhaust Gas Temperature Sensor Circuit High Bank 1 Sensor 1	The exhaust gas temperature sensor signal is used to indicate if the temperature is outside of acceptable limits (some emissions control functions are more efficient when the gas temperature is within specified limits). Excessive temperatures can rapidly cause deterioration of the catalyst and other emissions control components. High gas temperatures are often caused by faults in the engine operation e.g. misfires, air leaks, fuelling etc.	The voltage, frequency or value of the temperature sensor signal is either at full value (e.g. battery voltage with no signal or frequency) or, the signal exists but it is above the normal operating range (higher than the maximum operating limit). Possible fault in the temperature sensor or wiring e.g. short / open circuits, circuit resistance, incorrect sensor resistance or, fault with the reference / operating voltage; interference from other circuits can also affect sensor signals. Refer to list of sensor / actuator checks on page 19.
P0547	Exhaust Gas Temperature Sensor Circuit Bank 2 Sensor 1	The exhaust gas temperature sensor signal is used to indicate if the temperature is outside of acceptable limits (some emissions control functions are more efficient when the gas temperature is within specified limits). Excessive temperatures can rapidly cause deterioration of the catalyst and other emissions control components. High gas temperatures are often caused by faults in the engine operation e.g. misfires, air leaks, fuelling etc.	The voltage, frequency or value of the temperature sensor signal is incorrect (undefined fault). Sensor voltage / signal could be: out of normal operating range, constant value, non-existent, corrupt. Possible fault in the temperature sensor or wiring e.g. short / open circuits, circuit resistance, incorrect sensor resistance or, fault with the reference / operating voltage; interference from other circuits can also affect sensor signals. Refer to list of sensor / actuator checks on page 19.
P0548	Exhaust Gas Temperature Sensor Circuit Low Bank 2 Sensor 1	The exhaust gas temperature sensor signal is used to indicate if the temperature is outside of acceptable limits (some emissions control functions are more efficient when the gas temperature is within specified limits). Excessive temperatures can rapidly cause deterioration of the catalyst and other emissions control components. High gas temperatures are often caused by faults in the engine operation e.g. misfires, air leaks, fuelling etc.	The voltage, frequency or value of the temperature sensor signal is either "zero" (no signal) or, the signal exists but it is below the normal operating range (lower than the minimum operating limit). Possible fault in the temperature sensor or wiring e.g. short / open circuits, circuit resistance, incorrect sensor resistance or, fault with the reference / operating voltage; interference from other circuits can also affect sensor signals. Refer to list of sensor / actuator checks on page 19.
P0549	Exhaust Gas Temperature Sensor Circuit High Bank 2 Sensor 1	The exhaust gas temperature sensor signal is used to indicate if the temperature is outside of acceptable limits (some emissions control functions are more efficient when the gas temperature is within specified limits). Excessive temperatures can rapidly cause deterioration of the catalyst and other emissions control components. High gas temperatures are often caused by faults in the engine operation e.g. misfires, air leaks, fuelling etc.	The voltage, frequency or value of the temperature sensor signal is either at full value (e.g. battery voltage with no signal or frequency) or, the signal exists but it is above the normal operating range (higher than the maximum operating limit). Possible fault in the temperature sensor or wiring e.g. short / open circuits, circuit resistance, incorrect sensor resistance or, fault with the reference / operating voltage; interference from other circuits can also affect sensor signals. Refer to list of sensor / actuator checks on page 19.
P0550	Power Steering Pressure Sensor / Switch Circuit	The pressure in a power steering system can be monitored using either a pressure switch or a pressure sensor. The switch / sensor can be used to indicate power steering operation to the engine control unit to enable idle speed adjustment. For those systems using hydraulic pumps driven by electric motors (and possibly for systems with mechanical driven pump), the sensor can provide pressure information to enable pressure monitoring and control as necessary (speed of electric driven pump is altered to suit vehicle speed and steering angle thus requiring pressure monitoring) .	The voltage, frequency or value of the pressure sensor / switch signal is incorrect (undefined fault). Sensor / switch voltage / signal could be: out of normal operating range, constant value, non-existent, corrupt. Possible fault in the pressure sensor / switch or wiring e.g. short / open circuits, circuit resistance, incorrect sensor resistance (if applicable) or, fault with the reference / operating voltage; interference from other circuits can also affect sensor signals. Where a pressure switch is fitted, it is possible that the switch contacts are not opening / closing correctly or the switch contacts have a high resistance. Refer to list of sensor / actuator checks on page 19.
P0551	Power Steering Pressure Sensor / Switch Circuit Range / Performance	The pressure in a power steering system can be monitored using either a pressure switch or a pressure sensor. The switch / sensor can be used to indicate power steering operation to the engine control unit to enable idle speed adjustment. For those systems using hydraulic pumps driven by electric motors (and possibly for systems with mechanical driven pump), the sensor can provide pressure information to enable pressure monitoring and control as necessary (speed of electric driven pump is altered to suit vehicle speed and steering angle thus requiring pressure monitoring) .	The voltage, frequency or value of the pressure sensor / switch signal is within the normal operating range / tolerance but the signal is not plausible or is incorrect due to an undefined fault. The sensor / switch signal might not match the operating conditions or, the sensor signal is not changing / responding as expected to the changes in conditions. Possible fault in the pressure sensor / switch or wiring e.g. short / open circuits, circuit resistance, incorrect sensor resistance (if applicable) or, fault with the reference / operating voltage; interference from other circuits can also affect sensor signals. Where a pressure switch is fitted, it is possible that the switch contacts are not opening / closing correctly or the switch contacts have a high resistance. Refer to list of sensor / actuator checks on page 19. It is also possible that the pressure sensor is operating correctly but the pressure being measured by the sensor is unacceptable or incorrect for the operating conditions; if possible, use alternative method of checking pressure to establish whether the pressure indicated by the sensor is correct / incorrect.
P0552	Power Steering Pressure Sensor / Switch Circuit Low	The pressure in a power steering system can be monitored using either a pressure switch or a pressure sensor. The switch / sensor can be used to indicate power steering operation to the engine control unit to enable idle speed adjustment. For those systems using hydraulic pumps driven by electric motors (and possibly for systems with mechanical driven pump), the sensor can provide pressure information to enable pressure monitoring and control as necessary (speed of electric driven pump is altered to suit vehicle speed and steering angle thus requiring pressure monitoring) .	The voltage, frequency or value of the pressure sensor / switch signal is either "zero" (no signal) or, the signal exists but it is below the normal operating range (lower than the minimum operating limit). Possible fault in the pressure sensor / switch or wiring e.g. short / open circuits, circuit resistance, incorrect sensor resistance (if applicable) or, fault with the reference / operating voltage; interference from other circuits can also affect sensor / switch signals. Where a pressure switch is fitted, it is possible that the switch contacts are not opening / closing correctly or, the contacts are sticking (signal is remaining low when it should have risen to the high value); on some circuits, a high resistance on the contacts could also cause a low signal value to exist. Refer to list of sensor / actuator checks on page 19.
P0553	Power Steering Pressure Sensor / Switch Circuit High	The pressure in a power steering system can be monitored using either a pressure switch or a pressure sensor. The switch / sensor can be used to indicate power steering operation to the engine control unit to enable idle speed adjustment. For those systems using hydraulic pumps driven by electric motors (and possibly for systems with mechanical driven pump), the sensor can provide pressure information to enable pressure monitoring and control as necessary (speed of electric driven pump is altered to suit vehicle speed and steering angle thus requiring pressure monitoring) .	The voltage, frequency or value of the pressure sensor / switch signal is either at full value (e.g. battery voltage with no signal or frequency) or, the signal exists but it is above the normal operating range (higher than the maximum operating limit). Possible fault in the pressure sensor / switch or wiring e.g. short / open circuits, circuit resistance, incorrect sensor resistance (if applicable) or, fault with the reference / operating voltage; interference from other circuits can also affect sensor / switch signals. Where a pressure switch is fitted, it is possible that the switch contacts are not opening / closing correctly or, the contacts are sticking (signal is remaining high when it should have dropped to low value); on some circuits, a high resistance on the contacts could also cause a high signal value to exist. Refer to list of sensor / actuator checks on page 19.

NOTE 1. Check for other fault codes that could provide additional information.   **NOTE 2.** Communication between control units can pass via a CAN-Bus system; refer to Page 51 for CAN-Bus checks.
NOTE 3. If a fault cannot be located, it is also possible that the control unit is at fault.   **NOTE 4.** Refer to Page 53 for list of pages that show the ISO standard for component locations e.g. Sensor A - Bank 1.

code	Description		
P0554	Power Steering Pressure Sensor / Switch Circuit Intermittent	The pressure in a power steering system can be monitored using either a pressure switch or a pressure sensor. The switch / sensor can be used to indicate power steering operation to the engine control unit to enable idle speed adjustment. For those systems using hydraulic pumps driven by electric motors (and possibly for systems with mechanical driven pump), the sensor can provide pressure information to enable pressure monitoring and control as necessary (speed of electric driven pump is altered to suit vehicle speed and steering angle thus requiring pressure monitoring) .	The voltage, frequency or value of the pressure sensor / switch signal is intermittent (likely to be out of normal operating range / tolerance when intermittent fault is detected) or, the signal / voltage is erratic (e.g. signal changes are irregular / unpredictable). Possible fault with wiring or connections (including internal sensor / switch connections); also interference from other circuits can affect the signal. Where a pressure switch is fitted, it is possible that the switch contacts are not opening / closing correctly or the switch contacts have a high resistance. Refer to list of sensor / actuator checks on page 19. It is also possible that the pressure could be changing erratically, which could cause the fault code to be activated; if necessary, check pressure with a separate gauge to check for erratic pressure changes.
P0555	Brake Booster Pressure Sensor Circuit	A pressure / vacuum sensor can be fitted to monitor the vacuum at the brake servo (brake booster). Note that on engines with vacuum pumps (including some petrol engines) the pump operation may be dependent on the level of pressure / vacuum detected at the brake booster. Also note that on some petrol engines (with direct injection), the engine can operate during some conditions without throttle control (throttle held open), therefore vacuum is not always created in the intake system; if the pressure sensor indicates insufficient vacuum for brake booster operation, the throttle can then be positioned to momentarily produce vacuum in the intake system. The brake booster pressure sensor can be connected to the engine control unit or to the ABS or other control units (depending on the systems fitted to the vehicle).	The voltage, frequency or value of the pressure sensor signal is incorrect (undefined fault). Sensor voltage / signal could be: out of normal operating range, constant value, non-existent, corrupt. Possible fault in the pressure sensor or wiring e.g. short / open circuits, circuit resistance, incorrect sensor resistance (if applicable) or, fault with the reference / operating voltage; interference from other circuits can also affect sensor signals. Refer to list of sensor / actuator checks on page 19.
P0556	Brake Booster Pressure Sensor Circuit Range / Performance	A pressure / vacuum sensor can be fitted to monitor the vacuum at the brake servo (brake booster). Note that on engines with vacuum pumps (including some petrol engines) the pump operation may be dependent on the level of pressure / vacuum detected at the brake booster. Also note that on some petrol engines (with direct injection), the engine can operate during some conditions without throttle control (throttle held open), therefore vacuum is not always created in the intake system; if the pressure sensor indicates insufficient vacuum for brake booster operation, the throttle can then be positioned to momentarily produce vacuum in the intake system. The brake booster pressure sensor can be connected to the engine control unit or to the ABS or other control units (depending on the systems fitted to the vehicle).	The voltage, frequency or value of the pressure sensor signal is within the normal operating range / tolerance but the signal is not plausible or is incorrect due to an undefined fault. The sensor signal might not match the operating conditions indicated by other sensors or, the sensor signal is not changing / responding as expected to the changes in conditions. Possible fault in the pressure sensor or wiring e.g. short / open circuits, circuit resistance, incorrect sensor resistance (if applicable) or, fault with the reference / operating voltage; interference from other circuits can also affect sensor signals. Refer to list of sensor / actuator checks on page 19. Also check any air passages / pipes to the sensor for leaks / blockages etc. It is also possible that the pressure sensor is operating correctly but the pressure / vacuum being measured by the sensor is unacceptable or incorrect for the operating conditions; if possible, use alternative method of checking pressure / vacuum to establish if sensor system is operating correctly and whether the pressure / vacuum indicated by the sensor is correct / incorrect.
P0557	Brake Booster Pressure Sensor Circuit Low	A pressure / vacuum sensor can be fitted to monitor the vacuum at the brake servo (brake booster). Note that on engines with vacuum pumps (including some petrol engines) the pump operation may be dependent on the level of pressure / vacuum detected at the brake booster. Also note that on some petrol engines (with direct injection), the engine can operate during some conditions without throttle control (throttle held open), therefore vacuum is not always created in the intake system; if the pressure sensor indicates insufficient vacuum for brake booster operation, the throttle can then be positioned to momentarily produce vacuum in the intake system. The brake booster pressure sensor can be connected to the engine control unit or to the ABS or other control units (depending on the systems fitted to the vehicle).	The voltage, frequency or value of the pressure sensor signal is either "zero" (no signal) or, the signal exists but it is below the normal operating range (lower than the minimum operating limit). Possible fault in the pressure sensor or wiring e.g. short / open circuits, circuit resistance, incorrect sensor resistance (if applicable) or, fault with the reference / operating voltage; interference from other circuits can also affect sensor signals. Refer to list of sensor / actuator checks on page 19.
P0558	Brake Booster Pressure Sensor Circuit High	A pressure / vacuum sensor can be fitted to monitor the vacuum at the brake servo (brake booster). Note that on engines with vacuum pumps (including some petrol engines) the pump operation may be dependent on the level of pressure / vacuum detected at the brake booster. Also note that on some petrol engines (with direct injection), the engine can operate during some conditions without throttle control (throttle held open), therefore vacuum is not always created in the intake system; if the pressure sensor indicates insufficient vacuum for brake booster operation, the throttle can then be positioned to momentarily produce vacuum in the intake system. The brake booster pressure sensor can be connected to the engine control unit or to the ABS or other control units (depending on the systems fitted to the vehicle).	The voltage, frequency or value of the pressure sensor signal is either at full value (e.g. battery voltage with no signal or frequency) or, the signal exists but it is above the normal operating range (higher than the maximum operating limit). Possible fault in the pressure sensor or wiring e.g. short / open circuits, circuit resistance, incorrect sensor resistance (if applicable) or, fault with the reference / operating voltage; interference from other circuits can also affect sensor signals. Refer to list of sensor / actuator checks on page 19.

NOTE 1. Check for other fault codes that could provide additional information. NOTE 2. Communication between control units can pass via a CAN-Bus system; refer to Page 51 for CAN-Bus checks.

NOTE 3. If a fault cannot be located, it is also possible that the control unit is at fault. NOTE 4. Refer to Page 53 for list of pages that show the ISO standard for component locations e.g. Sensor A - Bank 1.

Fault code	EOBD / ISO Description	Component / System Description	Meaningful Description and Quick Check
P0559	Brake Booster Pressure Sensor Circuit Intermittent	A pressure / vacuum sensor can be fitted to monitor the vacuum at the brake servo (brake booster). Note that on engines with vacuum pumps (including some petrol engines) the pump operation may be dependent on the level of pressure / vacuum detected at the brake booster. Also note that on some petrol engines (with direct injection), the engine can operate during some conditions without throttle control (throttle held open), therefore vacuum is not always created in the intake system; if the pressure sensor indicates insufficient vacuum for brake booster operation, the throttle can then be positioned to momentarily produce vacuum in the intake system. The brake booster pressure sensor can be connected to the engine control unit or to the ABS or other control units (depending on the systems fitted to the vehicle).	The voltage, frequency or value of the pressure sensor signal is intermittent (likely to be out of normal operating range / tolerance when intermittent fault is detected) or, the signal / voltage is erratic (e.g. signal changes are irregular / unpredictable). Possible fault with wiring or connections (including internal sensor connections); also interference from other circuits can affect the signal. Refer to list of sensor / actuator checks on page 19. It is also possible that the pressure could be changing erratically, which could cause the fault code to be activated; if necessary, check pressure with a separate gauge to check for erratic pressure changes.
P0560	System Voltage	The control unit monitors system voltage (which is normally direct from the battery) with most actuators being provided with a power supply via a relay (power supply will therefore be battery voltage level). Note that there are some components where a regulated voltage is provided; it will therefore be necessary to refer to vehicle specific information to identify which components are provided with regulated voltage and whether the regulated voltage is provided by the control unit or other module. Other codes could indicate that system components also have a voltage related fault.	The fault code indicates that the system voltage is incorrect. Check for other related fault codes (if available) that could provide a more definitive fault description. Check battery / charging voltage (refer to fault codes P2502 - P2504 for additional information and checks relating to generator operation and voltage control). Check the input voltage at the control unit (direct from the battery). If the control unit voltage is different to the battery voltage, it could indicate a wiring related fault (check for high resistance). If fault occurs in a system where the voltage is regulated by the control unit or other module, refer to vehicle specific information, but note that wiring checks will still be applicable (short / open circuits, high resistances etc). Note that the fault could possibly be intermittent; it could therefore be necessary to clear the code and drive the vehicle or wiggle / move wiring etc to try and recreate the fault. Fault could be related to a cranking problem, check for excessive current draw or voltage drop during cranking.
P0561	System Voltage Unstable	The control unit monitors system voltage (which is normally direct from the battery) with most actuators being provided with a power supply via a relay (power supply will therefore be battery voltage level). Note that there are some components where a regulated voltage is provided; it will therefore be necessary to refer to vehicle specific information to identify which components are provided with regulated voltage and whether the regulated voltage is provided by the control unit or other module. Other codes could indicate that system components also have a voltage related fault.	The fault code indicates that the system voltage is incorrect (unstable or intermittent fault). Check battery and charging voltage (refer to fault codes P2502 - P2504 for additional information and checks relating to generator operation and voltage control). Check the input voltage available at the control unit (direct from the battery); if the voltage available at the control unit is different to the voltage at the battery, it would indicate a wiring related fault (check for high resistance, short / open circuits). If fault is occurring in a system where the voltage is regulated by the control unit or other module, refer to vehicle specific information, but note that wiring checks will still be applicable (short / open circuits and high resistances etc). Note that the fault could possibly be intermittent; it could therefore be necessary to clear the code and drive the vehicle or wiggle / move wiring etc to try and recreate the fault.
P0562	System Voltage Low	The control unit monitors system voltage (which is normally direct from the battery) with most actuators being provided with a power supply via a relay (power supply will therefore be battery voltage level). Note that there are some components where a regulated voltage is provided; it will therefore be necessary to refer to vehicle specific information to identify which components are provided with regulated voltage and whether the regulated voltage is provided by the control unit or other module. Other codes could indicate that system components also have a low system voltage.	The system voltage is either non existent (zero) or, below the normal operating range (out of tolerance) under certain conditions, which could be caused by a charging system or wiring fault. Note that the control unit could be trying to increase the voltage (by controlling alternator operation) but the voltage is not increasing as expected. Check battery / charging voltage (refer to fault codes P2502 - P2504 for additional information and checks relating to generator operation and voltage control). Check the input voltage available at the control unit (direct from the battery); if the voltage available at the control unit is different to the voltage at the battery, it would indicate a wiring related fault (check for high resistance, short / open circuits). If fault is occurring in a system where the voltage is regulated by the control unit or other module, refer to vehicle specific information, but note that wiring checks will still be applicable (short / open circuits and high resistances etc). Note that the fault could possibly be intermittent; it could therefore be necessary to clear the code and drive the vehicle or wiggle / move wiring etc to try and recreate the fault.
P0563	System Voltage High	The control unit monitors system voltage (which is normally direct from the battery) with most actuators being provided with a power supply via a relay (power supply will therefore be battery voltage level). Note that there are some components where a regulated voltage is provided; it will therefore be necessary to refer to vehicle specific information to identify which components are provided with regulated voltage and whether the regulated voltage is provided by the control unit or other module. Other codes could indicate that system components also have a high system voltage.	The system voltage is above the normal operating range (out of tolerance), which could be caused by a charging system or wiring fault. Note that the fault could indicate that the control unit is trying to reduce the voltage (by controlling alternator operation) but the voltage is not reducing as expected. Check battery and charging voltage (refer to fault codes P2502 - P2504 for additional information and checks relating to generator operation and voltage control). If fault is occurring in a system where the voltage is regulated by the control unit or other module, refer to vehicle specific information, but note that wiring checks will still be applicable (short / open circuits and high resistances etc). Note that the fault could possibly be intermittent; it could therefore be necessary to clear the code and drive the vehicle or wiggle / move wiring etc to try and recreate the fault.
P0564	Cruise Control Multi-Function Input "A" Circuit	The multi-function switch can allow all or most of the cruise control functions to be selected by the driver, although some systems make use of some separate switches. Switch selection / operation provides applicable control signals to the cruise control module (or other applicable control unit); the signals will be dependent on the functions selected by the driver. For many of the cruise control faults, the control unit uses a "Logic" to identify a fault e.g. a signal is indicating the system to accelerate the vehicle but the brakes are also being applied.	It may be necessary to refer to vehicle specific information to identify the specified input for the Multi-Function switch. The fault code indicates a "Circuit" related fault for the specified circuit; the fault is undefined. The fault could relate to the input voltage to the switch or, the output from the switch. The voltage or signal value could be: out of normal operating range, constant value, non-existent, corrupt. Possible fault in the wiring or switch e.g. short / open circuits, circuit resistance, incorrect sensor resistance or, fault with the reference / operating voltage. Note that on some systems, the fault could be detected by the Cruise Control Module, which then passes the information to the main control unit; however this information can be limited in detail, which will require further interrogation of the cruise control module and system.

NOTE 1. Check for other fault codes that could provide additional information.
NOTE 3. If a fault cannot be located, it is also possible that the control unit is at fault.

NOTE 2. Communication between control units can pass via a CAN-Bus system; refer to Page 51 for CAN-Bus checks.
NOTE 4. Refer to Page 53 for list of pages that show the ISO standard for component locations e.g. Sensor A - Bank 1.

code	Description		
P0565	Cruise Control "On" Signal	The multi-function switch can allow all or most of the cruise control functions to be selected by the driver, although some systems make use of some separate switches. Switch selection / operation provides applicable control signals to the cruise control module (or other applicable control unit); the signals will be dependent on the functions selected by the driver. For many of the cruise control faults, the control unit uses a "Logic" to identify a fault e.g. a signal is indicating the system to accelerate the vehicle but the brakes are also being applied.	It may be necessary to refer to vehicle specific information to identify the specified switch circuit. The fault code indicates that there is an undefined fault with the "ON" signal for the cruise control. The voltage or signal value could be: out of normal operating range, constant value, non-existent, corrupt. Possible fault in the switch or wiring e.g. short / open circuits, circuit / switch contact resistance or, fault with the switch input voltage. It is also possible that the "ON" signal is existing when other sensors or switches are indicating that the signal should not exist. Note that on some systems, the fault could be detected by the Cruise Control Module, which then passes the information to the main control unit; however this information can be limited in detail, which will require further interrogation of the cruise control module and system.
P0566	Cruise Control "Off" Signal	The multi-function switch can allow all or most of the cruise control functions to be selected by the driver, although some systems make use of some separate switches. Switch selection / operation provides applicable control signals to the cruise control module (or other applicable control unit); the signals will be dependent on the functions selected by the driver. For many of the cruise control faults, the control unit uses a "Logic" to identify a fault e.g. a signal is indicating the system to accelerate the vehicle but the brakes are also being applied.	It may be necessary to refer to vehicle specific information to identify the specified switch circuit. The fault code indicates that there is an undefined fault with the "OFF" signal for the cruise control. The voltage or signal value could be: out of normal operating range, constant value, non-existent, corrupt. Possible fault in the switch or wiring e.g. short / open circuits, circuit / switch contact resistance or, fault with the switch input voltage. It is also possible that the "OFF" signal is existing when other sensors or switches are indicating that the signal should not exist (e.g. when brakes are applied). Note that on some systems, the fault could be detected by the Cruise Control Module, which then passes the information to the main control unit; however this information can be limited in detail, which will require further interrogation of the cruise control module and system.
P0567	Cruise Control "Resume" Signal	The multi-function switch can allow all or most of the cruise control functions to be selected by the driver, although some systems make use of some separate switches. Switch selection / operation provides applicable control signals to the cruise control module (or other applicable control unit); the signals will be dependent on the functions selected by the driver. For many of the cruise control faults, the control unit uses a "Logic" to identify a fault e.g. a signal is indicating the system to accelerate the vehicle but the brakes are also being applied.	It may be necessary to refer to vehicle specific information to identify the specified switch circuit. The fault code indicates that there is an undefined fault with the "RESUME" signal for the cruise control. The voltage or signal value could be: out of normal operating range, constant value, non-existent, corrupt. Possible fault in the switch or wiring e.g. short / open circuits, circuit / switch contact resistance or, fault with the switch input voltage. It is also possible that the "RESUME" signal is existing when other sensors or switches are indicating that the signal should not exist (e.g. when brakes are applied). Note that on some systems, the fault could be detected by the Cruise Control Module, which then passes the information to the main control unit; however this information can be limited in detail, which will require further interrogation of the cruise control module and system.
P0568	Cruise Control "Set" Signal	The multi-function switch can allow all or most of the cruise control functions to be selected by the driver, although some systems make use of some separate switches. Switch selection / operation provides applicable control signals to the cruise control module (or other applicable control unit); the signals will be dependent on the functions selected by the driver. For many of the cruise control faults, the control unit uses a "Logic" to identify a fault e.g. a signal is indicating the system to accelerate the vehicle but the brakes are also being applied.	It may be necessary to refer to vehicle specific information to identify the specified switch circuit. The fault code indicates that there is an undefined fault with the "SET" signal for the cruise control. The voltage or signal value could be: out of normal operating range, constant value, non-existent, corrupt. Possible fault in the switch or wiring e.g. short / open circuits, circuit / switch contact resistance or, fault with the switch input voltage. It is also possible that the "SET" signal is existing when other sensors or switches are indicating that the signal should not exist (e.g. when brakes are applied). Note that on some systems, the fault could be detected by the Cruise Control Module, which then passes the information to the main control unit; however this information can be limited in detail, which will require further interrogation of the cruise control module and system.
P0569	Cruise Control "Coast" Signal	The multi-function switch can allow all or most of the cruise control functions to be selected by the driver, although some systems make use of some separate switches. Switch selection / operation provides applicable control signals to the cruise control module (or other applicable control unit); the signals will be dependent on the functions selected by the driver. For many of the cruise control faults, the control unit uses a "Logic" to identify a fault e.g. a signal is indicating the system to accelerate the vehicle but the brakes are also being applied.	It may be necessary to refer to vehicle specific information to identify the specified switch circuit. The fault code indicates that there is an undefined fault with the "COAST" signal for the cruise control. The voltage or signal value could be: out of normal operating range, constant value, non-existent, corrupt. Possible fault in the switch or wiring e.g. short / open circuits, circuit / switch contact resistance or, fault with the switch input voltage. It is also possible that the "COAST" signal is existing when other sensors or switches are indicating that the signal should not exist. Note that on some systems, the fault could be detected by the Cruise Control Module, which then passes the information to the main control unit; however this information can be limited in detail, which will require further interrogation of the cruise control module and system.
P056A	Cruise Control "Increase Distance" Signal	The multi-function switch can allow all or most of the cruise control functions to be selected by the driver, although some systems make use of some separate switches. Switch selection / operation provides applicable control signals to the cruise control module (or other applicable control unit); the signals will be dependent on the functions selected by the driver. For many of the cruise control faults, the control unit uses a "Logic" to identify a fault e.g. a signal is indicating the system to accelerate the vehicle but the brakes are also being applied.	It may be necessary to refer to vehicle specific information to identify the specified switch circuit. The fault code indicates that there is an undefined fault with the "INCREASE DISTANCE" signal for the cruise control. The voltage or signal value could be: out of normal operating range, constant value, non-existent, corrupt. Possible fault in the switch or wiring e.g. short / open circuits, circuit / switch contact resistance or, fault with the switch input voltage. It is also possible that the "INCREASE DISTANCE" signal is existing when other sensors or switches are indicating that the signal should not exist (e.g. when accelerator applied). Note that on some systems, the fault could be detected by the Cruise Control Module, which then passes the information to the main control unit; however this information can be limited in detail, which will require further interrogation of the cruise control module and system.
P056B	Cruise Control "Decrease Distance" Signal	The multi-function switch can allow all or most of the cruise control functions to be selected by the driver, although some systems make use of some separate switches. Switch selection / operation provides applicable control signals to the cruise control module (or other applicable control unit); the signals will be dependent on the functions selected by the driver. For many of the cruise control faults, the control unit uses a "Logic" to identify a fault e.g. a signal is indicating the system to accelerate the vehicle but the brakes are also being applied.	It may be necessary to refer to vehicle specific information to identify the specified switch circuit. The fault code indicates that there is an undefined fault with the "DECREASE DISTANCE" signal for the cruise control. The voltage or signal value could be: out of normal operating range, constant value, non-existent, corrupt. Possible fault in the switch or wiring e.g. short / open circuits, circuit / switch contact resistance or, fault with the switch input voltage. It is also possible that the "DECREASE DISTANCE" signal is existing when other sensors or switches are indicating that the signal should not exist (e.g. when brakes are applied). Note that on some systems, the fault could be detected by the Cruise Control Module, which then passes the information to the main control unit; however this information can be limited in detail, which will require further interrogation of the cruise control module and system.

Fault code	EOBD / ISO Description	Component / System Description	Meaningful Description and Quick Check
P0570	Cruise Control "Accelerate" Signal	The multi-function switch can allow all or most of the cruise control functions to be selected by the driver, although some systems make use of some separate switches. Switch selection / operation provides applicable control signals to the cruise control module (or other applicable control unit); the signals will be dependent on the functions selected by the driver. For many of the cruise control faults, the control unit uses a "Logic" to identify a fault e.g. a signal is indicating the system to accelerate the vehicle but the brakes are also being applied.	It may be necessary to refer to vehicle specific information to identify the specified switch circuit. The fault code indicates that there is an undefined fault with the "ACCELERATE" signal for the cruise control. The voltage or signal value could be: out of normal operating range, constant value, non-existent, corrupt. Possible fault in the switch or wiring e.g. short / open circuits, circuit / switch contact resistance, fault with the switch input voltage. It is also possible that the "ACCELERATE" signal is existing when other sensors or switches are indicating that the signal should not exist (e.g. when brakes are applied). Note that on some systems, the fault could be detected by the Cruise Control Module, which then passes the information to the main control unit; however this information can be limited in detail, which will require further interrogation of the cruise control module and system.
P0571	Brake Switch "A" Circuit	In addition to normal brake light operation, the brake switch signal can be used in the control of engine systems, vehicle stability systems, transmission and other vehicle systems. Where more than one brake switch is used, the signals can be passed to the different vehicle systems e.g. one for stability control / ABS (and other systems), and one for brake light operation. Note that some brake switches can be conventional type switches operated by brake pedal movement, but some types are "Pressure Switches" which operate due to brake line pressure (can be fitted into ABS modulator assembly).	Note. The switch assembly may contain more than one switch / sensor, refer to vehicle specific information to establish which switch is affected. The fault code indicates that the voltage or value of the switch signal is incorrect (undefined fault). Switch voltage / signal could be: out of normal operating range, constant value, non-existent, corrupt. Possible fault in the switch or wiring e.g. short / open circuits, circuit resistance or, fault with the reference / supply voltage; the switch contacts could also be faulty e.g. not opening / closing correctly or have a high contact resistance. Also check the mechanism operating the switches e.g. pedal and switch adjustment or, check that there is sufficient brake line pressure to operate switch (where applicable). Refer to list of sensor / actuator checks on page 19.
P0572	Brake Switch "A" Circuit Low	In addition to normal brake light operation, the brake switch signal can be used in the control of engine systems, vehicle stability systems, transmission and other vehicle systems. Where more than one brake switch is used, the signals can be passed to the different vehicle systems e.g. one for stability control / ABS (and other systems), and one for brake light operation. Note that some brake switches can be conventional type switches operated by brake pedal movement, but some types are "Pressure Switches" which operate due to brake line pressure (can be fitted into ABS modulator assembly).	Note. The switch assembly may contain more than one switch / sensor, refer to vehicle specific information to establish which switch is affected. The voltage, frequency or value of the switch signal is either "zero" (no signal) or, the signal exists but it is below the normal operating range (lower than the minimum operating limit). Possible fault in the switch or wiring e.g. short / open circuits, circuit resistance or, fault with the reference / supply voltage; the switch contacts could also be faulty e.g. not opening / closing correctly or have a high contact resistance. Also check the mechanism operating the switches e.g. pedal and switch adjustment or, check that there is sufficient brake line pressure to operate switch (where applicable). Refer to list of sensor / actuator checks on page 19.
P0573	Brake Switch "A" Circuit High	In addition to normal brake light operation, the brake switch signal can be used in the control of engine systems, vehicle stability systems, transmission and other vehicle systems. Where more than one brake switch is used, the signals can be passed to the different vehicle systems e.g. one for stability control / ABS (and other systems), and one for brake light operation. Note that some brake switches can be conventional type switches operated by brake pedal movement, but some types are "Pressure Switches" which operate due to brake line pressure (can be fitted into ABS modulator assembly).	Note. The switch assembly may contain more than one switch / sensor, refer to vehicle specific information to establish which switch is affected. The voltage, frequency or value of the switch signal is either at full value (e.g. battery voltage with no signal or frequency) or, the signal exists but it is above the normal operating range (higher than the maximum operating limit). Possible fault in the switch or wiring e.g. short / open circuits or, fault with the reference / operating voltage; the switch contacts could also be faulty e.g. not opening / closing correctly or have a high contact resistance. Also check the mechanism operating the switches e.g. pedal and switch adjustment or, check that there is sufficient brake line pressure to operate switch (where applicable). Refer to list of sensor / actuator checks on page 19.
P0574	Cruise Control System -Vehicle Speed Too High	Refer to vehicle specific information to indentify the cruise control system (components and operation). The cruise control system responds to inputs from the driver controls, including desired "cruise" speed. The control unit then monitors the actual speed, which is indicated by signals from a vehicle speed sensor. The cruise control module can be a separate item or from part of the main control unit.	The fault code indicates that the actual vehicle speed is higher than the selected / desired speed. The fault is undefined but can include problems with the cruise control module / control unit, wiring and connections from the driver control switches to the control unit. Also check for throttle operation (throttle control linkage or electric throttle actuation as applicable, and check for air leaks in the intake system or throttle body faults. Check for faulty switches / speed selector and also check for correct speed signal. Note that the speed signal could pass from the main control unit to the cruise control module. Check for faults and fault codes on other systems that could receive or provide a speed signal.
P0575	Cruise Control Input Circuit	The cruise control input circuit could relate to the communication circuit between the cruise control module and the main control unit (input signal), or a manufacturer could use the fault code to refer to a fault in the circuits between the cruise control module and the driver selection switches (multi-function switch etc). Refer to vehicle specific information to indentify the cruise control system (components, operation and wiring).	The fault code indicates an unidentified fault in the input signal, either between control units or from the driver selection switches. If possible check the "Live Data" from the control unit to try and identify any faulty / incorrect operating signals or, if the cruise control has a separate control module, check for any specific fault codes that might be available from the cruise control module. Check all wiring and connections (as applicable) for short / open circuits, check driver selection switch operation and check any communication circuits between the cruise control module and the main control unit.
P0576	Cruise Control Input Circuit Low	The cruise control input circuit could relate to the communication circuit between the cruise control module and the main control unit (input signal), or a manufacturer could use the fault code to refer to a fault in the circuits between the cruise control module and the driver selection switches (multi-function switch etc). Refer to vehicle specific information to indentify the cruise control system (components, operation and wiring).	The fault code indicates that the input signal (voltage frequency or signal value) is low (out of normal operating range). The signal could relate to either the communication between control units or from the driver selection switches. If possible check the "Live Data" from the control unit to try and identify any faulty / incorrect operating signals or, if the cruise control has a separate control module, check for any specific fault codes that might be available from the cruise control module. Check all wiring and connections (as applicable) for short / open circuits, check driver selection switch operation and check any communication circuits between the cruise control module and the main control unit. Note that the fault code could be activated because the control unit / module is expecting the signal to change (increase voltage or signal value) at specific times but the signal value remains low e.g. zero volts when it should rise to system voltage).

NOTE 1. Check for other fault codes that could provide additional information. **NOTE 2.** Communication between control units can pass via a CAN-Bus system; refer to Page 51 for CAN-Bus checks.
NOTE 3. If a fault cannot be located, it is also possible that the control unit is at fault. **NOTE 4.** Refer to Page 53 for list of pages that show the ISO standard for component locations e.g. Sensor A - Bank 1.

Fault code	OBD / JBD Description	Component / System Description	Meaningful Description and Quick Check
P0577	Cruise Control Input Circuit High	The cruise control input circuit could relate to the communication circuit between the cruise control module and the main control unit (input signal), or a manufacturer could use the fault code to refer to a fault in the circuits between the cruise control module and the driver selection switches (multi-function switch etc). Refer to vehicle specific information to indentify the cruise control system (components, operation and wiring).	The fault code indicates that the input signal (voltage frequency or signal value) is high (out of normal operating range). The signal could relate to either the communication between control units or from the driver selection switches. If possible check the "Live Data" from the control unit to try and identify any faulty / incorrect operating signals or, if the cruise control has a separate control module, check for any specific fault codes that might be available from the cruise control module. Check all wiring and connections (as applicable) for short / open circuits, check driver selection switch operation and check any communication circuits between the cruise control module and the main control unit. Note that the fault code could be activated because the control unit / module is expecting the signal to change (decrease voltage or signal value) at specific times but the signal value remains high e.g. system voltage when it should fall to zero volts.
P0578	Cruise Control Multi-Function Input "A" Circuit Stuck	The multi-function switch can allow all or most of the cruise control functions to be selected by the driver, although some systems make use of some separate switches. Switch selection / operation provides applicable control signals to the cruise control module (or other applicable control unit); the signals will be dependent on the functions selected by the driver. For many of the cruise control faults, the control unit uses a "Logic" to identify a fault e.g. a signal is indicating the system to accelerate the vehicle but the brakes are also being applied.	It may be necessary to refer to vehicle specific information to identify the specified switch circuit. The fault code indicates that the signal voltage or value on the specified circuit is "Stuck" e.g. stuck on or off; or if the signal is normally a changing value (e.g. increasing or decreasing voltage), the signal could be stuck at a fixed value. Possible fault in the switch or wiring e.g. short / open circuits, circuit / switch contact resistance or, fault with the switch input voltage. Note that on some systems, the fault could be detected by the Cruise Control Module, which then passes the information to the main control unit; however this information can be limited in detail, which will require further interrogation of the cruise control module and system.
P0579	Cruise Control Multi-Function Input "A" Circuit Range / Performance	The multi-function switch can allow all or most of the cruise control functions to be selected by the driver, although some systems make use of some separate switches. Switch selection / operation provides applicable control signals to the cruise control module (or other applicable control unit); the signals will be dependent on the functions selected by the driver. For many of the cruise control faults, the control unit uses a "Logic" to identify a fault e.g. a signal is indicating the system to accelerate the vehicle but the brakes are also being applied.	It may be necessary to refer to vehicle specific information to identify the specified switch circuit. The fault code indicates that the signal voltage or value on the specified circuit is within the normal operating range but is possibly incorrect for the operating conditions or, the signal is not changing as expected. Possible fault in the switch or wiring e.g. short / open circuits, circuit / switch contact resistance or, fault with the switch input voltage. Note that on some systems, the fault could be detected by the Cruise Control Module, which then passes the information to the main control unit; however this information can be limited in detail, which will require further interrogation of the cruise control module and system.
P0580	Cruise Control Multi-Function Input "A" Circuit Low	The multi-function switch can allow all or most of the cruise control functions to be selected by the driver, although some systems make use of some separate switches. Switch selection / operation provides applicable control signals to the cruise control module (or other applicable control unit); the signals will be dependent on the functions selected by the driver. For many of the cruise control faults, the control unit uses a "Logic" to identify a fault e.g. a signal is indicating the system to accelerate the vehicle but the brakes are also being applied.	It may be necessary to refer to vehicle specific information to identify the specified switch circuit. The fault code indicates that the signal voltage or value on the specified circuit is below the normal operating range. Possible fault in the switch or wiring e.g. short / open circuits, circuit / switch contact resistance or, fault with the switch input voltage. Note that on some systems, the fault could be detected by the Cruise Control Module, which then passes the information to the main control unit; however this information can be limited in detail, which will require further interrogation of the cruise control module and system.
P0581	Cruise Control Multi-Function Input "A" Circuit High	The multi-function switch can allow all or most of the cruise control functions to be selected by the driver, although some systems make use of some separate switches. Switch selection / operation provides applicable control signals to the cruise control module (or other applicable control unit); the signals will be dependent on the functions selected by the driver. For many of the cruise control faults, the control unit uses a "Logic" to identify a fault e.g. a signal is indicating the system to accelerate the vehicle but the brakes are also being applied.	It may be necessary to refer to vehicle specific information to identify the specified switch circuit. The fault code indicates that the signal voltage or value on the specified circuit is above the normal operating range. Possible fault in the switch or wiring e.g. short / open circuits, circuit / switch contact resistance or, fault with the switch input voltage. Note that on some systems, the fault could be detected by the Cruise Control Module, which then passes the information to the main control unit; however this information can be limited in detail, which will require further interrogation of the cruise control module and system.
P0582	Cruise Control Vacuum Control Circuit / Open	Due to the different designs of cruise control operation, it may be necessary to refer to vehicle specific information to identify components and system operation. The cruise control system can use a vacuum control to regulate the vacuum level passing to the cruise control speed control mechanism; the vacuum control will use an electric actuator (usually a solenoid) to operate a valve in the vacuum system.	The voltage, frequency or value of the control signal in the actuator circuit is incorrect. Signal could be:- out of normal operating range, constant value, non-existent, corrupt. The fault code indicates a possible "open circuit" but other undefined faults in the actuator or wiring could activate the same code e.g. actuator power supply, short circuit, high circuit resistance or incorrect actuator resistance. Note: the control unit could be providing the correct signal but it is being affected by the circuit fault. Refer to list of sensor / actuator checks on page 19.
P0583	Cruise Control Vacuum Control Circuit Low	Due to the different designs of cruise control operation, it may be necessary to refer to vehicle specific information to identify components and system operation. The cruise control system can use a vacuum control to regulate the vacuum level passing to the cruise control speed control mechanism; the vacuum control will use an electric actuator (usually a solenoid) to operate a valve in the vacuum system.	The voltage, frequency or value of the control signal in the actuator circuit is either "zero" (no signal) or, the signal exists but is below the normal operating range. Possible fault with actuator or wiring (e.g. actuator power supply, short / open circuit or resistance) that is causing a signal value to be lower than the minimum operating limit. Note: the control unit could be providing the correct signal but it is being affected by the circuit fault. Refer to list of sensor / actuator checks on page 19.
P0584	Cruise Control Vacuum Control Circuit High	Due to the different designs of cruise control operation, it may be necessary to refer to vehicle specific information to identify components and system operation. The cruise control system can use a vacuum control to regulate the vacuum level passing to the cruise control speed control mechanism; the vacuum control will use an electric actuator (usually a solenoid) to operate a valve in the vacuum system.	The voltage, frequency or value of the control signal in the actuator circuit is either at full value (e.g. battery voltage with no on / off signal) or, the signal exists but it is above the normal operating range. Possible fault with actuator or wiring (e.g. actuator power supply, short / open circuit or resistance) that is causing a signal value to be higher than the maximum operating limit. Note: the control unit could be providing the correct signal but it is being affected by the circuit fault. Refer to list of sensor / actuator checks on page 19.

NOTE 1. Check for other fault codes that could provide additional information.   **NOTE 2.** Communication between control units can pass via a CAN-Bus system; refer to Page 51 for CAN-Bus checks.

NOTE 3. If a fault cannot be located, it is also possible that the control unit is at fault.   **NOTE 4.** Refer to Page 53 for list of pages that show the ISO standard for component locations e.g. Sensor A - Bank 1.

Fault code	EOBD / ISO Description	Component / System Description	Meaningful Description and Quick Check
P0585	Cruise Control Multi-Function Input "A" / "B" Correlation	The multi-function switch can allow all or most of the cruise control functions to be selected by the drive, although some systems make use of some separate switches. Switch selection / operation provides applicable control signals to the cruise control module (or other applicable control unit); the signals will be dependent on the functions selected by the driver. For many of the cruise control faults, the control unit uses a "Logic" to identify a fault e.g. a signal is indicating the system to accelerate the vehicle but the brakes are also being applied.	It may be necessary to refer to vehicle specific information to identify the specified switch circuit. The fault code indicates that the signal voltage or value on the two specified input circuits are providing conflicting information i.e. the information in one circuit does not match or correspond to the information in the other circuit. It will be necessary to identify and perform checks on both switches / circuits. Possible fault in the switch or wiring e.g. short / open circuits, circuit / switch contact resistance or, fault with the switch input voltage. Note that on some systems, the fault could be detected by the Cruise Control Module, which then passes the information to the main control unit; however this information can be limited in detail, which will require further interrogation of the cruise control module and system.
P0586	Cruise Control Vent Control Circuit / Open	Due to the different designs of cruise control operation, it may be necessary to refer to vehicle specific information to identify components and system operation. For cruise control systems that make use of vacuum (applied to a vacuum servo unit) to move the throttle system (to hold / adjust speed), a vent valve is used to release the vacuum from the servo vacuum chamber; this would occur when the speed needs to be reduced or the cruise control disabled (under braking or when the driver switches off the system). The vent valve will be opened using an electric actuator e.g. a solenoid. Note that on some systems; the vent valve and the vacuum control valve (controlling vacuum input to servo unit) can be a combined unit (e.g. double action solenoid valve).	The voltage, frequency or value of the control signal in the actuator circuit is incorrect. Signal could be:- out of normal operating range, constant value, non-existent, corrupt. The fault code indicates a possible "open circuit" but other undefined faults in the actuator or wiring could activate the same code e.g. actuator power supply, short circuit, high circuit resistance or incorrect actuator resistance. Note: the control unit could be providing the correct signal but it is being affected by the circuit fault. Refer to list of sensor / actuator checks on page 19.
P0587	Cruise Control Vent Control Circuit Low	Due to the different designs of cruise control operation, it may be necessary to refer to vehicle specific information to identify components and system operation. For cruise control systems that make use of vacuum (applied to a vacuum servo unit) to move the throttle system (to hold / adjust speed), a vent valve is used to release the vacuum from the servo vacuum chamber; this would occur when the speed needs to be reduced or the cruise control disabled (under braking or when the driver switches off the system). The vent valve will be opened using an electric actuator e.g. a solenoid. Note that on some systems; the vent valve and the vacuum control valve (controlling vacuum input to servo unit) can be a combined unit (e.g. double action solenoid valve).	The voltage, frequency or value of the control signal in the actuator circuit is either "zero" (no signal) or, the signal exists but is below the normal operating range. Possible fault with actuator or wiring (e.g. actuator power supply, short / open circuit or resistance) that is causing a signal value to be lower than the minimum operating limit. Note: the control unit could be providing the correct signal but it is being affected by the circuit fault. Refer to list of sensor / actuator checks on page 19.
P0588	Cruise Control Vent Control Circuit High	Due to the different designs of cruise control operation, it may be necessary to refer to vehicle specific information to identify components and system operation. For cruise control systems that make use of vacuum (applied to a vacuum servo unit) to move the throttle system (to hold / adjust speed), a vent valve is used to release the vacuum from the servo vacuum chamber; this would occur when the speed needs to be reduced or the cruise control disabled (under braking or when the driver switches off the system). The vent valve will be opened using an electric actuator e.g. a solenoid. Note that on some systems; the vent valve and the vacuum control valve (controlling vacuum input to servo unit) can be a combined unit (e.g. double action solenoid valve).	The voltage, frequency or value of the control signal in the actuator circuit is either at full value (e.g. battery voltage with no on / off signal) or, the signal exists but it is above the normal operating range. Possible fault with actuator or wiring (e.g. actuator power supply, short / open circuit or resistance) that is causing a signal value to be higher than the maximum operating limit. Note: the control unit could be providing the correct signal but it is being affected by the circuit fault. Refer to list of sensor / actuator checks on page 19.
P0589	Cruise Control Multi-Function Input "B" Circuit	The multi-function switch can allow all or most of the cruise control functions to be selected by the driver, although some systems make use of some separate switches. Switch selection / operation provides applicable control signals to the cruise control module (or other applicable control unit); the signals will be dependent on the functions selected by the driver. For many of the cruise control faults, the control unit uses a "Logic" to identify a fault e.g. a signal is indicating the system to accelerate the vehicle but the brakes are also being applied.	It may be necessary to refer to vehicle specific information to identify the specified switch circuit. The fault code indicates that the signal voltage or value on the specified circuit is incorrect (undefined fault). Possible fault in the switch or wiring e.g. short / open circuits, circuit / switch contact resistance or, fault with the switch input voltage. Note that on some systems, the fault could be detected by the Cruise Control Module, which then passes the information to the main control unit; however this information can be limited in detail, which will require further interrogation of the cruise control module and system.
P0590	Cruise Control Multi-Function Input "B" Circuit Stuck	The multi-function switch can allow all or most of the cruise control functions to be selected by the driver, although some systems make use of some separate switches. Switch selection / operation provides applicable control signals to the cruise control module (or other applicable control unit); the signals will be dependent on the functions selected by the driver. For many of the cruise control faults, the control unit uses a "Logic" to identify a fault e.g. a signal is indicating the system to accelerate the vehicle but the brakes are also being applied.	It may be necessary to refer to vehicle specific information to identify the specified switch circuit. The fault code indicates that the signal voltage or value on the specified circuit is "Stuck" e.g. stuck on or off; or if the signal is normally a changing value (e.g. increasing or decreasing voltage), the signal could be stuck at a fixed value. Possible fault in the switch or wiring e.g. short / open circuits, circuit / switch contact resistance or, fault with the switch input voltage. Note that on some systems, the fault could be detected by the Cruise Control Module, which then passes the information to the main control unit; however this information can be limited in detail, which will require further interrogation of the cruise control module and system.

NOTE 1. Check for other fault codes that could provide additional information. NOTE 2. Communication between control units can pass via a CAN-Bus system; refer to Page 51 for CAN-Bus checks.

NOTE 3. If a fault cannot be located, it is also possible that the control unit is at fault. NOTE 4. Refer to Page 53 for list of pages that show the ISO standard for component locations e.g. Sensor A - Bank 1.

Fault code	Description	Component / System Description	Meaningful Description and Quick Check
P0591	Cruise Control Multi-Function Input "B" Circuit Range / Performance	The multi-function switch can allow all or most of the cruise control functions to be selected by the driver, although some systems make use of some separate switches. Switch selection / operation provides applicable control signals to the cruise control module (or other applicable control unit); the signals will be dependent on the functions selected by the driver. For many of the cruise control faults, the control unit uses a "Logic" to identify a fault e.g. a signal is indicating the system to accelerate the vehicle but the brakes are also being applied.	It may be necessary to refer to vehicle specific information to identify the specified switch circuit. The fault code indicates that the signal voltage or value on the specified circuit is within the normal operating range but is possibly incorrect for the operating conditions or, the signal is not changing as expected. Possible fault in the switch or wiring e.g. short / open circuits, circuit / switch contact resistance or, fault with the switch input voltage. Note that on some systems, the fault could be detected by the Cruise Control Module, which then passes the information to the main control unit; however this information can be limited in detail, which will require further interrogation of the cruise control module and system.
P0592	Cruise Control Multi-Function Input "B" Circuit Low	The multi-function switch can allow all or most of the cruise control functions to be selected by the driver, although some systems make use of some separate switches. Switch selection / operation provides applicable control signals to the cruise control module (or other applicable control unit); the signals will be dependent on the functions selected by the driver. For many of the cruise control faults, the control unit uses a "Logic" to identify a fault e.g. a signal is indicating the system to accelerate the vehicle but the brakes are also being applied.	It may be necessary to refer to vehicle specific information to identify the specified switch circuit. The fault code indicates that the signal voltage or value on the specified circuit is below the normal operating range. Possible fault in the switch or wiring e.g. short / open circuits, circuit / switch contact resistance or, fault with the switch input voltage. Note that on some systems, the fault could be detected by the Cruise Control Module, which then passes the information to the main control unit; however this information can be limited in detail, which will require further interrogation of the cruise control module and system.
P0593	Cruise Control Multi-Function Input "B" Circuit High	The multi-function switch can allow all or most of the cruise control functions to be selected by the driver, although some systems make use of some separate switches. Switch selection / operation provides applicable control signals to the cruise control module (or other applicable control unit); the signals will be dependent on the functions selected by the driver. For many of the cruise control faults, the control unit uses a "Logic" to identify a fault e.g. a signal is indicating the system to accelerate the vehicle but the brakes are also being applied.	It may be necessary to refer to vehicle specific information to identify the specified switch circuit. The fault code indicates that the signal voltage or value on the specified circuit is above the normal operating range. Possible fault in the switch or wiring e.g. short / open circuits, circuit / switch contact resistance or, fault with the switch input voltage. Note that on some systems, the fault could be detected by the Cruise Control Module, which then passes the information to the main control unit; however this information can be limited in detail, which will require further interrogation of the cruise control module and system.
P0594	Cruise Control Servo Control Circuit / Open	Due to the different designs of cruise control operation, it may be necessary to refer to vehicle specific information to identify components and system operation. For cruise control systems that make use of vacuum (applied to a vacuum servo unit) to move the throttle system (to hold / adjust speed), vacuum control valves can be used to regulate the vacuum being applied to the servo and a vent valve can also be used to release the vacuum from the servo. The valves will be opened using an electric actuator e.g. a solenoid. Note that on some systems, the vacuum control valve and the vent valve can be a combined unit (e.g. double action solenoid valve).	The voltage, frequency or value of the control signal in the actuator circuit is incorrect. Signal could be:- out of normal operating range, constant value, non-existent, corrupt. The fault code indicates a possible "open circuit" but other undefined faults in the actuator or wiring could activate the same code e.g. actuator power supply, short circuit, high circuit resistance or incorrect actuator resistance. Note: the control unit could be providing the correct signal but it is being affected by the circuit fault. Refer to list of sensor / actuator checks on page 19.
P0595	Cruise Control Servo Control Circuit Low	Due to the different designs of cruise control operation, it may be necessary to refer to vehicle specific information to identify components and system operation. For cruise control systems that make use of vacuum (applied to a vacuum servo unit) to move the throttle system (to hold / adjust speed), vacuum control valves can be used to regulate the vacuum being applied to the servo and a vent valve can also be used to release the vacuum from the servo. The valves will be opened using an electric actuator e.g. a solenoid. Note that on some systems, the vacuum control valve and the vent valve can be a combined unit (e.g. double action solenoid valve).	The voltage, frequency or value of the control signal in the actuator circuit is either "zero" (no signal) or, the signal exists but is below the normal operating range. Possible fault with actuator or wiring (e.g. actuator power supply, short / open circuit or resistance) that is causing a signal value to be lower than the minimum operating limit. Note: the control unit could be providing the correct signal but it is being affected by the circuit fault. Refer to list of sensor / actuator checks on page 19.
P0596	Cruise Control Servo Control Circuit High	Due to the different designs of cruise control operation, it may be necessary to refer to vehicle specific information to identify components and system operation. For cruise control systems that make use of vacuum (applied to a vacuum servo unit) to move the throttle system (to hold / adjust speed), vacuum control valves can be used to regulate the vacuum being applied to the servo and a vent valve can also be used to release the vacuum from the servo. The valves will be opened using an electric actuator e.g. a solenoid. Note that on some systems, the vacuum control valve and the vent valve can be a combined unit (e.g. double action solenoid valve).	The voltage, frequency or value of the control signal in the actuator circuit is either at full value (e.g. battery voltage with no on / off signal) or, the signal exists but is above the normal operating range. Possible fault with actuator or wiring (e.g. actuator power supply, short / open circuit or resistance) that is causing a signal value to be higher than the maximum operating limit. Note: the control unit could be providing the correct signal but it is being affected by the circuit fault. Refer to list of sensor / actuator checks on page 19.
P0597	Thermostat Heater Control Circuit / Open	A heating element can be incorporated into the coolant thermostat; the heater is used to regulate the coolant temperature more accurately and more quickly (depending on load and speed conditions). Refer to vehicle specific information if necessary to identify how the thermostat heater is controlled but note that it can be controlled directly by the control unit or, the control unit could operate a relay which in turn would control the thermostat heater.	Identify the control circuit and components for the thermostat heater system. The voltage, frequency or value of the control signal in the heater circuit is incorrect. Signal could be:- out of normal operating range, constant value, non-existent, corrupt. The fault code indicates a possible "open circuit" but other undefined faults in the heater or wiring could activate the same code e.g. heater power supply, short circuit, high circuit resistance or incorrect heater resistance. Note: the control unit could be providing the correct signal but it is being affected by a circuit fault. Refer to list of sensor / actuator checks on page 19.

Fault code	EOBD / ISO Description	Component / System Description	Meaningful Description and Quick Check
P0598	Thermostat Heater Control Circuit Low	A heating element can be incorporated into the coolant thermostat; the heater is used to regulate the coolant temperature more accurately and more quickly (depending on load and speed conditions). Refer to vehicle specific information if necessary to identify how the thermostat heater is controlled but note that it can be controlled directly by the control unit or, the control unit could operate a relay which in turn would control the thermostat heater.	Identify the control circuit and components for the thermostat heater system. The voltage, frequency or value of the control signal in the heater circuit is either "zero" (no signal) or, the signal exists but is below the normal operating range. Possible fault with heater or wiring (e.g. heater power supply, short / open circuit or resistance) that is causing a signal value to be lower than the minimum operating limit. Note: the control unit could be providing the correct signal but it is being affected by a circuit fault. Refer to list of sensor / actuator checks on page 19.
P0599	Thermostat Heater Control Circuit High	The heating element for the coolant thermostat (used to regulate the coolant temperature more accurately and more quickly, depending on load and speed conditions) can be controlled by the control unit but, the control unit could directly operate a relay which in turn would control the thermostat heater. Refer to vehicle specific information if necessary to identify how the thermostat heater is controlled.	Identify the control circuit and components for the thermostat heater system. The voltage, frequency or value of the control signal in the heater circuit is either at full value (e.g. battery voltage with no on / off signal) or, the signal exists but it is above the normal operating range. Possible fault with heater or wiring (e.g. heater power supply, short / open circuit or resistance) that is causing a signal value to be higher than the maximum operating limit. Note: the control unit could be providing the correct signal but it is being affected by a circuit fault. Refer to list of sensor / actuator checks on page 19.
P0600	Serial Communication Link	For most modern vehicles, there are a number of control units that communicate / share information (including fault related information). The fault code could relate to a communication problem between control units or, it could relate to the communication for the "Serial Link" to the diagnostic equipment (at the diagnostic connector).	If possible, check diagnostic information (fault codes and live data) using the diagnostic equipment. If the only available code is P0600 but no other information is accessible from the diagnostic connection (live data etc), this could indicate a diagnostic serial link fault. Note that it could be necessary to introduce a simple fault to the vehicle system to create an additional fault code that should be detectable). If the diagnostic link appears to be operating correctly, the fault code could therefore be indicating a communication fault between control units. If possible check the live data to try and identify and communication signals that are incorrect, and carry out appropriate checks.
P0601	Internal Control Module Memory Check Sum Error	The control unit monitors its internal memory and electronic operation as well as monitoring output signals to actuators and input signals from sensors. The fault indicates an internal control unit error (based on a mathematical check of specific items of data).	Control unit has detected a fault with one or more of its internal processor functions; this can be caused by the control unit hardware / software being faulty and requiring either reprogramming or replacement. Can relate to coding error when a system component is replaced; if necessary, carry out re-configuration process. Check for other fault codes that could provide additional information and rectify as applicable. If possible, clear code and drive the vehicle; if the code re-appears, this could indicate a control unit related fault; check for any software updates that might be applicable.
P0602	Control Module Programming Error	The control unit has identified a programming fault, which could relate to the fitting of a replacement part (might require re-configuration). It is also possible that a replacement control unit has been fitted and the configuration of the control unit has not been carried out or performed correctly (incorrect VIN number etc).	Check whether any replacement parts or a replacement control unit has been fitted; if necessary carry out re-configuration applicable for any new parts fitted. If a replacement control unit has been fitted, carry out the configuration / programming process to enable the control unit to recognise the vehicle.
P0603	Internal Control Module Keep Alive Memory (KAM) Error	The control unit can contain a memory function that retains operating data for the vehicle; the memory processor is usually supplied with a permanent power supply (even with engine and other systems switched off) to ensure data is retained in the memory.	Check for the "permanent" power supply to the control unit (this can be on a fused circuit therefore check for a blown fuse); also check the earth connections for the control unit. If the power supply and earth connections are good, clear the fault code and if possible drive the vehicle or run the engine; if the fault re-appears, it could indicate a control unit related fault.
P0604	Internal Control Module Random Access Memory (RAM) Error	The control unit can contain a memory function that enables other parts of the control unit to access information / data.	Although the fault is likely to relate to an internal control unit problem, but some basic checks should be carried out. Check the power supplies and earth circuits to the control unit. Clear the fault code and if possible drive the vehicle or run the engine; if the fault re-appears, it could indicate a control unit related fault.
P0605	Internal Control Module Read Only Memory (ROM) Error	The control unit can contain a memory function that enables other parts of the control unit to access specific information / data relating to the vehicle and vehicle system; this particular Read Only Memory is designed so that it can only be read or interrogated by other parts of the control unit, but the data cannot be changed or re-programmed during normal operation.	Although the fault is likely to relate to an internal control unit problem, some basic checks should be carried out. Check the power supplies and earth circuits to the control unit. Clear the fault code and if possible drive the vehicle or run the engine; if the fault re-appears, it could indicate a control unit related fault.
P0606	ECM / PCM Processor	The Engine Control Module / Powertrain Control Module (control unit) will contain one or more processors. The control unit monitors its own operation and provides a fault code if a fault is detected. For many of the faults, the control unit compares its output control signals (and output voltages to actuators where applicable) against the expected value (held in memory); if the actual signal or voltage differs from the expected value, this will result in a fault code being produced.	Although the fault is likely to relate to an internal control unit problem, some basic checks should be carried out. Check the power supplies and earth circuits to the control unit. Clear the fault code and if possible drive the vehicle or run the engine; if the fault re-appears, it could indicate a control unit related fault.
P0607	Control Module Performance	The control unit monitors its own operation and provides a fault code if a fault is detected. For many of the faults, the control unit compares its output control signals (and output voltages to actuators where applicable) against the expected value (held in memory); if the actual signal or voltage differs from the expected value, this will result in a fault code being produced.	The fault code indicates that the control module has detected an undefined fault, which is resulting in incorrect performance for one of its controlling functions. It may be necessary to refer to vehicle specific information to identify the module and its control functions. The possible faults include internal module components (processors, circuit boards etc), as well as possible faults with sensors / actuators and wiring connected to the module (faults with external component connected to the module can affect the control module performance). Check the power supplies and earth circuits to the control unit. Clear the fault code and if possible drive the vehicle or run the engine; if the fault re-appears, it could indicate a control unit related fault.

NOTE 1. Check for other fault codes that could provide additional information.
NOTE 3. If a fault cannot be located, it is also possible that the control unit is at fault.
NOTE 2. Communication between control units can pass via a CAN-Bus system; refer to Page 51 for CAN-Bus checks.
NOTE 4. Refer to Page 53 for list of pages that show the ISO standard for component locations e.g. Sensor A - Bank 1.

Code	Description		
P0608	Control Module VSS Output "A"	The control unit monitors its own operation and provides a fault code if a fault is detected. For many of the faults, the control unit compares its output control signals (and output voltages to actuators where applicable) against the expected value (held in memory); if the actual signal or voltage differs from the expected value, this will result in a fault code being produced.	The fault code indicates that the control unit has detected an undefined fault relating to the Vehicle Speed Sensor system and the output signal from the control unit; this could relate to a fault in the reference or supply voltage provided by the control unit to the speed sensor, but it could also relate to a speed signal being passed from the control unit to another control module. If possible, check the live data to try and identify any incorrect vehicle speed signals. Check vehicle speed sensor operation Refer to list of sensor / actuator checks on page 19. If applicable, check the speed sensor signal from the control unit to other control units that might use of the vehicle speed signal e.g. transmission control unit; check wiring and connections as applicable.
P0609	Control Module VSS Output "B"	The control unit monitors its own operation and provides a fault code if a fault is detected. For many of the faults, the control unit compares its output control signals (and output voltages to actuators where applicable) against the expected value (held in memory); if the actual signal or voltage differs from the expected value, this will result in a fault code being produced.	The fault code indicates that the control unit has detected an undefined fault relating to the Vehicle Speed Sensor system and the output signal from the control unit; this could relate to a fault in the reference or supply voltage provided by the control unit to the speed sensor, but it could also relate to a speed signal being passed from the control unit to another control module. If possible, check the live data to try and identify any incorrect vehicle speed signals. Check vehicle speed sensor operation Refer to list of sensor / actuator checks on page 19. If applicable, check the speed sensor signal from the control unit to other control units that might use of the vehicle speed signal e.g. transmission control unit; check wiring and connections as applicable.
P060A	Internal Control Module Monitoring Processor Performance	The control unit monitors its own operation and provides a fault code if a fault is detected. For many of the faults, the control unit compares its output control signals (and output voltages to actuators where applicable) against the expected value (held in memory); if the actual signal or voltage differs from the expected value, this will result in a fault code being produced.	The fault code indicates that the control module has detected an undefined fault, which is resulting in incorrect performance for one of its controlling functions. It may be necessary to refer to vehicle specific information to identify the module and its control functions. The possible faults include internal module components (processors, circuit boards etc), as well as possible faults with sensors / actuators and wiring connected to the module (faults with external component connected to the module can affect the control module performance). Check the power supplies and earth circuits to the control unit. Clear the fault code and if possible drive the vehicle or run the engine; if the fault re-appears, it could indicate a control unit related fault.
P060B	Internal Control Module A / D Processing Performance	The control unit processes analogue signals (usually from sensors) and converts the signals into a digital format.	The fault code indicates that there is a fault with the internal A / D process. The fault is unidentified but likely to be an internal control unit fault. Check the power supplies and earth circuits to the control unit (and other connections to the control unit) for short / open circuits. Clear the fault code and if possible drive the vehicle or run the engine; if the fault re-appears, it could indicate a control unit related fault.
P060C	Internal Control Module Main Processor Performance	The main control functions for the control unit are carried out by the main processor.	The fault code indicates that there is a fault with the control unit main processor. The fault is unidentified but likely to be an internal control unit fault. Check the power supplies and earth circuits to the control unit (and other connections to the control unit) for short / open circuits. Clear the fault code and if possible drive the vehicle or run the engine; if the fault re-appears, it could indicate a control unit related fault.
P060D	Internal Control Module Accelerator Pedal Position Performance	The control unit will manage some of the vehicle's systems based on information from various sensors e.g. accelerator position sensor signal, as well as using the operating data contained within the control unit memories.	The fault code indicates that there is a fault relating to the of the internal hardware or software that deals with the pedal position information. The fault is unidentified but likely to be an internal control unit fault but checks should be carried out on the pedal position sensor and circuit Refer to list of sensor / actuator checks on page 19. Check the power supplies and earth circuits to the control unit for short / open circuits. Clear the fault code and if possible drive the vehicle or run the engine; if the fault re-appears, it could indicate a control unit related fault.
P060E	Internal Control Module Throttle Position Performance	The control unit will manage some of the vehicle's systems based on information from various sensors e.g. throttle position sensor signal, as well as using the operating data contained within the control unit memories.	The fault code indicates that there is a fault relating to the of the internal hardware or software that deals with the throttle position information. The fault is unidentified but likely to be an internal control unit fault but checks should be carried out on the throttle position sensor and circuit Refer to list of sensor / actuator checks on page 19. Check the power supplies and earth circuits to the control unit for short / open circuits. Clear the fault code and if possible drive the vehicle or run the engine; if the fault re-appears, it could indicate a control unit related fault.
P060F	Internal Control Module Coolant Temperature Performance	The control unit will manage some of the vehicle's systems based on information from various sensors e.g. temperature sensor signal, as well as using the operating data contained within the control unit memories.	The fault code indicates that there is a fault relating to the of the internal hardware or software that deals with the coolant temperature information. The fault is unidentified but likely to be an internal control unit fault but checks should be carried out on the coolant temperature sensor and circuit Refer to list of sensor / actuator checks on page 19. Check the power supplies and earth circuits to the control unit for short / open circuits. Clear the fault code and if possible drive the vehicle or run the engine; if the fault re-appears, it could indicate a control unit related fault.
P0610	Control Module Vehicle Options Error	The control unit is configured / programmed with the vehicles identity (VIN), but it is also configured to match the options or accessories fitted to the vehicle e.g. air conditioning, auto transmission and other vehicle systems.	The fault code indicates a configuration fault e.g. the control unit has not been configured to match the vehicle or has not been configured correctly. The control unit is therefore not able to control or communicate with components or systems fitted to the vehicle. Check whether any replacement parts or a replacement control unit has been fitted; if necessary carry out re-configuration applicable for any new parts fitted. If a replacement control unit has been fitted, the control unit might require re-configuring; carry out the configuration / programming process to enable the control unit to recognise the vehicle or systems fitted.
P0611	Fuel Injector Control Module Performance	The fuel injector control module can monitor its own performance but, the main control unit will also monitor the injector control module operation and provide a fault code if a fault is detected.	The fault code indicates that the injector control module has an undefined internal fault. It may be necessary to refer to vehicle specific information to identify the module and its control functions. The possible faults include internal module components (processors, circuit boards etc), as well as possible faults with sensors / actuators and wiring connected to the module (faults with external components connected to the module can affect the control module performance). Check the power supplies and earth circuits to the control unit. Clear the fault code and if possible drive the vehicle or run the engine; if the fault re-appears, it could indicate a control unit related fault.
P0612	Fuel Injector Control Module Relay Control	The fuel injector control module can monitor its own performance but, the main control unit will also monitor the injector control module operation and provide a fault code if a fault is detected.	The fault code indicates that the injector control module has an undefined internal fault causing incorrect performance of the relay. It may be necessary to refer to vehicle specific information to identify the module, the relay and their control functions. The possible faults include internal module components (processors, circuit boards etc), but it is possible that a fault with the relay or with the wiring connecting the module with the relay). Check the relay for correct operation and check applicable wiring. Check the power supplies and earth circuits to the control unit. Clear the fault code and if possible drive the vehicle or run the engine; if the fault re-appears, it could indicate a control unit related fault.

NOTE 1. Check for other fault codes that could provide additional information.   **NOTE 2.** Communication between control units can pass via a CAN-Bus system; refer to Page 51 for CAN-Bus checks.

NOTE 3. If a fault cannot be located, it is also possible that the control unit is at fault.   **NOTE 4.** Refer to Page 53 for list of pages that show the ISO standard for component locations e.g. Sensor A - Bank 1.

Fault code	EOBD / ISO Description	Component / System Description	Meaningful Description and Quick Check
P0613	TCM Processor	The control Transmission Control Module monitors its own operation and provides a fault code if a fault is detected. Note that on many systems, when a fault is detected a limited fault related message or indication will be provided to the main control unit; it will therefore be necessary to interrogate the Transmission control unit for additional fault codes. For many of the faults, the control unit compares its output control signals (and output voltages to actuators where applicable) against the expected value (held in memory); if the actual signal or voltage differs from the expected value, this will result in a fault code being produced.	The fault code indicates an unidentified fault with the transmission control module internal processor(s). Although the fault is likely to relate to an internal transmission control unit problem, some basic checks should be carried out. Check the power supplies and earth circuits to the control unit. Check the communication wiring between the transmission control unit and the main control unit. Clear the fault code and if possible drive the vehicle or run the engine; if the fault re-appears, it could indicate a control unit related fault.
P0614	ECM / TCM Incompatible	The control unit is configured / programmed with the vehicles identity (VIN), but it is also configured to match the options or accessories fitted to the vehicle e.g. auto transmission. The transmission control unit will also be programmed to match the vehicle specification.	The fault code indicates the TCM and ECM are not compatible, which could be caused by a programming error e.g. one of the control units has not been configured to match the vehicle or has not been configured correctly. The control unit is therefore not able to control or communicate with components or systems fitted to the vehicle. Check whether any replacement parts or a replacement control units have been fitted; if necessary carry out re-configuration applicable for any new parts fitted. If a replacement control unit has been fitted, the control unit might require re-configuring; carry out the configuration / programming process to enable the control unit to recognise the vehicle or systems fitted. Clear the fault code and if possible drive the vehicle or run the engine; if the fault re-appears, it could indicate a control unit related fault.
P0615	Starter Relay Circuit	Also refer to P0512 for information on relay energising circuit. When the relay energising winding is operating (energising winding power and earth circuits connected during starting), the relay contacts should close and battery voltage should then pass to the starter motor solenoid. However, a number of different methods are used to wire and switch the relay, and it may be necessary to refer to vehicle specific information to identify the wiring and controls used for the relay. Note that the control unit often has a connection to the relay / starter circuit to monitor relay / starter operation	Refer to list of sensor / actuator checks on page 19 for notes on operation, expected voltages and relay checks. Different manufacturers could use the fault code to refer to different relay circuits. The fault code could refer to the power supply circuit (the power supply to the relay contacts and the circuit from the relay contacts to the starter). Alternatively, the fault code could refer to the relay energising winding circuit (power supply to the energising winding and the switched earth path through to the control unit). The voltage in the applicable relay circuit is incorrect (undefined fault). Check power supplies to relay contacts (possibly via a fused circuit) and to relay energising winding (might only be available when ignition switch is in start position). Check operation of energising winding (short / open circuit) and closing / opening of relay contacts. Check all wiring for short / open circuits. Also check "start inhibit" switch (if applicable), which can form part of the energising winding circuit on some vehicles.
P0616	Starter Relay Circuit Low	Also refer to P0512 for information on relay energising circuit. When the relay energising winding is operating (energising winding power and earth circuits connected during starting), the relay contacts should close and battery voltage should then pass to the starter motor solenoid. However, a number of different methods are used to wire and switch the relay, and it may be necessary to refer to vehicle specific information to identify the wiring and controls used for the relay. Note that the control unit often has a connection to the relay / starter circuit to monitor relay / starter operation	Refer to list of sensor / actuator checks on page 19 for notes on operation, expected voltages and relay checks. Different manufacturers could use the fault code to refer to different relay circuits. The fault code could refer to the power supply circuit (the power supply to the relay contacts and the circuit from the relay contacts to the starter). Alternatively, the fault code could refer to the relay energising winding circuit (power supply to the energising winding and the switched earth path through to the control unit). The voltage in the applicable relay circuit is low (likely to be zero volts) at a time when the voltage should be at battery voltage level. Check power supplies to relay contacts (usually via a fused circuit) and to relay energising winding (might only be available when ignition switch is in start position). Check operation of energising winding (short / open circuit) and closing / opening of relay contacts. Check all wiring for short open circuits. Also check "start inhibit" switch (if applicable), which can form part of the energising winding circuit on some vehicles.
P0617	Starter Relay Circuit High	Also refer to P0512 for information on relay energising circuit. When the relay energising winding is operating (energising winding power and earth circuits connected during starting), the relay contacts should close and battery voltage should then pass to the starter motor solenoid. However, a number of different methods are used to wire and switch the relay, and it may be necessary to refer to vehicle specific information to identify the wiring and controls used for the relay. Note that the control unit often has a connection to the relay / starter circuit to monitor relay / starter operation	Refer to list of sensor / actuator checks on page 19 for notes on operation, expected voltages and relay checks. Different manufacturers could use the fault code to refer to different relay circuits. The fault code could refer to the power supply circuit (the power supply to the relay contacts and the circuit from the relay contacts to the starter). Alternatively, the fault code could refer to the relay energising winding circuit (power supply to the energising winding and the switched earth path through to the control unit). The voltage in the relay circuit is high (likely to be battery voltage) when the relay circuit should be off (contacts open); it is possible that relay / starter circuits are switched "on" (starter motor operating) whilst the engine is running. Likely cause is relay contacts remaining closed (either due to relay fault or energising winding is being continuously activated); also possible short in wiring to another power supply or the energising winding earth path is shorted to earth. Check operation of relay, including closing / opening of relay contacts, and check operation of energising winding (short / open circuit).
P0618	Alternative Fuel Control Module KAM Error	The alternative fuel control module communicates with the main control unit and provides alternative control functions when the vehicle is operating on an alternative fuel. The operation alternative fuel control module is likely to be monitored by the main engine control unit (although some diagnostic capability could be contained with the alternative fuel control module).	The fault code indicates that the main control unit / alternative fuel control module has detected a fault with the internal memory (Keep Alive Memory). It may be necessary to refer to vehicle specific information to identify the module and its control functions. The KAM requires a permanent power supply to prevent the memory loss after the engine is switched off; check for the "permanent" power supply to the control unit (this can be on a fused circuit therefore check for a blown fuse). Other possible faults include internal module components (processors, circuit boards etc), but it is possible that a fault with sensors / actuators and wiring (connected to the module) can affect the control module operation. Check the power supplies and earth circuits to the control module and also check communication wiring to the main control unit. Clear the fault code and if possible drive the vehicle or run the engine; if the fault re-appears, it could indicate a control unit related fault. Note. It is possible that an electrical spike can erase some data or memory contained within the control unit.
P0619	Alternative Fuel Control Module RAM / ROM Error	The alternative fuel control module communicates with the main control unit and provides alternative control functions when the vehicle is operating on an alternative fuel. The operation of the alternative fuel control module is likely to be monitored by the main engine control unit (although some diagnostic capability could be contained with the alternative fuel control module).	The fault code indicates that the main control unit / alternative fuel control module has detected a fault with the internal memory (Random Access Memory or Read Only Memory). It may be necessary to refer to vehicle specific information to identify the module and its control functions. The possible faults include internal module components (processors, circuit boards etc), but it is possible that a fault with sensors / actuators and wiring (connected to the module) can affect the control module operation. Check the power supplies and earth circuits to the control module and also check communication wiring to the main control unit. Clear the fault code and if possible drive the vehicle or run the engine; if the fault re-appears, it could indicate a control unit related fault. Note. It is possible that an electrical spike can erase some data or memory contained within the control unit.

NOTE 1. Check for other fault codes that could provide additional information.

NOTE 2. Communication between control units can pass via a CAN-Bus system; refer to Page 51 for CAN-Bus checks.

NOTE 3. If a fault cannot be located, it is also possible that the control unit is at fault. NOTE 4. Refer to Page 53 for list of pages that show the ISO standard for component locations e.g. Sensor A - Bank 1.

Code	Description		
P061A	Internal Control Module Torque Performance	The control unit receives information from various sensors and vehicle systems (including throttle position selected by the driver); the information is then used to calculate the engine output (torque) to match the operating conditions and the performance request by the driver (throttle position).	The fault code indicates an internal fault with the processors etc that are calculating the torque that should be delivered by the engine. Although the fault is likely to relate to an internal control unit problem, some basic checks should be carried out. Check the power supplies and earth circuits to the control unit and check wiring / connections from the sensors and actuators to the control unit (check for short / open circuits etc. Check for other fault codes and if possible check live data to try and identify any other system faults). Clear the fault code and if possible drive the vehicle or run the engine; if the fault re-appears, it could indicate a control unit related fault.
P061B	Internal Control Module Torque Calculation Performance	The control unit receives information from various sensors and vehicle systems (including throttle position selected by the driver); the information is then used to calculate the engine output (torque) to match the operating conditions and the performance request by the driver (throttle position).	The fault code indicates an internal fault with the processors etc that are calculating the torque that should be delivered by the engine. Although the fault is likely to relate to an internal control unit problem, some basic checks should be carried out. Check the power supplies and earth circuits to the control unit and check wiring / connections from the sensors and actuators to the control unit (check for short / open circuits etc. Check for other fault codes and if possible check live data to try and identify any other system faults). Clear the fault code and if possible drive the vehicle or run the engine; if the fault re-appears, it could indicate a control unit related fault.
P061C	Internal Control Module Engine RPM Performance	The control unit will manage some of the vehicle's systems based on information from various sensors e.g. engine speed / rpm signal, as well as using the data contained within the control unit memories.	The fault code indicates that there is a fault relating to the internal hardware or software that deals with the engine speed information. The fault is unidentified but likely to be an internal control unit fault but checks should be carried out on the engine speed sensor and circuit Refer to list of sensor / actuator checks on page 19. Check the power supplies and earth circuits to the control unit for short / open circuits. Clear the fault code and if possible drive the vehicle or run the engine; if the fault re-appears, it could indicate a control unit related fault.
P061D	Internal Control Module Engine Air Mass Performance	The control unit will manage some of the vehicle's systems based on information from various sensors e.g. air mass sensor signal, as well as using the data contained within the control unit memories.	The fault code indicates that there is a fault relating to the internal hardware or software that deals with the intake air mass information. The fault is unidentified but likely to be an internal control unit fault but checks should be carried out on the air mass sensor and circuit Refer to list of sensor / actuator checks on page 19. Check the power supplies and earth circuits to the control unit for short / open circuits. Clear the fault code and if possible drive the vehicle or run the engine; if the fault re-appears, it could indicate a control unit related fault.
P061E	Internal Control Module Brake Signal Performance	The control unit will manage some of the vehicle's systems based on information from various sensors e.g. brake signal, as well as using the data contained within the control unit memories.	The fault code indicates that there is a fault relating to the internal hardware or software that deals with the brake system information. The fault is unidentified but likely to be an internal control unit fault but checks should be carried out on the brake system sensors and circuits Refer to list of sensor / actuator checks on page 19. Check the power supplies and earth circuits to the control unit for short / open circuits. Clear the fault code and if possible drive the vehicle or run the engine; if the fault re-appears, it could indicate a control unit related fault.
P061F	Internal Control Module Throttle Actuator Controller Performance	The control unit will manage some of the vehicle's systems e.g. throttle control, based on information from various sensors and the operating data contained within the control unit memories.	The fault code indicates that there is a fault relating to the internal hardware or software that deals with the throttle position control. The fault is unidentified but likely to be an internal control unit fault but checks should be carried out on the throttle control system (throttle control actuator and throttle mechanism), also check for other fault codes and live data that might help identify other system faults. Check the power supplies and earth circuits to the control unit for short / open circuits. Clear the fault code and if possible drive the vehicle or run the engine; if the fault re-appears, it could indicate a control unit related fault.
P0620	Generator Control Circuit	The generator can be connected to the main (power train) control unit to enable the control unit to monitor and / or to control generator output; the control unit can regulate the generator's field winding voltage / current, which in turn controls generator output.	It will be necessary to refer to vehicle specific information to identify the wiring / circuit used for the generator system. The fault code indicates that there is an undefined fault detected by the control unit in the circuits between the generator and the control unit. Note that the fault will be detected due to incorrect voltage / signal value in the circuit. Check wiring and connections between the generator and the control unit for short / open circuit. If the wiring and connections are good, it is possible that a charging fault exists. If charging output is low, also check condition and tension of generator drive belt / mechanism and battery condition.
P0621	Generator Lamp / L Terminal Circuit	The generator warning lamp can be controlled from the generator or, on many systems it is controlled by the engine control unit (possibly via the CAN-Bus network). The generator or control unit can provide an earth path or power to the warning light on the "L" circuit, but a signal could be provided to a separate module that in turn will turn on the light.	It will be necessary to refer to vehicle specific information to identify the wiring / circuit used for the generator warning light. The fault code indicates that the voltage, frequency or signal value in the "L" circuit is incorrect. The fault could be related to a wiring fault but, depending on the warning light system operation, the fault could be caused by a control module or CAN-Bus related fault. Check all applicable wiring for short / open circuits and check condition of warning light bulb (or lighting system).
P0622	Generator Field / F Terminal Circuit	The generator can be connected to the main (power train) control unit to enable the control unit to monitor and / or to control generator output; the control unit can regulate the generator's field winding voltage / current, which in turn controls generator output.	It will be necessary to refer to vehicle specific information to identify the wiring / circuit used for the generator system. The fault code indicates an undefined fault in the circuit between the control unit and the generator "Field" terminal. Check the wiring and connections between the generator "F" terminal and the control unit for short / open circuits. If the wiring and connections are good, it is possible that a fault exists in the generator.
P0623	Generator Lamp Control Circuit	The generator warning lamp can be controlled from the generator or, on many systems it is controlled by the engine control unit (possibly via the CAN-Bus network). The generator or control unit can provide an earth path or power to the warning light circuit, but a signal could be provided to a separate module that in turn will turn on the light.	It will be necessary to refer to vehicle specific information to identify the wiring / circuit used for the generator warning light. The fault code indicates that the voltage, frequency or signal value in the circuit is incorrect. The fault could be related to a wiring fault but, depending on the warning light system operation, the fault could be caused by a control module or CAN-Bus related fault. Check all applicable wiring for short / open circuits and check condition of warning light bulb (or lighting system).
P0624	Fuel Cap Lamp Control Circuit	The control unit monitors the removal / refitting of the fuel cap (via a fuel cap sensor); the control unit also illuminates a warning light if the cap is not refitted correctly. The warning light usually receives a power supply via a common power supply from the instrument cluster; the light will therefore be switched on via the control unit which will provide the earth path (note that different vehicles can use different wiring for the light).	If necessary, refer to vehicle specific information to identify the wiring for the warning light. The fault code indicates that there is an undefined fault with the warning light circuit. Check for short / open circuits in the wiring and for failure of the light bulb / LED. Check for a power supply and earth path for the warning light (as applicable depending on the type of wiring used).

NOTE 1. Check for other fault codes that could provide additional information.   **NOTE 2.** Communication between control units can pass via a CAN-Bus system; refer to Page 51 for CAN-Bus checks.   155

NOTE 3. If a fault cannot be located, it is also possible that the control unit is at fault.   **NOTE 4.** Refer to Page 53 for list of pages that show the ISO standard for component locations e.g. Sensor A - Bank 1.

Fault code	EOBD / ISO Description	Component / System Description	Meaningful Description and Quick Check
P0625	Generator Field / F Terminal Circuit Low	The generator can be connected to the main (power train) control unit to enable the control unit to monitor / control generator output; the control unit regulates the generator's field winding voltage / current, which in turn controls generator output.	It will be necessary to refer to vehicle specific information to identify the wiring / circuit used for the generator system. The fault code indicates that the voltage, frequency or signal value in the generator Field circuit is low. Check the wiring and connections between the generator "F" terminal and the control unit for short / open circuits. If the wiring and connections are good, it is possible that a fault exists in the generator / field windings.
P0626	Generator Field / F Terminal Circuit High	The generator can be connected to the main (power train) control unit to enable the control unit to monitor / control generator output; the control unit regulates the generator's field winding voltage / current, which in turn controls generator output.	It will be necessary to refer to vehicle specific information to identify the wiring / circuit used for the generator system. The fault code indicates that the voltage, frequency or signal value in the generator Field circuit is high. Check the wiring and connections between the generator "F" terminal and the control unit for short / open circuits. If the wiring and connections are good, it is possible that a fault exists in the generator / field windings.
P0627	Fuel Pump "A" Control Circuit / Open	Depending on the system design and wiring, the fuel pump control circuit can refer to the circuit used by the control unit to operate the fuel pump relay or, the circuit between the relay and the fuel pump. A manufacturer could in fact use the fault code to refer to both circuits.	It may be necessary to identify the wiring for the fuel pump and relay (as applicable) and identify how the relay is switched by the control unit (usually on the earth path for the relay energizing winding). The fault code indicates that there is a circuit fault (probably open circuit) in the wiring between the control unit and the relay or between the relay and the fuel pump. The voltage in the applicable circuit is likely to be: constant value at all times (e.g. battery voltage) or non-existent at all times. If the voltage is at zero at all times, this can be caused by open circuit in wiring (or relay energizing winding) or, no power supply to the energising winding (check back to power source). Check wiring between control unit and relay for short or open circuit and also check wiring between relay and fuel pump for short open circuit. Check relay operation (energising winding and contacts opening / closing) and check fuel pump motor (resistance / continuity). Refer to list of sensor / actuator checks on page 19. Also refer to fault codes P0230 - P0233 for additional information.
P0628	Fuel Pump "A" Control Circuit Low	Depending on the system design and wiring, the fuel pump control circuit can refer to the circuit used by the control unit to operate the fuel pump relay or, the circuit between the relay and the fuel pump. A manufacturer could in fact use the fault code to refer to both circuits.	It may be necessary to identify the wiring for the fuel pump and relay (as applicable) and identify how the relay is switched by the control unit (usually on the earth path for the relay energizing winding). The fault code indicates that there is a circuit fault in the wiring between the control unit and the relay or between the relay and the fuel pump. The voltage in the relay circuit is low (likely to be zero volts) when the relay should be providing power to the actuator (contacts closed). Check power supply to relay (usually via a fused circuit), check connection from relay to actuator (and monitoring wire to control unit if applicable). Check operation of relay, including closing / opening of relay contacts, and check operation of energising winding (short / open circuit). Check all wiring between control unit and relay, and between relay and pump for short / open circuit. Also check fuel pump motor (resistance / continuity). Refer to list of sensor / actuator checks on page 19. Also refer to fault codes P0230 - P0233 for additional information.
P0629	Fuel Pump "A" Control Circuit High	Depending on the system design and wiring, the fuel pump control circuit can refer to the circuit used by the control unit to operate the fuel pump relay or, the circuit between the relay and the fuel pump. A manufacturer could in fact use the fault code to refer to both circuits.	It may be necessary to identify the wiring for the fuel pump and relay (as applicable) and identify how the relay is switched by the control unit (usually on the earth path for the relay energizing winding). The fault code indicates that there is a circuit fault in the wiring between the control unit and the relay or between the relay and the fuel pump. The voltage in the relay circuit is high (likely to be battery volts) when the relay circuit should be off (contacts open); the check is probably detected by the control unit which monitors the voltage in the circuit (between relay and actuator). Likely cause is relay contacts remaining closed (either due to relay fault or energising winding is being continuously activated; also possible short in wiring to another power supply. Check all wiring between control unit and relay, and between relay and pump for short / open circuit. Also check fuel pump motor (resistance / continuity). Refer to list of sensor / actuator checks on page 19. Also refer to fault codes P0230 - P0233 for additional information.
P062A	Fuel Pump "A" Control Circuit Range / Performance	Depending on the system design and wiring, the fuel pump control circuit can refer to the circuit used by the control unit to operate the fuel pump relay or, the circuit between the relay and the fuel pump. A manufacturer could in fact use the fault code to refer to both circuits.	It may be necessary to identify the wiring for the fuel pump and relay (as applicable) and identify how the relay is switched by the control unit (usually on the earth path for the relay energizing winding). The fault code indicates that the voltage in the circuit is within the normal operating range / tolerance but it is likely that the voltage is not changing as expected with the change in the operating conditions. It is possible that the relay is not operating correctly at the applicable time or that the fuel pump is not operating correctly at the applicable time. Check all wiring between control unit and relay, and between relay and pump for short / open circuit. Also check fuel pump motor (resistance / continuity). Refer to list of sensor / actuator checks on page 19. Also refer to fault codes P0230 - P0233 for additional information.
P062B	Internal Control Module Fuel Injector Control Performance	The control unit will manage some of the vehicle's systems e.g. fuel injection, based on information from various sensors and using the operating data contained within the control unit memories. Note that for some systems, the main control unit could be controlling a separate fuel injector control module.	The fault code indicates that there is a fault relating to the internal hardware or software that deals with the fuel injector control. The fault is unidentified but likely to be an internal control unit fault, but checks should be carried out on the fuel injectors and wiring (also check wiring / connections to separate control module, if fitted). Check for other fault codes and live data that might help identify other system faults. Check the power supplies and earth circuits to the control unit for short / open circuits. Clear the fault code and if possible drive the vehicle or run the engine; if the fault re-appears, it could indicate a control unit related fault.
P062C	Internal Control Module Vehicle Speed Performance	The control unit will manage some of the vehicle's systems e.g. vehicle speed, based on information from various sensors and using the operating data contained within the control unit memories.	The fault code indicates that there is a fault relating to the of the internal hardware or software that deals with the vehicle speed control. The fault is unidentified but likely to be an internal control unit fault, but checks should be carried out on the components systems used to control vehicle speed e.g. throttle actuator motor. Check the power supplies and earth circuits to the control unit for short / open circuits. Clear the fault code and if possible drive the vehicle or run the engine; if the fault re-appears, it could indicate a control unit related fault.
P062D	Fuel Injector Driver Circuit Performance Bank 1	The control unit will manage some of the vehicle's systems e.g. fuel injection, based on information from various sensors and using the operating data contained within the control unit memories. The driver circuit refers to the electronic circuits / components used to switch the high current circuits for the fuel injectors. Note that for some systems, the main control unit could be controlling a separate fuel injector control module.	The fault code indicates that there is a fault relating to the internal hardware or software that deals with the fuel injector control. The fault is unidentified but likely to be an internal control unit fault, but checks should be carried out on the fuel injectors and wiring (also check wiring / connections to separate control module, if fitted). Check for other fault codes and live data that might help identify other system faults. Check the power supplies and earth circuits to the control unit for short / open circuits. Clear the fault code and if possible drive the vehicle or run the engine; if the fault re-appears, it could indicate a control unit related fault.

NOTE 1. Check for other fault codes that could provide additional information. **NOTE 2.** Communication between control units can pass via a CAN-Bus system; refer to Page 51 for CAN-Bus checks.

P062E	Fuel Injector Driver Circuit Performance Bank 2	The control unit will manage some of the vehicle's systems e.g. fuel injection, based on information from various sensors and using the operating data contained within the control unit memories. The driver circuit refers to the electronic circuits / components used to switch the high current circuits for the fuel injectors. Note that for some systems, the main control unit could be controlling a separate fuel injector control module.	The fault code indicates that there is a fault relating to the internal hardware or software that deals with the fuel injector control. The fault is unidentified but likely to be an internal control unit fault, but checks should be carried out on the fuel injectors and wiring (also check wiring / connections to separate control module, if fitted). Check for other fault codes and live data that might help identify other system faults. Check the power supplies and earth circuits to the control unit for short / open circuits. Clear the fault code and if possible drive the vehicle or run the engine; if the fault re-appears, it could indicate a control unit related fault.
P062F	Internal Control Module EEPROM Error	The control unit can contain a memory function that enables other parts of the control unit to access specific information / data relating to the vehicle and vehicle system.	Although the fault is likely to relate to an internal control unit problem, some basic checks should be carried out. Check the power supplies and earth circuits to the control unit. Clear the fault code and if possible drive the vehicle or run the engine; if the fault re-appears, it could indicate a control unit related fault. Note. It is possible that an electrical spike can erase some data or memory contained within the control unit.
P0630	VIN Not Programmed or Incompatible -ECM / PCM	The VIN number is programmed into the Engine / Powertrain Control Module during production and when a replacement control unit is fitted (or if the control unit has been re-programmed). If the VIN number cannot be programmed back into the control unit or, it has not been programmed as required after applicable repair tasks, the control unit will generate the fault code.	Refer to vehicle specific information to identify the process for re-programming the VIN code into the engine / powertrain control module; this may be possible using some diagnostic equipment. Note. It is possible that an electrical spike can erase some data or memory contained within the control unit e.g. the VIN number. The vehicle history should be checked to establish whether other faults have existed e.g. a faulty alternator that could have caused the problem. If applicable equipment is available, programme the control unit with the VIN number. If the control unit does not accept the VIN number, it could be that the number being programmed is incompatible with the control unit or there could be a control unit related fault.
P0631	VIN Not Programmed or Incompatible -TCM	The VIN number is programmed into the Transmission Control Module during production and when a replacement control unit is fitted (or if the control unit has been re-programmed). If the VIN number cannot be programmed back into the control unit or, it has not been programmed as required after applicable repair tasks, the control unit will generate the fault code.	Refer to vehicle specific information to identify the process for re-programming the VIN code into the transmission control unit; this may be possible using some diagnostic equipment. Note. It is possible that an electrical spike can erase some data or memory contained within the control unit e.g. the VIN number. The vehicle history should be checked to establish whether other faults have existed e.g. a faulty alternator that could have caused the problem. If applicable equipment is available, programme the control unit with the VIN number. If the control unit does not accept the VIN number, it could be that the number being programmed is incompatible with the control unit or there could be a control unit related fault.
P0632	Odometer Not Programmed -ECM / PCM	The Odometer systems records the vehicle miles / kilometres. The data is updated and stored within one or a number of control modules on the vehicle. If a replacement control unit / module is fitted, it can be necessary to programme in the miles / kilometres into the replacement module.	Refer to vehicle specific information to identify the process for re-programming the miles / kilometres information into the control unit; this may be possible using some diagnostic equipment. Note. It is possible that an electrical spike can erase some data or memory contained within the control unit.
P0633	Immobilizer Key Not Programmed -ECM / PCM	It may be necessary to refer to vehicle specific information to identify the operation and components involved for the immobilzer system.	The fault code indicates that the control unit / immobilzer system does not recognise the key (or other device) being used to enter / start the vehicle. The fault could be related to a control unit or key programming error (new key or control unit fitted). If possible, carry out reprogramming / learning process to enable the system to recognise the key. Note. It is possible that an electrical spike can erase some data or memory contained within the control unit.
P0634	PCM / ECM / TCM Internal Temperature Too High	A temperature sensor is located in the control unit (e.g. engine, transmission or other control unit), to monitor overheating of the control unit; this could be caused due to an excessive control unit work load as well as due to high ambient temperature (temperature in the region of the control unit location).	The fault code indicates that the internal temperature of the control unit is too high. Note that under these conditions, the control unit can limit the control functions it is performing, which will reduce the work load and current in the control thus reducing the temperature; this can cause restricted operation of the system being controlled by the control unit. Check air circulation for the control unit location, check cooling fan (if fitted). If it is however suspected that the temperature is not high and that there is a temperature sensor fault, refer to fault code P0666.
P0635	Power Steering Control Circuit	The power steering control circuit can refer to the control of an electric motor used to drive the hydraulic pump or, the control of actuators e.g. solenoids used to control hydraulic valves on systems with mechanically driven pumps; the power steering control circuit can also refer to the control of a fully electric power steering system (electrical assisted systems with a motor connected to the steering column. Refer to vehicle specific information to identify the type of power steering system, the components and the wiring.	Identify the type of power steering fitted to the vehicle and the actuators used to control the system (motor, solenoids etc). The fault code indicates that the control unit has identified a fault in the control circuit (between the control unit and the actuator being controlled). The voltage, frequency or value of the control signal in the circuit is incorrect (undefined fault). Signal could be:- out of normal operating range, constant value, non-existent, corrupt. Possible fault with actuator or wiring e.g. actuator power supply, short / open circuit, high circuit resistance or incorrect actuator resistance. Note: the control unit could be providing the correct signal but it is being affected by a circuit fault. Refer to list of sensor / actuator checks on page 19. If an electrical fault cannot be identified, check operation of actuators (motor solenoid etc).
P0636	Power Steering Control Circuit Low	The power steering control circuit can refer to the control of an electric motor used to drive the hydraulic pump or, the control of actuators e.g. solenoids used to control hydraulic valves on systems with mechanically driven pumps; the power steering control circuit can also refer to the control of a fully electric power steering system (electrical assisted systems with a motor connected to the steering column. Refer to vehicle specific information to identify the type of power steering system, the components and the wiring.	Identify the type of power steering fitted to the vehicle and the actuators used to control the system (motor, solenoids etc). The fault code indicates that the control unit has identified a fault in the control circuit (between the control unit and the actuator being controlled). The voltage, frequency or value of the control signal in the actuator circuit is either "zero" (no signal) or, the signal exists but is below the normal operating range. Possible fault with actuator or wiring (e.g. actuator power supply, short / open circuit or resistance) that is causing a signal value to be lower than the minimum operating limit. Note: the control unit could be providing the correct signal but it is being affected by a circuit fault. Refer to list of sensor / actuator checks on page 19.
P0637	Power Steering Control Circuit High	The power steering control circuit can refer to the control of an electric motor used to drive the hydraulic pump or, the control of actuators e.g. solenoids used to control hydraulic valves on systems with mechanically driven pumps; the power steering control circuit can also refer to the control of a fully electric power steering system (electrical assisted systems with a motor connected to the steering column. Refer to vehicle specific information to identify the type of power steering system, the components and the wiring.	Identify the type of power steering fitted to the vehicle and the actuators used to control the system (motor, solenoids etc). The fault code indicates that the control unit has identified a fault in the control circuit (between the control unit and the actuator being controlled). The voltage, frequency or value of the control signal in the actuator circuit is either at full value (e.g. battery voltage with no on / off signal) or, the signal exists but it is above the normal operating range. Possible fault with actuator or wiring (e.g. actuator power supply, short / open circuit or resistance) that is causing a signal value to be higher than the maximum operating limit. Note: the control unit could be providing the correct signal but it is being affected by a circuit fault. Refer to list of sensor / actuator checks on page 19.

NOTE 1. Check for other fault codes that could provide additional information.   **NOTE 2.** Communication between control units can pass via a CAN-Bus system; refer to Page 51 for CAN-Bus checks.
NOTE 3. If a fault cannot be located, it is also possible that the control unit is at fault.   **NOTE 4.** Refer to Page 53 for list of pages that show the ISO standard for component locations e.g. Sensor A - Bank 1.

Fault code	EOBD / ISO Description	Component / System Description	Meaningful Description and Quick Check
P0638	Throttle Actuator Control Range / Performance Bank 1	The throttle actuator is the motor that controls the throttle plate / butterfly opening angle. Note that the throttle actuator will probably form part of the throttle control assembly which can include a throttle position sensor as well as the control motor.	Note that the position of the motor can be regulated using a frequency / pulse width control signal or other electrical control (provided by the control unit on the earth or power circuit). It is advisable to refer to vehicle specific information to identify the type of control signal provided to the motor. The fault code could indicate that the throttle actuator is being provided with a specific control signal (by the control unit) but the actuator is not moving to the position expected by the control unit (position monitored by the throttle position sensor). The code could also indicate that the voltage, frequency or value of the control signal in the motor circuit is within the normal operating range / tolerance, but the signal might be incorrect for the operating conditions. Possible fault in motor or wiring e.g. short / open circuit or resistance, but also possible that the motor (or mechanism / system controlled by the motor) is not operating or moving correctly. Also check for contamination / restrictions that could affect airflow around the throttle butterfly. Refer to list of sensor / actuator checks on page 19.
P0639	Throttle Actuator Control Range / Performance Bank 2	The throttle actuator is the motor that controls the throttle plate / butterfly opening angle. Note that the throttle actuator will probably form part of the throttle control assembly which can include a throttle position sensor as well as the control motor.	Note that the position of the motor can be regulated using a frequency / pulse width control signal or other electrical control (provided by the control unit on the earth or power circuit). It is advisable to refer to vehicle specific information to identify the type of control signal provided to the motor. The fault code could indicate that the throttle actuator is being provided with a specific control signal (by the control unit) but the actuator is not moving to the position expected by the control unit (position monitored by the throttle position sensor). The code could also indicate that the voltage, frequency or value of the control signal in the motor circuit is within the normal operating range / tolerance, but the signal might be incorrect for the operating conditions. Possible fault in motor or wiring e.g. short / open circuit or resistance, but also possible that the motor (or mechanism / system controlled by the motor) is not operating or moving correctly. Also check for contamination / restrictions that could affect airflow around the throttle butterfly. Refer to list of sensor / actuator checks on page 19.
P063A	Generator Voltage Sense Circuit	The generator can be connected to the main (power train) control unit to enable the control unit to monitor / control generator output; the control unit regulates the generator's field winding voltage / current, which in turn controls generator output. A "sense" circuit (between the generator and the control unit) is used to enable the control unit to detect / monitor generator output / performance.	It will be necessary to refer to vehicle specific information to identify the wiring / circuit used for the generator system. The fault code indicates an undefined fault with "sense" circuit. Check the wiring and connections between the generator and the control unit for short / open circuits. If the wiring and connections are good, it is possible that a fault exists in the generator.
P063B	Generator Voltage Sense Circuit Range / Performance	The generator can be connected to the main (power train) control unit to enable the control unit to monitor / control generator output; the control unit regulates the generator's field winding voltage / current, which in turn controls generator output. A "sense" circuit (between the generator and the control unit) is used to enable the control unit to detect / monitor generator output / performance.	It will be necessary to refer to vehicle specific information to identify the wiring / circuit used for the generator system. The fault code indicates that the voltage, frequency or signal value on the sense circuit is not as expected for the operating conditions or, the output is not changing as expected when the control unit regulates the field winding voltage / current. Check the wiring and connections between the generator and the control unit for short / open circuits. If the wiring and connections are good, it is possible that a fault exists in the generator. Also check condition and tension of generator drive belt / mechanism and battery condition.
P063C	Generator Voltage Sense Circuit Low	The generator can be connected to the main (power train) control unit to enable the control unit to monitor / control generator output; the control unit regulates the generator's field winding voltage / current, which in turn controls generator output. A "sense" circuit (between the generator and the control unit) is used to enable the control unit to detect / monitor generator output / performance.	It will be necessary to refer to vehicle specific information to identify the wiring / circuit used for the generator system. The fault code indicates that the voltage in the sense circuit is low e.g. low generator output or low voltage in the circuit. Check the wiring and connections between the generator and the control unit for short / open circuits. If the wiring and connections are good, it is possible that a fault exists in the generator.
P063D	Generator Voltage Sense Circuit High	The generator can be connected to the main (power train) control unit to enable the control unit to monitor / control generator output; the control unit regulates the generator's field winding voltage / current, which in turn controls generator output. A "sense" circuit (between the generator and the control unit) is used to enable the control unit to detect / monitor generator output / performance.	It will be necessary to refer to vehicle specific information to identify the wiring / circuit used for the generator system. The fault code indicates that the voltage in the sense circuit (circuit used to detect generator output) is high e.g. high generator output or high voltage in the circuit. Check the wiring and connections between the generator and the control unit for short / open circuits. If the wiring and connections are good, it is possible that a fault exists in the generator.
P063E	Auto Configuration Throttle Input Not Present	The control unit can adapt (or re-configure the operating programme) when some new components are fitted to the system or, if the information from the sensor changes over time (re-learning of basic settings such as idle position signal values). Note that in many cases, the re-configuring must be carried out using appropriate equipment. The auto configuration process refers to the process where the control unit automatically learns / adapts to the new component.	The fault code indicates that the auto configure process is faulty e.g. a new throttle sensor has been fitted but the control unit has not adapted or auto re-configured to the new component; it is also possible that the operating information from the throttle sensor is not recognised (sensor information has changed over time and the control unit has not adapted). Check all wiring between the control unit and the throttle sensor for short / open circuits, check throttle position sensor operation (and throttle actuator operation if applicable). Clear the fault code and if possible drive the vehicle or run the engine; if the fault re-appears, it could indicate a control unit related fault.
P063F	Auto Configuration Engine Coolant Temperature Input Not Present	The control unit can adapt (or re-configure the operating programme) when some new components are fitted to the system or, if the information from the sensor changes over time (re-learning of basic settings such as idle position signal values). Note that in many cases, the re-configuring must be carried out using appropriate equipment. The auto configuration process refers to the process where the control unit automatically learns / adapts to the new component.	The fault code indicates that the auto configure process is faulty e.g. a new coolant temperature sensor has been fitted but the control unit has not adapted or auto re-configured to the new component; it is also possible that the operating information from the sensor is not recognised (sensor information has changed over time and the control unit has not adapted). Check all wiring between the control unit and temperature sensor for short / open circuits, check temperature sensor operation. Clear the fault code and if possible drive the vehicle or run the engine; if the fault re-appears, it could indicate a control unit related fault.
P0640	Intake Air Heater Control Circuit	Note that some heaters are simple electrical heating elements but some may use the heater element to ignite a dedicated fuel supply. The heater could be provided with a power supply via a relay, with the control unit switching the relay on / off; alternatively, a control unit / module could be providing a direct switching of the heater. Refer to vehicle specific information to identify the type of heater used and the wiring / control for the heater.	The heater circuit could refer to the circuit connecting the heater to a control unit / module or relay, but it could also refer to the circuit connecting the heater element to a power supply and earth path. The voltage, frequency or value of the control signal in the heater circuit is incorrect (undefined fault). Signal could be:- out of normal operating range, constant value, non-existent, corrupt. Possible fault with heater or wiring e.g. heater power supply, short / open circuit, high circuit resistance or incorrect heater resistance. Note: If the heater is directly controlled via a control unit / module, the control unit could be providing the correct control signal but it is being affected by a circuit fault. Refer to list of sensor / actuator checks on page 19.

NOTE 1. Check for other fault codes that could provide additional information. NOTE 2. Communication between control units can pass via a CAN-Bus system; refer to Page 51 for CAN-Bus checks.

P0641	Sensor Reference Voltage "A" Circuit / Open	The control unit provides a reference voltage to one or more system sensors (typically around 5 volts, although the voltage on some systems may differ); the reference voltage will have a tolerance range (typically in the range of 4.5 - 5.5 volts). Note that the reference voltage is sometimes referred to as a sensor power supply. Depending on the sensor type, the voltage is passed across the sensor resistance or an electronic circuit within the sensor, the sensor will then produce an output voltage or signal which can be compared with the original reference voltage level. On other sensor types, the reference voltage can form part of the sensor circuit and therefore the voltage will change as the sensor value changes. Refer to list of sensor / actuator checks on page 19 for additional information.	It may be necessary to refer to vehicle specific information to identify which is the specified reference voltage circuit. However, the voltage can be checked at various sensors (preferably with the engine running) to establish which sensors and circuit are affected. The fault code indicates that the control unit diagnostic system has detected that the reference voltage is incorrect. The exact fault is undefined but it is possible that the reference voltage is outside of the accepted tolerance due to either a short / open circuit or high resistance in the circuit to the applicable sensors. It is also possible that the control unit has an internal fault (the control unit can monitor the reference voltage and compare with an expected value). Also check Control unit power supply and earth connections.
P0642	Sensor Reference Voltage "A" Circuit Low	The control unit provides a reference voltage to one or more system sensors (typically around 5 volts, although the voltage on some systems may differ); the reference voltage will have a tolerance range (typically in the range of 4.5 - 5.5 volts). Note that the reference voltage is sometimes referred to as a sensor power supply. Depending on the sensor type, the voltage is passed across the sensor resistance or an electronic circuit within the sensor, the sensor will then produce an output voltage or signal which can be compared with the original reference voltage level. On other sensor types, the reference voltage can form part of the sensor circuit and therefore the voltage will change as the sensor value changes. Refer to list of sensor / actuator checks on page 19 for additional information.	It may be necessary to refer to vehicle specific information to identify which is the specified reference voltage circuit. However, the voltage can be checked at various sensors (preferably with the engine running) to establish which sensors and circuit are affected. The fault code indicates that the control unit diagnostic system has detected that the reference voltage is low; either "zero" (no signal) or, the signal exists but it is below the normal operating range (lower than the minimum operating limit). It is possible that a short circuit exists in the wiring or a sensor. It is also possible that the control unit has an internal fault (the control unit can monitor the reference voltage and compare with an expected value). Also check Control unit power supply and earth connections.
P0643	Sensor Reference Voltage "A" Circuit High	The control unit provides a reference voltage to one or more system sensors (typically around 5 volts, although the voltage on some systems may differ); the reference voltage will have a tolerance range (typically in the range of 4.5 - 5.5 volts). Note that the reference voltage is sometimes referred to as a sensor power supply. Depending on the sensor type, the voltage is passed across the sensor resistance or an electronic circuit within the sensor, the sensor will then produce an output voltage or signal which can be compared with the original reference voltage level. On other sensor types, the reference voltage can form part of the sensor circuit and therefore the voltage will change as the sensor value changes. Refer to list of sensor / actuator checks on page 19 for additional information.	It may be necessary to refer to vehicle specific information to identify which is the specified reference voltage circuit. However, the voltage can be checked at various sensors (preferably with the engine running) to establish which sensors and circuit are affected. The fault code indicates that the control unit diagnostic system has detected that the reference voltage is high; e.g. battery voltage with no signal or frequency or, the voltage exists but it is above the normal operating range (higher than the maximum operating limit). It is possible that there is a short to another power supply (in the wiring or a sensor). It is also possible that the control unit has an internal fault (the control unit can monitor the reference voltage and compare with an expected value). Also check Control unit power supply and earth connections.
P0644	Driver Display Serial Communication Circuit	The fault code is likely to refer to the communication circuit between the main control unit and the instrument display, which could possibly relate to information from the control unit to the instrumentation (fault warning light, driver information etc) or information passing from the instrumentation system to the control unit.	The fault code indicates a communication fault for the driver display. Identify the wiring and connectors for the circuit; check for short / open circuits, poor connections etc. If all wiring / connections appear to be good, it indicates a possible control unit fault but also check the instrumentation system for faults.
P0645	A / C Clutch Relay Control Circuit	The air condition "clutch Relay" can be switched on by the engine control unit or a separate control unit. It is normal practice for the control unit to switch the earth circuit for the relay i.e. the control unit will complete the earth path for the relay energising winding. The control unit can detect whether there is a power supply to the winding and is therefore able to detect if there is a fault in the circuit. Although the fault code is likely to relate to the energising winding circuit, some manufacturers may use the code to refer to the relay power output circuit (power supply to A / C clutch).	Refer to list of sensor / actuator checks on page 19 for notes on relay checks. Different manufacturers could use the fault code to refer to different relay circuits. The fault code could refer to the power supply circuit (the power supply to the relay contacts and the circuit from the relay contacts to the A / C clutch). Alternatively, the fault code could refer to the relay energising winding circuit (power supply to the energising winding and the switched earth path through to the control unit). The voltage in the applicable relay circuit is incorrect (undefined fault). Check power supplies to relay contacts (possibly via a fused circuit) and to relay energising winding. Check operation of energising winding (short / open circuit) and closing / opening of relay contacts. Check all wiring for short / open circuits.
P0646	A / C Clutch Relay Control Circuit Low	The air condition "clutch Relay" can be switched on by the engine control unit or a separate control unit. It is normal practice for the control unit to switch the earth circuit for the relay i.e. the control unit will complete the earth path for the relay energising winding. The control unit can detect whether there is a power supply to the winding and is therefore able to detect if there is a fault in the circuit. Although the fault code is likely to relate to the energising winding circuit, some manufacturers may use the code to refer to the relay power output circuit (power supply to A / C clutch).	Refer to list of sensor / actuator checks on page 19 for notes on relay checks. Different manufacturers could use the fault code to refer to different relay circuits. The fault code could refer to the power supply circuit (the power supply to the relay contacts and the circuit from the relay contacts to the A / C clutch). Alternatively, the fault code could refer to the relay energising winding circuit (power supply to the energising winding and the switched earth path through to the control unit). The voltage in the applicable relay circuit is low (likely to be zero volts) at a time when the voltage should be at battery voltage level. Check power supplies to relay contacts (usually via a fused circuit) and to relay energising winding. Check operation of energising winding (short / open circuit) and closing / opening of relay contacts. Check all wiring for short / open circuits.

NOTE 1. Check for other fault codes that could provide additional information.   **NOTE 2.** Communication between control units can pass via a CAN-Bus system; refer to Page 51 for CAN-Bus checks.
NOTE 3. If a fault cannot be located, it is also possible that the control unit is at fault.   **NOTE 4.** Refer to Page 53 for list of pages that show the ISO standard for component locations e.g. Sensor A - Bank 1.

159

Fault code	EOBD / ISO Description	Component / System Description	Meaningful Description and Quick Check
P0647	A / C Clutch Relay Control Circuit High	The air condition "clutch Relay" can be switched on by the engine control unit or a separate control unit. It is normal practice for the control unit to switch the earth circuit for the relay i.e. the control unit will complete the earth path for the relay energising winding. The control unit can detect whether there is a power supply to the winding and is therefore able to detect if there is a fault in the circuit. Although the fault code is likely to relate to the energising winding circuit, some manufacturers may use the code to refer to the relay power output circuit (power supply to A / C clutch).	Refer to list of sensor / actuator checks on page 19 for notes on relay checks. Different manufacturers could use the fault code to refer to different relay circuits. The fault code could refer to the power supply circuit (the power supply to the relay contacts and the circuit from the relay contacts to the A / C clutch). Alternatively, the fault code could refer to the relay energising winding circuit (power supply to the energising winding and the switched earth path through to the control unit). The voltage in the relay circuit is high (likely to be battery voltage) when the relay circuit should be off (contacts open). Likely cause is relay contacts remaining closed (either due to relay fault or energising winding is being continuously activated); also possible short in wiring to another power supply or the energising winding earth path is shorted to earth. Check operation of relay, including closing / opening of relay contacts, and check operation of energising winding (short / open circuit).
P0648	Immobilizer Lamp Control Circuit	It may be necessary to refer to vehicle specific information to identify the operation and components involved for the immobilzer system, and to identify the wiring for the immobilzer warning light circuit.	The fault code indicates that there is a fault detected with the immobilzer system warning light circuit; it does not recognise the key (or other device) being used to enter / start the vehicle. The fault is likely to be caused by a wiring / connection fault between the warning light and the control module or, a fault with the warning light bulb. Note that depending on the system design, the control module could be part of the engine management control unit or it could be a separate module.
P0649	Speed Control Lamp Control Circuit	The speed control light (cruise control indicator) is used to indicate to the driver that the speed / cruise control system is selected (on).	If necessary, refer to vehicle specific information to identify the wiring for the indicator light. The fault code indicates that there is an undefined fault with the indicator light circuit. Check for short / open circuits in the wiring and for failure of the light bulb / LED. Check for a power supply and earth path for the light (as applicable depending on the type of wiring used).
P0650	Malfunction Indicator Lamp (MIL) Control Circuit	The main control unit will usually switch the earth path for the Malfunction Indicator Light (MIL); the power supply usually being provided via a common power supply in the dashboard (note that different vehicles can use different wiring for the light).	Refer the vehicle specific information but note that the MIL light will normally be illuminated when the ignition is first switched on and the light should then go out after a brief period of time or after starting the engine (the light will remain illuminated if there are system faults / other fault codes). The fault code is indicating a fault with the Malfunction Indicator warning Light; this is an undefined fault with the MIL circuit or light. The fault code indicates that there is an undefined fault with the warning light circuit. Check for short / open circuits in the wiring and for failure of the light bulb / LED. Check for a power supply and earth path for the warning light (as applicable depending on the type of wiring used).
P0651	Sensor Reference Voltage "B" Circuit / Open	The control unit provides a reference voltage to one or more system sensors (typically around 5 volts, although the voltage on some systems may differ); the reference voltage will have a tolerance range (typically in the range of 4.5 - 5.5 volts). Note that the reference voltage is sometimes referred to as a sensor power supply. Depending on the sensor type, the voltage is passed across the sensor resistance or an electronic circuit within the sensor, the sensor will then produce an output voltage or signal which can be compared with the original reference voltage level. On other sensor types, the reference voltage can form part of the sensor circuit and therefore the voltage will change as the sensor value changes. Refer to list of sensor / actuator checks on page 19 for additional information.	It may be necessary to refer to vehicle specific information to identify which is the specified reference voltage circuit. However, the voltage can be checked at various sensors (preferably with the engine running) to establish which sensors and circuit are affected. The fault code indicates that the control unit diagnostic system has detected that the reference voltage is incorrect. The exact fault is undefined but it is possible that the reference voltage is outside of the accepted tolerance due to either a short / open circuit or high resistance in the circuit to the applicable sensors. It is also possible that the control unit has an internal fault (the control unit can monitor the reference voltage and compare with an expected value).
P0652	Sensor Reference Voltage "B" Circuit Low	The control unit provides a reference voltage to one or more system sensors (typically around 5 volts, although the voltage on some systems may differ); the reference voltage will have a tolerance range (typically in the range of 4.5 - 5.5 volts). Note that the reference voltage is sometimes referred to as a sensor power supply. Depending on the sensor type, the voltage is passed across the sensor resistance or an electronic circuit within the sensor, the sensor will then produce an output voltage or signal which can be compared with the original reference voltage level. On other sensor types, the reference voltage can form part of the sensor circuit and therefore the voltage will change as the sensor value changes. Refer to list of sensor / actuator checks on page 19 for additional information.	It may be necessary to refer to vehicle specific information to identify which is the specified reference voltage circuit. However, the voltage can be checked at various sensors (preferably with the engine running) to establish which sensors and circuit are affected. The fault code indicates that the control unit diagnostic system has detected that the reference voltage is low; either "zero" (no signal) or, the signal exists but it is below the normal operating range (lower than the minimum operating limit). It is possible that a short circuit exists in the wiring or a sensor. It is also possible that the control unit has an internal fault (the control unit can monitor the reference voltage and compare with an expected value).
P0653	Sensor Reference Voltage "B" Circuit High	The control unit provides a reference voltage to one or more system sensors (typically around 5 volts, although the voltage on some systems may differ); the reference voltage will have a tolerance range (typically in the range of 4.5 - 5.5 volts). Note that the reference voltage is sometimes referred to as a sensor power supply. Depending on the sensor type, the voltage is passed across the sensor resistance or an electronic circuit within the sensor, the sensor will then produce an output voltage or signal which can be compared with the original reference voltage level. On other sensor types, the reference voltage can form part of the sensor circuit and therefore the voltage will change as the sensor value changes. Refer to list of sensor / actuator checks on page 19 for additional information.	It may be necessary to refer to vehicle specific information to identify which is the specified reference voltage circuit. However, the voltage can be checked at various sensors (preferably with the engine running) to establish which sensors and circuit are affected. The fault code indicates that the control unit diagnostic system has detected that the reference voltage is high; e.g. battery voltage with no signal or frequency or, the voltage exists but it is above the normal operating range (higher than the maximum operating limit). It is possible that there is a short to another power supply (in the wiring or a sensor). It is also possible that the control unit has an internal fault (the control unit can monitor the reference voltage and compare with an expected value).

Code	Description		
P0654	Engine RPM Output Circuit	The code could be used to identify the speed signal passing from one control unit to another control unit e.g. from main Diesel pump control unit to the Diesel Glow Plug timing control unit; it will be necessary to establish which control units are applicable (this can be vehicle manufacturer specific). Note however that the code could be used by some manufacturers as a direct reference to the speed sensor used to detect engine speed e.g. crankshaft, camshaft or Diesel pump speed sensor. It will be necessary to identify whether code refers to the speed signal passing between control units or whether it is a direct reference to a speed sensor signal.	If the code is applicable to a speed sensor, identify the type of sensor e.g. Inductive, Hall effect etc; this will dictate the checks that are applicable. The voltage, frequency or value of the speed signal (either from a control unit or from a speed sensor) is incorrect (undefined fault). Signal value could be: out of normal operating range, constant value, non-existent, corrupt. If code is applicable to a signal from a control unit, there is a possible fault in the control unit wiring or CAN-Bus system (if applicable); also check the source of the speed signal input to the control unit (e.g. crankshaft sensor etc). If the code is applicable direct to a speed sensor, possible faults include: sensor or wiring e.g. short / open circuits, circuit resistance, incorrect sensor resistance (if applicable) or, fault with the reference / operating voltage; interference from other circuits can also affect sensor signals. Also check earth connections and cable screening. Refer to list of sensor / actuator checks on page 19.
P0655	Engine Hot Lamp Output Control Circuit	The Engine Hot warning light is used to indicate to the driver that the engine temperature is too high.	If necessary, refer to vehicle specific information to identify the wiring for the warning light. The fault code indicates that there is an undefined fault with the warning light circuit. Check for short / open circuits in the wiring and for failure of the light bulb / LED. Check for a power supply and earth path for the light (as applicable depending on the type of wiring used).
P0656	Fuel Level Output Circuit	The fuel level output circuit can refer to the signal provided by one control unit to another control unit e.g. from the main control unit to the instrument control module (or from the instrument control module to the main control unit).	The fault code indicates an undefined fault with the fuel level indication circuit (circuit passing fuel level information between control units). Check applicable wiring for short / open circuit. Note that it is unlikely that the fault code refers to the fuel level sensor, other fault codes are used to indicate fuel level sensor faults; however, if there are no faults detected on the wiring between control units, refer to fault codes P0460 - P0464 for fuel level sensor checks.
P0657	Actuator Supply Voltage "A" Circuit / Open	The control unit can monitor actuator supply voltages in those systems and circuits where the voltage supply is via relay or another control unit. The control unit can also provide the supply voltage for some actuators, as well as regulating the supply voltage (pulse width / duty cycle control); therefore the voltage can be regulated to enable more accurate control of the actuators. The control unit compares the actual voltage or regulated voltage provided to the actuator with the expected value (held in memory). It will be necessary to identify which actuators are affected and whether the supply voltage is provided by the main control unit, a relay or an independent control unit.	The fault code indicates an actuator(s) supply voltage fault, with a possible "open circuit" or other circuit fault. Identify the source of the voltage supply e.g. relay, main control unit or separate control unit. If the control unit detects a fault with an independent supply (e.g. relay or separate control unit), the main control unit could be connected to the supply using a "sensing wire". It will be necessary to check if the actuator(s) are receiving a supply, and check whether there is an open circuit in the sensing wire. The control unit could be monitoring the voltage in the actuator circuit via the earth path (earth path switched to earth by the control unit); if there is no supply voltage passing through the actuator into the earth circuit, the control unit can assess this as an open circuit. If the control unit is providing the supply voltage to an actuator, and there is an open circuit or circuit fault, this can also be detected by the control unit. In all cases, check for short / open circuits in the wiring and check for a short / open circuit or incorrect resistance with the actuator.
P0658	Actuator Supply Voltage "A" Circuit Low	The control unit can monitor actuator supply voltages in those systems and circuits where the voltage supply is via relay or another control unit. The control unit can also provide the supply voltage for some actuators, as well as regulating the supply voltage (pulse width / duty cycle control); therefore the voltage can be regulated to enable more accurate control of the actuators. The control unit compares the actual voltage or regulated voltage provided to the actuator with the expected value (held in memory). It will be necessary to identify which actuators are affected and whether the supply voltage is provided by the main control unit, a relay or an independent control unit.	The fault code indicates that the supply voltage to an actuator(s) is low (could be zero or below normal operating range). Identify the source of the voltage supply e.g. relay, main control unit or separate control unit. In all cases, check for short / open circuits in the wiring, high circuit resistances and check for an short / open circuit or incorrect resistance with the actuator. If the control unit detects a fault with an independent supply (e.g. relay or separate control unit), the main control unit could be detecting the fault via a "sensing wire". Check if the actuator(s) are receiving the correct supply (also check charging voltage), and check whether there is an open circuit or high resistance in the sensing wire. The control unit could be monitoring the actuator voltage via the earth path; earth path could be switched to earth by the control unit, and the control unit can therefore assess the available voltage via the earth circuit. If the control unit is providing the actuator supply voltage, a low voltage will be detected by the control unit (this could be a control unit or charging system fault).
P0659	Actuator Supply Voltage "A" Circuit High	The control unit can monitor actuator supply voltages in those systems and circuits where the voltage supply is via relay or another control unit. The control unit can also provide the supply voltage for some actuators, as well as regulating the supply voltage (pulse width / duty cycle control); therefore the voltage can be regulated to enable more accurate control of the actuators. The control unit compares the actual voltage or regulated voltage provided to the actuator with the expected value (held in memory). It will be necessary to identify which actuators are affected and whether the supply voltage is provided by the main control unit, a relay or an independent control unit.	The fault code indicates that the supply voltage to an actuator(s) is high; voltage could be at full value at all times e.g. battery voltage (with no signal or frequency for those actuators where the control unit provides a control signal on the power circuit). Identify the source of the voltage supply e.g. relay, main control unit or separate control unit, and identify the wiring circuit. In all cases, check for short / open circuits in the wiring, high circuit resistances and check for an short / open circuit or incorrect resistance with the actuator. For those actuators where the control unit switches the actuator earth path through to earth, it is possible that the earth path has a short to a power supply circuit (causing voltage to permanently exist in the circuit). If the control unit detects a fault with an independent power supply (e.g. relay or separate control unit), the main control unit could be detecting the fault via a "sensing wire". Check if the actuator(s) are receiving the correct supply (also check charging voltage).
P065A	Generator System Performance	The generator can be connected to the main (power train) control unit to enable the control unit to monitor and / or to control generator output; the control unit can regulate the generator's field winding voltage / current, which in turn controls generator output.	It will be necessary to refer to vehicle specific information to identify the wiring / circuit used for the generator warning light. The fault code indicates an undefined charging / battery system fault. For systems where the generator is not controlled by the main control unit, carry out generator / battery checks and check applicable wiring for short / open circuits and high resistances (including earth connections to battery, engine etc), also check generator drive belt or drive system. For systems where the main control unit is controlling the generator, it is possible that the control unit is providing a control signal to the generator field windings but, the generator output is not responding as expected. Check general operation of charging / battery system, check generator drive belt or drive system. Check wiring and connections between battery and generator for short / open circuits and high resistances. Check battery condition and earth circuit. Check wiring / connections between the generator and the control unit for short / open circuit or high resistance.
P065B	Generator Control Circuit Range / Performance	The generator can be connected to the main (power train) control unit to enable the control unit to monitor and / or to control generator output; the control unit can regulate the generator's field winding voltage / current, which in turn controls generator output.	It will be necessary to refer to vehicle specific information to identify the wiring / circuit used for the generator system. The fault code indicates that there is an undefined fault detected by the control unit in the circuits between the generator and the control unit. Note that the fault will be detected due to incorrect voltage / signal value in the circuit. Check wiring and connections between the generator and the control unit for short / open circuit. If the wiring and connections are good, it is possible that a charging fault exists. If charging output is low, also check condition and tension of generator drive belt / mechanism and battery condition.

NOTE 1. Check for other fault codes that could provide additional information. **NOTE 2.** Communication between control units can pass via a CAN-Bus system; refer to Page 51 for CAN-Bus checks.

NOTE 3. If a fault cannot be located, it is also possible that the control unit is at fault. **NOTE 4.** Refer to Page 53 for list of pages that show the ISO standard for component locations e.g. Sensor A - Bank 1.

Fault code	EOBD / ISO Description	Component / System Description	Meaningful Description and Quick Check
P0660	Intake Manifold Tuning Valve Control Circuit / Open Bank 1a	The intake manifold "Tuning" system (often referred to a "variable intake system") normally uses additional air valves (flaps or butterflies within the intake manifold) to divert intake air through longer or shorter intake paths; the intake tuning helps improve engine torque / efficiency at different engine speed / loads. The tuning valve can be moved using a vacuum operated mechanism (with the vacuum being regulated by a solenoid or other type of actuator) but it is possible that an electric actuator e.g. motor, could directly operate the mechanism. A position sensor / switch can be used to monitor the valve position.	The fault code will relate to the electric actuator (solenoid / motor etc) used to regulate the vacuum or operate the tuning valve mechanism. The voltage, frequency or value of the control signal in the actuator circuit is incorrect. Signal could be out of normal operating range, constant value, non-existent, corrupt. The fault code indicates a possible "open circuit" but other undefined faults in the actuator or wiring could activate the same code e.g. actuator power supply, short circuit, high circuit resistance or incorrect actuator resistance. Note: the control unit could be providing the correct signal but it is being affected by a circuit fault. Refer to list of sensor / actuator checks on page 19.
P0661	Intake Manifold Tuning Valve Control Circuit Low Bank 1a	The intake manifold "Tuning" system (often referred to a "variable intake system") normally uses additional air valves (flaps or butterflies within the intake manifold) to divert intake air through longer or shorter intake paths; the intake tuning helps improve engine torque / efficiency at different engine speed / loads. The tuning valve can be moved using a vacuum operated mechanism (with the vacuum being regulated by a solenoid or other type of actuator) but it is possible that an electric actuator e.g. motor, could directly operate the mechanism. A position sensor / switch can be used to monitor the valve position.	The fault code will relate to the electric actuator (solenoid / motor etc) used to regulate the vacuum or operate the tuning valve mechanism. The voltage, frequency or value of the control signal in the actuator circuit is either "zero" (no signal) or, the signal exists but is below the normal operating range. Possible fault with actuator or wiring (e.g. actuator power supply, short / open circuit or resistance) that is causing a signal value to be lower than the minimum operating limit. Note: the control unit could be providing the correct signal but it is being affected by a circuit fault. Refer to list of sensor / actuator checks on page 19.
P0662	Intake Manifold Tuning Valve Control Circuit High Bank 1a	The intake manifold "Tuning" system (often referred to a "variable intake system") normally uses additional air valves (flaps or butterflies within the intake manifold) to divert intake air through longer or shorter intake paths; the intake tuning helps improve engine torque / efficiency at different engine speed / loads. The tuning valve can be moved using a vacuum operated mechanism (with the vacuum being regulated by a solenoid or other type of actuator) but it is possible that an electric actuator e.g. motor, could directly operate the mechanism. A position sensor / switch can be used to monitor the valve position.	The fault code will relate to the electric actuator (solenoid / motor etc) used to regulate the vacuum or operate the tuning valve mechanism. The voltage, frequency or value of the control signal in the actuator circuit is either at full value (e.g. battery voltage with no on / off signal) or, the signal exists but it is above the normal operating range. Possible fault with actuator or wiring (e.g. actuator power supply, short / open circuit or resistance) that is causing a signal value to be higher than the maximum operating limit. Note: the control unit could be providing the correct signal but it is being affected by a circuit fault. Refer to list of sensor / actuator checks on page 19.
P0663	Intake Manifold Tuning Valve Control Circuit / Open Bank 2a	The intake manifold "Tuning" system (often referred to a "variable intake system") normally uses additional air valves (flaps or butterflies within the intake manifold) to divert intake air through longer or shorter intake paths; the intake tuning helps improve engine torque / efficiency at different engine speed / loads. The tuning valve can be moved using a vacuum operated mechanism (with the vacuum being regulated by a solenoid or other type of actuator) but it is possible that an electric actuator e.g. motor, could directly operate the mechanism. A position sensor / switch can be used to monitor the valve position.	The fault code will relate to the electric actuator (solenoid / motor etc) used to regulate the vacuum or operate the tuning valve mechanism. The voltage, frequency or value of the control signal in the actuator circuit is incorrect. Signal could be out of normal operating range, constant value, non-existent, corrupt. The fault code indicates a possible "open circuit" but other undefined faults in the actuator or wiring could activate the same code e.g. actuator power supply, short circuit, high circuit resistance or incorrect actuator resistance. Note: the control unit could be providing the correct signal but it is being affected by a circuit fault. Refer to list of sensor / actuator checks on page 19.
P0664	Intake Manifold Tuning Valve Control Circuit Low Bank 2a	The intake manifold "Tuning" system (often referred to a "variable intake system") normally uses additional air valves (flaps or butterflies within the intake manifold) to divert intake air through longer or shorter intake paths; the intake tuning helps improve engine torque / efficiency at different engine speed / loads. The tuning valve can be moved using a vacuum operated mechanism (with the vacuum being regulated by a solenoid or other type of actuator) but it is possible that an electric actuator e.g. motor, could directly operate the mechanism. A position sensor / switch can be used to monitor the valve position.	The fault code will relate to the electric actuator (solenoid / motor etc) used to regulate the vacuum or operate the tuning valve mechanism. The voltage, frequency or value of the control signal in the actuator circuit is either "zero" (no signal) or, the signal exists but is below the normal operating range. Possible fault with actuator or wiring (e.g. actuator power supply, short / open circuit or resistance) that is causing a signal value to be lower than the minimum operating limit. Note: the control unit could be providing the correct signal but it is being affected by a circuit fault. Refer to list of sensor / actuator checks on page 19.
P0665	Intake Manifold Tuning Valve Control Circuit High Bank 2a	The intake manifold "Tuning" system (often referred to a "variable intake system") normally uses additional air valves (flaps or butterflies within the intake manifold) to divert intake air through longer or shorter intake paths; the intake tuning helps improve engine torque / efficiency at different engine speed / loads. The tuning valve can be moved using a vacuum operated mechanism (with the vacuum being regulated by a solenoid or other type of actuator) but it is possible that an electric actuator e.g. motor, could directly operate the mechanism. A position sensor / switch can be used to monitor the valve position.	The fault code will relate to the electric actuator (solenoid / motor etc) used to regulate the vacuum or operate the tuning valve mechanism. The voltage, frequency or value of the control signal in the actuator circuit is either at full value (e.g. battery voltage with no on / off signal) or, the signal exists but it is above the normal operating range. Possible fault with actuator or wiring (e.g. actuator power supply, short / open circuit or resistance) that is causing a signal value to be higher than the maximum operating limit. Note: the control unit could be providing the correct signal but it is being affected by a circuit fault. Refer to list of sensor / actuator checks on page 19.
P0666	PCM / ECM / TCM Internal Temperature Sensor Circuit	A temperature sensor is located in the control unit (e.g. engine, transmission or other control unit), to monitor overheating of the control unit; the sensor is likely to be integrated into the control unit assembly.	The voltage, frequency or value of the temperature sensor signal is incorrect (undefined fault). The fault could be caused by a faulty sensor, poor internal connection or internal circuit board problem. Due to the fact that the sensor is an internal item (contained within the control unit), it is unlikely that checks could be easily carried out (other than by companies specialising in control unit faults / remanufacturing). Note that if the control unit detects an internal temperature related fault, it could reduce its performance (processing speed / power) to prevent overheating of the internal components.

NOTE 1. Check for other fault codes that could provide additional information.   **NOTE 2.** Communication between control units can pass via a CAN-Bus system; refer to Page 51 for CAN-Bus checks.

NOTE 3. If a fault cannot be located, it is also possible that the control unit is at fault   **NOTE 4.** Refer to Page 53 for list of pages that show the ISO standard for component locations e.g. Sensor A - Bank 1.

Code	Description		
P0667	PCM / ECM / TCM Internal Temperature Sensor Range / Performance	A temperature sensor is located in the control unit (e.g. engine, transmission or other control unit), to monitor overheating of the control unit; the sensor is likely to be integrated into the control unit assembly.	The voltage, frequency or value of the temperature sensor signal is within the normal operating range / tolerance but the signal is not plausible or is incorrect due to an undefined fault. The fault could be caused by a faulty sensor, poor internal connection or internal circuit board problem. Due to the fact that the sensor is an internal item (contained within the control unit), it is unlikely that checks could be easily carried out (other than by companies specialising in control unit faults / remanufacturing). Note that if the control unit detects an internal temperature related fault, it could reduce its performance (processing speed / power) to prevent overheating of the internal components.
P0668	PCM / ECM / TCM Internal Temperature Sensor Circuit Low	A temperature sensor is located in the control unit (e.g. engine, transmission or other control unit), to monitor overheating of the control unit; the sensor is likely to be integrated into the control unit assembly.	The fault code indicates that the temperature sensor signal value is below the normal operating range (out of tolerance) for this circuit. The fault could be caused by a faulty sensor, poor internal connection or internal circuit board problem. Due to the fact that the sensor is an internal item (contained within the control unit), it is unlikely that checks could be easily carried out (other than by companies specialising in control unit faults / remanufacturing). Note that if the control unit detects an internal temperature related fault, it could reduce its performance (processing speed / power) to prevent overheating of the internal components.
P0669	PCM / ECM / TCM Internal Temperature Sensor Circuit High	A temperature sensor is located in the control unit (e.g. engine, transmission or other control unit), to monitor overheating of the control unit; the sensor is likely to be integrated into the control unit assembly.	The fault code indicates that the signal value from the sensor is above the normal operating range (out of tolerance) for this circuit. The fault could be caused by a faulty sensor, poor internal connection or internal circuit board problem. Due to the fact that the sensor is an internal item (contained within the control unit), it is unlikely that checks could be easily carried out (other than by companies specialising in control unit faults / remanufacturing). Note that if the control unit detects an internal temperature related fault, it could reduce its performance (processing speed / power) to prevent overheating of the internal components.
P066A	Glow Plug 1 Control Circuit Low	Identify the type of glow plug control system. Glow plugs are usually controlled by a "Glow Plug Control Module" (can be integrated into the main Diesel control unit or, can be a separate unit which receives a control signal from the Diesel control unit). Depending on system design, full battery voltage can be applied to the glow plugs (possibly via a relay controlled by the control module) or, on many systems the circuit voltage / current is regulated by using a digital control signal (voltage / current is dependent on signal pulse width / duty cycle). Note that different types of glow plug can be specific to different types of control systems, and also note that the voltage / current level in the circuit can alter during the period of operation (including pre / post start conditions).	Refer to vehicle specific information for the glow plug control system and design of glow plug; the exact interpretation of the fault code will depend on the system design; however, it is likely that the fault code relates to the glow plug control signal on the circuit between the control module and the glow plugs. The voltage, frequency, current or value of the control signal in the control circuit is either "zero" (no signal / voltage) or, the signal / voltage exists but is below the normal operating range. Check the circuit between the control module and the glow plug for short / open circuit and high resistance. Check the resistance of the glow plugs (refer to vehicle specific information especially for ceramic element glow plugs). Check power supply and earth connections to control module. The control module could be providing the correct control signal but it is being affected by a circuit fault. Refer to list of sensor / actuator checks on page 19. Note that it is possible that a manufacturer could apply the code to the circuit between the Diesel control unit and the glow plug control module; if necessary refer to Fault Code P0383 for additional information.
P066B	Glow Plug 1 Control Circuit High	Identify the type of glow plug control system. Glow plugs are usually controlled by a "Glow Plug Control Module" (can be integrated into the main Diesel control unit or, can be a separate unit which receives a control signal from the Diesel control unit). Depending on system design, full battery voltage can be applied to the glow plugs (possibly via a relay controlled by the control module) or, on many systems the circuit voltage / current is regulated by using a digital control signal (voltage / current is dependent on signal pulse width / duty cycle). Note that different types of glow plug can be specific to different types of control systems, and also note that the voltage / current level in the circuit can alter during the period of operation (including pre / post start conditions).	Refer to vehicle specific information for the glow plug control system and design of glow plug; the exact interpretation of the fault code will depend on the system design; however, it is likely that the fault code relates to the glow plug control signal on the circuit between the control module and the glow plugs. The voltage, frequency, current or value of the control signal in the control circuit is either at full value (e.g. battery voltage with no on / off signal) or, the signal exists but it is above the normal operating range. Check the circuit between the control module and the glow plug for short / open circuit and high resistance. Check the resistance of the glow plugs (refer to vehicle specific information especially for ceramic element glow plugs). The control module could be providing the correct control signal but it is being affected by a circuit fault. Refer to list of sensor / actuator checks on page 19. Note that it is possible that a manufacturer could apply the code to the circuit between the Diesel control unit and the glow plug control module; if necessary refer to Fault Code P0384 for additional information.
P066C	Glow Plug 2 Control Circuit Low	Identify the type of glow plug control system. Glow plugs are usually controlled by a "Glow Plug Control Module" (can be integrated into the main Diesel control unit or, can be a separate unit which receives a control signal from the Diesel control unit). Depending on system design, full battery voltage can be applied to the glow plugs (possibly via a relay controlled by the control module) or, on many systems the circuit voltage / current is regulated by using a digital control signal (voltage / current is dependent on signal pulse width / duty cycle). Note that different types of glow plug can be specific to different types of control systems, and also note that the voltage / current level in the circuit can alter during the period of operation (including pre / post start conditions).	Refer to vehicle specific information for the glow plug control system and design of glow plug; the exact interpretation of the fault code will depend on the system design; however, it is likely that the fault code relates to the glow plug control signal on the circuit between the control module and the glow plugs. The voltage, frequency, current or value of the control signal in the control circuit is either "zero" (no signal / voltage) or, the signal / voltage exists but is below the normal operating range. Check the circuit between the control module and the glow plug for short / open circuit and high resistance. Check the resistance of the glow plugs (refer to vehicle specific information especially for ceramic element glow plugs). Check power supply and earth connections to control module. The control module could be providing the correct control signal but it is being affected by a circuit fault. Refer to list of sensor / actuator checks on page 19. Note that it is possible that a manufacturer could apply the code to the circuit between the Diesel control unit and the glow plug control module; if necessary refer to Fault Code P0383 for additional information.
P066D	Glow Plug 2 Control Circuit High	Identify the type of glow plug control system. Glow plugs are usually controlled by a "Glow Plug Control Module" (can be integrated into the main Diesel control unit or, can be a separate unit which receives a control signal from the Diesel control unit). Depending on system design, full battery voltage can be applied to the glow plugs (possibly via a relay controlled by the control module) or, on many systems the circuit voltage / current is regulated by using a digital control signal (voltage / current is dependent on signal pulse width / duty cycle). Note that different types of glow plug can be specific to different types of control systems, and also note that the voltage / current level in the circuit can alter during the period of operation (including pre / post start conditions).	Refer to vehicle specific information for the glow plug control system and design of glow plug; the exact interpretation of the fault code will depend on the system design; however, it is likely that the fault code relates to the glow plug control signal on the circuit between the control module and the glow plugs. The voltage, frequency, current or value of the control signal in the control circuit is either at full value (e.g. battery voltage with no on / off signal) or, the signal exists but it is above the normal operating range. Check the circuit between the control module and the glow plug for short / open circuit and high resistance. Check the resistance of the glow plugs (refer to vehicle specific information especially for ceramic element glow plugs). The control module could be providing the correct control signal but it is being affected by a circuit fault. Refer to list of sensor / actuator checks on page 19. Note that it is possible that a manufacturer could apply the code to the circuit between the Diesel control unit and the glow plug control module; if necessary refer to Fault Code P0384 for additional information.

NOTE 1. Check for other fault codes that could provide additional information.
NOTE 3. If a fault cannot be located, it is also possible that the control unit is at fault.

NOTE 2. Communication between control units can pass via a CAN-Bus system; refer to Page 51 for CAN-Bus checks.
NOTE 4. Refer to Page 53 for list of pages that show the ISO standard for component locations e.g. Sensor A - Bank 1.

Fault code	EOBD / ISO Description	Component / System Description	Meaningful Description and Quick Check
P066E	Glow Plug 3 Control Circuit Low	Identify the type of glow plug control system. Glow plugs are usually controlled by a "Glow Plug Control Module" (can be integrated into the main Diesel control unit or, can be a separate unit which receives a control signal from the Diesel control unit). Depending on system design, full battery voltage can be applied to the glow plugs (possibly via a relay controlled by the control module) or, on many systems the circuit voltage / current is regulated by using a digital control signal (voltage / current is dependent on signal pulse width / duty cycle). Note that different types of glow plug can be specific to different types of control systems, and also note that the voltage / current level in the circuit can alter during the period of operation (including pre / post start conditions).	Refer to vehicle specific information for the glow plug control system and design of glow plug; the exact interpretation of the fault code will depend on the system design; however, it is likely that the fault code relates to the glow plug control signal on the circuit between the control module and the glow plugs. The voltage, frequency, current or value of the control signal in the control circuit is either "zero" (no signal / voltage) or, the signal / voltage exists but is below the normal operating range. Check the circuit between the control module and the glow plug for short / open circuit and high resistance. Check the resistance of the glow plugs (refer to vehicle specific information especially for ceramic element glow plugs). Check power supply to control module. The control module could be providing the correct control signal but it is being affected by a circuit fault. Refer to list of sensor / actuator checks on page 19. Note that it is possible that a manufacturer could apply the code to the circuit between the Diesel control unit and the glow plug control module; if necessary refer to Fault Code P0383 for additional information.
P066F	Glow Plug 3 Control Circuit High	Identify the type of glow plug control system. Glow plugs are usually controlled by a "Glow Plug Control Module" (can be integrated into the main Diesel control unit or, can be a separate unit which receives a control signal from the Diesel control unit). Depending on system design, full battery voltage can be applied to the glow plugs (possibly via a relay controlled by the control module) or, on many systems the circuit voltage / current is regulated by using a digital control signal (voltage / current is dependent on signal pulse width / duty cycle). Note that different types of glow plug can be specific to different types of control systems, and also note that the voltage / current level in the circuit can alter during the period of operation (including pre / post start conditions).	Refer to vehicle specific information for the glow plug control system and design of glow plug; the exact interpretation of the fault code will depend on the system design; however, it is likely that the fault code relates to the glow plug control signal on the circuit between the control module and the glow plugs. The voltage, frequency, current or value of the control signal in the control circuit is either at full value (e.g. battery voltage with no on / off signal) or, the signal exists but it is above the normal operating range. Check the circuit between the control module and the glow plug for short / open circuit and high resistance. Check the resistance of the glow plugs (refer to vehicle specific information especially for ceramic element glow plugs). The control module could be providing the correct control signal but it is being affected by a circuit fault. Refer to list of sensor / actuator checks on page 19. Note that it is possible that a manufacturer could apply the code to the circuit between the Diesel control unit and the glow plug control module; if necessary refer to Fault Code P0384 for additional information.
P0670	Glow Plug Control Module Control Circuit / Open	The glow plugs are usually switched on / off by a "Glow Plug Control Module", which can be controlled by the main Diesel Engine control unit. The control module can be a separate item but it can form part of the main control unit. The Glow Plug Module will receive a control signal from the Diesel control unit, and the module will switch on / off the power supply to the glow plugs (the module can contain an internal relay / power stage, although it is possible to use an external relay). Note that some systems will switch groups of glow plugs e.g. 2, 3 or 4 glow plugs together.	Refer to vehicle specific information for the glow plug control system and design of glow plug. The exact interpretation of the fault code will depend on the system monitoring / diagnostic process and system design; it is however likely that the fault code relates to the circuit between the main Diesel control unit and the glow plug control module (the circuit carries the signal to the glow plug control module to instruct the module to switch on the glow plugs). The voltage, frequency or value of the control signal in the control circuit is incorrect (undefined fault but possibly caused by an open circuit). Note: The Diesel control unit could be providing the correct signal but it is being affected by a circuit fault. Check the circuit between the Diesel control unit and the glow plug control module for short / open circuit and high resistance. Refer to list of sensor / actuator checks on page 19. Note that it is possible that a manufacturer could apply the code to the circuit between the glow plug control module and the glow plugs; if necessary refer to Fault Code P0380 for additional information.
P0671	Cylinder 1 Glow Plug Circuit / Open	Identify the type of glow plug control system. Glow plugs are usually controlled by a "Glow Plug Control Module" (can be integrated into the main Diesel control unit or, can be a separate unit which receives a control signal from the Diesel control unit). Depending on system design, full battery voltage can be applied to the glow plugs (possibly via a relay controlled by the control module) or, on many systems the circuit voltage / current is regulated by using a digital control signal (voltage / current is dependent on signal pulse width / duty cycle). Note that different types of glow plug can be specific to different types of control systems, and also note that the voltage / current level in the circuit can alter during the period of operation (including pre / post start conditions).	Refer to vehicle specific information for the glow plug control system and design of glow plug; the exact interpretation of the fault code will depend on the system design; however, it is likely that the fault code relates to the glow plug control signal on the circuit between the control module and the glow plugs. The voltage, frequency, current or value of the control signal in the control circuit is either incorrect (undefined fault but possibly caused by an open circuit). Check the circuit between the control module and the glow plug for short / open circuit and high resistance. Check the resistance of the glow plugs (refer to vehicle specific information especially for ceramic element glow plugs). Check power supply and earth connections to control module. The control module could be providing the correct control signal but it is being affected by a circuit fault. Refer to list of sensor / actuator checks on page 19. Note that it is possible that a manufacturer could apply the code to the circuit between the Diesel control unit and the glow plug control module; if necessary refer to Fault Codes P0383 and P0384 for additional information.
P0672	Cylinder 2 Glow Plug Circuit / Open	Identify the type of glow plug control system. Glow plugs are usually controlled by a "Glow Plug Control Module" (can be integrated into the main Diesel control unit or, can be a separate unit which receives a control signal from the Diesel control unit). Depending on system design, full battery voltage can be applied to the glow plugs (possibly via a relay controlled by the control module) or, on many systems the circuit voltage / current is regulated by using a digital control signal (voltage / current is dependent on signal pulse width / duty cycle). Note that different types of glow plug can be specific to different types of control systems, and also note that the voltage / current level in the circuit can alter during the period of operation (including pre / post start conditions).	Refer to vehicle specific information for the glow plug control system and design of glow plug; the exact interpretation of the fault code will depend on the system design; however, it is likely that the fault code relates to the glow plug control signal on the circuit between the control module and the glow plugs. The voltage, frequency, current or value of the control signal in the control circuit is either incorrect (undefined fault but possibly caused by an open circuit). Check the circuit between the control module and the glow plug for short / open circuit and high resistance. Check the resistance of the glow plugs (refer to vehicle specific information especially for ceramic element glow plugs). Check power supply and earth connections to control module. The control module could be providing the correct control signal but it is being affected by a circuit fault. Refer to list of sensor / actuator checks on page 19. Note that it is possible that a manufacturer could apply the code to the circuit between the Diesel control unit and the glow plug control module; if necessary refer to Fault Codes P0383 and P0384 for additional information.
P0673	Cylinder 3 Glow Plug Circuit / Open	Identify the type of glow plug control system. Glow plugs are usually controlled by a "Glow Plug Control Module" (can be integrated into the main Diesel control unit or, can be a separate unit which receives a control signal from the Diesel control unit). Depending on system design, full battery voltage can be applied to the glow plugs (possibly via a relay controlled by the control module) or, on many systems the circuit voltage / current is regulated by using a digital control signal (voltage / current is dependent on signal pulse width / duty cycle). Note that different types of glow plug can be specific to different types of control systems, and also note that the voltage / current level in the circuit can alter during the period of operation (including pre / post start conditions).	Refer to vehicle specific information for the glow plug control system and design of glow plug; the exact interpretation of the fault code will depend on the system design; however, it is likely that the fault code relates to the glow plug control signal on the circuit between the control module and the glow plugs. The voltage, frequency, current or value of the control signal in the control circuit is either incorrect (undefined fault but possibly caused by an open circuit). Check the circuit between the control module and the glow plug for short / open circuit and high resistance. Check the resistance of the glow plugs (refer to vehicle specific information especially for ceramic element glow plugs). Check power supply and earth connections to control module. The control module could be providing the correct control signal but it is being affected by a circuit fault. Refer to list of sensor / actuator checks on page 19. Note that it is possible that a manufacturer could apply the code to the circuit between the Diesel control unit and the glow plug control module; if necessary refer to Fault Codes P0383 and P0384 for additional information.

NOTE 1. Check for other fault codes that could provide additional information. **NOTE 2.** Communication between control units can pass via a CAN-Bus system; refer to Page 51 for CAN-Bus checks.
NOTE 3. If a fault cannot be located, it is also possible that the control unit is at fault. **NOTE 4.** Refer to Page 53 for list of pages that show the ISO standard for component locations e.g. Sensor A - Bank 1.

Code	Description		
P0674	Cylinder 4 Glow Plug Circuit / Open	Identify the type of glow plug control system. Glow plugs are usually controlled by a "Glow Plug Control Module" (can be integrated into the main Diesel control unit or, can be a separate unit which receives a control signal from the Diesel control unit). Depending on system design, full battery voltage can be applied to the glow plugs (possibly via a relay controlled by the control module) or, on many systems the circuit voltage / current is regulated by using a digital control signal (voltage / current is dependent on signal pulse width / duty cycle). Note that different types of glow plug can be specific to different types of control systems, and also note that the voltage / current level in the circuit can alter during the period of operation (including pre / post start conditions).	Refer to vehicle specific information for the glow plug control system and design of glow plug; the exact interpretation of the fault code will depend on the system design; however, it is likely that the fault code relates to the glow plug control signal on the circuit between the control module and the glow plugs. The voltage, frequency, current or value of the control signal in the control circuit is either incorrect (undefined fault but possibly caused by an open circuit). Check the circuit between the control module and the glow plug for short / open circuit and high resistance. Check the resistance of the glow plugs (refer to vehicle specific information especially for ceramic element glow plugs). Check power supply and earth connections to control module. The control module could be providing the correct control signal but it is being affected by a circuit fault. Refer to list of sensor / actuator checks on page 19. Note that it is possible that a manufacturer could apply the code to the circuit between the Diesel control unit and the glow plug control module; if necessary refer to Fault Codes P0383 and P0384 for additional information.
P0675	Cylinder 5 Glow Plug Circuit / Open	Identify the type of glow plug control system. Glow plugs are usually controlled by a "Glow Plug Control Module" (can be integrated into the main Diesel control unit or, can be a separate unit which receives a control signal from the Diesel control unit). Depending on system design, full battery voltage can be applied to the glow plugs (possibly via a relay controlled by the control module) or, on many systems the circuit voltage / current is regulated by using a digital control signal (voltage / current is dependent on signal pulse width / duty cycle). Note that different types of glow plug can be specific to different types of control systems, and also note that the voltage / current level in the circuit can alter during the period of operation (including pre / post start conditions).	Refer to vehicle specific information for the glow plug control system and design of glow plug; the exact interpretation of the fault code will depend on the system design; however, it is likely that the fault code relates to the glow plug control signal on the circuit between the control module and the glow plugs. The voltage, frequency, current or value of the control signal in the control circuit is either incorrect (undefined fault but possibly caused by an open circuit). Check the circuit between the control module and the glow plug for short / open circuit and high resistance. Check the resistance of the glow plugs (refer to vehicle specific information especially for ceramic element glow plugs). Check power supply and earth connections to control module. The control module could be providing the correct control signal but it is being affected by a circuit fault. Refer to list of sensor / actuator checks on page 19. Note that it is possible that a manufacturer could apply the code to the circuit between the Diesel control unit and the glow plug control module; if necessary refer to Fault Codes P0383 and P0384 for additional information.
P0676	Cylinder 6 Glow Plug Circuit / Open	Identify the type of glow plug control system. Glow plugs are usually controlled by a "Glow Plug Control Module" (can be integrated into the main Diesel control unit or, can be a separate unit which receives a control signal from the Diesel control unit). Depending on system design, full battery voltage can be applied to the glow plugs (possibly via a relay controlled by the control module) or, on many systems the circuit voltage / current is regulated by using a digital control signal (voltage / current is dependent on signal pulse width / duty cycle). Note that different types of glow plug can be specific to different types of control systems, and also note that the voltage / current level in the circuit can alter during the period of operation (including pre / post start conditions).	Refer to vehicle specific information for the glow plug control system and design of glow plug; the exact interpretation of the fault code will depend on the system design; however, it is likely that the fault code relates to the glow plug control signal on the circuit between the control module and the glow plugs. The voltage, frequency, current or value of the control signal in the control circuit is either incorrect (undefined fault but possibly caused by an open circuit). Check the circuit between the control module and the glow plug for short / open circuit and high resistance. Check the resistance of the glow plugs (refer to vehicle specific information especially for ceramic element glow plugs). Check power supply and earth connections to control module. The control module could be providing the correct control signal but it is being affected by a circuit fault. Refer to list of sensor / actuator checks on page 19. Note that it is possible that a manufacturer could apply the code to the circuit between the Diesel control unit and the glow plug control module; if necessary refer to Fault Codes P0383 and P0384 for additional information.
P0677	Cylinder 7 Glow Plug Circuit / Open	Identify the type of glow plug control system. Glow plugs are usually controlled by a "Glow Plug Control Module" (can be integrated into the main Diesel control unit or, can be a separate unit which receives a control signal from the Diesel control unit). Depending on system design, full battery voltage can be applied to the glow plugs (possibly via a relay controlled by the control module) or, on many systems the circuit voltage / current is regulated by using a digital control signal (voltage / current is dependent on signal pulse width / duty cycle). Note that different types of glow plug can be specific to different types of control systems, and also note that the voltage / current level in the circuit can alter during the period of operation (including pre / post start conditions).	Refer to vehicle specific information for the glow plug control system and design of glow plug; the exact interpretation of the fault code will depend on the system design; however, it is likely that the fault code relates to the glow plug control signal on the circuit between the control module and the glow plugs. The voltage, frequency, current or value of the control signal in the control circuit is either incorrect (undefined fault but possibly caused by an open circuit). Check the circuit between the control module and the glow plug for short / open circuit and high resistance. Check the resistance of the glow plugs (refer to vehicle specific information especially for ceramic element glow plugs). Check power supply and earth connections to control module. The control module could be providing the correct control signal but it is being affected by a circuit fault. Refer to list of sensor / actuator checks on page 19. Note that it is possible that a manufacturer could apply the code to the circuit between the Diesel control unit and the glow plug control module; if necessary refer to Fault Codes P0383 and P0384 for additional information.
P0678	Cylinder 8 Glow Plug Circuit / Open	Identify the type of glow plug control system. Glow plugs are usually controlled by a "Glow Plug Control Module" (can be integrated into the main Diesel control unit or, can be a separate unit which receives a control signal from the Diesel control unit). Depending on system design, full battery voltage can be applied to the glow plugs (possibly via a relay controlled by the control module) or, on many systems the circuit voltage / current is regulated by using a digital control signal (voltage / current is dependent on signal pulse width / duty cycle). Note that different types of glow plug can be specific to different types of control systems, and also note that the voltage / current level in the circuit can alter during the period of operation (including pre / post start conditions).	Refer to vehicle specific information for the glow plug control system and design of glow plug; the exact interpretation of the fault code will depend on the system design; however, it is likely that the fault code relates to the glow plug control signal on the circuit between the control module and the glow plugs. The voltage, frequency, current or value of the control signal in the control circuit is either incorrect (undefined fault but possibly caused by an open circuit). Check the circuit between the control module and the glow plug for short / open circuit and high resistance. Check the resistance of the glow plugs (refer to vehicle specific information especially for ceramic element glow plugs). Check power supply and earth connections to control module. The control module could be providing the correct control signal but it is being affected by a circuit fault. Refer to list of sensor / actuator checks on page 19. Note that it is possible that a manufacturer could apply the code to the circuit between the Diesel control unit and the glow plug control module; if necessary refer to Fault Codes P0383 and P0384 for additional information.
P0679	Cylinder 9 Glow Plug Circuit / Open	Identify the type of glow plug control system. Glow plugs are usually controlled by a "Glow Plug Control Module" (can be integrated into the main Diesel control unit or, can be a separate unit which receives a control signal from the Diesel control unit). Depending on system design, full battery voltage can be applied to the glow plugs (possibly via a relay controlled by the control module) or, on many systems the circuit voltage / current is regulated by using a digital control signal (voltage / current is dependent on signal pulse width / duty cycle). Note that different types of glow plug can be specific to different types of control systems, and also note that the voltage / current level in the circuit can alter during the period of operation (including pre / post start conditions).	Refer to vehicle specific information for the glow plug control system and design of glow plug; the exact interpretation of the fault code will depend on the system design; however, it is likely that the fault code relates to the glow plug control signal on the circuit between the control module and the glow plugs. The voltage, frequency, current or value of the control signal in the control circuit is either incorrect (undefined fault but possibly caused by an open circuit). Check the circuit between the control module and the glow plug for short / open circuit and high resistance. Check the resistance of the glow plugs (refer to vehicle specific information especially for ceramic element glow plugs). Check power supply and earth connections to control module. The control module could be providing the correct control signal but it is being affected by a circuit fault. Refer to list of sensor / actuator checks on page 19. Note that it is possible that a manufacturer could apply the code to the circuit between the Diesel control unit and the glow plug control module; if necessary refer to Fault Codes P0383 and P0384 for additional information.

NOTE 1. Check for other fault codes that could provide additional information. **NOTE 2.** Communication between control units can pass via a CAN-Bus system; refer to Page 51 for CAN-Bus checks.
NOTE 3. If a fault cannot be located, it is also possible that the control unit is at fault. **NOTE 4.** Refer to Page 53 for list of pages that show the ISO standard for component locations e.g. Sensor A - Bank 1.
165

Fault code	EOBD / ISO Description	Component / System Description	Meaningful Description and Quick Check
P067A	Glow Plug 4 Control Circuit Low	The glow plugs are usually switched on / off by a "Glow Plug Control Module", which can be controlled by the main Diesel Engine control unit. The control module can be a separate item but it can form part of the main control unit. The Glow Plug Module will receive a control signal from the Diesel control unit, and the module will switch on / off the power supply to the glow plugs (the module can contain an internal relay / power stage, although it is possible to use an external relay).	Refer to vehicle specific information for the glow plug control system and design of glow plug; the exact interpretation of the fault code will depend on the system design; however, it is likely that the fault code relates to the glow plug control signal on the circuit between the control module and the glow plugs. The voltage, frequency, current or value of the control signal in the control circuit is either "zero" (no signal / voltage) or, the signal / voltage exists but is below the normal operating range. Check the circuit between the control module and the glow plug for short / open circuit and high resistance. Check the resistance of the glow plugs (refer to vehicle specific information especially for ceramic element glow plugs). Check power supply and earth connections to control module. The control module could be providing the correct control signal but it is being affected by a circuit fault. Refer to list of sensor / actuator checks on page 19. Note that it is possible that a manufacturer could apply the code to the circuit between the Diesel control unit and the glow plug control module; if necessary refer to Fault Code P0383 for additional information.
P067B	Glow Plug 4 Control Circuit High	Identify the type of glow plug control system. Glow plugs are usually controlled by a "Glow Plug Control Module" (can be integrated into the main Diesel control unit or, can be a separate unit which receives a control signal from the Diesel control unit). Depending on system design, full battery voltage can be applied to the glow plugs (possibly via a relay controlled by the control module) or, on many systems the circuit voltage / current is regulated by using a digital control signal (voltage / current is dependent on signal pulse width / duty cycle). Note that different types of glow plug can be specific to different types of control systems, and also note that the voltage / current level in the circuit can alter during the period of operation (including pre / post start conditions).	Refer to vehicle specific information for the glow plug control system and design of glow plug; the exact interpretation of the fault code will depend on the system design; however, it is likely that the fault code relates to the glow plug control signal on the circuit between the control module and the glow plugs. The voltage, frequency, current or value of the control signal in the control circuit is either at full value (e.g. battery voltage with no on / off signal) or, the signal exists but it is above the normal operating range. Check the circuit between the control module and the glow plug for short / open circuit and high resistance. Check the resistance of the glow plugs (refer to vehicle specific information especially for ceramic element glow plugs). The control module could be providing the correct control signal but it is being affected by a circuit fault. Refer to list of sensor / actuator checks on page 19. Note that it is possible that a manufacturer could apply the code to the circuit between the Diesel control unit and the glow plug control module; if necessary refer to Fault Code P0384 for additional information.
P067C	Glow Plug 5 Control Circuit Low	Identify the type of glow plug control system. Glow plugs are usually controlled by a "Glow Plug Control Module" (can be integrated into the main Diesel control unit or, can be a separate unit which receives a control signal from the Diesel control unit). Depending on system design, full battery voltage can be applied to the glow plugs (possibly via a relay controlled by the control module) or, on many systems the circuit voltage / current is regulated by using a digital control signal (voltage / current is dependent on signal pulse width / duty cycle). Note that different types of glow plug can be specific to different types of control systems, and also note that the voltage / current level in the circuit can alter during the period of operation (including pre / post start conditions).	Refer to vehicle specific information for the glow plug control system and design of glow plug; the exact interpretation of the fault code will depend on the system design; however, it is likely that the fault code relates to the glow plug control signal on the circuit between the control module and the glow plugs. The voltage, frequency, current or value of the control signal in the control circuit is either "zero" (no signal / voltage) or, the signal / voltage exists but is below the normal operating range. Check the circuit between the control module and the glow plug for short / open circuit and high resistance. Check the resistance of the glow plugs (refer to vehicle specific information especially for ceramic element glow plugs). Check power supply and earth connections to control module. The control module could be providing the correct control signal but it is being affected by a circuit fault. Refer to list of sensor / actuator checks on page 19. Note that it is possible that a manufacturer could apply the code to the circuit between the Diesel control unit and the glow plug control module; if necessary refer to Fault Code P0383 for additional information.
P067D	Glow Plug 5 Control Circuit High	Identify the type of glow plug control system. Glow plugs are usually controlled by a "Glow Plug Control Module" (can be integrated into the main Diesel control unit or, can be a separate unit which receives a control signal from the Diesel control unit). Depending on system design, full battery voltage can be applied to the glow plugs (possibly via a relay controlled by the control module) or, on many systems the circuit voltage / current is regulated by using a digital control signal (voltage / current is dependent on signal pulse width / duty cycle). Note that different types of glow plug can be specific to different types of control systems, and also note that the voltage / current level in the circuit can alter during the period of operation (including pre / post start conditions).	Refer to vehicle specific information for the glow plug control system and design of glow plug; the exact interpretation of the fault code will depend on the system design; however, it is likely that the fault code relates to the glow plug control signal on the circuit between the control module and the glow plugs. The voltage, frequency, current or value of the control signal in the control circuit is either at full value (e.g. battery voltage with no on / off signal) or, the signal exists but it is above the normal operating range. Check the circuit between the control module and the glow plug for short / open circuit and high resistance. Check the resistance of the glow plugs (refer to vehicle specific information especially for ceramic element glow plugs). The control module could be providing the correct control signal but it is being affected by a circuit fault. Refer to list of sensor / actuator checks on page 19. Note that it is possible that a manufacturer could apply the code to the circuit between the Diesel control unit and the glow plug control module; if necessary refer to Fault Code P0384 for additional information.
P067E	Glow Plug 6 Control Circuit Low	Identify the type of glow plug control system. Glow plugs are usually controlled by a "Glow Plug Control Module" (can be integrated into the main Diesel control unit or, can be a separate unit which receives a control signal from the Diesel control unit). Depending on system design, full battery voltage can be applied to the glow plugs (possibly via a relay controlled by the control module) or, on many systems the circuit voltage / current is regulated by using a digital control signal (voltage / current is dependent on signal pulse width / duty cycle). Note that different types of glow plug can be specific to different types of control systems, and also note that the voltage / current level in the circuit can alter during the period of operation (including pre / post start conditions).	Refer to vehicle specific information for the glow plug control system and design of glow plug; the exact interpretation of the fault code will depend on the system design; however, it is likely that the fault code relates to the glow plug control signal on the circuit between the control module and the glow plugs. The voltage, frequency, current or value of the control signal in the control circuit is either "zero" (no signal / voltage) or, the signal / voltage exists but is below the normal operating range. Check the circuit between the control module and the glow plug for short / open circuit and high resistance. Check the resistance of the glow plugs (refer to vehicle specific information especially for ceramic element glow plugs). Check power supply and earth connections to control module. The control module could be providing the correct control signal but it is being affected by a circuit fault. Refer to list of sensor / actuator checks on page 19. Note that it is possible that a manufacturer could apply the code to the circuit between the Diesel control unit and the glow plug control module; if necessary refer to Fault Code P0383 for additional information.
P067F	Glow Plug 6 Control Circuit High	Identify the type of glow plug control system. Glow plugs are usually controlled by a "Glow Plug Control Module" (can be integrated into the main Diesel control unit or, can be a separate unit which receives a control signal from the Diesel control unit). Depending on system design, full battery voltage can be applied to the glow plugs (possibly via a relay controlled by the control module) or, on many systems the circuit voltage / current is regulated by using a digital control signal (voltage / current is dependent on signal pulse width / duty cycle). Note that different types of glow plug can be specific to different types of control systems, and also note that the voltage / current level in the circuit can alter during the period of operation (including pre / post start conditions).	Refer to vehicle specific information for the glow plug control system and design of glow plug; the exact interpretation of the fault code will depend on the system design; however, it is likely that the fault code relates to the glow plug control signal on the circuit between the control module and the glow plugs. The voltage, frequency, current or value of the control signal in the control circuit is either at full value (e.g. battery voltage with no on / off signal) or, the signal exists but it is above the normal operating range. Check the circuit between the control module and the glow plug for short / open circuit and high resistance. Check the resistance of the glow plugs (refer to vehicle specific information especially for ceramic element glow plugs). The control module could be providing the correct control signal but it is being affected by a circuit fault. Refer to list of sensor / actuator checks on page 19. Note that it is possible that a manufacturer could apply the code to the circuit between the Diesel control unit and the glow plug control module; if necessary refer to Fault Code P0384 for additional information.

NOTE 1. Check for other fault codes that could provide additional information. **NOTE 2.** Communication between control units can pass via a CAN-Bus system; refer to Page 51 for CAN-Bus checks. 166
NOTE 3. If a fault cannot be located, it is also possible that the control unit is at fault. **NOTE 4.** Refer to Page 53 for list of pages that show the ISO standard for component locations e.g. Sensor A - Bank 1.

code	Description		
P0680	Cylinder 10 Glow Plug Circuit / Open	Identify the type of glow plug control system. Glow plugs are usually controlled by a "Glow Plug Control Module" (can be integrated into the main Diesel control unit or, can be a separate unit which receives a control signal from the Diesel control unit). Depending on system design, full battery voltage can be applied to the glow plugs (possibly via a relay controlled by the control module) or, on many systems the circuit voltage / current is regulated by using a digital control signal (voltage / current is dependent on signal pulse width / duty cycle). Note that different types of glow plug can be specific to different types of control systems, and also note that the voltage / current level in the circuit can alter during the period of operation (including pre / post start conditions).	Refer to vehicle specific information for the glow plug control system and design of glow plug; the exact interpretation of the fault code will depend on the system design; however, it is likely that the fault code relates to the glow plug control signal on the circuit between the control module and the glow plugs. The voltage, frequency, current or value of the control signal in the control circuit is either incorrect (undefined fault but possibly caused by an open circuit). Check the circuit between the control module and the glow plug for short / open circuit and high resistance. Check the resistance of the glow plugs (refer to vehicle specific information especially for ceramic element glow plugs). Check power supply and earth connections to control module. The control module could be providing the correct control signal but it is being affected by a circuit fault. Refer to list of sensor / actuator checks on page 19. Note that it is possible that a manufacturer could apply the code to the circuit between the Diesel control unit and the glow plug control module; if necessary refer to Fault Codes P0383 and P0384 for additional information.
P0681	Cylinder 11 Glow Plug Circuit / Open	Identify the type of glow plug control system. Glow plugs are usually controlled by a "Glow Plug Control Module" (can be integrated into the main Diesel control unit or, can be a separate unit which receives a control signal from the Diesel control unit). Depending on system design, full battery voltage can be applied to the glow plugs (possibly via a relay controlled by the control module) or, on many systems the circuit voltage / current is regulated by using a digital control signal (voltage / current is dependent on signal pulse width / duty cycle). Note that different types of glow plug can be specific to different types of control systems, and also note that the voltage / current level in the circuit can alter during the period of operation (including pre / post start conditions).	Refer to vehicle specific information for the glow plug control system and design of glow plug; the exact interpretation of the fault code will depend on the system design; however, it is likely that the fault code relates to the glow plug control signal on the circuit between the control module and the glow plugs. The voltage, frequency, current or value of the control signal in the control circuit is either incorrect (undefined fault but possibly caused by an open circuit). Check the circuit between the control module and the glow plug for short / open circuit and high resistance. Check the resistance of the glow plugs (refer to vehicle specific information especially for ceramic element glow plugs). Check power supply and earth connections to control module. The control module could be providing the correct control signal but it is being affected by a circuit fault. Refer to list of sensor / actuator checks on page 19. Note that it is possible that a manufacturer could apply the code to the circuit between the Diesel control unit and the glow plug control module; if necessary refer to Fault Codes P0383 and P0384 for additional information.
P0682	Cylinder 12 Glow Plug Circuit / Open	Identify the type of glow plug control system. Glow plugs are usually controlled by a "Glow Plug Control Module" (can be integrated into the main Diesel control unit or, can be a separate unit which receives a control signal from the Diesel control unit). Depending on system design, full battery voltage can be applied to the glow plugs (possibly via a relay controlled by the control module) or, on many systems the circuit voltage / current is regulated by using a digital control signal (voltage / current is dependent on signal pulse width / duty cycle). Note that different types of glow plug can be specific to different types of control systems, and also note that the voltage / current level in the circuit can alter during the period of operation (including pre / post start conditions).	Refer to vehicle specific information for the glow plug control system and design of glow plug; the exact interpretation of the fault code will depend on the system design; however, it is likely that the fault code relates to the glow plug control signal on the circuit between the control module and the glow plugs. The voltage, frequency, current or value of the control signal in the control circuit is either incorrect (undefined fault but possibly caused by an open circuit). Check the circuit between the control module and the glow plug for short / open circuit and high resistance. Check the resistance of the glow plugs (refer to vehicle specific information especially for ceramic element glow plugs). Check power supply and earth connections to control module. The control module could be providing the correct control signal but it is being affected by a circuit fault. Refer to list of sensor / actuator checks on page 19. Note that it is possible that a manufacturer could apply the code to the circuit between the Diesel control unit and the glow plug control module; if necessary refer to Fault Codes P0383 and P0384 for additional information.
P0683	Glow Plug Control Module to PCM Communication Circuit	The glow plugs are usually switched on / off by a "Glow Plug Control Module", which can be controlled by the main Diesel Engine control unit or the Powertrain Control Module (PCM). The Glow Plug Module will receive a control signal from the Diesel control unit / PCM, and the module will then switch / regulate the power supply to the glow plugs.	The fault code indicates that there is an undefined communication error between the Diesel engine control unit / PCM and the glow plug control module. The frequency or value of the communication signal is incorrect (undefined fault). The communication signal could be: out of normal operating range, constant value, non-existent, corrupt. The fault is likely to be related to a wiring / connection problem but note that the communication could be via a CAN-Bus network, Refer to list of sensor / actuator checks on page 19. Check the wiring and connections between the two modules for short / open circuits, high circuit resistance; interference from other circuits can also affect sensor signals.
P0684	Glow Plug Control Module to PCM Communication Circuit Range / Performance	The glow plugs are usually switched on / off by a "Glow Plug Control Module", which can be controlled by the main Diesel Engine control unit or the Powertrain Control Module (PCM). The Glow Plug Module will receive a control signal from the Diesel control unit / PCM, and the module will then switch / regulate the power supply to the glow plugs.	The fault code indicates that the communication between the Diesel engine control unit / PCM and the glow plug control module is not as expected or, the response to the signal is incorrect. The frequency or value of the communication signal is within the normal operating range / tolerance but the signal or, the response to the signal is not plausible or is incorrect due to an undefined fault. It is therefore possible that the glow plug system response differs from the expected / desired response (glow plug system not switching on or responding correctly). The fault could be related to a wiring / connection problem but note that the communication could be via a CAN-Bus network, Refer to list of sensor / actuator checks on page 19. Check the wiring and connections between the two modules for short / open circuits, high circuit resistance; interference from other circuits can also affect sensor signals. Also check that the glow plug control unit and glow plugs are responding / operating as expected (cold starts etc). Refer to list of sensor / actuator checks on page 19.
P0685	ECM / PCM Power Relay Control Circuit / Open	The main ECM / PCM (Engine / Power train Module) power relay will provide electrical power to most of the components on the system. It is normal practice for the control unit to switch the earth circuit for the relay i.e. the control unit will complete the earth path for the relay energising winding, which will cause the relay contacts to close and provide power to the system components. The control unit is therefore able to detect / sense whether there is a power supply to the energising winding circuit and if the circuit is good. A sense circuit can also be used, which connects the relay power output terminal to the control unit; this allows the control unit to detect that the relay has switched on the power output circuit.	Refer to list of sensor / actuator checks on page 19 for notes on relay checks. Different manufacturers could use the fault code to different relay circuits. The fault code could refer to the power supply circuit (the power supply to the relay contacts and the circuit from the relay contacts to the system components). Alternatively, the fault code could refer to the relay energising winding circuit (power supply to the energising winding and the switched earth path through to the control unit). The fault code indicates that the voltage on the applicable relay circuit is incorrect (undefined fault but possible open circuit). Check power supplies to relay contacts (usually via a fused circuit) and to relay energising winding. Check operation of energising winding (short / open circuit) and closing / opening of relay contacts. Check all wiring for short / open circuits.

NOTE 1. Check for other fault codes that could provide additional information.   NOTE 2. Communication between control units can pass via a CAN-Bus system; refer to Page 51 for CAN-Bus checks.   **167**
NOTE 3. If a fault cannot be located, it is also possible that the control unit is at fault.   NOTE 4. Refer to Page 53 for list of pages that show the ISO standard for component locations e.g. Sensor A - Bank 1.

Fault code	EOBD / ISO Description	Component / System Description	Meaningful Description and Quick Check
P0686	ECM / PCM Power Relay Control Circuit Low	The main ECM / PCM (Engine / Power train Module) power relay will provide electrical power to most of the components on the system. It is normal practice for the control unit to switch the earth circuit for the relay i.e. the control unit will complete the earth path for the relay energising winding, which will cause the relay contacts to close and provide power to the system components. The control unit is therefore able to detect / sense whether there is a power supply to the energising winding circuit and if the circuit is good. A sense circuit can also be used, which connects the relay power output terminal to the control unit; this allows the control unit to detect that the relay has switched on the power output circuit.	Refer to list of sensor / actuator checks on page 19 for notes on relay checks. Different manufacturers could use the fault code to refer to different relay circuits. The fault code could refer to the power supply circuit (the power supply to the relay contacts and the circuit from the relay contacts to the system components). Alternatively, the fault code could refer to the relay energising winding circuit (power supply to the energising winding and the switched earth path through to the control unit). The voltage in the applicable relay circuit is low (likely to be zero volts) at a time when the voltage should be at battery voltage level. Check power supplies to relay contacts (usually via a fused circuit) and to relay energising winding. Check operation of energising winding (short / open circuit) and closing / opening of relay contacts. Check all wiring for short / open circuits.
P0687	ECM / PCM Power Relay Control Circuit High	The main ECM / PCM (Engine / Power train Module) power relay will provide electrical power to most of the components on the system. It is normal practice for the control unit to switch the earth circuit for the relay i.e. the control unit will complete the earth path for the relay energising winding, which will cause the relay contacts to close and provide power to the system components. The control unit is therefore able to detect / sense whether there is a power supply to the energising winding circuit and if the circuit is good. A sense circuit can also be used, which connects the relay power output terminal to the control unit; this allows the control unit to detect that the relay has switched on the power output circuit.	Refer to list of sensor / actuator checks on page 19 for notes on relay checks. Different manufacturers could use the fault code to refer to different relay circuits. The fault code could refer to the power supply circuit (the power supply to the relay contacts and the circuit from the relay contacts to the system components). Alternatively, the fault code could refer to the relay energising winding circuit (power supply to the energising winding and the switched earth path through to the control unit). The voltage in the relay circuit is high (likely to be battery voltage) when the relay circuit should be off (contacts open). Likely cause is relay contacts remaining closed (either due to relay fault or energising winding is being continuously activated); also possible short in wiring to another power supply or the energising winding earth path is shorted to earth. Check operation of relay, including closing / opening of relay contacts, and check operation of energising winding (short / open circuit).
P0688	ECM / PCM Power Relay Sense Circuit / Open	The main ECM / PCM (Engine / Power train Module) power relay will provide electrical power to most of the components on the system. It is normal practice for the control unit to switch the earth circuit for the relay i.e. the control unit will complete the earth path for the relay energising winding, which will cause the relay contacts to close and provide power to the system components. The control unit is therefore able to detect / sense whether there is a power supply to the energising winding circuit and if the circuit is good. A sense circuit can also be used, which connects the relay power output terminal to the control unit; this allows the control unit to detect that the relay has switched on the power output circuit.	Refer to list of sensor / actuator checks on page 19 for notes on relay checks. Different manufacturers could use the fault code to refer to "sensing" of the different relay circuits. The fault code could refer to the power supply circuit (the power supply to the relay contacts and the circuit from the relay contacts to the system components). Alternatively, the fault code could refer to the relay energising winding circuit (power supply to the energising winding and the switched earth path through to the control unit). The fault code indicates that the voltage on the relay sense circuit is incorrect (undefined fault but possible open circuit). Check whether the system components are being provided with power from the relay; if the power is being provided, it indicates that the relay is operating but the sense circuit between the relay and the control unit is open circuit. If there is no power supply to the system components, check relay operation, including power supply to relay contacts and relay energising winding. Check for short / open circuit in all wiring and relay energising winding.
P0689	ECM / PCM Power Relay Sense Circuit Low	The main ECM / PCM (Engine / Power train Module) power relay will provide electrical power to most of the components on the system. It is normal practice for the control unit to switch the earth circuit for the relay i.e. the control unit will complete the earth path for the relay energising winding, which will cause the relay contacts to close and provide power to the system components. The control unit is therefore able to detect / sense whether there is a power supply to the energising winding circuit and if the circuit is good. A sense circuit can also be used, which connects the relay power output terminal to the control unit; this allows the control unit to detect that the relay has switched on the power output circuit.	Refer to list of sensor / actuator checks on page 19 for notes on relay checks. Different manufacturers could use the fault code to refer to "sensing" of the different relay circuits. The fault code could refer to the power supply circuit (the power supply to the relay contacts and the circuit from the relay contacts to the system components). Alternatively, the fault code could refer to the relay energising winding circuit (power supply to the energising winding and the switched earth path through to the control unit). The fault code indicates that the voltage on the relay sense circuit is low or non-existent. Check whether the system components are being provided with power from the relay; if the power is being provided, it indicates that the relay is operating but the sense circuit between the relay and the control unit is open circuit. If there is no power supply to the system components, check relay operation, including power supply to relay contacts and relay energising winding. Check for short / open circuit in all wiring and relay energising winding.
P068A	ECM / PCM Power Relay De-Energized Performance -Too Early	The main ECM / PCM (Engine / Power train Module) power relay will provide electrical power to most of the components on the system. It is normal practice for the control unit to switch the earth circuit for the relay i.e. the control unit will complete the earth path for the relay energising winding, which will cause the relay contacts to close and provide power to the system components. The control unit is therefore able to detect / sense whether there is a power supply to the energising winding circuit and if the circuit is good. A sense circuit can also be used, which connects the relay power output terminal to the control unit; this allows the control unit to detect that the relay has switched on the power output circuit.	The ECM / PCM relay can be controlled so that it De-Energises at pre-determined times i.e. a delayed switch off; this could occur at a pre-determined time e.g. after the engine is switched off to enable system components to reset for the next start or, for other control functions. Refer to list of sensor / actuator checks on page 19 for notes on relay checks, but it will be necessary to refer to vehicle specific information to identify when the relay should De-Energise and, how the process is controlled (control process could be an integrated part of the relay or it could be controlled by the control unit). The fault code indicates that the relay is De-Energising too early, which could prevent certain tasks or functions from occurring. The fault is likely to be related to the control of the relay De-Energising process (controlled either by the relay or the control unit). Check power supply and earth connections to the control unit.
P068B	ECM / PCM Power Relay De-Energized Performance -Too Late	The main ECM / PCM (Engine / Power train Module) power relay will provide electrical power to most of the components on the system. It is normal practice for the control unit to switch the earth circuit for the relay i.e. the control unit will complete the earth path for the relay energising winding, which will cause the relay contacts to close and provide power to the system components. The control unit is therefore able to detect / sense whether there is a power supply to the energising winding circuit and if the circuit is good. A sense circuit can also be used, which connects the relay power output terminal to the control unit; this allows the control unit to detect that the relay has switched on the power output circuit.	The ECM / PCM relay can be controlled so that it De-Energises at pre-determined times i.e. a delayed switch off; this could occur at a pre-determined time e.g. after the engine is switched off to enable systems components to reset for the next start or, for other control functions. Refer to list of sensor / actuator checks on page 19 for notes on relay checks, but it will be necessary to refer to vehicle specific information to identify when the relay should De-Energise and, how the process is controlled (control process could be an integrated part of the relay or it could be controlled by the control unit). The fault code indicates that the relay is De-Energising too late; the fault is likely to be related to the control of the relay De-Energising process (controlled either by the relay or the control unit). Check power supply and earth connections to the control unit.

NOTE 1. Check for other fault codes that could provide additional information. **NOTE 2.** Communication between control units can pass via a CAN-Bus system; refer to Page 51 for CAN-Bus checks.
NOTE 3. If a fault cannot be located, it is also possible that the control unit is at fault. **NOTE 4.** Refer to Page 53 for list of pages that show the ISO standard for component locations e.g. Sensor A - Bank 1.

code	Description		
P068C	Glow Plug 7 Control Circuit Low	Identify the type of glow plug control system. Glow plugs are usually controlled by a "Glow Plug Control Module" (can be integrated into the main Diesel control unit or, can be a separate unit which receives a control signal from the Diesel control unit). Depending on system design, full battery voltage can be applied to the glow plugs (possibly via a relay controlled by the control module) or, on many systems the circuit voltage / current is regulated by using a digital control signal (voltage / current is dependent on signal pulse width / duty cycle). Note that different types of glow plug can be specific to different types of control systems, and also note that the voltage / current level in the circuit can alter during the period of operation (including pre / post start conditions).	Refer to vehicle specific information for the glow plug control system and design of glow plug; the exact interpretation of the fault code will depend on the system design; however, it is likely that the fault code relates to the glow plug control signal on the circuit between the control module and the glow plugs. The voltage, frequency, current or value of the control signal in the control circuit is either "zero" (no signal / voltage) or, the signal / voltage exists but is below the normal operating range. Check the circuit between the control module and the glow plug for short / open circuit and high resistance. Check the resistance of the glow plugs (refer to vehicle specific information especially for ceramic element glow plugs). Check power supply and earth connections to control module. The control module could be providing the correct control signal but it is being affected by a circuit fault. Refer to list of sensor / actuator checks on page 19. Note that it is possible that a manufacturer could apply the code to the circuit between the Diesel control unit and the glow plug control module; if necessary refer to Fault Code P0383 for additional information.
P068D	Glow Plug 7 Control Circuit High	Identify the type of glow plug control system. Glow plugs are usually controlled by a "Glow Plug Control Module" (can be integrated into the main Diesel control unit or, can be a separate unit which receives a control signal from the Diesel control unit). Depending on system design, full battery voltage can be applied to the glow plugs (possibly via a relay controlled by the control module) or, on many systems the circuit voltage / current is regulated by using a digital control signal (voltage / current is dependent on signal pulse width / duty cycle). Note that different types of glow plug can be specific to different types of control systems, and also note that the voltage / current level in the circuit can alter during the period of operation (including pre / post start conditions).	Refer to vehicle specific information for the glow plug control system and design of glow plug; the exact interpretation of the fault code will depend on the system design; however, it is likely that the fault code relates to the glow plug control signal on the circuit between the control module and the glow plugs. The voltage, frequency, current or value of the control signal in the control circuit is either at full value (e.g. battery voltage with no on / off signal) or, the signal exists but it is above the normal operating range. Check the circuit between the control module and the glow plug for short / open circuit and high resistance. Check the resistance of the glow plugs (refer to vehicle specific information especially for ceramic element glow plugs). The control module could be providing the correct control signal but it is being affected by a circuit fault. Refer to list of sensor / actuator checks on page 19. Note that it is possible that a manufacturer could apply the code to the circuit between the Diesel control unit and the glow plug control module; if necessary refer to Fault Code P0384 for additional information.
P068E	Glow Plug 8 Control Circuit Low	Identify the type of glow plug control system. Glow plugs are usually controlled by a "Glow Plug Control Module" (can be integrated into the main Diesel control unit or, can be a separate unit which receives a control signal from the Diesel control unit). Depending on system design, full battery voltage can be applied to the glow plugs (possibly via a relay controlled by the control module) or, on many systems the circuit voltage / current is regulated by using a digital control signal (voltage / current is dependent on signal pulse width / duty cycle). Note that different types of glow plug can be specific to different types of control systems, and also note that the voltage / current level in the circuit can alter during the period of operation (including pre / post start conditions).	Refer to vehicle specific information for the glow plug control system and design of glow plug; the exact interpretation of the fault code will depend on the system design; however, it is likely that the fault code relates to the glow plug control signal on the circuit between the control module and the glow plugs. The voltage, frequency, current or value of the control signal in the control circuit is either "zero" (no signal / voltage) or, the signal / voltage exists but is below the normal operating range. Check the circuit between the control module and the glow plug for short / open circuit and high resistance. Check the resistance of the glow plugs (refer to vehicle specific information especially for ceramic element glow plugs). Check power supply and earth connections to control module. The control module could be providing the correct control signal but it is being affected by a circuit fault. Refer to list of sensor / actuator checks on page 19. Note that it is possible that a manufacturer could apply the code to the circuit between the Diesel control unit and the glow plug control module; if necessary refer to Fault Code P0383 for additional information.
P068F	Glow Plug 8 Control Circuit High	Identify the type of glow plug control system. Glow plugs are usually controlled by a "Glow Plug Control Module" (can be integrated into the main Diesel control unit or, can be a separate unit which receives a control signal from the Diesel control unit). Depending on system design, full battery voltage can be applied to the glow plugs (possibly via a relay controlled by the control module) or, on many systems the circuit voltage / current is regulated by using a digital control signal (voltage / current is dependent on signal pulse width / duty cycle). Note that different types of glow plug can be specific to different types of control systems, and also note that the voltage / current level in the circuit can alter during the period of operation (including pre / post start conditions).	Refer to vehicle specific information for the glow plug control system and design of glow plug; the exact interpretation of the fault code will depend on the system design; however, it is likely that the fault code relates to the glow plug control signal on the circuit between the control module and the glow plugs. The voltage, frequency, current or value of the control signal in the control circuit is either at full value (e.g. battery voltage with no on / off signal) or, the signal exists but it is above the normal operating range. Check the circuit between the control module and the glow plug for short / open circuit and high resistance. Check the resistance of the glow plugs (refer to vehicle specific information especially for ceramic element glow plugs). The control module could be providing the correct control signal but it is being affected by a circuit fault. Refer to list of sensor / actuator checks on page 19. Note that it is possible that a manufacturer could apply the code to the circuit between the Diesel control unit and the glow plug control module; if necessary refer to Fault Code P0384 for additional information.
P0690	ECM / PCM Power Relay Sense Circuit High	The main ECM / PCM (Engine / Power train Module) power relay will provide electrical power to most of the components on the system. It is normal practice for the control unit to switch the earth circuit for the relay i.e. the control unit will complete the earth path for the relay energising winding, which will cause the relay contacts to close and provide power to the system components. The control unit is therefore able to detect / sense whether there is a power supply to the energising winding circuit and if the circuit is good. A sense circuit can also be used, which connects the relay power output terminal to the control unit; this allows the control unit to detect that the relay has switched on the power output circuit.	Refer to list of sensor / actuator checks on page 19 for notes on relay checks. Different manufacturers could use the fault code to refer to "sensing" of the different relay circuits. The fault code could refer to the power supply circuit (the power supply to the relay contacts and the circuit from the relay contacts to the system components). Alternatively, the fault code could refer to the relay energising winding circuit (power supply to the energising winding and the switched earth path through to the control unit). The fault code indicates that the voltage on the relay sense circuit is high. It is likely that the relay is not switching off the power supply to the system components as expected e.g. when the engine is switched off (voltage will therefore exist in the sense circuit). It is likely that the relay is at fault but it is also possible that there is a short in the wiring to another power supply circuit or, a short on the energising winding earth circuit to another earth.
P0691	Fan 1 Control Circuit Low	It will be necessary to identify the type of control system for the cooling fans. The cooling fan control system can use the engine control unit, a separate control module, or the air conditioning control module to regulate fan motor speed (separate modules could also receive signals from the main control unit). The applicable control unit can provide a control signal to the fan motors, which is altered (duty cycle, voltage / current etc) to regulate the fan motor speed. Note that some systems used a resistor pack (different resistances used for different speed requirements) to regulate the fan speed. Also check for relays in the circuits, which can be used to provide power to the control units and / or the motors.	Note that the speed / power for some electric motors can be regulated using a frequency / pulse width control signal (provided by the control unit on the earth or power circuit). The voltage, frequency or value of the control signal in the motor circuit is either "zero" (no voltage or signal) or, the voltage / signal exists but is below the normal operating range. Possible fault with motor or wiring e.g. motor power supply (relay if fitted), short / open circuit or resistance, that is causing a signal value to be lower than the minimum operating limit (also check control resistances if fitted to the system). Note: the control unit or relay (if fitted) could be providing the correct voltage / signal but it is being affected by a circuit fault. Refer to list of sensor / actuator checks on page 19.

NOTE 1. Check for other fault codes that could provide additional information.   **NOTE 2.** Communication between control units can pass via a CAN-Bus system; refer to Page 51 for CAN-Bus checks.

NOTE 3. If a fault cannot be located, it is also possible that the control unit is at fault.   **NOTE 4.** Refer to Page 53 for list of pages that show the ISO standard for component locations e.g. Sensor A - Bank 1.

Fault code	EOBD / ISO Description	Component / System Description	Meaningful Description and Quick Check
P0692	Fan 1 Control Circuit High	It will be necessary to identify the type of control system for the cooling fans. The cooling fan control system can use the engine control unit, a separate control module, or the air conditioning control module to regulate fan motor speed (separate modules could also receive signals from the main control unit). The applicable control unit can provide a control signal to the fan motors, which is altered (duty cycle, voltage / current etc) to regulate the fan motor speed. Note that some systems used a resistor pack (different resistances used for different speed requirements) to regulate the fan speed. Also check for relays in the circuits, which can be used to provide power to the control units and / or the motors.	Note that the speed / power for some electric motors can be regulated using a frequency / pulse width control signal (provided by the control unit on the earth or power circuit). The voltage, frequency or value of the control signal in the motor circuit is either "zero" (no voltage or signal) or, the voltage / signal exists but is below the normal operating range. Possible fault with motor or wiring e.g. motor power supply (relay if fitted), short / open circuit or resistance, that is causing a signal value to be lower than the minimum operating limit (also check control resistances if fitted to the system). Note: the control unit or relay (if fitted) could be providing the correct voltage / signal but it is being affected by a circuit fault. Refer to list of sensor / actuator checks on page 19.
P0693	Fan 2 Control Circuit Low	It will be necessary to identify the type of control system for the cooling fans. The cooling fan control system can use the engine control unit, a separate control module, or the air conditioning control module to regulate fan motor speed (separate modules could also receive signals from the main control unit). The applicable control unit can provide a control signal to the fan motors, which is altered (duty cycle, voltage / current etc) to regulate the fan motor speed. Note that some systems used a resistor pack (different resistances used for different speed requirements) to regulate the fan speed. Also check for relays in the circuits, which can be used to provide power to the control units and / or the motors.	Note that the speed / power for some electric motors can be regulated using a frequency / pulse width control signal (provided by the control unit on the earth or power circuit). The voltage, frequency or value of the control signal in the motor circuit is either "zero" (no voltage or signal) or, the voltage / signal exists but is below the normal operating range. Possible fault with motor or wiring e.g. motor power supply (relay if fitted), short / open circuit or resistance, that is causing a signal value to be lower than the minimum operating limit (also check control resistances if fitted to the system). Note: the control unit or relay (if fitted) could be providing the correct voltage / signal but it is being affected by a circuit fault. Refer to list of sensor / actuator checks on page 19.
P0694	Fan 2 Control Circuit High	It will be necessary to identify the type of control system for the cooling fans. The cooling fan control system can use the engine control unit, a separate control module, or the air conditioning control module to regulate fan motor speed (separate modules could also receive signals from the main control unit). The applicable control unit can provide a control signal to the fan motors, which is altered (duty cycle, voltage / current etc) to regulate the fan motor speed. Note that some systems used a resistor pack (different resistances used for different speed requirements) to regulate the fan speed. Also check for relays in the circuits, which can be used to provide power to the control units and / or the motors.	Note that the speed / power for some electric motors can be regulated using a frequency / pulse width control signal (provided by the control unit on the earth or power circuit). The voltage, frequency or value of the control signal in the motor circuit is either "zero" (no voltage or signal) or, the voltage / signal exists but is below the normal operating range. Possible fault with motor or wiring e.g. motor power supply (relay if fitted), short / open circuit or resistance, that is causing a signal value to be lower than the minimum operating limit (also check control resistances if fitted to the system). Note: the control unit or relay (if fitted) could be providing the correct voltage / signal but it is being affected by a circuit fault. Refer to list of sensor / actuator checks on page 19.
P0695	Fan 3 Control Circuit Low	It will be necessary to identify the type of control system for the cooling fans. The cooling fan control system can use the engine control unit, a separate control module, or the air conditioning control module to regulate fan motor speed (separate modules could also receive signals from the main control unit). The applicable control unit can provide a control signal to the fan motors, which is altered (duty cycle, voltage / current etc) to regulate the fan motor speed. Note that some systems used a resistor pack (different resistances used for different speed requirements) to regulate the fan speed. Also check for relays in the circuits, which can be used to provide power to the control units and / or the motors.	Note that the speed / power for some electric motors can be regulated using a frequency / pulse width control signal (provided by the control unit on the earth or power circuit). The voltage, frequency or value of the control signal in the motor circuit is either "zero" (no voltage or signal) or, the voltage / signal exists but is below the normal operating range. Possible fault with motor or wiring e.g. motor power supply (relay if fitted), short / open circuit or resistance, that is causing a signal value to be lower than the minimum operating limit (also check control resistances if fitted to the system). Note: the control unit or relay (if fitted) could be providing the correct voltage / signal but it is being affected by a circuit fault. Refer to list of sensor / actuator checks on page 19.
P0696	Fan 3 Control Circuit High	It will be necessary to identify the type of control system for the cooling fans. The cooling fan control system can use the engine control unit, a separate control module, or the air conditioning control module to regulate fan motor speed (separate modules could also receive signals from the main control unit). The applicable control unit can provide a control signal to the fan motors, which is altered (duty cycle, voltage / current etc) to regulate the fan motor speed. Note that some systems used a resistor pack (different resistances used for different speed requirements) to regulate the fan speed. Also check for relays in the circuits, which can be used to provide power to the control units and / or the motors.	Note that the speed / power for some electric motors can be regulated using a frequency / pulse width control signal (provided by the control unit on the earth or power circuit). The voltage, frequency or value of the control signal in the motor circuit is either "zero" (no voltage or signal) or, the voltage / signal exists but is below the normal operating range. Possible fault with motor or wiring e.g. motor power supply (relay if fitted), short / open circuit or resistance, that is causing a signal value to be lower than the minimum operating limit (also check control resistances if fitted to the system). Note: the control unit or relay (if fitted) could be providing the correct voltage / signal but it is being affected by a circuit fault. Refer to list of sensor / actuator checks on page 19.

code	Description		
P0697	Sensor Reference Voltage "C" Circuit / Open	The control unit provides a reference voltage to one or more system sensors (typically around 5 volts, although the voltage on some systems may differ); the reference voltage will have a tolerance range (typically in the range of 4.5 - 5.5 volts). Note that the reference voltage is sometimes referred to as a sensor power supply. Depending on the sensor type, the voltage is passed across the sensor resistance or an electronic circuit within the sensor, the sensor will then produce an output voltage or signal which can be compared with the original reference voltage level. On other sensor types, the reference voltage can form part of the sensor circuit and therefore the voltage will change as the sensor value changes. Refer to list of sensor / actuator checks on page 19 for additional information.	It may be necessary to refer to vehicle specific information to identify which is the specified reference voltage circuit. However, the voltage can be checked at various sensors (preferably with the engine running) to establish which sensors and circuit are affected. The fault code indicates that the control unit diagnostic system has detected that the reference voltage is incorrect. The exact fault is undefined but it is possible that the reference voltage is outside of the accepted tolerance due to either a short / open circuit or high resistance in the circuit to the applicable sensors. It is also possible that the control unit has an internal fault (the control unit can monitor the reference voltage and compare with an expected value).
P0698	Sensor Reference Voltage "C" Circuit Low	The control unit provides a reference voltage to one or more system sensors (typically around 5 volts, although the voltage on some systems may differ); the reference voltage will have a tolerance range (typically in the range of 4.5 - 5.5 volts). Note that the reference voltage is sometimes referred to as a sensor power supply. Depending on the sensor type, the voltage is passed across the sensor resistance or an electronic circuit within the sensor, the sensor will then produce an output voltage or signal which can be compared with the original reference voltage level. On other sensor types, the reference voltage can form part of the sensor circuit and therefore the voltage will change as the sensor value changes. Refer to list of sensor / actuator checks on page 19 for additional information.	It may be necessary to refer to vehicle specific information to identify which is the specified reference voltage circuit. However, the voltage can be checked at various sensors (preferably with the engine running) to establish which sensors and circuit are affected. The fault code indicates that the control unit diagnostic system has detected that the reference voltage is low; either "zero" (no signal) or, the signal exists but it is below the normal operating range (lower than the minimum operating limit). It is possible that a short circuit exists in the wiring or a sensor. It is also possible that the control unit has an internal fault (the control unit can monitor the reference voltage and compare with an expected value).
P0699	Sensor Reference Voltage "C" Circuit High	The control unit provides a reference voltage to one or more system sensors (typically around 5 volts, although the voltage on some systems may differ); the reference voltage will have a tolerance range (typically in the range of 4.5 - 5.5 volts). Note that the reference voltage is sometimes referred to as a sensor power supply. Depending on the sensor type, the voltage is passed across the sensor resistance or an electronic circuit within the sensor, the sensor will then produce an output voltage or signal which can be compared with the original reference voltage level. On other sensor types, the reference voltage can form part of the sensor circuit and therefore the voltage will change as the sensor value changes. Refer to list of sensor / actuator checks on page 19 for additional information.	It may be necessary to refer to vehicle specific information to identify which is the specified reference voltage circuit. However, the voltage can be checked at various sensors (preferably with the engine running) to establish which sensors and circuit are affected. The fault code indicates that the control unit diagnostic system has detected that the reference voltage is high; e.g. battery voltage with no signal or frequency or, the voltage exists but it is above the normal operating range (higher than the maximum operating limit). It is possible that there is a short to another power supply (in the wiring or a sensor). It is also possible that the control unit has an internal fault (the control unit can monitor the reference voltage and compare with an expected value).
P069A	Glow Plug 9 Control Circuit Low	Identify the type of glow plug control system. Glow plugs are usually controlled by a "Glow Plug Control Module" (can be integrated into the main Diesel control unit or, can be a separate unit which receives a control signal from the Diesel control unit). Depending on system design, full battery voltage can be applied to the glow plugs (possibly via a relay controlled by the control module) or, on many systems the circuit voltage / current is regulated by using a digital control signal (voltage / current is dependent on signal pulse width / duty cycle). Note that different types of glow plug can be specific to different types of control systems, and also note that the voltage / current level in the circuit can alter during the period of operation (including pre / post start conditions).	Refer to vehicle specific information for the glow plug control system and design of glow plug; the exact interpretation of the fault code will depend on the system design; however, it is likely that the fault code relates to the glow plug control signal on the circuit between the control module and the glow plugs. The voltage, frequency, current or value of the control signal in the control circuit is either "zero" (no signal / voltage) or, the signal / voltage exists but is below the normal operating range. Check the circuit between the control module and the glow plug for short / open circuit and high resistance. Check the resistance of the glow plugs (refer to vehicle specific information especially for ceramic element glow plugs). Check power supply and earth connections to control module. The control module could be providing the correct control signal but it is being affected by a circuit fault. Refer to list of sensor / actuator checks on page 19. Note that it is possible that a manufacturer could apply the code to the circuit between the Diesel control unit and the glow plug control module; if necessary refer to Fault Code P0383 for additional information.
P069B	Glow Plug 9 Control Circuit High	Identify the type of glow plug control system. Glow plugs are usually controlled by a "Glow Plug Control Module" (can be integrated into the main Diesel control unit or, can be a separate unit which receives a control signal from the Diesel control unit). Depending on system design, full battery voltage can be applied to the glow plugs (possibly via a relay controlled by the control module) or, on many systems the circuit voltage / current is regulated by using a digital control signal (voltage / current is dependent on signal pulse width / duty cycle). Note that different types of glow plug can be specific to different types of control systems, and also note that the voltage / current level in the circuit can alter during the period of operation (including pre / post start conditions).	Refer to vehicle specific information for the glow plug control system and design of glow plug; the exact interpretation of the fault code will depend on the system design; however, it is likely that the fault code relates to the glow plug control signal on the circuit between the control module and the glow plugs. The voltage, frequency, current or value of the control signal in the control circuit is either at full value (e.g. battery voltage with no on / off signal) or, the signal exists but it is above the normal operating range. Check the circuit between the control module and the glow plug for short / open circuit and high resistance. Check the resistance of the glow plugs (refer to vehicle specific information especially for ceramic element glow plugs). The control module could be providing the correct control signal but it is being affected by a circuit fault. Refer to list of sensor / actuator checks on page 19. Note that it is possible that a manufacturer could apply the code to the circuit between the Diesel control unit and the glow plug control module; if necessary refer to Fault Code P0384 for additional information.

NOTE 1. Check for other fault codes that could provide additional information.

NOTE 3. If a fault cannot be located, it is also possible that the control unit is at fault.

NOTE 2. Communication between control units can pass via a CAN-Bus system; refer to Page 51 for CAN-Bus checks.

NOTE 4. Refer to Page 53 for list of pages that show the ISO standard for component locations e.g. Sensor A - Bank 1.

Fault code	EOBD / ISO Description	Component / System Description	Meaningful Description and Quick Check
P069C	Glow Plug 10 Control Circuit Low	Identify the type of glow plug control system. Glow plugs are usually controlled by a "Glow Plug Control Module" (can be integrated into the main Diesel control unit or, can be a separate unit which receives a control signal from the Diesel control unit). Depending on system design, full battery voltage can be applied to the glow plugs (possibly via a relay controlled by the control module) or, on many systems the circuit voltage / current is regulated by using a digital control signal (voltage / current is dependent on signal pulse width / duty cycle). Note that different types of glow plug can be specific to different types of control systems, and also note that the voltage / current level in the circuit can alter during the period of operation (including pre / post start conditions).	Refer to vehicle specific information for the glow plug control system and design of glow plug; the exact interpretation of the fault code will depend on the system design; however, it is likely that the fault code relates to the glow plug control signal on the circuit between the control module and the glow plugs. The voltage, frequency, current or value of the control signal in the control circuit is either "zero" (no signal / voltage) or, the signal / voltage exists but is below the normal operating range. Check the circuit between the control module and the glow plug for short / open circuit and high resistance. Check the resistance of the glow plugs (refer to vehicle specific information especially for ceramic element glow plugs). Check power supply and earth connections to control module. The control module could be providing the correct control signal but it is being affected by a circuit fault. Refer to list of sensor / actuator checks on page 19. Note that it is possible that a manufacturer could apply the code to the circuit between the Diesel control unit and the glow plug control module; if necessary refer to Fault Code P0383 for additional information.
P069D	Glow Plug 10 Control Circuit High	Identify the type of glow plug control system. Glow plugs are usually controlled by a "Glow Plug Control Module" (can be integrated into the main Diesel control unit or, can be a separate unit which receives a control signal from the Diesel control unit). Depending on system design, full battery voltage can be applied to the glow plugs (possibly via a relay controlled by the control module) or, on many systems the circuit voltage / current is regulated by using a digital control signal (voltage / current is dependent on signal pulse width / duty cycle). Note that different types of glow plug can be specific to different types of control systems, and also note that the voltage / current level in the circuit can alter during the period of operation (including pre / post start conditions).	Refer to vehicle specific information for the glow plug control system and design of glow plug; the exact interpretation of the fault code will depend on the system design; however, it is likely that the fault code relates to the glow plug control signal on the circuit between the control module and the glow plugs. The voltage, frequency, current or value of the control signal in the control circuit is either at full value (e.g. battery voltage with no on / off signal) or, the signal exists but it is above the normal operating range. Check the circuit between the control module and the glow plug for short / open circuit and high resistance. Check the resistance of the glow plugs (refer to vehicle specific information especially for ceramic element glow plugs). The control module could be providing the correct control signal but it is being affected by a circuit fault. Refer to list of sensor / actuator checks on page 19. Note that it is possible that a manufacturer could apply the code to the circuit between the Diesel control unit and the glow plug control module; if necessary refer to Fault Code P0384 for additional information.
P069E	ISO / SAE reserved		Not used, reserved for future allocation.
P069F	ISO / SAE reserved		Not used, reserved for future allocation.
P0700	Transmission Control System (MIL Request)	The transmission control unit can directly switch on the warning light when a transmission fault occurs. On other systems, the transmission control unit can pass information to the main control unit if a transmission fault occurs; this information can be limited to requesting the MIL (Malfunction Indicator Light) is switched on.	It may be necessary to identify the method of switching the transmission warning light e.g. direct from the transmission control unit or, via the main control unit. The fault code indicates that there is a fault with the transmission MIL request system. Initially check warning light bulb and wiring back to the applicable control unit. Depending on the system design, the fault could also be related to a wiring / connection fault between the between the transmission control unit and the main control unit.
P0701	Transmission Control System Range / Performance	Fault codes for the transmission control system will usually be detected by the transmission control unit but, depending on the transmission system and the vehicle age, the EOBD code retrieved via the standard EOBD diagnostic plug, might only indicate that a fault exists (without a clear fault definition). If possible, check for additional codes from the transmission control unit (possibly only manufacturer specific codes).	The fault code indicates that there is an undefined fault within the transmission control system; the fault may be detected due to a transmission control unit failure or, because the transmission is not responding correctly to the control signals provided by the transmission control unit. Check for other fault codes that might provide additional information. Check all wiring between transmission control unit and the transmission system components (sensors and actuators). If possible check the "live data" values for the transmission components to identify any control signal or sensor errors. Check power supplies and earths for the control unit and for the transmission system components.
P0702	Transmission Control System Electrical	Fault codes for the transmission control system will usually be detected by the transmission control unit but, depending on the transmission system and the vehicle age, the EOBD code retrieved via the standard EOBD diagnostic plug, might only indicate that a fault exists (without a clear fault definition). If possible, check for additional codes from the transmission control unit (possibly only manufacturer specific codes).	The fault code indicates that there is an undefined electrical fault within the transmission control system; the fault may be detected due to a transmission control unit failure or, because one or more of the control signals and / or sensor signals are incorrect. Check for other fault codes that might provide additional information. Check all wiring between transmission control unit and the transmission system components (sensors and actuators). If possible check the "live data" values for the transmission components to identify any control signal or sensor errors. Check power supplies and earths for the control unit and for the transmission system components.
P0703	Brake Switch "B" Circuit	In addition to normal brake light operation, the brake switch signal can be used in the control of engine systems, vehicle stability systems, transmission and other vehicle systems. Where more than one brake switch is used, the signals can be passed to the different vehicle systems e.g. one for stability control / ABS (and other systems), and one for brake light operation. Note that some brake switches can be conventional type switches operated by brake pedal movement, but some types are "Pressure Switches" which operate due to brake line pressure (can be fitted into ABS modulator assembly).	Note. The switch assembly may contain more than one switch / sensor; refer to vehicle specific information to establish which switch is affected. The fault code indicates that the voltage or value of the switch signal is incorrect (undefined fault). Switch voltage / signal could be: out of normal operating range, constant value, non-existent, corrupt. Possible fault in the switch or wiring e.g. short / open circuits, circuit resistance, fault with the reference / supply voltage; the switch contacts could also be faulty e.g. not opening / closing correctly or have a high contact resistance. Also check the mechanism operating the switches e.g. pedal and switch adjustment or, check that there is sufficient brake line pressure to operate switch (where applicable). Refer to list of sensor / actuator checks on page 19.
P0704	Clutch Switch Input Circuit	The clutch switch is used to indicate the operation of the clutch mechanism. The switch can be operated by clutch pedal movement or, on other systems; the switch could be operated by the clutch operating mechanism. Note that the code could be used for automated clutch systems (electronically controlled clutch actuation). Also note that the clutch switch fault code is classified as a Transmission code but, the switch could be providing a signal to other vehicle systems / control units.	The voltage, frequency or value of the switch signal is within the normal operating range / tolerance but the signal is not plausible or is incorrect due to an undefined fault. The switch signal might not match the operating conditions indicated by other sensors or, the switch signal is not changing / responding as expected. Possible fault in the switch or wiring e.g. short / open circuits, circuit resistance or, fault with the reference / operating voltage; also check switch contacts to ensure that they open / close and check continuity of switch contacts (ensure that there is not a high resistance). Check mechanism / linkage used to operate switch, and check base settings for the switch if applicable. Refer to list of sensor / actuator checks on page 19.
P0705	Transmission Range Sensor "A" Circuit (PRNDL Input)	Manufacturers generally refer to the range sensor as being the sensor that indicates the selection of P - R - N - 1 -2 -3 - 4 etc. More than one sensor can be used for back-up or for indicating additional information (depending on the system); one sensor could be located on the selector gate with the other sensor being located on the selector linkage or transmission housing.	The voltage, frequency or value of the sensor signal is incorrect (undefined fault). Sensor voltage / signal could be: out of normal operating range, constant value, non-existent, corrupt. Possible fault in the sensor or wiring e.g. short / open circuits, circuit resistance, incorrect sensor resistance or, fault with the reference / operating voltage; interference from other circuits can also affect sensor signals. Refer to list of sensor / actuator checks on page 19.

NOTE 1. Check for other fault codes that could provide additional information. **NOTE 2.** Communication between control units can pass via a CAN-Bus system; refer to Page 51 for CAN-Bus checks.

NOTE 3. If a fault cannot be located, it is also possible that the control unit is at fault. **NOTE 4.** Refer to Page 53 for list of pages that show the ISO standard for component locations e.g. Sensor A - Bank 1.

code	Description		
P0706	Transmission Range Sensor "A" Circuit Range / Performance	Manufacturers generally refer to the range sensor as being the sensor that indicates the selection of P - R - N - 1 -2 - 3 - 4 etc. More than one sensor can be used for back-up or for indicating additional information (depending on the system); one sensor could be located on the selector gate with the other sensor being located on the selector linkage or transmission housing.	The voltage, frequency or value of the sensor signal is within the normal operating range / tolerance but the signal is not plausible or is incorrect due to an undefined fault. The sensor signal might not match the operating conditions indicated by other sensors or, the sensor signal is not changing / responding as expected. Possible fault in the sensor or wiring e.g. short / open circuits, circuit resistance, incorrect sensor resistance or, fault with the reference / operating voltage; interference from other circuits can also affect sensor signals. Refer to list of sensor / actuator checks on page 19. It is also possible that the sensor is operating correctly but the position / gear being indicated by the sensor is unacceptable or incorrect (not as expected by the control unit with regard to information from other sensors). Check operation of range selector and mechanism; also check for any base settings that could be applicable to the sensor and selector.
P0707	Transmission Range Sensor "A" Circuit Low	Manufacturers generally refer to the range sensor as being the sensor that indicates the selection of P - R - N - 1 -2 - 3 - 4 etc. More than one sensor can be used for back-up or for indicating additional information (depending on the system); one sensor could be located on the selector gate with the other sensor being located on the selector linkage or transmission housing.	The voltage, frequency or value of the sensor signal is either "zero" (no signal) or, the signal exists but it is below the normal operating range (lower than the minimum operating limit). Possible fault in the sensor or wiring e.g. short / open circuits, circuit resistance, incorrect sensor resistance or, fault with the reference / operating voltage; interference from other circuits can also affect sensor signals. Refer to list of sensor / actuator checks on page 19. It is also possible that the control unit is detecting "no change" in the signal when the range selector is moved due to a fault in the mechanism connecting the sensor to the selector.
P0708	Transmission Range Sensor "A" Circuit High	Manufacturers generally refer to the range sensor as being the sensor that indicates the selection of P - R - N - 1 -2 - 3 - 4 etc. More than one sensor can be used for back-up or for indicating additional information (depending on the system); one sensor could be located on the selector gate with the other sensor being located on the selector linkage or transmission housing.	The voltage, frequency or value of the sensor signal is either at full value (e.g. battery voltage with no signal or frequency) or, the signal exists but it is above the normal operating range (higher than the maximum operating limit). Possible fault in the sensor or wiring e.g. short / open circuits, circuit resistance, incorrect sensor resistance or, fault with the reference / operating voltage; interference from other circuits can also affect sensor signals. Refer to list of sensor / actuator checks on page 19. It is also possible that the control unit is detecting "no change" in the signal when the range selector is moved due to a fault in the mechanism connecting the sensor to the selector.
P0709	Transmission Range Sensor "A" Circuit Intermittent	Manufacturers generally refer to the range sensor as being the sensor that indicates the selection of P - R - N - 1 -2 - 3 - 4 etc. More than one sensor can be used for back-up or for indicating additional information (depending on the system); one sensor could be located on the selector gate with the other sensor being located on the selector linkage or transmission housing.	The voltage, frequency or value of the sensor signal is intermittent (likely to be out of normal operating range / tolerance when intermittent fault is detected) or, the signal / voltage is erratic (e.g. signal changes are irregular / unpredictable). Possible fault with wiring or connections (including internal sensor connections); also interference from other circuits can affect the signal. Refer to list of sensor / actuator checks on page 19. It is also possible that the control unit is detecting "no change" in the signal when the range selector is moved due to a fault in the mechanism connecting the sensor to the selector.
P070A	Transmission Fluid Level Sensor Circuit	The code identifies an electrical related fault in the "Transmission" fluid level sensor but, it might be advisable to check oil level. Note that oil level sensors can be simple on / off devices that respond when the oil level is above or below pre-set limits.	The voltage, frequency or value of the level sensor signal is incorrect (undefined fault). Sensor voltage / signal could be: out of normal operating range, constant value, non-existent, corrupt. Possible fault in the level sensor or wiring e.g. short / open circuits, circuit resistance, incorrect sensor resistance or, fault with the reference / operating voltage; interference from other circuits can also affect sensor signals. Refer to list of sensor / actuator checks on page 19.
P070B	Transmission Fluid Level Sensor Circuit Range / Performance	The code identifies an electrical related fault in the "Transmission" fluid level sensor but, it might be advisable to check oil level. Note that oil level sensors can be simple on / off devices that respond when the oil level is above or below pre-set limits.	The voltage, frequency or value of the level sensor signal is within the normal operating range / tolerance but the signal is not plausible or is incorrect due to an undefined fault. Possible fault in the level sensor or wiring e.g. short / open circuits, circuit resistance, incorrect sensor resistance or, fault with the reference / operating voltage; interference from other circuits can also affect sensor signals. Refer to list of sensor / actuator checks on page 19. It is also possible that the level sensor is operating correctly but the level being measured by the sensor is unacceptable or incorrect.
P070C	Transmission Fluid Level Sensor Circuit Low	The code identifies an electrical related fault in the "Transmission" fluid level sensor but, it might be advisable to check oil level. Note that oil level sensors can be simple on / off devices that respond when the oil level is above or below pre-set limits.	The voltage, frequency or value of the level sensor signal is either "zero" (no signal) or, the signal exists but it is below the normal operating range (lower than the minimum operating limit). Possible fault in the level sensor or wiring e.g. short / open circuits, circuit resistance, incorrect sensor resistance or, fault with the reference / operating voltage; interference from other circuits can also affect sensor signals. Refer to list of sensor / actuator checks on page 19.
P070D	Transmission Fluid Level Sensor Circuit High	The code identifies an electrical related fault in the "Transmission" fluid level sensor but, it might be advisable to check oil level. Note that oil level sensors can be simple on / off devices that respond when the oil level is above or below pre-set limits.	The voltage, frequency or value of the level sensor signal is either at full value (e.g. battery voltage with no signal or frequency) or, the signal exists but it is above the normal operating range (higher than the maximum operating limit). Possible fault in the level sensor or wiring e.g. short / open circuits, circuit resistance, incorrect sensor resistance or, fault with the reference / operating voltage; interference from other circuits can also affect sensor signals. Refer to list of sensor / actuator checks on page 19.
P070E	Transmission Fluid Level Sensor Circuit intermittent / Erratic	The code identifies an electrical related fault in the "Transmission" fluid level sensor but, it might be advisable to check oil level. Note that oil level sensors can be simple on / off devices that respond when the oil level is above or below pre-set limits.	The voltage, frequency or value of the level sensor signal is intermittent (likely to be out of normal operating range / tolerance when intermittent fault is detected) or, the signal / voltage is erratic (e.g. signal changes are irregular / unpredictable). Possible fault with wiring or connections (including internal sensor connections); also interference from other circuits can affect the signal. Refer to list of sensor / actuator checks on page 19.
P070F	Transmission Fluid Level Too Low	The code identifies an electrical related fault in the "Transmission" fluid level sensor but, it might be advisable to check oil level. Note that oil level sensors can be simple on / off devices that respond when the oil level is above or below pre-set limits.	The fault code relates to the oil level sensor signal which is indicating that oil level is too low. The sensor signal is within normal operating range / tolerance but the indicated oil level is below specified values. If the oil level is not too low, it is possible that the sensor signal is incorrect; refer to fault code P070B for additional information.
P0710	Transmission Fluid Temperature Sensor "A" Circuit	The information from the transmission fluid temperature sensor can be used to alter transmission operation during cold running / warm up and also influence the gear change strategy; under certain conditions engine control could be influenced by transmission oil temperature. Excessive fluid temperatures will also be indicated by the temperature sensor.	The code identifies an electrical related fault in the fluid temperature sensor but, it might be advisable to check oil cooling, oil level and oil condition / grade. The voltage, frequency or value of the temperature sensor signal is incorrect (undefined fault). Sensor voltage / signal could be: out of normal operating range, constant value, non-existent, corrupt. Possible fault in the temperature sensor or wiring e.g. short / open circuits, circuit resistance, incorrect sensor resistance or, fault with the reference / operating voltage; interference from other circuits can also affect sensor signals. Refer to list of sensor / actuator checks on page 19.

NOTE 1. Check for other fault codes that could provide additional information.   **NOTE 2.** Communication between control units can pass via a CAN-Bus system; refer to Page 51 for CAN-Bus checks.   **173**

NOTE 3. If a fault cannot be located, it is also possible that the control unit is at fault.   **NOTE 4.** Refer to Page 53 for list of pages that show the ISO standard for component locations e.g. Sensor A - Bank 1.

Fault code	EOBD / ISO Description	Component / System Description	Meaningful Description and Quick Check
P0711	Transmission Fluid Temperature Sensor "A" Circuit Range / Performance	The information from the transmission fluid temperature sensor can be used to alter transmission operation during cold running / warm up and also influence the gear change strategy; under certain conditions engine control could be influenced by transmission oil temperature. Excessive fluid temperatures will also be indicated by the temperature sensor.	The code identifies an electrical related fault in the fluid temperature sensor but, it might be advisable to check oil cooling, oil level and oil condition / grade. The voltage, frequency or value of the temperature sensor signal is within the normal operating range / tolerance but the signal is not plausible or is incorrect due to an undefined fault. The sensor signal might not match the operating conditions indicated by other sensors or, the sensor signal is not changing / responding as expected to the changes in conditions. Possible fault in the temperature sensor or wiring e.g. short / open circuits, circuit resistance, incorrect sensor resistance or, fault with the reference / operating voltage; interference from other circuits can also affect sensor signals. Refer to list of sensor / actuator checks on page 19. It is also possible that the sensor is operating correctly but the oil temperature is unacceptable or incorrect. If it is suspected that the temperature is incorrect, check transmission fluid cooling (intercooler etc) and check for correct oil, oil condition and quantity.
P0712	Transmission Fluid Temperature Sensor "A" Circuit Low	The information from the transmission fluid temperature sensor can be used to alter transmission operation during cold running / warm up and also influence the gear change strategy; under certain conditions engine control could be influenced by transmission oil temperature. Excessive fluid temperatures will also be indicated by the temperature sensor.	The code identifies an electrical related fault in the fluid temperature sensor but, it might be advisable to check oil cooling, oil level and oil condition / grade. The voltage, frequency or value of the temperature sensor signal is either "zero" (no signal) or the signal exists but it is below the normal operating range (lower than the minimum operating limit). Possible fault in the temperature sensor or wiring e.g. short / open circuits, circuit resistance, incorrect sensor resistance or, fault with the reference / operating voltage; interference from other circuits can also affect sensor signals. Refer to list of sensor / actuator checks on page 19.
P0713	Transmission Fluid Temperature Sensor "A" Circuit High	The information from the transmission fluid temperature sensor can be used to alter transmission operation during cold running / warm up and also influence the gear change strategy; under certain conditions engine control could be influenced by transmission oil temperature. Excessive fluid temperatures will also be indicated by the temperature sensor.	The code identifies an electrical related fault in the fluid temperature sensor but, it might be advisable to check oil cooling, oil level and oil condition / grade. The voltage, frequency or value of the temperature sensor signal is either at full value (e.g. battery voltage with no signal or frequency) or, the signal exists but it is above the normal operating range (higher than the maximum operating limit). Possible fault in the temperature sensor or wiring e.g. short / open circuits, circuit resistance, incorrect sensor resistance or, fault with the reference / operating voltage; interference from other circuits can also affect sensor signals. Refer to list of sensor / actuator checks on page 19.
P0714	Transmission Fluid Temperature Sensor "A" Circuit Intermittent	The information from the transmission fluid temperature sensor can be used to alter transmission operation during cold running / warm up and also influence the gear change strategy; under certain conditions engine control could be influenced by transmission oil temperature. Excessive fluid temperatures will also be indicated by the temperature sensor.	The code identifies an electrical related fault in the fluid temperature sensor but, it might be advisable to check oil cooling, oil level and oil condition / grade. The voltage, frequency or value of the temperature sensor signal is intermittent (likely to be out of normal operating range / tolerance when intermittent fault is detected) or, the signal / voltage is erratic (e.g. signal changes are irregular / unpredictable). Possible fault with wiring or connections (including internal sensor connections); also interference from other circuits can affect the signal. Refer to list of sensor / actuator checks on page 19. Check for any other system faults that could cause rapid or erratic change in fluid temperature.
P0715	Input / Turbine Speed Sensor "A" Circuit	Transmission input / output speeds indicate the gear ratio (indicates requested gear has been fully selected, without slip on the gear selection clutches used in auto transmission). The code is likely to relate to the speed sensor for the gearbox input shaft (connected to the "output" side of the torque converter, which the code refers to as the turbine). Note: some manufacturers refer to the turbine as being the "input" side of the torque converter. If a manufacturer has allocated this code to a torque converter "input speed" fault, this could refer to the engine speed signal (passed from the engine control unit or the engine speed sensor). Refer to torque converter fault codes P0740-P0756, P2756-P2770 and also to P0725.	Identify what is providing the speed signal to the transmission system e.g. an engine speed sensor, the engine control unit, or a specific sensor fitted to the transmission. Identify the type of sensor e.g. Inductive, Hall effect etc; this will dictate the checks that are applicable. The voltage, frequency or value of the speed sensor signal is incorrect (undefined fault). Sensor voltage / signal value could be: out of normal operating range, constant value, non-existent, corrupt. Possible fault in the speed sensor or wiring e.g. short / open circuits, circuit resistance, incorrect sensor resistance (if applicable) or, fault with the reference / operating voltage; interference from other circuits can also affect sensor signals. Also check earth connections and cable screening. Refer to list of sensor / actuator checks on page 19.
P0716	Input / Turbine Speed Sensor "A" Circuit Range / Performance	Transmission input / output speeds indicate the gear ratio (indicates requested gear has been fully selected, without slip on the gear selection clutches used in auto transmission). The code is likely to relate to the speed sensor for the gearbox input shaft (connected to the "output" side of the torque converter, which the code refers to as the turbine). Note: some manufacturers refer to the turbine as being the "input" side of the torque converter. If a manufacturer has allocated this code to a torque converter "input speed" fault, this could refer to the engine speed signal (passed from the engine control unit or the engine speed sensor). Refer to torque converter fault codes P0740-P0756, P2756-P2770 and also to P0725.	Identify what is providing the speed signal to the transmission system e.g. an engine speed sensor, the engine control unit, or a specific sensor fitted to the transmission. Identify the type of sensor e.g. Inductive, Hall effect etc. The voltage, frequency or value of the speed sensor signal is within the normal operating range / tolerance but the signal is not plausible or is incorrect due to an undefined fault. It is possible that the speed being measured by the sensor is unacceptable or incorrect for the operating conditions or, the sensor signal is not changing / responding as expected to the changes in conditions; if possible, check whether the indicated speed is correct. Possible fault in the speed sensor or wiring e.g. short / open circuits, circuit resistance, incorrect sensor resistance or, fault with the reference / operating voltage; interference from other circuits can also affect sensor signals. Refer to list of sensor / actuator checks on page 19. Check for other transmission related faults. Also check condition of reference teeth on trigger disc / rotor and ensure rotor is turning with shaft. Note that on some applications, the fault code will be activated if the speed exceeds specified limits.
P0717	Input / Turbine Speed Sensor "A" Circuit No Signal	Transmission input / output speeds indicate the gear ratio (indicates requested gear has been fully selected, without slip on the gear selection clutches used in auto transmission). The code is likely to relate to the speed sensor for the gearbox input shaft (connected to the "output" side of the torque converter, which the code refers to as the turbine). Note: some manufacturers refer to the turbine as being the "input" side of the torque converter. If a manufacturer has allocated this code to a torque converter "input speed" fault, this could refer to the engine speed signal (passed from the engine control unit or the engine speed sensor). Refer to torque converter fault codes P0740-P0756, P2756-P2770 and also to P0725.	Identify what is providing the speed signal to the transmission system e.g. an engine speed sensor, the engine control unit, or a specific sensor fitted to the transmission. Identify the type of sensor e.g. Inductive, Hall effect etc; this will dictate the checks that are applicable. The voltage, frequency or value of the speed sensor signal is incorrect (undefined fault resulting in no signal being detected). Sensor voltage / signal value could be: constant value or non-existent. Possible fault in the speed sensor or wiring e.g. short / open circuits, circuit resistance, incorrect sensor resistance (if applicable) or, fault with the reference / operating voltage; interference from other circuits can also affect sensor signals Also check earth connections and cable screening. Refer to list of sensor / actuator checks on page 19. Also check condition of reference teeth on trigger disc / rotor and ensure rotor is turning with shaft.

NOTE 1. Check for other fault codes that could provide additional information.   NOTE 2. Communication between control units can pass via a CAN-Bus system; refer to Page 51 for CAN-Bus checks.

NOTE 3. If a fault cannot be located, it is also possible that the control unit is at fault.   NOTE 4. Refer to Page 53 for list of pages that show the ISO standard for component locations e.g. Sensor A - Bank 1.

code	Description		
P0718	Input / Turbine Speed Sensor "A" Circuit Intermittent	Transmission input / output speeds indicate the gear ratio (indicates requested gear has been fully selected, without slip on the gear selection clutches used in auto transmission). The code is likely to relate to the speed sensor for the gearbox input shaft (connected to the "output" side of the torque converter, which the code refers to as the turbine). Note: some manufacturers refer to the turbine as being the "input" side of the torque converter. If a manufacturer has allocated this code to a torque converter "input speed" fault, this could refer to the engine speed signal (passed from the engine control unit or the engine speed sensor). Refer to torque converter fault codes P0740-P0756, P2756-P2770 and also to P0725.	Identify what is providing the speed signal to the transmission system e.g. an engine speed sensor, the engine control unit, or a specific sensor fitted to the transmission. Identify the type of sensor e.g. Inductive, Hall effect etc; this will dictate the checks that are applicable. The voltage, frequency or value of the speed sensor signal is intermittent (likely to be out of normal operating range / tolerance when intermittent fault is detected) or, the signal / voltage is erratic (e.g. signal changes are irregular / unpredictable). Possible fault with wiring or connections (including internal sensor connections); also interference from other circuits can affect the signal. Refer to list of sensor / actuator checks on page 19. Also check condition of reference teeth on trigger disc / rotor and ensure rotor is turning with shaft.
P0719	Brake Switch "B" Circuit Low	In addition to normal brake light operation, the brake switch signal can be used in the control of engine systems, vehicle stability systems, transmission and other vehicle systems. Where more than one brake switch is used, the signals can be passed to the different vehicle systems e.g. one for stability control / ABS (and other systems), and one for brake light operation. Note that some brake switches can be conventional type switches operated by brake pedal movement, but some types are "Pressure Switches" which operate due to brake line pressure (can be fitted into ABS modulator assembly).	Note. The switch assembly may contain more than one switch / sensor; refer to vehicle specific information to establish which switch is affected. The voltage, frequency or value of the switch signal is either "zero" (no signal) or, the signal exists but it is below the normal operating range (lower than the minimum operating limit). Possible fault in the switch or wiring e.g. short / open circuits, circuit resistance or, fault with the reference / supply voltage; the switch contacts could also be faulty e.g. not opening / closing correctly or have a high contact resistance. Also check the mechanism operating the switches e.g. pedal and switch adjustment or, check that there is sufficient brake line pressure to operate switch (where applicable). Refer to list of sensor / actuator checks on page 19.
P071A	Transmission Mode Switch "A" Circuit	The mode switch refers to a switch used to select different operating modes e.g. sport / economy or other available modes on the transmission system.	The voltage, frequency or value of the switch signal is incorrect (undefined fault). Switch voltage / signal could be: out of normal operating range, constant value, non-existent, corrupt. Possible fault in the switch or wiring e.g. short / open circuits, circuit resistance or, fault with the reference / operating voltage; the switch contacts could also be faulty e.g. not opening / closing correctly or have a high contact resistance. Refer to list of sensor / actuator checks on page 19. Also check mechanism / linkage that operates the switch.
P071B	Transmission Mode Switch "A" Circuit Low	The mode switch refers to a switch used to select different operating modes e.g. sport / economy or other available modes on the transmission system.	The voltage, frequency or value of the switch signal is either "zero" (no signal) or, the signal exists but it is below the normal operating range (lower than the minimum operating limit). Possible fault in the switch or wiring e.g. short / open circuits, circuit resistance or, fault with the reference / operating voltage; the switch contacts could also be faulty e.g. not opening / closing correctly or have a high contact resistance. Refer to list of sensor / actuator checks on page 19. Also check mechanism / linkage that operates the switch.
P071C	Transmission Mode Switch "A" Circuit High	The mode switch refers to a switch used to select different operating modes e.g. sport / economy or other available modes on the transmission system.	The voltage, frequency or value of the switch signal is either at full value (e.g. battery voltage with no signal or frequency) or, the signal exists but it is above the normal operating range (higher than the maximum operating limit). Possible fault in the switch or wiring e.g. short / open circuits, circuit resistance or, fault with the reference / operating voltage; the switch contacts could also be faulty e.g. not opening / closing correctly or have a high contact resistance. Refer to list of sensor / actuator checks on page 19. Also check mechanism / linkage that operates the switch.
P071D	Transmission Mode Switch "B" Circuit	The mode switch refers to a switch used to select different operating modes e.g. sport / economy or other available modes on the transmission system.	The voltage, frequency or value of the switch signal is incorrect (undefined fault). Switch voltage / signal could be: out of normal operating range, constant value, non-existent, corrupt. Possible fault in the switch or wiring e.g. short / open circuits, circuit resistance or, fault with the reference / operating voltage; the switch contacts could also be faulty e.g. not opening / closing correctly or have a high contact resistance. Refer to list of sensor / actuator checks on page 19. Also check mechanism / linkage that operates the switch.
P071E	Transmission Mode Switch "B" Circuit Low	The mode switch refers to a switch used to select different operating modes e.g. sport / economy or other available modes on the transmission system.	The voltage, frequency or value of the switch signal is either "zero" (no signal) or, the signal exists but it is below the normal operating range (lower than the minimum operating limit). Possible fault in the switch or wiring e.g. short / open circuits, circuit resistance or, fault with the reference / operating voltage; the switch contacts could also be faulty e.g. not opening / closing correctly or have a high contact resistance. Refer to list of sensor / actuator checks on page 19. Also check mechanism / linkage that operates the switch.
P071F	Transmission Mode Switch "B" Circuit High	The mode switch refers to a switch used to select different operating modes e.g. sport / economy or other available modes on the transmission system.	The voltage, frequency or value of the switch signal is either at full value (e.g. battery voltage with no signal or frequency) or, the signal exists but it is above the normal operating range (higher than the maximum operating limit). Possible fault in the switch or wiring e.g. short / open circuits, circuit resistance or, fault with the reference / operating voltage; the switch contacts could also be faulty e.g. not opening / closing correctly or have a high contact resistance. Refer to list of sensor / actuator checks on page 19. Also check mechanism / linkage that operates the switch.
P0720	Output Speed Sensor Circuit	Transmission input / output speeds indicate the gear ratio (indicates requested gear has been fully selected, without slip on the gear selection clutches used in auto transmission). The transmission output speed signal can also be used to indicate vehicle speed.	Identify the type of sensor e.g. Inductive, Hall effect etc; this will dictate the checks that are applicable. The voltage, frequency or value of the speed sensor signal is incorrect (undefined fault). Sensor voltage / signal value could be: out of normal operating range, constant value, non-existent, corrupt. Possible fault in the speed sensor or wiring e.g. short / open circuits, circuit resistance, incorrect sensor resistance (if applicable) or, fault with the reference / operating voltage; interference from other circuits can also affect sensor signals. Also check earth connections and cable screening. Refer to list of sensor / actuator checks on page 19.

NOTE 1. Check for other fault codes that could provide additional information. **NOTE 2.** Communication between control units can pass via a CAN-Bus system; refer to Page 51 for CAN-Bus checks.

NOTE 3. If a fault cannot be located, it is also possible that the control unit is at fault. **NOTE 4.** Refer to Page 53 for list of pages that show the ISO standard for component locations e.g. Sensor A - Bank 1.

Fault code	EOBD / ISO Description	Component / System Description	Meaningful Description and Quick Check
P0721	Output Speed Sensor Circuit Range / Performance	Transmission input / output speeds indicate the gear ratio (indicates requested gear has been fully selected, without slip on the gear selection clutches used in auto transmission). The transmission output speed signal could also be used to indicate vehicle speed.	Identify the type of sensor e.g. Inductive, Hall effect etc; this will dictate the checks that are applicable. The voltage, frequency or value of the speed sensor signal is within the normal operating range / tolerance but the signal is not plausible or is incorrect due to an undefined fault. The sensor signal might not match the operating conditions indicated by other sensors or, the sensor signal is not changing / responding as expected to the changes in conditions. Possible fault in the speed sensor or wiring e.g. short / open circuits, circuit resistance, incorrect sensor resistance or, fault with the reference / operating voltage; interference from other circuits can also affect sensor signals. Refer to list of sensor / actuator checks on page 19. It is also possible that the speed sensor is operating correctly but the speed being measured by the sensor is unacceptable or incorrect for the operating conditions; check for other transmission related faults. Also check condition of reference teeth on trigger disc / rotor and ensure rotor is turning with shaft. Note that on some applications, the fault code will be activated if the speed exceeds specified limits.
P0722	Output Speed Sensor Circuit No Signal	Transmission input / output speeds indicate the gear ratio (indicates requested gear has been fully selected, without slip on the gear selection clutches used in auto transmission). The transmission output speed signal could also be used to indicate vehicle speed.	Identify the type of sensor e.g. Inductive, Hall effect etc; this will dictate the checks that are applicable. The voltage, frequency or value of the speed sensor signal is incorrect (undefined fault resulting in no signal being detected). Sensor voltage / signal value could be: constant value or non-existent. Possible fault in the speed sensor or wiring e.g. short / open circuits, circuit resistance, incorrect sensor resistance (if applicable) or, fault with the reference / operating voltage; interference from other circuits can also affect sensor signals Also check earth connections and cable screening. Refer to list of sensor / actuator checks on page 19. Also check condition of reference teeth on trigger disc / rotor and ensure rotor is turning with shaft.
P0723	Output Speed Sensor Circuit Intermittent	Transmission input / output speeds indicate the gear ratio (indicates requested gear has been selected, without slip on the gear selection clutches used in auto transmission). The transmission output speed signal could also be used to indicate vehicle speed.	Identify the type of sensor e.g. Inductive, Hall effect etc; this will dictate the checks that are applicable. The voltage, frequency or value of the speed sensor signal is intermittent (likely to be out of normal operating range / tolerance when intermittent fault is detected) or, the signal / voltage is erratic (e.g. signal changes are irregular / unpredictable). Possible fault with wiring or connections (including internal sensor connections); also interference from other circuits can affect the signal. Refer to list of sensor / actuator checks on page 19. Also check condition of reference teeth on trigger disc / rotor and ensure rotor is turning with shaft.
P0724	Brake Switch "B" Circuit High	In addition to normal brake light operation, the brake switch signal can be used in the control of engine systems, vehicle stability systems, transmission and other vehicle systems. Where more than one brake switch is used, the signals can be passed to the different vehicle systems e.g. one for stability control / ABS (and other systems), and one for brake light operation. Note that some brake switches can be conventional type switches operated by brake pedal movement, but some types are "Pressure Switches" which operate due to brake line pressure (can be fitted into ABS modulator assembly).	Note. The switch assembly may contain more than one switch / sensor; refer to vehicle specific information to establish which switch is affected. The voltage, frequency or value of the switch signal is either at full value (e.g. battery voltage with no signal or frequency) or, the signal exists but it is above the normal operating range (higher than the maximum operating limit). Possible fault in the switch or wiring e.g. short / open circuits or, fault with the reference / operating voltage; the switch contacts could also be faulty e.g. not opening / closing correctly or have a high contact resistance. Also check the mechanism operating the switches e.g. pedal and switch adjustment or, check that there is sufficient brake line pressure to operate switch (where applicable). Refer to list of sensor / actuator checks on page 19.
P0725	Engine Speed Input Circuit	The fault code is categorised as a transmission related code but, the speed signal can be provided direct from the engine control unit to the transmission control unit (often via the CAN-Bus system); the speed signal will therefore originate at the engine speed sensor e.g. crankshaft / camshaft speed sensor or from Diesel pump speed sensor. It could be possible that a speed sensor is fitted that initially passes the signal to the transmission control unit, which then passes a speed signal to the engine control unit. It will be necessary to identify how the speed signal is passed to transmission control unit.	The voltage, frequency or value of the speed / sensor signal is incorrect (undefined fault). Signal value could be: out of normal operating range, constant value, non-existent, corrupt. If the speed signal is passed from the engine control unit, check wiring and connections (short / open circuits etc) or check for a CAN-Bus fault. Refer to list of sensor / actuator checks on page 19. If there is no detectable communication fault between the control units or, the speed signal is direct from a speed sensor, check speed sensor operation. Identify the type of sensor e.g. Inductive, Hall effect etc; this will dictate the checks that are applicable. Possible fault in the speed sensor or wiring e.g. short / open circuits, circuit resistance, incorrect sensor resistance (if applicable) or, fault with the reference / operating voltage; interference from other circuits can also affect sensor signals. Also check earth connections and cable screening. Refer to list of sensor / actuator checks on page 19.
P0726	Engine Speed Input Circuit Range / Performance	The fault code is categorised as a transmission related code but, the speed signal can be provided direct from the engine control unit to the transmission control unit (often via the CAN-Bus system); the speed signal will therefore originate at the engine speed sensor e.g. crankshaft / camshaft speed sensor or from Diesel pump speed sensor. It could be possible that a speed sensor is fitted that initially passes the signal to the transmission control unit, which then passes a speed signal to the engine control unit. It will be necessary to identify how the speed signal is passed to transmission control unit.	The voltage, frequency or value of the speed signal is within the normal operating range / tolerance but the signal is not plausible or is incorrect. The signal might not match the operating conditions or, the signal is not changing as expected. If the speed signal is passed from the engine control unit, check wiring and connections (short / open circuits etc) or check for a CAN-Bus fault. Refer to list of sensor / actuator checks on page 19. If there is no detectable communication fault between the control units or, the speed signal is direct from a speed sensor, check speed sensor operation. Identify the type of sensor e.g. Inductive, Hall effect etc; this will dictate the checks that are applicable. Possible fault in the speed sensor or wiring e.g. short / open circuits, circuit resistance, incorrect sensor resistance (if applicable) or, fault with the reference / operating voltage; interference from other circuits can also affect sensor signals. Also check earth connections and cable screening. Refer to list of sensor / actuator checks on page 19.
P0727	Engine Speed Input Circuit No Signal	The fault code is categorised as a transmission related code but, the speed signal can be provided direct from the engine control unit to the transmission control unit (often via the CAN-Bus system); the speed signal will therefore originate at the engine speed sensor e.g. crankshaft / camshaft speed sensor or from Diesel pump speed sensor. It could be possible that a speed sensor is fitted that initially passes the signal to the transmission control unit, which then passes a speed signal to the engine control unit. It will be necessary to identify how the speed signal is passed to transmission control unit.	The voltage, frequency or value of the speed sensor signal is incorrect (undefined fault resulting in no signal being detected). Sensor voltage / signal value could be: constant value or non-existent. If the speed signal is passed from the engine control unit, check wiring and connections (short / open circuits etc) or check for a CAN-Bus fault. Refer to list of sensor / actuator checks on page 19. If there is no detectable communication fault between the control units or, the speed signal is direct from a speed sensor, check speed sensor operation. Identify the type of sensor e.g. Inductive, Hall effect etc; this will dictate the checks that are applicable. Possible fault in the speed sensor or wiring e.g. short / open circuits, circuit resistance, incorrect sensor resistance (if applicable) or, fault with the reference / operating voltage; interference from other circuits can also affect sensor signals. Also check earth connections and cable screening. Refer to list of sensor / actuator checks on page 19.

NOTE 1. Check for other fault codes that could provide additional information. **NOTE 2.** Communication between control units can pass via a CAN-Bus system; refer to Page 51 for CAN-Bus checks.

NOTE 3. If a fault cannot be located, it is also possible that the control unit is at fault. **NOTE 4.** Refer to Page 53 for list of pages that show the ISO standard for component locations e.g. Sensor A - Bank 1.

Code	Description		
P0728	Engine Speed Input Circuit Intermittent	The fault code is categorised as a transmission related code but, the speed signal can be provided direct from the engine control unit to the transmission control unit (often via the CAN-Bus system); the speed signal will therefore originate at the engine speed sensor e.g. crankshaft / camshaft speed sensor or from Diesel pump speed sensor. It could be possible that a speed sensor is fitted that initially passes the signal to the transmission control unit, which then passes a speed signal to the engine control unit. It will be necessary to identify how the speed signal is passed to transmission control unit.	The voltage, frequency or value of the speed sensor signal is intermittent (likely to be out of normal operating range / tolerance when intermittent fault is detected) or, the signal / voltage is erratic (e.g. signal changes are irregular / unpredictable). If the speed signal is passed from the engine control unit, check wiring and connections (short / open circuits etc) or check for a CAN-Bus fault. Refer to list of sensor / actuator checks on page 19. If there is no detectable communication fault between the control units or, the speed signal is direct from a speed sensor, check speed sensor operation. Identify the type of sensor e.g. Inductive, Hall effect etc; this will dictate the checks that are applicable. Possible fault in the speed sensor or wiring e.g. short / open circuits, circuit resistance, incorrect sensor resistance (if applicable) or, fault with the reference / operating voltage; interference from other circuits can also affect sensor signals. Also check earth connections and cable screening. Refer to list of sensor / actuator checks on page 19.
P0729	Gear 6 Incorrect Ratio	The fault does not necessarily indicate that the wrong gear has been selected, it normally indicates that the gear has been selected but, due to a fault within the transmission (such as a slipping transmission clutch / brake band) the ratio is incorrect. The control unit monitors the speeds of the transmission input and output shafts; when each gear is selected, the input and output speeds should correspond to the selected gear ratio e.g. if the gear ratio is 4 to 1 then the input speed should be 4 times faster than the output speed. If the input and output speeds do not match the selected ratio (for the identified gear), this will be regarded as a fault (likely to be a mechanical / hydraulic pressure fault).	Refer to vehicle specific information to identify the method of changing / gear in the transmission and the components used e.g. shift solenoid valves, pressure control valves etc. Check transmission control system for additional fault codes. Fault is likely to be related to a problem with the gear selection mechanism for the identified gear e.g. a slipping clutch band or brake; note that this could be a mechanical fault within the transmission or a hydraulic pressure control problem. Check for correct operation of gear shift solenoid valves and pressure control valves, also check control pressures. If solenoid and control valves are not operating correctly, check wiring to power supply and to control unit, also check control signal from control unit to valves. Refer to vehicle specific information to identify the method of changing / gear in the transmission and the components used e.g. shift solenoid valves, pressure control valves etc.
P0730	Incorrect Gear Ratio	The fault does not necessarily indicate that the wrong gear has been selected, it normally indicates that the gear has been selected but, due to a fault within the transmission (such as a slipping transmission clutch / brake band) the ratio is incorrect. The control unit monitors the speeds of the transmission input and output shafts; when each gear is selected, the input and output speeds should correspond to the selected gear ratio e.g. if the gear ratio is 4 to 1 then the input speed should be 4 times faster than the output speed. If the input and output speeds do not match the selected ratio (for the identified gear), this will be regarded as a fault (likely to be a mechanical / hydraulic pressure fault).	Refer to vehicle specific information to identify the method of changing / gear in the transmission and the components used e.g. shift solenoid valves, pressure control valves etc. Check transmission control system for additional fault codes. Fault is likely to be related to a problem with the gear selection mechanism e.g. a slipping clutch band or brake; note that this could be a mechanical fault within the transmission or a hydraulic pressure control problem. Check for correct operation of gear shift solenoid valves and pressure control valves, also check control pressures. If solenoid and control valves are not operating correctly, check wiring to power supply and to control unit, also check control signal from control unit to valves. Refer to vehicle specific information to identify the method of changing / gear in the transmission and the components used e.g. shift solenoid valves, pressure control valves etc.
P0731	Gear 1 Incorrect Ratio	The fault does not necessarily indicate that the wrong gear has been selected, it normally indicates that the gear has been selected but, due to a fault within the transmission (such as a slipping transmission clutch / brake band) the ratio is incorrect. The control unit monitors the speeds of the transmission input and output shafts; when each gear is selected, the input and output speeds should correspond to the selected gear ratio e.g. if the gear ratio is 4 to 1 then the input speed should be 4 times faster than the output speed. If the input and output speeds do not match the selected ratio (for the identified gear), this will be regarded as a fault (likely to be a mechanical / hydraulic pressure fault).	Refer to vehicle specific information to identify the method of changing gear in the transmission and the components used e.g. shift solenoid valves, pressure control valves etc. Check transmission control system for additional fault codes. Fault is likely to be related to a problem with the gear selection mechanism for the identified gear e.g. a slipping clutch band or brake; note that this could be a mechanical fault within the transmission or a hydraulic pressure control problem. Check for correct operation of gear shift solenoid valves and pressure control valves, also check control pressures. If solenoid and control valves are not operating correctly, check wiring to power supply and to control unit, also check control signal from control unit to valves. Refer to vehicle specific information to identify the method of changing / gear in the transmission and the components used e.g. shift solenoid valves, pressure control valves etc.
P0732	Gear 2 Incorrect Ratio	The fault does not necessarily indicate that the wrong gear has been selected, it normally indicates that the gear has been selected but, due to a fault within the transmission (such as a slipping transmission clutch / brake band) the ratio is incorrect. The control unit monitors the speeds of the transmission input and output shafts; when each gear is selected, the input and output speeds should correspond to the selected gear ratio e.g. if the gear ratio is 4 to 1 then the input speed should be 4 times faster than the output speed. If the input and output speeds do not match the selected ratio (for the identified gear), this will be regarded as a fault (likely to be a mechanical / hydraulic pressure fault).	Refer to vehicle specific information to identify the method of changing gear in the transmission and the components used e.g. shift solenoid valves, pressure control valves etc. Check transmission control system for additional fault codes. Fault is likely to be related to a problem with the gear selection mechanism for the identified gear e.g. a slipping clutch band or brake; note that this could be a mechanical fault within the transmission or a hydraulic pressure control problem. Check for correct operation of gear shift solenoid valves and pressure control valves, also check control pressures. If solenoid and control valves are not operating correctly, check wiring to power supply and to control unit, also check control signal from control unit to valves. Refer to vehicle specific information to identify the method of changing / gear in the transmission and the components used e.g. shift solenoid valves, pressure control valves etc.
P0733	Gear 3 Incorrect Ratio	The fault does not necessarily indicate that the wrong gear has been selected, it normally indicates that the gear has been selected but, due to a fault within the transmission (such as a slipping transmission clutch / brake band) the ratio is incorrect. The control unit monitors the speeds of the transmission input and output shafts; when each gear is selected, the input and output speeds should correspond to the selected gear ratio e.g. if the gear ratio is 4 to 1 then the input speed should be 4 times faster than the output speed. If the input and output speeds do not match the selected ratio (for the identified gear), this will be regarded as a fault (likely to be a mechanical / hydraulic pressure fault).	Refer to vehicle specific information to identify the method of changing gear in the transmission and the components used e.g. shift solenoid valves, pressure control valves etc. Check transmission control system for additional fault codes. Fault is likely to be related to a problem with the gear selection mechanism for the identified gear e.g. a slipping clutch band or brake; note that this could be a mechanical fault within the transmission or a hydraulic pressure control problem. Check for correct operation of gear shift solenoid valves and pressure control valves, also check control pressures. If solenoid and control valves are not operating correctly, check wiring to power supply and to control unit, also check control signal from control unit to valves. Refer to vehicle specific information to identify the method of changing / gear in the transmission and the components used e.g. shift solenoid valves, pressure control valves etc.

NOTE 1. Check for other fault codes that could provide additional information.   **NOTE 2.** Communication between control units can pass via a CAN-Bus system; refer to Page 51 for CAN-Bus checks.

NOTE 3. If a fault cannot be located, it is also possible that the control unit is at fault.   **NOTE 4.** Refer to Page 53 for list of pages that show the ISO standard for component locations e.g. Sensor A - Bank 1.

Fault code	EOBD / ISO Description	Component / System Description	Meaningful Description and Quick Check
P0734	Gear 4 Incorrect Ratio	The fault does not necessarily indicate that the wrong gear has been selected, it normally indicates that the gear has been selected but, due to a fault within the transmission (such as a slipping transmission clutch / brake band) the ratio is incorrect. The control unit monitors the speeds of the transmission input and output shafts; when each gear is selected, the input and output speeds should correspond to the selected gear ratio e.g. if the gear ratio is 4 to 1 then the input speed should be 4 times faster than the output speed. If the input and output speeds do not match the selected ratio (for the identified gear), this will be regarded as a fault (likely to be a mechanical / hydraulic pressure fault).	Refer to vehicle specific information to identify the method of changing gear in the transmission and the components used e.g. shift solenoid valves, pressure control valves etc. Check transmission control system for additional fault codes. Fault is likely to be related to a problem with the gear selection mechanism for the identified gear e.g. a slipping clutch band or brake; note that this could be a mechanical fault within the transmission or a hydraulic pressure control problem. Check for correct operation of gear shift solenoid valves and pressure control valves, also check control pressures. If solenoid and control valves are not operating correctly, check wiring to power supply and to control unit, also check control signal from control unit to valves. Refer to vehicle specific information to identify the method of changing / gear in the transmission and the components used e.g. shift solenoid valves, pressure control valves etc.
P0735	Gear 5 Incorrect Ratio	The fault does not necessarily indicate that the wrong gear has been selected, it normally indicates that the gear has been selected but, due to a fault within the transmission (such as a slipping transmission clutch / brake band) the ratio is incorrect. The control unit monitors the speeds of the transmission input and output shafts; when each gear is selected, the input and output speeds should correspond to the selected gear ratio e.g. if the gear ratio is 4 to 1 then the input speed should be 4 times faster than the output speed. If the input and output speeds do not match the selected ratio (for the identified gear), this will be regarded as a fault (likely to be a mechanical / hydraulic pressure fault).	Refer to vehicle specific information to identify the method of changing gear in the transmission and the components used e.g. shift solenoid valves, pressure control valves etc. Check transmission control system for additional fault codes. Fault is likely to be related to a problem with the gear selection mechanism for the identified gear e.g. a slipping clutch band or brake; note that this could be a mechanical fault within the transmission or a hydraulic pressure control problem. Check for correct operation of gear shift solenoid valves and pressure control valves, also check control pressures. If solenoid and control valves are not operating correctly, check wiring to power supply and to control unit, also check control signal from control unit to valves. Refer to vehicle specific information to identify the method of changing / gear in the transmission and the components used e.g. shift solenoid valves, pressure control valves etc.
P0736	Reverse Incorrect Ratio	The fault code indicates that the gear has probably been selected but, due to a fault within the transmission (such as a slipping transmission clutch / brake band) the ratio is incorrect. The control unit monitors the speeds of the transmission input and output shafts; when each gear is selected, the input and output speeds should correspond to the selected gear ratio e.g. if the gear ratio is 4 to 1 then the input speed should be 4 times faster than the output speed. If the input and output speeds do not match the selected ratio (for the identified gear), this will be regarded as a fault (likely to be a mechanical / hydraulic pressure fault).	Refer to vehicle specific information to identify the method of changing gear in the transmission and the components used e.g. shift solenoid valves, pressure control valves etc. Check transmission control system for additional fault codes. Fault is likely to be related to a problem with the gear selection mechanism e.g. a slipping clutch or brake band; note that this could be a mechanical fault within the transmission or a hydraulic pressure control problem. Check for correct operation of gear shift solenoid valves and pressure control valves (check valves are not sticking), also check control pressures. If solenoid and control valves are not operating correctly, check wiring to power supply and to control unit, also check control signal from control unit to valves. Refer to vehicle specific information to identify the method of changing / gear in the transmission and the components used e.g. shift solenoid valves, pressure control valves etc.
P0737	TCM Engine Speed Output Circuit	The fault code is likely to relate to the speed signal passing between the transmission control unit and another control unit e.g. engine control unit. Other fault codes are used to identify engine speed sensor faults, therefore check for other codes that could indicate an engine speed sensor problem	The fault code indicates that there is an engine speed circuit fault or an incorrect speed signal which could be due to a wiring / connection fault or due to a control unit fault (transmission control unit or other control unit / system connected to the speed signal circuit). The voltage, frequency or value of the signal is incorrect (undefined fault). Signal could be: out of normal operating range, constant value, non-existent, corrupt or, the speed signal could be incorrect for the operating conditions. Possible fault in the wiring between the control units e.g. short / open circuits or high circuit resistance; interference from other circuits can also affect signals. If there is not a detectable fault in the wiring, it is possible there is an engine speed sensor fault; Refer to list of sensor / actuator checks on page 19.
P0738	TCM Engine Speed Output Circuit Low	The fault code is likely to relate to the speed signal passing between the transmission control unit and another control unit e.g. engine control unit. Other fault codes are used to identify engine speed sensor faults, therefore check for other codes that could indicate an engine speed sensor problem	The fault code indicates that there is an engine speed circuit fault or an incorrect speed signal which could be due to a wiring / connection fault or due to a control unit fault (transmission control unit or other control unit / system connected to the speed signal circuit). The voltage, frequency or value of the sensor signal is either "zero" (no signal) or, the signal exists but it is below the normal operating range (lower than the minimum operating limit). Possible fault in the wiring between the control units e.g. short / open circuits or high circuit resistance; interference from other circuits can also affect signals. If there is not a detectable fault in the wiring, it is possible there is an engine speed sensor fault; Refer to list of sensor / actuator checks on page 19.
P0739	TCM Engine Speed Output Circuit High	The fault code is likely to relate to the speed signal passing between the transmission control unit and another control unit e.g. engine control unit. Other fault codes are used to identify engine speed sensor faults, therefore check for other codes that could indicate an engine speed sensor problem	The fault code indicates that there is an engine speed circuit fault or an incorrect speed signal which could be due to a wiring / connection fault or due to a control unit fault (transmission control unit or other control unit / system connected to the speed signal circuit). The voltage, frequency or value of the sensor signal is either at full value (e.g. battery voltage with no signal or frequency) or, the signal exists but it is above the normal operating range (higher than the maximum operating limit). Possible fault in the wiring between the control units e.g. short / open circuits; interference from other circuits can also affect signals. If there is not a detectable fault in the wiring, it is possible there is an engine speed sensor fault; Refer to list of sensor / actuator checks on page 19.
P0740	Torque Converter Clutch Circuit / Open	The torque converter clutch is used to effectively lock the input and output sides of the converter during certain driving conditions; the clutch is usually operated by hydraulic pressure controlled by a clutch actuator (usually a solenoid valve). The control unit monitors the converter input speed (engine rpm) and the output speed to determine the amount of slip.	The voltage, frequency or value of the control signal in the actuator circuit is incorrect. Signal could be out of normal operating range, constant value, non-existent, corrupt. The fault code indicates a possible "open circuit" but other undefined faults in the actuator or wiring could activate the same code e.g. actuator power supply, short circuit, high circuit resistance or incorrect actuator resistance. Note: the control unit could be providing the correct signal but it is being affected by the circuit fault. Refer to list of sensor / actuator checks on page 19.
P0741	Torque Converter Clutch Circuit Performance / Stuck Off	The torque converter clutch is used to effectively lock the input and output sides of the converter during certain driving conditions; the clutch is usually operated by hydraulic pressure controlled by a clutch actuator (usually a solenoid valve). The control unit monitors the converter input speed (engine rpm) and the output speed to determine the amount of slip.	The control unit is providing a control signal to the actuator (signal is within the normal operating range / tolerance) but the control unit has detected that the actuator / clutch is stuck in a fixed (off) position and is not responding to the control signal. The fault could be related to an actuator / wiring fault. Refer to list of sensor / actuator checks on page 19. However, the fault could be detected because of incorrect or no response of the torque converter clutch; it is likely that the control signal is requesting that the actuator should be "on" (there should be very little or zero slip in the torque converter) but the torque converter clutch appears to remain "off" (allowing slip to exist in the torque converter); the assessment is that the converter clutch is stuck "off".

NOTE 1. Check for other fault codes that could provide additional information.   **NOTE 2.** Communication between control units can pass via a CAN-Bus system; refer to Page 51 for CAN-Bus checks.   178

NOTE 3. If a fault cannot be located, it is also possible that the control unit is at fault.   **NOTE 4.** Refer to Page 53 for list of pages that show the ISO standard for component locations e.g. Sensor A - Bank 1.

code	Description		
P0742	Torque Converter Clutch Circuit Stuck On	The torque converter clutch is used to effectively lock the input and output sides of the converter during certain driving conditions; the clutch is usually operated by hydraulic pressure controlled by a clutch actuator (usually a solenoid valve). The control unit monitors the converter input speed (engine rpm) and the output speed to determine the amount of slip.	The control unit is providing a control signal to the actuator (signal is within the normal operating range / tolerance) but the control unit has detected that the actuator / clutch is stuck in a fixed (off) position and is not responding to the control signal. The fault could be related to an actuator / wiring fault. Refer to list of sensor / actuator checks on page 19. However, the fault could be detected because of incorrect or no response of the torque converter clutch; it is likely that the control signal is requesting that the actuator should be "off" (allowing slip to exist in the torque converter) but the torque converter clutch appears to remain "on" (there is very little or zero slip); the assessment is that the converter clutch is stuck on.
P0743	Torque Converter Clutch Circuit Electrical	The torque converter clutch is used to effectively lock the input and output sides of the converter during certain driving conditions; the clutch is usually operated by hydraulic pressure controlled by a clutch actuator (usually a solenoid valve). The control unit monitors the converter input speed (engine rpm) and the output speed to determine the amount of slip.	The voltage, frequency or value of the control signal in the actuator circuit is incorrect (undefined electrical fault). Signal could be out of normal operating range, constant value, non-existent, corrupt. Possible fault with actuator or wiring e.g. actuator power supply, short / open circuit, high circuit resistance or incorrect actuator resistance. Note: the control unit could be providing the correct signal but it is being affected by the circuit fault. Refer to list of sensor / actuator checks on page 19.
P0744	Torque Converter Clutch Circuit Intermittent	The torque converter clutch is used to effectively lock the input and output sides of the converter during certain driving conditions; the clutch is usually operated by hydraulic pressure controlled by a clutch actuator (usually a solenoid valve). The control unit monitors the converter input speed (engine rpm) and the output speed to determine the amount of slip.	The voltage, frequency or value of the control signal in the actuator circuit is intermittent (likely to be out of normal operating range / tolerance when intermittent fault is detected) or, the signal is erratic e.g. signal changes are irregular / unpredictable. Possible fault with actuator or wiring (e.g. internal / external connections, broken wiring) that is causing intermittent or erratic signal to exist. Note: the control unit could be providing the correct signal but it is being affected by the circuit fault. Refer to list of sensor / actuator checks on page 19.
P0745	Pressure Control Solenoid "A"	Pressure control solenoids are used in the transmission to regulate the pressure in the hydraulic circuits (usually for gear shift control). The pressure control solenoid can also influence / control the actual gear shift. Refer to vehicle specific information to establish which gear(s) are applicable to the identified solenoid.	The control unit has detected an undefined fault with the pressure control solenoid system; the fault could be detected due to an electrical problem with the solenoid / wiring (e.g. solenoid power supply, short / open circuit or resistance). Refer to list of sensor / actuator checks on page 19. The fault could also be detected because sensors are detecting incorrect gear shift operation; check operation of any position sensors (if applicable), check hydraulic pressures and gear selection mechanism.
P0746	Pressure Control Solenoid "A" Performance / Stuck Off	Pressure control solenoids are used in the transmission to regulate the pressure in the hydraulic circuits (usually for gear shift control). The pressure control solenoid can also influence / control the actual gear shift. Refer to vehicle specific information to establish which gear(s) are applicable to the identified solenoid.	The control unit is providing a control signal to the solenoid (signal is within the normal operating range / tolerance) but the control unit has detected that the solenoid is stuck in a fixed (off) position and is not responding to the control signal. The fault could be related to a solenoid / wiring fault. Refer to list of sensor / actuator checks on page 19. Note that the control unit could be detecting the fault because of "a lack of response / no response" from the system or mechanism being controlled by the solenoid; check system (as well as the solenoid) for correct operation, movement and response. A position / pressure sensor (if fitted) could be indicating the "stuck position"; also check sensor operation.
P0747	Pressure Control Solenoid "A" Stuck On	Pressure control solenoids are used in the transmission to regulate the pressure in the hydraulic circuits (usually for gear shift control). The pressure control solenoid can also influence / control the actual gear shift. Refer to vehicle specific information to establish which gear(s) are applicable to the identified solenoid.	The control unit is providing a control signal to the solenoid (signal is within the normal operating range / tolerance) but the control unit has detected that the solenoid is stuck in a fixed (on) position and is not responding to the control signal. The fault could be related to a solenoid / wiring fault. Refer to list of sensor / actuator checks on page 19. Note that the control unit could be detecting the fault because of "a lack of response / no response" from the system or mechanism being controlled by the solenoid; check system (as well as the solenoid) for correct operation, movement and response. A position / pressure sensor (if fitted) could be indicating the "stuck position"; also check sensor operation.
P0748	Pressure Control Solenoid "A" Electrical	Pressure control solenoids are used in the transmission to regulate the pressure in the hydraulic circuits (usually for gear shift control). The pressure control solenoid can also influence / control the actual gear shift. Refer to vehicle specific information to establish which gear(s) are applicable to the identified solenoid.	The voltage, frequency or value of the control signal in the solenoid circuit is incorrect (undefined electrical fault). Signal could be out of normal operating range, constant value, non-existent, corrupt. Possible fault with solenoid or wiring e.g. solenoid power supply, short / open circuit, high circuit resistance or incorrect solenoid resistance. Note: the control unit could be providing the correct signal but it is being affected by a circuit fault. Refer to list of sensor / actuator checks on page 19.
P0749	Pressure Control Solenoid "A" Intermittent	Pressure control solenoids are used in the transmission to regulate the pressure in the hydraulic circuits (usually for gear shift control). The pressure control solenoid can also influence / control the actual gear shift. Refer to vehicle specific information to establish which gear(s) are applicable to the identified solenoid.	The voltage, frequency or value of the control signal in the solenoid circuit is intermittent (likely to be out of normal operating range / tolerance when intermittent fault is detected) or, the signal is erratic e.g. signal changes are irregular / unpredictable. Possible fault with solenoid or wiring (e.g. internal / external connections, broken wiring) that is causing intermittent or erratic signal to exist. Note: the control unit could be providing the correct signal but it is being affected by a circuit fault. Refer to list of sensor / actuator checks on page 19.
P0750	Shift Solenoid "A"	The shift solenoid is used to control a hydraulic circuit in the transmission, which enables the applicable gear(s) to be selected / deselected. Refer to vehicle specific information to establish which gear(s) are applicable to the identified solenoid.	The control unit has detected an undefined fault with the shift solenoid system; the fault could be detected due to an electrical problem with the solenoid / wiring (e.g. solenoid power supply, short / open circuit or resistance). Refer to list of sensor / actuator checks on page 19. The fault could also be detected because sensors are detecting incorrect gear shift operation; check operation of any position sensors (if applicable), check hydraulic pressures and gear selection mechanism. Note that gear engagement can be detected by transmission speed sensors which monitor the input and output speeds thus allowing the control unit to calculate if the gear is fully engaged and the time taken to engage a gear (slip in the friction element of the gear engagement system can cause slow engagement or incomplete engagement).
P0751	Shift Solenoid "A" Performance / Stuck Off	The shift solenoid is used to control a hydraulic circuit in the transmission, which enables the applicable gear(s) to be selected / deselected. Refer to vehicle specific information to establish which gear(s) are applicable to the identified solenoid.	The control unit is providing a control signal to the solenoid (signal is within the normal operating range / tolerance) but the control unit has detected that the solenoid is stuck in a fixed (off) position and is not responding to the control signal. The fault could be related to a solenoid / wiring fault. Refer to list of sensor / actuator checks on page 19. Note that the control unit could be detecting the fault because of "a lack of response / no response" from the system or mechanism being controlled by the solenoid; check system (as well as the solenoid) for correct operation, movement and response. A position sensor (if fitted) could be indicating the "stuck position"; also check position sensor operation. Note that gear engagement can be detected by transmission speed sensors which monitor the input and output speeds thus allowing the control unit to calculate if the gear is fully engaged and the time taken to engage a gear (slip in the friction element of the gear engagement system can cause slow engagement or incomplete engagement).

NOTE 1. Check for other fault codes that could provide additional information. NOTE 2. Communication between control units can pass via a CAN-Bus system; refer to Page 51 for CAN-Bus checks.

NOTE 3. If a fault cannot be located, it is also possible that the control unit is at fault. NOTE 4. Refer to Page 53 for list of pages that show the ISO standard for component locations e.g. Sensor A - Bank 1.

Fault code	EOBD / ISO Description	Component / System Description	Meaningful Description and Quick Check
P0752	Shift Solenoid "A" Stuck On	The shift solenoid is used to control a hydraulic circuit in the transmission, which enables the applicable gear(s) to be selected / deselected. Refer to vehicle specific information to establish which gear(s) are applicable to the identified solenoid.	The control unit is providing a control signal to the solenoid (signal is within the normal operating range / tolerance) but the control unit has detected that the solenoid is stuck in a fixed (on) position and is not responding to the control signal. The fault could be related to a solenoid / wiring fault. Refer to list of sensor / actuator checks on page 19. Note that the control unit could be detecting the fault because of "a lack of response / no response" from the system or mechanism being controlled by the solenoid; check system (as well as the solenoid) for correct operation, movement and response. A position sensor (if fitted) could be indicating the "stuck position"; also check position sensor operation.
P0753	Shift Solenoid "A" Electrical	The shift solenoid is used to control a hydraulic circuit in the transmission, which enables the applicable gear(s) to be selected / deselected. Refer to vehicle specific information to establish which gear(s) are applicable to the identified solenoid.	The voltage, frequency or value of the control signal in the solenoid circuit is incorrect (undefined electrical fault). Signal could be out of normal operating range, constant value, non-existent, corrupt. Possible fault with solenoid or wiring e.g. solenoid power supply, short / open circuit, high circuit resistance or incorrect solenoid resistance. Note: the control unit could be providing the correct signal but it is being affected by the circuit fault. Refer to list of sensor / actuator checks on page 19.
P0754	Shift Solenoid "A" Intermittent	The shift solenoid is used to control a hydraulic circuit in the transmission, which enables the applicable gear(s) to be selected / deselected. Refer to vehicle specific information to establish which gear(s) are applicable to the identified solenoid.	The voltage, frequency or value of the control signal in the solenoid circuit is intermittent (likely to be out of normal operating range / tolerance when intermittent fault is detected) or, the signal is erratic e.g. signal changes are irregular / unpredictable. Possible fault with solenoid or wiring (e.g. internal / external connections, broken wiring) that is causing intermittent or erratic signal to exist. Note: the control unit could be providing the correct signal but it is being affected by a circuit fault. Refer to list of sensor / actuator checks on page 19.
P0755	Shift Solenoid "B"	The shift solenoid is used to control a hydraulic circuit in the transmission, which enables the applicable gear(s) to be selected / deselected. Refer to vehicle specific information to establish which gear(s) are applicable to the identified solenoid.	The control unit has detected an undefined fault with the shift solenoid system; the fault could be detected due to an electrical problem with the solenoid / wiring (e.g. solenoid power supply, short / open circuit or resistance). Refer to list of sensor / actuator checks on page 19. The fault could also be detected because sensors are detecting incorrect gear shift operation; check operation of any position sensors (if applicable), check hydraulic pressures and gear selection mechanism. Note that gear engagement can be detected by transmission speed sensors which monitor the input and output speeds thus allowing the control unit to calculate if the gear is fully engaged and the time taken to engage a gear (slip in the friction element of the gear engagement system can cause slow engagement or incomplete engagement).
P0756	Shift Solenoid "B" Performance / Stuck Off	The shift solenoid is used to control a hydraulic circuit in the transmission, which enables the applicable gear(s) to be selected / deselected. Refer to vehicle specific information to establish which gear(s) are applicable to the identified solenoid.	The control unit is providing a control signal to the solenoid (signal is within the normal operating range / tolerance) but the control unit has detected that the solenoid is stuck in a fixed (off) position and is not responding to the control signal. The fault could be related to a solenoid / wiring fault. Refer to list of sensor / actuator checks on page 19. Note that the control unit could be detecting the fault because of "a lack of response / no response" from the system or mechanism being controlled by the solenoid; check system (as well as the solenoid) for correct operation, movement and response. A position sensor (if fitted) could be indicating the "stuck position"; also check position sensor operation. Note that gear engagement can be detected by transmission speed sensors which monitor the input and output speeds thus allowing the control unit to calculate if the gear is fully engaged and the time taken to engage a gear (slip in the friction element of the gear engagement system can cause slow engagement or incomplete engagement).
P0757	Shift Solenoid "B" Stuck On	The shift solenoid is used to control a hydraulic circuit in the transmission, which enables the applicable gear(s) to be selected / deselected. Refer to vehicle specific information to establish which gear(s) are applicable to the identified solenoid.	The control unit is providing a control signal to the solenoid (signal is within the normal operating range / tolerance) but the control unit has detected that the solenoid is stuck in a fixed (on) position and is not responding to the control signal. The fault could be related to a solenoid / wiring fault. Refer to list of sensor / actuator checks on page 19. Note that the control unit could be detecting the fault because of "a lack of response / no response" from the system or mechanism being controlled by the solenoid; check system (as well as the solenoid) for correct operation, movement and response. A position sensor (if fitted) could be indicating the "stuck position"; also check position sensor operation.
P0758	Shift Solenoid "B" Electrical	The shift solenoid is used to control a hydraulic circuit in the transmission, which enables the applicable gear(s) to be selected / deselected. Refer to vehicle specific information to establish which gear(s) are applicable to the identified solenoid.	The voltage, frequency or value of the control signal in the solenoid circuit is incorrect (undefined electrical fault). Signal could be out of normal operating range, constant value, non-existent, corrupt. Possible fault with solenoid or wiring e.g. solenoid power supply, short / open circuit, high circuit resistance or incorrect solenoid resistance. Note: the control unit could be providing the correct signal but it is being affected by the circuit fault. Refer to list of sensor / actuator checks on page 19.
P0759	Shift Solenoid "B" Intermittent	The shift solenoid is used to control a hydraulic circuit in the transmission, which enables the applicable gear(s) to be selected / deselected. Refer to vehicle specific information to establish which gear(s) are applicable to the identified solenoid.	The voltage, frequency or value of the control signal in the solenoid circuit is intermittent (likely to be out of normal operating range / tolerance when intermittent fault is detected) or, the signal is erratic e.g. signal changes are irregular / unpredictable. Possible fault with solenoid or wiring (e.g. internal / external connections, broken wiring) that is causing intermittent or erratic signal to exist. Note: the control unit could be providing the correct signal but it is being affected by a circuit fault. Refer to list of sensor / actuator checks on page 19.
P075A	Shift Solenoid "G"	The shift solenoid is used to control a hydraulic circuit in the transmission, which enables the applicable gear(s) to be selected / deselected. Refer to vehicle specific information to establish which gear(s) are applicable to the identified solenoid.	The control unit has detected an undefined fault with the shift solenoid system; the fault could be detected due to an electrical problem with the solenoid / wiring (e.g. solenoid power supply, short / open circuit or resistance). Refer to list of sensor / actuator checks on page 19. The fault could also be detected because sensors are detecting incorrect gear shift operation; check operation of any position sensors (if applicable), check hydraulic pressures and gear selection mechanism. Note that gear engagement can be detected by transmission speed sensors which monitor the input and output speeds thus allowing the control unit to calculate if the gear is fully engaged and the time taken to engage a gear (slip in the friction element of the gear engagement system can cause slow engagement or incomplete engagement).

NOTE 1. Check for other fault codes that could provide additional information. **NOTE 2.** Communication between control units can pass via a CAN-Bus system; refer to Page 51 for CAN-Bus checks.
NOTE 3. If a fault cannot be located, it is also possible that the control unit is at fault. **NOTE 4.** Refer to Page 53 for list of pages that show the ISO standard for component locations e.g. Sensor A - Bank 1.

code	Description		
P075B	Shift Solenoid "G" Performance / Stuck Off	The shift solenoid is used to control a hydraulic circuit in the transmission, which enables the applicable gear(s) to be selected / deselected. Refer to vehicle specific information to establish which gear(s) are applicable to the identified solenoid.	The control unit is providing a control signal to the solenoid (signal is within the normal operating range / tolerance) but the control unit has detected that the solenoid is stuck in a fixed (off) position and is not responding to the control signal. The fault could be related to a solenoid / wiring fault. Refer to list of sensor / actuator checks on page 19. Note that the control unit could be detecting the fault because of "a lack of response / no response" from the system or mechanism being controlled by the solenoid; check system (as well as the solenoid) for correct operation, movement and response. A position sensor (if fitted) could be indicating the "stuck position"; also check position sensor operation. Note that gear engagement can be detected by transmission speed sensors which monitor the input and output speeds thus allowing the control unit to calculate if the gear is fully engaged and the time taken to engage a gear (slip in the friction element of the gear engagement system can cause slow engagement or incomplete engagement).
P075C	Shift Solenoid "G" Stuck On	The shift solenoid is used to control a hydraulic circuit in the transmission, which enables the applicable gear(s) to be selected / deselected. Refer to vehicle specific information to establish which gear(s) are applicable to the identified solenoid.	The control unit is providing a control signal to the solenoid (signal is within the normal operating range / tolerance) but the control unit has detected that the solenoid is stuck in a fixed (on) position and is not responding to the control signal. The fault could be related to a solenoid / wiring fault. Refer to list of sensor / actuator checks on page 19. Note that the control unit could be detecting the fault because of "a lack of response / no response" from the system or mechanism being controlled by the solenoid; check system (as well as the solenoid) for correct operation, movement and response. A position sensor (if fitted) could be indicating the "stuck position"; also check position sensor operation. Note that gear engagement can be detected by transmission speed sensors which monitor the input and output speeds thus allowing the control unit to calculate if the gear is fully engaged and the time taken to engage a gear (slip in the friction element of the gear engagement system can cause slow engagement or incomplete engagement).
P075D	Shift Solenoid "G" Electrical	The shift solenoid is used to control a hydraulic circuit in the transmission, which enables the applicable gear(s) to be selected / deselected. Refer to vehicle specific information to establish which gear(s) are applicable to the identified solenoid.	The voltage, frequency or value of the control signal in the solenoid circuit is incorrect (undefined electrical fault). Signal could be out of normal operating range, constant value, non-existent, corrupt. Possible fault with solenoid or wiring e.g. solenoid power supply, short / open circuit, high circuit resistance or incorrect solenoid resistance. Note: the control unit could be providing the correct signal but it is being affected by the circuit fault. Refer to list of sensor / actuator checks on page 19.
P075E	Shift Solenoid "G" Intermittent	The shift solenoid is used to control a hydraulic circuit in the transmission, which enables the applicable gear(s) to be selected / deselected. Refer to vehicle specific information to establish which gear(s) are applicable to the identified solenoid.	The voltage, frequency or value of the control signal in the solenoid circuit is intermittent (likely to be out of normal operating range / tolerance when intermittent fault is detected) or, the signal is erratic e.g. signal changes are irregular / unpredictable. Possible fault with solenoid or wiring (e.g. internal / external connections, broken wiring) that is causing intermittent or erratic signal to exist. Note: the control unit could be providing the correct signal but it is being affected by a circuit fault. Refer to list of sensor / actuator checks on page 19.
P075F		ISO / SAE reserved	Not used, reserved for future allocation.
P0760	Shift Solenoid "C"	The shift solenoid is used to control a hydraulic circuit in the transmission, which enables the applicable gear(s) to be selected / deselected. Refer to vehicle specific information to establish which gear(s) are applicable to the identified solenoid.	The control unit has detected an undefined fault with the shift solenoid system; the fault could be detected due to an electrical problem with the solenoid / wiring (e.g. solenoid power supply, short / open circuit or resistance). Refer to list of sensor / actuator checks on page 19. The fault could also be detected because sensors are detecting incorrect gear shift operation; check operation of any position sensors (if applicable), check hydraulic pressures and gear selection mechanism. Note that gear engagement can be detected by transmission speed sensors which monitor the input and output speeds thus allowing the control unit to calculate if the gear is fully engaged and the time taken to engage a gear (slip in the friction element of the gear engagement system can cause slow engagement or incomplete engagement).
P0761	Shift Solenoid "C" Performance / Stuck Off	The shift solenoid is used to control a hydraulic circuit in the transmission, which enables the applicable gear(s) to be selected / deselected. Refer to vehicle specific information to establish which gear(s) are applicable to the identified solenoid.	The control unit is providing a control signal to the solenoid (signal is within the normal operating range / tolerance) but the control unit has detected that the solenoid is stuck in a fixed (off) position and is not responding to the control signal. The fault could be related to a solenoid / wiring fault. Refer to list of sensor / actuator checks on page 19. Note that the control unit could be detecting the fault because of "a lack of response / no response" from the system or mechanism being controlled by the solenoid; check system (as well as the solenoid) for correct operation, movement and response. A position sensor (if fitted) could be indicating the "stuck position"; also check position sensor operation. Note that gear engagement can be detected by transmission speed sensors which monitor the input and output speeds thus allowing the control unit to calculate if the gear is fully engaged and the time taken to engage a gear (slip in the friction element of the gear engagement system can cause slow engagement or incomplete engagement).
P0762	Shift Solenoid "C" Stuck On	The shift solenoid is used to control a hydraulic circuit in the transmission, which enables the applicable gear(s) to be selected / deselected. Refer to vehicle specific information to establish which gear(s) are applicable to the identified solenoid.	The control unit is providing a control signal to the solenoid (signal is within the normal operating range / tolerance) but the control unit has detected that the solenoid is stuck in a fixed (on) position and is not responding to the control signal. The fault could be related to a solenoid / wiring fault. Refer to list of sensor / actuator checks on page 19. Note that the control unit could be detecting the fault because of "a lack of response / no response" from the system or mechanism being controlled by the solenoid; check system (as well as the solenoid) for correct operation, movement and response. A position sensor (if fitted) could be indicating the "stuck position"; also check position sensor operation.
P0763	Shift Solenoid "C" Electrical	The shift solenoid is used to control a hydraulic circuit in the transmission, which enables the applicable gear(s) to be selected / deselected. Refer to vehicle specific information to establish which gear(s) are applicable to the identified solenoid.	The voltage, frequency or value of the control signal in the solenoid circuit is incorrect (undefined electrical fault). Signal could be:- out of normal operating range, constant value, non-existent, corrupt. Possible fault with solenoid or wiring e.g. solenoid power supply, short / open circuit, high circuit resistance or incorrect solenoid resistance. Note: the control unit could be providing the correct signal but it is being affected by the circuit fault. Refer to list of sensor / actuator checks on page 19.
P0764	Shift Solenoid "C" Intermittent	The shift solenoid is used to control a hydraulic circuit in the transmission, which enables the applicable gear(s) to be selected / deselected. Refer to vehicle specific information to establish which gear(s) are applicable to the identified solenoid.	The voltage, frequency or value of the control signal in the solenoid circuit is intermittent (likely to be out of normal operating range / tolerance when intermittent fault is detected) or, the signal is erratic e.g. signal changes are irregular / unpredictable. Possible fault with solenoid or wiring (e.g. internal / external connections, broken wiring) that is causing intermittent or erratic signal to exist. Note: the control unit could be providing the correct signal but it is being affected by a circuit fault. Refer to list of sensor / actuator checks on page 19.

NOTE 1. Check for other fault codes that could provide additional information.   **NOTE 2.** Communication between control units can pass via a CAN-Bus system; refer to Page 51 for CAN-Bus checks.
NOTE 3. If a fault cannot be located, it is also possible that the control unit is at fault.   **NOTE 4.** Refer to Page 53 for list of pages that show the ISO standard for component locations e.g. Sensor A - Bank 1.

Fault code	EOBD / ISO Description	Component / System Description	Meaningful Description and Quick Check
P0765	Shift Solenoid "D"	The shift solenoid is used to control a hydraulic circuit in the transmission, which enables the applicable gear(s) to be selected / deselected. Refer to vehicle specific information to establish which gear(s) are applicable to the identified solenoid.	The control unit has detected an undefined fault with the shift solenoid system; the fault could be detected due to an electrical problem with the solenoid / wiring (e.g. solenoid power supply, short / open circuit or resistance). Refer to list of sensor / actuator checks on page 19. The fault could also be detected because sensors are detecting incorrect gear shift operation; check operation of any position sensors (if applicable), check hydraulic pressures and gear selection mechanism. Note that gear engagement can be detected by transmission speed sensors which monitor the input and output speeds thus allowing the control unit to calculate if the gear is fully engaged and the time taken to engage a gear (slip in the friction element of the gear engagement system can cause slow engagement or incomplete engagement).
P0766	Shift Solenoid "D" Performance / Stuck Off	The shift solenoid is used to control a hydraulic circuit in the transmission, which enables the applicable gear(s) to be selected / deselected. Refer to vehicle specific information to establish which gear(s) are applicable to the identified solenoid.	The control unit is providing a control signal to the solenoid (signal is within the normal operating range / tolerance) but the control unit has detected that the solenoid is stuck in a fixed (off) position and is not responding to the control signal. The fault could be related to a solenoid / wiring fault. Refer to list of sensor / actuator checks on page 19. Note that the control unit could be detecting the fault because of "a lack of response / no response" from the system or mechanism being controlled by the solenoid; check system (as well as the solenoid) for correct operation, movement and response. A position sensor (if fitted) could be indicating the "stuck position"; also check position sensor operation. Note that gear engagement can be detected by transmission speed sensors which monitor the input and output speeds thus allowing the control unit to calculate if the gear is fully engaged and the time taken to engage a gear (slip in the friction element of the gear engagement system can cause slow engagement or incomplete engagement).
P0767	Shift Solenoid "D" Stuck On	The shift solenoid is used to control a hydraulic circuit in the transmission, which enables the applicable gear(s) to be selected / deselected. Refer to vehicle specific information to establish which gear(s) are applicable to the identified solenoid.	The control unit is providing a control signal to the solenoid (signal is within the normal operating range / tolerance) but the control unit has detected that the solenoid is stuck in a fixed (on) position and is not responding to the control signal. The fault could be related to a solenoid / wiring fault. Refer to list of sensor / actuator checks on page 19. Note that the control unit could be detecting the fault because of "a lack of response / no response" from the system or mechanism being controlled by the solenoid; check system (as well as the solenoid) for correct operation, movement and response. A position sensor (if fitted) could be indicating the "stuck position"; also check position sensor operation.
P0768	Shift Solenoid "D" Electrical	The shift solenoid is used to control a hydraulic circuit in the transmission, which enables the applicable gear(s) to be selected / deselected. Refer to vehicle specific information to establish which gear(s) are applicable to the identified solenoid.	The voltage, frequency or value of the control signal in the solenoid circuit is incorrect (undefined electrical fault). Signal could be:- out of normal operating range, constant value, non-existent, corrupt. Possible fault with solenoid or wiring e.g. solenoid power supply, short / open circuit, high circuit resistance or incorrect solenoid resistance. Note: the control unit could be providing the correct signal but it is being affected by the circuit fault. Refer to list of sensor / actuator checks on page 19.
P0769	Shift Solenoid "D" Intermittent	The shift solenoid is used to control a hydraulic circuit in the transmission, which enables the applicable gear(s) to be selected / deselected. Refer to vehicle specific information to establish which gear(s) are applicable to the identified solenoid.	The voltage, frequency or value of the control signal in the solenoid circuit is intermittent (likely to be out of normal operating range / tolerance when intermittent fault is detected) or, the signal is erratic e.g. signal changes are irregular / unpredictable. Possible fault with solenoid or wiring (e.g. internal / external connections, broken wiring) that is causing intermittent or erratic signal to exist. Note: the control unit could be providing the correct signal but it is being affected by a circuit fault. Refer to list of sensor / actuator checks on page 19.
P076A	Shift Solenoid "H"	The shift solenoid is used to control a hydraulic circuit in the transmission, which enables the applicable gear(s) to be selected / deselected. Refer to vehicle specific information to establish which gear(s) are applicable to the identified solenoid.	The control unit has detected an undefined fault with the shift solenoid system; the fault could be detected due to an electrical problem with the solenoid / wiring (e.g. solenoid power supply, short / open circuit or resistance). Refer to list of sensor / actuator checks on page 19. The fault could also be detected because sensors are detecting incorrect gear shift operation; check operation of any position sensors (if applicable), check hydraulic pressures and gear selection mechanism. Note that gear engagement can be detected by transmission speed sensors which monitor the input and output speeds thus allowing the control unit to calculate if the gear is fully engaged and the time taken to engage a gear (slip in the friction element of the gear engagement system can cause slow engagement or incomplete engagement).
P076B	Shift Solenoid "H" Performance / Stuck Off	The shift solenoid is used to control a hydraulic circuit in the transmission, which enables the applicable gear(s) to be selected / deselected. Refer to vehicle specific information to establish which gear(s) are applicable to the identified solenoid.	The control unit is providing a control signal to the solenoid (signal is within the normal operating range / tolerance) but the control unit has detected that the solenoid is stuck in a fixed (off) position and is not responding to the control signal. The fault could be related to a solenoid / wiring fault. Refer to list of sensor / actuator checks on page 19. Note that the control unit could be detecting the fault because of "a lack of response / no response" from the system or mechanism being controlled by the solenoid; check system (as well as the solenoid) for correct operation, movement and response. A position sensor (if fitted) could be indicating the "stuck position"; also check position sensor operation. Note that gear engagement can be detected by transmission speed sensors which monitor the input and output speeds thus allowing the control unit to calculate if the gear is fully engaged and the time taken to engage a gear (slip in the friction element of the gear engagement system can cause slow engagement or incomplete engagement).
P076C	Shift Solenoid "H" Stuck On	The shift solenoid is used to control a hydraulic circuit in the transmission, which enables the applicable gear(s) to be selected / deselected. Refer to vehicle specific information to establish which gear(s) are applicable to the identified solenoid.	The control unit is providing a control signal to the solenoid (signal is within the normal operating range / tolerance) but the control unit has detected that the solenoid is stuck in a fixed (on) position and is not responding to the control signal. The fault could be related to a solenoid / wiring fault. Refer to list of sensor / actuator checks on page 19. Note that the control unit could be detecting the fault because of "a lack of response / no response" from the system or mechanism being controlled by the solenoid; check system (as well as the solenoid) for correct operation, movement and response. A position sensor (if fitted) could be indicating the "stuck position"; also check position sensor operation.
P076D	Shift Solenoid "H" Electrical	The shift solenoid is used to control a hydraulic circuit in the transmission, which enables the applicable gear(s) to be selected / deselected. Refer to vehicle specific information to establish which gear(s) are applicable to the identified solenoid.	The voltage, frequency or value of the control signal in the solenoid circuit is incorrect (undefined electrical fault). Signal could be:- out of normal operating range, constant value, non-existent, corrupt. Possible fault with solenoid or wiring e.g. solenoid power supply, short / open circuit, high circuit resistance or incorrect solenoid resistance. Note: the control unit could be providing the correct signal but it is being affected by the circuit fault. Refer to list of sensor / actuator checks on page 19.

NOTE 1. Check for other fault codes that could provide additional information. **NOTE 2.** Communication between control units can pass via a CAN-Bus system; refer to Page 51 for CAN-Bus checks.

NOTE 3. If a fault cannot be located, it is also possible that the control unit is at fault. **NOTE 4.** Refer to Page 53 for list of pages that show the ISO standard for component locations e.g. Sensor A - Bank 1.

code	Description		
P076E	Shift Solenoid "H" Intermittent	The shift solenoid is used to control a hydraulic circuit in the transmission, which enables the applicable gear(s) to be selected / deselected. Refer to vehicle specific information to establish which gear(s) are applicable to the identified solenoid.	The voltage, frequency or value of the control signal in the solenoid circuit is intermittent (likely to be out of normal operating range / tolerance when intermittent fault is detected) or, the signal is erratic e.g. signal changes are irregular / unpredictable. Possible fault with solenoid or wiring (e.g. internal / external connections, broken wiring) that is causing intermittent or erratic signal to exist. Note: the control unit could be providing the correct signal but it is being affected by a circuit fault. Refer to list of sensor / actuator checks on page 19.
P076F	Gear 7 Incorrect Ratio	The fault does not necessarily indicate that the wrong gear has been selected, it normally indicates that the gear has been selected but, due to a fault within the transmission (such as a slipping transmission clutch / brake band) the ratio is incorrect. The control unit monitors the speeds of the transmission input and output shafts; when each gear is selected, the input and output speeds should correspond to the selected gear ratio e.g. if the gear ratio is 4 to 1 then the input speed should be 4 times faster than the output speed. If the input and output speeds do not match the selected ratio (for the identified gear), this will be regarded as a fault (likely to be a mechanical / hydraulic pressure fault).	Refer to vehicle specific information to identify the method of changing / gear in the transmission and the components used e.g. shift solenoid valves, pressure control valves etc. Check transmission control system for additional fault codes. Fault is likely to be related to a problem with the gear selection mechanism for the identified gear e.g. a slipping clutch band or brake; note that this could be a mechanical fault within the transmission or a hydraulic pressure control problem. Check for correct operation of gear shift solenoid valves and pressure control valves, also check control pressures. If solenoid and control valves are not operating correctly, check wiring to power supply and to control unit, also check control signal from control unit to valves. Refer to vehicle specific information to identify the method of changing / gear in the transmission and the components used e.g. shift solenoid valves, pressure control valves etc.
P0770	Shift Solenoid "E"	The shift solenoid is used to control a hydraulic circuit in the transmission, which enables the applicable gear(s) to be selected / deselected. Refer to vehicle specific information to establish which gear(s) are applicable to the identified solenoid.	The control unit has detected an undefined fault with the shift solenoid system; the fault could be detected due to an electrical problem with the solenoid / wiring (e.g. solenoid power supply, short / open circuit or resistance). Refer to list of sensor / actuator checks on page 19. The fault could also be detected because sensors are detecting incorrect gear shift operation; check operation of any position sensors (if applicable), check hydraulic pressures and gear selection mechanism. Note that gear engagement can be detected by transmission speed sensors which monitor the input and output speeds thus allowing the control unit to calculate if the gear is fully engaged and the time taken to engage a gear (slip in the friction element of the gear engagement system can cause slow engagement or incomplete engagement).
P0771	Shift Solenoid "E" Performance / Stuck Off	The shift solenoid is used to control a hydraulic circuit in the transmission, which enables the applicable gear(s) to be selected / deselected. Refer to vehicle specific information to establish which gear(s) are applicable to the identified solenoid.	The control unit is providing a control signal to the solenoid (signal is within the normal operating range / tolerance) but the control unit has detected that the solenoid is stuck in a fixed (off) position and is not responding to the control signal. The fault could be related to a solenoid / wiring fault. Refer to list of sensor / actuator checks on page 19. Note that the control unit could be detecting the fault because of "a lack of response / no response" from the system or mechanism being controlled by the solenoid; check system (as well as the solenoid) for correct operation, movement and response. A position sensor (if fitted) could be indicating the "stuck position"; also check position sensor operation. Note that gear engagement can be detected by transmission speed sensors which monitor the input and output speeds thus allowing the control unit to calculate if the gear is fully engaged and the time taken to engage a gear (slip in the friction element of the gear engagement system can cause slow engagement or incomplete engagement).
P0772	Shift Solenoid "E" Stuck On	The shift solenoid is used to control a hydraulic circuit in the transmission, which enables the applicable gear(s) to be selected / deselected. Refer to vehicle specific information to establish which gear(s) are applicable to the identified solenoid.	The control unit is providing a control signal to the solenoid (signal is within the normal operating range / tolerance) but the control unit has detected that the solenoid is stuck in a fixed (on) position and is not responding to the control signal. The fault could be related to a solenoid / wiring fault. Refer to list of sensor / actuator checks on page 19. Note that the control unit could be detecting the fault because of "a lack of response / no response" from the system or mechanism being controlled by the solenoid; check system (as well as the solenoid) for correct operation, movement and response. A position sensor (if fitted) could be indicating the "stuck position"; also check position sensor operation.
P0773	Shift Solenoid "E" Electrical	The shift solenoid is used to control a hydraulic circuit in the transmission, which enables the applicable gear(s) to be selected / deselected. Refer to vehicle specific information to establish which gear(s) are applicable to the identified solenoid.	The voltage, frequency or value of the control signal in the solenoid circuit is incorrect (undefined electrical fault). Signal could be:- out of normal operating range, constant value, non-existent, corrupt. Possible fault with solenoid or wiring e.g. solenoid power supply, short / open circuit, high circuit resistance or incorrect solenoid resistance. Note: the control unit could be providing the correct signal but it is being affected by the circuit fault. Refer to list of sensor / actuator checks on page 19.
P0774	Shift Solenoid "E" Intermittent	The shift solenoid is used to control a hydraulic circuit in the transmission, which enables the applicable gear(s) to be selected / deselected. Refer to vehicle specific information to establish which gear(s) are applicable to the identified solenoid.	The voltage, frequency or value of the control signal in the solenoid circuit is intermittent (likely to be out of normal operating range / tolerance when intermittent fault is detected) or, the signal is erratic e.g. signal changes are irregular / unpredictable. Possible fault with solenoid or wiring (e.g. internal / external connections, broken wiring) that is causing intermittent or erratic signal to exist. Note: the control unit could be providing the correct signal but it is being affected by a circuit fault. Refer to list of sensor / actuator checks on page 19.
P0775	Pressure Control Solenoid "B"	Pressure control solenoids are used in the transmission to regulate the pressure in the hydraulic circuits (usually for gear shift control). The pressure control solenoid can also influence / control the actual gear shift. Refer to vehicle specific information to establish which gear(s) are applicable to the identified solenoid.	The control unit has detected an undefined fault with the pressure control solenoid system; the fault could be detected due to an electrical problem with the solenoid / wiring (e.g. solenoid power supply, short / open circuit or resistance). Refer to list of sensor / actuator checks on page 19. The fault could also be detected because sensors are detecting incorrect gear shift operation; check operation of any position sensors (if applicable), check hydraulic pressures and gear selection mechanism.
P0776	Pressure Control Solenoid "B" Performance / Stuck Off	Pressure control solenoids are used in the transmission to regulate the pressure in the hydraulic circuits (usually for gear shift control). The pressure control solenoid can also influence / control the actual gear shift. Refer to vehicle specific information to establish which gear(s) are applicable to the identified solenoid.	The control unit is providing a control signal to the solenoid (signal is within the normal operating range / tolerance) but the control unit has detected that the solenoid is stuck in a fixed (off) position and is not responding to the control signal. The fault could be related to a solenoid / wiring fault. Refer to list of sensor / actuator checks on page 19. Note that the control unit could be detecting the fault because of "a lack of response / no response" from the system or mechanism being controlled by the solenoid; check system (as well as the solenoid) for correct operation, movement and response. A position / pressure sensor (if fitted) could be indicating the "stuck position"; also check sensor operation.

NOTE 1. Check for other fault codes that could provide additional information.   **NOTE 2.** Communication between control units can pass via a CAN-Bus system; refer to Page 51 for CAN-Bus checks.

NOTE 3. If a fault cannot be located, it is also possible that the control unit is at fault.   **NOTE 4.** Refer to Page 53 for list of pages that show the ISO standard for component locations e.g. Sensor A - Bank 1.

Fault code	EOBD / ISO Description	Component / System Description	Meaningful Description and Quick Check
P0777	Pressure Control Solenoid "B" Stuck On	Pressure control solenoids are used in the transmission to regulate the pressure in the hydraulic circuits (usually for gear shift control). The pressure control solenoid can also influence / control the actual gear shift. Refer to vehicle specific information to establish which gear(s) are applicable to the identified solenoid.	The control unit is providing a control signal to the solenoid (signal is within the normal operating range / tolerance) but the control unit has detected that the solenoid is stuck in a fixed (on) position and is not responding to the control signal. The fault could be related to a solenoid / wiring fault. Refer to list of sensor / actuator checks on page 19. Note that the control unit could be detecting the fault because of "a lack of response / no response" from the system or mechanism being controlled by the solenoid; check system (as well as the solenoid) for correct operation, movement and response. A position / pressure sensor (if fitted) could be indicating the "stuck position"; also check sensor operation.
P0778	Pressure Control Solenoid "B" Electrical	Pressure control solenoids are used in the transmission to regulate the pressure in the hydraulic circuits (usually for gear shift control). The pressure control solenoid can also influence / control the actual gear shift. Refer to vehicle specific information to establish which gear(s) are applicable to the identified solenoid.	The voltage, frequency or value of the control signal in the solenoid circuit is incorrect (undefined electrical fault). Signal could be:- out of normal operating range, constant value, non-existent, corrupt. Possible fault with solenoid or wiring e.g. solenoid power supply, short / open circuit, high circuit resistance or incorrect solenoid resistance. Note: the control unit could be providing the correct signal but it is being affected by a circuit fault. Refer to list of sensor / actuator checks on page 19.
P0779	Pressure Control Solenoid "B" Intermittent	Pressure control solenoids are used in the transmission to regulate the pressure in the hydraulic circuits (usually for gear shift control). The pressure control solenoid can also influence / control the actual gear shift. Refer to vehicle specific information to establish which gear(s) are applicable to the identified solenoid.	The voltage, frequency or value of the control signal in the solenoid circuit is intermittent (likely to be out of normal operating range / tolerance when intermittent fault is detected) or, the signal is erratic e.g. signal changes are irregular / unpredictable. Possible fault with solenoid or wiring (e.g. internal / external connections, broken wiring) that is causing intermittent or erratic signal to exist. Note: the control unit could be providing the correct signal but it is being affected by a circuit fault. Refer to list of sensor / actuator checks on page 19.
P0780	Shift Error	The control unit monitors the input speed and output speed in the transmission to establish that the requested gear has been selected or is being selected (gear is requested by control unit).	The fault code indicates that there is an unidentified error when shifting / gear changing (between unidentified gears). The fault could be identified due to information from the input and output speed sensors or, due to information from other sensors e.g. position sensor on the gear shift mechanism. It will be necessary to identify the components fitted to the transmission, but checks will need to be carried out on actuators and sensors (as applicable to the particular transmission system). Refer to list of sensor / actuator checks on page 19. Note that it is also possible that the fault code has been activated due to a mechanical gear selection fault e.g. gear change clutch failure or other component fault; also check transmission hydraulic system for pressures etc.
P0781	1-2 Shift	The control unit monitors the input speed and output speed in the transmission to establish that the requested gear has been selected or is being selected (gear is requested by control unit). Some systems may use a position sensor to identify movement of the gear selection mechanism.	The fault code indicates that the identified gear(s) requested by the control unit is not selected. The fault could be on the up shift or the downshift. Identify system operation and components. Check actuators and control signals e.g. shift / solenoid valves etc. Check any applicable position, speed or pressure sensors and sensor signals. Refer to list of sensor / actuator checks on page 19. Note that it is also possible that the fault code has been activated due to a mechanical gear selection fault e.g. gear change clutch failure or other component fault; also check transmission hydraulic system for pressures etc.
P0782	2-3 Shift	The control unit monitors the input speed and output speed in the transmission to establish that the requested gear has been selected or is being selected (gear is requested by control unit). Some systems may use a position sensor to identify movement of the gear selection mechanism.	The fault code indicates that the identified gear(s) requested by the control unit is not selected. The fault could be on the up shift or the downshift. Identify system operation and components. Check actuators and control signals e.g. shift / solenoid valves etc. Check any applicable position, speed or pressure sensors and sensor signals. Refer to list of sensor / actuator checks on page 19. Note that it is also possible that the fault code has been activated due to a mechanical gear selection fault e.g. gear change clutch failure or other component fault; also check transmission hydraulic system for pressures etc.
P0783	3-4 Shift	The control unit monitors the input speed and output speed in the transmission to establish that the requested gear has been selected or is being selected (gear is requested by control unit). Some systems may use a position sensor to identify movement of the gear selection mechanism.	The fault code indicates that the identified gear(s) requested by the control unit is not selected. The fault could be on the up shift or the downshift. Identify system operation and components. Check actuators and control signals e.g. shift / solenoid valves etc. Check any applicable position, speed or pressure sensors and sensor signals. Refer to list of sensor / actuator checks on page 19. Note that it is also possible that the fault code has been activated due to a mechanical gear selection fault e.g. gear change clutch failure or other component fault; also check transmission hydraulic system for pressures etc.
P0784	4-5 Shift	The control unit monitors the input speed and output speed in the transmission to establish that the requested gear has been selected or is being selected (gear is requested by control unit). Some systems may use a position sensor to identify movement of the gear selection mechanism.	The fault code indicates that the identified gear(s) requested by the control unit is not selected. The fault could be on the up shift or the downshift. Identify system operation and components. Check actuators and control signals e.g. shift / solenoid valves etc. Check any applicable position, speed or pressure sensors and sensor signals. Refer to list of sensor / actuator checks on page 19. Note that it is also possible that the fault code has been activated due to a mechanical gear selection fault e.g. gear change clutch failure or other component fault; also check transmission hydraulic system for pressures etc.
P0785	Shift / Timing Solenoid	The shift solenoid is used to control a hydraulic circuit in the transmission, which enables the applicable gear(s) to be selected / deselected. The Solenoid can be referred to as the Shift / Timing solenoid due to the control of the time to engage the gear (e.g. control of hydraulic pressure acting on the friction element on auto transmission gear engagement system).	The control unit has detected an undefined fault with the shift / timing control solenoid system; the fault could be detected due to an electrical problem with the solenoid / wiring (e.g. solenoid power supply, short / open circuit or resistance). Refer to list of sensor / actuator checks on page 19. The fault could be detected due to information from the input and output speed sensors (or position sensors that indicate linkage / mechanism position); check operation of any speed / position sensors (as applicable), check hydraulic pressures and gear selection mechanism. Note that gear engagement can be detected by transmission speed sensors which monitor the input and output speeds thus allowing the control unit to calculate if the gear is fully engaged and the time taken to engage a gear (slip in the friction element of the gear engagement system can cause slow engagement or incomplete engagement).
P0786	Shift / Timing Solenoid Range / Performance	The shift solenoid is used to control a hydraulic circuit in the transmission, which enables the applicable gear(s) to be selected / deselected. The Solenoid can be referred to as the Shift / Timing solenoid due to the control of the time to engage the gear (e.g. control of hydraulic pressure acting on the friction element on auto transmission gear engagement system).	The voltage, frequency or value of the control signal in the solenoid circuit is within the normal operating range / tolerance, but the signal might be incorrect for the operating conditions. It is also possible that the solenoid response (or response of system controlled by the solenoid) differs from the expected / desired response (incorrect response to the control signal provided by the control unit). Possible fault in solenoid or wiring, but also possible that the solenoid (or mechanism / system controlled by the solenoid) is not operating or moving correctly. Note: the control unit could be providing the correct signal but it is being affected by a circuit fault. Refer to list of sensor / actuator checks on page 19. Note that the fault code could also be activated due to mechanical or hydraulic faults e.g. restriction in the movement of the solenoid, a mechanical fault with the gear selection / deselection mechanism, a fault with the hydraulic pressure or valves.

NOTE 1. Check for other fault codes that could provide additional information. NOTE 2. Communication between control units can pass via a CAN-Bus system; refer to Page 51 for CAN-Bus checks. 184
NOTE 3. If a fault cannot be located, it is also possible that the control unit is at fault. NOTE 4. Refer to Page 53 for list of pages that show the ISO standard for component locations e.g. Sensor A - Bank 1.

Code	Description		
P0787	Shift / Timing Solenoid Low	The shift solenoid is used to control a hydraulic circuit in the transmission, which enables the applicable gear(s) to be selected / deselected. The Solenoid can be referred to as the Shift / Timing solenoid due to the control of the time to engage the gear (e.g. control of hydraulic pressure acting on the friction element on auto transmission gear engagement system).	The voltage, frequency or value of the control signal in the solenoid circuit is either "zero" (no signal) or, the signal exists but is below the normal operating range. Possible fault with solenoid or wiring (e.g. solenoid power supply, short / open circuit or resistance) that is causing a signal value to be lower than the minimum operating limit. Note: the control unit could be providing the correct signal but it is being affected by a circuit fault. Refer to list of sensor / actuator checks on page 19.
P0788	Shift / Timing Solenoid High	The shift solenoid is used to control a hydraulic circuit in the transmission, which enables the applicable gear(s) to be selected / deselected. The Solenoid can be referred to as the Shift / Timing solenoid due to the control of the time to engage the gear (e.g. control of hydraulic pressure acting on the friction element on auto transmission gear engagement system).	The voltage, frequency or value of the control signal in the solenoid circuit is either at full value (e.g. battery voltage with no on / off signal) or, the signal exists but it is above the normal operating range. Possible fault with solenoid or wiring (e.g. solenoid power supply, short / open circuit or resistance) that is causing a signal value to be higher than the maximum operating limit. Note: the control unit could be providing the correct signal but it is being affected by the circuit fault. Refer to list of sensor / actuator checks on page 19.
P0789	Shift / Timing Solenoid Intermittent	The shift solenoid is used to control a hydraulic circuit in the transmission, which enables the applicable gear(s) to be selected / deselected. The Solenoid can be referred to as the Shift / Timing solenoid due to the control of the time to engage the gear (e.g. control of hydraulic pressure acting on the friction element on auto transmission gear engagement system).	The voltage, frequency or value of the control signal in the solenoid circuit is intermittent (likely to be out of normal operating range / tolerance when intermittent fault is detected) or, the signal is erratic e.g. signal changes are irregular / unpredictable. Possible fault with solenoid or wiring (e.g. internal / external connections, broken wiring) that is causing intermittent or erratic signal to exist. Note: the control unit could be providing the correct signal but it is being affected by a circuit fault. Refer to list of sensor / actuator checks on page 19.
P0790	Normal / Performance Switch Circuit	The switch used to select different operating modes e.g. sport / performance / economy or other available modes on the transmission system.	The voltage, frequency or value of the switch signal is incorrect (undefined fault). Switch voltage / signal could be: out of normal operating range, constant value, non-existent, corrupt. Possible fault in the switch or wiring e.g. short / open circuits, circuit resistance or, fault with the reference / operating voltage; the switch contacts could also be faulty e.g. not opening / closing correctly or have a high contact resistance. Refer to list of sensor / actuator checks on page 19. Also check mechanism / linkage that operates the switch.
P0791	Intermediate Shaft Speed Sensor "A" Circuit	Depending on the type of transmission, the information from the transmission intermediate shaft speed sensor can be used to indicate correct selection of gears, timing (speed) for gear changes and for any slip in the transmission system (clutches etc).	Identify the type of sensor e.g. Inductive, Hall effect etc; this will dictate the checks that are applicable. The voltage, frequency or value of the speed sensor signal is incorrect (undefined fault). Sensor voltage / signal value could be: out of normal operating range, constant value, non-existent, corrupt. Possible fault in the speed sensor or wiring e.g. short / open circuits, circuit resistance, incorrect sensor resistance (if applicable) or, fault with the reference / operating voltage; interference from other circuits can also affect sensor signals Also check earth connections and cable screening. Refer to list of sensor / actuator checks on page 19.
P0792	Intermediate Shaft Speed Sensor "A" Circuit Range / Performance	Depending on the type of transmission, the information from the transmission intermediate shaft speed sensor can be used to indicate correct selection of gears, timing (speed) for gear changes and for any slip in the transmission system (clutches etc).	Identify the type of sensor e.g. Inductive, Hall effect etc; this will dictate the checks that are applicable. The voltage, frequency or value of the speed sensor signal is within the normal operating range / tolerance but the signal is not plausible or is incorrect due to an undefined fault. The sensor signal might not match the operating conditions indicated by other sensors or, the sensor signal is not changing / responding as expected to the changes in conditions. Possible fault in the speed sensor or wiring e.g. short / open circuits, circuit resistance, incorrect sensor resistance or, fault with the reference / operating voltage; interference from other circuits can also affect sensor signals. Refer to list of sensor / actuator checks on page 19. It is also possible that the speed sensor is operating correctly but the speed being measured by the sensor is unacceptable or incorrect for the operating conditions (e.g. the requested gear is not engaged causing incorrect shaft speed). Also check condition of reference teeth on trigger disc / rotor and ensure rotor is turning with shaft.
P0793	Intermediate Shaft Speed Sensor "A" Circuit No Signal	Depending on the type of transmission, the information from the transmission intermediate shaft speed sensor can be used to indicate correct selection of gears, timing (speed) for gear changes and for any slip in the transmission system (clutches etc).	Identify the type of sensor e.g. Inductive, Hall effect etc; this will dictate the checks that are applicable. The voltage, frequency or value of the speed sensor signal is incorrect (undefined fault resulting in no signal being detected). Sensor voltage / signal value could be: constant value or non-existent. Possible fault in the speed sensor or wiring e.g. short / open circuits, circuit resistance, incorrect sensor resistance (if applicable) or, fault with the reference / operating voltage; interference from other circuits can also affect sensor signals Also check earth connections and cable screening. Refer to list of sensor / actuator checks on page 19. Check to ensure that the trigger rotor is turning with the shaft.
P0794	Intermediate Shaft Speed Sensor "A" Circuit Intermittent	Depending on the type of transmission, the information from the transmission intermediate shaft speed sensor can be used to indicate correct selection of gears, timing (speed) for gear changes and for any slip in the transmission system (clutches etc).	Identify the type of sensor e.g. Inductive, Hall effect etc; this will dictate the checks that are applicable. The voltage, frequency or value of the speed sensor signal is intermittent (likely to be out of normal operating range / tolerance when intermittent fault is detected) or, the signal / voltage is erratic (e.g. signal changes are irregular / unpredictable). Possible fault with wiring or connections (including internal sensor connections); also interference from other circuits can affect the signal. Refer to list of sensor / actuator checks on page 19. Also check condition of reference teeth on trigger disc / rotor and ensure rotor is turning with shaft.
P0795	Pressure Control Solenoid "C"	Pressure control solenoids are used in the transmission to regulate the pressure in the hydraulic circuits (usually for gear shift control). The pressure control solenoid can also influence / control the actual gear shift. Refer to vehicle specific information to establish which gear(s) are applicable to the identified solenoid.	The control unit has detected an undefined fault with the pressure control solenoid system; the fault could be detected due to an electrical problem with the solenoid / wiring (e.g. solenoid power supply, short / open circuit or resistance). Refer to list of sensor / actuator checks on page 19. The fault could also be detected because sensors are detecting incorrect gear shift operation; check operation of any position sensors (if applicable), check hydraulic pressures and gear selection mechanism.
P0796	Pressure Control Solenoid "C" Performance / Stuck Off	Pressure control solenoids are used in the transmission to regulate the pressure in the hydraulic circuits (usually for gear shift control). The pressure control solenoid can also influence / control the actual gear shift. Refer to vehicle specific information to establish which gear(s) are applicable to the identified solenoid.	The control unit is providing a control signal to the solenoid (signal is within the normal operating range / tolerance) but the control unit has detected that the solenoid is stuck in a fixed (off) position and is not responding to the control signal. The fault could be related to a solenoid / wiring fault. Refer to list of sensor / actuator checks on page 19. Note that the control unit could be detecting the fault because of "a lack of response / no response" from the system or mechanism being controlled by the solenoid; check system (as well as the solenoid) for correct operation, movement and response. A position / pressure sensor (if fitted) could be indicating the "stuck position"; also check sensor operation.

NOTE 1. Check for other fault codes that could provide additional information. **NOTE 2.** Communication between control units can pass via a CAN-Bus system; refer to Page 51 for CAN-Bus checks.

NOTE 3. If a fault cannot be located, it is also possible that the control unit is at fault. **NOTE 4.** Refer to Page 53 for list of pages that show the ISO standard for component locations e.g. Sensor A - Bank 1.

Fault code	EOBD / ISO Description	Component / System Description	Meaningful Description and Quick Check
P0797	Pressure Control Solenoid "C" Stuck On	Pressure control solenoids are used in the transmission to regulate the pressure in the hydraulic circuits (usually for gear shift control). The pressure control solenoid can also influence / control the actual gear shift. Refer to vehicle specific information to establish which gear(s) are applicable to the identified solenoid.	The control unit is providing a control signal to the solenoid (signal is within the normal operating range / tolerance) but the control unit has detected that the solenoid is stuck in a fixed (on) position and is not responding to the control signal. The fault could be related to a solenoid / wiring fault. Refer to list of sensor / actuator checks on page 19. Note that the control unit could be detecting the fault because of "a lack of response / no response" from the system or mechanism being controlled by the solenoid; check system (as well as the solenoid) for correct operation, movement and response. A position / pressure sensor (if fitted) could be indicating the "stuck position"; also check sensor operation.
P0798	Pressure Control Solenoid "C" Electrical	Pressure control solenoids are used in the transmission to regulate the pressure in the hydraulic circuits (usually for gear shift control). The pressure control solenoid can also influence / control the actual gear shift. Refer to vehicle specific information to establish which gear(s) are applicable to the identified solenoid.	The voltage, frequency or value of the control signal in the solenoid circuit is incorrect (undefined electrical fault). Signal could be:- out of normal operating range, constant value, non-existent, corrupt. Possible fault with solenoid or wiring e.g. solenoid power supply, short / open circuit, high circuit resistance or incorrect solenoid resistance. Note: the control unit could be providing the correct signal but it is being affected by a circuit fault. Refer to list of sensor / actuator checks on page 19.
P0799	Pressure Control Solenoid "C" Intermittent	Pressure control solenoids are used in the transmission to regulate the pressure in the hydraulic circuits (usually for gear shift control). The pressure control solenoid can also influence / control the actual gear shift. Refer to vehicle specific information to establish which gear(s) are applicable to the identified solenoid.	The voltage, frequency or value of the control signal in the solenoid circuit is intermittent (likely to be out of normal operating range / tolerance when intermittent fault is detected) or, the signal is erratic e.g. signal changes are irregular / unpredictable. Possible fault with solenoid or wiring (e.g. internal / external connections, broken wiring) that is causing intermittent or erratic signal to exist. Note: the control unit could be providing the correct signal but it is being affected by a circuit fault. Refer to list of sensor / actuator checks on page 19.
P0800	Transfer Case Control System (MIL Request)	The control unit for the transfer case (typically the power transfer box used on 4-wheel drive systems) can pass information to the main control unit if a fault occurs; this information can be limited to requesting that the main control unit should switch on the MIL (Malfunction Indicator Light).	The fault code could indicate that there is a transfer case fault and that the MIL (Malfunction Indicator Light) should be switched on by the main control unit; it will therefore be necessary to investigate the transfer case fault. However, it is also possible that the fault code indicates that there is a fault with the transfer case MIL request circuit i.e. the communication circuit between the transfer case control unit and the main control unit. The fault code could be related to a wiring / connection fault between the between the transfer case control unit and the main control unit. Check wiring for short / open circuits and connector fault.
P0801	Reverse Inhibit Control Circuit	It may be necessary to refer to vehicle specific information to identify the design / wiring for the inhibit system. The system can be used to prevent engine starting when reverse gear is selected but, could also be used to prevent engagement of reverse gear under certain conditions.	The fault code indicates an undefined fault with the reverse gear inhibit circuit. The fault could be detected due to the voltage in the circuit remaining constant e.g. permanently at full circuit voltage or permanently at zero vaults. Check wiring for short / open circuits, check reverse gear switch / sensor operation (note that there could be a switch sensor on the gear lever assembly or on the gear selection mechanism). Check for a power supply / reference voltage to any applicable switches / sensors e.g. reverse gear indicator switch / sensor. Also note that the transmission control unit can send an appropriate "Reverse Gear" selected signal to the engine control unit or starter circuit; the fault code could possible be used to indicate a fault in this circuit.
P0802	Transmission Control System MIL Request Circuit / Open	The transmission control unit can directly switch on the warning light when a transmission fault occurs. On other systems, the transmission control unit can pass information to the main control unit if a transmission fault occurs; this information can be limited to requesting the MIL (Malfunction Indicator Light) is switched on.	The fault code is likely to relate to the circuit between the transmission control unit and the main control unit; this circuit carries the signal that requests the main control unit to illuminate the MIL. The request would be initiated because of a transmission fault, which will require further investigation. The fault could be detected because the request signal is out of normal operating range, constant value, non-existent, corrupt. The fault code indicates a possible "open circuit" but other undefined faults in the wiring could activate the same code e.g. short circuit or high circuit resistance. Note: the transmission control unit could be providing the correct "request" signal but it is being affected by a circuit fault. It is also possible the transmission or main control units are faulty.
P0803	Upshift / Skip Shift Solenoid Control Circuit	The Upshift / Skip Shift system is generally applicable to vehicles sold in US markets. The system causes the transmission to jump or skip gears (usually from 1st gear to 4th gear) during low load and low speed conditions e.g. town driving; the system is designed to reduces fuel consumption under the specified conditions. A solenoid (fitted to the transmission system) is used to activate the Upshift / Skip Shift process.	The voltage, frequency or value of the control signal in the solenoid circuit is incorrect (undefined fault). Signal could be:- out of normal operating range, constant value, non-existent, corrupt. Possible fault with solenoid or wiring e.g. solenoid power supply, short / open circuit, high circuit resistance or incorrect solenoid resistance. Note: the control unit could be providing the correct signal but it is being affected by the circuit fault. Refer to list of sensor / actuator checks on page 19.
P0804	Upshift / Skip Shift Lamp Control Circuit	The Upshift / Skip Shift system is generally applicable to vehicles sold in US markets. The system causes the transmission to jump or skip gears (usually from 1st gear to 4th gear) during low load and low speed conditions e.g. town driving; the system is designed to reduce fuel consumption under the specified conditions.	An indicator light is used to inform the driver that the Upshift / Skip Shift system is active (speed, load and temperature conditions allow the system to operate). The fault code indicates that there is fault in the indicator lamp circuit. The fault could be caused by a short / open circuit or a lamp failure.
P0805	Clutch Position Sensor Circuit	The information from the clutch position sensor can be used to assist in accurate control / positioning of the clutch engagement / disengagement.	It may be necessary to identify the system design and components used (sensor type etc) to allow selection of applicable sensor checks. Refer to list of sensor / actuator checks on page 19. The voltage, frequency or value of the sensor signal is incorrect (undefined fault). Sensor voltage / signal could be: out of normal operating range, constant value, non-existent, corrupt. Possible fault in the sensor or wiring e.g. short / open circuits, circuit resistance, incorrect sensor resistance or, fault with the reference / operating voltage; interference from other circuits can also affect sensor signals.
P0806	Clutch Position Sensor Circuit Range / Performance	The information from the clutch position sensor can be used to assist in accurate control / positioning of the clutch engagement / disengagement.	It may be necessary to identify the system design and components used (sensor type etc) to allow selection of applicable sensor checks. Refer to list of sensor / actuator checks on page 19. The voltage, frequency or value of the sensor signal is within the normal operating range / tolerance but the signal is not plausible or is incorrect due to an undefined fault. The sensor signal might not be changing / responding as expected to the changes in conditions. Possible fault in the sensor or wiring e.g. short / open circuits, circuit resistance, incorrect sensor resistance or, fault with the reference / operating voltage; interference from other circuits can also affect sensor signals. It is also possible that the sensor is operating correctly but the clutch movement / position being detected by the sensor is unacceptable or incorrect for the operating conditions.

NOTE 1. Check for other fault codes that could provide additional information. NOTE 2. Communication between control units can pass via a CAN-Bus system; refer to Page 51 for CAN-Bus checks.

NOTE 3. If a fault cannot be located, it is also possible that the control unit is at fault. NOTE 4. Refer to Page 53 for list of pages that show the ISO standard for component locations e.g. Sensor A - Bank 1.

code	Description		
P0807	Clutch Position Sensor Circuit Low	The information from the clutch position sensor can be used to assist in accurate control / positioning of the clutch engagement / disengagement.	It may be necessary to identify the system design and components used (sensor type etc) to allow selection of applicable sensor checks. Refer to list of sensor / actuator checks on page 19. The voltage, frequency or value of the sensor signal is either "zero" (no signal) or, the signal exists but it is below the normal operating range (lower than the minimum operating limit). Possible fault in the sensor or wiring e.g. short / open circuits, circuit resistance, incorrect sensor resistance or, fault with the reference / operating voltage; interference from other circuits can also affect sensor signals
P0808	Clutch Position Sensor Circuit High	The information from the clutch position sensor can be used to assist in accurate control / positioning of the clutch engagement / disengagement.	It may be necessary to identify the system design and components used (sensor type etc) to allow selection of applicable sensor checks. Refer to list of sensor / actuator checks on page 19. The voltage, frequency or value of the sensor signal is either at full value (e.g. battery voltage with no signal or frequency) or, the signal exists but it is above the normal operating range (higher than the maximum operating limit). Possible fault in the sensor or wiring e.g. short / open circuits, circuit resistance, incorrect sensor resistance or, fault with the reference / operating voltage; interference from other circuits can also affect sensor signals.
P0809	Clutch Position Sensor Circuit Intermittent	The information from the clutch position sensor can be used to assist in accurate control / positioning of the clutch engagement / disengagement.	It may be necessary to identify the system design and components used (sensor type etc) to allow selection of applicable sensor checks. Refer to list of sensor / actuator checks on page 19. The voltage, frequency or value of the sensor signal is intermittent (likely to be out of normal operating range / tolerance when intermittent fault is detected) or, the signal / voltage is erratic (e.g. signal changes are irregular / unpredictable). Possible fault with wiring or connections (including internal sensor connections); also interference from other circuits can affect the signal.
P080A	Clutch Position Not Learned	A control module (likely to be the transmission control module) is unable to learn the clutch position, which is indicated by a clutch position sensor. The control module can require the position information to enable effective control of the clutch.	The fault code indicates that the control module is unable to learn the position of the clutch; this could be caused by a fault in the clutch position sensor / wiring e.g. short / open circuit Refer to list of sensor / actuator checks on page 19. The fault could also be caused by clutch actuation faults (actuation mechanism or a mechanical fault) or, the control unit could be faulty. On some vehicle, there can be a set up procedure which enables the control unit to "learn" the base setting e.g. after fitting of a new clutch; it will therefore be necessary check for any specific procedures.
P080B	Upshift / Skip Shift Solenoid Control Circuit Range / Performance	The Upshift / Skip Shift system is generally applicable to vehicles sold in US markets. The system causes the transmission to jump or skip gears (usually from 1st gear to 4th gear) during low load and low speed conditions e.g. town driving; the system is designed to reduces fuel consumption under the specified conditions. A solenoid (fitted to the transmission system) is used to activate the Upshift / Skip Shift process.	The voltage, frequency or value of the control signal in the solenoid circuit is within the normal operating range / tolerance, but the signal might be incorrect for the operating conditions. It is also possible that the solenoid response (or response of system controlled by the solenoid) differs from the expected / desired response (incorrect response to the control signal provided by the control unit). Possible fault in solenoid or wiring, but also possible that the solenoid (or mechanism / system controlled by the solenoid) is not operating or moving correctly. Note: the control unit could be providing the correct signal but it is being affected by a circuit fault. Refer to list of sensor / actuator checks on page 19.
P080C	Upshift / Skip Shift Solenoid Control Circuit Low	The Upshift / Skip Shift system is generally applicable to vehicles sold in US markets. The system causes the transmission to jump or skip gears (usually from 1st gear to 4th gear) during low load and low speed conditions e.g. town driving; the system is designed to reduces fuel consumption under the specified conditions. A solenoid (fitted to the transmission system) is used to activate the Upshift / Skip Shift process.	The voltage, frequency or value of the control signal in the solenoid circuit is either "zero" (no signal) or, the signal exists but is below the normal operating range. Possible fault with solenoid or wiring (e.g. solenoid power supply, short / open circuit or resistance) that is causing a signal value to be lower than the minimum operating limit. Note: the control unit could be providing the correct signal but it is being affected by a circuit fault. Refer to list of sensor / actuator checks on page 19.
P080D	Upshift / Skip Shift Solenoid Control Circuit High	The Upshift / Skip Shift system is generally applicable to vehicles sold in US markets. The system causes the transmission to jump or skip gears (usually from 1st gear to 4th gear) during low load and low speed conditions e.g. town driving; the system is designed to reduces fuel consumption under the specified conditions. A solenoid (fitted to the transmission system) is used to activate the Upshift / Skip Shift process.	The voltage, frequency or value of the control signal in the solenoid circuit is either at full value (e.g. battery voltage with no on / off signal) or, the signal exists but it is above the normal operating range. Possible fault with solenoid or wiring (e.g. solenoid power supply, short / open circuit or resistance) that is causing a signal value to be higher than the maximum operating limit. Note: the control unit could be providing the correct signal but it is being affected by a circuit fault. Refer to list of sensor / actuator checks on page 19.
P080E		ISO / SAE reserved	Not used, reserved for future allocation.
P080F		ISO / SAE reserved	Not used, reserved for future allocation.
P0810	Clutch Position Control Error	The clutch position can be controlled by an electronic system. The clutch engagement and disengagement will therefore be dependent on driving conditions, operating conditions as well as performance request from the driver (throttle position, transmission mode selection etc).	It will be necessary to identify the operation of the clutch system and identify sensors, actuators and wiring. The fault code indicates an undefined fault with the control of the clutch position. The fault could be related to a clutch actuator problem (e.g. solenoid controlling hydraulic circuit), a position sensor or a mechanical fault (e.g. conventional clutch wear). Check clutch actuation mechanism for smooth operation and check electric actuator / wiring for short / open circuits etc. Also check for any base setting procedures for the actuation system. Check operation of clutch position sensors and check wiring for short / open circuit; Refer to list of sensor / actuator checks on page 19.
P0811	Excessive Clutch "A" Slippage	Clutch slip can be detected by comparing selected speed signals e.g. engine speed compared with transmission input shaft speed. Also refer to fault code P0810.	The fault could be activated due to conventional wear in the clutch, but it could also be activated due to control problems with an electronic controlled system e.g. slow clutch engagement etc. Check clutch actuation mechanism for smooth operation and check electric actuator / wiring for short / open circuits etc; Refer to list of sensor / actuator checks on page 19. Also check for any base setting procedures for the actuation system. Check for mechanical wear in the clutch and actuation system. Check operation of clutch position sensors and check wiring for short / open circuit; Refer to list of sensor / actuator checks on page 19.
P0812	Reverse Input Circuit	The fault code relates to the reverse switch system (typically used on automatic transmission systems), the switch can be used to operate reverse lights but can also be used for transmission control and other functions.	The fault code indicates that the signal in the reverse switch circuit is incorrect; this could indicate that the reverse switch is "ON" when other sensors or gear selector position sensors are indicating that reverse gear is not selected. Alternatively, the switch could be "OFF" when other sensors are indicating that reverse gear is selected. Check operating of reverse switch and applicable wiring. Check for short / open circuits and check operating of switch contacts. Check for correct adjustment of switch and check that switch operates at the correct time.

NOTE 1. Check for other fault codes that could provide additional information.   **NOTE 2.** Communication between control units can pass via a CAN-Bus system; refer to Page 51 for CAN-Bus checks.

NOTE 3. If a fault cannot be located, it is also possible that the control unit is at fault.   **NOTE 4.** Refer to Page 53 for list of pages that show the ISO standard for component locations e.g. Sensor A - Bank 1.

Fault code	EOBD / ISO Description	Component / System Description	Meaningful Description and Quick Check
P0813	Reverse Output Circuit	The fault code relates to the reverse switch system (typically used on automatic transmission systems), the switch can be used to operate reverse lights but can also be used for transmission control and other functions.	The fault code indicates that the signal in the reverse switch circuit is incorrect; this could indicate that the reverse switch is "ON" when other sensors or gear selector position sensors are indicating that reverse gear is not selected. Alternatively, the switch could be "OFF" when other sensors are indicating that reverse gear is selected. Check operating of reverse switch and applicable wiring. Check for short / open circuits and check operating of switch contacts. Check for correct adjustment of switch and check that switch operates at the correct time.
P0814	Transmission Range Display Circuit	The range display fault codes usually relate to the gear range selection driver display (PRNDL etc).	Control unit has detected a fault with the Transmission range display system. The fault could be caused by a light / lamp / display unit fault or the associated wiring between the display and the transmission control module (or the range selector lever switches / sensor). Check wiring and connections for short / open circuits.
P0815	Upshift Switch Circuit	The Upshift switch is likely to refer to the system that enables manual control of gear changes on an auto transmission or possibly on an electronically controlled manual gearbox. The manual controls could be up or down shift paddles / buttons (on steering column or steering wheel), or it could be a specific movement of the main gear lever.	The fault code indicates a circuit related fault with the Upshift control. The fault could possibly be detected due to the voltage in the circuit indicating permanent switch operation i.e. voltage remaining permanently high or low. It is also possible that the fault is detected during a system self check (possibly by passing current through the circuit). Possible fault in the switch or wiring e.g. short / open circuits, circuit resistance or, fault with the reference / operating voltage (provided by the control module); the switch contacts could also be faulty e.g. not opening / closing correctly or have a high contact resistance.
P0816	Downshift Switch Circuit	The Downshift switch is likely to refer to the system that enables manual control of gear changes on an auto transmission or possibly on an electronically controlled manual gearbox. The manual controls could be up or down shift paddles / buttons (on steering column or steering wheel), or it could be a specific movement of the main gear lever.	The fault code indicates a circuit related fault with the Downshift control. The fault could possibly be detected due to the voltage in the circuit indicating permanent switch operation i.e. voltage remaining permanently high or low. It is also possible that the fault is detected during a system self check (possibly by passing current through the circuit). Possible fault in the switch or wiring e.g. short / open circuits, circuit resistance or, fault with the reference / operating voltage (provided by the control module); the switch contacts could also be faulty e.g. not opening / closing correctly or have a high contact resistance.
P0817	Starter Disable Circuit / Open	The system can be used to prevent engine starting under certain conditions e.g. when a gear is selected or a door is open. It may be necessary to refer to vehicle specific information to identify the design / wiring for the starter disable system, and identify what conditions will prevent starter operation.	The fault code indicates a fault with the starter disable circuit (possibly an open circuit). The fault could be detected because the voltage or signal value of the switch / sensor signal is remaining constant e.g. permanently at full circuit voltage or permanently at zero vaults. Possible fault in the switch / sensor or wiring e.g. short / open circuits, circuit resistance, incorrect sensor resistance or, fault with the reference / operating voltage. Check gear switch / sensor operation (to ensure that neutral or park is being indicated), check operation of other sensors and systems that could affect starter operation; also check if the immobilzer system is influencing the starter disable circuit. Refer to list of sensor / actuator checks on page 19. The fault could be detected due to a permanent "Low or High" voltage or signal in the circuit but, the control unit is expecting the voltage to change with the changes in conditions. Also note that the transmission control unit can send an appropriate signal to the engine control unit or starter circuit to prevent starter / engine operation; the fault code could possible be used to indicate a fault in this circuit.
P0818	Driveline Disconnect Switch Input Circuit	The Driveline switch is used to indicate whether the drive is selected / deselected; this driveline selection switch will normally relate to a 4-wheel drive High / Low box that allows driveline disconnect.	The voltage or value of the switch signal is incorrect (undefined fault). Switch voltage / signal could be: out of normal operating range, constant value, non-existent, corrupt. Possible fault in the switch or wiring e.g. short / open circuits, circuit resistance or, fault with the reference / operating voltage; the switch contacts could also be faulty e.g. not opening / closing correctly or have a high contact resistance. Refer to list of sensor / actuator checks on page 19.
P0819	Up and Down Shift Switch to Transmission Range Correlation	The transmission shift control levers / selection switches include an "up / down" shift selection system as well as the range selection (PRNDL etc). There can be some form of "Lock out" or control system to prevent or enable manual shift selection; the lock out could be based on the range selector position.	The fault code indicates that there is conflicting information / signals from the range selection system and the up / down shift switch. It is possible that there is a fault on the range selection switch / sensor system or on the up / down shift switch. Identify the applicable switches / sensors and check the operation, wiring (short / open circuits etc) and base settings / adjustments for switches / sensors; Refer to list of sensor / actuator checks on page 19.
P081A	Starter Disable Circuit Low	The system can be used to prevent engine starting under certain conditions e.g. when a gear is selected or a door is open. It may be necessary to refer to vehicle specific information to identify the design / wiring for the starter disable system, and identify what conditions will prevent starter operation.	The fault could be detected because the voltage or signal value of the switch / sensor signal (in the starter disable circuit) is either "zero" (no signal) or, the signal exists but it is below the normal operating range (lower than the minimum operating limit). Possible fault in the switch / sensor or wiring e.g. short / open circuits, circuit resistance, incorrect sensor resistance or, fault with the reference / operating voltage. Check gear switch / sensor operation (to ensure that neutral or park is being indicated), check operation of other sensors and systems that could affect starter operation; also check if the immobilzer system is influencing the starter disable circuit. Refer to list of sensor / actuator checks on page 19. The fault could be detected due to a permanent "Low" voltage or signal in the circuit but, the control unit is expecting the voltage to change with the changes in conditions. Also note that the transmission control unit can send an appropriate signal to the engine control unit or starter circuit to prevent starter / engine operation; the fault code could possible be used to indicate a fault in this circuit.
P081B	Starter Disable Circuit High	The system can be used to prevent engine starting under certain conditions e.g. when a gear is selected or a door is open. It may be necessary to refer to vehicle specific information to identify the design / wiring for the starter disable system, and identify what conditions will prevent starter operation.	The fault could be detected because the voltage or signal value of the switch / sensor signal (in the starter disable circuit) is either at full value (e.g. battery voltage with no signal) or, the signal exists but it is above the normal operating range (higher than the maximum operating limit). Possible fault in the switch / sensor or wiring e.g. short / open circuits, circuit resistance, incorrect sensor resistance or, fault with the reference / operating voltage. Check gear switch / sensor operation (to ensure that neutral or park is being indicated), check operation of other sensors and systems that could affect starter operation; also check if the immobilzer system is influencing the starter disable circuit. Refer to list of sensor / actuator checks on page 19. The fault could be detected due to a permanent "High" voltage or signal in the circuit but, the control unit is expecting the voltage to change with the changes in conditions. Also note that the transmission control unit can send an appropriate signal to the engine control unit or starter circuit to prevent starter / engine operation; the fault code could possible be used to indicate a fault in this circuit.
P081C	Park Input Circuit	It may be necessary to refer to vehicle specific information to identify the design / wiring for the Park Input circuit. The Park Input circuit could relate to the signal passing from the transmission control system to the engine control system; the signal will indicate that the gear lever is placed into the Park position or that the gearbox is in "Park".	The fault code indicates that there is an undefined fault in the Park input circuit. Although it may be necessary to identify the system and wiring, check the operation of any switches / sensors used to indicate the "Park" selection or position; check all applicable wiring for short / open circuits and check for correct reference / supply voltage to applicable switches / sensors. Check the applicable communication circuits between the transmission control system and the engine control system (note that communication could be via the CAN-Bus network).

NOTE 1. Check for other fault codes that could provide additional information. **NOTE 2.** Communication between control units can pass via a CAN-Bus system; refer to Page 51 for CAN-Bus checks.

NOTE 3. If a fault cannot be located, it is also possible that the control unit is at fault. NOTE 4. Refer to Page 53 for list of pages that show the ISO standard for component locations e.g. Sensor A - Bank 1.

Code	Description		
P081D	Neutral Input Circuit	It may be necessary to refer to vehicle specific information to identify the design / wiring for the Neutral Input circuit. The Neutral Input circuit could relate to the signal passing from the transmission control system to the engine control system; the signal will indicate that the gear lever is placed into the Neutral position or that the gearbox is in "Neutral".	The fault code indicates that there is an undefined fault in the Neutral input circuit. Although it may be necessary to identify the system and wiring, check the operation of any switches / sensors used to indicate the "Neutral" selection or position; check all applicable wiring for short / open circuits and check for correct reference / supply voltage to applicable switches / sensors. Check the applicable communication circuits between the transmission control system and the engine control system (note that communication could be via the CAN-Bus network).
P081E	Excessive Clutch "B" Slippage	Clutch slip can be detected by comparing selected speed signals e.g. engine speed compared with transmission input shaft speed. Also refer to fault code P0810.	The fault could be activated due to conventional wear in the clutch, but it could also be activated due to control problems with an electronic controlled system e.g. slow clutch engagement etc. Check clutch actuation mechanism for smooth operation and check electric actuator / wiring for short / open circuits etc; Refer to list of sensor / actuator checks on page 19. Also check for any base setting procedures for the actuation system. Check for mechanical wear in the clutch and actuation system. Check operation of clutch position sensors and check wiring for short / open circuit; Refer to list of sensor / actuator checks on page 19.
P0820	Gear Lever X-Y Position Sensor Circuit	A sensor can be used to indicate the movement of the gear lever in two directions on two axis (typically left / right and Forward / backward). A single sensor assembly can be used which can detect / monitor movement in both directions or, it is possible to use two or four separate sensors. Note that on some systems, the terms X and Y refer to two gear shift mechanisms (actuating two sets of gear clusters in the transmission); in these cases, the sensor will monitor the position of the actuator and / or gear shift mechanism. If necessary, refer to vehicle specific information to identify the application of the fault code and the components used.	The voltage or value of the position sensor signal is incorrect (undefined fault). Sensor voltage / signal value could be: out of normal operating range, constant value, non-existent, corrupt. Possible fault in the position sensor or wiring e.g. short / open circuits, circuit resistance, incorrect sensor resistance (if applicable) or, fault with the reference / operating voltage; interference from other circuits can also affect sensor signals. Also check earth connections and cable screening. Refer to list of sensor / actuator checks on page 19. Also check that the sensor and gear lever are set / adjusted correctly (base setting); it could be possible that the sensor is indicating an incorrect position or not operating / responding correctly to the movement of the gear lever.
P0821	Gear Lever X Position Circuit	A sensor can be used to indicate the movement of the gear lever in two directions on two axis (typically left / right and Forward / backward). A single sensor assembly can be used which can detect / monitor movement in both directions or, it is possible to use two or four separate sensors. Note that on some systems, the terms X and Y refer to two gear shift mechanisms (actuating two sets of gear clusters in the transmission); in these cases, the sensor will monitor the position of the actuator and / or gear shift mechanism. If necessary, refer to vehicle specific information to identify the application of the fault code and the components used.	The voltage or value of the position sensor signal is incorrect (undefined fault). Sensor voltage / signal value could be: out of normal operating range, constant value, non-existent, corrupt. Possible fault in the position sensor or wiring e.g. short / open circuits, circuit resistance, incorrect sensor resistance (if applicable) or, fault with the reference / operating voltage; interference from other circuits can also affect sensor signals. Also check earth connections and cable screening. Refer to list of sensor / actuator checks on page 19. Also check that the sensor and gear lever are set / adjusted correctly (base setting); it could be possible that the sensor is indicating an incorrect position or not operating / responding correctly to the movement of the gear lever.
P0822	Gear Lever Y Position Circuit	A sensor can be used to indicate the movement of the gear lever in two directions on two axis (typically left / right and Forward / backward). A single sensor assembly can be used which can detect / monitor movement in both directions or, it is possible to use two or four separate sensors. Note that on some systems, the terms X and Y refer to two gear shift mechanisms (actuating two sets of gear clusters in the transmission); in these cases, the sensor will monitor the position of the actuator and / or gear shift mechanism. If necessary, refer to vehicle specific information to identify the application of the fault code and the components used.	The voltage or value of the position sensor signal is incorrect (undefined fault). Sensor voltage / signal value could be: out of normal operating range, constant value, non-existent, corrupt. Possible fault in the position sensor or wiring e.g. short / open circuits, circuit resistance, incorrect sensor resistance (if applicable) or, fault with the reference / operating voltage; interference from other circuits can also affect sensor signals. Also check earth connections and cable screening. Refer to list of sensor / actuator checks on page 19. Also check that the sensor and gear lever are set / adjusted correctly (base setting); it could be possible that the sensor is indicating an incorrect position or not operating / responding correctly to the movement of the gear lever.
P0823	Gear Lever X Position Circuit Intermittent	A sensor can be used to indicate the movement of the gear lever in two directions on two axis (typically left / right and Forward / backward). A single sensor assembly can be used which can detect / monitor movement in both directions or, it is possible to use two or four separate sensors. Note that on some systems, the terms X and Y refer to two gear shift mechanisms (actuating two sets of gear clusters in the transmission); in these cases, the sensor will monitor the position of the actuator and / or gear shift mechanism. If necessary, refer to vehicle specific information to identify the application of the fault code and the components used.	The voltage or value of the position sensor signal is intermittent (likely to be out of normal operating range / tolerance when intermittent fault is detected) or, the signal / voltage is erratic (e.g. signal changes are irregular / unpredictable). Possible fault with wiring or connections (including internal sensor connections); also interference from other circuits can affect the signal. Refer to list of sensor / actuator checks on page 19. Also check that the sensor and gear lever are set / adjusted correctly (base setting); if the sensor is very close the limit of the setting, the sensor be detecting small movements of the gear lever and the signal could alter in and out of the correct value.
P0824	Gear Lever Y Position Circuit Intermittent	A sensor can be used to indicate the movement of the gear lever in two directions on two axis (typically left / right and Forward / backward). A single sensor assembly can be used which can detect / monitor movement in both directions or, it is possible to use two or four separate sensors. Note that on some systems, the terms X and Y refer to two gear shift mechanisms (actuating two sets of gear clusters in the transmission); in these cases, the sensor will monitor the position of the actuator and / or gear shift mechanism. If necessary, refer to vehicle specific information to identify the application of the fault code and the components used.	The voltage or value of the position sensor signal is intermittent (likely to be out of normal operating range / tolerance when intermittent fault is detected) or, the signal / voltage is erratic (e.g. signal changes are irregular / unpredictable). Possible fault with wiring or connections (including internal sensor connections); also interference from other circuits can affect the signal. Refer to list of sensor / actuator checks on page 19. Also check that the sensor and gear lever are set / adjusted correctly (base setting); if the sensor is very close the limit of the setting, the sensor be detecting small movements of the gear lever and the signal could alter in and out of the correct value.

NOTE 1. Check for other fault codes that could provide additional information. NOTE 2. Communication between control units can pass via a CAN-Bus system; refer to Page 51 for CAN-Bus checks.

NOTE 3. If a fault cannot be located, it is also possible that the control unit is at fault. NOTE 4. Refer to Page 53 for list of pages that show the ISO standard for component locations e.g. Sensor A - Bank 1.

Fault code	EOBD / ISO Description	Component / System Description	Meaningful Description and Quick Check
P0825	Gear Lever Push-Pull Switch (Shift Anticipate)	Different manufacturers use different systems to select transmission mode and range; it may therefore be necessary to refer to vehicle specific information to identify the gear selection system fitted to the vehicle. Some vehicles may be fitted with a specific "Push-Pull" switch to select certain transmission functions e.g. range selection, upshift / downshift, reverse gear selection etc.	The fault code indicates an unidentified fault with the Push-Pull switch. It may be necessary to identify the switch and its function. The fault could be caused due to the switch voltage / signal being out of normal operating range, constant value, non-existent, corrupt. Possible fault in the switch or wiring e.g. short / open circuits, circuit resistance or, fault with the reference / operating voltage; the switch contacts (if applicable) could also be faulty e.g. not opening / closing correctly or have a high contact resistance.
P0826	Up and Down Shift Switch Circuit	The switch will indicate movement / position of the Up and Down lever or controls for the gear shift / changes.	The voltage or value of the switch signal is incorrect (undefined fault). Switch voltage / signal could be: out of normal operating range, constant value, non-existent, corrupt. Possible fault in the switch or wiring e.g. short / open circuits, circuit resistance or, fault with the reference / operating voltage; the switch contacts could also be faulty e.g. not opening / closing correctly or have a high contact resistance. Refer to list of sensor / actuator checks on page 19. Also check that the switch and gear lever are set / adjusted correctly (base setting); it could be possible that the switch is indicating an incorrect position or not operating / responding correctly to the movement of the gear lever / controls
P0827	Up and Down Shift Switch Circuit Low	The switch will indicate movement / position of the Up and Down lever or controls for the gear shift / changes.	The voltage or value of the switch signal is either "zero" (no signal) or, the signal exists but it is below the normal operating range (lower than the minimum operating limit). Possible fault in the switch or wiring e.g. short / open circuits, circuit resistance or, fault with the reference / operating voltage; the switch contacts could also be faulty e.g. not opening / closing correctly or have a high contact resistance. Refer to list of sensor / actuator checks on page 19. Also check that the switch and gear lever are set / adjusted correctly (base setting); it could be possible that the switch is indicating an incorrect position or not operating / responding correctly to the movement of the gear lever / controls
P0828	Up and Down Shift Switch Circuit High	The switch will indicate movement / position of the Up and Down lever or controls for the gear shift / changes.	The voltage or value of the switch signal is either at full value (e.g. battery voltage with no signal or frequency) or, the signal exists but it is above the normal operating range (higher than the maximum operating limit). Possible fault in the switch or wiring e.g. short / open circuits, circuit resistance or, fault with the reference / operating voltage; the switch contacts could also be faulty e.g. not opening / closing correctly or have a high contact resistance. Refer to list of sensor / actuator checks on page 19. Also check that the switch and gear lever are set / adjusted correctly (base setting); it could be possible that the switch is indicating an incorrect position or not operating / responding correctly to the movement of the gear lever / controls.
P0829	5-6 Shift	The control unit monitors the input speed and output speed in the transmission to establish that the requested gear has been selected or is being selected (gear is requested by control unit). Some systems may use a position sensor to identify movement of the gear selection mechanism.	The fault code indicates that the identified gear(s) requested by the control unit is not selected. The fault could be on the up shift or the downshift. Identify system operation and components. Check actuators and control signals e.g. shift / solenoid valves etc. Check any applicable position, speed or pressure sensors and sensor signals. Refer to list of sensor / actuator checks on page 19. Note that it is also possible that the fault code has been activated due to a mechanical gear selection fault e.g. gear change clutch failure or other component fault; also check transmission hydraulic system for pressures etc.
P0830	Clutch Pedal Switch "A" Circuit	Note. Although this code can be related to a transmission problem, the clutch switch could also be providing a signal to other systems. Clutch switches can be used for various control functions e.g. transmission, cruise control, engine control, vehicle stability systems etc (note that some vehicles require the clutch pedal to be depressed prior to engine starts). Some vehicles can use a switch to indicate that the pedal is released (clutch engaged) and they can be used to indicate that the clutch is depressed (clutch disengaged).	The voltage or value of the switch signal is incorrect (undefined fault). Switch voltage / signal could be: out of normal operating range, constant value, non-existent, corrupt. Possible fault in the switch or wiring e.g. short / open circuits, circuit resistance or, fault with the reference / operating voltage; the switch contacts could also be faulty e.g. not opening / closing correctly or have a high contact resistance. Refer to list of sensor / actuator checks on page 19. Also check clutch switch operating mechanism (pedal / linkage) to ensure switch is being operated correctly. Check for any base settings for the switch position.
P0831	Clutch Pedal Switch "A" Circuit Low	Note. Although this code can be related to a transmission problem, the clutch switch could also be providing a signal to other systems. Clutch switches can be used for various control functions e.g. transmission, cruise control, engine control, vehicle stability systems etc (note that some vehicles require the clutch pedal to be depressed prior to engine starts). Some vehicles can use a switch to indicate that the pedal is released (clutch engaged) and they can be used to indicate that the clutch is depressed (clutch disengaged).	The voltage or value of the switch signal is either "zero" (no signal) or, the signal exists but it is below the normal operating range (lower than the minimum operating limit). Possible fault in the switch or wiring e.g. short / open circuits, circuit resistance or, fault with the reference / operating voltage; the switch contacts could also be faulty e.g. not opening / closing correctly or have a high contact resistance. Refer to list of sensor / actuator checks on page 19. Note that the fault code can be activated for systems with position switches, if the switch output remains at the low value e.g. zero volts and does not rise to the high value e.g. battery voltage (the control unit will detect no switch signal change but it has detected change in engine operating conditions). Also check clutch switch operating mechanism (pedal / linkage) to ensure switch is being operated correctly. Check for any base settings for the switch position.
P0832	Clutch Pedal Switch "A" Circuit High	Note. Although this code can be related to a transmission problem, the clutch switch could also be providing a signal to other systems. Clutch switches can be used for various control functions e.g. transmission, cruise control, engine control, vehicle stability systems etc (note that some vehicles require the clutch pedal to be depressed prior to engine starts). Some vehicles can use a switch to indicate that the pedal is released (clutch engaged) and they can be used to indicate that the clutch is depressed (clutch disengaged).	The voltage or value of the switch signal is either at full value (e.g. battery voltage with no on / off signal) or, the signal exists but it is above the normal operating range (higher than the maximum operating limit). Possible fault in the switch or wiring e.g. short / open circuits, circuit resistance or, fault with the reference / operating voltage; the switch contacts could also be faulty e.g. not opening / closing correctly or have a high contact resistance. Refer to list of sensor / actuator checks on page 19. Note that the fault code can be activated for systems with position switches, if the switch output remains at the high value e.g. battery voltage and does not fall to the low value e.g. zero volts (the control unit will detect no switch signal change but it has detected change in operating conditions). Also check clutch switch operating mechanism (pedal / linkage) to ensure switch is being operated correctly. Check for any base settings for the switch position.
P0833	Clutch Pedal Switch "B" Circuit	Note. Although this code can be related to a transmission problem, the clutch switch could also be providing a signal to other systems. Clutch switches can be used for various control functions e.g. transmission, cruise control, engine control, vehicle stability systems etc (note that some vehicles require the clutch pedal to be depressed prior to engine starts). Some vehicles can use a switch to indicate that the pedal is released (clutch engaged) and they can be used to indicate that the clutch is depressed (clutch disengaged).	The voltage or value of the switch signal is incorrect (undefined fault). Switch voltage / signal could be: out of normal operating range, constant value, non-existent, corrupt. Possible fault in the switch or wiring e.g. short / open circuits, circuit resistance or, fault with the reference / operating voltage; the switch contacts could also be faulty e.g. not opening / closing correctly or have a high contact resistance. Refer to list of sensor / actuator checks on page 19. Also check clutch switch operating mechanism (pedal / linkage) to ensure switch is being operated correctly. Check for any base settings for the switch position.

NOTE 1. Check for other fault codes that could provide additional information. **NOTE 2.** Communication between control units can pass via a CAN-Bus system; refer to Page 51 for CAN-Bus checks. 190

NOTE 3. If a fault cannot be located, it is also possible that the control unit is at fault **NOTE 4.** Refer to Page 53 for list of pages that show the ISO standard for component locations e.g. Sensor A - Bank 1.

Code	Description		
P0834	Clutch Pedal Switch "B" Circuit Low	Note. Although this code can be related to a transmission problem, the clutch switch could also be providing a signal to other systems. Clutch switches can be used for various control functions e.g. transmission, cruise control, engine control, vehicle stability systems etc (note that some vehicles require the clutch pedal to be depressed prior to engine starts). Some vehicles can use a switch to indicate that the pedal is released (clutch engaged) and they can be used to indicate that the clutch is depressed (clutch disengaged).	The voltage or value of the switch signal is either "zero" (no signal) or, the signal exists but it is below the normal operating range (lower than the minimum operating limit). Possible fault in the switch or wiring e.g. short / open circuits, circuit resistance or, fault with the reference / operating voltage; the switch contacts could also be faulty e.g. not opening / closing correctly or have a high contact resistance. Refer to list of sensor / actuator checks on page 19. Note that the fault code can be activated for systems with position switches, if the switch output remains at the low value e.g. zero volts and does not rise to the high value e.g. battery voltage (the control unit will detect no switch signal change but it has detected change in engine operating conditions). Also check clutch switch operating mechanism (pedal / linkage) to ensure switch is being operated correctly. Check for any base settings for the switch position.
P0835	Clutch Pedal Switch "B" Circuit High	Note. Although this code can be related to a transmission problem, the clutch switch could also be providing a signal to other systems. Clutch switches can be used for various control functions e.g. transmission, cruise control, engine control, vehicle stability systems etc (note that some vehicles require the clutch pedal to be depressed prior to engine starts). Some vehicles can use a switch to indicate that the pedal is released (clutch engaged) and they can be used to indicate that the clutch is depressed (clutch disengaged).	The voltage or value of the switch signal is either at full value (e.g. battery voltage with no on / off signal) or, the signal exists but it is above the normal operating range (higher than the maximum operating limit). Possible fault in the switch or wiring e.g. short / open circuits, circuit resistance or, fault with the reference / operating voltage; the switch contacts could also be faulty e.g. not opening / closing correctly or have a high contact resistance. Refer to list of sensor / actuator checks on page 19. Note that the fault code can be activated for systems with position switches, if the switch output remains at the high value e.g. battery voltage and does not fall to the low value e.g. zero volts (the control unit will detect no switch signal change but it has detected change in operating conditions). Also check clutch switch operating mechanism (pedal / linkage) to ensure switch is being operated correctly. Check for any base settings for the switch position.
P0836	Four Wheel Drive (4WD) Switch Circuit	The switch can be used to select / indicate the different functions of the Four Wheel Drive operation e.g. two-wheel or four-wheel drive.	The voltage or value of the switch signal is incorrect (undefined fault). Switch voltage / signal could be: out of normal operating range, constant value, non-existent, corrupt. Possible fault in the switch or wiring e.g. short / open circuits, circuit resistance or, fault with the reference / operating voltage; the switch contacts could also be faulty e.g. not opening / closing correctly or have a high contact resistance. Refer to list of sensor / actuator checks on page 19. Also check that the switch is set / adjusted correctly (base setting); it could be possible that the switch is indicating an incorrect position or not operating / responding correctly.
P0837	Four Wheel Drive (4WD) Switch Circuit Range / Performance	The switch can be used to select / indicate the different functions of the Four Wheel Drive operation e.g. two-wheel or four-wheel drive.	The voltage value of the switch signal is within the normal operating range / tolerance but the signal is not plausible or is incorrect due to an undefined fault. The switch signal might not match the operating conditions indicated by other sensors or, the switch signal is not changing / responding as expected. Possible fault in the switch or wiring e.g. short / open circuits, circuit resistance or, fault with the reference / operating voltage; also check switch contacts to ensure that they open / close and check continuity of switch contacts (ensure that there is not a high resistance). Refer to list of sensor / actuator checks on page 19. Also check that the switch is set / adjusted correctly (base setting); it could be possible that the switch is indicating an incorrect position or not operating / responding correctly.
P0838	Four Wheel Drive (4WD) Switch Circuit Low	The switch can be used to select / indicate the different functions of the Four Wheel Drive operation e.g. two-wheel or four-wheel drive.	The voltage or value of the switch signal is either "zero" (no signal) or, the signal exists but it is below the normal operating range (lower than the minimum operating limit). Possible fault in the switch or wiring e.g. short / open circuits, circuit resistance or, fault with the reference / operating voltage; the switch contacts could also be faulty e.g. not opening / closing correctly or have a high contact resistance. Refer to list of sensor / actuator checks on page 19. Also check that the switch is set / adjusted correctly (base setting); it could be possible that the switch is indicating an incorrect position or not operating / responding correctly.
P0839	Four Wheel Drive (4WD) Switch Circuit High	The switch can be used to select / indicate the different functions of the Four Wheel Drive operation e.g. two-wheel or four-wheel drive.	The voltage or value of the switch signal is either at full value (e.g. battery voltage with no signal or frequency) or, the signal exists but it is above the normal operating range (higher than the maximum operating limit). Possible fault in the switch or wiring e.g. short / open circuits, circuit resistance or, fault with the reference / operating voltage; the switch contacts could also be faulty e.g. not opening / closing correctly or have a high contact resistance. Refer to list of sensor / actuator checks on page 19. Also check that the switch is set / adjusted correctly (base setting); it could be possible that the switch is indicating an incorrect position or not operating / responding correctly.
P083A	Transmission Fluid Pressure Sensor / Switch "G" Circuit	Transmission systems can make use of fluid pressure to control various functions, including gear changes. Fluid pressure sensors are therefore used to monitor pressures and changes in pressure when functions such as gear changes take place. It may be necessary to refer to vehicle specific information to identify the applicable pressure circuit (and its function) for the identified pressure sensor.	The voltage, frequency or value of the pressure sensor signal is incorrect (undefined fault). Sensor voltage / signal could be: out of normal operating range, constant value, non-existent, corrupt. Possible fault in the pressure sensor or wiring e.g. short / open circuits, circuit resistance, incorrect sensor resistance (if applicable) or, fault with the reference / operating voltage; interference from other circuits can also affect sensor signals. Where a pressure switch is fitted, it is possible that the switch contacts are not opening / closing correctly or the switch contacts have a high resistance. Refer to list of sensor / actuator checks on page 19.
P083B	Transmission Fluid Pressure Sensor / Switch "G" Circuit Range / Performance	Transmission systems can make use of fluid pressure to control various functions, including gear changes. Fluid pressure sensors are therefore used to monitor pressures and changes in pressure when functions such as gear changes take place. It may be necessary to refer to vehicle specific information to identify the applicable pressure circuit (and its function) for the identified pressure sensor.	The voltage, frequency or value of the pressure sensor signal is within the normal operating range / tolerance but the signal is not plausible or is incorrect due to an undefined fault. The sensor signal might not match the operating conditions indicated by other sensors or, the sensor signal is not changing / responding as expected. Possible fault in the pressure sensor or wiring e.g. short / open circuits, circuit resistance, incorrect sensor resistance (if applicable) or, fault with the reference / operating voltage; interference from other circuits can also affect sensor signals. Where a pressure switch is fitted, it is possible that the switch contacts are not opening / closing correctly or the switch contacts have a high resistance. Refer to list of sensor / actuator checks on page 19. It is also possible that the pressure sensor is operating correctly but the pressure being measured by the sensor is unacceptable or incorrect for the operating conditions; if possible, use alternative method of checking pressure to establish if sensor system is operating correctly and whether the pressure indicated by the sensor is correct / incorrect.

NOTE 1. Check for other fault codes that could provide additional information. NOTE 2. Communication between control units can pass via a CAN-Bus system; refer to Page 51 for CAN-Bus checks. 191
NOTE 3. If a fault cannot be located, it is also possible that the control unit is at fault. NOTE 4. Refer to Page 53 for list of pages that show the ISO standard for component locations e.g. Sensor A - Bank 1.

Fault code	EOBD / ISO Description	Component / System Description	Meaningful Description and Quick Check
P083C	Transmission Fluid Pressure Sensor / Switch "G" Circuit Low	Transmission systems can make use of fluid pressure to control various functions, including gear changes. Fluid pressure sensors are therefore used to monitor pressures and changes in pressure when functions such as gear changes take place. It may be necessary to refer to vehicle specific information to identify the applicable pressure circuit (and its function) for the identified pressure sensor.	The voltage, frequency or value of the pressure sensor signal is either "zero" (no signal) or, the signal exists but it is below the normal operating range (lower than the minimum operating limit). Possible fault in the pressure sensor or wiring e.g. short / open circuits, circuit resistance, incorrect sensor resistance (if applicable) or, fault with the reference / operating voltage; interference from other circuits can also affect sensor signals. Where a pressure switch is fitted, it is possible that the switch contacts are not opening / closing correctly or, the contacts are sticking (signal is remaining low when it should have risen to the high value); on some circuits, a high resistance on the contacts could also cause a low signal value to exist. Refer to list of sensor / actuator checks on page 19.
P083D	Transmission Fluid Pressure Sensor / Switch "G" Circuit High	Transmission systems can make use of fluid pressure to control various functions, including gear changes. Fluid pressure sensors are therefore used to monitor pressures and changes in pressure when functions such as gear changes take place. It may be necessary to refer to vehicle specific information to identify the applicable pressure circuit (and its function) for the identified pressure sensor.	The voltage, frequency or value of the pressure sensor signal is either at full value (e.g. battery voltage with no signal or frequency) or, the signal exists but it is above the normal operating range (higher than the maximum operating limit). Possible fault in the pressure sensor or wiring e.g. short / open circuits, circuit resistance, incorrect sensor resistance (if applicable) or, fault with the reference / operating voltage; interference from other circuits can also affect sensor signals. Where a pressure switch is fitted, it is possible that the switch contacts are not opening / closing correctly or, the contacts are sticking (signal is remaining high when it should have dropped to low value); on some circuits, a high resistance on the contacts could also cause a high signal value to exist. Refer to list of sensor / actuator checks on page 19.
P083E	Transmission Fluid Pressure Sensor / Switch "G" Circuit Intermittent	Transmission systems can make use of fluid pressure to control various functions, including gear changes. Fluid pressure sensors are therefore used to monitor pressures and changes in pressure when functions such as gear changes take place. It may be necessary to refer to vehicle specific information to identify the applicable pressure circuit (and its function) for the identified pressure sensor.	The voltage, frequency or value of the pressure sensor signal is intermittent (likely to be out of normal operating range / tolerance when intermittent fault is detected) or, the signal / voltage is erratic (e.g. signal changes are irregular / unpredictable). Possible fault with wiring or connections (including internal sensor connections); also interference from other circuits can affect the signal. Where a pressure switch is fitted, it is possible that the switch contacts are not opening / closing correctly or the switch contacts have a high resistance. Refer to list of sensor / actuator checks on page 19. It is also possible that the pressure could be changing erratically, which could cause the fault code to be activated; if necessary, check pressure with a separate gauge to check for erratic pressure changes.
P083F	Clutch Pedal Switch "A" / "B" Correlation	Clutch switches can be used for various control functions e.g. transmission, cruise control, engine control, vehicle stability systems etc (note that some vehicles require the clutch pedal to be depressed prior to engine starts). Some vehicles can use a switch to indicate that the pedal is released (clutch engaged) and they can be used to indicate that the clutch is depressed (clutch disengaged). Note that on some applications, with the clutch in either the depressed or released position, one switch could be "OFF" whilst the other switch could be "ON".	The signals from the two clutch switches are providing conflicting or implausible information e.g. one switch is indicating that the pedal is depressed but the other switch is indicating pedal is released. Check the operation of the switches to establish if one of the switches is faulty or if there is a wiring fault; check for short or open circuits in the wiring, check that switch contacts are closing / opening correctly and check for high resistance at switch contacts. Check for reference voltage / power supply to the switches. Also check the mechanism operating the switches e.g. pedal and switch adjustment. Refer to list of sensor / actuator checks on page 19.
P0840	Transmission Fluid Pressure Sensor / Switch "A" Circuit	Transmission systems can make use of fluid pressure to control various functions, including gear changes. Fluid pressure sensors are therefore used to monitor pressures and changes in pressure when functions such as gear changes take place. It may be necessary to refer to vehicle specific information to identify the applicable pressure circuit (and its function) for the identified pressure sensor.	The voltage, frequency or value of the pressure sensor signal is incorrect (undefined fault). Sensor voltage / signal could be: out of normal operating range, constant value, non-existent, corrupt. Possible fault in the pressure sensor or wiring e.g. short / open circuits, circuit resistance, incorrect sensor resistance (if applicable) or, fault with the reference / operating voltage; interference from other circuits can also affect sensor signals. Where a pressure switch is fitted, it is possible that the switch contacts are not opening / closing correctly or the switch contacts have a high resistance. Refer to list of sensor / actuator checks on page 19.
P0841	Transmission Fluid Pressure Sensor / Switch "A" Circuit Range / Performance	Transmission systems can make use of fluid pressure to control various functions, including gear changes. Fluid pressure sensors are therefore used to monitor pressures and changes in pressure when functions such as gear changes take place. It may be necessary to refer to vehicle specific information to identify the applicable pressure circuit (and its function) for the identified pressure sensor.	The voltage, frequency or value of the pressure sensor signal is within the normal operating range / tolerance but the signal is not plausible or is incorrect due to an undefined fault. The sensor signal might not match the operating conditions indicated by other sensors or, the sensor signal is not changing / responding as expected. Possible fault in the pressure sensor or wiring e.g. short / open circuits, circuit resistance, incorrect sensor resistance (if applicable) or, fault with the reference / operating voltage; interference from other circuits can also affect sensor signals. Where a pressure switch is fitted, it is possible that the switch contacts are not opening / closing correctly or the switch contacts have a high resistance. Refer to list of sensor / actuator checks on page 19. It is also possible that the pressure sensor is operating correctly but the pressure being measured by the sensor is unacceptable or incorrect for the operating conditions; if possible, use alternative method of checking pressure to establish if sensor system is operating correctly and whether the pressure indicated by the sensor is correct / incorrect.
P0842	Transmission Fluid Pressure Sensor / Switch "A" Circuit Low	Transmission systems can make use of fluid pressure to control various functions, including gear changes. Fluid pressure sensors are therefore used to monitor pressures and changes in pressure when functions such as gear changes take place. It may be necessary to refer to vehicle specific information to identify the applicable pressure circuit (and its function) for the identified pressure sensor.	The voltage, frequency or value of the pressure sensor signal is either "zero" (no signal) or, the signal exists but it is below the normal operating range (lower than the minimum operating limit). Possible fault in the pressure sensor or wiring e.g. short / open circuits, circuit resistance, incorrect sensor resistance (if applicable) or, fault with the reference / operating voltage; interference from other circuits can also affect sensor signals. Where a pressure switch is fitted, it is possible that the switch contacts are not opening / closing correctly or, the contacts are sticking (signal is remaining low when it should have risen to the high value); on some circuits, a high resistance on the contacts could also cause a low signal value to exist. Refer to list of sensor / actuator checks on page 19.
P0843	Transmission Fluid Pressure Sensor / Switch "A" Circuit High	Transmission systems can make use of fluid pressure to control various functions, including gear changes. Fluid pressure sensors are therefore used to monitor pressures and changes in pressure when functions such as gear changes take place. It may be necessary to refer to vehicle specific information to identify the applicable pressure circuit (and its function) for the identified pressure sensor.	The voltage, frequency or value of the pressure sensor signal is either at full value (e.g. battery voltage with no signal or frequency) or, the signal exists but it is above the normal operating range (higher than the maximum operating limit). Possible fault in the pressure sensor or wiring e.g. short / open circuits, circuit resistance, incorrect sensor resistance (if applicable) or, fault with the reference / operating voltage; interference from other circuits can also affect sensor signals. Where a pressure switch is fitted, it is possible that the switch contacts are not opening / closing correctly or, the contacts are sticking (signal is remaining high when it should have dropped to low value); on some circuits, a high resistance on the contacts could also cause a high signal value to exist. Refer to list of sensor / actuator checks on page 19.

NOTE 1. Check for other fault codes that could provide additional information. **NOTE 2.** Communication between control units can pass via a CAN-Bus system; refer to Page 51 for CAN-Bus checks.
NOTE 3. If a fault cannot be located, it is also possible that the control unit is at fault. **NOTE 4.** Refer to Page 53 for list of pages that show the ISO standard for component locations e.g. Sensor A - Bank 1.

Code	Description		
P0844	Transmission Fluid Pressure Sensor / Switch "A" Circuit Intermittent	Transmission systems can make use of fluid pressure to control various functions, including gear changes. Fluid pressure sensors are therefore used to monitor pressures and changes in pressure when functions such as gear changes take place. It may be necessary to refer to vehicle specific information to identify the applicable pressure circuit (and its function) for the identified pressure sensor.	The voltage, frequency or value of the pressure sensor signal is intermittent (likely to be out of normal operating range / tolerance when intermittent fault is detected) or, the signal / voltage is erratic (e.g. signal changes are irregular / unpredictable). Possible fault with wiring or connections (including internal sensor connections); also interference from other circuits can affect the signal. Where a pressure switch is fitted, it is possible that the switch contacts are not opening / closing correctly or the switch contacts have a high resistance. Refer to list of sensor / actuator checks on page 19. It is also possible that the pressure could be changing erratically, which could cause the fault code to be activated; if necessary, check pressure with a separate gauge to check for erratic pressure changes.
P0845	Transmission Fluid Pressure Sensor / Switch "B" Circuit	Transmission systems can make use of fluid pressure to control various functions, including gear changes. Fluid pressure sensors are therefore used to monitor pressures and changes in pressure when functions such as gear changes take place. It may be necessary to refer to vehicle specific information to identify the applicable pressure circuit (and its function) for the identified pressure sensor.	The voltage, frequency or value of the pressure sensor signal is incorrect (undefined fault). Sensor voltage / signal could be: out of normal operating range, constant value, non-existent, corrupt. Possible fault in the pressure sensor or wiring e.g. short / open circuits, circuit resistance, incorrect sensor resistance (if applicable) or, fault with the reference / operating voltage; interference from other circuits can also affect sensor signals. Where a pressure switch is fitted, it is possible that the switch contacts are not opening / closing correctly or the switch contacts have a high resistance. Refer to list of sensor / actuator checks on page 19.
P0846	Transmission Fluid Pressure Sensor / Switch "B" Circuit Range / Performance	Transmission systems can make use of fluid pressure to control various functions, including gear changes. Fluid pressure sensors are therefore used to monitor pressures and changes in pressure when functions such as gear changes take place. It may be necessary to refer to vehicle specific information to identify the applicable pressure circuit (and its function) for the identified pressure sensor.	The voltage, frequency or value of the pressure sensor signal is within the normal operating range / tolerance but the signal is not plausible or is incorrect due to an undefined fault. The sensor signal might not match the operating conditions indicated by other sensors or, the sensor signal is not changing / responding as expected. Possible fault in the pressure sensor or wiring e.g. short / open circuits, circuit resistance, incorrect sensor resistance (if applicable) or, fault with the reference / operating voltage; interference from other circuits can also affect sensor signals. Where a pressure switch is fitted, it is possible that the switch contacts are not opening / closing correctly or the switch contacts have a high resistance. Refer to list of sensor / actuator checks on page 19. It is also possible that the pressure sensor is operating correctly but the pressure being measured by the sensor is unacceptable or incorrect for the operating conditions; if possible, use alternative method of checking pressure to establish if sensor system is operating correctly and whether the pressure indicated by the sensor is correct / incorrect.
P0847	Transmission Fluid Pressure Sensor / Switch "B" Circuit Low	Transmission systems can make use of fluid pressure to control various functions, including gear changes. Fluid pressure sensors are therefore used to monitor pressures and changes in pressure when functions such as gear changes take place. It may be necessary to refer to vehicle specific information to identify the applicable pressure circuit (and its function) for the identified pressure sensor.	The voltage, frequency or value of the pressure sensor signal is either "zero" (no signal) or, the signal exists but it is below the normal operating range (lower than the minimum operating limit). Possible fault in the pressure sensor or wiring e.g. short / open circuits, circuit resistance, incorrect sensor resistance (if applicable) or, fault with the reference / operating voltage; interference from other circuits can also affect sensor signals. Where a pressure switch is fitted, it is possible that the switch contacts are not opening / closing correctly or, the contacts are sticking (signal is remaining low when it should have risen to the high value); on some circuits, a high resistance on the contacts could also cause a low signal value to exist. Refer to list of sensor / actuator checks on page 19.
P0848	Transmission Fluid Pressure Sensor / Switch "B" Circuit High	Transmission systems can make use of fluid pressure to control various functions, including gear changes. Fluid pressure sensors are therefore used to monitor pressures and changes in pressure when functions such as gear changes take place. It may be necessary to refer to vehicle specific information to identify the applicable pressure circuit (and its function) for the identified pressure sensor.	The voltage, frequency or value of the pressure sensor signal is either at full value (e.g. battery voltage with no signal or frequency) or, the signal exists but it is above the normal operating range (higher than the maximum operating limit). Possible fault in the pressure sensor or wiring e.g. short / open circuits, circuit resistance, incorrect sensor resistance (if applicable) or, fault with the reference / operating voltage; interference from other circuits can also affect sensor signals. Where a pressure switch is fitted, it is possible that the switch contacts are not opening / closing correctly or, the contacts are sticking (signal is remaining high when it should have dropped to low value); on some circuits, a high resistance on the contacts could also cause a high signal value to exist. Refer to list of sensor / actuator checks on page 19.
P0849	Transmission Fluid Pressure Sensor / Switch "B" Circuit Intermittent	Transmission systems can make use of fluid pressure to control various functions, including gear changes. Fluid pressure sensors are therefore used to monitor pressures and changes in pressure when functions such as gear changes take place. It may be necessary to refer to vehicle specific information to identify the applicable pressure circuit (and its function) for the identified pressure sensor.	The voltage, frequency or value of the pressure sensor signal is intermittent (likely to be out of normal operating range / tolerance when intermittent fault is detected) or, the signal / voltage is erratic (e.g. signal changes are irregular / unpredictable). Possible fault with wiring or connections (including internal sensor connections); also interference from other circuits can affect the signal. Where a pressure switch is fitted, it is possible that the switch contacts are not opening / closing correctly or the switch contacts have a high resistance. Refer to list of sensor / actuator checks on page 19. It is also possible that the pressure could be changing erratically, which could cause the fault code to be activated; if necessary, check pressure with a separate gauge to check for erratic pressure changes.
P084A	Transmission Fluid Pressure Sensor / Switch "H" Circuit	Transmission systems can make use of fluid pressure to control various functions, including gear changes. Fluid pressure sensors are therefore used to monitor pressures and changes in pressure when functions such as gear changes take place. It may be necessary to refer to vehicle specific information to identify the applicable pressure circuit (and its function) for the identified pressure sensor.	The voltage, frequency or value of the pressure sensor signal is incorrect (undefined fault). Sensor voltage / signal could be: out of normal operating range, constant value, non-existent, corrupt. Possible fault in the pressure sensor or wiring e.g. short / open circuits, circuit resistance, incorrect sensor resistance (if applicable) or, fault with the reference / operating voltage; interference from other circuits can also affect sensor signals. Where a pressure switch is fitted, it is possible that the switch contacts are not opening / closing correctly or the switch contacts have a high resistance. Refer to list of sensor / actuator checks on page 19.
P084B	Transmission Fluid Pressure Sensor / Switch "H" Circuit Range / Performance	Transmission systems can make use of fluid pressure to control various functions, including gear changes. Fluid pressure sensors are therefore used to monitor pressures and changes in pressure when functions such as gear changes take place. It may be necessary to refer to vehicle specific information to identify the applicable pressure circuit (and its function) for the identified pressure sensor.	The voltage, frequency or value of the pressure sensor signal is within the normal operating range / tolerance but the signal is not plausible or is incorrect due to an undefined fault. The sensor signal might not match the operating conditions indicated by other sensors or, the sensor signal is not changing / responding as expected. Possible fault in the pressure sensor or wiring e.g. short / open circuits, circuit resistance, incorrect sensor resistance (if applicable) or, fault with the reference / operating voltage; interference from other circuits can also affect sensor signals. Where a pressure switch is fitted, it is possible that the switch contacts are not opening / closing correctly or the switch contacts have a high resistance. Refer to list of sensor / actuator checks on page 19. It is also possible that the pressure sensor is operating correctly but the pressure being measured by the sensor is unacceptable or incorrect for the operating conditions; if possible, use alternative method of checking pressure to establish if sensor system is operating correctly and whether the pressure indicated by the sensor is correct / incorrect.

NOTE 1. Check for other fault codes that could provide additional information. **NOTE 2.** Communication between control units can pass via a CAN-Bus system; refer to Page 51 for CAN-Bus checks.

NOTE 3. If a fault cannot be located, it is also possible that the control unit is at fault. **NOTE 4.** Refer to Page 53 for list of pages that show the ISO standard for component locations e.g. Sensor A - Bank 1.

Fault code	EOBD / ISO Description	Component / System Description	Meaningful Description and Quick Check
P084C	Transmission Fluid Pressure Sensor / Switch "H" Circuit Low	Transmission systems can make use of fluid pressure to control various functions, including gear changes. Fluid pressure sensors are therefore used to monitor pressures and changes in pressure when functions such as gear changes take place. It may be necessary to refer to vehicle specific information to identify the applicable pressure circuit (and its function) for the identified pressure sensor.	The voltage, frequency or value of the pressure sensor signal is either "zero" (no signal) or, the signal exists but it is below the normal operating range (lower than the minimum operating limit). Possible fault in the pressure sensor or wiring e.g. short / open circuits, circuit resistance, incorrect sensor resistance (if applicable) or, fault with the reference / operating voltage; interference from other circuits can also affect sensor signals. Where a pressure switch is fitted, it is possible that the switch contacts are not opening / closing correctly or, the contacts are sticking (signal is remaining low when it should have risen to the high value); on some circuits, a high resistance on the contacts could also cause a low signal value to exist. Refer to list of sensor / actuator checks on page 19.
P084D	Transmission Fluid Pressure Sensor / Switch "H" Circuit High	Transmission systems can make use of fluid pressure to control various functions, including gear changes. Fluid pressure sensors are therefore used to monitor pressures and changes in pressure when functions such as gear changes take place. It may be necessary to refer to vehicle specific information to identify the applicable pressure circuit (and its function) for the identified pressure sensor.	The voltage, frequency or value of the pressure sensor signal is either at full value (e.g. battery voltage with no signal or frequency) or, the signal exists but it is above the normal operating range (higher than the maximum operating limit). Possible fault in the pressure sensor or wiring e.g. short / open circuits, circuit resistance, incorrect sensor resistance (if applicable) or, fault with the reference / operating voltage; interference from other circuits can also affect sensor signals. Where a pressure switch is fitted, it is possible that the switch contacts are not opening / closing correctly or, the contacts are sticking (signal is remaining high when it should have dropped to low value); on some circuits, a high resistance on the contacts could also cause a high signal value to exist. Refer to list of sensor / actuator checks on page 19.
P084E	Transmission Fluid Pressure Sensor / Switch "H" Circuit Intermittent	Transmission systems can make use of fluid pressure to control various functions, including gear changes. Fluid pressure sensors are therefore used to monitor pressures and changes in pressure when functions such as gear changes take place. It may be necessary to refer to vehicle specific information to identify the applicable pressure circuit (and its function) for the identified pressure sensor.	The voltage, frequency or value of the pressure sensor signal is intermittent (likely to be out of normal operating range / tolerance when intermittent fault is detected) or, the signal / voltage is erratic (e.g. signal changes are irregular / unpredictable). Possible fault with wiring or connections (including internal sensor connections); also interference from other circuits can affect the signal. Where a pressure switch is fitted, it is possible that the switch contacts are not opening / closing correctly or the switch contacts have a high resistance. Refer to list of sensor / actuator checks on page 19. It is also possible that the pressure could be changing erratically, which could cause the fault code to be activated; if necessary, check pressure with a separate gauge to check for erratic pressure changes.
P084F	ISO / SAE reserved		Not used, reserved for future allocation.
P0850	Park / Neutral Switch Input Circuit	The park / neutral switch is used to indicate that the driver has selected Park or Neutral (using the gear selection lever); the signal will be passed to the transmission control unit but it can also be passed to other systems.	The voltage or value of the switch signal is incorrect (undefined fault). Switch voltage / signal could be: out of normal operating range, constant value, non-existent, corrupt. Possible fault in the switch or wiring e.g. short / open circuits, circuit resistance or, fault with the reference / operating voltage; the switch contacts could also be faulty e.g. not opening / closing correctly or have a high contact resistance. Refer to list of sensor / actuator checks on page 19. Also check switch operating mechanism (gear lever / linkage) to ensure switch is being operated correctly. Check for any base settings for the switch position.
P0851	Park / Neutral Switch Input Circuit Low	The park / neutral switch is used to indicate that the driver has selected Park or Neutral (using the gear selection lever); the signal will be passed to the transmission control unit but it can also be passed to other systems.	The voltage or value of the switch signal is either "zero" (no signal) or, the signal exists but it is below the normal operating range (lower than the minimum operating limit). Possible fault in the switch or wiring e.g. short / open circuits, circuit resistance or, fault with the reference / operating voltage; the switch contacts could also be faulty e.g. not opening / closing correctly or have a high contact resistance. Refer to list of sensor / actuator checks on page 19. Note that the fault code can be activated if the switch output remains at the low value e.g. zero volts and does not rise to the high value e.g. battery voltage (the control unit will detect no switch signal change but it has detected change in operating conditions). Also check switch operating mechanism (gear lever / linkage) to ensure switch is being operated correctly. Check for any base settings for the switch position.
P0852	Park / Neutral Switch Input Circuit High	The park / neutral switch is used to indicate that the driver has selected Park or Neutral (using the gear selection lever); the signal will be passed to the transmission control unit but it can also be passed to other systems.	The voltage or value of the switch signal is either at full value (e.g. battery voltage with no on / off signal) or, the signal exists but it is above the normal operating range (higher than the maximum operating limit). Possible fault in the switch or wiring e.g. short / open circuits, circuit resistance or, fault with the reference / operating voltage; the switch contacts could also be faulty e.g. not opening / closing correctly or have a high contact resistance. Refer to list of sensor / actuator checks on page 19. Note that the fault code can be activated if the switch output remains at the high value e.g. battery voltage and does not fall to the low value e.g. zero volts (the control unit will detect no switch signal change but it has detected change in operating conditions). Also check switch operating mechanism (gear lever / linkage) to ensure switch is being operated correctly. Check for any base settings for the switch position.
P0853	Drive Switch Input Circuit	The "Drive" switch is used to indicate that the driver has selected Drive for the transmission (using the gear selection lever); the signal will be passed to the transmission control unit but it can also be passed to other systems.	The voltage or value of the switch signal is incorrect (undefined fault). Switch voltage / signal could be: out of normal operating range, constant value, non-existent, corrupt. Possible fault in the switch or wiring e.g. short / open circuits, circuit resistance or, fault with the reference / operating voltage; the switch contacts could also be faulty e.g. not opening / closing correctly or have a high contact resistance. Refer to list of sensor / actuator checks on page 19. Also check switch operating mechanism (gear lever / linkage) to ensure switch is being operated correctly. Check for any base settings for the switch position.
P0854	Drive Switch Input Circuit Low	The "Drive" switch is used to indicate that the driver has selected Drive for the transmission (using the gear selection lever); the signal will be passed to the transmission control unit but it can also be passed to other systems.	The voltage or value of the switch signal is either "zero" (no signal) or, the signal exists but it is below the normal operating range (lower than the minimum operating limit). Possible fault in the switch or wiring e.g. short / open circuits, circuit resistance or, fault with the reference / operating voltage; the switch contacts could also be faulty e.g. not opening / closing correctly or have a high contact resistance. Refer to list of sensor / actuator checks on page 19. Note that the fault code can be activated if the switch output remains at the low value e.g. zero volts and does not rise to the high value e.g. battery voltage (the control unit will detect no switch signal change but it has detected change in operating conditions). Also check switch operating mechanism (gear lever / linkage) to ensure switch is being operated correctly. Check for any base settings for the switch position.

code	Description		
P0855	Drive Switch Input Circuit High	The "Drive" switch is used to indicate that the driver has selected Drive for the transmission (using the gear selection lever); the signal will be passed to the transmission control unit but it can also be passed to other systems.	The voltage or value of the switch signal is either at full value (e.g. battery voltage with no on / off signal) or, the signal exists but it is above the normal operating range (higher than the maximum operating limit). Possible fault in the switch or wiring e.g. short / open circuits, circuit resistance or, fault with the reference / operating voltage; the switch contacts could also be faulty e.g. not opening / closing correctly or have a high contact resistance. Refer to list of sensor / actuator checks on page 19. Note that the fault code can be activated for systems with position switches, if the switch output remains at the high value e.g. battery voltage and does not fall to the low value e.g. zero volts (the control unit will detect no switch signal change but it has detected change in operating conditions). Also check switch operating mechanism (gear lever / linkage) to ensure switch is being operated correctly. Check for any base settings for the switch position.
P0856	Traction Control Input Signal	The traction control system control module / unit will communicate with the engine control unit to enable control of engine torque, and other for other control functions; information can pass both ways between the control units.	The fault code indicates that there is a fault with the signal passing between the traction control and main control units. The voltage, frequency or value of the signal is incorrect (undefined fault). The voltage / signal could be: out of normal operating range, constant value, non-existent, corrupt. Possible fault in the wiring e.g. short / open circuits, circuit resistance, incorrect sensor resistance; interference from other circuits can also affect sensor signals. It is also possible that there is a fault with one of the control units.
P0857	Traction Control Input Signal Range / Performance	The traction control system control module / unit will communicate with the engine control unit to enable control of engine torque, and other for other control functions; information can pass both ways between the control units.	The fault code indicates that there is a fault with the signal passing between the traction control and main control units. The voltage, frequency or value of the signal is within the normal operating range / tolerance but the signal is not plausible or is incorrect due to an undefined fault. The signal might not match the operating conditions indicated by other sensors or, the sensor signal is not changing / responding as expected to the changes in operating conditions. Possible fault in the wiring e.g. short / open circuits, circuit resistance, incorrect sensor resistance; interference from other circuits can also affect signals. It is also possible that there is a fault with one of the control units.
P0858	Traction Control Input Signal Low	The traction control system control module / unit will communicate with the engine control unit to enable control of engine torque, and other for other control functions; information can pass both ways between the control units.	The fault code indicates that there is a fault with the signal passing between the traction control and main control units. The voltage, frequency or value of the signal is either "zero" (zero volts or no signal) or, the signal exists but it is below the normal operating range (voltage, frequency or signal value is lower than the minimum operating limit). Possible fault in the sensor or wiring e.g. short / open circuits, circuit resistance; interference from other circuits can also affect sensor signals. It is also possible that there is a fault with one of the control units.
P0859	Traction Control Input Signal High	The traction control system control module / unit will communicate with the engine control unit to enable control of engine torque, and other for other control functions; information can pass both ways between the control units.	The fault code indicates that there is a fault with the signal passing between the traction control and main control units. The voltage, frequency or value of the signal is either at full value (e.g. battery voltage with no signal or frequency) or, the signal exists but it is above the normal operating range (voltage, frequency or signal value is higher than the maximum operating limit). Possible fault in the sensor or wiring e.g. short / open circuits, circuit resistance; interference from other circuits can also affect sensor signals. It is also possible that there is a fault with one of the control units.
P085A	Gear Shift Module "B" Communication Circuit	The fault code refers to one of the gear shift module communication circuits. It may be necessary to refer to vehicle specific information to identify the module and communication circuits. The communication circuit could relate to the circuits carrying information between the gear shift module and other control units.	The voltage, frequency or value of the communication signal is incorrect (undefined fault). Voltage / signal could be: out of normal operating range, constant value, non-existent, corrupt. Possible fault in the wiring e.g. short / open circuits, circuit resistance; interference from other circuits can also affect signals. If there are no obvious wiring connections faults, it is possible that there is a fault with the control module.
P085B	Gear Shift Module "B" Communication Circuit Low	The fault code refers to one of the gear shift module communication circuits. It may be necessary to refer to vehicle specific information to identify the module and communication circuits. The communication circuit could relate to the circuits carrying information between the gear shift module and other control units.	The voltage, frequency or value of the communication signal is either "zero" (no signal) or, the signal exists but it is below the normal operating range (lower than the minimum operating limit). Possible fault in the wiring e.g. short / open circuits, circuit resistance; interference from other circuits can also affect signals. If there are no obvious wiring connections faults, it is possible that there is a fault with the control module.
P085C	Gear Shift Module "B" Communication Circuit High	The fault code refers to one of the gear shift module communication circuits. It may be necessary to refer to vehicle specific information to identify the module and communication circuits. The communication circuit could relate to the circuits carrying information between the gear shift module and other control units.	The voltage, frequency or value of the communication signal is either at full value (e.g. battery voltage with no signal or frequency) or, the signal exists but it is above the normal operating range (higher than the maximum operating limit). Possible fault in the wiring e.g. short / open circuits, circuit resistance; interference from other circuits can also affect signals. If there are no obvious wiring connections faults, it is possible that there is a fault with the control module.
P0860	Gear Shift Module "A" Communication Circuit	The fault code refers to one of the gear shift module communication circuits. It may be necessary to refer to vehicle specific information to identify the module and communication circuits. The communication circuit could relate to the circuits carrying information between the gear shift module and other control units.	The voltage, frequency or value of the communication signal is incorrect (undefined fault). Voltage / signal could be: out of normal operating range, constant value, non-existent, corrupt. Possible fault in the wiring e.g. short / open circuits, circuit resistance; interference from other circuits can also affect signals. If there are no obvious wiring connections faults, it is possible that there is a fault with the control module.
P0861	Gear Shift Module "A" Communication Circuit Low	The fault code refers to one of the gear shift module communication circuits. It may be necessary to refer to vehicle specific information to identify the module and communication circuits. The communication circuit could relate to the circuits carrying information between the gear shift module and other control units.	The voltage, frequency or value of the communication signal is either "zero" (no signal) or, the signal exists but it is below the normal operating range (lower than the minimum operating limit). Possible fault in the wiring e.g. short / open circuits, circuit resistance; interference from other circuits can also affect signals. If there are no obvious wiring connections faults, it is possible that there is a fault with the control module.
P0862	Gear Shift Module "A" Communication Circuit High	The fault code refers to one of the gear shift module communication circuits. It may be necessary to refer to vehicle specific information to identify the module and communication circuits. The communication circuit could relate to the circuits carrying information between the gear shift module and other control units.	The voltage, frequency or value of the communication signal is either at full value (e.g. battery voltage with no signal or frequency) or, the signal exists but it is above the normal operating range (higher than the maximum operating limit). Possible fault in the wiring e.g. short / open circuits, circuit resistance; interference from other circuits can also affect signals. If there are no obvious wiring connections faults, it is possible that there is a fault with the control module.

NOTE 1. Check for other fault codes that could provide additional information. **NOTE 2.** Communication between control units can pass via a CAN-Bus system; refer to Page 51 for CAN-Bus checks.

NOTE 3. If a fault cannot be located, it is also possible that the control unit is at fault. **NOTE 4.** Refer to Page 53 for list of pages that show the ISO standard for component locations e.g. Sensor A - Bank 1.

Fault code	EOBD / ISO Description	Component / System Description	Meaningful Description and Quick Check
P0863	TCM Communication Circuit	The Transmission Control Module will communicate with other control units (sharing information between control units or passing fault related information). The communication circuit may form part of a CAN-Bus network	The fault code indicates a fault with the communication signal / circuit from the TCM to the engine control unit (or to other control units). The voltage, frequency or value of the signal is incorrect (undefined fault). Voltage / signal could be: out of normal operating range, constant value, non-existent, corrupt. Possible fault in the wiring e.g. short / open circuits, circuit resistance; interference from other circuits can also affect signals. If there are no faults in the wiring / connections, it could indicate a possible fault with the TCM or other control unit.
P0864	TCM Communication Circuit Range / Performance	The Transmission Control Module will communicate with other control units (sharing information between control units or passing fault related information). The communication circuit may form part of a CAN-Bus network	The fault code indicates a fault with the communication signal / circuit from the TCM to the engine control unit (or to other control units). The voltage, frequency or value of the signal is within the normal operating range / tolerance but the signal is not plausible or is incorrect due to an undefined fault. The signal might not match the operating conditions indicated by other sensors or, the signal is not changing / responding as expected to the changes in conditions. Possible fault in the wiring e.g. short / open circuits, circuit resistance; interference from other circuits can also affect signals. If there are no faults in the wiring / connections, it could indicate a possible fault with the TCM or other control unit.
P0865	TCM Communication Circuit Low	The Transmission Control Module will communicate with other control units (sharing information between control units or passing fault related information). The communication circuit may form part of a CAN-Bus network	The fault code indicates a fault with the communication signal / circuit from the TCM to the engine control unit (or to other control units). The voltage, frequency or value of the signal is either "zero" (no signal) or, the signal exists but it is below the normal operating range (lower than the minimum operating limit). Possible fault in the wiring e.g. short / open circuits, circuit resistance; interference from other circuits can also affect signals. If there are no faults in the wiring / connections, it could indicate a possible fault with the TCM or other control unit.
P0866	TCM Communication Circuit High	The Transmission Control Module will communicate with other control units (sharing information between control units or passing fault related information). The communication circuit may form part of a CAN-Bus network	The fault code indicates a fault with the communication signal / circuit from the TCM to the engine control unit (or to other control units). The voltage, frequency or value of the signal is either at full value (e.g. battery voltage with no signal or frequency) or, the signal exists but it is above the normal operating range (higher than the maximum operating limit). Possible fault in the wiring e.g. short / open circuits, circuit resistance; interference from other circuits can also affect signals. If there are no faults in the wiring / connections, it could indicate a possible fault with the TCM or other control unit.
P0867	Transmission Fluid Pressure	Transmission systems can make use of fluid pressure to control various functions, including gear changes. Fluid pressure sensors are therefore used to monitor pressures and changes in pressure when functions such as gear changes take place. It may be necessary to refer to vehicle specific information to identify the applicable pressure circuit (and its function).	The voltage, frequency or value of the pressure sensor signal is within the normal operating range / tolerance but is indicating an unacceptable pressure value. The control unit has not detected an electrical pressure sensor fault, therefore the assessment is that the pressure is incorrect for the operating conditions (indicated by other sensors) or, the pressure does not match values expected by the control unit. If possible, use alternative method of checking pressure to establish if sensor system is operating correctly and whether the pressure indicated by the sensor is correct / incorrect. If the pressure is incorrect, refer to vehicle specific information to identify system operation and components and check as applicable e.g. pressure pump, pressure regulator / pressure valves, filters oil passages etc. If the oil pressure appears to be correct, it could indicate a pressure sensor fault. Also check oil quantity, grade and condition. Refer to list of sensor / actuator checks on page 19.
P0868	Transmission Fluid Pressure Low	Transmission systems can make use of fluid pressure to control various functions, including gear changes. Fluid pressure sensors are therefore used to monitor pressures and changes in pressure when functions such as gear changes take place. It may be necessary to refer to vehicle specific information to identify the applicable pressure circuit (and its function).	The voltage, frequency or value of the pressure sensor signal is within the normal operating range / tolerance but is indicating a low pressure value. The control unit has not detected an electrical pressure sensor fault, therefore the assessment is that the pressure is too low for the operating conditions (indicated by other sensors) or, the pressure does not match values expected by the control unit. If possible, use alternative method of checking pressure to establish if sensor system is operating correctly and whether the pressure indicated by the sensor is correct / incorrect. If the pressure is incorrect, refer to vehicle specific information to identify system operation and components and check as applicable e.g. pressure pump, pressure regulator / pressure valves, filters oil passages etc. If the oil pressure appears to be correct, it could indicate a pressure sensor fault. Also check oil quantity, grade and condition. Refer to list of sensor / actuator checks on page 19.
P0869	Transmission Fluid Pressure High	Transmission systems can make use of fluid pressure to control various functions, including gear changes. Fluid pressure sensors are therefore used to monitor pressures and changes in pressure when functions such as gear changes take place. It may be necessary to refer to vehicle specific information to identify the applicable pressure circuit (and its function) .	The voltage, frequency or value of the pressure sensor signal is within the normal operating range / tolerance but is indicating a high pressure value. The control unit has not detected an electrical pressure sensor fault, therefore the assessment is that the pressure is too high for the operating conditions (indicated by other sensors) or, the pressure does not match values expected by the control unit. If possible, use alternative method of checking pressure to establish if sensor system is operating correctly and whether the pressure indicated by the sensor is correct / incorrect. If the pressure is incorrect, refer to vehicle specific information to identify system operation and components and check as applicable e.g. pressure pump, pressure regulator / pressure valves, filters oil passages etc. If the oil pressure appears to be correct, it could indicate a pressure sensor fault. Also check oil quantity, grade and condition. Refer to list of sensor / actuator checks on page 19.
P0870	Transmission Fluid Pressure Sensor / Switch "C" Circuit	Transmission systems can make use of fluid pressure to control various functions, including gear changes. Fluid pressure sensors are therefore used to monitor pressures and changes in pressure when functions such as gear changes take place. It may be necessary to refer to vehicle specific information to identify the applicable pressure circuit (and its function).	The voltage, frequency or value of the pressure sensor signal is incorrect (undefined fault). Sensor voltage / signal could be: out of normal operating range, constant value, non-existent, corrupt. Possible fault in the pressure sensor or wiring e.g. short / open circuits, circuit resistance, incorrect sensor resistance (if applicable) or, fault with the reference / operating voltage; interference from other circuits can also affect sensor signals. Where a pressure switch is fitted, it is possible that the switch contacts are not opening / closing correctly or the switch contacts have a high resistance. Refer to list of sensor / actuator checks on page 19.
P0871	Transmission Fluid Pressure Sensor / Switch "C" Circuit Range / Performance	Transmission systems can make use of fluid pressure to control various functions, including gear changes. Fluid pressure sensors are therefore used to monitor pressures and changes in pressure when functions such as gear changes take place. It may be necessary to refer to vehicle specific information to identify the applicable pressure circuit (and its function) for the identified pressure sensor.	The voltage, frequency or value of the pressure sensor signal is within the normal operating range / tolerance but the signal is not plausible or is incorrect due to an undefined fault. The sensor signal might not match the operating conditions indicated by other sensors or, the sensor signal is not changing / responding as expected. Possible fault in the pressure sensor or wiring e.g. short / open circuits, circuit resistance, incorrect sensor resistance (if applicable) or, fault with the reference / operating voltage; interference from other circuits can also affect sensor signals. Where a pressure switch is fitted, it is possible that the switch contacts are not opening / closing correctly or the switch contacts have a high resistance. Refer to list of sensor / actuator checks on page 19. It is also possible that the pressure sensor is operating correctly but the pressure being measured by the sensor is unacceptable or incorrect for the operating conditions; if possible, use alternative method of checking pressure to establish if sensor system is operating correctly and whether the pressure indicated by the sensor is correct / incorrect.

NOTE 1. Check for other fault codes that could provide additional information.   **NOTE 2.** Communication between control units can pass via a CAN-Bus system; refer to Page 51 for CAN-Bus checks.
NOTE 3. If a fault cannot be located, it is also possible that the control unit is at fault.   **NOTE 4.** Refer to Page 53 for list of pages that show the ISO standard for component locations e.g. Sensor A - Bank 1.

code	Description		
P0872	Transmission Fluid Pressure Sensor / Switch "C" Circuit Low	Transmission systems can make use of fluid pressure to control various functions, including gear changes. Fluid pressure sensors are therefore used to monitor pressures and changes in pressure when functions such as gear changes take place. It may be necessary to refer to vehicle specific information to identify the applicable pressure circuit (and its function) for the identified pressure sensor.	The voltage, frequency or value of the pressure sensor signal is either "zero" (no signal) or, the signal exists but it is below the normal operating range (lower than the minimum operating limit). Possible fault in the pressure sensor or wiring e.g. short / open circuits, circuit resistance, incorrect sensor resistance (if applicable) or, fault with the reference / operating voltage; interference from other circuits can also affect sensor signals. Where a pressure switch is fitted, it is possible that the switch contacts are not opening / closing correctly or, the contacts are sticking (signal is remaining low when it should have risen to the high value); on some circuits, a high resistance on the contacts could also cause a low signal value to exist. Refer to list of sensor / actuator checks on page 19.
P0873	Transmission Fluid Pressure Sensor / Switch "C" Circuit High	Transmission systems can make use of fluid pressure to control various functions, including gear changes. Fluid pressure sensors are therefore used to monitor pressures and changes in pressure when functions such as gear changes take place. It may be necessary to refer to vehicle specific information to identify the applicable pressure circuit (and its function) for the identified pressure sensor.	The voltage, frequency or value of the pressure sensor signal is either at full value (e.g. battery voltage with no signal or frequency) or, the signal exists but it is above the normal operating range (higher than the maximum operating limit). Possible fault in the pressure sensor or wiring e.g. short / open circuits, circuit resistance, incorrect sensor resistance (if applicable) or, fault with the reference / operating voltage; interference from other circuits can also affect sensor signals. Where a pressure switch is fitted, it is possible that the switch contacts are not opening / closing correctly or, the contacts are sticking (signal is remaining high when it should have dropped to low value); on some circuits, a high resistance on the contacts could also cause a high signal value to exist. Refer to list of sensor / actuator checks on page 19.
P0874	Transmission Fluid Pressure Sensor / Switch "C" Circuit Intermittent	Transmission systems can make use of fluid pressure to control various functions, including gear changes. Fluid pressure sensors are therefore used to monitor pressures and changes in pressure when functions such as gear changes take place. It may be necessary to refer to vehicle specific information to identify the applicable pressure circuit (and its function) for the identified pressure sensor.	The voltage, frequency or value of the pressure sensor signal is intermittent (likely to be out of normal operating range / tolerance when intermittent fault is detected) or, the signal / voltage is erratic (e.g. signal changes are irregular / unpredictable). Possible fault with wiring or connections (including internal sensor connections); also interference from other circuits can affect the signal. Where a pressure switch is fitted, it is possible that the switch contacts are not opening / closing correctly or the switch contacts have a high resistance. Refer to list of sensor / actuator checks on page 19. It is also possible that the pressure could be changing erratically, which could cause the fault code to be activated; if necessary, check pressure with a separate gauge to check for erratic pressure changes.
P0875	Transmission Fluid Pressure Sensor / Switch "D" Circuit	Transmission systems can make use of fluid pressure to control various functions, including gear changes. Fluid pressure sensors are therefore used to monitor pressures and changes in pressure when functions such as gear changes take place. It may be necessary to refer to vehicle specific information to identify the applicable pressure circuit (and its function) for the identified pressure sensor.	The voltage, frequency or value of the pressure sensor signal is incorrect (undefined fault). Sensor voltage / signal could be: out of normal operating range, constant value, non-existent, corrupt. Possible fault in the pressure sensor or wiring e.g. short / open circuits, circuit resistance, incorrect sensor resistance (if applicable) or, fault with the reference / operating voltage; interference from other circuits can also affect sensor signals. Where a pressure switch is fitted, it is possible that the switch contacts are not opening / closing correctly or the switch contacts have a high resistance. Refer to list of sensor / actuator checks on page 19.
P0876	Transmission Fluid Pressure Sensor / Switch "D" Circuit Range / Performance	Transmission systems can make use of fluid pressure to control various functions, including gear changes. Fluid pressure sensors are therefore used to monitor pressures and changes in pressure when functions such as gear changes take place. It may be necessary to refer to vehicle specific information to identify the applicable pressure circuit (and its function) for the identified pressure sensor.	The voltage, frequency or value of the pressure sensor signal is within the normal operating range / tolerance but the signal is not plausible or is incorrect due to an undefined fault. The sensor signal might not match the operating conditions indicated by other sensors or, the sensor signal is not changing / responding as expected. Possible fault in the pressure sensor or wiring e.g. short / open circuits, circuit resistance, incorrect sensor resistance (if applicable) or, fault with the reference / operating voltage; interference from other circuits can also affect sensor signals. Where a pressure switch is fitted, it is possible that the switch contacts are not opening / closing correctly or the switch contacts have a high resistance. Refer to list of sensor / actuator checks on page 19. It is also possible that the pressure sensor is operating correctly but the pressure being measured by the sensor is unacceptable or incorrect for the operating conditions; if possible, use alternative method of checking pressure to establish if sensor system is operating correctly and whether the pressure indicated by the sensor is correct / incorrect.
P0877	Transmission Fluid Pressure Sensor / Switch "D" Circuit Low	Transmission systems can make use of fluid pressure to control various functions, including gear changes. Fluid pressure sensors are therefore used to monitor pressures and changes in pressure when functions such as gear changes take place. It may be necessary to refer to vehicle specific information to identify the applicable pressure circuit (and its function) for the identified pressure sensor.	The voltage, frequency or value of the pressure sensor signal is either "zero" (no signal) or, the signal exists but it is below the normal operating range (lower than the minimum operating limit). Possible fault in the pressure sensor or wiring e.g. short / open circuits, circuit resistance, incorrect sensor resistance (if applicable) or, fault with the reference / operating voltage; interference from other circuits can also affect sensor signals. Where a pressure switch is fitted, it is possible that the switch contacts are not opening / closing correctly or, the contacts are sticking (signal is remaining low when it should have risen to the high value); on some circuits, a high resistance on the contacts could also cause a low signal value to exist. Refer to list of sensor / actuator checks on page 19.
P0878	Transmission Fluid Pressure Sensor / Switch "D" Circuit High	Transmission systems can make use of fluid pressure to control various functions, including gear changes. Fluid pressure sensors are therefore used to monitor pressures and changes in pressure when functions such as gear changes take place. It may be necessary to refer to vehicle specific information to identify the applicable pressure circuit (and its function) for the identified pressure sensor.	The voltage, frequency or value of the pressure sensor signal is either at full value (e.g. battery voltage with no signal or frequency) or, the signal exists but it is above the normal operating range (higher than the maximum operating limit). Possible fault in the pressure sensor or wiring e.g. short / open circuits, circuit resistance, incorrect sensor resistance (if applicable) or, fault with the reference / operating voltage; interference from other circuits can also affect sensor signals. Where a pressure switch is fitted, it is possible that the switch contacts are not opening / closing correctly or, the contacts are sticking (signal is remaining high when it should have dropped to low value); on some circuits, a high resistance on the contacts could also cause a high signal value to exist. Refer to list of sensor / actuator checks on page 19.
P0879	Transmission Fluid Pressure Sensor / Switch "D" Circuit Intermittent	Transmission systems can make use of fluid pressure to control various functions, including gear changes. Fluid pressure sensors are therefore used to monitor pressures and changes in pressure when functions such as gear changes take place. It may be necessary to refer to vehicle specific information to identify the applicable pressure circuit (and its function) for the identified pressure sensor.	The voltage, frequency or value of the pressure sensor signal is intermittent (likely to be out of normal operating range / tolerance when intermittent fault is detected) or, the signal / voltage is erratic (e.g. signal changes are irregular / unpredictable). Possible fault with wiring or connections (including internal sensor connections); also interference from other circuits can affect the signal. Where a pressure switch is fitted, it is possible that the switch contacts are not opening / closing correctly or the switch contacts have a high resistance. Refer to list of sensor / actuator checks on page 19. It is also possible that the pressure could be changing erratically, which could cause the fault code to be activated; if necessary, check pressure with a separate gauge to check for erratic pressure changes.

NOTE 1. Check for other fault codes that could provide additional information.   **NOTE 2.** Communication between control units can pass via a CAN-Bus system; refer to Page 51 for CAN-Bus checks.

NOTE 3. If a fault cannot be located, it is also possible that the control unit is at fault.   **NOTE 4.** Refer to Page 53 for list of pages that show the ISO standard for component locations e.g. Sensor A - Bank 1.

Fault code	EOBD / ISO Description	Component / System Description	Meaningful Description and Quick Check
P0880	TCM Power Input Signal	The Transmission Control Module can be provided with one or more power supplies. Note that the power supplies can be fed via a relay or a power control module and that a fuse can be included in the circuit. Also note that some systems may retain a power supply after the ignition is switched off; this is used to prevent loss of data from part of the internal memory system (Keep Alive Memory or KAM). The TCM can also function as the power supply control for some actuators and provide sensor reference or operating voltage to some sensors.	The fault code indicates that there is a fault with the power input to the TCM (undefined fault). The voltage could be out of normal operating range, constant but incorrect value, non-existent, corrupt. Possible fault in the wiring e.g. short / open circuits or circuit resistance. Check power supply from source e.g. from ignition switch or from the battery via a relay and / or fuse; check relay operation if applicable Refer to list of sensor / actuator checks on page 19. Note that a relay could be switched by another control unit, which could also require testing. If power supplies are being provided to the TCM but the fault code re-appears after clearing the code, it could indicate a that the TCM is at fault.
P0881	TCM Power Input Signal Range / Performance	The Transmission Control Module can be provided with one or more power supplies. Note that the power supplies can be fed via a relay or a power control module and that a fuse can be included in the circuit. Also note that some systems may retain a power supply after the ignition is switched off; this is used to prevent loss of data from part of the internal memory system (Keep Alive Memory or KAM). The TCM can also function as the power supply control for some actuators and provide sensor reference or operating voltage to some sensors.	The fault code indicates that there is a fault with the power input to the TCM (undefined fault). The voltage is probably within the normal operating range / tolerance but the voltage is not plausible or is incorrect due to an undefined fault. The voltage might not match the operating conditions (e.g. low voltage but charging voltage is correct) or, the voltage is not changing / responding as expected. Possible fault in the wiring e.g. short / open circuits or circuit resistance. Check power supply from source e.g. from ignition switch or from the battery via a relay and / or fuse; check relay operation if applicable Refer to list of sensor / actuator checks on page 19. Note that a relay could be switched by another control unit, which could also require testing. If power supplies are being provided to the TCM but the fault code re-appears after clearing the code, it could indicate a that the TCM is at fault.
P0882	TCM Power Input Signal Low	The Transmission Control Module can be provided with one or more power supplies. Note that the power supplies can be fed via a relay or a power control module and that a fuse can be included in the circuit. Also note that some systems may retain a power supply after the ignition is switched off; this is used to prevent loss of data from part of the internal memory system (Keep Alive Memory or KAM). The TCM can also function as the power supply control for some actuators and provide sensor reference or operating voltage to some sensors.	The fault code indicates that there is a fault with the power input to the TCM. The voltage could be either "zero" (no supply) or, the voltage exists but it is below the normal operating range (lower than the minimum operating limit). Note that the fault code could be activated because the power supply is OFF when it should be switched ON; this could be caused by a faulty relay, fuse etc. Possible fault in the wiring e.g. short / open circuits or circuit resistance; check actual voltage level in the circuit to ensure that the supply voltage is not significantly lower than the charging / battery voltage. Check power supply from source e.g. from ignition switch or from the battery via a relay and / or fuse to identify whether the power supplies are being switched on / off at the correct times; check relay operation if applicable, Refer to list of sensor / actuator checks on page 19. Note that a relay could be switched by another control unit, which could also require testing. If power supplies are being provided to the TCM but the fault code re-appears after clearing the code, it could indicate a that the TCM is at fault.
P0883	TCM Power Input Signal High	The Transmission Control Module can be provided with one or more power supplies. Note that the power supplies can be fed via a relay or a power control module and that a fuse can be included in the circuit. Also note that some systems may retain a power supply after the ignition is switched off; this is used to prevent loss of data from part of the internal memory system (Keep Alive Memory or KAM). The TCM can also function as the power supply control for some actuators and provide sensor reference or operating voltage to some sensors.	The fault code indicates that there is a fault with the power input to the TCM. The voltage could be either at full value (e.g. battery voltage exists at all times and is not switching off at the correct times) or, the voltage exists but it is above the normal operating range (this could be a charging fault). Note that the fault code could be activated because the power supply is ON when it should be switched OFF; this could be caused by a faulty relay, fuse etc. Possible fault in the wiring e.g. short / open circuits or circuit resistance. Check power supply from source e.g. from ignition switch or from the battery via a relay and / or fuse to identify whether the power supplies are being switched on / off at the correct times; check relay operation if applicable, Refer to list of sensor / actuator checks on page 19. Note that a relay could be switched by another control unit, which could also require testing. If power supplies to the TCM are correct but the fault code re-appears after clearing the code, it could indicate that the TCM is at fault.
P0884	TCM Power Input Signal Intermittent	The Transmission Control Module can be provided with one or more power supplies. Note that the power supplies can be fed via a relay or a power control module and that a fuse can be included in the circuit. Also note that some systems may retain a power supply after the ignition is switched off; this is used to prevent loss of data from part of the internal memory system (Keep Alive Memory or KAM). The TCM can also function as the power supply control for some actuators and provide sensor reference or operating voltage to some sensors.	The fault code indicates that there is a fault with the power input to the TCM. The power supply could be intermittent (likely to be out of normal operating range / tolerance when intermittent fault is detected) or, the voltage level is erratic (e.g. rapid changes to the voltage level). Possible fault in the wiring e.g. intermittent connections, intermittent short / open circuits. Check power supply from source e.g. from ignition switch or from the battery via a relay and / or fuse; check relay operation if applicable, Refer to list of sensor / actuator checks on page 19. Note that a relay could be switched by another control unit, which could also require testing. If power supplies to the TCM appear to be stable but the fault code re-appears after clearing the code, it could indicate a that the TCM is at fault.
P0885	TCM Power Relay Control Circuit / Open	The TCM (Transmission Control Module) power relay will provide electrical power to most of the components on the transmission system. It is normal practice for the control unit to switch the earth circuit for the relay i.e. the control unit will complete the earth path for the relay energising winding, which will cause the relay contacts to close and provide power to the system components. The control unit is therefore able to detect / sense whether there is a power supply to the energising winding circuit and if the circuit is good. A sense circuit can also be used which connects the relay power output terminal to the control unit; this allows the control unit to detect that the relay has switched on the power output circuit.	Refer to list of sensor / actuator checks on page 19 for notes on relay checks. Different manufacturers could use the fault code to refer to different relay circuits. The fault code could refer to the power supply circuit (the power supply to the relay contacts and the circuit from the relay contacts to the system components). Alternatively, the fault code could refer to the relay energising winding circuit (power supply to the energising winding and the switched earth path through to the control unit). The fault code indicates that the voltage on the applicable relay circuit is incorrect (undefined fault but possible open circuit). Check power supplies to relay contacts (usually via a fused circuit) and to relay energising winding. Check operation of energising winding (short / open circuit) and closing / opening of relay contacts. Check all wiring for short / open circuits.

NOTE 1. Check for other fault codes that could provide additional information. NOTE 2. Communication between control units can pass via a CAN-Bus system; refer to Page 51 for CAN-Bus checks. 198

NOTE 3. If a fault cannot be located, it is also possible that the control unit is at fault. NOTE 4. Refer to Page 53 for list of pages that show the ISO standard for component locations e.g. Sensor A - Bank 1.

code	Description		
P0886	TCM Power Relay Control Circuit Low	The TCM (Transmission Control Module) power relay will provide electrical power to most of the components on the transmission system. It is normal practice for the control unit to switch the earth circuit for the relay i.e. the control unit will complete the earth path for the relay energising winding, which will cause the relay contacts to close and provide power to the system components. The control unit is therefore able to detect / sense whether there is a power supply to the energising winding circuit and if the circuit is good. A sense circuit can also be used which connects the relay power output terminal to the control unit; this allows the control unit to detect that the relay has switched on the power output circuit.	Refer to list of sensor / actuator checks on page 19 for notes on relay checks. Different manufacturers could use the fault code to refer to different relay circuits. The fault code could refer to the power supply circuit (the power supply to the relay contacts and the circuit from the relay contacts to the system components). Alternatively, the fault code could refer to the relay energising winding circuit (power supply to the energising winding and the switched earth path through to the control unit). The voltage in the applicable relay circuit is low (likely to be zero volts) at a time when the voltage should be at battery voltage level. Check power supplies to relay contacts (usually via a fused circuit) and to relay energising winding. Check operation of energising winding (short / open circuit) and closing / opening of relay contacts. Check all wiring for short / open circuits.
P0887	TCM Power Relay Control Circuit High	The TCM (Transmission Control Module) power relay will provide electrical power to most of the components on the transmission system. It is normal practice for the control unit to switch the earth circuit for the relay i.e. the control unit will complete the earth path for the relay energising winding, which will cause the relay contacts to close and provide power to the system components. The control unit is therefore able to detect / sense whether there is a power supply to the energising winding circuit and if the circuit is good. A sense circuit can also be used which connects the relay power output terminal to the control unit; this allows the control unit to detect that the relay has switched on the power output circuit.	Refer to list of sensor / actuator checks on page 19 for notes on relay checks. Different manufacturers could use the fault code to refer to different relay circuits. The fault code could refer to the power supply circuit (the power supply to the relay contacts and the circuit from the relay contacts to the system components). Alternatively, the fault code could refer to the relay energising winding circuit (power supply to the energising winding and the switched earth path through to the control unit). The voltage in the relay circuit is high (likely to be battery voltage) when the relay circuit should be off (contacts open). Likely cause is relay contacts remaining closed (either due to relay fault or energising winding is being continuously activated); also possible short in wiring to another power supply or the energising winding earth path is shorted to earth. Check operation of relay, including closing / opening of relay contacts, and check operation of energising winding (short / open circuit). If the relay is not operating correctly, this could be caused by a control unit fault (not switching the relay energising circuit).
P0888	TCM Power Relay Sense Circuit	The TCM (Transmission Control Module) power relay will provide electrical power to most of the components on the transmission system. It is normal practice for the control unit to switch the earth circuit for the relay i.e. the control unit will complete the earth path for the relay energising winding, which will cause the relay contacts to close and provide power to the system components. The control unit is therefore able to detect / sense whether there is a power supply to the energising winding circuit and if the circuit is good. A sense circuit can also be used which connects the relay power output terminal to the control unit; this allows the control unit to detect that the relay has switched on the power output circuit.	Refer to list of sensor / actuator checks on page 19 for notes on relay checks. Different manufacturers could use the fault code to refer to "sensing" of the different relay circuits. The fault code could refer to the power supply circuit (the power supply to the relay contacts and the circuit from the relay contacts to the system components). Alternatively, the fault code could refer to the relay energising winding circuit (power supply to the energising winding and the switched earth path through to the control unit). The fault code indicates that the voltage on the relay sense circuit is incorrect (undefined fault). Check whether the system components are being provided with power from the relay; if the power is being provided, it indicates that the relay is operating but the sense circuit between the relay and the control unit is open circuit. If there is no power supply to the system components, check relay operation, including power supply to relay contacts and relay energising winding. Check for short / open circuit in all wiring and relay energising winding.
P0889	TCM Power Relay Sense Circuit Range / Performance	The TCM (Transmission Control Module) power relay will provide electrical power to most of the components on the transmission system. It is normal practice for the control unit to switch the earth circuit for the relay i.e. the control unit will complete the earth path for the relay energising winding, which will cause the relay contacts to close and provide power to the system components. The control unit is therefore able to detect / sense whether there is a power supply to the energising winding circuit and if the circuit is good. A sense circuit can also be used which connects the relay power output terminal to the control unit; this allows the control unit to detect that the relay has switched on the power output circuit.	Refer to list of sensor / actuator checks on page 19 for notes on relay checks. Different manufacturers could use the fault code to refer to "sensing" of the different relay circuits. The fault code could refer to the power supply circuit (the power supply to the relay contacts and the circuit from the relay contacts to the system components). Alternatively, the fault code could refer to the relay energising winding circuit (power supply to the energising winding and the switched earth path through to the control unit). The fault code indicates that the voltage on the relay sense circuit is incorrect or, the voltage is not as expected under certain operating conditions (e.g. circuit is on when it should be off or, off when it should be on). The fault could be related to relay operation e.g. contacts stuck closed or open, wiring faults (short / open circuit) or possible control unit fault (switching relay at incorrect times). Check all relay wiring and operation, but also check "sense" circuit wiring.
P0890	TCM Power Relay Sense Circuit Low	The TCM (Transmission Control Module) power relay will provide electrical power to most of the components on the transmission system. It is normal practice for the control unit to switch the earth circuit for the relay i.e. the control unit will complete the earth path for the relay energising winding, which will cause the relay contacts to close and provide power to the system components. The control unit is therefore able to detect / sense whether there is a power supply to the energising winding circuit and if the circuit is good. A sense circuit can also be used which connects the relay power output terminal to the control unit; this allows the control unit to detect that the relay has switched on the power output circuit.	Refer to list of sensor / actuator checks on page 19 for notes on relay checks. Different manufacturers could use the fault code to refer to "sensing" of the different relay circuits. The fault code could refer to the power supply circuit (the power supply to the relay contacts and the circuit from the relay contacts to the system components). Alternatively, the fault code could refer to the relay energising winding circuit (power supply to the energising winding and the switched earth path through to the control unit). The fault code indicates that the voltage on the relay sense circuit is low or non-existent. Check whether the system components are being provided with power from the relay; if the power is being provided, it indicates that the relay is operating but the sense circuit between the relay and the control unit is open circuit. If there is no power supply to the system components, check relay operation, including power supply to relay contacts and relay energising winding. Check for short / open circuit in all wiring and relay energising winding.
P0891	TCM Power Relay Sense Circuit High	The TCM (Transmission Control Module) power relay will provide electrical power to most of the components on the transmission system. It is normal practice for the control unit to switch the earth circuit for the relay i.e. the control unit will complete the earth path for the relay energising winding, which will cause the relay contacts to close and provide power to the system components. The control unit is therefore able to detect / sense whether there is a power supply to the energising winding circuit and if the circuit is good. A sense circuit can also be used which connects the relay power output terminal to the control unit; this allows the control unit to detect that the relay has switched on the power output circuit.	Refer to list of sensor / actuator checks on page 19 for notes on relay checks. Different manufacturers could use the fault code to refer to "sensing" of the different relay circuits. The fault code could refer to the power supply circuit (the power supply to the relay contacts and the circuit from the relay contacts to the system components). Alternatively, the fault code could refer to the relay energising winding circuit (power supply to the energising winding and the switched earth path through to the control unit). The fault code indicates that the voltage on the relay sense circuit is high. It is likely that the relay is not switching off the power supply to the system components as expected e.g. when the engine is switched off (voltage will therefore exist in the sense circuit). It is likely that the relay is at fault but it is also possible that there is a short in the wiring to another power supply circuit or, a short on the energising winding earth circuit to another earth.

NOTE 1. Check for other fault codes that could provide additional information.   **NOTE 2.** Communication between control units can pass via a CAN-Bus system; refer to Page 51 for CAN-Bus checks.   **199**
NOTE 3. If a fault cannot be located, it is also possible that the control unit is at fault.   **NOTE 4.** Refer to Page 53 for list of pages that show the ISO standard for component locations e.g. Sensor A - Bank 1.

Fault code	EOBD / ISO Description	Component / System Description	Meaningful Description and Quick Check
P0892	TCM Power Relay Sense Circuit Intermittent	The TCM (Transmission Control Module) power relay will provide electrical power to most of the components on the transmission system. It is normal practice for the control unit to switch the earth circuit for the relay i.e. the control unit will complete the earth path for the relay energising winding, which will cause the relay contacts to close and provide power to the system components. The control unit is therefore able to detect / sense whether there is a power supply to the energising winding circuit and if the circuit is good. A sense circuit can also be used which connects the relay power output terminal to the control unit; this allows the control unit to detect that the relay has switched on the power output circuit.	Refer to list of sensor / actuator checks on page 19 for notes on relay checks. Different manufacturers could use the fault code to refer to "sensing" of the different relay circuits. The fault code could refer to the power supply circuit (the power supply to the relay contacts and the circuit from the relay contacts to the system components). Alternatively, the fault code could refer to the relay energising winding circuit (power supply to the energising winding and the switched earth path through to the control unit). The fault code indicates that the voltage on the relay sense circuit is in intermittent or erratic. The fault could be caused by loose connections or breaks in the wiring, but also check operation of relay contacts to ensure that they not opening / closing erratically.
P0893	Multiple Gears Engaged	Depending on the transmission design, there are various methods of detecting multiple gear engagement. Systems may use a position sensors to identify movement of the gear selection mechanism (both external selection mechanisms and internal engagement mechanisms). Which ever system is used to detect gear engagement, the information will be passed from the sensors to the transmission control system, and an appropriate signal can also be passed to the engine control system (this signal could be a simple message indicating that a transmission fault exists).	The fault code indicates that an undefined fault has resulted in multiple gear engagement. Depending on the transmission type and design, the fault could be caused by a number of possible electrical or mechanical failures including: a failure of a gear selection mechanism to disengage a gear (electrical or mechanical fault), faulty gear engagement sensors, or possibly a transmission control module fault. If possible, interrogate the transmission control module for additional fault codes that might indicate a specific component / electrical fault. Check operation and wiring for all applicable sensors and gear selection actuators (check for short / open circuits etc).
P0894	Transmission Component Slipping	The control unit monitors the speeds of the transmission input and output shafts; when each gear is selected, the input and output speeds should correspond to the selected gear ratio e.g. if the gear ratio is 4 to 1 then the input speed should be 4 times faster than the output speed. If the input and output speeds do not match the selected ratio (for the identified gear); this will be regarded as a fault.	The fault code indicates that there is "Slip" in one of the transmission system components; the fault could be caused by a worn friction element in the gear change mechanism e.g. engagement clutch but, it could also be caused by a fault with the engagement system (mechanical or a hydraulic pressure fault) that is preventing correct operation of the engagement system thus resulting in "Slip" of the friction element of the gear engagement system. Check operation of shift solenoids / pressure control valves and check system pressures. Check for full movement of any engagement mechanisms.
P0895	Shift Time Too Short	Different transmission systems could use different methods of monitoring gear engagement e.g. for many transmission systems, the control unit monitors the speeds of the transmission input and output shafts; when each gear is selected, the input and output speeds are therefore able to indicate whether gears have been fully engaged thus enable engagement time to be monitored. On some systems, position sensors could also be used to monitor movement of engagement mechanisms.	Identify the type of transmission and gear engagement system and investigate as applicable. The fault code indicates that gear engagement time is too short / quick for the conditions indicated by other sensors. Note that the expected shift time is stored in the control unit, which will alter hydraulic pressures to adjust shift time if necessary; if the system cannot restore shift time to expected value, a fault code will be activated. It is therefore possible that the gear change / shift system is operating correctly but the shift speed is not correct for the driving / operating conditions e.g. shift speed can change with temperature, load and mode selection (such as sport or economy). Check for other fault codes that could indicate other related faults. The fault code could be caused by a transmission control module fault or actuation system fault e.g. hydraulic pressures, shift / pressure control solenoids or mechanical fault (including friction elements of the gear shift mechanism).
P0896	Shift Time Too Long	Different transmission systems could use different methods of monitoring gear engagement e.g. for many transmission systems, the control unit monitors the speeds of the transmission input and output shafts; when each gear is selected, the input and output speeds are therefore able to indicate whether gears have been fully engaged thus enable engagement time to be monitored. On some systems, position sensors could also be used to monitor movement of engagement mechanisms.	Identify the type of transmission and gear engagement system and investigate as applicable. The fault code indicates that gear engagement time is too long / slow for the conditions indicated by other sensors. Note that the expected shift time is stored in the control unit, which will alter hydraulic pressures to adjust shift time if necessary; if the system cannot restore shift time to expected value, a fault code will be activated. It is therefore possible that the gear change / shift system is operating correctly but the shift speed is not correct for the driving / operating conditions e.g. shift speed can change with temperature, load and mode selection (such as sport or economy). Check for other fault codes that could indicate other related faults. The fault code could be caused by a transmission control module fault or actuation system fault e.g. hydraulic pressures, shift / pressure control solenoids or mechanical fault (including friction elements of the gear shift mechanism).
P0897	Transmission Fluid Deteriorated	The transmission control unit monitors information from various sensors e.g. transmission / torque converter input / output shaft speed sensors; this information can be used to detect faults such as transmission fluid condition as well as mechanical system faults.	During the process of locking and unlocking the torque converter the transmission control unit detects "shudder" or rough engagement; this symptom can be caused a deterioration of the hydraulic transmission fluid. Check fluid condition / level and grade. Change fluid if applicable. Note that some vehicles may make use of a sensor to monitor condition of the hydraulic fluid; refer to vehicle specific information if necessary.
P0898	Transmission Control System MIL Request Circuit Low	The transmission control unit can directly switch on the warning light when a transmission fault occurs. On other systems, the transmission control unit can pass information to the main control unit if a transmission fault occurs; this information can be limited to requesting the MIL (Malfunction Indicator Light) is switched on.	The fault code is likely to relate to the circuit between the transmission control unit and the main control unit; this circuit carries the signal that requests the main control unit to illuminate the MIL. The request would be initiated because of a transmission fault, which will require further investigation. The fault could be detected because the request signal (voltage / frequency) in the actuator circuit is either "zero" (no signal) or, the signal exists but is below the normal operating range. Possible fault with actuator or wiring (e.g. short / open circuit or high resistance) that is causing a signal value to be lower than the minimum operating limit. Note: the control unit could be providing the correct signal but it is being affected by a circuit fault. It is also possible the transmission or main control units are faulty.
P0899	Transmission Control System MIL Request Circuit High	The transmission control unit can directly switch on the warning light when a transmission fault occurs. On other systems, the transmission control unit can pass information to the main control unit if a transmission fault occurs; this information can be limited to requesting the MIL (Malfunction Indicator Light) is switched on.	The fault code is likely to relate to the circuit between the transmission control unit and the main control unit; this circuit carries the signal that requests the main control unit to illuminate the MIL. The request would be initiated because of a transmission fault, which will require further investigation. The fault could be detected because the request signal (voltage / frequency) exists but it is above the normal operating range or the signal / voltage exists at an incorrect time. Possible fault with actuator or wiring (e.g. short / open circuit) that is causing a signal voltage / value to be higher than the maximum operating limit. Note: the control unit could be providing the correct signal but it is being affected by a circuit fault. It is also possible the transmission or main control units are faulty.

NOTE 1. Check for other fault codes that could provide additional information.   **NOTE 2.** Communication between control units can pass via a CAN-Bus system; refer to Page 51 for CAN-Bus checks.

NOTE 3. If a fault cannot be located, it is also possible that the control unit is at fault.   **NOTE 4.** Refer to Page 53 for list of pages that show the ISO standard for component locations e.g. Sensor A - Bank 1.

code	Description		
P0900	Clutch Actuator Circuit / Open	An electric actuator can be used to control the operation of the clutch. Depending on the system design, the actuator could control a hydraulic pressure system. Refer to vehicle specific information to identify the type of clutch operation and the type of actuator e.g. solenoid.	The voltage, frequency or value of the control signal in the actuator circuit is incorrect. Signal could be:- out of normal operating range, constant value, non-existent, corrupt. The fault code indicates a possible "open circuit" but other undefined faults in the actuator or wiring could activate the same code e.g. actuator power supply, short circuit, high circuit resistance or incorrect actuator resistance. Note: the control unit could be providing the correct signal but it is being affected by the circuit fault. Refer to list of sensor / actuator checks on page 19.
P0901	Clutch Actuator Circuit Range / Performance	An electric actuator can be used to control the operation of the clutch. Depending on the system design, the actuator could control a hydraulic pressure system. Refer to vehicle specific information to identify the type of clutch operation and the type of actuator e.g. solenoid.	The voltage, frequency or value of the control signal in the actuator circuit is within the normal operating range / tolerance, but the signal might be incorrect for the operating conditions. It is also possible that the actuator response (or response of system controlled by the actuator) differs from the expected / desired response (incorrect response for the control signal provided by the control unit). Possible fault in actuator or wiring, but also possible that the actuator (or mechanism / system controlled by the actuator) is not operating or moving correctly. Note: the control unit could be providing the correct signal but it is being affected by the circuit fault. Refer to list of sensor / actuator checks on page 19. It is also possible that the actuator and clutch system are operating correctly but a clutch position sensor is providing an incorrect potion signal; check for any positions sensors and check operation of the system and ensure that the sensor is correctly indicating clutch operation.
P0902	Clutch Actuator Circuit Low	An electric actuator can be used to control the operation of the clutch. Depending on the system design, the actuator could control a hydraulic pressure system. Refer to vehicle specific information to identify the type of clutch operation and the type of actuator e.g. solenoid.	The voltage, frequency or value of the control signal in the actuator circuit is either "zero" (no signal) or, the signal exists but is below the normal operating range. Possible fault with actuator or wiring (e.g. actuator power supply, short / open circuit or resistance) that is causing a signal value to be lower than the minimum operating limit. Note: the control unit could be providing the correct signal but it is being affected by the circuit fault. Refer to list of sensor / actuator checks on page 19.
P0903	Clutch Actuator Circuit High	An electric actuator can be used to control the operation of the clutch. Depending on the system design, the actuator could control a hydraulic pressure system. Refer to vehicle specific information to identify the type of clutch operation and the type of actuator e.g. solenoid.	The voltage, frequency or value of the control signal in the actuator circuit is either at full value (e.g. battery voltage with no on / off signal) or, the signal exists but it is above the normal operating range. Possible fault with actuator or wiring (e.g. actuator power supply, short / open circuit or resistance) that is causing a signal value to be higher than the maximum operating limit. Note: the control unit could be providing the correct signal but it is being affected by the circuit fault. Refer to list of sensor / actuator checks on page 19.
P0904	Gate Select Position Circuit	A switch / sensor can be used to indicate the movement of the gear lever into the different "Gates" (usually used to select different gear change performance or enable manual selection of gear changes). However, a sensor / switch can also be used to monitor the position of the gear engagement mechanism (and therefore respond to movement of the engagement mechanism when different gates are selected). It may be necessary to refer to vehicle specific information to identify the system and components fitted to the vehicle.	The voltage or value of the switch / sensor signal is incorrect (undefined fault). Switch / sensor voltage / signal could be: out of normal operating range, constant value, non-existent, corrupt. Possible fault in the switch / sensor or wiring e.g. short / open circuits, circuit resistance, incorrect sensor resistance (if applicable) or, fault with the reference / operating voltage; interference from other circuits can also affect switch / sensor signals. Where a switch is fitted, it is possible that the switch contacts are not opening / closing correctly or the switch contacts have a high resistance. Refer to list of sensor / actuator checks on page 19. Also check that the switch / sensor and gear lever are set / adjusted correctly (base setting); it could be possible that the switch / sensor is indicating an incorrect position or not operating / responding correctly to the movement of the gear lever.
P0905	Gate Select Position Circuit Range / Performance	A switch / sensor can be used to indicate the movement of the gear lever into the different "Gates" (usually used to select different gear change performance or enable manual selection of gear changes). However, a sensor / switch can also be used to monitor the position of the gear engagement mechanism (and therefore respond to movement of the engagement mechanism when different gates are selected). It may be necessary to refer to vehicle specific information to identify the system and components fitted to the vehicle.	The voltage or value of the switch / sensor signal is within the normal operating range / tolerance but the signal is not plausible or is incorrect due to an undefined fault. The switch / sensor signal might not match the operating conditions or, the switch / sensor signal is not changing / responding as expected. Possible fault in the switch / sensor or wiring e.g. short / open circuits, circuit resistance, incorrect sensor resistance (if applicable) or, fault with the reference / operating voltage; interference from other circuits can also affect switch / sensor signals. Where a switch is fitted, it is possible that the switch contacts are not opening / closing correctly or the switch contacts have a high resistance. Refer to list of sensor / actuator checks on page 19. Also check that the switch / sensor and gear lever are set / adjusted correctly (base setting) and not indicating an incorrect gear or range; it could be possible that the switch / sensor is indicating an incorrect position or not operating / responding correctly to the movement of the gear lever.
P0906	Gate Select Position Circuit Low	A switch / sensor can be used to indicate the movement of the gear lever into the different "Gates" (usually used to select different gear change performance or enable manual selection of gear changes). However, a sensor / switch can also be used to monitor the position of the gear engagement mechanism (and therefore respond to movement of the engagement mechanism when different gates are selected). It may be necessary to refer to vehicle specific information to identify the system and components fitted to the vehicle.	The voltage or value of the switch / sensor signal is either "zero" (no signal) or, the signal exists but it is below the normal operating range (lower than the minimum operating limit). Possible fault in the switch / sensor or wiring e.g. short / open circuits, circuit resistance, incorrect sensor resistance (if applicable) or, fault with the reference / operating voltage; interference from other circuits can also affect switch / sensor signals. Where a switch is fitted, it is possible that the switch contacts are not opening / closing correctly or, the contacts are sticking (signal is remaining low when it should have risen to the high value); on some circuits, a high resistance on the contacts could also cause a low signal value to exist. Refer to list of sensor / actuator checks on page 19. Also check that the switch / sensor and gear lever are set / adjusted correctly (base setting); it could be possible that the switch / sensor is indicating an incorrect position or not operating / responding correctly to the movement of the gear lever.
P0907	Gate Select Position Circuit High	A switch / sensor can be used to indicate the movement of the gear lever into the different "Gates" (usually used to select different gear change performance or enable manual selection of gear changes). However, a sensor / switch can also be used to monitor the position of the gear engagement mechanism (and therefore respond to movement of the engagement mechanism when different gates are selected). It may be necessary to refer to vehicle specific information to identify the system and components fitted to the vehicle.	The voltage or value of the switch / sensor signal is either at full value (e.g. battery voltage with no signal or frequency) or, the signal exists but it is above the normal operating range (higher than the maximum operating limit). Possible fault in the switch / sensor or wiring e.g. short / open circuits, circuit resistance, incorrect sensor resistance (if applicable) or, fault with the reference / operating voltage; interference from other circuits can also affect switch / sensor signals. Where a switch is fitted, it is possible that the switch contacts are not opening / closing correctly or, the contacts are sticking (signal is remaining high when it should have dropped to low value); on some circuits, a high resistance on the contacts could also cause a high signal value to exist. Refer to list of sensor / actuator checks on page 19. Also check that the switch / sensor and gear lever are set / adjusted correctly (base setting); it could be possible that the switch / sensor is indicating an incorrect position or not operating / responding correctly to the movement of the gear lever.

NOTE 1. Check for other fault codes that could provide additional information.   NOTE 2. Communication between control units can pass via a CAN-Bus system; refer to Page 51 for CAN-Bus checks.
NOTE 3. If a fault cannot be located, it is also possible that the control unit is at fault.   NOTE 4. Refer to Page 53 for list of pages that show the ISO standard for component locations e.g. Sensor A - Bank 1.

Fault code	EOBD / ISO Description	Component / System Description	Meaningful Description and Quick Check
P0908	Gate Select Position Circuit Intermittent	A switch / sensor can be used to indicate the movement of the gear lever into the different "Gates" (usually used to select different gear change performance or enable manual selection of gear changes). However, a sensor / switch can also be used to monitor the position of the gear engagement mechanism (and therefore respond to movement of the engagement mechanism when different gates are selected). It may be necessary to refer to vehicle specific information to identify the system and components fitted to the vehicle.	The voltage or value of the switch / sensor signal is intermittent (likely to be out of normal operating range / tolerance when intermittent fault is detected) or, the signal / voltage is erratic (e.g. signal changes are irregular / unpredictable). Possible fault with wiring or connections (including internal switch / sensor connections); also interference from other circuits can affect the signal. Where a switch is fitted, it is possible that the switch contacts are not opening / closing correctly or the switch contacts have a high resistance. Refer to list of sensor / actuator checks on page 19. Also check that the switch / sensor and gear lever are set / adjusted correctly (base setting); if the switch / sensor is very close the limit of the setting, it could be detecting small movements of the gear lever and the signal could alter in and out of the correct value.
P0909	Gate Select Control Error	The driver can use the gear / shift lever to select a "Gate" e.g. manual or auto gate; the "Gate Select Actuator" is likely to refer to the electric actuator that is then used to activate the engagement mechanism for the selected transmission range / mode of operation. Depending on system design, the actuator could be a motor or a solenoid. It may be necessary to refer to vehicle specific information to identify the system and components fitted to the vehicle.	The fault code indicates an undefined fault relating to the Gear select system and control. Refer to code P0904 - P0908 for information relating to checking the Gate select position switch / sensor and refer to codes P0910 - P0913 for information relating to the actuator. If the switch / sensor operation and actuator operation are correct, the fault could relate to the control unit operation or a transmission fault.
P0910	Gate Select Actuator Circuit / Open	The driver can use the gear / shift lever to select a "Gate" e.g. manual or auto gate; the "Gate Select Actuator" is likely to refer to the electric actuator that is then used to activate the engagement mechanism for the selected transmission range / mode of operation. Depending on system design, the actuator could be a motor or a solenoid. It may be necessary to refer to vehicle specific information to identify the system and components fitted to the vehicle.	The voltage, frequency or value of the control signal in the actuator circuit is incorrect. Signal could be:- out of normal operating range, constant value, non-existent, corrupt. The fault code indicates a possible "open circuit" but other undefined faults in the actuator or wiring could activate the same code e.g. actuator power supply, short circuit, high circuit resistance or incorrect actuator resistance. Note: the control unit could be providing the correct signal but it is being affected by the circuit fault. Refer to list of sensor / actuator checks on page 19.
P0911	Gate Select Actuator Circuit Range / Performance	The driver can use the gear / shift lever to select a "Gate" e.g. manual or auto gate; the "Gate Select Actuator" is likely to refer to the electric actuator that is then used to activate the engagement mechanism for the selected transmission range / mode of operation. Depending on system design, the actuator could be a motor or a solenoid. It may be necessary to refer to vehicle specific information to identify the system and components fitted to the vehicle.	The voltage, frequency or value of the control signal in the actuator circuit is within the normal operating range / tolerance, but the signal might be incorrect for the operating conditions. It is also possible that the actuator response (or response of system controlled by the actuator) differs from the expected / desired response (incorrect response for the control signal provided by the control unit). Possible fault in actuator or wiring, but also possible that the actuator (or mechanism / system controlled by the actuator) is not operating or moving correctly. Note: the control unit could be providing the correct signal but it is being affected by the circuit fault. Refer to list of sensor / actuator checks on page 19. Also check that the actuator operation is not being restricted by a mechanism fault (seized or worn components) and that the actuator / mechanism is set / adjusted correctly.
P0912	Gate Select Actuator Circuit Low	The driver can use the gear / shift lever to select a "Gate" e.g. manual or auto gate; the "Gate Select Actuator" is likely to refer to the electric actuator that is then used to activate the engagement mechanism for the selected transmission range / mode of operation. Depending on system design, the actuator could be a motor or a solenoid. It may be necessary to refer to vehicle specific information to identify the system and components fitted to the vehicle.	The voltage, frequency or value of the control signal in the actuator circuit is either "zero" (no signal) or, the signal exists but is below the normal operating range. Possible fault with actuator or wiring (e.g. actuator power supply, short / open circuit or resistance) that is causing a signal value to be lower than the minimum operating limit. Note: the control unit could be providing the correct signal but it is being affected by the circuit fault. Refer to list of sensor / actuator checks on page 19.
P0913	Gate Select Actuator Circuit High	The driver can use the gear / shift lever to select a "Gate" e.g. manual or auto gate; the "Gate Select Actuator" is likely to refer to the electric actuator that is then used to activate the engagement mechanism for the selected transmission range / mode of operation. Depending on system design, the actuator could be a motor or a solenoid. It may be necessary to refer to vehicle specific information to identify the system and components fitted to the vehicle.	The voltage, frequency or value of the control signal in the actuator circuit is either at full value (e.g. battery voltage with no on / off signal) or, the signal exists but it is above the normal operating range. Possible fault with actuator or wiring (e.g. actuator power supply, short / open circuit or resistance) that is causing a signal value to be higher than the maximum operating limit. Note: the control unit could be providing the correct signal but it is being affected by the circuit fault. Refer to list of sensor / actuator checks on page 19.
P0914	Gear Shift Position Circuit	A switch / sensor can be used to indicate the different gear selection. Depending on the type of transmission, the switch / sensor could respond to gear lever / control movement or to the movement of the gear engagement mechanism. A manufacturer could also use the fault code to refer to the circuit that carries the gear shift information from the Transmission control unit to the engine control unit (refer to fault codes P0863 - P0868 for additional information on this type of fault).	The voltage or value of the switch / sensor signal is incorrect (undefined fault). Switch / sensor voltage / signal could be: out of normal operating range, constant value, non-existent, corrupt. Possible fault in the switch / sensor or wiring e.g. short / open circuits, circuit resistance, incorrect sensor resistance (if applicable), fault with the reference / operating voltage; interference from other circuits can also affect switch / sensor signals. Where a switch is fitted, it is possible that the switch contacts are not opening / closing correctly or the switch contacts have a high resistance. Refer to list of sensor / actuator checks on page 19. Also check that the switch / sensor and gear lever (or gear selector controls) are set / adjusted correctly (base setting); it could be possible that the switch / sensor is indicating an incorrect position or not operating / responding correctly to the movement of the gear lever.
P0915	Gear Shift Position Circuit Range / Performance	A switch / sensor can be used to indicate the different gear selection. Depending on the type of transmission, the switch / sensor could respond to gear lever / control movement or to the movement of the gear engagement mechanism. A manufacturer could also use the fault code to refer to the circuit that carries the gear shift information from the Transmission control unit to the engine control unit (refer to fault codes P0863 - P0868 for additional information on this type of fault).	The voltage or value of the switch / sensor signal is within the normal operating range / tolerance but the signal is not plausible or is incorrect due to an undefined fault. The switch / sensor signal might not match the operating conditions or, the switch / sensor signal is not changing / responding as expected. Possible fault in the switch / sensor or wiring e.g. short / open circuits, circuit resistance, incorrect sensor resistance (if applicable) or, fault with the reference / operating voltage; interference from other circuits can also affect switch / sensor signals. Where a switch is fitted, it is possible that the switch contacts are not opening / closing correctly or the switch contacts have a high resistance. Refer to list of sensor / actuator checks on page 19. Also check that the switch / sensor and gear lever (or gear selector controls) are set / adjusted correctly (base setting); it could be possible that the switch / sensor is indicating an incorrect position or not operating / responding correctly to the movement of the gear lever.

NOTE 1. Check for other fault codes that could provide additional information. **NOTE 2.** Communication between control units can pass via a CAN-Bus system; refer to Page 51 for CAN-Bus checks.
NOTE 3. If a fault cannot be located, it is also possible that the control unit is at fault. **NOTE 4.** Refer to Page 53 for list of pages that show the ISO standard for component locations e.g. Sensor A - Bank 1.

code	Description		
P0916	Gear Shift Position Circuit Low	A switch / sensor can be used to indicate the different gear selection. Depending on the type of transmission, the switch / sensor could respond to gear lever / control movement or to the movement of the gear engagement mechanism. A manufacturer could also use the fault code to refer to the circuit that carries the gear shift information from the Transmission control unit to the engine control unit (refer to fault codes P0863 - P0868 for additional information on this type of fault).	The voltage or value of the switch / sensor signal is either "zero" (no signal) or, the signal exists but it is below the normal operating range (lower than the minimum operating limit). Possible fault in the switch / sensor or wiring e.g. short / open circuits, circuit resistance, incorrect sensor resistance (if applicable) or, fault with the reference / operating voltage; interference from other circuits can also affect switch / sensor signals. Where a switch is fitted, it is possible that the switch contacts are not opening / closing correctly or, the contacts are sticking (signal is remaining low when it should have risen to the high value); on some circuits, a high resistance on the contacts could also cause a low signal value to exist. Refer to list of sensor / actuator checks on page 19. Also check that the switch / sensor and gear lever (or gear selector controls) are set / adjusted correctly (base setting); it could be possible that the sensor is indicating an incorrect position or not operating / responding correctly to the movement of the gear lever.
P0917	Gear Shift Position Circuit High	A switch / sensor can be used to indicate the different gear selection. Depending on the type of transmission, the switch / sensor could respond to gear lever / control movement or to the movement of the gear engagement mechanism. A manufacturer could also use the fault code to refer to the circuit that carries the gear shift information from the Transmission control unit to the engine control unit (refer to fault codes P0863 - P0868 for additional information on this type of fault).	The voltage or value of the switch / sensor signal is either at full value (e.g. battery voltage with no signal or frequency) or, the signal exists but it is above the normal operating range (higher than the maximum operating limit). Possible fault in the switch / sensor or wiring e.g. short / open circuits, circuit resistance, incorrect sensor resistance (if applicable) or, fault with the reference / operating voltage; interference from other circuits can also affect switch / sensor signals. Where a switch is fitted, it is possible that the switch contacts are not opening / closing correctly or, the contacts are sticking (signal is remaining high when it should have dropped to low value); on some circuits, a high resistance on the contacts could also cause a high signal value to exist. Refer to list of sensor / actuator checks on page 19. Also check that the switch / sensor and gear lever (or gear selector controls) are set / adjusted correctly (base setting); it could be possible that the switch / sensor is indicating an incorrect position or not operating / responding correctly to the movement of the gear lever.
P0918	Gear Shift Position Circuit Intermittent	A switch / sensor can be used to indicate the different gear selection. Depending on the type of transmission, the switch / sensor could respond to gear lever / control movement or to the movement of the gear engagement mechanism. A manufacturer could also use the fault code to refer to the circuit that carries the gear shift information from the Transmission control unit to the engine control unit (refer to fault codes P0863 - P0868 for additional information on this type of fault).	The voltage or value of the switch / sensor signal is intermittent (likely to be out of normal operating range / tolerance when intermittent fault is detected) or, the signal / voltage is erratic (e.g. signal changes are irregular / unpredictable). Possible fault with wiring or connections (including internal switch / sensor connections); also interference from other circuits can affect the signal. Where a switch is fitted, it is possible that the switch contacts are not opening / closing correctly or the switch contacts have a high resistance. Refer to list of sensor / actuator checks on page 19. Also check that the switch / sensor and gear lever (or gear selector controls) are set / adjusted correctly (base setting); if the switch / sensor is very close to the limit of the setting, it could be detecting small movements of the gear lever and the signal could alter in and out of the correct value.
P0919	Gear Shift Position Control Error	A switch / sensor can be used to indicate the movement of the gear lever into the different "Gates" and sensors can be used to monitor the position of the gear engagement mechanism (and therefore respond to movement of the engagement mechanism when different gates are selected). Actuators can then be used to operate the gear shift engagement mechanism. It may be necessary to refer to vehicle specific information to identify the system and components fitted to the vehicle.	The fault code indicates an undefined fault relating to the Gear Shift position control. It is likely that the control unit is detecting an incorrect operation of the gear shift actuators / mechanism; this could be detected due to incorrect or implausible information from the shift mechanism position sensors or, it is possible that the shift mechanism position does not correspond to or match the gear lever position. Refer to code P0914 - P0918 for information relating to checking the gear shift position sensors and refer to codes P0920 - P0927 for information relating to the actuators. If the switch / sensor operation and actuator operation are correct, the fault could relate to the control unit operation or a transmission fault.
P0920	Gear Shift Forward Actuator Circuit / Open	Depending on transmission system design and type of transmission control, the transmission can be fitted with forward and reverse selection actuators. Note however that the "Forward" and "Reverse" terms can refer to the movement of the gear shift actuating mechanism rather than the selection of the forward and reverse gears or, they can refer to the location of the gear clusters in the transmission case (e.g. the forward gear clusters correspond to gears 1, 3 and 5, and the reverse clusters correspond to gears 2, 4 and 6) . Refer to manufacturer specific information if necessary to identify the application of the fault code, the actuator, actuator control and any associated mechanism.	The voltage, frequency or value of the control signal in the actuator circuit is incorrect. Signal could be:- out of normal operating range, constant value, non-existent, corrupt. The fault code indicates a possible "open circuit" but other undefined faults in the actuator or wiring could activate the same code e.g. actuator power supply, short circuit, high circuit resistance or incorrect actuator resistance. Note: the control unit could be providing the correct signal but it is being affected by the circuit fault. Refer to list of sensor / actuator checks on page 19.
P0921	Gear Shift Forward Actuator Circuit Range / Performance	Depending on transmission system design and type of transmission control, the transmission can be fitted with forward and reverse selection actuators. Note however that the "Forward" and "Reverse" terms can refer to the movement of the gear shift actuating mechanism rather than the selection of the forward and reverse gears or, they can refer to the location of the gear clusters in the transmission case (e.g. the forward gear clusters correspond to gears 1, 3 and 5, and the reverse clusters correspond to gears 2, 4 and 6) . Refer to manufacturer specific information if necessary to identify the application of the fault code, the actuator, actuator control and any associated mechanism.	The voltage, frequency or value of the control signal in the actuator circuit is within the normal operating range / tolerance, but the signal might be incorrect for the operating conditions. It is also possible that the actuator response (or response of system controlled by the actuator) differs from the expected / desired response (incorrect response for the control signal provided by the control unit). Possible fault in actuator or wiring, but also possible that the actuator (or mechanism / system controlled by the actuator) is not operating or moving correctly. Note: the control unit could be providing the correct signal but it is being affected by the circuit fault. Refer to list of sensor / actuator checks on page 19. Also check that the gear shift mechanism is free to operate correctly and is not worn; also check for correct operating of any gear shift mechanism position sensors.
P0922	Gear Shift Forward Actuator Circuit Low	Depending on transmission system design and type of transmission control, the transmission can be fitted with forward and reverse selection actuators. Note however that the "Forward" and "Reverse" terms can refer to the movement of the gear shift actuating mechanism rather than the selection of the forward and reverse gears or, they can refer to the location of the gear clusters in the transmission case (e.g. the forward gear clusters correspond to gears 1, 3 and 5, and the reverse clusters correspond to gears 2, 4 and 6) . Refer to manufacturer specific information if necessary to identify the application of the fault code, the actuator, actuator control and any associated mechanism.	The voltage, frequency or value of the control signal in the actuator circuit is either "zero" (no signal) or, the signal exists but is below the normal operating range. Possible fault with actuator or wiring (e.g. actuator power supply, short / open circuit or resistance) that is causing a signal value to be lower than the minimum operating limit. Note: the control unit could be providing the correct signal but it is being affected by the circuit fault. Refer to list of sensor / actuator checks on page 19.

Fault code	EOBD / ISO Description	Component / System Description	Meaningful Description and Quick Check
P0923	Gear Shift Forward Actuator Circuit High	Depending on transmission system design and type of transmission control, the transmission can be fitted with forward and reverse selection actuators. Note however that the "Forward" and "Reverse" terms can refer to the movement of the gear shift actuating mechanism rather than the selection of the forward and reverse gears or, they can refer to the location of the gear clusters in the transmission case (e.g. the forward gear clusters correspond to gears 1, 3 and 5, and the reverse clusters correspond to gears 2, 4 and 6) . Refer to manufacturer specific information if necessary to identify the application of the fault code, the actuator, actuator control and any associated mechanism.	The voltage, frequency or value of the control signal in the actuator circuit is either at full value (e.g. battery voltage with no on / off signal) or, the signal exists but it is above the normal operating range. Possible fault with actuator or wiring (e.g. actuator power supply, short / open circuit or resistance) that is causing a signal value to be higher than the maximum operating limit. Note: the control unit could be providing the correct signal but it is being affected by the circuit fault. Refer to list of sensor / actuator checks on page 19.
P0924	Gear Shift Reverse Actuator Circuit / Open	Depending on transmission system design and type of transmission control, the transmission can be fitted with forward and reverse selection actuators. Note however that the "Forward" and "Reverse" terms can refer to the movement of the gear shift actuating mechanism rather than the selection of the forward and reverse gears or, they can refer to the location of the gear clusters in the transmission case (e.g. the forward gear clusters correspond to gears 1, 3 and 5, and the reverse clusters correspond to gears 2, 4 and 6) . Refer to manufacturer specific information if necessary to identify the application of the fault code, the actuator, actuator control and any associated mechanism.	The voltage, frequency or value of the control signal in the actuator circuit is incorrect. Signal could be:- out of normal operating range, constant value, non-existent, corrupt. The fault code indicates a possible "open circuit" but other undefined faults in the actuator or wiring could activate the same code e.g. actuator power supply, short circuit, high circuit resistance or incorrect actuator resistance. Note: the control unit could be providing the correct signal but it is being affected by the circuit fault. Refer to list of sensor / actuator checks on page 19.
P0925	Gear Shift Reverse Actuator Circuit Range / Performance	Depending on transmission system design and type of transmission control, the transmission can be fitted with forward and reverse selection actuators. Note however that the "Forward" and "Reverse" terms can refer to the movement of the gear shift actuating mechanism rather than the selection of the forward and reverse gears or, they can refer to the location of the gear clusters in the transmission case (e.g. the forward gear clusters correspond to gears 1, 3 and 5, and the reverse clusters correspond to gears 2, 4 and 6) . Refer to manufacturer specific information if necessary to identify the application of the fault code, the actuator, actuator control and any associated mechanism.	The voltage, frequency or value of the control signal in the actuator circuit is within the normal operating range / tolerance, but the signal might be incorrect for the operating conditions. It is also possible that the actuator response (or response of system controlled by the actuator) differs from the expected / desired response (incorrect response for the control signal provided by the control unit). Possible fault in actuator or wiring, but also possible that the actuator (or mechanism / system controlled by the actuator) is not operating or moving correctly. Note: the control unit could be providing the correct signal but it is being affected by the circuit fault. Refer to list of sensor / actuator checks on page 19. Also check that the gear shift mechanism is free to operate correctly and is not worn; also check for correct operating of any gear shift mechanism position sensors.
P0926	Gear Shift Reverse Actuator Circuit Low	Depending on transmission system design and type of transmission control, the transmission can be fitted with forward and reverse selection actuators. Note however that the "Forward" and "Reverse" terms can refer to the movement of the gear shift actuating mechanism rather than the selection of the forward and reverse gears or, they can refer to the location of the gear clusters in the transmission case (e.g. the forward gear clusters correspond to gears 1, 3 and 5, and the reverse clusters correspond to gears 2, 4 and 6) . Refer to manufacturer specific information if necessary to identify the application of the fault code, the actuator, actuator control and any associated mechanism.	The voltage, frequency or value of the control signal in the actuator circuit is either "zero" (no signal) or, the signal exists but is below the normal operating range. Possible fault with actuator or wiring (e.g. actuator power supply, short / open circuit or resistance) that is causing a signal value to be lower than the minimum operating limit. Note: the control unit could be providing the correct signal but it is being affected by the circuit fault. Refer to list of sensor / actuator checks on page 19.
P0927	Gear Shift Reverse Actuator Circuit High	Depending on transmission system design and type of transmission control, the transmission can be fitted with forward and reverse selection actuators. Note however that the "Forward" and "Reverse" terms can refer to the movement of the gear shift actuating mechanism rather than the selection of the forward and reverse gears or, they can refer to the location of the gear clusters in the transmission case (e.g. the forward gear clusters correspond to gears 1, 3 and 5, and the reverse clusters correspond to gears 2, 4 and 6) . Refer to manufacturer specific information if necessary to identify the application of the fault code, the actuator, actuator control and any associated mechanism.	The voltage, frequency or value of the control signal in the actuator circuit is either at full value (e.g. battery voltage with no on / off signal) or, the signal exists but it is above the normal operating range. Possible fault with actuator or wiring (e.g. actuator power supply, short / open circuit or resistance) that is causing a signal value to be higher than the maximum operating limit. Note: the control unit could be providing the correct signal but it is being affected by the circuit fault. Refer to list of sensor / actuator checks on page 19.
P0928	Gear Shift Lock Solenoid Control Circuit / Open	A mechanism can be used to lock the gear selector (lever or other controls) to prevent the gear lever from being moved e.g. the lock system can be used to prevent gear selection without the brake pedal being depressed. A solenoid is normally used to actuate the locking mechanism.	The voltage, frequency or value of the control signal in the solenoid circuit is incorrect. Signal could be:- out of normal operating range, constant value, non-existent, corrupt. The fault code indicates a possible "open circuit" but other undefined faults in the solenoid or wiring could activate the same code e.g. solenoid power supply, short circuit, high circuit resistance or incorrect solenoid resistance. Note: the control unit could be providing the correct signal but it is being affected by the circuit fault. Refer to list of sensor / actuator checks on page 19.
P0929	Gear Shift Lock Solenoid Control Circuit Range / Performance	A mechanism can be used to lock the gear selector (lever or other controls) to prevent the gear lever from being moved e.g. the lock system can be used to prevent gear selection without the brake pedal being depressed. A solenoid is normally used to actuate the locking mechanism.	The voltage, frequency or value of the control signal in the solenoid circuit is within the normal operating range / tolerance, but the signal might be incorrect for the operating conditions. It is also possible that the solenoid response (or response of mechanism actuated by the solenoid) differs from the expected / desired response (incorrect response to the control signal provided by the control unit). Possible fault in solenoid or wiring, but also possible that the solenoid (or mechanism / system controlled by the solenoid) is not operating or moving correctly. Note that the control unit could be providing the correct signal but it is being affected by a circuit fault. Refer to list of sensor / actuator checks on page 19. Also check for free movement of the solenoid and the locking mechanism and check that the locking mechanism can engage correctly with the gear lever / shift mechanism.

NOTE 1. Check for other fault codes that could provide additional information. NOTE 2. Communication between control units can pass via a CAN-Bus system; refer to Page 51 for CAN-Bus checks.

NOTE 3. If a fault cannot be located, it is also possible that the control unit is at fault. NOTE 4. Refer to Page 53 for list of pages that show the ISO standard for component locations e.g. Sensor A - Bank 1.

Code	Description		
P0930	Gear Shift Lock Solenoid Control Circuit Low	A mechanism can be used to lock the gear selector (lever or other controls) to prevent the gear lever from being moved e.g. the lock system can be used to prevent gear selection without the brake pedal being depressed. A solenoid is normally used to actuate the locking mechanism.	The voltage, frequency or value of the control signal in the solenoid circuit is either "zero" (no signal) or, the signal exists but is below the normal operating range. Possible fault with solenoid or wiring (e.g. solenoid power supply, short / open circuit or resistance) that is causing a signal value to be lower than the minimum operating limit. Note: the control unit could be providing the correct signal but it is being affected by a circuit fault. Refer to list of sensor / actuator checks on page 19.
P0931	Gear Shift Lock Solenoid Control Circuit High	A mechanism can be used to lock the gear selector (lever or other controls) to prevent the gear lever from being moved e.g. the lock system can be used to prevent gear selection without the brake pedal being depressed. A solenoid is normally used to actuate the locking mechanism.	The voltage, frequency or value of the control signal in the solenoid circuit is either at full value (e.g. battery voltage with no on / off signal) or, the signal exists but it is above the normal operating range. Possible fault with solenoid or wiring (e.g. solenoid power supply, short / open circuit or resistance) that is causing a signal value to be higher than the maximum operating limit. Note: the control unit could be providing the correct signal but it is being affected by the circuit fault. Refer to list of sensor / actuator checks on page 19.
P0932	Hydraulic Pressure Sensor Circuit	Transmission systems can make use of hydraulic pressure to control / actuate various functions, including gear changes (and main clutch operation on electro / hydraulic controlled clutches). A pressure sensor is therefore used to monitor the system pressure and changes in pressure within the system. It may be necessary to refer to vehicle specific information to identify the function of the pressure circuit, which will depend on the system design.	The voltage, frequency or value of the pressure sensor signal is incorrect (undefined fault). Sensor voltage / signal could be: out of normal operating range, constant value, non-existent, corrupt. Possible fault in the pressure sensor or wiring e.g. short / open circuits, circuit resistance, incorrect sensor resistance (if applicable) or, fault with the reference / operating voltage; interference from other circuits can also affect sensor signals. Refer to list of sensor / actuator checks on page 19.
P0933	Hydraulic Pressure Sensor Range / Performance	Transmission systems can make use of hydraulic pressure to control / actuate various functions, including gear changes (and main clutch operation on electro / hydraulic controlled clutches). A pressure sensor is therefore used to monitor the system pressure and changes in pressure within the system. It may be necessary to refer to vehicle specific information to identify the function of the pressure circuit, which will depend on the system design.	The voltage, frequency or value of the pressure sensor signal is within the normal operating range / tolerance but the signal is not plausible or is incorrect due to an undefined fault. The sensor signal might not match the operating conditions indicated by other sensors or, the sensor signal is not changing / responding as expected to the changes in conditions. Possible fault in the pressure sensor or wiring e.g. short / open circuits, circuit resistance, incorrect sensor resistance (if applicable) or, fault with the reference / operating voltage; interference from other circuits can also affect sensor signals. Refer to list of sensor / actuator checks on page 19. It is also possible that the pressure sensor is operating correctly but the pressure being measured by the sensor is unacceptable or incorrect for the operating conditions; if possible, use alternative method of checking pressure to establish if sensor system is operating correctly and whether the pressure indicated by the sensor is correct / incorrect.
P0934	Hydraulic Pressure Sensor Circuit Low	Transmission systems can make use of hydraulic pressure to control / actuate various functions, including gear changes (and main clutch operation on electro / hydraulic controlled clutches). A pressure sensor is therefore used to monitor the system pressure and changes in pressure within the system. It may be necessary to refer to vehicle specific information to identify the function of the pressure circuit, which will depend on the system design.	The voltage, frequency or value of the pressure sensor signal is either "zero" (no signal) or, the signal exists but it is below the normal operating range (lower than the minimum operating limit). Possible fault in the pressure sensor or wiring e.g. short / open circuits, circuit resistance, incorrect sensor resistance (if applicable) or, fault with the reference / operating voltage; interference from other circuits can also affect sensor signals. Refer to list of sensor / actuator checks on page 19.
P0935	Hydraulic Pressure Sensor Circuit High	Transmission systems can make use of hydraulic pressure to control / actuate various functions, including gear changes (and main clutch operation on electro / hydraulic controlled clutches). A pressure sensor is therefore used to monitor the system pressure and changes in pressure within the system. It may be necessary to refer to vehicle specific information to identify the function of the pressure circuit, which will depend on the system design.	The voltage, frequency or value of the pressure sensor signal is either at full value (e.g. battery voltage with no signal or frequency) or, the signal exists but it is above the normal operating range (higher than the maximum operating limit). Possible fault in the pressure sensor or wiring e.g. short / open circuits, circuit resistance, incorrect sensor resistance (if applicable) or, fault with the reference / operating voltage; interference from other circuits can also affect sensor signals. Refer to list of sensor / actuator checks on page 19.
P0936	Hydraulic Pressure Sensor Circuit Intermittent	Transmission systems can make use of hydraulic pressure to control / actuate various functions, including gear changes (and main clutch operation on electro / hydraulic controlled clutches). A pressure sensor is therefore used to monitor the system pressure and changes in pressure within the system. It may be necessary to refer to vehicle specific information to identify the function of the pressure circuit, which will depend on the system design.	The voltage, frequency or value of the pressure sensor signal is intermittent (likely to be out of normal operating range / tolerance when intermittent fault is detected) or, the signal / voltage is erratic (e.g. signal changes are irregular / unpredictable). Possible fault with wiring or connections (including internal sensor connections); also interference from other circuits can affect the signal. Refer to list of sensor / actuator checks on page 19. It is also possible that the pressure could be changing erratically, which could cause the fault code to be activated; if necessary, check pressure with a separate gauge to check for erratic pressure changes.
P0937	Hydraulic Oil Temperature Sensor Circuit	The information from the transmission fluid temperature sensor can be used to alter transmission operation during cold running / warm up and also influence the gear change strategy; under certain conditions engine control could be influenced by transmission oil temperature. Excessive fluid temperatures will also be indicated by the temperature sensor.	The code identifies an electrical related fault in the fluid temperature sensor but, it might be advisable to check oil cooling and oil condition / grade. The voltage, frequency or value of the temperature sensor signal is incorrect (undefined fault). Sensor voltage / signal could be: out of normal operating range, constant value, non-existent, corrupt. Possible fault in the temperature sensor or wiring e.g. short / open circuits, circuit resistance, incorrect sensor resistance or, fault with the reference / operating voltage; interference from other circuits can also affect sensor signals. Refer to list of sensor / actuator checks on page 19.
P0938	Hydraulic Oil Temperature Sensor Range / Performance	The information from the transmission fluid temperature sensor can be used to alter transmission operation during cold running / warm up and also influence the gear change strategy; under certain conditions engine control could be influenced by transmission oil temperature. Excessive fluid temperatures will also be indicated by the temperature sensor.	The code identifies an electrical related fault in the fluid temperature sensor but, it might be advisable to check oil cooling and oil condition / grade. The voltage, frequency or value of the temperature sensor signal is within the normal operating range / tolerance but the signal is not plausible or is incorrect due to an undefined fault. The sensor signal might not match the operating conditions indicated by other sensors or, the sensor signal is not changing / responding as expected to the changes in conditions. Possible fault in the temperature sensor or wiring e.g. short / open circuits, circuit resistance, incorrect sensor resistance or, fault with the reference / operating voltage; interference from other circuits can also affect sensor signals. Refer to list of sensor / actuator checks on page 19. It is also possible that the sensor is operating correctly but the oil temperature is unacceptable or incorrect. If it is suspected that the temperature is incorrect, check transmission fluid cooling (intercooler etc) and check for correct oil, oil condition and quantity.

NOTE 1. Check for other fault codes that could provide additional information.   **NOTE 2.** Communication between control units can pass via a CAN-Bus system; refer to Page 51 for CAN-Bus checks.

NOTE 3. If a fault cannot be located, it is also possible that the control unit is at fault.   **NOTE 4.** Refer to Page 53 for list of pages that show the ISO standard for component locations e.g. Sensor A - Bank 1.

Fault code	EOBD / ISO Description	Component / System Description	Meaningful Description and Quick Check
P0939	Hydraulic Oil Temperature Sensor Circuit Low	The information from the transmission fluid temperature sensor can.be used to alter transmission operation during cold running / warm up and also influence the gear change strategy; under certain conditions engine control could be influenced by transmission oil temperature. Excessive fluid temperatures will also be indicated by the temperature sensor.	The code identifies an electrical related fault in the fluid temperature sensor but, it might be advisable to check oil cooling, oil quantity and oil condition / grade. The voltage, frequency or value of the temperature sensor signal is either "zero" (no signal) or, the signal exists but it is below the normal operating range (lower than the minimum operating limit). Possible fault in the temperature sensor or wiring e.g. short / open circuits, circuit resistance, incorrect sensor resistance or, fault with the reference / operating voltage; interference from other circuits can also affect sensor signals. Refer to list of sensor / actuator checks on page 19.
P0940	Hydraulic Oil Temperature Sensor Circuit High	The information from the transmission fluid temperature sensor can be used to alter transmission operation during cold running / warm up and also influence the gear change strategy; under certain conditions engine control could be influenced by transmission oil temperature. Excessive fluid temperatures will also be indicated by the temperature sensor.	The code identifies an electrical related fault in the fluid temperature sensor but, it might be advisable to check oil cooling, oil quantity and oil condition / grade. The voltage, frequency or value of the temperature sensor signal is either at full value (e.g. battery voltage with no signal or frequency) or, the signal exists but it is above the normal operating range (higher than the maximum operating limit). Possible fault in the temperature sensor or wiring e.g. short / open circuits, circuit resistance, incorrect sensor resistance or, fault with the reference / operating voltage; interference from other circuits can also affect sensor signals. Refer to list of sensor / actuator checks on page 19.
P0941	Hydraulic Oil Temperature Sensor Circuit Intermittent	The information from the transmission fluid temperature sensor can be used to alter transmission operation during cold running / warm up and also influence the gear change strategy; under certain conditions engine control could be influenced by transmission oil temperature. Excessive fluid temperatures will also be indicated by the temperature sensor.	The code identifies an electrical related fault in the fluid temperature sensor but, it might be advisable to check oil cooling, oil quantity and oil condition / grade. The voltage, frequency or value of the temperature sensor signal is either at full value (e.g. battery voltage with no signal or frequency) or, the signal exists but it is above the normal operating range (higher than the maximum operating limit). Possible fault in the temperature sensor or wiring e.g. short / open circuits, circuit resistance, incorrect sensor resistance or, fault with the reference / operating voltage; interference from other circuits can also affect sensor signals. Refer to list of sensor / actuator checks on page 19.
P0942	Hydraulic Pressure Unit	Transmission systems can make use of hydraulic pressure systems / units to provide regulated pressure for various transmission related functions; the functions will depend on system design but can include: gear changes, clutch operation etc. Fluid pressure sensors are usually fitted to monitor pressures and changes in pressure when functions such as gear changes take place. It may be necessary to refer to vehicle specific information to identify the applicable system components and hydraulic pressure circuit (and its function).	The fault code indicates that there is an undefined fault with the hydraulic pressure unit (can contain the pump and other hydraulic devices such as pressure regulator, pressure sensor etc). It will be necessary to identify the system components and carry out a full range of systems checks. It is possible that the pressure (detected by a pressure sensor) is incorrect for the operating conditions or, the pressure does not match values expected by the control unit. It is also possible that specific tasks are not being carried out efficiently (e.g. a gear change etc). If possible, use alternative method of checking pressure to establish if the pressure indicated by the sensor is correct / incorrect. If the pressure is incorrect, refer to vehicle specific information to identify system operation and components and check as applicable e.g. pressure pump, pressure regulator / pressure valves, filters oil passages etc. If the pressure appears to be correct, it could indicate a pressure sensor fault. Refer to list of sensor / actuator checks on page 19.
P0943	Hydraulic Pressure Unit Cycling Period Too Short	Transmission systems can make use of hydraulic pressure systems / units to provide regulated pressure for various transmission related functions; the functions will depend on system design but can include: gear changes, clutch operation etc. Fluid pressure sensors are usually fitted to monitor pressures and changes in pressure when functions such as gear changes take place. It may be necessary to refer to vehicle specific information to identify the applicable system components and hydraulic pressure circuit (and its function).	It will be necessary to identify the system components and carry out a full range of systems checks, note that the hydraulic pressure unit (can contain the pump and other hydraulic devices such as pressure regulator, pressure sensor etc). The fault code indicates that the "cycling period" for the hydraulic unit is too short; this is likely to relate to the operation of the pressure pump, which might be switching on for an insufficient length of time and therefore not creating sufficient volume / pressure in the system. The problem could relate to a pressure sensor fault (providing incorrect pressure information) or a pump control system fault. It will be necessary to refer to vehicle specific information to identify the control system for the pump and also Refer to list of sensor / actuator checks on page 19. Also check for pressure loss due to leaks etc.
P0944	Hydraulic Pressure Unit Loss of Pressure	Transmission systems can make use of hydraulic pressure systems / units to provide regulated pressure for various transmission related functions; the functions will depend on system design but can include: gear changes, clutch operation etc. Fluid pressure sensors are usually fitted to monitor pressures and changes in pressure when functions such as gear changes take place. It may be necessary to refer to vehicle specific information to identify the applicable system components and hydraulic pressure circuit (and its function).	It may necessary to identify the system components and carry out a full range of systems checks, note that the hydraulic pressure unit (can contain the pump and other hydraulic devices such as pressure regulator, pressure sensor etc). The fault code indicates that there is a pressure loss in the hydraulic unit, which is probably caused by a fluid leak or a faulty pressure regulator.. Depending on construction / location of the system it might be difficult to identify the location of any internal fluid leaks. However, if possible, check the hydraulic pressure using a separate pressure gauge and check for pressure loss. If there is a pressure loss, check operation of pressure regulation system and check for fluid leaks. If there is no pressure loss, it could indicate a possible pressure sensor fault (Refer to list of sensor / actuator checks on page 19).
P0945	Hydraulic Pump Relay Circuit / Open	The transmission Hydraulic Pump Relay can be switched on by the transmission control unit or could be switched on a simple electrical circuit when the ignition is switched on. It is normal practice on systems with electronic control units for the control unit to switch the earth circuit for the relay i.e. the control unit will complete the earth path for the relay energising winding. The control unit can detect whether there is a power supply to the winding and is therefore able to detect if there is a fault in the circuit.	Refer to list of sensor / actuator checks on page 19 for notes on relay checks. Different manufacturers could use the fault code to refer to different relay circuits. The fault code could refer to the power supply circuit (the power supply to the relay contacts and the circuit from the relay contacts to the Hydraulic pump). Alternatively, the fault code could refer to the relay energising winding circuit (power supply to the energising winding and the switched earth path through to the control unit). The voltage in the applicable relay circuit is incorrect (undefined fault but possibly an open circuit). Check power supplies to relay contacts (possibly via a fused circuit) and to relay energising winding. Check operation of energising winding (short / open circuit) and closing / opening of relay contacts. Check all wiring for short / open circuits.
P0946	Hydraulic Pump Relay Circuit Range / Performance	The transmission Hydraulic Pump Relay can be switched on by the transmission control unit or could be switched on a simple electrical circuit when the ignition is switched on. It is normal practice on systems with electronic control units for the control unit to switch the earth circuit for the relay i.e. the control unit will complete the earth path for the relay energising winding. The control unit can detect whether there is a power supply to the winding and is therefore able to detect if there is a fault in the circuit. Although the fault code is likely to relate to the relay energising winding circuit, some manufacturers may use the code to refer to the relay power output circuit (power supply to hydraulic pump).	Refer to list of sensor / actuator checks on page 19 for notes on relay checks. Different manufacturers could use the fault code to refer to the different relay circuits. The fault code could refer to the power supply circuit (the power supply to the relay contacts and the circuit from the relay contacts to the hydraulic pump). Alternatively, the fault code could refer to the relay energising winding circuit (power supply to the energising winding and the switched earth path through to the control unit). The fault code indicates that the voltage on the applicable relay circuit is incorrect or, the voltage is not as expected under certain operating conditions (e.g. circuit is on when it should be off or, off when it should be on). The fault could be related to relay operation e.g. contacts stuck closed or open, wiring faults (short / open circuit) or possible control unit fault (switching relay at incorrect times).

NOTE 1. Check for other fault codes that could provide additional information. **NOTE 2.** Communication between control units can pass via a CAN-Bus system; refer to Page 51 for CAN-Bus checks. **206**

NOTE 3. If a fault cannot be located, it is also possible that the control unit is at fault **NOTE 4.** Refer to Page 53 for list of pages that show the ISO standard for component locations e.g. Sensor A - Bank 1.

Code	Description		
P0947	Hydraulic Pump Relay Circuit Low	The transmission Hydraulic Pump Relay can be switched on by the transmission control unit or could be switched on a simple electrical circuit when the ignition is switched on. It is normal practice on systems with electronic control units for the control unit to switch the earth circuit for the relay i.e. the control unit will complete the earth path for the relay energising winding. The control unit can detect whether there is a power supply to the winding and is therefore able to detect if there is a fault in the circuit. Although the fault code is likely to relate to the relay energising winding circuit, some manufacturers may use the code to refer to the relay power output circuit (power supply to hydraulic pump).	Refer to list of sensor / actuator checks on page 19 for notes on relay checks. Different manufacturers could use the fault code to refer to different relay circuits. The fault code could refer to the power supply circuit (the power supply to the relay contacts and the circuit from the relay contacts to the hydraulic pump). Alternatively, the fault code could refer to the relay energising winding circuit (power supply to the energising winding and the switched earth path through to the control unit). The voltage in the applicable relay circuit is low (likely to be zero volts) at a time when the voltage should be at battery voltage level. Check power supplies to relay contacts (usually via a fused circuit) and to relay energising winding. Check operation of energising winding (short / open circuit) and closing / opening of relay contacts. Check all wiring for short / open circuits.
P0948	Hydraulic Pump Relay Circuit High	The transmission Hydraulic Pump Relay can be switched on by the transmission control unit or could be switched on a simple electrical circuit when the ignition is switched on. It is normal practice on systems with electronic control units for the control unit to switch the earth circuit for the relay i.e. the control unit will complete the earth path for the relay energising winding. The control unit can detect whether there is a power supply to the winding and is therefore able to detect if there is a fault in the circuit. Although the fault code is likely to relate to the relay energising winding circuit, some manufacturers may use the code to refer to the relay power output circuit (power supply to hydraulic pump).	Refer to list of sensor / actuator checks on page 19 for notes on relay checks. Different manufacturers could use the fault code to refer to different relay circuits. The fault code could refer to the power supply circuit (the power supply to the relay contacts and the circuit from the relay contacts to the hydraulic pump). Alternatively, the fault code could refer to the relay energising winding circuit (power supply to the energising winding and the switched earth path through to the control unit). The voltage in the relay circuit is high (likely to be battery voltage) when the relay circuit should be off (contacts open). Likely cause is relay contacts remaining closed (either due to relay fault or energising winding is being continuously activated); also possible short in wiring to another power supply or the energising winding earth path is shorted to earth. Check operation of relay, including closing / opening of relay contacts, and check operation of energising winding (short / open circuit).
P0949	Auto Shift Manual Adaptive Learning Not Complete	The auto shift manual control refers to the system that allows manual or auto control of gear changes on an auto transmission or possibly on an electronically controlled manual gearbox. The manual controls could be up or down shift paddles / buttons (on steering column or steering wheel), or it could be a specific movement of the main gear lever.	The adaptive learning process enables the control module to adapt the operation of the system based on changes detected by the control module; this can include changes in operating conditions (temperature etc), changes in driving style, but can also be related to changes in component operation e.g. if an actuator requires a slightly different control signal to complete a task such as changing to the next gear. For this fault code, it is possible that the control module is unable to adapt (learn or re-learn) to changes that have occurred. It is possible that fitting a replacement component could have caused the fault (also check if there is a re-initialisation process for the module). The fault code indicates that there is a fault with the adaptive learning process; this could relate to manual or auto shift process. Check for other fault codes, check all wiring and connections to the control unit (from the sensors, actuators and selection switches (driver controls).
P0950	Auto Shift Manual Control Circuit	The auto shift manual control refers to the system that allows manual control of gear changes on an auto transmission or possibly on an electronically controlled manual gearbox. The manual controls could be up or down shift paddles / buttons (on steering column or steering wheel), or it could be a specific movement of the main gear lever.	The control unit has detected a fault with the manual gear selection control circuits e.g. gear change paddle / button switch circuits or gear lever position switch circuits (for manual up / down shifts). The voltage or value of the switch signal is incorrect (undefined fault). Switch voltage / signal could be: out of normal operating range, constant value, non-existent, corrupt. Possible fault in the switch or wiring e.g. short / open circuits, circuit resistance or, fault with the reference / operating voltage; the switch contacts could also be faulty e.g. not opening / closing correctly or have a high contact resistance. The fault code could be activated if one of the gear change switches is permanently ON or OFF or, if both up and down shift switches are on at the same time. If there does not appear to be any faults with the switches and wiring, it possibly indicates a control module fault.
P0951	Auto Shift Manual Control Circuit Range / Performance	The auto shift manual control refers to the system that allows manual control of gear changes on an auto transmission or possibly on an electronically controlled manual gearbox. The manual controls could be up or down shift paddles / buttons (on steering column or steering wheel), or it could be a specific movement of the main gear lever.	The control unit has detected a fault with the manual gear selection control circuits e.g. gear change paddle / button switch circuits or gear lever position switch circuits (for manual up / down shifts). The voltage of the switch signal is within the normal operating range / tolerance but the signal is not plausible or is incorrect due to an undefined fault. The switch signal might not match the operating conditions indicated by other sensors or, the switch signal is not changing / responding as expected. Possible fault in the switch or wiring e.g. short / open circuits, circuit resistance or, fault with the reference / operating voltage; the switch contacts could also be faulty e.g. not opening / closing correctly or have a high contact resistance. The fault code could be activated if one of the gear change switches is permanently ON or OFF or, if both up and down shift switches are on at the same time. If there does not appear to be any faults with the switches and wiring, it possibly indicates a control module fault.
P0952	Auto Shift Manual Control Circuit Low	The auto shift manual control refers to the system that allows manual control of gear changes on an auto transmission or possibly on an electronically controlled manual gearbox. The manual controls could be up or down shift paddles / buttons (on steering column or steering wheel), or it could be a specific movement of the main gear lever.	The control unit has detected a fault with the manual gear selection control circuits e.g. gear change paddle / button switch circuits or gear lever position switch circuits (for manual up / down shifts). The voltage is likely to be "zero" (with no voltage change). Possible fault in the switch or wiring e.g. short / open circuits, circuit resistance or, fault with the reference / operating voltage; the switch contacts could also be faulty e.g. not opening / closing correctly or have a high contact resistance. The fault code could be activated if one of the gear change switches is permanently ON or OFF or, if both up and down shift switches are on at the same time. If there does not appear to be any faults with the switches and wiring, it possibly indicates a control module fault.
P0953	Auto Shift Manual Control Circuit High	The auto shift manual control refers to the system that allows manual control of gear changes on an auto transmission or possibly on an electronically controlled manual gearbox. The manual controls could be up or down shift paddles / buttons (on steering column or steering wheel), or it could be a specific movement of the main gear lever.	The control unit has detected a fault with the manual gear selection control circuits e.g. gear change paddle / button switch circuits or gear lever position switch circuits (for manual up / down shifts). The voltage is likely to be at full value (e.g. battery or reference voltage with no change). Possible fault in the switch or wiring e.g. short / open circuits, circuit resistance or, fault with the reference / operating voltage; the switch contacts could also be faulty e.g. not opening / closing correctly or have a high contact resistance. The fault code could be activated if one of the gear change switches is permanently ON or OFF or, if both up and down shift switches are on at the same time. If there does not appear to be any faults with the switches and wiring, it possibly indicates a control module fault.

NOTE 1. Check for other fault codes that could provide additional information. NOTE 2. Communication between control units can pass via a CAN-Bus system; refer to Page 51 for CAN-Bus checks.

NOTE 3. If a fault cannot be located, it is also possible that the control unit is at fault. NOTE 4. Refer to Page 53 for list of pages that show the ISO standard for component locations e.g. Sensor A - Bank 1.

Fault code	EOBD / ISO Description	Component / System Description	Meaningful Description and Quick Check
P0954	Auto Shift Manual Control Circuit Intermittent	The auto shift manual control refers to the system that allows manual control of gear changes on an auto transmission or possibly on an electronically controlled manual gearbox. The manual controls could be up or down shift paddles / buttons (on steering column or steering wheel), or it could be a specific movement of the main gear lever.	The control unit has detected a fault with the manual gear selection control circuits e.g. gear change paddle / button switch circuits or gear lever position switch circuits (for manual up / down shifts). The voltage is likely to be intermittent (likely to be out of normal operating range / tolerance when intermittent fault is detected) or, the signal / voltage is erratic (e.g. signal changes are irregular / unpredictable). Possible fault with wiring or connections (including internal switch connections); the switch contacts could also be faulty e.g. not opening / closing correctly or have a high contact resistance.
P0955	Auto Shift Manual Mode Circuit	The auto shift manual control refers to the system that allows manual control of gear changes on an auto transmission or possibly on an electronically controlled manual gearbox. The fault code is likely to refer to the mode selector switch operation, but also refer to fault codes P0950 - P0954 for additional information on manual change switches.	The voltage or value of the switch signal is incorrect (undefined fault). Switch voltage / signal could be: out of normal operating range, constant value, non-existent, corrupt. Possible fault in the switch or wiring e.g. short / open circuits, circuit resistance or, fault with the reference / operating voltage; the switch contacts could also be faulty e.g. not opening / closing correctly or have a high contact resistance. Refer to list of sensor / actuator checks on page 19.
P0956	Auto Shift Manual Mode Circuit Range / Performance	The auto shift manual control refers to the system that allows manual control of gear changes on an auto transmission or possibly on an electronically controlled manual gearbox. The fault code is likely to refer to the mode selector switch operation, but also refer to fault codes P0950 - P0954 for additional information on manual change switches.	The voltage or value of the switch signal is within the normal operating range / tolerance but the signal is not plausible or is incorrect due to an undefined fault. The switch signal might not match the operating conditions indicated by other sensors or, the switch signal is not changing / responding as expected. Possible fault in the switch or wiring e.g. short / open circuits, circuit resistance or, fault with the reference / operating voltage; also check switch contacts to ensure that they open / close and check continuity of switch contacts (ensure that there is not a high resistance). Refer to list of sensor / actuator checks on page 19. It is also possible that the switch is operating correctly but the position being indicated by the switch is unacceptable or incorrect for the operating conditions indicated by other sensors or, is not as expected by the control unit; check mechanism / linkage used to operate switch, and check base settings for the switch if applicable.
P0957	Auto Shift Manual Mode Circuit Low	The auto shift manual control refers to the system that allows manual control of gear changes on an auto transmission or possibly on an electronically controlled manual gearbox. The fault code is likely to refer to the mode selector switch operation, but also refer to fault codes P0950 - P0954 for additional information on manual change switches.	The voltage or value of the switch signal is either "zero" (no signal) or, the signal exists but it is below the normal operating range (lower than the minimum operating limit). Possible fault in the switch or wiring e.g. short / open circuits, circuit resistance or, fault with the reference / operating voltage; the switch contacts could also be faulty e.g. not opening / closing correctly or have a high contact resistance. Refer to list of sensor / actuator checks on page 19. Note that the fault code can be activated if the switch output remains at the low value e.g. zero volts and does not rise to the high value e.g. battery voltage (the control unit will detect no switch signal change but it has detected change in engine operating conditions).
P0958	Auto Shift Manual Mode Circuit High	The auto shift manual control refers to the system that allows manual control of gear changes on an auto transmission or possibly on an electronically controlled manual gearbox. The fault code is likely to refer to the mode selector switch operation, but also refer to fault codes P0950 - P0954 for additional information on manual change switches.	The voltage or value of the switch signal is either at full value (e.g. battery voltage with no signal or frequency) or, the signal exists but it is above the normal operating range (higher than the maximum operating limit). Possible fault in the switch or wiring e.g. short / open circuits, circuit resistance or, fault with the reference / operating voltage; the switch contacts could also be faulty e.g. not opening / closing correctly or have a high contact resistance. Refer to list of sensor / actuator checks on page 19. Note that the fault code can be activated if the switch output remains at the high value e.g. battery voltage and does not fall to the low value e.g. zero volts (the control unit will detect no switch signal change but it has detected change in engine operating conditions).
P0959	Auto Shift Manual Mode Circuit Intermittent	The auto shift manual control refers to the system that allows manual control of gear changes on an auto transmission or possibly on an electronically controlled manual gearbox. The fault code is likely to refer to the mode selector switch operation, but also refer to fault codes P0950 - P0954 for additional information on manual change switches.	The voltage or value of the switch signal is intermittent (likely to be out of normal operating range / tolerance when intermittent fault is detected) or, the signal / voltage is erratic (e.g. signal changes are irregular / unpredictable). Possible fault with wiring or connections (including internal switch connections); the switch contacts could also be faulty e.g. not opening / closing correctly or have a high contact resistance. Refer to list of sensor / actuator checks on page 19.
P0960	Pressure Control Solenoid "A" Control Circuit / Open	Pressure control solenoids are used in the transmission to regulate the pressure in the hydraulic circuits (usually for gear shift control). The pressure control solenoid can also influence / control the actual gear shift. Refer to vehicle specific information to establish which gear(s) are applicable to the identified solenoid.	The voltage, frequency or value of the control signal in the solenoid circuit is incorrect. Signal could be:- out of normal operating range, constant value, non-existent, corrupt. The fault code indicates a possible "open circuit" but other undefined faults in the solenoid or wiring could activate the same code e.g. solenoid power supply, short circuit, high circuit resistance or incorrect solenoid resistance. Note: the control unit could be providing the correct signal but it is being affected by the circuit fault. Refer to list of sensor / actuator checks on page 19.
P0961	Pressure Control Solenoid "A" Control Circuit Range / Performance	Pressure control solenoids are used in the transmission to regulate the pressure in the hydraulic circuits (usually for gear shift control). The pressure control solenoid can also influence / control the actual gear shift. Refer to vehicle specific information to establish which gear(s) are applicable to the identified solenoid.	The voltage, frequency or value of the control signal in the solenoid circuit is within the normal operating range / tolerance, but the signal might be incorrect for the operating conditions. It is also possible that the solenoid response (or response of system controlled by the solenoid) differs from the expected / desired response (incorrect response to the control signal provided by the control unit). Possible fault in solenoid or wiring, but also possible that the solenoid (or mechanism / system controlled by the solenoid) is not operating or moving correctly. Note: the control unit could be providing the correct signal but it is being affected by a circuit fault. Refer to list of sensor / actuator checks on page 19. Note that the fault code could also be activated due to mechanical or hydraulic faults e.g. restriction in the movement of the solenoid, a mechanical fault with the gear selection / deselection mechanism, a fault with the hydraulic pressure or valves.
P0962	Pressure Control Solenoid "A" Control Circuit Low	Pressure control solenoids are used in the transmission to regulate the pressure in the hydraulic circuits (usually for gear shift control). The pressure control solenoid can also influence / control the actual gear shift. Refer to vehicle specific information to establish which gear(s) are applicable to the identified solenoid.	The voltage, frequency or value of the control signal in the solenoid circuit is either "zero" (no signal) or, the signal exists but is below the normal operating range. Possible fault with solenoid or wiring (e.g. solenoid power supply, short / open circuit or resistance) that is causing a signal value to be lower than the minimum operating limit. Note: the control unit could be providing the correct signal but it is being affected by a circuit fault. Refer to list of sensor / actuator checks on page 19.
P0963	Pressure Control Solenoid "A" Control Circuit High	Pressure control solenoids are used in the transmission to regulate the pressure in the hydraulic circuits (usually for gear shift control). The pressure control solenoid can also influence / control the actual gear shift. Refer to vehicle specific information to establish which gear(s) are applicable to the identified solenoid.	The voltage, frequency or value of the control signal in the solenoid circuit is either at full value (e.g. battery voltage with no on / off signal) or, the signal exists but it is above the normal operating range. Possible fault with solenoid or wiring (e.g. solenoid power supply, short / open circuit or resistance) that is causing a signal value to be higher than the maximum operating limit. Note: the control unit could be providing the correct signal but it is being affected by a circuit fault. Refer to list of sensor / actuator checks on page 19.

NOTE 1. Check for other fault codes that could provide additional information.

NOTE 2. Communication between control units can pass via a CAN-Bus system; refer to Page 51 for CAN-Bus checks.

NOTE 3. If a fault cannot be located, it is also possible that the control unit is at fault

NOTE 4. Refer to Page 53 for list of pages that show the ISO standard for component locations e.g. Sensor A - Bank 1.

208

Code	Fault Location	Description	Probable Cause / Notes
P0964	Pressure Control Solenoid "B" Control Circuit / Open	Pressure control solenoids are used in the transmission to regulate the pressure in the hydraulic circuits (usually for gear shift control). The pressure control solenoid can also influence / control the actual gear shift. Refer to vehicle specific information to establish which gear(s) are applicable to the identified solenoid.	The voltage, frequency or value of the control signal in the solenoid circuit is incorrect. Signal could be:- out of normal operating range, constant value, non-existent, corrupt. The fault code indicates a possible "open circuit" but other undefined faults in the solenoid or wiring could activate the same code e.g. solenoid power supply, short circuit, high circuit resistance or incorrect solenoid resistance. Note: the control unit could be providing the correct signal but it is being affected by the circuit fault. Refer to list of sensor / actuator checks on page 19.
P0965	Pressure Control Solenoid "B" Control Circuit Range / Performance	Pressure control solenoids are used in the transmission to regulate the pressure in the hydraulic circuits (usually for gear shift control). The pressure control solenoid can also influence / control the actual gear shift. Refer to vehicle specific information to establish which gear(s) are applicable to the identified solenoid.	The voltage, frequency or value of the control signal in the solenoid circuit is within the normal operating range / tolerance, but the signal might be incorrect for the operating conditions. It is also possible that the solenoid response (or response of system controlled by the solenoid) differs from the expected / desired response (incorrect response to the control signal provided by the control unit). Possible fault in solenoid or wiring, but also possible that the solenoid (or mechanism / system controlled by the solenoid) is not operating or moving correctly. Note: the control unit could be providing the correct signal but it is being affected by a circuit fault. Refer to list of sensor / actuator checks on page 19. Note that the fault code could also be activated due to mechanical or hydraulic faults e.g. restriction in the movement of the solenoid, a mechanical fault with the gear selection / deselection mechanism, a fault with the hydraulic pressure or valves.
P0966	Pressure Control Solenoid "B" Control Circuit Low	Pressure control solenoids are used in the transmission to regulate the pressure in the hydraulic circuits (usually for gear shift control). The pressure control solenoid can also influence / control the actual gear shift. Refer to vehicle specific information to establish which gear(s) are applicable to the identified solenoid.	The voltage, frequency or value of the control signal in the solenoid circuit is either "zero" (no signal) or, the signal exists but is below the normal operating range. Possible fault with solenoid or wiring (e.g. solenoid power supply, short / open circuit or resistance) that is causing a signal value to be lower than the minimum operating limit. Note: the control unit could be providing the correct signal but it is being affected by a circuit fault. Refer to list of sensor / actuator checks on page 19.
P0967	Pressure Control Solenoid "B" Control Circuit High	Pressure control solenoids are used in the transmission to regulate the pressure in the hydraulic circuits (usually for gear shift control). The pressure control solenoid can also influence / control the actual gear shift. Refer to vehicle specific information to establish which gear(s) are applicable to the identified solenoid.	The voltage, frequency or value of the control signal in the solenoid circuit is either at full value (e.g. battery voltage with no on / off signal) or, the signal exists but it is above the normal operating range. Possible fault with solenoid or wiring (e.g. solenoid power supply, short / open circuit or resistance) that is causing a signal value to be higher than the maximum operating limit. Note: the control unit could be providing the correct signal but it is being affected by a circuit fault. Refer to list of sensor / actuator checks on page 19.
P0968	Pressure Control Solenoid "C" Control Circuit / Open	Pressure control solenoids are used in the transmission to regulate the pressure in the hydraulic circuits (usually for gear shift control). The pressure control solenoid can also influence / control the actual gear shift. Refer to vehicle specific information to establish which gear(s) are applicable to the identified solenoid.	The voltage, frequency or value of the control signal in the solenoid circuit is incorrect. Signal could be:- out of normal operating range, constant value, non-existent, corrupt. The fault code indicates a possible "open circuit" but other undefined faults in the solenoid or wiring could activate the same code e.g. solenoid power supply, short circuit, high circuit resistance or incorrect solenoid resistance. Note: the control unit could be providing the correct signal but it is being affected by the circuit fault. Refer to list of sensor / actuator checks on page 19.
P0969	Pressure Control Solenoid "C" Control Circuit Range / Performance	Pressure control solenoids are used in the transmission to regulate the pressure in the hydraulic circuits (usually for gear shift control). The pressure control solenoid can also influence / control the actual gear shift. Refer to vehicle specific information to establish which gear(s) are applicable to the identified solenoid.	The voltage, frequency or value of the control signal in the solenoid circuit is within the normal operating range / tolerance, but the signal might be incorrect for the operating conditions. It is also possible that the solenoid response (or response of system controlled by the solenoid) differs from the expected / desired response, but also possible that the solenoid (or mechanism / system controlled by the solenoid) is not operating or moving correctly. Note: the control unit could be providing the correct signal but it is being affected by a circuit fault. Refer to list of sensor / actuator checks on page 19. Note that the fault code could also be activated due to mechanical or hydraulic faults e.g. restriction in the movement of the solenoid, a mechanical fault with the gear selection / deselection mechanism, a fault with the hydraulic pressure or valves.
P0970	Pressure Control Solenoid "C" Control Circuit Low	Pressure control solenoids are used in the transmission to regulate the pressure in the hydraulic circuits (usually for gear shift control). The pressure control solenoid can also influence / control the actual gear shift. Refer to vehicle specific information to establish which gear(s) are applicable to the identified solenoid.	The voltage, frequency or value of the control signal in the solenoid circuit is either "zero" (no signal) or, the signal exists but is below the normal operating range. Possible fault with solenoid or wiring (e.g. solenoid power supply, short / open circuit or resistance) that is causing a signal value to be lower than the minimum operating limit. Note: the control unit could be providing the correct signal but it is being affected by a circuit fault. Refer to list of sensor / actuator checks on page 19.
P0971	Pressure Control Solenoid "C" Control Circuit High	Pressure control solenoids are used in the transmission to regulate the pressure in the hydraulic circuits (usually for gear shift control). The pressure control solenoid can also influence / control the actual gear shift. Refer to vehicle specific information to establish which gear(s) are applicable to the identified solenoid.	The voltage, frequency or value of the control signal in the solenoid circuit is either at full value (e.g. battery voltage with no on / off signal) or, the signal exists but it is above the normal operating range. Possible fault with solenoid or wiring (e.g. solenoid power supply, short / open circuit or resistance) that is causing a signal value to be higher than the maximum operating limit. Note: the control unit could be providing the correct signal but it is being affected by a circuit fault. Refer to list of sensor / actuator checks on page 19.
P0972	Shift Solenoid "A" Control Circuit Range / Performance	The shift solenoid is used to control a hydraulic circuit in the transmission, which enables the applicable gear(s) to be selected / deselected. Refer to vehicle specific information to establish which gear(s) are applicable to the identified solenoid.	The voltage, frequency or value of the control signal in the solenoid circuit is within the normal operating range / tolerance, but the signal might be incorrect for the operating conditions. It is also possible that the solenoid response (or response of system controlled by the solenoid) differs from the expected / desired response (incorrect response to the control signal provided by the control unit). Possible fault in solenoid or wiring, but also possible that the solenoid (or mechanism / system controlled by the solenoid) is not operating or moving correctly. Note: the control unit could be providing the correct signal but it is being affected by a circuit fault. Refer to list of sensor / actuator checks on page 19. Note that the fault code could also be activated due to mechanical or hydraulic faults e.g. restriction in the movement of the solenoid, a mechanical fault with the gear selection / deselection mechanism, a fault with the hydraulic pressure or valves.
P0973	Shift Solenoid "A" Control Circuit Low	The shift solenoid is used to control a hydraulic circuit in the transmission, which enables the applicable gear(s) to be selected / deselected. Refer to vehicle specific information to establish which gear(s) are applicable to the identified solenoid.	The voltage, frequency or value of the control signal in the solenoid circuit is either "zero" (no signal) or, the signal exists but is below the normal operating range. Possible fault with solenoid or wiring (e.g. solenoid power supply, short / open circuit or resistance) that is causing a signal value to be lower than the minimum operating limit. Note: the control unit could be providing the correct signal but it is being affected by a circuit fault. Refer to list of sensor / actuator checks on page 19.

NOTE 1. Check for other fault codes that could provide additional information. **NOTE 2.** Communication between control units can pass via a CAN-Bus system; refer to Page 51 for CAN-Bus checks.

NOTE 3. If a fault cannot be located, it is also possible that the control unit is at fault. **NOTE 4.** Refer to Page 53 for list of pages that show the ISO standard for component locations e.g. Sensor A - Bank 1.

Fault code	EOBD / ISO Description	Component / System Description	Meaningful Description and Quick Check
P0974	Shift Solenoid "A" Control Circuit High	The shift solenoid is used to control a hydraulic circuit in the transmission, which enables the applicable gear(s) to be selected / deselected. Refer to vehicle specific information to establish which gear(s) are applicable to the identified solenoid.	The voltage, frequency or value of the control signal in the solenoid circuit is either at full value (e.g. battery voltage with no on / off signal) or, the signal exists but it is above the normal operating range. Possible fault with solenoid or wiring (e.g. solenoid power supply, short / open circuit or resistance) that is causing a signal value to be higher than the maximum operating limit. Note: the control unit could be providing the correct signal but it is being affected by the circuit fault. Refer to list of sensor / actuator checks on page 19.
P0975	Shift Solenoid "B" Control Circuit Range / Performance	The shift solenoid is used to control a hydraulic circuit in the transmission, which enables the applicable gear(s) to be selected / deselected. Refer to vehicle specific information to establish which gear(s) are applicable to the identified solenoid.	The voltage, frequency or value of the control signal in the solenoid circuit is within the normal operating range / tolerance, but the signal might be incorrect for the operating conditions. It is also possible that the solenoid response (or response of system controlled by the solenoid) differs from the expected / desired response (incorrect response to the control signal provided by the control unit). Possible fault in solenoid or wiring, but also possible that the solenoid (or mechanism / system controlled by the solenoid) is not operating or moving correctly. Note: the control unit could be providing the correct signal but it is being affected by a circuit fault. Refer to list of sensor / actuator checks on page 19. Note that the fault code could also be activated due to mechanical or hydraulic faults e.g. restriction in the movement of the solenoid, a mechanical fault with the gear selection / deselection mechanism, a fault with the hydraulic pressure or valves
P0976	Shift Solenoid "B" Control Circuit Low	The shift solenoid is used to control a hydraulic circuit in the transmission, which enables the applicable gear(s) to be selected / deselected. Refer to vehicle specific information to establish which gear(s) are applicable to the identified solenoid.	The voltage, frequency or value of the control signal in the solenoid circuit is either "zero" (no signal) or, the signal exists but is below the normal operating range. Possible fault with solenoid or wiring (e.g. solenoid power supply, short / open circuit or resistance) that is causing a signal value to be lower than the minimum operating limit. Note: the control unit could be providing the correct signal but it is being affected by a circuit fault. Refer to list of sensor / actuator checks on page 19.
P0977	Shift Solenoid "B" Control Circuit High	The shift solenoid is used to control a hydraulic circuit in the transmission, which enables the applicable gear(s) to be selected / deselected. Refer to vehicle specific information to establish which gear(s) are applicable to the identified solenoid.	The voltage, frequency or value of the control signal in the solenoid circuit is either at full value (e.g. battery voltage with no on / off signal) or, the signal exists but it is above the normal operating range. Possible fault with solenoid or wiring (e.g. solenoid power supply, short / open circuit or resistance) that is causing a signal value to be higher than the maximum operating limit. Note: the control unit could be providing the correct signal but it is being affected by the circuit fault. Refer to list of sensor / actuator checks on page 19.
P0978	Shift Solenoid "C" Control Circuit Range / Performance	The shift solenoid is used to control a hydraulic circuit in the transmission, which enables the applicable gear(s) to be selected / deselected. Refer to vehicle specific information to establish which gear(s) are applicable to the identified solenoid.	The voltage, frequency or value of the control signal in the solenoid circuit is within the normal operating range / tolerance, but the signal might be incorrect for the operating conditions. It is also possible that the solenoid response (or response of system controlled by the solenoid) differs from the expected / desired response (incorrect response to the control signal provided by the control unit). Possible fault in solenoid or wiring, but also possible that the solenoid (or mechanism / system controlled by the solenoid) is not operating or moving correctly. Note: the control unit could be providing the correct signal but it is being affected by a circuit fault. Refer to list of sensor / actuator checks on page 19. Note that the fault code could also be activated due to mechanical or hydraulic faults e.g. restriction in the movement of the solenoid, a mechanical fault with the gear selection / deselection mechanism, a fault with the hydraulic pressure or valves.
P0979	Shift Solenoid "C" Control Circuit Low	The shift solenoid is used to control a hydraulic circuit in the transmission, which enables the applicable gear(s) to be selected / deselected. Refer to vehicle specific information to establish which gear(s) are applicable to the identified solenoid.	The voltage, frequency or value of the control signal in the solenoid circuit is either "zero" (no signal) or, the signal exists but is below the normal operating range. Possible fault with solenoid or wiring (e.g. solenoid power supply, short / open circuit or resistance) that is causing a signal value to be lower than the minimum operating limit. Note: the control unit could be providing the correct signal but it is being affected by a circuit fault. Refer to list of sensor / actuator checks on page 19.
P0980	Shift Solenoid "C" Control Circuit High	The shift solenoid is used to control a hydraulic circuit in the transmission, which enables the applicable gear(s) to be selected / deselected. Refer to vehicle specific information to establish which gear(s) are applicable to the identified solenoid.	The voltage, frequency or value of the control signal in the solenoid circuit is either at full value (e.g. battery voltage with no on / off signal) or, the signal exists but it is above the normal operating range. Possible fault with solenoid or wiring (e.g. solenoid power supply, short / open circuit or resistance) that is causing a signal value to be higher than the maximum operating limit. Note: the control unit could be providing the correct signal but it is being affected by the circuit fault. Refer to list of sensor / actuator checks on page 19.
P0981	Shift Solenoid "D" Control Circuit Range / Performance	The shift solenoid is used to control a hydraulic circuit in the transmission, which enables the applicable gear(s) to be selected / deselected. Refer to vehicle specific information to establish which gear(s) are applicable to the identified solenoid.	The voltage, frequency or value of the control signal in the solenoid circuit is within the normal operating range / tolerance, but the signal might be incorrect for the operating conditions. It is also possible that the solenoid response (or response of system controlled by the solenoid) differs from the expected / desired response (incorrect response to the control signal provided by the control unit). Possible fault in solenoid or wiring, but also possible that the solenoid (or mechanism / system controlled by the solenoid) is not operating or moving correctly. Note: the control unit could be providing the correct signal but it is being affected by a circuit fault. Refer to list of sensor / actuator checks on page 19. Note that the fault code could also be activated due to mechanical or hydraulic faults e.g. restriction in the movement of the solenoid, a mechanical fault with the gear selection / deselection mechanism, a fault with the hydraulic pressure or valves.
P0982	Shift Solenoid "D" Control Circuit Low	The shift solenoid is used to control a hydraulic circuit in the transmission, which enables the applicable gear(s) to be selected / deselected. Refer to vehicle specific information to establish which gear(s) are applicable to the identified solenoid.	The voltage, frequency or value of the control signal in the solenoid circuit is either "zero" (no signal) or, the signal exists but is below the normal operating range. Possible fault with solenoid or wiring (e.g. solenoid power supply, short / open circuit or resistance) that is causing a signal value to be lower than the minimum operating limit. Note: the control unit could be providing the correct signal but it is being affected by a circuit fault. Refer to list of sensor / actuator checks on page 19.
P0983	Shift Solenoid "D" Control Circuit High	The shift solenoid is used to control a hydraulic circuit in the transmission, which enables the applicable gear(s) to be selected / deselected. Refer to vehicle specific information to establish which gear(s) are applicable to the identified solenoid.	The voltage, frequency or value of the control signal in the solenoid circuit is either at full value (e.g. battery voltage with no on / off signal) or, the signal exists but it is above the normal operating range. Possible fault with solenoid or wiring (e.g. solenoid power supply, short / open circuit or resistance) that is causing a signal value to be higher than the maximum operating limit. Note: the control unit could be providing the correct signal but it is being affected by the circuit fault. Refer to list of sensor / actuator checks on page 19.

NOTE 1. Check for other fault codes that could provide additional information.   **NOTE 2.** Communication between control units can pass via a CAN-Bus system; refer to Page 51 for CAN-Bus checks.

NOTE 4. Refer to Page 53 for list of pages that show the ISO standard for component locations e.g. Sensor A - Bank 1.

Code	Fault Location	Description	Probable Cause
P0984	Shift Solenoid "E" Control Circuit Range / Performance	The shift solenoid is used to control a hydraulic circuit in the transmission, which enables the applicable gear(s) to be selected / deselected. Refer to vehicle specific information to establish which gear(s) are applicable to the identified solenoid.	The voltage, frequency or value of the control signal in the solenoid circuit is within the normal operating range / tolerance, but the signal might be incorrect for the operating conditions. It is also possible that the solenoid response (or response of system controlled by the solenoid) differs from the expected / desired response (incorrect response to the control signal provided by the control unit). Possible fault in solenoid or wiring, but also possible that the solenoid (or mechanism / system controlled by the solenoid) is not operating or moving correctly. Note: the control unit could be providing the correct signal but it is being affected by a circuit fault. Refer to list of sensor / actuator checks on page 19. Note that the fault code could also be activated due to mechanical or hydraulic faults e.g. restriction in the movement of the solenoid, a mechanical fault with the gear selection / deselection mechanism, a fault with the hydraulic pressure or valves.
P0985	Shift Solenoid "E" Control Circuit Low	The shift solenoid is used to control a hydraulic circuit in the transmission, which enables the applicable gear(s) to be selected / deselected. Refer to vehicle specific information to establish which gear(s) are applicable to the identified solenoid.	The voltage, frequency or value of the control signal in the solenoid circuit is either "zero" (no signal) or, the signal exists but is below the normal operating range. Possible fault with solenoid or wiring (e.g. solenoid power supply, short / open circuit or resistance) that is causing a signal value to be lower than the minimum operating limit. Note: the control unit could be providing the correct signal but it is being affected by a circuit fault. Refer to list of sensor / actuator checks on page 19.
P0986	Shift Solenoid "E" Control Circuit High	The shift solenoid is used to control a hydraulic circuit in the transmission, which enables the applicable gear(s) to be selected / deselected. Refer to vehicle specific information to establish which gear(s) are applicable to the identified solenoid.	The voltage, frequency or value of the control signal in the solenoid circuit is either at full value (e.g. battery voltage with no on / off signal) or, the signal exists but it is above the normal operating range. Possible fault with solenoid or wiring (e.g. solenoid power supply, short / open circuit or resistance) that is causing a signal value to be higher than the maximum operating limit. Note: the control unit could be providing the correct signal but it is being affected by the circuit fault. Refer to list of sensor / actuator checks on page 19.
P0987	Transmission Fluid Pressure Sensor / Switch "E" Circuit	Transmission systems can make use of fluid pressure to control various functions, including gear changes. Fluid pressure sensors are therefore used to monitor pressures and changes in pressure when functions such as gear changes take place. It may be necessary to refer to vehicle specific information to identify the applicable pressure circuit (and its function) for the identified pressure sensor.	The voltage, frequency or value of the pressure sensor signal is incorrect (undefined fault). Sensor voltage / signal could be: out of normal operating range, constant value, non-existent, corrupt. Possible fault in the pressure sensor or wiring e.g. short / open circuits, circuit resistance, incorrect sensor resistance (if applicable) or, fault with the reference / operating voltage; interference from other circuits can also affect sensor signals. Where a pressure switch is fitted, it is possible that the switch contacts are not opening / closing correctly or the switch contacts have a high resistance. Refer to list of sensor / actuator checks on page 19.
P0988	Transmission Fluid Pressure Sensor / Switch "E" Circuit Range / Performance	Transmission systems can make use of fluid pressure to control various functions, including gear changes. Fluid pressure sensors are therefore used to monitor pressures and changes in pressure when functions such as gear changes take place. It may be necessary to refer to vehicle specific information to identify the applicable pressure circuit (and its function) for the identified pressure sensor.	The voltage, frequency or value of the pressure sensor signal is within the normal operating range / tolerance but the signal is not plausible or is incorrect due to an undefined fault. The sensor signal might not match the operating conditions indicated by other sensors or, the sensor signal is not changing / responding as expected. Possible fault in the pressure sensor or wiring e.g. short / open circuits, circuit resistance, incorrect sensor resistance (if applicable) or, fault with the reference / operating voltage; interference from other circuits can also affect sensor signals. Where a pressure switch is fitted, it is possible that the switch contacts are not opening / closing correctly or the switch contacts have a high resistance. Refer to list of sensor / actuator checks on page 19. It is also possible that the pressure sensor is operating correctly but the pressure being measured by the sensor is unacceptable or incorrect for the operating conditions; if possible, use alternative method of checking pressure to establish if sensor system is operating correctly and whether the pressure indicated by the sensor is correct / incorrect.
P0989	Transmission Fluid Pressure Sensor / Switch "E" Circuit Low	Transmission systems can make use of fluid pressure to control various functions, including gear changes. Fluid pressure sensors are therefore used to monitor pressures and changes in pressure when functions such as gear changes take place. It may be necessary to refer to vehicle specific information to identify the applicable pressure circuit (and its function) for the identified pressure sensor.	The voltage, frequency or value of the pressure sensor signal is either "zero" (no signal) or, the signal exists but it is below the normal operating range (lower than the minimum operating limit). Possible fault in the pressure sensor or wiring e.g. short / open circuits, circuit resistance, incorrect sensor resistance (if applicable) or, fault with the reference / operating voltage; interference from other circuits can also affect sensor signals. Where a pressure switch is fitted, it is possible that the switch contacts are not opening / closing correctly or, the contacts are sticking (signal is remaining low when it should have risen to the high value); on some circuits, a high resistance on the contacts could also cause a low signal value to exist. Refer to list of sensor / actuator checks on page 19.
P0990	Transmission Fluid Pressure Sensor / Switch "E" Circuit High	Transmission systems can make use of fluid pressure to control various functions, including gear changes. Fluid pressure sensors are therefore used to monitor pressures and changes in pressure when functions such as gear changes take place. It may be necessary to refer to vehicle specific information to identify the applicable pressure circuit (and its function) for the identified pressure sensor.	The voltage, frequency or value of the pressure sensor signal is either at full value (e.g. battery voltage with no signal or frequency) or, the signal exists but it is above the normal operating range (higher than the maximum operating limit). Possible fault in the pressure sensor or wiring e.g. short / open circuits, circuit resistance, incorrect sensor resistance (if applicable) or, fault with the reference / operating voltage; interference from other circuits can also affect sensor signals. Where a pressure switch is fitted, it is possible that the switch contacts are not opening / closing correctly or, the contacts are sticking (signal is remaining high when it should have dropped to low value); on some circuits, a high resistance on the contacts could also cause a high signal value to exist. Refer to list of sensor / actuator checks on page 19.
P0991	Transmission Fluid Pressure Sensor / Switch "E" Circuit Intermittent	Transmission systems can make use of fluid pressure to control various functions, including gear changes. Fluid pressure sensors are therefore used to monitor pressures and changes in pressure when functions such as gear changes take place. It may be necessary to refer to vehicle specific information to identify the applicable pressure circuit (and its function) for the identified pressure sensor.	The voltage, frequency or value of the pressure sensor signal is intermittent (likely to be out of normal operating range / tolerance when intermittent fault is detected) or, the signal / voltage is erratic (e.g. signal changes are irregular / unpredictable). Possible fault with wiring or connections (including internal sensor connections); also interference from other circuits can affect the signal. Where a pressure switch is fitted, it is possible that the switch contacts are not opening / closing correctly or the switch contacts have a high resistance. Refer to list of sensor / actuator checks on page 19. It is also possible that the pressure could be changing erratically, which could cause the fault code to be activated; if necessary, check pressure with a separate gauge to check for erratic pressure changes.
P0992	Transmission Fluid Pressure Sensor / Switch "F" Circuit	Transmission systems can make use of fluid pressure to control various functions, including gear changes. Fluid pressure sensors are therefore used to monitor pressures and changes in pressure when functions such as gear changes take place. It may be necessary to refer to vehicle specific information to identify the applicable pressure circuit (and its function) for the identified pressure sensor.	The voltage, frequency or value of the pressure sensor signal is incorrect (undefined fault). Sensor voltage / signal could be: out of normal operating range, constant value, non-existent, corrupt. Possible fault in the pressure sensor or wiring e.g. short / open circuits, circuit resistance, incorrect sensor resistance (if applicable) or, fault with the reference / operating voltage; interference from other circuits can also affect sensor signals. Where a pressure switch is fitted, it is possible that the switch contacts are not opening / closing correctly or the switch contacts have a high resistance. Refer to list of sensor / actuator checks on page 19.

NOTE 1. Check for other fault codes that could provide additional information. **NOTE 2.** Communication between control units can pass via a CAN-Bus system; refer to Page 51 for CAN-Bus checks.

NOTE 3. If a fault cannot be located, it is also possible that the control unit is at fault. **NOTE 4.** Refer to Page 53 for list of pages that show the ISO standard for component locations e.g. Sensor A - Bank 1.

Fault code	EOBD / ISO Description	Component / System Description	Meaningful Description and Quick Check
P0993	Transmission Fluid Pressure Sensor / Switch "F" Circuit Range / Performance	Transmission systems can make use of fluid pressure to control various functions, including gear changes. Fluid pressure sensors are therefore used to monitor pressures and changes in pressure when functions such as gear changes take place. It may be necessary to refer to vehicle specific information to identify the applicable pressure circuit (and its function) for the identified pressure sensor.	The voltage, frequency or value of the pressure sensor signal is within the normal operating range / tolerance but the signal is not plausible or is incorrect due to an undefined fault. The sensor signal might not match the operating conditions indicated by other sensors or, the sensor signal is not changing / responding as expected. Possible fault in the pressure sensor or wiring e.g. short / open circuits, circuit resistance, incorrect sensor resistance (if applicable) or, fault with the reference / operating voltage; interference from other circuits can also affect sensor signals. Where a pressure switch is fitted, it is possible that the switch contacts are not opening / closing correctly or the switch contacts have a high resistance. Refer to list of sensor / actuator checks on page 19. It is also possible that the pressure sensor is operating correctly but the pressure being measured by the sensor is unacceptable or incorrect for the operating conditions; if possible, use alternative method of checking pressure to establish if sensor system is operating correctly and whether the pressure indicated by the sensor is correct / incorrect.
P0994	Transmission Fluid Pressure Sensor / Switch "F" Circuit Low	Transmission systems can make use of fluid pressure to control various functions, including gear changes. Fluid pressure sensors are therefore used to monitor pressures and changes in pressure when functions such as gear changes take place. It may be necessary to refer to vehicle specific information to identify the applicable pressure circuit (and its function) for the identified pressure sensor.	The voltage, frequency or value of the pressure sensor signal is either "zero" (no signal) or, the signal exists but it is below the normal operating range (lower than the minimum operating limit). Possible fault in the pressure sensor or wiring e.g. short / open circuits, circuit resistance, incorrect sensor resistance (if applicable) or, fault with the reference / operating voltage; interference from other circuits can also affect sensor signals. Where a pressure switch is fitted, it is possible that the switch contacts are not opening / closing correctly or, the contacts are sticking (signal is remaining low when it should have risen to the high value); on some circuits, a high resistance on the contacts could also cause a low signal value to exist. Refer to list of sensor / actuator checks on page 19.
P0995	Transmission Fluid Pressure Sensor / Switch "F" Circuit High	Transmission systems can make use of fluid pressure to control various functions, including gear changes. Fluid pressure sensors are therefore used to monitor pressures and changes in pressure when functions such as gear changes take place. It may be necessary to refer to vehicle specific information to identify the applicable pressure circuit (and its function) for the identified pressure sensor.	The voltage, frequency or value of the pressure sensor signal is either at full value (e.g. battery voltage with no signal or frequency) or, the signal exists but it is above the normal operating range (higher than the maximum operating limit). Possible fault in the pressure sensor or wiring e.g. short / open circuits, circuit resistance, incorrect sensor resistance (if applicable) or, fault with the reference / operating voltage; interference from other circuits can also affect sensor signals. Where a pressure switch is fitted, it is possible that the switch contacts are not opening / closing correctly or, the contacts are sticking (signal is remaining high when it should have dropped to low value); on some circuits, a high resistance on the contacts could also cause a high signal value to exist. Refer to list of sensor / actuator checks on page 19.
P0996	Transmission Fluid Pressure Sensor / Switch "F" Circuit Intermittent	Transmission systems can make use of fluid pressure to control various functions, including gear changes. Fluid pressure sensors are therefore used to monitor pressures and changes in pressure when functions such as gear changes take place. It may be necessary to refer to vehicle specific information to identify the applicable pressure circuit (and its function) for the identified pressure sensor.	The voltage, frequency or value of the pressure sensor signal is intermittent (likely to be out of normal operating range / tolerance when intermittent fault is detected) or, the signal / voltage is erratic (e.g. signal changes are irregular / unpredictable). Possible fault with wiring or connections (including internal sensor connections); also interference from other circuits can affect the signal. Where a pressure switch is fitted, it is possible that the switch contacts are not opening / closing correctly or the switch contacts have a high resistance. Refer to list of sensor / actuator checks on page 19. It is also possible that the pressure could be changing erratically, which could cause the fault code to be activated; if necessary, check pressure with a separate gauge to check for erratic pressure changes.
P0997	Shift Solenoid "F" Control Circuit Range / Performance	The shift solenoid is used to control a hydraulic circuit in the transmission, which enables the applicable gear(s) to be selected / deselected. Refer to vehicle specific information to establish which gear(s) are applicable to the identified solenoid.	The voltage, frequency or value of the pressure sensor signal is within the normal operating range / tolerance but the signal is not plausible or is incorrect due to an undefined fault. The sensor signal might not match the operating conditions indicated by other sensors or, the sensor signal is not changing / responding as expected to the changes in conditions. Possible fault in the pressure sensor or wiring e.g. short / open circuits, circuit resistance, incorrect sensor resistance (if applicable) or, fault with the reference / operating voltage; interference from other circuits can also affect sensor signals. Where a pressure switch is fitted, it is possible that the switch contacts are not opening / closing correctly or the switch contacts have a high resistance. Refer to list of sensor / actuator checks on page 19. It is also possible that the pressure sensor is operating correctly but the pressure being measured by the sensor is unacceptable or incorrect for the operating conditions; if possible, use alternative method of checking pressure to establish if sensor system is operating correctly and whether the pressure indicated by the sensor is correct / incorrect.
P0998	Shift Solenoid "F" Control Circuit Low	The shift solenoid is used to control a hydraulic circuit in the transmission, which enables the applicable gear(s) to be selected / deselected. Refer to vehicle specific information to establish which gear(s) are applicable to the identified solenoid.	The voltage, frequency or value of the pressure sensor signal is within the normal operating range / tolerance but the signal is not plausible or is incorrect due to an undefined fault. The sensor signal might not match the operating conditions indicated by other sensors or, the sensor signal is not changing / responding as expected to the changes in conditions. Possible fault in the pressure sensor or wiring e.g. short / open circuits, circuit resistance, incorrect sensor resistance (if applicable) or, fault with the reference / operating voltage; interference from other circuits can also affect sensor signals. Where a pressure switch is fitted, it is possible that the switch contacts are not opening / closing correctly or the switch contacts have a high resistance. Refer to list of sensor / actuator checks on page 19. It is also possible that the pressure sensor is operating correctly but the pressure being measured by the sensor is unacceptable or incorrect for the operating conditions; if possible, use alternative method of checking pressure to establish if sensor system is operating correctly and whether the pressure indicated by the sensor is correct / incorrect.
P0999	Shift Solenoid "F" Control Circuit High	The shift solenoid is used to control a hydraulic circuit in the transmission, which enables the applicable gear(s) to be selected / deselected. Refer to vehicle specific information to establish which gear(s) are applicable to the identified solenoid.	The voltage, frequency or value of the pressure sensor signal is within the normal operating range / tolerance but the signal is not plausible or is incorrect due to an undefined fault. The sensor signal might not match the operating conditions indicated by other sensors or, the sensor signal is not changing / responding as expected to the changes in conditions. Possible fault in the pressure sensor or wiring e.g. short / open circuits, circuit resistance, incorrect sensor resistance (if applicable) or, fault with the reference / operating voltage; interference from other circuits can also affect sensor signals. Where a pressure switch is fitted, it is possible that the switch contacts are not opening / closing correctly or the switch contacts have a high resistance. Refer to list of sensor / actuator checks on page 19. It is also possible that the pressure sensor is operating correctly but the pressure being measured by the sensor is unacceptable or incorrect for the operating conditions; if possible, use alternative method of checking pressure to establish if sensor system is operating correctly and whether the pressure indicated by the sensor is correct / incorrect.

NOTE 1. Check for other fault codes that could provide additional information. **NOTE 2.** Communication between control units can pass via a CAN-Bus system; refer to Page 51 for CAN-Bus checks.

Code	Description		
P099A	Shift Solenoid "G" Control Circuit Range / Performance	The shift solenoid is used to control a hydraulic circuit in the transmission, which enables the applicable gear(s) to be selected / deselected. Refer to vehicle specific information to establish which gear(s) are applicable to the identified solenoid.	The voltage, frequency or value of the control signal in the solenoid circuit is within the normal operating range / tolerance, but the signal might be incorrect for the operating conditions. It is also possible that the solenoid response (or response of system controlled by the solenoid) differs from the expected / desired response (incorrect response to the control signal provided by the control unit). Possible fault in solenoid or wiring, but also possible that the solenoid (or mechanism / system controlled by the solenoid) is not operating or moving correctly. Note: the control unit could be providing the correct signal but it is being affected by a circuit fault. Refer to list of sensor / actuator checks on page 19. Note that the fault code could also be activated due to mechanical or hydraulic faults e.g. restriction in the movement of the solenoid, a mechanical fault with the gear selection / deselection mechanism, a fault with the hydraulic pressure or valves.
P099B	Shift Solenoid "G" Control Circuit Low	The shift solenoid is used to control a hydraulic circuit in the transmission, which enables the applicable gear(s) to be selected / deselected. Refer to vehicle specific information to establish which gear(s) are applicable to the identified solenoid.	The voltage, frequency or value of the control signal in the solenoid circuit is either "zero" (no signal) or, the signal exists but is below the normal operating range. Possible fault with solenoid or wiring (e.g. solenoid power supply, short / open circuit or resistance) that is causing a signal value to be lower than the minimum operating limit. Note: the control unit could be providing the correct signal but it is being affected by a circuit fault. Refer to list of sensor / actuator checks on page 19.
P099C	Shift Solenoid "G" Control Circuit High	The shift solenoid is used to control a hydraulic circuit in the transmission, which enables the applicable gear(s) to be selected / deselected. Refer to vehicle specific information to establish which gear(s) are applicable to the identified solenoid.	The voltage, frequency or value of the control signal in the solenoid circuit is either at full value (e.g. battery voltage with no on / off signal) or, the signal exists but it is above the normal operating range. Possible fault with solenoid or wiring (e.g. solenoid power supply, short / open circuit or resistance) that is causing a signal value to be higher than the maximum operating limit. Note: the control unit could be providing the correct signal but it is being affected by the circuit fault. Refer to list of sensor / actuator checks on page 19.
P099D	Shift Solenoid "H" Control Circuit Range / Performance	The shift solenoid is used to control a hydraulic circuit in the transmission, which enables the applicable gear(s) to be selected / deselected. Refer to vehicle specific information to establish which gear(s) are applicable to the identified solenoid.	The voltage, frequency or value of the control signal in the solenoid circuit is within the normal operating range / tolerance, but the signal might be incorrect for the operating conditions. It is also possible that the solenoid response (or response of system controlled by the solenoid) differs from the expected / desired response (incorrect response to the control signal provided by the control unit). Possible fault in solenoid or wiring, but also possible that the solenoid (or mechanism / system controlled by the solenoid) is not operating or moving correctly. Note: the control unit could be providing the correct signal but it is being affected by a circuit fault. Refer to list of sensor / actuator checks on page 19. Note that the fault code could also be activated due to mechanical or hydraulic faults e.g. restriction in the movement of the solenoid, a mechanical fault with the gear selection / deselection mechanism, a fault with the hydraulic pressure or valves.
P099E	Shift Solenoid "H" Control Circuit Low	The shift solenoid is used to control a hydraulic circuit in the transmission, which enables the applicable gear(s) to be selected / deselected. Refer to vehicle specific information to establish which gear(s) are applicable to the identified solenoid.	The voltage, frequency or value of the control signal in the solenoid circuit is either "zero" (no signal) or, the signal exists but is below the normal operating range. Possible fault with solenoid or wiring (e.g. solenoid power supply, short / open circuit or resistance) that is causing a signal value to be lower than the minimum operating limit. Note: the control unit could be providing the correct signal but it is being affected by a circuit fault. Refer to list of sensor / actuator checks on page 19.
P099F	Shift Solenoid "H" Control Circuit High	The shift solenoid is used to control a hydraulic circuit in the transmission, which enables the applicable gear(s) to be selected / deselected. Refer to vehicle specific information to establish which gear(s) are applicable to the identified solenoid.	The voltage, frequency or value of the control signal in the solenoid circuit is either at full value (e.g. battery voltage with no on / off signal) or, the signal exists but it is above the normal operating range. Possible fault with solenoid or wiring (e.g. solenoid power supply, short / open circuit or resistance) that is causing a signal value to be higher than the maximum operating limit. Note: the control unit could be providing the correct signal but it is being affected by the circuit fault. Refer to list of sensor / actuator checks on page 19.
P2000	NOx Trap Efficiency Below Threshold Bank 1	A NOx trap can be used to reduce the emissions of NOx (Oxides of Nitrogen). The trap, which forms part of a normal 3-way catalytic converter, can be located close to the engine (as used on some Diesel engines) or it can be located after the normal 3-way catalyst (as used on some petrol engines). Note that the trap can be referred to as a "NOx accumulator". The operation generally relies on the trap / accumulator capturing and storing NOx for a short period, and then by very briefly providing a slightly enriched fuel mixture, this allows the NOx to be converted to less harmful gases and released from the trap.	The operation and efficiency of a NOx Trap / accumulator can be controlled and monitored using temperature and NOx / oxygen sensors. Systems can rely on a NOx / oxygen sensor value and / or air temperature values to determine if the NOx trap efficiency is acceptable; the sensor information can also be used to determine when the short process for converting and releasing the NOx is complete. It may be necessary to refer to vehicle specific information to identify any specific checks or replacement procedures. Also check signals from temperature sensors, NOx / oxygen sensors, to ensure that the information from the sensors is correct.
P2001	NOx Trap Efficiency Below Threshold Bank 2	A NOx trap can be used to reduce the emissions of NOx (Oxides of Nitrogen). The trap, which forms part of a normal 3-way catalytic converter, can be located close to the engine (as used on some Diesel engines) or it can be located after the normal 3-way catalyst (as used on some petrol engines). Note that the trap can be referred to as a "NOx accumulator". The operation generally relies on the trap / accumulator capturing and storing NOx for a short period, and then by very briefly providing a slightly enriched fuel mixture, this allows the NOx to be converted to less harmful gases and released from the trap.	The operation and efficiency of a NOx Trap / accumulator can be controlled and monitored using temperature and NOx / oxygen sensors. Systems can rely on a NOx / oxygen sensor value and / or air temperature values to determine if the NOx trap efficiency is acceptable; the sensor information can also be used to determine when the short process for converting and releasing the NOx is complete. It may be necessary to refer to vehicle specific information to identify any specific checks or replacement procedures. Also check signals from temperature sensors, NOx / oxygen sensors, to ensure that the information from the sensors is correct.
P2002	Diesel Particulate Filter Efficiency Below Threshold Bank 1	The Diesel particulate filter can be checked using a pressure difference measurement (differential exhaust pressures from the input to the output of the filter). If the particulate filter is becoming blocked then the input pressure will be significantly higher than the output pressure. Systems may also use a soot / ash sensor or exhaust gas temperature measurements to detect whether the filter system is operating efficiently.	The fault code indicates that the particulate filter efficiency is below acceptable limits. The fault is likely to be detected due to an unacceptable pressure difference across the filter (measured by differential pressure sensor or pressure sensors); on some systems, the soot / ash sensor may also indicate unacceptable levels of soot passing from the filter or an exhaust gas temperature sensor signal can be used for the calculation. Whichever method is used to detect filter efficiency, the fault code would indicate that the filter might require cleaning or replacement; refer to vehicle specific information regarding cleaning or replacement frequency. If the particulate filter is not blocked or inefficient, it may be necessary to check sensor operation. Refer to list of sensor / actuator checks on page 19.

NOTE 1. Check for other fault codes that could provide additional information. **NOTE 2.** Communication between control units can pass via a CAN-Bus system; refer to Page 51 for CAN-Bus checks.

NOTE 3. If a fault cannot be located, it is also possible that the control unit is at fault. **NOTE 4.** Refer to Page 53 for list of pages that show the ISO standard for component locations e.g. Sensor A - Bank 1.

Fault code	EOBD / ISO Description	Component / System Description	Meaningful Description and Quick Check
P2003	Diesel Particulate Filter Efficiency Below Threshold Bank 2	The Diesel particulate filter can be checked using a pressure difference measurement (differential exhaust pressures from the input to the output of the filter). If the particulate filter is becoming blocked then the input pressure will be significantly higher than the output pressure. Systems may also use a soot / ash sensor or exhaust gas temperature measurements to detect whether the filter system is operating efficiently.	The fault code indicates that the particulate filter efficiency is below acceptable limits. The fault is likely to be detected due to an unacceptable pressure difference across the filter (measured by differential pressure sensor or pressure sensors); on some systems, the soot / ash sensor may also indicate unacceptable levels of soot passing from the filter or an exhaust gas temperature sensor signal can be used for the calculation. Whichever method is used to detect filter efficiency, the fault code would indicate that the filter might require cleaning or replacement; refer to vehicle specific information regarding cleaning or replacement frequency. If the particulate filter is not blocked or inefficient, it may be necessary to check sensor operation. Refer to list of sensor / actuator checks on page 19.
P2004	Intake Manifold Runner Control Stuck Open Bank 1	The intake manifold runner control system is normally used to change the motion of the intake airflow within the intake system. The manifold runner control movement could be actuated by engine vacuum (regulated by a solenoid controlled valve) or, there could be a motor directly altering the valve position. The solenoid or motor is referred to as the actuator.	The control unit is providing a control signal to the actuator but the control unit has detected that the actuator is stuck in the open position and is not responding to the control signal. The fault could be related to an actuator / wiring fault e.g. power supply, short / open circuit, incorrect actuator resistance. Refer to list of sensor / actuator checks on page 19. Note that the control unit could be detecting no response from the manifold runner system; check system (as well as the actuator) for correct operation, movement and response. Check vacuum, hoses, linkage etc. A position sensor (if fitted) would be indicating the "stuck position"; also check position sensor operation.
P2005	Intake Manifold Runner Control Stuck Open Bank 2	The intake manifold runner control system is normally used to change the motion of the intake airflow within the intake system. The manifold runner control movement could be actuated by engine vacuum (regulated by a solenoid controlled valve) or, there could be a motor directly altering the valve position. The solenoid or motor is referred to as the actuator.	The control unit is providing a control signal to the actuator but the control unit has detected that the actuator is stuck in the open position and is not responding to the control signal. The fault could be related to an actuator / wiring fault e.g. power supply, short / open circuit, incorrect actuator resistance. Refer to list of sensor / actuator checks on page 19. Note that the control unit could be detecting no response from the manifold runner system; check system (as well as the actuator) for correct operation, movement and response. Check vacuum, hoses, linkage etc. A position sensor (if fitted) would be indicating the "stuck position"; also check position sensor operation.
P2006	Intake Manifold Runner Control Stuck Closed Bank 1	The intake manifold runner control system is normally used to change the motion of the intake airflow within the intake system. The manifold runner control movement could be actuated by engine vacuum (regulated by a solenoid controlled valve) or, there could be a motor directly altering the valve position. The solenoid or motor is referred to as the actuator.	The control unit is providing a control signal to the actuator but the control unit has detected that the actuator is stuck in the closed position and is not responding to the control signal. The fault could be related to an actuator / wiring fault e.g. power supply, short / open circuit, incorrect actuator resistance. Refer to list of sensor / actuator checks on page 19. Note that the control unit could be detecting no response from the manifold runner system; check system (as well as the actuator) for correct operation, movement and response. Check vacuum, hoses, linkage etc. A position sensor (if fitted) would be indicating the "stuck position"; also check position sensor operation.
P2007	Intake Manifold Runner Control Stuck Closed Bank 2	The intake manifold runner control system is normally used to change the motion of the intake airflow within the intake system. The manifold runner control movement could be actuated by engine vacuum (regulated by a solenoid controlled valve) or, there could be a motor directly altering the valve position. The solenoid or motor is referred to as the actuator.	The control unit is providing a control signal to the actuator but the control unit has detected that the actuator is stuck in the closed position and is not responding to the control signal. The fault could be related to an actuator / wiring fault e.g. power supply, short / open circuit, incorrect actuator resistance. Refer to list of sensor / actuator checks on page 19. Note that the control unit could be detecting no response from the manifold runner system; check system (as well as the actuator) for correct operation, movement and response. Check vacuum, hoses, linkage etc. A position sensor (if fitted) would be indicating the "stuck position"; also check position sensor operation.
P2008	Intake Manifold Runner Control Circuit / Open Bank 1	The intake manifold runner control system is normally used to change the motion of the intake airflow within the intake system; a valve will be used to affect the motion. The manifold runner control valve could be actuated by engine vacuum (regulated by a solenoid controlled valve) or, there could be a motor directly altering the valve position.	The voltage, frequency or value of the control signal in the actuator circuit is incorrect. Signal could be:- out of normal operating range, constant value, non-existent, corrupt. The fault code indicates a possible "open circuit" but other undefined faults in the actuator or wiring could activate the same code e.g. actuator power supply, short circuit, high circuit resistance or incorrect actuator resistance. Note: the control unit could be providing the correct signal but it is being affected by the circuit fault. Refer to list of sensor / actuator checks on page 19.
P2009	Intake Manifold Runner Control Circuit Low Bank 1	The intake manifold runner control system is normally used to change the motion of the intake airflow within the intake system; a valve will be used to affect the motion. The manifold runner control valve could be actuated by engine vacuum (regulated by a solenoid controlled valve) or, there could be a motor directly altering the valve position.	The voltage, frequency or value of the control signal in the actuator circuit is either "zero" (no signal) or, the signal exists but is below the normal operating range. Possible fault with actuator or wiring (e.g. actuator power supply, short / open circuit or resistance) that is causing a signal value to be lower than the minimum operating limit. Note: the control unit could be providing the correct signal but it is being affected by the circuit fault. Refer to list of sensor / actuator checks on page 19.
P2010	Intake Manifold Runner Control Circuit High Bank 1	The intake manifold runner control system is normally used to change the motion of the intake airflow within the intake system; a valve will be used to affect the motion. The manifold runner control valve could be actuated by engine vacuum (regulated by a solenoid controlled valve) or, there could be a motor directly altering the valve position.	The voltage, frequency or value of the control signal in the actuator circuit is either at full value (e.g. battery voltage with no on / off signal) or, the signal exists but it is above the normal operating range. Possible fault with actuator or wiring (e.g. actuator power supply, short / open circuit or resistance) that is causing a signal value to be higher than the maximum operating limit. Note: the control unit could be providing the correct signal but it is being affected by the circuit fault. Refer to list of sensor / actuator checks on page 19.
P2011	Intake Manifold Runner Control Circuit / Open Bank 2	The intake manifold runner control system is normally used to change the motion of the intake airflow within the intake system; a valve will be used to affect the motion. The manifold runner control valve could be actuated by engine vacuum (regulated by a solenoid controlled valve) or, there could be a motor directly altering the valve position.	The voltage, frequency or value of the control signal in the actuator circuit is incorrect. Signal could be:- out of normal operating range, constant value, non-existent, corrupt. The fault code indicates a possible "open circuit" but other undefined faults in the actuator or wiring could activate the same code e.g. actuator power supply, short circuit, high circuit resistance or incorrect actuator resistance. Note: the control unit could be providing the correct signal but it is being affected by the circuit fault. Refer to list of sensor / actuator checks on page 19.
P2012	Intake Manifold Runner Control Circuit Low Bank 2	The intake manifold runner control system is normally used to change the motion of the intake airflow within the intake system; a valve will be used to affect the motion. The manifold runner control valve could be actuated by engine vacuum (regulated by a solenoid controlled valve) or, there could be a motor directly altering the valve position.	The voltage, frequency or value of the control signal in the actuator circuit is either "zero" (no signal) or, the signal exists but is below the normal operating range. Possible fault with actuator or wiring (e.g. actuator power supply, short / open circuit or resistance) that is causing a signal value to be lower than the minimum operating limit. Note: the control unit could be providing the correct signal but it is being affected by the circuit fault. Refer to list of sensor / actuator checks on page 19.

NOTE 1. Check for other fault codes that could provide additional information. **NOTE 2.** Communication between control units can pass via a CAN-Bus system; refer to Page 51 for CAN-Bus checks. 214

NOTE 3. If a fault cannot be located, it is also possible that the control unit is at fault. NOTE 4. Refer to Page 53 for list of pages that show the ISO standard for component locations e.g. Sensor A - Bank 1.

Code	Description		
P2013	Intake Manifold Runner Control Circuit High Bank 2	The intake manifold runner control system is normally used to change the motion of the intake airflow within the intake system; a valve will be used to affect the motion. The manifold runner control valve could be actuated by engine vacuum (regulated by a solenoid controlled valve) or, there could be a motor directly altering the valve position.	The voltage, frequency or value of the control signal in the actuator circuit is either at full value (e.g. battery voltage with no on / off signal) or, the signal exists but it is above the normal operating range. Possible fault with actuator or wiring (e.g. actuator power supply, short / open circuit or resistance) that is causing a signal value to be higher than the maximum operating limit. Note: the control unit could be providing the correct signal but it is being affected by the circuit fault. Refer to list of sensor / actuator checks on page 19.
P2014	Intake Manifold Runner Position Sensor / Switch Circuit Bank 1	The intake manifold runner control system is normally used to change the motion (e.g. swirl / turbulence) of the intake airflow within the intake system; a valve / butterfly system can be fitted into the intake manifold to affect the motion. The valve / butterfly can be moved using a vacuum operated mechanism, with the vacuum being regulated by an electrical actuator e.g. solenoid. A sensor can be used to monitor the position of the valve / mechanism.	The voltage, frequency or value of the position sensor signal is incorrect (undefined fault). Sensor voltage / signal value could be: out of normal operating range, constant value, non-existent, corrupt. Possible fault in the position sensor or wiring e.g. short / open circuits, circuit resistance, incorrect sensor resistance (if applicable) or, fault with the reference / operating voltage; interference from other circuits can also affect sensor signals. Also check earth connections and cable screening. Refer to list of sensor / actuator checks on page 19.
P2015	Intake Manifold Runner Position Sensor / Switch Circuit Range / Performance Bank 1	The intake manifold runner control system is normally used to change the motion (e.g. swirl / turbulence) of the intake airflow within the intake system; a valve / butterfly system can be fitted into the intake manifold to affect the motion. The valve / butterfly can be moved using a vacuum operated mechanism, with the vacuum being regulated by an electrical actuator e.g. solenoid. A sensor can be used to monitor the position of the valve / mechanism.	The voltage, frequency or value of the position sensor signal is within the normal operating range / tolerance but the signal is not plausible or is incorrect due to an undefined fault. The sensor signal might not match the operating conditions indicated by other sensors or, the sensor signal is not changing / responding as expected to the requested changes in the intake runner position. Possible fault in the position sensor or wiring e.g. short / open circuits, circuit resistance, incorrect sensor resistance or, fault with the reference / operating voltage; interference from other circuits can also affect sensor signals. Refer to list of sensor / actuator checks on page 19. It is also possible that the sensor is operating correctly but the position being detected by the sensor is unacceptable or incorrect for the operating conditions. Check for freedom of movement for the runner control mechanism. Check for operation of vacuum actuator (check for vacuum leaks / blockages etc.), check solenoid control (or other actuator as applicable).
P2016	Intake Manifold Runner Position Sensor / Switch Circuit Low Bank 1	The intake manifold runner control system is normally used to change the motion (e.g. swirl / turbulence) of the intake airflow within the intake system; a valve / butterfly system can be fitted into the intake manifold to affect the motion. The valve / butterfly can be moved using a vacuum operated mechanism, with the vacuum being regulated by an electrical actuator e.g. solenoid. A sensor can be used to monitor the position of the valve / mechanism.	The voltage, frequency or value of the position sensor signal is either "zero" (no signal) or, the signal exists but it is below the normal operating range (lower than the minimum operating limit). Possible fault in the position sensor or wiring e.g. short / open circuits, circuit resistance, incorrect sensor resistance or, fault with the reference / operating voltage; interference from other circuits can also affect sensor signals. Refer to list of sensor / actuator checks on page 19.
P2017	Intake Manifold Runner Position Sensor / Switch Circuit High Bank 1	The intake manifold runner control system is normally used to change the motion (e.g. swirl / turbulence) of the intake airflow within the intake system; a valve / butterfly system can be fitted into the intake manifold to affect the motion. The valve / butterfly can be moved using a vacuum operated mechanism, with the vacuum being regulated by an electrical actuator e.g. solenoid. A sensor can be used to monitor the position of the valve / mechanism.	The voltage, frequency or value of the position sensor signal is either at full value (e.g. battery voltage with no signal or frequency) or, the signal exists but it is above the normal operating range (higher than the maximum operating limit). Possible fault in the position sensor or wiring e.g. short / open circuits, circuit resistance, incorrect sensor resistance or, fault with the reference / operating voltage; interference from other circuits can also affect sensor signals. Refer to list of sensor / actuator checks on page 19.
P2018	Intake Manifold Runner Position Sensor / Switch Circuit Intermittent Bank 1	The intake manifold runner control system is normally used to change the motion (e.g. swirl / turbulence) of the intake airflow within the intake system; a valve / butterfly system can be fitted into the intake manifold to affect the motion. The valve / butterfly can be moved using a vacuum operated mechanism, with the vacuum being regulated by an electrical actuator e.g. solenoid. A sensor can be used to monitor the position of the valve / mechanism.	The voltage, frequency or value of the position sensor signal is intermittent (likely to be out of normal operating range / tolerance when intermittent fault is detected) or, the signal / voltage is erratic (e.g. signal changes are irregular / unpredictable). Possible fault with wiring or connections (including internal sensor connections); also interference from other circuits can affect the signal. Refer to list of sensor / actuator checks on page 19. Check operation of runner control mechanism to ensure that it is not moving erratically (can be caused by vacuum / solenoid control problems).
P2019	Intake Manifold Runner Position Sensor / Switch Circuit Bank 2	The intake manifold runner control system is normally used to change the motion (e.g. swirl / turbulence) of the intake airflow within the intake system; a valve / butterfly system can be fitted into the intake manifold to affect the motion. The valve / butterfly can be moved using a vacuum operated mechanism, with the vacuum being regulated by an electrical actuator e.g. solenoid. A sensor can be used to monitor the position of the valve / mechanism.	The voltage, frequency or value of the position sensor signal is incorrect (undefined fault). Sensor voltage / signal value could be: out of normal operating range, constant value, non-existent, corrupt. Possible fault in the position sensor or wiring e.g. short / open circuits, circuit resistance, incorrect sensor resistance (if applicable) or, fault with the reference / operating voltage; interference from other circuits can also affect sensor signals. Also check earth connections and cable screening. Refer to list of sensor / actuator checks on page 19.
P2020	Intake Manifold Runner Position Sensor / Switch Circuit Range / Performance Bank 2	The intake manifold runner control system is normally used to change the motion (e.g. swirl / turbulence) of the intake airflow within the intake system; a valve / butterfly system can be fitted into the intake manifold to affect the motion. The valve / butterfly can be moved using a vacuum operated mechanism, with the vacuum being regulated by an electrical actuator e.g. solenoid. A sensor can be used to monitor the position of the valve / mechanism.	The voltage, frequency or value of the position sensor signal is within the normal operating range / tolerance but the signal is not plausible or is incorrect due to an undefined fault. The sensor signal might not match the operating conditions indicated by other sensors or, the sensor signal is not changing / responding as expected to the requested changes in the intake runner position. Possible fault in the position sensor or wiring e.g. short / open circuits, circuit resistance, incorrect sensor resistance or, fault with the reference / operating voltage; interference from other circuits can also affect sensor signals. Refer to list of sensor / actuator checks on page 19. It is also possible that the sensor is operating correctly but the position being detected by the sensor is unacceptable or incorrect for the operating conditions. Check for freedom of movement for the runner control mechanism. Check for operation of vacuum actuator (check for vacuum leaks / blockages etc.), check solenoid control (or other actuator as applicable).

NOTE 1. Check for other fault codes that could provide additional information.
NOTE 3. If a fault cannot be located, it is also possible that the control unit is at fault.
NOTE 2. Communication between control units can pass via a CAN-Bus system; refer to Page 51 for CAN-Bus checks.
NOTE 4. Refer to Page 53 for list of pages that show the ISO standard for component locations e.g. Sensor A - Bank 1.

Fault code	EOBD / ISO Description	Component / System Description	Meaningful Description and Quick Check
P2021	Intake Manifold Runner Position Sensor / Switch Circuit Low Bank 2	The intake manifold runner control system is normally used to change the motion (e.g. swirl / turbulence) of the intake airflow within the intake system; a valve / butterfly system can be fitted into the intake manifold to affect the motion. The valve / butterfly can be moved using a vacuum operated mechanism, with the vacuum being regulated by an electrical actuator e.g. solenoid. A sensor can be used to monitor the position of the valve / mechanism.	The voltage, frequency or value of the position sensor signal is either "zero" (no signal) or, the signal exists but it is below the normal operating range (lower than the minimum operating limit). Possible fault in the position sensor or wiring e.g. short / open circuits, circuit resistance, incorrect sensor resistance or, fault with the reference / operating voltage; interference from other circuits can also affect sensor signals. Refer to list of sensor / actuator checks on page 19.
P2022	Intake Manifold Runner Position Sensor / Switch Circuit High Bank 2	The intake manifold runner control system is normally used to change the motion (e.g. swirl / turbulence) of the intake airflow within the intake system; a valve / butterfly system can be fitted into the intake manifold to affect the motion. The valve / butterfly can be moved using a vacuum operated mechanism, with the vacuum being regulated by an electrical actuator e.g. solenoid. A sensor can be used to monitor the position of the valve / mechanism.	The voltage, frequency or value of the position sensor signal is either at full value (e.g. battery voltage with no signal or frequency) or, the signal exists but it is above the normal operating range (higher than the maximum operating limit). Possible fault in the position sensor or wiring e.g. short / open circuits, circuit resistance, incorrect sensor resistance or, fault with the reference / operating voltage; interference from other circuits can also affect sensor signals. Refer to list of sensor / actuator checks on page 19.
P2023	Intake Manifold Runner Position Sensor / Switch Circuit Intermittent Bank 2	The intake manifold runner control system is normally used to change the motion (e.g. swirl / turbulence) of the intake airflow within the intake system; a valve / butterfly system can be fitted into the intake manifold to affect the motion. The valve / butterfly can be moved using a vacuum operated mechanism, with the vacuum being regulated by an electrical actuator e.g. solenoid. A sensor can be used to monitor the position of the valve / mechanism.	The voltage, frequency or value of the position sensor signal is intermittent (likely to be out of normal operating range / tolerance when intermittent fault is detected) or, the signal / voltage is erratic (e.g. signal changes are irregular / unpredictable). Possible fault with wiring or connections (including internal sensor connections); also interference from other circuits can affect the signal. Refer to list of sensor / actuator checks on page 19. Check operation of runner control mechanism to ensure that it is not moving erratically (can be caused by vacuum / solenoid control problems).
P2024	Evaporative Emissions (EVAP) Fuel Vapour Temperature Sensor Circuit	EVAP systems make use of a canister to collect fuel vapours and then release them (under controlled conditions) to the engine intake system. A temperature sensor is used to monitor fuel vapour temperature in the EVAP system; refer to vehicle specific information for location.	The voltage of the temperature sensor signal is incorrect (undefined fault). Sensor voltage / signal could be: out of normal operating range, constant value, non-existent, corrupt. Possible fault in the temperature sensor or wiring e.g. short / open circuits, circuit resistance, incorrect sensor resistance or, fault with the reference / operating voltage; interference from other circuits can also affect sensor signals. Refer to list of sensor / actuator checks on page 19.
P2025	Evaporative Emissions (EVAP) Fuel Vapour Temperature Sensor Performance	EVAP systems make use of a canister to collect fuel vapours and then release them (under controlled conditions) to the engine intake system. A temperature sensor is used to monitor fuel vapour temperature in the EVAP system; refer to vehicle specific information for location.	The voltage of the temperature sensor signal is incorrect (undefined fault). Sensor voltage / signal could be: out of normal operating range, constant value, non-existent, corrupt. Possible fault in the temperature sensor or wiring e.g. short / open circuits, circuit resistance, incorrect sensor resistance or, fault with the reference / operating voltage; interference from other circuits can also affect sensor signals. Refer to list of sensor / actuator checks on page 19. It is also possible that the sensor is operating correctly but the temperature being measured is incorrect for the operating conditions; if possible check temperature with a separate thermometer and if necessary identify reason for incorrect temperature.
P2026	Evaporative Emissions (EVAP) Fuel Vapour Temperature Sensor Circuit Low Voltage	EVAP systems make use of a canister to collect fuel vapours and then release them (under controlled conditions) to the engine intake system. A temperature sensor is used to monitor fuel vapour temperature in the EVAP system; refer to vehicle specific information for location.	The voltage value of the temperature sensor signal is either "zero" (no signal) or, the signal exists but it is below the normal operating range (lower than the minimum operating limit). Possible fault in the temperature sensor or wiring e.g. short / open circuits, circuit resistance, incorrect sensor resistance or, fault with the reference / operating voltage; interference from other circuits can also affect sensor signals. Refer to list of sensor / actuator checks on page 19.
P2027	Evaporative Emissions (EVAP) Fuel Vapour Temperature Sensor Circuit High Voltage	EVAP systems make use of a canister to collect fuel vapours and then release them (under controlled conditions) to the engine intake system. A temperature sensor is used to monitor fuel vapour temperature in the EVAP system; refer to vehicle specific information for location.	The voltage of the temperature sensor signal is either at full value (e.g. battery voltage with no signal or frequency) or, the signal exists but it is above the normal operating range (higher than the maximum operating limit). Possible fault in the temperature sensor or wiring e.g. short / open circuits, circuit resistance, incorrect sensor resistance or, fault with the reference / operating voltage; interference from other circuits can also affect sensor signals. Refer to list of sensor / actuator checks on page 19.
P2028	Evaporative Emissions (EVAP) Fuel Vapour Temperature Sensor Circuit Intermittent	EVAP systems make use of a canister to collect fuel vapours and then release them (under controlled conditions) to the engine intake system. A temperature sensor is used to monitor fuel vapour temperature in the EVAP system; refer to vehicle specific information for location.	The voltage of the temperature sensor signal is either at full value (e.g. battery voltage with no signal or frequency) or, the signal exists but it is above the normal operating range (higher than the maximum operating limit). Possible fault in the temperature sensor or wiring e.g. short / open circuits, circuit resistance, incorrect sensor resistance or, fault with the reference / operating voltage; interference from other circuits can also affect sensor signals. Refer to list of sensor / actuator checks on page 19.
P2029	Fuel Fired Heater Disabled	A fuel fired heater is fitted to some vehicles to improve the warm up time for the cooling system (and for the vehicle interior); the heater is used after start up when the ambient and coolant temperatures are low. Coolant passes into the heater assembly and is therefore rapidly heated, coolant can then pass through heat exchangers (radiators) to heat the vehicle interior. The fuel fired heaters can use a fuel pump, fuel injector, glow plug / igniter, temperature sensor etc and be controlled by an independent control unit (possibly linked to climate control system).	The fault code indicates that the fuel fired heater system has been disabled, this is likely to be caused by a heater system fault, which has caused the control unit to disable the heater system operation. It may be necessary to identify the components and system fitted to the vehicle and refer to any specific information for the system. Note that the heater system will probably pass a limited level of fault detection information to the main engine control unit, it will therefore be necessary to investigate the heater system for faults (identify which control module controls the heater system e.g. body or climate control module, and check for fault codes from the control module). If the Fuel Fired Heater system has a fault that has required the system to be disabled, this could be detected because of information from the heater temperature sensor (indicating poor or no heating operation) or, the coolant temperature is not increasing as expected during the period when the heater is operating. However, it is possible that the fault represents a possible dangerous condition (potential fuel leaks etc) and the control unit therefore disables the system.

NOTE 1. Check for other fault codes that could provide additional information. **NOTE 2.** Communication between control units can pass via a CAN-Bus system; refer to Page 51 for CAN-Bus checks. 216

NOTE 3. If a fault cannot be located, it is also possible that the control unit is at fault **NOTE 4.** Refer to Page 53 for list of pages that show the ISO standard for component locations e.g. Sensor A - Bank 1.

P2030	Fuel Fired Heater Performance	A fuel fired heater is fitted to some vehicles to improve the warm up time for the cooling system (and for the vehicle interior); the heater is used after start up when the ambient and coolant temperatures are low. Coolant passes into the heater assembly and is therefore rapidly heated, coolant can then pass through heat exchangers (radiators) to heat the vehicle interior. The fuel fired heaters can use a fuel pump, fuel injector, temperature sensor etc and be controlled by an independent control unit (possibly linked to climate control system).	It may be necessary to identify the components and system fitted to the vehicle and refer to any specific information for the system. Note that the heater system will probably pass a limited level of fault detection information to the main engine control unit, it will therefore be necessary to investigate the heater system for faults (identify which control module controls the heater system e.g. body or climate control module, and check for fault codes from the control module). The fault code indicates that the Fuel Fired Heater system is not operating as expected; this is likely to be detected either due to information from the heater temperature sensor (indicating poor or no heating operation) or, the coolant temperature is not increasing as expected during the period when the heater is operating.
P2031	Exhaust Gas Temperature Sensor Circuit Bank 1 Sensor 2	The exhaust gas temperature sensor signal is used to indicate if the temperature is outside of acceptable limits (some emissions control functions are more efficient when the gas temperature is within specified limits). Excessive temperatures can rapidly cause deterioration of the catalyst and other emissions control components. High gas temperatures are often caused by faults in the engine operation e.g. misfires, air leaks, fuelling etc.	The voltage, frequency or value of the temperature sensor signal is incorrect (undefined fault). Sensor voltage / signal could be: out of normal operating range, constant value, non-existent, corrupt. Possible fault in the temperature sensor or wiring e.g. short / open circuits, circuit resistance, incorrect sensor resistance or, fault with the reference / operating voltage; interference from other circuits can also affect sensor signals. Refer to list of sensor / actuator checks on page 19.
P2032	Exhaust Gas Temperature Sensor Circuit Low Bank 1 Sensor 2	The exhaust gas temperature sensor signal is used to indicate if the temperature is outside of acceptable limits (some emissions control functions are more efficient when the gas temperature is within specified limits). Excessive temperatures can rapidly cause deterioration of the catalyst and other emissions control components. High gas temperatures are often caused by faults in the engine operation e.g. misfires, air leaks, fuelling etc.	The voltage, frequency or value of the temperature sensor signal is either "zero" (no signal) or, the signal exists but it is below the normal operating range (lower than the minimum operating limit). Possible fault in the temperature sensor or wiring e.g. short / open circuits, circuit resistance, incorrect sensor resistance or, fault with the reference / operating voltage; interference from other circuits can also affect sensor signals. Refer to list of sensor / actuator checks on page 19.
P2033	Exhaust Gas Temperature Sensor Circuit High Bank 1 Sensor 2	The exhaust gas temperature sensor signal is used to indicate if the temperature is outside of acceptable limits (some emissions control functions are more efficient when the gas temperature is within specified limits). Excessive temperatures can rapidly cause deterioration of the catalyst and other emissions control components. High gas temperatures are often caused by faults in the engine operation e.g. misfires, air leaks, fuelling etc.	The voltage, frequency or value of the temperature sensor signal is either at full value (e.g. battery voltage with no signal or frequency) or, the signal exists but it is above the normal operating range (higher than the maximum operating limit). Possible fault in the temperature sensor or wiring e.g. short / open circuits, circuit resistance, incorrect sensor resistance or, fault with the reference / operating voltage; interference from other circuits can also affect sensor signals. Refer to list of sensor / actuator checks on page 19.
P2034	Exhaust Gas Temperature Sensor Circuit Bank 2 Sensor 2	The exhaust gas temperature sensor signal is used to indicate if the temperature is outside of acceptable limits (some emissions control functions are more efficient when the gas temperature is within specified limits). Excessive temperatures can rapidly cause deterioration of the catalyst and other emissions control components. High gas temperatures are often caused by faults in the engine operation e.g. misfires, air leaks, fuelling etc.	The voltage, frequency or value of the temperature sensor signal is incorrect (undefined fault). Sensor voltage / signal could be: out of normal operating range, constant value, non-existent, corrupt. Possible fault in the temperature sensor or wiring e.g. short / open circuits, circuit resistance, incorrect sensor resistance or, fault with the reference / operating voltage; interference from other circuits can also affect sensor signals. Refer to list of sensor / actuator checks on page 19.
P2035	Exhaust Gas Temperature Sensor Circuit Low Bank 2 Sensor 2	The exhaust gas temperature sensor signal is used to indicate if the temperature is outside of acceptable limits (some emissions control functions are more efficient when the gas temperature is within specified limits). Excessive temperatures can rapidly cause deterioration of the catalyst and other emissions control components. High gas temperatures are often caused by faults in the engine operation e.g. misfires, air leaks, fuelling etc.	The voltage, frequency or value of the temperature sensor signal is either "zero" (no signal) or, the signal exists but it is below the normal operating range (lower than the minimum operating limit). Possible fault in the temperature sensor or wiring e.g. short / open circuits, circuit resistance, incorrect sensor resistance or, fault with the reference / operating voltage; interference from other circuits can also affect sensor signals. Refer to list of sensor / actuator checks on page 19.
P2036	Exhaust Gas Temperature Sensor Circuit High Bank 2 Sensor 2	The exhaust gas temperature sensor signal is used to indicate if the temperature is outside of acceptable limits (some emissions control functions are more efficient when the gas temperature is within specified limits). Excessive temperatures can rapidly cause deterioration of the catalyst and other emissions control components. High gas temperatures are often caused by faults in the engine operation e.g. misfires, air leaks, fuelling etc.	The voltage, frequency or value of the temperature sensor signal is either at full value (e.g. battery voltage with no signal or frequency) or, the signal exists but it is above the normal operating range (higher than the maximum operating limit). Possible fault in the temperature sensor or wiring e.g. short / open circuits, circuit resistance, incorrect sensor resistance or, fault with the reference / operating voltage; interference from other circuits can also affect sensor signals. Refer to list of sensor / actuator checks on page 19.
P2037	Reductant Injection Air Pressure Sensor Circuit	A Reductant is a chemical that is stored in a separate tank and is added either to the fuel or to the exhaust gas (depending on system design). The Reductant combines with the Diesel soot particles, which reduces the temperature at which the soot will combust; this allows the soot particles to combust more easily within the particle filter (during the particle filter regeneration process). On some systems, a small air pump and metering assembly mixes the chemical with the air and injects the mixture into the exhaust (via a nozzle / injector). A sensor monitors injection air pressure.	The voltage or value of the pressure sensor signal is incorrect (undefined fault). Sensor voltage / signal could be: out of normal operating range, constant value, non-existent, corrupt. Possible fault in the pressure sensor or wiring e.g. short / open circuits, circuit resistance, incorrect sensor resistance (if applicable) or, fault with the reference / operating voltage; interference from other circuits can also affect sensor signals. Refer to list of sensor / actuator checks on page 19.
P2038	Reductant Injection Air Pressure Sensor Circuit Range / Performance	A Reductant is a chemical that is stored in a separate tank and is added either to the fuel or to the exhaust gas (depending on system design). The Reductant combines with the Diesel soot particles, which reduces the temperature at which the soot will combust; this allows the soot particles to combust more easily within the particle filter (during the particle filter regeneration process). On some systems, a small air pump and metering assembly mixes the chemical with the air and injects the mixture into the exhaust (via a nozzle / injector). A sensor monitors injection air pressure.	The voltage or value of the pressure sensor signal is within the normal operating range / tolerance but the signal is not plausible or is incorrect due to an undefined fault. The sensor signal might not match the operating conditions indicated by other sensors or, the sensor signal is not changing / responding as expected to the changes in conditions. Possible fault in the pressure sensor or wiring e.g. short / open circuits, circuit resistance, incorrect sensor resistance (if applicable) or, fault with the reference / operating voltage; interference from other circuits can also affect sensor signals. Refer to list of sensor / actuator checks on page 19. Also check any air passages / pipes to the sensor for leaks / blockages etc. It is also possible that the pressure sensor is operating correctly but the air pressure being measured by the sensor is unacceptable or incorrect for the operating conditions; if possible check pressure with a separate gauge, also check air pump operation.

NOTE 1. Check for other fault codes that could provide additional information. NOTE 2. Communication between control units can pass via a CAN-Bus system; refer to Page 51 for CAN-Bus checks.

NOTE 3. If a fault cannot be located, it is also possible that the control unit is at fault. NOTE 4. Refer to Page 53 for list of pages that show the ISO standard for component locations e.g. Sensor A - Bank 1.

Fault code	EOBD / ISO Description	Component / System Description	Meaningful Description and Quick Check
P2039	Reductant Injection Air Pressure Sensor Circuit Low	A Reductant is a chemical that is stored in a separate tank and is added either to the fuel or to the exhaust gas (depending on system design). The Reductant combines with the Diesel soot particles, which reduces the temperature at which the soot will combust; this allows the soot particles to combust more easily within the particle filter (during the particle filter regeneration process). On some systems, a small air pump and metering assembly mixes the chemical with the air and injects the mixture into the exhaust (via a nozzle / injector). A sensor monitors injection pressure.	The voltage or value of the pressure sensor signal is either "zero" (no signal) or, the signal exists but it is below the normal operating range (lower than the minimum operating limit). Possible fault in the pressure sensor or wiring e.g. short / open circuits, circuit resistance, incorrect sensor resistance (if applicable) or, fault with the reference / operating voltage; interference from other circuits can also affect sensor signals. Refer to list of sensor / actuator checks on page 19.
P203A	Reductant Level Sensor Circuit	A Reductant is a chemical that is stored in a separate tank and is added either to the fuel or to the exhaust gas (depending on system design). The Reductant combines with the Diesel soot particles, which reduces the temperature at which the soot will combust; this allows the soot particles to combust more easily within the particle filter (during the particle filter regeneration process). A sensor can be used to monitor the level of the Reductant in the tank.	The voltage or value of the level sensor signal is incorrect (undefined fault). Sensor voltage / signal could be: out of normal operating range, constant value, non-existent, corrupt. Possible fault in the level sensor or wiring e.g. short / open circuits, circuit resistance, incorrect sensor resistance or, fault with the reference / operating voltage; interference from other circuits can also affect sensor signals. Refer to list of sensor / actuator checks on page 19.
P203B	Reductant Level Sensor Circuit Range / Performance	A Reductant is a chemical that is stored in a separate tank and is added either to the fuel or to the exhaust gas (depending on system design). The Reductant combines with the Diesel soot particles, which reduces the temperature at which the soot will combust; this allows the soot particles to combust more easily within the particle filter (during the particle filter regeneration process). A sensor can be used to monitor the level of the Reductant in the tank.	The voltage or value of the level sensor signal is within the normal operating range / tolerance but the signal is not plausible or is incorrect due to an undefined fault. The sensor signal might not be changing / responding as expected. Possible fault in the level sensor or wiring e.g. short / open circuits, circuit resistance, incorrect sensor resistance or, fault with the reference / operating voltage; interference from other circuits can also affect sensor signals. Refer to list of sensor / actuator checks on page 19. It is also possible that the level sensor is operating correctly but the level being measured by the sensor is unacceptable or incorrect.
P203C	Reductant Level Sensor Circuit Low	A Reductant is a chemical that is stored in a separate tank and is added either to the fuel or to the exhaust gas (depending on system design). The Reductant combines with the Diesel soot particles, which reduces the temperature at which the soot will combust; this allows the soot particles to combust more easily within the particle filter (during the particle filter regeneration process). A sensor can be used to monitor the level of the Reductant in the tank.	The voltage or value of the level sensor signal is either "zero" (no signal) or, the signal exists but it is below the normal operating range (lower than the minimum operating limit). Possible fault in the level sensor or wiring e.g. short / open circuits, circuit resistance, incorrect sensor resistance or, fault with the reference / operating voltage; interference from other circuits can also affect sensor signals. Refer to list of sensor / actuator checks on page 19.
P203D	Reductant Level Sensor Circuit High	A Reductant is a chemical that is stored in a separate tank and is added either to the fuel or to the exhaust gas (depending on system design). The Reductant combines with the Diesel soot particles, which reduces the temperature at which the soot will combust; this allows the soot particles to combust more easily within the particle filter (during the particle filter regeneration process). A sensor can be used to monitor the level of the Reductant in the tank.	The voltage or value of the level sensor signal is either at full value (e.g. battery voltage with no signal or frequency) or, the signal exists but it is above the normal operating range (higher than the maximum operating limit). Possible fault in the level sensor or wiring e.g. short / open circuits, circuit resistance, incorrect sensor resistance or, fault with the reference / operating voltage; interference from other circuits can also affect sensor signals. Refer to list of sensor / actuator checks on page 19.
P203E	Reductant Level Sensor Circuit Intermittent / Erratic	A Reductant is a chemical that is stored in a separate tank and is added either to the fuel or to the exhaust gas (depending on system design). The Reductant combines with the Diesel soot particles, which reduces the temperature at which the soot will combust; this allows the soot particles to combust more easily within the particle filter (during the particle filter regeneration process). A sensor can be used to monitor the level of the Reductant in the tank.	The voltage or value of the level sensor signal is intermittent (likely to be out of normal operating range / tolerance when intermittent fault is detected) or, the signal / voltage is erratic (e.g. signal changes are irregular / unpredictable). Possible fault with wiring or connections (including internal sensor connections); also interference from other circuits can affect the signal. Refer to list of sensor / actuator checks on page 19.
P203F	Reductant Level Low	A Reductant is a chemical that is stored in a separate tank and is added either to the fuel or to the exhaust gas (depending on system design). The Reductant combines with the Diesel soot particles, which reduces the temperature at which the soot will combust; this allows the soot particles to combust more easily within the particle filter (during the particle filter regeneration process). A sensor can be used to monitor the level of the Reductant in the tank.	The level sensor is indicating that there is a low level of the Reductant chemical in the Reductant tank. Add Reductant as necessary. If the fault code remains, it could indicate a level sensor fault (refer to Fault code P203B)
P2040	Reductant Injection Air Pressure Sensor Circuit High	A Reductant is a chemical that is stored in a separate tank and is added either to the fuel or to the exhaust gas (depending on system design). The Reductant combines with the Diesel soot particles, which reduces the temperature at which the soot will combust; this allows the soot particles to combust more easily within the particle filter (during the particle filter regeneration process). On some systems, a small air pump and metering assembly mixes the chemical with the air and injects the mixture into the exhaust (via a nozzle / injector). A sensor monitors injection air pressure.	The voltage or value of the pressure sensor signal is either at full value (e.g. battery voltage with no signal or frequency) or, the signal exists but it is above the normal operating range (higher than the maximum operating limit). Possible fault in the pressure sensor or wiring e.g. short / open circuits, circuit resistance, incorrect sensor resistance (if applicable) or, fault with the reference / operating voltage; interference from other circuits can also affect sensor signals. Refer to list of sensor / actuator checks on page 19.

Code	Description		
P2041	Reductant Injection Air Pressure Sensor Circuit Intermittent	A Reductant is a chemical that is stored in a separate tank and is added either to the fuel or to the exhaust gas (depending on system design). The Reductant combines with the Diesel soot particles, which reduces the temperature at which the soot will combust; this allows the soot particles to combust more easily within the particle filter (during the particle filter regeneration process). On some systems, a small air pump and metering assembly mixes the chemical with the air and injects the mixture into the exhaust (via a nozzle / injector). A sensor monitors injection air pressure.	The voltage or value of the pressure sensor signal is intermittent (likely to be out of normal operating range / tolerance when intermittent fault is detected) or, the signal / voltage is erratic (e.g. signal changes are irregular / unpredictable). Possible fault with wiring or connections (including internal sensor connections); also interference from other circuits can affect the signal. Refer to list of sensor / actuator checks on page 19. It is also possible that the pressure could be changing erratically, which could cause the fault code to be activated; if necessary, check pressure with a separate gauge to check for erratic pressure changes.
P2042	Reductant Temperature Sensor Circuit	A Reductant is a chemical that is stored in a separate tank and is added either to the fuel or to the exhaust gas (depending on system design). The Reductant combines with the Diesel soot particles, which reduces the temperature at which the soot will combust; this allows the soot particles to combust more easily within the particle filter (during the particle filter regeneration process). On some systems, a small air pump and metering assembly mixes the chemical with the air and injects the mixture into the exhaust (via a nozzle / injector). A sensor can be used to monitor the Reductant temperature.	The voltage or value of the temperature sensor signal is incorrect (undefined fault). Sensor voltage / signal could be: out of normal operating range, constant value, non-existent, corrupt. Possible fault in the temperature sensor or wiring e.g. short / open circuits, circuit resistance, incorrect sensor resistance or, fault with the reference / operating voltage; interference from other circuits can also affect sensor signals. Refer to list of sensor / actuator checks on page 19.
P2043	Reductant Temperature Sensor Circuit Range / Performance	A Reductant is a chemical that is stored in a separate tank and is added either to the fuel or to the exhaust gas (depending on system design). The Reductant combines with the Diesel soot particles, which reduces the temperature at which the soot will combust; this allows the soot particles to combust more easily within the particle filter (during the particle filter regeneration process). A sensor can be used to monitor the Reductant temperature.	The voltage or value of the temperature sensor signal is within the normal operating range / tolerance but the signal is not plausible or is incorrect due to an undefined fault. The sensor signal might not match the operating conditions indicated by other sensors or, the sensor signal is not changing / responding as expected to the changes in conditions. Possible fault in the temperature sensor or wiring e.g. short / open circuits, circuit resistance, incorrect sensor resistance or, fault with the reference / operating voltage; interference from other circuits can also affect sensor signals. Refer to list of sensor / actuator checks on page 19. Note that the Reductant temperature could also be unacceptable due to ambient temperatures or due to a system fault.
P2044	Reductant Temperature Sensor Circuit Low	A Reductant is a chemical that is stored in a separate tank and is added either to the fuel or to the exhaust gas (depending on system design). The Reductant combines with the Diesel soot particles, which reduces the temperature at which the soot will combust; this allows the soot particles to combust more easily within the particle filter (during the particle filter regeneration process). A sensor can be used to monitor the Reductant temperature.	The voltage or value of the temperature sensor signal is either "zero" (no signal) or, the signal exists but it is below the normal operating range (lower than the minimum operating limit). Possible fault in the temperature sensor or wiring e.g. short / open circuits, circuit resistance, incorrect sensor resistance or, fault with the reference / operating voltage; interference from other circuits can also affect sensor signals. Refer to list of sensor / actuator checks on page 19.
P2045	Reductant Temperature Sensor Circuit High	A Reductant is a chemical that is stored in a separate tank and is added either to the fuel or to the exhaust gas (depending on system design). The Reductant combines with the Diesel soot particles, which reduces the temperature at which the soot will combust; this allows the soot particles to combust more easily within the particle filter (during the particle filter regeneration process). A sensor can be used to monitor the Reductant temperature.	The voltage or value of the temperature sensor signal is either at full value (e.g. battery voltage with no signal or frequency) or, the signal exists but it is above the normal operating range (higher than the maximum operating limit). Possible fault in the temperature sensor or wiring e.g. short / open circuits, circuit resistance, incorrect sensor resistance or, fault with the reference / operating voltage; interference from other circuits can also affect sensor signals. Refer to list of sensor / actuator checks on page 19.
P2046	Reductant Temperature Sensor Circuit Intermittent	A Reductant is a chemical that is stored in a separate tank and is added either to the fuel or to the exhaust gas (depending on system design). The Reductant combines with the Diesel soot particles, which reduces the temperature at which the soot will combust; this allows the soot particles to combust more easily within the particle filter (during the particle filter regeneration process). A sensor can be used to monitor the Reductant temperature.	The voltage or value of the temperature sensor signal is intermittent (likely to be out of normal operating range / tolerance when intermittent fault is detected) or, the signal / voltage is erratic (e.g. signal changes are irregular / unpredictable). Possible fault with wiring or connections (including internal sensor connections); also interference from other circuits can affect the signal. Check the sensor signal, using "live data" or other test equipment and wiggle wiring / connections to try and recreate the fault. Refer to list of sensor / actuator checks on page 19.
P2047	Reductant Injector Circuit / Open Bank 1 Unit 1	A Reductant is a chemical that is stored in a separate tank and is added either to the fuel or to the exhaust gas (depending on system design). The Reductant combines with the Diesel soot particles, which reduces the temperature at which the soot will combust; this allows the soot particles to combust more easily within the particle filter (during the particle filter regeneration process). An injector (controlled by the main Diesel control unit or a separate control module) controls the delivery of the Reductant from the metering unit into the exhaust system (close to the particle filter).	Identify the type of injector and the wiring / control modules used on the system. The voltage, frequency or value of the control signal in the injector circuit is incorrect. Signal could be:- out of normal operating range, constant value, non-existent, corrupt. The fault code indicates a possible "open circuit" but other undefined faults in the injector or wiring could activate the same code e.g. injector power supply, short circuit, high circuit resistance or incorrect injector resistance (solenoid type injectors). Note: the control unit could be providing the correct signal but it is being affected by a circuit fault. Identify the type of injector and the wiring / control modules used on the system. Refer to list of sensor / actuator checks on page 19.
P2048	Reductant Injector Circuit Low Bank 1 Unit 1	A Reductant is a chemical that is stored in a separate tank and is added either to the fuel or to the exhaust gas (depending on system design). The Reductant combines with the Diesel soot particles, which reduces the temperature at which the soot will combust; this allows the soot particles to combust more easily within the particle filter (during the particle filter regeneration process). An injector (controlled by the main Diesel control unit or a separate control module) controls the delivery of the Reductant from the metering unit into the exhaust system (close to the particle filter).	Identify the type of injector and the wiring / control modules used on the system. The voltage, frequency or value of the control signal in the injector circuit is either "zero" (no signal) or, the signal exists but is below the normal operating range. Possible fault with injector or wiring (e.g. injector power supply, short / open circuit or resistance) that is causing a signal value to be lower than the minimum operating limit. Note: the control unit could be providing the correct signal but it is being affected by a circuit fault. Refer to list of sensor / actuator checks on page 19.

NOTE 1. Check for other fault codes that could provide additional information.
NOTE 3. If a fault cannot be located, it is also possible that the control unit is at fault.
NOTE 2. Communication between control units can pass via a CAN-Bus system; refer to Page 51 for CAN-Bus checks.
NOTE 4. Refer to Page 53 for list of pages that show the ISO standard for component locations e.g. Sensor A - Bank 1.

Fault code	EOBD / ISO Description	Component / System Description	Meaningful Description and Quick Check
P2049	Reductant Injector Circuit High Bank 1 Unit 1	A Reductant is a chemical that is stored in a separate tank and is added either to the fuel or to the exhaust gas (depending on system design). The Reductant combines with the Diesel soot particles, which reduces the temperature at which the soot will combust; this allows the soot particles to combust more easily within the particle filter (during the particle filter regeneration process). An injector (controlled by the main Diesel control unit or a separate control module) controls the delivery of the Reductant from the metering unit into the exhaust system (close to the particle filter).	Identify the type of injector and the wiring / control modules used on the system. The voltage, frequency or value of the control signal in the injector circuit is either at full value (e.g. battery voltage with no on / off signal) or, the signal exists but it is above the normal operating range. Possible fault with injector or wiring (e.g. injector power supply, short / open circuit or resistance) that is causing a signal value to be higher than the maximum operating limit. Note: the control unit could be providing the correct signal but it is being affected by a circuit fault. Refer to list of sensor / actuator checks on page 19. It is possible that the when the control unit completes the earth path for the injector, the voltage on the earth circuit remains at supply voltage level e.g. battery voltage instead of dropping to zero (or close to zero); this could be caused by a open circuit, poor earth for the control unit or control unit fault.
P204A	Reductant Pressure Sensor Circuit	A Reductant is a chemical that is stored in a separate tank and is added either to the fuel or to the exhaust gas (depending on system design). The Reductant combines with the Diesel soot particles, which reduces the temperature at which the soot will combust; this allows the soot particles to combust more easily within the particle filter (during the particle filter regeneration process). A sensor can be used to monitor the pressure of Reductant within the system.	The voltage or value of the pressure sensor signal is incorrect (undefined fault). Sensor voltage / signal could be: out of normal operating range, constant value, non-existent, corrupt. Possible fault in the pressure sensor or wiring e.g. short / open circuits, circuit resistance, incorrect sensor resistance (if applicable) or, fault with the reference / operating voltage; interference from other circuits can also affect sensor signals. Refer to list of sensor / actuator checks on page 19.
P204B	Reductant Pressure Sensor Circuit Range / Performance	A Reductant is a chemical that is stored in a separate tank and is added either to the fuel or to the exhaust gas (depending on system design). The Reductant combines with the Diesel soot particles, which reduces the temperature at which the soot will combust; this allows the soot particles to combust more easily within the particle filter (during the particle filter regeneration process). A sensor can be used to monitor the pressure of Reductant within the system.	The voltage or value of the pressure sensor signal is within the normal operating range / tolerance but the signal is not plausible or is incorrect due to an undefined fault. The sensor signal might not match the operating conditions indicated by other sensors or, the sensor signal is not changing / responding as expected to the changes in conditions. Possible fault in the pressure sensor or wiring e.g. short / open circuits, circuit resistance, incorrect sensor resistance (if applicable) or, fault with the reference / operating voltage; interference from other circuits can also affect sensor signals. Refer to list of sensor / actuator checks on page 19. Also check any air passages / pipes to the sensor for leaks / blockages etc. It is also possible that the pressure sensor is operating correctly but the pressure being measured by the sensor is unacceptable or incorrect for the operating conditions; if possible, use alternative method of checking pressure to establish if sensor system is operating correctly and whether the pressure indicated by the sensor is correct / incorrect.
P204C	Reductant Pressure Sensor Circuit Low	A Reductant is a chemical that is stored in a separate tank and is added either to the fuel or to the exhaust gas (depending on system design). The Reductant combines with the Diesel soot particles, which reduces the temperature at which the soot will combust; this allows the soot particles to combust more easily within the particle filter (during the particle filter regeneration process). A sensor can be used to monitor the pressure of Reductant within the system.	The voltage or value of the pressure sensor signal is either "zero" (no signal) or, the signal exists but it is below the normal operating range (lower than the minimum operating limit). Possible fault in the pressure sensor or wiring e.g. short / open circuits, circuit resistance, incorrect sensor resistance (if applicable) or, fault with the reference / operating voltage; interference from other circuits can also affect sensor signals. Refer to list of sensor / actuator checks on page 19.
P204D	Reductant Pressure Sensor Circuit High	A Reductant is a chemical that is stored in a separate tank and is added either to the fuel or to the exhaust gas (depending on system design). The Reductant combines with the Diesel soot particles, which reduces the temperature at which the soot will combust; this allows the soot particles to combust more easily within the particle filter (during the particle filter regeneration process). A sensor can be used to monitor the pressure of Reductant within the system.	The voltage or value of the pressure sensor signal is either at full value (e.g. battery voltage with no signal or frequency) or, the signal exists but it is above the normal operating range (higher than the maximum operating limit). Possible fault in the pressure sensor or wiring e.g. short / open circuits, circuit resistance, incorrect sensor resistance (if applicable) or, fault with the reference / operating voltage; interference from other circuits can also affect sensor signals. Refer to list of sensor / actuator checks on page 19.
P204E	Reductant Pressure Sensor Circuit Intermittent / Erratic	A Reductant is a chemical that is stored in a separate tank and is added either to the fuel or to the exhaust gas (depending on system design). The Reductant combines with the Diesel soot particles, which reduces the temperature at which the soot will combust; this allows the soot particles to combust more easily within the particle filter (during the particle filter regeneration process). A sensor can be used to monitor the pressure of Reductant within the system.	The voltage or value of the pressure sensor signal is intermittent (likely to be out of normal operating range / tolerance when intermittent fault is detected) or, the signal / voltage is erratic (e.g. signal changes are irregular / unpredictable). Possible fault with wiring or connections (including internal sensor connections); also interference from other circuits can affect the signal. Refer to list of sensor / actuator checks on page 19. It is also possible that the pressure could be changing erratically, which could cause the fault code to be activated; if necessary, check pressure with a separate gauge to check for erratic pressure changes.
P204F		ISO / SAE reserved	Not used, reserved for future allocation.
P2050	Reductant Injector Circuit / Open Bank 2 Unit 1	A Reductant is a chemical that is stored in a separate tank and is added either to the fuel or to the exhaust gas (depending on system design). The Reductant combines with the Diesel soot particles, which reduces the temperature at which the soot will combust; this allows the soot particles to combust more easily within the particle filter (during the particle filter regeneration process). An injector (controlled by the main Diesel control unit or a separate control module) controls the delivery of the Reductant from the metering unit into the exhaust system (close to the particle filter).	Identify the type of injector and the wiring / control modules used on the system. The voltage, frequency or value of the control signal in the injector circuit is incorrect. Signal could be:- out of normal operating range, constant value, non-existent, corrupt. The fault code indicates a possible "open circuit" but other undefined faults in the injector or wiring could activate the same code e.g. injector power supply, short circuit, high circuit resistance or incorrect injector resistance (solenoid type injectors). Note: the control unit could be providing the correct signal but it is being affected by a circuit fault. Identify the type of injector and the wiring / control modules used on the system. Refer to list of sensor / actuator checks on page 19.
P2051	Reductant Injector Circuit Low Bank 2 Unit 1	A Reductant is a chemical that is stored in a separate tank and is added either to the fuel or to the exhaust gas (depending on system design). The Reductant combines with the Diesel soot particles, which reduces the temperature at which the soot will combust; this allows the soot particles to combust more easily within the particle filter (during the particle filter regeneration process). An injector (controlled by the main Diesel control unit or a separate control module) controls the delivery of the Reductant from the metering unit into the exhaust system (close to the particle filter).	Identify the type of injector and the wiring / control modules used on the system. The voltage, frequency or value of the control signal in the injector circuit is either "zero" (no signal) or, the signal exists but is below the normal operating range. Possible fault with injector or wiring (e.g. injector power supply, short / open circuit or resistance) that is causing a signal value to be lower than the minimum operating limit. Note: the control unit could be providing the correct signal but it is being affected by a circuit fault. Refer to list of sensor / actuator checks on page 19.

Code	Fault Location	Description	Probable Cause
P2052	Reductant Injector Circuit High Bank 2 Unit 1	A Reductant is a chemical that is stored in a separate tank and is added either to the fuel or to the exhaust gas (depending on system design). The Reductant combines with the Diesel soot particles, which reduces the temperature at which the soot will combust; this allows the soot particles to combust more easily within the particle filter (during the particle filter regeneration process). An injector (controlled by the main Diesel control unit or a separate control module) controls the delivery of the Reductant from the metering unit into the exhaust system (close to the particle filter).	Identify the type of injector and the wiring / control modules used on the system. The voltage, frequency or value of the control signal in the injector circuit is either at full value (e.g. battery voltage with no on / off signal) or, the signal exists but it is above the normal operating range. Possible fault with injector or wiring (e.g. injector power supply, short / open circuit or resistance) that is causing a signal value to be higher than the maximum operating limit. Note: the control unit could be providing the correct signal but it is being affected by a circuit fault. Refer to list of sensor / actuator checks on page 19. It is possible that the when the control unit completes the earth path for the injector, the voltage on the earth circuit remains at supply voltage level e.g. battery voltage instead of dropping to zero (or close to zero); this could be caused by a open circuit, poor earth for the control unit or control unit fault.
P2053	Reductant Injector Circuit / Open Bank 1 Unit 2	A Reductant is a chemical that is stored in a separate tank and is added either to the fuel or to the exhaust gas (depending on system design). The Reductant combines with the Diesel soot particles, which reduces the temperature at which the soot will combust; this allows the soot particles to combust more easily within the particle filter (during the particle filter regeneration process). An injector (controlled by the main Diesel control unit or a separate control module) controls the delivery of the Reductant from the metering unit into the exhaust system (close to the particle filter).	Identify the type of injector and the wiring / control modules used on the system. The voltage, frequency or value of the control signal in the injector circuit is incorrect. Signal could be:- out of normal operating range, constant value, non-existent, corrupt. The fault code indicates a possible "open circuit" but other undefined faults in the injector or wiring could activate the same code e.g. injector power supply, short circuit, high circuit resistance or incorrect injector resistance (solenoid type injectors). Note: the control unit could be providing the correct signal but it is being affected by a circuit fault. Identify the type of injector and the wiring / control modules used on the system. Refer to list of sensor / actuator checks on page 19.
P2054	Reductant Injector Circuit Low Bank 1 Unit 2	A Reductant is a chemical that is stored in a separate tank and is added either to the fuel or to the exhaust gas (depending on system design). The Reductant combines with the Diesel soot particles, which reduces the temperature at which the soot will combust; this allows the soot particles to combust more easily within the particle filter (during the particle filter regeneration process). An injector (controlled by the main Diesel control unit or a separate control module) controls the delivery of the Reductant from the metering unit into the exhaust system (close to the particle filter).	Identify the type of injector and the wiring / control modules used on the system. The voltage, frequency or value of the control signal in the injector circuit is either "zero" (no signal) or, the signal exists but is below the normal operating range. Possible fault with injector or wiring (e.g. injector power supply, short / open circuit or resistance) that is causing a signal value to be lower than the minimum operating limit. Note: the control unit could be providing the correct signal but it is being affected by a circuit fault. Refer to list of sensor / actuator checks on page 19.
P2055	Reductant Injector Circuit High Bank 1 Unit 2	A Reductant is a chemical that is stored in a separate tank and is added either to the fuel or to the exhaust gas (depending on system design). The Reductant combines with the Diesel soot particles, which reduces the temperature at which the soot will combust; this allows the soot particles to combust more easily within the particle filter (during the particle filter regeneration process). An injector (controlled by the main Diesel control unit or a separate control module) controls the delivery of the Reductant from the metering unit into the exhaust system (close to the particle filter).	Identify the type of injector and the wiring / control modules used on the system. The voltage, frequency or value of the control signal in the injector circuit is either at full value (e.g. battery voltage with no on / off signal), the signal exists but it is above the normal operating range. Possible fault with injector or wiring (e.g. injector power supply, short / open circuit or resistance) that is causing a signal value to be higher than the maximum operating limit. Note: the control unit could be providing the correct signal but it is being affected by a circuit fault. Refer to list of sensor / actuator checks on page 19. It is possible that the when the control unit completes the earth path for the injector, the voltage on the earth circuit remains at supply voltage level e.g. battery voltage instead of dropping to zero (or close to zero); this could be caused by a open circuit, poor earth for the control unit or control unit fault.
P2056	Reductant Injector Circuit / Open Bank 2 Unit 2	A Reductant is a chemical that is stored in a separate tank and is added either to the fuel or to the exhaust gas (depending on system design). The Reductant combines with the Diesel soot particles, which reduces the temperature at which the soot will combust; this allows the soot particles to combust more easily within the particle filter (during the particle filter regeneration process). An injector (controlled by the main Diesel control unit or a separate control module) controls the delivery of the Reductant from the metering unit into the exhaust system (close to the particle filter).	Identify the type of injector and the wiring / control modules used on the system. The voltage, frequency or value of the control signal in the injector circuit is incorrect. Signal could be:- out of normal operating range, constant value, non-existent, corrupt. The fault code indicates a possible "open circuit" but other undefined faults in the injector or wiring could activate the same code e.g. injector power supply, short circuit, high circuit resistance or incorrect injector resistance (solenoid type injectors). Note: the control unit could be providing the correct signal but it is being affected by a circuit fault. Identify the type of injector and the wiring / control modules used on the system. Refer to list of sensor / actuator checks on page 19.
P2057	Reductant Injector Circuit Low Bank 2 Unit 2	A Reductant is a chemical that is stored in a separate tank and is added either to the fuel or to the exhaust gas (depending on system design). The Reductant combines with the Diesel soot particles, which reduces the temperature at which the soot will combust; this allows the soot particles to combust more easily within the particle filter (during the particle filter regeneration process). An injector (controlled by the main Diesel control unit or a separate control module) controls the delivery of the Reductant from the metering unit into the exhaust system (close to the particle filter).	Identify the type of injector and the wiring / control modules used on the system. The voltage, frequency or value of the control signal in the injector circuit is either "zero" (no signal) or, the signal exists but is below the normal operating range. Possible fault with injector or wiring (e.g. injector power supply, short / open circuit or resistance) that is causing a signal value to be lower than the minimum operating limit. Note: the control unit could be providing the correct signal but it is being affected by a circuit fault. Refer to list of sensor / actuator checks on page 19.
P2058	Reductant Injector Circuit High Bank 2 Unit 2	A Reductant is a chemical that is stored in a separate tank and is added either to the fuel or to the exhaust gas (depending on system design). The Reductant combines with the Diesel soot particles, which reduces the temperature at which the soot will combust; this allows the soot particles to combust more easily within the particle filter (during the particle filter regeneration process). An injector (controlled by the main Diesel control unit or a separate control module) controls the delivery of the Reductant from the metering unit into the exhaust system (close to the particle filter).	Identify the type of injector and the wiring / control modules used on the system. The voltage, frequency or value of the control signal in the injector circuit is either at full value (e.g. battery voltage with no on / off signal) or, the signal exists but it is above the normal operating range. Possible fault with injector or wiring (e.g. injector power supply, short / open circuit or resistance) that is causing a signal value to be higher than the maximum operating limit. Note: the control unit could be providing the correct signal but it is being affected by a circuit fault. Refer to list of sensor / actuator checks on page 19. It is possible that the when the control unit completes the earth path for the injector, the voltage on the earth circuit remains at supply voltage level e.g. battery voltage instead of dropping to zero (or close to zero); this could be caused by a open circuit, poor earth for the control unit or control unit fault.

NOTE 1. Check for other fault codes that could provide additional information. **NOTE 2.** Communication between control units can pass via a CAN-Bus system; refer to Page 51 for CAN-Bus checks.

NOTE 3. If a fault cannot be located, it is also possible that the control unit is at fault. **NOTE 4.** Refer to Page 53 for list of pages that show the ISO standard for component locations e.g. Sensor A - Bank 1.

Fault code	EOBD / ISO Description	Component / System Description	Meaningful Description and Quick Check
P2059	Reductant Injection Air Pump Control Circuit / Open	A Reductant is a chemical that is stored in a separate tank and is added either to the fuel or to the exhaust gas (depending on system design). The Reductant combines with the Diesel soot particles, which reduces the temperature at which the soot will combust; this allows the soot particles to combust more easily within the particle filter (during the particle filter regeneration process). On some systems, a small air pump and metering assembly mixes the chemical with the air and injects the mixture into the exhaust (via a nozzle / injector).	The pump will be driven by an actuator (probably an electric motor). The voltage, frequency or value of the control signal in the actuator circuit is incorrect. Signal could be:- out of normal operating range, constant value, non-existent, corrupt. The fault code indicates a possible "open circuit" in the wiring to the pump actuator, but other undefined faults in the actuator or wiring could activate the same code e.g. actuator power supply, short circuit, high circuit resistance or incorrect actuator resistance. Note: the control unit could be providing the correct control signal but it is being affected by a circuit fault. Refer to list of sensor / actuator checks on page 19.
P205A	Reductant Tank Temperature Sensor Circuit	A Reductant is a chemical that is stored in a separate tank and is added either to the fuel or to the exhaust gas (depending on system design). The Reductant combines with the Diesel soot particles, which reduces the temperature at which the soot will combust; this allows the soot particles to combust more easily within the particle filter (during the particle filter regeneration process). A sensor can be used to monitor the temperature of the Reductant in the tank.	The voltage or value of the temperature sensor signal is incorrect (undefined fault). Sensor voltage / signal could be: out of normal operating range, constant value, non-existent, corrupt. Possible fault in the temperature sensor or wiring e.g. short / open circuits, circuit resistance, incorrect sensor resistance or, fault with the reference / operating voltage; interference from other circuits can also affect sensor signals. Refer to list of sensor / actuator checks on page 19.
P205B	Reductant Tank Temperature Sensor Circuit Range / Performance	A Reductant is a chemical that is stored in a separate tank and is added either to the fuel or to the exhaust gas (depending on system design). The Reductant combines with the Diesel soot particles, which reduces the temperature at which the soot will combust; this allows the soot particles to combust more easily within the particle filter (during the particle filter regeneration process). A sensor can be used to monitor the temperature of the Reductant in the tank.	The voltage or value of the temperature sensor signal is within the normal operating range / tolerance but the signal is not plausible or is incorrect due to an undefined fault. The sensor signal might not match the operating conditions indicated by other sensors or, the sensor signal is not changing / responding as expected to the changes in conditions. Possible fault in the temperature sensor or wiring e.g. short / open circuits, circuit resistance, incorrect sensor resistance or, fault with the reference / operating voltage; interference from other circuits can also affect sensor signals. Refer to list of sensor / actuator checks on page 19. It is also possible that the temperature sensor is operating correctly but the temperature being measured by the sensor is unacceptable or incorrect for the operating conditions; if possible, use alternative method of checking temperature to establish if sensor system is operating correctly and whether temperature is correct / incorrect.
P205C	Reductant Tank Temperature Sensor Circuit Low	A Reductant is a chemical that is stored in a separate tank and is added either to the fuel or to the exhaust gas (depending on system design). The Reductant combines with the Diesel soot particles, which reduces the temperature at which the soot will combust; this allows the soot particles to combust more easily within the particle filter (during the particle filter regeneration process). A sensor can be used to monitor the temperature of the Reductant in the tank.	The voltage or value of the temperature sensor signal is either "zero" (no signal) or, the signal exists but it is below the normal operating range (lower than the minimum operating limit). Possible fault in the temperature sensor or wiring e.g. short / open circuits, circuit resistance, incorrect sensor resistance or, fault with the reference / operating voltage; interference from other circuits can also affect sensor signals. Refer to list of sensor / actuator checks on page 19.
P205D	Reductant Tank Temperature Sensor Circuit High	A Reductant is a chemical that is stored in a separate tank and is added either to the fuel or to the exhaust gas (depending on system design). The Reductant combines with the Diesel soot particles, which reduces the temperature at which the soot will combust; this allows the soot particles to combust more easily within the particle filter (during the particle filter regeneration process). A sensor can be used to monitor the temperature of the Reductant in the tank.	The voltage or value of the temperature sensor signal is either at full value (e.g. battery voltage with no signal or frequency) or, the signal exists but it is above the normal operating range (higher than the maximum operating limit). Possible fault in the temperature sensor or wiring e.g. short / open circuits, circuit resistance, incorrect sensor resistance or, fault with the reference / operating voltage; interference from other circuits can also affect sensor signals. Refer to list of sensor / actuator checks on page 19.
P205E	Reductant Tank Temperature Sensor Circuit Intermittent / Erratic	A Reductant is a chemical that is stored in a separate tank and is added either to the fuel or to the exhaust gas (depending on system design). The Reductant combines with the Diesel soot particles, which reduces the temperature at which the soot will combust; this allows the soot particles to combust more easily within the particle filter (during the particle filter regeneration process). A sensor can be used to monitor the temperature of the Reductant in the tank.	The voltage value of the temperature sensor signal is intermittent (likely to be out of normal operating range / tolerance when intermittent fault is detected) or, the signal / voltage is erratic (e.g. signal changes are irregular / unpredictable). Possible fault with wiring or connections (including internal sensor connections); also interference from other circuits can affect the signal. Check the sensor signal, using "live data" or other test equipment and wiggle wiring / connections to try and recreate the fault. Refer to list of sensor / actuator checks on page 19.
P205F	ISO / SAE reserved		Not used, reserved for future allocation.
P2060	Reductant Injection Air Pump Control Circuit Low	A Reductant is a chemical that is stored in a separate tank and is added either to the fuel or to the exhaust gas (depending on system design). The Reductant combines with the Diesel soot particles, which reduces the temperature at which the soot will combust; this allows the soot particles to combust more easily within the particle filter (during the particle filter regeneration process). On some systems, a small air pump and metering assembly mixes the chemical with the air and injects the mixture into the exhaust (via a nozzle / injector).	The voltage, frequency or value of the control signal in the actuator circuit is either "zero" (no signal) or, the signal exists but is below the normal operating range. Possible fault with actuator or wiring (e.g. actuator power supply, short / open circuit or resistance) that is causing a control signal value to be lower than the minimum operating limit. Note: the control unit could be providing the correct signal but it is being affected by a circuit fault. Refer to list of sensor / actuator checks on page 19.
P2061	Reductant Injection Air Pump Control Circuit High	A Reductant is a chemical that is stored in a separate tank and is added either to the fuel or to the exhaust gas (depending on system design). The Reductant combines with the Diesel soot particles, which reduces the temperature at which the soot will combust; this allows the soot particles to combust more easily within the particle filter (during the particle filter regeneration process). On some systems, a small air pump and metering assembly mixes the chemical with the air and injects the mixture into the exhaust (via a nozzle / injector).	The voltage, frequency or value of the control signal in the actuator circuit is either at full value (e.g. battery voltage with no on / off signal) or, the signal exists but it is above the normal operating range. Possible fault with actuator or wiring (e.g. actuator power supply, short / open circuit or resistance) that is causing a control signal value to be higher than the maximum operating limit. Note: the control unit could be providing the correct signal but it is being affected by a circuit fault. Refer to list of sensor / actuator checks on page 19.

NOTE 1. Check for other fault codes that could provide additional information.   **NOTE 2.** Communication between control units can pass via a CAN-Bus system; refer to Page 51 for CAN-Bus checks.   222

NOTE 3. If a fault cannot be located, it is also possible that the control unit is at fault.   **NOTE 4.** Refer to Page 53 for list of pages that show the ISO standard for component locations e.g. Sensor A - Bank 1.

Code	Fault Location	Description	Probable Cause
P2062	Reductant / Regeneration Supply Control Circuit / Open	A Reductant is a chemical that is stored in a separate tank and is added either to the fuel or to the exhaust gas (depending on system design). The Reductant combines with the Diesel soot particles, which reduces the temperature at which the soot will combust; this allows the soot particles to combust more easily within the particle filter (during the particle filter regeneration process). Different systems provide different methods of delivering the reductant to the system; it may be necessary to identify the reductant supply process	The reductant supply could be controlled via a pump or other electrical actuator e.g. solenoid valve. The voltage, frequency or value of the control signal in the actuator circuit is incorrect. Signal could be:- out of normal operating range, constant value, non-existent, corrupt. The fault code indicates a possible "open circuit" but other undefined faults in the actuator or wiring could activate the same code e.g. actuator power supply, short circuit, high circuit resistance or incorrect actuator resistance. Note: the control unit could be providing the correct signal but it is being affected by a circuit fault. Refer to list of sensor / actuator checks on page 19.
P2063	Reductant Supply Control Circuit Low	A Reductant is a chemical that is stored in a separate tank and is added either to the fuel or to the exhaust gas (depending on system design). The Reductant combines with the Diesel soot particles, which reduces the temperature at which the soot will combust; this allows the soot particles to combust more easily within the particle filter (during the particle filter regeneration process). Different systems provide different methods of delivering the reductant to the system; it may be necessary to identify the reductant supply process	The reductant supply could be controlled via a pump or other electrical actuator e.g. solenoid valve. The voltage, frequency or value of the control signal in the actuator circuit is either "zero" (no signal) or, the signal exists but is below the normal operating range. Possible fault with actuator or wiring (e.g. actuator power supply, short / open circuit or resistance) that is causing a signal value to be lower than the minimum operating limit. Note: the control unit could be providing the correct signal but it is being affected by a circuit fault. Refer to list of sensor / actuator checks on page 19.
P2064	Reductant Supply Control Circuit High	A Reductant is a chemical that is stored in a separate tank and is added either to the fuel or to the exhaust gas (depending on system design). The Reductant combines with the Diesel soot particles, which reduces the temperature at which the soot will combust; this allows the soot particles to combust more easily within the particle filter (during the particle filter regeneration process). Different systems provide different methods of delivering the reductant to the system; it may be necessary to identify the reductant supply process	The reductant supply could be controlled via a pump or other electrical actuator e.g. solenoid valve. The voltage, frequency or value of the control signal in the actuator circuit is either at full value (e.g. battery voltage with no on / off signal) or, the signal exists but it is above the normal operating range. Possible fault with actuator or wiring (e.g. actuator power supply, short / open circuit or resistance) that is causing a signal value to be higher than the maximum operating limit. Note: the control unit could be providing the correct signal but it is being affected by a circuit fault. Refer to list of sensor / actuator checks on page 19.
P2065	Fuel Level Sensor "B" Circuit	Note that there could be more than one fuel level sensor; one sensor being used for the fuel gauge and the second providing a signal to the control unit for other purposes such as assisting Evaporative Emission control calculations. It is therefore possible that a fault has been detected due to the difference in the signals between the two sensors (the control unit may be able to assess which sensor is at fault due to fuel used compared with distance travelled).	The voltage, frequency or value of the level sensor signal is incorrect (undefined fault). Sensor voltage / signal could be: out of normal operating range, constant value, non-existent, corrupt. Possible fault in the level sensor or wiring e.g. short / open circuits, circuit resistance, incorrect sensor resistance or, fault with the reference / operating voltage; interference from other circuits can also affect sensor signals. Refer to list of sensor / actuator checks on page 19.
P2066	Fuel Level Sensor "B" Performance	Note that there could be more than one fuel level sensor; one sensor being used for the fuel gauge and the second providing a signal to the control unit for other purposes such as assisting Evaporative Emission control calculations. It is therefore possible that a fault has been detected due to the difference in the signals between the two sensors (the control unit may be able to assess which sensor is at fault due to fuel used compared with distance travelled).	The voltage, frequency or value of the level sensor signal is incorrect (undefined fault). Sensor voltage / signal could be: out of normal operating range, constant value, non-existent, corrupt. Possible fault in the level sensor or wiring e.g. short / open circuits, circuit resistance, incorrect sensor resistance or, fault with the reference / operating voltage; interference from other circuits can also affect sensor signals. Refer to list of sensor / actuator checks on page 19. It is also possible that the sensor is operating correctly but the measurement being made by the sensor is unacceptable or incorrect (does not match the values expected / calculated by the control unit).
P2067	Fuel Level Sensor "B" Circuit Low	Note that there could be more than one fuel level sensor; one sensor being used for the fuel gauge and the second providing a signal to the control unit for other purposes such as assisting Evaporative Emission control calculations. It is therefore possible that a fault has been detected due to the difference in the signals between the two sensors (the control unit may be able to assess which sensor is at fault due to fuel used compared with distance travelled).	The voltage, frequency or value of the level sensor signal is either "zero" (no signal) or, the signal exists but it is below the normal operating range (lower than the minimum operating limit). Possible fault in the level sensor or wiring e.g. short / open circuits, circuit resistance, incorrect sensor resistance or, fault with the reference / operating voltage; interference from other circuits can also affect sensor signals. Refer to list of sensor / actuator checks on page 19.
P2068	Fuel Level Sensor "B" Circuit High	Note that there could be more than one fuel level sensor; one sensor being used for the fuel gauge and the second providing a signal to the control unit for other purposes such as assisting Evaporative Emission control calculations. It is therefore possible that a fault has been detected due to the difference in the signals between the two sensors (the control unit may be able to assess which sensor is at fault due to fuel used compared with distance travelled).	The voltage, frequency or value of the level sensor signal is either at full value (e.g. battery voltage with no signal or frequency) or, the signal exists but it is above the normal operating range (higher than the maximum operating limit). Possible fault in the level sensor or wiring e.g. short / open circuits, circuit resistance, incorrect sensor resistance or, fault with the reference / operating voltage; interference from other circuits can also affect sensor signals. Refer to list of sensor / actuator checks on page 19.
P2069	Fuel Level Sensor "B" Circuit Intermittent	Note that there could be more than one fuel level sensor; one sensor being used for the fuel gauge and the second providing a signal to the control unit for other purposes such as assisting Evaporative Emission control calculations. It is therefore possible that a fault has been detected due to the difference in the signals between the two sensors (the control unit may be able to assess which sensor is at fault due to fuel used compared with distance travelled).	The voltage, frequency or value of the level sensor signal is intermittent (likely to be out of normal operating range / tolerance when intermittent fault is detected) or, the signal / voltage is erratic (e.g. signal changes are irregular / unpredictable). Possible fault with wiring or connections (including internal sensor connections); also interference from other circuits can affect the signal. Refer to list of sensor / actuator checks on page 19.
P206A		ISO / SAE reserved	Not used, reserved for future allocation.
P206B		ISO / SAE reserved	Not used, reserved for future allocation.
P206C		ISO / SAE reserved	Not used, reserved for future allocation.
P206D		ISO / SAE reserved	Not used, reserved for future allocation.

NOTE 1. Check for other fault codes that could provide additional information. **NOTE 2.** Communication between control units can pass via a CAN-Bus system; refer to Page 51 for CAN-Bus checks.

NOTE 3. If a fault cannot be located, it is also possible that the control unit is at fault. **NOTE 4.** Refer to Page 53 for list of pages that show the ISO standard for component locations e.g. Sensor A - Bank 1.

Fault code	EOBD / ISO Description	Component / System Description	Meaningful Description and Quick Check
P206E	Intake Manifold Tuning (IMT) Valve Stuck Open Bank 2	The intake manifold "Tuning" system (often referred to a "variable intake system") normally uses additional air valves (flaps or butterflies within the intake manifold) to divert intake air through longer or shorter intake paths; the intake tuning helps improve engine torque / efficiency at different engine speed / loads. The tuning valve can be moved using a vacuum operated mechanism (with the vacuum being regulated by a solenoid or other type of actuator) but it is possible that an electric actuator e.g. motor, could directly operate the mechanism. A position sensor / switch can be used to monitor the valve position.	The control unit is providing a control signal to the actuator but the control unit has detected that the actuator / tuning valve mechanism is stuck in the open position (not responding to the control signal). The fault could be related to an actuator / wiring fault e.g. power supply, short / open circuit, incorrect actuator / circuit resistance. Refer to list of sensor / actuator checks on page 19. Note that the control unit could be detecting the fault because of "a lack of response / no response" from the system or mechanism being controlled by the actuator; check system (as well as the actuator) for correct operation, movement and response. For vacuum operated systems, check for leaks blockages in vacuum pipes and check movement of mechanism when vacuum is applied. A position sensor (if fitted) would be indicating the "stuck position"; also check position sensor operation (refer to fault code P2076).
P206F	Intake Manifold Tuning (IMT) Valve Stuck Closed Bank 2	The intake manifold "Tuning" system (often referred to a "variable intake system") normally uses additional air valves (flaps or butterflies within the intake manifold) to divert intake air through longer or shorter intake paths; the intake tuning helps improve engine torque / efficiency at different engine speed / loads. The tuning valve can be moved using a vacuum operated mechanism (with the vacuum being regulated by a solenoid or other type of actuator) but it is possible that an electric actuator e.g. motor, could directly operate the mechanism. A position sensor / switch can be used to monitor the valve position.	The control unit is providing a control signal to the actuator but the control unit has detected that the actuator / tuning valve mechanism is stuck in the closed position (not responding to the control signal). The fault could be related to an actuator / wiring fault e.g. power supply, short / open circuit, incorrect actuator / circuit resistance. Refer to list of sensor / actuator checks on page 19. Note that the control unit could be detecting the fault because of "a lack of response / no response" from the system or mechanism being controlled by the actuator; check system (as well as the actuator) for correct operation, movement and response. For vacuum operated systems, check for leaks blockages in vacuum pipes and check movement of mechanism when vacuum is applied. A position sensor (if fitted) would be indicating the "stuck position"; also check position sensor operation (refer to fault code P2076).
P2070	Intake Manifold Tuning (IMT) Valve Stuck Open Bank 1	The intake manifold "Tuning" system (often referred to a "variable intake system") normally uses additional air valves (flaps or butterflies within the intake manifold) to divert intake air through longer or shorter intake paths; the intake tuning helps improve engine torque / efficiency at different engine speed / loads. The tuning valve can be moved using a vacuum operated mechanism (with the vacuum being regulated by a solenoid or other type of actuator) but it is possible that an electric actuator e.g. motor, could directly operate the mechanism. A position sensor / switch can be used to monitor the valve position.	The control unit is providing a control signal to the actuator but the control unit has detected that the actuator / tuning valve mechanism is stuck in the open position (not responding to the control signal). The fault could be related to an actuator / wiring fault e.g. power supply, short / open circuit, incorrect actuator / circuit resistance. Refer to list of sensor / actuator checks on page 19. Note that the control unit could be detecting the fault because of "a lack of response / no response" from the system or mechanism being controlled by the actuator; check system (as well as the actuator) for correct operation, movement and response. For vacuum operated systems, check for leaks blockages in vacuum pipes and check movement of mechanism when vacuum is applied. A position sensor (if fitted) would be indicating the "stuck position"; also check position sensor operation (refer to fault code P2076).
P2071	Intake Manifold Tuning (IMT) Valve Stuck Closed Bank 1	The intake manifold "Tuning" system (often referred to a "variable intake system") normally uses additional air valves (flaps or butterflies within the intake manifold) to divert intake air through longer or shorter intake paths; the intake tuning helps improve engine torque / efficiency at different engine speed / loads. The tuning valve can be moved using a vacuum operated mechanism (with the vacuum being regulated by a solenoid or other type of actuator) but it is possible that an electric actuator e.g. motor, could directly operate the mechanism. A position sensor / switch can be used to monitor the valve position.	The control unit is providing a control signal to the actuator but the control unit has detected that the actuator / tuning valve mechanism is stuck in the closed position (not responding to the control signal). The fault could be related to an actuator / wiring fault e.g. power supply, short / open circuit, incorrect actuator / circuit resistance. Refer to list of sensor / actuator checks on page 19. Note that the control unit could be detecting the fault because of "a lack of response / no response" from the system or mechanism being controlled by the actuator; check system (as well as the actuator) for correct operation, movement and response. For vacuum operated systems, check for leaks blockages in vacuum pipes and check movement of mechanism when vacuum is applied. A position sensor (if fitted) would be indicating the "stuck position"; also check position sensor operation (refer to fault code P2076).
P2072	Throttle Actuator Control System -Ice Blockage	The throttle actuator is the motor that controls the throttle plate / butterfly opening angle. Note that the throttle actuator will probably form part of the throttle control assembly which can include a throttle position sensor as well as the control motor.	The fault code is indicating that there is a likely Ice Blockage in the throttle actuator system. The fault is likely to be detected due to information from other sensors (including ambient / intake air temperature). It is possible that the throttle actuator (e.g. motor) requires an abnormal position setting to maintain a controlled speed e.g. idle. Check for signs of icing, but also check for signs of blockage or severe contamination within the throttle body. If there are no signs of icing or other throttle actuator / throttle body fault, it is possible that the throttle actuator control module or main control unit has a fault.
P2073	Manifold Absolute Pressure / Mass Air Flow -Throttle Position Correlation at Idle	The MAP (Manifold Absolute Pressure) sensor or MAF (Mass Airflow) sensor (depending on which sensor is fitted) is used in conjunction with the Throttle position sensor to indicate load and airflow to the control unit.	The signals from the different sensors are providing conflicting or implausible information at idle e.g. MAP or MAF sensor or indicates high airflow but Throttle Position sensor indicates throttle is in "closed / idle" position. It is possible that one of the sensors is operating incorrectly, therefore check each sensor for a fault with the sensor / wiring e.g. short / open circuits, circuit resistance, incorrect sensor resistance or, reference / operating voltage fault. Refer to list of sensor / actuator checks on page 19. Also check for intake and exhaust air leaks / blockages or leak / blockage in MAP sensor air pipe (if applicable).
P2074	Manifold Absolute Pressure / Mass Air Flow -Throttle Position Correlation at Higher Load	The MAP (Manifold Absolute Pressure) sensor or MAF (Mass Airflow) sensor (depending on which sensor is fitted) is used in conjunction with the Throttle position sensor to indicate load and airflow to the control unit.	The signals from the different sensors are providing conflicting or implausible information at "higher load" e.g. MAP or MAF sensor indicates high airflow but Throttle Position sensor indicates throttle is in "almost closed " position. It is possible that one of the sensors is operating incorrectly, therefore check each sensor for a fault with the sensor / wiring e.g. short / open circuits, circuit resistance, incorrect sensor resistance or, reference / operating voltage fault. Refer to list of sensor / actuator checks on page 19. Also check for intake and exhaust air leaks / blockages or leak / blockage in MAP sensor air pipe (if applicable).

NOTE 1. Check for other fault codes that could provide additional information.

NOTE 2. Communication between control units can pass via a CAN-Bus system; refer to Page 51 for CAN-Bus checks.

NOTE 3. If a fault cannot be located, it is also possible that the control unit is at fault.　NOTE 4. Refer to Page 53 for list of pages that show the ISO standard for component locations e.g. Sensor A - Bank 1.

P2075	Intake Manifold Tuning (IMT) Valve Position Sensor / Switch Circuit Bank 1	The intake manifold "Tuning" system (often referred to a "variable intake system") normally uses additional air valves (flaps or butterflies within the intake manifold) to divert intake air through longer or shorter intake paths; the intake tuning helps improve engine torque / efficiency at different engine speed / loads. The tuning valve can be moved using a vacuum operated mechanism (with the vacuum being regulated by a solenoid or other type of actuator) but it is possible that an electric actuator e.g. motor, could directly operate the mechanism. A position sensor / switch can be used to monitor the valve position.	The voltage, frequency or value of the position sensor signal is incorrect (undefined fault). Sensor voltage / signal value could be: out of normal operating range, constant value, non-existent, corrupt. Possible fault in the position sensor or wiring e.g. short / open circuits, circuit resistance, incorrect sensor resistance (if applicable) or, fault with the reference / operating voltage; interference from other circuits can also affect sensor signals. Also check earth connections and cable screening. Refer to list of sensor / actuator checks on page 19.
P2076	Intake Manifold Tuning (IMT) Valve Position Sensor / Switch Circuit Range / Performance Bank 1	The intake manifold "Tuning" system (often referred to a "variable intake system") normally uses additional air valves (flaps or butterflies within the intake manifold) to divert intake air through longer or shorter intake paths; the intake tuning helps improve engine torque / efficiency at different engine speed / loads. The tuning valve can be moved using a vacuum operated mechanism (with the vacuum being regulated by a solenoid or other type of actuator) but it is possible that an electric actuator e.g. motor, could directly operate the mechanism. A position sensor / switch can be used to monitor the valve position.	The voltage, frequency or value of the position sensor signal is within the normal operating range / tolerance but the signal is not plausible or is incorrect due to an undefined fault. The sensor signal might not match the operating conditions indicated by other sensors or, the sensor signal is not changing / responding as expected to the requested changes in the intake tuning valve position. Possible fault in the position sensor or wiring e.g. short / open circuits, circuit resistance, incorrect sensor resistance or, fault with the reference / operating voltage; interference from other circuits can also affect sensor signals. Refer to list of sensor / actuator checks on page 19. It is also possible that the sensor is operating correctly but the position being detected by the sensor is unacceptable or incorrect for the operating conditions. Check for freedom of movement for the tuning valve control mechanism. Check for operation of vacuum actuator (check for vacuum leaks / blockages etc.), check solenoid control (or other actuator as applicable).
P2077	Intake Manifold Tuning (IMT) Valve Position Sensor / Switch Circuit Low Bank 1	The intake manifold "Tuning" system (often referred to a "variable intake system") normally uses additional air valves (flaps or butterflies within the intake manifold) to divert intake air through longer or shorter intake paths; the intake tuning helps improve engine torque / efficiency at different engine speed / loads. The tuning valve can be moved using a vacuum operated mechanism (with the vacuum being regulated by a solenoid or other type of actuator) but it is possible that an electric actuator e.g. motor, could directly operate the mechanism. A position sensor / switch can be used to monitor the valve position.	The voltage, frequency or value of the position sensor signal is either "zero" (no signal) or, the signal exists but it is below the normal operating range (lower than the minimum operating limit). Possible fault in the position sensor or wiring e.g. short / open circuits, circuit resistance, incorrect sensor resistance or, fault with the reference / operating voltage; interference from other circuits can also affect sensor signals. Refer to list of sensor / actuator checks on page 19.
P2078	Intake Manifold Tuning (IMT) Valve Position Sensor / Switch Circuit High Bank 1	The intake manifold "Tuning" system (often referred to a "variable intake system") normally uses additional air valves (flaps or butterflies within the intake manifold) to divert intake air through longer or shorter intake paths; the intake tuning helps improve engine torque / efficiency at different engine speed / loads. The tuning valve can be moved using a vacuum operated mechanism (with the vacuum being regulated by a solenoid or other type of actuator) but it is possible that an electric actuator e.g. motor, could directly operate the mechanism. A position sensor / switch can be used to monitor the valve position.	The voltage, frequency or value of the position sensor signal is either at full value (e.g. battery voltage with no signal or frequency) or, the signal exists but it is above the normal operating range (higher than the maximum operating limit). Possible fault in the position sensor or wiring e.g. short / open circuits, circuit resistance, incorrect sensor resistance or, fault with the reference / operating voltage; interference from other circuits can also affect sensor signals. Refer to list of sensor / actuator checks on page 19.
P2079	Intake Manifold Tuning (IMT) Valve Position Sensor / Switch Circuit Intermittent Bank 1	The intake manifold "Tuning" system (often referred to a "variable intake system") normally uses additional air valves (flaps or butterflies within the intake manifold) to divert intake air through longer or shorter intake paths; the intake tuning helps improve engine torque / efficiency at different engine speed / loads. The tuning valve can be moved using a vacuum operated mechanism (with the vacuum being regulated by a solenoid or other type of actuator) but it is possible that an electric actuator e.g. motor, could directly operate the mechanism. A position sensor / switch can be used to monitor the valve position.	The voltage, frequency or value of the position sensor signal is intermittent (likely to be out of normal operating range / tolerance when intermittent fault is detected) or, the signal / voltage is erratic (e.g. signal changes are irregular / unpredictable). Possible fault with wiring or connections (including internal sensor connections); also interference from other circuits can affect the signal. Refer to list of sensor / actuator checks on page 19. Check operation of tuning valve control mechanism to ensure that it is not moving erratically (can be caused by vacuum / solenoid control problems).
P207A	Intake Manifold Tuning (IMT) Valve Position Sensor / Switch Circuit Bank 2	The intake manifold "Tuning" system (often referred to a "variable intake system") normally uses additional air valves (flaps or butterflies within the intake manifold) to divert intake air through longer or shorter intake paths; the intake tuning helps improve engine torque / efficiency at different engine speed / loads. The tuning valve can be moved using a vacuum operated mechanism (with the vacuum being regulated by a solenoid or other type of actuator) but it is possible that an electric actuator e.g. motor, could directly operate the mechanism. A position sensor / switch can be used to monitor the valve position.	The voltage, frequency or value of the position sensor signal is incorrect (undefined fault). Sensor voltage / signal value could be: out of normal operating range, constant value, non-existent, corrupt. Possible fault in the position sensor or wiring e.g. short / open circuits, circuit resistance, incorrect sensor resistance (if applicable) or, fault with the reference / operating voltage; interference from other circuits can also affect sensor signals. Also check earth connections and cable screening. Refer to list of sensor / actuator checks on page 19.

Fault code	EOBD / ISO Description	Component / System Description	Meaningful Description and Quick Check
P207B	Intake Manifold Tuning (IMT) Valve Position Sensor / Switch Circuit Range / Performance Bank 2	The intake manifold "Tuning" system (often referred to a "variable intake system") normally uses additional air valves (flaps or butterflies within the intake manifold) to divert intake air through longer or shorter intake paths; the intake tuning helps improve engine torque / efficiency at different engine speed / loads. The tuning valve can be moved using a vacuum operated mechanism (with the vacuum being regulated by a solenoid or other type of actuator) but it is possible that an electric actuator e.g. motor, could directly operate the mechanism. A position sensor / switch can be used to monitor the valve position.	The voltage, frequency or value of the position sensor signal is within the normal operating range / tolerance but the signal is not plausible or is incorrect due to an undefined fault. The sensor signal might not match the operating conditions indicated by other sensors or, the sensor signal is not changing / responding as expected to the requested changes in the intake tuning valve position. Possible fault in the position sensor or wiring e.g. short / open circuits, circuit resistance, incorrect sensor resistance or, fault with the reference / operating voltage; interference from other circuits can also affect sensor signals. Refer to list of sensor / actuator checks on page 19. It is also possible that the sensor is operating correctly but the position being detected by the sensor is unacceptable or incorrect for the operating conditions. Check for freedom of movement for the tuning valve control mechanism. Check for operation of vacuum actuator (check for vacuum leaks / blockages etc.), check solenoid control (or other actuator as applicable).
P207C	Intake Manifold Tuning (IMT) Valve Position Sensor / Switch Circuit Low Bank 2	The intake manifold "Tuning" system (often referred to a "variable intake system") normally uses additional air valves (flaps or butterflies within the intake manifold) to divert intake air through longer or shorter intake paths; the intake tuning helps improve engine torque / efficiency at different engine speed / loads. The tuning valve can be moved using a vacuum operated mechanism (with the vacuum being regulated by a solenoid or other type of actuator) but it is possible that an electric actuator e.g. motor, could directly operate the mechanism. A position sensor / switch can be used to monitor the valve position.	The voltage, frequency or value of the position sensor signal is either "zero" (no signal) or, the signal exists but it is below the normal operating range (lower than the minimum operating limit). Possible fault in the position sensor or wiring e.g. short / open circuits, circuit resistance, incorrect sensor resistance or, fault with the reference / operating voltage; interference from other circuits can also affect sensor signals. Refer to list of sensor / actuator checks on page 19.
P207D	Intake Manifold Tuning (IMT) Valve Position Sensor / Switch Circuit High Bank 2	The intake manifold "Tuning" system (often referred to a "variable intake system") normally uses additional air valves (flaps or butterflies within the intake manifold) to divert intake air through longer or shorter intake paths; the intake tuning helps improve engine torque / efficiency at different engine speed / loads. The tuning valve can be moved using a vacuum operated mechanism (with the vacuum being regulated by a solenoid or other type of actuator) but it is possible that an electric actuator e.g. motor, could directly operate the mechanism. A position sensor / switch can be used to monitor the valve position.	The voltage, frequency or value of the position sensor signal is either at full value (e.g. battery voltage with no signal or frequency) or, the signal exists but it is above the normal operating range (higher than the maximum operating limit). Possible fault in the position sensor or wiring e.g. short / open circuits, circuit resistance, incorrect sensor resistance or, fault with the reference / operating voltage; interference from other circuits can also affect sensor signals. Refer to list of sensor / actuator checks on page 19.
P207E	Intake Manifold Tuning (IMT) Valve Position Sensor / Switch Circuit Intermittent Bank 2	The intake manifold "Tuning" system (often referred to a "variable intake system") normally uses additional air valves (flaps or butterflies within the intake manifold) to divert intake air through longer or shorter intake paths; the intake tuning helps improve engine torque / efficiency at different engine speed / loads. The tuning valve can be moved using a vacuum operated mechanism (with the vacuum being regulated by a solenoid or other type of actuator) but it is possible that an electric actuator e.g. motor, could directly operate the mechanism. A position sensor / switch can be used to monitor the valve position.	The voltage, frequency or value of the position sensor signal is intermittent (likely to be out of normal operating range / tolerance when intermittent fault is detected) or, the signal / voltage is erratic (e.g. signal changes are irregular / unpredictable). Possible fault with wiring or connections (including internal sensor connections); also interference from other circuits can affect the signal. Refer to list of sensor / actuator checks on page 19. Check operation of tuning valve control mechanism to ensure that it is not moving erratically (can be caused by vacuum / solenoid control problems).
P207F	ISO / SAE reserved		Not used, reserved for future allocation.
P2080	Exhaust Gas Temperature Sensor Circuit Range / Performance Bank 1 Sensor 1	The exhaust gas temperature sensor signal is used to indicate if the temperature is outside of acceptable limits (some emissions control functions are more efficient when the gas temperature is within specified limits). Excessive temperatures can rapidly cause deterioration of the catalyst and other emissions control components. High gas temperatures are often caused by faults in the engine operation e.g. misfires, air leaks, fuelling etc.	The voltage, frequency or value of the temperature sensor signal is within the normal operating range / tolerance but the signal is not plausible or is incorrect due to an undefined fault. The sensor signal might not match the operating conditions indicated by other sensors or, the sensor signal is not changing / responding as expected to the changes in conditions. Possible fault in the temperature sensor or wiring e.g. short / open circuits, circuit resistance, incorrect sensor resistance or, fault with the reference / operating voltage; interference from other circuits can also affect sensor signals. Refer to list of sensor / actuator checks on page 19. It is also possible that the temperature sensor is operating correctly but the temperature being measured by the sensor is unacceptable; check for exhaust air leaks, fuelling (mixture), misfires, secondary air injection faults, or other faults causing high oxygen and / or high fuel content in the exhaust.
P2081	Exhaust Gas Temperature Sensor Circuit Intermittent Bank 1 Sensor 1	The exhaust gas temperature sensor signal is used to indicate if the temperature is outside of acceptable limits (some emissions control functions are more efficient when the gas temperature is within specified limits). Excessive temperatures can rapidly cause deterioration of the catalyst and other emissions control components. High gas temperatures are often caused by faults in the engine operation e.g. misfires, air leaks, fuelling etc.	The voltage, frequency or value of the temperature sensor signal is intermittent (likely to be out of normal operating range / tolerance when intermittent fault is detected) or, the signal / voltage is erratic (e.g. signal changes are irregular / unpredictable). Possible fault with wiring or connections (including internal sensor connections); also interference from other circuits can affect the signal. Refer to list of sensor / actuator checks on page 19. Although the problem is likely to be a sensor / electrical fault, also check for other faults could cause rapid / erratic changes in the exhaust temperature.
P2082	Exhaust Gas Temperature Sensor Circuit Range / Performance Bank 2 Sensor 1	The exhaust gas temperature sensor signal is used to indicate if the temperature is outside of acceptable limits (some emissions control functions are more efficient when the gas temperature is within specified limits). Excessive temperatures can rapidly cause deterioration of the catalyst and other emissions control components. High gas temperatures are often caused by faults in the engine operation e.g. misfires, air leaks, fuelling etc.	The voltage, frequency or value of the temperature sensor signal is within the normal operating range / tolerance but the signal is not plausible or is incorrect due to an undefined fault. The sensor signal might not match the operating conditions indicated by other sensors or, the sensor signal is not changing / responding as expected to the changes in conditions. Possible fault in the temperature sensor or wiring e.g. short / open circuits, circuit resistance, incorrect sensor resistance or, fault with the reference / operating voltage; interference from other circuits can also affect sensor signals. Refer to list of sensor / actuator checks on page 19. It is also possible that the temperature sensor is operating correctly but the temperature being measured by the sensor is unacceptable; check for exhaust air leaks, fuelling (mixture), misfires, secondary air injection faults, or other faults causing high oxygen and / or high fuel content in the exhaust.

NOTE 1. Check for other fault codes that could provide additional information.
NOTE 2. Communication between control units can pass via a CAN-Bus system; refer to Page 51 for CAN-Bus checks.
NOTE 3. If a fault cannot be located, it is also possible that the control unit is at fault.
NOTE 4. Refer to Page 53 for list of pages that show the ISO standard for component locations e.g. Sensor A - Bank 1.

Code	Description		
P2083	Exhaust Gas Temperature Sensor Circuit Intermittent Bank 2 Sensor 1	The exhaust gas temperature sensor signal is used to indicate if the temperature is outside of acceptable limits (some emissions control functions are more efficient when the gas temperature is within specified limits). Excessive temperatures can rapidly cause deterioration of the catalyst and other emissions control components. High gas temperatures are often caused by faults in the engine operation e.g. misfires, air leaks, fuelling etc.	The voltage, frequency or value of the temperature sensor signal is intermittent (likely to be out of normal operating range / tolerance when intermittent fault is detected) or, the signal / voltage is erratic (e.g. signal changes are irregular / unpredictable). Possible fault with wiring or connections (including internal sensor connections); also interference from other circuits can affect the signal. Refer to list of sensor / actuator checks on page 19. Although the problem is likely to be a sensor / electrical fault, also check for other faults could cause rapid / erratic changes in the exhaust temperature.
P2084	Exhaust Gas Temperature Sensor Circuit Range / Performance Bank 1 Sensor 2	The exhaust gas temperature sensor signal is used to indicate if the temperature is outside of acceptable limits (some emissions control functions are more efficient when the gas temperature is within specified limits). Excessive temperatures can rapidly cause deterioration of the catalyst and other emissions control components. High gas temperatures are often caused by faults in the engine operation e.g. misfires, air leaks, fuelling etc.	The voltage, frequency or value of the temperature sensor signal is within the normal operating range / tolerance but the signal is not plausible or is incorrect due to an undefined fault. The sensor signal might not match the operating conditions indicated by other sensors or, the sensor signal is not changing / responding as expected to the changes in conditions. Possible fault in the temperature sensor or wiring e.g. short / open circuits, circuit resistance, incorrect sensor resistance or, fault with the reference / operating voltage; interference from other circuits can also affect sensor signals. Refer to list of sensor / actuator checks on page 19. It is also possible that the temperature sensor is operating correctly but the temperature being measured by the sensor is unacceptable; check for exhaust air leaks, fuelling (mixture), misfires, secondary air injection faults, or other faults causing high oxygen and / or high fuel content in the exhaust.
P2085	Exhaust Gas Temperature Sensor Circuit Intermittent Bank 1 Sensor 2	The exhaust gas temperature sensor signal is used to indicate if the temperature is outside of acceptable limits (some emissions control functions are more efficient when the gas temperature is within specified limits). Excessive temperatures can rapidly cause deterioration of the catalyst and other emissions control components. High gas temperatures are often caused by faults in the engine operation e.g. misfires, air leaks, fuelling etc.	The voltage, frequency or value of the temperature sensor signal is intermittent (likely to be out of normal operating range / tolerance when intermittent fault is detected) or, the signal / voltage is erratic (e.g. signal changes are irregular / unpredictable). Possible fault with wiring or connections (including internal sensor connections); also interference from other circuits can affect the signal. Refer to list of sensor / actuator checks on page 19. Although the problem is likely to be a sensor / electrical fault, also check for other faults could cause rapid / erratic changes in the exhaust temperature.
P2086	Exhaust Gas Temperature Sensor Circuit Range / Performance Bank 2 Sensor 2	The exhaust gas temperature sensor signal is used to indicate if the temperature is outside of acceptable limits (some emissions control functions are more efficient when the gas temperature is within specified limits). Excessive temperatures can rapidly cause deterioration of the catalyst and other emissions control components. High gas temperatures are often caused by faults in the engine operation e.g. misfires, air leaks, fuelling etc.	The voltage, frequency or value of the temperature sensor signal is within the normal operating range / tolerance but the signal is not plausible or is incorrect due to an undefined fault. The sensor signal might not match the operating conditions indicated by other sensors or, the sensor signal is not changing / responding as expected to the changes in conditions. Possible fault in the temperature sensor or wiring e.g. short / open circuits, circuit resistance, incorrect sensor resistance or, fault with the reference / operating voltage; interference from other circuits can also affect sensor signals. Refer to list of sensor / actuator checks on page 19. It is also possible that the temperature sensor is operating correctly but the temperature being measured by the sensor is unacceptable; check for exhaust air leaks, fuelling (mixture), misfires, secondary air injection faults, or other faults causing high oxygen and / or high fuel content in the exhaust.
P2087	Exhaust Gas Temperature Sensor Circuit Intermittent Bank 2 Sensor 2	The exhaust gas temperature sensor signal is used to indicate if the temperature is outside of acceptable limits (some emissions control functions are more efficient when the gas temperature is within specified limits). Excessive temperatures can rapidly cause deterioration of the catalyst and other emissions control components. High gas temperatures are often caused by faults in the engine operation e.g. misfires, air leaks, fuelling etc.	The voltage, frequency or value of the temperature sensor signal is intermittent (likely to be out of normal operating range / tolerance when intermittent fault is detected) or, the signal / voltage is erratic (e.g. signal changes are irregular / unpredictable). Possible fault with wiring or connections (including internal sensor connections); also interference from other circuits can affect the signal. Refer to list of sensor / actuator checks on page 19. Although the problem is likely to be a sensor / electrical fault, also check for other faults could cause rapid / erratic changes in the exhaust temperature.
P2088	"A" Camshaft Position Actuator Control Circuit Low Bank 1	Refers to operation of variable valve timing system, which is used to alter camshaft position / valve timing; the control unit will control the actuator (e.g. solenoid) which then regulates oil pressure or other mechanism to alter camshaft position.	The voltage, frequency or value of the control signal in the actuator circuit is either "zero" (no signal) or, the signal exists but is below the normal operating range. Possible fault with actuator or wiring (e.g. actuator power supply, short / open circuit or resistance) that is causing a signal value to be lower than the minimum operating limit. Note: the control unit could be providing the correct signal but it is being affected by the circuit fault. Refer to list of sensor / actuator checks on page 19.
P2089	"A" Camshaft Position Actuator Control Circuit High Bank 1	Refers to operation of variable valve timing system, which is used to alter camshaft position / valve timing; the control unit will control the actuator (e.g. solenoid) which then regulates oil pressure or other mechanism to alter camshaft position.	The voltage, frequency or value of the control signal in the actuator circuit is either at full value (e.g. battery voltage with no on / off signal) or, the signal exists but it is above the normal operating range. Possible fault with actuator or wiring (e.g. actuator power supply, short / open circuit or resistance) that is causing a signal value to be higher than the maximum operating limit. Note: the control unit could be providing the correct signal but it is being affected by the circuit fault. Refer to list of sensor / actuator checks on page 19.
P2090	"B" Camshaft Position Actuator Control Circuit Low Bank 1	Refers to operation of variable valve timing system, which is used to alter camshaft position / valve timing; the control unit will control the actuator (e.g. solenoid) which then regulates oil pressure or other mechanism to alter camshaft position.	The voltage, frequency or value of the control signal in the actuator circuit is either "zero" (no signal) or, the signal exists but is below the normal operating range. Possible fault with actuator or wiring (e.g. actuator power supply, short / open circuit or resistance) that is causing a signal value to be lower than the minimum operating limit. Note: the control unit could be providing the correct signal but it is being affected by the circuit fault. Refer to list of sensor / actuator checks on page 19.
P2091	"B" Camshaft Position Actuator Control Circuit High Bank 1	Refers to operation of variable valve timing system, which is used to alter camshaft position / valve timing; the control unit will control the actuator (e.g. solenoid) which then regulates oil pressure or other mechanism to alter camshaft position.	The voltage, frequency or value of the control signal in the actuator circuit is either at full value (e.g. battery voltage with no on / off signal) or, the signal exists but it is above the normal operating range. Possible fault with actuator or wiring (e.g. actuator power supply, short / open circuit or resistance) that is causing a signal value to be higher than the maximum operating limit. Note: the control unit could be providing the correct signal but it is being affected by the circuit fault. Refer to list of sensor / actuator checks on page 19.
P2092	"A" Camshaft Position Actuator Control Circuit Low Bank 2	Refers to operation of variable valve timing system, which is used to alter camshaft position / valve timing; the control unit will control the actuator (e.g. solenoid) which then regulates oil pressure or other mechanism to alter camshaft position.	The voltage, frequency or value of the control signal in the actuator circuit is either "zero" (no signal) or, the signal exists but is below the normal operating range. Possible fault with actuator or wiring (e.g. actuator power supply, short / open circuit or resistance) that is causing a signal value to be lower than the minimum operating limit. Note: the control unit could be providing the correct signal but it is being affected by the circuit fault. Refer to list of sensor / actuator checks on page 19.

NOTE 1. Check for other fault codes that could provide additional information. **NOTE 2.** Communication between control units can pass via a CAN-Bus system; refer to Page 51 for CAN-Bus checks.
NOTE 3. If a fault cannot be located, it is also possible that the control unit is at fault. **NOTE 4.** Refer to Page 53 for list of pages that show the ISO standard for component locations e.g. Sensor A - Bank 1.

Fault code	EOBD / ISO Description	Component / System Description	Meaningful Description and Quick Check
P2093	"A" Camshaft Position Actuator Control Circuit High Bank 2	Refers to operation of variable valve timing system, which is used to alter camshaft position / valve timing; the control unit will control the actuator (e.g. solenoid) which then regulates oil pressure or other mechanism to alter camshaft position.	The voltage, frequency or value of the control signal in the actuator circuit is either at full value (e.g. battery voltage with no on / off signal) or, the signal exists but it is above the normal operating range. Possible fault with actuator or wiring (e.g. actuator power supply, short / open circuit or resistance) that is causing a signal value to be higher than the maximum operating limit. Note: the control unit could be providing the correct signal but it is being affected by the circuit fault. Refer to list of sensor / actuator checks on page 19.
P2094	"B" Camshaft Position Actuator Control Circuit Low Bank 2	Refers to operation of variable valve timing system, which is used to alter camshaft position / valve timing; the control unit will control the actuator (e.g. solenoid) which then regulates oil pressure or other mechanism to alter camshaft position.	The voltage, frequency or value of the control signal in the actuator circuit is either "zero" (no signal) or, the signal exists but is below the normal operating range. Possible fault with actuator or wiring (e.g. actuator power supply, short / open circuit or resistance) that is causing a signal value to be lower than the minimum operating limit. Note: the control unit could be providing the correct signal but it is being affected by the circuit fault. Refer to list of sensor / actuator checks on page 19.
P2095	"B" Camshaft Position Actuator Control Circuit High Bank 2	Refers to operation of variable valve timing system, which is used to alter camshaft position / valve timing; the control unit will control the actuator (e.g. solenoid) which then regulates oil pressure or other mechanism to alter camshaft position.	The voltage, frequency or value of the control signal in the actuator circuit is either at full value (e.g. battery voltage with no on / off signal) or, the signal exists but it is above the normal operating range. Possible fault with actuator or wiring (e.g. actuator power supply, short / open circuit or resistance) that is causing a signal value to be higher than the maximum operating limit. Note: the control unit could be providing the correct signal but it is being affected by the circuit fault. Refer to list of sensor / actuator checks on page 19.
P2096	Post Catalyst Fuel Trim System Too Lean Bank 1	Note: Many O2 / Lambda sensor related fault codes are activated due to faults in other systems (engine, fuelling, misfires etc). Fuel trim refers to the monitoring and control of the air / fuel ratio. The control unit monitors the Oxygen / Lambda sensor signal (indicates exhaust gas oxygen content). The control unit adjusts the fuelling to maintain the oxygen content within defined operating limits (helps efficient catalytic converter operation). If the control unit applies a fuelling adjustment, but this adjustment does not alter the oxygen content as expected, this can be regarded as a "fuel trim" fault. Fuel trim can apply to Pre-cat oxygen measurements (adjusting to a pre-defined air / fuel ratio), but it can also apply to Post-cat measurements (measuring after the catalyst indicates if there is a need for any fine tuning of the air / fuel ratio to maximise catalyst operation).	The fault code indicates that there is a post-cat fuel trim problem (biased towards a lean mixture, but the normal minor adjustments are not correcting the fault). The fault could be caused by a sensor / wiring fault; check wiring between sensor and control unit for short / open circuits and high resistance. It is also possible that the exhaust gas oxygen content is too high. Check signals for the pre and post-cat sensors under various operating conditions (load, constant rpm etc). If the pre-cat sensor signal is correct but post-cat sensor signal is still "lean", it could indicate a sensor fault but, it could also be caused by an inefficient catalyst or a leak into the catalyst (or close to the post-cat sensor). If pre-cat sensor signal is also incorrect, the fault could be caused by an engine / fuelling fault, or other fault causing incorrect oxygen content in the exhaust gas e.g. exhaust leak. For a detailed list of checks, Refer to list of sensor / actuator checks on page 19. Different types of O2 / Lambda sensors (e.g. Zirconia or Titania / wideband etc) provide different signals and values, it will be necessary to identify the type of sensor fitted to the vehicle.
P2097	Post Catalyst Fuel Trim System Too Rich Bank 1	Note: Many O2 / Lambda sensor related fault codes are activated due to faults in other systems (engine, fuelling, misfires etc). Fuel trim refers to the monitoring and control of the air / fuel ratio. The control unit monitors the Oxygen / Lambda sensor signal (indicates exhaust gas oxygen content). The control unit adjusts the fuelling to maintain the oxygen content within defined operating limits (helps efficient catalytic converter operation). If the control unit applies a fuelling adjustment, but this adjustment does not alter the oxygen content as expected, this can be regarded as a "fuel trim" fault. Fuel trim can apply to Pre-cat oxygen measurements (adjusting to a pre-defined air / fuel ratio), but it can also apply to Post-cat measurements (measuring after the catalyst indicates if there is a need for any fine tuning of the air / fuel ratio to maximise catalyst operation).	The fault code indicates that there is a post-cat fuel trim problem (biased towards a rich mixture, but the normal minor adjustments are not correcting the fault). The fault could be caused by a sensor / wiring fault; check wiring between sensor and control unit for short / open circuits and high resistance. It is also possible that the exhaust gas oxygen content is too low (rich mixture). Check signals for the pre and post-cat sensors under various operating conditions (load, constant rpm etc). If the pre-cat sensor signal is correct but post-cat sensor signal is still "rich", it could indicate a sensor fault. If pre-cat sensor signal is also incorrect, the fault could be caused by an engine / fuelling fault, or other fault causing a low oxygen content in the exhaust gas (rich mixture). For a detailed list of checks, Refer to list of sensor / actuator checks on page 19. Different types of O2 / Lambda sensors (e.g. Zirconia or Titania / wideband etc) provide different signals and values, it will be necessary to identify the type of sensor fitted to the vehicle.
P2098	Post Catalyst Fuel Trim System Too Lean Bank 2	Note: Many O2 / Lambda sensor related fault codes are activated due to faults in other systems (engine, fuelling, misfires etc). Fuel trim refers to the monitoring and control of the air / fuel ratio. The control unit monitors the Oxygen / Lambda sensor signal (indicates exhaust gas oxygen content). The control unit adjusts the fuelling to maintain the oxygen content within defined operating limits (helps efficient catalytic converter operation). If the control unit applies a fuelling adjustment, but this adjustment does not alter the oxygen content as expected, this can be regarded as a "fuel trim" fault. Fuel trim can apply to Pre-cat oxygen measurements (adjusting to a pre-defined air / fuel ratio), but it can also apply to Post-cat measurements (measuring after the catalyst indicates if there is a need for any fine tuning of the air / fuel ratio to maximise catalyst operation).	The fault code indicates that there is a post-cat fuel trim problem (biased towards a lean mixture, but the normal minor adjustments are not correcting the fault). The fault could be caused by a sensor / wiring fault; check wiring between sensor and control unit for short / open circuits and high resistance. It is also possible that the exhaust gas oxygen content is too high. Check signals for the pre and post-cat sensors under various operating conditions (load, constant rpm etc). If the pre-cat sensor signal is correct but post-cat sensor signal is still "lean", it could indicate a sensor fault but, it could also be caused by an inefficient catalyst or a leak into the catalyst (or close to the post-cat sensor). If pre-cat sensor signal is also incorrect, the fault could be caused by an engine / fuelling fault, or other fault causing incorrect oxygen content in the exhaust gas e.g. exhaust leak. For a detailed list of checks, Refer to list of sensor / actuator checks on page 19. Different types of O2 / Lambda sensors (e.g. Zirconia or Titania / wideband etc) provide different signals and values, it will be necessary to identify the type of sensor fitted to the vehicle.

NOTE 1. Check for other fault codes that could provide additional information.
NOTE 2. Communication between control units can pass via a CAN-Bus system; refer to Page 51 for CAN-Bus checks.
NOTE 3. If a fault cannot be located, it is also possible that the control unit is at fault.
NOTE 4. Refer to Page 53 for list of pages that show the ISO standard for component locations e.g. Sensor A - Bank 1.

code	Description		
P2099	Post Catalyst Fuel Trim System Too Rich Bank 2	Note: Many O2 / Lambda sensor related fault codes are activated due to faults in other systems (engine, fuelling, misfires etc). Fuel trim refers to the monitoring and control of the air / fuel ratio. The control unit monitors the Oxygen / Lambda sensor signal (indicates exhaust gas oxygen content). The control unit adjusts the fuelling to maintain the oxygen content within defined operating limits (helps efficient catalytic converter operation). If the control unit applies a fuelling adjustment, but this adjustment does not alter the oxygen content as expected, this can be regarded as a "fuel trim" fault. Fuel trim can apply to Pre-cat oxygen measurements (adjusting to a pre-defined air / fuel ratio), but it can also apply to Post-cat measurements (measuring after the catalyst indicates if there is a need for any fine tuning of the air / fuel ratio to maximise catalyst operation).	The fault code indicates that there is a post-cat fuel trim problem (biased towards a rich mixture, but the normal minor adjustments are not correcting the fault). The fault could be caused by a sensor / wiring fault; check wiring between sensor and control unit for short / open circuits and high resistance. It is also possible that the exhaust gas oxygen content is too low (rich mixture). Check signals for the pre and post-cat sensors under various operating conditions (load, constant rpm etc). If the pre-cat sensor signal is correct but post-cat sensor signal is still "rich", it could indicate a sensor fault. If pre-cat sensor signal is also incorrect, the fault could be caused by an engine / fuelling fault, or other fault causing a low oxygen content in the exhaust gas (rich mixture). For a detailed list of checks, Refer to list of sensor / actuator checks on page 19. Different types of O2 / Lambda sensors (e.g. Zirconia or Titania / wideband etc) provide different signals and values, it will be necessary to identify the type of sensor fitted to the vehicle.
P2100	Throttle Actuator Control Motor Circuit / Open	The throttle actuator is the motor that controls the throttle plate / butterfly opening angle. Note that the throttle actuator will probably form part of the throttle control assembly which can include a throttle position sensor as well as the control motor.	Note that the position of the motor can be regulated using a frequency / pulse width control signal or other electrical control (provided by the control unit on the earth or power circuit). It is advisable to refer to vehicle specific information to identify the type of control signal provided to the motor. The voltage, frequency or value of the control signal in the throttle motor circuit is incorrect. Signal could be:- out of normal operating range, constant value, non-existent, corrupt. The fault code indicates a possible "open circuit" but other undefined faults in the motor or wiring could activate the same code e.g. motor power supply, short circuit, high circuit resistance or incorrect motor resistance. Note: the control unit could be providing the correct signal but it is being affected by a circuit fault. Identify the type of motor control, control signal and the wiring used on the system. Refer to list of sensor / actuator checks on page 19.
P2101	Throttle Actuator Control Motor Circuit Range / Performance	The throttle actuator is the motor that controls the throttle plate / butterfly opening angle. Note that the throttle actuator will probably form part of the throttle control assembly which can include a throttle position sensor as well as the control motor.	Note that the position of the motor can be regulated using a frequency / pulse width control signal or other electrical control (provided by the control unit on the earth or power circuit). It is advisable to refer to vehicle specific information to identify the type of control signal provided to the motor. The fault code could indicate that the throttle actuator is being provided with a specific control signal (by the control unit) but the actuator is not moving to the position expected by the control unit (position monitored by the throttle position sensor). The code could also indicate that the voltage, frequency or value of the control signal in the motor circuit is within the normal operating range / tolerance, but the signal might be incorrect for the operating conditions. Possible fault in motor or wiring e.g. short / open circuit or resistance, but also possible that the motor (or mechanism / system controlled by the motor) is not operating or moving correctly. Also check for contamination / restrictions that could affect airflow around the throttle butterfly. Refer to list of sensor / actuator checks on page 19.
P2102	Throttle Actuator Control Motor Circuit Low	The throttle actuator is the motor that controls the throttle plate / butterfly opening angle. Note that the throttle actuator will probably form part of the throttle control assembly which can include a throttle position sensor as well as the control motor.	Note that the position of the motor can be regulated using a frequency / pulse width control signal or other electrical control (provided by the control unit on the earth or power circuit). It is advisable to refer to vehicle specific information to identify the type of control signal provided to the motor. The voltage, frequency or value of the control signal in the motor circuit is either "zero" (no voltage or signal) or, the voltage / signal exists but is below the normal operating range. Possible fault with motor or wiring e.g. motor power supply, short / open circuit or resistance that is causing a signal value to be lower than the minimum operating limit. Note: the control unit could be providing the correct voltage / signal but it is being affected by a circuit fault. Refer to list of sensor / actuator checks on page 19.
P2103	Throttle Actuator Control Motor Circuit High	The throttle actuator is the motor that controls the throttle plate / butterfly opening angle. Note that the throttle actuator will probably form part of the throttle control assembly which can include a throttle position sensor as well as the control motor.	Note that the position of the motor can be regulated using a frequency / pulse width control signal or other electrical control (provided by the control unit on the earth or power circuit). It is advisable to refer to vehicle specific information to identify the type of control signal provided to the motor. The voltage, frequency or value of the voltage / control signal in the motor circuit is either at full value (e.g. battery voltage with no on / off control signal) or, the voltage / signal exists but it is above the normal operating range. Possible fault with motor or wiring e.g. motor power supply (relay if fitted), short / open circuit or resistance that is causing a voltage / signal value to be higher than the maximum operating limit. Note: the control unit could be providing the correct voltage / signal but it is being affected by a circuit fault. Refer to list of sensor / actuator checks on page 19.
P2104	Throttle Actuator Control System - Forced Idle	The throttle actuator is the motor that controls the throttle plate / butterfly opening angle. Note that the throttle actuator will probably form part of the throttle control assembly which can include a throttle position sensor as well as the control motor.	The fault code indicates that the control unit has implemented a "Forced Idle" control mode; although this could be related to a throttle actuator problem, it can also be caused by a fault in an unrelated engine system (or other system) that could damage the engine / vehicle (or is classified as being unsafe for normal driving). Refer to fault code P2101 for additional information and checks relating to the throttle control motor.
P2105	Throttle Actuator Control System - Forced Engine Shutdown	The throttle actuator is the motor that controls the throttle plate / butterfly opening angle. Note that the throttle actuator will probably form part of the throttle control assembly which can include a throttle position sensor as well as the control motor.	The fault code indicates that the control unit has implemented a "Forced Engine shutdown" control mode; although this could be related to a throttle actuator problem, it can also be caused by a fault in an unrelated engine system (or other system) that could damage the engine / vehicle (or is classified as being unsafe for normal driving). Refer to fault code P2101 for additional information and checks relating to the throttle control motor.
P2106	Throttle Actuator Control System - Forced Limited Power	The throttle actuator is the motor that controls the throttle plate / butterfly opening angle. Note that the throttle actuator will probably form part of the throttle control assembly which can include a throttle position sensor as well as the control motor.	The fault code indicates that the control unit has implemented a "Forced Limited power" control mode; although this could be related to a throttle actuator problem, but can also be caused due to incorrect information from other related sensors e.g. throttle position sensor, rpm sensor etc. The fault could also be caused by a fault in an unrelated engine system (or other system) that could damage the engine / vehicle (or is classified as being unsafe for normal driving). Check for other fault codes. Refer to fault code P2101 for additional information and checks relating to the throttle control motor.

NOTE 1. Check for other fault codes that could provide additional information.   **NOTE 2.** Communication between control units can pass via a CAN-Bus system; refer to Page 51 for CAN-Bus checks.   **229**

NOTE 3. If a fault cannot be located, it is also possible that the control unit is at fault.   **NOTE 4.** Refer to Page 53 for list of pages that show the ISO standard for component locations e.g. Sensor A - Bank 1.

Fault code	EOBD / ISO Description	Component / System Description	Meaningful Description and Quick Check
P2107	Throttle Actuator Control Module Processor	The throttle actuator is the motor that controls the throttle plate / butterfly opening angle. Note that the throttle actuator will probably form part of the throttle control assembly which can include a throttle position sensor as well as the control motor.	The fault code indicates a fault with the internal processor for the throttle control module (which could be combined with the engine control unit or a separate module). Check the power supplies and earth circuits to the control module and check for short / open circuits. Clear the fault code and if possible drive the vehicle or run the engine; if the fault re-appears, it could indicate a control module related fault.
P2108	Throttle Actuator Control Module Performance	The throttle actuator is the motor that controls the throttle plate / butterfly opening angle. Note that the throttle actuator will probably form part of the throttle control assembly which can include a throttle position sensor as well as the control motor.	The fault code indicates an undefined performance problem with the throttle control motor. The fault is likely to be detected due to a positioning problem with the throttle butterfly / plate e.g. the control unit is providing a specific control signal to the throttle control motor but, the response to the control signal is not as expected. The incorrect response could be that the engine speed / power does not alter as expected or, the throttle position sensor signal indicates that the throttle is not moving to the desired position. Check all wiring to the throttle control system (including throttle position sensor). Check for short / open circuits, check control signal to the motor (check live data if possible). Check for freedom of movement for throttle motor and butterfly assembly.
P2109	Throttle / Pedal Position Sensor "A" Minimum Stop Performance	Note that the throttle actuator will probably form part of the throttle control assembly, which can include a throttle position sensor as well as the control motor (in some cases a separate throttle pedal sensor can be fitted). The minimum stop relates to the position of the throttle motor (therefore the throttle butterfly / plate), which is "learnt" by the control unit for the closed throttle position. In some cases it can relate to the fixed "stop" position (maximum closure of throttle) or, it could relate to the position used for pre-starts or start up.	The fault code indicates a performance problem relating to the minimum stop position; the position (indicated by a throttle position sensor) can be "learnt" by the control unit each time the ignition is switched on. The fault is likely to be caused due to an incorrect / unexpected throttle position signal (higher or lower than the expected value) when the throttle motor is positioned by the control unit. The cause of the fault is undefined but can be caused by a throttle motor fault, a restriction to the movement of the throttle motor or butterfly, a throttle position sensor fault or, a fault with the learning process. The fault could also occur if a new throttle component has been fitted (sensor or throttle motor) or, work has been carried out on the throttle body / components. Checks will therefore need to be carried out on the sensor motor and wiring; Refer to list of sensor / actuator checks on page 19.
P2110	Throttle Actuator Control System - Forced Limited RPM	The throttle actuator is the motor that controls the throttle plate / butterfly opening angle. Note that the throttle actuator will probably form part of the throttle control assembly which can include a throttle position sensor as well as the control motor.	The fault code indicates that the control unit has implemented a "Forced Limited RPM" control mode; although this could be related to a throttle actuator problem, but can also be caused due to incorrect information from other related sensors e.g. throttle position sensor, rpm sensor etc. The fault could also be caused by a fault in an unrelated engine system (or other system) that could damage the engine / vehicle (or is classified as being unsafe for normal driving). Check for other fault codes. Refer to fault code P2101 for additional information and checks relating to the throttle control motor.
P2111	Throttle Actuator Control System - Stuck Open	The throttle actuator is the motor that controls the throttle plate / butterfly opening angle. Note that the throttle actuator will probably form part of the throttle control assembly which can include a throttle position sensor as well as the control motor.	The fault code indicates that the control unit has identified that the throttle plate is stuck in an open position, but the control unit would be passing a signal to the throttle motor to close the throttle. The control unit will receive throttle position information from the throttle position sensor, and other information could be indicating high airflow or other applicable information. Refer to fault code P2101 for additional information and checks relating to the throttle control motor.
P2112	Throttle Actuator Control System - Stuck Closed	The throttle actuator is the motor that controls the throttle plate / butterfly opening angle. Note that the throttle actuator will probably form part of the throttle control assembly which can include a throttle position sensor as well as the control motor.	The fault code indicates that the control unit has identified that the throttle plate is stuck in a closed position, but the control unit would be passing a signal to the throttle motor to open the throttle. The control unit will receive throttle position information from the throttle position sensor and other information could be indicating low airflow or other applicable information. Refer to fault code P2101 for additional information and checks relating to the throttle control motor.
P2113	Throttle / Pedal Position Sensor "B" Minimum Stop Performance	Note that the throttle actuator will probably form part of the throttle control assembly, which can include a throttle position sensor as well as the control motor (in some cases a separate throttle pedal sensor can be fitted). The minimum stop relates to the position of the throttle motor (therefore the throttle butterfly / plate), which is "learnt" by the control unit for the closed throttle position. In some cases it can relate to the fixed "stop" position (maximum closure of throttle) or, it could relate to the position used for pre-starts or start up.	The fault code indicates a performance problem relating to the minimum stop position; the position (indicated by a throttle position sensor) can be "learnt" by the control unit each time the ignition is switched on. The fault is likely to be caused due to an incorrect / unexpected throttle position signal (higher or lower than the expected value) when the throttle motor is positioned by the control unit. The cause of the fault is undefined but can be caused by a throttle motor fault, a restriction to the movement of the throttle motor or butterfly, a throttle position sensor fault or, a fault with the learning process. The fault could also occur if a new throttle component has been fitted (sensor or throttle motor) or, work has been carried out on the throttle body / components. Checks will therefore need to be carried out on the sensor motor and wiring; Refer to list of sensor / actuator checks on page 19.
P2114	Throttle / Pedal Position Sensor "C" Minimum Stop Performance	Note that the throttle actuator will probably form part of the throttle control assembly, which can include a throttle position sensor as well as the control motor (in some cases a separate throttle pedal sensor can be fitted). The minimum stop relates to the position of the throttle motor (therefore the throttle butterfly / plate), which is "learnt" by the control unit for the closed throttle position. In some cases it can relate to the fixed "stop" position (maximum closure of throttle) or, it could relate to the position used for pre-starts or start up.	The fault code indicates a performance problem relating to the minimum stop position; the position (indicated by a throttle position sensor) can be "learnt" by the control unit each time the ignition is switched on. The fault is likely to be caused due to an incorrect / unexpected throttle position signal (higher or lower than the expected value) when the throttle motor is positioned by the control unit. The cause of the fault is undefined but can be caused by a throttle motor fault, a restriction to the movement of the throttle motor or butterfly, a throttle position sensor fault or, a fault with the learning process. The fault could also occur if a new throttle component has been fitted (sensor or throttle motor) or, work has been carried out on the throttle body / components. Checks will therefore need to be carried out on the sensor motor and wiring; Refer to list of sensor / actuator checks on page 19.
P2115	Throttle / Pedal Position Sensor "D" Minimum Stop Performance	Note that the throttle actuator will probably form part of the throttle control assembly, which can include a throttle position sensor as well as the control motor (in some cases a separate throttle pedal sensor can be fitted). The minimum stop relates to the position of the throttle motor (therefore the throttle butterfly / plate), which is "learnt" by the control unit for the closed throttle position. In some cases it can relate to the fixed "stop" position (maximum closure of throttle) or, it could relate to the position used for pre-starts or start up.	The fault code indicates a performance problem relating to the minimum stop position; the position (indicated by a throttle position sensor) can be "learnt" by the control unit each time the ignition is switched on. The fault is likely to be caused due to an incorrect / unexpected throttle position signal (higher or lower than the expected value) when the throttle motor is positioned by the control unit. The cause of the fault is undefined but can be caused by a throttle motor fault, a restriction to the movement of the throttle motor or butterfly, a throttle position sensor fault or, a fault with the learning process. The fault could also occur if a new throttle component has been fitted (sensor or throttle motor) or, work has been carried out on the throttle body / components. Checks will therefore need to be carried out on the sensor motor and wiring; Refer to list of sensor / actuator checks on page 19.

NOTE 1. Check for other fault codes that could provide additional information. **NOTE 2.** Communication between control units can pass via a CAN-Bus system; refer to Page 51 for CAN-Bus checks.

NOTE 3. If a fault cannot be located, it is also possible that the control unit is at fault. NOTE 4. Refer to Page 53 for list of pages that show the ISO standard for component locations e.g. Sensor A - Bank 1.

230

code	Description		
P2116	Throttle / Pedal Position Sensor "E" Minimum Stop Performance	Note that the throttle actuator will probably form part of the throttle control assembly, which can include a throttle position sensor as well as the control motor (in some cases a separate throttle pedal sensor can be fitted). The minimum stop relates to the position of the throttle motor (therefore the throttle butterfly / plate), which is "learned" by the control unit for the closed throttle position. In some cases it can relate to the fixed "stop" position (maximum closure of throttle) or, it could relate to the position used for pre-starts or start up.	The fault code indicates a performance problem relating to the minimum stop position; the position (indicated by a throttle position sensor) can be "learned" by the control unit each time the ignition is switched on. The fault is likely to be caused due to an incorrect / unexpected throttle position signal (higher or lower than the expected value) when the throttle motor is positioned by the control unit. The cause of the fault is undefined but can be caused by a throttle motor fault, a restriction to the movement of the throttle motor or butterfly, a throttle position sensor fault or, a fault with the learning process. The fault could also occur if a new throttle component has been fitted (sensor or throttle motor) or, work has been carried out on the throttle body / components. Checks will therefore need to be carried out on the sensor motor and wiring; Refer to list of sensor / actuator checks on page 19.
P2117	Throttle / Pedal Position Sensor "F" Minimum Stop Performance	Note that the throttle actuator will probably form part of the throttle control assembly, which can include a throttle position sensor as well as the control motor (in some cases a separate throttle pedal sensor can be fitted). The minimum stop relates to the position of the throttle motor (therefore the throttle butterfly / plate), which is "learned" by the control unit for the closed throttle position. In some cases it can relate to the fixed "stop" position (maximum closure of throttle) or, it could relate to the position used for pre-starts or start up.	The fault code indicates a performance problem relating to the minimum stop position; the position (indicated by a throttle position sensor) can be "learned" by the control unit each time the ignition is switched on. The fault is likely to be caused due to an incorrect / unexpected throttle position signal (higher or lower than the expected value) when the throttle motor is positioned by the control unit. The cause of the fault is undefined but can be caused by a throttle motor fault, a restriction to the movement of the throttle motor or butterfly, a throttle position sensor fault or, a fault with the learning process. The fault could also occur if a new throttle component has been fitted (sensor or throttle motor) or, work has been carried out on the throttle body / components. Checks will therefore need to be carried out on the sensor motor and wiring; Refer to list of sensor / actuator checks on page 19.
P2118	Throttle Actuator Control Motor Current Range / Performance	The throttle actuator is the motor that controls the throttle plate / butterfly opening angle. The fault code indicates that the control unit has detected that the power supply voltage / current for the actuator motor circuit is incorrect or unstable. It is also possible that the control signal provided by the control unit is assessed as being incorrect, which could be caused by a circuit fault or other faults that influence the control motor operation.	The voltage, frequency or value of the control signal in the actuator circuit is within the normal operating range / tolerance, but the signal might be incorrect for the operating conditions. It is also possible that the actuator response (or response of system controlled by the actuator) differs from the expected / desired response (incorrect response for the control signal provided by the control unit). Possible fault in actuator or wiring, but also possible that the actuator (or mechanism / system controlled by the actuator) is not operating or moving correctly. Note: the control unit could be providing the correct signal but it is being affected by the circuit fault. Refer to list of sensor / actuator checks on page 19.
P2119	Throttle Actuator Control Throttle Body Range / Performance	The throttle actuator is the motor that controls the throttle plate / butterfly opening angle; the opening angle and movement of the throttle plate is then monitored by the throttle / accelerator position sensors. The actuator motor, the throttle plate and the position sensor are normally combined into the throttle body assembly. The fault code refers to a throttle body performance problem; this code definition covers a number of potential faults that can occur, which can include faults relating to the actuator motor, the position sensor as well as the control unit. Also, faults can relate to the ability of the system to adapt to other problems e.g. if an intake air leak occurs which might affect the idle airflow and speed thus require the control unit to learn a new control signal value to achieve the correct idle speed.	The voltage, frequency or value of the control signal in the actuator circuit is within the normal operating range / tolerance, but the signal might be incorrect for the operating conditions. It is also possible that the actuator response (or response of system controlled by the actuator) differs from the expected / desired response (incorrect response for the control signal provided by the control unit). Possible fault in actuator or wiring, but also possible that the actuator (or mechanism / system controlled by the actuator) is not operating or moving correctly. Note: the control unit could be providing the correct signal but it is being affected by the circuit fault. Refer to list of sensor / actuator checks on page 19.
P2120	Throttle / Pedal Position Sensor / Switch "D" Circuit	The throttle / pedal position sensor / switch could be located on the throttle body, throttle cable / linkage or on the throttle pedal. If a switch is fitted, it could be used to indicate "Throttle Closed", "Throttle Open" or both positions. If a position sensor is fitted, it can indicate all angles of throttle opening as well as the speed of throttle opening (rate of change). Systems can be fitted with combinations of sensors and switches, which can be separate or combined into a single unit.	The sensor assembly may contain more than one switch or sensor, refer to vehicle specific information to establish which sensor or switch is affected. The voltage, frequency or value of the pedal position sensor / switch signal is incorrect (undefined fault). Sensor voltage / signal could be: out of normal operating range, constant value, non-existent, corrupt. Possible fault in the pedal position sensor / switch or wiring e.g. short / open circuits, circuit resistance, incorrect sensor resistance or, fault with the reference / operating voltage; the switch contacts (if applicable) could also be faulty e.g. not opening / closing correctly or have a high contact resistance. Refer to list of sensor / actuator checks on page 19.
P2121	Throttle / Pedal Position Sensor / Switch "D" Circuit Range / Performance	The throttle / pedal position sensor / switch could be located on the throttle body, throttle cable / linkage or on the throttle pedal. If a switch is fitted, it could be used to indicate "Throttle Closed", "Throttle Open" or both positions. If a position sensor is fitted, it can indicate all angles of throttle opening as well as the speed of throttle opening (rate of change). Systems can be fitted with combinations of sensors and switches, which can be separate or combined into a single unit.	The sensor assembly may contain more than one switch or sensor, refer to vehicle specific information to establish which sensor or switch is affected. The voltage, frequency or value of the pedal position sensor / switch signal is within the normal operating range / tolerance but the signal is not plausible or is incorrect due to an undefined fault. The sensor / switch signal might not match the operating conditions indicated by other sensors or, the sensor / switch signal is not changing / responding as expected. Possible fault in the pedal position sensor / switch or wiring e.g. short / open circuits, circuit resistance, incorrect sensor resistance or, fault with the reference / operating voltage; also check switch contacts (if applicable) to ensure that they open / close and check continuity of switch contacts (ensure that there is not a high resistance). Refer to list of sensor / actuator checks on page 19. It is also possible that the pedal position sensor / switch is operating correctly but the position being indicated by the sensor is unacceptable or incorrect for the operating conditions indicated by other sensors or, is not as expected by the control unit; check mechanism / linkage used to operate switch / sensor, and check base settings for throttle mechanism and for the sensor / switch.
P2122	Throttle / Pedal Position Sensor / Switch "D" Circuit Low	The throttle / pedal position sensor / switch could be located on the throttle body, throttle cable / linkage or on the throttle pedal. If a switch is fitted, it could be used to indicate "Throttle Closed", "Throttle Open" or both positions. If a position sensor is fitted, it can indicate all angles of throttle opening as well as the speed of throttle opening (rate of change). Systems can be fitted with combinations of sensors and switches, which can be separate or combined into a single unit.	Note. The sensor assembly may contain more than one switch or sensor, refer to vehicle specific information to establish which sensor or switch is affected. The voltage, frequency or value of the pedal position sensor / switch signal is either "zero" (no signal) or, the signal exists but it is below the normal operating range (lower than the minimum operating limit). Possible fault in the pedal position sensor / switch or wiring e.g. short / open circuits, circuit resistance, incorrect sensor resistance or, fault with the reference / operating voltage; the switch contacts (if applicable) could also be faulty e.g. not opening / closing correctly or have a high contact resistance. Refer to list of sensor / actuator checks on page 19. Note that the fault code can be activated for systems with position switches, if the switch output remains at the low value e.g. zero volts and does not rise to the high value e.g. battery voltage (the control unit will detect no switch signal change but it has detected change in engine operating conditions).

NOTE 1. Check for other fault codes that could provide additional information.
NOTE 2. Communication between control units can pass via a CAN-Bus system; refer to Page 51 for CAN-Bus checks.
NOTE 3. If a fault cannot be located, it is also possible that the control unit is at fault.
NOTE 4. Refer to Page 53 for list of pages that show the ISO standard for component locations e.g. Sensor A - Bank 1.

Fault code	EOBD / ISO Description	Component / System Description	Meaningful Description and Quick Check
P2123	Throttle / Pedal Position Sensor / Switch "D" Circuit High	The throttle / pedal position sensor / switch could be located on the throttle body, throttle cable / linkage or on the throttle pedal. If a switch is fitted, it could be used to indicate "Throttle Closed", "Throttle Open" or both positions. If a position sensor is fitted, it can indicate all angles of throttle opening as well as the speed of throttle opening (rate of change). Systems can be fitted with combinations of sensors and switches, which can be separate or combined into a single unit.	The sensor assembly may contain more than one switch or sensor, refer to vehicle specific information to establish which sensor or switch is affected. The voltage, frequency or value of the pedal position sensor / switch signal is either at full value (e.g. battery voltage with no signal or frequency) or, the signal exists but it is above the normal operating range (higher than the maximum operating limit). Possible fault in the pedal position sensor / switch or wiring e.g. short / open circuits, circuit resistance, incorrect sensor resistance or, fault with the reference / operating voltage; the switch contacts (if applicable) could also be faulty e.g. not opening / closing correctly or have a high contact resistance. Refer to list of sensor / actuator checks on page 19. Note that the fault code can be activated for systems with position switches, if the switch output remains at the high value e.g. battery voltage and does not fall to the low value e.g. zero volts (the control unit will detect no switch signal change but it has detected change in engine operating conditions).
P2124	Throttle / Pedal Position Sensor / Switch "D" Circuit Intermittent	The throttle / pedal position sensor / switch could be located on the throttle body, throttle cable / linkage or on the throttle pedal. If a switch is fitted, it could be used to indicate "Throttle Closed", "Throttle Open" or both positions. If a position sensor is fitted, it can indicate all angles of throttle opening as well as the speed of throttle opening (rate of change). Systems can be fitted with combinations of sensors and switches, which can be separate or combined into a single unit.	The sensor assembly may contain more than one switch or sensor, refer to vehicle specific information to establish which sensor or switch is affected. The voltage, frequency or value of the pedal position sensor / switch signal is intermittent (likely to be out of normal operating range / tolerance when intermittent fault is detected) or, the signal / voltage is erratic (e.g. signal changes are irregular / unpredictable). Possible fault with wiring or connections (including internal sensor / switch connections); the switch contacts (if applicable) could also be faulty e.g. not opening / closing correctly or have a high contact resistance. Refer to list of sensor / actuator checks on page 19.
P2125	Throttle / Pedal Position Sensor / Switch "E" Circuit	The throttle / pedal position sensor / switch could be located on the throttle body, throttle cable / linkage or on the throttle pedal. If a switch is fitted, it could be used to indicate "Throttle Closed", "Throttle Open" or both positions. If a position sensor is fitted, it can indicate all angles of throttle opening as well as the speed of throttle opening (rate of change). Systems can be fitted with combinations of sensors and switches, which can be separate or combined into a single unit.	The sensor assembly may contain more than one switch or sensor, refer to vehicle specific information to establish which sensor or switch is affected. The voltage, frequency or value of the pedal position sensor / switch signal is incorrect (undefined fault). Sensor voltage / signal could be: out of normal operating range, constant value, non-existent, corrupt. Possible fault in the pedal position sensor / switch or wiring e.g. short / open circuits, circuit resistance, incorrect sensor resistance or, fault with the reference / operating voltage; the switch contacts (if applicable) could also be faulty e.g. not opening / closing correctly or have a high contact resistance. Refer to list of sensor / actuator checks on page 19.
P2126	Throttle / Pedal Position Sensor / Switch "E" Circuit Range / Performance	The throttle / pedal position sensor / switch could be located on the throttle body, throttle cable / linkage or on the throttle pedal. If a switch is fitted, it could be used to indicate "Throttle Closed", "Throttle Open" or both positions. If a position sensor is fitted, it can indicate all angles of throttle opening as well as the speed of throttle opening (rate of change). Systems can be fitted with combinations of sensors and switches, which can be separate or combined into a single unit.	The sensor assembly may contain more than one switch or sensor, refer to vehicle specific information to establish which sensor or switch is affected. The voltage, frequency or value of the pedal position sensor / switch signal is within the normal operating range / tolerance but the signal is not plausible or is incorrect due to an undefined fault. The sensor / switch signal might not match the operating conditions indicated by other sensors or, the sensor / switch signal is not changing / responding as expected. Possible fault in the pedal position sensor / switch or wiring e.g. short / open circuits, circuit resistance, incorrect sensor resistance or, fault with the reference / operating voltage; also check switch contacts (if applicable) to ensure that they open / close and check continuity of switch contacts (ensure that there is not a high resistance). Refer to list of sensor / actuator checks on page 19. It is also possible that the pedal position sensor / switch is operating correctly but the position being indicated by the sensor is unacceptable or incorrect for the operating conditions indicated by other sensors or, is not as expected by the control unit; check mechanism / linkage used to operate switch / sensor, and check base settings for throttle mechanism and for the sensor / switch.
P2127	Throttle / Pedal Position Sensor / Switch "E" Circuit Low	The throttle / pedal position sensor / switch could be located on the throttle body, throttle cable / linkage or on the throttle pedal. If a switch is fitted, it could be used to indicate "Throttle Closed", "Throttle Open" or both positions. If a position sensor is fitted, it can indicate all angles of throttle opening as well as the speed of throttle opening (rate of change). Systems can be fitted with combinations of sensors and switches, which can be separate or combined into a single unit.	Note. The sensor assembly may contain more than one switch or sensor, refer to vehicle specific information to establish which sensor or switch is affected. The voltage, frequency or value of the pedal position sensor / switch signal is either "zero" (no signal) or, the signal exists but it is below the normal operating range (lower than the minimum operating limit). Possible fault in the pedal position sensor / switch or wiring e.g. short / open circuits, circuit resistance, incorrect sensor resistance or, fault with the reference / operating voltage; the switch contacts (if applicable) could also be faulty e.g. not opening / closing correctly or have a high contact resistance. Refer to list of sensor / actuator checks on page 19. Note that the fault code can be activated for systems with position switches, if the switch output remains at the low value e.g. zero volts and does not rise to the high value e.g. battery voltage (the control unit will detect no switch signal change but it has detected change in engine operating conditions).
P2128	Throttle / Pedal Position Sensor / Switch "E" Circuit High	The throttle / pedal position sensor / switch could be located on the throttle body, throttle cable / linkage or on the throttle pedal. If a switch is fitted, it could be used to indicate "Throttle Closed", "Throttle Open" or both positions. If a position sensor is fitted, it can indicate all angles of throttle opening as well as the speed of throttle opening (rate of change). Systems can be fitted with combinations of sensors and switches, which can be separate or combined into a single unit.	The sensor assembly may contain more than one switch or sensor, refer to vehicle specific information to establish which sensor or switch is affected. The voltage, frequency or value of the pedal position sensor / switch signal is either at full value (e.g. battery voltage with no signal or frequency) or, the signal exists but it is above the normal operating range (higher than the maximum operating limit). Possible fault in the pedal position sensor / switch or wiring e.g. short / open circuits, circuit resistance, incorrect sensor resistance or, fault with the reference / operating voltage; the switch contacts (if applicable) could also be faulty e.g. not opening / closing correctly or have a high contact resistance. Refer to list of sensor / actuator checks on page 19. Note that the fault code can be activated for systems with position switches, if the switch output remains at the high value e.g. battery voltage and does not fall to the low value e.g. zero volts (the control unit will detect no switch signal change but it has detected change in engine operating conditions).
P2129	Throttle / Pedal Position Sensor / Switch "E" Circuit Intermittent	The throttle / pedal position sensor / switch could be located on the throttle body, throttle cable / linkage or on the throttle pedal. If a switch is fitted, it could be used to indicate "Throttle Closed", "Throttle Open" or both positions. If a position sensor is fitted, it can indicate all angles of throttle opening as well as the speed of throttle opening (rate of change). Systems can be fitted with combinations of sensors and switches, which can be separate or combined into a single unit.	The sensor assembly may contain more than one switch or sensor, refer to vehicle specific information to establish which sensor or switch is affected. The voltage, frequency or value of the pedal position sensor / switch signal is intermittent (likely to be out of normal operating range / tolerance when intermittent fault is detected) or, the signal / voltage is erratic (e.g. signal changes are irregular / unpredictable). Possible fault with wiring or connections (including internal sensor / switch connections); the switch contacts (if applicable) could also be faulty e.g. not opening / closing correctly or have a high contact resistance. Refer to list of sensor / actuator checks on page 19.

NOTE 1. Check for other fault codes that could provide additional information. **NOTE 2.** Communication between control units can pass via a CAN-Bus system; refer to Page 51 for CAN-Bus checks. 232

NOTE 3. If a fault cannot be located, it is also possible that the control unit is at fault. **NOTE 4.** Refer to Page 53 for list of pages that show the ISO standard for component locations e.g. Sensor A - Bank 1.

Code	Fault Location	Description	Probable Cause
P2130	Throttle / Pedal Position Sensor / Switch "F" Circuit	The throttle / pedal position sensor / switch could be located on the throttle body, throttle cable / linkage or on the throttle pedal. If a switch is fitted, it could be used to indicate "Throttle Closed", "Throttle Open" or both positions. If a position sensor is fitted, it can indicate all angles of throttle opening as well as the speed of throttle opening (rate of change). Systems can be fitted with combinations of sensors and switches, which can be separate or combined into a single unit.	The sensor assembly may contain more than one switch or sensor, refer to vehicle specific information to establish which sensor or switch is affected. The voltage, frequency or value of the pedal position sensor / switch signal is incorrect (undefined fault). Sensor voltage / signal could be: out of normal operating range, constant value, non-existent, corrupt. Possible fault in the pedal sensor / switch or wiring e.g. short / open circuits, circuit resistance, incorrect sensor resistance or, fault with the reference / operating voltage; the switch contacts (if applicable) could also be faulty e.g. not opening / closing correctly or have a high contact resistance. Refer to list of sensor / actuator checks on page 19.
P2131	Throttle / Pedal Position Sensor / Switch "F" Circuit Range / Performance	The throttle / pedal position sensor / switch could be located on the throttle body, throttle cable / linkage or on the throttle pedal. If a switch is fitted, it could be used to indicate "Throttle Closed", "Throttle Open" or both positions. If a position sensor is fitted, it can indicate all angles of throttle opening as well as the speed of throttle opening (rate of change). Systems can be fitted with combinations of sensors and switches, which can be separate or combined into a single unit.	The sensor assembly may contain more than one switch or sensor, refer to vehicle specific information to establish which sensor or switch is affected. The voltage, frequency or value of the pedal position sensor / switch signal is within the normal operating range / tolerance but the signal is not plausible or is incorrect due to an undefined fault. The sensor / switch signal might not match the operating conditions indicated by other sensors or, the sensor / switch signal is not changing / responding as expected. Possible fault in the pedal position sensor / switch or wiring e.g. short / open circuits, circuit resistance, incorrect sensor resistance or, fault with the reference / operating voltage; also check switch contacts (if applicable) to ensure that they open / close and check continuity of switch contacts (ensure that there is not a high resistance). Refer to list of sensor / actuator checks on page 19. It is also possible that the pedal position sensor / switch is operating correctly but the position being indicated by the sensor is unacceptable or incorrect for the operating conditions indicated by other sensors or, is not as expected by the control unit; check mechanism / linkage used to operate switch / sensor, and check base settings for throttle mechanism and for the sensor / switch.
P2132	Throttle / Pedal Position Sensor / Switch "F" Circuit Low	The throttle / pedal position sensor / switch could be located on the throttle body, throttle cable / linkage or on the throttle pedal. If a switch is fitted, it could be used to indicate "Throttle Closed", "Throttle Open" or both positions. If a position sensor is fitted, it can indicate all angles of throttle opening as well as the speed of throttle opening (rate of change). Systems can be fitted with combinations of sensors and switches, which can be separate or combined into a single unit.	Note. The sensor assembly may contain more than one switch or sensor, refer to vehicle specific information to establish which sensor or switch is affected. The voltage, frequency or value of the pedal position sensor / switch signal is either "zero" (no signal) or, the signal exists but it is below the normal operating range (lower than the minimum operating limit). Possible fault in the pedal position sensor / switch or wiring e.g. short / open circuits, circuit resistance, incorrect sensor resistance or, fault with the reference / operating voltage; the switch contacts (if applicable) could also be faulty e.g. not opening / closing correctly or have a high contact resistance. Refer to list of sensor / actuator checks on page 19. Note that the fault code can be activated for systems with position switches, if the switch output remains at the low value e.g. zero volts and does not rise to the high value e.g. battery voltage (the control unit will detect no switch signal change but it has detected change in engine operating conditions).
P2133	Throttle / Pedal Position Sensor / Switch "F" Circuit High	The throttle / pedal position sensor / switch could be located on the throttle body, throttle cable / linkage or on the throttle pedal. If a switch is fitted, it could be used to indicate "Throttle Closed", "Throttle Open" or both positions. If a position sensor is fitted, it can indicate all angles of throttle opening as well as the speed of throttle opening (rate of change). Systems can be fitted with combinations of sensors and switches, which can be separate or combined into a single unit.	The sensor assembly may contain more than one switch or sensor, refer to vehicle specific information to establish which sensor or switch is affected. The voltage, frequency or value of the pedal position sensor / switch signal is either at full value (e.g. battery voltage with no signal or frequency) or, the signal exists but it is above the normal operating range (higher than the maximum operating limit). Possible fault in the pedal position sensor / switch or wiring e.g. short / open circuits, circuit resistance, incorrect sensor resistance or, fault with the reference / operating voltage; the switch contacts (if applicable) could also be faulty e.g. not opening / closing correctly or have a high contact resistance. Refer to list of sensor / actuator checks on page 19. Note that the fault code can be activated for systems with position switches, if the switch output remains at the high value e.g. battery voltage and does not fall to the low value e.g. zero volts (the control unit will detect no switch signal change but it has detected change in engine operating conditions).
P2134	Throttle / Pedal Position Sensor / Switch "F" Circuit Intermittent	The throttle / pedal position sensor / switch could be located on the throttle body, throttle cable / linkage or on the throttle pedal. If a switch is fitted, it could be used to indicate "Throttle Closed", "Throttle Open" or both positions. If a position sensor is fitted, it can indicate all angles of throttle opening as well as the speed of throttle opening (rate of change). Systems can be fitted with combinations of sensors and switches, which can be separate or combined into a single unit.	The sensor assembly may contain more than one switch or sensor, refer to vehicle specific information to establish which sensor or switch is affected. The voltage, frequency or value of the pedal position sensor / switch signal is intermittent (likely to be out of normal operating range / tolerance when intermittent fault is detected) or, the signal / voltage is erratic (e.g. signal changes are irregular / unpredictable). Possible fault with wiring or connections (including internal sensor / switch connections); the switch contacts (if applicable) could also be faulty e.g. not opening / closing correctly or have a high contact resistance. Refer to list of sensor / actuator checks on page 19.
P2135	Throttle / Pedal Position Sensor / Switch "A" / "B" Voltage Correlation	A number of different throttle position sensors / switches can be used; The two sensors / switches may be combined into one assembly e.g. throttle body or pedal box, or the sensors / switches may be two separate assemblies, refer to vehicle specific information to establish which sensor or switch is affected. Different sensors / switches can also be used on the different intake systems for "V" engines.	The signals from the different sensors are providing conflicting or implausible information e.g. information from one sensor does not match the information from the other sensor. It is possible that one of the sensors / switches is operating incorrectly, therefore check each sensor for a fault with the sensor / wiring e.g. short / open circuits, circuit resistance, incorrect sensor resistance, faulty switch contacts or, reference / operating voltage fault. Refer to list of sensor / actuator checks on page 19. If the different sensors / switches are used on a "V" engine, check the base setting for the sensors, check throttle actuator operation (incorrect electronic throttle movement will affect throttle position sensor signal); also ensure that both throttles open at the same rate / time.
P2136	Throttle / Pedal Position Sensor / Switch "A" / "C" Voltage Correlation	A number of different throttle position sensors / switches can be used; The two sensors / switches may be combined into one assembly e.g. throttle body or pedal box, or the sensors / switches may be two separate assemblies, refer to vehicle specific information to establish which sensor or switch is affected. Different sensors / switches can also be used on the different intake systems for "V" engines.	The signals from the different sensors are providing conflicting or implausible information e.g. information from one sensor does not match the information from the other sensor. It is possible that one of the sensors / switches is operating incorrectly, therefore check each sensor for a fault with the sensor / wiring e.g. short / open circuits, circuit resistance, incorrect sensor resistance, faulty switch contacts or, reference / operating voltage fault. Refer to list of sensor / actuator checks on page 19. If the different sensors / switches are used on a "V" engine, check the base setting for the sensors, check throttle actuator operation (incorrect electronic throttle movement will affect throttle position sensor signal); also ensure that both throttles open at the same rate / time.
P2137	Throttle / Pedal Position Sensor / Switch "B" / "C" Voltage Correlation	A number of different throttle position sensors / switches can be used; The two sensors / switches may be combined into one assembly e.g. throttle body or pedal box, or the sensors / switches may be two separate assemblies, refer to vehicle specific information to establish which sensor or switch is affected. Different sensors / switches can also be used on the different intake systems for "V" engines.	The signals from the different sensors are providing conflicting or implausible information e.g. information from one sensor does not match the information from the other sensor. It is possible that one of the sensors / switches is operating incorrectly, therefore check each sensor for a fault with the sensor / wiring e.g. short / open circuits, circuit resistance, incorrect sensor resistance, faulty switch contacts or, reference / operating voltage fault. Refer to list of sensor / actuator checks on page 19. If the different sensors / switches are used on a "V" engine, check the base setting for the sensors, check throttle actuator operation (incorrect electronic throttle movement will affect throttle position sensor signal); also ensure that both throttles open at the same rate / time.

NOTE 1. Check for other fault codes that could provide additional information. **NOTE 2.** Communication between control units can pass via a CAN-Bus system; refer to Page 51 for CAN-Bus checks.
NOTE 3. If a fault cannot be located, it is also possible that the control unit is at fault. **NOTE 4.** Refer to Page 53 for list of pages that show the ISO standard for component locations e.g. Sensor A - Bank 1.

Fault code	EOBD / ISO Description	Component / System Description	Meaningful Description and Quick Check
P2138	Throttle / Pedal Position Sensor / Switch "D" / "E" Voltage Correlation	A number of different throttle position sensors / switches can be used; The two sensors / switches may be combined into one assembly e.g. throttle body or pedal box, or the sensors / switches may be two separate assemblies, refer to vehicle specific information to establish which sensor or switch is affected. Different sensors / switches can also be used on the different intake systems for "V" engines.	The signals from the different sensors are providing conflicting or implausible information e.g. information from one sensor does not match the information from the other sensor. It is possible that one of the sensors / switches is operating incorrectly, therefore check each sensor for a fault with the sensor / wiring e.g. short / open circuits, circuit resistance, incorrect sensor resistance, faulty switch contacts or, reference / operating voltage fault. Refer to list of sensor / actuator checks on page 19. If the different sensors / switches are used on a "V" engine, check the base setting for the sensors, check throttle actuator operation (incorrect electronic throttle movement will affect throttle position sensor signal); also ensure that both throttles open at the same rate / time.
P2139	Throttle / Pedal Position Sensor / Switch "D" / "F" Voltage Correlation	A number of different throttle position sensors / switches can be used; The two sensors / switches may be combined into one assembly e.g. throttle body or pedal box, or the sensors / switches may be two separate assemblies, refer to vehicle specific information to establish which sensor or switch is affected. Different sensors / switches can also be used on the different intake systems for "V" engines.	The signals from the different sensors are providing conflicting or implausible information e.g. information from one sensor does not match the information from the other sensor. It is possible that one of the sensors / switches is operating incorrectly, therefore check each sensor for a fault with the sensor / wiring e.g. short / open circuits, circuit resistance, incorrect sensor resistance, faulty switch contacts or, reference / operating voltage fault. Refer to list of sensor / actuator checks on page 19. If the different sensors / switches are used on a "V" engine, check the base setting for the sensors, check throttle actuator operation (incorrect electronic throttle movement will affect throttle position sensor signal); also ensure that both throttles open at the same rate / time.
P213A	Exhaust Gas Recirculation Throttle Control Circuit "B" / Open	The EGR throttle can be fitted to Diesel engines and to some petrol engines (usually Direct injection engines where under certain driving conditions, the normal throttle is held fully open). The EGR throttle can be located in the intake system and, will be closed off when the EGR valve is opened; partially closing the EGR throttle reduces intake air volume (air volume is replaced by EGR gases) and it also creates a negative pressure / vacuum in the intake to assist in drawing in the EGR gases. The EGR throttle can be controlled by an electric actuator (usually a motor).	The voltage, frequency or value of the control signal in the actuator circuit is incorrect. Signal could be:- out of normal operating range, constant value, non-existent, corrupt. The fault code indicates a possible "open circuit" but other undefined faults in the actuator or wiring could activate the same code e.g. actuator power supply, short circuit, high circuit resistance or incorrect actuator resistance. Note: the control unit could be providing the correct signal but it is being affected by the circuit fault. Refer to list of sensor / actuator checks on page 19.
P213B	Exhaust Gas Recirculation Throttle Control Circuit "B" Range / Performance	The EGR throttle can be fitted to Diesel engines and to some petrol engines (usually Direct injection engines where under certain driving conditions, the normal throttle is held fully open). The EGR throttle can be located in the intake system and, will be closed off when the EGR valve is opened; partially closing the EGR throttle reduces intake air volume (air volume is replaced by EGR gases) and it also creates a negative pressure / vacuum in the intake to assist in drawing in the EGR gases. The EGR throttle can be controlled by an electric actuator (usually a motor).	The voltage, frequency or value of the control signal in the actuator circuit is within the normal operating range / tolerance, but the signal might be incorrect for the operating conditions. It is also possible that the actuator response (or response of system controlled by the actuator) differs from the expected / desired response (incorrect response for the control signal provided by the control unit). Possible fault in actuator or wiring, but also possible that the actuator (or mechanism / system controlled by the actuator) is not operating or moving correctly. Note: the control unit could be providing the correct signal but it is being affected by the circuit fault. Refer to list of sensor / actuator checks on page 19.
P213C	Exhaust Gas Recirculation Throttle Control Circuit "B" Low	The EGR throttle can be fitted to Diesel engines and to some petrol engines (usually Direct injection engines where under certain driving conditions, the normal throttle is held fully open). The EGR throttle can be located in the intake system and, will be closed off when the EGR valve is opened; partially closing the EGR throttle reduces intake air volume (air volume is replaced by EGR gases) and it also creates a negative pressure / vacuum in the intake to assist in drawing in the EGR gases. The EGR throttle can be controlled by an electric actuator (usually a motor).	The voltage, frequency or value of the control signal in the actuator circuit is either "zero" (no signal) or, the signal exists but is below the normal operating range. Possible fault with actuator or wiring (e.g. actuator power supply, short / open circuit or resistance) that is causing a signal value to be lower than the minimum operating limit. Note: the control unit could be providing the correct signal but it is being affected by the circuit fault. Refer to list of sensor / actuator checks on page 19.
P213D	Exhaust Gas Recirculation Throttle Control Circuit "B" High	The EGR throttle can be fitted to Diesel engines and to some petrol engines (usually Direct injection engines where under certain driving conditions, the normal throttle is held fully open). The EGR throttle can be located in the intake system and, will be closed off when the EGR valve is opened; partially closing the EGR throttle reduces intake air volume (air volume is replaced by EGR gases) and it also creates a negative pressure / vacuum in the intake to assist in drawing in the EGR gases. The EGR throttle can be controlled by an electric actuator (usually a motor).	The voltage, frequency or value of the control signal in the actuator circuit is either at full value (e.g. battery voltage with no on / off signal) or, the signal exists but it is above the normal operating range. Possible fault with actuator or wiring (e.g. actuator power supply, short / open circuit or resistance) that is causing a signal value to be higher than the maximum operating limit. Note: the control unit could be providing the correct signal but it is being affected by the circuit fault. Refer to list of sensor / actuator checks on page 19.
P2140	Throttle / Pedal Position Sensor / Switch "E" / "F" Voltage Correlation	A number of different throttle position sensors / switches can be used; The two sensors / switches may be combined into one assembly e.g. throttle body or pedal box, or the sensors / switches may be two separate assemblies, refer to vehicle specific information to establish which sensor or switch is affected. Different sensors / switches can also be used on the different intake systems for "V" engines.	The signals from the different sensors are providing conflicting or implausible information e.g. information from one sensor does not match the information from the other sensor. It is possible that one of the sensors / switches is operating incorrectly, therefore check each sensor for a fault with the sensor / wiring e.g. short / open circuits, circuit resistance, incorrect sensor resistance, faulty switch contacts or, reference / operating voltage fault. Refer to list of sensor / actuator checks on page 19. If the different sensors / switches are used on a "V" engine, check the base setting for the sensors, check throttle actuator operation (incorrect electronic throttle movement will affect throttle position sensor signal); also ensure that both throttles open at the same rate / time.
P2141	Exhaust Gas Recirculation Throttle Control Circuit "A" Low	The EGR throttle can be fitted to Diesel engines and to some petrol engines (usually Direct injection engines where under certain driving conditions, the normal throttle is held fully open). The EGR throttle can be located in the intake system and, will be closed off when the EGR valve is opened; partially closing the EGR throttle reduces intake air volume (air volume is replaced by EGR gases) and it also creates a negative pressure / vacuum in the intake to assist in drawing in the EGR gases. The EGR throttle can be controlled by an electric actuator (usually a motor).	The voltage, frequency or value of the control signal in the actuator circuit is either "zero" (no signal) or, the signal exists but is below the normal operating range. Possible fault with actuator or wiring (e.g. actuator power supply, short / open circuit or resistance) that is causing a signal value to be lower than the minimum operating limit. Note: the control unit could be providing the correct signal but it is being affected by the circuit fault. Refer to list of sensor / actuator checks on page 19.

NOTE 1. Check for other fault codes that could provide additional information. **NOTE 2.** Communication between control units can pass via a CAN-Bus system; refer to Page 51 for CAN-Bus checks. 234

NOTE 3. If a fault code is located, it is also possible that the control unit is at fault. NOTE 4. Refer to Page 53 for list of pages that show the ISO standard for component locations e.g. Sensor A - Bank 1.

Code	Description		
P2142	Exhaust Gas Recirculation Throttle Control Circuit "A" High	The EGR throttle can be fitted to Diesel engines and to some petrol engines (usually Direct injection engines where under certain driving conditions, the normal throttle is held fully open). The EGR throttle can be located in the intake system and, will be closed off when the EGR valve is opened; partially closing the EGR throttle reduces intake air volume (air volume is replaced by EGR gases) and it also creates a negative pressure / vacuum in the intake to assist in drawing in the EGR gases. The EGR throttle can be controlled by an electric actuator (usually a motor).	The voltage, frequency or value of the control signal in the actuator circuit is either at full value (e.g. battery voltage with no on / off signal) or, the signal exists but it is above the normal operating range. Possible fault with actuator or wiring (e.g. actuator power supply, short / open circuit or resistance) that is causing a signal value to be higher than the maximum operating limit. Note: the control unit could be providing the correct signal but it is being affected by the circuit fault. Refer to list of sensor / actuator checks on page 19.
P2143	Exhaust Gas Recirculation Vent Control Circuit / Open	On EGR systems where vacuum is applied to an actuator to open the EGR valve, the vacuum being applied can be regulated by a valve (usually a solenoid valve) and, the vacuum can then be vented to allow the EGR valve to close; the vent valve will be operated by an electric actuator (usually a solenoid). Note that on some systems, the vent valve and the vacuum control valve (controlling vacuum input to the EGR) can be a combined unit (e.g. double action solenoid valve).	The voltage, frequency or value of the control signal in the actuator circuit is incorrect. Signal could be:- out of normal operating range, constant value, non-existent, corrupt. The fault code indicates a possible "open circuit" but other undefined faults in the actuator or wiring could activate the same code e.g. actuator power supply, short circuit, high circuit resistance or incorrect actuator resistance. Note: the control unit could be providing the correct signal but it is being affected by the circuit fault. Refer to list of sensor / actuator checks on page 19.
P2144	Exhaust Gas Recirculation Vent Control Circuit Low	On EGR systems where vacuum is applied to an actuator to open the EGR valve, the vacuum being applied can be regulated by a valve (usually a solenoid valve) and, the vacuum can then be vented to allow the EGR valve to close; the vent valve will be operated by an electric actuator (usually a solenoid). Note that on some systems, the vent valve and the vacuum control valve (controlling vacuum input to the EGR) can be a combined unit (e.g. double action solenoid valve).	The voltage, frequency or value of the control signal in the actuator circuit is either "zero" (no signal) or, the signal exists but is below the normal operating range. Possible fault with actuator or wiring (e.g. actuator power supply, short / open circuit or resistance) that is causing a signal value to be lower than the minimum operating limit. Note: the control unit could be providing the correct signal but it is being affected by the circuit fault. Refer to list of sensor / actuator checks on page 19.
P2145	Exhaust Gas Recirculation Vent Control Circuit High	On EGR systems where vacuum is applied to an actuator to open the EGR valve, the vacuum being applied can be regulated by a valve (usually a solenoid valve) and, the vacuum can then be vented to allow the EGR valve to close; the vent valve will be operated by an electric actuator (usually a solenoid). Note that on some systems, the vent valve and the vacuum control valve (controlling vacuum input to the EGR) can be a combined unit (e.g. double action solenoid valve).	The voltage, frequency or value of the control signal in the actuator circuit is either at full value (e.g. battery voltage with no on / off signal) or, the signal exists but it is above the normal operating range. Possible fault with actuator or wiring (e.g. actuator power supply, short / open circuit or resistance) that is causing a signal value to be higher than the maximum operating limit. Note: the control unit could be providing the correct signal but it is being affected by a circuit fault. Refer to list of sensor / actuator checks on page 19.
P2146	Fuel Injector Group "A" Supply Voltage Circuit / Open	Fuel injectors can receive supply voltage via a relay or on some systems, the voltage could be provided via a an injector control module (separate or combined with the main control unit). Injectors can be grouped with regard to the voltage supply. The control unit can detect the supply voltage fault either because it is directly providing the supply or, there is a connection from the relay / injector control module to enable monitoring of the supply voltage. Refer to vehicle specific information to identify the specified injector group and to identify the wiring / source of the power supply (relay, control module etc).	The fault code is indicating a fault with the power supply to the injector group; the fault could be an open circuit in the wiring or is undefined. Check voltage at injector (engine cranking or running); if the voltage is incorrect or non existent, check wiring back to power source (relay, control module etc) and check for short / open circuit or high resistance. If there is no power supply to the injectors and the power source is a relay, it may be necessary to check relay operation (input voltage, closing of relay contacts etc), also check for a short circuit that may have caused a fuse to blow. If the power source is from a control module or control unit, it may be necessary to carry out module checks. Note that on some systems, the control unit is able to detect the supply voltage fault because the control unit is switching the earth path for the injectors (the control unit can detect the voltage when the earth path is switched off); on these systems, it will also be necessary to check for short / open circuits in the injector earth path (between injector and control unit).
P2147	Fuel Injector Group "A" Supply Voltage Circuit Low	Fuel injectors can receive supply voltage via a relay or on some systems, the voltage could be provided via a an injector control module (separate or combined with the main control unit). Injectors can be grouped with regard to the voltage supply. The control unit can detect the supply voltage fault either because it is directly providing the supply or, there is a connection from the relay / injector control module to enable monitoring of the supply voltage.	The fault code is indicating a fault with the power supply to the injector group is low; voltage could be non-existent or below the normal operating range. Check voltage at injector (engine cranking or running); if the voltage is incorrect or non existent, check wiring back to power source (relay, control module etc) and check for short / open circuit or high resistance. If there is no power supply to the injectors and the power source is a relay, it may be necessary to check relay operation (input voltage, closing of relay contacts etc) also check for a short circuit that may have caused a fuse to blow. If the power source is from a control module or control unit, it may be necessary to carry out module checks. Note that on some systems, the control unit is able to detect the supply voltage fault because the control unit is switching the earth path for the injectors (the control unit can detect the voltage when the earth path is switched off); on these systems, it will also be necessary to check for short / open circuits in the injector earth path (between injector and control unit).
P2148	Fuel Injector Group "A" Supply Voltage Circuit High	Fuel injectors can receive supply voltage via a relay or on some systems, the voltage could be provided via a an injector control module (separate or combined with the main control unit). Injectors can be grouped with regard to the voltage supply. The control unit can detect the supply voltage fault either because it is directly providing the supply or, there is a connection from the relay / injector control module to enable monitoring of the supply voltage.	The fault code is indicating that the power supply to the injector group is high; voltage could be above the normal operating range. Note that a high voltage could be caused by an overcharging generator system, check charging system and rectify as necessary. It is also possible that the fault has been activated because voltage exists at the injectors when the circuit should be switched off (e.g. ignition switched off or engine not running). Check for short circuits in the injector wiring to other power supplies. Check relay operation (if applicable) to ensure relay contacts open correctly. If the power source is from a control module or control unit, it may be necessary to carry out module checks. Note that on some systems, the control unit is able to detect the supply voltage fault because the control unit is switching the earth path for the injectors (the control unit can detect the voltage when the earth path is switched off); on these systems, it will also be necessary to check for short circuits in the injector earth path to a power supply.

NOTE 1. Check for other fault codes that could provide additional information. NOTE 2. Communication between control units can pass via a CAN-Bus system; refer to Page 51 for CAN-Bus checks.

NOTE 3. If a fault cannot be located, it is also possible that the control unit is at fault. NOTE 4. Refer to Page 53 for list of pages that show the ISO standard for component locations e.g. Sensor A - Bank 1.

Fault code	EOBD / ISO Description	Component / System Description	Meaningful Description and Quick Check
P2149	Fuel Injector Group "B" Supply Voltage Circuit / Open	Fuel injectors can receive supply voltage via a relay or on some systems, the voltage could be provided via a an injector control module (separate or combined with the main control unit). Injectors can be grouped with regard to the voltage supply. The control unit can detect the supply voltage fault either because it is directly providing the supply or, there is a connection from the relay / injector control module to enable monitoring of the supply voltage.	The fault code is indicating a fault with the power supply to the injector group; the fault could be an open circuit in the wiring or is undefined. Check voltage at injector (engine cranking or running); if the voltage is incorrect or non existent, check wiring back to power source (relay, control module etc) and check for short / open circuit or high resistance. If there is no power supply to the injectors and the power source is a relay, it may be necessary to check relay operation (input voltage, closing of relay contacts etc), also check for a short circuit that may have caused a fuse to blow. If the power source is from a control module or control unit, it may be necessary to carry out module checks. Note that on some systems, the control unit is able to detect the supply voltage fault because the control unit is switching the earth path for the injectors (the control unit can detect the voltage when the earth path is switched off); on these systems, it will also be necessary to check for short / open circuits in the injector earth path (between injector and control unit).
P2150	Fuel Injector Group "B" Supply Voltage Circuit Low	Fuel injectors can receive supply voltage via a relay or on some systems, the voltage could be provided via a an injector control module (separate or combined with the main control unit). Injectors can be grouped with regard to the voltage supply. The control unit can detect the supply voltage fault either because it is directly providing the supply or, there is a connection from the relay / injector control module to enable monitoring of the supply voltage.	The fault code is indicating a fault with the power supply to the injector group is low; voltage could be non-existent or below the normal operating range. Check voltage at injector (engine cranking or running); if the voltage is incorrect or non existent, check wiring back to power source (relay, control module etc) and check for short / open circuit or high resistance. If there is no power supply to the injectors and the power source is a relay, it may be necessary to check relay operation (input voltage, closing of relay contacts etc) also check for a short circuit that may have caused a fuse to blow. If the power source is from a control module or control unit, it may be necessary to carry out module checks. Note that on some systems, the control unit is able to detect the supply voltage fault because the control unit is switching the earth path for the injectors (the control unit can detect the voltage when the earth path is switched off); on these systems, it will also be necessary to check for short / open circuits in the injector earth path (between injector and control unit).
P2151	Fuel Injector Group "B" Supply Voltage Circuit High	Fuel injectors can receive supply voltage via a relay or on some systems, the voltage could be provided via a an injector control module (separate or combined with the main control unit). Injectors can be grouped with regard to the voltage supply. The control unit can detect the supply voltage fault either because it is directly providing the supply or, there is a connection from the relay / injector control module to enable monitoring of the supply voltage.	The fault code is indicating that the power supply to the injector group is high; voltage could be above the normal operating range. Note that a high voltage could be caused by an overcharging generator system, check charging system and rectify as necessary. It is also possible that the fault has been activated because voltage exists at the injectors when the circuit should be switched off (e.g. ignition switched off or engine not running). Check for short circuits in the injector wiring to other power supplies. Check relay operation (if applicable) to ensure relay contacts open correctly. If the power source is from a control module or control unit, it may be necessary to carry out module checks. Note that on some systems, the control unit is able to detect the supply voltage fault because the control unit is switching the earth path for the injectors (the control unit can detect the voltage when the earth path is switched off); on these systems, it will also be necessary to check for short circuits in the injector earth path to a power supply.
P2152	Fuel Injector Group "C" Supply Voltage Circuit / Open	Fuel injectors can receive supply voltage via a relay or on some systems, the voltage could be provided via a an injector control module (separate or combined with the main control unit). Injectors can be grouped with regard to the voltage supply. The control unit can detect the supply voltage fault either because it is directly providing the supply or, there is a connection from the relay / injector control module to enable monitoring of the supply voltage.	The fault code is indicating a fault with the power supply to the injector group; the fault could be an open circuit in the wiring or is undefined. Check voltage at injector (engine cranking or running); if the voltage is incorrect or non existent, check wiring back to power source (relay, control module etc) and check for short / open circuit or high resistance. If there is no power supply to the injectors and the power source is a relay, it may be necessary to check relay operation (input voltage, closing of relay contacts etc), also check for a short circuit that may have caused a fuse to blow. If the power source is from a control module or control unit, it may be necessary to carry out module checks. Note that on some systems, the control unit is able to detect the supply voltage fault because the control unit is switching the earth path for the injectors (the control unit can detect the voltage when the earth path is switched off); on these systems, it will also be necessary to check for short / open circuits in the injector earth path (between injector and control unit).
P2153	Fuel Injector Group "C" Supply Voltage Circuit Low	Fuel injectors can receive supply voltage via a relay or on some systems, the voltage could be provided via a an injector control module (separate or combined with the main control unit). Injectors can be grouped with regard to the voltage supply. The control unit can detect the supply voltage fault either because it is directly providing the supply or, there is a connection from the relay / injector control module to enable monitoring of the supply voltage.	The fault code is indicating a fault with the power supply to the injector group is low; voltage could be non-existent or below the normal operating range. Check voltage at injector (engine cranking or running); if the voltage is incorrect or non existent, check wiring back to power source (relay, control module etc) and check for short / open circuit or high resistance. If there is no power supply to the injectors and the power source is a relay, it may be necessary to check relay operation (input voltage, closing of relay contacts etc) also check for a short circuit that may have caused a fuse to blow. If the power source is from a control module or control unit, it may be necessary to carry out module checks. Note that on some systems, the control unit is able to detect the supply voltage fault because the control unit is switching the earth path for the injectors (the control unit can detect the voltage when the earth path is switched off); on these systems, it will also be necessary to check for short / open circuits in the injector earth path (between injector and control unit).
P2154	Fuel Injector Group "C" Supply Voltage Circuit High	Fuel injectors can receive supply voltage via a relay or on some systems, the voltage could be provided via a an injector control module (separate or combined with the main control unit). Injectors can be grouped with regard to the voltage supply. The control unit can detect the supply voltage fault either because it is directly providing the supply or, there is a connection from the relay / injector control module to enable monitoring of the supply voltage.	The fault code is indicating that the power supply to the injector group is high; voltage could be above the normal operating range. Note that a high voltage could be caused by an overcharging generator system, check charging system and rectify as necessary. It is also possible that the fault has been activated because voltage exists at the injectors when the circuit should be switched off (e.g. ignition switched off or engine not running). Check for short circuits in the injector wiring to other power supplies. Check relay operation (if applicable) to ensure relay contacts open correctly. If the power source is from a control module or control unit, it may be necessary to carry out module checks. Note that on some systems, the control unit is able to detect the supply voltage fault because the control unit is switching the earth path for the injectors (the control unit can detect the voltage when the earth path is switched off); on these systems, it will also be necessary to check for short circuits in the injector earth path to a power supply.

NOTE 1. Check for other fault codes that could provide additional information. **NOTE 2.** Communication between control units can pass via a CAN-Bus system; refer to Page 51 for CAN-Bus checks.

236

NOTE 4. Refer to Page 53 for list of pages that show the ISO standard for component locations e.g. Sensor A - Bank 1.

Code	Fault	Description	Diagnosis
P2155	Fuel Injector Group "D" Supply Voltage Circuit / Open	Fuel injectors can receive supply voltage via a relay or on some systems, the voltage could be provided via a an injector control module (separate or combined with the main control unit). Injectors can be grouped with regard to the voltage supply. The control unit can detect the supply voltage fault either because it is directly providing the supply or, there is a connection from the relay / injector control module to enable monitoring of the supply voltage.	The fault code is indicating a fault with the power supply to the injector group; the fault could be an open circuit in the wiring or is undefined. Check voltage at injector (engine cranking or running); if the voltage is incorrect or non existent, check wiring back to power source (relay, control module etc) and check for short / open circuit or high resistance. If there is no power supply to the injectors and the power source is a relay, it may be necessary to check relay operation (input voltage, closing of relay contacts etc), also check for a short circuit that may have caused a fuse to blow. If the power source is from a control module or control unit, it may be necessary to carry out module checks. Note that on some systems, the control unit is able to detect the supply voltage fault because the control unit is switching the earth path for the injectors (the control unit can detect the voltage when the earth path is switched off); on these systems, it will also be necessary to check for short / open circuits in the injector earth path (between injector and control unit).
P2156	Fuel Injector Group "D" Supply Voltage Circuit Low	Fuel injectors can receive supply voltage via a relay or on some systems, the voltage could be provided via a an injector control module (separate or combined with the main control unit). Injectors can be grouped with regard to the voltage supply. The control unit can detect the supply voltage fault either because it is directly providing the supply or, there is a connection from the relay / injector control module to enable monitoring of the supply voltage.	The fault code is indicating a fault with the power supply to the injector group is low; voltage could be non-existent or below the normal operating range. Check voltage at injector (engine cranking or running); if the voltage is incorrect or non existent, check wiring back to power source (relay, control module etc) and check for short / open circuit or high resistance. If there is no power supply to the injectors and the power source is a relay, it may be necessary to check relay operation (input voltage, closing of relay contacts etc) also check for a short circuit that may have caused a fuse to blow. If the power source is from a control module or control unit, it may be necessary to carry out module checks. Note that on some systems, the control unit is able to detect the supply voltage fault because the control unit is switching the earth path for the injectors (the control unit can detect the voltage when the earth path is switched off); on these systems, it will also be necessary to check for short / open circuits in the injector earth path (between injector and control unit).
P2157	Fuel Injector Group "D" Supply Voltage Circuit High	Fuel injectors can receive supply voltage via a relay or on some systems, the voltage could be provided via a an injector control module (separate or combined with the main control unit). Injectors can be grouped with regard to the voltage supply. The control unit can detect the supply voltage fault either because it is directly providing the supply or, there is a connection from the relay / injector control module to enable monitoring of the supply voltage.	The fault code is indicating that the power supply to the injector group is high; voltage could be above the normal operating range. Note that a high voltage could be caused by an overcharging generator system, check charging system and rectify as necessary. It is also possible that the fault has been activated because voltage exists at the injectors when the circuit should be switched off (e.g. ignition switched off or engine not running). Check for short circuits in the injector wiring to other power supplies. Check relay operation (if applicable) to ensure relay contacts open correctly. If the power source is from a control module or control unit, it may be necessary to carry out module checks. Note that on some systems, the control unit is able to detect the supply voltage fault because the control unit is switching the earth path for the injectors (the control unit can detect the voltage when the earth path is switched off); on these systems, it will also be necessary to check for short circuits in the injector earth path to a power supply.
P2158	Vehicle Speed Sensor "B"	Note that the vehicle speed sensor could be an independent device (transmission or drive train located) or, the vehicle speed could be indicated by a wheel speed sensor (ABS); refer to vehicle specific information if necessary. Identify the type of sensor e.g. Inductive, Hall effect etc; this will dictate the checks that are applicable.	The voltage, frequency or value of the speed sensor signal is incorrect (undefined fault). Sensor voltage / signal value could be: out of normal operating range, constant value, non-existent, corrupt. Possible fault in the speed sensor or wiring e.g. short / open circuits, circuit resistance, incorrect sensor resistance (if applicable) or, fault with the reference / operating voltage; interference from other circuits can also affect sensor signals Also check earth connections and cable screening. Refer to list of sensor / actuator checks on page 19. Note, because the fault is undefined, it is also possible that the speed sensor is operating correctly but the speed being measured by the sensor is unacceptable or incorrect for the operating conditions; if possible, use alternative method of checking speed to establish if sensor system is operating correctly and whether speed is correct / incorrect. Also check condition of reference teeth on trigger disc / rotor and ensure rotor is turning with shaft.
P2159	Vehicle Speed Sensor "B" Range / Performance	Note that the vehicle speed sensor could be an independent device (transmission or drive train located) or, the vehicle speed could be indicated by a wheel speed sensor (ABS); refer to vehicle specific information if necessary. Identify the type of sensor e.g. Inductive, Hall effect etc; this will dictate the checks that are applicable.	The voltage, frequency or value of the speed sensor signal is within the normal operating range / tolerance but the signal is not plausible or is incorrect due to an undefined fault. The sensor signal might not match the operating conditions indicated by other sensors or, the sensor signal is not changing / responding as expected to the changes in conditions. Possible fault in the speed sensor or wiring e.g. short / open circuits, circuit resistance, incorrect sensor resistance or, fault with the reference / operating voltage; interference from other circuits can also affect sensor signals. Refer to list of sensor / actuator checks on page 19. It is also possible that the speed sensor is operating correctly but the speed being measured by the sensor is unacceptable or incorrect for the operating conditions; if possible, use alternative method of checking speed to establish if sensor system is operating correctly and whether speed is correct / incorrect. Also check condition of reference teeth on trigger disc / rotor and ensure rotor is turning with shaft.
P215A	Vehicle Speed - Wheel Speed Correlation	Note that the vehicle speed sensor could be an independent device (transmission or drive train located) or, the vehicle speed could be indicated by a wheel speed sensor (ABS); refer to vehicle specific information if necessary. Identify the type of sensor e.g. Inductive, Hall effect etc; this will dictate the checks that are applicable.	Note that if one wheel speed sensor is used for vehicle speed, the conflict could be with another wheel speed sensor. The signals from the two applicable sensors are providing conflicting or implausible information e.g. the two sensors are indicating different vehicle speeds. It is possible that one of the sensors is operating incorrectly, therefore check each sensor for a fault with the sensor / wiring e.g. short / open circuits, circuit resistance, incorrect sensor resistance or reference / operating voltage fault. Refer to list of sensor / actuator checks on page 19.
P215B	Vehicle Speed - Output Shaft Speed Correlation	Note that the vehicle speed sensor could be an independent device (transmission or drive train located) or, the vehicle speed could be indicated by a wheel speed sensor (ABS); refer to vehicle specific information if necessary. Identify the type of sensor e.g. Inductive, Hall effect etc; this will dictate the checks that are applicable.	The signals from the two applicable sensors are providing conflicting or implausible information e.g. the two sensors are indicating different vehicle speeds. It is possible that one of the sensors is operating incorrectly, therefore check each sensor for a fault with the sensor / wiring e.g. short / open circuits, circuit resistance, incorrect sensor resistance or reference / operating voltage fault. Refer to list of sensor / actuator checks on page 19.
P2160	Vehicle Speed Sensor "B" Circuit Low	Note that the vehicle speed sensor could be an independent device (transmission or drive train located) or, the vehicle speed could be indicated by a wheel speed sensor (ABS); refer to vehicle specific information if necessary. Identify the type of sensor e.g. Inductive, Hall effect etc; this will dictate the checks that are applicable.	The voltage, frequency or value of the speed sensor signal is either "zero" (no signal) or, the signal exists but it is below the normal operating range (lower than the minimum operating limit). Possible fault in the speed sensor or wiring e.g. short / open circuits, circuit resistance, incorrect sensor resistance or, fault with the reference / operating voltage; interference from other circuits can also affect sensor signals. Refer to list of sensor / actuator checks on page 19.

NOTE 1. Check for other fault codes that could provide additional information.

NOTE 2. Communication between control units can pass via a CAN-Bus system; refer to Page 51 for CAN-Bus checks.

NOTE 3. If a fault cannot be located, it is also possible that the control unit is at fault.

NOTE 4. Refer to Page 53 for list of pages that show the ISO standard for component locations e.g. Sensor A - Bank 1.

Fault code	EOBD / ISO Description	Component / System Description	Meaningful Description and Quick Check
P2161	Vehicle Speed Sensor "B" Intermittent / Erratic	Note that the vehicle speed sensor could be an independent device (transmission or drive train located) or, the vehicle speed could be indicated by a wheel speed sensor (ABS); refer to vehicle specific information if necessary. Identify the type of sensor e.g. Inductive, Hall effect etc; this will dictate the checks that are applicable.	The voltage, frequency or value of the speed sensor signal is intermittent (likely to be out of normal operating range / tolerance when intermittent fault is detected) or, the signal / voltage is erratic (e.g. signal changes are irregular / unpredictable). Possible fault with wiring or connections (including internal sensor connections); also interference from other circuits can affect the signal. Refer to list of sensor / actuator checks on page 19. Also check condition of reference teeth on trigger disc / rotor and ensure rotor is turning with shaft.
P2162	Vehicle Speed Sensor "A" / "B" Correlation	Note that the vehicle speed sensor could be an independent device (transmission or drive train located) or, the vehicle speed could be indicated by a wheel speed sensor (ABS); refer to vehicle specific information if necessary. Identify the type of sensor e.g. Inductive, Hall effect etc; this will dictate the checks that are applicable.	Identify location and type of sensors A and B. The signals from the two applicable sensors are providing conflicting or implausible information e.g. the two sensors are indicating different vehicle speeds. It is possible that one of the sensors is operating incorrectly, therefore check each sensor for a fault with the sensor / wiring e.g. short / open circuits, circuit resistance, incorrect sensor resistance or reference / operating voltage fault. Refer to list of sensor / actuator checks on page 19.
P2163	Throttle / Pedal Position Sensor "A" Maximum Stop Performance	Note that the throttle actuator will probably form part of the throttle control assembly, which can include one or more throttle position sensors as well as the throttle motor (in some cases a separate throttle pedal sensor can be fitted). The maximum stop relates to the position of the throttle motor (therefore the throttle butterfly / plate), which is "learned" by the control unit for the open or maximum throttle position.	The fault code indicates a performance problem relating to the maximum stop position; the position (indicated by a throttle position sensor) can be "learned" by the control unit. The fault is likely to be caused due to an incorrect / unexpected throttle position signal (higher or lower than the expected value) when the throttle motor is positioned by the control unit. The cause of the fault is undefined but can be caused by a throttle motor fault, a restriction to the movement of the throttle motor or butterfly, a throttle position sensor fault or, a fault with the learning process. The fault could also occur if a new throttle component has been fitted (sensor or throttle motor) or, work has been carried out on the throttle body / components. Checks will therefore need to be carried out on the sensor motor and wiring; Refer to list of sensor / actuator checks on page 19.
P2164	Throttle / Pedal Position Sensor "B" Maximum Stop Performance	Note that the throttle actuator will probably form part of the throttle control assembly, which can include one or more throttle position sensors as well as the throttle motor (in some cases a separate throttle pedal sensor can be fitted). The maximum stop relates to the position of the throttle motor (therefore the throttle butterfly / plate), which is "learned" by the control unit for the open or maximum throttle position.	The fault code indicates a performance problem relating to the maximum stop position; the position (indicated by a throttle position sensor) can be "learned" by the control unit. The fault is likely to be caused due to an incorrect / unexpected throttle position signal (higher or lower than the expected value) when the throttle motor is positioned by the control unit. The cause of the fault is undefined but can be caused by a throttle motor fault, a restriction to the movement of the throttle motor or butterfly, a throttle position sensor fault or, a fault with the learning process. The fault could also occur if a new throttle component has been fitted (sensor or throttle motor) or, work has been carried out on the throttle body / components. Checks will therefore need to be carried out on the sensor motor and wiring; Refer to list of sensor / actuator checks on page 19.
P2165	Throttle / Pedal Position Sensor "C" Maximum Stop Performance	Note that the throttle actuator will probably form part of the throttle control assembly, which can include one or more throttle position sensors as well as the throttle motor (in some cases a separate throttle pedal sensor can be fitted). The maximum stop relates to the position of the throttle motor (therefore the throttle butterfly / plate), which is "learned" by the control unit for the open or maximum throttle position.	The fault code indicates a performance problem relating to the maximum stop position; the position (indicated by a throttle position sensor) can be "learned" by the control unit. The fault is likely to be caused due to an incorrect / unexpected throttle position signal (higher or lower than the expected value) when the throttle motor is positioned by the control unit. The cause of the fault is undefined but can be caused by a throttle motor fault, a restriction to the movement of the throttle motor or butterfly, a throttle position sensor fault or, a fault with the learning process. The fault could also occur if a new throttle component has been fitted (sensor or throttle motor) or, work has been carried out on the throttle body / components. Checks will therefore need to be carried out on the sensor motor and wiring; Refer to list of sensor / actuator checks on page 19.
P2166	Throttle / Pedal Position Sensor "D" Maximum Stop Performance	Note that the throttle actuator will probably form part of the throttle control assembly, which can include one or more throttle position sensors as well as the throttle motor (in some cases a separate throttle pedal sensor can be fitted). The maximum stop relates to the position of the throttle motor (therefore the throttle butterfly / plate), which is "learned" by the control unit for the open or maximum throttle position.	The fault code indicates a performance problem relating to the maximum stop position; the position (indicated by a throttle position sensor) can be "learned" by the control unit. The fault is likely to be caused due to an incorrect / unexpected throttle position signal (higher or lower than the expected value) when the throttle motor is positioned by the control unit. The cause of the fault is undefined but can be caused by a throttle motor fault, a restriction to the movement of the throttle motor or butterfly, a throttle position sensor fault or, a fault with the learning process. The fault could also occur if a new throttle component has been fitted (sensor or throttle motor) or, work has been carried out on the throttle body / components. Checks will therefore need to be carried out on the sensor motor and wiring; Refer to list of sensor / actuator checks on page 19.
P2167	Throttle / Pedal Position Sensor "E" Maximum Stop Performance	Note that the throttle actuator will probably form part of the throttle control assembly, which can include one or more throttle position sensors as well as the throttle motor (in some cases a separate throttle pedal sensor can be fitted). The maximum stop relates to the position of the throttle motor (therefore the throttle butterfly / plate), which is "learned" by the control unit for the open or maximum throttle position.	The fault code indicates a performance problem relating to the maximum stop position; the position (indicated by a throttle position sensor) can be "learned" by the control unit. The fault is likely to be caused due to an incorrect / unexpected throttle position signal (higher or lower than the expected value) when the throttle motor is positioned by the control unit. The cause of the fault is undefined but can be caused by a throttle motor fault, a restriction to the movement of the throttle motor or butterfly, a throttle position sensor fault or, a fault with the learning process. The fault could also occur if a new throttle component has been fitted (sensor or throttle motor) or, work has been carried out on the throttle body / components. Checks will therefore need to be carried out on the sensor motor and wiring; Refer to list of sensor / actuator checks on page 19.
P2168	Throttle / Pedal Position Sensor "F" Maximum Stop Performance	Note that the throttle actuator will probably form part of the throttle control assembly, which can include one or more throttle position sensors as well as the throttle motor (in some cases a separate throttle pedal sensor can be fitted). The maximum stop relates to the position of the throttle motor (therefore the throttle butterfly / plate), which is "learned" by the control unit for the open or maximum throttle position.	The fault code indicates a performance problem relating to the maximum stop position; the position (indicated by a throttle position sensor) can be "learned" by the control unit. The fault is likely to be caused due to an incorrect / unexpected throttle position signal (higher or lower than the expected value) when the throttle motor is positioned by the control unit. The cause of the fault is undefined but can be caused by a throttle motor fault, a restriction to the movement of the throttle motor or butterfly, a throttle position sensor fault or, a fault with the learning process. The fault could also occur if a new throttle component has been fitted (sensor or throttle motor) or, work has been carried out on the throttle body / components. Checks will therefore need to be carried out on the sensor motor and wiring; Refer to list of sensor / actuator checks on page 19.
P2169	Exhaust Pressure Regulator Vent Solenoid Control Circuit / Open	Refer to fault code P0476 for additional information on exhaust pressure control systems, which may form part of the same or similar functions. The exhaust pressure regulator can be a valve / flap in the exhaust system which is opened / closed by a vacuum operated mechanism. The vacuum will be applied to operate the mechanism in one direction e.g. to close the valve, and the vacuum will then be vented to allow the mechanism to open the valve. The vacuum control and venting process can be controlled using a solenoid operated valve.	The voltage, frequency or value of the control signal in the solenoid circuit is incorrect. Signal could be:- out of normal operating range, constant value, non-existent, corrupt. The fault code indicates a possible "open circuit" but other undefined faults in the solenoid or wiring could activate the same code e.g. solenoid power supply, short circuit, high circuit resistance or incorrect solenoid resistance. Note: the control unit could be providing the correct signal but it is being affected by the circuit fault. Refer to list of sensor / actuator checks on page 19.

NOTE 1. Check for other fault codes that could provide additional information. **NOTE 2.** Communication between control units can pass via a CAN-Bus system; refer to Page 51 for CAN-Bus checks.

NOTE 3. If a fault cannot be located, it is also possible that the control unit is at fault. **NOTE 4.** Refer to Page 53 for list of pages that show the ISO standard for component locations e.g. Sensor A - Bank 1.

P2170	Exhaust Pressure Regulator Vent Solenoid Control Circuit Low	Refer to fault code P0476 for additional information on exhaust pressure control systems, which may form part of the same or similar functions. The exhaust pressure regulator can be a valve / flap in the exhaust system which is opened / closed by a vacuum operated mechanism. The vacuum will be applied to operate the mechanism in one direction e.g. to close the valve, and the vacuum will then be vented to allow the mechanism to open the valve. The vacuum control and venting process can be controlled using a solenoid operated valve.	The voltage, frequency or value of the control signal in the solenoid circuit is either "zero" (no signal) or, the signal exists but is below the normal operating range. Possible fault with solenoid or wiring (e.g. solenoid power supply, short / open circuit or resistance) that is causing a signal value to be lower than the minimum operating limit. Note: the control unit could be providing the correct signal but it is being affected by a circuit fault. Refer to list of sensor / actuator checks on page 19.
P2171	Exhaust Pressure Regulator Vent Solenoid Control Circuit High	Refer to fault code P0476 for additional information on exhaust pressure control systems, which may form part of the same or similar functions. The exhaust pressure regulator can be a valve / flap in the exhaust system which is opened / closed by a vacuum operated mechanism. The vacuum will be applied to operate the mechanism in one direction e.g. to close the valve, and the vacuum will then be vented to allow the mechanism to open the valve. The vacuum control and venting process can be controlled using a solenoid operated valve.	The voltage, frequency or value of the control signal in the solenoid circuit is either at full value (e.g. battery voltage with no on / off signal) or, the signal exists but it is above the normal operating range. Possible fault with solenoid or wiring (e.g. solenoid power supply, short / open circuit or resistance) that is causing a signal value to be higher than the maximum operating limit. Note: the control unit could be providing the correct signal but it is being affected by a circuit fault. Refer to list of sensor / actuator checks on page 19.
P2172	Throttle Actuator Control System - Sudden High Airflow Detected	The fault code is probably activated due to conflicting information from the throttle position sensor signal and the MAP (Manifold Absolute Pressure) sensor signal e.g. throttle position could correspond to a "low load" condition (throttle closed) but the MAP sensor corresponds to a "high load" condition. The control unit therefore calculates that an air leak exists in the intake system (assuming that the control unit has not detected any other intake component faults e.g. MAP sensor or throttle position sensor).	The fault code indicates that the throttle position sensor and the MAP or airflow sensors are not providing consistent information relating to the engine load conditions. A leak in the intake system can result in a MAP sensor signal that is not plausible or is inconsistent with information from other sensors. Check for intake system related fault codes e.g. MAP sensor or throttle position sensor; if necessary Refer to list of sensor / actuator checks on page 19. If there are no other faults identified then it is likely that there is an air leak in the intake system.
P2173	Throttle Actuator Control System -High Airflow Detected	The fault code is probably activated due to conflicting information from the throttle position sensor signal and the MAP (Manifold Absolute Pressure) sensor signal e.g. throttle position could correspond to a "low load" condition (throttle closed) but the MAP sensor corresponds to a "high load" condition. The control unit therefore calculates that an air leak exists in the intake system (assuming that the control unit has not detected any other intake component faults e.g. MAP sensor or throttle position sensor).	The fault code indicates that the throttle position sensor and the MAP or airflow sensors are not providing consistent information relating to the engine load conditions. A leak in the intake system can result in a MAP sensor signal that is not plausible or is inconsistent with information from other sensors. Check for intake system related fault codes e.g. MAP sensor or throttle position sensor; if necessary Refer to list of sensor / actuator checks on page 19. If there are no other faults identified then it is likely that there is an air leak in the intake system.
P2174	Throttle Actuator Control System - Sudden Low Airflow Detected	The fault code is probably activated due to conflicting information from the throttle position sensor signal and the MAP (Manifold Absolute Pressure) sensor signal e.g. throttle position could correspond to a "high load" condition (throttle open) but the MAP sensor corresponds to a "low load" condition. The control unit therefore calculates that there could be an air flow restriction in the intake system (assuming that the control unit has not detected any other intake component faults e.g. MAP sensor or throttle position sensor).	The fault code indicates that the throttle position sensor and the MAP or airflow sensors are not providing consistent information relating to the engine load conditions. A blockage or restriction in the intake system can result in a MAP sensor signal that is not plausible or is inconsistent with information from other sensors. Check for intake system related fault codes e.g. MAP sensor or throttle position sensor; if necessary Refer to list of sensor / actuator checks on page 19. If there are no other faults identified then it is possible that there is an intake system restriction or an exhaust system / catalytic converter blockage.
P2175	Throttle Actuator Control System -Low Airflow Detected	The fault code is probably activated due to conflicting information from the throttle position sensor signal and the MAP (Manifold Absolute Pressure) sensor signal e.g. throttle position could correspond to a "high load" condition (throttle open) but the MAP sensor corresponds to a "low load" condition. The control unit therefore calculates that there could be an air flow restriction in the intake system (assuming that the control unit has not detected any other intake component faults e.g. MAP sensor or throttle position sensor).	The fault code indicates that the throttle position sensor and the MAP or airflow sensors are not providing consistent information relating to the engine load conditions. A blockage or restriction in the intake system can result in a MAP sensor signal that is not plausible or is inconsistent with information from other sensors. Check for intake system related fault codes e.g. MAP sensor or throttle position sensor; if necessary Refer to list of sensor / actuator checks on page 19. If there are no other faults identified then it is possible that there is an intake system restriction or an exhaust system / catalytic converter blockage.
P2176	Throttle Actuator Control System -Idle Position Not Learned	Note that the throttle actuator will probably form part of the throttle control assembly, which can include throttle position sensors as well as the throttle motor. The idle position for the throttle motor (therefore the throttle butterfly / plate), is "learnt" by the control unit, which monitors the throttle position sensor signal (often when the ignition is switched on and the throttle motor is operated to achieve idle position).	The fault code indicates that the control unit is unable to learn the idle position for the throttle butterfly / plate. The fault can be caused by an incorrect / inconsistent or unexpected throttle position signal (higher or lower than the expected value) when the throttle motor is positioned at idle by the control unit. The fault can be caused by a throttle motor fault, a restriction to the movement of the throttle motor or butterfly, a throttle position sensor fault or, a fault with the learning process. The fault could also occur if a new throttle component has been fitted (sensor or throttle motor) or, work has been carried out on the throttle body / components. Checks will therefore need to be carried out on the sensor, motor and wiring; Refer to list of sensor / actuator checks on page 19.
P2177	System Too Lean Off Idle Bank 1	Note: Many O2 / Lambda sensor related fault codes are activated due to faults in other systems (engine, fuelling, misfires etc). The control unit is able to detect the ratio of the air / fuel mixture by monitoring the information from different sensors. The O2 (Oxygen / Lambda) sensors (primarily the pre-cat sensor) will indicate the oxygen content in the exhaust gas, which is an indicator of air / fuel ratio. The control unit will also monitor other sensors to identify the operating conditions (e.g. heavy or light load etc) and (using the O2 sensor signal), assess whether the air / fuel ratio is correct. Note that faulty engine operation e.g. misfires and exhaust air leaks can cause increased oxygen in the exhaust gas which will affect the oxygen sensor signal.	For a detailed list of checks and identification of the sensor location, Refer to list of sensor / actuator checks on page 19. Different types of O2 / Lambda sensors (e.g. Zirconia or Titania / wideband etc) provide different signals and values, it will be necessary to identify the type of sensor fitted to the vehicle. The fault code indicates a lean mixture exists when the engine is running "off" idle speed (this could refer to the transition from idle to throttle just opening). The fault could be caused by a genuine lean mixture but, the fault code could be activated due to an Oxygen / Lambda sensor fault or an exhaust air leak (can cause high oxygen content in the exhaust gas). Check sensor signal under different operating conditions (idle, load, etc) with particular reference to when throttle is opened from the idle position. If the sensor appears to be operating correctly but the signal value is indicating a genuine lean mixture, this could indicate an engine related fault e.g. incorrect mixtures, fuel pressure, misfires etc; also check operation and base setting of any idle or throttle position switches / sensors, and check operation of load sensors (MAP, MAF as applicable).

NOTE 1. Check for other fault codes that could provide additional information.　**NOTE 2.** Communication between control units can pass via a CAN-Bus system; refer to Page 51 for CAN-Bus checks.
NOTE 3. If a fault cannot be located, it is also possible that the control unit is at fault.　**NOTE 4.** Refer to Page 53 for list of pages that show the ISO standard for component locations e.g. Sensor A - Bank 1.

Fault code	EOBD / ISO Description	Component / System Description	Meaningful Description and Quick Check
P2178	System Too Rich Off Idle Bank 1	Note: Many O2 / Lambda sensor related fault codes are activated due to faults in other systems (engine, fuelling, misfires etc). The control unit is able to detect the ratio of the air / fuel mixture by monitoring the information from different sensors. The O2 (Oxygen / Lambda) sensors (primarily the pre-cat sensor) will indicate the oxygen content in the exhaust gas, which is an indicator of air / fuel ratio. The control unit will also monitor other sensors to identify the operating conditions (e.g. heavy or light load etc) and (using the O2 sensor signal), assess whether the air / fuel ratio is correct. Note that faulty engine operation e.g. misfires and exhaust air leaks can cause increased oxygen in the exhaust gas which will affect the oxygen sensor signal.	For a detailed list of checks and identification of the sensor location, Refer to list of sensor / actuator checks on page 19. Different types of O2 / Lambda sensors (e.g. Zirconia or Titania / wideband etc) provide different signals and values, it will be necessary to identify the type of sensor fitted to the vehicle. The fault code indicates a rich mixture exists when the engine is running "off" idle speed (this could refer to the transition from idle to throttle just opening). The fault is likely to be caused by a genuine rich mixture but, the fault code could be activated due to an Oxygen / Lambda sensor fault. Check sensor signal under different operating conditions (idle, load, etc) with particular reference to when the throttle is opened from the idle position. If the sensor appears to be operating correctly but the signal value is indicating a genuine rich mixture, this could indicate an engine related fault e.g. incorrect mixtures, fuel pressure, leaking injectors etc; also check operation and base setting of any idle or throttle position switches / sensors, and check operation of load sensors (MAP, MAF as applicable).
P2179	System Too Lean Off Idle Bank 2	Note: Many O2 / Lambda sensor related fault codes are activated due to faults in other systems (engine, fuelling, misfires etc). The control unit is able to detect the ratio of the air / fuel mixture by monitoring the information from different sensors. The O2 (Oxygen / Lambda) sensors (primarily the pre-cat sensor) will indicate the oxygen content in the exhaust gas, which is an indicator of air / fuel ratio. The control unit will also monitor other sensors to identify the operating conditions (e.g. heavy or light load etc) and (using the O2 sensor signal), assess whether the air / fuel ratio is correct. Note that faulty engine operation e.g. misfires and exhaust air leaks can cause increased oxygen in the exhaust gas which will affect the oxygen sensor signal.	For a detailed list of checks and identification of the sensor location, Refer to list of sensor / actuator checks on page 19. Different types of O2 / Lambda sensors (e.g. Zirconia or Titania / wideband etc) provide different signals and values, it will be necessary to identify the type of sensor fitted to the vehicle. The fault code indicates a lean mixture exists when the engine is running "off" idle speed (this could refer to the transition from idle to throttle just opening). The fault could be caused by a genuine lean mixture but, the fault code could be activated due to an Oxygen / Lambda sensor fault or an exhaust air leak (can cause high oxygen content in the exhaust gas). Check sensor signal under different operating conditions (idle, load, etc) with particular reference to when the throttle is opened from the idle position. If the sensor appears to be operating correctly but the signal value is indicating a genuine lean mixture, this could indicate an engine related fault e.g. incorrect mixtures, fuel pressure, misfires etc; also check operation and base setting of any idle or throttle position switches / sensors, and check operation of load sensors (MAP, MAF as applicable).
P2180	System Too Rich Off Idle Bank 2	Note: Many O2 / Lambda sensor related fault codes are activated due to faults in other systems (engine, fuelling, misfires etc). The control unit is able to detect the ratio of the air / fuel mixture by monitoring the information from different sensors. The O2 (Oxygen / Lambda) sensors (primarily the pre-cat sensor) will indicate the oxygen content in the exhaust gas, which is an indicator of air / fuel ratio. The control unit will also monitor other sensors to identify the operating conditions (e.g. heavy or light load etc) and (using the O2 sensor signal), assess whether the air / fuel ratio is correct. Note that faulty engine operation e.g. misfires and exhaust air leaks can cause increased oxygen in the exhaust gas which will affect the oxygen sensor signal.	For a detailed list of checks and identification of the sensor location, Refer to list of sensor / actuator checks on page 19. Different types of O2 / Lambda sensors (e.g. Zirconia or Titania / wideband etc) provide different signals and values, it will be necessary to identify the type of sensor fitted to the vehicle. The fault code indicates a rich mixture exists when the engine is running "off" idle speed (this could refer to the transition from idle to throttle just opening). The fault is likely to be caused by a genuine rich mixture but, the fault code could be activated due to an Oxygen / Lambda sensor fault. Check sensor signal under different operating conditions (idle, load, etc) with particular reference to when the throttle is opened from the idle position. If the sensor appears to be operating correctly but the signal value is indicating a genuine rich mixture, this could indicate an engine related fault e.g. incorrect mixtures, fuel pressure, leaking injectors etc; also check operation and base setting of any idle or throttle position switches / sensors, and check operation of load sensors (MAP, MAF as applicable).
P2181	Cooling System Performance	The operation of the cooing system will be monitored by temperature sensor(s). The control unit will expect responses to the coolant temperature e.g. expected temperature change within a certain time after a cold start. Also, if the control unit operates cooling fans, thermostat heater or electric water pump, the control unit will expect a specific temperature change response.	The fault code indicates that the cooling system (temperature) is not responding as expected to the operating conditions or, the temperature change response is not as expected e.g. the control unit could expect a change in temperature when the cooling fans, electric water pump or electric thermostat are operated. Carry out all routine cooling system checks (coolant level, thermostat operation, pressure cap etc). Also identify any parts of the cooling system that are controlled by a control unit e.g. cooling fans, electric water pump etc; check operation of these components or systems. Also check operation of temperature sensors (note that where more than one temperature sensor is used, differing signals could cause the fault code to be activated).
P2182	Engine Coolant Temperature Sensor 2 Circuit	The coolant temperature sensor signal is used for control of fuel, ignition and emissions control. The sensor information can be used for cooling fan control as well as for other vehicle systems. If there is more than one sensor, it can provide independent measurement for banks on a "V" engine or, it could provide information to different systems e.g. engine management or cooling fan control.	The voltage, frequency or value of the temperature sensor signal is incorrect (undefined fault). Sensor voltage / signal could be: out of normal operating range, constant value, non-existent, corrupt. Possible fault in the temperature sensor or wiring e.g. short / open circuits, circuit resistance, incorrect sensor resistance or, fault with the reference / operating voltage; interference from other circuits can also affect sensor signals. Refer to list of sensor / actuator checks on page 19.
P2183	Engine Coolant Temperature Sensor 2 Circuit Range / Performance	The coolant temperature sensor signal is used for control of fuel, ignition and emissions control. The sensor information can be used for cooling fan control as well as for other vehicle systems. If there is more than one sensor, it can provide independent measurement for banks on a "V" engine or, it could provide information to different systems e.g. engine management or cooling fan control.	The voltage, frequency or value of the temperature sensor signal is within the normal operating range / tolerance but the signal is not plausible or is incorrect due to an undefined fault. The sensor signal might not match the operating conditions indicated by other sensors or, the sensor signal is not changing / responding as expected to the changes in conditions. Possible fault in the temperature sensor or wiring e.g. short / open circuits, circuit resistance, incorrect sensor resistance or, fault with the reference / operating voltage; interference from other circuits can also affect sensor signals. Refer to list of sensor / actuator checks on page 19. It is also possible that the temperature sensor is operating correctly but the temperature being measured by the sensor is unacceptable or incorrect for the operating conditions e.g. cooling system fault causing unacceptable coolant temperatures; if possible, use alternative method of checking temperature to establish if sensor system is operating correctly and whether temperature is correct / incorrect..
P2184	Engine Coolant Temperature Sensor 2 Circuit Low	The coolant temperature sensor signal is used for control of fuel, ignition and emissions control. The sensor information can be used for cooling fan control as well as for other vehicle systems. If there is more than one sensor, it can provide independent measurement for banks on a "V" engine or, it could provide information to different systems e.g. engine management or cooling fan control.	The voltage, frequency or value of the temperature sensor signal is either "zero" (no signal) or, the signal exists but it is below the normal operating range (lower than the minimum operating limit). Possible fault in the temperature sensor or wiring e.g. short / open circuits, circuit resistance, incorrect sensor resistance or, fault with the reference / operating voltage; interference from other circuits can also affect sensor signals. Refer to list of sensor / actuator checks on page 19.

NOTE 1. Check for other fault codes that could provide additional information.   **NOTE 2.** Communication between control units can pass via a CAN-Bus system; refer to Page 51 for CAN-Bus checks.   240

NOTE 4. Refer to Page 53 for list of pages that show the ISO standard for component locations e.g. Sensor A - Bank 1.

Code	Fault Location	Description	Diagnosis / Possible Causes
P2185	Engine Coolant Temperature Sensor 2 Circuit High	The coolant temperature sensor signal is used for control of fuel, ignition and emissions control. The sensor information can be used for cooling fan control as well as for other vehicle systems. If there is more than one sensor, it can provide independent measurement for banks on a "V" engine or, it could provide information to different systems e.g. engine management or cooling fan control.	The voltage, frequency or value of the temperature sensor signal is either at full value (e.g. battery voltage with no signal or frequency) or, the signal exists but it is above the normal operating range (higher than the maximum operating limit). Possible fault in the temperature sensor or wiring e.g. short / open circuits, circuit resistance, incorrect sensor resistance or, fault with the reference / operating voltage; interference from other circuits can also affect sensor signals. Refer to list of sensor / actuator checks on page 19.
P2186	Engine Coolant Temperature Sensor 2 Circuit Intermittent / Erratic	The coolant temperature sensor signal is used for control of fuel, ignition and emissions control. The sensor information can be used for cooling fan control as well as for other vehicle systems. If there is more than one sensor, it can provide independent measurement for banks on a "V" engine or, it could provide information to different systems e.g. engine management or cooling fan control.	The voltage, frequency or value of the temperature sensor signal is intermittent (likely to be out of normal operating range / tolerance when intermittent fault is detected) or, the signal / voltage is erratic (e.g. signal changes are irregular / unpredictable). Possible fault with wiring or connections (including internal sensor connections); also interference from other circuits can affect the signal. Check the sensor signal, using "live data" or other test equipment and wiggle wiring / connections to try and recreate the fault. Refer to list of sensor / actuator checks on page 19.
P2187	System Too Lean at Idle Bank 1	Note: Many O2 / Lambda sensor related fault codes are activated due to faults in other systems (engine, fuelling, misfires etc). The control unit is able to detect the ratio of the air / fuel mixture by monitoring the information from different sensors. The O2 (Oxygen / Lambda) sensors (primarily the pre-cat sensor) will indicate the oxygen content in the exhaust gas, which is an indicator of air / fuel ratio. The control unit will also monitor other sensors to identify the operating conditions (e.g. heavy or light load etc) and (using the O2 sensor signal), assess whether the air / fuel ratio is correct. Note that faulty engine operation e.g. misfires and exhaust air leaks can cause increased oxygen in the exhaust gas which will affect the oxygen sensor signal.	Refer to list of sensor / actuator checks on page 19. Different types of O2 / Lambda sensors (e.g. Zirconia or Titania / wideband etc) provide different signals and values, it will be necessary to identify the type of sensor fitted to the vehicle. The fault code indicates a lean mixture exists at idle speed. The fault could be caused by a genuine lean mixture or, the fault code could be activated due to an Oxygen / Lambda sensor fault; also note that exhaust air leaks can cause a high oxygen content to exist in the exhaust gas, which will cause a "lean mixture" sensor signal. Check sensor signal and signal response under different operating conditions (idle, load, etc). If the sensor appears to be operating correctly but the signal value is indicating a genuine lean mixture at idle, this could indicate an engine related fault e.g. incorrect mixtures, misfires, fuel pressure, air leaks or other engine / engine system problems; also check operation and base setting of any idle or throttle position switches / sensors.
P2188	System Too Rich at Idle Bank 1	Note: Many O2 / Lambda sensor related fault codes are activated due to faults in other systems (engine, fuelling, misfires etc). The control unit is able to detect the ratio of the air / fuel mixture by monitoring the information from different sensors. The O2 (Oxygen / Lambda) sensors (primarily the pre-cat sensor) will indicate the oxygen content in the exhaust gas, which is an indicator of air / fuel ratio. The control unit will also monitor other sensors to identify the operating conditions (e.g. heavy or light load etc) and (using the O2 sensor signal), assess whether the air / fuel ratio is correct. Note that faulty engine operation e.g. misfires and exhaust air leaks can cause increased oxygen in the exhaust gas which will affect the oxygen sensor signal.	Refer to list of sensor / actuator checks on page 19. Different types of O2 / Lambda sensors (e.g. Zirconia or Titania / wideband etc) provide different signals and values, it will be necessary to identify the type of sensor fitted to the vehicle. The fault code indicates a rich mixture exists at idle speed. The fault could be caused by a genuine rich mixture or, the fault code could be activated due to an Oxygen / Lambda sensor fault. Check sensor signal and signal response under different operating conditions (idle, load, etc). If the sensor appears to be operating correctly but the signal value is indicating a genuine rich mixture at idle, this could indicate an engine related fault e.g. incorrect mixtures, fuel pressure, leaking injectors or other engine / engine system problems; also check operation and base setting of any idle or throttle position switches / sensors.
P2189	System Too Lean at Idle Bank 2	Note: Many O2 / Lambda sensor related fault codes are activated due to faults in other systems (engine, fuelling, misfires etc). The control unit is able to detect the ratio of the air / fuel mixture by monitoring the information from different sensors. The O2 (Oxygen / Lambda) sensors (primarily the pre-cat sensor) will indicate the oxygen content in the exhaust gas, which is an indicator of air / fuel ratio. The control unit will also monitor other sensors to identify the operating conditions (e.g. heavy or light load etc) and (using the O2 sensor signal), assess whether the air / fuel ratio is correct. Note that faulty engine operation e.g. misfires and exhaust air leaks can cause increased oxygen in the exhaust gas which will affect the oxygen sensor signal.	Refer to list of sensor / actuator checks on page 19. Different types of O2 / Lambda sensors (e.g. Zirconia or Titania / wideband etc) provide different signals and values, it will be necessary to identify the type of sensor fitted to the vehicle. The fault code indicates a lean mixture exists at idle speed. The fault could be caused by a genuine lean mixture or, the fault code could be activated due to an Oxygen / Lambda sensor fault; also note that exhaust air leaks can cause a high oxygen content to exist in the exhaust gas, which will cause a "lean mixture" sensor signal. Check sensor signal and signal response under different operating conditions (idle, load, etc). If the sensor appears to be operating correctly but the signal value is indicating a genuine lean mixture at idle, this could indicate an engine related fault e.g. incorrect mixtures, misfires, fuel pressure, air leaks or other engine / engine system problems; also check operation and base setting of any idle or throttle position switches / sensors.
P2190	System Too Rich at Idle Bank 2	Note: Many O2 / Lambda sensor related fault codes are activated due to faults in other systems (engine, fuelling, misfires etc). The control unit is able to detect the ratio of the air / fuel mixture by monitoring the information from different sensors. The O2 (Oxygen / Lambda) sensors (primarily the pre-cat sensor) will indicate the oxygen content in the exhaust gas, which is an indicator of air / fuel ratio. The control unit will also monitor other sensors to identify the operating conditions (e.g. heavy or light load etc) and (using the O2 sensor signal), assess whether the air / fuel ratio is correct. Note that faulty engine operation e.g. misfires and exhaust air leaks can cause increased oxygen in the exhaust gas which will affect the oxygen sensor signal.	Refer to list of sensor / actuator checks on page 19. Different types of O2 / Lambda sensors (e.g. Zirconia or Titania / wideband etc) provide different signals and values, it will be necessary to identify the type of sensor fitted to the vehicle. The fault code indicates a rich mixture exists at idle speed. The fault could be caused by a genuine rich mixture or, the fault code could be activated due to an Oxygen / Lambda sensor fault. Check sensor signal and signal response under different operating conditions (idle, load, etc). If the sensor appears to be operating correctly but the signal value is indicating a genuine rich mixture at idle, this could indicate an engine related fault e.g. incorrect mixtures, fuel pressure, leaking injectors or other engine / engine system problems; also check operation and base setting of any idle or throttle position switches / sensors.

NOTE 1. Check for other fault codes that could provide additional information.   **NOTE 2.** Communication between control units can pass via a CAN-Bus system; refer to Page 51 for CAN-Bus checks.

NOTE 3. If a fault cannot be located, it is also possible that the control unit is at fault.   **NOTE 4.** Refer to Page 53 for list of pages that show the ISO standard for component locations e.g. Sensor A - Bank 1.

Fault code	EOBD / ISO Description	Component / System Description	Meaningful Description and Quick Check
P2191	System Too Lean at Higher Load Bank 1	Note: Many O2 / Lambda sensor related fault codes are activated due to faults in other systems (engine, fuelling, misfires etc). The control unit is able to detect the ratio of the air / fuel mixture by monitoring the information from different sensors. The O2 (Oxygen / Lambda) sensors (primarily the pre-cat sensor) will indicate the oxygen content in the exhaust gas, which is an indicator of air / fuel ratio. The control unit will also monitor other sensors to identify the operating conditions (e.g. heavy or light load etc) and (using the O2 sensor signal), assess whether the air / fuel ratio is correct. Note that faulty engine operation e.g. misfires and exhaust air leaks can cause increased oxygen in the exhaust gas which will affect the oxygen sensor signal.	Refer to list of sensor / actuator checks on page 19. Different types of O2 / Lambda sensors (e.g. Zirconia or Titania / wideband etc) provide different signals and values, it will be necessary to identify the type of sensor fitted to the vehicle. The fault code indicates a lean mixture exists at higher load conditions. The fault could be caused by a genuine lean mixture but, the fault code could be activated due to an Oxygen / Lambda sensor fault or an exhaust air leak (can cause high oxygen content in the exhaust gas). Check sensor signal and signal response under different operating conditions (idle, load, etc) with particular reference to the signal under higher load conditions. If the sensor appears to be operating correctly but the signal value is indicating a genuine lean mixture, this could indicate an engine related fault e.g. incorrect mixtures, fuel pressure, misfires etc, but check operation of load sensors e.g. throttle position, MAP, MAF (as applicable).
P2192	System Too Rich at Higher Load Bank 1	Note: Many O2 / Lambda sensor related fault codes are activated due to faults in other systems (engine, fuelling, misfires etc). The control unit is able to detect the ratio of the air / fuel mixture by monitoring the information from different sensors. The O2 (Oxygen / Lambda) sensors (primarily the pre-cat sensor) will indicate the oxygen content in the exhaust gas, which is an indicator of air / fuel ratio. The control unit will also monitor other sensors to identify the operating conditions (e.g. heavy or light load etc) and (using the O2 sensor signal), assess whether the air / fuel ratio is correct. Note that faulty engine operation e.g. misfires and exhaust air leaks can cause increased oxygen in the exhaust gas which will affect the oxygen sensor signal.	Refer to list of sensor / actuator checks on page 19. Different types of O2 / Lambda sensors (e.g. Zirconia or Titania / wideband etc) provide different signals and values, it will be necessary to identify the type of sensor fitted to the vehicle. The fault code indicates a rich mixture exists at higher load conditions. The fault could be caused by a genuine rich mixture but, the fault code could be activated due to an Oxygen / Lambda sensor fault or an exhaust air leak. Check sensor signal and signal response under different operating conditions (idle, load, etc) with particular reference to the signal under high load conditions. If the sensor appears to be operating correctly but the signal value is indicating a genuine rich mixture, this could indicate an engine related fault e.g. incorrect mixtures, fuel pressure, leaking injectors etc, but check operation of load sensors e.g. throttle position, MAP, MAF (as applicable).
P2193	System Too Lean at Higher Load Bank 2	Note: Many O2 / Lambda sensor related fault codes are activated due to faults in other systems (engine, fuelling, misfires etc). The control unit is able to detect the ratio of the air / fuel mixture by monitoring the information from different sensors. The O2 (Oxygen / Lambda) sensors (primarily the pre-cat sensor) will indicate the oxygen content in the exhaust gas, which is an indicator of air / fuel ratio. The control unit will also monitor other sensors to identify the operating conditions (e.g. heavy or light load etc) and (using the O2 sensor signal), assess whether the air / fuel ratio is correct. Note that faulty engine operation e.g. misfires and exhaust air leaks can cause increased oxygen in the exhaust gas which will affect the oxygen sensor signal.	Refer to list of sensor / actuator checks on page 19. Different types of O2 / Lambda sensors (e.g. Zirconia or Titania / wideband etc) provide different signals and values, it will be necessary to identify the type of sensor fitted to the vehicle. The fault code indicates a lean mixture exists at higher load conditions. The fault could be caused by a genuine lean mixture but, the fault code could be activated due to an Oxygen / Lambda sensor fault or an exhaust air leak (can cause high oxygen content in the exhaust gas). Check sensor signal and signal response under different operating conditions (idle, load, etc) with particular reference to the signal under higher load conditions. If the sensor appears to be operating correctly but the signal value is indicating a genuine lean mixture, this could indicate an engine related fault e.g. incorrect mixtures, fuel pressure, misfires etc, but check operation of load sensors e.g. throttle position, MAP, MAF (as applicable).
P2194	System Too Rich at Higher Load Bank 2	Note: Many O2 / Lambda sensor related fault codes are activated due to faults in other systems (engine, fuelling, misfires etc). The control unit is able to detect the ratio of the air / fuel mixture by monitoring the information from different sensors. The O2 (Oxygen / Lambda) sensors (primarily the pre-cat sensor) will indicate the oxygen content in the exhaust gas, which is an indicator of air / fuel ratio. The control unit will also monitor other sensors to identify the operating conditions (e.g. heavy or light load etc) and (using the O2 sensor signal), assess whether the air / fuel ratio is correct. Note that faulty engine operation e.g. misfires and exhaust air leaks can cause increased oxygen in the exhaust gas which will affect the oxygen sensor signal.	Refer to list of sensor / actuator checks on page 19. Different types of O2 / Lambda sensors (e.g. Zirconia or Titania / wideband etc) provide different signals and values, it will be necessary to identify the type of sensor fitted to the vehicle. The fault code indicates a rich mixture exists at higher load conditions. The fault could be caused by a genuine rich mixture but, the fault code could be activated due to an Oxygen / Lambda sensor fault or an exhaust air leak. Check sensor signal and signal response under different operating conditions (idle, load, etc) with particular reference to the signal under high load conditions. If the sensor appears to be operating correctly but the signal value is indicating a genuine rich mixture, this could indicate an engine related fault e.g. incorrect mixtures, fuel pressure, leaking injectors etc, but check operation of load sensors e.g. throttle position, MAP, MAF (as applicable).
P2195	O2 Sensor Signal Biased / Stuck Lean Bank 1 Sensor 1	Note: Many O2 / Lambda sensor related fault codes are activated due to faults in other systems (engine, fuelling, misfires etc). The control unit is able to detect the ratio of the air / fuel mixture by monitoring the information from different sensors. The O2 (Oxygen / Lambda) sensors (primarily the pre-cat sensor) will indicate the oxygen content in the exhaust gas, which is an indicator of air / fuel ratio. The control unit will also monitor other sensors to identify the operating conditions (e.g. heavy or light load etc) and (using the O2 sensor signal), assess whether the air / fuel ratio is correct. Note that faulty engine operation e.g. misfires and exhaust air leaks can cause increased oxygen in the exhaust gas which will affect the oxygen sensor signal.	Refer to list of sensor / actuator checks on page 19. Different types of O2 / Lambda sensors (e.g. Zirconia or Titania / wideband etc) provide different signals and values, it will be necessary to identify the type of sensor fitted to the vehicle. The fault code indicates that the signal from the Oxygen / Lambda sensor is stuck or biased towards a lean mixture value, possibly during all operating conditions; the control unit will be applying a control signal to the injectors to provide a richer mixture but the Oxygen / Lambda sensor signal is remaining lean. The fault could be caused by a genuine lean mixture but, the fault code could be activated due to an Oxygen / Lambda sensor fault or an exhaust air leak (can cause high oxygen content in the exhaust gas). Check sensor signal and signal response under different operating conditions (idle, load, etc). If the sensor appears to be operating correctly but the signal value is indicating a genuine lean mixture, this could indicate an engine related fault e.g. incorrect mixtures, fuel pressure, misfires etc. but check operation of load sensors e.g. throttle position, MAP, MAF (as applicable).
P2196	O2 Sensor Signal Biased / Stuck Rich Bank 1 Sensor 1	Note: Many O2 / Lambda sensor related fault codes are activated due to faults in other systems (engine, fuelling, misfires etc). The control unit is able to detect the ratio of the air / fuel mixture by monitoring the information from different sensors. The O2 (Oxygen / Lambda) sensors (primarily the pre-cat sensor) will indicate the oxygen content in the exhaust gas, which is an indicator of air / fuel ratio. The control unit will also monitor other sensors to identify the operating conditions (e.g. heavy or light load etc) and (using the O2 sensor signal), assess whether the air / fuel ratio is correct. Note that faulty engine operation e.g. misfires and exhaust air leaks can cause increased oxygen in the exhaust gas which will affect the oxygen sensor signal.	Refer to list of sensor / actuator checks on page 19. Different types of O2 / Lambda sensors (e.g. Zirconia or Titania / wideband etc) provide different signals and values, it will be necessary to identify the type of sensor fitted to the vehicle. The fault code indicates that the signal from the Oxygen / Lambda sensor is stuck or biased towards a rich mixture value, possibly during all operating conditions; the control unit will be applying a control signal to the injectors to provide a leaner mixture but the Oxygen / Lambda sensor signal is remaining rich. The fault could be caused by a genuine rich mixture but, the fault code could be activated due to an Oxygen / Lambda sensor fault. Check sensor signal and signal response under different operating conditions (idle, load, etc). If the sensor appears to be operating correctly but the signal value is indicating a genuine rich mixture, this could indicate an engine related fault e.g. incorrect mixtures, fuel pressure, leaking injector etc. but check operation of load sensors e.g. throttle position, MAP, MAF (as applicable).

NOTE 1. Check for other fault codes that could provide additional information.   NOTE 2. Communication between control units can pass via a CAN-Bus system; refer to Page 51 for CAN-Bus checks.   242

NOTE 3. If a fault cannot be located, it is also possible that the control unit is at fault   NOTE 4. Refer to Page 53 for list of pages that show the ISO standard for component locations e.g. Sensor A - Bank 1.

Code	Description		
P2197	O2 Sensor Signal Biased / Stuck Lean Bank 2 Sensor 1	Note: Many O2 / Lambda sensor related fault codes are activated due to faults in other systems (engine, fuelling, misfires etc). The control unit is able to detect the ratio of the air / fuel mixture by monitoring the information from different sensors. The O2 (Oxygen / Lambda) sensors (primarily the pre-cat sensor) will indicate the oxygen content in the exhaust gas, which is an indicator of air / fuel ratio. The control unit will also monitor other sensors to identify the operating conditions (e.g. heavy or light load etc) and (using the O2 sensor signal), assess whether the air / fuel ratio is correct. Note that faulty engine operation e.g. misfires and exhaust air leaks can cause increased oxygen in the exhaust gas which will affect the oxygen sensor signal.	Refer to list of sensor / actuator checks on page 19. Different types of O2 / Lambda sensors (e.g. Zirconia or Titania / wideband etc) provide different signals and values, it will be necessary to identify the type of sensor fitted to the vehicle. The fault code indicates that the signal from the Oxygen / Lambda sensor is stuck or biased towards a lean mixture value, possibly during all operating conditions; the control unit will be applying a control signal to the injectors to provide a richer mixture but the Oxygen / Lambda sensor signal is remaining lean. The fault could be caused by a genuine lean mixture but, the fault code could be activated due to an Oxygen / Lambda sensor fault or an exhaust air leak (can cause high oxygen content in the exhaust gas). Check sensor signal and signal response under different operating conditions (idle, load, etc). If the sensor appears to be operating correctly but the signal value is indicating a genuine lean mixture, this could indicate an engine related fault e.g. incorrect mixtures, fuel pressure, misfires etc. but check operation of load sensors e.g. throttle position, MAP, MAF (as applicable).
P2198	O2 Sensor Signal Biased / Stuck Rich Bank 2 Sensor 1	Note: Many O2 / Lambda sensor related fault codes are activated due to faults in other systems (engine, fuelling, misfires etc). The control unit is able to detect the ratio of the air / fuel mixture by monitoring the information from different sensors. The O2 (Oxygen / Lambda) sensors (primarily the pre-cat sensor) will indicate the oxygen content in the exhaust gas, which is an indicator of air / fuel ratio. The control unit will also monitor other sensors to identify the operating conditions (e.g. heavy or light load etc) and (using the O2 sensor signal), assess whether the air / fuel ratio is correct. Note that faulty engine operation e.g. misfires and exhaust air leaks can cause increased oxygen in the exhaust gas which will affect the oxygen sensor signal.	Refer to list of sensor / actuator checks on page 19. Different types of O2 / Lambda sensors (e.g. Zirconia or Titania / wideband etc) provide different signals and values, it will be necessary to identify the type of sensor fitted to the vehicle. The fault code indicates that the signal from the Oxygen / Lambda sensor is stuck or biased towards a rich mixture value, possibly during all operating conditions; the control unit will be applying a control signal to the injectors to provide a leaner mixture but the Oxygen / Lambda sensor signal is remaining rich. The fault could be caused by a genuine rich mixture but, the fault code could be activated due to an Oxygen / Lambda sensor fault. Check sensor signal and signal response under different operating conditions (idle, load, etc). If the sensor appears to be operating correctly but the signal value is indicating a genuine rich mixture, this could indicate an engine related fault e.g. incorrect mixtures, fuel pressure, leaking injector etc. but check operation of load sensors e.g. throttle position, MAP, MAF (as applicable).
P2199	Intake Air Temperature Sensor 1 / 2 Correlation	The intake air temperature sensor can be used for fine tuning of fuel, ignition and emissions control. On turbo / supercharged engines, the intake air temperature can form part of the calculation for controlling boost pressure or controlling airflow through the intercooler.	The signals from the different sensors are providing conflicting or implausible information i.e. Sensor 1 indicates a different temperature value to sensor 2. It is possible that one of the sensor signals is incorrect; if possible check the operation of each sensor and note that with the engine off (ignition on) the sensors should indicate the same temperature (check air temperature with independent thermometer to identify which sensor is correct). Check the sensors and wiring for faults e.g. short / open circuits, circuit resistance, incorrect sensor resistance or reference / operating voltage fault; interference from other circuits can affect sensor signals. Refer to list of sensor / actuator checks on page 19.
P2200	NOx Sensor Circuit Bank 1	NOx sensors can be combined into the post-cat Lambda sensor; the sensor (which monitors oxygen / NOx content) will be located downstream of the rear catalyst, which functions as a NOx accumulator / converter (stores and then converts the NOx into less harmful gasses). The NOx sensor monitors oxygen / NOx content and, when the NOx accumulator carries out the conversion process, the oxygen / NOx content is monitored by the NOx sensor to identify when the process is complete. A separate electronic module can form part of the NOx sensor circuit, the module enhances the operation of the sensor and a signal is passed from the module to the main engine control unit. For identification of the sensor type, location and additional information,	Note that if a separate NOx control module is fitted, the fault could be related to the circuit carrying the signal from the module to the main engine control unit. The fault code indicates that the voltage, frequency or value of the sensor signal is incorrect (undefined fault). The signal value could be: out of normal operating range, constant value, non-existent, corrupt. The problem could be related to a sensor or wiring fault; check wiring between sensor and control unit (or separate module) for short / open circuits and high resistance (also check wiring from module to engine control unit). Check sensor signal or Live Data to identify whether sensor signal is correct. If the sensor appears to be operating correctly but the signal value is incorrect, this could indicate that the oxygen / NOx content in the exhaust gas is incorrect for the operating conditions (causing sensor signal to be out of normal operating range); an incorrect signal could be caused by inefficient NOx accumulator operation. Refer to list of sensor / actuator checks on page 19.
P2201	NOx Sensor Circuit Range / Performance Bank 1	NOx sensors can be combined into the post-cat Lambda sensor; the sensor (which monitors oxygen / NOx content) will be located downstream of the rear catalyst, which functions as a NOx accumulator / converter (stores and then converts the NOx into less harmful gasses). The NOx sensor monitors oxygen / NOx content and, when the NOx accumulator carries out the conversion process, the oxygen / NOx content is monitored by the NOx sensor to identify when the process is complete. A separate electronic module can form part of the NOx sensor circuit, the module enhances the operation of the sensor and a signal is passed from the module to the main engine control unit. For identification of the sensor type, location and additional information,	Note that if a separate NOx control module is fitted, the fault could be related to the circuit carrying the signal from the module to the main engine control unit. The fault code indicates that the voltage, frequency or value of the sensor signal is incorrect (undefined fault). The signal value could be: out of normal operating range, constant value, non-existent, corrupt. The problem could be related to a sensor or wiring fault; check wiring between sensor and control unit (or separate module) for short / open circuits and high resistance (also check wiring from module to engine control unit). Check sensor signal or Live Data to identify whether sensor signal is correct. If the sensor appears to be operating correctly but the signal value is incorrect, this could indicate that the oxygen / NOx content in the exhaust gas is incorrect for the operating conditions (causing sensor signal to be out of normal operating range); an incorrect signal could be caused by inefficient NOx accumulator operation. Refer to list of sensor / actuator checks on page 19.
P2202	NOx Sensor Circuit Low Bank 1	NOx sensors can be combined into the post-cat Lambda sensor; the sensor (which monitors oxygen / NOx content) will be located downstream of the rear catalyst, which functions as a NOx accumulator / converter (stores and then converts the NOx into less harmful gasses). The NOx sensor monitors oxygen / NOx content and, when the NOx accumulator carries out the conversion process, the oxygen / NOx content is monitored by the NOx sensor to identify when the process is complete. A separate electronic module can form part of the NOx sensor circuit, the module enhances the operation of the sensor and a signal is passed from the module to the main engine control unit. For identification of the sensor type, location and additional information,	Note that if a separate NOx control module is fitted, the fault could be related to the circuit carrying the signal from the module to the main engine control unit. The fault code indicates that the voltage of the sensor signal is low (possibly out of normal operating range). The problem could be related to a sensor or wiring fault; check wiring between sensor and control unit (or separate module) for short / open circuits and high resistance (also check wiring from module to engine control unit). Check sensor signal or Live Data to identify whether sensor signal is correct. If the sensor appears to be operating correctly but the signal value is incorrect, this could indicate that the oxygen / NOx content in the exhaust gas is incorrect for the operating conditions (causing sensor signal to be out of normal operating range); an incorrect signal could be caused by inefficient NOx accumulator operation. Refer to list of sensor / actuator checks on page 19.

NOTE 1. Check for other fault codes that could provide additional information.
NOTE 2. Communication between control units can pass via a CAN-Bus system; refer to Page 51 for CAN-Bus checks.
NOTE 3. If a fault cannot be located, it is also possible that the control unit is at fault.
NOTE 4. Refer to Page 53 for list of pages that show the ISO standard for component locations e.g. Sensor A - Bank 1.

Fault code	EOBD / ISO Description	Component / System Description	Meaningful Description and Quick Check
P2203	NOx Sensor Circuit High Bank 1	NOx sensors can be combined into the post-cat Lambda sensor; the sensor (which monitors oxygen / NOx content) will be located downstream of the rear catalyst, which functions as a NOx accumulator / converter (stores and then converts the NOx into less harmful gasses). The NOx sensor monitors oxygen / NOx content and, when the NOx accumulator carries out the conversion process, the oxygen / NOx content is monitored by the NOx sensor to identify when the process is complete. A separate electronic module can form part of the NOx sensor circuit, the module enhances the operation of the sensor and a signal is passed from the module to the main engine control unit. For identification of the sensor type, location and additional information,	Note that if a separate NOx control module is fitted, the fault could be related to the circuit carrying the signal from the module to the main engine control unit. The fault code indicates that the voltage of the sensor signal is high (possibly out of normal operating range). The problem could be related to a sensor or wiring fault; check wiring between sensor and control unit (or separate module) for short / open circuits and high resistance (also check wiring from module to engine control unit). Check sensor signal or Live Data to identify whether sensor signal is correct. If the sensor appears to be operating correctly but the signal value is incorrect, this could indicate that the oxygen / NOx content in the exhaust gas is incorrect for the operating conditions (causing sensor signal to be out of normal operating range); an incorrect signal could be caused by inefficient NOx accumulator operation. Refer to list of sensor / actuator checks on page 19.
P2204	NOx Sensor Circuit Intermittent Bank 1	NOx sensors can be combined into the post-cat Lambda sensor; the sensor (which monitors oxygen / NOx content) will be located downstream of the rear catalyst, which functions as a NOx accumulator / converter (stores and then converts the NOx into less harmful gasses). The NOx sensor monitors oxygen / NOx content and, when the NOx accumulator carries out the conversion process, the oxygen / NOx content is monitored by the NOx sensor to identify when the process is complete. A separate electronic module can form part of the NOx sensor circuit, the module enhances the operation of the sensor and a signal is passed from the module to the main engine control unit. For identification of the sensor type, location and additional information,	Note that if a separate NOx control module is fitted, the fault could be related to the circuit carrying the signal from the module to the main engine control unit. The fault code indicates that the voltage, frequency or value of the sensor signal is intermittent (likely to be out of normal operating range / tolerance when intermittent fault is detected) or, the signal / voltage is erratic (e.g. signal changes are irregular / unpredictable). The problem could be related to a sensor or wiring fault; check wiring between sensor and control unit (or separate module) for short / open circuits and high resistance (also check wiring from module to engine control unit). Check sensor signal or Live Data and try to recreate the fault by moving all applicable wiring and connections. Refer to list of sensor / actuator checks on page 19.
P2205	NOx Sensor Heater Control Circuit / Open Bank 1	The heater for the NOx sensor is used to heat up and provide stable temperature for the sensor (primarily after cold starts). The heating element can receive a power supply via a relay (often the main system relay) and the heater earth path can be switched to earth via the control unit (this is the heater control circuit), the control unit is therefore able to control the operation of the heating element by controlling the earth circuit. Note that some systems may also provide the power supply to the heating element via the control unit. It may be necessary to refer to vehicle specific information to identify the sensor and the wiring.	The fault code can relate to the earth path from the heater element to the control unit (which can be the control circuit for the heater), but note that in some cases (depending on the wiring for the heater and the manufacturers interpretation of the code), the code could relate to the heater power supply circuit. The fault code indicates that there is an undefined fault in the sensor heater control circuit, which is causing an incorrect voltage / current in the circuit; this is likely to be: non-existent, constant value (but incorrect for the operating conditions), corrupt or, is out of normal operating range (out of tolerance). The fault could be caused by a faulty sensor heater element (open circuit or high resistance) or a wiring fault (short / open circuit). Also check heater supply voltage (could be provided via a relay or from a control unit). Refer to list of sensor / actuator checks on page 19.
P2206	NOx Sensor Heater Control Circuit Low Bank 1	The heater for the NOx sensor is used to heat up and provide stable temperature for the sensor (primarily after cold starts). The heating element can receive a power supply via a relay (often the main system relay) and the heater earth path can be switched to earth via the control unit (this is the heater control circuit), the control unit is therefore able to control the operation of the heating element by controlling the earth circuit. Note that some systems may also provide the power supply to the heating element via the control unit. It may be necessary to refer to vehicle specific information to identify the sensor and the wiring.	The fault code can relate to the earth path from the heater element to the control unit (which can be the control circuit for the heater) but, depending on the wiring for the heater and the manufacturer's interpretation of the code, the code could relate to the heater power supply circuit. The fault code indicates that the voltage / current in the sensor heater control circuit is low e.g. zero volts, when the control unit is expecting to detect a high voltage e.g. battery volts (when the control unit is not completing the earth circuit for the heater, battery voltage would normally be detected in this circuit). The fault is likely to be caused by an open circuit in the wiring or heater element (or short to earth); it also possible that there is no power supply (either from the relay or control unit, as applicable). Check power supply and wiring to heater, check continuity / resistance of heater element (as applicable), and check wiring through to the control unit for short / open circuit to heater. Refer to list of sensor / actuator checks on page 19.
P2207	NOx Sensor Heater Control Circuit High Bank 1	The heater for the NOx sensor is used to heat up and provide stable temperature for the sensor (primarily after cold starts). The heating element can receive a power supply via a relay (often the main system relay) and the heater earth path can be switched to earth via the control unit (this is the heater control circuit), the control unit is therefore able to control the operation of the heating element by controlling the earth circuit. Note that some systems may also provide the power supply to the heating element via the control unit. It may be necessary to refer to vehicle specific information to identify the sensor and the wiring.	The fault code can relate to the earth path from the heater element to the control unit (which can be the control circuit for the heater) but, depending on the wiring for the heater and the manufacturer's interpretation of the code, the code could relate to the heater power supply circuit. The fault code indicates that the voltage / current in the sensor heater control circuit is high e.g. battery voltage when it should be low e.g. zero volts (it would normally be zero / close to zero, when the control unit completes the earth circuit). The fault could also be caused by a short from the heater earth circuit to a another circuit (e.g. power supply). Check all applicable wiring for short / open circuits, and check for good connections between the heater and the control unit. Refer to list of sensor / actuator checks on page 19.
P2208	NOx Sensor Heater Sense Circuit Bank 1	The heater for the NOx sensor is used to heat up and provide stable temperature for the sensor (primarily after cold starts). The heating element can receive a power supply via a relay (often the main system relay) and the heater earth path can be switched to earth via the control unit (this is the heater control circuit), the control unit is therefore able to control the operation of the heating element by controlling the earth circuit. Note that some systems may also provide the power supply to the heating element via the control unit. It may be necessary to refer to vehicle specific information to identify the sensor and the wiring.	The fault code refers to the heater "sense" circuit. Depending on the type of sensor and wring for the heater, some systems can make use of a sense wire or circuit to monitor the heater operation. The fault is likely to relate to the sense circuit; check the wiring for a sense circuit between the sensor and the control unit, but note that a manufacturer could also use the code to refer to the heater control circuit wire as the sense circuit (it may therefore be necessary to carry out heater control circuit checks - also refer to fault code P2205). The fault code indicates that there is an undefined fault in the sense circuit (possible open circuit), which is causing an incorrect voltage / current in the circuit; this is likely to be: non-existent, constant value (but incorrect for the operating conditions), corrupt or, is out of normal operating range (out of tolerance). The fault could be caused by faulty sensor wiring (short / open circuit or high circuit resistance). Refer to list of sensor / actuator checks on page 19.

NOTE 1. Check for other fault codes that could provide additional information.   **NOTE 2.** Communication between control units can pass via a CAN-Bus system; refer to Page 51 for CAN-Bus checks.   **244**

NOTE 4. Refer to Page 53 for list of pages that show the ISO standard for component locations e.g. Sensor A - Bank 1.

P2209	NOx Sensor Heater Sense Circuit Range / Performance Bank 1	The heater for the NOx sensor is used to heat up and provide stable temperature for the sensor (primarily after cold starts). The heating element can receive a power supply via a relay (often the main system relay) and the heater earth path can be switched to earth via the control unit (this is the heater control circuit), the control unit is therefore able to control the operation of the heating element by controlling the earth circuit. Note that some systems may also provide the power supply to the heating element via the control unit. It may be necessary to refer to vehicle specific information to identify the sensor and the wiring.	The fault code refers to the heater "sense" circuit. Depending on the type of sensor and wring for the heater, some systems can make use of a sense wire or circuit to monitor the heater operation. The fault is likely to relate to the sense circuit; check the wiring for a sense circuit between the sensor and the control unit, but note that the sense circuit signal will be dependent on the heater circuit / operation and, a manufacturer could also use the code to refer to the heater control circuit wire as the sense circuit (it may therefore be necessary to carry out heater control circuit checks - also refer to fault code P2207). The fault code indicates that the voltage / current in the sense circuit is within the normal operating range / tolerance, but the signal might be incorrect for the operating conditions. Check all applicable wiring for short / open circuits, and check for good connections between the sensor and the control unit. Refer to list of sensor / actuator checks on page 19.
P2210	NOx Sensor Heater Sense Circuit Low Bank 1	The heater for the NOx sensor is used to heat up and provide stable temperature for the sensor (primarily after cold starts). The heating element can receive a power supply via a relay (often the main system relay) and the heater earth path can be switched to earth via the control unit (this is the heater control circuit), the control unit is therefore able to control the operation of the heating element by controlling the earth circuit. Note that some systems may also provide the power supply to the heating element via the control unit. It may be necessary to refer to vehicle specific information to identify the sensor and the wiring.	The fault code refers to the heater "sense" circuit. Depending on the type of sensor and wring for the heater, some systems can make use of a sense wire or circuit to monitor the heater operation. The fault is likely to relate to the sense circuit; check the wiring for a sense circuit between the sensor and the control unit, but note that a manufacturer could also use the code to refer to the heater control circuit wire as the sense circuit (it may therefore be necessary to carry out heater control circuit checks - also refer to fault code P2206). The fault code indicates that the voltage / current in the sense circuit is low e.g. zero volts, when the control unit is expecting to detect a high voltage e.g. battery volts. The fault is likely to be caused by an open circuit in the wiring or short to earth. Check all applicable wiring for short / open circuits, and check for good connections between the sensor and the control unit. Refer to list of sensor / actuator checks on page 19.
P2211	NOx Sensor Heater Sense Circuit High Bank 1	The heater for the NOx sensor is used to heat up and provide stable temperature for the sensor (primarily after cold starts). The heating element can receive a power supply via a relay (often the main system relay) and the heater earth path can be switched to earth via the control unit (this is the heater control circuit), the control unit is therefore able to control the operation of the heating element by controlling the earth circuit. Note that some systems may also provide the power supply to the heating element via the control unit. It may be necessary to refer to vehicle specific information to identify the sensor and the wiring.	The fault code refers to the heater "sense" circuit. Depending on the type of sensor and wring for the heater, some systems can make use of a sense wire or circuit to monitor the heater operation. The fault is likely to relate to the sense circuit; check the wiring for a sense circuit between the sensor and the control unit, but note that a manufacturer could also use the code to refer to the heater control circuit wire as the sense circuit (it may therefore be necessary to carry out heater control circuit checks - also refer to fault code P2207). The fault code indicates that the voltage / current in the sense circuit is high e.g. battery voltage when it should be low e.g. zero volts. The fault could be caused by a short from the sense circuit to a another circuit (e.g. power supply). Check all applicable wiring for short / open circuits, and check for good connections between the sensor and the control unit. Refer to list of sensor / actuator checks on page 19.
P2212	NOx Sensor Heater Sense Circuit Intermittent Bank 1	The heater for the NOx sensor is used to heat up and provide stable temperature for the sensor (primarily after cold starts). The heating element can receive a power supply via a relay (often the main system relay) and the heater earth path can be switched to earth via the control unit (this is the heater control circuit), the control unit is therefore able to control the operation of the heating element by controlling the earth circuit. Note that some systems may also provide the power supply to the heating element via the control unit. It may be necessary to refer to vehicle specific information to identify the sensor and the wiring.	The fault code refers to the heater "sense" circuit. Depending on the type of sensor and wring for the heater, some systems can make use of a sense wire or circuit to monitor the heater operation. The fault is likely to relate to the sense circuit; check the wiring for a sense circuit between the sensor and the control unit, but note that a manufacturer could also use the code to refer to the heater control circuit wire as the sense circuit (it may therefore be necessary to carry out heater control circuit checks - also refer to fault code P2209). The fault code indicates that the voltage / current in the sense circuit is intermittent (likely to be out of normal operating range / tolerance when intermittent fault is detected) or, the signal / voltage is erratic (signal changes are irregular / unpredictable). Possible fault with wiring or connections (including internal sensor connections). Check all applicable wiring between the sensor and the control unit for intermittent short / open circuits. Refer to list of sensor / actuator checks on page 19.
P2213	NOx Sensor Circuit Bank 2	NOx sensors can be combined into the post-cat Lambda sensor; the sensor (which monitors oxygen / NOx content) will be located downstream of the rear catalyst, which functions as a NOx accumulator / converter (stores and then converts the NOx into less harmful gasses). The NOx sensor monitors oxygen / NOx content and, when the NOx accumulator carries out the conversion process, the oxygen / NOx content is monitored by the NOx sensor to identify when the process is complete. A separate electronic module can form part of the NOx sensor circuit, the module enhances the operation of the sensor and a signal is passed from the module to the main engine control unit. For identification of the sensor type, location and additional information,	Note that if a separate NOx control module is fitted, the fault could be related to the circuit carrying the signal from the module to the main engine control unit. The fault code indicates that the voltage, frequency or value of the sensor signal is incorrect (undefined fault). The signal value could be: out of normal operating range, constant value, non-existent, corrupt. The problem could be related to a sensor or wiring fault; check wiring between sensor and control unit (or separate module) for short / open circuits and high resistance (also check wiring from module to engine control unit). Check sensor signal or Live Data to identify whether sensor signal is correct. If the sensor appears to be operating correctly but the signal value is incorrect, this could indicate that the oxygen / NOx content in the exhaust gas is incorrect for the operating conditions (causing sensor signal to be out of normal operating range); an incorrect signal could be caused by inefficient NOx accumulator operation. Refer to list of sensor / actuator checks on page 19.
P2214	NOx Sensor Circuit Range / Performance Bank 2	NOx sensors can be combined into the post-cat Lambda sensor; the sensor (which monitors oxygen / NOx content) will be located downstream of the rear catalyst, which functions as a NOx accumulator / converter (stores and then converts the NOx into less harmful gasses). The NOx sensor monitors oxygen / NOx content and, when the NOx accumulator carries out the conversion process, the oxygen / NOx content is monitored by the NOx sensor to identify when the process is complete. A separate electronic module can form part of the NOx sensor circuit, the module enhances the operation of the sensor and a signal is passed from the module to the main engine control unit. For identification of the sensor type, location and additional information,	Note that if a separate NOx control module is fitted, the fault could be related to the circuit carrying the signal from the module to the main engine control unit. The fault code indicates that the voltage, frequency or value of the sensor signal is incorrect (undefined fault). The signal value could be: out of normal operating range, constant value, non-existent, corrupt. The problem could be related to a sensor or wiring fault; check wiring between sensor and control unit (or separate module) for short / open circuits and high resistance (also check wiring from module to engine control unit). Check sensor signal or Live Data to identify whether sensor signal is correct. If the sensor appears to be operating correctly but the signal value is incorrect, this could indicate that the oxygen / NOx content in the exhaust gas is incorrect for the operating conditions (causing sensor signal to be out of normal operating range); an incorrect signal could be caused by inefficient NOx accumulator operation. Refer to list of sensor / actuator checks on page 19.

NOTE 1. Check for other fault codes that could provide additional information.   **NOTE 2.** Communication between control units can pass via a CAN-Bus system; refer to Page 51 for CAN-Bus checks.
NOTE 3. If a fault cannot be located, it is also possible that the control unit is at fault.   **NOTE 4.** Refer to Page 53 for list of pages that show the ISO standard for component locations e.g. Sensor A - Bank 1.

245

Fault code	EOBD / ISO Description	Component / System Description	Meaningful Description and Quick Check
P2215	NOx Sensor Circuit Low Bank 2	NOx sensors can be combined into the post-cat Lambda sensor; the sensor (which monitors oxygen / NOx content) will be located downstream of the rear catalyst, which functions as a NOx accumulator / converter (stores and then converts the NOx into less harmful gasses). The NOx sensor monitors oxygen / NOx content and, when the NOx accumulator carries out the conversion process, the oxygen / NOx content is monitored by the NOx sensor to identify when the process is complete. A separate electronic module can form part of the NOx sensor circuit, the module enhances the operation of the sensor and a signal is passed from the module to the main engine control unit. For identification of the sensor type, location and additional information,	Note that if a separate NOx control module is fitted, the fault could be related to the circuit carrying the signal from the module to the main engine control unit. The fault code indicates that the voltage of the sensor signal is low (possibly out of normal operating range). The problem could be related to a sensor or wiring fault; check wiring between sensor and control unit (or separate module) for short / open circuits and high resistance (also check wiring from module to engine control unit). Check sensor signal or Live Data to identify whether sensor signal is correct. If the sensor appears to be operating correctly but the signal value is incorrect, this could indicate that the oxygen / NOx content in the exhaust gas is incorrect for the operating conditions (causing sensor signal to be out of normal operating range); an incorrect signal could be caused by inefficient NOx accumulator operation. Refer to list of sensor / actuator checks on page 19.
P2216	NOx Sensor Circuit High Bank 2	NOx sensors can be combined into the post-cat Lambda sensor; the sensor (which monitors oxygen / NOx content) will be located downstream of the rear catalyst, which functions as a NOx accumulator / converter (stores and then converts the NOx into less harmful gasses). The NOx sensor monitors oxygen / NOx content and, when the NOx accumulator carries out the conversion process, the oxygen / NOx content is monitored by the NOx sensor to identify when the process is complete. A separate electronic module can form part of the NOx sensor circuit, the module enhances the operation of the sensor and a signal is passed from the module to the main engine control unit. For identification of the sensor type, location and additional information,	Note that if a separate NOx control module is fitted, the fault could be related to the circuit carrying the signal from the module to the main engine control unit. The fault code indicates that the voltage of the sensor signal is high (possibly out of normal operating range). The problem could be related to a sensor or wiring fault; check wiring between sensor and control unit (or separate module) for short / open circuits and high resistance (also check wiring from module to engine control unit). Check sensor signal or Live Data to identify whether sensor signal is correct. If the sensor appears to be operating correctly but the signal value is incorrect, this could indicate that the oxygen / NOx content in the exhaust gas is incorrect for the operating conditions (causing sensor signal to be out of normal operating range); an incorrect signal could be caused by inefficient NOx accumulator operation. Refer to list of sensor / actuator checks on page 19.
P2217	NOx Sensor Circuit Intermittent Bank 2	NOx sensors can be combined into the post-cat Lambda sensor; the sensor (which monitors oxygen / NOx content) will be located downstream of the rear catalyst, which functions as a NOx accumulator / converter (stores and then converts the NOx into less harmful gasses). The NOx sensor monitors oxygen / NOx content and, when the NOx accumulator carries out the conversion process, the oxygen / NOx content is monitored by the NOx sensor to identify when the process is complete. A separate electronic module can form part of the NOx sensor circuit, the module enhances the operation of the sensor and a signal is passed from the module to the main engine control unit. For identification of the sensor type, location and additional information,	Note that if a separate NOx control module is fitted, the fault could be related to the circuit carrying the signal from the module to the main engine control unit. The fault code indicates that the voltage, frequency or value of the sensor signal is intermittent (likely to be out of normal operating range / tolerance when intermittent fault is detected) or, the signal / voltage is erratic (e.g. signal changes are irregular / unpredictable). The problem could be related to a sensor or wiring fault; check wiring between sensor and control unit (or separate module) for short / open circuits and high resistance (also check wiring from module to engine control unit). Check sensor signal or Live Data and try to recreate the fault by moving all applicable wiring and connections. Refer to list of sensor / actuator checks on page 19.
P2218	NOx Sensor Heater Control Circuit / Open Bank 2	The heater for the NOx sensor is used to heat up and provide stable temperature for the sensor (primarily after cold starts). The heating element can receive a power supply via a relay (often the main system relay) and the heater earth path can be switched to earth via the control unit (this is the heater control circuit), the control unit is therefore able to control the operation of the heating element by controlling the earth circuit. Note that some systems may also provide the power supply to the heating element via the control unit. It may be necessary to refer to vehicle specific information to identify the sensor and the wiring.	The fault code can relate to the earth path from the heater element to the control unit (which can be the control circuit for the heater), but note that in some cases (depending on the wiring for the heater and the manufacturers interpretation of the code), the code could relate to the heater power supply circuit. The fault code indicates that there is an undefined fault in the sensor heater control circuit, which is causing an incorrect voltage / current in the circuit; this is likely to be: non-existent, constant value (but incorrect for the operating conditions), corrupt or, is out of normal operating range (out of tolerance). The fault could be caused by a faulty sensor heater element (open circuit or high resistance) or a wiring fault (short / open circuit). Also check heater supply voltage (could be provided via a relay or from a control unit). Refer to list of sensor / actuator checks on page 19.
P2219	NOx Sensor Heater Control Circuit Low Bank 2	The heater for the NOx sensor is used to heat up and provide stable temperature for the sensor (primarily after cold starts). The heating element can receive a power supply via a relay (often the main system relay) and the heater earth path can be switched to earth via the control unit (this is the heater control circuit), the control unit is therefore able to control the operation of the heating element by controlling the earth circuit. Note that some systems may also provide the power supply to the heating element via the control unit. It may be necessary to refer to vehicle specific information to identify the sensor and the wiring.	The fault code can relate to the earth path from the heater element to the control unit (which can be the control circuit for the heater) but, depending on the wiring for the heater and the manufacturer's interpretation of the code, the code could relate to the heater power supply circuit. The fault code indicates that the voltage / current in the sensor heater control circuit is low e.g. zero volts, when the control unit is expecting to detect a high voltage e.g. battery volts (when the control unit is not completing the earth circuit for the heater, battery voltage would normally be detected in this circuit). The fault is likely to be caused by an open circuit in the wiring or heater element (or short to earth); it also possible that there is no power supply (either from the relay or control unit, as applicable). Check power supply and wiring to heater, check continuity / resistance of heater element (as applicable), and check wiring through to the control unit for short / open circuit to heater. Refer to list of sensor / actuator checks on page 19.
P2220	NOx Sensor Heater Control Circuit High Bank 2	The heater for the NOx sensor is used to heat up and provide stable temperature for the sensor (primarily after cold starts). The heating element can receive a power supply via a relay (often the main system relay) and the heater earth path can be switched to earth via the control unit (this is the heater control circuit), the control unit is therefore able to control the operation of the heating element by controlling the earth circuit. Note that some systems may also provide the power supply to the heating element via the control unit. It may be necessary to refer to vehicle specific information to identify the sensor and the wiring.	The fault code can relate to the earth path from the heater element to the control unit (which can be the control circuit for the heater) but, depending on the wiring for the heater and the manufacturer's interpretation of the code, the code could relate to the heater power supply circuit. The fault code indicates that the voltage / current in the sensor heater control circuit is high e.g. battery voltage when it should be low e.g. zero volts (it would normally be zero / close to zero, when the control unit completes the earth circuit). The fault could also be caused by a short from the heater earth circuit to a another circuit (e.g. power supply). Check all applicable wiring for short / open circuits, and check for good connections between the heater and the control unit. Refer to list of sensor / actuator checks on page 19.

NOTE 1. Check for other fault codes that could provide additional information. **NOTE 2.** Communication between control units can pass via a CAN-Bus system; refer to Page 51 for CAN-Bus checks. 246

NOTE 3. If a fault cannot be located, it is also possible that the control unit is at fault NOTE 4. Refer to Page 53 for list of pages that show the ISO standard for component locations e.g. Sensor A - Bank 1.

code	Description		
P2221	NOx Sensor Heater Sense Circuit Bank 2	The heater for the NOx sensor is used to heat up and provide stable temperature for the sensor (primarily after cold starts). The heating element can receive a power supply via a relay (often the main system relay) and the heater earth path can be switched to earth via the control unit (this is the heater control circuit), the control unit is therefore able to control the operation of the heating element by controlling the earth circuit. Note that some systems may also provide the power supply to the heating element via the control unit. It may be necessary to refer to vehicle specific information to identify the sensor and the wiring.	The fault code refers to the heater "sense" circuit. Depending on the type of sensor and wring for the heater, some systems can make use of a sense wire or circuit to monitor the heater operation. The fault is likely to relate to the sense circuit; check the wiring for a sense circuit between the sensor and the control unit, but note that a manufacturer could also use the code to refer to the heater control circuit wire as the sense circuit (it may therefore be necessary to carry out heater control circuit checks - also refer to fault code P2205). The fault code indicates that there is an undefined fault in the sense circuit (possible open circuit), which is causing an incorrect voltage / current in the circuit; this is likely to be: non-existent, constant value (but incorrect for the operating conditions), corrupt or, is out of normal operating range (out of tolerance). The fault could be caused by faulty sensor wiring (short / open circuit or high circuit resistance). Refer to list of sensor / actuator checks on page 19.
P2222	NOx Sensor Heater Sense Circuit Range / Performance Bank 2	The heater for the NOx sensor is used to heat up and provide stable temperature for the sensor (primarily after cold starts). The heating element can receive a power supply via a relay (often the main system relay) and the heater earth path can be switched to earth via the control unit (this is the heater control circuit), the control unit is therefore able to control the operation of the heating element by controlling the earth circuit. Note that some systems may also provide the power supply to the heating element via the control unit. It may be necessary to refer to vehicle specific information to identify the sensor and the wiring.	The fault code refers to the heater "sense" circuit. Depending on the type of sensor and wring for the heater, some systems can make use of a sense wire or circuit to monitor the heater operation. The fault is likely to relate to the sense circuit; check the wiring for a sense circuit between the sensor and the control unit, but note that the sense circuit signal will be dependent on the heater circuit / operation and, a manufacturer could also use the code to refer to the heater control circuit wire as the sense circuit (it may therefore be necessary to carry out heater control circuit checks - also refer to fault code P2207). The fault code indicates that the voltage / current in the sense circuit is within the normal operating range / tolerance, but the signal might be incorrect for the operating conditions. Check all applicable wiring for short / open circuits, and check for good connections between the sensor and the control unit. Refer to list of sensor / actuator checks on page 19.
P2223	NOx Sensor Heater Sense Circuit Low Bank 2	The heater for the NOx sensor is used to heat up and provide stable temperature for the sensor (primarily after cold starts). The heating element can receive a power supply via a relay (often the main system relay) and the heater earth path can be switched to earth via the control unit (this is the heater control circuit), the control unit is therefore able to control the operation of the heating element by controlling the earth circuit. Note that some systems may also provide the power supply to the heating element via the control unit. It may be necessary to refer to vehicle specific information to identify the sensor and the wiring.	The fault code refers to the heater "sense" circuit. Depending on the type of sensor and wring for the heater, some systems can make use of a sense wire or circuit to monitor the heater operation. The fault is likely to relate to the sense circuit; check the wiring for a sense circuit between the sensor and the control unit, but note that a manufacturer could also use the code to refer to the heater control circuit wire as the sense circuit (it may therefore be necessary to carry out heater control circuit checks - also refer to fault code P2206). The fault code indicates that the voltage / current in the sense circuit is low e.g. zero volts, when the control unit is expecting to detect a high voltage e.g. battery volts. The fault is likely to be caused by an open circuit in the wiring or short to earth. Check all applicable wiring for short / open circuits, and check for good connections between the sensor and the control unit. Refer to list of sensor / actuator checks on page 19.
P2224	NOx Sensor Heater Sense Circuit High Bank 2	The heater for the NOx sensor is used to heat up and provide stable temperature for the sensor (primarily after cold starts). The heating element can receive a power supply via a relay (often the main system relay) and the heater earth path can be switched to earth via the control unit (this is the heater control circuit), the control unit is therefore able to control the operation of the heating element by controlling the earth circuit. Note that some systems may also provide the power supply to the heating element via the control unit. It may be necessary to refer to vehicle specific information to identify the sensor and the wiring.	The fault code refers to the heater "sense" circuit. Depending on the type of sensor and wring for the heater, some systems can make use of a sense wire or circuit to monitor the heater operation. The fault is likely to relate to the sense circuit; check the wiring for a sense circuit between the sensor and the control unit, but note that a manufacturer could also use the code to refer to the heater control circuit wire as the sense circuit (it may therefore be necessary to carry out heater control circuit checks - also refer to fault code P2207). The fault code indicates that the voltage / current in the sense circuit is high e.g. battery voltage when it should be low e.g. zero volts. The fault could be caused by a short from the sense circuit to a another circuit (e.g. power supply). Check all applicable wiring for short / open circuits, and check for good connections between the sensor and the control unit. Refer to list of sensor / actuator checks on page 19.
P2225	NOx Sensor Heater Sense Circuit Intermittent Bank 2	The heater for the NOx sensor is used to heat up and provide stable temperature for the sensor (primarily after cold starts). The heating element can receive a power supply via a relay (often the main system relay) and the heater earth path can be switched to earth via the control unit (this is the heater control circuit), the control unit is therefore able to control the operation of the heating element by controlling the earth circuit. Note that some systems may also provide the power supply to the heating element via the control unit. It may be necessary to refer to vehicle specific information to identify the sensor and the wiring.	The fault code refers to the heater "sense" circuit. Depending on the type of sensor and wring for the heater, some systems can make use of a sense wire or circuit to monitor the heater operation. The fault is likely to relate to the sense circuit; check the wiring for a sense circuit between the sensor and the control unit, but note that a manufacturer could also use the code to refer to the heater control circuit wire as the sense circuit (it may therefore be necessary to carry out heater control circuit checks - also refer to fault code P2209). The fault code indicates that the voltage / current in the sense circuit is intermittent (likely to be out of normal operating range / tolerance when intermittent fault is detected) or, the signal / voltage is erratic (signal changes are irregular / unpredictable). Possible fault with wiring or connections (including internal sensor connections). Check all applicable wiring between the sensor and the control unit for intermittent short / open circuits. Refer to list of sensor / actuator checks on page 19.
P2226	Barometric Pressure Circuit	The barometric / atmospheric air pressure sensor signal can be used for fine tuning of the fuel, ignition and emissions control systems. Barometric pressure can also be used in the calculations for boost pressure control on turbo / supercharger systems.	The voltage, frequency or value of the pressure sensor signal is incorrect (undefined fault). Sensor voltage / signal could be: out of normal operating range, constant value, non-existent, corrupt. Possible fault in the pressure sensor or wiring e.g. short / open circuits, circuit resistance, incorrect sensor resistance (if applicable) or, fault with the reference / operating voltage; interference from other circuits can also affect sensor signals. Refer to list of sensor / actuator checks on page 19.
P2227	Barometric Pressure Circuit Range / Performance	The barometric / atmospheric air pressure sensor signal can be used for fine tuning of the fuel, ignition and emissions control systems. Barometric pressure can also be used in the calculations for boost pressure control on turbo / supercharger systems.	The voltage, frequency or value of the pressure sensor signal is within the normal operating range / tolerance but the signal is not plausible or is incorrect due to an undefined fault. The sensor signal might not match the operating conditions indicated by other sensors or, the sensor signal is not changing / responding as expected to the changes in conditions. Possible fault in the pressure sensor or wiring e.g. short / open circuits, circuit resistance, incorrect sensor resistance (if applicable) or, fault with the reference / operating voltage; interference from other circuits can also affect sensor signals. Refer to list of sensor / actuator checks on page 19. Also check any air passages / pipes to the sensor for leaks / blockages etc. It is also possible that the pressure sensor is operating correctly but the pressure being measured by the sensor is unacceptable or incorrect for the operating conditions; if possible, use alternative method of checking pressure to establish if sensor system is operating correctly and whether the pressure indicated by the sensor is correct / incorrect.

NOTE 1. Check for other fault codes that could provide additional information. **NOTE 2.** Communication between control units can pass via a CAN-Bus system; refer to Page 51 for CAN-Bus checks.

NOTE 3. If a fault cannot be located, it is also possible that the control unit is at fault. **NOTE 4.** Refer to Page 53 for list of pages that show the ISO standard for component locations e.g. Sensor A - Bank 1.

Fault code	EOBD / ISO Description	Component / System Description	Meaningful Description and Quick Check
P2228	Barometric Pressure Circuit Low	The barometric / atmospheric air pressure sensor signal can be used for fine tuning of the fuel, ignition and emissions control systems. Barometric pressure can also be used in the calculations for boost pressure control on turbo / supercharger systems.	The voltage, frequency or value of the pressure sensor signal is either "zero" (no signal) or, the signal exists but it is below the normal operating range (lower than the minimum operating limit). Possible fault in the pressure sensor or wiring e.g. short / open circuits, circuit resistance, incorrect sensor resistance (if applicable) or, fault with the reference / operating voltage; interference from other circuits can also affect sensor signals. Refer to list of sensor / actuator checks on page 19.
P2229	Barometric Pressure Circuit High	The barometric / atmospheric air pressure sensor signal can be used for fine tuning of the fuel, ignition and emissions control systems. Barometric pressure can also be used in the calculations for boost pressure control on turbo / supercharger systems.	The voltage, frequency or value of the pressure sensor signal is either at full value (e.g. battery voltage with no signal or frequency) or, the signal exists but it is above the normal operating range (higher than the maximum operating limit). Possible fault in the pressure sensor or wiring e.g. short / open circuits, circuit resistance, incorrect sensor resistance (if applicable) or, fault with the reference / operating voltage; interference from other circuits can also affect sensor signals. Refer to list of sensor / actuator checks on page 19.
P2230	Barometric Pressure Circuit Intermittent	The barometric / atmospheric air pressure sensor signal can be used for fine tuning of the fuel, ignition and emissions control systems. Barometric pressure can also be used in the calculations for boost pressure control on turbo / supercharger systems.	The voltage, frequency or value of the pressure sensor signal is intermittent (likely to be out of normal operating range / tolerance when intermittent fault is detected) or, the signal / voltage is erratic (e.g. signal changes are irregular / unpredictable). Possible fault with wiring or connections (including internal sensor connections); also interference from other circuits can affect the signal. Refer to list of sensor / actuator checks on page 19.
P2231	O2 Sensor Signal Circuit Shorted to Heater Circuit Bank 1 Sensor 1	Note: Many O2 / Lambda sensor related fault codes are activated due to faults in other systems (engine, fuelling, misfires etc). O2 (Oxygen / Lambda) sensors detect the exhaust gas oxygen content, which is also an indicator of air / fuel ratio (rich / lean). Oxygen content is indicated by a "Lambda" value. The air / fuel ratio (and therefore oxygen content) is controlled by the engine control unit to maximise efficiency of catalytic converters and reduce emissions of harmful gases.	For a detailed list of checks and identification of the sensor location, Refer to list of sensor / actuator checks on page 19. Different types of O2 / Lambda sensors (e.g. Zirconia or Titania / wideband etc) provide different signals and values, it will be necessary to identify the type of sensor fitted to the vehicle. The fault code indicates that there is a likely short circuit between the O2 (oxygen / Lambda) sensor signal wire and the sensor heater circuit. It is likely that the supply voltage for the heater (which is typically at battery voltage level) is being detected on the signal circuit. Check for a short between the signal wire and the heater circuit (also check for a short to other circuits). It is possible that the fault could affect operation of other systems e.g. fuelling etc; therefore check for other fault codes.
P2232	O2 Sensor Signal Circuit Shorted to Heater Circuit Bank 1 Sensor 2	Note: Many O2 / Lambda sensor related fault codes are activated due to faults in other systems (engine, fuelling, misfires etc). O2 (Oxygen / Lambda) sensors detect the exhaust gas oxygen content, which is also an indicator of air / fuel ratio (rich / lean). Oxygen content is indicated by a "Lambda" value. The air / fuel ratio (and therefore oxygen content) is controlled by the engine control unit to maximise efficiency of catalytic converters and reduce emissions of harmful gases.	For a detailed list of checks and identification of the sensor location, Refer to list of sensor / actuator checks on page 19. Different types of O2 / Lambda sensors (e.g. Zirconia or Titania / wideband etc) provide different signals and values, it will be necessary to identify the type of sensor fitted to the vehicle. The fault code indicates that there is a likely short circuit between the O2 (oxygen / Lambda) sensor signal wire and the sensor heater circuit. It is likely that the supply voltage for the heater (which is typically at battery voltage level) is being detected on the signal circuit. Check for a short between the signal wire and the heater circuit (also check for a short to other circuits). It is possible that the fault could affect operation of other systems e.g. fuelling etc; therefore check for other fault codes.
P2233	O2 Sensor Signal Circuit Shorted to Heater Circuit Bank 1 Sensor 3	Note: Many O2 / Lambda sensor related fault codes are activated due to faults in other systems (engine, fuelling, misfires etc). O2 (Oxygen / Lambda) sensors detect the exhaust gas oxygen content, which is also an indicator of air / fuel ratio (rich / lean). Oxygen content is indicated by a "Lambda" value. The air / fuel ratio (and therefore oxygen content) is controlled by the engine control unit to maximise efficiency of catalytic converters and reduce emissions of harmful gases.	For a detailed list of checks and identification of the sensor location, Refer to list of sensor / actuator checks on page 19. Different types of O2 / Lambda sensors (e.g. Zirconia or Titania / wideband etc) provide different signals and values, it will be necessary to identify the type of sensor fitted to the vehicle. The fault code indicates that there is a likely short circuit between the O2 (oxygen / Lambda) sensor signal wire and the sensor heater circuit. It is likely that the supply voltage for the heater (which is typically at battery voltage level) is being detected on the signal circuit. Check for a short between the signal wire and the heater circuit (also check for a short to other circuits). It is possible that the fault could affect operation of other systems e.g. fuelling etc; therefore check for other fault codes.
P2234	O2 Sensor Signal Circuit Shorted to Heater Circuit Bank 2 Sensor 1	Note: Many O2 / Lambda sensor related fault codes are activated due to faults in other systems (engine, fuelling, misfires etc). O2 (Oxygen / Lambda) sensors detect the exhaust gas oxygen content, which is also an indicator of air / fuel ratio (rich / lean). Oxygen content is indicated by a "Lambda" value. The air / fuel ratio (and therefore oxygen content) is controlled by the engine control unit to maximise efficiency of catalytic converters and reduce emissions of harmful gases.	For a detailed list of checks and identification of the sensor location, Refer to list of sensor / actuator checks on page 19. Different types of O2 / Lambda sensors (e.g. Zirconia or Titania / wideband etc) provide different signals and values, it will be necessary to identify the type of sensor fitted to the vehicle. The fault code indicates that there is a likely short circuit between the O2 (oxygen / Lambda) sensor signal wire and the sensor heater circuit. It is likely that the supply voltage for the heater (which is typically at battery voltage level) is being detected on the signal circuit. Check for a short between the signal wire and the heater circuit (also check for a short to other circuits). It is possible that the fault could affect operation of other systems e.g. fuelling etc; therefore check for other fault codes.
P2235	O2 Sensor Signal Circuit Shorted to Heater Circuit Bank 2 Sensor 2	Note: Many O2 / Lambda sensor related fault codes are activated due to faults in other systems (engine, fuelling, misfires etc). O2 (Oxygen / Lambda) sensors detect the exhaust gas oxygen content, which is also an indicator of air / fuel ratio (rich / lean). Oxygen content is indicated by a "Lambda" value. The air / fuel ratio (and therefore oxygen content) is controlled by the engine control unit to maximise efficiency of catalytic converters and reduce emissions of harmful gases.	For a detailed list of checks and identification of the sensor location, Refer to list of sensor / actuator checks on page 19. Different types of O2 / Lambda sensors (e.g. Zirconia or Titania / wideband etc) provide different signals and values, it will be necessary to identify the type of sensor fitted to the vehicle. The fault code indicates that there is a likely short circuit between the O2 (oxygen / Lambda) sensor signal wire and the sensor heater circuit. It is likely that the supply voltage for the heater (which is typically at battery voltage level) is being detected on the signal circuit. Check for a short between the signal wire and the heater circuit (also check for a short to other circuits). It is possible that the fault could affect operation of other systems e.g. fuelling etc; therefore check for other fault codes.
P2236	O2 Sensor Signal Circuit Shorted to Heater Circuit Bank 2 Sensor 3	Note: Many O2 / Lambda sensor related fault codes are activated due to faults in other systems (engine, fuelling, misfires etc). O2 (Oxygen / Lambda) sensors detect the exhaust gas oxygen content, which is also an indicator of air / fuel ratio (rich / lean). Oxygen content is indicated by a "Lambda" value. The air / fuel ratio (and therefore oxygen content) is controlled by the engine control unit to maximise efficiency of catalytic converters and reduce emissions of harmful gases.	For a detailed list of checks and identification of the sensor location, Refer to list of sensor / actuator checks on page 19. Different types of O2 / Lambda sensors (e.g. Zirconia or Titania / wideband etc) provide different signals and values, it will be necessary to identify the type of sensor fitted to the vehicle. The fault code indicates that there is a likely short circuit between the O2 (oxygen / Lambda) sensor signal wire and the sensor heater circuit. It is likely that the supply voltage for the heater (which is typically at battery voltage level) is being detected on the signal circuit. Check for a short between the signal wire and the heater circuit (also check for a short to other circuits). It is possible that the fault could affect operation of other systems e.g. fuelling etc; therefore check for other fault codes.

NOTE 1. Check for other fault codes that could provide additional information.

NOTE 2. Communication between control units can pass via a CAN-Bus system; refer to Page 51 for CAN-Bus checks.

NOTE 3. If a fault cannot be located, it is also possible that the control unit is at fault.

NOTE 4. Refer to Page 53 for list of pages that show the ISO standard for component locations e.g. Sensor A - Bank 1.

Code	Description	Note (general)	Detailed checks
P2237	O2 Sensor Positive Current Control Circuit / Open Bank 1 Sensor 1	Note: Many O2 / Lambda sensor related fault codes are activated due to faults in other systems (engine, fuelling, misfires etc). O2 (Oxygen / Lambda) sensors detect the exhaust gas oxygen content (an indicator of air / fuel ratio). Oxygen content is indicated by a "Lambda" value. The control unit alters the fuelling to achieve the optimum Lambda value to enable good efficiency of catalytic converters (reduce emissions of harmful gases). For the operation of "broadband" sensors, the measurement process involves applying positive or negative current (pumping current) to one part of the sensing element; the level of positive or negative current being applied is an indicator of the oxygen content.	For a detailed list of checks and identification of the sensor location, Refer to list of sensor / actuator checks on page 19. The fault code indicates that the there is a possibly open circuit / fault in the circuit used to apply current to the measuring element of the Oxygen / Lambda sensor. Due to the fact that the fault code refers to a circuit / open circuit fault, the problem is most likely to be related to a sensor or wiring fault; check wiring between sensor and control unit for short / open circuits and high resistance. Note that the sensor could also be contaminated or faulty. Check sensor signal and signal response under different operating conditions (load, constant rpm etc) and Refer to list of sensor / actuator checks on page 19. If the sensor does however appear to be operating correctly but the signal value is incorrect, this could indicate an engine / fuelling fault, or other fault causing incorrect oxygen content in the exhaust gas.
P2238	O2 Sensor Positive Current Control Circuit Low Bank 1 Sensor 1	Note: Many O2 / Lambda sensor related fault codes are activated due to faults in other systems (engine, fuelling, misfires etc). O2 (Oxygen / Lambda) sensors detect the exhaust gas oxygen content (an indicator of air / fuel ratio). Oxygen content is indicated by a "Lambda" value. The control unit alters the fuelling to achieve the optimum Lambda value to enable good efficiency of catalytic converters (reduce emissions of harmful gases). For the operation of "broadband" sensors, the measurement process involves applying positive or negative current (pumping current) to one part of the sensing element; the level of positive or negative current being applied is an indicator of the oxygen content.	For a detailed list of checks and identification of the sensor location, Refer to list of sensor / actuator checks on page 19. The fault code indicates that the current (being applied to the measuring element of the Oxygen / Lambda sensor) is low; the current is likely to be below the normal operating range. The problem is most likely to be related to a sensor or wiring fault; check wiring between sensor and control unit for short / open circuits and high resistance. Note that the sensor could also be contaminated or faulty. Check sensor signal and signal response under different operating conditions (load, constant rpm etc) and Refer to list of sensor / actuator checks on page 19. If the sensor does however appear to be operating correctly but the current value remains incorrect, this could indicate an engine / fuelling fault, or other fault causing incorrect oxygen content in the exhaust gas.
P2239	O2 Sensor Positive Current Control Circuit High Bank 1 Sensor 1	Note: Many O2 / Lambda sensor related fault codes are activated due to faults in other systems (engine, fuelling, misfires etc). O2 (Oxygen / Lambda) sensors detect the exhaust gas oxygen content (an indicator of air / fuel ratio). Oxygen content is indicated by a "Lambda" value. The control unit alters the fuelling to achieve the optimum Lambda value to enable good efficiency of catalytic converters (reduce emissions of harmful gases). For the operation of "broadband" sensors, the measurement process involves applying positive or negative current (pumping current) to one part of the sensing element; the level of positive or negative current being applied is an indicator of the oxygen content.	For a detailed list of checks and identification of the sensor location, Refer to list of sensor / actuator checks on page 19. The fault code indicates that the current (being applied to the measuring element of the Oxygen / Lambda sensor) is high; the current is likely to be above the normal operating range. The problem is most likely to be related to a sensor or wiring fault; check wiring between sensor and control unit for short / open circuits and high resistance. Note that the sensor could also be contaminated or faulty. Check sensor signal and signal response under different operating conditions (load, constant rpm etc) and Refer to list of sensor / actuator checks on page 19. If the sensor does however appear to be operating correctly but the current value remains incorrect, this could indicate an engine / fuelling fault, or other fault causing incorrect oxygen content in the exhaust gas.
P2240	O2 Sensor Positive Current Control Circuit / Open Bank 2 Sensor 1	Note: Many O2 / Lambda sensor related fault codes are activated due to faults in other systems (engine, fuelling, misfires etc). O2 (Oxygen / Lambda) sensors detect the exhaust gas oxygen content (an indicator of air / fuel ratio). Oxygen content is indicated by a "Lambda" value. The control unit alters the fuelling to achieve the optimum Lambda value to enable good efficiency of catalytic converters (reduce emissions of harmful gases). For the operation of "broadband" sensors, the measurement process involves applying positive or negative current (pumping current) to one part of the sensing element; the level of positive or negative current being applied is an indicator of the oxygen content.	For a detailed list of checks and identification of the sensor location, Refer to list of sensor / actuator checks on page 19. The fault code indicates that the there is a possibly open circuit / fault in the circuit used to apply current to the measuring element of the Oxygen / Lambda sensor. Due to the fact that the fault code refers to a circuit / open circuit fault, the problem is most likely to be related to a sensor or wiring fault; check wiring between sensor and control unit for short / open circuits and high resistance. Note that the sensor could also be contaminated or faulty. Check sensor signal and signal response under different operating conditions (load, constant rpm etc) and Refer to list of sensor / actuator checks on page 19. If the sensor does however appear to be operating correctly but the signal value is incorrect, this could indicate an engine / fuelling fault, or other fault causing incorrect oxygen content in the exhaust gas.
P2241	O2 Sensor Positive Current Control Circuit Low Bank 2 Sensor 1	Note: Many O2 / Lambda sensor related fault codes are activated due to faults in other systems (engine, fuelling, misfires etc). O2 (Oxygen / Lambda) sensors detect the exhaust gas oxygen content (an indicator of air / fuel ratio). Oxygen content is indicated by a "Lambda" value. The control unit alters the fuelling to achieve the optimum Lambda value to enable good efficiency of catalytic converters (reduce emissions of harmful gases). For the operation of "broadband" sensors, the measurement process involves applying positive or negative current (pumping current) to one part of the sensing element; the level of positive or negative current being applied is an indicator of the oxygen content.	For a detailed list of checks and identification of the sensor location, Refer to list of sensor / actuator checks on page 19. The fault code indicates that the current (being applied to the measuring element of the Oxygen / Lambda sensor) is low; the current is likely to be below the normal operating range. The problem is most likely to be related to a sensor or wiring fault; check wiring between sensor and control unit for short / open circuits and high resistance. Note that the sensor could also be contaminated or faulty. Check sensor signal and signal response under different operating conditions (load, constant rpm etc) and Refer to list of sensor / actuator checks on page 19. If the sensor does however appear to be operating correctly but the current value remains incorrect, this could indicate an engine / fuelling fault, or other fault causing incorrect oxygen content in the exhaust gas.
P2242	O2 Sensor Positive Current Control Circuit High Bank 2 Sensor 1	Note: Many O2 / Lambda sensor related fault codes are activated due to faults in other systems (engine, fuelling, misfires etc). O2 (Oxygen / Lambda) sensors detect the exhaust gas oxygen content (an indicator of air / fuel ratio). Oxygen content is indicated by a "Lambda" value. The control unit alters the fuelling to achieve the optimum Lambda value to enable good efficiency of catalytic converters (reduce emissions of harmful gases). For the operation of "broadband" sensors, the measurement process involves applying positive or negative current (pumping current) to one part of the sensing element; the level of positive or negative current being applied is an indicator of the oxygen content.	For a detailed list of checks and identification of the sensor location, Refer to list of sensor / actuator checks on page 19. The fault code indicates that the current (being applied to the measuring element of the Oxygen / Lambda sensor) is high; the current is likely to be above the normal operating range. The problem is most likely to be related to a sensor or wiring fault; check wiring between sensor and control unit for short / open circuits and high resistance. Note that the sensor could also be contaminated or faulty. Check sensor signal and signal response under different operating conditions (load, constant rpm etc) and Refer to list of sensor / actuator checks on page 19. If the sensor does however appear to be operating correctly but the current value remains incorrect, this could indicate an engine / fuelling fault, or other fault causing incorrect oxygen content in the exhaust gas.

NOTE 1. Check for other fault codes that could provide additional information. NOTE 2. Communication between control units can pass via a CAN-Bus system; refer to Page 51 for CAN-Bus checks.

NOTE 3. If a fault cannot be located, it is also possible that the control unit is at fault. NOTE 4. Refer to Page 53 for list of pages that show the ISO standard for component locations e.g. Sensor A - Bank 1.

Fault code	EOBD / ISO Description	Component / System Description	Meaningful Description and Quick Check
P2243	O2 Sensor Reference Voltage Circuit / Open Bank 1 Sensor 1	Note: Many O2 / Lambda sensor related fault codes are activated due to faults in other systems (engine, fuelling, misfires etc). O2 (Oxygen / Lambda) sensors detect the exhaust gas oxygen content (an indicator of air / fuel ratio). Oxygen content is indicated by a "Lambda" value. The control unit alters the fuelling to achieve the optimum Lambda value to enable good efficiency of catalytic converters (reduce emissions of harmful gases).	For a detailed list of checks and identification of the sensor location, Refer to list of sensor / actuator checks on page 19. The fault code refers to a "Reference Voltage", which can be applied to two different types of Oxygen / Lambda sensors. It will therefore be necessary to identify which type of sensor is fitted to the vehicle i.e. Titania or Broadband. For both types of sensor, the fault code indicates that the reference voltage (or circuit carrying the reference voltage) is incorrect and, in both cases the fault could be related to a sensor / wiring fault; check wiring between sensor and control unit for short / open circuits and high resistance. Note that the sensor could also be contaminated or faulty. In both cases it is also possible that the oxygen content in the exhaust gas is incorrect for the operating conditions, which could indicate that the sensor might be operating correctly. If the sensor does appear to be operating correctly the fault could therefore be caused by an engine / fuelling fault, or other fault causing incorrect oxygen content in the exhaust gas.
P2244	O2 Sensor Reference Voltage Performance Bank 1 Sensor 1	Note: Many O2 / Lambda sensor related fault codes are activated due to faults in other systems (engine, fuelling, misfires etc). O2 (Oxygen / Lambda) sensors detect the exhaust gas oxygen content (an indicator of air / fuel ratio). Oxygen content is indicated by a "Lambda" value. The control unit alters the fuelling to achieve the optimum Lambda value to enable good efficiency of catalytic converters (reduce emissions of harmful gases).	For a detailed list of checks and identification of the sensor location, Refer to list of sensor / actuator checks on page 19. The fault code refers to a "Reference Voltage", which can be applied to two different types of Oxygen / Lambda sensors. It will therefore be necessary to identify which type of sensor is fitted to the vehicle i.e. Titania or Broadband. For both types of sensor, the fault code indicates that the reference voltage (or circuit carrying the reference voltage) is incorrect and, in both cases the fault could be related to a sensor / wiring fault; check wiring between sensor and control unit for short / open circuits and high resistance. Note that the sensor could also be contaminated or faulty. In both cases it is also possible that the oxygen content in the exhaust gas is incorrect for the operating conditions, which could indicate that the sensor might be operating correctly. If the sensor does appear to be operating correctly the fault could therefore be caused by an engine / fuelling fault, or other fault causing incorrect oxygen content in the exhaust gas.
P2245	O2 Sensor Reference Voltage Circuit Low Bank 1 Sensor 1	Note: Many O2 / Lambda sensor related fault codes are activated due to faults in other systems (engine, fuelling, misfires etc). O2 (Oxygen / Lambda) sensors detect the exhaust gas oxygen content (an indicator of air / fuel ratio). Oxygen content is indicated by a "Lambda" value. The control unit alters the fuelling to achieve the optimum Lambda value to enable good efficiency of catalytic converters (reduce emissions of harmful gases).	For a detailed list of checks and identification of the sensor location, Refer to list of sensor / actuator checks on page 19. The fault code refers to a "Reference Voltage", which can be applied to two different types of Oxygen / Lambda sensors. It will therefore be necessary to identify which type of sensor is fitted to the vehicle i.e. Titania or Broadband. For both types of sensor, the fault code indicates that the reference voltage (or circuit carrying the reference voltage) is incorrect and, in both cases the fault could be related to a sensor / wiring fault; check wiring between sensor and control unit for short / open circuits and high resistance. Note that the sensor could also be contaminated or faulty. In both cases it is also possible that the oxygen content in the exhaust gas is incorrect for the operating conditions, which could indicate that the sensor might be operating correctly. If the sensor does appear to be operating correctly the fault could therefore be caused by an engine / fuelling fault, or other fault causing incorrect oxygen content in the exhaust gas.
P2246	O2 Sensor Reference Voltage Circuit High Bank 1 Sensor 1	Note: Many O2 / Lambda sensor related fault codes are activated due to faults in other systems (engine, fuelling, misfires etc). O2 (Oxygen / Lambda) sensors detect the exhaust gas oxygen content (an indicator of air / fuel ratio). Oxygen content is indicated by a "Lambda" value. The control unit alters the fuelling to achieve the optimum Lambda value to enable good efficiency of catalytic converters (reduce emissions of harmful gases).	For a detailed list of checks and identification of the sensor location, Refer to list of sensor / actuator checks on page 19. The fault code refers to a "Reference Voltage", which can be applied to two different types of Oxygen / Lambda sensors. It will therefore be necessary to identify which type of sensor is fitted to the vehicle i.e. Titania or Broadband. For both types of sensor, the fault code indicates that the reference voltage (or circuit carrying the reference voltage) is incorrect and, in both cases the fault could be related to a sensor / wiring fault; check wiring between sensor and control unit for short / open circuits and high resistance. Note that the sensor could also be contaminated or faulty. In both cases it is also possible that the oxygen content in the exhaust gas is incorrect for the operating conditions, which could indicate that the sensor might be operating correctly. If the sensor does appear to be operating correctly the fault could therefore be caused by an engine / fuelling fault, or other fault causing incorrect oxygen content in the exhaust gas.
P2247	O2 Sensor Reference Voltage Circuit / Open Bank 2 Sensor 1	Note: Many O2 / Lambda sensor related fault codes are activated due to faults in other systems (engine, fuelling, misfires etc). O2 (Oxygen / Lambda) sensors detect the exhaust gas oxygen content (an indicator of air / fuel ratio). Oxygen content is indicated by a "Lambda" value. The control unit alters the fuelling to achieve the optimum Lambda value to enable good efficiency of catalytic converters (reduce emissions of harmful gases).	For a detailed list of checks and identification of the sensor location, Refer to list of sensor / actuator checks on page 19. The fault code refers to a "Reference Voltage", which can be applied to two different types of Oxygen / Lambda sensors. It will therefore be necessary to identify which type of sensor is fitted to the vehicle i.e. Titania or Broadband. For both types of sensor, the fault code indicates that the reference voltage (or circuit carrying the reference voltage) is incorrect and, in both cases the fault could be related to a sensor / wiring fault; check wiring between sensor and control unit for short / open circuits and high resistance. Note that the sensor could also be contaminated or faulty. In both cases it is also possible that the oxygen content in the exhaust gas is incorrect for the operating conditions, which could indicate that the sensor might be operating correctly. If the sensor does appear to be operating correctly the fault could therefore be caused by an engine / fuelling fault, or other fault causing incorrect oxygen content in the exhaust gas.
P2248	O2 Sensor Reference Voltage Performance Bank 2 Sensor 1	Note: Many O2 / Lambda sensor related fault codes are activated due to faults in other systems (engine, fuelling, misfires etc). O2 (Oxygen / Lambda) sensors detect the exhaust gas oxygen content (an indicator of air / fuel ratio). Oxygen content is indicated by a "Lambda" value. The control unit alters the fuelling to achieve the optimum Lambda value to enable good efficiency of catalytic converters (reduce emissions of harmful gases).	For a detailed list of checks and identification of the sensor location, Refer to list of sensor / actuator checks on page 19. The fault code refers to a "Reference Voltage", which can be applied to two different types of Oxygen / Lambda sensors. It will therefore be necessary to identify which type of sensor is fitted to the vehicle i.e. Titania or Broadband. For both types of sensor, the fault code indicates that the reference voltage (or circuit carrying the reference voltage) is incorrect and, in both cases the fault could be related to a sensor / wiring fault; check wiring between sensor and control unit for short / open circuits and high resistance. Note that the sensor could also be contaminated or faulty. In both cases it is also possible that the oxygen content in the exhaust gas is incorrect for the operating conditions, which could indicate that the sensor might be operating correctly. If the sensor does appear to be operating correctly the fault could therefore be caused by an engine / fuelling fault, or other fault causing incorrect oxygen content in the exhaust gas.

NOTE 1. Check for other fault codes that could provide additional information. **NOTE 2.** Communication between control units can pass via a CAN-Bus system; refer to Page 51 for CAN-Bus checks.

250

Code	Description	Note	Detail
P2249	O2 Sensor Reference Voltage Circuit Low Bank 2 Sensor 1	Note: Many O2 / Lambda sensor related fault codes are activated due to faults in other systems (engine, fuelling, misfires etc). O2 (Oxygen / Lambda) sensors detect the exhaust gas oxygen content (an indicator of air / fuel ratio). Oxygen content is indicated by a "Lambda" value. The control unit alters the fuelling to achieve the optimum Lambda value to enable good efficiency of catalytic converters (reduce emissions of harmful gases).	For a detailed list of checks and identification of the sensor location, Refer to list of sensor / actuator checks on page 19. The fault code refers to a "Reference Voltage", which can be applied to two different types of Oxygen / Lambda sensors. It will therefore be necessary to identify which type of sensor is fitted to the vehicle i.e. Titania or Broadband. For both types of sensor, the fault code indicates that the reference voltage (or circuit carrying the reference voltage) is incorrect and, in both cases the fault could be related to a sensor / wiring fault; check wiring between sensor and control unit for short / open circuits and high resistance. Note that the sensor could also be contaminated or faulty. In both cases it is also possible that the oxygen content in the exhaust gas is incorrect for the operating conditions, which could indicate that the sensor might be operating correctly. If the sensor does appear to be operating correctly the fault could therefore be caused by an engine / fuelling fault, or other fault causing incorrect oxygen content in the exhaust gas.
P2250	O2 Sensor Reference Voltage Circuit High Bank 2 Sensor 1	Note: Many O2 / Lambda sensor related fault codes are activated due to faults in other systems (engine, fuelling, misfires etc). O2 (Oxygen / Lambda) sensors detect the exhaust gas oxygen content (an indicator of air / fuel ratio). Oxygen content is indicated by a "Lambda" value. The control unit alters the fuelling to achieve the optimum Lambda value to enable good efficiency of catalytic converters (reduce emissions of harmful gases).	For a detailed list of checks and identification of the sensor location, Refer to list of sensor / actuator checks on page 19. The fault code refers to a "Reference Voltage", which can be applied to two different types of Oxygen / Lambda sensors. It will therefore be necessary to identify which type of sensor is fitted to the vehicle i.e. Titania or Broadband. For both types of sensor, the fault code indicates that the reference voltage (or circuit carrying the reference voltage) is incorrect and, in both cases the fault could be related to a sensor / wiring fault; check wiring between sensor and control unit for short / open circuits and high resistance. Note that the sensor could also be contaminated or faulty. In both cases it is also possible that the oxygen content in the exhaust gas is incorrect for the operating conditions, which could indicate that the sensor might be operating correctly. If the sensor does appear to be operating correctly the fault could therefore be caused by an engine / fuelling fault, or other fault causing incorrect oxygen content in the exhaust gas
P2251	O2 Sensor Negative Current Control Circuit / Open Bank 1 Sensor 1	Note: Many O2 / Lambda sensor related fault codes are activated due to faults in other systems (engine, fuelling, misfires etc). O2 (Oxygen / Lambda) sensors detect the exhaust gas oxygen content (an indicator of air / fuel ratio). Oxygen content is indicated by a "Lambda" value. The control unit alters the fuelling to achieve the optimum Lambda value to enable good efficiency of catalytic converters (reduce emissions of harmful gases). For the operation of "broadband" sensors, the measurement process involves applying positive or negative current (pumping current) to one part of the sensing element; the level of positive or negative current being applied is an indicator of the oxygen content.	For a detailed list of checks and identification of the sensor location, Refer to list of sensor / actuator checks on page 19. The fault code indicates that the there is a possibly open circuit / fault in the circuit used to apply current to the measuring element of the Oxygen / Lambda sensor. Due to the fact that the fault code refers to a circuit / open circuit fault, the problem is most likely to be related to a sensor or wiring fault; check wiring between sensor and control unit for short / open circuits and high resistance. Note that the sensor could also be contaminated or faulty. Check sensor signal and signal response under different operating conditions (load, constant rpm etc) and Refer to list of sensor / actuator checks on page 19. If the sensor does however appear to be operating correctly but the signal value is incorrect, this could indicate an engine / fuelling fault, or other fault causing incorrect oxygen content in the exhaust gas.
P2252	O2 Sensor Negative Current Control Circuit Low Bank 1 Sensor 1	Note: Many O2 / Lambda sensor related fault codes are activated due to faults in other systems (engine, fuelling, misfires etc). O2 (Oxygen / Lambda) sensors detect the exhaust gas oxygen content (an indicator of air / fuel ratio). Oxygen content is indicated by a "Lambda" value. The control unit alters the fuelling to achieve the optimum Lambda value to enable good efficiency of catalytic converters (reduce emissions of harmful gases). For the operation of "broadband" sensors, the measurement process involves applying positive or negative current (pumping current) to one part of the sensing element; the level of positive or negative current being applied is an indicator of the oxygen content.	For a detailed list of checks and identification of the sensor location, Refer to list of sensor / actuator checks on page 19. The fault code indicates that the current (being applied to the measuring element of the Oxygen / Lambda sensor) is low; the current is likely to be below the normal operating range. The problem is most likely to be related to a sensor or wiring fault; check wiring between sensor and control unit for short / open circuits and high resistance. Note that the sensor could also be contaminated or faulty. Check sensor signal and signal response under different operating conditions (load, constant rpm etc) and Refer to list of sensor / actuator checks on page 19. If the sensor does however appear to be operating correctly but the current value remains incorrect, this could indicate an engine / fuelling fault, or other fault causing incorrect oxygen content in the exhaust gas.
P2253	O2 Sensor Negative Current Control Circuit High Bank 1 Sensor 1	Note: Many O2 / Lambda sensor related fault codes are activated due to faults in other systems (engine, fuelling, misfires etc). O2 (Oxygen / Lambda) sensors detect the exhaust gas oxygen content (an indicator of air / fuel ratio). Oxygen content is indicated by a "Lambda" value. The control unit alters the fuelling to achieve the optimum Lambda value to enable good efficiency of catalytic converters (reduce emissions of harmful gases). For the operation of "broadband" sensors, the measurement process involves applying positive or negative current (pumping current) to one part of the sensing element; the level of positive or negative current being applied is an indicator of the oxygen content.	For a detailed list of checks and identification of the sensor location, Refer to list of sensor / actuator checks on page 19. The fault code indicates that the current (being applied to the measuring element of the Oxygen / Lambda sensor) is high; the current is likely to be above the normal operating range. The problem is most likely to be related to a sensor or wiring fault; check wiring between sensor and control unit for short / open circuits and high resistance. Note that the sensor could also be contaminated or faulty. Check sensor signal and signal response under different operating conditions (load, constant rpm etc) and Refer to list of sensor / actuator checks on page 19. If the sensor does however appear to be operating correctly but the current value remains incorrect, this could indicate an engine / fuelling fault, or other fault causing incorrect oxygen content in the exhaust gas.
P2254	O2 Sensor Negative Current Control Circuit / Open Bank 2 Sensor 1	Note: Many O2 / Lambda sensor related fault codes are activated due to faults in other systems (engine, fuelling, misfires etc). O2 (Oxygen / Lambda) sensors detect the exhaust gas oxygen content (an indicator of air / fuel ratio). Oxygen content is indicated by a "Lambda" value. The control unit alters the fuelling to achieve the optimum Lambda value to enable good efficiency of catalytic converters (reduce emissions of harmful gases). For the operation of "broadband" sensors, the measurement process involves applying positive or negative current (pumping current) to one part of the sensing element; the level of positive or negative current being applied is an indicator of the oxygen content.	For a detailed list of checks and identification of the sensor location, Refer to list of sensor / actuator checks on page 19. The fault code indicates that the there is a possibly open circuit / fault in the circuit used to apply current to the measuring element of the Oxygen / Lambda sensor. Due to the fact that the fault code refers to a circuit / open circuit fault, the problem is most likely to be related to a sensor or wiring fault; check wiring between sensor and control unit for short / open circuits and high resistance. Note that the sensor could also be contaminated or faulty. Check sensor signal and signal response under different operating conditions (load, constant rpm etc) and Refer to list of sensor / actuator checks on page 19. If the sensor does however appear to be operating correctly but the signal value is incorrect, this could indicate an engine / fuelling fault, or other fault causing incorrect oxygen content in the exhaust gas.

Fault code	EOBD / ISO Description	Component / System Description	Meaningful Description and Quick Check
P2255	O2 Sensor Negative Current Control Circuit Low Bank 2 Sensor 1	Note: Many O2 / Lambda sensor related fault codes are activated due to faults in other systems (engine, fuelling, misfires etc). O2 (Oxygen / Lambda) sensors detect the exhaust gas oxygen content (an indicator of air / fuel ratio). Oxygen content is indicated by a "Lambda" value. The control unit alters the fuelling to achieve the optimum Lambda value to enable good efficiency of catalytic converters (reduce emissions of harmful gases). For the operation of "broadband" sensors, the measurement process involves applying positive or negative current (pumping current) to one part of the sensing element; the level of positive or negative current being applied is an indicator of the oxygen content.	For a detailed list of checks and identification of the sensor location, Refer to list of sensor / actuator checks on page 19. The fault code indicates that the current (being applied to the measuring element of the Oxygen / Lambda sensor) is low; the current is likely to be below the normal operating range. The problem is most likely to be related to a sensor or wiring fault; check wiring between sensor and control unit for short / open circuits and high resistance. Note that the sensor could also be contaminated or faulty. Check sensor signal and signal response under different operating conditions (load, constant rpm etc) and Refer to list of sensor / actuator checks on page 19. If the sensor does however appear to be operating correctly but the current value remains incorrect, this could indicate an engine / fuelling fault, or other fault causing incorrect oxygen content in the exhaust gas.
P2256	O2 Sensor Negative Current Control Circuit High Bank 2 Sensor 1	Note: Many O2 / Lambda sensor related fault codes are activated due to faults in other systems (engine, fuelling, misfires etc). O2 (Oxygen / Lambda) sensors detect the exhaust gas oxygen content (an indicator of air / fuel ratio). Oxygen content is indicated by a "Lambda" value. The control unit alters the fuelling to achieve the optimum Lambda value to enable good efficiency of catalytic converters (reduce emissions of harmful gases). For the operation of "broadband" sensors, the measurement process involves applying positive or negative current (pumping current) to one part of the sensing element; the level of positive or negative current being applied is an indicator of the oxygen content.	For a detailed list of checks and identification of the sensor location, Refer to list of sensor / actuator checks on page 19. The fault code indicates that the current (being applied to the measuring element of the Oxygen / Lambda sensor) is high; the current is likely to be above the normal operating range. The problem is most likely to be related to a sensor or wiring fault; check wiring between sensor and control unit for short / open circuits and high resistance. Note that the sensor could also be contaminated or faulty. Check sensor signal and signal response under different operating conditions (load, constant rpm etc) and Refer to list of sensor / actuator checks on page 19. If the sensor does however appear to be operating correctly but the current value remains incorrect, this could indicate an engine / fuelling fault, or other fault causing incorrect oxygen content in the exhaust gas.
P2257	Secondary Air Injection System Control "A" Circuit Low	After cold starts, the secondary air injection system pumps air (oxygen) into the exhaust gas (which is rich with HC and CO), this helps the HC / CO to combust (combustion heat in exhaust gas helps to heat the catalyst); the additional air can be detected by the Lambda (oxygen) sensor although a flow / pressure sensor can be used. Note that some systems (pulse air systems) use negative exhaust pressure pulses to draw air into the exhaust system instead of a pump. Refer to vehicle specific information to establish the components fitted to the system.	The voltage, frequency or value of the control signal in the actuator circuit is either "zero" (no signal) or, the signal exists but is below the normal operating range. Possible fault with actuator or wiring (e.g. actuator power supply, short / open circuit or resistance) that is causing a signal value to be lower than the minimum operating limit. Note: the control unit could be providing the correct signal but it is being affected by the circuit fault. Refer to list of sensor / actuator checks on page 19.
P2258	Secondary Air Injection System Control "A" Circuit High	After cold starts, the secondary air injection system pumps air (oxygen) into the exhaust gas (which is rich with HC and CO), this helps the HC / CO to combust (combustion heat in exhaust gas helps to heat the catalyst); the additional air can be detected by the Lambda (oxygen) sensor although a flow / pressure sensor can be used. Note that some systems (pulse air systems) use negative exhaust pressure pulses to draw air into the exhaust system instead of a pump. Refer to vehicle specific information to establish the components fitted to the system.	The voltage, frequency or value of the control signal in the actuator circuit is either at full value (e.g. battery voltage with no on / off signal) or, the signal exists but it is above the normal operating range. Possible fault with actuator or wiring (e.g. actuator power supply, short / open circuit or resistance) that is causing a signal value to be higher than the maximum operating limit. Note: the control unit could be providing the correct signal but it is being affected by the circuit fault. Refer to list of sensor / actuator checks on page 19.
P2259	Secondary Air Injection System Control "B" Circuit Low	After cold starts, the secondary air injection system pumps air (oxygen) into the exhaust gas (which is rich with HC and CO), this helps the HC / CO to combust (combustion heat in exhaust gas helps to heat the catalyst); the additional air can be detected by the Lambda (oxygen) sensor although a flow / pressure sensor can be used. Note that some systems (pulse air systems) use negative exhaust pressure pulses to draw air into the exhaust system instead of a pump. Refer to vehicle specific information to establish the components fitted to the system.	The voltage, frequency or value of the control signal in the actuator circuit is either "zero" (no signal) or, the signal exists but is below the normal operating range. Possible fault with actuator or wiring (e.g. actuator power supply, short / open circuit or resistance) that is causing a signal value to be lower than the minimum operating limit. Note: the control unit could be providing the correct signal but it is being affected by the circuit fault. Refer to list of sensor / actuator checks on page 19.
P2260	Secondary Air Injection System Control "B" Circuit High	After cold starts, the secondary air injection system pumps air (oxygen) into the exhaust gas (which is rich with HC and CO), this helps the HC / CO to combust (combustion heat in exhaust gas helps to heat the catalyst); the additional air can be detected by the Lambda (oxygen) sensor although a flow / pressure sensor can be used. Note that some systems (pulse air systems) use negative exhaust pressure pulses to draw air into the exhaust system instead of a pump. Refer to vehicle specific information to establish the components fitted to the system.	The voltage, frequency or value of the control signal in the actuator circuit is either at full value (e.g. battery voltage with no on / off signal) or, the signal exists but it is above the normal operating range. Possible fault with actuator or wiring (e.g. actuator power supply, short / open circuit or resistance) that is causing a signal value to be higher than the maximum operating limit. Note: the control unit could be providing the correct signal but it is being affected by the circuit fault. Refer to list of sensor / actuator checks on page 19.
P2261	Turbocharger / Supercharger Bypass Valve -Mechanical	Turbo / supercharger systems have different methods of regulating boost pressure e.g. wastegate or variable geometry control. Some systems (blow through systems where the turbo / supercharger is located before / upstream of the throttle body), have a bypass valve (blow off / dump valve) which opens to relieve pressure build up and surging in the intake system during closed throttle deceleration. The bypass valve actuator is often a solenoid valve which controls the vacuum used to open / close the bypass valve.	The fault code indicates that the control unit detects a fault with the bypass valve operation (possibly detected due to high intake pressure) but, the control unit does not detect any electrical faults in the applicable actuators or sensor circuits. It is therefore assessed that a mechanical fault exists in the by pass valve or associated control mechanism. If checks need to be carried out on the electrical actuator and wiring, Refer to list of sensor / actuator checks on page 19.

NOTE 1. Check for other fault codes that could provide additional information.   **NOTE 2.** Communication between control units can pass via a CAN-Bus system; refer to Page 51 for CAN-Bus checks.

NOTE 3. If a fault cannot be located, it is also possible that the control unit is at fault   NOTE 4. Refer to Page 53 for list of pages that show the ISO standard for component locations e.g. Sensor A - Bank 1.

Code	Fault	Description	Diagnosis
P2262	Turbocharger / Supercharger Boost Pressure Not Detected -Mechanical	Different types of turbo / supercharger systems can be fitted to produce additional engine performance. For this fault code, it is likely that the system uses a variable geometry turbocharger with a mechanism to control the geometry of the turbine blades or nozzle. It may be necessary to identify the system and components used.	The control unit detects incorrect or no boost pressure but, the control unit does not detect any electrical faults in the applicable actuators or sensor circuits. It is therefore assessed that a mechanical fault exists in the turbo / supercharger or associated control mechanism. The control unit is possibly making the assessment because of information from a number of sensors e.g. boost pressure, engine speed / load, throttle position etc. Check boost control system (and boost pressure sensor) e.g. wastegate or variable geometry system, and check turbo / supercharger operation.
P2263	Turbocharger / Supercharger Boost System Performance	Different types of turbo / supercharger systems can be fitted to produce additional engine performance. For this fault code, it may be necessary to identify the system and components used.	The control unit detects an undefined fault with the turbo / supercharger boost system. The fault is likely to be related to incorrect boost pressure at a specific driving condition (or at various conditions). The control unit is possibly making the assessment because of information from a number of sensors e.g. boost pressure, engine speed / load, throttle position etc. The fault could caused by an electrical problem (e.g. boost pressure sensor or boost control actuator). Refer to list of sensor / actuator checks on page 19. Also check boost control system (and boost pressure sensor) e.g. wastegate or variable geometry system, and check turbo / supercharger operation.
P2264	Water in Fuel Sensor Circuit	A sensor can be used to detect water in the fuel system; this could be a sensor detecting the presence of water in the fuel or a sensor that detects the level of water in a water trap; on Diesel engines, both types are usually located in the base of the Diesel fuel filter	If necessary, identify the type of sensor fitted to the vehicle and the wiring for the sensor. The voltage, frequency or value of the sensor signal is incorrect (undefined fault). Sensor voltage / signal could be: out of normal operating range, constant value, non-existent, corrupt. Possible fault in the sensor or wiring e.g. short / open circuits, circuit resistance, incorrect sensor resistance or, fault with the reference / operating voltage; interference from other circuits can also affect sensor signals. Refer to list of sensor / actuator checks on page 19. If the sensor / switch contains a set of contacts, it is also possible that the contacts are stuck open / stuck closed, or contacts have a high resistance. Also check for excess water / moisture in water trap
P2265	Water in Fuel Sensor Circuit Range / Performance	A sensor can be used to detect water in the fuel system; this could be a sensor detecting the presence of water in the fuel or a sensor that detects the level of water in a water trap; on Diesel engines, both types are usually located in the base of the Diesel fuel filter	If necessary, identify the type of sensor fitted to the vehicle and the wiring for the sensor. The voltage, frequency or value of the sensor signal is within the normal operating range / tolerance but the signal is not plausible or is incorrect due to an undefined fault. The sensor signal might not be changing / responding as expected. Possible fault in the sensor or wiring e.g. short / open circuits, circuit resistance, incorrect sensor resistance or, fault with the reference / operating voltage; interference from other circuits can also affect sensor signals. Refer to list of sensor / actuator checks on page 19. If the sensor / switch contains a set of contacts, it is also possible that the contacts are stuck open or stuck closed (depends on how the sensor is wired) or contacts have a high resistance. It is also possible that the sensor is operating correctly but the information being detected by the sensor is unacceptable or incorrect i.e. water content too high. Check for excessive water in the water trap (drain if necessary) and also refer to fault codes P2266 and P2267 for a guide to other checks.
P2266	Water in Fuel Sensor Circuit Low	A sensor can be used to detect water in the fuel system; this could be a sensor detecting the presence of water in the fuel or a sensor that detects the level of water in a water trap; on Diesel engines, both types are usually located in the base of the Diesel fuel filter	If necessary, identify the type of sensor fitted to the vehicle and the wiring for the sensor. The voltage, frequency or value of the level sensor signal is either "zero" (no signal) or, the signal exists but it is below the normal operating range (lower than the minimum operating limit). Possible fault in the sensor or wiring e.g. short / open circuits, circuit resistance, incorrect sensor resistance or, fault with the reference / operating voltage; interference from other circuits can also affect sensor signals. Refer to list of sensor / actuator checks on page 19. Note that the fault code could be activated because the signal from the sensor does not increase as expected e.g. sensor signal stuck at a low value. If the sensor / switch contains a set of contacts, it is also possible that the contacts are stuck open or stuck closed (depends on how the sensor is wired) or contacts have a high resistance.
P2267	Water in Fuel Sensor Circuit High	A sensor can be used to detect water in the fuel system; this could be a sensor detecting the presence of water in the fuel or a sensor that detects the level of water in a water trap; on Diesel engines, both types are usually located in the base of the Diesel fuel filter	If necessary, identify the type of sensor fitted to the vehicle and the wiring for the sensor. The voltage, frequency or value of the sensor signal is either at full value (e.g. battery voltage with no signal or frequency) or, the signal exists but it is above the normal operating range (higher than the maximum operating limit). Possible fault in the sensor or wiring e.g. short / open circuits, circuit resistance, incorrect sensor resistance or, fault with the reference / operating voltage; interference from other circuits can also affect sensor signals. Refer to list of sensor / actuator checks on page 19. Note that the fault code could be activated because the signal from the sensor does not decrease as expected e.g. sensor signal stuck at a high value. If the sensor / switch contains a set of contacts, it is also possible that the contacts are stuck open or stuck closed (depends on how the sensor is wired) or contacts have a high resistance.
P2268	Water in Fuel Sensor Circuit Intermittent	A sensor can be used to detect water in the fuel system; this could be a sensor detecting the presence of water in the fuel or a sensor that detects the level of water in a water trap; on Diesel engines, both types are usually located in the base of the Diesel fuel filter	If necessary, identify the type of sensor fitted to the vehicle and the wiring for the sensor. The voltage, frequency or value of the sensor signal is intermittent (likely to be out of normal operating range / tolerance when intermittent fault is detected) or, the signal / voltage is erratic (e.g. signal changes are irregular / unpredictable). Possible fault with wiring or connections (including internal sensor connections); also interference from other circuits can affect the signal. If the sensor / switch contains a set of contacts, it is also possible that the contacts are intermittently opening / closing. Refer to list of sensor / actuator checks on page 19.
P2269	Water in Fuel Condition	A sensor can be used to detect water in the fuel system; this could be a sensor detecting the presence of water in the fuel or a sensor that detects the level of water in a water trap; on Diesel engines, both types are usually located in the base of the Diesel fuel filter.	The fault code is activated because the sensor signal indicates a high water content in the fuel. If applicable, use the water drain facility (usually located on the fuel filter base) to allow the water to drain from the water trap. If the sensor signal appears to be incorrect i.e. the water content is not high, it may be necessary to check the sensor operation. If necessary, identify the type of sensor fitted to the vehicle and the wiring for the sensor.

NOTE 1. Check for other fault codes that could provide additional information.
NOTE 2. Communication between control units can pass via a CAN-Bus system; refer to Page 51 for CAN-Bus checks.
NOTE 3. If a fault cannot be located, it is also possible that the control unit is at fault.
NOTE 4. Refer to Page 53 for list of pages that show the ISO standard for component locations e.g. Sensor A - Bank 1.

Fault code	EOBD / ISO Description	Component / System Description	Meaningful Description and Quick Check
P2270	O2 Sensor Signal Stuck Lean Bank 1 Sensor 2	Note: Many O2 / Lambda sensor related fault codes are activated due to faults in other systems (engine, fuelling, misfires etc). The control unit is able to detect the ratio of the air / fuel mixture by monitoring the information from different sensors. The O2 (Oxygen / Lambda) sensors (primarily the pre-cat sensor) will indicate the oxygen content in the exhaust gas, which is an indicator of air / fuel ratio. The control unit will also monitor other sensors to identify the operating conditions (e.g. heavy or light load etc) and (using the O2 sensor signal), assess whether the air / fuel ratio is correct. Note that faulty engine operation e.g. misfires and exhaust air leaks can cause increased oxygen in the exhaust gas which will affect the oxygen sensor signal.	For a detailed list of checks and identification of the sensor location, Refer to list of sensor / actuator checks on page 19. Different types of O2 / Lambda sensors (e.g. Zirconia or Titania / wideband etc) provide different signals and values, it will be necessary to identify the type of sensor fitted to the vehicle. The fault code indicates that the signal from the Oxygen / Lambda sensor is stuck or biased towards a lean mixture value, possibly during all operating conditions; the control unit could be applying a control signal to the injectors to provide a richer mixture but, the Oxygen / Lambda sensor signal is remaining lean. Note that this sensor is a post-cat sensor and if the fuel mixture is lean, it is likely that the pre-cat sensor will also be stuck lean (check for correct pre-cat sensor signal); check for an inefficient catalytic converter or air leak between the pre and post- cat sensors. The fault could be caused by a genuine lean mixture but, the fault code could be activated due to an Oxygen / Lambda sensor fault. Check sensor signal and signal response under different operating conditions (idle, load, etc). If the sensor appears to be operating correctly but the signal value is indicating a genuine lean mixture, this could indicate an engine related fault e.g. incorrect mixtures, fuel pressure, misfires etc. If the sensor appears to be at fault, Refer to list of sensor / actuator checks on page 19.
P2271	O2 Sensor Signal Stuck Rich Bank 1 Sensor 2	Note: Many O2 / Lambda sensor related fault codes are activated due to faults in other systems (engine, fuelling, misfires etc). The control unit is able to detect the ratio of the air / fuel mixture by monitoring the information from different sensors. The O2 (Oxygen / Lambda) sensors (primarily the pre-cat sensor) will indicate the oxygen content in the exhaust gas, which is an indicator of air / fuel ratio. The control unit will also monitor other sensors to identify the operating conditions (e.g. heavy or light load etc) and (using the O2 sensor signal), assess whether the air / fuel ratio is correct. Note that faulty engine operation e.g. misfires and exhaust air leaks can cause increased oxygen in the exhaust gas which will affect the oxygen sensor signal.	For a detailed list of checks and identification of the sensor location, Refer to list of sensor / actuator checks on page 19. Different types of O2 / Lambda sensors (e.g. Zirconia or Titania / wideband etc) provide different signals and values, it will be necessary to identify the type of sensor fitted to the vehicle. The fault code indicates that the signal from the Oxygen / Lambda sensor is stuck or biased towards a rich mixture value, possibly during all operating conditions; the control unit will be applying a control signal to the injectors to provide a leaner mixture but, the Oxygen / Lambda sensor signal is remaining rich. Note that this sensor is a post-cat sensor and if the fuel mixture is rich, it is likely that the pre-cat sensor will also be stuck rich (check for correct pre-cat sensor signal). The fault could be caused by a genuine rich mixture but, the fault code could be activated due to an Oxygen / Lambda sensor fault. Check sensor signal and signal response under different operating conditions (idle, load, etc). If the sensor appears to be operating correctly but the signal value is indicating a genuine rich mixture, this could indicate an engine related fault e.g. incorrect mixtures, fuel pressure, leaking injector etc.
P2272	O2 Sensor Signal Stuck Lean Bank 2 Sensor 2	Note: Many O2 / Lambda sensor related fault codes are activated due to faults in other systems (engine, fuelling, misfires etc). The control unit is able to detect the ratio of the air / fuel mixture by monitoring the information from different sensors. The O2 (Oxygen / Lambda) sensors (primarily the pre-cat sensor) will indicate the oxygen content in the exhaust gas, which is an indicator of air / fuel ratio. The control unit will also monitor other sensors to identify the operating conditions (e.g. heavy or light load etc) and (using the O2 sensor signal), assess whether the air / fuel ratio is correct. Note that faulty engine operation e.g. misfires and exhaust air leaks can cause increased oxygen in the exhaust gas which will affect the oxygen sensor signal.	For a detailed list of checks and identification of the sensor location, Refer to list of sensor / actuator checks on page 19. Different types of O2 / Lambda sensors (e.g. Zirconia or Titania / wideband etc) provide different signals and values, it will be necessary to identify the type of sensor fitted to the vehicle. The fault code indicates that the signal from the Oxygen / Lambda sensor is stuck or biased towards a lean mixture value, possibly during all operating conditions; the control unit could be applying a control signal to the injectors to provide a richer mixture but, the Oxygen / Lambda sensor signal is remaining lean. Note that this sensor is a post-cat sensor and if the fuel mixture is lean, it is likely that the pre-cat sensor will also be stuck lean (check for correct pre-cat sensor signal); check for an inefficient catalytic converter or air leak between the pre and post- cat sensors. The fault could be caused by a genuine lean mixture but, the fault code could be activated due to an Oxygen / Lambda sensor fault. Check sensor signal and signal response under different operating conditions (idle, load, etc). If the sensor appears to be operating correctly but the signal value is indicating a genuine lean mixture, this could indicate an engine related fault e.g. incorrect mixtures, fuel pressure, misfires etc. If the sensor appears to be at fault, Refer to list of sensor / actuator checks on page 19.
P2273	O2 Sensor Signal Stuck Rich Bank 2 Sensor 2	Note: Many O2 / Lambda sensor related fault codes are activated due to faults in other systems (engine, fuelling, misfires etc). The control unit is able to detect the ratio of the air / fuel mixture by monitoring the information from different sensors. The O2 (Oxygen / Lambda) sensors (primarily the pre-cat sensor) will indicate the oxygen content in the exhaust gas, which is an indicator of air / fuel ratio. The control unit will also monitor other sensors to identify the operating conditions (e.g. heavy or light load etc) and (using the O2 sensor signal), assess whether the air / fuel ratio is correct. Note that faulty engine operation e.g. misfires and exhaust air leaks can cause increased oxygen in the exhaust gas which will affect the oxygen sensor signal.	For a detailed list of checks and identification of the sensor location, Refer to list of sensor / actuator checks on page 19. Different types of O2 / Lambda sensors (e.g. Zirconia or Titania / wideband etc) provide different signals and values, it will be necessary to identify the type of sensor fitted to the vehicle. The fault code indicates that the signal from the Oxygen / Lambda sensor is stuck or biased towards a rich mixture value, possibly during all operating conditions; the control unit will be applying a control signal to the injectors to provide a leaner mixture but the Oxygen / Lambda sensor signal is remaining rich. Note that this sensor is a post-cat sensor and if the fuel mixture is rich, it is likely that the pre-cat sensor will also be stuck rich (check for correct pre-cat sensor signal). The fault could be caused by a genuine rich mixture but, the fault code could be activated due to an Oxygen / Lambda sensor fault. Check sensor signal and signal response under different operating conditions (idle, load, etc). If the sensor appears to be operating correctly but the signal value is indicating a genuine rich mixture, this could indicate an engine related fault e.g. incorrect mixtures, fuel pressure, leaking injector etc.
P2274	O2 Sensor Signal Stuck Lean Bank 1 Sensor 3	Note: Many O2 / Lambda sensor related fault codes are activated due to faults in other systems (engine, fuelling, misfires etc). The control unit is able to detect the ratio of the air / fuel mixture by monitoring the information from different sensors. The O2 (Oxygen / Lambda) sensors (primarily the pre-cat sensor) will indicate the oxygen content in the exhaust gas, which is an indicator of air / fuel ratio. The control unit will also monitor other sensors to identify the operating conditions (e.g. heavy or light load etc) and (using the O2 sensor signal), assess whether the air / fuel ratio is correct. Note that faulty engine operation e.g. misfires and exhaust air leaks can cause increased oxygen in the exhaust gas which will affect the oxygen sensor signal.	For a detailed list of checks and identification of the sensor location, Refer to list of sensor / actuator checks on page 19. Different types of O2 / Lambda sensors (e.g. Zirconia or Titania / wideband etc) provide different signals and values, it will be necessary to identify the type of sensor fitted to the vehicle. The fault code indicates that the signal from the Oxygen / Lambda sensor is stuck or biased towards a lean mixture value, possibly during all operating conditions; the control unit could be applying a control signal to the injectors to provide a richer mixture but, the Oxygen / Lambda sensor signal is remaining lean. Note that this sensor is a post-cat sensor (possibly a post "NOx accumulator" sensor) and if the fuel mixture is lean, it is likely that the pre-cat sensor will also be stuck lean (check for correct pre-cat sensor signals); check for an inefficient catalytic converter or air leak between the pre and post-cat sensors. The fault could be caused by a genuine lean mixture but, the fault code could be activated due to an Oxygen / Lambda sensor fault. Check sensor signal and signal response under different operating conditions (idle, load, etc). If the sensor appears to be operating correctly but the signal value is indicating a genuine lean mixture or excess oxygen in post cat exhaust gas, this could indicate an engine related fault e.g. incorrect mixtures, fuel pressure, misfires etc or, a catalytic converter fault. If the sensor appears to be at fault, Refer to list of sensor / actuator checks on page 19.

Code	Description	Note	Details
P2275	O2 Sensor Signal Stuck Rich Bank 1 Sensor 3	Note: Many O2 / Lambda sensor related fault codes are activated due to faults in other systems (engine, fuelling, misfires etc). The control unit is able to detect the ratio of the air / fuel mixture by monitoring the information from different sensors. The O2 (Oxygen / Lambda) sensors (primarily the pre-cat sensor) will indicate the oxygen content in the exhaust gas, which is an indicator of air / fuel ratio. The control unit will also monitor other sensors to identify the operating conditions (e.g. heavy or light load etc) and (using the O2 sensor signal), assess whether the air / fuel ratio is correct. Note that faulty engine operation e.g. misfires and exhaust air leaks can cause increased oxygen in the exhaust gas which will affect the oxygen sensor signal.	For a detailed list of checks and identification of the sensor location, Refer to list of sensor / actuator checks on page 19. Different types of O2 / Lambda sensors (e.g. Zirconia or Titania / wideband etc) provide different signals and values, it will be necessary to identify the type of sensor fitted to the vehicle. The fault code indicates that the signal from the Oxygen / Lambda sensor is stuck or biased towards a rich mixture value, possibly during all operating conditions; the control unit will be applying a control signal to the injectors to provide a leaner mixture but the Oxygen / Lambda sensor signal is remaining rich. Note that this sensor is a post-cat sensor (possibly a post "NOx accumulator" sensor) and if the fuel mixture is rich, it is likely that the pre-cat sensor will also be stuck rich (check for correct pre-cat sensor signals); The fault could be caused by a genuine rich mixture but, the fault code could be activated due to an Oxygen / Lambda sensor fault. Check sensor signal and signal response under different operating conditions (idle, load, etc). If the sensor appears to be operating correctly but the signal value is indicating a genuine rich mixture, this could indicate an engine related fault e.g. incorrect mixtures, fuel pressure, leaking injector etc.
P2276	O2 Sensor Signal Stuck Lean Bank 2 Sensor 3	Note: Many O2 / Lambda sensor related fault codes are activated due to faults in other systems (engine, fuelling, misfires etc). The control unit is able to detect the ratio of the air / fuel mixture by monitoring the information from different sensors. The O2 (Oxygen / Lambda) sensors (primarily the pre-cat sensor) will indicate the oxygen content in the exhaust gas, which is an indicator of air / fuel ratio. The control unit will also monitor other sensors to identify the operating conditions (e.g. heavy or light load etc) and (using the O2 sensor signal), assess whether the air / fuel ratio is correct. Note that faulty engine operation e.g. misfires and exhaust air leaks can cause increased oxygen in the exhaust gas which will affect the oxygen sensor signal.	For a detailed list of checks and identification of the sensor location, Refer to list of sensor / actuator checks on page 19. Different types of O2 / Lambda sensors (e.g. Zirconia or Titania / wideband etc) provide different signals and values, it will be necessary to identify the type of sensor fitted to the vehicle. The fault code indicates that the signal from the Oxygen / Lambda sensor is stuck or biased towards a lean mixture value, possibly during all operating conditions; the control unit could be applying a control signal to the injectors to provide a richer mixture but, the Oxygen / Lambda sensor signal is remaining lean. Note that this sensor is a post-cat sensor (possibly a post "NOx accumulator" sensor) and if the fuel mixture is lean, it is likely that the pre-cat sensor will also be stuck lean (check for correct pre-cat sensor signals); check for an inefficient catalytic converter or air leak between the pre and post-cat sensors. The fault could be caused by a genuine lean mixture but, the fault code could be activated due to an Oxygen / Lambda sensor fault. Check sensor signal and signal response under different operating conditions (idle, load, etc). If the sensor appears to be operating correctly but the signal value is indicating a genuine lean mixture or excess oxygen in post cat exhaust gas, this could indicate an engine related fault e.g. incorrect mixtures, fuel pressure, misfires etc or, a catalytic converter fault. If the sensor appears to be at fault, Refer to list of sensor / actuator checks on page 19.
P2277	O2 Sensor Signal Stuck Rich Bank 2 Sensor 3	Note: Many O2 / Lambda sensor related fault codes are activated due to faults in other systems (engine, fuelling, misfires etc). The control unit is able to detect the ratio of the air / fuel mixture by monitoring the information from different sensors. The O2 (Oxygen / Lambda) sensors (primarily the pre-cat sensor) will indicate the oxygen content in the exhaust gas, which is an indicator of air / fuel ratio. The control unit will also monitor other sensors to identify the operating conditions (e.g. heavy or light load etc) and (using the O2 sensor signal), assess whether the air / fuel ratio is correct. Note that faulty engine operation e.g. misfires and exhaust air leaks can cause increased oxygen in the exhaust gas which will affect the oxygen sensor signal.	For a detailed list of checks and identification of the sensor location, Refer to list of sensor / actuator checks on page 19. Different types of O2 / Lambda sensors (e.g. Zirconia or Titania / wideband etc) provide different signals and values, it will be necessary to identify the type of sensor fitted to the vehicle. The fault code indicates that the signal from the Oxygen / Lambda sensor is stuck or biased towards a rich mixture value, possibly during all operating conditions; the control unit will be applying a control signal to the injectors to provide a leaner mixture but the Oxygen / Lambda sensor signal is remaining rich. Note that this sensor is a post-cat sensor (possibly a post "NOx accumulator" sensor) and if the fuel mixture is rich, it is likely that the pre-cat sensor will also be stuck rich (check for correct pre-cat sensor signals); The fault could be caused by a genuine rich mixture but, the fault code could be activated due to an Oxygen / Lambda sensor fault. Check sensor signal and signal response under different operating conditions (idle, load, etc). If the sensor appears to be operating correctly but the signal value is indicating a genuine rich mixture, this could indicate an engine related fault e.g. incorrect mixtures, fuel pressure, leaking injector etc.
P2278	O2 Sensor Signals Swapped Bank 1 Sensor 3 / Bank 2 Sensor 3	Note: Many O2 / Lambda sensor related fault codes are activated due to faults in other systems (engine, fuelling, misfires etc). O2 (Oxygen / Lambda) sensors detect the exhaust gas oxygen content, which is also an indicator of air / fuel ratio (rich / lean). Oxygen content is indicated by a "Lambda" value. The air / fuel ratio (and therefore oxygen content) is controlled by the engine control unit to maximise efficiency of catalytic converters and reduce emissions of harmful gases.	The control unit has been able to identify that the signals from the two identified O2 / Lambda sensors are swapped i.e. signal from one sensor is swapped with the signal from the other sensor. It is likely that the harness connector plugs for the two identified sensors have been connected incorrectly (the connector plugs are most likely identical). Locate the wiring harness for the sensors and reconnect correctly. Also check Lambda sensor wiring and connections in case any wiring repairs have been carried out that could have resulted in the signal wires for the two sensors being swapped. For information on location and identification of the sensors, Refer to list of sensor / actuator checks on page 19.
P2279	Intake Air System Leak	An intake air leak can be detected by the control unit, which monitors the signals from different sensors that indicate operating conditions in the intake system e.g. throttle position, MAP (Manifold Absolute Pressure) sensor or MAF (Mass Air Flow) sensor. The control unit can also monitor information from other sensor to calculate whether a leak exists e.g. the Oxygen / Lambda sensor or engine rpm..	The fault indicates that an intake system leak exists but, the code does not identify the area of the leak e.g. upstream or down stream of the throttle butterfly / plate. If possible, check the Live Data to obtain an indication of sensor signals e.g. is the air flow low and throttle closed but engine rpm high (air leak between throttle butterfly and engine); by analysing the information, it can help to identify the location of a leak. Visually check for leaks and if necessary use leak detection equipment to locate the fault. Note that there is a possibility that one of the sensor signals is incorrect, but because most systems will make use of a number of sensors signals, the control unit is usually able to identify intake leaks / restrictions as opposed to sensor faults. If however a leak / restriction is not located, sensor checks should be carried out.
P2280	Air Flow Restriction / Air Leak Between Air Filter and MAF	An intake air leak / restriction can be detected by the control unit, which monitors the signals from different sensors that indicate operating conditions in the intake system e.g. throttle position or MAF (Mass Air Flow) sensor. Information from other sensors can also be used to identify intake restrictions / leaks.	The fault indicates that an intake system restriction or leak exists between the air filter and the MAF sensor. Visually check for leaks / restrictions (including air filter and collapsing intake pipes) and if necessary use leak detection equipment to locate the fault. If the restriction / leak cannot be located in the defined area, check the Live Data (if possible) to obtain an indication of sensor signals e.g. is the air flow low and throttle closed but engine rpm high (air leak between throttle butterfly and engine); by analysing the information, it can help to identify the location of a leak. Note that there is a possibility that one of the sensor signals is incorrect, but because most systems will make use of a number of sensors signals, the control unit is usually able to identify intake leaks / restrictions as opposed to sensor faults. If however a leak / restriction is not located, sensor checks should be carried out.
P2281	Air Leak Between MAF and Throttle Body	An intake air leak can be detected by the control unit, which monitors the signals from different sensors that indicate operating conditions in the intake system e.g. throttle position or MAF (Mass Air Flow) sensor. The control unit can also monitor information from other sensor to calculate whether a leak exists e.g. the Oxygen / Lambda sensor or engine rpm..	The fault indicates that an intake system leak exists between the MAF sensor and throttle body / butterfly; this is likely to be detected due to a low MAF signal value compared to the throttle position (the Oxygen / Lambda sensor will also indicate a weak mixture). If possible, check the Live Data to obtain an indication of sensor signals e.g. is the air flow signal low but throttle is open. Visually check for leaks and if necessary use leak detection equipment to locate the fault. Note that there is a possibility that one of the sensor signals is incorrect, but because most systems will make use of a number of sensors signals, the control unit is usually able to identify intake leaks as opposed to sensor faults. If however a leak is not located, sensor checks should be carried out.

NOTE 1. Check for other fault codes that could provide additional information.
NOTE 3. If a fault cannot be located, it is also possible that the control unit is at fault.
NOTE 2. Communication between control units can pass via a CAN-Bus system; refer to Page 51 for CAN-Bus checks.
NOTE 4. Refer to Page 53 for list of pages that show the ISO standard for component locations e.g. Sensor A - Bank 1.

Fault code	EOBD / ISO Description	Component / System Description	Meaningful Description and Quick Check
P2282	Air Leak Between Throttle Body and Intake Valves	An intake air leak can be detected by the control unit, which monitors the signals from different sensors that indicate operating conditions in the intake system e.g. throttle position, MAP (Manifold Absolute Pressure) sensor or MAF (Mass Air Flow) sensor. The control unit can also monitor information from other sensor to calculate whether a leak exists e.g. the Oxygen / Lambda sensor or engine rpm..	The fault indicates that an intake system leak exists between the throttle body / butterfly and the intake valves; this is likely to be detected due to a small throttle opening but a higher then expected engine rpm and, possible a weak mixture signal from the Oxygen / Lambda sensor. Visually check for leaks and check for faults with other systems connected to the intake system e.g. PCV (crankcase ventilation), if necessary use leak detection equipment to locate the fault. If a leak cannot be easily located, check the Live Data (if possible) to obtain an indication of sensor signals e.g. is the air flow signal low but throttle is open. Note that there is a possibility that one of the sensor signals is incorrect, but because most systems will make use of a number of sensors signals, the control unit is usually able to identify intake leaks as opposed to sensor faults. If however a leak is not located, sensor checks should be carried out. Refer to list of sensor / actuator checks on page 19.
P2283	Injector Control Pressure Sensor Circuit	On some types of diesel engine (typically US engines) engine oil is passed to a high pressure pump, which then passes oil at high pressure (in the region of 3000 - 4000 psi) to "unit type fuel injectors" (located in the cylinder head). A solenoid at each injector, controls the flow of the high pressure oil through the injector; when the high pressure oil is allowed to flow into the injector, it acts on a plunger, which pressurises the fuel thus causing the injector to open (and fuel will be injected into the engine). The high pressure engine oil system is referred to as the "control pressure circuit". The control pressure can be altered, depending on the operating conditions.	The pressure sensor monitors the oil pressure in the "Control Pressure" circuit. The fault code indicates a circuit related fault. The voltage, frequency or value of the pressure sensor signal is incorrect (undefined fault). Sensor voltage / signal could be: out of normal operating range, constant value, non-existent, corrupt. Possible fault in the pressure sensor or wiring e.g. short / open circuits, circuit resistance, incorrect sensor resistance (if applicable) or, fault with the reference / operating voltage; interference from other circuits can also affect sensor signals. Refer to list of sensor / actuator checks on page 19.
P2284	Injector Control Pressure Sensor Circuit Range / Performance	On some types of diesel engine (typically US engines) engine oil is passed to a high pressure pump, which then passes oil at high pressure (in the region of 3000 - 4000 psi) to "unit type fuel injectors" (located in the cylinder head). A solenoid at each injector, controls the flow of the high pressure oil through the injector; when the high pressure oil is allowed to flow into the injector, it acts on a plunger, which pressurises the fuel thus causing the injector to open (and fuel will be injected into the engine). The high pressure engine oil system is referred to as the "control pressure circuit". The control pressure can be altered, depending on the operating conditions.	The pressure sensor monitors the oil pressure in the "Control Pressure" circuit. The voltage, frequency or value of the pressure sensor signal is within the normal operating range / tolerance but the signal is not plausible or is incorrect due to an undefined fault. The sensor signal might not match the operating conditions or, the sensor signal is not changing / responding as expected. Possible fault in the pressure sensor or wiring e.g. short / open circuits, circuit resistance, incorrect sensor resistance (if applicable) or, fault with the reference / operating voltage; interference from other circuits can also affect sensor signals. Refer to list of sensor / actuator checks on page 19. Also check any air passages / pipes to the sensor for leaks / blockages etc. It is also possible that the pressure sensor is operating correctly but the pressure being measured by the sensor is unacceptable or incorrect for the operating conditions; if possible, use alternative method of checking pressure to establish if sensor system is operating correctly and whether the pressure indicated by the sensor is correct / incorrect.
P2285	Injector Control Pressure Sensor Circuit Low	On some types of diesel engine (typically US engines) engine oil is passed to a high pressure pump, which then passes oil at high pressure (in the region of 3000 - 4000 psi) to "unit type fuel injectors" (located in the cylinder head). A solenoid at each injector, controls the flow of the high pressure oil through the injector; when the high pressure oil is allowed to flow into the injector, it acts on a plunger, which pressurises the fuel thus causing the injector to open (and fuel will be injected into the engine). The high pressure engine oil system is referred to as the "control pressure circuit". The control pressure can be altered, depending on the operating conditions.	The pressure sensor monitors the oil pressure in the "Control Pressure" circuit. The voltage, frequency or value of the pressure sensor signal is either "zero" (no signal) or, the signal exists but it is below the normal operating range (lower than the minimum operating limit). Possible fault in the pressure sensor or wiring e.g. short / open circuits, circuit resistance, incorrect sensor resistance (if applicable) or, fault with the reference / operating voltage; interference from other circuits can also affect sensor signals. Refer to list of sensor / actuator checks on page 19.
P2286	Injector Control Pressure Sensor Circuit High	On some types of diesel engine (typically US engines) engine oil is passed to a high pressure pump, which then passes oil at high pressure (in the region of 3000 - 4000 psi) to "unit type fuel injectors" (located in the cylinder head). A solenoid at each injector, controls the flow of the high pressure oil through the injector; when the high pressure oil is allowed to flow into the injector, it acts on a plunger, which pressurises the fuel thus causing the injector to open (and fuel will be injected into the engine). The high pressure engine oil system is referred to as the "control pressure circuit". The control pressure can be altered, depending on the operating conditions.	The pressure sensor monitors the oil pressure in the "Control Pressure" circuit. The voltage, frequency or value of the pressure sensor signal is either at full value (e.g. battery voltage with no signal or frequency) or, the signal exists but it is above the normal operating range (higher than the maximum operating limit). Possible fault in the pressure sensor or wiring e.g. short / open circuits, circuit resistance, incorrect sensor resistance (if applicable) or, fault with the reference / operating voltage; interference from other circuits can also affect sensor signals. Refer to list of sensor / actuator checks on page 19.
P2287	Injector Control Pressure Sensor Circuit Intermittent	On some types of diesel engine (typically US engines) engine oil is passed to a high pressure pump, which then passes oil at high pressure (in the region of 3000 - 4000 psi) to "unit type fuel injectors" (located in the cylinder head). A solenoid at each injector, controls the flow of the high pressure oil through the injector; when the high pressure oil is allowed to flow into the injector, it acts on a plunger, which pressurises the fuel thus causing the injector to open (and fuel will be injected into the engine). The high pressure engine oil system is referred to as the "control pressure circuit". The control pressure can be altered, depending on the operating conditions.	The pressure sensor monitors the oil pressure in the "Control Pressure" circuit. The voltage, frequency or value of the pressure sensor signal is intermittent (likely to be out of normal operating range / tolerance when intermittent fault is detected) or, the signal / voltage is erratic (e.g. signal changes are irregular / unpredictable). Possible fault with wiring or connections (including internal sensor connections); also interference from other circuits can affect the signal. Refer to list of sensor / actuator checks on page 19. It is also possible that the pressure could be changing erratically, which could cause the fault code to be activated; if necessary, check pressure with a separate gauge to check for erratic pressure changes.

Code	Fault Location	Description	Diagnosis / Possible Cause
P2288	Injector Control Pressure Too High	On some types of diesel engine (typically US engines) engine oil is passed to a high pressure pump, which then passes oil at high pressure (in the region of 3000 - 4000 psi) to "unit type fuel injectors" (located in the cylinder head). A solenoid at each injector, controls the flow of the high pressure oil through the injector; when the high pressure oil is allowed to flow into the injector, it acts on a plunger, which pressurises the fuel thus causing the injector to open (and fuel will be injected into the engine). The high pressure engine oil system is referred to as the "control pressure circuit". The control pressure can be altered, depending on the operating conditions.	It may be necessary to refer to vehicle specific information for correct pressure specifications etc. The fault code indicates that the injector control pressure (used to actuate the injectors) is higher than the maximum limit, or higher than the value normally expected under the operating conditions. A control pressure regulator (probably located in the high pressure pump) should regulate the pressure according to operating conditions (temperature, engine speed, etc). Check the pressure using independent pressure gauge to establish of the pressure is too high or if the pressure sensor is incorrect. If the pressure is high, check pressure regulator operation (check electrical connections to pressure regulator and to pressure sensor), check for blockages in the high pressure line and also on the drain line (through the pump and back to the engine oil sump).
P2289	Injector Control Pressure Too High - Engine Off	On some types of diesel engine (typically US engines) engine oil is passed to a high pressure pump, which then passes oil at high pressure (in the region of 3000 - 4000 psi) to "unit type fuel injectors" (located in the cylinder head). A solenoid at each injector, controls the flow of the high pressure oil through the injector; when the high pressure oil is allowed to flow into the injector, it acts on a plunger, which pressurises the fuel thus causing the injector to open (and fuel will be injected into the engine). The high pressure engine oil system is referred to as the "control pressure circuit". The control pressure can be altered, depending on the operating conditions.	It may be necessary to refer to vehicle specific information for correct pressure specifications etc. The fault code indicates that the injector control pressure (used to actuate the injectors) is higher than the value normally expected under the "engine off" conditions. A control pressure regulator (probably located in the high pressure pump) should regulate the pressure according to operating conditions (temperature, engine speed, etc). Check the pressure using independent pressure gauge to establish of the pressure is too high or if the pressure sensor is incorrect. If the pressure is high, check pressure regulator operation (check electrical connections to pressure regulator and to pressure sensor), check for blockages in the high pressure line and also on the drain line (through the pump and back to the engine oil sump).
P2290	Injector Control Pressure Too Low	On some types of diesel engine (typically US engines) engine oil is passed to a high pressure pump, which then passes oil at high pressure (in the region of 3000 - 4000 psi) to "unit type fuel injectors" (located in the cylinder head). A solenoid at each injector, controls the flow of the high pressure oil through the injector; when the high pressure oil is allowed to flow into the injector, it acts on a plunger, which pressurises the fuel thus causing the injector to open (and fuel will be injected into the engine). The high pressure engine oil system is referred to as the "control pressure circuit". The control pressure can be altered, depending on the operating conditions.	It may be necessary to refer to vehicle specific information for correct pressure specifications etc. The fault code indicates that the injector control pressure (used to actuate the injectors) is lower than the minimum limit, or lower than the value normally expected under the operating conditions. A control pressure regulator (probably located in the high pressure pump) should regulate the pressure according to operating conditions (temperature, engine speed, etc). Check the pressure using independent pressure gauge to establish of the pressure is too low or if the pressure sensor is incorrect. If the pressure is low, check pressure regulator operation (check electrical connections to pressure regulator and to pressure sensor), check for blockages in the high pressure line. The fault could also be caused by restriction in the oil supply to the high pressure pump e.g. blocked filter etc (or lack of oil), or there could be a pump fault.
P2291	Injector Control Pressure Too Low - Engine Cranking	On some types of diesel engine (typically US engines) engine oil is passed to a high pressure pump, which then passes oil at high pressure (in the region of 3000 - 4000 psi) to "unit type fuel injectors" (located in the cylinder head). A solenoid at each injector, controls the flow of the high pressure oil through the injector; when the high pressure oil is allowed to flow into the injector, it acts on a plunger, which pressurises the fuel thus causing the injector to open (and fuel will be injected into the engine). The high pressure engine oil system is referred to as the "control pressure circuit". The control pressure can be altered, depending on the operating conditions.	It may be necessary to refer to vehicle specific information for correct pressure specifications etc. The fault code indicates that the injector control pressure (used to actuate the injectors) is lower than the value normally expected during cranking conditions. A control pressure regulator (probably located in the high pressure pump) should regulate the pressure according to operating conditions (temperature, engine speed, etc). Note that during cranking, the oil feed to the high pressure pump can be from a small reservoir / gallery in the engine / cylinder head; it will necessary to ensure that oil is able to reach the high pressure pump. Check the pressure using independent pressure gauge to establish of the pressure is too low or if the pressure sensor is incorrect. If the pressure is low, check pressure regulator operation (check electrical connections to pressure regulator and to pressure sensor), check for blockages in the high pressure line. The fault could also be caused by restriction in the oil supply to the high pressure pump e.g. blocked filter etc (or lack of oil / incorrect oil), or there could be a pump fault.
P2292	Injector Control Pressure Erratic	On some types of diesel engine (typically US engines) engine oil is passed to a high pressure pump, which then passes oil at high pressure (in the region of 3000 - 4000 psi) to "unit type fuel injectors" (located in the cylinder head). A solenoid at each injector, controls the flow of the high pressure oil through the injector; when the high pressure oil is allowed to flow into the injector, it acts on a plunger, which pressurises the fuel thus causing the injector to open (and fuel will be injected into the engine). The high pressure engine oil system is referred to as the "control pressure circuit". The control pressure can be altered, depending on the operating conditions.	It may be necessary to refer to vehicle specific information for correct pressure specifications etc. The fault code indicates that the injector control pressure (used to actuate the injectors) is erratic or fluctuating. A control pressure regulator (probably located in the high pressure pump) should regulate the pressure according to operating conditions (temperature, engine speed, etc). Note that during cranking, the oil feed to the high pressure pump can be from a small reservoir / gallery in the engine / cylinder head; it will necessary to ensure that oil is able to reach the high pressure pump. Check the pressure using independent pressure gauge to establish of the pressure is erratic or whether the pressure sensor is incorrect. If the pressure is erratic, check pressure regulator operation, check for blockages in the high pressure line. The fault could also be caused by restriction in the oil supply to the high pressure pump e.g. blocked filter etc (or lack of oil), or there could be a pump fault. Also check electrical connections to pressure regulator and to pressure sensor.
P2293	Fuel Pressure Regulator 2 Performance	The fuel pressure regulator is used to control / maintain the fuel system pressure at the specified value. Regulators are normally controlled by an electrical actuator (usually a solenoid) and pressure sensors are normally used to monitor fuel pressure. Refer to vehicle specific information to identify regulator 1 and 2 (if more than one regulator is fitted).	The control unit has detected an undefined fault / performance problem with the fuel pressure regulator / fuel pressure system. Note that the fault does not indicate an electrical fault with the regulator / wiring but checks should be made to ensure that the wiring and regulator are not at fault. Refer to list of sensor / actuator checks on page 19. The fault could be detected because the fuel pressure sensor is detecting incorrect pressure under some or all operating conditions. Check fuel pressure if possible with a separate gauge and if necessary check operation of pressure sensor and fuel system.
P2294	Fuel Pressure Regulator 2 Control Circuit	The fuel pressure regulator is used to control / maintain the fuel system pressure at the specified value. Regulators are normally controlled by an electrical actuator (usually a solenoid). Refer to vehicle specific information to identify regulator 1 and 2 (if more than one regulator is fitted).	The voltage, frequency or value of the control signal in the regulator circuit is incorrect (undefined fault). Signal could be:- out of normal operating range, constant value, non-existent, corrupt. Possible fault with regulator or wiring e.g. regulator power supply, short / open circuit, high circuit resistance or incorrect regulator resistance. Note: the control unit could be providing the correct signal but it is being affected by the circuit fault. Refer to list of sensor / actuator checks on page 19.
P2295	Fuel Pressure Regulator 2 Control Circuit Low	The fuel pressure regulator is used to control / maintain the fuel system pressure at the specified value. Regulators are normally controlled by an electrical actuator (usually a solenoid). Refer to vehicle specific information to identify regulator 1 and 2 (if more than one regulator is fitted).	The voltage, frequency or value of the control signal in the regulator circuit is either "zero" (no signal) or, the signal exists but is below the normal operating range. Possible fault with regulator or wiring (e.g. regulator power supply, short / open circuit or resistance) that is causing a signal value to be lower than the minimum operating limit. Note: the control unit could be providing the correct signal but it is being affected by the circuit fault. Refer to list of sensor / actuator checks on page 19.

NOTE 1. Check for other fault codes that could provide additional information.
NOTE 3. If a fault cannot be located, it is also possible that the control unit is at fault.
NOTE 2. Communication between control units can pass via a CAN-Bus system; refer to Page 51 for CAN-Bus checks.
NOTE 4. Refer to Page 53 for list of pages that show the ISO standard for component locations e.g. Sensor A - Bank 1.

Fault code	EOBD / ISO Description	Component / System Description	Meaningful Description and Quick Check
P2296	Fuel Pressure Regulator 2 Control Circuit High	The fuel pressure regulator is used to control / maintain the fuel system pressure at the specified value. Regulators are normally controlled by an electrical actuator (usually a solenoid). Refer to vehicle specific information to identify regulator 1 and 2 (if more than one regulator is fitted).	The voltage, frequency or value of the control signal in the regulator circuit is either at full value (e.g. battery voltage with no on / off signal) or, the signal exists but it is above the normal operating range. Possible fault with regulator or wiring (e.g. regulator power supply, short / open circuit or resistance) that is causing a signal value to be higher than the maximum operating limit. Note: the control unit could be providing the correct signal but it is being affected by the circuit fault. Refer to list of sensor / actuator checks on page 19.
P2297	O2 Sensor Out of Range During Deceleration Bank 1 Sensor 1	Note: Many O2 / Lambda sensor related fault codes are activated due to faults in other systems (engine, fuelling, misfires etc). O2 (Oxygen / Lambda) sensors detect the exhaust gas oxygen content, which is also an indicator of air / fuel ratio (rich / lean). Oxygen content is indicated by a "Lambda" value. The air / fuel ratio (and therefore oxygen content) is controlled by the engine control unit to maximise efficiency of catalytic converters and reduce emissions of harmful gases.	For a detailed list of checks and identification of the sensor location, Refer to list of sensor / actuator checks on page 19. Different types of O2 / Lambda sensors (e.g. Zirconia or Titania / wideband etc) provide different signals and values, it will be necessary to identify the type of sensor fitted to the vehicle. The fault code indicates that the voltage of the O2 / Lambda sensor signal is outside of the normal operating range during deceleration conditions. It is possible that the problem is caused by a fuelling system related fault, with particular reference to operation and base setting of throttle position switches / sensors and load sensors (MAP / MAF etc); checks should also be carried out on the Oxygen / Lambda sensor. Note that O2 / Lambda sensor Performance fault codes are frequently activated due to the exhaust gas oxygen content being incorrect for the operating conditions; this can be caused by incorrect mixtures, misfires, air leaks or other engine related problems. It is also possible that a failing catalytic converter can activate the Performance fault code for post-cat Lambda sensors.
P2298	O2 Sensor Out of Range During Deceleration Bank 2 Sensor 1	Note: Many O2 / Lambda sensor related fault codes are activated due to faults in other systems (engine, fuelling, misfires etc). O2 (Oxygen / Lambda) sensors detect the exhaust gas oxygen content, which is also an indicator of air / fuel ratio (rich / lean). Oxygen content is indicated by a "Lambda" value. The air / fuel ratio (and therefore oxygen content) is controlled by the engine control unit to maximise efficiency of catalytic converters and reduce emissions of harmful gases.	For a detailed list of checks and identification of the sensor location, Refer to list of sensor / actuator checks on page 19. Different types of O2 / Lambda sensors (e.g. Zirconia or Titania / wideband etc) provide different signals and values, it will be necessary to identify the type of sensor fitted to the vehicle. The fault code indicates that the voltage of the O2 / Lambda sensor signal is outside of the normal operating range during deceleration conditions. It is possible that the problem is caused by a fuelling system related fault, with particular reference to operation and base setting of throttle position switches / sensors and load sensors (MAP / MAF etc); checks should also be carried out on the Oxygen / Lambda sensor. Note that O2 / Lambda sensor Performance fault codes are frequently activated due to the exhaust gas oxygen content being incorrect for the operating conditions; this can be caused by incorrect mixtures, misfires, air leaks or other engine related problems. It is also possible that a failing catalytic converter can activate the Performance fault code for post-cat Lambda sensors.
P2299	Brake Pedal Position / Accelerator Pedal Position Incompatible	The signals from the brake switch and the accelerator pedal position sensor can be used in the control of engine systems, vehicle stability systems and transmission systems.	The signals from the brake switch and the accelerator position sensor / switch are providing conflicting or implausible information e.g. the brake switch could be indicating that the brake pedal is being depressed but the accelerator pedal sensor could be indicating that the vehicle is accelerating. It is possible that one of the sensors / switches is operating incorrectly, therefore check each sensor for a fault with the sensor / wiring e.g. short / open circuits, circuit resistance, incorrect sensor resistance or, reference / operating voltage fault. Refer to list of sensor / actuator checks on page 19. Check that the operation of the brake pedal and throttle pedal is not preventing correct operation of the sensors / switches.
P2300	Ignition Coil "A" Primary Control Circuit Low	The control unit (or separate ignition module if fitted) normally "switches on / off" the current to the ignition coil(s) on the coil earth circuit; this circuit combined with the coil power supply circuit is referred to as the "primary circuit". The control unit monitors the voltage, current and / or dwell time (coil charge time) in the primary circuit; if the control unit detects a measurement that is outside of expected values, this will be assessed as a coil circuit fault. Refer to vehicle specific information to identify the specified coil.	Check operation of other coils (if applicable) to establish whether the fault affects all coils. The voltage (or other value monitored by the control unit) in the coil primary circuit is either "zero" (no signal) or, the signal exists but is below the normal operating range. It is likely that the fault has been detected because the voltage in the switching circuit does not increase to supply voltage level (normally battery voltage) when the control unit / module breaks / opens the circuit (primary circuit switched off). The fault could be caused by a short to earth or a high resistance / poor connection in the wiring or coil.
P2301	Ignition Coil "A" Primary Control Circuit High	The control unit (or separate ignition module if fitted) normally "switches on / off" the current to the ignition coil(s) on the coil earth circuit; this circuit combined with the coil power supply circuit is referred to as the "primary circuit". The control unit monitors the voltage, current and / or dwell time (coil charge time) in the primary circuit; if the control unit detects a measurement that is outside of expected values, this will be assessed as a coil circuit fault. Refer to vehicle specific information to indentify the specified coil.	Check operation of other coils (if applicable) to establish whether the fault affects all coils. The voltage (or other value monitored by the control unit) in the coil primary switching circuit is either at full value (e.g. battery voltage with no on / off switching signal) or, the signal exists but it is above the normal operating range. It is likely that the fault has been detected because the voltage has not reduced to zero (or close to zero) on the earth circuit when the control unit / module completes the earth circuit (primary circuit switched on). The fault could be caused by a short to positive (power supply) in the wiring or coil. It is also possible that the control unit or ignition module is not switching the circuit through to earth; check for a good earth connection for the control unit or module (as applicable).
P2302	Ignition Coil "A" Secondary Circuit	Some ignition systems have made use of sensors (and other sophisticated electrical checks) to directly monitor the voltages in the secondary circuit; however, the control unit can monitor the voltage and other values in the primary circuit, which can be influenced by secondary circuit performance and voltage. If the control unit detects a measurement that is outside of expected values, this will be assessed as a fault. Refer to vehicle specific information to indentify the specified coil.	The fault code does not define the exact fault, but the control unit will have detected an ignition related problem that is likely to be caused by a secondary circuit fault; note that the spark plug forms part of the secondary circuit. Due to the different types of coils and secondary circuits, checks will be applicable to the system. Initial checks should include secondary circuit connections as applicable (e.g. coil leads where fitted, coil to plug connections etc). Check for poor insulation on secondary circuit (arcing or tracking of high voltage). Check the secondary winding resistance (refer to vehicle specific information).
P2303	Ignition Coil "B" Primary Control Circuit Low	The control unit (or separate ignition module if fitted) normally "switches on / off" the current to the ignition coil(s) on the coil earth circuit; this circuit combined with the coil power supply circuit is referred to as the "primary circuit". The control unit monitors the voltage, current and / or dwell time (coil charge time) in the primary circuit; if the control unit detects a measurement that is outside of expected values, this will be assessed as a coil circuit fault. Refer to vehicle specific information to indentify the specified coil.	Check operation of other coils (if applicable) to establish whether the fault affects all coils. The voltage (or other value monitored by the control unit) in the coil primary circuit is either "zero" (no signal) or, the signal exists but is below the normal operating range. It is likely that the fault has been detected because the voltage in the switching circuit does not increase to supply voltage level (normally battery voltage) when the control unit / module breaks / opens the circuit (primary circuit switched off). The fault could be caused by a short to earth or a high resistance / poor connection in the wiring or coil.

NOTE 1. Check for other fault codes that could provide additional information. **NOTE 2.** Communication between control units can pass via a CAN-Bus system; refer to Page 51 for CAN-Bus checks. 258

NOTE 4 Refer to Page 53 for list of pages that show the ISO standard for component locations e.g. Sensor A - Bank 1.

Code	Description		
P2304	Ignition Coil "B" Primary Control Circuit High	The control unit (or separate ignition module if fitted) normally "switches on / off" the current to the ignition coil(s) on the coil earth circuit; this circuit combined with the coil power supply circuit is referred to as the "primary circuit". The control unit monitors the voltage, current and / or dwell time (coil charge time) in the primary circuit; if the control unit detects a measurement that is outside of expected values, this will be assessed as a coil circuit fault. Refer to vehicle specific information to indentify the specified coil.	Check operation of other coils (if applicable) to establish whether the fault affects all coils. The voltage (or other value monitored by the control unit) in the coil primary switching circuit is either at full value (e.g. battery voltage with no on / off switching signal) or, the signal exists but it is above the normal operating range. It is likely that the fault has been detected because the voltage has not reduced to zero (or close to zero) on the earth circuit when the control unit / module completes the earth circuit (primary circuit switched on). The fault could be caused by a short to positive (power supply) in the wiring or coil. It is also possible that the control unit or ignition module is not switching the circuit through to earth; check for a good earth connection for the control unit or module (as applicable).
P2305	Ignition Coil "B" Secondary Circuit	Some ignition systems have made use of sensors (and other sophisticated electrical checks) to directly monitor the voltages in the secondary circuit; however, the control unit can monitor the voltage and other values in the primary circuit, which can be influenced by secondary circuit performance and voltage. If the control unit detects a measurement that is outside of expected values, this will be assessed as a fault. Refer to vehicle specific information to indentify the specified coil.	The fault code does not define the exact fault, but the control unit will have detected an ignition related problem that is likely to be caused by a secondary circuit fault; note that the spark plug forms part of the secondary circuit. Due to the different types of coils and secondary circuits, checks will be applicable to the system. Initial checks should include secondary circuit connections as applicable (e.g. coil leads where fitted, coil to plug connections etc). Check for poor insulation on secondary circuit (arcing or tracking of high voltage). Check the secondary winding resistance (refer to vehicle specific information).
P2306	Ignition Coil "C" Primary Control Circuit Low	The control unit (or separate ignition module if fitted) normally "switches on / off" the current to the ignition coil(s) on the coil earth circuit; this circuit combined with the coil power supply circuit is referred to as the "primary circuit". The control unit monitors the voltage, current and / or dwell time (coil charge time) in the primary circuit; if the control unit detects a measurement that is outside of expected values, this will be assessed as a coil circuit fault. Refer to vehicle specific information to indentify the specified coil.	Check operation of other coils (if applicable) to establish whether the fault affects all coils. The voltage (or other value monitored by the control unit) in the coil primary circuit is either "zero" (no signal) or, the signal exists but is below the normal operating range. It is likely that the fault has been detected because the voltage in the switching circuit does not increase to supply voltage level (normally battery voltage) when the control unit / module breaks / opens the circuit (primary circuit switched off). The fault could be caused by a short to earth or a high resistance / poor connection in the wiring or coil.
P2307	Ignition Coil "C" Primary Control Circuit High	The control unit (or separate ignition module if fitted) normally "switches on / off" the current to the ignition coil(s) on the coil earth circuit; this circuit combined with the coil power supply circuit is referred to as the "primary circuit". The control unit monitors the voltage, current and / or dwell time (coil charge time) in the primary circuit; if the control unit detects a measurement that is outside of expected values, this will be assessed as a coil circuit fault. Refer to vehicle specific information to indentify the specified coil.	Check operation of other coils (if applicable) to establish whether the fault affects all coils. The voltage (or other value monitored by the control unit) in the coil primary switching circuit is either at full value (e.g. battery voltage with no on / off switching signal) or, the signal exists but it is above the normal operating range. It is likely that the fault has been detected because the voltage has not reduced to zero (or close to zero) on the earth circuit when the control unit / module completes the earth circuit (primary circuit switched on). The fault could be caused by a short to positive (power supply) in the wiring or coil. It is also possible that the control unit or ignition module is not switching the circuit through to earth; check for a good earth connection for the control unit or module (as applicable).
P2308	Ignition Coil "C" Secondary Circuit	Some ignition systems have made use of sensors (and other sophisticated electrical checks) to directly monitor the voltages in the secondary circuit; however, the control unit can monitor the voltage and other values in the primary circuit, which can be influenced by secondary circuit performance and voltage. If the control unit detects a measurement that is outside of expected values, this will be assessed as a fault. Refer to vehicle specific information to indentify the specified coil.	The fault code does not define the exact fault, but the control unit will have detected an ignition related problem that is likely to be caused by a secondary circuit fault; note that the spark plug forms part of the secondary circuit. Due to the different types of coils and secondary circuits, checks will be applicable to the system. Initial checks should include secondary circuit connections as applicable (e.g. coil leads where fitted, coil to plug connections etc). Check for poor insulation on secondary circuit (arcing or tracking of high voltage). Check the secondary winding resistance (refer to vehicle specific information).
P2309	Ignition Coil "D" Primary Control Circuit Low	The control unit (or separate ignition module if fitted) normally "switches on / off" the current to the ignition coil(s) on the coil earth circuit; this circuit combined with the coil power supply circuit is referred to as the "primary circuit". The control unit monitors the voltage, current and / or dwell time (coil charge time) in the primary circuit; if the control unit detects a measurement that is outside of expected values, this will be assessed as a coil circuit fault. Refer to vehicle specific information to indentify the specified coil.	Check operation of other coils (if applicable) to establish whether the fault affects all coils. The voltage (or other value monitored by the control unit) in the coil primary circuit is either "zero" (no signal) or, the signal exists but is below the normal operating range. It is likely that the fault has been detected because the voltage in the switching circuit does not increase to supply voltage level (normally battery voltage) when the control unit / module breaks / opens the circuit (primary circuit switched off). The fault could be caused by a short to earth or a high resistance / poor connection in the wiring or coil.
P2310	Ignition Coil "D" Primary Control Circuit High	The control unit (or separate ignition module if fitted) normally "switches on / off" the current to the ignition coil(s) on the coil earth circuit; this circuit combined with the coil power supply circuit is referred to as the "primary circuit". The control unit monitors the voltage, current and / or dwell time (coil charge time) in the primary circuit; if the control unit detects a measurement that is outside of expected values, this will be assessed as a coil circuit fault. Refer to vehicle specific information to indentify the specified coil.	Check operation of other coils (if applicable) to establish whether the fault affects all coils. The voltage (or other value monitored by the control unit) in the coil primary switching circuit is either at full value (e.g. battery voltage with no on / off switching signal) or, the signal exists but it is above the normal operating range. It is likely that the fault has been detected because the voltage has not reduced to zero (or close to zero) on the earth circuit when the control unit / module completes the earth circuit (primary circuit switched on). The fault could be caused by a short to positive (power supply) in the wiring or coil. It is also possible that the control unit or ignition module is not switching the circuit through to earth; check for a good earth connection for the control unit or module (as applicable).
P2311	Ignition Coil "D" Secondary Circuit	Some ignition systems have made use of sensors (and other sophisticated electrical checks) to directly monitor the voltages in the secondary circuit; however, the control unit can monitor the voltage and other values in the primary circuit, which can be influenced by secondary circuit performance and voltage. If the control unit detects a measurement that is outside of expected values, this will be assessed as a fault. Refer to vehicle specific information to indentify the specified coil.	The fault code does not define the exact fault, but the control unit will have detected an ignition related problem that is likely to be caused by a secondary circuit fault; note that the spark plug forms part of the secondary circuit. Due to the different types of coils and secondary circuits, checks will be applicable to the system. Initial checks should include secondary circuit connections as applicable (e.g. coil leads where fitted, coil to plug connections etc). Check for poor insulation on secondary circuit (arcing or tracking of high voltage). Check the secondary winding resistance (refer to vehicle specific information).

NOTE 1. Check for other fault codes that could provide additional information.   **NOTE 2.** Communication between control units can pass via a CAN-Bus system; refer to Page 51 for CAN-Bus checks.

NOTE 3. If a fault cannot be located, it is also possible that the control unit is at fault.   **NOTE 4.** Refer to Page 53 for list of pages that show the ISO standard for component locations e.g. Sensor A - Bank 1.

Fault code	EOBD / ISO Description	Component / System Description	Meaningful Description and Quick Check
P2312	Ignition Coil "E" Primary Control Circuit Low	The control unit (or separate ignition module if fitted) normally "switches on / off" the current to the ignition coil(s) on the coil earth circuit; this circuit combined with the coil power supply circuit is referred to as the "primary circuit". The control unit monitors the voltage, current and / or dwell time (coil charge time) in the primary circuit; if the control unit detects a measurement that is outside of expected values, this will be assessed as a coil circuit fault. Refer to vehicle specific information to indentify the specified coil.	Check operation of other coils (if applicable) to establish whether the fault affects all coils. The voltage (or other value monitored by the control unit) in the coil primary circuit is either "zero" (no signal) or, the signal exists but is below the normal operating range. It is likely that the fault has been detected because the voltage in the switching circuit does not increase to supply voltage level (normally battery voltage) when the control unit / module breaks / opens the circuit (primary circuit switched off). The fault could be caused by a short to earth or a high resistance / poor connection in the wiring or coil.
P2313	Ignition Coil "E" Primary Control Circuit High	The control unit (or separate ignition module if fitted) normally "switches on / off" the current to the ignition coil(s) on the coil earth circuit; this circuit combined with the coil power supply circuit is referred to as the "primary circuit". The control unit monitors the voltage, current and / or dwell time (coil charge time) in the primary circuit; if the control unit detects a measurement that is outside of expected values, this will be assessed as a coil circuit fault. Refer to vehicle specific information to indentify the specified coil.	Check operation of other coils (if applicable) to establish whether the fault affects all coils. The voltage (or other value monitored by the control unit) in the coil primary switching circuit is either at full value (e.g. battery voltage with no on / off switching signal) or, the signal exists but it is above the normal operating range. It is likely that the fault has been detected because the voltage has not reduced to zero (or close to zero) on the earth circuit when the control unit / module completes the earth circuit (primary circuit switched on). The fault could be caused by a short to positive (power supply) in the wiring or coil. It is also possible that the control unit or ignition module is not switching the circuit through to earth; check for a good earth connection for the control unit or module (as applicable).
P2314	Ignition Coil "E" Secondary Circuit	Some ignition systems have made use of sensors (and other sophisticated electrical checks) to directly monitor the voltages in the secondary circuit; however, the control unit can monitor the voltage and other values in the primary circuit, which can be influenced by secondary circuit performance and voltage. If the control unit detects a measurement that is outside of expected values, this will be assessed as a fault. Refer to vehicle specific information to indentify the specified coil.	The fault code does not define the exact fault, but the control unit will have detected an ignition related problem that is likely to be caused by a secondary circuit fault; note that the spark plug forms part of the secondary circuit. Due to the different types of coils and secondary circuits, checks will be applicable to the system. Initial checks should include secondary circuit connections as applicable (e.g. coil leads where fitted, coil to plug connections etc). Check for poor insulation on secondary circuit (arcing or tracking of high voltage). Check the secondary winding resistance (refer to vehicle specific information).
P2315	Ignition Coil "F" Primary Control Circuit Low	The control unit (or separate ignition module if fitted) normally "switches on / off" the current to the ignition coil(s) on the coil earth circuit; this circuit combined with the coil power supply circuit is referred to as the "primary circuit". The control unit monitors the voltage, current and / or dwell time (coil charge time) in the primary circuit; if the control unit detects a measurement that is outside of expected values, this will be assessed as a coil circuit fault. Refer to vehicle specific information to indentify the specified coil.	Check operation of other coils (if applicable) to establish whether the fault affects all coils. The voltage (or other value monitored by the control unit) in the coil primary circuit is either "zero" (no signal) or, the signal exists but is below the normal operating range. It is likely that the fault has been detected because the voltage in the switching circuit does not increase to supply voltage level (normally battery voltage) when the control unit / module breaks / opens the circuit (primary circuit switched off). The fault could be caused by a short to earth or a high resistance / poor connection in the wiring or coil.
P2316	Ignition Coil "F" Primary Control Circuit High	The control unit (or separate ignition module if fitted) normally "switches on / off" the current to the ignition coil(s) on the coil earth circuit; this circuit combined with the coil power supply circuit is referred to as the "primary circuit". The control unit monitors the voltage, current and / or dwell time (coil charge time) in the primary circuit; if the control unit detects a measurement that is outside of expected values, this will be assessed as a coil circuit fault. Refer to vehicle specific information to indentify the specified coil.	Check operation of other coils (if applicable) to establish whether the fault affects all coils. The voltage (or other value monitored by the control unit) in the coil primary switching circuit is either at full value (e.g. battery voltage with no on / off switching signal) or, the signal exists but it is above the normal operating range. It is likely that the fault has been detected because the voltage has not reduced to zero (or close to zero) on the earth circuit when the control unit / module completes the earth circuit (primary circuit switched on). The fault could be caused by a short to positive (power supply) in the wiring or coil. It is also possible that the control unit or ignition module is not switching the circuit through to earth; check for a good earth connection for the control unit or module (as applicable).
P2317	Ignition Coil "F" Secondary Circuit	Some ignition systems have made use of sensors (and other sophisticated electrical checks) to directly monitor the voltages in the secondary circuit; however, the control unit can monitor the voltage and other values in the primary circuit, which can be influenced by secondary circuit performance and voltage. If the control unit detects a measurement that is outside of expected values, this will be assessed as a fault. Refer to vehicle specific information to indentify the specified coil.	The fault code does not define the exact fault, but the control unit will have detected an ignition related problem that is likely to be caused by a secondary circuit fault; note that the spark plug forms part of the secondary circuit. Due to the different types of coils and secondary circuits, checks will be applicable to the system. Initial checks should include secondary circuit connections as applicable (e.g. coil leads where fitted, coil to plug connections etc). Check for poor insulation on secondary circuit (arcing or tracking of high voltage). Check the secondary winding resistance (refer to vehicle specific information).
P2318	Ignition Coil "G" Primary Control Circuit Low	The control unit (or separate ignition module if fitted) normally "switches on / off" the current to the ignition coil(s) on the coil earth circuit; this circuit combined with the coil power supply circuit is referred to as the "primary circuit". The control unit monitors the voltage, current and / or dwell time (coil charge time) in the primary circuit; if the control unit detects a measurement that is outside of expected values, this will be assessed as a coil circuit fault. Refer to vehicle specific information to indentify the specified coil.	Check operation of other coils (if applicable) to establish whether the fault affects all coils. The voltage (or other value monitored by the control unit) in the coil primary circuit is either "zero" (no signal) or, the signal exists but is below the normal operating range. It is likely that the fault has been detected because the voltage in the switching circuit does not increase to supply voltage level (normally battery voltage) when the control unit / module breaks / opens the circuit (primary circuit switched off). The fault could be caused by a short to earth or a high resistance / poor connection in the wiring or coil.
P2319	Ignition Coil "G" Primary Control Circuit High	The control unit (or separate ignition module if fitted) normally "switches on / off" the current to the ignition coil(s) on the coil earth circuit; this circuit combined with the coil power supply circuit is referred to as the "primary circuit". The control unit monitors the voltage, current and / or dwell time (coil charge time) in the primary circuit; if the control unit detects a measurement that is outside of expected values, this will be assessed as a coil circuit fault. Refer to vehicle specific information to indentify the specified coil.	Check operation of other coils (if applicable) to establish whether the fault affects all coils. The voltage (or other value monitored by the control unit) in the coil primary switching circuit is either at full value (e.g. battery voltage with no on / off switching signal) or, the signal exists but it is above the normal operating range. It is likely that the fault has been detected because the voltage has not reduced to zero (or close to zero) on the earth circuit when the control unit / module completes the earth circuit (primary circuit switched on). The fault could be caused by a short to positive (power supply) in the wiring or coil. It is also possible that the control unit or ignition module is not switching the circuit through to earth; check for a good earth connection for the control unit or module (as applicable).

NOTE 1. Check for other fault codes that could provide additional information.　　NOTE 2. Communication between control units can pass via a CAN-Bus system; refer to Page 51 for CAN-Bus checks.

NOTE 3. If a fault cannot be located, it is also possible that the control unit is at fault　　NOTE 4. Refer to Page 53 for list of pages that show the ISO standard for component locations e.g. Sensor A - Bank 1.

Code	Description		
P2320	Ignition Coil "G" Secondary Circuit	Some ignition systems have made use of sensors (and other sophisticated electrical checks) to directly monitor the voltages in the secondary circuit; however, the control unit can monitor the voltage and other values in the primary circuit, which can be influenced by secondary circuit performance and voltage. If the control unit detects a measurement that is outside of expected values, this will be assessed as a fault. Refer to vehicle specific information to indentify the specified coil.	The fault code does not define the exact fault, but the control unit will have detected an ignition related problem that is likely to be caused by a secondary circuit fault; note that the spark plug forms part of the secondary circuit. Due to the different types of coils and secondary circuits, checks will be applicable to the system. Initial checks should include secondary circuit connections as applicable (e.g. coil leads where fitted, coil to plug connections etc). Check for poor insulation on secondary circuit (arcing or tracking of high voltage). Check the secondary winding resistance (refer to vehicle specific information).
P2321	Ignition Coil "H" Primary Control Circuit Low	The control unit (or separate ignition module if fitted) normally "switches on / off" the current to the ignition coil(s) on the coil earth circuit; this circuit combined with the coil power supply circuit is referred to as the "primary circuit". The control unit monitors the voltage, current and / or dwell time (coil charge time) in the primary circuit; if the control unit detects a measurement that is outside of expected values, this will be assessed as a coil circuit fault. Refer to vehicle specific information to indentify the specified coil.	Check operation of other coils (if applicable) to establish whether the fault affects all coils. The voltage (or other value monitored by the control unit) in the coil primary circuit is either "zero" (no signal) or, the signal exists but is below the normal operating range. It is likely that the fault has been detected because the voltage in the switching circuit does not increase to supply voltage level (normally battery voltage) when the control unit / module breaks / opens the circuit (primary circuit switched off). The fault could be caused by a short to earth or a high resistance / poor connection in the wiring or coil.
P2322	Ignition Coil "H" Primary Control Circuit High	The control unit (or separate ignition module if fitted) normally "switches on / off" the current to the ignition coil(s) on the coil earth circuit; this circuit combined with the coil power supply circuit is referred to as the "primary circuit". The control unit monitors the voltage, current and / or dwell time (coil charge time) in the primary circuit; if the control unit detects a measurement that is outside of expected values, this will be assessed as a coil circuit fault. Refer to vehicle specific information to indentify the specified coil.	Check operation of other coils (if applicable) to establish whether the fault affects all coils. The voltage (or other value monitored by the control unit) in the coil primary switching circuit is either at full value (e.g. battery voltage with no on / off switching signal) or, the signal exists but it is above the normal operating range. It is likely that the fault has been detected because the voltage has not reduced to zero (or close to zero) on the earth circuit when the control unit / module completes the earth circuit (primary circuit switched on). The fault could be caused by a short to positive (power supply) in the wiring or coil. It is also possible that the control unit or ignition module is not switching the circuit through to earth; check for a good earth connection for the control unit or module (as applicable).
P2323	Ignition Coil "H" Secondary Circuit	Some ignition systems have made use of sensors (and other sophisticated electrical checks) to directly monitor the voltages in the secondary circuit; however, the control unit can monitor the voltage and other values in the primary circuit, which can be influenced by secondary circuit performance and voltage. If the control unit detects a measurement that is outside of expected values, this will be assessed as a fault. Refer to vehicle specific information to indentify the specified coil.	The fault code does not define the exact fault, but the control unit will have detected an ignition related problem that is likely to be caused by a secondary circuit fault; note that the spark plug forms part of the secondary circuit. Due to the different types of coils and secondary circuits, checks will be applicable to the system. Initial checks should include secondary circuit connections as applicable (e.g. coil leads where fitted, coil to plug connections etc). Check for poor insulation on secondary circuit (arcing or tracking of high voltage). Check the secondary winding resistance (refer to vehicle specific information).
P2324	Ignition Coil "I" Primary Control Circuit Low	The control unit (or separate ignition module if fitted) normally "switches on / off" the current to the ignition coil(s) on the coil earth circuit; this circuit combined with the coil power supply circuit is referred to as the "primary circuit". The control unit monitors the voltage, current and / or dwell time (coil charge time) in the primary circuit; if the control unit detects a measurement that is outside of expected values, this will be assessed as a coil circuit fault. Refer to vehicle specific information to indentify the specified coil.	Check operation of other coils (if applicable) to establish whether the fault affects all coils. The voltage (or other value monitored by the control unit) in the coil primary circuit is either "zero" (no signal) or, the signal exists but is below the normal operating range. It is likely that the fault has been detected because the voltage in the switching circuit does not increase to supply voltage level (normally battery voltage) when the control unit / module breaks / opens the circuit (primary circuit switched off). The fault could be caused by a short to earth or a high resistance / poor connection in the wiring or coil.
P2325	Ignition Coil "I" Primary Control Circuit High	The control unit (or separate ignition module if fitted) normally "switches on / off" the current to the ignition coil(s) on the coil earth circuit; this circuit combined with the coil power supply circuit is referred to as the "primary circuit". The control unit monitors the voltage, current and / or dwell time (coil charge time) in the primary circuit; if the control unit detects a measurement that is outside of expected values, this will be assessed as a coil circuit fault. Refer to vehicle specific information to indentify the specified coil.	Check operation of other coils (if applicable) to establish whether the fault affects all coils. The voltage (or other value monitored by the control unit) in the coil primary switching circuit is either at full value (e.g. battery voltage with no on / off switching signal) or, the signal exists but it is above the normal operating range. It is likely that the fault has been detected because the voltage has not reduced to zero (or close to zero) on the earth circuit when the control unit / module completes the earth circuit (primary circuit switched on). The fault could be caused by a short to positive (power supply) in the wiring or coil. It is also possible that the control unit or ignition module is not switching the circuit through to earth; check for a good earth connection for the control unit or module (as applicable).
P2326	Ignition Coil "I" Secondary Circuit	Some ignition systems have made use of sensors (and other sophisticated electrical checks) to directly monitor the voltages in the secondary circuit; however, the control unit can monitor the voltage and other values in the primary circuit, which can be influenced by secondary circuit performance and voltage. If the control unit detects a measurement that is outside of expected values, this will be assessed as a fault. Refer to vehicle specific information to indentify the specified coil.	The fault code does not define the exact fault, but the control unit will have detected an ignition related problem that is likely to be caused by a secondary circuit fault; note that the spark plug forms part of the secondary circuit. Due to the different types of coils and secondary circuits, checks will be applicable to the system. Initial checks should include secondary circuit connections as applicable (e.g. coil leads where fitted, coil to plug connections etc). Check for poor insulation on secondary circuit (arcing or tracking of high voltage). Check the secondary winding resistance (refer to vehicle specific information).
P2327	Ignition Coil "J" Primary Control Circuit Low	The control unit (or separate ignition module if fitted) normally "switches on / off" the current to the ignition coil(s) on the coil earth circuit; this circuit combined with the coil power supply circuit is referred to as the "primary circuit". The control unit monitors the voltage, current and / or dwell time (coil charge time) in the primary circuit; if the control unit detects a measurement that is outside of expected values, this will be assessed as a coil circuit fault. Refer to vehicle specific information to indentify the specified coil.	Check operation of other coils (if applicable) to establish whether the fault affects all coils. The voltage (or other value monitored by the control unit) in the coil primary circuit is either "zero" (no signal) or, the signal exists but is below the normal operating range. It is likely that the fault has been detected because the voltage in the switching circuit does not increase to supply voltage level (normally battery voltage) when the control unit / module breaks / opens the circuit (primary circuit switched off). The fault could be caused by a short to earth or a high resistance / poor connection in the wiring or coil.

NOTE 1. Check for other fault codes that could provide additional information.   **NOTE 2.** Communication between control units can pass via a CAN-Bus system; refer to Page 51 for CAN-Bus checks.

NOTE 3. If a fault cannot be located, it is also possible that the control unit is at fault.   **NOTE 4.** Refer to Page 53 for list of pages that show the ISO standard for component locations e.g. Sensor A - Bank 1.

Fault code	EOBD / ISO Description	Component / System Description	Meaningful Description and Quick Check
P2328	Ignition Coil "J" Primary Control Circuit High	The control unit (or separate ignition module if fitted) normally "switches on / off" the current to the ignition coil(s) on the coil earth circuit; this circuit combined with the coil power supply circuit is referred to as the "primary circuit". The control unit monitors the voltage, current and / or dwell time (coil charge time) in the primary circuit; if the control unit detects a measurement that is outside of expected values, this will be assessed as a coil circuit fault. Refer to vehicle specific information to indentify the specified coil.	Check operation of other coils (if applicable) to establish whether the fault affects all coils. The voltage (or other value monitored by the control unit) in the coil primary switching circuit is either at full value (e.g. battery voltage with no on / off switching signal) or, the signal exists but it is above the normal operating range. It is likely that the fault has been detected because the voltage has not reduced to zero (or close to zero) on the earth circuit when the control unit / module completes the earth circuit (primary circuit switched on). The fault could be caused by a short to positive (power supply) in the wiring or coil. It is also possible that the control unit or ignition module is not switching the circuit through to earth; check for a good earth connection for the control unit or module (as applicable)
P2329	Ignition Coil "J" Secondary Circuit	Some ignition systems have made use of sensors (and other sophisticated electrical checks) to directly monitor the voltages in the secondary circuit; however, the control unit can monitor the voltage and other values in the primary circuit, which can be influenced by secondary circuit performance and voltage. If the control unit detects a measurement that is outside of expected values, this will be assessed as a fault. Refer to vehicle specific information to indentify the specified coil.	The fault code does not define the exact fault, but the control unit will have detected an ignition related problem that is likely to be caused by a secondary circuit fault; note that the spark plug forms part of the secondary circuit. Due to the different types of coils and secondary circuits, checks will be applicable to the system. Initial checks should include secondary circuit connections as applicable (e.g. coil leads where fitted, coil to plug connections etc). Check for poor insulation on secondary circuit (arcing or tracking of high voltage). Check the secondary winding resistance (refer to vehicle specific information).
P2330	Ignition Coil "K" Primary Control Circuit Low	The control unit (or separate ignition module if fitted) normally "switches on / off" the current to the ignition coil(s) on the coil earth circuit; this circuit combined with the coil power supply circuit is referred to as the "primary circuit". The control unit monitors the voltage, current and / or dwell time (coil charge time) in the primary circuit; if the control unit detects a measurement that is outside of expected values, this will be assessed as a coil circuit fault. Refer to vehicle specific information to indentify the specified coil.	Check operation of other coils (if applicable) to establish whether the fault affects all coils. The voltage (or other value monitored by the control unit) in the coil primary circuit is either "zero" (no signal) or, the signal exists but is below the normal operating range. It is likely that the fault has been detected because the voltage in the switching circuit does not increase to supply voltage level (normally battery voltage) when the control unit / module breaks / opens the circuit (primary circuit switched off). The fault could be caused by a short to earth or a high resistance / poor connection in the wiring or coil.
P2331	Ignition Coil "K" Primary Control Circuit High	The control unit (or separate ignition module if fitted) normally "switches on / off" the current to the ignition coil(s) on the coil earth circuit; this circuit combined with the coil power supply circuit is referred to as the "primary circuit". The control unit monitors the voltage, current and / or dwell time (coil charge time) in the primary circuit; if the control unit detects a measurement that is outside of expected values, this will be assessed as a coil circuit fault. Refer to vehicle specific information to indentify the specified coil.	Check operation of other coils (if applicable) to establish whether the fault affects all coils. The voltage (or other value monitored by the control unit) in the coil primary switching circuit is either at full value (e.g. battery voltage with no on / off switching signal) or, the signal exists but it is above the normal operating range. It is likely that the fault has been detected because the voltage has not reduced to zero (or close to zero) on the earth circuit when the control unit / module completes the earth circuit (primary circuit switched on). The fault could be caused by a short to positive (power supply) in the wiring or coil. It is also possible that the control unit or ignition module is not switching the circuit through to earth; check for a good earth connection for the control unit or module (as applicable).
P2332	Ignition Coil "K" Secondary Circuit	Some ignition systems have made use of sensors (and other sophisticated electrical checks) to directly monitor the voltages in the secondary circuit; however, the control unit can monitor the voltage and other values in the primary circuit, which can be influenced by secondary circuit performance and voltage. If the control unit detects a measurement that is outside of expected values, this will be assessed as a fault. Refer to vehicle specific information to indentify the specified coil.	The fault code does not define the exact fault, but the control unit will have detected an ignition related problem that is likely to be caused by a secondary circuit fault; note that the spark plug forms part of the secondary circuit. Due to the different types of coils and secondary circuits, checks will be applicable to the system. Initial checks should include secondary circuit connections as applicable (e.g. coil leads where fitted, coil to plug connections etc). Check for poor insulation on secondary circuit (arcing or tracking of high voltage). Check the secondary winding resistance (refer to vehicle specific information).
P2333	Ignition Coil "L" Primary Control Circuit Low	The control unit (or separate ignition module if fitted) normally "switches on / off" the current to the ignition coil(s) on the coil earth circuit; this circuit combined with the coil power supply circuit is referred to as the "primary circuit". The control unit monitors the voltage, current and / or dwell time (coil charge time) in the primary circuit; if the control unit detects a measurement that is outside of expected values, this will be assessed as a coil circuit fault. Refer to vehicle specific information to indentify the specified coil.	Check operation of other coils (if applicable) to establish whether the fault affects all coils. The voltage (or other value monitored by the control unit) in the coil primary circuit is either "zero" (no signal) or, the signal exists but is below the normal operating range. It is likely that the fault has been detected because the voltage in the switching circuit does not increase to supply voltage level (normally battery voltage) when the control unit / module breaks / opens the circuit (primary circuit switched off). The fault could be caused by a short to earth or a high resistance / poor connection in the wiring or coil.
P2334	Ignition Coil "L" Primary Control Circuit High	The control unit (or separate ignition module if fitted) normally "switches on / off" the current to the ignition coil(s) on the coil earth circuit; this circuit combined with the coil power supply circuit is referred to as the "primary circuit". The control unit monitors the voltage, current and / or dwell time (coil charge time) in the primary circuit; if the control unit detects a measurement that is outside of expected values, this will be assessed as a coil circuit fault. Refer to vehicle specific information to indentify the specified coil.	Check operation of other coils (if applicable) to establish whether the fault affects all coils. The voltage (or other value monitored by the control unit) in the coil primary switching circuit is either at full value (e.g. battery voltage with no on / off switching signal) or, the signal exists but it is above the normal operating range. It is likely that the fault has been detected because the voltage has not reduced to zero (or close to zero) on the earth circuit when the control unit / module completes the earth circuit (primary circuit switched on). The fault could be caused by a short to positive (power supply) in the wiring or coil. It is also possible that the control unit or ignition module is not switching the circuit through to earth; check for a good earth connection for the control unit or module (as applicable).
P2335	Ignition Coil "L" Secondary Circuit	Some ignition systems have made use of sensors (and other sophisticated electrical checks) to directly monitor the voltages in the secondary circuit; however, the control unit can monitor the voltage and other values in the primary circuit, which can be influenced by secondary circuit performance and voltage. If the control unit detects a measurement that is outside of expected values, this will be assessed as a fault. Refer to vehicle specific information to indentify the specified coil.	The fault code does not define the exact fault, but the control unit will have detected an ignition related problem that is likely to be caused by a secondary circuit fault; note that the spark plug forms part of the secondary circuit. Due to the different types of coils and secondary circuits, checks will be applicable to the system. Initial checks should include secondary circuit connections as applicable (e.g. coil leads where fitted, coil to plug connections etc). Check for poor insulation on secondary circuit (arcing or tracking of high voltage). Check the secondary winding resistance (refer to vehicle specific information).

NOTE 1. Check for other fault codes that could provide additional information.

NOTE 2. Communication between control units can pass via a CAN-Bus system; refer to Page 51 for CAN-Bus checks.

NOTE 3. If a fault cannot be located, it is also possible that the control unit is at fault

NOTE 4. Refer to Page 53 for list of pages that show the ISO standard for component locations e.g. Sensor A - Bank 1.

Code	Description		
P2336	Cylinder 1 Above Knock Threshold	Knock sensors (usually fitted to the cylinder block or cylinder head) are used to detect combustion knock. Depending on the engine and systems, the knock system (usually integrated into the main control unit) will then alter engine operation e.g. alter ignition timing, fuelling or reduce boost pressure, to reduce / eliminate the knock.	The knock control system has identified a "knock condition" on the specified cylinder; however, corrective action by the control unit e.g. timing or fuelling changes, has not corrected the problem i.e. the knock condition continues or, exceeds the limit of knock that could normally be corrected / eliminated. Note, check for other fault codes relating to a similar fault on other or all cylinders. If the fault only relates to one cylinder, this would indicate a likely cylinder specific problem e.g. injector fault, combustion chamber condition etc. If fault codes indicate other cylinders, this could indicate a cylinder bank fault (if applicable) or a more general fuelling or ignition problem. If all cylinders are affected, it could also indicate a knock sensor related fault or, mechanical or internal engine problems could also cause knock (also, if applicable, excessive boost pressure or intake temperature).
P2337	Cylinder 2 Above Knock Threshold	Knock sensors (usually fitted to the cylinder block or cylinder head) are used to detect combustion knock. Depending on the engine and systems, the knock system (usually integrated into the main control unit) will then alter engine operation e.g. alter ignition timing, fuelling or reduce boost pressure, to reduce / eliminate the knock.	The knock control system has identified a "knock condition" on the specified cylinder; however, corrective action by the control unit e.g. timing or fuelling changes, has not corrected the problem i.e. the knock condition continues or, exceeds the limit of knock that could normally be corrected / eliminated. Note, check for other fault codes relating to a similar fault on other or all cylinders. If the fault only relates to one cylinder, this would indicate a likely cylinder specific problem e.g. injector fault, combustion chamber condition etc. If fault codes indicate other cylinders, this could indicate a cylinder bank fault (if applicable) or a more general fuelling or ignition problem. If all cylinders are affected, it could also indicate a knock sensor related fault or, mechanical or internal engine problems could also cause knock (also, if applicable, excessive boost pressure or intake temperature).
P2338	Cylinder 3 Above Knock Threshold	Knock sensors (usually fitted to the cylinder block or cylinder head) are used to detect combustion knock. Depending on the engine and systems, the knock system (usually integrated into the main control unit) will then alter engine operation e.g. alter ignition timing, fuelling or reduce boost pressure, to reduce / eliminate the knock.	The knock control system has identified a "knock condition" on the specified cylinder; however, corrective action by the control unit e.g. timing or fuelling changes, has not corrected the problem i.e. the knock condition continues or, exceeds the limit of knock that could normally be corrected / eliminated. Note, check for other fault codes relating to a similar fault on other or all cylinders. If the fault only relates to one cylinder, this would indicate a likely cylinder specific problem e.g. injector fault, combustion chamber condition etc. If fault codes indicate other cylinders, this could indicate a cylinder bank fault (if applicable) or a more general fuelling or ignition problem. If all cylinders are affected, it could also indicate a knock sensor related fault or, mechanical or internal engine problems could also cause knock (also, if applicable, excessive boost pressure or intake temperature).
P2339	Cylinder 4 Above Knock Threshold	Knock sensors (usually fitted to the cylinder block or cylinder head) are used to detect combustion knock. Depending on the engine and systems, the knock system (usually integrated into the main control unit) will then alter engine operation e.g. alter ignition timing, fuelling or reduce boost pressure, to reduce / eliminate the knock.	The knock control system has identified a "knock condition" on the specified cylinder; however, corrective action by the control unit e.g. timing or fuelling changes, has not corrected the problem i.e. the knock condition continues or, exceeds the limit of knock that could normally be corrected / eliminated. Note, check for other fault codes relating to a similar fault on other or all cylinders. If the fault only relates to one cylinder, this would indicate a likely cylinder specific problem e.g. injector fault, combustion chamber condition etc. If fault codes indicate other cylinders, this could indicate a cylinder bank fault (if applicable) or a more general fuelling or ignition problem. If all cylinders are affected, it could also indicate a knock sensor related fault or, mechanical or internal engine problems could also cause knock (also, if applicable, excessive boost pressure or intake temperature).
P2340	Cylinder 5 Above Knock Threshold	Knock sensors (usually fitted to the cylinder block or cylinder head) are used to detect combustion knock. Depending on the engine and systems, the knock system (usually integrated into the main control unit) will then alter engine operation e.g. alter ignition timing, fuelling or reduce boost pressure, to reduce / eliminate the knock.	The knock control system has identified a "knock condition" on the specified cylinder; however, corrective action by the control unit e.g. timing or fuelling changes, has not corrected the problem i.e. the knock condition continues or, exceeds the limit of knock that could normally be corrected / eliminated. Note, check for other fault codes relating to a similar fault on other or all cylinders. If the fault only relates to one cylinder, this would indicate a likely cylinder specific problem e.g. injector fault, combustion chamber condition etc. If fault codes indicate other cylinders, this could indicate a cylinder bank fault (if applicable) or a more general fuelling or ignition problem. If all cylinders are affected, it could also indicate a knock sensor related fault or, mechanical or internal engine problems could also cause knock (also, if applicable, excessive boost pressure or intake temperature).
P2341	Cylinder 6 Above Knock Threshold	Knock sensors (usually fitted to the cylinder block or cylinder head) are used to detect combustion knock. Depending on the engine and systems, the knock system (usually integrated into the main control unit) will then alter engine operation e.g. alter ignition timing, fuelling or reduce boost pressure, to reduce / eliminate the knock.	The knock control system has identified a "knock condition" on the specified cylinder; however, corrective action by the control unit e.g. timing or fuelling changes, has not corrected the problem i.e. the knock condition continues or, exceeds the limit of knock that could normally be corrected / eliminated. Note, check for other fault codes relating to a similar fault on other or all cylinders. If the fault only relates to one cylinder, this would indicate a likely cylinder specific problem e.g. injector fault, combustion chamber condition etc. If fault codes indicate other cylinders, this could indicate a cylinder bank fault (if applicable) or a more general fuelling or ignition problem. If all cylinders are affected, it could also indicate a knock sensor related fault or, mechanical or internal engine problems could also cause knock (also, if applicable, excessive boost pressure or intake temperature).
P2342	Cylinder 7 Above Knock Threshold	Knock sensors (usually fitted to the cylinder block or cylinder head) are used to detect combustion knock. Depending on the engine and systems, the knock system (usually integrated into the main control unit) will then alter engine operation e.g. alter ignition timing, fuelling or reduce boost pressure, to reduce / eliminate the knock.	The knock control system has identified a "knock condition" on the specified cylinder; however, corrective action by the control unit e.g. timing or fuelling changes, has not corrected the problem i.e. the knock condition continues or, exceeds the limit of knock that could normally be corrected / eliminated. Note, check for other fault codes relating to a similar fault on other or all cylinders. If the fault only relates to one cylinder, this would indicate a likely cylinder specific problem e.g. injector fault, combustion chamber condition etc. If fault codes indicate other cylinders, this could indicate a cylinder bank fault (if applicable) or a more general fuelling or ignition problem. If all cylinders are affected, it could also indicate a knock sensor related fault or, mechanical or internal engine problems could also cause knock (also, if applicable, excessive boost pressure or intake temperature).
P2343	Cylinder 8 Above Knock Threshold	Knock sensors (usually fitted to the cylinder block or cylinder head) are used to detect combustion knock. Depending on the engine and systems, the knock system (usually integrated into the main control unit) will then alter engine operation e.g. alter ignition timing, fuelling or reduce boost pressure, to reduce / eliminate the knock.	The knock control system has identified a "knock condition" on the specified cylinder; however, corrective action by the control unit e.g. timing or fuelling changes, has not corrected the problem i.e. the knock condition continues or, exceeds the limit of knock that could normally be corrected / eliminated. Note, check for other fault codes relating to a similar fault on other or all cylinders. If the fault only relates to one cylinder, this would indicate a likely cylinder specific problem e.g. injector fault, combustion chamber condition etc. If fault codes indicate other cylinders, this could indicate a cylinder bank fault (if applicable) or a more general fuelling or ignition problem. If all cylinders are affected, it could also indicate a knock sensor related fault or, mechanical or internal engine problems could also cause knock (also, if applicable, excessive boost pressure or intake temperature).

NOTE 1. Check for other fault codes that could provide additional information.

NOTE 3. If a fault cannot be located, it is also possible that the control unit is at fault.

NOTE 2. Communication between control units can pass via a CAN-Bus system; refer to Page 51 for CAN-Bus checks.

NOTE 4. Refer to Page 53 for list of pages that show the ISO standard for component locations e.g. Sensor A - Bank 1.

Fault code	EOBD / ISO Description	Component / System Description	Meaningful Description and Quick Check
P2344	Cylinder 9 Above Knock Threshold	Knock sensors (usually fitted to the cylinder block or cylinder head) are used to detect combustion knock. Depending on the engine and systems, the knock system (usually integrated into the main control unit) will then alter engine operation e.g. alter ignition timing, fuelling or reduce boost pressure, to reduce / eliminate the knock.	The knock control system has identified a "knock condition" on the specified cylinder; however, corrective action by the control unit e.g. timing or fuelling changes, has not corrected the problem i.e. the knock condition continues or, exceeds the limit of knock that could normally be corrected / eliminated. Note, check for other fault codes relating to a similar fault on other or all cylinders. If the fault only relates to one cylinder, this would indicate a likely cylinder specific problem e.g. injector fault, combustion chamber condition etc. If fault codes indicate other cylinders, this could indicate a cylinder bank fault (if applicable) or a more general fuelling or ignition problem. If all cylinders are affected, it could also indicate a knock sensor related fault or, mechanical or internal engine problems could also cause knock (also, if applicable, excessive boost pressure or intake temperature).
P2345	Cylinder 10 Above Knock Threshold	Knock sensors (usually fitted to the cylinder block or cylinder head) are used to detect combustion knock. Depending on the engine and systems, the knock system (usually integrated into the main control unit) will then alter engine operation e.g. alter ignition timing, fuelling or reduce boost pressure, to reduce / eliminate the knock.	The knock control system has identified a "knock condition" on the specified cylinder; however, corrective action by the control unit e.g. timing or fuelling changes, has not corrected the problem i.e. the knock condition continues or, exceeds the limit of knock that could normally be corrected / eliminated. Note, check for other fault codes relating to a similar fault on other or all cylinders. If the fault only relates to one cylinder, this would indicate a likely cylinder specific problem e.g. injector fault, combustion chamber condition etc. If fault codes indicate other cylinders, this could indicate a cylinder bank fault (if applicable) or a more general fuelling or ignition problem. If all cylinders are affected, it could also indicate a knock sensor related fault or, mechanical or internal engine problems could also cause knock (also, if applicable, excessive boost pressure or intake temperature).
P2346	Cylinder 11 Above Knock Threshold	Knock sensors (usually fitted to the cylinder block or cylinder head) are used to detect combustion knock. Depending on the engine and systems, the knock system (usually integrated into the main control unit) will then alter engine operation e.g. alter ignition timing, fuelling or reduce boost pressure, to reduce / eliminate the knock.	The knock control system has identified a "knock condition" on the specified cylinder; however, corrective action by the control unit e.g. timing or fuelling changes, has not corrected the problem i.e. the knock condition continues or, exceeds the limit of knock that could normally be corrected / eliminated. Note, check for other fault codes relating to a similar fault on other or all cylinders. If the fault only relates to one cylinder, this would indicate a likely cylinder specific problem e.g. injector fault, combustion chamber condition etc. If fault codes indicate other cylinders, this could indicate a cylinder bank fault (if applicable) or a more general fuelling or ignition problem. If all cylinders are affected, it could also indicate a knock sensor related fault or, mechanical or internal engine problems could also cause knock (also, if applicable, excessive boost pressure or intake temperature).
P2347	Cylinder 12 Above Knock Threshold	Knock sensors (usually fitted to the cylinder block or cylinder head) are used to detect combustion knock. Depending on the engine and systems, the knock system (usually integrated into the main control unit) will then alter engine operation e.g. alter ignition timing, fuelling or reduce boost pressure, to reduce / eliminate the knock.	The knock control system has identified a "knock condition" on the specified cylinder; however, corrective action by the control unit e.g. timing or fuelling changes, has not corrected the problem i.e. the knock condition continues or, exceeds the limit of knock that could normally be corrected / eliminated. Note, check for other fault codes relating to a similar fault on other or all cylinders. If the fault only relates to one cylinder, this would indicate a likely cylinder specific problem e.g. injector fault, combustion chamber condition etc. If fault codes indicate other cylinders, this could indicate a cylinder bank fault (if applicable) or a more general fuelling or ignition problem. If all cylinders are affected, it could also indicate a knock sensor related fault or, mechanical or internal engine problems could also cause knock (also, if applicable, excessive boost pressure or intake temperature).
P2400	Evaporative Emission System Leak Detection Pump Control Circuit / Open	EVAP systems collect fuel vapours and then release them (under controlled conditions) to the engine intake. On some systems, a pump provides pressure / vacuum (via a calibrated orifice) to the fuel tank and EVAP system for leak detection. Some pumps are driven by a motor and, the current required by the motor to achieve or maintain the pressure / vacuum level is "learnt" by the control unit. The control unit will check the current during each leak test and, if the current value is different from the "learnt" value, this will indicate a possible leak. Other systems use engine vacuum to drive a pump element; the application of vacuum is pulsed (to create the pumping action) using a solenoid controlled valve; on these systems, a switch / sensor indicates pump action, which stops when the system is pressurised.	The fault code refers to the electric actuator (e.g. electric motor) that is used to operate the pump. The voltage, frequency or value of the control signal in the actuator circuit is incorrect. Signal could be:- out of normal operating range, constant value, non-existent, corrupt. The fault code indicates a possible "open circuit" but other undefined faults in the actuator or wiring could activate the same code e.g. actuator power supply, short circuit, high circuit resistance or incorrect actuator resistance. Note: the control unit could be providing the correct signal but it is being affected by a circuit fault. Refer to list of sensor / actuator checks on page 19.
P2401	Evaporative Emission System Leak Detection Pump Control Circuit Low	EVAP systems collect fuel vapours and then release them (under controlled conditions) to the engine intake. On some systems, a pump provides pressure / vacuum (via a calibrated orifice) to the fuel tank and EVAP system for leak detection. Some pumps are driven by a motor and, the current required by the motor to achieve or maintain the pressure / vacuum level is "learnt" by the control unit. The control unit will check the current during each leak test and, if the current value is different from the "learnt" value, this will indicate a possible leak. Other systems use engine vacuum to drive a pump element; the application of vacuum is pulsed (to create the pumping action) using a solenoid controlled valve; on these systems, a switch / sensor indicates pump action, which stops when the system is pressurised.	The fault code refers to the electric actuator (e.g. electric motor) that is used to operate the pump. The voltage, frequency or value of the control signal in the actuator circuit is either "zero" (no signal) or, the signal exists but is below the normal operating range. Possible fault with actuator or wiring (e.g. actuator power supply, short / open circuit or resistance) that is causing a signal value to be lower than the minimum operating limit. Note: the control unit could be providing the correct signal but it is being affected by a circuit fault. Refer to list of sensor / actuator checks on page 19.

NOTE 1. Check for other fault codes that could provide additional information. **NOTE 2.** Communication between control units can pass via a CAN-Bus system; refer to Page 51 for CAN-Bus checks.
NOTE 3. If a fault cannot be located, it is also possible that the control unit is at fault. **NOTE 4.** Refer to Page 53 for list of pages that show the ISO standard for component locations e.g. Sensor A - Bank 1.

264

P2402	Evaporative Emission System Leak Detection Pump Control Circuit High	EVAP systems collect fuel vapours and then release them (under controlled conditions) to the engine intake. On some systems, a pump provides pressure / vacuum (via a calibrated orifice) to the fuel tank and EVAP system for leak detection. Some pumps are driven by a motor and, the current required by the motor to achieve or maintain the pressure / vacuum level is "learnt" by the control unit. The control unit will check the current during each leak test and, if the current value is different from the "learnt" value, this will indicate a possible leak. Other systems use engine vacuum to drive a pump element; the application of vacuum is pulsed (to create the pumping action) using a solenoid controlled valve; on these systems, a switch / sensor indicates pump action, which stops when the system is pressurised.	The fault code refers to the electric actuator (e.g. electric motor) that is used to operate the pump. The voltage, frequency or value of the control signal in the actuator circuit is either at full value (e.g. battery voltage with no on / off signal) or, the signal exists but it is above the normal operating range. Possible fault with actuator or wiring (e.g. actuator power supply, short / open circuit or resistance) that is causing a signal value to be higher than the maximum operating limit. Note: the control unit could be providing the correct signal but it is being affected by a circuit fault. Refer to list of sensor / actuator checks on page 19.
P2403	Evaporative Emission System Leak Detection Pump Sense Circuit / Open	EVAP systems collect fuel vapours and then release them (under controlled conditions) to the engine intake. On some systems, a pump provides pressure / vacuum (via a calibrated orifice) to the fuel tank and EVAP system for leak detection. Some pumps are driven by a motor and, the current required by the motor to achieve or maintain the pressure / vacuum level is "learnt" by the control unit. The control unit will check the current during each leak test and, if the current value is different from the "learnt" value, this will indicate a possible leak. Other systems use engine vacuum to drive a pump element; the application of vacuum is pulsed (to create the pumping action) using a solenoid controlled valve; on these systems, a switch / sensor indicates pump action, which stops when the system is pressurised.	The pump "sense" circuit is likely to refer to the circuit that connects the pump switch / sensor to the control unit; the signal from the switch indicates pump operation and on the vacuum operated pumps, it can indicate when the pump has pressurised the system to the correct value. The voltage or value of the switch signal is incorrect (undefined fault but possibly an open circuit). Switch voltage / signal could be: out of normal operating range, constant value, non-existent, corrupt. Possible fault in the switch / sensor or wiring e.g. short / open circuits, circuit resistance or, fault with the reference / operating voltage; if applicable, the switch contacts could also be faulty e.g. not opening / closing correctly or have a high contact resistance. The sensor could be a magnetic / reed switch type sensor, therefore check operation of sensor as applicable. Refer to list of sensor / actuator checks on page 19.
P2404	Evaporative Emission System Leak Detection Pump Sense Circuit Range / Performance	EVAP systems collect fuel vapours and then release them (under controlled conditions) to the engine intake. On some systems, a pump provides pressure / vacuum (via a calibrated orifice) to the fuel tank and EVAP system for leak detection. Some pumps are driven by a motor and, the current required by the motor to achieve or maintain the pressure / vacuum level is "learnt" by the control unit. The control unit will check the current during each leak test and, if the current value is different from the "learnt" value, this will indicate a possible leak. Other systems use engine vacuum to drive a pump element; the application of vacuum is pulsed (to create the pumping action) using a solenoid controlled valve; on these systems, a switch / sensor indicates pump action, which stops when the system is pressurised.	The pump "sense" circuit is likely to refer to the circuit that connects the pump switch / sensor to the control unit; the signal from the switch / sensor indicates pump operation and on the vacuum operated pumps, it can indicate when the pump has pressurised the system to the correct value. The voltage, frequency or value of the switch signal is within the normal operating range / tolerance but the signal is not plausible or is incorrect due to an undefined fault. The switch signal might not match the operating conditions indicated by other sensors or, the switch signal is not changing / responding as expected. Possible fault in the switch or wiring e.g. short / open circuits, circuit resistance or, fault with the reference / operating voltage; if applicable, also check switch contacts to ensure that they open / close and check continuity of switch contacts (ensure that there is not a high resistance). The sensor could be a magnetic / reed switch type sensor, therefore check operation of sensor as applicable. Refer to list of sensor / actuator checks on page 19.
P2405	Evaporative Emission System Leak Detection Pump Sense Circuit Low	EVAP systems collect fuel vapours and then release them (under controlled conditions) to the engine intake. On some systems, a pump provides pressure / vacuum (via a calibrated orifice) to the fuel tank and EVAP system for leak detection. Some pumps are driven by a motor and, the current required by the motor to achieve or maintain the pressure / vacuum level is "learnt" by the control unit. The control unit will check the current during each leak test and, if the current value is different from the "learnt" value, this will indicate a possible leak. Other systems use engine vacuum to drive a pump element; the application of vacuum is pulsed (to create the pumping action) using a solenoid controlled valve; on these systems, a switch / sensor indicates pump action, which stops when the system is pressurised.	The pump "sense" circuit is likely to refer to the circuit that connects the pump switch / sensor to the control unit; the signal from the switch / sensor indicates pump operation and on the vacuum operated pumps, it can indicate when the pump has pressurised the system to the correct value. The voltage, frequency or value of the switch signal is either "zero" (no signal) or, the signal exists but it is below the normal operating range (lower than the minimum operating limit). Possible fault in the switch or wiring e.g. short / open circuits, circuit resistance or, fault with the reference / operating voltage; if applicable, the switch contacts could also be faulty e.g. not opening / closing correctly or have a high contact resistance. The sensor could be a magnetic / reed switch type sensor, therefore check operation of sensor as applicable. Refer to list of sensor / actuator checks on page 19. Note that the fault code can be activated if the switch output remains at the low value e.g. zero volts and does not rise to the high value e.g. battery voltage (the control unit will detect no switch signal change but it has detected change in engine operating conditions).
P2406	Evaporative Emission System Leak Detection Pump Sense Circuit High	EVAP systems collect fuel vapours and then release them (under controlled conditions) to the engine intake. On some systems, a pump provides pressure / vacuum (via a calibrated orifice) to the fuel tank and EVAP system for leak detection. Some pumps are driven by a motor and, the current required by the motor to achieve or maintain the pressure / vacuum level is "learnt" by the control unit. The control unit will check the current during each leak test and, if the current value is different from the "learnt" value, this will indicate a possible leak. Other systems use engine vacuum to drive a pump element; the application of vacuum is pulsed (to create the pumping action) using a solenoid controlled valve; on these systems, a switch / sensor indicates pump action, which stops when the system is pressurised.	The pump "sense" circuit is likely to refer to the circuit that connects the pump switch / sensor to the control unit; the signal from the switch / sensor indicates pump operation and on the vacuum operated pumps, it can indicate when the pump has pressurised the system to the correct value. The voltage, frequency or value of the switch signal is either at full value (e.g. battery voltage with no signal or frequency) or, the signal exists but it is above the normal operating range (higher than the maximum operating limit). Possible fault in the switch or wiring e.g. short / open circuits, circuit resistance or, fault with the reference / operating voltage; if applicable, the switch contacts could also be faulty e.g. not opening / closing correctly or have a high contact resistance. The sensor could be a magnetic / reed switch type sensor, therefore check operation of sensor as applicable. Refer to list of sensor / actuator checks on page 19. Note that the fault code can be activated if the switch output remains at the high value e.g. battery voltage and does not fall to the low value e.g. zero volts (the control unit will detect no switch signal change but it has detected change in engine operating conditions).

NOTE 1. Check for other fault codes that could provide additional information.   **NOTE 2.** Communication between control units can pass via a CAN-Bus system; refer to Page 51 for CAN-Bus checks.
NOTE 3. If a fault cannot be located, it is also possible that the control unit is at fault.   **NOTE 4.** Refer to Page 53 for list of pages that show the ISO standard for component locations e.g. Sensor A - Bank 1.

265

Fault code	EOBD / ISO Description	Component / System Description	Meaningful Description and Quick Check
P2407	Evaporative Emission System Leak Detection Pump Sense Circuit Intermittent / Erratic	EVAP systems collect fuel vapours and then release them (under controlled conditions) to the engine intake. On some systems, a pump provides pressure / vacuum (via a calibrated orifice) to the fuel tank and EVAP system for leak detection. Some pumps are driven by a motor and, the current required by the motor to achieve or maintain the pressure / vacuum level is "learnt" by the control unit. The control unit will check the current during each leak test and, if the current value is different from the "learnt" value, this will indicate a possible leak. Other systems use engine vacuum to drive a pump element; the application of vacuum is pulsed (to create the pumping action) using a solenoid controlled valve; on these systems, a switch / sensor indicates pump action, which stops when the system is pressurised.	The pump "sense" circuit is likely to refer to the circuit that connects the pump switch / sensor to the control unit; the signal from the switch / sensor indicates pump operation and on the vacuum operated pumps, it can indicate when the pump has pressurised the system to the correct value. The voltage, frequency or value of the switch signal is intermittent (likely to be out of normal operating range / tolerance when intermittent fault is detected) or, the signal / voltage is erratic (e.g. signal changes are irregular / unpredictable). Possible fault with wiring or connections (including internal switch connections); the switch contacts could also be faulty e.g. not opening / closing correctly or have a high contact resistance. Refer to list of sensor / actuator checks on page 19.
P2408	Fuel Cap Sensor / Switch Circuit	On some vehicles, the control unit monitors the removal / refitting of the fuel cap (via a fuel cap sensor); in most cases, this information can be used on vehicles that require a fuel additive to be added to the fuel system (either into the main fuel tank or into a separate tank). When the cap is removed / replaced, this indicates that fuel has probably been added to the main fuel tank and therefore, the fuel additive should also be topped up. The sensor is usually a simple magnetically operated switch (although other sensor types could be used). For petrol engine vehicles, an incorrectly fitted fuel cap will allow fuel vapour to be released to the atmosphere. On some petrol vehicles, if the fuel cap is not fitted / closed correctly, this can affect the leak detection process on the EVAP system.	The voltage or value of the sensor signal is incorrect (undefined fault). Sensor voltage / signal value could be: out of normal operating range, constant value, non-existent, corrupt. Possible fault in the sensor or wiring e.g. short / open circuits, circuit resistance, incorrect sensor resistance (if applicable) or, fault with the reference / operating voltage; interference from other circuits can also affect sensor signals. Note that sensor may be a simple switch, therefore the switch contacts could have a high resistance or the contacts may not be opening / closing correctly. Refer to list of sensor / actuator checks on page 19.
P2409	Fuel Cap Sensor / Switch Circuit Range / Performance	On some vehicles, the control unit monitors the removal / refitting of the fuel cap (via a fuel cap sensor); in most cases, this information can be used on vehicles that require a fuel additive to be added to the fuel system (either into the main fuel tank or into a separate tank). When the cap is removed / replaced, this indicates that fuel has probably been added to the main fuel tank and therefore, the fuel additive should also be topped up. The sensor is usually a simple magnetically operated switch (although other sensor types could be used). For petrol engine vehicles, an incorrectly fitted fuel cap will allow fuel vapour to be released to the atmosphere. On some petrol vehicles, if the fuel cap is not fitted / closed correctly, this can affect the leak detection process on the EVAP system.	The voltage or value of the sensor signal is within the normal operating range / tolerance but the signal is not plausible or is incorrect due to an undefined fault. The sensor signal might not be changing / responding as expected e.g. the signal has not changed although the distance travelled by the vehicle would indicate that fuel must have been added (fuel cap must have been removed). Possible fault in the sensor or wiring e.g. short / open circuits, circuit resistance, incorrect sensor resistance or, fault with the reference / operating voltage; interference from other circuits can also affect sensor signals. Refer to list of sensor / actuator checks on page 19. Note that sensor may be a simple switch, therefore the switch contacts could have a high resistance or the contacts may not be opening / closing correctly. It is also possible that the sensor is operating correctly but the fuel cap is not fitted correctly or the incorrect cap has been fitted.
P240A	Evaporative Emission System Leak Detection Pump Heater Control Circuit / Open	EVAP systems make use of a canister to collect fuel vapours and then release them (under controlled conditions) to the engine intake system. On some EVAP systems, a pump is used to provide pressure / vacuum to the fuel tank and EVAP system for leak detection (engine off); the pump applies the pressure / vacuum via a calibrated orifice. Some systems can use a heater to prevent icing within the pump and calibrated orifice (refer to P2400 for additional information on the Leak Detection Pump systems.	It may be necessary to refer to vehicle specific information to identify the heater circuit e.g. power supply, relay and heater control. The voltage or value of the control signal in the heater circuit is incorrect. The voltage or value could be:- out of normal operating range, constant value, non-existent, corrupt. The fault code indicates a possible "open circuit" but other undefined faults in the heater or wiring could activate the same code e.g. actuator power supply, short circuit, high circuit resistance or incorrect heater resistance. Note: the control unit could be providing the correct signal but it is being affected by a circuit fault.
P240B	Evaporative Emission System Leak Detection Pump Heater Control Circuit Low	EVAP systems make use of a canister to collect fuel vapours and then release them (under controlled conditions) to the engine intake system. On some EVAP systems, a pump is used to provide pressure / vacuum to the fuel tank and EVAP system for leak detection (engine off); the pump applies the pressure / vacuum via a calibrated orifice. Some systems can use a heater to prevent icing within the pump and calibrated orifice (refer to P2400 for additional information on the Leak Detection Pump systems.	It may be necessary to refer to vehicle specific information to identify the heater circuit e.g. power supply, relay and heater control. The voltage or value of the control signal in the heater circuit is either "zero" (no signal) or, the signal exists but is below the normal operating range. Possible fault with actuator or wiring (e.g. actuator power supply, short / open circuit or resistance) that is causing a signal value to be lower than the minimum operating limit. Note: the control unit could be providing the correct signal but it is being affected by a circuit fault.
P240C	Evaporative Emission System Leak Detection Pump Heater Control Circuit High	EVAP systems make use of a canister to collect fuel vapours and then release them (under controlled conditions) to the engine intake system. On some EVAP systems, a pump is used to provide pressure / vacuum to the fuel tank and EVAP system for leak detection (engine off); the pump applies the pressure / vacuum via a calibrated orifice. Some systems can use a heater to prevent icing within the pump and calibrated orifice (refer to P2400 for additional information on the Leak Detection Pump systems.	It may be necessary to refer to vehicle specific information to identify the heater circuit e.g. power supply, relay and heater control. The voltage or value of the control signal in the heater circuit is either at full value (e.g. battery voltage with no on / off signal) or, the signal exists but it is above the normal operating range. Possible fault with actuator or wiring (e.g. actuator power supply, short / open circuit or resistance) that is causing a signal value to be higher than the maximum operating limit. Note: the control unit could be providing the correct signal but it is being affected by a circuit fault.
P240D		ISO / SAE reserved	Not used, reserved for future allocation.
P240E		ISO / SAE reserved	Not used, reserved for future allocation.
P240F		ISO / SAE reserved	Not used, reserved for future allocation.

NOTE 1. Check for other fault codes that could provide additional information. **NOTE 2.** Communication between control units can pass via a CAN-Bus system; refer to Page 51 for CAN-Bus checks.

Code	Fault Location	Description	Probable Cause
P2410	Fuel Cap Sensor / Switch Circuit Low	On some vehicles, the control unit monitors the removal / refitting of the fuel cap (via a fuel cap sensor); in most cases, this information can be used on vehicles that require a fuel additive to be added to the fuel system (either into the main fuel tank or into a separate tank). When the cap is removed / replaced, this indicates that fuel has probably been added to the main fuel tank and therefore, the fuel additive should also be topped up. The sensor is usually a simple magnetically operated switch (although other sensor types could be used). For petrol engine vehicles, an incorrectly fitted fuel cap will allow fuel vapour to be released to the atmosphere. On some petrol vehicles, if the fuel cap is not fitted / closed correctly, this can affect the leak detection process on the EVAP system.	The voltage or value of the position sensor signal is either "zero" (no signal) or, the signal exists but it is below the normal operating range (lower than the minimum operating limit). Possible fault in the sensor or wiring e.g. short / open circuits, circuit resistance, incorrect sensor resistance or, fault with the reference / operating voltage; interference from other circuits can also affect sensor signals. Refer to list of sensor / actuator checks on page 19. Note that sensor may be a simple switch, therefore the switch contacts could have a high resistance or the contacts may not be opening / closing correctly. It is also possible that the sensor is operating correctly but the fuel cap is not fitted correctly or the incorrect cap has been fitted, therefore the sensor signal is not changing.
P2411	Fuel Cap Sensor / Switch Circuit High	On some vehicles, the control unit monitors the removal / refitting of the fuel cap (via a fuel cap sensor); in most cases, this information can be used on vehicles that require a fuel additive to be added to the fuel system (either into the main fuel tank or into a separate tank). When the cap is removed / replaced, this indicates that fuel has probably been added to the main fuel tank and therefore, the fuel additive should also be topped up. The sensor is usually a simple magnetically operated switch (although other sensor types could be used). For petrol engine vehicles, an incorrectly fitted fuel cap will allow fuel vapour to be released to the atmosphere. On some petrol vehicles, if the fuel cap is not fitted / closed correctly, this can affect the leak detection process on the EVAP system.	The voltage or value of the position sensor signal is either at full value (e.g. battery voltage with no signal or frequency) or, the signal exists but it is above the normal operating range (higher than the maximum operating limit). Possible fault in the sensor or wiring e.g. short / open circuits, circuit resistance, incorrect sensor resistance or, fault with the reference / operating voltage; interference from other circuits can also affect sensor signals. Refer to list of sensor / actuator checks on page 19. Note that sensor may be a simple switch, therefore the switch contacts could have a high resistance or the contacts may not be opening / closing correctly. It is also possible that the sensor is operating correctly but the fuel cap is not fitted correctly or the incorrect cap has been fitted, therefore the sensor signal is not changing.
P2412	Fuel Cap Sensor / Switch Circuit Intermittent / Erratic	On some vehicles, the control unit monitors the removal / refitting of the fuel cap (via a fuel cap sensor); in most cases, this information can be used on vehicles that require a fuel additive to be added to the fuel system (either into the main fuel tank or into a separate tank). When the cap is removed / replaced, this indicates that fuel has probably been added to the main fuel tank and therefore, the fuel additive should also be topped up. The sensor is usually a simple magnetically operated switch (although other sensor types could be used). For petrol engine vehicles, an incorrectly fitted fuel cap will allow fuel vapour to be released to the atmosphere. On some petrol vehicles, if the fuel cap is not fitted / closed correctly, this can affect the leak detection process on the EVAP system.	The voltage or value of the position sensor signal is intermittent (likely to be out of normal operating range / tolerance when intermittent fault is detected) or, the signal / voltage is erratic (e.g. signal changes are irregular / unpredictable). Possible fault with wiring or connections (including internal sensor connections); also interference from other circuits can affect the signal. Refer to list of sensor / actuator checks on page 19. Note that sensor may be a simple switch, therefore the switch contacts may not be opening / closing correctly. It is also possible that the sensor is operating correctly but the fuel cap is not fitted correctly thus causing the signal to be erratic.
P2413	Exhaust Gas Recirculation System Performance	The measurement of gas flow within an Exhaust Gas Recirculation (EGR) system is often a calculated value; the calculation will be based on information provided by other sensors e.g. intake airflow sensor, throttle position sensor, temperature sensor etc. It is also possible to make use of an EGR valve lift / movement sensor to enable the control unit to assess the movement and position of the EGR valve. Also refer to additional information attached to fault code P0404.	The fault code indicates that the EGR gas flow is calculated or measured as being incorrect for the operating conditions i.e. the flow is not as expected or requested by the control unit. The checks required will be depend on the system design; refer to vehicle specific information to identify the type of system. Checks should however be made on the operation of the EGR valve; check that the valve opens / closes correctly (use vacuum pump to test operation if applicable), check for contamination of valve and ports. Check EGR valve wiring and check condition of vacuum hoses and pipes carrying the exhaust gases. Where applicable check operation of other applicable sensors; other related fault codes (if available) may provide a more definitive fault description or might indicate a fault in another related component.
P2414	O2 Sensor Exhaust Sample Error Bank 1 Sensor 1	There are different possible applications for this fault code. However, the fault code is likely to refer to the exhaust gas sample, which is then used by a broadband oxygen / Lambda sensor as a reference value. If the sample is not at the correct value i.e. Lambda = 1, the sensor then provides an internal "pumping current" to alter the oxygen content in the sample (to achieve Lambda 1).	The fault code indicates that the exhaust gas sample is incorrect and it is likely that the sensor is unable to alter the sample (oxygen correction) to achieve Lambda 1. The fault could be caused by a fuelling fault, exhaust air leak or any other fault that could cause the oxygen content in the exhaust gas to be excessively high or low. The fault could also be caused by a sensor fault. Refer to list of sensor / actuator checks on page 19.
P2415	O2 Sensor Exhaust Sample Error Bank 2 Sensor 1	There are different possible applications for this fault code. However, the fault code is likely to refer to the exhaust gas sample, which is then used by a broadband oxygen / Lambda sensor as a reference value. If the sample is not at the correct value i.e. Lambda = 1, the sensor then provides an internal "pumping current" to alter the oxygen content in the sample (to achieve Lambda 1).	The fault code indicates that the exhaust gas sample is incorrect and it is likely that the sensor is unable to alter the sample (oxygen correction) to achieve Lambda 1. The fault could be caused by a fuelling fault, exhaust air leak or any other fault that could cause the oxygen content in the exhaust gas to be excessively high or low. The fault could also be caused by a sensor fault. Refer to list of sensor / actuator checks on page 19.
P2416	O2 Sensor Signals Swapped Bank 1 Sensor 2 / Bank 1 Sensor 3	Note: Many O2 / Lambda sensor related fault codes are activated due to faults in other systems (engine, fuelling, misfires etc). O2 (Oxygen / Lambda) sensors detect the exhaust gas oxygen content, which is also an indicator of air / fuel ratio (rich / lean). Oxygen content is indicated by a "Lambda" value. The air / fuel ratio (and therefore oxygen content) is controlled by the engine control unit to maximise efficiency of catalytic converters and reduce emissions of harmful gases.	The control unit has been able to identify that the signals from the two identified O2 / Lambda sensors are swapped i.e. signal from one sensor is swapped with the signal from the other sensor. It is likely that the harness connector plugs for the two identified sensors have been connected incorrectly (the connector plugs are most likely identical). Locate the wiring harness for the sensors and reconnect correctly. Also check Lambda sensor wiring and connections in case any wiring repairs have been carried out that could have resulted in the signal wires for the two sensors being swapped. For information on location and identification of the sensors, Refer to list of sensor / actuator checks on page 19.

NOTE 1. Check for other fault codes that could provide additional information. **NOTE 2.** Communication between control units can pass via a CAN-Bus system; refer to Page 51 for CAN-Bus checks.

NOTE 3. If a fault cannot be located, it is also possible that the control unit is at fault. **NOTE 4.** Refer to Page 53 for list of pages that show the ISO standard for component locations e.g. Sensor A - Bank 1.

Fault code	EOBD / ISO Description	Component / System Description	Meaningful Description and Quick Check
P2417	O2 Sensor Signals Swapped Bank 2 Sensor 2 / Bank 2 Sensor 3	Note: Many O2 / Lambda sensor related fault codes are activated due to faults in other systems (engine, fuelling, misfires etc). O2 (Oxygen / Lambda) sensors detect the exhaust gas oxygen content, which is also an indicator of air / fuel ratio (rich / lean). Oxygen content is indicated by a "Lambda" value. The air / fuel ratio (and therefore oxygen content) is controlled by the engine control unit to maximise efficiency of catalytic converters and reduce emissions of harmful gases.	The control unit has been able to identify that the signals from the two identified O2 / Lambda sensors are swapped i.e. signal from one sensor is swapped with the signal from the other sensor. It is likely that the harness connector plugs for the two identified sensors have been connected incorrectly (the connector plugs are most likely identical). Locate the wiring harness for the sensors and reconnect correctly. Also check Lambda sensor wiring and connections in case any wiring repairs have been carried out that could have resulted in the signal wires for the two sensors being swapped. For information on location and identification of the sensors, Refer to list of sensor / actuator checks on page 19.
P2418	Evaporative Emission System Switching Valve Control Circuit / Open	EVAP systems make use of a canister to collect fuel vapours and then release them (under controlled conditions) to the engine intake system. It may be necessary to refer to vehicle specific information to identify the system design and components used. Some vehicle manufacturers may use the term "switching valve" as an alternative to "vent valve". On some applications, a vent valve is contained within a pump module (contains the leak test pressure / vacuum pump). Refer to vehicle specific information to identify the valve and location, Refer to fault code P2407 for other information on the pump. Refer to fault code P0448 for additional information on vent valves.	The fault code refers to the electric actuator that is used to operate the switching valve. The voltage, frequency or value of the control signal in the actuator circuit is incorrect. Signal could be:- out of normal operating range, constant value, non-existent, corrupt. The fault code indicates a possible "open circuit" but other undefined faults in the actuator or wiring could activate the same code e.g. actuator power supply, short circuit, high circuit resistance or incorrect actuator resistance. Note: the control unit could be providing the correct signal but it is being affected by a circuit fault. Refer to list of sensor / actuator checks on page 19.
P2419	Evaporative Emission System Switching Valve Control Circuit Low	EVAP systems make use of a canister to collect fuel vapours and then release them (under controlled conditions) to the engine intake system. It may be necessary to refer to vehicle specific information to identify the system design and components used. Some vehicle manufacturers may use the term "switching valve" as an alternative to "vent valve". On some applications, a vent valve is contained within a pump module (contains the leak test pressure / vacuum pump). Refer to vehicle specific information to identify the valve and location, Refer to fault code P2407 for other information on the pump. Refer to fault code P0448 for additional information on vent valves.	The fault code refers to the electric actuator that is used to operate the switching valve. The voltage, frequency or value of the control signal in the actuator circuit is either "zero" (no signal) or, the signal exists but is below the normal operating range. Possible fault with actuator or wiring (e.g. actuator power supply, short / open circuit or resistance) that is causing a signal value to be lower than the minimum operating limit. Note: the control unit could be providing the correct signal but it is being affected by a circuit fault. Refer to list of sensor / actuator checks on page 19.
P2420	Evaporative Emission System Switching Valve Control Circuit High	EVAP systems make use of a canister to collect fuel vapours and then release them (under controlled conditions) to the engine intake system. It may be necessary to refer to vehicle specific information to identify the system design and components used. Some vehicle manufacturers may use the term "switching valve" as an alternative to "vent valve". On some applications, a vent valve is contained within a pump module (contains the leak test pressure / vacuum pump). Refer to vehicle specific information to identify the valve and location, Refer to fault code P2407 for other information on the pump. Refer to fault code P0448 for additional information on vent valves.	The fault code refers to the electric actuator that is used to operate the switching valve. The voltage, frequency or value of the control signal in the actuator circuit is either at full value (e.g. battery voltage with no on / off signal) or, the signal exists but it is above the normal operating range. Possible fault with actuator or wiring (e.g. actuator power supply, short / open circuit or resistance) that is causing a signal value to be higher than the maximum operating limit. Note: the control unit could be providing the correct signal but it is being affected by a circuit fault. Refer to list of sensor / actuator checks on page 19.
P2421	Evaporative Emission System Vent Valve Stuck Open	EVAP systems make use of a canister to collect fuel vapours and then release them (under controlled conditions) to the engine intake system. The vent control valve is operated by an electric actuator (usually a solenoid), which is provided by a control signal from the control unit. The valve allows air to enter the canister (when vapours are being drawn out of the canister) but, the valve is closed during the leak test cycle. Note that "Purge Valves" have also been referred to as "Vent Valves" on some vehicles and documentation (refer to P0443 for information on purge valves).	The control unit is providing a control signal to the actuator but the control unit has detected that the actuator is stuck in the open position and is not responding to the control signal. The fault could be related to an actuator / wiring fault e.g. power supply, short / open circuit, incorrect actuator resistance. Refer to list of sensor / actuator checks on page 19. Note that the control unit could be detecting the fault because of "a lack of response / no response" from the vent valve; check vent valve (as well as the actuator) for correct operation, movement and response.
P2422	Evaporative Emission System Vent Valve Stuck Closed	EVAP systems make use of a canister to collect fuel vapours and then release them (under controlled conditions) to the engine intake system. The vent control valve is operated by an electric actuator (usually a solenoid), which is provided by a control signal from the control unit. The valve allows air to enter the canister (when vapours are being drawn out of the canister) but, the valve is closed during the leak test cycle. Note that "Purge Valves" have also been referred to as "Vent Valves" on some vehicles and documentation (refer to P0443 for information on purge valves).	The control unit is providing a control signal to the actuator but the control unit has detected that the actuator is stuck in the closed position and is not responding to the control signal. The fault could be related to an actuator / wiring fault e.g. power supply, short / open circuit, incorrect actuator resistance. Refer to list of sensor / actuator checks on page 19. Note that the control unit could be detecting the fault because of "a lack of response / no response" from the vent valve; check vent valve (as well as the actuator) for correct operation, movement and response.
P2423	HC Adsorption Catalyst Efficiency Below Threshold Bank 1	HC adsorption catalysts are used to adsorb the unburned hydrocarbons (HC) during the warm up period before the main catalysts are at working temperature. The adsorption catalyst stores the HC and then releases it when the main catalyst is able to operate efficiently (up to operating temperature and therefore able to convert the HC into less harmful gases); note that the adsorption catalyst can be closely coupled or integrated into a normal / main catalyst, and a valve can be used to control when the HC is released from the adsorption catalyst into the main catalyst.	The fault code indicates that the efficiency of the HC adsorption catalyst is below an acceptable level. Due to the different arrangements for catalytic converters and Lambda sensors it may be advisable to refer to applicable vehicle specific information to identify the types of catalytic converters and Lambda sensors fitted to the vehicle. The control unit can monitor information from sensors e.g. Lambda / oxygen sensor, to assess the efficiency of the adsorption catalyst. The fault can be caused by a deterioration in the adsorption catalyst, but it is possible that fuelling faults or other system faults could affect the efficiency; check for other fault codes and correct as necessary.

NOTE 1. Check for other fault codes that could provide additional information.   **NOTE 2.** Communication between control units can pass via a CAN-Bus system; refer to Page 51 for CAN-Bus checks.   268

NOTE 3. If a fault ... located, it is also possible that the control unit is at fault.   NOTE 4. Refer to Page 53 for list of pages that show the ISO standard for component locations e.g. Sensor A - Bank 1.

Code	Description	System Description	Fault Description
P2424	HC Adsorption Catalyst Efficiency Below Threshold Bank 2	HC adsorption catalysts are used to adsorb the unburned hydrocarbons (HC) during the warm up period before the main catalysts are at working temperature. The adsorption catalyst stores the HC and then releases it when the main catalyst is able to operate efficiently; note that the adsorption catalyst can be closely coupled or integrated into a normal / main catalyst.	The fault code indicates that the efficiency of the HC adsorption catalyst is below an acceptable level. Due to the different arrangements for catalytic converters and Lambda sensors it may be advisable to refer to applicable vehicle specific information to identify the types of catalytic converters and Lambda sensors fitted to the vehicle. The control unit can monitor information from sensors e.g. Lambda / oxygen sensor, to assess the efficiency of the adsorption catalyst. The fault can be caused by a deterioration in the adsorption catalyst, but it is possible that fuelling faults or other system faults could affect the efficiency; check for other fault codes and correct as necessary.
P2425	Exhaust Gas Recirculation Cooling Valve Control Circuit / Open	The Exhaust Gas Recirculation (EGR) system can be fitted with a cooler (fed by the normal engine coolant) to reduce the exhaust gas temperature. A temperature sensor can be used to monitor the exhaust gas outlet temperature from the cooler to ensure effective cooling. A valve can be used to control the flow of exhaust gas through the cooler; on Diesel engines, the valve can form part of the EGR throttle assembly.	Identify the type of valve actuator e.g. motor, solenoid etc. The voltage, frequency or value of the control signal (provided by the control unit) in the valve circuit is incorrect (undefined problem). Signal could be: out of normal operating range, constant value, non-existent, corrupt. The code indicates a possible "open circuit" but other faults could activate the same fault code e.g. short circuit, high resistance, interference from other circuits. Note: the control unit could be providing the correct signal but it is being affected by a circuit fault. Identify the type of valve and the wiring used on the system. Refer to list of sensor / actuator checks on page 19.
P2426	Exhaust Gas Recirculation Cooling Valve Control Circuit Low	The Exhaust Gas Recirculation (EGR) system can be fitted with a cooler (fed by the normal engine coolant) to reduce the exhaust gas temperature. A temperature sensor can be used to monitor the exhaust gas outlet temperature from the cooler to ensure effective cooling. A valve can be used to control the flow of exhaust gas through the cooler; on Diesel engines, the valve can form part of the EGR throttle assembly.	Identify the type of valve actuator e.g. motor, solenoid etc. The voltage, frequency or value of the control signal (provided by the control unit) in the valve circuit is either "zero" (no signal) or, the signal exists but is below the normal operating range. Possible fault with valve / wiring (short / open circuit or resistance) that is causing a signal value to be lower than the minimum operating limit. Refer to list of sensor / actuator checks on page 19.
P2427	Exhaust Gas Recirculation Cooling Valve Control Circuit High	The Exhaust Gas Recirculation (EGR) system can be fitted with a cooler (fed by the normal engine coolant) to reduce the exhaust gas temperature. A temperature sensor can be used to monitor the exhaust gas outlet temperature from the cooler to ensure effective cooling. A valve can be used to control the flow of exhaust gas through the cooler; on Diesel engines, the valve can form part of the EGR throttle assembly.	Identify the type of valve actuator e.g. motor, solenoid etc. The voltage, frequency or value of the control signal (provided by the control unit) in the valve circuit is either at full value (e.g. battery voltage with no on / off signal) or, the signal exists but it is above the normal operating range. Possible fault with valve / wiring (short / open circuit or resistance) that is causing a signal value to be higher than the maximum operating limit. Refer to list of sensor / actuator checks on page 19.
P2428	Exhaust Gas Temperature Too High Bank 1	The exhaust gas temperature sensor signal is used to indicate if the temperature is outside of acceptable limits (some emissions control functions are more efficient when the gas temperature is within specified limits). Excessive temperatures can rapidly cause deterioration of the catalyst and other emissions control components. High gas temperatures are often caused by faults in the engine operation e.g. misfires, air leaks, fuelling etc.	The voltage, frequency or value of the temperature sensor signal is within the normal operating range / tolerance but is indicating a high temperature value. The control unit has not detected an electrical fault, therefore the assessment is that the temperature is too high for the operating conditions (indicated by other sensors) or, the temperature does not match values expected by the control unit. High temperatures are often caused by high levels of oxygen and / or fuel in the exhaust gas; check for exhaust air leaks, fuelling (mixture), misfires, secondary air injection faults, or other faults causing high oxygen / high fuel content in the exhaust gas. If the sensor does appear to be faulty, check temperature sensor / wiring e.g. short / open circuits, circuit resistance, incorrect sensor resistance or, fault with the reference / operating voltage; interference from other circuits can also affect sensor signals. Refer to list of sensor / actuator checks on page 19.
P2429	Exhaust Gas Temperature Too High Bank 2	The exhaust gas temperature sensor signal is used to indicate if the temperature is outside of acceptable limits (some emissions control functions are more efficient when the gas temperature is within specified limits). Excessive temperatures can rapidly cause deterioration of the catalyst and other emissions control components. High gas temperatures are often caused by faults in the engine operation e.g. misfires, air leaks, fuelling etc.	The voltage, frequency or value of the temperature sensor signal is within the normal operating range / tolerance but is indicating a high temperature value. The control unit has not detected an electrical fault, therefore the assessment is that the temperature is too high for the operating conditions (indicated by other sensors) or, the temperature does not match values expected by the control unit. High temperatures are often caused by high levels of oxygen and / or fuel in the exhaust gas; check for exhaust air leaks, fuelling (mixture), misfires, secondary air injection faults, or other faults causing high oxygen / high fuel content in the exhaust gas. If the sensor does appear to be faulty, check temperature sensor / wiring e.g. short / open circuits, circuit resistance, incorrect sensor resistance or, fault with the reference / operating voltage; interference from other circuits can also affect sensor signals. Refer to list of sensor / actuator checks on page 19.
P242A	Exhaust Gas Temperature Sensor Circuit Bank 1 Sensor 3	The exhaust gas temperature sensor signal is used to indicate if the temperature is outside of acceptable limits (some emissions control functions are more efficient when the gas temperature is within specified limits). Excessive temperatures can rapidly cause deterioration of the catalyst and other emissions control components. High gas temperatures are often caused by faults in the engine operation e.g. misfires, air leaks, fuelling etc.	The voltage, frequency or value of the temperature sensor signal is incorrect (undefined fault). Sensor voltage / signal could be: out of normal operating range, constant value, non-existent, corrupt. Possible fault in the temperature sensor or wiring e.g. short / open circuits, circuit resistance, incorrect sensor resistance or, fault with the reference / operating voltage; interference from other circuits can also affect sensor signals. Refer to list of sensor / actuator checks on page 19.
P242B	Exhaust Gas Temperature Sensor Circuit Range / Performance Bank 1 Sensor 3	The exhaust gas temperature sensor signal is used to indicate if the temperature is outside of acceptable limits (some emissions control functions are more efficient when the gas temperature is within specified limits). Excessive temperatures can rapidly cause deterioration of the catalyst and other emissions control components. High gas temperatures are often caused by faults in the engine operation e.g. misfires, air leaks, fuelling etc.	The voltage, frequency or value of the temperature sensor signal is within the normal operating range / tolerance but the signal is not plausible or is incorrect due to an undefined fault. The sensor signal might not match the operating conditions indicated by other sensors or, the sensor signal is not changing / responding as expected to the changes in conditions. Possible fault in the temperature sensor or wiring e.g. short / open circuits, circuit resistance, incorrect sensor resistance or, fault with the reference / operating voltage; interference from other circuits can also affect sensor signals. Refer to list of sensor / actuator checks on page 19. It is also possible that the temperature sensor is operating correctly but the temperature being measured by the sensor is unacceptable; check for exhaust air leaks, fuelling (mixture), misfires, secondary air injection faults, or other faults causing high oxygen and / or high fuel content in the exhaust.

Fault code	EOBD / ISO Description	Component / System Description	Meaningful Description and Quick Check
P242C	Exhaust Gas Temperature Sensor Circuit Low Bank 1 Sensor 3	The exhaust gas temperature sensor signal is used to indicate if the temperature is outside of acceptable limits (some emissions control functions are more efficient when the gas temperature is within specified limits). Excessive temperatures can rapidly cause deterioration of the catalyst and other emissions control components. High gas temperatures are often caused by faults in the engine operation e.g. misfires, air leaks, fuelling etc.	The voltage, frequency or value of the temperature sensor signal is either "zero" (no signal) or, the signal exists but it is below the normal operating range (lower than the minimum operating limit). Possible fault in the temperature sensor or wiring e.g. short / open circuits, circuit resistance, incorrect sensor resistance or, fault with the reference / operating voltage; interference from other circuits can also affect sensor signals. Refer to list of sensor / actuator checks on page 19.
P242D	Exhaust Gas Temperature Sensor Circuit High Bank 1 Sensor 3	The exhaust gas temperature sensor signal is used to indicate if the temperature is outside of acceptable limits (some emissions control functions are more efficient when the gas temperature is within specified limits). Excessive temperatures can rapidly cause deterioration of the catalyst and other emissions control components. High gas temperatures are often caused by faults in the engine operation e.g. misfires, air leaks, fuelling etc.	The voltage, frequency or value of the temperature sensor signal is either at full value (e.g. battery voltage with no signal or frequency) or, the signal exists but it is above the normal operating range (higher than the maximum operating limit). Possible fault in the temperature sensor or wiring e.g. short / open circuits, circuit resistance, incorrect sensor resistance or, fault with the reference / operating voltage; interference from other circuits can also affect sensor signals. Refer to list of sensor / actuator checks on page 19.
P242E	Exhaust Gas Temperature Sensor Circuit Intermittent / Erratic Bank 1 Sensor 3	The exhaust gas temperature sensor signal is used to indicate if the temperature is outside of acceptable limits (some emissions control functions are more efficient when the gas temperature is within specified limits). Excessive temperatures can rapidly cause deterioration of the catalyst and other emissions control components. High gas temperatures are often caused by faults in the engine operation e.g. misfires, air leaks, fuelling etc.	The voltage, frequency or value of the temperature sensor signal is intermittent (likely to be out of normal operating range / tolerance when intermittent fault is detected) or, the signal / voltage is erratic (e.g. signal changes are irregular / unpredictable). Possible fault with wiring or connections (including internal sensor connections); also interference from other circuits can affect the signal. Refer to list of sensor / actuator checks on page 19. Although the problem is likely to be a sensor / electrical fault, also check for other faults could cause rapid / erratic changes in the exhaust temperature.
P242F	Diesel Particulate Filter Restriction -Ash Accumulation	The Diesel particulate filter can be checked using a pressure difference measurement (differential exhaust pressures from the input to the output of the filter). If the particulate filter is becoming blocked then the input pressure will be significantly higher than the output pressure. Systems may also use a soot or ash sensor to detect whether the filter system is operating efficiently.	The information from the pressure and / or temperature sensors (or ash / soot sensor) is used to assess when regeneration is required. The fault code indicates that there is a restriction or accumulation of ash in the filter (which should normally be cleared during the particulate filter regeneration process). It is therefore possible that the filter is not efficient / faulty, or the regeneration process is not effective (also refer to fault codes P2458 and P2459 for additional information). For those systems using an Reductant additive (added to the fuel tank or injected into the exhaust gas), there could be a fault with the Reductant system; refer to fault codes P2037 to P2064 (including P203A-F, P204A-E and P205A-E).
P2430	Secondary Air Injection System Air Flow / Pressure Sensor Circuit Bank 1	After cold starts, the secondary air injection system pumps air (oxygen) into the exhaust gas (which is rich with HC and CO), this helps the HC / CO to combust (combustion heat in exhaust gas helps to heat the catalyst); the additional air can be detected in a number of ways including using a Lambda (oxygen) sensor (detecting oxygen level) or a flow / pressure sensor. Some systems direct air from the main intake system and then use the main airflow sensor to detect increase in air flow. Note that some systems (pulse air systems) use negative exhaust pressure pulses to draw air into the exhaust system instead of an air pump. If additional air is not detected or quantity is incorrect, this is regarded as a secondary air system fault. It may be necessary to refer to vehicle specific information to identify the system and components used.	The voltage or value of the flow / pressure sensor signal is incorrect (undefined fault). Sensor voltage / signal could be: out of normal operating range, constant value, non-existent, corrupt. Possible fault in the sensor or wiring e.g. short / open circuits, circuit resistance, incorrect sensor resistance or, fault with the reference / operating voltage; interference from other circuits can also affect sensor signals. Refer to list of sensor / actuator checks on page 19.
P2431	Secondary Air Injection System Air Flow / Pressure Sensor Circuit Range / Performance Bank 1	After cold starts, the secondary air injection system pumps air (oxygen) into the exhaust gas (which is rich with HC and CO), this helps the HC / CO to combust (combustion heat in exhaust gas helps to heat the catalyst); the additional air can be detected in a number of ways including using a Lambda (oxygen) sensor (detecting oxygen level) or a flow / pressure sensor. Some systems direct air from the main intake system and then use the main airflow sensor to detect increase in air flow. Note that some systems (pulse air systems) use negative exhaust pressure pulses to draw air into the exhaust system instead of an air pump. If additional air is not detected or quantity is incorrect, this is regarded as a secondary air system fault. It may be necessary to refer to vehicle specific information to identify the system and components used.	The voltage or value of the flow / pressure sensor signal is within the normal operating range / tolerance but the signal is not plausible or is incorrect due to an undefined fault. The sensor signal might not match the operating conditions indicated by other sensors or, the sensor signal is not changing / responding as expected to the changes in conditions. Possible fault in the sensor or wiring e.g. short / open circuits, circuit resistance, incorrect sensor resistance or, fault with the reference / operating voltage; interference from other circuits can also affect sensor signals. Refer to list of sensor / actuator checks on page 19. Also check for any blockages / leaks at the sensor or sensor pipes. It is also possible that the sensor is operating correctly but the flow / pressure being detected by the sensor is unacceptable or incorrect for the operating conditions.
P2432	Secondary Air Injection System Air Flow / Pressure Sensor Circuit Low Bank 1	After cold starts, the secondary air injection system pumps air (oxygen) into the exhaust gas (which is rich with HC and CO), this helps the HC / CO to combust (combustion heat in exhaust gas helps to heat the catalyst); the additional air can be detected in a number of ways including using a Lambda (oxygen) sensor (detecting oxygen level) or a flow / pressure sensor. Some systems direct air from the main intake system and then use the main airflow sensor to detect increase in air flow. Note that some systems (pulse air systems) use negative exhaust pressure pulses to draw air into the exhaust system instead of an air pump. If additional air is not detected or quantity is incorrect, this is regarded as a secondary air system fault. It may be necessary to refer to vehicle specific information to identify the system and components used.	The voltage or value of the flow / pressure sensor signal is either "zero" (no signal) or, the signal exists but it is below the normal operating range (lower than the minimum operating limit). Possible fault in the sensor or wiring e.g. short / open circuits, circuit resistance, incorrect sensor resistance or, fault with the reference / operating voltage; interference from other circuits can also affect sensor signals. Refer to list of sensor / actuator checks on page 19.

P2433	Secondary Air Injection System Air Flow / Pressure Sensor Circuit High Bank 1	After cold starts, the secondary air injection system pumps air (oxygen) into the exhaust gas (which is rich with HC and CO), this helps the HC / CO to combust (combustion heat in exhaust gas helps to heat the catalyst); the additional air can be detected in a number of ways including using a Lambda (oxygen) sensor (detecting oxygen level) or a flow / pressure sensor. Some systems direct air from the main intake system and then use the main airflow sensor to detect increase in air flow. Note that some systems (pulse air systems) use negative exhaust pressure pulses to draw air into the exhaust system instead of an air pump. If additional air is not detected or quantity is incorrect, this is regarded as a secondary air system fault. It may be necessary to refer to vehicle specific information to identify the system and components used.	The voltage or value of the flow / pressure sensor signal is either at full value (e.g. battery voltage with no signal or frequency) or, the signal exists but it is above the normal operating range (higher than the maximum operating limit). Possible fault in the sensor or wiring e.g. short / open circuits, circuit resistance, incorrect sensor resistance or, fault with the reference / operating voltage; interference from other circuits can also affect sensor signals. Refer to list of sensor / actuator checks on page 19.
P2434	Secondary Air Injection System Air Flow / Pressure Sensor Circuit Intermittent / Erratic Bank 1	After cold starts, the secondary air injection system pumps air (oxygen) into the exhaust gas (which is rich with HC and CO), this helps the HC / CO to combust (combustion heat in exhaust gas helps to heat the catalyst); the additional air can be detected in a number of ways including using a Lambda (oxygen) sensor (detecting oxygen level) or a flow / pressure sensor. Some systems direct air from the main intake system and then use the main airflow sensor to detect increase in air flow. Note that some systems (pulse air systems) use negative exhaust pressure pulses to draw air into the exhaust system instead of an air pump. If additional air is not detected or quantity is incorrect, this is regarded as a secondary air system fault. It may be necessary to refer to vehicle specific information to identify the system and components used.	The voltage or value of the flow / pressure sensor signal is intermittent (likely to be out of normal operating range / tolerance when intermittent fault is detected) or, the signal / voltage is erratic (e.g. signal changes are irregular / unpredictable). Possible fault with wiring or connections (including internal sensor connections); also interference from other circuits can affect the signal. Refer to list of sensor / actuator checks on page 19. It is also possible that the flow / pressure being detected is fluctuating or erratic. If possible use a separate pressure gauge to check the secondary air pressure. Also check for any blockages / leaks at the sensor or sensor pipes.
P2435	Secondary Air Injection System Air Flow / Pressure Sensor Circuit Bank 2	After cold starts, the secondary air injection system pumps air (oxygen) into the exhaust gas (which is rich with HC and CO), this helps the HC / CO to combust (combustion heat in exhaust gas helps to heat the catalyst); the additional air can be detected in a number of ways including using a Lambda (oxygen) sensor (detecting oxygen level) or a flow / pressure sensor. Some systems direct air from the main intake system and then use the main airflow sensor to detect increase in air flow. Note that some systems (pulse air systems) use negative exhaust pressure pulses to draw air into the exhaust system instead of an air pump. If additional air is not detected or quantity is incorrect, this is regarded as a secondary air system fault. It may be necessary to refer to vehicle specific information to identify the system and components used.	The voltage or value of the flow / pressure sensor signal is incorrect (undefined fault). Sensor voltage / signal could be: out of normal operating range, constant value, non-existent, corrupt. Possible fault in the sensor or wiring e.g. short / open circuits, circuit resistance, incorrect sensor resistance or, fault with the reference / operating voltage; interference from other circuits can also affect sensor signals. Refer to list of sensor / actuator checks on page 19.
P2436	Secondary Air Injection System Air Flow / Pressure Sensor Circuit Range / Performance Bank 2	After cold starts, the secondary air injection system pumps air (oxygen) into the exhaust gas (which is rich with HC and CO), this helps the HC / CO to combust (combustion heat in exhaust gas helps to heat the catalyst); the additional air can be detected in a number of ways including using a Lambda (oxygen) sensor (detecting oxygen level) or a flow / pressure sensor. Some systems direct air from the main intake system and then use the main airflow sensor to detect increase in air flow. Note that some systems (pulse air systems) use negative exhaust pressure pulses to draw air into the exhaust system instead of an air pump. If additional air is not detected or quantity is incorrect, this is regarded as a secondary air system fault. It may be necessary to refer to vehicle specific information to identify the system and components used.	The voltage or value of the flow / pressure sensor signal is within the normal operating range / tolerance but the signal is not plausible or is incorrect due to an undefined fault. The sensor signal might not match the operating conditions indicated by other sensors or, the sensor signal is not changing / responding as expected to the changes in conditions. Possible fault in the sensor or wiring e.g. short / open circuits, circuit resistance, incorrect sensor resistance or, fault with the reference / operating voltage; interference from other circuits can also affect sensor signals. Refer to list of sensor / actuator checks on page 19. Also check for any blockages / leaks at the sensor or sensor pipes. It is also possible that the sensor is operating correctly but the flow / pressure being detected by the sensor is unacceptable or incorrect for the operating conditions.
P2437	Secondary Air Injection System Air Flow / Pressure Sensor Circuit Low Bank 2	After cold starts, the secondary air injection system pumps air (oxygen) into the exhaust gas (which is rich with HC and CO), this helps the HC / CO to combust (combustion heat in exhaust gas helps to heat the catalyst); the additional air can be detected in a number of ways including using a Lambda (oxygen) sensor (detecting oxygen level) or a flow / pressure sensor. Some systems direct air from the main intake system and then use the main airflow sensor to detect increase in air flow. Note that some systems (pulse air systems) use negative exhaust pressure pulses to draw air into the exhaust system instead of an air pump. If additional air is not detected or quantity is incorrect, this is regarded as a secondary air system fault. It may be necessary to refer to vehicle specific information to identify the system and components used.	The voltage or value of the flow / pressure sensor signal is either "zero" (no signal) or, the signal exists but it is below the normal operating range (lower than the minimum operating limit). Possible fault in the sensor or wiring e.g. short / open circuits, circuit resistance, incorrect sensor resistance or, fault with the reference / operating voltage; interference from other circuits can also affect sensor signals. Refer to list of sensor / actuator checks on page 19.

Fault code	EOBD / ISO Description	Component / System Description	Meaningful Description and Quick Check
P2438	Secondary Air Injection System Air Flow / Pressure Sensor Circuit High Bank 2	After cold starts, the secondary air injection system pumps air (oxygen) into the exhaust gas (which is rich with HC and CO), this helps the HC / CO to combust (combustion heat in exhaust gas helps to heat the catalyst); the additional air can be detected in a number of ways including using a Lambda (oxygen) sensor (detecting oxygen level) or a flow / pressure sensor. Some systems direct air from the main intake system and then use the main airflow sensor to detect increase in air flow. Note that some systems (pulse air systems) use negative exhaust pressure pulses to draw air into the exhaust system instead of an air pump. If additional air is not detected or quantity is incorrect, this is regarded as a secondary air system fault. It may be necessary to refer to vehicle specific information to identify the system and components used.	The voltage or value of the flow / pressure sensor signal is either at full value (e.g. battery voltage with no signal or frequency) or, the signal exists but it is above the normal operating range (higher than the maximum operating limit). Possible fault in the sensor or wiring e.g. short / open circuits, circuit resistance, incorrect sensor resistance or, fault with the reference / operating voltage; interference from other circuits can also affect sensor signals. Refer to list of sensor / actuator checks on page 19.
P2439	Secondary Air Injection System Air Flow / Pressure Sensor Circuit Intermittent / Erratic Bank 2	After cold starts, the secondary air injection system pumps air (oxygen) into the exhaust gas (which is rich with HC and CO), this helps the HC / CO to combust (combustion heat in exhaust gas helps to heat the catalyst); the additional air can be detected in a number of ways including using a Lambda (oxygen) sensor (detecting oxygen level) or a flow / pressure sensor. Some systems direct air from the main intake system and then use the main airflow sensor to detect increase in air flow. Note that some systems (pulse air systems) use negative exhaust pressure pulses to draw air into the exhaust system instead of an air pump. If additional air is not detected or quantity is incorrect, this is regarded as a secondary air system fault. It may be necessary to refer to vehicle specific information to identify the system and components used.	The voltage or value of the flow / pressure sensor signal is intermittent (likely to be out of normal operating range / tolerance when intermittent fault is detected) or, the signal / voltage is erratic (e.g. signal changes are irregular / unpredictable). Possible fault with wiring or connections (including internal sensor connections); also interference from other circuits can affect the signal. Refer to list of sensor / actuator checks on page 19. It is also possible that the flow / pressure being detected is fluctuating or erratic. If possible use a separate pressure gauge to check the secondary air pressure. Also check for any blockages / leaks at the sensor or sensor pipes.
P2440	Secondary Air Injection System Switching Valve Stuck Open Bank 1	After cold starts, the secondary air injection system pumps air (oxygen) into the exhaust gas (which is rich with HC and CO), the additional air helps the HC / CO to combust (combustion heat in exhaust gas helps to heat the catalyst). The switching valve is operated by an electric actuator (usually a solenoid), and is used to regulate the vacuum that opens the secondary air valve (which allows the air into the exhaust gas).	The control unit is providing a control signal to the actuator but the control unit has detected that the actuator is stuck in the open position and is not responding to the control signal. The fault could be related to an actuator / wiring fault e.g. power supply, short / open circuit, incorrect actuator resistance. Refer to list of sensor / actuator checks on page 19. Note that the control unit could be detecting the fault because of "a lack of response / no response" from the system or mechanism being controlled by the actuator; check system (as well as the actuator) for correct operation, movement and response. Check that the valve is closing effectively (no leaks) and that there are no leaks allowing air to enter the exhaust system. Many systems rely on the Lambda sensor to register the increase in oxygen in the exhaust gas (when air injection occurs), it is therefore possible that a Lambda sensor fault could be providing incorrect oxygen content information. Also check vacuum pipes for blockages / leaks.
P2441	Secondary Air Injection System Switching Valve Stuck Closed Bank 1	After cold starts, the secondary air injection system pumps air (oxygen) into the exhaust gas (which is rich with HC and CO), the additional air helps the HC / CO to combust (combustion heat in exhaust gas helps to heat the catalyst). The switching valve is operated by an electric actuator (usually a solenoid), and is used to regulate the vacuum that opens the secondary air valve (which allows the air into the exhaust gas).	The control unit is providing a control signal to the actuator but the control unit has detected that the actuator is stuck in the closed position and is not responding to the control signal. The fault could be related to an actuator / wiring fault e.g. power supply, short / open circuit, incorrect actuator resistance. Refer to list of sensor / actuator checks on page 19. Note that the control unit could be detecting the fault because of "a lack of response / no response" from the system or mechanism being controlled by the actuator; check system (as well as the actuator) for correct operation and blockages in the air pipes or valve. Many systems rely on the Lambda sensor to register the increase in oxygen in the exhaust gas (when air injection occurs), it is therefore possible that a Lambda sensor fault could be providing incorrect oxygen content information. Also check vacuum pipes for blockages / leaks.
P2442	Secondary Air Injection System Switching Valve Stuck Open Bank 2	After cold starts, the secondary air injection system pumps air (oxygen) into the exhaust gas (which is rich with HC and CO), the additional air helps the HC / CO to combust (combustion heat in exhaust gas helps to heat the catalyst). The switching valve is operated by an electric actuator (usually a solenoid), and is used to regulate the vacuum that opens the secondary air valve (which allows the air into the exhaust gas).	The control unit is providing a control signal to the actuator but the control unit has detected that the actuator is stuck in the open position and is not responding to the control signal. The fault could be related to an actuator / wiring fault e.g. power supply, short / open circuit, incorrect actuator resistance. Refer to list of sensor / actuator checks on page 19. Note that the control unit could be detecting the fault because of "a lack of response / no response" from the system or mechanism being controlled by the actuator; check system (as well as the actuator) for correct operation, movement and response. Check that the valve is closing effectively (no leaks) and that there are no leaks allowing air to enter the exhaust system. Many systems rely on the Lambda sensor to register the increase in oxygen in the exhaust gas (when air injection occurs), it is therefore possible that a Lambda sensor fault could be providing incorrect oxygen content information. Also check vacuum pipes for blockages / leaks.
P2443	Secondary Air Injection System Switching Valve Stuck Closed Bank 2	After cold starts, the secondary air injection system pumps air (oxygen) into the exhaust gas (which is rich with HC and CO), the additional air helps the HC / CO to combust (combustion heat in exhaust gas helps to heat the catalyst). The switching valve is operated by an electric actuator (usually a solenoid), and is used to regulate the vacuum that opens the secondary air valve (which allows the air into the exhaust gas).	The control unit is providing a control signal to the actuator but the control unit has detected that the actuator is stuck in the closed position and is not responding to the control signal. The fault could be related to an actuator / wiring fault e.g. power supply, short / open circuit, incorrect actuator resistance. Refer to list of sensor / actuator checks on page 19. Note that the control unit could be detecting the fault because of "a lack of response / no response" from the system or mechanism being controlled by the actuator; check system (as well as the actuator) for correct operation and blockages in the air pipes or valve. Many systems rely on the Lambda sensor to register the increase in oxygen in the exhaust gas (when air injection occurs), it is therefore possible that a Lambda sensor fault could be providing incorrect oxygen content information. Also check vacuum pipes for blockages / leaks.

Code	Fault Location	Description	Probable Cause
P2444	Secondary Air Injection System Pump Stuck On Bank 1	After cold starts, the secondary air injection system pumps air (oxygen) into the exhaust gas (which is rich with HC and CO), this helps the HC / CO to combust (combustion heat in exhaust gas helps to heat the catalyst. The additional air can be detected by the Lambda (oxygen) sensor although a flow / pressure sensor can be used. The air pump delivers the secondary air to the exhaust manifold (via a secondary air valve). The pump is driven by an electric actuator (usually a motor).	The fault could be detected due to incorrect oxygen content in the exhaust gas or, on some systems the secondary airflow / pressure sensor is indicating incorrect flow / pressure. The exact fault could depend on the type of control system and wiring for the pump. The fault code indicates that the pump is stuck in a fixed (on) position; this could be caused by a wiring fault (e.g. short circuit to another power supply) or relay stuck on (if applicable); Refer to list of sensor / actuator checks on page 19. Note that the control unit could be detecting the fault because of "a lack of response / no response" from the secondary air injection system when the control unit switches off the control signal; check secondary air system (as well as the pump / wiring) for correct operation, movement and response. Check for incorrect operation of system valves and check for leaks / blockages of secondary air system hoses / pipes etc. Also refer to fault code P2448 for additional information.
P2445	Secondary Air Injection System Pump Stuck Off Bank 1	After cold starts, the secondary air injection system pumps air (oxygen) into the exhaust gas (which is rich with HC and CO), this helps the HC / CO to combust (combustion heat in exhaust gas helps to heat the catalyst. The additional air can be detected by the Lambda (oxygen) sensor although a flow / pressure sensor can be used. The air pump delivers the secondary air to the exhaust manifold (via a secondary air valve). The pump is driven by an electric actuator (usually a motor).	The fault could be detected due to incorrect oxygen content in the exhaust gas or, on some systems the secondary airflow / pressure sensor is indicating incorrect flow / pressure. The control unit is providing a control signal to the actuator (signal is within the normal operating range / tolerance) but the control unit has detected that the actuator is stuck in a fixed (off) position and is not responding to the control signal. The fault could be related to an actuator / wiring fault, Refer to list of sensor / actuator checks on page 19. Note that the control unit could be detecting the fault because of "a lack of response / no response" from the secondary air injection system; check secondary air system (as well as the actuator) for correct operation, movement and response. Check for incorrect operation of system valves and check for leaks / blockages of secondary air system hoses / pipes etc. Also refer to fault code P2448 for additional information.
P2446	Secondary Air Injection System Pump Stuck On Bank 2	After cold starts, the secondary air injection system pumps air (oxygen) into the exhaust gas (which is rich with HC and CO), this helps the HC / CO to combust (combustion heat in exhaust gas helps to heat the catalyst. The additional air can be detected by the Lambda (oxygen) sensor although a flow / pressure sensor can be used. The air pump delivers the secondary air to the exhaust manifold (via a secondary air valve). The pump is driven by an electric actuator (usually a motor).	The fault could be detected due to incorrect oxygen content in the exhaust gas or, on some systems the secondary airflow / pressure sensor is indicating incorrect flow / pressure. The exact fault could depend on the type of control system and wiring for the pump. The fault code indicates that the pump is stuck in a fixed (on) position; this could be caused by a wiring fault (e.g. short circuit to another power supply) or relay stuck on (if applicable); Refer to list of sensor / actuator checks on page 19. Note that the control unit could be detecting the fault because of "a lack of response / no response" from the secondary air injection system when the control unit switches off the control signal; check secondary air system (as well as the pump / wiring) for correct operation, movement and response. Check for incorrect operation of system valves and check for leaks / blockages of secondary air system hoses / pipes etc. Also refer to fault code P2448 for additional information.
P2447	Secondary Air Injection System Pump Stuck Off Bank 2	After cold starts, the secondary air injection system pumps air (oxygen) into the exhaust gas (which is rich with HC and CO), this helps the HC / CO to combust (combustion heat in exhaust gas helps to heat the catalyst. The additional air can be detected by the Lambda (oxygen) sensor although a flow / pressure sensor can be used. The air pump delivers the secondary air to the exhaust manifold (via a secondary air valve). The pump is driven by an electric actuator (usually a motor).	The fault could be detected due to incorrect oxygen content in the exhaust gas or, on some systems the secondary airflow / pressure sensor is indicating incorrect flow / pressure. The control unit is providing a control signal to the actuator (signal is within the normal operating range / tolerance) but the control unit has detected that the actuator is stuck in a fixed (off) position and is not responding to the control signal. The fault could be related to an actuator / wiring fault, Refer to list of sensor / actuator checks on page 19. Note that the control unit could be detecting the fault because of "a lack of response / no response" from the secondary air injection system; check secondary air system (as well as the actuator) for correct operation, movement and response. Check for incorrect operation of system valves and check for leaks / blockages of secondary air system hoses / pipes etc. Also refer to fault code P2448 for additional information.
P2448	Secondary Air Injection System High Air Flow Bank 1	After cold starts, the secondary air injection system pumps air (oxygen) into the exhaust gas (which is rich with HC and CO), this helps the HC / CO to combust (combustion heat in exhaust gas helps to heat the catalyst); the additional air can be detected in a number of ways including using a Lambda (oxygen) sensor (detecting oxygen level) or a flow / pressure sensor. Some systems direct air from the main intake system and then use the main airflow sensor to detect increase in air flow. Note that some systems (pulse air systems) use negative exhaust pressure pulses to draw air into the exhaust system instead of an air pump. If additional air is not detected or quantity is incorrect, this is regarded as a secondary air system fault. It may be necessary to refer to vehicle specific information to identify the system and components used.	The fault code indicates that the air flow provided by the secondary air system (into the exhaust system) is higher than the expected value; the flow could be a calculated value based on the Lambda sensor signal or could be detected by the secondary air "flow / pressure" sensor. Check operation of the secondary air system; depending on the system design, this could involve checks on the secondary air control valves and switching valve control systems (vacuum control, vacuum pipes etc) and air pump. Check that the secondary air valve is closing properly and check that the switching valve is closing off the vacuum supply to the air valve. Check all applicable wiring and connections for short / open circuits and high resistance. Also check for correct supply voltage to applicable secondary air system components. Check that the non-return valve (in the air pipe or combined with switching valve) is operating correctly. Check for other fault codes e.g. flow / pressure sensor (if fitted) and oxygen / Lambda sensor fault codes that might indicate if the oxygen / Lambda sensor is at fault.
P2449	Secondary Air Injection System High Air Flow Bank 2	After cold starts, the secondary air injection system pumps air (oxygen) into the exhaust gas (which is rich with HC and CO), this helps the HC / CO to combust (combustion heat in exhaust gas helps to heat the catalyst); the additional air can be detected in a number of ways including using a Lambda (oxygen) sensor (detecting oxygen level) or a flow / pressure sensor. Some systems direct air from the main intake system and then use the main airflow sensor to detect increase in air flow. Note that some systems (pulse air systems) use negative exhaust pressure pulses to draw air into the exhaust system instead of an air pump. If additional air is not detected or quantity is incorrect, this is regarded as a secondary air system fault. It may be necessary to refer to vehicle specific information to identify the system and components used.	The fault code indicates that the air flow provided by the secondary air system (into the exhaust system) is higher than the expected value; the flow could be a calculated value based on the Lambda sensor signal or could be detected by the secondary air "flow / pressure" sensor. Check operation of the secondary air system; depending on the system design, this could involve checks on the secondary air control valves and switching valve control systems (vacuum control, vacuum pipes etc) and air pump. Check that the secondary air valve is closing properly and check that the switching valve is closing off the vacuum supply to the air valve. Check all applicable wiring and connections for short / open circuits and high resistance. Also check for correct supply voltage to applicable secondary air system components. Check that the non-return valve (in the air pipe or combined with switching valve) is operating correctly. Check for other fault codes e.g. flow / pressure sensor (if fitted) and oxygen / Lambda sensor fault codes that might indicate if the oxygen / Lambda sensor is at fault.

NOTE 1. Check for other fault codes that could provide additional information.
NOTE 3. If a fault cannot be located, it is also possible that the control unit is at fault.

NOTE 2. Communication between control units can pass via a CAN-Bus system; refer to Page 51 for CAN-Bus checks.
NOTE 4. Refer to Page 53 for list of pages that show the ISO standard for component locations e.g. Sensor A - Bank 1.

Fault code	EOBD / ISO Description	Component / System Description	Meaningful Description and Quick Check
P2450	Evaporative Emission System Switching Valve Performance / Stuck Open	EVAP systems make use of a canister to collect fuel vapours and then release them (under controlled conditions) to the engine intake system. It may be necessary to refer to vehicle specific information to identify the system design and components used. Some vehicle manufacturers may use the term "switching valve" as an alternative to "vent valve". On some applications, a vent valve is contained within a pump module (contains the leak test pressure / vacuum pump). Refer to vehicle specific information to identify the valve and location, Refer to fault code P2407 for other information on the pump. Refer to fault code P0448 for additional information on vent valves.	The fault code refers to the electric actuator that is used to operate the switching valve. The control unit is providing a control signal to the actuator (signal is within the normal operating range / tolerance) but the control unit has detected that the actuator is stuck in a fixed (open) position and is not responding to the control signal. The fault could be related to an actuator / wiring fault. Refer to list of sensor / actuator checks on page 19. Note that the control unit could be detecting the fault because of "a lack of response / no response" from the switching valve; check switching valve (as well as the actuator) for correct operation, movement and response. Also check for blockages or leaks in the hoses / pipes etc. in the EVAP system.
P2451	Evaporative Emission System Switching Valve Stuck Closed	EVAP systems make use of a canister to collect fuel vapours and then release them (under controlled conditions) to the engine intake system. It may be necessary to refer to vehicle specific information to identify the system design and components used. Some vehicle manufacturers may use the term "switching valve" as an alternative to "vent valve". On some applications, a vent valve is contained within a pump module (contains the leak test pressure / vacuum pump). Refer to vehicle specific information to identify the valve and location, Refer to fault code P2407 for other information on the pump. Refer to fault code P0448 for additional information on vent valves.	The fault code refers to the electric actuator that is used to operate the switching valve. The control unit is providing a control signal to the actuator (signal is within the normal operating range / tolerance) but the control unit has detected that the actuator is stuck in a fixed (closed) position and is not responding to the control signal. The fault could be related to an actuator / wiring fault. Refer to list of sensor / actuator checks on page 19. Note that the control unit could be detecting the fault because of "a lack of response / no response" from the switching valve; check switching valve (as well as the actuator) for correct operation, movement and response. Also check for blockages or leaks in the hoses / pipes etc. in the EVAP system.
P2452	Diesel Particulate Filter Differential Pressure Sensor Circuit	The Diesel particulate filter can be checked using a pressure difference measurement (differential exhaust pressures from the input to the output of the filter). If the particulate filter is becoming blocked then the input pressure will be significantly higher than the output pressure. Systems may also use a soot or ash sensor to detect whether the filter system is operating efficiently.	The voltage, frequency or value of the pressure sensor signal is incorrect (undefined fault). Sensor voltage / signal could be: out of normal operating range, constant value, non-existent, corrupt. Possible fault in the pressure sensor or wiring e.g. short / open circuits, circuit resistance, incorrect sensor resistance (if applicable) or, fault with the reference / operating voltage; interference from other circuits can also affect sensor signals. Refer to list of sensor / actuator checks on page 19.
P2453	Diesel Particulate Filter Differential Pressure Sensor Circuit Range / Performance	The Diesel particulate filter can be checked using a pressure difference measurement (differential exhaust pressures from the input to the output of the filter). If the particulate filter is becoming blocked then the input pressure will be significantly higher than the output pressure. Systems may also use a soot or ash sensor to detect whether the filter system is operating efficiently.	The voltage, frequency or value of the pressure sensor signal is within the normal operating range / tolerance but the signal is not plausible or is incorrect due to an undefined fault. It is possible that the pressure sensor is operating correctly but the differential pressure being measured by the sensor is unacceptable or incorrect for the operating conditions, or, the sensor signal is not changing / responding as expected to the changes in conditions. Possible fault in the pressure sensor or wiring e.g. short / open circuits, circuit resistance, incorrect sensor resistance (if applicable) or, fault with the reference / operating voltage; interference from other circuits can also affect sensor signals. Refer to list of sensor / actuator checks on page 19. Also check any air passages / pipes to the sensor for leaks / blockages etc. It is also possible that the particulate filter is becoming blocked causing incorrect differential pressures.
P2454	Diesel Particulate Filter Differential Pressure Sensor Circuit Low	The differential pressure sensors monitors the exhaust pressures at the input and output of the particulate filter to establish if the filter is efficient.	The voltage, frequency or value of the pressure sensor signal is either "zero" (no signal) or, the signal exists but it is below the normal operating range (lower than the minimum operating limit). Possible fault in the pressure sensor or wiring e.g. short / open circuits, circuit resistance, incorrect sensor resistance (if applicable) or, fault with the reference / operating voltage; interference from other circuits can also affect sensor signals. Refer to list of sensor / actuator checks on page 19.
P2455	Diesel Particulate Filter Differential Pressure Sensor Circuit High	The Diesel particulate filter can be checked using a pressure difference measurement (differential exhaust pressures from the input to the output of the filter). If the particulate filter is becoming blocked then the input pressure will be significantly higher than the output pressure. Systems may also use a soot or ash sensor to detect whether the filter system is operating efficiently.	The voltage, frequency or value of the pressure sensor signal is either at full value (e.g. battery voltage with no signal or frequency) or, the signal exists but it is above the normal operating range (higher than the maximum operating limit). Possible fault in the pressure sensor or wiring e.g. short / open circuits, circuit resistance, incorrect sensor resistance (if applicable) or, fault with the reference / operating voltage; interference from other circuits can also affect sensor signals. Refer to list of sensor / actuator checks on page 19.
P2456	Diesel Particulate Filter Differential Pressure Sensor Circuit Intermittent / Erratic	The Diesel particulate filter can be checked using a pressure difference measurement (differential exhaust pressures from the input to the output of the filter). If the particulate filter is becoming blocked then the input pressure will be significantly higher than the output pressure. Systems may also use a soot or ash sensor to detect whether the filter system is operating efficiently.	The voltage, frequency or value of the pressure sensor signal is intermittent (likely to be out of normal operating range / tolerance when intermittent fault is detected) or, the signal / voltage is erratic (e.g. signal changes are irregular / unpredictable). Possible fault with wiring or connections (including internal sensor connections); also interference from other circuits can affect the signal. Refer to list of sensor / actuator checks on page 19. It is also possible that the differential pressure could be changing erratically, which could cause the fault code to be activated; check the leaks in any connecting pipes and also particulate filter system for correct operation
P2457	Exhaust Gas Recirculation Cooling System Performance	The Exhaust Gas Recirculation (EGR) system can be fitted with a cooler (fed by the normal engine coolant) to reduce the exhaust gas temperature. A temperature sensor can be used to monitor the exhaust gas outlet temperature from the cooler to ensure effective cooling. A valve can be used to control the flow of exhaust gas through the cooler or through a by-pass; on Diesel engines, the valve can form part of the EGR throttle assembly.	The fault code is indicating that there is an unidentified fault with the performance of the EGR gas cooling. The fault can be detected due to high exhaust gas temperatures (EGR gases exiting the cooler system). The fault could be caused by a cooling system fault (e.g. low coolant, incorrect cooling fan operation etc). If a cooling valve is fitted (see fault code P2427), it is possible that the valve is not operating correctly. Also check operation of exhaust gas temperature sensors.

NOTE 1. Check for other fault codes that could provide additional information.   NOTE 2. Communication between control units can pass via a CAN-Bus system; refer to Page 51 for CAN-Bus checks.   274

NOTE 3. If a fault cannot be located, it is also possible that the control unit is at fault.   NOTE 4. Refer to Page 53 for list of pages that show the ISO standard for component locations e.g. Sensor A - Bank 1.

	Description		
P2458	Diesel Particulate Filter Regeneration Duration	The Diesel particulate filter can be checked using a pressure difference measurement (differential exhaust pressures from the input to the output of the filter). If the particulate filter is becoming blocked then the input pressure will be significantly higher than the output pressure. Systems may also use a soot / ash sensor or exhaust gas temperature measurements to detect whether the filter system is operating efficiently.	The information from the pressure and / or temperature sensors is used to assess when regeneration is required; the normal regeneration duration and frequency values are stored in the control unit memory. The control unit will implement engine control changes e.g. late injection timing, to create higher exhaust gas temperatures; the higher temperatures are necessary for the regeneration process (burning off particulates). The fault code can be activated if the normal "duration" of the regeneration process does not successfully reduce the level of particulates in the filter (sensor readings could indicate further regeneration is required). The fault could be due to an inefficient particulate filter or, possibly the process to increase the exhaust gas temperature is not effective. For those systems using an Reductant additive (added to the fuel tank or injected into the exhaust gas), there could be a fault with the Reductant system; refer to fault codes P2037 to P2064 (including P203A-F, P204A-E and P205A-E).
P2459	Diesel Particulate Filter Regeneration Frequency	The Diesel particulate filter can be checked using a pressure difference measurement (differential exhaust pressures from the input to the output of the filter). If the particulate filter is becoming blocked then the input pressure will be significantly higher than the output pressure. Systems may also use a soot / ash sensor or exhaust gas temperature measurements to detect whether the filter system is operating efficiently.	The information from the pressure and / or temperature sensors is used to assess when regeneration is required; the normal regeneration duration and frequency values are stored in the control unit memory. The control unit will implement engine control changes e.g. late injection timing, to create higher exhaust gas temperatures; the higher temperatures are necessary for the regeneration process (burning off particulates). The fault code can be activated if it is necessary to increase the normal "frequency" of the regeneration process i.e. the particulate filter is becoming saturated with particulates more frequently than expected thus requiring an increased frequency for the regeneration process. The fault could be due to an inefficient particulate filter or, possibly the process to increase the exhaust gas temperature is not effective. For those systems using an Reductant additive (added to the fuel tank or injected into the exhaust gas), there could be a fault with the Reductant system; refer to fault codes P2037 to P2064 (including P203A-F, P204A-E and P205A-E). Also check operation of exhaust differential pressure sensor (or individual pressure sensors) and temperature sensors.
P245A	Exhaust Gas Recirculation Cooler Bypass Control Circuit / Open	The Exhaust Gas Recirculation (EGR) system can be fitted with a cooler (fed by the normal engine coolant) to reduce the exhaust gas temperature. A temperature sensor can be used to monitor the exhaust gas outlet temperature from the cooler to ensure effective cooling. A valve can be used to control the flow of exhaust gas through the cooler or through a by-pass; on Diesel engines, the valve can form part of the EGR throttle assembly.	Identify the type of actuator used for the by-pass control. The voltage, frequency or value of the control signal in the actuator circuit is incorrect. Signal could be:- out of normal operating range, constant value, non-existent, corrupt. The fault code indicates a possible "open circuit" but other undefined faults in the actuator or wiring could activate the same code e.g. actuator power supply, short circuit, high circuit resistance or incorrect actuator resistance. Note: the control unit could be providing the correct signal but it is being affected by the circuit fault. Refer to list of sensor / actuator checks on page 19.
P245B	Exhaust Gas Recirculation Cooler Bypass Control Circuit Range / Performance	The Exhaust Gas Recirculation (EGR) system can be fitted with a cooler (fed by the normal engine coolant) to reduce the exhaust gas temperature. A temperature sensor can be used to monitor the exhaust gas outlet temperature from the cooler to ensure effective cooling. A valve can be used to control the flow of exhaust gas through the cooler or through a by-pass; on Diesel engines, the valve can form part of the EGR throttle assembly.	Identify the type of actuator used for the by-pass control. The voltage, frequency or value of the control signal in the actuator circuit is within the normal operating range / tolerance, but the signal might be incorrect for the operating conditions. It is also possible that the actuator response (or response of system controlled by the actuator) differs from the expected / desired response (incorrect response for the control signal provided by the control unit). Possible fault in actuator or wiring, but also possible that the actuator (or mechanism / system controlled by the actuator) is not operating or moving correctly. Note: the control unit could be providing the correct signal but it is being affected by the circuit fault. Refer to list of sensor / actuator checks on page 19.
P245C	Exhaust Gas Recirculation Cooler Bypass Control Circuit Low	The Exhaust Gas Recirculation (EGR) system can be fitted with a cooler (fed by the normal engine coolant) to reduce the exhaust gas temperature. A temperature sensor can be used to monitor the exhaust gas outlet temperature from the cooler to ensure effective cooling. A valve can be used to control the flow of exhaust gas through the cooler or through a by-pass; on Diesel engines, the valve can form part of the EGR throttle assembly.	Identify the type of actuator used for the by-pass control. The voltage, frequency or value of the control signal in the actuator circuit is either "zero" (no signal) or, the signal exists but is below the normal operating range. Possible fault with actuator or wiring (e.g. actuator power supply, short / open circuit or resistance) that is causing a signal value to be lower than the minimum operating limit. Note: the control unit could be providing the correct signal but it is being affected by the circuit fault. Refer to list of sensor / actuator checks on page 19.
P245D	Exhaust Gas Recirculation Cooler Bypass Control Circuit High	The Exhaust Gas Recirculation (EGR) system can be fitted with a cooler (fed by the normal engine coolant) to reduce the exhaust gas temperature. A temperature sensor can be used to monitor the exhaust gas outlet temperature from the cooler to ensure effective cooling. A valve can be used to control the flow of exhaust gas through the cooler or through a by-pass; on Diesel engines, the valve can form part of the EGR throttle assembly.	Identify the type of actuator used for the by-pass control. The voltage, frequency or value of the control signal in the actuator circuit is either at full value (e.g. battery voltage with no on / off signal) or, the signal exists but is above the normal operating range. Possible fault with actuator or wiring (e.g. actuator power supply, short / open circuit or resistance) that is causing a signal value to be higher than the maximum operating limit. Note: the control unit could be providing the correct signal but it is being affected by the circuit fault. Refer to list of sensor / actuator checks on page 19.
P245E		ISO / SAE reserved	Not used, reserved for future allocation.
P245F		ISO / SAE reserved	Not used, reserved for future allocation.
P2500	Generator Lamp / L–Terminal Circuit Low	The generator warning lamp can be controlled from the generator or, on many systems it is controlled by the engine control unit (possibly via the CAN-Bus network). The generator or control unit can provide an earth path or power to the warning light on the "L" circuit, but a signal could be provided to a separate module that in turn will turn on the light.	It will be necessary to refer to vehicle specific information to identify the wiring / circuit used for the generator warning light. The fault code indicates that the voltage, frequency or signal value in the "L" circuit is low. The fault could be related to a wiring fault but, depending on the warning light system operation, the fault could be caused by a control module or CAN-Bus related fault. Check all applicable wiring for short / open circuits and check condition of warning light bulb (or lighting system).
P2501	Generator Lamp / L–Terminal Circuit High	The generator warning lamp can be controlled from the generator or, on many systems it is controlled by the engine control unit (possibly via the CAN-Bus network). The generator or control unit can provide an earth path or power to the warning light on the "L" circuit, but a signal could be provided to a separate module that in turn will turn on the light.	It will be necessary to refer to vehicle specific information to identify the wiring / circuit used for the generator warning light. The fault code indicates that the voltage, frequency or signal value in the "L" circuit is high. The fault could be related to a wiring fault but, depending on the warning light system operation, the fault could be caused by a control module or CAN-Bus related fault. Check all applicable wiring for short / open circuits and check condition of warning light bulb (or lighting system).

NOTE 1. Check for other fault codes that could provide additional information.
NOTE 3. If a fault cannot be located, it is also possible that the control unit is at fault.
NOTE 2. Communication between control units can pass via a CAN-Bus system; refer to Page 51 for CAN-Bus checks.
NOTE 4. Refer to Page 53 for list of pages that show the ISO standard for component locations e.g. Sensor A - Bank 1.

Fault code	EOBD / ISO Description	Component / System Description	Meaningful Description and Quick Check
P2502	Charging System Voltage	The main engine / powertrain control unit will be able to detect charging system voltage due to the fact that the control unit and components controlled by the control unit receive a power supply via the battery / charging system. However, it is becoming standard practice for the generator to be directly connected to the main control unit to enable the control unit to monitor and control generator output. Control units can regulate the generator field winding voltage / current, which in turn controls generator output.	The fault code indicates an undefined charging / battery system fault. For those systems where the generator is not controlled by the main control unit, carry out generator / battery checks and check applicable wiring for short / open circuits and high resistances (including earth connections to battery, engine etc), also check generator drive belt or drive system. For those systems where the main control unit is controlling the generator, it is possible that the control unit is providing a control signal to the generator field windings but, the generator output is not responding as expected. Check charging / battery system against specification for correct values and operation. Check generator drive belt or drive system. Check wiring and connections between battery and generator for short / open circuits and high resistances. Check battery condition and earth circuit. Check wiring / connections between the generator and the control unit for short / open circuit or high resistance (faulty wiring / connections could result in incorrect information passing from the generator to the control unit).
P2503	Charging System Voltage Low	The main engine / power train control unit will be able to detect charging system voltage due to the fact that the control unit and components controlled by the control unit receive a power supply via the battery / charging system. However, it is becoming standard practice for the generator to be directly connected to the main control unit to enable the control unit to monitor and control generator output. Control units can regulate the generator field winding voltage / current, which in turn controls generator output.	The fault code indicates a low voltage is being produced by the charging / battery system. For those systems where the generator is not controlled by the main control unit, carry out generator and battery checks and check applicable wiring for short / open circuits and high resistances (including earth connections to battery, engine etc), also check generator drive belt or drive system. For those systems where the main control unit is controlling the generator, it is possible that the control unit is providing a control signal to the generator field windings but, the generator output is lower than expected. Check charging / battery system against specification for correct values and operation. Check generator drive belt or drive system. Check wiring and connections between battery and generator for short / open circuits and high resistances. Check battery condition and earth circuit. Check wiring / connections between the generator and the control unit for short / open circuit or high resistance (faulty wiring / connections could result in incorrect information passing from the generator to the control unit).
P2504	Charging System Voltage High	The main engine / power train control unit will be able to detect charging system voltage due to the fact that the control unit and components controlled by the control unit receive a power supply via the battery / charging system. However, it is becoming standard practice for the generator to be directly connected to the main control unit to enable the control unit to monitor and control generator output. Control units can regulate the generator field winding voltage / current, which in turn controls generator output.	It will be necessary to refer to vehicle specific information to identify the wiring / circuit used for the generator warning light. The fault code indicates an undefined charging / battery system fault. For systems where the generator is not controlled by the main control unit, carry out generator / battery checks and check applicable wiring for short / open circuits and high resistances (including earth connections to battery, engine etc), also check generator drive belt or drive system. For systems where the main control unit is controlling the generator, it is possible that the control unit is providing a control signal to the generator field windings but, the generator output is not responding as expected. Check general operation of charging / battery system, check generator drive belt or drive system. Check wiring and connections between battery and generator for short / open circuits and high resistances. Check battery condition and earth circuit. Check wiring / connections between the generator and the control unit for short / open circuit or high resistance.
P2505	ECM / PCM Power Input Signal	The Engine Control Module / Powertrain Control Module can be provided with one or more power supplies. Note that the power supplies can be fed via a relay or a power control module and that a fuse can be included in the circuit. Also note that some systems may retain a power supply after the ignition is switched off; this is used to prevent loss of data from part of the internal memory system (Keep Alive Memory or KAM). The control module can also function as the power supply control for some actuators and provide sensor reference or operating voltage to some sensors.	The fault code indicates that there is a fault with the power input to the ECM / PCM (undefined fault). The voltage could be out of normal operating range, constant but incorrect value, non-existent, corrupt. Possible fault in the wiring e.g. short / open circuits or circuit resistance. Check power supply from source e.g. from ignition switch or from the battery via a relay and / or fuse; check relay operation if applicable Refer to list of sensor / actuator checks on page 19. Note that a relay could be switched by a power module, which could also require testing. If power supplies are being provided to the ECM / PCM but the fault code re-appears after clearing the code, it could indicate that the ECM / PCM is at fault.
P2506	ECM / PCM Power Input Signal Range / Performance	The Engine Control Module / Powertrain Control Module can be provided with one or more power supplies. Note that the power supplies can be fed via a relay or a power control module and that a fuse can be included in the circuit. Also note that some systems may retain a power supply after the ignition is switched off; this is used to prevent loss of data from part of the internal memory system (Keep Alive Memory or KAM). The control module can also function as the power supply control for some actuators and provide sensor reference or operating voltage to some sensors.	The fault code indicates that there is a fault with the power input to the ECM / PCM (undefined fault). The voltage is probably within the normal operating range / tolerance but the voltage is not plausible or is incorrect due to an undefined fault. The voltage might not match the operating conditions (e.g. low voltage but charging voltage is correct) or, the voltage is not changing / responding as expected. Possible fault in the wiring e.g. short / open circuits or circuit resistance. Check power supply from source e.g. from ignition switch or from the battery via a relay and / or fuse; check relay operation if applicable Refer to list of sensor / actuator checks on page 19. Note that a relay could be switched by a power module, which could also require testing. If power supplies are being provided to the ECM / PCM but the fault code re-appears after clearing the code, it could indicate that the ECM / PCM is at fault.
P2507	ECM / PCM Power Input Signal Low	The Engine Control Module / Powertrain Control Module can be provided with one or more power supplies. Note that the power supplies can be fed via a relay or a power control module and that a fuse can be included in the circuit. Also note that some systems may retain a power supply after the ignition is switched off; this is used to prevent loss of data from part of the internal memory system (Keep Alive Memory or KAM). The control module can also function as the power supply control for some actuators and provide sensor reference or operating voltage to some sensors.	The fault code indicates that there is a fault with the power input to the ECM / PCM. The voltage could be either "zero" (no supply) or, the voltage exists but it is below the normal operating range (lower than the minimum operating limit). Note that the fault code could be activated because the power supply is OFF when it should be switched ON; this could be caused by a faulty relay, fuse etc. Possible fault in the wiring e.g. short / open circuits or circuit resistance; check actual voltage level in the circuit to ensure that the supply voltage is not significantly lower than the charging / battery voltage. Check power supply from source e.g. from ignition switch or from the battery via a relay and / or fuse to identify whether the power supplies are being switched on / off at the correct times; check relay operation if applicable, Refer to list of sensor / actuator checks on page 19. Note that a relay could be switched by a power module, which could also require testing. If power supplies are being provided to the ECM / PCM but the fault code re-appears after clearing the code, it could indicate that the ECM / PCM is at fault.

NOTE 1. Check for other fault codes that could provide additional information.   **NOTE 2.** Communication between control units can pass via a CAN-Bus system; refer to Page 51 for CAN-Bus checks.   276

NOTE 4. Refer to Page 53 for list of pages that show the ISO standard for component locations e.g. Sensor A - Bank 1.

P2508	ECM / PCM Power Input Signal High	The Engine Control Module / Powertrain Control Module can be provided with one or more power supplies. Note that the power supplies can be fed via a relay or a power control module and that a fuse can be included in the circuit. Also note that some systems may retain a power supply after the ignition is switched off; this is used to prevent loss of data from part of the internal memory system (Keep Alive Memory or KAM). The control module can also function as the power supply control for some actuators and provide sensor reference or operating voltage to some sensors.	The fault code indicates that there is a fault with the power input to the ECM / PCM. The voltage could be either at full value (e.g. battery voltage exists at all times and is not switching off at the correct times) or, the voltage exists but it is above the normal operating range (this could be a charging fault). Note that the fault code could be activated because the power supply is ON when it should be switched OFF; this could be caused by a faulty relay, fuse etc. Possible fault in the wiring e.g. short / open circuits or circuit resistance. Check power supply from source e.g. from ignition switch or from the battery via a relay and / or fuse to identify whether the power supplies are being switched on / off at the correct times; check relay operation if applicable, Refer to list of sensor / actuator checks on page 19. Note that a relay could be switched by a power module, which could also require testing. If power supplies to the ECM / PCM are correct but the fault code re-appears after clearing the code, it could indicate that the ECM / PCM is at fault.
P2509	ECM / PCM Power Input Signal Intermittent	The Engine Control Module / Powertrain Control Module can be provided with one or more power supplies. Note that the power supplies can be fed via a relay or a power control module and that a fuse can be included in the circuit. Also note that some systems may retain a power supply after the ignition is switched off; this is used to prevent loss of data from part of the internal memory system (Keep Alive Memory or KAM). The control module can also function as the power supply control for some actuators and provide sensor reference or operating voltage to some sensors.	The fault code indicates that there is a fault with the power input to the ECM / PCM. The power supply could be intermittent (likely to be out of normal operating range / tolerance when intermittent fault is detected) or, the voltage level is erratic (e.g. rapid changes to the voltage level). Possible fault in the wiring e.g. intermittent connections, intermittent short / open circuits. Check power supply from source e.g. from ignition switch or from the battery via a relay and / or fuse; check relay operation if applicable, Refer to list of sensor / actuator checks on page 19. Note that a relay could be switched by a power module, which could also require testing. If power supplies to the ECM / PCM appear to be stable but the fault code re-appears after clearing the code, it could indicate that the ECM / PCM is at fault.
P250A	Engine Oil Level Sensor Circuit	Engine oil level sensors can be part of the oil level dipstick or, they can be separate items. Note that oil level sensors can be simple on / off devices that respond when the oil level is above or below pre-set limits.	The voltage, frequency or value of the level sensor signal is incorrect (undefined fault). Sensor voltage / signal could be: out of normal operating range, constant value, non-existent, corrupt. Possible fault in the level sensor or wiring e.g. short / open circuits, circuit resistance, incorrect sensor resistance or, fault with the reference / operating voltage; interference from other circuits can also affect sensor signals. Refer to list of sensor / actuator checks on page 19.
P250B	Engine Oil Level Sensor Circuit Range / Performance	Engine oil level sensors can be part of the oil level dipstick or, they can be separate items. Note that oil level sensors can be simple on / off devices that respond when the oil level is above or below pre-set limits.	The voltage, frequency or value of the level sensor signal is within the normal operating range / tolerance but the signal is not plausible or is incorrect due to an undefined fault. Possible fault in the level sensor or wiring e.g. short / open circuits, circuit resistance, incorrect sensor resistance or, fault with the reference / operating voltage; interference from other circuits can also affect sensor signals. Refer to list of sensor / actuator checks on page 19. It is also possible that the level sensor is operating correctly but the level being measured by the sensor is unacceptable or incorrect.
P250C	Engine Oil Level Sensor Circuit Low	Engine oil level sensors can be part of the oil level dipstick or, they can be separate items. Note that oil level sensors can be simple on / off devices that respond when the oil level is above or below pre-set limits.	The voltage, frequency or value of the level sensor signal is either "zero" (no signal) or, the signal exists but it is below the normal operating range (lower than the minimum operating limit). Possible fault in the level sensor or wiring e.g. short / open circuits, circuit resistance, incorrect sensor resistance or, fault with the reference / operating voltage; interference from other circuits can also affect sensor signals. Refer to list of sensor / actuator checks on page 19.
P250D	Engine Oil Level Sensor Circuit High	Engine oil level sensors can be part of the oil level dipstick or, they can be separate items. Note that oil level sensors can be simple on / off devices that respond when the oil level is above or below pre-set limits.	The voltage, frequency or value of the level sensor signal is either at full value (e.g. battery voltage with no signal or frequency) or, the signal exists but it is above the normal operating range (higher than the maximum operating limit). Possible fault in the level sensor or wiring e.g. short / open circuits, circuit resistance, incorrect sensor resistance or, fault with the reference / operating voltage; interference from other circuits can also affect sensor signals. Refer to list of sensor / actuator checks on page 19.
P250E	Engine Oil Level Sensor Circuit Intermittent / Erratic	Engine oil level sensors can be part of the oil level dipstick or, they can be separate items. Note that oil level sensors can be simple on / off devices that respond when the oil level is above or below pre-set limits.	The voltage, frequency or value of the level sensor signal is intermittent (likely to be out of normal operating range / tolerance when intermittent fault is detected) or, the signal / voltage is erratic (e.g. signal changes are irregular / unpredictable). Possible fault with wiring or connections (including internal sensor connections); also interference from other circuits can affect the signal. Refer to list of sensor / actuator checks on page 19.
P250F	Engine Oil Level Too Low	Engine oil level sensors can be part of the oil level dipstick or, they can be separate items. Note that oil level sensors can be simple on / off devices that respond when the oil level is above or below pre-set limits.	The fault code relates to the oil level sensor signal which is indicating that oil level is too low. The sensor signal is within normal operating range / tolerance but the indicated oil level is below specified values. If the oil level is not too low, it is possible that the sensor signal is incorrect; refer to fault code P250B for additional information.
P2510	ECM / PCM Power Relay Sense Circuit Range / Performance	The main ECM / PCM (Engine / Power train Module) power relay will provide electrical power to most of the components on the system. It is normal practice for the control unit to switch the earth circuit for the relay i.e. the control unit will complete the earth path for the relay energising winding, which will cause the relay contacts to close and provide power to the system components. The control unit is therefore able to detect / sense whether there is a power supply to the energising winding circuit and if the circuit is good. A sense circuit can also be used, which connects the relay power output terminal to the control unit; this allows the control unit to detect that the relay has switched on the power output circuit.	Refer to list of sensor / actuator checks on page 19 for notes on relay checks. Different manufacturers could use the fault code to refer to "sensing" of the different relay circuits. The fault code could refer to the power supply circuit (the power supply to the relay contacts and the circuit from the relay contacts to the system components). Alternatively, the fault code could refer to the relay energising winding circuit (power supply to the energising winding and the switched earth path through to the control unit). The fault code indicates that the voltage on the relay sense circuit is incorrect or, the voltage is not as expected under certain operating conditions (e.g. circuit is on when it should be off, or off when it should be on). The fault could be related to relay operation e.g. contacts stuck closed or open, wiring faults (short / open circuit) or possible control unit fault (switching relay at incorrect times). Check all relay wiring and operation, but also check "sense" circuit wiring.

NOTE 1. Check for other fault codes that could provide additional information.   **NOTE 2.** Communication between control units can pass via a CAN-Bus system; refer to Page 51 for CAN-Bus checks.   277

NOTE 3. If a fault cannot be located, it is also possible that the control unit is at fault.   **NOTE 4.** Refer to Page 53 for list of pages that show the ISO standard for component locations e.g. Sensor A - Bank 1.

Fault code	EOBD / ISO Description	Component / System Description	Meaningful Description and Quick Check
P2511	ECM / PCM Power Relay Sense Circuit Intermittent	The main ECM / PCM (Engine / Power train Module) power relay will provide electrical power to most of the components on the system. It is normal practice for the control unit to switch the earth circuit for the relay i.e. the control unit will complete the earth path for the relay energising winding, which will cause the relay contacts to close and provide power to the system components. The control unit is therefore able to detect / sense whether there is a power supply to the energising winding circuit and if the circuit is good. A sense circuit can also be used, which connects the relay power output terminal to the control unit; this allows the control unit to detect that the relay has switched on the power output circuit.	Refer to list of sensor / actuator checks on page 19 for notes on relay checks. Different manufacturers could use the fault code to refer to "sensing" of the different relay circuits. The fault code could refer to the power supply circuit (the power supply to the relay contacts and the circuit from the relay contacts to the system components). Alternatively, the fault code could refer to the relay energising winding circuit (power supply to the energising winding and the switched earth path through to the control unit). The fault code indicates that the voltage on the relay sense circuit is in intermittent or erratic. The fault could be caused by loose connections or breaks in the wiring, but also check operation of relay contacts to ensure that they not opening / closing erratically.
P2512	Event Data Recorder Request Circuit / Open	Event data recorders are used to capture data on specific events e.g. the data will be captured at the time of an accident (legal requirement in some countries).	The fault code indicates that there is a fault in the signal from the Event Recorder due to a circuit related fault (possibly open circuit). Refer to vehicle specific information to identify the communication path (e.g. via the CAN-Bus network or direct to the control unit) and to identify an specific test procedures. Check the applicable wiring for connection faults and for short / open circuits; also check operation of Data Recorder if possible.
P2513	Event Data Recorder Request Circuit Low	Event data recorders are used to capture data on specific events e.g. the data will be captured at the time of an accident (legal requirement in some countries).	The fault code indicates that there is a fault in the signal from the Event Recorder due to a circuit related fault which is causing a low value signal from the Data recorder. Refer to vehicle specific information to identify the communication path (e.g. via the CAN-Bus network or direct to the control unit) and to identify an specific test procedures. Check the applicable wiring for connection faults and for short / open circuits; also check operation of Data Recorder if possible.
P2514	Event Data Recorder Request Circuit High	Event data recorders are used to capture data on specific events e.g. the data will be captured at the time of an accident (legal requirement in some countries).	The fault code indicates that there is a fault in the signal from the Event Recorder due to a circuit related fault which is causing a high value signal from the Data recorder. Refer to vehicle specific information to identify the communication path (e.g. via the CAN-Bus network or direct to the control unit) and to identify an specific test procedures. Check the applicable wiring for connection faults and for short / open circuits; also check operation of Data Recorder if possible.
P2515	A / C Refrigerant Pressure Sensor "B" Circuit	The information from the pressure sensor can assist in control of the air conditioning system. The information can be also used by the main control unit to indicate the load demand of the air conditioning system, which allows the control unit to regulate the idle speed, as well as for safety and other control purposes. It is normal practice to fit pressure sensors on the "High Pressure Side" of the system but it may be necessary to refer to vehicle specific information to identify location / function of the sensor.	The voltage, frequency or value of the pressure sensor signal is incorrect (undefined fault). Sensor voltage / signal could be: out of normal operating range, constant value, non-existent, corrupt. Possible fault in the pressure sensor or wiring e.g. short / open circuits, circuit resistance, incorrect sensor resistance (if applicable) or, fault with the reference / operating voltage; interference from other circuits can also affect sensor signals. Refer to list of sensor / actuator checks on page 19.
P2516	A / C Refrigerant Pressure Sensor "B" Circuit Range / Performance	The information from the pressure sensor can assist in control of the air conditioning system. The information can be also used by the main control unit to indicate the load demand of the air conditioning system, which allows the control unit to regulate the idle speed, as well as for safety and other control purposes. It is normal practice to fit pressure sensors on the "High Pressure Side" of the system but it may be necessary to refer to vehicle specific information to identify location / function of the sensor.	The voltage, frequency or value of the pressure sensor signal is within the normal operating range / tolerance but the signal is not plausible or is incorrect due to an undefined fault. The sensor signal might not match the operating conditions indicated by other sensors or, the sensor signal is not changing / responding as expected to the changes in conditions. Possible fault in the pressure sensor or wiring e.g. short / open circuits, circuit resistance, incorrect sensor resistance (if applicable) or, fault with the reference / operating voltage; interference from other circuits can also affect sensor signals. Refer to list of sensor / actuator checks on page 19. It is also possible that the pressure sensor is operating correctly but the refrigerant pressure is unacceptable or incorrect for the operating conditions; if possible, use alternative method of checking pressure to establish if sensor system is operating correctly and whether the pressure indicated by the sensor is correct / incorrect.
P2517	A / C Refrigerant Pressure Sensor "B" Circuit Low	The information from the pressure sensor can assist in control of the air conditioning system. The information can be also used by the main control unit to indicate the load demand of the air conditioning system, which allows the control unit to regulate the idle speed, as well as for safety and other control purposes. It is normal practice to fit pressure sensors on the "High Pressure Side" of the system but it may be necessary to refer to vehicle specific information to identify location / function of the sensor.	The voltage, frequency or value of the pressure sensor signal is either "zero" (no signal) or, the signal exists but it is below the normal operating range (lower than the minimum operating limit). Possible fault in the pressure sensor or wiring e.g. short / open circuits, circuit resistance, incorrect sensor resistance (if applicable) or, fault with the reference / operating voltage; interference from other circuits can also affect sensor signals. Refer to list of sensor / actuator checks on page 19.
P2518	A / C Refrigerant Pressure Sensor "B" Circuit High	The information from the pressure sensor can assist in control of the air conditioning system. The information can be also used by the main control unit to indicate the load demand of the air conditioning system, which allows the control unit to regulate the idle speed, as well as for safety and other control purposes. It is normal practice to fit pressure sensors on the "High Pressure Side" of the system but it may be necessary to refer to vehicle specific information to identify location / function of the sensor.	The voltage, frequency or value of the pressure sensor signal is either at full value (e.g. battery voltage with no signal or frequency) or, the signal exists but it is above the normal operating range (higher than the maximum operating limit). Possible fault in the pressure sensor or wiring e.g. short / open circuits, circuit resistance, incorrect sensor resistance (if applicable) or, fault with the reference / operating voltage; interference from other circuits can also affect sensor signals. Refer to list of sensor / actuator checks on page 19.
P2519	A / C Request "A" Circuit	It may be necessary to refer to vehicle specific information to identify the A / C request circuits, their function and the applicable control units. In many cases, the request circuit can be used where one module requests an action from another module e.g. under certain operating conditions there could be a requirement to switch off the air conditioning compressor; this request can pass from the main control unit to the air conditioning control module.	The fault code indicates that there is an unidentified fault in the A / C request circuit. The fault can be detected because the voltage, frequency or value of the signal in the circuit is incorrect (undefined fault). The frequency or value of the signal could be out of normal operating range, constant value, non-existent, corrupt. Possible fault in the sensor or wiring e.g. short / open circuits, circuit resistance or interference from other circuits can also affect sensor signals. Note that the request circuit could pass through the CAN-Bus network.

NOTE 1. Check for other fault codes that could provide additional information. **NOTE 2.** Communication between control units can pass via a CAN-Bus system; refer to Page 51 for CAN-Bus checks. 278

NOTE 3. If a fault cannot be located, it is also possible that the control unit is at fault. NOTE 4. Refer to Page 53 for list of pages that show the ISO standard for component locations e.g. Sensor A - Bank 1.

Code	Description		
P251A	PTO Enable Switch Circuit / Open	The PTO (Power Take Off) unit can be controlled by the main engine control unit or a separate control module (which will communicate with the main control unit). The enable switch allows the operator to select PTO operation.	The voltage or value of the switch signal is incorrect (undefined fault). Switch voltage / signal could be: out of normal operating range, constant value, non-existent, corrupt. Possible fault in the switch or wiring e.g. short / open circuits, circuit resistance or, fault with the reference / operating voltage; the switch contacts could also be faulty e.g. not opening / closing correctly or have a high contact resistance. Refer to list of sensor / actuator checks on page 19.
P251B	PTO Enable Switch Circuit Low	The PTO (Power Take Off) unit can be controlled by the main engine control unit or a separate control module (which will communicate with the main control unit). The enable switch allows the operator to select PTO operation.	The voltage or value of the switch signal is either "zero" (no signal) or, the signal exists but it is below the normal operating range (lower than the minimum operating limit). Possible fault in the switch or wiring e.g. short / open circuits, circuit resistance or, fault with the reference / operating voltage; the switch contacts could also be faulty e.g. not opening / closing correctly or have a high contact resistance. Refer to list of sensor / actuator checks on page 19. Note that the fault code can be activated if the switch output remains at the low value e.g. zero volts and does not rise to the high value e.g. battery voltage (the control unit will detect no switch signal change but it has detected change in operating conditions).
P251C	PTO Enable Switch Circuit High	The PTO (Power Take Off) unit can be controlled by the main engine control unit or a separate control module (which will communicate with the main control unit). The enable switch allows the operator to select PTO operation.	The voltage or value of the switch signal is either at full value (e.g. battery voltage with no signal or frequency) or, the signal exists but it is above the normal operating range (higher than the maximum operating limit). Possible fault in the switch or wiring e.g. short / open circuits, circuit resistance or, fault with the reference / operating voltage; the switch contacts could also be faulty e.g. not opening / closing correctly or have a high contact resistance. Refer to list of sensor / actuator checks on page 19. Note that the fault code can be activated if the switch output remains at the high value e.g. battery voltage and does not fall to the low value e.g. zero volts (the control unit will detect no switch signal change but it has detected change in operating conditions).
P251D	PTO Engine Shutdown Circuit / Open	The PTO (Power Take Off) unit can be controlled by the main engine control unit or a separate control module (which will communicate with the main control unit). A shutdown system can be fitted to switch off the engine if a fault or an unsafe operation exists.	Refer to vehicle specific information to identify the operation of the "Shutdown" system. The fault code indicates that there is a fault (possibly an open circuit) in the circuit carrying the "Shutdown" signal to the engine control unit (or other device used to switch off the engine). Check all applicable wiring and connections for short / open circuits. Also check operation of PTO sensors or operation of separate PTO control module (if fitted).
P251E	PTO Engine Shutdown Circuit Low	The PTO (Power Take Off) unit can be controlled by the main engine control unit or a separate control module (which will communicate with the main control unit).	Refer to vehicle specific information to identify the operation of the "Shutdown" system. The fault code indicates that there is a fault in the circuit carrying the "Shutdown" signal to the engine control unit (or other device used to switch off the engine). The voltage or value of the signal is either "zero" (no signal) or, the signal exists but it is below the normal operating range (lower than the minimum operating limit). Possible fault in the wiring e.g. short / open circuits, circuit resistance. Check all applicable wiring and connections for short / open circuits. Also check operation of PTO sensors or operation of separate PTO control module (if fitted).
P251F	PTO Engine Shutdown Circuit High	The PTO (Power Take Off) unit can be controlled by the main engine control unit or a separate control module (which will communicate with the main control unit).	Refer to vehicle specific information to identify the operation of the "Shutdown" system. The fault code indicates that there is a fault in the circuit carrying the "Shutdown" signal to the engine control unit (or other device used to switch off the engine). The voltage or value of the signal is either at full value (e.g. battery voltage with no signal or frequency) or, the signal exists but it is above the normal operating range (higher than the maximum operating limit). Possible fault in the wiring e.g. short / open circuits, circuit resistance. Check all applicable wiring and connections for short / open circuits. Also check operation of PTO sensors or operation of separate PTO control module (if fitted).
P2520	A / C Request "A" Circuit Low	It may be necessary to refer to vehicle specific information to identify the A / C request circuits, their function and the applicable control units. In many cases, the request circuit can be used where one module requests an action from another module e.g. under certain operating conditions there could be a requirement to switch off the air conditioning compressor; this request can pass from the main control unit to the air conditioning control module.	The fault code indicates that there is an unidentified fault in the A / C request circuit. The fault can be detected because the voltage, frequency or value of the sensor signal is either "zero" (no signal) or, the signal exists but it is below the normal operating range (lower than the minimum operating limit). Possible fault in the sensor or wiring e.g. short / open circuits, circuit resistance or interference from other circuits can also affect sensor signals. Note that the request circuit could pass through the CAN-Bus network.
P2521	A / C Request "A" Circuit High	It may be necessary to refer to vehicle specific information to identify the A / C request circuits, their function and the applicable control units. In many cases, the request circuit can be used where one module requests an action from another module e.g. under certain operating conditions there could be a requirement to switch off the air conditioning compressor; this request can pass from the main control unit to the air conditioning control module.	The fault code indicates that there is an unidentified fault in the A / C request circuit. The fault can be detected because the voltage, frequency or value of the sensor signal is either at full value (e.g. battery voltage with no signal or frequency) or, the signal exists but it is above the normal operating range (higher than the maximum operating limit). Possible fault in the sensor or wiring e.g. short / open circuits, circuit resistance or interference from other circuits can also affect sensor signals. Note that the request circuit could pass through the CAN-Bus network.
P2522	A / C Request "B" Circuit	It may be necessary to refer to vehicle specific information to identify the A / C request circuits, their function and the applicable control units. In many cases, the request circuit can be used where one module requests an action from another module e.g. under certain operating conditions there could be a requirement to switch off the air conditioning compressor; this request can pass from the main control unit to the air conditioning control module.	The fault code indicates that there is an unidentified fault in the A / C request circuit. The fault can be detected because the voltage, frequency or value of the signal in the circuit is incorrect (undefined fault). The frequency or value of the signal could be out of normal operating range, constant value, non-existent, corrupt. Possible fault in the sensor or wiring e.g. short / open circuits, circuit resistance or interference from other circuits can also affect sensor signals. Note that the request circuit could pass through the CAN-Bus network.
P2523	A / C Request "B" Circuit Low	It may be necessary to refer to vehicle specific information to identify the A / C request circuits, their function and the applicable control units. In many cases, the request circuit can be used where one module requests an action from another module e.g. under certain operating conditions there could be a requirement to switch off the air conditioning compressor; this request can pass from the main control unit to the air conditioning control module.	The fault code indicates that there is an unidentified fault in the A / C request circuit. The fault can be detected because the voltage, frequency or value of the sensor signal is either "zero" (no signal) or, the signal exists but it is below the normal operating range (lower than the minimum operating limit). Possible fault in the sensor or wiring e.g. short / open circuits, circuit resistance or interference from other circuits can also affect sensor signals. Note that the request circuit could pass through the CAN-Bus network.

NOTE 1. Check for other fault codes that could provide additional information.   **NOTE 2.** Communication between control units can pass via a CAN-Bus system; refer to Page 51 for CAN-Bus checks.

NOTE 3. If a fault cannot be located, it is also possible that the control unit is at fault.   **NOTE 4.** Refer to Page 53 for list of pages that show the ISO standard for component locations e.g. Sensor A - Bank 1.

Fault code	EOBD / ISO Description	Component / System Description	Meaningful Description and Quick Check
P2524	A / C Request "B" Circuit High	It may be necessary to refer to vehicle specific information to identify the A / C request circuits, their function and the applicable control units. In many cases, the request circuit can be used where one module requests an action from another module e.g. under certain operating conditions there could be a requirement to switch off the air conditioning compressor; this request can pass from the main control unit to the air conditioning control module.	The fault code indicates that there is an unidentified fault in the A / C request circuit. The fault can be detected because the voltage, frequency or value of the sensor signal is either at full value (e.g. battery voltage with no signal or frequency) or, the signal exists but it is above the normal operating range (higher than the maximum operating limit). Possible fault in the sensor or wiring e.g. short / open circuits, circuit resistance or interference from other circuits can also affect sensor signals. Note that the request circuit could pass through the CAN-Bus network.
P2525	Vacuum Reservoir Pressure Sensor Circuit	The vacuum reservoir, provides a vacuum supply to various vacuum operated systems. The reservoirs (and a vacuum pump) are usually fitted to vehicles with diesel engines. Some petrol engines (usually direct injection engines) that operate without throttle control during some operating phases, also use reservoirs to store vacuum (due to the lack of vacuum when the engine is not being throttled).	The voltage, frequency or value of the vacuum / pressure sensor signal is incorrect (undefined fault). Sensor voltage / signal could be: out of normal operating range, constant value, non-existent, corrupt. Possible fault in the sensor or wiring e.g. short / open circuits, circuit resistance, incorrect sensor resistance (if applicable) or, fault with the reference / operating voltage; interference from other circuits can also affect sensor signals. Refer to list of sensor / actuator checks on page 19.
P2526	Vacuum Reservoir Pressure Sensor Circuit Range / Performance	The vacuum reservoir, provides a vacuum supply to various vacuum operated systems. The reservoirs (and a vacuum pump) are usually fitted to vehicles with diesel engines. Some petrol engines (usually direct injection engines) that operate without throttle control during some operating phases, also use reservoirs to store vacuum (due to the lack of vacuum when the engine is not being throttled).	The voltage, frequency or value of the vacuum / pressure sensor signal is within the normal operating range / tolerance but the signal is not plausible or is incorrect due to an undefined fault. The sensor signal might not be changing / responding as expected. Possible fault in the sensor or wiring e.g. short / open circuits, circuit resistance, incorrect sensor resistance (if applicable) or, fault with the reference / operating voltage; interference from other circuits can also affect sensor signals. Refer to list of sensor / actuator checks on page 19. Also check any air passages / pipes to the sensor for leaks / blockages etc. It is also possible that the sensor is operating correctly but the vacuum / pressure being measured by the sensor is unacceptable or incorrect for the operating conditions; if possible, use alternative method of checking vacuum to establish if sensor system is operating correctly and whether the vacuum indicated by the sensor is correct / incorrect.
P2527	Vacuum Reservoir Pressure Sensor Circuit Low	The vacuum reservoir, provides a vacuum supply to various vacuum operated systems. The reservoirs (and a vacuum pump) are usually fitted to vehicles with diesel engines. Some petrol engines (usually direct injection engines) that operate without throttle control during some operating phases, also use reservoirs to store vacuum (due to the lack of vacuum when the engine is not being throttled).	The voltage, frequency or value of the vacuum / pressure sensor signal is either "zero" (no signal) or, the signal exists but it is below the normal operating range (lower than the minimum operating limit). Possible fault in the sensor or wiring e.g. short / open circuits, circuit resistance, incorrect sensor resistance (if applicable) or, fault with the reference / operating voltage; interference from other circuits can also affect sensor signals. Refer to list of sensor / actuator checks on page 19.
P2528	Vacuum Reservoir Pressure Sensor Circuit High	The vacuum reservoir, provides a vacuum supply to various vacuum operated systems. The reservoirs (and a vacuum pump) are usually fitted to vehicles with diesel engines. Some petrol engines (usually direct injection engines) that operate without throttle control during some operating phases, also use reservoirs to store vacuum (due to the lack of vacuum when the engine is not being throttled).	The voltage, frequency or value of the vacuum / pressure sensor signal is either at full value (e.g. battery voltage with no signal or frequency) or, the signal exists but it is above the normal operating range (higher than the maximum operating limit). Possible fault in the sensor or wiring e.g. short / open circuits, circuit resistance, incorrect sensor resistance (if applicable) or, fault with the reference / operating voltage; interference from other circuits can also affect sensor signals. Refer to list of sensor / actuator checks on page 19.
P2529	Vacuum Reservoir Pressure Sensor Circuit Intermittent	The vacuum reservoir, provides a vacuum supply to various vacuum operated systems. The reservoirs (and a vacuum pump) are usually fitted to vehicles with diesel engines. Some petrol engines (usually direct injection engines) that operate without throttle control during some operating phases, also use reservoirs to store vacuum (due to the lack of vacuum when the engine is not being throttled).	The voltage, frequency or value of the vacuum / pressure sensor signal is intermittent (likely to be out of normal operating range / tolerance when intermittent fault is detected) or, the signal / voltage is erratic (e.g. signal changes are irregular / unpredictable). Possible fault with wiring or connections (including internal sensor connections); also interference from other circuits can affect the signal. Refer to list of sensor / actuator checks on page 19. It is also possible that the pressure could be changing erratically, which could cause the fault code to be activated; if necessary, check pressure with a separate gauge to check for erratic pressure changes / leaks etc.
P252A	Engine Oil Quality Sensor Circuit	A sensor can be used to monitor oil quality. Refer to vehicle specific information to identify the exact operation and the wiring for the sensor.	The voltage, frequency or value of the sensor signal is incorrect (undefined fault). Sensor voltage / signal could be: out of normal operating range, constant value, non-existent, corrupt. Possible fault in the sensor or wiring e.g. short / open circuits, circuit resistance, incorrect sensor resistance or, fault with the reference / operating voltage; interference from other circuits can also affect sensor signals. Refer to list of sensor / actuator checks on page 19.
P252B	Engine Oil Quality Sensor Circuit Range / Performance	A sensor can be used to monitor oil quality. Refer to vehicle specific information to identify the exact operation and the wiring for the sensor.	The voltage, frequency or value of the sensor signal is within the normal operating range / tolerance but the signal is not plausible or is incorrect due to an undefined fault. The sensor signal might not match the operating conditions indicated by other sensors or, the sensor signal is not changing / responding as expected to the changes in conditions. Possible fault in the sensor or wiring e.g. short / open circuits, circuit resistance, incorrect sensor resistance or, fault with the reference / operating voltage; interference from other circuits can also affect sensor signals. Refer to list of sensor / actuator checks on page 19. It is also possible that the sensor is operating correctly but the oil quality is unacceptable.
P252C	Engine Oil Quality Sensor Circuit Low	A sensor can be used to monitor oil quality. Refer to vehicle specific information to identify the exact operation and the wiring for the sensor.	The voltage, frequency or value of the sensor signal is either "zero" (no signal) or, the signal exists but it is below the normal operating range (lower than the minimum operating limit). Possible fault in the sensor or wiring e.g. short / open circuits, circuit resistance, incorrect sensor resistance or, fault with the reference / operating voltage; interference from other circuits can also affect sensor signals. Refer to list of sensor / actuator checks on page 19.
P252D	Engine Oil Quality Sensor Circuit High	A sensor can be used to monitor oil quality. Refer to vehicle specific information to identify the exact operation and the wiring for the sensor.	The voltage, frequency or value of the sensor signal is either at full value (e.g. battery voltage with no signal or frequency) or, the signal exists but it is above the normal operating range (higher than the maximum operating limit). Possible fault in the sensor or wiring e.g. short / open circuits, circuit resistance, incorrect sensor resistance or, fault with the reference / operating voltage; interference from other circuits can also affect sensor signals. Refer to list of sensor / actuator checks on page 19.

NOTE 1. Check for other fault codes that could provide additional information. **NOTE 2.** Communication between control units can pass via a CAN-Bus system; refer to Page 51 for CAN-Bus checks.

NOTE 3. If a fault cannot be located, it is also possible that the control unit is at fault. **NOTE 4.** Refer to Page 53 for list of pages that show the ISO standard for component locations e.g. Sensor A - Bank 1.

280

Code	Fault Location	General Description	Specific Description
P252E	Engine Oil Quality Circuit Intermittent / Erratic	A sensor can be used to monitor oil quality. Refer to vehicle specific information to identify the exact operation and the wiring for the sensor.	The voltage, frequency or value of the sensor signal is intermittent (likely to be out of normal operating range / tolerance when intermittent fault is detected) or, the signal / voltage is erratic (e.g. signal changes are irregular / unpredictable). Possible fault with wiring or connections (including internal sensor connections); also interference from other circuits can affect the signal. Refer to list of sensor / actuator checks on page 19.
P252F	Engine Oil Level Too High	Engine oil level sensors can be part of the oil level dipstick or, they can be separate items. Note that oil level sensors can be simple on / off devices that respond when the oil level is above or below pre-set limits.	The fault code relates to the oil level sensor signal which is indicating that oil level is too high. The sensor signal is within normal operating range / tolerance but the indicated oil level is above specified values. If the oil level is not too high, it is possible that the sensor signal is incorrect; refer to fault code P250B for additional information.
P2530	Ignition Switch Run Position Circuit	For all ignition switch related faults, checks should be made on the input supply to the switch (which could be direct from the battery on a fused circuit). Ignition switch faults often occur due to poor connection across the applicable switch contacts, which can be due to general wear of the switch assembly.	The fault code identifies an undefined fault in the circuit when the switch is in the run position. The output voltage from the switch could be out of normal operating range (typically a low voltage) or the voltage could be non-existent. Note that voltage checks at the switch may not be accurate if there is no electrical load being applied to the circuit. Check battery voltage (at battery terminals). Check voltage at the switch "run" position output terminal (with an electrical load being applied to the circuit i.e. all systems running). If the voltage at the switch is different to the battery voltage, check supply voltage to the switch and if incorrect, check fuses and check wiring between battery and switch for short / open circuits and high resistance. If input voltage to ignition switch is good but output voltage is incorrect, check ignition switch contacts for poor connection / high resistance. Note that the fault could be intermittent and it may be necessary to wiggle the switch assembly to try and recreate the fault.
P2531	Ignition Switch Run Position Circuit Low	For all ignition switch related faults, checks should be made on the input supply to the switch (which could be direct from the battery on a fused circuit). Ignition switch faults often occur due to poor connection across the applicable switch contacts, which can be due to general wear of the switch assembly.	The fault code identifies that the output voltage at the ignition switch run position is low. The output voltage could be below normal operating range or the voltage could be non-existent. Note that voltage checks at the switch may not be accurate if there is no electrical load being applied to the circuit. Check battery voltage (at battery terminals). Check voltage at the switch "run" position output terminal (with an electrical load being applied to the circuit i.e. all systems running). If the voltage at the switch is different to the battery voltage, check supply voltage to the switch and if incorrect, check fuses and check wiring between battery and switch for short / open circuits and high resistance. If input voltage to ignition switch is good but output voltage is incorrect, check ignition switch contacts for poor connection / high resistance. Note that the fault could be intermittent and it may be necessary to wiggle the switch assembly to try and recreate the fault.
P2532	Ignition Switch Run Position Circuit High	For all ignition switch related faults, checks should be made on the input supply to the switch (which could be direct from the battery on a fused circuit). Ignition switch faults often occur due to poor connection across the applicable switch contacts, which can be due to general wear of the switch assembly.	The fault code identifies that the output voltage at the ignition switch "run" position is high. The output voltage could be above normal operating range (check charging voltage) or the voltage could exist when the circuit should be off (switch not in run position). Check voltage at the switch "run" position output terminal (with an electrical load being applied to the circuit i.e. all systems running) and then check the voltage with the switch in the accessory / auxiliary position or with the switch in the off position; note that the voltage should be zero at the "run" terminal with the switch in the accessory or off position (the fault could be intermittent at it may be necessary to wiggle the switch assembly to try and recreate the fault). Note that the fault could be intermittent and it may be necessary to wiggle the switch assembly to try and recreate the fault.
P2533	Ignition Switch Run / Start Position Circuit	For all ignition switch related faults, checks should be made on the input supply to the switch (which could be direct from the battery on a fused circuit). Ignition switch faults often occur due to poor connection across the applicable switch contacts, which can be due to general wear of the switch assembly.	The fault code identifies an undefined fault in the circuit when the switch is in the run and / or start position. The output voltage from the switch could be out of normal operating range (typically a low voltage) or the voltage could be non-existent. Note that voltage checks at the switch may not be accurate if there is no electrical load being applied to the circuit. Check battery voltage (at battery terminals) then check voltage at the switch "run" and / or "start" position output terminal (with an electrical load being applied to the circuit i.e. all systems running or engine starting). If the voltage at the switch is different to the battery voltage, check supply voltage to the switch and if incorrect, check fuses and check wiring between battery and switch for short / open circuits and high resistance. If input voltage to ignition switch is good but output voltage is incorrect, check ignition switch contacts for poor connection / high resistance. Note that the fault could be intermittent and it may be necessary to wiggle the switch assembly to try and recreate the fault.
P2534	Ignition Switch Run / Start Position Circuit Low	For all ignition switch related faults, checks should be made on the input supply to the switch (which could be direct from the battery on a fused circuit). Ignition switch faults often occur due to poor connection across the applicable switch contacts, which can be due to general wear of the switch assembly.	The fault code identifies that the output voltage at the ignition switch run and / or start position is low. The output voltage could be below normal operating range or the voltage could be non-existent. Note that voltage checks at the switch may not be accurate if there is no electrical load being applied to the circuit. Check battery voltage (at battery terminals). Check voltage at the switch "run" and "start" position output terminal (with an electrical load being applied to the circuit i.e. all systems running or engine starting). If the voltage at the switch is different to the battery voltage, check supply voltage to the switch and if incorrect, check fuses and check wiring between battery and switch for short / open circuits and high resistance. If input voltage to ignition switch is good but output voltage is incorrect, check ignition switch contacts for poor connection / high resistance. Note that the fault could be intermittent and it may be necessary to wiggle the switch assembly to try and recreate the fault.
P2535	Ignition Switch Run / Start Position Circuit High	For all ignition switch related faults, checks should be made on the input supply to the switch (which could be direct from the battery on a fused circuit). Ignition switch faults often occur due to poor connection across the applicable switch contacts, which can be due to general wear of the switch assembly.	The fault code identifies that the output voltage at the ignition switch "run" and / or "start" position is high. The output voltage could be above normal operating range (check charging voltage) or the voltage could exist when the circuit should be off (switch not in run or start position). Check voltage at the switch "run" and / or "start" position output terminal (with an electrical load being applied to the circuit i.e. all systems running and / or engine starting) and then check the voltage with the switch in the accessory / auxiliary position or with the switch in the off position; note that the voltage should be zero at the "run" and "start" terminal with the switch in the accessory or off position. Note that the fault could be intermittent and it may be necessary to wiggle the switch assembly to try and recreate the fault.

Fault code	EOBD / ISO Description	Component / System Description	Meaningful Description and Quick Check
P2536	Ignition Switch Accessory Position Circuit	For all ignition switch related faults, checks should be made on the input supply to the switch (which could be direct from the battery on a fused circuit). Ignition switch faults often occur due to poor connection across the applicable switch contacts, which can be due to general wear of the switch assembly.	The fault code identifies an undefined fault in the circuit when the switch is in the "accessory" / "auxiliary" position. The output voltage from the switch could be out of normal operating range (typically a low voltage) or the voltage could be non-existent. Note that voltage checks at the switch may not be accurate if there is no electrical load being applied to the circuit. Check battery voltage (at battery terminals). Check voltage at the switch "accessory" position output terminal (with an electrical load being applied to the circuit e.g. any high load devices that operate in the accessory position). If the voltage at the switch is different to the battery voltage, check supply voltage to the switch and if incorrect, check fuses and check wiring between battery and switch for short / open circuits and high resistance. If input voltage to ignition switch is good but output voltage is incorrect, check ignition switch contacts for poor connection / high resistance. Note that the fault could be intermittent and it may be necessary to wiggle the switch assembly to try and recreate the fault.
P2537	Ignition Switch Accessory Position Circuit Low	For all ignition switch related faults, checks should be made on the input supply to the switch (which could be direct from the battery on a fused circuit). Ignition switch faults often occur due to poor connection across the applicable switch contacts, which can be due to general wear of the switch assembly.	The fault code identifies that the output voltage at the ignition switch "accessory" / "auxiliary" position is low. The output voltage could be below normal operating range or the voltage could be non-existent. Note that voltage checks at the switch may not be accurate if there is no electrical load being applied to the circuit. Check battery voltage (at battery terminals). Check voltage at the switch "accessory" position output terminal (with an electrical load being applied to the circuit e.g. any high load devices that operate in the accessory position). If the voltage at the switch is different to the battery voltage, check supply voltage to the switch and if incorrect, check fuses and check wiring between battery and switch for short / open circuits and high resistance. If input voltage to ignition switch is good but output voltage is incorrect, check ignition switch contacts for poor connection / high resistance. Note that the fault could be intermittent and it may be necessary to wiggle the switch assembly to try and recreate the fault.
P2538	Ignition Switch Accessory Position Circuit High	For all ignition switch related faults, checks should be made on the input supply to the switch (which could be direct from the battery on a fused circuit). Ignition switch faults often occur due to poor connection across the applicable switch contacts, which can be due to general wear of the switch assembly.	The fault code identifies that the output voltage at the ignition switch "accessory" / "auxiliary" position is high. The output voltage could be above normal operating range (check charging voltage) or the voltage could exist when the circuit should be off (switch not in accessory position). Check voltage at the switch "accessory" position output terminal (with an electrical load being applied to the circuit e.g. any high load devices that operate in the accessory position) and then check the voltage with the switch in the off position; note that the voltage should be zero at the "accessory" terminal with the switch in the off position. Note that the fault could be intermittent and it may be necessary to wiggle the switch assembly to try and recreate the fault.
P2539	Low Pressure Fuel System Sensor Circuit	A sensor can be used to monitor the pressure in the "Low" pressure side of the fuel system. If the pressure is low, this can indicate a low pressure pump fault or blockage in the system e.g. filter etc.	The voltage, frequency or value of the pressure sensor signal is incorrect (undefined fault). Sensor voltage / signal could be: out of normal operating range, constant value, non-existent, corrupt. Possible fault in the pressure sensor or wiring e.g. short / open circuits, circuit resistance, incorrect sensor resistance (if applicable) or, fault with the reference / operating voltage; interference from other circuits can also affect sensor signals. Refer to list of sensor / actuator checks on page 19.
P253A	PTO Sense Circuit / Open	The PTO (Power Take Off) unit can be controlled by the main engine control unit or a separate control module (which will communicate with the main control unit). A sense or monitoring system can be used to provide operating information to the engine control unit e.g. PTO load information etc. .	It will be necessary to refer to vehicle specific information to identify the sense / monitoring system and wiring. The fault code indicates that there is a fault (possibly an open circuit) in the sense / monitoring circuit. Check all applicable wiring and connections for short / open circuits. Also check operation of PTO sensors or operation of separate PTO control module (if fitted).
P253B	PTO Sense Circuit Range / Performance	The PTO (Power Take Off) unit can be controlled by the main engine control unit or a separate control module (which will communicate with the main control unit). A sense or monitoring system can be used to provide operating information to the engine control unit e.g. PTO load information etc. .	It will be necessary to refer to vehicle specific information to identify the sense / monitoring system and wiring. The fault code indicates that there is a fault in the sense / monitoring circuit. Check all applicable wiring and connections for short / open circuits. Also check operation of PTO sensors or operation of separate PTO control module (if fitted).
P253C	PTO Sense Circuit Low	The PTO (Power Take Off) unit can be controlled by the main engine control unit or a separate control module (which will communicate with the main control unit). A sense or monitoring system can be used to provide operating information to the engine control unit e.g. PTO load information etc. .	It will be necessary to refer to vehicle specific information to identify the sense / monitoring system and wiring. The fault code indicates that there is a fault in the sense / monitoring circuit. The voltage or value of the signal is either "zero" (no signal) or, the signal exists but it is below the normal operating range (lower than the minimum operating limit). Possible fault in the wiring e.g. short / open circuits, circuit resistance. Check all applicable wiring and connections for short / open circuits. Also check operation of PTO sensors or operation of separate PTO control module (if fitted).
P253D	PTO Sense Circuit High	The PTO (Power Take Off) unit can be controlled by the main engine control unit or a separate control module (which will communicate with the main control unit). A sense or monitoring system can be used to provide operating information to the engine control unit e.g. PTO load information etc. .	It will be necessary to refer to vehicle specific information to identify the sense / monitoring system and wiring. The fault code indicates that there is a fault in the sense / monitoring circuit. The voltage or value of the signal is either at full value (e.g. battery voltage with no signal or frequency) or, the signal exists but it is above the normal operating range (higher than the maximum operating limit). Possible fault in the wiring e.g. short / open circuits, circuit resistance. Check all applicable wiring and connections for short / open circuits. Also check operation of PTO sensors or operation of separate PTO control module (if fitted).
P253E	PTO Sense Circuit Intermittent / Erratic	The PTO (Power Take Off) unit can be controlled by the main engine control unit or a separate control module (which will communicate with the main control unit). A sense or monitoring system can be used to provide operating information to the engine control unit e.g. PTO load information etc. .	It will be necessary to refer to vehicle specific information to identify the sense / monitoring system and wiring. The fault code indicates that there is a fault in the sense / monitoring circuit. The voltage value of the signal is intermittent (likely to be out of normal operating range / tolerance when intermittent fault is detected) or, the signal / voltage is erratic (e.g. signal changes are irregular / unpredictable). Possible fault in the wiring e.g. short / open circuits, circuit resistance. Check all applicable wiring and connections for short / open circuits. Also check operation of PTO sensors or operation of separate PTO control module (if fitted).
P253F	Engine Oil Deteriorated	A sensor can be used to monitor oil quality. Refer to vehicle specific information to identify the exact operation and the wiring for the sensor.	The sensor is indicating that the oil quality is unacceptable. The control unit has not detected a sensor related fault and therefore the oil condition should be checked. If however the oil quality appears to be good, refer to P252B for information on sensor checks

NOTE 1. Check for other fault codes that could provide additional information. **NOTE 2.** Communication between control units can pass via a CAN-Bus system; refer to Page 51 for CAN-Bus checks. 282

NOTE 4. Refer to Page 53 for list of pages that show the ISO standard for component locations e.g. Sensor A - Bank 1.

P2540	Low Pressure Fuel System Sensor Circuit Range / Performance	A sensor can be used to monitor the pressure in the "Low" pressure side of the fuel system. If the pressure is low, this can indicate a low pressure pump fault or blockage in the system e.g. filter etc.	The voltage, frequency or value of the pressure sensor signal is within the normal operating range / tolerance but the signal is not plausible or is incorrect due to an undefined fault. The sensor signal might not match the operating conditions indicated by other sensors or, the sensor signal is not changing / responding as expected. Possible fault in the pressure sensor or wiring e.g. short / open circuits, circuit resistance, incorrect sensor resistance (if applicable) or, fault with the reference / operating voltage; interference from other circuits can also affect sensor signals. Refer to list of sensor / actuator checks on page 19. Also check pipes to the sensor for leaks / blockages etc. It is also possible that the pressure sensor is operating correctly but the pressure being measured by the sensor is unacceptable or incorrect for the operating conditions; if possible, use alternative method of checking pressure to establish if sensor system is operating correctly and whether the pressure indicated by the sensor is correct / incorrect.
P2541	Low Pressure Fuel System Sensor Circuit Low	A sensor can be used to monitor the pressure in the "Low" pressure side of the fuel system. If the pressure is low, this can indicate a low pressure pump fault or blockage in the system e.g. filter etc.	The voltage, frequency or value of the pressure sensor signal is either "zero" (no signal) or, the signal exists but it is below the normal operating range (lower than the minimum operating limit). Possible fault in the pressure sensor or wiring e.g. short / open circuits, circuit resistance, incorrect sensor resistance (if applicable) or, fault with the reference / operating voltage; interference from other circuits can also affect sensor signals. Refer to list of sensor / actuator checks on page 19.
P2542	Low Pressure Fuel System Sensor Circuit High	A sensor can be used to monitor the pressure in the "Low" pressure side of the fuel system. If the pressure is low, this can indicate a low pressure pump fault or blockage in the system e.g. filter etc.	The voltage, frequency or value of the pressure sensor signal is either at full value (e.g. battery voltage with no signal or frequency) or, the signal exists but it is above the normal operating range (higher than the maximum operating limit). Possible fault in the pressure sensor or wiring e.g. short / open circuits, circuit resistance, incorrect sensor resistance (if applicable) or, fault with the reference / operating voltage; interference from other circuits can also affect sensor signals. Refer to list of sensor / actuator checks on page 19.
P2543	Low Pressure Fuel System Sensor Circuit Intermittent	A sensor can be used to monitor the pressure in the "Low" pressure side of the fuel system. If the pressure is low, this can indicate a low pressure pump fault or blockage in the system e.g. filter etc.	The voltage, frequency or value of the pressure sensor signal is intermittent (likely to be out of normal operating range / tolerance when intermittent fault is detected) or, the signal / voltage is erratic (e.g. signal changes are irregular / unpredictable). Possible fault with wiring or connections (including internal sensor connections); also interference from other circuits can affect the signal. Refer to list of sensor / actuator checks on page 19. It is also possible that the pressure could be changing erratically, which could cause the fault code to be activated; if necessary, check pressure with a separate gauge to check for erratic pressure changes.
P2544	Torque Management Request Input Signal "A"	Different vehicle systems can pass a request (from the control module to the engine control unit) for a change in engine torque to match operating conditions e.g. the transmission can request a torque reduction during gear changes (changing up). It may be necessary to identify which input circuit (to the engine control unit is being referred to (input A, B etc). Note that the torque request can pass through a CAN-Bus network.	The fault code could be detected due to an incorrect "torque request" signal value. The voltage, frequency or value of the signal is incorrect (undefined fault). The voltage / signal could be: out of normal operating range, constant value, non-existent, corrupt. Possible fault in the wiring e.g. short / open circuits, circuit resistance, incorrect sensor resistance or, fault with the reference / operating voltage; interference from other circuits can also affect sensor signals. Also check for CAN-Bus fault codes (if applicable). The fault could also be due to a control module / engine control unit fault.
P2545	Torque Management Request Input Signal "A" Range / Performance	Different vehicle systems can pass a request (from the control module to the engine control unit) for a change in engine torque to match operating conditions e.g. the transmission can request a torque reduction during gear changes (changing up). It may be necessary to identify which input circuit (to the engine control unit is being referred to (input A, B etc). Note that the torque request can pass through a CAN-Bus network.	The fault code could be detected due to an incorrect "torque request" signal value. The voltage, frequency or value of the signal is within the normal operating range / tolerance but the signal is not plausible or is incorrect due to an undefined fault. The signal might not match the operating conditions indicated by other sensors or, the signal is not changing / responding as expected to the changes in conditions. Possible fault in the wiring e.g. short / open circuits, circuit resistance, incorrect sensor resistance or, fault with the reference / operating voltage; interference from other circuits can also affect sensor signals. Also check for CAN-Bus fault codes (if applicable). The fault could also be due to a control module / engine control unit fault.
P2546	Torque Management Request Input Signal "A" Low	Different vehicle systems can pass a request (from the control module to the engine control unit) for a change in engine torque to match operating conditions e.g. the transmission can request a torque reduction during gear changes (changing up). It may be necessary to identify which input circuit (to the engine control unit is being referred to (input A, B etc). Note that the torque request can pass through a CAN-Bus network.	The fault code could be detected due to an incorrect "torque request" signal value. The voltage, frequency or value of the signal is either "zero" (no signal) or, the signal exists but it is below the normal operating range (lower than the minimum operating limit). Possible fault in the wiring e.g. short / open circuits, circuit resistance, incorrect sensor resistance or, fault with the reference / operating voltage; interference from other circuits can also affect sensor signals. Also check for CAN-Bus fault codes (if applicable). The fault could also be due to a control module / engine control unit fault.
P2547	Torque Management Request Input Signal "A" High	Different vehicle systems can pass a request (from the control module to the engine control unit) for a change in engine torque to match operating conditions e.g. the transmission can request a torque reduction during gear changes (changing up). It may be necessary to identify which input circuit (to the engine control unit is being referred to (input A, B etc). Note that the torque request can pass through a CAN-Bus network.	The fault code could be detected due to an incorrect "torque request" signal value. The voltage, frequency or value of the signal is either at full value (e.g. battery voltage with no signal or frequency) or, the signal exists but it is above the normal operating range (higher than the maximum operating limit). Possible fault in the wiring e.g. short / open circuits, circuit resistance, incorrect sensor resistance or, fault with the reference / operating voltage; interference from other circuits can also affect sensor signals. Also check for CAN-Bus fault codes (if applicable). The fault could also be due to a control module / engine control unit fault.
P2548	Torque Management Request Input Signal "B"	Different vehicle systems can pass a request (from the control module to the engine control unit) for a change in engine torque to match operating conditions e.g. the transmission can request a torque reduction during gear changes (changing up). It may be necessary to identify which input circuit (to the engine control unit is being referred to (input A, B etc). Note that the torque request can pass through a CAN-Bus network.	The fault code could be detected due to an incorrect "torque request" signal value. The voltage, frequency or value of the signal is incorrect (undefined fault). The voltage / signal could be: out of normal operating range, constant value, non-existent, corrupt. Possible fault in the wiring e.g. short / open circuits, circuit resistance, incorrect sensor resistance or, fault with the reference / operating voltage; interference from other circuits can also affect sensor signals. Also check for CAN-Bus fault codes (if applicable). The fault could also be due to a control module / engine control unit fault.
P2549	Torque Management Request Input Signal "B" Range / Performance	Different vehicle systems can pass a request (from the control module to the engine control unit) for a change in engine torque to match operating conditions e.g. the transmission can request a torque reduction during gear changes (changing up). It may be necessary to identify which input circuit (to the engine control unit is being referred to (input A, B etc). Note that the torque request can pass through a CAN-Bus network.	The fault code could be detected due to an incorrect "torque request" signal value. The voltage, frequency or value of the signal is within the normal operating range / tolerance but the signal is not plausible or is incorrect due to an undefined fault. The signal might not match the operating conditions indicated by other sensors or, the signal is not changing / responding as expected to the changes in conditions. Possible fault in the wiring e.g. short / open circuits, circuit resistance, incorrect sensor resistance or, fault with the reference / operating voltage; interference from other circuits can also affect sensor signals. Also check for CAN-Bus fault codes (if applicable). The fault could also be due to a control module / engine control unit fault.

NOTE 1. Check for other fault codes that could provide additional information.

NOTE 2. Communication between control units can pass via a CAN-Bus system; refer to Page 51 for CAN-Bus checks.

NOTE 3. If a fault cannot be located, it is also possible that the control unit is at fault.

NOTE 4. Refer to Page 53 for list of pages that show the ISO standard for component locations e.g. Sensor A - Bank 1.

283

Fault code	EOBD / ISO Description	Component / System Description	Meaningful Description and Quick Check
P254A	PTO Speed Selector Sensor / Switch 1 Circuit / Open	The PTO (Power Take Off) unit can be controlled by the main engine control unit or a separate control module (which will communicate with the main control unit). A speed selection system can be used to select the desired PTO operating speed.	It will be necessary to refer to vehicle specific information to identify the Speed selector switch / sensor and speed control system / wiring. The switch / sensor signal could be: out of normal operating range, constant value, non-existent, corrupt. Possible fault in the switch / sensor or wiring e.g. short / open circuits, circuit resistance or, fault with the reference / operating voltage; the switch contacts (if applicable) could also be faulty e.g. not opening / closing correctly or have a high contact resistance. Refer to list of sensor / actuator checks on page 19. Also check operation of PTO sensors or operation of separate PTO control module (if fitted).
P254B	PTO Speed Selector Sensor / Switch 1 Range / Performance	The PTO (Power Take Off) unit can be controlled by the main engine control unit or a separate control module (which will communicate with the main control unit). A speed selection system can be used to select the desired PTO operating speed.	It will be necessary to refer to vehicle specific information to identify the Speed selector switch / sensor and speed control system / wiring. The switch / sensor signal could be within the normal operating range / tolerance but the signal is not plausible or is incorrect due to an undefined fault. The switch / sensor signal might not match the operating conditions indicated by other sensors or, the switch / sensor signal is not changing / responding as expected. Possible fault in the switch / sensor or wiring e.g. short / open circuits, circuit resistance or, fault with the reference / operating voltage; the switch contacts (if applicable) could also be faulty e.g. not opening / closing correctly or have a high contact resistance. Refer to list of sensor / actuator checks on page 19. Also check operation of PTO sensors or operation of separate PTO control module (if fitted).
P254C	PTO Speed Selector Sensor / Switch 1 Circuit Low	The PTO (Power Take Off) unit can be controlled by the main engine control unit or a separate control module (which will communicate with the main control unit). A speed selection system can be used to select the desired PTO operating speed.	It will be necessary to refer to vehicle specific information to identify the Speed selector switch / sensor and speed control system / wiring. The switch / sensor signal could be "zero" (no signal) or, the signal exists but it is below the normal operating range (lower than the minimum operating limit). The switch / sensor signal might not match the operating conditions indicated by other sensors or, the switch / sensor signal is not changing / responding as expected. Possible fault in the switch / sensor or wiring e.g. short / open circuits, circuit resistance or, fault with the reference / operating voltage; the switch contacts (if applicable) could also be faulty e.g. not opening / closing correctly or have a high contact resistance. Refer to list of sensor / actuator checks on page 19. Also check operation of PTO sensors or operation of separate PTO control module (if fitted).
P254D	PTO Speed Selector Sensor / Switch 1 Circuit High	The PTO (Power Take Off) unit can be controlled by the main engine control unit or a separate control module (which will communicate with the main control unit). A speed selection system can be used to select the desired PTO operating speed.	It will be necessary to refer to vehicle specific information to identify the Speed selector switch / sensor and speed control system / wiring. The switch / sensor signal could be at full value (e.g. battery voltage with no signal or frequency) or, the signal exists but it is above the normal operating range (higher than the maximum operating limit). The switch / sensor signal might not match the operating conditions indicated by other sensors or, the switch / sensor signal is not changing / responding as expected. Possible fault in the switch / sensor or wiring e.g. short / open circuits, circuit resistance or, fault with the reference / operating voltage; the switch contacts (if applicable) could also be faulty e.g. not opening / closing correctly or have a high contact resistance. Refer to list of sensor / actuator checks on page 19. Also check operation of PTO sensors or operation of separate PTO control module (if fitted).
P254E	PTO Speed Selector Sensor / Switch 1 Circuit Intermittent / Erratic	The PTO (Power Take Off) unit can be controlled by the main engine control unit or a separate control module (which will communicate with the main control unit). A speed selection system can be used to select the desired PTO operating speed.	It will be necessary to refer to vehicle specific information to identify the Speed selector switch / sensor and speed control system / wiring. The switch / sensor signal could is intermittent (likely to be out of normal operating range / tolerance when intermittent fault is detected) or, the signal / voltage is erratic (e.g. signal changes are irregular / unpredictable). Possible fault in the switch / sensor or wiring e.g. loose connections; the switch contacts (if applicable) could also be faulty e.g. not opening / closing correctly or have a high contact resistance. Refer to list of sensor / actuator checks on page 19. Also check operation of PTO sensors or operation of separate PTO control module (if fitted).
P254F	Engine Hood Switch Circuit	The switch detects whether the engine hood / bonnet is shut / fully closed. Note the switch might not contain mechanical "Contacts" but could be a magnetic or other type of switch. If necessary, refer to applicable information to identify the switch operation.	The voltage or value of the switch signal is incorrect (undefined fault). Switch voltage / signal could be: out of normal operating range, constant value, non-existent, corrupt. Possible fault in the switch or wiring e.g. short / open circuits, circuit resistance or, fault with the reference / operating voltage. If applicable, check switch contacts to ensure that they open / close and check continuity of switch contacts (ensure that there is not a high resistance). Refer to list of sensor / actuator checks on page 19. Check mechanism / linkage used to operate switch, and check base settings for the switch if applicable.
P2550	Torque Management Request Input Signal "B" Low	Different vehicle systems can pass a request (from the control module to the engine control unit) for a change in engine torque to match operating conditions e.g. the transmission can request a torque reduction during gear changes (changing up). It may be necessary to identify which input circuit (to the engine control unit is being referred to (input A, B etc). Note that the torque request can pass through a CAN-Bus network.	The fault code could be detected due to an incorrect "torque request" signal value. The voltage, frequency or value of the signal is either "zero" (no signal) or, the signal exists but it is below the normal operating range (lower than the minimum operating limit). Possible fault in the wiring e.g. short / open circuits, circuit resistance, incorrect sensor resistance or, fault with the reference / operating voltage; interference from other circuits can also affect sensor signals. Also check for CAN-Bus fault codes (if applicable). The fault could also be due to a control module / engine control unit fault.
P2551	Torque Management Request Input Signal "B" High	Different vehicle systems can pass a request (from the control module to the engine control unit) for a change in engine torque to match operating conditions e.g. the transmission can request a torque reduction during gear changes (changing up). It may be necessary to identify which input circuit (to the engine control unit is being referred to (input A, B etc). Note that the torque request can pass through a CAN-Bus network.	The fault code could be detected due to an incorrect "torque request" signal value. The voltage, frequency or value of the signal is either at full value (e.g. battery voltage with no signal or frequency) or, the signal exists but it is above the normal operating range (higher than the maximum operating limit). Possible fault in the wiring e.g. short / open circuits, circuit resistance, incorrect sensor resistance or, fault with the reference / operating voltage; interference from other circuits can also affect sensor signals. Also check for CAN-Bus fault codes (if applicable). The fault could also be due to a control module / engine control unit fault.
P2552	Throttle / Fuel Inhibit Circuit	Note that different vehicle manufacturers have applied this code in different ways, it may therefore be necessary to refer to vehicle specific information. In most cases, the fault code refers to the circuit which carries an instruction from a control module or, transmission or engine control, to a separate fuel or throttle control module; the instruction is to reduce / cut off the fuel or close the throttle (this can be due to a fault or operating conditions requiring inhibiting of fuel or throttle).	The fault code indicates that there is an unidentified fault in the circuit carrying the "Inhibit Instruction". The voltage, frequency or value of the signal in the circuit is incorrect (undefined fault). Voltage / signal could be: out of normal operating range, constant value, non-existent, corrupt. Possible fault in the wiring e.g. short / open circuits or circuit resistance. It is also possible that one of the applicable control module is at fault. Also check whether the Throttle / Fuel Inhibit signal passes through the CAN-Bus network.

NOTE 1. Check for other fault codes that could provide additional information.   **NOTE 2.** Communication between control units can pass via a CAN-Bus system; refer to Page 51 for CAN-Bus checks.   284

NOTE 3. If a fault cannot be located, it is also possible that the control unit is at fault.   NOTE 4. Refer to Page 53 for list of pages that show the ISO standard for component locations e.g. Sensor A - Bank 1.

Code	Fault Location	Description	Probable Cause
P2553	Throttle / Fuel Inhibit Circuit Range / Performance	Note that different vehicle manufacturers have applied this code in different ways, it may therefore be necessary to refer to vehicle specific information. In most cases, the fault code refers to the circuit which carries an instruction from a control module e.g. transmission or engine control, to a separate fuel or throttle control module; the instruction is to reduce / cut off the fuel or close the throttle (this can be due to a fault or operating conditions requiring inhibiting of fuel or throttle).	The fault code indicates that there is an unidentified fault in the circuit carrying the "Inhibit Instruction". The voltage, frequency or value of the signal is within the normal operating range / tolerance but the signal is not plausible or is incorrect due to an undefined fault. The signal might not match the operating conditions indicated by other sensors or, the signal is not changing / responding as expected to the changes in conditions. Possible fault in the wiring e.g. short / open circuits or circuit resistance. It is also possible that one of the applicable control module is at fault. Also check whether the Throttle / Fuel Inhibit signal passes through the CAN-Bus network.
P2554	Throttle / Fuel Inhibit Circuit Low	Note that different vehicle manufacturers have applied this code in different ways, it may therefore be necessary to refer to vehicle specific information. In most cases, the fault code refers to the circuit which carries an instruction from a control module e.g. transmission or engine control, to a separate fuel or throttle control module; the instruction is to reduce / cut off the fuel or close the throttle (this can be due to a fault or operating conditions requiring inhibiting of fuel or throttle).	The fault code indicates that there is an unidentified fault in the circuit carrying the "Inhibit Instruction". The voltage, frequency or value of the signal is either "zero" (no signal) or, the signal exists but it is below the normal operating range (lower than the minimum operating limit). Possible fault in the wiring e.g. short / open circuits or circuit resistance. It is also possible that one of the applicable control module is at fault. Also check whether the Throttle / Fuel Inhibit signal passes through the CAN-Bus network.
P2555	Throttle / Fuel Inhibit Circuit High	Note that different vehicle manufacturers have applied this code in different ways, it may therefore be necessary to refer to vehicle specific information. In most cases, the fault code refers to the circuit which carries an instruction from a control module e.g. transmission or engine control, to a separate fuel or throttle control module; the instruction is to reduce / cut off the fuel or close the throttle (this can be due to a fault or operating conditions requiring inhibiting of fuel or throttle).	The fault code indicates that there is an unidentified fault in the circuit carrying the "Inhibit Instruction". The voltage, frequency or value of the signal is either at full value (e.g. battery voltage with no signal or frequency) or, the signal exists but it is above the normal operating range (higher than the maximum operating limit). Possible fault in the wiring e.g. short / open circuits or circuit resistance. It is also possible that one of the applicable control module is at fault. Also check whether the Throttle / Fuel Inhibit signal passes through the CAN-Bus network.
P2556	Engine Coolant Level Sensor / Switch Circuit	Engine coolant level sensors can be simple on / off devices that respond when the coolant level is above or below pre-set limits.	The voltage, frequency or value of the level sensor signal is incorrect (undefined fault). Sensor voltage / signal could be: out of normal operating range, constant value, non-existent, corrupt. Possible fault in the level sensor or wiring e.g. short / open circuits, circuit resistance, incorrect sensor resistance or, fault with the reference / operating voltage; interference from other circuits can also affect sensor signals. Refer to list of sensor / actuator checks on page 19.
P2557	Engine Coolant Level Sensor / Switch Circuit Range / Performance	Engine coolant level sensors can be simple on / off devices that respond when the coolant level is above or below pre-set limits.	The voltage, frequency or value of the level sensor signal is within the normal operating range / tolerance but the signal is not plausible or is incorrect due to an undefined fault. The sensor signal might not match the operating conditions indicated by other sensors or, the sensor signal is not changing / responding as expected to the changes in conditions. Possible fault in the level sensor or wiring e.g. short / open circuits, circuit resistance, incorrect sensor resistance or, fault with the reference / operating voltage; interference from other circuits can also affect sensor signals. Refer to list of sensor / actuator checks on page 19. It is also possible that the level sensor is operating correctly but the level being measured by the sensor is unacceptable or incorrect.
P2558	Engine Coolant Level Sensor / Switch Circuit Low	Engine coolant level sensors can be simple on / off devices that respond when the coolant level is above or below pre-set limits.	The voltage, frequency or value of the level sensor signal is either "zero" (no signal) or, the signal exists but it is below the normal operating range (lower than the minimum operating limit). Possible fault in the level sensor or wiring e.g. short / open circuits, circuit resistance, incorrect sensor resistance or, fault with the reference / operating voltage; interference from other circuits can also affect sensor signals. Refer to list of sensor / actuator checks on page 19.
P2559	Engine Coolant Level Sensor / Switch Circuit High	Engine coolant level sensors can be simple on / off devices that respond when the coolant level is above or below pre-set limits.	The voltage, frequency or value of the level sensor signal is either at full value (e.g. battery voltage with no signal or frequency) or, the signal exists but it is above the normal operating range (higher than the maximum operating limit). Possible fault in the level sensor or wiring e.g. short / open circuits, circuit resistance, incorrect sensor resistance or, fault with the reference / operating voltage; interference from other circuits can also affect sensor signals. Refer to list of sensor / actuator checks on page 19.
P255A	PTO Speed Selector Sensor / Switch 2 Circuit / Open	The PTO (Power Take Off) unit can be controlled by the main engine control unit or a separate control module (which will communicate with the main control unit). A speed selection system can be used to select the desired PTO operating speed.	It will be necessary to refer to vehicle specific information to identify the Speed selector switch / sensor and speed control system / wiring. The switch / sensor signal could be: out of normal operating range, constant value, non-existent, corrupt. Possible fault in the switch / sensor or wiring e.g. short / open circuits, circuit resistance or, fault with the reference / operating voltage; the switch contacts (if applicable) could also be faulty e.g. not opening / closing correctly or have a high contact resistance. Refer to list of sensor / actuator checks on page 19. Also check operation of PTO sensors or operation of separate PTO control module (if fitted).
P255B	PTO Speed Selector Sensor / Switch 2 Range / Performance	The PTO (Power Take Off) unit can be controlled by the main engine control unit or a separate control module (which will communicate with the main control unit). A speed selection system can be used to select the desired PTO operating speed.	It will be necessary to refer to vehicle specific information to identify the Speed selector switch / sensor and speed control system / wiring. The switch / sensor signal could be within the normal operating range / tolerance but the signal is not plausible or is incorrect due to an undefined fault. The switch / sensor signal might not match the operating conditions indicated by other sensors or, the switch / sensor signal is not changing / responding as expected. Possible fault in the switch / sensor or wiring e.g. short / open circuits, circuit resistance or, fault with the reference / operating voltage; the switch contacts (if applicable) could also be faulty e.g. not opening / closing correctly or have a high contact resistance. Refer to list of sensor / actuator checks on page 19. Also check operation of PTO sensors or operation of separate PTO control module (if fitted).
P255C	PTO Speed Selector Sensor / Switch 2 Circuit Low	The PTO (Power Take Off) unit can be controlled by the main engine control unit or a separate control module (which will communicate with the main control unit). A speed selection system can be used to select the desired PTO operating speed.	It will be necessary to refer to vehicle specific information to identify the Speed selector switch / sensor and speed control system / wiring. The switch / sensor signal could be "zero" (no signal) or, the signal exists but it is below the normal operating range (lower than the minimum operating limit). The switch / sensor signal might not match the operating conditions indicated by other sensors or, the switch / sensor signal is not changing / responding as expected. Possible fault in the switch / sensor or wiring e.g. short / open circuits, circuit resistance or, fault with the reference / operating voltage; the switch contacts (if applicable) could also be faulty e.g. not opening / closing correctly or have a high contact resistance. Refer to list of sensor / actuator checks on page 19. Also check operation of PTO sensors or operation of separate PTO control module (if fitted).

NOTE 1. Check for other fault codes that could provide additional information.
NOTE 2. Communication between control units can pass via a CAN-Bus system; refer to Page 51 for CAN-Bus checks.
NOTE 3. If a fault cannot be located, it is also possible that the control unit is at fault.
NOTE 4. Refer to Page 53 for list of pages that show the ISO standard for component locations e.g. Sensor A - Bank 1.

Fault code	EOBD / ISO Description	Component / System Description	Meaningful Description and Quick Check
P255D	PTO Speed Selector Sensor / Switch 2 Circuit High	The PTO (Power Take Off) unit can be controlled by the main engine control unit or a separate control module (which will communicate with the main control unit). A speed selection system can be used to select the desired PTO operating speed.	It will be necessary to refer to vehicle specific information to identify the Speed selector switch / sensor and speed control system / wiring. The switch / sensor signal could be at full value (e.g. battery voltage with no signal or frequency) or, the signal exists but it is above the normal operating range (higher than the maximum operating limit). The switch / sensor signal might not match the operating conditions indicated by other sensors or, the switch / sensor signal is not changing / responding as expected. Possible fault in the switch / sensor or wiring e.g. short / open circuits, circuit resistance or, fault with the reference / operating voltage; the switch contacts (if applicable) could also be faulty e.g. not opening / closing correctly or have a high contact resistance. Refer to list of sensor / actuator checks on page 19. Also check operation of PTO sensors or operation of separate PTO control module (if fitted).
P255E	PTO Speed Selector Sensor / Switch 2 Circuit Intermittent / Erratic	The PTO (Power Take Off) unit can be controlled by the main engine control unit or a separate control module (which will communicate with the main control unit). A speed selection system can be used to select the desired PTO operating speed.	It will be necessary to refer to vehicle specific information to identify the Speed selector switch / sensor and speed control system / wiring. The switch / sensor signal could is intermittent (likely to be out of normal operating range / tolerance when intermittent fault is detected) or, the signal / voltage is erratic (e.g. signal changes are irregular / unpredictable). Possible fault in the switch / sensor or wiring e.g. loose connections; the switch contacts (if applicable) could also be faulty e.g. not opening / closing correctly or have a high contact resistance. Refer to list of sensor / actuator checks on page 19. Also check operation of PTO sensors or operation of separate PTO control module (if fitted).
P255F	ISO / SAE reserved		Not used, reserved for future allocation.
P2560	Engine Coolant Level Low	Engine coolant level sensors can be simple on / off devices that respond when the coolant level is above or below pre-set limits.	The fault code relates to the coolant level sensor signal which is indicating that coolant level is too low. The sensor signal is within normal operating range / tolerance but the indicated coolant level is below specified values. If the coolant level is not too low, it is possible that the sensor signal is incorrect; refer to fault code P2557 for additional information.
P2561	A / C Control Module Requested MIL Illumination	If the Air Conditioning system control module detect a fault in the air conditioning system, it can pass a request to the engine control unit to illuminate the MIL (Malfunction Indicator Lamp).	The fault code Indicates that the air conditioning control module has detected an undefined system fault, and has requested that the MIL (Malfunction Indicator Lamp) should be illuminated. If possible, check for air conditioning system fault codes (via the air conditioning control module if applicable) or investigate the air conditioning system for faults.
P2562	Turbocharger Boost Control Position Sensor "A" Circuit	The boost control position sensor is generally used on turbochargers that have: Variable geometry, Variable vane / nozzle turbochargers. The geometry change of the turbine blades or nozzle blades regulates boost pressure instead of a boost pressure wastegate. The sensor detects the position of the variable geometry mechanism to establish the correct operation and position of the variable geometry system.	The voltage, frequency or value of the position sensor signal is incorrect (undefined fault). Sensor voltage / signal value could be: out of normal operating range, constant value, non-existent, corrupt. Possible fault in the position sensor or wiring e.g. short / open circuits, circuit resistance, incorrect sensor resistance (if applicable) or, fault with the reference / operating voltage; interference from other circuits can also affect sensor signals. Also check earth connections and cable screening. Refer to list of sensor / actuator checks on page 19.
P2563	Turbocharger Boost Control Position Sensor "A" Circuit Range / Performance	The boost control position sensor is generally used on turbochargers that have: Variable geometry, Variable vane / nozzle turbochargers. The geometry change of the turbine blades or nozzle blades regulates boost pressure instead of a boost pressure wastegate. The sensor detects the position of the variable geometry mechanism to establish the correct operation and position of the variable geometry system.	The voltage, frequency or value of the position sensor signal is within the normal operating range / tolerance but the signal is not plausible or is incorrect due to an undefined fault. The sensor signal might not match the operating conditions indicated by other sensors or, the sensor signal is not changing / responding as expected to the changes in boost control position (as requested by the control unit). Possible fault in the position sensor or wiring e.g. short / open circuits, circuit resistance, incorrect sensor resistance or, fault with the reference / operating voltage; interference from other circuits can also affect sensor signals. Refer to list of sensor / actuator checks on page 19. It is also possible that the sensor is operating correctly but the position being detected by the sensor is unacceptable or incorrect for the operating conditions. Check operation of boost control system and check for freedom of movement of operating mechanism.
P2564	Turbocharger Boost Control Position Sensor "A" Circuit Low	The boost control position sensor is generally used on turbochargers that have: Variable geometry, Variable vane / nozzle turbochargers. The geometry change of the turbine blades or nozzle blades regulates boost pressure instead of a boost pressure wastegate. The sensor detects the position of the variable geometry mechanism to establish the correct operation and position of the variable geometry system.	The voltage, frequency or value of the position sensor signal is either "zero" (no signal) or, the signal exists but it is below the normal operating range (lower than the minimum operating limit). Possible fault in the position sensor or wiring e.g. short / open circuits, circuit resistance, incorrect sensor resistance or, fault with the reference / operating voltage; interference from other circuits can also affect sensor signals. Refer to list of sensor / actuator checks on page 19.
P2565	Turbocharger Boost Control Position Sensor "A" Circuit High	The boost control position sensor is generally used on turbochargers that have: Variable geometry, Variable vane / nozzle turbochargers. The geometry change of the turbine blades or nozzle blades regulates boost pressure instead of a boost pressure wastegate. The sensor detects the position of the variable geometry mechanism to establish the correct operation and position of the variable geometry system.	The voltage, frequency or value of the position sensor signal is either at full value (e.g. battery voltage with no signal or frequency) or, the signal exists but it is above the normal operating range (higher than the maximum operating limit). Possible fault in the position sensor or wiring e.g. short / open circuits, circuit resistance, incorrect sensor resistance or, fault with the reference / operating voltage; interference from other circuits can also affect sensor signals. Refer to list of sensor / actuator checks on page 19.
P2566	Turbocharger Boost Control Position Sensor "A" Circuit Intermittent	The boost control position sensor is generally used on turbochargers that have: Variable geometry, Variable vane / nozzle turbochargers. The geometry change of the turbine blades or nozzle blades regulates boost pressure instead of a boost pressure wastegate. The sensor detects the position of the variable geometry mechanism to establish the correct operation and position of the variable geometry system.	The voltage, frequency or value of the position sensor signal is intermittent (likely to be out of normal operating range / tolerance when intermittent fault is detected) or, the signal / voltage is erratic (e.g. signal changes are irregular / unpredictable). Possible fault with wiring or connections (including internal sensor connections); also interference from other circuits can affect the signal. Refer to list of sensor / actuator checks on page 19. Also check for erratic operation of boost control system that could cause position sensor signal to be erratic.

NOTE 1. Check for other fault codes that could provide additional information. NOTE 2. Communication between control units can pass via a CAN-Bus system; refer to Page 51 for CAN-Bus checks.

P2567	Direct Ozone Reduction Catalyst Temperature Sensor Circuit	The Ozone Reduction Catalyst forms part of the cooling system radiator. The radiator is treated with a catalytic material and under certain operating conditions the heat of the radiator and the catalytic material promote the reduction of Ozone in the atmosphere. Sensors are used (including temperature sensors) to monitor the operation of the catalyst and the deterioration of the catalyst efficiency.	The voltage or value of the temperature sensor signal is incorrect (undefined fault). Sensor voltage / signal could be: out of normal operating range, constant value, non-existent, corrupt. Possible fault in the temperature sensor or wiring e.g. short / open circuits, circuit resistance, incorrect sensor resistance or, fault with the reference / operating voltage; interference from other circuits can also affect sensor signals. Refer to list of sensor / actuator checks on page 19. The code refers to a sensor / circuit fault but if the sensor does appear to be operating correctly, check for other faults that could affect catalyst temperatures.
P2568	Direct Ozone Reduction Catalyst Temperature Sensor Circuit Range / Performance	The Ozone Reduction Catalyst forms part of the cooling system radiator. The radiator is treated with a catalytic material and under certain operating conditions the heat of the radiator and the catalytic material promote the reduction of Ozone in the atmosphere. Sensors are used (including temperature sensors) to monitor the operation of the catalyst and the deterioration of the catalyst efficiency.	The voltage or value of the temperature sensor signal is within the normal operating range / tolerance but the signal is not plausible or is incorrect due to an undefined fault. The sensor signal might not match the operating conditions indicated by other sensors or, the sensor signal is not changing / responding as expected to the changes in conditions. Possible fault in the temperature sensor or wiring e.g. short / open circuits, circuit resistance, incorrect sensor resistance or, fault with the reference / operating voltage; interference from other circuits can also affect sensor signals. Refer to list of sensor / actuator checks on page 19. It is also possible that the temperature sensor is operating correctly but the catalyst / radiator temperature is unacceptable or incorrect for the operating conditions. Check for faults that could affect catalyst temperatures e.g. high exhaust gas temperatures caused by engine faults.
P2569	Direct Ozone Reduction Catalyst Temperature Sensor Circuit Low	The Ozone Reduction Catalyst forms part of the cooling system radiator. The radiator is treated with a catalytic material and under certain operating conditions the heat of the radiator and the catalytic material promote the reduction of Ozone in the atmosphere. Sensors are used (including temperature sensors) to monitor the operation of the catalyst and the deterioration of the catalyst efficiency.	The voltage or value of the temperature sensor signal is either "zero" (no signal) or, the signal exists but it is below the normal operating range (lower than the minimum operating limit). Possible fault in the temperature sensor or wiring e.g. short / open circuits, circuit resistance, incorrect sensor resistance or, fault with the reference / operating voltage; interference from other circuits can also affect sensor signals. Refer to list of sensor / actuator checks on page 19.
P256A	Engine Idle Speed Selector Sensor / Switch Circuit / Open	An idle speed selector system can be used where vehicles are equipped with a PTO (Power Take Off) system or hydraulic pumps etc. The selector enables a high idle speed to be selected where there is a requirement for more engine power.	The voltage or value of the switch signal is incorrect (possibly caused by an open circuit). Switch voltage / signal could be: out of normal operating range, constant value, non-existent, corrupt. Possible fault in the switch or wiring e.g. short / open circuits, circuit resistance or, fault with the reference / operating voltage; the switch contacts could also be faulty e.g. not opening / closing correctly or have a high contact resistance. Refer to list of sensor / actuator checks on page 19.
P256B	Engine Idle Speed Selector Sensor / Switch Range / Performance	An idle speed selector system can be used where vehicles are equipped with a PTO (Power Take Off) system or hydraulic pumps etc. The selector enables a high idle speed to be selected where there is a requirement for more engine power.	The voltage, frequency or value of the switch signal is within the normal operating range / tolerance but the signal is not plausible or is incorrect due to an undefined fault. The switch signal might not match the operating conditions indicated by other sensors or, the switch signal is not changing / responding as expected. Possible fault in the switch or wiring e.g. short / open circuits, circuit resistance or, fault with the reference / operating voltage; also check switch contacts to ensure that they open / close and check continuity of switch contacts (ensure that there is not a high resistance). Refer to list of sensor / actuator checks on page 19.
P256C	Engine Idle Speed Selector Sensor / Switch Circuit Low	An idle speed selector system can be used where vehicles are equipped with a PTO (Power Take Off) system or hydraulic pumps etc. The selector enables a high idle speed to be selected where there is a requirement for more engine power.	The voltage, frequency or value of the switch signal is either "zero" (no signal) or, the signal exists but it is below the normal operating range (lower than the minimum operating limit). Possible fault in the switch or wiring e.g. short / open circuits, circuit resistance or, fault with the reference / operating voltage; the switch contacts could also be faulty e.g. not opening / closing correctly or have a high contact resistance. Refer to list of sensor / actuator checks on page 19.
P256D	Engine Idle Speed Selector Sensor / Switch Circuit High	An idle speed selector system can be used where vehicles are equipped with a PTO (Power Take Off) system or hydraulic pumps etc. The selector enables a high idle speed to be selected where there is a requirement for more engine power.	The voltage, frequency or value of the switch signal is either at full value (e.g. battery voltage with no signal or frequency) or, the signal exists but it is above the normal operating range (higher than the maximum operating limit). Possible fault in the switch or wiring e.g. short / open circuits, circuit resistance or, fault with the reference / operating voltage; the switch contacts could also be faulty e.g. not opening / closing correctly or have a high contact resistance. Refer to list of sensor / actuator checks on page 19.
P256E	Engine Idle Speed Selector Sensor / Switch Circuit Intermittent / Erratic	An idle speed selector system can be used where vehicles are equipped with a PTO (Power Take Off) system or hydraulic pumps etc. The selector enables a high idle speed to be selected where there is a requirement for more engine power.	The voltage, frequency or value of the switch signal is intermittent (likely to be out of normal operating range / tolerance when intermittent fault is detected) or, the signal / voltage is erratic (possible faulty / intermittent connection). Possible fault with wiring or connections (including internal switch connections); the switch contacts could also be faulty e.g. not opening / closing correctly or have a high contact resistance. Refer to list of sensor / actuator checks on page 19.
P256F	ISO / SAE reserved		Not used, reserved for future allocation.
P2570	Direct Ozone Reduction Catalyst Temperature Sensor Circuit High	The Ozone Reduction Catalyst forms part of the cooling system radiator. The radiator is treated with a catalytic material and under certain operating conditions the heat of the radiator and the catalytic material promote the reduction of Ozone in the atmosphere. Sensors are used (including temperature sensors) to monitor the operation of the catalyst and the deterioration of the catalyst efficiency.	The voltage or value of the temperature sensor signal is either at full value (e.g. battery voltage with no signal or frequency) or, the signal exists but it is above the normal operating range (higher than the maximum operating limit). Possible fault in the temperature sensor or wiring e.g. short / open circuits, circuit resistance, incorrect sensor resistance or, fault with the reference / operating voltage; interference from other circuits can also affect sensor signals. Refer to list of sensor / actuator checks on page 19.
P2571	Direct Ozone Reduction Catalyst Temperature Sensor Circuit Intermittent / Erratic	The Ozone Reduction Catalyst forms part of the cooling system radiator. The radiator is treated with a catalytic material and under certain operating conditions the heat of the radiator and the catalytic material promote the reduction of Ozone in the atmosphere. Sensors are used (including temperature sensors) to monitor the operation of the catalyst and the deterioration of the catalyst efficiency.	The voltage or value of the temperature sensor signal is intermittent (likely to be out of normal operating range / tolerance when intermittent fault is detected) or, the signal / voltage is erratic (e.g. signal changes are irregular / unpredictable). Possible fault with wiring or connections (including internal sensor connections); also interference from other circuits can affect the signal. Refer to list of sensor / actuator checks on page 19.

NOTE 1. Check for other fault codes that could provide additional information.   **NOTE 2.** Communication between control units can pass via a CAN-Bus system; refer to Page 51 for CAN-Bus checks.

NOTE 3. If a fault cannot be located, it is also possible that the control unit is at fault.   **NOTE 4.** Refer to Page 53 for list of pages that show the ISO standard for component locations e.g. Sensor A - Bank 1.

Fault code	EOBD / ISO Description	Component / System Description	Meaningful Description and Quick Check
P2572	Direct Ozone Reduction Catalyst Deterioration Sensor Circuit	The Ozone Reduction Catalyst forms part of the cooling system radiator. The radiator is treated with a catalytic material and under certain operating conditions the heat of the radiator and the catalytic material promote the reduction of Ozone in the atmosphere. Sensors are used to monitor the operation of the catalyst and the deterioration of the catalyst efficiency. It may be necessary to refer to vehicle specific information to identify the type of sensor used to monitor the catalyst operation / deterioration.	Identify the type of sensor fitted to the Ozone Reduction Catalyst. Note that the Ozone reduction process is affected by temperature (coolant and ambient), airflow (including effects from vehicle speed, cooling fan operation and Air / Con system operation). The fault code indicates that the voltage or value of the sensor signal is incorrect (undefined fault). Sensor voltage / signal could be: out of normal operating range, constant value, non-existent, corrupt. Possible fault in the sensor or wiring e.g. short / open circuits, circuit resistance, incorrect sensor resistance or, fault with the reference / operating voltage; interference from other circuits can also affect sensor signals
P2573	Direct Ozone Reduction Catalyst Deterioration Sensor Circuit Range / Performance	The Ozone Reduction Catalyst forms part of the cooling system radiator. The radiator is treated with a catalytic material and under certain operating conditions the heat of the radiator and the catalytic material promote the reduction of Ozone in the atmosphere. Sensors are used to monitor the operation of the catalyst and the deterioration of the catalyst efficiency. It may be necessary to refer to vehicle specific information to identify the type of sensor used to monitor the catalyst operation / deterioration.	Identify the type of sensor fitted to the Ozone Reduction Catalyst. Note that the Ozone reduction process is affected by temperature (coolant and ambient), airflow (including effects from vehicle speed, cooling fan operation and Air / Con system operation). The fault code indicates that the voltage or value of the sensor signal is within the normal operating range / tolerance but the signal is not plausible or is incorrect due to an undefined fault. The sensor signal might not match the operating conditions indicated by other sensors or, the sensor signal is not changing / responding as expected to the changes in conditions. Possible fault in the sensor or wiring e.g. short / open circuits, circuit resistance, incorrect sensor resistance or, fault with the reference / operating voltage; interference from other circuits can also affect sensor signals. For this code, it is also possible that the sensor is operating correctly but the information provided by the sensor is unacceptable or incorrect for the operating conditions.
P2574	Direct Ozone Reduction Catalyst Deterioration Sensor Circuit Low	The Ozone Reduction Catalyst forms part of the cooling system radiator. The radiator is treated with a catalytic material and under certain operating conditions the heat of the radiator and the catalytic material promote the reduction of Ozone in the atmosphere. Sensors are used to monitor the operation of the catalyst and the deterioration of the catalyst efficiency. It may be necessary to refer to vehicle specific information to identify the type of sensor used to monitor the catalyst operation / deterioration.	Identify the type of sensor fitted to the Ozone Reduction Catalyst. Note that the Ozone reduction process is affected by temperature (coolant and ambient), airflow (including effects from vehicle speed, cooling fan operation and Air / Con system operation). The fault code indicates that the voltage or value of the sensor signal is either "zero" (no signal) or, the signal exists but it is below the normal operating range (lower than the minimum operating limit). Possible fault in the sensor or wiring e.g. short / open circuits, circuit resistance, incorrect sensor resistance or, fault with the reference / operating voltage; interference from other circuits can also affect sensor signals
P2575	Direct Ozone Reduction Catalyst Deterioration Sensor Circuit High	The Ozone Reduction Catalyst forms part of the cooling system radiator. The radiator is treated with a catalytic material and under certain operating conditions the heat of the radiator and the catalytic material promote the reduction of Ozone in the atmosphere. Sensors are used to monitor the operation of the catalyst and the deterioration of the catalyst efficiency. It may be necessary to refer to vehicle specific information to identify the type of sensor used to monitor the catalyst operation / deterioration.	Identify the type of sensor fitted to the Ozone Reduction Catalyst. Note that the Ozone reduction process is affected by temperature (coolant and ambient), airflow (including effects from vehicle speed, cooling fan operation and Air / Con system operation). The fault code indicates that the voltage or value of the sensor signal is either at full value (e.g. battery voltage with no signal or frequency) or, the signal exists but it is above the normal operating range (higher than the maximum operating limit). Possible fault in the sensor or wiring e.g. short / open circuits, circuit resistance, incorrect sensor resistance or, fault with the reference / operating voltage; interference from other circuits can also affect sensor signals
P2576	Direct Ozone Reduction Catalyst Deterioration Sensor Circuit Intermittent / Erratic	The Ozone Reduction Catalyst forms part of the cooling system radiator. The radiator is treated with a catalytic material and under certain operating conditions the heat of the radiator and the catalytic material promote the reduction of Ozone in the atmosphere. Sensors are used to monitor the operation of the catalyst and the deterioration of the catalyst efficiency. It may be necessary to refer to vehicle specific information to identify the type of sensor used to monitor the catalyst operation / deterioration.	Identify the type of sensor fitted to the Ozone Reduction Catalyst. Note that the Ozone reduction process is affected by temperature (coolant and ambient), airflow (including effects from vehicle speed, cooling fan operation and Air / Con system operation). The fault code indicates that the voltage or value of the sensor signal is intermittent (likely to be out of normal operating range / tolerance when intermittent fault is detected) or, the signal / voltage is erratic (e.g. signal changes are irregular / unpredictable). Possible fault with wiring or connections (including internal sensor connections), also check sensor and wiring e.g. short / open circuits, circuit resistance, incorrect sensor resistance or, fault with the reference / operating voltage; interference from other circuits can also affect sensor signals.
P2577	Direct Ozone Reduction Catalyst Efficiency Below Threshold	The Ozone Reduction Catalyst forms part of the cooling system radiator. The radiator is treated with a catalytic material and under certain operating conditions the heat of the radiator and the catalytic material promote the reduction of Ozone in the atmosphere. Sensors are used to monitor the operation of the catalyst and the deterioration of the catalyst efficiency. It may be necessary to refer to vehicle specific information to identify the type of sensor used to monitor the catalyst operation / deterioration.	Identify the type of sensor fitted to the Ozone Reduction Catalyst. Note that the Ozone reduction process is affected by temperature (coolant and ambient), airflow (including effects from vehicle speed, cooling fan operation and Air / Con system operation). The fault code indicates that the efficiency of the Ozone reduction catalyst is below the acceptable limits / threshold. The fault could be caused by deterioration of the catalyst material or could be caused by a radiator / cooling system fault (restricted airflow, incorrect temperature in the radiator etc). It is also possible that the sensor used to monitor the Ozone reduction process is faulty (refer to fault codes P2572 - P2576).
P2578	Turbocharger Speed Sensor Circuit	Speed sensors can be used to detect the turbine speed on turbo / supercharger systems.	Identify the type of sensor e.g. Inductive, Hall effect etc; this will dictate the checks that are applicable. The voltage, frequency or value of the speed sensor signal is incorrect (undefined fault). Sensor voltage / signal value could be: out of normal operating range, constant value, non-existent, corrupt. Possible fault in the speed sensor or wiring e.g. short / open circuits, circuit resistance, incorrect sensor resistance (if applicable) or, fault with the reference / operating voltage; interference from other circuits can also affect sensor signals. Also check earth connections and cable screening. Refer to list of sensor / actuator checks on page 19.
P2579	Turbocharger Speed Sensor Circuit Range / Performance	Speed sensors can be used to detect the turbine speed on turbo / supercharger systems.	Identify the type of sensor e.g. Inductive, Hall effect etc; this will dictate the checks that are applicable. The voltage, frequency or value of the speed sensor signal is within the normal operating range / tolerance but the signal is not plausible or is incorrect due to an undefined fault. It is possible that the speed sensor is operating correctly but the speed being measured by the sensor is unacceptable or incorrect for the operating conditions; if possible, use alternative method of checking speed to establish if sensor system is operating correctly and whether speed is correct / incorrect. It is also possible that the sensor signal is not changing / responding as expected to the expected or actual changes in speed. Possible fault in the speed sensor or wiring e.g. short / open circuits, circuit resistance, incorrect sensor resistance or, fault with the reference / operating voltage; interference from other circuits can also affect sensor signals. Refer to list of sensor / actuator checks on page 19. Also check condition of reference teeth on trigger disc / rotor and ensure rotor is turning with shaft.

Code	Fault Location	Description	Probable Cause / Checks
P257A	Vacuum Reservoir Control Circuit Open	A vacuum reservoir can be used on vehicles with petrol and Diesel engines, where it is not effective or possible to rely on the engine as a vacuum source. A control system can be used to regulate the vacuum level; the system can use valves controlled by electric actuators e.g. solenoid operated valve.	The voltage, frequency or value of the control signal in the actuator circuit is incorrect. Signal could be:- out of normal operating range, constant value, non-existent, corrupt. The fault code indicates a possible "open circuit" but other undefined faults in the actuator or wiring could activate the same code e.g. actuator power supply, short circuit, high circuit resistance or incorrect actuator resistance. Note: the control unit could be providing the correct signal but it is being affected by the circuit fault. Refer to list of sensor / actuator checks on page 19.
P257B	Vacuum Reservoir Control Circuit Low	A vacuum reservoir can be used on vehicles with petrol and Diesel engines, where it is not effective or possible to rely on the engine as a vacuum source. A control system can be used to regulate the vacuum level; the system can use valves controlled by electric actuators e.g. solenoid operated valve.	The voltage, frequency or value of the control signal in the actuator circuit is either "zero" (no signal) or, the signal exists but is below the normal operating range. Possible fault with actuator or wiring (e.g. actuator power supply, short / open circuit or resistance) that is causing a signal value to be lower than the minimum operating limit. Note: the control unit could be providing the correct signal but it is being affected by the circuit fault. Refer to list of sensor / actuator checks on page 19.
P257C	Vacuum Reservoir Control Circuit High	A vacuum reservoir can be used on vehicles with petrol and Diesel engines, where it is not effective or possible to rely on the engine as a vacuum source. A control system can be used to regulate the vacuum level; the system can use valves controlled by electric actuators e.g. solenoid operated valve.	The voltage, frequency or value of the control signal in the actuator circuit is either at full value (e.g. battery voltage with no on / off signal) or, the signal exists but it is above the normal operating range. Possible fault with actuator or wiring (e.g. actuator power supply, short / open circuit or resistance) that is causing a signal value to be higher than the maximum operating limit. Note: the control unit could be providing the correct signal but it is being affected by the circuit fault. Refer to list of sensor / actuator checks on page 19.
P257D	Engine Hood Switch Circuit Range / Performance	The switch detects whether the engine hood / bonnet is shut / fully closed. Note the switch might not contain mechanical "Contacts" but could be a magnetic or other type of switch. If necessary, refer to applicable information to identify the switch operation.	The voltage or value of the switch signal is within the normal operating range / tolerance but the signal is not plausible or is incorrect due to an undefined fault. The switch signal might not be changing / responding as expected. Possible fault in the switch or wiring e.g. short / open circuits, circuit resistance or, fault with the reference / operating voltage. If applicable, check switch contacts to ensure that they open / close and check continuity of switch contacts (ensure that there is not a high resistance). Refer to list of sensor / actuator checks on page 19. Check mechanism / linkage used to operate switch, and check base settings for the switch if applicable.
P257E	Engine Hood Switch Circuit Low	The switch detects whether the engine hood / bonnet is shut / fully closed. Note the switch might not contain mechanical "Contacts" but could be a magnetic or other type of switch. If necessary, refer to applicable information to identify the switch operation.	The voltage or value of the switch signal is either "zero" (no signal) and does not change or, the signal exists but it is below the normal operating range (lower than the minimum operating limit). Possible fault in the switch or wiring e.g. short / open circuits, circuit resistance or, fault with the reference / operating voltage. If applicable, check switch contacts to ensure that they open / close and check continuity of switch contacts (ensure that there is not a high resistance). Refer to list of sensor / actuator checks on page 19. Check mechanism / linkage used to operate switch, and check base settings for the switch if applicable.
P257F	Engine Hood Switch Circuit High	The switch detects whether the engine hood / bonnet is shut / fully closed. Note the switch might not contain mechanical "Contacts" but could be a magnetic or other type of switch. If necessary, refer to applicable information to identify the switch operation.	The voltage or value of the switch signal is either at full value (e.g. battery voltage with no signal) and does change or, the signal exists but it is above the normal operating range (higher than the maximum operating limit). Possible fault in the switch or wiring e.g. short / open circuits, circuit resistance or, fault with the reference / operating voltage. If applicable, check switch contacts to ensure that they open / close and check continuity of switch contacts (ensure that there is not a high resistance). Refer to list of sensor / actuator checks on page 19. Check mechanism / linkage used to operate switch, and check base settings for the switch if applicable.
P2580	Turbocharger Speed Sensor Circuit Low	Speed sensors can be used to detect the turbine speed on turbo / supercharger systems.	Identify the type of sensor e.g. Inductive, Hall effect etc; this will dictate the checks that are applicable. The voltage, frequency or value of the speed sensor signal is either "zero" (no signal) or, the signal exists but it is below the normal operating range (lower than the minimum operating limit). Possible fault in the speed sensor or wiring e.g. short / open circuits, circuit resistance, incorrect sensor resistance or, fault with the reference / operating voltage; interference from other circuits can also affect sensor signals. Refer to list of sensor / actuator checks on page 19.
P2581	Turbocharger Speed Sensor Circuit High	Speed sensors can be used to detect the turbine / impeller speed on turbo / supercharger systems.	Identify the type of sensor e.g. Inductive, Hall effect etc; this will dictate the checks that are applicable. The voltage, frequency or value of the speed sensor signal is either at full value (e.g. battery voltage with no signal or frequency) or, the signal exists but it is above the normal operating range (higher than the maximum operating limit). Possible fault in the speed sensor or wiring e.g. short / open circuits, circuit resistance, incorrect sensor resistance or, fault with the reference / operating voltage; interference from other circuits can also affect sensor signals. Refer to list of sensor / actuator checks on page 19.
P2582	Turbocharger Speed Sensor Circuit Intermittent	Speed sensors can be used to detect the turbine speed on turbo / supercharger systems.	Identify the type of sensor e.g. Inductive, Hall effect etc; this will dictate the checks that are applicable. The voltage, frequency or value of the speed sensor signal is intermittent (likely to be out of normal operating range / tolerance when intermittent fault is detected) or, the signal / voltage is erratic (e.g. signal changes are irregular / unpredictable). Possible fault with wiring or connections (including internal sensor connections); also interference from other circuits can affect the signal. Refer to list of sensor / actuator checks on page 19. Also check condition of reference teeth on trigger disc / rotor and ensure rotor is turning with shaft.
P2583	Cruise Control Front Distance Range Sensor	Some cruise control systems make use of a distance control system to enable the vehicle to maintain a pre-determined distance from the vehicle in front. A sensor is used to detect the distance to the vehicle in front, which enables the cruise control system to reduce the vehicle speed and re-accelerate as necessary (to maintain the correct distance).	It may be necessary to refer to vehicle specific information to identify the sensor and wiring. The fault code indicates that there is an unidentified fault with the range sensor. Basic checks can include checking the wiring and connections for short / open circuits, and checking for the correct reference / supply voltage to the sensor. It is likely that reference will be required to vehicle specific information to identify any specific sensor checks. Note that on some systems, the fault could be detected by the Cruise Control Module, which then passes the information to the main control unit; however this information can be limited in detail, which will require further interrogation of the cruise control module and system.

NOTE 1. Check for other fault codes that could provide additional information.
NOTE 3. If a fault cannot be located, it is also possible that the control unit is at fault.
NOTE 2. Communication between control units can pass via a CAN-Bus system; refer to Page 51 for CAN-Bus checks.
NOTE 4. Refer to Page 53 for list of pages that show the ISO standard for component locations e.g. Sensor A - Bank 1.

Fault code	EOBD / ISO Description	Component / System Description	Meaningful Description and Quick Check
P2584	Fuel Additive Control Module Requested MIL Illumination	The fuel additive system relates to the additive used for the regeneration process of the Diesel Particulate filter. The additive (often referred to as a "Reductant") is stored in a separate tank and is added (using controlled quantities) to the fuel in the tank after refuelling. The Reductant combines with the Diesel soot particles, which reduces the temperature at which the soot will combust; this allows the soot particles to combust more easily within the particle filter (during the particle filter regeneration process).	The fault code relates to the additive control module (which controls the additive passing into the fuel tank); the control module communicates with the main engine control unit (or could be combined into the control unit). If a fault exists with the operation of the additive system (that could affect the particulate regeneration process), the additive control module can request the MIL (Malfunction indicator Lamp) to be illuminated. The fault code indicates that the additive control module has detected an undefined fault and requested illumination of the MIL; If possible check for fault codes from the additive control module or carry out further investigation of the system. Also Refer to fault code P2585. Note that the fault could relate to the calculation of the required additive quantity; the quantity required is based on how much fuel has been added to the tank there could be a calculation error.
P2585	Fuel Additive Control Module Warning Lamp Request	The fuel additive system relates to the additive used for the regeneration process of the Diesel Particulate filter. The additive (often referred to as a "Reductant") is stored in a separate tank and is added (using controlled quantities) to the fuel in the tank after refuelling. The Reductant combines with the Diesel soot particles, which reduces the temperature at which the soot will combust; this allows the soot particles to combust more easily within the particle filter (during the particle filter regeneration process).	The fault code relates to the additive control module (which controls the additive passing into the fuel tank); the control module communicates with the main engine control unit (or could be combined into the control unit). The additive control module can request the warning light to be illuminated if there is a system problem; this could be related to low additive level rather than to a specific fault (also Refer to fault code P2584). The fault code indicates that the additive control module has detected an undefined problem and requested illumination of the warning light; check additive level and general operation of additive system. If possible check for fault codes from the additive control module. Note that the fault could relate to the calculation of the required additive quantity; the quantity required is based on how much fuel has been added to the tank there could be a calculation error.
P2586	Turbocharger Boost Control Position Sensor "B" Circuit	The boost control position sensor is generally used on turbochargers that have: Variable geometry, Variable vane / nozzle turbochargers. The geometry change of the turbine blades or nozzle blades regulates boost pressure instead of a boost pressure wastegate. The sensor detects the position of the variable geometry mechanism to establish the correct operation and position of the variable geometry system.	The voltage, frequency or value of the position sensor signal is incorrect (undefined fault). Sensor voltage / signal value could be: out of normal operating range, constant value, non-existent, corrupt. Possible fault in the position sensor or wiring e.g. short / open circuits, circuit resistance, incorrect sensor resistance (if applicable) or, fault with the reference / operating voltage; interference from other circuits can also affect sensor signals. Also check earth connections and cable screening. Refer to list of sensor / actuator checks on page 19.
P2587	Turbocharger Boost Control Position Sensor "B" Circuit Range / Performance	The boost control position sensor is generally used on turbochargers that have: Variable geometry, Variable vane / nozzle turbochargers. The geometry change of the turbine blades or nozzle blades regulates boost pressure instead of a boost pressure wastegate. The sensor detects the position of the variable geometry mechanism to establish the correct operation and position of the variable geometry system.	The voltage, frequency or value of the position sensor signal is within the normal operating range / tolerance but the signal is not plausible or is incorrect due to an undefined fault. The sensor signal might not match the operating conditions indicated by other sensors or, the sensor signal is not changing / responding as expected to the changes in boost control position (as requested by the control unit). Possible fault in the position sensor or wiring e.g. short / open circuits, circuit resistance, incorrect sensor resistance or, fault with the reference / operating voltage; interference from other circuits can also affect sensor signals. Refer to list of sensor / actuator checks on page 19. It is also possible that the sensor is operating correctly but the position being detected by the sensor is unacceptable or incorrect for the operating conditions. Check operation of boost control system and check for freedom of movement of operating mechanism
P2588	Turbocharger Boost Control Position Sensor "B" Circuit Low	The boost control position sensor is generally used on turbochargers that have: Variable geometry, Variable vane / nozzle turbochargers. The geometry change of the turbine blades or nozzle blades regulates boost pressure instead of a boost pressure wastegate. The sensor detects the position of the variable geometry mechanism to establish the correct operation and position of the variable geometry system.	The voltage, frequency or value of the position sensor signal is either "zero" (no signal) or, the signal exists but it is below the normal operating range (lower than the minimum operating limit). Possible fault in the position sensor or wiring e.g. short / open circuits, circuit resistance, incorrect sensor resistance or, fault with the reference / operating voltage; interference from other circuits can also affect sensor signals. Refer to list of sensor / actuator checks on page 19.
P2589	Turbocharger Boost Control Position Sensor "B" Circuit High	The boost control position sensor is generally used on turbochargers that have: Variable geometry, Variable vane / nozzle turbochargers. The geometry change of the turbine blades or nozzle blades regulates boost pressure instead of a boost pressure wastegate. The sensor detects the position of the variable geometry mechanism to establish the correct operation and position of the variable geometry system.	The voltage, frequency or value of the position sensor signal is either at full value (e.g. battery voltage with no signal or frequency) or, the signal exists but it is above the normal operating range (higher than the maximum operating limit). Possible fault in the position sensor or wiring e.g. short / open circuits, circuit resistance, incorrect sensor resistance or, fault with the reference / operating voltage; interference from other circuits can also affect sensor signals. Refer to list of sensor / actuator checks on page 19.
P2590	Turbocharger Boost Control Position Sensor "B" Circuit Intermittent / Erratic	The boost control position sensor is generally used on turbochargers that have: Variable geometry, Variable vane / nozzle turbochargers. The geometry change of the turbine blades or nozzle blades regulates boost pressure instead of a boost pressure wastegate. The sensor detects the position of the variable geometry mechanism to establish the correct operation and position of the variable geometry system.	The voltage, frequency or value of the position sensor signal is intermittent (likely to be out of normal operating range / tolerance when intermittent fault is detected) or, the signal / voltage is erratic (e.g. signal changes are irregular / unpredictable). Possible fault with wiring or connections (including internal sensor connections); also interference from other circuits can affect the signal. Refer to list of sensor / actuator checks on page 19. Also check for erratic operation of boost control system that could cause position sensor signal to be erratic.
P2600	Coolant Pump Control Circuit / Open	The fault code refers to the control circuit for the pump motor. It may be necessary to identify the control system, which could be a separate control module or the pump could be controlled by the engine control unit. Also identify whether the control circuit contains relays / fuses and whether the control module directly controls the pump or whether the control module controls relays that in turn switch the pump on / off.	The fault code identifies a fault on the pump control circuit. It may be necessary to identify the control circuit and components. The fault is likely to be between the pump and the control module or, between the pump and a relay (as applicable to the type of control system). The voltage, frequency or value of the control signal in the pump circuit is incorrect. Signal could be:- out of normal operating range, constant value, non-existent, corrupt. The fault code indicates a possible "open circuit" but other undefined faults in the pump motor or wiring could activate the same code e.g. power supply, short circuit, high circuit resistance or incorrect actuator resistance. Note: the control unit could be providing the correct signal but it is being affected by a circuit fault. Refer to list of sensor / actuator checks on page 19.

Code	Fault Location	Description (middle)	Description (right)
P2601	Coolant Pump Control Circuit Range / Performance	The fault code refers to the control circuit for the pump motor. It may be necessary to identify the control system, which could be a separate control module or the pump could be controlled by the engine control unit. Also identify whether the control circuit contains relays / fuses and whether the control module directly controls the pump or whether the control module controls relays that in turn switch the pump on / off.	The fault code identifies a fault on the pump control circuit. It may be necessary to identify the control circuit and components. The fault is likely to be between the pump and the control module or, between the pump and a relay (as applicable to the type of control system). The voltage, frequency or value of the control signal in the actuator circuit is within the normal operating range / tolerance, but the signal might be incorrect for the operating conditions. It is also possible that the pump response differs from the expected / desired response (incorrect response to the control signal). Possible fault in actuator or wiring, but also possible that the actuator / motor (or pump assembly) is not operating or moving correctly. Note: the control unit could be providing the correct signal but it is being affected by a circuit fault. Refer to list of sensor / actuator checks on page 19. It is possible that the pump is operating correctly but the cooling system temperature does not change as expected when the pump operates or is switched off; therefore cooling system checks may also be required.
P2602	Coolant Pump Control Circuit Low	The fault code refers to the control circuit for the pump motor. It may be necessary to identify the control system, which could be a separate control module or the pump could be controlled by the engine control unit. Also identify whether the control circuit contains relays / fuses and whether the control module directly controls the pump or whether the control module controls relays that in turn switch the pump on / off.	The fault code identifies a fault on the pump control circuit. It may be necessary to identify the control circuit and components. The fault is likely to be between the pump and the control module or, between the pump and a relay (as applicable to the type of control system). The voltage, frequency or value of the control signal in the actuator circuit is either "zero" (no signal) or, the signal exists but is below the normal operating range. Possible fault with the pump motor or wiring (e.g. power supply, short / open circuit or resistance) that is causing a signal value to be lower than the minimum operating limit. Note: the control unit could be providing the correct signal but it is being affected by a circuit fault. Refer to list of sensor / actuator checks on page 19.
P2603	Coolant Pump Control Circuit High	The fault code refers to the control circuit for the pump motor. It may be necessary to identify the control system, which could be a separate control module or the pump could be controlled by the engine control unit. Also identify whether the control circuit contains relays / fuses and whether the control module directly controls the pump or whether the control module controls relays that in turn switch the pump on / off.	The fault code identifies a fault on the pump control circuit. It may be necessary to identify the control circuit and components. The fault is likely to be between the pump and the control module or, between the pump and a relay (as applicable to the type of control system). The voltage, frequency or value of the control signal in the actuator circuit is either at full value (e.g. battery voltage with no on / off signal) or, the signal exists but it is above the normal operating range. Possible fault with pump motor or wiring (e.g. power supply, short / open circuit or resistance) that is causing a signal value to be higher than the maximum operating limit. Note: the control unit could be providing the correct signal but it is being affected by a circuit fault. Refer to list of sensor / actuator checks on page 19.
P2604	Intake Air Heater "A" Circuit Range / Performance	Note that some heaters are simple electrical heating elements but some may use the heater element to ignite a dedicated fuel supply. The heater could be provided with a power supply via a relay, with the control unit switching the relay on / off; alternatively, a control unit / module could be providing a direct switching of the heater. Refer to vehicle specific information to identify the type of heater used and the wiring / control for the heater.	The heater circuit is likely to refer to the circuit connecting the heater element to a power supply and earth path. The voltage, frequency or value of the signal in the heater circuit is within the normal operating range / tolerance, but the heater response differs from the expected / desired response. Possible fault with heater or wiring e.g. heater power supply, short / open circuit, high circuit resistance or incorrect heater resistance. Note: If the heater is directly controlled via a control unit / module, the control unit could be providing the correct control signal but it is being affected by a circuit fault. Refer to list of sensor / actuator checks on page 19.
P2605	Intake Air Heater "A" Circuit / Open	Note that some heaters are simple electrical heating elements but some may use the heater element to ignite a dedicated fuel supply. The heater could be provided with a power supply via a relay, with the control unit switching the relay on / off; alternatively, a control unit / module could be providing a direct switching of the heater. Refer to vehicle specific information to identify the type of heater used and the wiring / control for the heater.	The heater circuit is likely to refer to the circuit connecting the heater element to a power supply and earth path. The voltage, frequency or value of the control signal in the heater circuit is incorrect (undefined fault but possibly caused by an open circuit). Signal could be:- non-existent, out of normal operating range, constant value, corrupt. Possible fault with heater or wiring e.g. heater power supply, short / open circuit, high circuit resistance or incorrect heater resistance. Note: If the heater is directly controlled via a control unit / module, the control unit could be providing the correct control signal but it is being affected by a circuit fault. Refer to list of sensor / actuator checks on page 19.
P2606	Intake Air Heater "B" Circuit Range / Performance	Note that some heaters are simple electrical heating elements but some may use the heater element to ignite a dedicated fuel supply. The heater could be provided with a power supply via a relay, with the control unit switching the relay on / off; alternatively, a control unit / module could be providing a direct switching of the heater. Refer to vehicle specific information to identify the type of heater used and the wiring / control for the heater.	The heater circuit is likely to refer to the circuit connecting the heater element to a power supply and earth path. The voltage, frequency or value of the signal in the heater circuit is within the normal operating range / tolerance, but the heater response differs from the expected / desired response. Possible fault with heater or wiring e.g. heater power supply, short / open circuit, high circuit resistance or incorrect heater resistance. Note: If the heater is directly controlled via a control unit / module, the control unit could be providing the correct control signal but it is being affected by a circuit fault. Refer to list of sensor / actuator checks on page 19.
P2607	Intake Air Heater "B" Circuit Low	Note that some heaters are simple electrical heating elements but some may use the heater element to ignite a dedicated fuel supply. The heater could be provided with a power supply via a relay, with the control unit switching the relay on / off; alternatively, a control unit / module could be providing a direct switching of the heater. Refer to vehicle specific information to identify the type of heater used and the wiring / control for the heater.	The heater circuit is likely to refer to the circuit connecting the heater element to a power supply and earth path. The voltage, frequency or value of the signal in the heater circuit is either "zero" (no signal) or, the signal exists but is below the normal operating range. Possible fault with heater or wiring (e.g. heater power supply, short / open circuit or resistance) that is causing a signal value to be lower than the minimum operating limit. Note: If the heater is directly controlled via a control unit / module, the control unit could be providing the correct control signal but it is being affected by a circuit fault. Refer to list of sensor / actuator checks on page 19.
P2608	Intake Air Heater "B" Circuit High	Note that some heaters are simple electrical heating elements but some may use the heater element to ignite a dedicated fuel supply. The heater could be provided with a power supply via a relay, with the control unit switching the relay on / off; alternatively, a control unit / module could be providing a direct switching of the heater. Refer to vehicle specific information to identify the type of heater used and the wiring / control for the heater.	The heater circuit is likely to refer to the circuit connecting the heater element to a power supply and earth path. The voltage, frequency or value of the control in the heater circuit is either at full value (e.g. battery voltage with no on / off signal) or, the signal exists but it is above the normal operating range. Possible fault with heater or wiring (e.g. heater power supply, short / open circuit or resistance) that is causing a signal value to be higher than the maximum operating limit. Note: If the heater is directly controlled via a control unit / module, the control unit could be providing the correct control signal but it is being affected by a circuit fault. Refer to list of sensor / actuator checks on page 19.
P2609	Intake Air Heater System Performance	Note that some heaters are simple electrical heating elements but some may use the heater element to ignite a dedicated fuel supply. The heater could be provided with a power supply via a relay, with the control unit switching the relay on / off; alternatively, a control unit / module could be providing a direct switching of the heater. Refer to vehicle specific information to identify the type of heater used and the wiring / control for the heater.	The fault code indicates that the heater system is not operating as expected; the problem could be caused by a wiring related fault or a control system fault. Note that the system can include a relay, control module or main control unit (as applicable to the system) as well as the heater. Check heater and wiring e.g. heater power supply, short / open circuit, high circuit resistance or incorrect heater resistance. Note: If the heater is directly controlled via a control unit / module, the control unit could be providing the correct control signal but it is being affected by a circuit fault.

NOTE 1. Check for other fault codes that could provide additional information. **NOTE 2.** Communication between control units can pass via a CAN-Bus system; refer to Page 51 for CAN-Bus checks.
NOTE 3. If a fault cannot be located, it is also possible that the control unit is at fault. **NOTE 4.** Refer to Page 53 for list of pages that show the ISO standard for component locations e.g. Sensor A - Bank 1.

Fault code	EOBD / ISO Description	Component / System Description	Meaningful Description and Quick Check
P260A	PTO Control Circuit / Open	The PTO (Power Take Off) unit can be controlled by the main engine control unit or a separate control module (which will communicate with the main control unit).	The fault code will probably relate to the circuit providing the control signal from the control unit / module to the power take off unit; however the fault could also be located in the communication circuit between the main control unit and the PTO control module. The voltage, frequency or value of the control signal in the circuit is incorrect. Signal could be:- out of normal operating range, constant value, non-existent, corrupt. The fault code indicates a possible "open circuit" but other undefined faults in the actuator or wiring could activate the same code e.g. power supply, earth connection, short circuit, or high circuit resistance. Note: the control unit could be providing the correct signal but it is being affected by a circuit fault.
P260B	PTO Control Circuit Low	The PTO (Power Take Off) unit can be controlled by the main engine control unit or a separate control module (which will communicate with the main control unit).	The fault code will probably relate to the circuit providing the control signal from the control unit / module to the power take off unit; however the fault could also be located in the communication circuit between the main control unit and the PTO control module. The voltage, frequency or value of the control signal in the circuit is either "zero" (no signal) or, the signal exists but is below the normal operating range. Possible fault with wiring (e.g. power supply, earth connection, short / open circuit or circuit resistance) that is causing a signal value to be lower than the minimum operating limit. Note: the control unit could be providing the correct signal but it is being affected by a circuit fault.
P260C	PTO Control Circuit High	The PTO (Power Take Off) unit can be controlled by the main engine control unit or a separate control module (which will communicate with the main control unit).	The fault code will probably relate to the circuit providing the control signal from the control unit / module to the power take off unit; however the fault could also be located in the communication circuit between the main control unit and the PTO control module. The voltage, frequency or value of the control signal in the actuator circuit is either at full value (e.g. battery voltage with no on / off signal) or, the signal exists but it is above the normal operating range. Possible fault with wiring (e.g. power supply, earth connection, short / open circuit or circuit resistance) that is causing a signal value to be lower than the minimum operating limit. Note: the control unit could be providing the correct signal but it is being affected by a circuit fault.
P260D	PTO Engaged Lamp Control Circuit	When the PTO (Power Take Off) is engaged (operating) an Indicator Lamp can be used to provide a warning that the PTO is operating.	The fault code indicates an that there is an undefined fault with the PTO lamp / lamp circuit. Check wiring and connections between the lamp and control unit, check for earth / power to the lamp and check lamp bulb.
P260E		ISO / SAE reserved	Not used, reserved for future allocation.
P260F	Evaporative System Monitoring Processor Performance	EVAP systems make use of a canister to collect fuel vapours and then release them (under controlled conditions) to the engine intake system. The engine control unit will have an internal processor to monitor the operation of the EVAP system.	The control unit has detected an internal fault with the processor used to control the monitoring of the EVAP system. The fault could be related to one of the components or associated wiring of the EVAP system but it is also possible that the processor in the control unit is faulty. If necessary, refer to vehicle specific information relating to re-programming / re-configuring the control unit and for any information on processor faults.
P2610	ECM / PCM Internal Engine Off Timer Performance	The internal timer can be used to control some systems, after the engine has been switched off e.g. some EVAP system leak detection processes can be activated a few hours after the engine has been switched off.	The fault code indicates a problem with the timer system. The fault is possibly related to an internal fault within the PCM / ECM (Power Train / Engine Control Module). Refer to vehicle specific information to identify what functions are activated by the internal timer system, and check for correct operation. Also check PCM / ECM for correct power supply and earth connections.
P2611	A / C Refrigerant Distribution Valve Control Circuit / Open	Some large passenger vehicles can be fitted with air conditioning systems that use more than one evaporator; these systems can also use a distribution valve to control the flow of refrigerant through the evaporators.	The fault code indicates a fault (possibly an open circuit) with the actuator controlling the distribution valve. The voltage, frequency or value of the control signal in the actuator circuit is incorrect. Signal could be out of normal operating range, constant value, non-existent, corrupt. The fault code indicates a possible "open circuit" but other undefined faults in the actuator or wiring could activate the same code e.g. actuator power supply, short circuit, high circuit resistance or incorrect actuator resistance. Note: the control unit could be providing the correct signal but it is being affected by a circuit fault. Refer to list of sensor / actuator checks on page 19.
P2612	A / C Refrigerant Distribution Valve Control Circuit Low	Some large passenger vehicles can be fitted with air conditioning systems that use more than one evaporator; these systems can also use a distribution valve to control the flow of refrigerant through the evaporators.	The fault code indicates a fault with the actuator controlling the distribution valve. The voltage, frequency or value of the control signal in the actuator circuit is either "zero" (no signal) or, the signal exists but is below the normal operating range. Possible fault with actuator or wiring (e.g. actuator power supply, short / open circuit or resistance) that is causing a signal value to be lower than the minimum operating limit. Note: the control unit could be providing the correct signal but it is being affected by a circuit fault. Note: the control unit could be providing the correct signal but it is being affected by a circuit fault. Refer to list of sensor / actuator checks on page 19.
P2613	A / C Refrigerant Distribution Valve Control Circuit High	Some large passenger vehicles can be fitted with air conditioning systems that use more than one evaporator; these systems can also use a distribution valve to control the flow of refrigerant through the evaporators.	The fault code indicates a fault with the actuator controlling the distribution valve. The voltage, frequency or value of the control signal in the actuator circuit is either at full value (e.g. battery voltage with no on / off signal) or, the signal exists but it is above the normal operating range. Possible fault with actuator or wiring (e.g. actuator power supply, short / open circuit or resistance) that is causing a signal value to be higher than the maximum operating limit. Note: the control unit could be providing the correct signal but it is being affected by a circuit fault. Note: the control unit could be providing the correct signal but it is being affected by a circuit fault. Refer to list of sensor / actuator checks on page 19.
P2614	Camshaft Position Signal Output Circuit / Open	The sensor is used to detect camshaft position (angular position relative to crankshaft position); sensor information can be used to enable accurate control of ignition timing, fuel injection timing and variable valve timing.	Identify the type of sensor e.g. Inductive, Hall effect etc; this will dictate the checks that are applicable. The voltage, frequency or value of the position / speed sensor signal is incorrect or non-existent; undefined fault but possible open circuit. Sensor voltage / signal value could be: out of normal operating range, constant value, non-existent, corrupt. Possible fault in the position / speed sensor or wiring e.g. short / open circuits, circuit resistance, incorrect sensor resistance (if applicable) or, fault with the reference / operating voltage; interference from other circuits can also affect sensor signals. Also check earth connections and cable screening. Refer to list of sensor / actuator checks on page 19. Note that some manufacturers have used this code to refer to a correlation between the Camshaft and Crankshaft sensor signals (refer to fault code P0016 for additional information).
P2615	Camshaft Position Signal Output Circuit Low	The sensor is used to detect camshaft position (angular position relative to crankshaft position); sensor information can be used to enable accurate control of ignition timing, fuel injection timing and variable valve timing.	Identify the type of sensor e.g. Inductive, Hall effect etc; this will dictate the checks that are applicable. The voltage, frequency or value of the position / speed sensor signal is either "zero" (no signal) or, the signal exists but is below the normal operating range (lower than the minimum operating limit). Possible fault in the position / speed sensor or wiring e.g. short / open circuits, circuit resistance, incorrect sensor resistance or, fault with the reference / operating voltage; interference from other circuits can also affect sensor signals. Refer to list of sensor / actuator checks on page 19.

NOTE 1. Check for other fault codes that could provide additional information.   **NOTE 2.** Communication between control units can pass via a CAN-Bus system; refer to Page 51 for CAN-Bus checks.

P2616	Camshaft Position Signal Output Circuit High	The sensor is used to detect camshaft position (angular position relative to crankshaft position); sensor information can be used to enable accurate control of ignition timing, fuel injection timing and variable valve timing.	Identify the type of sensor e.g. Inductive, Hall effect etc; this will dictate the checks that are applicable. The voltage, frequency or value of the position / speed sensor signal is either at full value (e.g. battery voltage with no signal or frequency) or, the signal exists but it is above the normal operating range (higher than the maximum operating limit). Possible fault in the position / speed sensor or wiring e.g. short / open circuits, circuit resistance, incorrect sensor resistance or, fault with the reference / operating voltage; interference from other circuits can also affect sensor signals. Refer to list of sensor / actuator checks on page 19.
P2617	Crankshaft Position Signal Output Circuit / Open	The crankshaft position sensor provides position (and usually speed information) to the control unit. The information is used in calculations for fuelling, ignition timing, variable valve timing and emissions control functions; the information is also used for other vehicle systems such as transmission, stability control systems etc.	Identify the type of sensor e.g. Inductive, Hall effect etc; this will dictate the checks that are applicable. The voltage, frequency or value of the position / speed sensor signal is incorrect or non-existent; undefined fault but possible open circuit. Sensor voltage / signal value could be: out of normal operating range, constant value, non-existent, corrupt. Possible fault in the position / speed sensor or wiring e.g. short / open circuits, circuit resistance, incorrect sensor resistance (if applicable) or, fault with the reference / operating voltage; interference from other circuits can also affect sensor signals. Also check earth connections and cable screening. Refer to list of sensor / actuator checks on page 19. Note that some manufacturers have used this code to refer to a correlation between the Camshaft and Crankshaft sensor signals (refer to fault code P0016 for additional information).
P2618	Crankshaft Position Signal Output Circuit Low	The crankshaft position sensor provides position (and usually speed information) to the control unit. The information is used in calculations for fuelling, ignition timing, variable valve timing and emissions control functions; the information is also used for other vehicle systems such as transmission, stability control systems etc.	Identify the type of sensor e.g. Inductive, Hall effect etc; this will dictate the checks that are applicable. The voltage, frequency or value of the position / speed sensor signal is either "zero" (no signal) or, the signal exists but it is below the normal operating range (lower than the minimum operating limit). Possible fault in the position / speed sensor or wiring e.g. short / open circuits, circuit resistance, incorrect sensor resistance or, fault with the reference / operating voltage; interference from other circuits can also affect sensor signals. Refer to list of sensor / actuator checks on page 19.
P2619	Crankshaft Position Signal Output Circuit High	The crankshaft position sensor provides position (and usually speed information) to the control unit. The information is used in calculations for fuelling, ignition timing, variable valve timing and emissions control functions; the information is also used for other vehicle systems such as transmission, stability control systems etc.	Identify the type of sensor e.g. Inductive, Hall effect etc; this will dictate the checks that are applicable. The voltage, frequency or value of the position / speed sensor signal is either at full value (e.g. battery voltage with no signal or frequency) or, the signal exists but it is above the normal operating range (higher than the maximum operating limit). Possible fault in the position / speed sensor or wiring e.g. short / open circuits, circuit resistance, incorrect sensor resistance or, fault with the reference / operating voltage; interference from other circuits can also affect sensor signals. Refer to list of sensor / actuator checks on page 19.
P2620	Throttle Position Output Circuit / Open	The fault code relates to the information provided by the throttle position sensor but, because the signal from the sensor could pass to the CAN-Bus system (or from one control unit to another), the code could be referring to a problem in the circuit carrying the signal between different control units (or between modules in the CAN-Bus). If necessary refer to vehicle specific information to identify which control units receive the throttle position signal.	Note. The sensor assembly may contain more than one switch or sensor, refer to vehicle specific information to establish which sensor or switch is affected. The voltage, frequency or value of the throttle position sensor signal (or signal being passed between control units) is incorrect (undefined fault but possible open circuit). Sensor voltage / signal could be: out of normal operating range, constant value, non-existent, corrupt. Possible fault in the position sensor or wiring e.g. short / open circuits, circuit resistance, incorrect sensor resistance, or fault with the reference / operating voltage; the switch contacts (if applicable) could also be faulty e.g. not opening / closing correctly or have a high contact resistance. Refer to list of sensor / actuator checks on page 19. Check any applicable circuits that could be carrying the sensor signal between control units or modules (e.g. engine, transmission Can-Bus etc). If applicable, refer to CAN-Bus notes in Chapter 7.
P2621	Throttle Position Output Circuit Low	The fault code relates to the information provided by the throttle position sensor but, because the signal from the sensor could pass to the CAN-Bus system (or from one control unit to another), the code could be referring to a problem in the circuit carrying the signal between different control units (or between modules in the CAN-Bus). If necessary refer to vehicle specific information to identify which control units receive the throttle position signal.	Note. The sensor assembly may contain more than one switch or sensor, refer to vehicle specific information to establish which sensor or switch is affected. The voltage, frequency or value of the throttle position sensor signal (or signal being passed between control units) is either "zero" (no signal) or, the signal exists but it is below the normal operating range (lower than the minimum operating limit). Possible fault in the position sensor or wiring e.g. short / open circuits, circuit resistance, incorrect sensor resistance or, fault with the reference / operating voltage; the switch contacts (if applicable) could also be faulty e.g. not opening / closing correctly or have a high contact resistance. Refer to list of sensor / actuator checks on page 19. Check any applicable circuits that could be carrying the sensor signal between control units or modules (e.g. engine, transmission Can-Bus etc). If applicable, refer to CAN-Bus notes in Chapter 7.
P2622	Throttle Position Output Circuit High	The fault code relates to the information provided by the throttle position sensor but, because the signal from the sensor could pass to the CAN-Bus system (or from one control unit to another), the code could be referring to a problem in the circuit carrying the signal between different control units (or between modules in the CAN-Bus). If necessary refer to vehicle specific information to identify which control units receive the throttle position signal.	Note. The sensor assembly may contain more than one switch or sensor, refer to vehicle specific information to establish which sensor or switch is affected. The voltage, frequency or value of the throttle position sensor signal (or signal being passed between control units) is either at full value (e.g. battery voltage with no signal or frequency) or, the signal exists but it is above the normal operating range (higher than the maximum operating limit). Possible fault in the position sensor or wiring e.g. short / open circuits, circuit resistance, incorrect sensor resistance or, fault with the reference / operating voltage; the switch contacts (if applicable) could also be faulty e.g. not opening / closing correctly. Refer to list of sensor / actuator checks on page 19. Check any applicable circuits that could be carrying the sensor signal between control units or modules (e.g. engine, transmission Can-Bus etc). If applicable, refer to CAN-Bus notes in Chapter 7.
P2623	Injector Control Pressure Regulator Circuit / Open	On some types of diesel engine (typically US engines) engine oil is passed to a high pressure pump, which then passes oil at high pressure (in the region of 3000 - 4000 psi) to "unit type fuel injectors" (located in the cylinder head). A solenoid at each injector, controls the flow of the high pressure oil through the injector; when the high pressure oil is allowed to flow into the injector, it acts on a plunger, which pressurises the fuel thus causing the injector to open (and fuel will be injected into the engine). The high pressure engine oil system is referred to as the "control pressure circuit". The control pressure can be altered, depending on the operating conditions using a pressure regulator.	The voltage, frequency or value of the control signal in the regulator circuit is incorrect. Signal could be:- out of normal operating range, constant value, non-existent, corrupt. The fault code indicates a possible "open circuit" but other undefined faults in the regulator or wiring could activate the same code e.g. regulator power supply, short circuit, high circuit resistance or incorrect regulator resistance. Note: the control unit could be providing the correct signal but it is being affected by the circuit fault. Refer to list of sensor / actuator checks on page 19.

NOTE 1. Check for other fault codes that could provide additional information. **NOTE 2.** Communication between control units can pass via a CAN-Bus system; refer to Page 51 for CAN-Bus checks.

NOTE 3. If a fault cannot be located, it is also possible that the control unit is at fault. **NOTE 4.** Refer to Page 53 for list of pages that show the ISO standard for component locations e.g. Sensor A - Bank 1.

Fault code	EOBD / ISO Description	Component / System Description	Meaningful Description and Quick Check
P2624	Injector Control Pressure Regulator Circuit Low	On some types of diesel engine (typically US engines) engine oil is passed to a high pressure pump, which then passes oil at high pressure (in the region of 3000 - 4000 psi) to "unit type fuel injectors" (located in the cylinder head). A solenoid at each injector, controls the flow of the high pressure oil through the injector; when the high pressure oil is allowed to flow into the injector, it acts on a plunger, which pressurises the fuel thus causing the injector to open (and fuel will be injected into the engine). The high pressure engine oil system is referred to as the "control pressure circuit". The control pressure can be altered, depending on the operating conditions using a pressure regulator.	The voltage, frequency or value of the control signal in the regulator circuit is either "zero" (no signal) or, the signal exists but is below the normal operating range. Possible fault with regulator or wiring (e.g. regulator power supply, short / open circuit or resistance) that is causing a signal value to be lower than the minimum operating limit. Note: the control unit could be providing the correct signal but it is being affected by the circuit fault. Refer to list of sensor / actuator checks on page 19.
P2625	Injector Control Pressure Regulator Circuit High	On some types of diesel engine (typically US engines) engine oil is passed to a high pressure pump, which then passes oil at high pressure (in the region of 3000 - 4000 psi) to "unit type fuel injectors" (located in the cylinder head). A solenoid at each injector, controls the flow of the high pressure oil through the injector; when the high pressure oil is allowed to flow into the injector, it acts on a plunger, which pressurises the fuel thus causing the injector to open (and fuel will be injected into the engine). The high pressure engine oil system is referred to as the "control pressure circuit". The control pressure can be altered, depending on the operating conditions using a pressure regulator.	The voltage, frequency or value of the control signal in the regulator circuit is either at full value (e.g. battery voltage with no on / off signal) or, the signal exists but it is above the normal operating range. Possible fault with regulator or wiring (e.g. regulator power supply, short / open circuit or resistance) that is causing a signal value to be higher than the maximum operating limit. Note: the control unit could be providing the correct signal but it is being affected by the circuit fault. Refer to list of sensor / actuator checks on page 19.
P2626	O2 Sensor Pumping Current Trim Circuit / Open Bank 1 Sensor 1	Note: Many O2 / Lambda sensor related fault codes are activated due to faults in other systems (engine, fuelling, misfires etc). O2 (Oxygen / Lambda) sensors detect the exhaust gas oxygen content (an indicator of air / fuel ratio). Oxygen content is indicated by a "Lambda" value. The control unit alters the fuelling to achieve the optimum Lambda value to enable good efficiency of catalytic converters (reduce emissions of harmful gases). For the operation of "broadband" sensors, the measurement process involves applying positive or negative current (pumping current) to one part of the sensing element; the level of positive or negative current being applied is an indicator of the oxygen content.	For a detailed list of checks and identification of the sensor location, Refer to list of sensor / actuator checks on page 19. The fault code indicates that the there is a possibly open circuit / fault in the circuit used to apply current to the measuring element of the Oxygen / Lambda sensor. Due to the fact that the fault code refers to a circuit / open circuit fault, the problem is most likely to be related to a sensor or wiring fault; check wiring between sensor and control unit for short / open circuits and high resistance. Note that the sensor could also be contaminated or faulty. Check sensor signal and signal response under different operating conditions (load, constant rpm etc) and Refer to list of sensor / actuator checks on page 19. If the sensor does however appear to be operating correctly but the signal value is incorrect, this could indicate an engine / fuelling fault, or other fault causing incorrect oxygen content in the exhaust gas.
P2627	O2 Sensor Pumping Current Trim Circuit Low Bank 1 Sensor 1	Note: Many O2 / Lambda sensor related fault codes are activated due to faults in other systems (engine, fuelling, misfires etc). O2 (Oxygen / Lambda) sensors detect the exhaust gas oxygen content (an indicator of air / fuel ratio). Oxygen content is indicated by a "Lambda" value. The control unit alters the fuelling to achieve the optimum Lambda value to enable good efficiency of catalytic converters (reduce emissions of harmful gases). For the operation of "broadband" sensors, the measurement process involves applying positive or negative current (pumping current) to one part of the sensing element; the level of positive or negative current being applied is an indicator of the oxygen content.	For a detailed list of checks and identification of the sensor location, Refer to list of sensor / actuator checks on page 19. The fault code indicates that the current (being applied to the measuring element of the Oxygen / Lambda sensor) is low; the current is likely to be below the normal operating range. The problem is most likely to be related to a sensor or wiring fault; check wiring between sensor and control unit for short / open circuits and high resistance. Note that the sensor could also be contaminated or faulty. Check sensor signal and signal response under different operating conditions (load, constant rpm etc) and Refer to list of sensor / actuator checks on page 19. If the sensor does however appear to be operating correctly but the current value remains incorrect, this could indicate an engine / fuelling fault, or other fault causing incorrect oxygen content in the exhaust gas.
P2628	O2 Sensor Pumping Current Trim Circuit High Bank 1 Sensor 1	Note: Many O2 / Lambda sensor related fault codes are activated due to faults in other systems (engine, fuelling, misfires etc). O2 (Oxygen / Lambda) sensors detect the exhaust gas oxygen content (an indicator of air / fuel ratio). Oxygen content is indicated by a "Lambda" value. The control unit alters the fuelling to achieve the optimum Lambda value to enable good efficiency of catalytic converters (reduce emissions of harmful gases). For the operation of "broadband" sensors, the measurement process involves applying positive or negative current (pumping current) to one part of the sensing element; the level of positive or negative current being applied is an indicator of the oxygen content.	For a detailed list of checks and identification of the sensor location, Refer to list of sensor / actuator checks on page 19. The fault code indicates that the current (being applied to the measuring element of the Oxygen / Lambda sensor) is high; the current is likely to be above the normal operating range. The problem is most likely to be related to a sensor or wiring fault; check wiring between sensor and control unit for short / open circuits and high resistance. Note that the sensor could also be contaminated or faulty. Check sensor signal and signal response under different operating conditions (load, constant rpm etc) and Refer to list of sensor / actuator checks on page 19. If the sensor does however appear to be operating correctly but the current value remains incorrect, this could indicate an engine / fuelling fault, or other fault causing incorrect oxygen content in the exhaust gas.
P2629	O2 Sensor Pumping Current Trim Circuit / Open Bank 2 Sensor 1	Note: Many O2 / Lambda sensor related fault codes are activated due to faults in other systems (engine, fuelling, misfires etc). O2 (Oxygen / Lambda) sensors detect the exhaust gas oxygen content (an indicator of air / fuel ratio). Oxygen content is indicated by a "Lambda" value. The control unit alters the fuelling to achieve the optimum Lambda value to enable good efficiency of catalytic converters (reduce emissions of harmful gases). For the operation of "broadband" sensors, the measurement process involves applying positive or negative current (pumping current) to one part of the sensing element; the level of positive or negative current being applied is an indicator of the oxygen content.	For a detailed list of checks and identification of the sensor location, Refer to list of sensor / actuator checks on page 19. The fault code indicates that the there is a possibly open circuit / fault in the circuit used to apply current to the measuring element of the Oxygen / Lambda sensor. Due to the fact that the fault code refers to a circuit / open circuit fault, the problem is most likely to be related to a sensor or wiring fault; check wiring between sensor and control unit for short / open circuits and high resistance. Note that the sensor could also be contaminated or faulty. Check sensor signal and signal response under different operating conditions (load, constant rpm etc) and Refer to list of sensor / actuator checks on page 19. If the sensor does however appear to be operating correctly but the signal value is incorrect, this could indicate an engine / fuelling fault, or other fault causing incorrect oxygen content in the exhaust gas.

NOTE 1. Check for other fault codes that could provide additional information. **NOTE 2.** Communication between control units can pass via a CAN-Bus system; refer to Page 51 for CAN-Bus checks. 294

NOTE 3. If a fault cannot be located, it is also possible that the control unit is at fault. NOTE 4. Refer to Page 52 for list of pages that show the ISO standard for component locations e.g. Sensor A - Bank 1.

code	Description		
P2630	O2 Sensor Pumping Current Trim Circuit Low Bank 2 Sensor 1	Note: Many O2 / Lambda sensor related fault codes are activated due to faults in other systems (engine, fuelling, misfires etc). O2 (Oxygen / Lambda) sensors detect the exhaust gas oxygen content (an indicator of air / fuel ratio). Oxygen content is indicated by a "Lambda" value. The control unit alters the fuelling to achieve the optimum Lambda value to enable good efficiency of catalytic converters (reduce emissions of harmful gases). For the operation of "broadband" sensors, the measurement process involves applying positive or negative current (pumping current) to one part of the sensing element; the level of positive or negative current being applied is an indicator of the oxygen content.	For a detailed list of checks and identification of the sensor location, Refer to list of sensor / actuator checks on page 19. The fault code indicates that the current (being applied to the measuring element of the Oxygen / Lambda sensor) is low; the current is likely to be below the normal operating range. The problem is most likely to be related to a sensor or wiring fault; check wiring between sensor and control unit for short / open circuits and high resistance. Note that the sensor could also be contaminated or faulty. Check sensor signal and signal response under different operating conditions (load, constant rpm etc) and Refer to list of sensor / actuator checks on page 19. If the sensor does however appear to be operating correctly but the current value remains incorrect, this could indicate an engine / fuelling fault, or other fault causing incorrect oxygen content in the exhaust gas.
P2631	O2 Sensor Pumping Current Trim Circuit High Bank 2 Sensor 1	Note: Many O2 / Lambda sensor related fault codes are activated due to faults in other systems (engine, fuelling, misfires etc). O2 (Oxygen / Lambda) sensors detect the exhaust gas oxygen content (an indicator of air / fuel ratio). Oxygen content is indicated by a "Lambda" value. The control unit alters the fuelling to achieve the optimum Lambda value to enable good efficiency of catalytic converters (reduce emissions of harmful gases). For the operation of "broadband" sensors, the measurement process involves applying positive or negative current (pumping current) to one part of the sensing element; the level of positive or negative current being applied is an indicator of the oxygen content.	For a detailed list of checks and identification of the sensor location, Refer to list of sensor / actuator checks on page 19. The fault code indicates that the current (being applied to the measuring element of the Oxygen / Lambda sensor) is high; the current is likely to be above the normal operating range. The problem is most likely to be related to a sensor or wiring fault; check wiring between sensor and control unit for short / open circuits and high resistance. Note that the sensor could also be contaminated or faulty. Check sensor signal and signal response under different operating conditions (load, constant rpm etc) and Refer to list of sensor / actuator checks on page 19. If the sensor does however appear to be operating correctly but the current value remains incorrect, this could indicate an engine / fuelling fault, or other fault causing incorrect oxygen content in the exhaust gas.
P2632	Fuel Pump "B" Control Circuit / Open	Electric fuel pumps can be provided with a power supply from the pump relay. Some systems can use two power supply circuits from the relay; one circuit provides battery voltage (pump runs at high speed for high fuel demand / starting). The second circuit contains a resistor to reduce the voltage / current (pump runs at lower speeds for low fuel demand); the selection of the two circuits is made by the control unit, which controls the relay operation. Note that the speed / power for some electric motors can be regulated using a frequency / pulse width control signal (provided by the control unit on the earth or power circuit).	Identify whether the pump motor speed is controlled using a relay or other type of control. Wiring checks should include circuits to the relay (including resistor if fitted) and connections to the control unit. The voltage, frequency or value of the control signal in the pump motor circuit is incorrect. Signal could be:- out of normal operating range, constant value, non-existent, corrupt. The fault code indicates a possible "open circuit" but other undefined faults in the motor or wiring could activate the same code e.g. motor power supply, short circuit, high circuit resistance or incorrect motor resistance. Note: the control unit or relay could be providing the correct signal / voltage but it is being affected by a circuit fault. Identify the type of motor control (if applicable) and the wiring used on the system. Refer to list of sensor / actuator checks on page 19.
P2633	Fuel Pump "B" Control Circuit Low	Electric fuel pumps can be provided with a power supply from the pump relay. Some systems can use two power supply circuits from the relay; one circuit provides battery voltage (pump runs at high speed for high fuel demand / starting). The second circuit contains a resistor to reduce the voltage / current (pump runs at lower speeds for low fuel demand); the selection of the two circuits is made by the control unit, which controls the relay operation. Note that the speed / power for some electric motors can be regulated using a frequency / pulse width control signal (provided by the control unit on the earth or power circuit).	Identify whether the pump motor speed is controlled using a relay or other type of control. Wiring checks should include circuits to the relay (including resistor if fitted) and connections to the control unit. The voltage, frequency or value of the control signal in the motor circuit is either "zero" (no voltage or signal) or, the voltage / signal exists but is below the normal operating range. Possible fault with motor or wiring (e.g. motor power supply, short / open circuit or resistance) that is causing a voltage / signal value to be lower than the minimum operating limit. Note: the control unit / relay could be providing the correct voltage signal but it is being affected by a circuit fault. Refer to list of sensor / actuator checks on page 19.
P2634	Fuel Pump "B" Control Circuit High	Electric fuel pumps can be provided with a power supply from the pump relay. Some systems can use two power supply circuits from the relay; one circuit provides battery voltage (pump runs at high speed for high fuel demand / starting). The second circuit contains a resistor to reduce the voltage / current (pump runs at lower speeds for low fuel demand); the selection of the two circuits is made by the control unit, which controls the relay operation. Note that the speed / power for some electric motors can be regulated using a frequency / pulse width control signal (provided by the control unit on the earth or power circuit).	Identify whether the pump motor speed is controlled using a relay or other type of control. Wiring checks should include circuits to the relay (including resistor if fitted) and connections to the control unit. The voltage, frequency or value of the control signal in the motor circuit is either at full value (e.g. battery voltage with no on / off signal) or, the voltage / signal exists but it is above the normal operating range. Possible fault with motor or wiring (e.g. motor power supply, short / open circuit or resistance) that is causing a voltage / signal value to be higher than the maximum operating limit. Note: the control unit / relay could be providing the correct voltage / signal but it is being affected by a circuit fault. Refer to list of sensor / actuator checks on page 19.
P2635	Fuel Pump "A" Low Flow / Performance	Refer to vehicle specific information to identify Pump A and B, and also to identify the wiring and power supply. The pump could be a primary (low pressure) pump. Note that some pumps can be controlled for speed and delivery volume, often by using normal and low voltage supplies (can be switched via a relay with more than one output circuit, one of which will pass through a resistance). Some systems can make use of a Pump Controller.	The fault code indicates a performance related problem with the fuel pump which can be related to a "low Flow" problem. The fault could be detected due to a low pressure signal from a pressure sensor, although other types of sensor could be used to monitor fuel flow; refer to vehicle specific information to identify the system components and sensors, and carry out applicable checks. Check for blockages in the system e.g. blocked filter, restrictions in pipes etc. Check power supply and earth connections to the pump.
P2636	Fuel Pump "B" Low Flow / Performance	Refer to vehicle specific information to identify Pump A and B, and also to identify the wiring and power supply. The pump could be a primary (low pressure) pump. Note that some pumps can be controlled for speed and delivery volume, often by using normal and low voltage supplies (can be switched via a relay with more than one output circuit, one of which will pass through a resistance). Some systems can make use of a Pump Controller.	The fault code indicates a performance related problem with the fuel pump which can be related to a "low Flow" problem. The fault could be detected due to a low pressure signal from a pressure sensor, although other types of sensor could be used to monitor fuel flow; refer to vehicle specific information to identify the system components and sensors, and carry out applicable checks. Check for blockages in the system e.g. blocked filter, restrictions in pipes etc. Check power supply and earth connections to the pump.

NOTE 1. Check for other fault codes that could provide additional information.   **NOTE 2.** Communication between control units can pass via a CAN-Bus system; refer to Page 51 for CAN-Bus checks.

NOTE 3. If a fault cannot be located, it is also possible that the control unit is at fault.   **NOTE 4.** Refer to Page 53 for list of pages that show the ISO standard for component locations e.g. Sensor A - Bank 1.

Fault code	EOBD / ISO Description	Component / System Description	Meaningful Description and Quick Check
P2637	Torque Management Feedback Signal "A"	A torque management signal refers to the signal sent from the engine control unit to other vehicle systems e.g. transmission control unit (note that signal can be sent via the CAN-Bus network). The signal is an indication of the torque being requested / delivered by the engine; the torque level is based on driver request (throttle position) as well as temperature and other information. The torque signal is used by the transmission to determine gear selection, shift speed etc. but can be used by other systems for control functions e.g. traction control etc. The fault code indicates that the torque management signal is invalid (not plausible or is incorrect).	The fault code indicates that the torque management signal is invalid. The fault could be related to a wiring / connection fault between the engine and transmission control units, but note that the signal could pass through the CAN-Bus network; check for a Can-Bus related fault code. It is also possible that the torque signal information is invalid because of a "Calculation" fault within the engine control unit (could be caused by incorrect sensor information passing to the engine control unit or a control unit fault). Check for other fault codes relating to engine control / transmission control systems and check for fault codes relating to the CAN-Bus system; check all applicable wiring and connections. Also refer to Fault codes P2638 to P2640.
P2638	Torque Management Feedback Signal "A" Range / Performance	A torque management signal refers to the signal sent from the engine control unit to other vehicle systems e.g. transmission control unit (note that signal can be sent via the CAN-Bus network). The signal is an indication of the torque being requested / delivered by the engine; the torque level is based on driver request (throttle position) as well as temperature and other information. The torque signal is used by the transmission to determine gear selection, shift speed etc. but can be used by other systems for control functions e.g. traction control etc. The fault code indicates that the torque management signal is invalid (not plausible or is incorrect).	The fault code indicates that voltage, frequency or value of the torque management signal is within the normal operating range / tolerance but the signal is not plausible or is incorrect due to an undefined fault. The torque signal might not match the operating conditions indicated by other sensors or, the signal is not changing / responding as expected. Possible fault in the wiring e.g. short / open circuits, circuit resistance; interference from other circuits can also affect sensor signals. The fault could be related to a wiring / connection fault between the engine and transmission control units, but note that the signal could pass through the CAN-Bus network; check for a Can-Bus related fault code. It is also possible that the torque signal information is invalid because of a "Calculation" fault within the engine control unit (could be caused by incorrect sensor information passing to the engine control unit or a control unit fault). Check for other fault codes relating to engine / transmission and CAN-Bus systems.
P2639	Torque Management Feedback Signal "A" Low	A torque management signal refers to the signal sent from the engine control unit to other vehicle systems e.g. transmission control unit (note that signal can be sent via the CAN-Bus network). The signal is an indication of the torque being requested / delivered by the engine; the torque level is based on driver request (throttle position) as well as temperature and other information. The torque signal is used by the transmission to determine gear selection, shift speed etc. but can be used by other systems for control functions e.g. traction control etc. The fault code indicates that the torque management signal is invalid (not plausible or is incorrect).	The fault code indicates that voltage, frequency or value of the torque management signal is either "zero" (no signal) or, the signal exists but it is below the normal operating range (lower than the minimum operating limit). Possible fault in the wiring e.g. short / open circuits, circuit resistance; interference from other circuits can also affect sensor signals. The fault could be related to a wiring / connection fault between the engine and transmission control units, but note that the signal could pass through the CAN-Bus network; check for a Can-Bus related fault code. Check for other fault codes relating to engine control / transmission control systems and check for fault codes relating to the CAN-Bus system; check all applicable wiring and connections.
P2640	Torque Management Feedback Signal "A" High	A torque management signal refers to the signal sent from the engine control unit to other vehicle systems e.g. transmission control unit (note that signal can be sent via the CAN-Bus network). The signal is an indication of the torque being requested / delivered by the engine; the torque level is based on driver request (throttle position) as well as temperature and other information. The torque signal is used by the transmission to determine gear selection, shift speed etc. but can be used by other systems for control functions e.g. traction control etc. The fault code indicates that the torque management signal is invalid (not plausible or is incorrect).	The fault code indicates that voltage, frequency or value of the torque management signal is either at full value (e.g. battery voltage with no signal or frequency) or, the signal exists but it is above the normal operating range (higher than the maximum operating limit). Possible fault in the wiring e.g. short / open circuits, circuit resistance; interference from other circuits can also affect sensor signals. The fault could be related to a wiring / connection fault between the engine and transmission control units, but note that the signal could pass through the CAN-Bus network; check for a Can-Bus related fault code. Check for other fault codes relating to engine control / transmission control systems and check for fault codes relating to the CAN-Bus system; check all applicable wiring and connections.
P2641	Torque Management Feedback Signal "B"	A torque management signal refers to the signal sent from the engine control unit to other vehicle systems e.g. transmission control unit (note that signal can be sent via the CAN-Bus network). The signal is an indication of the torque being requested / delivered by the engine; the torque level is based on driver request (throttle position) as well as temperature and other information. The torque signal is used by the transmission to determine gear selection, shift speed etc. but can be used by other systems for control functions e.g. traction control etc. The fault code indicates that the torque management signal is invalid (not plausible or is incorrect).	The fault code indicates that the torque management signal is invalid. The fault could be related to a wiring / connection fault between the engine and transmission control units, but note that the signal could pass through the CAN-Bus network; check for a Can-Bus related fault code. It is also possible that the torque signal information is invalid because of a "Calculation" fault within the engine control unit (could be caused by incorrect sensor information passing to the engine control unit or a control unit fault). Check for other fault codes relating to engine control / transmission control systems and check for fault codes relating to the CAN-Bus system; check all applicable wiring and connections. Also refer to Fault codes P2638 to P2640.
P2642	Torque Management Feedback Signal "B" Range / Performance	A torque management signal refers to the signal sent from the engine control unit to other vehicle systems e.g. transmission control unit (note that signal can be sent via the CAN-Bus network). The signal is an indication of the torque being requested / delivered by the engine; the torque level is based on driver request (throttle position) as well as temperature and other information. The torque signal is used by the transmission to determine gear selection, shift speed etc. but can be used by other systems for control functions e.g. traction control etc. The fault code indicates that the torque management signal is invalid (not plausible or is incorrect).	The fault code indicates that voltage, frequency or value of the torque management signal is within the normal operating range / tolerance but the signal is not plausible or is incorrect due to an undefined fault. The torque signal might not match the operating conditions indicated by other sensors or, the signal is not changing / responding as expected. Possible fault in the wiring e.g. short / open circuits, circuit resistance; interference from other circuits can also affect sensor signals. The fault could be related to a wiring / connection fault between the engine and transmission control units, but note that the signal could pass through the CAN-Bus network; check for a Can-Bus related fault code. It is also possible that the torque signal information is invalid because of a "Calculation" fault within the engine control unit (could be caused by incorrect sensor information passing to the engine control unit or a control unit fault). Check for other fault codes relating to engine / transmission and CAN-Bus systems.

Code	Fault Location	Description	Diagnosis
P2643	Torque Management Feedback Signal "B" Low	A torque management signal refers to the signal sent from the engine control unit to other vehicle systems e.g. transmission control unit (note that signal can be sent via the CAN-Bus network). The signal is an indication of the torque being requested / delivered by the engine; the torque level is based on driver request (throttle position) as well as temperature and other information. The torque signal is used by the transmission to determine gear selection, shift speed etc. but can be used by other systems for control functions e.g. traction control etc. The fault code indicates that the torque management signal is invalid (not plausible or is incorrect).	The fault code indicates that voltage, frequency or value of the torque management signal is either "zero" (no signal) or, the signal exists but it is below the normal operating range (lower than the minimum operating limit). Possible fault in the wiring e.g. short / open circuits, circuit resistance; interference from other circuits can also affect sensor signals. The fault could be related to a wiring / connection fault between the engine and transmission control units, but note that the signal could pass through the CAN-Bus network; check for a Can-Bus related fault code. Check for other fault codes relating to engine control / transmission control systems and check for fault codes relating to the CAN-Bus system; check all applicable wiring and connections.
P2644	Torque Management Feedback Signal "B" High	A torque management signal refers to the signal sent from the engine control unit to other vehicle systems e.g. transmission control unit (note that signal can be sent via the CAN-Bus network). The signal is an indication of the torque being requested / delivered by the engine; the torque level is based on driver request (throttle position) as well as temperature and other information. The torque signal is used by the transmission to determine gear selection, shift speed etc. but can be used by other systems for control functions e.g. traction control etc. The fault code indicates that the torque management signal is invalid (not plausible or is incorrect).	The fault code indicates that voltage, frequency or value of the torque management signal is either at full value (e.g. battery voltage with no signal or frequency) or, the signal exists but it is above the normal operating range (higher than the maximum operating limit). Possible fault in the wiring e.g. short / open circuits, circuit resistance; interference from other circuits can also affect sensor signals. The fault could be related to a wiring / connection fault between the engine and transmission control units, but note that the signal could pass through the CAN-Bus network; check for a Can-Bus related fault code. Check for other fault codes relating to engine control / transmission control systems and check for fault codes relating to the CAN-Bus system; check all applicable wiring and connections.
P2645	"A" Rocker Arm Actuator Control Circuit / Open Bank 1	Rocker arm actuators are used on some variable valve timing / lift systems. Systems can use oil pressure to engage / disengage the rocker arm mechanisms, which when activated enables changes in valve timing / lift. A solenoid operated valve can be used to control the application of the oil (under pressure) to the mechanism. Oil pressure sensors and position sensors can be used to monitor / indicate rocker arm actuation.	The voltage, frequency or value of the control signal in the actuator circuit is incorrect. Signal could be:- out of normal operating range, constant value, non-existent, corrupt. The fault code indicates a possible "open circuit" but other undefined faults in the actuator or wiring could activate the same code e.g. actuator power supply, short circuit, high circuit resistance or incorrect actuator resistance. Note: the control unit could be providing the correct signal but it is being affected by the circuit fault. Refer to list of sensor / actuator checks on page 19.
P2646	"A" Rocker Arm Actuator System Performance or Stuck Off Bank 1	Rocker arm actuators are used on some variable valve timing / lift systems. Systems can use oil pressure to engage / disengage the rocker arm mechanisms, which when activated enables changes in valve timing / lift. A solenoid operated valve can be used to control the application of the oil (under pressure) to the mechanism. Oil pressure sensors and position sensors can be used to monitor / indicate rocker arm actuation.	The control unit is providing a control signal to the actuator (signal is within the normal operating range / tolerance) but the control unit has detected that the actuator is stuck in a fixed (off) position and is not responding to the control signal. The fault could be related to an actuator / wiring fault. Refer to list of sensor / actuator checks on page 19. Note that the control unit could be detecting "no response" from the system or mechanism being controlled by the actuator; check system (as well as the actuator) for correct operation, movement and response. A position sensor (if fitted) would be indicating the "stuck position"; also check position sensor operation. Also check for correct oil pressure (if applicable).
P2647	"A" Rocker Arm Actuator System Stuck On Bank 1	Rocker arm actuators are used on some variable valve timing / lift systems. Systems can use oil pressure to engage / disengage the rocker arm mechanisms, which when activated enables changes in valve timing / lift. A solenoid operated valve can be used to control the application of the oil (under pressure) to the mechanism. Oil pressure sensors and position sensors can be used to monitor / indicate rocker arm actuation.	The control unit is providing a control signal to the actuator (signal is within the normal operating range / tolerance) but the control unit has detected that the actuator is stuck in a fixed (on) position and is not responding to the control signal. The fault could be related to an actuator / wiring fault. Refer to list of sensor / actuator checks on page 19. Note that the control unit could be detecting "no response" from the system or mechanism being controlled by the actuator; check system (as well as the actuator) for correct operation, movement and response. A position sensor (if fitted) would be indicating the "stuck position"; also check position sensor operation. Also check for correct oil pressure (if applicable).
P2648	"A" Rocker Arm Actuator Control Circuit Low Bank 1	Rocker arm actuators are used on some variable valve timing / lift systems. Systems can use oil pressure to engage / disengage the rocker arm mechanisms, which when activated enables changes in valve timing / lift. A solenoid operated valve can be used to control the application of the oil (under pressure) to the mechanism. Oil pressure sensors and position sensors can be used to monitor / indicate rocker arm actuation.	The voltage, frequency or value of the control signal in the actuator circuit is either "zero" (no signal) or, the signal exists but is below the normal operating range. Possible fault with actuator or wiring (e.g. actuator power supply, short / open circuit or resistance) that is causing a signal value to be lower than the minimum operating limit. Note: the control unit could be providing the correct signal but it is being affected by the circuit fault. Refer to list of sensor / actuator checks on page 19.
P2649	"A" Rocker Arm Actuator Control Circuit High Bank 1	Rocker arm actuators are used on some variable valve timing / lift systems. Systems can use oil pressure to engage / disengage the rocker arm mechanisms, which when activated enables changes in valve timing / lift. A solenoid operated valve can be used to control the application of the oil (under pressure) to the mechanism. Oil pressure sensors and position sensors can be used to monitor / indicate rocker arm actuation.	The voltage, frequency or value of the control signal in the actuator circuit is either at full value (e.g. battery voltage with no on / off signal) or, the signal exists but it is above the normal operating range. Possible fault with actuator or wiring (e.g. actuator power supply, short / open circuit or resistance) that is causing a signal value to be higher than the maximum operating limit. Note: the control unit could be providing the correct signal but it is being affected by the circuit fault. Refer to list of sensor / actuator checks on page 19.
P264A	"A" Rocker Arm Actuator Position Sensor Circuit Bank 1	Rocker arm actuators are used on some variable valve timing / lift systems. Systems can use oil pressure to engage / disengage the rocker arm mechanisms, which when activated enables changes in valve timing / lift. A solenoid operated valve can be used to control the application of the oil (under pressure) to the mechanism. Oil pressure sensors and position sensors can be used to monitor / indicate rocker arm actuation. It is possible that some manufacturers could apply the code to "Cylinder Deactivation" systems (that use the rocker arm as part of the deactivation mechanism).	Identify the type of position sensor fitted (e.g. Hall effect), which can dictate the specific checks that should be carried out. Note that some systems can use an oil pressure sensor to indicate that the Rocker Arm Actuator mechanism has operated / moved. The voltage or value of the position sensor signal is incorrect (undefined fault). Sensor voltage / signal value could be: out of normal operating range, constant value, non-existent, corrupt. Possible fault in the position sensor or wiring e.g. short / open circuits, circuit resistance, incorrect sensor resistance (if applicable) or, fault with the reference / operating voltage; interference from other circuits can also affect sensor signals. Also check earth connections and cable screening. Refer to list of sensor / actuator checks on page 19.

NOTE 1. Check for other fault codes that could provide additional information.
NOTE 2. Communication between control units can pass via a CAN-Bus system; refer to Page 51 for CAN-Bus checks.
NOTE 3. If a fault cannot be located, it is also possible that the control unit is at fault.
NOTE 4. Refer to Page 53 for list of pages that show the ISO standard for component locations e.g. Sensor A - Bank 1.

Fault code	EOBD / ISO Description	Component / System Description	Meaningful Description and Quick Check
P264B	"A" Rocker Arm Actuator Position Sensor Circuit Range / Performance Bank 1	Rocker arm actuators are used on some variable valve timing / lift systems. Systems can use oil pressure to engage / disengage the rocker arm mechanisms, which when activated enables changes in valve timing / lift. A solenoid operated valve can be used to control the application of the oil (under pressure) to the mechanism. Oil pressure sensors and position sensors can be used to monitor / indicate rocker arm actuation. It is possible that some manufacturers could apply the code to "Cylinder Deactivation" systems (that use the rocker arm as part of the deactivation mechanism).	Identify the type of position sensor fitted (e.g. Hall effect), which can dictate the specific checks that should be carried out. Note that some systems can use an oil pressure sensor to indicate that the Rocker Arm Actuator mechanism has operated / moved. The voltage or value of the position sensor signal is within the normal operating range / tolerance but the signal is not plausible or is incorrect due to an undefined fault. The sensor signal might not match the operating conditions or, the sensor signal is not changing / responding as expected to the changes in conditions. Possible fault in the position sensor or wiring e.g. short / open circuits, circuit resistance, incorrect sensor resistance or, fault with the reference / operating voltage; interference from other circuits can also affect sensor signals. Refer to list of sensor / actuator checks on page 19. It is also possible that the sensor is operating correctly but the position / pressure or activation detected by the sensor is unacceptable for the operating conditions. Check Rocker Arm Actuator operation / movement to ensure correct operation.
P264C	"A" Rocker Arm Actuator Position Sensor Circuit Low Bank 1	Rocker arm actuators are used on some variable valve timing / lift systems. Systems can use oil pressure to engage / disengage the rocker arm mechanisms, which when activated enables changes in valve timing / lift. A solenoid operated valve can be used to control the application of the oil (under pressure) to the mechanism. Oil pressure sensors and position sensors can be used to monitor / indicate rocker arm actuation. It is possible that some manufacturers could apply the code to "Cylinder Deactivation" systems (that use the rocker arm as part of the deactivation mechanism).	Identify the type of position sensor fitted (e.g. Hall effect), which can dictate the specific checks that should be carried out. Note that some systems can use an oil pressure sensor to indicate that the Rocker Arm Actuator mechanism has operated / moved. The voltage or value of the position sensor signal is either "zero" (no signal) or, the signal exists but it is below the normal operating range (lower than the minimum operating limit). Possible fault in the position sensor or wiring e.g. short / open circuits, circuit resistance, incorrect sensor resistance or, fault with the reference / operating voltage; interference from other circuits can also affect sensor signals. Refer to list of sensor / actuator checks on page 19.
P264D	"A" Rocker Arm Actuator Position Sensor Circuit High Bank 1	Rocker arm actuators are used on some variable valve timing / lift systems. Systems can use oil pressure to engage / disengage the rocker arm mechanisms, which when activated enables changes in valve timing / lift. A solenoid operated valve can be used to control the application of the oil (under pressure) to the mechanism. Oil pressure sensors and position sensors can be used to monitor / indicate rocker arm actuation. It is possible that some manufacturers could apply the code to "Cylinder Deactivation" systems (that use the rocker arm as part of the deactivation mechanism).	Identify the type of position sensor fitted (e.g. Hall effect), which can dictate the specific checks that should be carried out. Note that some systems can use an oil pressure sensor to indicate that the Rocker Arm Actuator mechanism has operated / moved. The voltage or value of the position sensor signal is either at full value (e.g. battery voltage with no signal) or, the signal exists but it is above the normal operating range (higher than the maximum operating limit). Possible fault in the position sensor or wiring e.g. short / open circuits, circuit resistance, incorrect sensor resistance or, fault with the reference / operating voltage; interference from other circuits can also affect sensor signals. Refer to list of sensor / actuator checks on page 19.
P264E	"A" Rocker Arm Actuator Position Sensor Circuit Intermittent / Erratic Bank 1	Rocker arm actuators are used on some variable valve timing / lift systems. Systems can use oil pressure to engage / disengage the rocker arm mechanisms, which when activated enables changes in valve timing / lift. A solenoid operated valve can be used to control the application of the oil (under pressure) to the mechanism. Oil pressure sensors and position sensors can be used to monitor / indicate rocker arm actuation. It is possible that some manufacturers could apply the code to "Cylinder Deactivation" systems (that use the rocker arm as part of the deactivation mechanism).	Identify the type of position sensor fitted (e.g. Hall effect), which can dictate the specific checks that should be carried out. Note that some systems can use an oil pressure sensor to indicate that the Rocker Arm Actuator mechanism has operated / moved. The voltage or value of the position sensor signal is intermittent (likely to be out of normal operating range / tolerance when intermittent fault is detected) or, the signal / voltage is erratic (e.g. signal changes are irregular / unpredictable). Possible fault with wiring or connections (including internal sensor connections); also interference from other circuits can affect the signal. Refer to list of sensor / actuator checks on page 19. Also check for erratic operation of the Rocker Arm Actuation system.
P264F	ISO / SAE reserved		Not used, reserved for future allocation.
P2650	"B" Rocker Arm Actuator Control Circuit / Open Bank 1	Rocker arm actuators are used on some variable valve timing / lift systems. Systems can use oil pressure to engage / disengage the rocker arm mechanisms, which when activated enables changes in valve timing / lift. A solenoid operated valve can be used to control the application of the oil (under pressure) to the mechanism. Oil pressure sensors and position sensors can be used to monitor / indicate rocker arm actuation.	The voltage, frequency or value of the control signal in the actuator circuit is incorrect. Signal could be:- out of normal operating range, constant value, non-existent, corrupt. The fault code indicates a possible "open circuit" but other undefined faults in the actuator or wiring could activate the same code e.g. actuator power supply, short circuit, high circuit resistance or incorrect actuator resistance. Note: the control unit could be providing the correct signal but it is being affected by the circuit fault. Refer to list of sensor / actuator checks on page 19.
P2651	"B" Rocker Arm Actuator System Performance / Stuck Off Bank 1	Rocker arm actuators are used on some variable valve timing / lift systems. Systems can use oil pressure to engage / disengage the rocker arm mechanisms, which when activated enables changes in valve timing / lift. A solenoid operated valve can be used to control the application of the oil (under pressure) to the mechanism. Oil pressure sensors and position sensors can be used to monitor / indicate rocker arm actuation.	The control unit is providing a control signal to the actuator (signal is within the normal operating range / tolerance) but the control unit has detected that the actuator is stuck in a fixed (off) position and is not responding to the control signal. The fault could be related to an actuator / wiring fault. Refer to list of sensor / actuator checks on page 19. Note that the control unit could be detecting "no response" from the system or mechanism being controlled by the actuator; check system (as well as the actuator) for correct operation, movement and response. A position sensor (if fitted) would be indicating the "stuck position"; also check position sensor operation. Also check for correct oil pressure (if applicable).
P2652	"B" Rocker Arm Actuator System Stuck On Bank 1	Rocker arm actuators are used on some variable valve timing / lift systems. Systems can use oil pressure to engage / disengage the rocker arm mechanisms, which when activated enables changes in valve timing / lift. A solenoid operated valve can be used to control the application of the oil (under pressure) to the mechanism. Oil pressure sensors and position sensors can be used to monitor / indicate rocker arm actuation.	The control unit is providing a control signal to the actuator (signal is within the normal operating range / tolerance) but the control unit has detected that the actuator is stuck in a fixed (on) position and is not responding to the control signal. The fault could be related to an actuator / wiring fault. Refer to list of sensor / actuator checks on page 19. Note that the control unit could be detecting "no response" from the system or mechanism being controlled by the actuator; check system (as well as the actuator) for correct operation, movement and response. A position sensor (if fitted) would be indicating the "stuck position"; also check position sensor operation. Also check for correct oil pressure (if applicable).
P2653	"B" Rocker Arm Actuator Control Circuit Low Bank 1	Rocker arm actuators are used on some variable valve timing / lift systems. Systems can use oil pressure to engage / disengage the rocker arm mechanisms, which when activated enables changes in valve timing / lift. A solenoid operated valve can be used to control the application of the oil (under pressure) to the mechanism. Oil pressure sensors and position sensors can be used to monitor / indicate rocker arm actuation.	The voltage, frequency or value of the control signal in the actuator circuit is either "zero" (no signal) or, the signal exists but is below the normal operating range. Possible fault with actuator or wiring (e.g. actuator power supply, short / open circuit or resistance) that is causing a signal value to be lower than the minimum operating limit. Note: the control unit could be providing the correct signal but it is being affected by the circuit fault. Refer to list of sensor / actuator checks on page 19.

NOTE 1. Check for other fault codes that could provide additional information. **NOTE 2.** Communication between control units can pass via a CAN-Bus system; refer to Page 51 for CAN-Bus checks.

P2654	"B" Rocker Arm Actuator Control Circuit High Bank 1	Rocker arm actuators are used on some variable valve timing / lift systems. Systems can use oil pressure to engage / disengage the rocker arm mechanisms, which when activated enables changes in valve timing / lift. A solenoid operated valve can be used to control the application of the oil (under pressure) to the mechanism. Oil pressure sensors and position sensors can be used to monitor / indicate rocker arm actuation.	The voltage, frequency or value of the control signal in the actuator circuit is either at full value (e.g. battery voltage with no on / off signal) or, the signal exists but it is above the normal operating range. Possible fault with actuator or wiring (e.g. actuator power supply, short / open circuit or resistance) that is causing a signal value to be higher than the maximum operating limit. Note: the control unit could be providing the correct signal but it is being affected by the circuit fault. Refer to list of sensor / actuator checks on page 19.
P2655	"A" Rocker Arm Actuator Control Circuit / Open Bank 2	Rocker arm actuators are used on some variable valve timing / lift systems. Systems can use oil pressure to engage / disengage the rocker arm mechanisms, which when activated enables changes in valve timing / lift. A solenoid operated valve can be used to control the application of the oil (under pressure) to the mechanism. Oil pressure sensors and position sensors can be used to monitor / indicate rocker arm actuation.	The voltage, frequency or value of the control signal in the actuator circuit is incorrect (undefined fault). Signal could be:- out of normal operating range, constant value, non-existent, corrupt. Possible fault with actuator or wiring e.g. actuator power supply, short / open circuit, high circuit resistance or incorrect actuator resistance. Note: the control unit could be providing the correct signal but it is being affected by the circuit fault. Refer to list of sensor / actuator checks on page 19.
P2656	"A" Rocker Arm Actuator System Performance or Stuck Off Bank 2	Rocker arm actuators are used on some variable valve timing / lift systems. Systems can use oil pressure to engage / disengage the rocker arm mechanisms, which when activated enables changes in valve timing / lift. A solenoid operated valve can be used to control the application of the oil (under pressure) to the mechanism. Oil pressure sensors and position sensors can be used to monitor / indicate rocker arm actuation.	The control unit is providing a control signal to the actuator (signal is within the normal operating range / tolerance) but the control unit has detected that the actuator is stuck in a fixed (off) position and is not responding to the control signal. The fault could be related to an actuator / wiring fault. Refer to list of sensor / actuator checks on page 19. Note that the control unit could be detecting "no response" from the system or mechanism being controlled by the actuator; check system (as well as the actuator) for correct operation, movement and response. A position sensor (if fitted) would be indicating the "stuck position"; also check position sensor operation. Also check for correct oil pressure (if applicable).
P2657	"A" Rocker Arm Actuator System Stuck On Bank 2	Rocker arm actuators are used on some variable valve timing / lift systems. Systems can use oil pressure to engage / disengage the rocker arm mechanisms, which when activated enables changes in valve timing / lift. A solenoid operated valve can be used to control the application of the oil (under pressure) to the mechanism. Oil pressure sensors and position sensors can be used to monitor / indicate rocker arm actuation.	The control unit is providing a control signal to the actuator (signal is within the normal operating range / tolerance) but the control unit has detected that the actuator is stuck in a fixed (on) position and is not responding to the control signal. The fault could be related to an actuator / wiring fault. Refer to list of sensor / actuator checks on page 19. Note that the control unit could be detecting "no response" from the system or mechanism being controlled by the actuator; check system (as well as the actuator) for correct operation, movement and response. A position sensor (if fitted) would be indicating the "stuck position"; also check position sensor operation. Also check for correct oil pressure (if applicable).
P2658	"A" Rocker Arm Actuator Control Circuit Low Bank 2	Rocker arm actuators are used on some variable valve timing / lift systems. Systems can use oil pressure to engage / disengage the rocker arm mechanisms, which when activated enables changes in valve timing / lift. A solenoid operated valve can be used to control the application of the oil (under pressure) to the mechanism. Oil pressure sensors and position sensors can be used to monitor / indicate rocker arm actuation.	The voltage, frequency or value of the control signal in the actuator circuit is either "zero" (no signal) or, the signal exists but is below the normal operating range. Possible fault with actuator or wiring (e.g. actuator power supply, short / open circuit or resistance) that is causing a signal value to be lower than the minimum operating limit. Note: the control unit could be providing the correct signal but it is being affected by the circuit fault. Refer to list of sensor / actuator checks on page 19.
P2659	"A" Rocker Arm Actuator Control Circuit High Bank 2	Rocker arm actuators are used on some variable valve timing / lift systems. Systems can use oil pressure to engage / disengage the rocker arm mechanisms, which when activated enables changes in valve timing / lift. A solenoid operated valve can be used to control the application of the oil (under pressure) to the mechanism. Oil pressure sensors and position sensors can be used to monitor / indicate rocker arm actuation.	The voltage, frequency or value of the control signal in the actuator circuit is either at full value (e.g. battery voltage with no on / off signal) or, the signal exists but it is above the normal operating range. Possible fault with actuator or wiring (e.g. actuator power supply, short / open circuit or resistance) that is causing a signal value to be higher than the maximum operating limit. Note: the control unit could be providing the correct signal but it is being affected by the circuit fault. Refer to list of sensor / actuator checks on page 19.
P265A	"B" Rocker Arm Actuator Position Sensor Circuit Bank 1	Rocker arm actuators are used on some variable valve timing / lift systems. Systems can use oil pressure to engage / disengage the rocker arm mechanisms, which when activated enables changes in valve timing / lift. A solenoid operated valve can be used to control the application of the oil (under pressure) to the mechanism. Oil pressure sensors and position sensors can be used to monitor / indicate rocker arm actuation. It is possible that some manufacturers could apply the code to "Cylinder Deactivation" systems (that use the rocker arm as part of the deactivation mechanism).	Identify the type of position sensor fitted (e.g. Hall effect), which can dictate the specific checks that should be carried out. Note that some systems can use an oil pressure sensor to indicate that the Rocker Arm Actuator mechanism has operated / moved. The voltage or value of the position sensor signal is incorrect (undefined fault). Sensor voltage / signal value could be: out of normal operating range, constant value, non-existent, corrupt. Possible fault in the position sensor or wiring e.g. short / open circuits, circuit resistance, incorrect sensor resistance (if applicable) or, fault with the reference / operating voltage; interference from other circuits can also affect sensor signals. Also check earth connections and cable screening. Refer to list of sensor / actuator checks on page 19.
P265B	"B" Rocker Arm Actuator Position Sensor Circuit Range / Performance Bank 1	Rocker arm actuators are used on some variable valve timing / lift systems. Systems can use oil pressure to engage / disengage the rocker arm mechanisms, which when activated enables changes in valve timing / lift. A solenoid operated valve can be used to control the application of the oil (under pressure) to the mechanism. Oil pressure sensors and position sensors can be used to monitor / indicate rocker arm actuation. It is possible that some manufacturers could apply the code to "Cylinder Deactivation" systems (that use the rocker arm as part of the deactivation mechanism).	Identify the type of position sensor fitted (e.g. Hall effect), which can dictate the specific checks that should be carried out. Note that some systems can use an oil pressure sensor to indicate that the Rocker Arm Actuator mechanism has operated / moved. The voltage or value of the position sensor signal is within the normal operating range / tolerance but the signal is not plausible or is incorrect due to an undefined fault. The sensor signal might not match the operating conditions or, the sensor signal is not changing / responding as expected to the changes in conditions. Possible fault in the position sensor or wiring e.g. short / open circuits, circuit resistance, incorrect sensor resistance or, fault with the reference / operating voltage; interference from other circuits can also affect sensor signals. Refer to list of sensor / actuator checks on page 19. It is also possible that the sensor is operating correctly but the position / pressure or activation detected by the sensor is unacceptable for the operating conditions. Check Rocker Arm Actuator operation / movement to ensure correct operation.

NOTE 1. Check for other fault codes that could provide additional information.
NOTE 2. Communication between control units can pass via a CAN-Bus system; refer to Page 51 for CAN-Bus checks.
NOTE 3. If a fault cannot be located, it is also possible that the control unit is at fault.
NOTE 4. Refer to Page 53 for list of pages that show the ISO standard for component locations e.g. Sensor A - Bank 1.

Fault code	EOBD / ISO Description	Component / System Description	Meaningful Description and Quick Check
P265C	"B" Rocker Arm Actuator Position Sensor Circuit Low Bank 1	Rocker arm actuators are used on some variable valve timing / lift systems. Systems can use oil pressure to engage / disengage the rocker arm mechanisms, which when activated enables changes in valve timing / lift. A solenoid operated valve can be used to control the application of the oil (under pressure) to the mechanism. Oil pressure sensors and position sensors can be used to monitor / indicate rocker arm actuation. It is possible that some manufacturers could apply the code to "Cylinder Deactivation" systems (that use the rocker arm as part of the deactivation mechanism).	Identify the type of position sensor fitted (e.g. Hall effect), which can dictate the specific checks that should be carried out. Note that some systems can use an oil pressure sensor to indicate that the Rocker Arm Actuator mechanism has operated / moved. The voltage or value of the position sensor signal is either "zero" (no signal) or, the signal exists but it is below the normal operating range (lower than the minimum operating limit). Possible fault in the position sensor or wiring e.g. short / open circuits, circuit resistance, incorrect sensor resistance or, fault with the reference / operating voltage; interference from other circuits can also affect sensor signals. Refer to list of sensor / actuator checks on page 19.
P265D	"B" Rocker Arm Actuator Position Sensor Circuit High Bank 1	Rocker arm actuators are used on some variable valve timing / lift systems. Systems can use oil pressure to engage / disengage the rocker arm mechanisms, which when activated enables changes in valve timing / lift. A solenoid operated valve can be used to control the application of the oil (under pressure) to the mechanism. Oil pressure sensors and position sensors can be used to monitor / indicate rocker arm actuation. It is possible that some manufacturers could apply the code to "Cylinder Deactivation" systems (that use the rocker arm as part of the deactivation mechanism).	Identify the type of position sensor fitted (e.g. Hall effect), which can dictate the specific checks that should be carried out. Note that some systems can use an oil pressure sensor to indicate that the Rocker Arm Actuator mechanism has operated / moved. The voltage or value of the position sensor signal is either at full value (e.g. battery voltage with no signal) or, the signal exists but it is above the normal operating range (higher than the maximum operating limit). Possible fault in the position sensor or wiring e.g. short / open circuits, circuit resistance, incorrect sensor resistance or, fault with the reference / operating voltage; interference from other circuits can also affect sensor signals. Refer to list of sensor / actuator checks on page 19.
P265E	"B" Rocker Arm Actuator Position Sensor Circuit Intermittent / Erratic Bank 1	Rocker arm actuators are used on some variable valve timing / lift systems. Systems can use oil pressure to engage / disengage the rocker arm mechanisms, which when activated enables changes in valve timing / lift. A solenoid operated valve can be used to control the application of the oil (under pressure) to the mechanism. Oil pressure sensors and position sensors can be used to monitor / indicate rocker arm actuation. It is possible that some manufacturers could apply the code to "Cylinder Deactivation" systems (that use the rocker arm as part of the deactivation mechanism).	Identify the type of position sensor fitted (e.g. Hall effect), which can dictate the specific checks that should be carried out. Note that some systems can use an oil pressure sensor to indicate that the Rocker Arm Actuator mechanism has operated / moved. The voltage or value of the position sensor signal is intermittent (likely to be out of normal operating range / tolerance when intermittent fault is detected) or, the signal / voltage is erratic (e.g. signal changes are irregular / unpredictable). Possible fault with wiring or connections (including internal sensor connections); also interference from other circuits can affect the signal. Refer to list of sensor / actuator checks on page 19. Also check for erratic operation of the Rocker Arm Actuation system.
P265F	ISO / SAE reserved		Not used, reserved for future allocation.
P2660	"B" Rocker Arm Actuator Control Circuit / Open Bank 2	Rocker arm actuators are used on some variable valve timing / lift systems. Systems can use oil pressure to engage / disengage the rocker arm mechanisms, which when activated enables changes in valve timing / lift. A solenoid operated valve can be used to control the application of the oil (under pressure) to the mechanism. Oil pressure sensors and position sensors can be used to monitor / indicate rocker arm actuation.	The voltage, frequency or value of the control signal in the actuator circuit is incorrect. Signal could be:- out of normal operating range, constant value, non-existent, corrupt. The fault code indicates a possible "open circuit" but other undefined faults in the actuator or wiring could activate the same code e.g. actuator power supply, short circuit, high circuit resistance or incorrect actuator resistance. Note: the control unit could be providing the correct signal but it is being affected by the circuit fault. Refer to list of sensor / actuator checks on page 19.
P2661	"B" Rocker Arm Actuator System Performance / Stuck Off Bank 2	Rocker arm actuators are used on some variable valve timing / lift systems. Systems can use oil pressure to engage / disengage the rocker arm mechanisms, which when activated enables changes in valve timing / lift. A solenoid operated valve can be used to control the application of the oil (under pressure) to the mechanism. Oil pressure sensors and position sensors can be used to monitor / indicate rocker arm actuation.	The control unit is providing a control signal to the actuator (signal is within the normal operating range / tolerance) but the control unit has detected that the actuator is stuck in a fixed (off) position and is not responding to the control signal. The fault could be related to an actuator / wiring fault. Refer to list of sensor / actuator checks on page 19. Note that the control unit could be detecting "no response" from the system or mechanism being controlled by the actuator; check system (as well as the actuator) for correct operation, movement and response. A position sensor (if fitted) would be indicating the "stuck position"; also check position sensor operation. Also check for correct oil pressure (if applicable).
P2662	"B" Rocker Arm Actuator System Stuck On Bank 2	Rocker arm actuators are used on some variable valve timing / lift systems. Systems can use oil pressure to engage / disengage the rocker arm mechanisms, which when activated enables changes in valve timing / lift. A solenoid operated valve can be used to control the application of the oil (under pressure) to the mechanism. Oil pressure sensors and position sensors can be used to monitor / indicate rocker arm actuation.	The control unit is providing a control signal to the actuator (signal is within the normal operating range / tolerance) but the control unit has detected that the actuator is stuck in a fixed (on) position and is not responding to the control signal. The fault could be related to an actuator / wiring fault. Refer to list of sensor / actuator checks on page 19. Note that the control unit could be detecting "no response" from the system or mechanism being controlled by the actuator; check system (as well as the actuator) for correct operation, movement and response. A position sensor (if fitted) would be indicating the "stuck position"; also check position sensor operation. Also check for correct oil pressure (if applicable).
P2663	"B" Rocker Arm Actuator Control Circuit Low Bank 2	Rocker arm actuators are used on some variable valve timing / lift systems. Systems can use oil pressure to engage / disengage the rocker arm mechanisms, which when activated enables changes in valve timing / lift. A solenoid operated valve can be used to control the application of the oil (under pressure) to the mechanism. Oil pressure sensors and position sensors can be used to monitor / indicate rocker arm actuation.	The voltage, frequency or value of the control signal in the actuator circuit is either "zero" (no signal) or, the signal exists but is below the normal operating range. Possible fault with actuator or wiring (e.g. actuator power supply, short / open circuit or resistance) that is causing a signal value to be lower than the minimum operating limit. Note: the control unit could be providing the correct signal but it is being affected by the circuit fault. Refer to list of sensor / actuator checks on page 19.
P2664	"B" Rocker Arm Actuator Control Circuit High Bank 2	Rocker arm actuators are used on some variable valve timing / lift systems. Systems can use oil pressure to engage / disengage the rocker arm mechanisms, which when activated enables changes in valve timing / lift. A solenoid operated valve can be used to control the application of the oil (under pressure) to the mechanism. Oil pressure sensors and position sensors can be used to monitor / indicate rocker arm actuation.	The voltage, frequency or value of the control signal in the actuator circuit is either at full value (e.g. battery voltage with no on / off signal) or, the signal exists but it is above the normal operating range. Possible fault with actuator or wiring (e.g. actuator power supply, short / open circuit or resistance) that is causing a signal value to be higher than the maximum operating limit. Note: the control unit could be providing the correct signal but it is being affected by the circuit fault. Refer to list of sensor / actuator checks on page 19.

NOTE 1. Check for other fault codes that could provide additional information.   **NOTE 2.** Communication between control units can pass via a CAN-Bus system; refer to Page 51 for CAN-Bus checks.

NOTE 4. Refer to Page 53 for list of pages that show the ISO standard for component locations e.g. Sensor A - Bank 1.

300

Code		Description	
P2665	Fuel Shutoff Valve "B" Control Circuit / Open	A fuel shut off valve is usually located in a fuel injection pump. The valve is operated by an electrical actuator (usually a solenoid) and is used to switch off the fuel supply when engine is stopped. Note, some systems move the fuel quantity control (in the diesel pump) to the "zero" fuel delivery position. Note that different fuel systems may use a different type of shut off valve.	Identify the type of shutoff valve; the control signal from the control unit could be a simple "on or off" (on the power supply circuit). On some "fuel quantity" control systems a frequency / duty cycle signal could be provided. The voltage or value of the control signal in the actuator circuit is incorrect. Signal / voltage could be out of normal operating range, constant value, non-existent, corrupt. The fault code indicates a possible "open circuit" but other undefined faults in the actuator or wiring could activate the same code e.g. actuator power supply, short circuit, high circuit resistance or incorrect actuator resistance. Note: the control unit could be providing the correct signal but it is being affected by a circuit fault. Refer to list of sensor / actuator checks on page 19.
P2666	Fuel Shutoff Valve "B" Control Circuit Low	A fuel shut off valve is usually located in a fuel injection pump. The valve is operated by an electrical actuator (usually a solenoid) and is used to switch off the fuel supply when engine is stopped. Note, some systems move the fuel quantity control (in the diesel pump) to the "zero" fuel delivery position. Note that different fuel systems may use a different type of shut off valve.	Identify the type of shutoff valve; the control signal from the control unit could be a simple "on or off" (on the power supply circuit). On some "fuel quantity" control systems a frequency / duty cycle signal could be provided. The voltage or value of the control signal in the actuator circuit is incorrect. Signal / voltage could be out of normal operating range, constant value, non-existent, corrupt. The fault code indicates a possible "open circuit" but other undefined faults in the actuator or wiring could activate the same code e.g. actuator power supply, short circuit, high circuit resistance or incorrect actuator resistance. Note: the control unit could be providing the correct signal but it is being affected by a circuit fault. Refer to list of sensor / actuator checks on page 19.
P2667	Fuel Shutoff Valve "B" Control Circuit High	A fuel shut off valve is usually located in a fuel injection pump. The valve is operated by an electrical actuator (usually a solenoid) and is used to switch off the fuel supply when engine is stopped. Note, some systems move the fuel quantity control (in the diesel pump) to the "zero" fuel delivery position. Note that different fuel systems may use a different type of shut off valve.	I Identify the type of shutoff valve; the control signal from the control unit could be a simple "on or off" (on the power supply circuit). On some "fuel quantity" control systems a frequency / duty cycle signal could be provided. The voltage or value of the control signal in the actuator circuit is incorrect. Signal / voltage could be out of normal operating range, constant value, non-existent, corrupt. The fault code indicates a possible "open circuit" but other undefined faults in the actuator or wiring could activate the same code e.g. actuator power supply, short circuit, high circuit resistance or incorrect actuator resistance. Note: the control unit could be providing the correct signal but it is being affected by a circuit fault. Refer to list of sensor / actuator checks on page 19.
P2668	Fuel Mode Indicator Lamp Control Circuit	The fuel mode indicator Lamp provide the driver with an indication that a particular fuel mode is selected e.g. Bio-Fuel operation.	The fault code indicates a fault with the indicator lamp / circuit. Check the wiring and connections for short / open circuits and check the lamp. If applicable, check wiring / connections to the Fuel Mode selector switch.
P2669	Actuator Supply Voltage "B" Circuit / Open	The control unit can monitor actuator supply voltages in those systems and circuits where the voltage supply is via relay or another control unit. The control unit can also provide the supply voltage for some actuators, as well as regulating the supply voltage (pulse width / duty cycle control), therefore the voltage can be regulated to enable more accurate control of the actuators. The control unit compares the actual voltage or regulated voltage provided to the actuator with the expected value (held in memory). It will be necessary to identify which actuators are affected and whether the supply voltage is provided by the main control unit, a relay or an independent control unit.	The fault code indicates an actuator(s) supply voltage fault, with a possible "open circuit" or other circuit fault. Identify the source of the voltage supply e.g. relay, main control unit or separate control unit. If the control unit detects a fault with an independent supply (e.g. relay or separate control unit), the main control unit could be connected to the supply using a "sensing wire". It will be necessary to check if the actuator(s) are receiving a supply, and check whether there is an open circuit in the sensing wire. The control unit could be monitoring the voltage in the actuator circuit via the earth path (earth path switched to earth by the control unit); if there is no supply voltage passing through the actuator into the earth circuit, the control unit can assess this as an open circuit. If the control unit is providing the supply voltage to an actuator, and there is an open circuit or circuit fault, this can also be detected by the control unit. In all cases, check for short / open circuits in the wiring and check for an short / open circuit or incorrect resistance with the actuator.
P266A	"A" Rocker Arm Actuator Position Sensor Circuit Bank 2	Rocker arm actuators are used on some variable valve timing / lift systems. Systems can use oil pressure to engage / disengage the rocker arm mechanisms, which when activated enables changes in valve timing / lift. A solenoid operated valve can be used to control the application of the oil (under pressure) to the mechanism. Oil pressure sensors and position sensors can be used to monitor / indicate rocker arm actuation. It is possible that some manufacturers could apply the code to "Cylinder Deactivation" systems (that use the rocker arm as part of the deactivation mechanism).	Identify the type of position sensor fitted (e.g. Hall effect), which can dictate the specific checks that should be carried out. Note that some systems can use an oil pressure sensor to indicate that the Rocker Arm Actuator mechanism has operated / moved. The voltage or value of the position sensor signal is incorrect (undefined fault). Sensor voltage / signal value could be: out of normal operating range, constant value, non-existent, corrupt. Possible fault in the position sensor or wiring e.g. short / open circuits, circuit resistance, incorrect sensor resistance (if applicable) or, fault with the reference / operating voltage; interference from other circuits can also affect sensor signals. Also check earth connections and cable screening. Refer to list of sensor / actuator checks on page 19.
P266B	"A" Rocker Arm Actuator Position Sensor Circuit Range / Performance Bank 2	Rocker arm actuators are used on some variable valve timing / lift systems. Systems can use oil pressure to engage / disengage the rocker arm mechanisms, which when activated enables changes in valve timing / lift. A solenoid operated valve can be used to control the application of the oil (under pressure) to the mechanism. Oil pressure sensors and position sensors can be used to monitor / indicate rocker arm actuation. It is possible that some manufacturers could apply the code to "Cylinder Deactivation" systems (that use the rocker arm as part of the deactivation mechanism).	Identify the type of position sensor fitted (e.g. Hall effect), which can dictate the specific checks that should be carried out. Note that some systems can use an oil pressure sensor to indicate that the Rocker Arm Actuator mechanism has operated / moved. The voltage or value of the position sensor signal is within the normal operating range / tolerance but the signal is not plausible or is incorrect due to an undefined fault. The sensor signal might not match the operating conditions or, the sensor signal is not changing / responding as expected to the changes in conditions. Possible fault in the position sensor or wiring e.g. short / open circuits, circuit resistance, incorrect sensor resistance or, fault with the reference / operating voltage; interference from other circuits can also affect sensor signals. Refer to list of sensor / actuator checks on page 19. It is also possible that the sensor is operating correctly but the position / pressure or activation detected by the sensor is unacceptable for the operating conditions. Check Rocker Arm Actuator operation / movement to ensure correct operation.
P266C	"A" Rocker Arm Actuator Position Sensor Circuit Low Bank 2	Rocker arm actuators are used on some variable valve timing / lift systems. Systems can use oil pressure to engage / disengage the rocker arm mechanisms, which when activated enables changes in valve timing / lift. A solenoid operated valve can be used to control the application of the oil (under pressure) to the mechanism. Oil pressure sensors and position sensors can be used to monitor / indicate rocker arm actuation. It is possible that some manufacturers could apply the code to "Cylinder Deactivation" systems (that use the rocker arm as part of the deactivation mechanism).	Identify the type of position sensor fitted (e.g. Hall effect), which can dictate the specific checks that should be carried out. Note that some systems can use an oil pressure sensor to indicate that the Rocker Arm Actuator mechanism has operated / moved. The voltage or value of the position sensor signal is either "zero" (no signal) or, the signal exists but it is below the normal operating range (lower than the minimum operating limit). Possible fault in the position sensor or wiring e.g. short / open circuits, circuit resistance, incorrect sensor resistance or, fault with the reference / operating voltage; interference from other circuits can also affect sensor signals. Refer to list of sensor / actuator checks on page 19.

NOTE 1. Check for other fault codes that could provide additional information.   **NOTE 2.** Communication between control units can pass via a CAN-Bus system; refer to Page 51 for CAN-Bus checks.

NOTE 3. If a fault cannot be located, it is also possible that the control unit is at fault.   **NOTE 4.** Refer to Page 53 for list of pages that show the ISO standard for component locations e.g. Sensor A - Bank 1.

Fault code	EOBD / ISO Description	Component / System Description	Meaningful Description and Quick Check
P266D	"A" Rocker Arm Actuator Position Sensor Circuit High Bank 2	Rocker arm actuators are used on some variable valve timing / lift systems. Systems can use oil pressure to engage / disengage the rocker arm mechanisms, which when activated enables changes in valve timing / lift. A solenoid operated valve can be used to control the application of the oil (under pressure) to the mechanism. Oil pressure sensors and position sensors can be used to monitor / indicate rocker arm actuation. It is possible that some manufacturers could apply the code to "Cylinder Deactivation" systems (that use the rocker arm as part of the deactivation mechanism).	Identify the type of position sensor fitted (e.g. Hall effect), which can dictate the specific checks that should be carried out. Note that some systems can use an oil pressure sensor to indicate that the Rocker Arm Actuator mechanism has operated / moved. The voltage or value of the position sensor signal is either at full value (e.g. battery voltage with no signal) or, the signal exists but it is above the normal operating range (higher than the maximum operating limit). Possible fault in the position sensor or wiring e.g. short / open circuits, circuit resistance, incorrect sensor resistance or, fault with the reference / operating voltage; interference from other circuits can also affect sensor signals. Refer to list of sensor / actuator checks on page 19.
P266E	"A" Rocker Arm Actuator Position Sensor Circuit Intermittent / Erratic Bank 2	Rocker arm actuators are used on some variable valve timing / lift systems. Systems can use oil pressure to engage / disengage the rocker arm mechanisms, which when activated enables changes in valve timing / lift. A solenoid operated valve can be used to control the application of the oil (under pressure) to the mechanism. Oil pressure sensors and position sensors can be used to monitor / indicate rocker arm actuation. It is possible that some manufacturers could apply the code to "Cylinder Deactivation" systems (that use the rocker arm as part of the deactivation mechanism).	Identify the type of position sensor fitted (e.g. Hall effect), which can dictate the specific checks that should be carried out. Note that some systems can use an oil pressure sensor to indicate that the Rocker Arm Actuator mechanism has operated / moved. The voltage or value of the position sensor signal is intermittent (likely to be out of normal operating range / tolerance when intermittent fault is detected) or, the signal / voltage is erratic (e.g. signal changes are irregular / unpredictable). Possible fault with wiring or connections (including internal sensor connections); also interference from other circuits can affect the signal. Refer to list of sensor / actuator checks on page 19. Also check for erratic operation of the Rocker Arm Actuation system.
P266F	ISO / SAE reserved		Not used, reserved for future allocation.
P2670	Actuator Supply Voltage "B" Circuit Low	The control unit can monitor actuator supply voltages in those systems and circuits where the voltage supply is via relay or another control unit. The control unit can also provide the supply voltage for some actuators, as well as regulating the supply voltage (pulse width / duty cycle control), therefore the voltage can be regulated to enable more accurate control of the actuators. The control unit compares the actual voltage or regulated voltage provided to the actuator with the expected value (held in memory). It will be necessary to identify which actuators are affected and whether the supply voltage is provided by the main control unit, a relay or an independent control unit.	The fault code indicates that the supply voltage to an actuator(s) is low (could be zero or below normal operating range). Identify the source of the voltage supply e.g. relay, main control unit or separate control unit. In all cases, check for short / open circuits in the wiring, high circuit resistances and check for an short / open circuit or incorrect resistance with the actuator. If the control unit detects a fault with an independent supply (e.g. relay or separate control unit), the main control unit could be detecting the fault via a "sensing wire". Check if the actuator(s) are receiving the correct supply (also check charging voltage), and check whether there is an open circuit or high resistance in the sensing wire. The control unit could be monitoring the actuator voltage via the earth path; earth path could be switched to earth by the control unit, and the control unit can therefore assess the available voltage via the earth circuit. If the control unit is providing the actuator supply voltage, a low voltage will be detected by the control unit (this could be a control unit or charging system fault).
P2671	Actuator Supply Voltage "B" Circuit High	The control unit can monitor actuator supply voltages in those systems and circuits where the voltage supply is via relay or another control unit. The control unit can also provide the supply voltage for some actuators, as well as regulating the supply voltage (pulse width / duty cycle control), therefore the voltage can be regulated to enable more accurate control of the actuators. The control unit compares the actual voltage or regulated voltage provided to the actuator with the expected value (held in memory). It will be necessary to identify which actuators are affected and whether the supply voltage is provided by the main control unit, a relay or an independent control unit.	The fault code indicates that the supply voltage to an actuator(s) is high; voltage could be at full value at all times e.g. battery voltage (with no signal or frequency for those actuators where the control unit provides a control signal on the power circuit). Identify the source of the voltage supply e.g. relay, main control unit or separate control unit, and identify the wiring circuit. In all cases, check for short / open circuits in the wiring, high circuit resistances and check for an short / open circuit or incorrect resistance with the actuator. For those actuators where the control unit switches the actuator earth path through to earth, it is possible that the earth path has a short to a power supply circuit (causing voltage to permanently exist in the circuit). If the control unit detects a fault with an independent power supply (e.g. relay or separate control unit), the main control unit could be detecting the fault via a "sensing wire". Check if the actuator(s) are receiving the correct supply (also check charging voltage).
P2672	Injection Pump Timing Offset	Injection pump timing offset usually refers to the injection timing that is theoretically being delivered by the Diesel injection pump compared with the actual timing. The Diesel pump control unit will be monitoring crankshaft / camshaft position and / or Diesel pump shaft position and, it will receive an injection timing signal (possibly from an injector motion sensor / lift sensor). If the actual timing is incorrect and the control of the pump timing system does not correct the timing, this can be regarded as a timing offset fault.	It may be necessary to refer to vehicle specific information to identify specific fuel pump / system checks and control unit re-programming / configuration process. The fault code indicates a fault (undefined) with the timing offset. Check pump timing base setting, check for any wear or slackness in the pump drive (belt tension etc). Check operation of pump timing mechanism (control solenoids etc) and check for correct operation of timing sensor e.g. injector motion sensor. Faults with the fuelling system (pump or sensors / actuators) could result in loss of configuration data from the control unit. It is possible that the fitting of new components (including the control unit) or, maintenance on the system has resulted in loss of timing offset information / data, it may therefore be necessary on some systems to re-programme / re-configure the Diesel control unit. Note that a separate electronic module can be located on the pump (or remote), the module can be used to provide the high current control signals to the pump actuators (solenoids etc); it is therefore possible that there is a fault in the communication between the Diesel control unit and the separate module.
P2673	Injection Pump Timing Calibration Not Learned	Injection pump timing calibration usually refers to the programme in the Diesel control unit which controls the Diesel timing (as applicable to the specific vehicle and the operating conditions). The Diesel pump control unit will be monitoring crankshaft / camshaft position and / or Diesel pump shaft position and, it will receive an injection timing signal (possibly from an injector motion sensor / lift sensor). If the control unit cannot detect or learn the actual timing via the timing sensor signal, this could activate the fault code.	It may be necessary to refer to vehicle specific information to identify specific fuel pump / system checks and control unit re-programming / configuration process. The fault could be caused by a control unit problem (including wiring / connections to the fuel system and sensors), but checks should be carried out on the fuel system operation and components. Check pump timing base setting, check for any wear or slackness in the pump drive (belt tension etc). Check operation of pump timing mechanism (control solenoids etc) and check for correct operation of timing sensor e.g. injector motion sensor. Faults with the fuelling system (pump or sensors / actuators) could result in loss of configuration data from the control unit. It is possible that the fitting of new components (including the control unit) or, maintenance on the system has resulted in loss of timing information / data, it may therefore be necessary on some systems to re-programme / re-configure the Diesel control unit. Note that a separate electronic module can be located on the pump (or remote), the module can be used to provide the high current control signals to the pump actuators (solenoids etc); it is therefore possible that there is a fault in the communication between the Diesel control unit and the separate module.

NOTE 1. Check for other fault codes that could provide additional information.   **NOTE 2.** Communication between control units can pass via a CAN-Bus system; refer to Page 51 for CAN-Bus checks.   302

NOTE 3. If a fault cannot be located, it is also possible that the control unit is at fault   NOTE 4. Refer to Page 53 for list of pages that show the ISO standard for component locations e.g. Sensor A - Bank 1.

Code	Fault Description	Description	Checks / Notes
P2674	Injection Pump Fuel Calibration Not Learned	Injection pump fuel calibration usually refers to the fuelling control that is theoretically being delivered by the Diesel injection pump for the specific vehicle and / or operating conditions. The Diesel pump control unit will be monitoring various sensors to enable correct fuelling control.	It may be necessary to refer to vehicle specific information to identify specific fuel pump / system checks and control unit re-programming / configuration process. The fault could be caused by a control unit problem (including wiring / connections to the fuel system and sensors), but checks should be carried out on the fuel system operation and components. Faults with the fuelling system (pump or sensors / actuators) could result in loss of configuration data from the control unit. It is possible that the fitting of new components (including the control unit) or, maintenance on the system has resulted in loss of timing offset information / data, it may therefore be necessary on some systems to re-programme / re-configure the Diesel control unit. Note that a separate electronic module can be located on the pump (or remote), the module can be used to provide the high current control signals to the pump actuators (solenoids etc); it is therefore possible that there is a fault in the communication between the Diesel control unit and the separate module.
P2675	Air Cleaner Inlet Control Circuit / Open	The air passing to the air cleaner can be fresh ambient air or, it can be passed from an alternative inlet that could collect the air from a warm location e.g. close to the exhaust system. A diverter system e.g. a flap in the inlet pipe, can be controlled using vacuum, which in turn is controlled by an electric actuator (usually a solenoid).	The voltage, frequency or value of the control signal in the actuator circuit is incorrect. Signal could be:- out of normal operating range, constant value, non-existent, corrupt. The fault code indicates a possible "open circuit" but other undefined faults in the actuator or wiring could activate the same code e.g. actuator power supply, short circuit, high circuit resistance or incorrect actuator resistance. Note: the control unit could be providing the correct signal but it is being affected by the circuit fault. Refer to list of sensor / actuator checks on page 19.
P2676	Air Cleaner Inlet Control Circuit Low	The air passing to the air cleaner can be fresh ambient air or, it can be passed from an alternative inlet that could collect the air from a warm location e.g. close to the exhaust system. A diverter system e.g. a flap in the inlet pipe, can be controlled using vacuum, which in turn is controlled by an electric actuator (usually a solenoid).	The voltage, frequency or value of the control signal in the actuator circuit is either "zero" (no signal) or, the signal exists but is below the normal operating range. Possible fault with actuator or wiring (e.g. actuator power supply, short / open circuit or resistance) that is causing a signal value to be lower than the minimum operating limit. Note: the control unit could be providing the correct signal but it is being affected by the circuit fault. Refer to list of sensor / actuator checks on page 19.
P2677	Air Cleaner Inlet Control Circuit High	The air passing to the air cleaner can be fresh ambient air or, it can be passed from an alternative inlet that could collect the air from a warm location e.g. close to the exhaust system. A diverter system e.g. a flap in the inlet pipe, can be controlled using vacuum, which in turn is controlled by an electric actuator (usually a solenoid).	The voltage, frequency or value of the control signal in the actuator circuit is either at full value (e.g. battery voltage with no on / off signal) or, the signal exists but it is above the normal operating range. Possible fault with actuator or wiring (e.g. actuator power supply, short / open circuit or resistance) that is causing a signal value to be higher than the maximum operating limit. Note: the control unit could be providing the correct signal but it is being affected by the circuit fault. Refer to list of sensor / actuator checks on page 19.
P2678	Coolant Degassing Valve Control Circuit / Open	A "degassing tank" is a US term referring to the expansion / header tank. Some system can use a valve to control coolant flow to aid the degassing / aeration	Identify the degassing valve and valve actuator e.g. solenoid. The voltage, frequency or value of the control signal (provided by the control unit) in the valve circuit is incorrect (undefined problem). Signal could be: out of normal operating range, constant value, non-existent, corrupt. The code indicates a possible "open circuit" but other faults could activate the same fault code e.g. short circuit, high resistance, interference from other circuits. Note: the control unit could be providing the correct signal but it is being affected by a circuit fault. Identify the type of valve and the wiring used on the system. Refer to list of sensor / actuator checks on page 19.
P2679	Coolant Degassing Valve Control Circuit Low	A "degassing tank" is a US term referring to the expansion / header tank. Some system can use a valve to control coolant flow to aid the degassing / aeration	Identify the degassing valve and valve actuator e.g. solenoid. The voltage, frequency or value of the control signal (provided by the control unit) in the valve circuit is either "zero" (no signal) or, the signal exists but is below the normal operating range. The code indicates a possible "open circuit" but other faults could activate the same fault code e.g. short circuit, high resistance, interference from other circuits. Note: the control unit could be providing the correct signal but it is being affected by a circuit fault. Identify the type of valve and the wiring used on the system. Refer to list of sensor / actuator checks on page 19.
P267A	"B" Rocker Arm Actuator Position Sensor Circuit Bank 2	Rocker arm actuators are used on some variable valve timing / lift systems. Systems can use oil pressure to engage / disengage the rocker arm mechanisms, which when activated enables changes in valve timing / lift. A solenoid operated valve can be used to control the application of the oil (under pressure) to the mechanism. Oil pressure sensors and position sensors can be used to monitor / indicate rocker arm actuation. It is possible that some manufacturers could apply the code to "Cylinder Deactivation" systems (that use the rocker arm as part of the deactivation mechanism).	Identify the type of position sensor fitted (e.g. Hall effect), which can dictate the specific checks that should be carried out. Note that some systems can use an oil pressure sensor to indicate that the Rocker Arm Actuator mechanism has operated / moved. The voltage or value of the position sensor signal is incorrect (undefined fault). Sensor voltage / signal value could be: out of normal operating range, constant value, non-existent, corrupt. Possible fault in the position sensor or wiring e.g. short / open circuits, circuit resistance, incorrect sensor resistance (if applicable) or, fault with the reference / operating voltage; interference from other circuits can also affect sensor signals. Also check earth connections and cable screening. Refer to list of sensor / actuator checks on page 19.
P267B	"B" Rocker Arm Actuator Position Sensor Circuit Range / Performance Bank 2	Rocker arm actuators are used on some variable valve timing / lift systems. Systems can use oil pressure to engage / disengage the rocker arm mechanisms, which when activated enables changes in valve timing / lift. A solenoid operated valve can be used to control the application of the oil (under pressure) to the mechanism. Oil pressure sensors and position sensors can be used to monitor / indicate rocker arm actuation. It is possible that some manufacturers could apply the code to "Cylinder Deactivation" systems (that use the rocker arm as part of the deactivation mechanism).	Identify the type of position sensor fitted (e.g. Hall effect), which can dictate the specific checks that should be carried out. Note that some systems can use an oil pressure sensor to indicate that the Rocker Arm Actuator mechanism has operated / moved. The voltage or value of the position sensor signal is within the normal operating range / tolerance but the signal is not plausible or is incorrect due to an undefined fault. The sensor signal might not match the operating conditions or, the sensor signal is not changing / responding as expected to the changes in conditions. Possible fault in the position sensor or wiring e.g. short / open circuits, circuit resistance, incorrect sensor resistance or, fault with the reference / operating voltage; interference from other circuits can also affect sensor signals. Refer to list of sensor / actuator checks on page 19. It is also possible that the sensor is operating correctly but the position / pressure or activation detected by the sensor is unacceptable for the operating conditions. Check Rocker Arm Actuator operation / movement to ensure correct operation.

Fault code	EOBD / ISO Description	Component / System Description	Meaningful Description and Quick Check
P267C	"B" Rocker Arm Actuator Position Sensor Circuit Low Bank 2	Rocker arm actuators are used on some variable valve timing / lift systems. Systems can use oil pressure to engage / disengage the rocker arm mechanisms, which when activated enables changes in valve timing / lift. A solenoid operated valve can be used to control the application of the oil (under pressure) to the mechanism. Oil pressure sensors and position sensors can be used to monitor / indicate rocker arm actuation. It is possible that some manufacturers could apply the code to "Cylinder Deactivation" systems (that use the rocker arm as part of the deactivation mechanism).	Identify the type of position sensor fitted (e.g. Hall effect), which can dictate the specific checks that should be carried out. Note that some systems can use an oil pressure sensor to indicate that the Rocker Arm Actuator mechanism has operated / moved. The voltage or value of the position sensor signal is either "zero" (no signal) or, the signal exists but it is below the normal operating range (lower than the minimum operating limit). Possible fault in the position sensor or wiring e.g. short / open circuits, circuit resistance, incorrect sensor resistance or, fault with the reference / operating voltage; interference from other circuits can also affect sensor signals. Refer to list of sensor / actuator checks on page 19.
P267D	"B" Rocker Arm Actuator Position Sensor Circuit High Bank 2	Rocker arm actuators are used on some variable valve timing / lift systems. Systems can use oil pressure to engage / disengage the rocker arm mechanisms, which when activated enables changes in valve timing / lift. A solenoid operated valve can be used to control the application of the oil (under pressure) to the mechanism. Oil pressure sensors and position sensors can be used to monitor / indicate rocker arm actuation. It is possible that some manufacturers could apply the code to "Cylinder Deactivation" systems (that use the rocker arm as part of the deactivation mechanism).	Identify the type of position sensor fitted (e.g. Hall effect), which can dictate the specific checks that should be carried out. Note that some systems can use an oil pressure sensor to indicate that the Rocker Arm Actuator mechanism has operated / moved. The voltage or value of the position sensor signal is either at full value (e.g. battery voltage with no signal) or, the signal exists but it is above the normal operating range (higher than the maximum operating limit). Possible fault in the position sensor or wiring e.g. short / open circuits, circuit resistance, incorrect sensor resistance or, fault with the reference / operating voltage; interference from other circuits can also affect sensor signals. Refer to list of sensor / actuator checks on page 19.
P267E	"B" Rocker Arm Actuator Position Sensor Circuit Intermittent / Erratic Bank 2	Rocker arm actuators are used on some variable valve timing / lift systems. Systems can use oil pressure to engage / disengage the rocker arm mechanisms, which when activated enables changes in valve timing / lift. A solenoid operated valve can be used to control the application of the oil (under pressure) to the mechanism. Oil pressure sensors and position sensors can be used to monitor / indicate rocker arm actuation. It is possible that some manufacturers could apply the code to "Cylinder Deactivation" systems (that use the rocker arm as part of the deactivation mechanism).	Identify the type of position sensor fitted (e.g. Hall effect), which can dictate the specific checks that should be carried out. Note that some systems can use an oil pressure sensor to indicate that the Rocker Arm Actuator mechanism has operated / moved. The voltage or value of the position sensor signal is intermittent (likely to be out of normal operating range / tolerance when intermittent fault is detected) or, the signal / voltage is erratic (e.g. signal changes are irregular / unpredictable). Possible fault with wiring or connections (including internal sensor connections); also interference from other circuits can affect the signal. Refer to list of sensor / actuator checks on page 19. Also check for erratic operation of the Rocker Arm Actuation system.
P267F	ISO / SAE reserved		Not used, reserved for future allocation.
P2680	Coolant Degassing Valve Control Circuit High	A "degassing tank" is a US term referring to the expansion / header tank. Some system can use a valve to control coolant flow to aid the degassing / aeration	Identify the degassing valve and valve actuator e.g. solenoid. The voltage, frequency or value of the control signal (provided by the control unit) in the valve circuit is either at full value (e.g. battery voltage with no on / off signal) or, the signal exists but it is above the normal operating range. The code indicates a possible "open circuit" but other faults could activate the same fault code e.g. short circuit, high resistance, interference from other circuits. Note: the control unit could be providing the correct signal but it is being affected by a circuit fault. Identify the type of valve and the wiring used on the system. Refer to list of sensor / actuator checks on page 19.
P2681	Engine Coolant Bypass Valve Control Circuit / Open	The coolant system can include a by-pass system, which often is used as part of the de-gassing system (Refer to fault codes P02678 - 2680); note however a manufacturer could use the fault code to apply to other coolant by-pass systems e.g. EGR coolant by-pass.	Identify the by-pass valve and valve actuator e.g. solenoid. The voltage, frequency or value of the control signal (provided by the control unit) in the valve circuit is incorrect (undefined problem). Signal could be: out of normal operating range, constant value, non-existent, corrupt. The code indicates a possible "open circuit" but other faults could activate the same fault code e.g. short circuit, high resistance, interference from other circuits. Note: the control unit could be providing the correct signal but it is being affected by a circuit fault. Identify the type of valve and the wiring used on the system. Refer to list of sensor / actuator checks on page 19.
P2682	Engine Coolant Bypass Valve Control Circuit Low	The coolant system can include a by-pass system, which often is used as part of the de-gassing system (Refer to fault codes P02678 - 2680); note however a manufacturer could use the fault code to apply to other coolant by-pass systems e.g. EGR coolant by-pass.	Identify the by-pass valve and valve actuator e.g. solenoid. The voltage, frequency or value of the control signal (provided by the control unit) in the valve circuit is either "zero" (no signal) or, the signal exists but is below the normal operating range. The code indicates a possible "open circuit" but other faults could activate the same fault code e.g. short circuit, high resistance, interference from other circuits. Note: the control unit could be providing the correct signal but it is being affected by a circuit fault. Identify the type of valve and the wiring used on the system. Refer to list of sensor / actuator checks on page 19.
P2683	Engine Coolant Bypass Valve Control Circuit High	The coolant system can include a by-pass system, which often is used as part of the de-gassing system (Refer to fault codes P02678 - 2680); note however a manufacturer could use the fault code to apply to other coolant by-pass systems e.g. EGR coolant by-pass.	Identify the by-pass valve and valve actuator e.g. solenoid. The voltage, frequency or value of the control signal (provided by the control unit) in the valve circuit is either at full value (e.g. battery voltage with no on / off signal) or, the signal exists but it is above the normal operating range. The code indicates a possible "open circuit" but other faults could activate the same fault code e.g. short circuit, high resistance, interference from other circuits. Note: the control unit could be providing the correct signal but it is being affected by a circuit fault. Identify the type of valve and the wiring used on the system. Refer to list of sensor / actuator checks on page 19.
P2684	Actuator Supply Voltage "C" Circuit / Open	The control unit can monitor actuator supply voltages in those systems and circuits where the voltage supply is via relay or another control unit. The control unit can also provide the supply voltage for some actuators, as well as regulating the supply voltage (pulse width / duty cycle control), therefore the voltage can be regulated to enable more accurate control of the actuators. The control unit compares the actual voltage or regulated voltage provided to the actuator with the expected value (held in memory). It will be necessary to identify which actuators are affected and whether the supply voltage is provided by the main control unit, a relay or an independent control unit.	The fault code indicates an actuator(s) supply voltage fault, with a possible "open circuit" or other circuit fault. Identify the source of the voltage supply e.g. relay, main control unit or separate control unit. If the control unit detects a fault with an independent supply (e.g. relay or separate control unit), the main control unit could be connected to the supply using a "sensing wire". It will be necessary to check if the actuator(s) are receiving a supply, and check whether there is an open circuit in the sensing wire. The control unit could be monitoring the voltage in the actuator circuit via the earth path (earth path switched to earth by the control unit); if there is no supply voltage passing through the actuator into the earth circuit, the control unit can assess this as an open circuit. If the control unit is providing the supply voltage to an actuator, and there is an open circuit or circuit fault, this can also be detected by the control unit. In all cases, check for short / open circuits in the wiring and check for an short / open circuit or incorrect resistance with the actuator.

Code	Description	System Operation	Fault Diagnosis
P2685	Actuator Supply Voltage "C" Circuit Low	The control unit can monitor actuator supply voltages in those systems and circuits where the voltage supply is via relay or another control unit. The control unit can also provide the supply voltage for some actuators, as well as regulating the supply voltage (pulse width / duty cycle control), therefore the voltage can be regulated to enable more accurate control of the actuators. The control unit compares the actual voltage or regulated voltage provided to the actuator with the expected value (held in memory). It will be necessary to identify which actuators are affected and whether the supply voltage is provided by the main control unit, a relay or an independent control unit.	The fault code indicates that the supply voltage to an actuator(s) is high; voltage could be at full value at all times e.g. battery voltage (with no signal or frequency for those actuators where the control unit provides a control signal on the power circuit). Identify the source of the voltage supply e.g. relay, main control unit or separate control unit, and identify the wiring circuit. In all cases, check for short / open circuits in the wiring, high circuit resistances and check for a short / open circuit or incorrect resistance with the actuator. For those actuators where the control unit switches the actuator earth path through to earth, it is possible that the earth path has a short to a power supply circuit (causing voltage to permanently exist in the circuit). If the control unit detects a fault with an independent power supply (e.g. relay or separate control unit), the main control unit could be detecting the fault via a "sensing wire". Check if the actuator(s) are receiving the correct supply (also check charging voltage).
P2686	Actuator Supply Voltage "C" Circuit High	The control unit can monitor actuator supply voltages in those systems and circuits where the voltage supply is via relay or another control unit. The control unit can also provide the supply voltage for some actuators, as well as regulating the supply voltage (pulse width / duty cycle control), therefore the voltage can be regulated to enable more accurate control of the actuators. The control unit compares the actual voltage or regulated voltage provided to the actuator with the expected value (held in memory). It will be necessary to identify which actuators are affected and whether the supply voltage is provided by the main control unit, a relay or an independent control unit.	The fault code indicates that the supply voltage to an actuator(s) is high; voltage could be at full value at all times e.g. battery voltage (with no signal or frequency for those actuators where the control unit provides a control signal on the power circuit). Identify the source of the voltage supply e.g. relay, main control unit or separate control unit, and identify the wiring circuit. In all cases, check for short / open circuits in the wiring, high circuit resistances and check for a short / open circuit or incorrect resistance with the actuator. For those actuators where the control unit switches the actuator earth path through to earth, it is possible that the earth path has a short to a power supply circuit (causing voltage to permanently exist in the circuit). If the control unit detects a fault with an independent power supply (e.g. relay or separate control unit), the main control unit could be detecting the fault via a "sensing wire". Check if the actuator(s) are receiving the correct supply (also check charging voltage).
P2687	Fuel Supply Heater Control Circuit / Open	A fuel heater can be fitted (often in the Diesel fuel filter) to increase the temperature of the fuel, which can prevent low temperature "waxing". Heaters can receive a power supply via a relay, which can be controlled by the Diesel / engine control unit.	It may be necessary to identify the system wiring. The fault code indicates a fault in the Heater control circuit, which could relate to the wiring between the heater and the control module. If a relay is used in the circuit, the fault could relate to the wiring between the heater and the relay or between the relay and the control unit. The fault code indicates a possible "open circuit" but other undefined faults in the actuator or wiring could activate the same code e.g. power supply to heater or relay, short circuit in heater or relay wiring, high circuit resistance or faulty relay operation. Check all applicable wiring for short / open circuits, check relay operation and check power supply to relay or heater. Also check the earth path for the heater for short / open circuit and high resistance. The fault code could be activated because the relay contacts are stuck open / closed (circuit voltage is not changing as expected).
P2688	Fuel Supply Heater Control Circuit Low	A fuel heater can be fitted (often in the Diesel fuel filter) to increase the temperature of the fuel, which can prevent low temperature "waxing". Heaters can receive a power supply via a relay, which can be controlled by the Diesel / engine control unit.	It may be necessary to identify the system wiring. The fault code indicates a fault in the Heater control circuit, which could relate to the wiring between the heater and the control module. If a relay is used in the circuit, the fault could relate to the wiring between the heater and the relay or between the relay and the control unit. The fault code indicates that the voltage / control signal value in the control circuit is either "zero" (no signal) or, the signal exists but it is below the normal operating range (lower than the minimum operating limit). The fault could be caused by a fault with the power supply to heater or relay, short circuit in heater or relay wiring, high circuit resistance or faulty relay operation. Check all applicable wiring for short / open circuits, check relay operation and check power supply to relay or heater. Also check the earth path for the heater for short / open circuit and high resistance. The fault code could be activated because the relay contacts are stuck open / closed (circuit voltage is not changing as expected).
P2689	Fuel Supply Heater Control Circuit High	A fuel heater can be fitted (often in the Diesel fuel filter) to increase the temperature of the fuel, which can prevent low temperature "waxing". Heaters can receive a power supply via a relay, which can be controlled by the Diesel / engine control unit.	It may be necessary to identify the system wiring. The fault code indicates a fault in the Heater control circuit, which could relate to the wiring between the heater and the control module. If a relay is used in the circuit, the fault could relate to the wiring between the heater and the relay or between the relay and the control unit. The fault code indicates that the voltage / control signal value in the control circuit is either at full value (e.g. battery voltage with no signal or frequency) or, the signal exists but it is above the normal operating range (higher than the maximum operating limit). The fault could be caused by a fault with the power supply to heater or relay, short circuit in heater or relay wiring, high circuit resistance or faulty relay operation. Check all applicable wiring for short / open circuits, check relay operation and check power supply to relay or heater. Also check the earth path for the heater for short / open circuit and high resistance. The fault code could be activated because the relay contacts are stuck open / closed (circuit voltage is not changing as expected).
P268A		ISO / SAE reserved	Not used, reserved for future allocation.
P268B		ISO / SAE reserved	Not used, reserved for future allocation.
P268C		ISO / SAE reserved	Not used, reserved for future allocation.
P268D		ISO / SAE reserved	Not used, reserved for future allocation.
P268E		ISO / SAE reserved	Not used, reserved for future allocation.
P268F		ISO / SAE reserved	Not used, reserved for future allocation.
P2700	Transmission Friction Element "A" Apply Time Range / Performance	The control unit monitors the input speed and output speed in the transmission to establish that the requested gear has been selected or is being selected (gear is requested by control unit). It is therefore possible for the control unit to detect the time taken for the gear change system to complete the engagement of a selected gear. Gear change time is dependent on the electro / hydraulic actuation of the friction element.	The fault code indicates that "engagement time" for the gear requested by the control unit is not correct; the fault is related to the operation of the identified friction element e.g. gear selection clutch or brake band. Refer to vehicle specific information to identify the applicable friction element. The fault could be related to an electrical / electronic fault e.g. control unit, shift or pressure control solenoids. The fault could also be caused by a hydraulic pressure problem or, due to a fault with the friction element and / or mechanical problem with the gear engagement system.

NOTE 1. Check for other fault codes that could provide additional information. **NOTE 2.** Communication between control units can pass via a CAN-Bus system; refer to Page 51 for CAN-Bus checks.
NOTE 3. If a fault cannot be located, it is also possible that the control unit is at fault. **NOTE 4.** Refer to Page 53 for list of pages that show the ISO standard for component locations e.g. Sensor A - Bank 1.

Fault code	EOBD / ISO Description	Component / System Description	Meaningful Description and Quick Check
P2701	Transmission Friction Element "B" Apply Time Range / Performance	The control unit monitors the input speed and output speed in the transmission to establish that the requested gear has been selected or is being selected (gear is requested by control unit). It is therefore possible for the control unit to detect the time taken for the gear change system to complete the engagement of a selected gear. Gear change time is dependent on the electro / hydraulic actuation of the friction element.	The fault code indicates that "engagement time" for the gear requested by the control unit is not correct; the fault is related to the operation of the identified friction element e.g. gear selection clutch or brake band. Refer to vehicle specific information to identify the applicable friction element. The fault could be related to an electrical / electronic fault e.g. control unit, shift or pressure control solenoids. The fault could also be caused by a hydraulic pressure problem or, due to a fault with the friction element and / or mechanical problem with the gear engagement system.
P2702	Transmission Friction Element "C" Apply Time Range / Performance	The control unit monitors the input speed and output speed in the transmission to establish that the requested gear has been selected or is being selected (gear is requested by control unit). It is therefore possible for the control unit to detect the time taken for the gear change system to complete the engagement of a selected gear. Gear change time is dependent on the electro / hydraulic actuation of the friction element.	The fault code indicates that "engagement time" for the gear requested by the control unit is not correct; the fault is related to the operation of the identified friction element e.g. gear selection clutch or brake band. Refer to vehicle specific information to identify the applicable friction element. The fault could be related to an electrical / electronic fault e.g. control unit, shift or pressure control solenoids. The fault could also be caused by a hydraulic pressure problem or, due to a fault with the friction element and / or mechanical problem with the gear engagement system.
P2703	Transmission Friction Element "D" Apply Time Range / Performance	The control unit monitors the input speed and output speed in the transmission to establish that the requested gear has been selected or is being selected (gear is requested by control unit). It is therefore possible for the control unit to detect the time taken for the gear change system to complete the engagement of a selected gear. Gear change time is dependent on the electro / hydraulic actuation of the friction element.	The fault code indicates that "engagement time" for the gear requested by the control unit is not correct; the fault is related to the operation of the identified friction element e.g. gear selection clutch or brake band. Refer to vehicle specific information to identify the applicable friction element. The fault could be related to an electrical / electronic fault e.g. control unit, shift or pressure control solenoids. The fault could also be caused by a hydraulic pressure problem or, due to a fault with the friction element and / or mechanical problem with the gear engagement system.
P2704	Transmission Friction Element "E" Apply Time Range / Performance	The control unit monitors the input speed and output speed in the transmission to establish that the requested gear has been selected or is being selected (gear is requested by control unit). It is therefore possible for the control unit to detect the time taken for the gear change system to complete the engagement of a selected gear. Gear change time is dependent on the electro / hydraulic actuation of the friction element.	The fault code indicates that "engagement time" for the gear requested by the control unit is not correct; the fault is related to the operation of the identified friction element e.g. gear selection clutch or brake band. Refer to vehicle specific information to identify the applicable friction element. The fault could be related to an electrical / electronic fault e.g. control unit, shift or pressure control solenoids. The fault could also be caused by a hydraulic pressure problem or, due to a fault with the friction element and / or mechanical problem with the gear engagement system.
P2705	Transmission Friction Element "F" Apply Time Range / Performance	The control unit monitors the input speed and output speed in the transmission to establish that the requested gear has been selected or is being selected (gear is requested by control unit). It is therefore possible for the control unit to detect the time taken for the gear change system to complete the engagement of a selected gear. Gear change time is dependent on the electro / hydraulic actuation of the friction element.	The fault code indicates that "engagement time" for the gear requested by the control unit is not correct; the fault is related to the operation of the identified friction element e.g. gear selection clutch or brake band. Refer to vehicle specific information to identify the applicable friction element. The fault could be related to an electrical / electronic fault e.g. control unit, shift or pressure control solenoids. The fault could also be caused by a hydraulic pressure problem or, due to a fault with the friction element and / or mechanical problem with the gear engagement system.
P2706	Shift Solenoid "F"	The shift solenoid is used to control a hydraulic circuit in the transmission, which enables the applicable gear(s) to be selected / deselected. Refer to vehicle specific information to establish which gear(s) are applicable to the identified solenoid.	The control unit has detected an undefined fault with the shift solenoid system; the fault could be detected due to an electrical problem with the solenoid / wiring (e.g. solenoid power supply, short / open circuit or resistance). Refer to list of sensor / actuator checks on page 19. The fault could also be detected because sensors are detecting incorrect gear shift operation; check operation of any position sensors (if applicable), check hydraulic pressures and gear selection mechanism. Note that gear engagement can be detected by transmission speed sensors which monitor the input and output speeds thus allowing the control unit to calculate if the gear is fully engaged and the time taken to engage a gear (slip in the friction element of the gear engagement system can cause slow engagement or incomplete engagement).
P2707	Shift Solenoid "F" Performance / Stuck Off	The shift solenoid is used to control a hydraulic circuit in the transmission, which enables the applicable gear(s) to be selected / deselected. Refer to vehicle specific information to establish which gear(s) are applicable to the identified solenoid.	The control unit is providing a control signal to the solenoid (signal is within the normal operating range / tolerance) but the control unit has detected that the solenoid is stuck in a fixed (off) position and is not responding to the control signal. The fault could be related to a solenoid / wiring fault. Refer to list of sensor / actuator checks on page 19. Note that the control unit could be detecting the fault because of "a lack of response / no response" from the system or mechanism being controlled by the solenoid; check system (as well as the solenoid) for correct operation, movement and response. A position sensor (if fitted) could be indicating the "stuck position"; also check position sensor operation. Note that gear engagement can be detected by transmission speed sensors which monitor the input and output speeds thus allowing the control unit to calculate if the gear is fully engaged and the time taken to engage a gear (slip in the friction element of the gear engagement system can cause slow engagement or incomplete engagement)
P2708	Shift Solenoid "F" Stuck On	The shift solenoid is used to control a hydraulic circuit in the transmission, which enables the applicable gear(s) to be selected / deselected. Refer to vehicle specific information to establish which gear(s) are applicable to the identified solenoid.	The control unit is providing a control signal to the solenoid (signal is within the normal operating range / tolerance) but the control unit has detected that the solenoid is stuck in a fixed (on) position and is not responding to the control signal. The fault could be related to a solenoid / wiring fault. Refer to list of sensor / actuator checks on page 19. Note that the control unit could be detecting the fault because of "a lack of response / no response" from the system or mechanism being controlled by the solenoid; check system (as well as the solenoid) for correct operation, movement and response. A position sensor (if fitted) could be indicating the "stuck position"; also check position sensor operation.
P2709	Shift Solenoid "F" Electrical	The shift solenoid is used to control a hydraulic circuit in the transmission, which enables the applicable gear(s) to be selected / deselected. Refer to vehicle specific information to establish which gear(s) are applicable to the identified solenoid.	The voltage, frequency or value of the control signal in the solenoid circuit is incorrect (undefined electrical fault). Signal could be:- out of normal operating range, constant value, non-existent, corrupt. Possible fault with solenoid or wiring e.g. solenoid power supply, short / open circuit, high circuit resistance or incorrect solenoid resistance. Note: the control unit could be providing the correct signal but it is being affected by the circuit fault. Refer to list of sensor / actuator checks on page 19.

NOTE 1. Check for other fault codes that could provide additional information.

NOTE 2. Communication between control units can pass via a CAN-Bus system; refer to Page 51 for CAN-Bus checks.

NOTE 3. If a fault cannot be located, it is also possible that the control unit is at fault.

NOTE 4. Refer to Page 53 for list of pages that show the ISO standard for component locations e.g. Sensor A - Bank 1.

306

Code	Fault Location	Description	Possible Cause / Diagnosis
P2710	Shift Solenoid "F" Intermittent	The shift solenoid is used to control a hydraulic circuit in the transmission, which enables the applicable gear(s) to be selected / deselected. Refer to vehicle specific information to establish which gear(s) are applicable to the identified solenoid.	The voltage, frequency or value of the control signal in the solenoid circuit is intermittent (likely to be out of normal operating range / tolerance when intermittent fault is detected) or, the signal is erratic e.g. signal changes are irregular / unpredictable. Possible fault with solenoid or wiring (e.g. internal / external connections, broken wiring) that is causing intermittent or erratic signal to exist. Note: the control unit could be providing the correct signal but it is being affected by a circuit fault. Refer to list of sensor / actuator checks on page 19.
P2711	Unexpected Mechanical Gear Disengagement	The fault code is indicating that a gear has disengaged without a "disengagement command" from the transmission control unit (or without driver selection). It may be necessary to refer to vehicle specific information to identify the components and selection mechanism fitted to the vehicle.	The fault code is indicating that the problem is related to the "mechanical" gear disengagement, which would indicate that the control unit has not requested disengagement; the fault is therefore likely to relate to the gear selection / engagement mechanism or a wiring fault (that has caused the selection engagement mechanism to disengage a gear). Depending on transmission type and design, the gear engagement mechanism could include hydraulic systems, shift / pressure control valves as well as mechanical systems. Check for other fault codes and also refer to vehicle specific information to help identify the possible component or system faults.
P2712	Hydraulic Power Unit Leakage	The fault code indicates that the control unit has detected a hydraulic leak; the leak is likely to be identified due to pressure loss in the system or because pump operation is required frequently / continuously to maintain pressure.	Identify the system components and pressure monitoring system. Check for obvious leaks and rectify. If there are no leaks detected, check pressure with a separate gauge. If the pressure is correct, it indicates a possible pressure sensor fault (where fitted), or the fault could be caused by faulty hydraulic pump. Check all applicable sensor and pump wiring for short / open circuits etc.
P2713	Pressure Control Solenoid "D"	Pressure control solenoids are used in the transmission to regulate the pressure in the hydraulic circuits (usually for gear shift control). The pressure control solenoid can also influence / control the actual gear shift. Refer to vehicle specific information to establish which gear(s) are applicable to the identified solenoid.	The control unit has detected an undefined fault with the pressure control solenoid system; the fault could be detected due to an electrical problem with the solenoid / wiring (e.g. solenoid power supply, short / open circuit or resistance). Refer to list of sensor / actuator checks on page 19. The fault could also be detected because sensors are detecting incorrect gear shift operation; check operation of any position sensors (if applicable), check hydraulic pressures and gear selection mechanism.
P2714	Pressure Control Solenoid "D" Performance / Stuck Off	Pressure control solenoids are used in the transmission to regulate the pressure in the hydraulic circuits (usually for gear shift control). The pressure control solenoid can also influence / control the actual gear shift. Refer to vehicle specific information to establish which gear(s) are applicable to the identified solenoid.	The control unit is providing a control signal to the solenoid (signal is within the normal operating range / tolerance) but the control unit has detected that the solenoid is stuck in a fixed (off) position and is not responding to the control signal. The fault could be related to a solenoid / wiring fault. Refer to list of sensor / actuator checks on page 19. Note that the control unit could be detecting the fault because of "a lack of response / no response" from the system or mechanism being controlled by the solenoid; check system (as well as the solenoid) for correct operation, movement and response. A position / pressure sensor (if fitted) could be indicating the "stuck position"; also check sensor operation.
P2715	Pressure Control Solenoid "D" Stuck On	Pressure control solenoids are used in the transmission to regulate the pressure in the hydraulic circuits (usually for gear shift control). The pressure control solenoid can also influence / control the actual gear shift. Refer to vehicle specific information to establish which gear(s) are applicable to the identified solenoid.	The control unit is providing a control signal to the solenoid (signal is within the normal operating range / tolerance) but the control unit has detected that the solenoid is stuck in a fixed (on) position and is not responding to the control signal. The fault could be related to a solenoid / wiring fault. Refer to list of sensor / actuator checks on page 19. Note that the control unit could be detecting the fault because of "a lack of response / no response" from the system or mechanism being controlled by the solenoid; check system (as well as the solenoid) for correct operation, movement and response. A position / pressure sensor (if fitted) could be indicating the "stuck position"; also check sensor operation.
P2716	Pressure Control Solenoid "D" Electrical	Pressure control solenoids are used in the transmission to regulate the pressure in the hydraulic circuits (usually for gear shift control). The pressure control solenoid can also influence / control the actual gear shift. Refer to vehicle specific information to establish which gear(s) are applicable to the identified solenoid.	The voltage, frequency or value of the control signal in the solenoid circuit is incorrect (undefined electrical fault). Signal could be:- out of normal operating range, constant value, non-existent, corrupt. Possible fault with solenoid or wiring e.g. solenoid power supply, short / open circuit, high circuit resistance or incorrect solenoid resistance. Note: the control unit could be providing the correct signal but it is being affected by a circuit fault. Refer to list of sensor / actuator checks on page 19.
P2717	Pressure Control Solenoid "D" Intermittent	Pressure control solenoids are used in the transmission to regulate the pressure in the hydraulic circuits (usually for gear shift control). The pressure control solenoid can also influence / control the actual gear shift. Refer to vehicle specific information to establish which gear(s) are applicable to the identified solenoid.	The voltage, frequency or value of the control signal in the solenoid circuit is intermittent (likely to be out of normal operating range / tolerance when intermittent fault is detected) or, the signal is erratic e.g. signal changes are irregular / unpredictable. Possible fault with solenoid or wiring (e.g. internal / external connections, broken wiring) that is causing intermittent or erratic signal to exist. Note: the control unit could be providing the correct signal but it is being affected by a circuit fault. Refer to list of sensor / actuator checks on page 19.
P2718	Pressure Control Solenoid "D" Control Circuit / Open	Pressure control solenoids are used in the transmission to regulate the pressure in the hydraulic circuits (usually for gear shift control). The pressure control solenoid can also influence / control the actual gear shift. Refer to vehicle specific information to establish which gear(s) are applicable to the identified solenoid.	The voltage, frequency or value of the control signal in the solenoid circuit is incorrect. Signal could be:- out of normal operating range, constant value, non-existent, corrupt. The fault code indicates a possible "open circuit" but other undefined faults in the solenoid or wiring could activate the same code e.g. solenoid power supply, short circuit, high circuit resistance or incorrect solenoid resistance. Note: the control unit could be providing the correct signal but it is being affected by the circuit fault. Refer to list of sensor / actuator checks on page 19.
P2719	Pressure Control Solenoid "D" Control Circuit Range / Performance	Pressure control solenoids are used in the transmission to regulate the pressure in the hydraulic circuits (usually for gear shift control). The pressure control solenoid can also influence / control the actual gear shift. Refer to vehicle specific information to establish which gear(s) are applicable to the identified solenoid.	The voltage, frequency or value of the control signal in the solenoid circuit is within the normal operating range / tolerance, but the signal might be incorrect for the operating conditions. It is also possible that the solenoid response (or response of system controlled by the solenoid) differs from the expected / desired response (incorrect response to the control signal provided by the control unit). Possible fault in solenoid or wiring, but also possible that the solenoid (or mechanism / system controlled by the solenoid) is not operating or moving correctly. Note: the control unit could be providing the correct signal but it is being affected by a circuit fault. Refer to list of sensor / actuator checks on page 19. Note that the fault code could also be activated due to mechanical or hydraulic faults e.g. restriction in the movement of the solenoid, a mechanical fault with the gear selection / deselection mechanism, a fault with the hydraulic pressure or valves.

NOTE 1. Check for other fault codes that could provide additional information.   **NOTE 2.** Communication between control units can pass via a CAN-Bus system; refer to Page 51 for CAN-Bus checks.   **307**

NOTE 3. If a fault cannot be located, it is also possible that the control unit is at fault.   **NOTE 4.** Refer to Page 53 for list of pages that show the ISO standard for component locations e.g. Sensor A - Bank 1.

Fault code	EOBD / ISO Description	Component / System Description	Meaningful Description and Quick Check
P2720	Pressure Control Solenoid "D" Control Circuit Low	Pressure control solenoids are used in the transmission to regulate the pressure in the hydraulic circuits (usually for gear shift control). The pressure control solenoid can also influence / control the actual gear shift. Refer to vehicle specific information to establish which gear(s) are applicable to the identified solenoid.	The voltage, frequency or value of the control signal in the solenoid circuit is either "zero" (no signal) or, the signal exists but is below the normal operating range. Possible fault with solenoid or wiring (e.g. solenoid power supply, short / open circuit or resistance) that is causing a signal value to be lower than the minimum operating limit. Note: the control unit could be providing the correct signal but it is being affected by a circuit fault. Refer to list of sensor / actuator checks on page 19.
P2721	Pressure Control Solenoid "D" Control Circuit High	Pressure control solenoids are used in the transmission to regulate the pressure in the hydraulic circuits (usually for gear shift control). The pressure control solenoid can also influence / control the actual gear shift. Refer to vehicle specific information to establish which gear(s) are applicable to the identified solenoid.	The voltage, frequency or value of the control signal in the solenoid circuit is either at full value (e.g. battery voltage with no on / off signal) or, the signal exists but it is above the normal operating range. Possible fault with solenoid or wiring (e.g. solenoid power supply, short / open circuit or resistance) that is causing a signal value to be higher than the maximum operating limit. Note: the control unit could be providing the correct signal but it is being affected by a circuit fault. Refer to list of sensor / actuator checks on page 19.
P2722	Pressure Control Solenoid "E"	Pressure control solenoids are used in the transmission to regulate the pressure in the hydraulic circuits (usually for gear shift control). The pressure control solenoid can also influence / control the actual gear shift. Refer to vehicle specific information to establish which gear(s) are applicable to the identified solenoid.	The control unit has detected an undefined fault with the pressure control solenoid system; the fault could be detected due to an electrical problem with the solenoid / wiring (e.g. solenoid power supply, short / open circuit or resistance). Refer to list of sensor / actuator checks on page 19. The fault could also be detected because sensors are detecting incorrect gear shift operation; check operation of any position sensors (if applicable), check hydraulic pressures and gear selection mechanism.
P2723	Pressure Control Solenoid "E" Performance / Stuck Off	Pressure control solenoids are used in the transmission to regulate the pressure in the hydraulic circuits (usually for gear shift control). The pressure control solenoid can also influence / control the actual gear shift. Refer to vehicle specific information to establish which gear(s) are applicable to the identified solenoid.	The control unit is providing a control signal to the solenoid (signal is within the normal operating range / tolerance) but the control unit has detected that the solenoid is stuck in a fixed (off) position and is not responding to the control signal. The fault could be related to a solenoid / wiring fault. Refer to list of sensor / actuator checks on page 19. Note that the control unit could be detecting the fault because of "a lack of response / no response" from the system or mechanism being controlled by the solenoid; check system (as well as the solenoid) for correct operation, movement and response. A position / pressure sensor (if fitted) could be indicating the "stuck position"; also check sensor operation.
P2724	Pressure Control Solenoid "E" Stuck On	Pressure control solenoids are used in the transmission to regulate the pressure in the hydraulic circuits (usually for gear shift control). The pressure control solenoid can also influence / control the actual gear shift. Refer to vehicle specific information to establish which gear(s) are applicable to the identified solenoid.	The control unit is providing a control signal to the solenoid (signal is within the normal operating range / tolerance) but the control unit has detected that the solenoid is stuck in a fixed (on) position and is not responding to the control signal. The fault could be related to a solenoid / wiring fault. Refer to list of sensor / actuator checks on page 19. Note that the control unit could be detecting the fault because of "a lack of response / no response" from the system or mechanism being controlled by the solenoid; check system (as well as the solenoid) for correct operation, movement and response. A position / pressure sensor (if fitted) could be indicating the "stuck position"; also check sensor operation.
P2725	Pressure Control Solenoid "E" Electrical	Pressure control solenoids are used in the transmission to regulate the pressure in the hydraulic circuits (usually for gear shift control). The pressure control solenoid can also influence / control the actual gear shift. Refer to vehicle specific information to establish which gear(s) are applicable to the identified solenoid.	The voltage, frequency or value of the control signal in the solenoid circuit is incorrect (undefined electrical fault). Signal could be:- out of normal operating range, constant value, non-existent, corrupt. Possible fault with solenoid or wiring e.g. solenoid power supply, short / open circuit, high circuit resistance or incorrect solenoid resistance. Note: the control unit could be providing the correct signal but it is being affected by a circuit fault. Refer to list of sensor / actuator checks on page 19.
P2726	Pressure Control Solenoid "E" Intermittent	Pressure control solenoids are used in the transmission to regulate the pressure in the hydraulic circuits (usually for gear shift control). The pressure control solenoid can also influence / control the actual gear shift. Refer to vehicle specific information to establish which gear(s) are applicable to the identified solenoid.	The voltage, frequency or value of the control signal in the solenoid circuit is intermittent (likely to be out of normal operating range / tolerance when intermittent fault is detected) or, the signal is erratic e.g. signal changes are irregular / unpredictable. Possible fault with solenoid or wiring (e.g. internal / external connections, broken wiring) that is causing intermittent or erratic signal to exist. Note: the control unit could be providing the correct signal but it is being affected by a circuit fault. Refer to list of sensor / actuator checks on page 19.
P2727	Pressure Control Solenoid "E" Control Circuit / Open	Pressure control solenoids are used in the transmission to regulate the pressure in the hydraulic circuits (usually for gear shift control). The pressure control solenoid can also influence / control the actual gear shift. Refer to vehicle specific information to establish which gear(s) are applicable to the identified solenoid.	The voltage, frequency or value of the control signal in the solenoid circuit is incorrect. Signal could be:- out of normal operating range, constant value, non-existent, corrupt. The fault code indicates a possible "open circuit" but other undefined faults in the solenoid or wiring could activate the same code e.g. solenoid power supply, short circuit, high circuit resistance or incorrect solenoid resistance. Note: the control unit could be providing the correct signal but it is being affected by the circuit fault. Refer to list of sensor / actuator checks on page 19.
P2728	Pressure Control Solenoid "E" Control Circuit Range / Performance	Pressure control solenoids are used in the transmission to regulate the pressure in the hydraulic circuits (usually for gear shift control). The pressure control solenoid can also influence / control the actual gear shift. Refer to vehicle specific information to establish which gear(s) are applicable to the identified solenoid.	The voltage, frequency or value of the control signal in the solenoid circuit is within the normal operating range / tolerance, but the signal might be incorrect for the operating conditions. It is also possible that the solenoid response (or response of system controlled by the solenoid) differs from the expected / desired response (incorrect response to the control signal provided by the control unit). Possible fault in solenoid or wiring, but also possible that the solenoid (or mechanism / system controlled by the solenoid) is not operating or moving correctly. Note: the control unit could be providing the correct signal but it is being affected by a circuit fault. Refer to list of sensor / actuator checks on page 19. Note that the fault code could also be activated due to mechanical or hydraulic faults e.g. restriction in the movement of the solenoid, a mechanical fault with the gear selection / deselection mechanism, a fault with the hydraulic pressure or valves.
P2729	Pressure Control Solenoid "E" Control Circuit Low	Pressure control solenoids are used in the transmission to regulate the pressure in the hydraulic circuits (usually for gear shift control). The pressure control solenoid can also influence / control the actual gear shift. Refer to vehicle specific information to establish which gear(s) are applicable to the identified solenoid.	The voltage, frequency or value of the control signal in the solenoid circuit is either "zero" (no signal) or, the signal exists but is below the normal operating range. Possible fault with solenoid or wiring (e.g. solenoid power supply, short / open circuit or resistance) that is causing a signal value to be lower than the minimum operating limit. Note: the control unit could be providing the correct signal but it is being affected by a circuit fault. Refer to list of sensor / actuator checks on page 19.

NOTE 1. Check for other fault codes that could provide additional information. **NOTE 2.** Communication between control units can pass via a CAN-Bus system; refer to Page 51 for CAN-Bus checks.

NOTE 4. Refer to Page 52 for list of pages that show the ISO standard for component locations e.g. Sensor A - Bank 1.

P2730	Pressure Control Solenoid "E" Control Circuit High	Pressure control solenoids are used in the transmission to regulate the pressure in the hydraulic circuits (usually for gear shift control). The pressure control solenoid can also influence / control the actual gear shift. Refer to vehicle specific information to establish which gear(s) are applicable to the identified solenoid.	The voltage, frequency or value of the control signal in the solenoid circuit is either at full value (e.g. battery voltage with no on / off signal) or, the signal exists but it is above the normal operating range. Possible fault with solenoid or wiring (e.g. solenoid power supply, short / open circuit or resistance) that is causing a signal value to be higher than the maximum operating limit. Note: the control unit could be providing the correct signal but it is being affected by a circuit fault. Refer to list of sensor / actuator checks on page 19.
P2731	Pressure Control Solenoid "F"	Pressure control solenoids are used in the transmission to regulate the pressure in the hydraulic circuits (usually for gear shift control). The pressure control solenoid can also influence / control the actual gear shift. Refer to vehicle specific information to establish which gear(s) are applicable to the identified solenoid.	The control unit has detected an undefined fault with the pressure control solenoid system; the fault could be detected due to an electrical problem with the solenoid / wiring (e.g. solenoid power supply, short / open circuit or resistance). Refer to list of sensor / actuator checks on page 19. The fault could also be detected because sensors are detecting incorrect gear shift operation; check operation of any position sensors (if applicable), check hydraulic pressures and gear selection mechanism.
P2732	Pressure Control Solenoid "F" Performance / Stuck Off	Pressure control solenoids are used in the transmission to regulate the pressure in the hydraulic circuits (usually for gear shift control). The pressure control solenoid can also influence / control the actual gear shift. Refer to vehicle specific information to establish which gear(s) are applicable to the identified solenoid.	The control unit is providing a control signal to the solenoid (signal is within the normal operating range / tolerance) but the control unit has detected that the solenoid is stuck in a fixed (off) position and is not responding to the control signal. The fault could be related to a solenoid / wiring fault. Refer to list of sensor / actuator checks on page 19. Note that the control unit could be detecting the fault because of "a lack of response / no response" from the system or mechanism being controlled by the solenoid; check system (as well as the solenoid) for correct operation, movement and response. A position / pressure sensor (if fitted) could be indicating the "stuck position"; also check sensor operation.
P2733	Pressure Control Solenoid "F" Stuck On	Pressure control solenoids are used in the transmission to regulate the pressure in the hydraulic circuits (usually for gear shift control). The pressure control solenoid can also influence / control the actual gear shift. Refer to vehicle specific information to establish which gear(s) are applicable to the identified solenoid.	The control unit is providing a control signal to the solenoid (signal is within the normal operating range / tolerance) but the control unit has detected that the solenoid is stuck in a fixed (on) position and is not responding to the control signal. The fault could be related to a solenoid / wiring fault. Refer to list of sensor / actuator checks on page 19. Note that the control unit could be detecting the fault because of "a lack of response / no response" from the system or mechanism being controlled by the solenoid; check system (as well as the solenoid) for correct operation, movement and response. A position / pressure sensor (if fitted) could be indicating the "stuck position"; also check sensor operation.
P2734	Pressure Control Solenoid "F" Electrical	Pressure control solenoids are used in the transmission to regulate the pressure in the hydraulic circuits (usually for gear shift control). The pressure control solenoid can also influence / control the actual gear shift. Refer to vehicle specific information to establish which gear(s) are applicable to the identified solenoid.	The voltage, frequency or value of the control signal in the solenoid circuit is incorrect (undefined electrical fault). Signal could be:- out of normal operating range, constant value, non-existent, corrupt. Possible fault with solenoid or wiring e.g. solenoid power supply, short / open circuit, high circuit resistance or incorrect solenoid resistance. Note: the control unit could be providing the correct signal but it is being affected by a circuit fault. Refer to list of sensor / actuator checks on page 19.
P2735	Pressure Control Solenoid "F" Intermittent	Pressure control solenoids are used in the transmission to regulate the pressure in the hydraulic circuits (usually for gear shift control). The pressure control solenoid can also influence / control the actual gear shift. Refer to vehicle specific information to establish which gear(s) are applicable to the identified solenoid.	The voltage, frequency or value of the control signal in the solenoid circuit is intermittent (likely to be out of normal operating range / tolerance when intermittent fault is detected) or, the signal is erratic e.g. signal changes are irregular / unpredictable. Possible fault with solenoid or wiring (e.g. internal / external connections, broken wiring) that is causing intermittent or erratic signal to exist. Note: the control unit could be providing the correct signal but it is being affected by a circuit fault. Refer to list of sensor / actuator checks on page 19.
P2736	Pressure Control Solenoid "F" Control Circuit / Open	Pressure control solenoids are used in the transmission to regulate the pressure in the hydraulic circuits (usually for gear shift control). The pressure control solenoid can also influence / control the actual gear shift. Refer to vehicle specific information to establish which gear(s) are applicable to the identified solenoid.	The voltage, frequency or value of the control signal in the solenoid circuit is incorrect. Signal could be:- out of normal operating range, constant value, non-existent, corrupt. The fault code indicates a possible "open circuit" but other undefined faults in the solenoid or wiring could activate the same code e.g. solenoid power supply, short circuit, high circuit resistance or incorrect solenoid resistance. Note: the control unit could be providing the correct signal but it is being affected by the circuit fault. Refer to list of sensor / actuator checks on page 19.
P2737	Pressure Control Solenoid "F" Control Circuit Range / Performance	Pressure control solenoids are used in the transmission to regulate the pressure in the hydraulic circuits (usually for gear shift control). The pressure control solenoid can also influence / control the actual gear shift. Refer to vehicle specific information to establish which gear(s) are applicable to the identified solenoid.	The voltage, frequency or value of the control signal in the solenoid circuit is within the normal operating range / tolerance, but the signal might be incorrect for the operating conditions. It is also possible that the solenoid response (or response of system controlled by the solenoid) differs from the expected / desired response (incorrect response to the control signal provided by the control unit). Possible fault in solenoid or wiring, but also possible that the solenoid (or mechanism / system controlled by the solenoid) is not operating or moving correctly. Note: the control unit could be providing the correct signal but it is being affected by a circuit fault. Refer to list of sensor / actuator checks on page 19. Note that the fault code could also be activated due to mechanical or hydraulic faults e.g. restriction in the movement of the solenoid, a mechanical fault with the gear selection / deselection mechanism, a fault with the hydraulic pressure or valves.
P2738	Pressure Control Solenoid "F" Control Circuit Low	Pressure control solenoids are used in the transmission to regulate the pressure in the hydraulic circuits (usually for gear shift control). The pressure control solenoid can also influence / control the actual gear shift. Refer to vehicle specific information to establish which gear(s) are applicable to the identified solenoid.	The voltage, frequency or value of the control signal in the solenoid circuit is either "zero" (no signal) or, the signal exists but is below the normal operating range. Possible fault with solenoid or wiring (e.g. solenoid power supply, short / open circuit or resistance) that is causing a signal value to be lower than the minimum operating limit. Note: the control unit could be providing the correct signal but it is being affected by a circuit fault. Refer to list of sensor / actuator checks on page 19.
P2739	Pressure Control Solenoid "F" Control Circuit High	Pressure control solenoids are used in the transmission to regulate the pressure in the hydraulic circuits (usually for gear shift control). The pressure control solenoid can also influence / control the actual gear shift. Refer to vehicle specific information to establish which gear(s) are applicable to the identified solenoid.	The voltage, frequency or value of the control signal in the solenoid circuit is either at full value (e.g. battery voltage with no on / off signal) or, the signal exists but it is above the normal operating range. Possible fault with solenoid or wiring (e.g. solenoid power supply, short / open circuit or resistance) that is causing a signal value to be higher than the maximum operating limit. Note: the control unit could be providing the correct signal but it is being affected by a circuit fault. Refer to list of sensor / actuator checks on page 19.

NOTE 1. Check for other fault codes that could provide additional information.   **NOTE 2.** Communication between control units can pass via a CAN-Bus system; refer to Page 51 for CAN-Bus checks.   **309**

NOTE 3. If a fault cannot be located, it is also possible that the control unit is at fault.   **NOTE 4.** Refer to Page 53 for list of pages that show the ISO standard for component locations e.g. Sensor A - Bank 1.

Fault code	EOBD / ISO Description	Component / System Description	Meaningful Description and Quick Check
P273A	Transmission Friction Element "G" Apply Time Range / Performance	The control unit monitors the input speed and output speed in the transmission to establish that the requested gear has been selected or is being selected (gear is requested by control unit). It is therefore possible for the control unit to detect the time taken for the gear change system to complete the engagement of a selected gear. Gear change time is dependent on the electro / hydraulic actuation of the friction element.	The fault code indicates that "engagement time" for the gear requested by the control unit is not correct; the fault is related to the operation of the identified friction element e.g. gear selection clutch or brake band. Refer to vehicle specific information to identify the applicable friction element. The fault could be related to an electrical / electronic fault e.g. control unit, shift or pressure control solenoids. The fault could also be caused by a hydraulic pressure problem or, due to a fault with the friction element and / or mechanical problem with the gear engagement system.
P273B	Transmission Friction Element "H" Apply Time Range / Performance	The control unit monitors the input speed and output speed in the transmission to establish that the requested gear has been selected or is being selected (gear is requested by control unit). It is therefore possible for the control unit to detect the time taken for the gear change system to complete the engagement of a selected gear. Gear change time is dependent on the electro / hydraulic actuation of the friction element.	The fault code indicates that "engagement time" for the gear requested by the control unit is not correct; the fault is related to the operation of the identified friction element e.g. gear selection clutch or brake band. Refer to vehicle specific information to identify the applicable friction element. The fault could be related to an electrical / electronic fault e.g. control unit, shift or pressure control solenoids. The fault could also be caused by a hydraulic pressure problem or, due to a fault with the friction element and / or mechanical problem with the gear engagement system.
P273C		ISO / SAE reserved	Not used, reserved for future allocation.
P273D		ISO / SAE reserved	Not used, reserved for future allocation.
P273E		ISO / SAE reserved	Not used, reserved for future allocation.
P273F		ISO / SAE reserved	Not used, reserved for future allocation.
P2740	Transmission Fluid Temperature Sensor "B" Circuit	The information from the transmission fluid temperature sensor can be used to alter transmission operation during cold running / warm up and also influence the gear change strategy; under certain conditions engine control could be influenced by transmission oil temperature. Excessive fluid temperatures will also be indicated by the temperature sensor.	The code identifies an electrical related fault in the fluid temperature sensor but, it might be advisable to check oil cooling and oil condition / grade. The voltage, frequency or value of the temperature sensor signal is incorrect (undefined fault). Sensor voltage / signal could be: out of normal operating range, constant value, non-existent, corrupt. Possible fault in the temperature sensor or wiring e.g. short / open circuits, circuit resistance, incorrect sensor resistance or, fault with the reference / operating voltage; interference from other circuits can also affect sensor signals. Refer to list of sensor / actuator checks on page 19.
P2741	Transmission Fluid Temperature Sensor "B" Circuit Range / Performance	The information from the transmission fluid temperature sensor can be used to alter transmission operation during cold running / warm up and also influence the gear change strategy; under certain conditions engine control could be influenced by transmission oil temperature. Excessive fluid temperatures will also be indicated by the temperature sensor.	The code identifies an electrical related fault in the fluid temperature sensor but, it might be advisable to check oil cooling and oil condition / grade. The voltage, frequency or value of the temperature sensor signal is within the normal operating range / tolerance but the signal is not plausible or is incorrect due to an undefined fault. The sensor signal might not match the operating conditions indicated by other sensors or, the sensor signal is not changing / responding as expected to the changes in conditions. Possible fault in the temperature sensor or wiring e.g. short / open circuits, circuit resistance, incorrect sensor resistance or, fault with the reference / operating voltage; interference from other circuits can also affect sensor signals. Refer to list of sensor / actuator checks on page 19. It is also possible that the sensor is operating correctly but the oil temperature is unacceptable or incorrect. If it is suspected that the temperature is incorrect, check transmission fluid cooling (intercooler etc) and check for correct oil, oil condition and quantity
P2742	Transmission Fluid Temperature Sensor "B" Circuit Low	The information from the transmission fluid temperature sensor can be used to alter transmission operation during cold running / warm up and also influence the gear change strategy; under certain conditions engine control could be influenced by transmission oil temperature. Excessive fluid temperatures will also be indicated by the temperature sensor.	The code identifies an electrical related fault in the fluid temperature sensor but, it might be advisable to check oil cooling and oil condition / grade. The voltage, frequency or value of the temperature sensor signal is either "zero" (no signal) or, the signal exists but it is below the normal operating range (lower than the minimum operating limit). Possible fault in the temperature sensor or wiring e.g. short / open circuits, circuit resistance, incorrect sensor resistance, fault with the reference / operating voltage; interference from other circuits can also affect sensor signals. Refer to list of sensor / actuator checks on page 19.
P2743	Transmission Fluid Temperature Sensor "B" Circuit High	The information from the transmission fluid temperature sensor can be used to alter transmission operation during cold running / warm up and also influence the gear change strategy; under certain conditions engine control could be influenced by transmission oil temperature. Excessive fluid temperatures will also be indicated by the temperature sensor.	The code identifies an electrical related fault in the fluid temperature sensor but, it might be advisable to check oil cooling and oil condition / grade. The voltage, frequency or value of the temperature sensor signal is either at full value (e.g. battery voltage with no signal or frequency) or, the signal exists but it is above the normal operating range (higher than the maximum operating limit). Possible fault in the temperature sensor or wiring e.g. short / open circuits, circuit resistance, incorrect sensor resistance or, fault with the reference / operating voltage; interference from other circuits can also affect sensor signals. Refer to list of sensor / actuator checks on page 19.
P2744	Transmission Fluid Temperature Sensor "B" Circuit Intermittent	The information from the transmission fluid temperature sensor can be used to alter transmission operation during cold running / warm up and also influence the gear change strategy; under certain conditions engine control could be influenced by transmission oil temperature. Excessive fluid temperatures will also be indicated by the temperature sensor.	The code identifies an electrical related fault in the fluid temperature sensor but, it might be advisable to check oil cooling and oil condition / grade. The voltage, frequency or value of the temperature sensor signal is either at full value (e.g. battery voltage with no signal or frequency) or, the signal exists but it is above the normal operating range (higher than the maximum operating limit). Possible fault in the temperature sensor or wiring e.g. short / open circuits, circuit resistance, incorrect sensor resistance or, fault with the reference / operating voltage; interference from other circuits can also affect sensor signals. Refer to list of sensor / actuator checks on page 19.
P2745	Intermediate Shaft Speed Sensor "B" Circuit	Depending on the type of transmission, the information from the transmission intermediate shaft speed sensor can be used to indicate correct selection of gears, timing (speed) for gear changes and for any slip in the transmission system (clutches etc).	Identify the type of sensor e.g. Inductive, Hall effect etc; this will dictate the checks that are applicable. The voltage, frequency or value of the speed sensor signal is incorrect (undefined fault). Sensor voltage / signal value could be: out of normal operating range, constant value, non-existent, corrupt. Possible fault in the speed sensor or wiring e.g. short / open circuits, circuit resistance, incorrect sensor resistance (if applicable) or, fault with the reference / operating voltage; interference from other circuits can also affect sensor signals Also check earth connections and cable screening. Refer to list of sensor / actuator checks on page 19.

NOTE 1. Check for other fault codes that could provide additional information. **NOTE 2.** Communication between control units can pass via a CAN-Bus system; refer to Page 51 for CAN-Bus checks. 310

NOTE 3. If a fault cannot be located, it is also possible that the control unit is at fault. NOTE 4. Refer to Page 53 for list of pages that show the ISO standard for component locations e.g. Sensor A - Bank 1.

Code	Description		
P2746	Intermediate Shaft Speed Sensor "B" Circuit Range / Performance	Depending on the type of transmission, the information from the transmission intermediate shaft speed sensor can be used to indicate correct selection of gears, timing (speed) for gear changes and for any slip in the transmission system (clutches etc).	Identify the type of sensor e.g. Inductive, Hall effect etc; this will dictate the checks that are applicable. The voltage, frequency or value of the speed sensor signal is within the normal operating range / tolerance but the signal is not plausible or is incorrect due to an undefined fault. The sensor signal might not match the operating conditions indicated by other sensors or, the sensor signal is not changing / responding as expected to the changes in conditions. Possible fault in the speed sensor or wiring e.g. short / open circuits, circuit resistance, incorrect sensor resistance or, fault with the reference / operating voltage; interference from other circuits can also affect sensor signals. Refer to list of sensor / actuator checks on page 19. It is also possible that the speed sensor is operating correctly but the speed being measured by the sensor is unacceptable or incorrect for the operating conditions (e.g. the requested gear is not engaged causing incorrect shaft speed). Also check condition of reference teeth on trigger disc / rotor and ensure rotor is turning with shaft.
P2747	Intermediate Shaft Speed Sensor "B" Circuit No Signal	Depending on the type of transmission, the information from the transmission intermediate shaft speed sensor can be used to indicate correct selection of gears, timing (speed) for gear changes and for any slip in the transmission system (clutches etc).	Identify the type of sensor e.g. Inductive, Hall effect etc; this will dictate the checks that are applicable. The voltage, frequency or value of the speed sensor signal is incorrect (undefined fault resulting in no signal being detected). Sensor voltage / signal value could be: constant value or non-existent. Possible fault in the speed sensor or wiring e.g. short / open circuits, circuit resistance, incorrect sensor resistance (if applicable) or, fault with the reference / operating voltage; interference from other circuits can also affect sensor signals Also check earth connections and cable screening. Refer to list of sensor / actuator checks on page 19. Check to ensure that the trigger rotor is turning with the shaft.
P2748	Intermediate Shaft Speed Sensor "B" Circuit Intermittent	Depending on the type of transmission, the information from the transmission intermediate shaft speed sensor can be used to indicate correct selection of gears, timing (speed) for gear changes and for any slip in the transmission system (clutches etc).	Identify the type of sensor e.g. Inductive, Hall effect etc; this will dictate the checks that are applicable. The voltage, frequency or value of the speed sensor signal is intermittent (likely to be out of normal operating range / tolerance when intermittent fault is detected) or, the signal / voltage is erratic (e.g. signal changes are irregular / unpredictable). Possible fault with wiring or connections (including internal sensor connections); also interference from other circuits can affect the signal. Refer to list of sensor / actuator checks on page 19. Also check condition of reference teeth on trigger disc / rotor and ensure rotor is turning with shaft.
P2749	Intermediate Shaft Speed Sensor "C" Circuit	Depending on the type of transmission, the information from the transmission intermediate shaft speed sensor can be used to indicate correct selection of gears, timing (speed) for gear changes and for any slip in the transmission system (clutches etc).	Identify the type of sensor e.g. Inductive, Hall effect etc; this will dictate the checks that are applicable. The voltage, frequency or value of the speed sensor signal is incorrect (undefined fault). Sensor voltage / signal value could be: out of normal operating range, constant value, non-existent, corrupt. Possible fault in the speed sensor or wiring e.g. short / open circuits, circuit resistance, incorrect sensor resistance (if applicable) or, fault with the reference / operating voltage; interference from other circuits can also affect sensor signals Also check earth connections and cable screening. Refer to list of sensor / actuator checks on page 19.
P2750	Intermediate Shaft Speed Sensor "C" Circuit Range / Performance	Depending on the type of transmission, the information from the transmission intermediate shaft speed sensor can be used to indicate correct selection of gears, timing (speed) for gear changes and for any slip in the transmission system (clutches etc).	Identify the type of sensor e.g. Inductive, Hall effect etc; this will dictate the checks that are applicable. The voltage, frequency or value of the speed sensor signal is within the normal operating range / tolerance but the signal is not plausible or is incorrect due to an undefined fault. The sensor signal might not match the operating conditions indicated by other sensors or, the sensor signal is not changing / responding as expected to the changes in conditions. Possible fault in the speed sensor or wiring e.g. short / open circuits, circuit resistance, incorrect sensor resistance or, fault with the reference / operating voltage; interference from other circuits can also affect sensor signals. Refer to list of sensor / actuator checks on page 19. It is also possible that the speed sensor is operating correctly but the speed being measured by the sensor is unacceptable or incorrect for the operating conditions (e.g. the requested gear is not engaged causing incorrect shaft speed). Also check condition of reference teeth on trigger disc / rotor and ensure rotor is turning with shaft.
P2751	Intermediate Shaft Speed Sensor "C" Circuit No Signal	Depending on the type of transmission, the information from the transmission intermediate shaft speed sensor can be used to indicate correct selection of gears, timing (speed) for gear changes and for any slip in the transmission system (clutches etc).	Identify the type of sensor e.g. Inductive, Hall effect etc; this will dictate the checks that are applicable. The voltage, frequency or value of the speed sensor signal is incorrect (undefined fault resulting in no signal being detected). Sensor voltage / signal value could be: constant value or non-existent. Possible fault in the speed sensor or wiring e.g. short / open circuits, circuit resistance, incorrect sensor resistance (if applicable) or, fault with the reference / operating voltage; interference from other circuits can also affect sensor signals Also check earth connections and cable screening. Refer to list of sensor / actuator checks on page 19. Check to ensure that the trigger rotor is turning with the shaft.
P2752	Intermediate Shaft Speed Sensor "C" Circuit Intermittent	Depending on the type of transmission, the information from the transmission intermediate shaft speed sensor can be used to indicate correct selection of gears, timing (speed) for gear changes and for any slip in the transmission system (clutches etc).	Identify the type of sensor e.g. Inductive, Hall effect etc; this will dictate the checks that are applicable. The voltage, frequency or value of the speed sensor signal is intermittent (likely to be out of normal operating range / tolerance when intermittent fault is detected) or, the signal / voltage is erratic (e.g. signal changes are irregular / unpredictable). Possible fault with wiring or connections (including internal sensor connections); also interference from other circuits can affect the signal. Refer to list of sensor / actuator checks on page 19. Also check condition of reference teeth on trigger disc / rotor and ensure rotor is turning with shaft.
P2753	Transmission Fluid Cooler Control Circuit / Open	The transmission system can be fitted with a cooler to maintain the temperature of the transmission fluid at the correct operating temperature. A control system can be used to regulate the transmission fluid flow through the cooler e.g. to ensure that the fluid is not cooled excessively after a cold start.	Identify the cooler control circuit actuator e.g. solenoid. The voltage, frequency or value of the control signal (provided by the control unit) in the valve circuit is incorrect (undefined problem). Signal could be: out of normal operating range, constant value, non-existent, corrupt. The code indicates a possible "open circuit" but other faults could activate the same fault code e.g. short circuit, high resistance, interference from other circuits. Note: the control unit could be providing the correct signal but it is being affected by a circuit fault. Identify the type of valve and the wiring used on the system. Refer to list of sensor / actuator checks on page 19.
P2754	Transmission Fluid Cooler Control Circuit Low	The transmission system can be fitted with a cooler to maintain the temperature of the transmission fluid at the correct operating temperature. A control system can be used to regulate the transmission fluid flow through the cooler e.g. to ensure that the fluid is not cooled excessively after a cold start.	Identify the cooler control circuit actuator e.g. solenoid. The voltage, frequency or value of the control signal (provided by the control unit) in the valve circuit is either "zero" (no signal) or, the signal exists but is below the normal operating range. The code indicates a possible "open circuit" but other faults could activate the same fault code e.g. short circuit, high resistance, interference from other circuits. Note: the control unit could be providing the correct signal but it is being affected by a circuit fault. Identify the type of valve and the wiring used on the system. Refer to list of sensor / actuator checks on page 19.

NOTE 1. Check for other fault codes that could provide additional information. NOTE 2. Communication between control units can pass via a CAN-Bus system; refer to Page 51 for CAN-Bus checks.

NOTE 3. If a fault cannot be located, it is also possible that the control unit is at fault. NOTE 4. Refer to Page 53 for list of pages that show the ISO standard for component locations e.g. Sensor A - Bank 1.

Fault code	EOBD / ISO Description	Component / System Description	Meaningful Description and Quick Check
P2755	Transmission Fluid Cooler Control Circuit High	The transmission system can be fitted with a cooler to maintain the temperature of the transmission fluid at the correct operating temperature. A control system can be used to regulate the transmission fluid flow through the cooler e.g. to ensure that the fluid is not cooled excessively after a cold start.	Identify the cooler control circuit actuator e.g. solenoid. The voltage, frequency or value of the control signal (provided by the control unit) in the valve circuit is either at full value (e.g. battery voltage with no on / off signal) or, the signal exists but it is above the normal operating range. The code indicates a possible "open circuit" but other faults could activate the same fault code e.g. short circuit, high resistance, interference from other circuits. Note: the control unit could be providing the correct signal but it is being affected by a circuit fault. Identify the type of valve and the wiring used on the system. Refer to list of sensor / actuator checks on page 19.
P2756	Torque Converter Clutch Pressure Control Solenoid	The torque converter clutch is used to effectively lock the input and output sides of the converter during certain driving conditions; the clutch is usually operated by hydraulic pressure controlled by a clutch actuator (usually a solenoid valve); a pressure control solenoid can also be used to regulate the hydraulic pressure in the clutch actuation hydraulic circuit. The control unit monitors the converter input speed (engine rpm) and the output speed to determine the amount of slip.	The control unit has detected an undefined fault with the pressure control solenoid system; the fault could be detected due to an electrical problem with the solenoid / wiring (e.g. solenoid power supply, short / open circuit or resistance). Refer to list of sensor / actuator checks on page 19. Note the fault could be detected because the input and output speed sensors are indicating that the clutch lock-up operation is incorrect or, there is "a lack of response / no response" from the mechanism being controlled by the solenoid; check system (as well as the solenoid) for correct operation, movement and response. Also check input / output speed sensors if necessary.
P2757	Torque Converter Clutch Pressure Control Solenoid Control Circuit Performance / Stuck Off	The torque converter clutch is used to effectively lock the input and output sides of the converter during certain driving conditions; the clutch is usually operated by hydraulic pressure controlled by a clutch actuator (usually a solenoid valve); a pressure control solenoid can also be used to regulate the hydraulic pressure in the clutch actuation hydraulic circuit. The control unit monitors the converter input speed (engine rpm) and the output speed to determine the amount of slip.	The control unit is providing a control signal to the solenoid (signal is within the normal operating range / tolerance) but the control unit has detected that the solenoid is stuck in a fixed (off) position and is not responding to the control signal. The fault could be related to a solenoid / wiring fault. Refer to list of sensor / actuator checks on page 19. Note the fault could be detected because the input and output speed sensors are indicating that the clutch lock-up operation is incorrect or, there is "a lack of response / no response" from the mechanism being controlled by the solenoid; check system (as well as the solenoid) for correct operation, movement and response. Also check input / output speed sensors if necessary.
P2758	Torque Converter Clutch Pressure Control Solenoid Control Circuit Stuck On	The torque converter clutch is used to effectively lock the input and output sides of the converter during certain driving conditions; the clutch is usually operated by hydraulic pressure controlled by a clutch actuator (usually a solenoid valve); a pressure control solenoid can also be used to regulate the hydraulic pressure in the clutch actuation hydraulic circuit. The control unit monitors the converter input speed (engine rpm) and the output speed to determine the amount of slip.	The control unit is providing a control signal to the solenoid (signal is within the normal operating range / tolerance) but the control unit has detected that the solenoid is stuck in a fixed (on) position and is not responding to the control signal. The fault could be related to a solenoid / wiring fault. Refer to list of sensor / actuator checks on page 19. Note the fault could be detected because the input and output speed sensors are indicating that the clutch lock-up operation is incorrect or, there is "a lack of response / no response" from the mechanism being controlled by the solenoid; check system (as well as the solenoid) for correct operation, movement and response. Also check input / output speed sensors if necessary.
P2759	Torque Converter Clutch Pressure Control Solenoid Control Circuit Electrical	The torque converter clutch is used to effectively lock the input and output sides of the converter during certain driving conditions; the clutch is usually operated by hydraulic pressure controlled by a clutch actuator (usually a solenoid valve); a pressure control solenoid can also be used to regulate the hydraulic pressure in the clutch actuation hydraulic circuit.	The voltage, frequency or value of the control signal in the solenoid circuit is incorrect (undefined electrical fault). Signal could be:- out of normal operating range, constant value, non-existent, corrupt. Possible fault with solenoid or wiring e.g. solenoid power supply, short / open circuit, high circuit resistance or incorrect solenoid resistance. Note: the control unit could be providing the correct signal but it is being affected by a circuit fault. Refer to list of sensor / actuator checks on page 19.
P2760	Torque Converter Clutch Pressure Control Solenoid Control Circuit Intermittent	The torque converter clutch is used to effectively lock the input and output sides of the converter during certain driving conditions; the clutch is usually operated by hydraulic pressure controlled by a clutch actuator (usually a solenoid valve); a pressure control solenoid can also be used to regulate the hydraulic pressure in the clutch actuation hydraulic circuit.	The voltage, frequency or value of the control signal in the solenoid circuit is intermittent (likely to be out of normal operating range / tolerance when intermittent fault is detected) or, the signal is erratic e.g. signal changes are irregular / unpredictable. Possible fault with solenoid or wiring (e.g. internal / external connections, broken wiring) that is causing intermittent or erratic signal to exist. Note: the control unit could be providing the correct signal but it is being affected by a circuit fault. Refer to list of sensor / actuator checks on page 19.
P2761	Torque Converter Clutch Pressure Control Solenoid Control Circuit / Open	The torque converter clutch is used to effectively lock the input and output sides of the converter during certain driving conditions; the clutch is usually operated by hydraulic pressure controlled by a clutch actuator (usually a solenoid valve); a pressure control solenoid can also be used to regulate the hydraulic pressure in the clutch actuation hydraulic circuit.	The voltage, frequency or value of the control signal in the solenoid circuit is incorrect. Signal could be:- out of normal operating range, constant value, non-existent, corrupt. The fault code indicates a possible "open circuit" but other undefined faults in the solenoid or wiring could activate the same code e.g. solenoid power supply, short circuit, high circuit resistance or incorrect solenoid resistance. Note: the control unit could be providing the correct signal but it is being affected by the circuit fault. Refer to list of sensor / actuator checks on page 19.
P2762	Torque Converter Clutch Pressure Control Solenoid Control Circuit Range / Performance	The torque converter clutch is used to effectively lock the input and output sides of the converter during certain driving conditions; the clutch is usually operated by hydraulic pressure controlled by a clutch actuator (usually a solenoid valve); a pressure control solenoid can also be used to regulate the hydraulic pressure in the clutch actuation hydraulic circuit.	The voltage, frequency or value of the control signal in the solenoid circuit is within the normal operating range / tolerance, but the signal might be incorrect for the operating conditions. It is also possible that the solenoid response (or response of system controlled by the solenoid) differs from the expected / desired response (incorrect response to the control signal provided by the control unit). Possible fault in solenoid or wiring, but also possible that the solenoid (or mechanism / system controlled by the solenoid) is not operating or moving correctly. Note: the control unit could be providing the correct signal but it is being affected by a circuit fault. Refer to list of sensor / actuator checks on page 19.
P2763	Torque Converter Clutch Pressure Control Solenoid Control Circuit High	The torque converter clutch is used to effectively lock the input and output sides of the converter during certain driving conditions; the clutch is usually operated by hydraulic pressure controlled by a clutch actuator (usually a solenoid valve); a pressure control solenoid can also be used to regulate the hydraulic pressure in the clutch actuation hydraulic circuit.	The voltage, frequency or value of the control signal in the solenoid circuit is either at full value (e.g. battery voltage with no on / off signal) or, the signal exists but it is above the normal operating range. Possible fault with solenoid or wiring (e.g. solenoid power supply, short / open circuit or resistance) that is causing a signal value to be higher than the maximum operating limit. Note: the control unit could be providing the correct signal but it is being affected by a circuit fault. Refer to list of sensor / actuator checks on page 19.
P2764	Torque Converter Clutch Pressure Control Solenoid Control Circuit Low	The torque converter clutch is used to effectively lock the input and output sides of the converter during certain driving conditions; the clutch is usually operated by hydraulic pressure controlled by a clutch actuator (usually a solenoid valve); a pressure control solenoid can also be used to regulate the hydraulic pressure in the clutch actuation hydraulic circuit.	The voltage, frequency or value of the control signal in the solenoid circuit is either "zero" (no signal) or, the signal exists but is below the normal operating range. Possible fault with solenoid or wiring (e.g. solenoid power supply, short / open circuit or resistance) that is causing a signal value to be lower than the minimum operating limit. Note: the control unit could be providing the correct signal but it is being affected by a circuit fault. Refer to list of sensor / actuator checks on page 19.

NOTE 1. Check for other fault codes that could provide additional information. **NOTE 2.** Communication between control units can pass via a CAN-Bus system; refer to Page 51 for CAN-Bus checks.

NOTE 3. If a fault cannot be located, it is also possible that the control unit is at fault. **NOTE 4.** Refer to Page 53 for list of pages that show the ISO standard for component locations e.g. Sensor A - Bank 1.

P2765	Input / Turbine Speed Sensor "B" Circuit	Transmission input / output speeds indicate the gear ratio (indicates requested gear has been fully selected, without slip on the gear selection clutches used in auto transmission). The code is likely to relate to the speed sensor for the gearbox input shaft (connected to the "output" side of the torque converter, which the code refers to as the turbine). Note: some manufacturers refer to the turbine as being the "input" side of the torque converter. If a manufacturer has allocated this code to a torque converter "input speed" fault, this could refer to the engine speed signal (passed from the engine control unit or the engine speed sensor). Refer to torque converter fault codes P0740-P0756, P2756-P2770 and also to P0725.	Identify the type of sensor e.g. Inductive, Hall effect etc; this will dictate the checks that are applicable. The voltage, frequency or value of the speed sensor signal is incorrect (undefined fault). Sensor voltage / signal value could be: out of normal operating range, constant value, non-existent, corrupt. Possible fault in the speed sensor or wiring e.g. short / open circuits, circuit resistance, incorrect sensor resistance (if applicable) or, fault with the reference / operating voltage; interference from other circuits can also affect sensor signals. Also check earth connections and cable screening. Refer to list of sensor / actuator checks on page 19.
P2766	Input / Turbine Speed Sensor "B" Circuit Range / Performance	Transmission input / output speeds indicate the gear ratio (indicates requested gear has been fully selected, without slip on the gear selection clutches used in auto transmission). The code is likely to relate to the speed sensor for the gearbox input shaft (connected to the "output" side of the torque converter, which the code refers to as the turbine). Note: some manufacturers refer to the turbine as being the "input" side of the torque converter. If a manufacturer has allocated this code to a torque converter "input speed" fault, this could refer to the engine speed signal (passed from the engine control unit or the engine speed sensor). Refer to torque converter fault codes P0740-P0756, P2756-P2770 and also to P0725.	Identify the type of sensor e.g. Inductive, Hall effect etc; this will dictate the checks that are applicable. The voltage, frequency or value of the speed sensor signal is within the normal operating range / tolerance but the signal is not plausible or is incorrect due to an undefined fault. The sensor signal might not match the operating conditions indicated by other sensors or, the sensor signal is not changing / responding as expected to the changes in conditions. Possible fault in the speed sensor or wiring e.g. short / open circuits, circuit resistance, incorrect sensor resistance or, fault with the reference / operating voltage; interference from other circuits can also affect sensor signals. Refer to list of sensor / actuator checks on page 19. It is also possible that the speed sensor is operating correctly but the speed being measured by the sensor is unacceptable or incorrect for the operating conditions; check for other transmission related faults. Also check condition of reference teeth on trigger disc / rotor and ensure rotor is turning with shaft.
P2767	Input / Turbine Speed Sensor "B" Circuit No Signal	Transmission input / output speeds indicate the gear ratio (indicates requested gear has been fully selected, without slip on the gear selection clutches used in auto transmission). The code is likely to relate to the speed sensor for the gearbox input shaft (connected to the "output" side of the torque converter, which the code refers to as the turbine). Note: some manufacturers refer to the turbine as being the "input" side of the torque converter. If a manufacturer has allocated this code to a torque converter "input speed" fault, this could refer to the engine speed signal (passed from the engine control unit or the engine speed sensor). Refer to torque converter fault codes P0740-P0756, P2756-P2770 and also to P0725.	Identify the type of sensor e.g. Inductive, Hall effect etc; this will dictate the checks that are applicable. The voltage, frequency or value of the speed sensor signal is incorrect (undefined fault resulting in no signal being detected). Sensor voltage / signal value could be: constant value or non-existent. Possible fault in the speed sensor or wiring e.g. short / open circuits, circuit resistance, incorrect sensor resistance (if applicable) or, fault with the reference / operating voltage; interference from other circuits can also affect sensor signals Also check earth connections and cable screening. Refer to list of sensor / actuator checks on page 19. Also check condition of reference teeth on trigger disc / rotor and ensure rotor is turning with shaft.
P2768	Input / Turbine Speed Sensor "B" Circuit Intermittent	Transmission input / output speeds indicate the gear ratio (indicates requested gear has been fully selected, without slip on the gear selection clutches used in auto transmission). The code is likely to relate to the speed sensor for the gearbox input shaft (connected to the "output" side of the torque converter, which the code refers to as the turbine). Note: some manufacturers refer to the turbine as being the "input" side of the torque converter. If a manufacturer has allocated this code to a torque converter "input speed" fault, this could refer to the engine speed signal (passed from the engine control unit or the engine speed sensor). Refer to torque converter fault codes P0740-P0756, P2756-P2770 and also to P0725.	Identify the type of sensor e.g. Inductive, Hall effect etc; this will dictate the checks that are applicable. The voltage, frequency or value of the speed sensor signal is intermittent (likely to be out of normal operating range / tolerance when intermittent fault is detected) or, the signal / voltage is erratic (e.g. signal changes are irregular / unpredictable). Possible fault with wiring or connections (including internal sensor connections); also interference from other circuits can affect the signal. Refer to list of sensor / actuator checks on page 19. Also check condition of reference teeth on trigger disc / rotor and ensure rotor is turning with shaft.
P2769	Torque Converter Clutch Circuit Low	The torque converter clutch is used to effectively lock the input and output sides of the converter during certain driving conditions; the clutch is usually operated by hydraulic pressure controlled by a clutch actuator (usually a solenoid valve); a pressure control solenoid can also be used to regulate the hydraulic pressure in the clutch actuation hydraulic circuit.	The voltage, frequency or value of the control signal in the clutch actuator circuit is either "zero" (no signal) or, the signal exists but is below the normal operating range. Possible fault with actuator or wiring (e.g. power supply, short / open circuit or resistance) that is causing a signal value to be lower than the minimum operating limit. Note: the control unit could be providing the correct signal but it is being affected by a circuit fault. Refer to list of sensor / actuator checks on page 19.
P2770	Torque Converter Clutch Circuit High	The torque converter clutch is used to effectively lock the input and output sides of the converter during certain driving conditions; the clutch is usually operated by hydraulic pressure controlled by a clutch actuator (usually a solenoid valve); a pressure control solenoid can also be used to regulate the hydraulic pressure in the clutch actuation hydraulic circuit.	The voltage, frequency or value of the control signal in the actuator circuit is either at full value (e.g. battery voltage with no on / off signal) or, the signal exists but it is above the normal operating range. Possible fault with actuator or wiring (e.g. power supply, short / open circuit or resistance) that is causing a signal value to be higher than the maximum operating limit. Note: the control unit could be providing the correct signal but it is being affected by a circuit fault. Refer to list of sensor / actuator checks on page 19.
P2771	Four Wheel Drive (4WD) Low Switch Circuit	The switch can be used to select / indicate the "Low" range for the Four Wheel Drive operation.	The voltage or value of the switch signal is incorrect (undefined fault). Switch voltage / signal could be: out of normal operating range, constant value, non-existent, corrupt. Possible fault in the switch or wiring e.g. short / open circuits, circuit resistance or, fault with the reference / operating voltage; the switch contacts could also be faulty e.g. not opening / closing correctly or have a high contact resistance. Refer to list of sensor / actuator checks on page 19. Also check that the switch is set / adjusted correctly (base setting); it could be possible that the switch is indicating an incorrect position or not operating / responding correctly.

Fault code	EOBD / ISO Description	Component / System Description	Meaningful Description and Quick Check
P2772	Four Wheel Drive (4WD) Low Switch Circuit Range / Performance	The switch can be used to select / indicate the "Low" range for the Four Wheel Drive operation.	The voltage value of the switch signal is within the normal operating range / tolerance but the signal is not plausible or is incorrect due to an undefined fault. The switch signal might not match the operating conditions indicated by other sensors or, the switch signal is not changing / responding as expected. Possible fault in the switch or wiring e.g. short / open circuits, circuit resistance or, fault with the reference / operating voltage; also check switch contacts to ensure that they open / close and check continuity of switch contacts (ensure that there is not a high resistance). Refer to list of sensor / actuator checks on page 19. Also check that the switch is set / adjusted correctly (base setting); it could be possible that the switch is indicating an incorrect position or not operating / responding correctly.
P2773	Four Wheel Drive (4WD) Low Switch Circuit Low	The switch can be used to select / indicate the "Low" range for the Four Wheel Drive operation.	The voltage or value of the switch signal is either "zero" (no signal) or, the signal exists but it is below the normal operating range (lower than the minimum operating limit). Possible fault in the switch or wiring e.g. short / open circuits, circuit resistance or, fault with the reference / operating voltage; the switch contacts could also be faulty e.g. not opening / closing correctly or have a high contact resistance. Refer to list of sensor / actuator checks on page 19. Also check that the switch is set / adjusted correctly (base setting); it could be possible that the switch is indicating an incorrect position or not operating / responding correctly.
P2774	Four Wheel Drive (4WD) Low Switch Circuit High	The switch can be used to select / indicate the "Low" range for the Four Wheel Drive operation.	The voltage or value of the switch signal is either at full value (e.g. battery voltage with no signal or frequency) or, the signal exists but it is above the normal operating range (higher than the maximum operating limit). Possible fault in the switch or wiring e.g. short / open circuits, circuit resistance or, fault with the reference / operating voltage; the switch contacts could also be faulty e.g. not opening / closing correctly or have a high contact resistance. Refer to list of sensor / actuator checks on page 19. Also check that the switch is set / adjusted correctly (base setting); it could be possible that the switch is indicating an incorrect position or not operating / responding correctly.
P2775	Upshift Switch Circuit Range / Performance	The Upshift switch is likely to refer to the system that enables manual control of gear changes on an auto transmission or possibly on an electronically controlled manual gearbox. The manual controls could be up or down shift paddles / buttons (on steering column or steering wheel), or it could be a specific movement of the main gear lever.	The fault could possibly be detected due to the voltage in the Upshift switch circuit indicating permanent switch operation i.e. voltage remaining permanently high or low. It is also possible that the fault is detected during a system self check. The switch signal value is within the normal operating range / tolerance but the signal might not match the operating conditions indicated by other sensors or, the switch signal is not changing / responding as expected. Possible fault in the switch or wiring e.g. short / open circuits, circuit resistance or, fault with the reference / operating voltage; also check switch contacts to ensure that they open / close and check continuity of switch contacts (ensure that there is not a high resistance). Check mechanism / linkage used to operate switch, and check base settings for the switch if applicable. Refer to list of sensor / actuator checks on page 19.
P2776	Upshift Switch Circuit Low	The Upshift switch is likely to refer to the system that enables manual control of gear changes on an auto transmission or possibly on an electronically controlled manual gearbox. The manual controls could be up or down shift paddles / buttons (on steering column or steering wheel), or it could be a specific movement of the main gear lever.	The fault could possibly be detected due to the voltage in the Upshift switch circuit being permanently "off" i.e. voltage remaining permanently low. The switch signal is either "zero" (no voltage / signal) or, the signal exists but it is below the normal operating range (lower than the minimum operating limit). Possible fault in the switch or wiring e.g. short / open circuits, circuit resistance or, fault with the reference / operating voltage; the switch contacts could also be faulty e.g. not opening / closing correctly or have a high contact resistance. Refer to list of sensor / actuator checks on page 19.
P2777	Upshift Switch Circuit High	The Upshift switch is likely to refer to the system that enables manual control of gear changes on an auto transmission or possibly on an electronically controlled manual gearbox. The manual controls could be up or down shift paddles / buttons (on steering column or steering wheel), or it could be a specific movement of the main gear lever.	The fault could possibly be detected due to the voltage in the Upshift switch circuit being permanently "off" i.e. voltage remaining permanently High. The switch signal is either at full value (e.g. battery voltage with no signal or frequency) or, the signal exists but it is above the normal operating range (higher than the maximum operating limit). Possible fault in the switch or wiring e.g. short / open circuits, circuit resistance or, fault with the reference / operating voltage; the switch contacts could also be faulty e.g. not opening / closing correctly or have a high contact resistance. Refer to list of sensor / actuator checks on page 19.
P2778	Upshift Switch Circuit Intermittent / Erratic	The Upshift switch is likely to refer to the system that enables manual control of gear changes on an auto transmission or possibly on an electronically controlled manual gearbox. The manual controls could be up or down shift paddles / buttons (on steering column or steering wheel), or it could be a specific movement of the main gear lever.	The fault could possibly be detected due to the voltage in the Upshift switch circuit being intermittent or erratic (probably when switch is on but could possibly be an intermittent voltage occurring when switch is off). The switch signal is intermittent (likely to be out of normal operating range / tolerance when intermittent fault is detected) or, the signal / voltage is erratic. Possible fault with wiring or connections (including internal switch connections); the switch contacts (if applicable) could also be faulty e.g. not opening / closing correctly or have a high contact resistance. Refer to list of sensor / actuator checks on page 19.
P2779	Downshift Switch Circuit Range / Performance	The Downshift switch is likely to refer to the system that enables manual control of gear changes on an auto transmission or possibly on an electronically controlled manual gearbox. The manual controls could be up or down shift paddles / buttons (on steering column or steering wheel), or it could be a specific movement of the main gear lever.	The fault could possibly be detected due to the voltage in the Upshift switch circuit indicating permanent switch operation i.e. voltage remaining permanently high or low. It is also possible that the fault is detected during a system self check. The switch signal value is within the normal operating range / tolerance but the signal might not match the operating conditions indicated by other sensors or, the switch signal is not changing / responding as expected. Possible fault in the switch or wiring e.g. short / open circuits, circuit resistance or, fault with the reference / operating voltage; also check switch contacts to ensure that they open / close and check continuity of switch contacts (ensure that there is not a high resistance). Check mechanism / linkage used to operate switch, and check base settings for the switch if applicable. Refer to list of sensor / actuator checks on page 19.
P2780	Downshift Switch Circuit Low	The Downshift switch is likely to refer to the system that enables manual control of gear changes on an auto transmission or possibly on an electronically controlled manual gearbox. The manual controls could be up or down shift paddles / buttons (on steering column or steering wheel), or it could be a specific movement of the main gear lever.	The fault could possibly be detected due to the voltage in the Upshift switch circuit being permanently "off" i.e. voltage remaining permanently low. The switch signal is either "zero" (no voltage / signal) or, the signal exists but it is below the normal operating range (lower than the minimum operating limit). Possible fault in the switch or wiring e.g. short / open circuits, circuit resistance or, fault with the reference / operating voltage; the switch contacts could also be faulty e.g. not opening / closing correctly or have a high contact resistance. Refer to list of sensor / actuator checks on page 19.
P2781	Downshift Switch Circuit High	The Downshift switch is likely to refer to the system that enables manual control of gear changes on an auto transmission or possibly on an electronically controlled manual gearbox. The manual controls could be up or down shift paddles / buttons (on steering column or steering wheel), or it could be a specific movement of the main gear lever.	The fault could possibly be detected due to the voltage in the Upshift switch circuit being permanently "off" i.e. voltage remaining permanently High. The switch signal is either at full value (e.g. battery voltage with no signal or frequency) or, the signal exists but it is above the normal operating range (higher than the maximum operating limit). Possible fault in the switch or wiring e.g. short / open circuits, circuit resistance or, fault with the reference / operating voltage; the switch contacts could also be faulty e.g. not opening / closing correctly or have a high contact resistance. Refer to list of sensor / actuator checks on page 19.

NOTE 1. Check for other fault codes that could provide additional information.   **NOTE 2.** Communication between control units can pass via a CAN-Bus system; refer to Page 51 for CAN-Bus checks.   314

NOTE 3. If a fault is not cleared / it is also possible that the control unit is at fault   NOTE 4. Refer to Page 53 for list of pages that show the ISO standard for component locations e.g. Sensor A - Bank 1.

Code	Fault Location	Description	Fault Details / Diagnosis
P2782	Downshift Switch Circuit Intermittent / Erratic	The Downshift switch is likely to refer to the system that enables manual control of gear changes on an auto transmission or possibly on an electronically controlled manual gearbox. The manual controls could be up or down shift paddles / buttons (on steering column or steering wheel), or it could be a specific movement of the main gear lever.	The fault could possibly be detected due to the voltage in the Upshift switch circuit being intermittent or erratic (probably when switch is on but could possibly be an intermittent voltage occurring when switch is off). The switch signal is intermittent (likely to be out of normal operating range / tolerance when intermittent fault is detected) or, the signal / voltage is erratic. Possible fault with wiring or connections (including internal switch connections); the switch contacts (if applicable) could also be faulty e.g. not opening / closing correctly or have a high contact resistance. Refer to list of sensor / actuator checks on page 19.
P2783	Torque Converter Temperature Too High	A temperature sensor can be used to monitor the torque converter temperature.	The fault code could is likely to be activated if the torque converter temperature exceeds the maximum acceptable limit. The fault could be temporary (due to unusual driving conditions or high ambient temperature etc). However the fault could also be caused due to transmission fluid cooler faults, oil deteriorated / low / incorrect or transmission / torque converter fault. If there are no obvious faults, clear the fault code and if the code re-appears carry out further investigations. If necessary, also check operation of temperature sensor, Refer to list of sensor / actuator checks on page 19.
P2784	Input / Turbine Speed Sensor "A" / "B" Correlation	Transmission systems can be fitted with a number of speed sensors to indicate input / output shaft speeds, and torque converter turbine speed.	If necessary, refer to vehicle specific information to identify the A and B speed sensors and to identify the type of speed sensors (Hall effect, inductive etc). The fault code indicates that the two speed sensors are providing conflicting speed information. Initially carry out speed sensor and wiring checks to establish whether one of the sensors is faulty; Refer to list of sensor / actuator checks on page 19. If the speed sensors appear to be operating correctly, it could indicate a transmission / torque converter related fault; the transmission system type and design will dictate what checks will need to be carried out.
P2785	Clutch Actuator Temperature Too High	A temperature sensor can be used to monitor the clutch actuator temperature (usually on electronically controlled systems). Depending on system design, the fault could relate to the electrical actuator e.g. a solenoid valve or to other components of the clutch actuation system.	The fault code could is likely to be activated if the clutch actuator temperature exceeds the maximum acceptable limit. The fault could be temporary (due to unusual driving conditions / excessive load or high ambient temperature etc). However the fault could also be caused due to worn / faulty clutch or clutch actuation system components. If there are no obvious faults, clear the fault code and if the code re-appears carry out further investigations. If necessary, also check operation of temperature sensor, Refer to list of sensor / actuator checks on page 19.
P2786	Gear Shift Actuator Temperature Too High	A temperature sensor can be used to monitor the temperature of the gear shift actuator e.g. solenoid / motor.	The fault code could is likely to be activated if the gear shift actuator temperature exceeds the maximum acceptable limit. The fault could be temporary (due to unusual driving conditions / excessive use or high ambient temperature etc). However the fault could also be caused due to worn or faulty gear shift actuation system components. If there are no obvious faults, clear the fault code and if the code re-appears carry out further investigations. If necessary, also check operation of temperature sensor, Refer to list of sensor / actuator checks on page 19.
P2787	Clutch Temperature Too High	A temperature sensor could be used to monitor the temperature of the clutch (usually on electronically controlled systems). However, systems also use a "Calculated" temperature value which is based on: Ambient Temperature, Time since engine start, the amount of torque being transmitted through the clutch and the amount of clutch slip. A high clutch temperature is therefore most likely to occur due to operating conditions and / or wear on some components (also check for correct system operation e.g. clutch engagement).	The fault code could is likely to be activated if the clutch temperature exceeds the maximum acceptable limit. The fault could be temporary due to a period where abnormal driving conditions were causing the clutch to overheat e.g. continuous stop / start traffic going uphill or, too many aggressive / "launch controlled" starts; it is therefore possible that the clutch is not faulty but it may have been briefly exposed to aggressive use. However the fault could also be caused due to worn / faulty clutch or faulty operation of the clutch control / engagement system. If there are no obvious faults, clear the fault code and if the code re-appears carry out further investigations. If necessary, also check operation of temperature sensor (ambient or clutch as applicable). Refer to list of sensor / actuator checks on page 19.
P2788	Auto Shift Manual Adaptive Learning at Limit	The auto shift manual control refers to the system that allows manual or auto control of gear changes on an auto transmission or possibly on an electronically controlled manual gearbox. The manual controls could be up or down shift paddles / buttons (on steering column or steering wheel), or it could be a specific movement of the main gear lever.	The adaptive learning process enables the control module to adapt the operation of the system based on changes detected by the control module; this can include changes in operating conditions (temperature etc), changes in driving style, but can also be related to changes in component operation e.g. if an actuator requires a slightly different control signal to complete a task such as changing to the next gear. For this fault code, it is possible that the control module is unable to adapt (learn or re-learn) to changes that have occurred; although the adaptive learning process has been functioning, it has reached the limit of its ability to control / adapt to the changes. It is possible that fitting a replacement component could have caused the fault (also check if there is a re-initialisation process for the module). The fault code indicates that there is a fault with the adaptive learning process; this could relate to manual or auto shift process. Check for other fault codes, check all wiring and connections to the control unit (from the sensors, actuators and selection switches (driver controls).
P2789	Clutch Adaptive Learning at Limit	The fault code relates to the a system of electronic control for clutch operation. The clutch adaptive learning process refers to the system where the transmission control module learns what value of control signal is required to position / control the clutch movement a given amount, as well as learning the stop positions (clutch engaged / disengaged). A clutch position / movement sensor provides information back to the control unit to support the learning and clutch operation process.	The adaptive learning process enables the control module to adapt the operation of the clutch based on changes detected by the control module; this can include changes in operating conditions (temperature etc), changes in driving style, but can also be related to changes in component operation due to wear etc. For this fault code, it is possible that the control module is unable to adapt (learn or re-learn) to changes that have occurred; although the adaptive learning process has been functioning, it has reached the limit of its ability to control / adapt to the changes. The problem could be caused by a fault in the clutch position sensor / wiring e.g. short / open circuit; Refer to list of sensor / actuator checks on page 19. The fault could also be caused by clutch actuation faults (actuation mechanism or a mechanical fault) as well as clutch wear or, the control unit could be faulty. On some vehicles, there can be a set up procedure which enables the control unit to "learn" the base setting e.g. after fitting of a new clutch; it will therefore be necessary check for any specific procedures.
P278A	Kick Down Switch Circuit	Auto transmission systems can use a switch to indicate full open throttle position; the switch signal is used to help calculate whether a lower gear should be selected to achieve the desired vehicle performance. The switch could be connected to the throttle pedal or connected to the throttle linkage / mechanism. The switch could form part of a throttle position sensor assembly which could contain more than one switch or sensor, refer to vehicle specific information to establish which sensor or switch is affected. Note that some systems may use a vacuum operated switch (activated by changes in intake manifold vacuum).	The voltage or value of the switch signal is incorrect (undefined fault). The switch voltage / signal could be: out of normal operating range, constant value, non-existent, corrupt. Possible fault in switch or wiring e.g. short / open circuits, circuit resistance, incorrect sensor resistance or, fault with the reference / operating voltage; the switch contacts (if applicable) could also be faulty e.g. not opening / closing correctly or have a high contact resistance. Refer to list of sensor / actuator checks on page 19.

NOTE 1. Check for other fault codes that could provide additional information. NOTE 2. Communication between control units can pass via a CAN-Bus system; refer to Page 51 for CAN-Bus checks.

NOTE 3. If a fault cannot be located, it is also possible that the control unit is at fault. NOTE 4. Refer to Page 53 for list of pages that show the ISO standard for component locations e.g. Sensor A - Bank 1.

Fault code	EOBD / ISO Description	Component / System Description	Meaningful Description and Quick Check
P278B	Kick Down Switch Circuit Range / Performance	Auto transmission systems can use a switch to indicate full open throttle position; the switch signal is used to help calculate whether a lower gear should be selected to achieve the desired vehicle performance. The switch could be connected to the throttle pedal or connected to the throttle linkage / mechanism. The switch could form part of a throttle position sensor assembly which could contain more than one switch or sensor, refer to vehicle specific information to establish which sensor or switch is affected. Note that some systems may use a vacuum operated switch (activated by changes in intake manifold vacuum).	The voltage or value of the switch signal is within the normal operating range / tolerance but the signal is not plausible or is incorrect due to an undefined fault or, the switch signal is not changing / responding as expected. Possible fault in the switch or wiring e.g. short / open circuits, circuit resistance, incorrect sensor resistance or, fault with the reference / operating voltage; also check switch contacts (if applicable) to ensure that they open / close and check continuity of switch contacts (ensure that there is not a high resistance). Refer to list of sensor / actuator checks on page 19. It is also possible that the switch is operating correctly but the position being indicated by the switch is unacceptable or incorrect for the operating conditions indicated by other sensors or, is not as expected by the control unit; check mechanism / linkage used to operate switch, and check base settings the switch and throttle mechanism.
P278C	Kick Down Switch Circuit Low	Auto transmission systems can use a switch to indicate full open throttle position; the switch signal is used to help calculate whether a lower gear should be selected to achieve the desired vehicle performance. The switch could be connected to the throttle pedal or connected to the throttle linkage / mechanism. The switch could form part of a throttle position sensor assembly which could contain more than one switch or sensor, refer to vehicle specific information to establish which sensor or switch is affected. Note that some systems may use a vacuum operated switch (activated by changes in intake manifold vacuum).	The voltage or value of the switch signal is either "zero" (no signal) or, the signal exists but it is below the normal operating range (lower than the minimum operating limit). Possible fault in the switch or wiring e.g. short / open circuits, circuit resistance, incorrect sensor resistance or, fault with the reference / operating voltage; the switch contacts (if applicable) could also be faulty e.g. not opening / closing correctly or have a high contact resistance. Refer to list of sensor / actuator checks on page 19. Note that the fault code can be activated if the switch output remains at the low value e.g. zero volts and does not rise to the high value e.g. battery voltage (the control unit will detect no switch signal change but it has detected change in engine operating conditions or detected conflicting signals from other throttle position sensors).
P278D	Kick Down Switch Circuit High	Auto transmission systems can use a switch to indicate full open throttle position; the switch signal is used to help calculate whether a lower gear should be selected to achieve the desired vehicle performance. The switch could be connected to the throttle pedal or connected to the throttle linkage / mechanism. The switch could form part of a throttle position sensor assembly which could contain more than one switch or sensor, refer to vehicle specific information to establish which sensor or switch is affected. Note that some systems may use a vacuum operated switch (activated by changes in intake manifold vacuum).	The voltage or value of the switch signal is either at full value (e.g. battery voltage with no signal or frequency) or, the signal exists but it is above the normal operating range (higher than the maximum operating limit). Possible fault in the switch or wiring e.g. short / open circuits, circuit resistance, incorrect sensor resistance or, fault with the reference / operating voltage; the switch contacts (if applicable) could also be faulty e.g. not opening / closing correctly or have a high contact resistance. Refer to list of sensor / actuator checks on page 19. Note that the fault code can be activated if the switch output remains at the high value e.g. battery voltage and does not fall to the low value e.g. zero volts (the control unit will detect no switch signal change but it has detected change in engine operating conditions or detected conflicting signals from other throttle position sensors).
P278E	Kick Down Switch Circuit Intermittent / Erratic	Auto transmission systems can use a switch to indicate full open throttle position; the switch signal is used to help calculate whether a lower gear should be selected to achieve the desired vehicle performance. The switch could be connected to the throttle pedal or connected to the throttle linkage / mechanism. The switch could form part of a throttle position sensor assembly which could contain more than one switch or sensor, refer to vehicle specific information to establish which sensor or switch is affected. Note that some systems may use a vacuum operated switch (activated by changes in intake manifold vacuum).	The voltage or value of the switch signal is intermittent (likely to be out of normal operating range / tolerance when intermittent fault is detected) or, the signal / voltage is erratic (e.g. signal changes are irregular / unpredictable). Possible fault with wiring or connections (including internal sensor / switch connections); the switch contacts (if applicable) could also be faulty e.g. not opening / closing correctly or have a high contact resistance. Refer to list of sensor / actuator checks on page 19.
P278F		SAE / ISO reserved	Not used, reserved for future allocation.
P2790	Gate Select Direction Circuit	Different manufacturers use different systems to select transmission mode and range; it may therefore be necessary to refer to vehicle specific information to identify the components and gear selection system fitted to the vehicle. The Gate Select Direction is likely to refer to the sensor that monitors the movement of the gear lever / selector between the gates, and will possibly indicate movement of the lever within the gate.	Identify the components fitted to the gear selection system and locate the sensors / switches used to monitor the movement of the gear lever / selector (if possible refer to vehicle specific information to identify the Gate Select Direction sensor / switch). The fault code indicates that there is an unidentified fault with the signal in the circuit. The voltage, frequency or value of the sensor signal is either at full value (e.g. battery voltage with no signal or frequency) or, the signal exists but it is above the normal operating range (higher than the maximum operating limit). Possible fault in the sensor or wiring e.g. short / open circuits, circuit resistance, incorrect sensor resistance or, fault with the reference / operating voltage; interference from other circuits can also affect sensor signals. Refer to list of sensor / actuator checks on page 19. Also check that the sensor / switch is set correctly (base position setting) and that any mechanism operating the sensor / switch is acting on the switch / sensor correctly.
P2791	Gate Select Direction Circuit Low	Different manufacturers use different systems to select transmission mode and range; it may therefore be necessary to refer to vehicle specific information to identify the components and gear selection system fitted to the vehicle. The Gate Select Direction is likely to refer to the sensor that monitors the movement of the gear lever / selector between the gates, and will possibly indicate movement of the lever within the gate.	Identify the components fitted to the gear selection system and locate the sensors / switches used to monitor the movement of the gear lever / selector (if possible refer to vehicle specific information to identify the Gate Select Direction sensor / switch). The fault code indicates that there is an unidentified fault with the signal in the circuit. The voltage or value of the sensor signal is either "zero" (no signal) or, the signal exists but it is below the normal operating range (lower than the minimum operating limit). Possible fault in the sensor or wiring e.g. short / open circuits, circuit resistance, incorrect sensor resistance or, fault with the reference / operating voltage; interference from other circuits can also affect sensor signals. Refer to list of sensor / actuator checks on page 19. Also check that the sensor is set correctly (base position setting) and that any mechanism operating the sensor / switch is acting on the switch / sensor correctly.

NOTE 1. Check for other fault codes that could provide additional information.   **NOTE 2.** Communication between control units can pass via a CAN-Bus system; refer to Page 51 for CAN-Bus checks.   316

NOTE 4. Refer to Page 53 for list of pages that show the ISO standard for component locations e.g. Sensor A - Bank 1.

P2792	Gate Select Direction Circuit High	Different manufacturers use different systems to select transmission mode and range; it may therefore be necessary to refer to vehicle specific information to identify the components and gear selection system fitted to the vehicle. The Gate Select Direction is likely to refer to the sensor that monitors the movement of the gear lever / selector between the gates, and will possibly indicate movement of the lever within the gate.	Identify the components fitted to the gear selection system and locate the sensors / switches used to monitor the movement of the gear lever / selector (if possible refer to vehicle specific information to identify the Gate Select Direction sensor / switch). The fault code indicates that there is an unidentified fault with the signal in the circuit. The voltage, frequency or value of the sensor signal is either at full value (e.g. battery voltage with no signal or frequency) or, the signal exists but it is above the normal operating range (higher than the maximum operating limit). Possible fault in the sensor or wiring e.g. short / open circuits, circuit resistance, incorrect sensor resistance or, fault with the reference / operating voltage; interference from other circuits can also affect sensor signals. Refer to list of sensor / actuator checks on page 19. Also check that the sensor is set correctly (base position setting) and that any mechanism operating the sensor / switch is acting on the switch / sensor correctly.
P2793	Gear Shift Direction Circuit	Different manufacturers use different systems to select transmission mode and range; it may therefore be necessary to refer to vehicle specific information to identify the components and gear selection system fitted to the vehicle. The Gear Shift Direction is likely to refer to a sensor / switch that monitors the movement of the gear lever / selector; the sensor / switch could monitor movement between the gates or possibly monitor movement to select up or down shifts.	Identify the components fitted to the gear selection system and locate the sensors / switches used to monitor the movement of the gear lever / selector (if possible refer to vehicle specific information to identify the Gear Shift Direction sensor / switch). The fault code indicates that there is an unidentified fault with the signal in the circuit. The voltage, frequency or value of the sensor signal is either at full value (e.g. battery voltage with no signal or frequency) or, the signal exists but it is above the normal operating range (higher than the maximum operating limit). Possible fault in the sensor or wiring e.g. short / open circuits, circuit resistance, incorrect sensor resistance or, fault with the reference / operating voltage; interference from other circuits can also affect sensor signals. Refer to list of sensor / actuator checks on page 19. Also check that the sensor / switch is set correctly (base position setting) and that any mechanism operating the sensor / switch is acting on the switch / sensor correctly.
P2794	Gear Shift Direction Circuit Low	Different manufacturers use different systems to select transmission mode and range; it may therefore be necessary to refer to vehicle specific information to identify the components and gear selection system fitted to the vehicle. The Gear Shift Direction is likely to refer to a sensor / switch that monitors the movement of the gear lever / selector; the sensor / switch could monitor movement between the gates or possibly monitor movement to select up or down shifts.	Identify the components fitted to the gear selection system and locate the sensors / switches used to monitor the movement of the gear lever / selector (if possible refer to vehicle specific information to identify the Gear Shift Direction sensor / switch). The fault code indicates that there is an unidentified fault with the signal in the circuit. The voltage or value of the sensor signal is either "zero" (no signal) or, the signal exists but it is below the normal operating range (lower than the minimum operating limit). Possible fault in the sensor or wiring e.g. short / open circuits, circuit resistance, incorrect sensor resistance or, fault with the reference / operating voltage; interference from other circuits can also affect sensor signals. Refer to list of sensor / actuator checks on page 19. Also check that the sensor is set correctly (base position setting) and that any mechanism operating the sensor / switch is acting on the switch / sensor correctly.
P2795	Gear Shift Direction Circuit High	Different manufacturers use different systems to select transmission mode and range; it may therefore be necessary to refer to vehicle specific information to identify the components and gear selection system fitted to the vehicle. The Gear Shift Direction is likely to refer to a sensor / switch that monitors the movement of the gear lever / selector; the sensor / switch could monitor movement between the gates or possibly monitor movement to select up or down shifts.	Identify the components fitted to the gear selection system and locate the sensors / switches used to monitor the movement of the gear lever / selector (if possible refer to vehicle specific information to identify the Gear Shift Direction sensor / switch). The fault code indicates that there is an unidentified fault with the signal in the circuit. The voltage, frequency or value of the sensor signal is either at full value (e.g. battery voltage with no signal or frequency) or, the signal exists but it is above the normal operating range (higher than the maximum operating limit). Possible fault in the sensor or wiring e.g. short / open circuits, circuit resistance, incorrect sensor resistance or, fault with the reference / operating voltage; interference from other circuits can also affect sensor signals. Refer to list of sensor / actuator checks on page 19. Also check that the sensor is set correctly (base position setting) and that any mechanism operating the sensor / switch is acting on the switch / sensor correctly.
P2796	Auxiliary Transmission Fluid Pump Control Circuit / Open	The auxiliary transmission pump provides pressurised fluid to the auxiliary transmission system; the pump operation can be controlled by a separate module or by the main control unit. It may be necessary to identify the wiring and components for the pump system e.g. relay, power supply and control.	The voltage, frequency or value of the control signal in the pump circuit is incorrect. Signal could be:- out of normal operating range, constant value, non-existent, corrupt. The fault code indicates a possible "open circuit" but other undefined faults in the actuator or wiring could activate the same code e.g. actuator power supply, short circuit, high circuit resistance or incorrect actuator resistance. Note: the control unit could be providing the correct signal but it is being affected by a circuit fault. Refer to list of sensor / actuator checks on page 19.
P2797	Auxiliary Transmission Fluid Pump Performance	The auxiliary transmission pump provides pressurised fluid to the auxiliary transmission system; the pump operation can be controlled by a separate module or by the main control unit. It may be necessary to identify the wiring and components for the pump system e.g. relay, power supply and control.	The fault code indicates a performance problem with the fluid pump; this could be caused by an electrical fault or a fluid / hydraulic fault. Note that pressure switches / sensors can be used to monitor system pressure; it is therefore possible that the fault code has been activated due to lack of system pressure. Check for fluid leaks and correct as necessary. If there are no obvious leaks, check the pressure with a separate gauge. If the pressure is correct it could indicate a pressure switch / sensor fault. If the pressure is low or the volume of fluid delivered by the pump is low, check pump operation. Check for high resistances in the pump wiring, or if the pump is not operating, check for short / open circuits and power supply / earth faults.
P2798	Auxiliary Transmission Fluid Pump Control Circuit Low	The auxiliary transmission pump provides pressurised fluid to the auxiliary transmission system; the pump operation can be controlled by a separate module or by the main control unit. It may be necessary to identify the wiring and components for the pump system e.g. relay, power supply and control.	The fault code refers to a fault in the pump / motor circuit. The voltage, frequency or value of the control signal in the pump circuit is either "zero" (no signal) or, the signal exists but is below the normal operating range. Possible fault with pump / motor or wiring (e.g. pump power supply, short / open circuit or resistance) that is causing a signal value to be lower than the minimum operating limit. Note: the control unit could be providing the correct control signal but it is being affected by a circuit fault. Refer to list of sensor / actuator checks on page 19.
P2799	Auxiliary Transmission Fluid Pump Control Circuit High	The auxiliary transmission pump provides pressurised fluid to the auxiliary transmission system; the pump operation can be controlled by a separate module or by the main control unit. It may be necessary to identify the wiring and components for the pump system e.g. relay, power supply and control.	The voltage, frequency or value of the control signal in the pump / motor circuit is either at full value (e.g. battery voltage with no on / off signal) or, the signal exists but it is above the normal operating range. Possible fault with pump / motor or wiring (e.g. pump power supply, short / open circuit or resistance) that is causing a signal value to be higher than the maximum operating limit. Note: the control unit could be providing the correct control signal but it is being affected by a circuit fault. Refer to list of sensor / actuator checks on page 19.
P279A		ISO / SAE reserved	Not used, reserved for future allocation.
P279B		ISO / SAE reserved	Not used, reserved for future allocation.
P279C		ISO / SAE reserved	Not used, reserved for future allocation.
P279D		ISO / SAE reserved	Not used, reserved for future allocation.
P279E		ISO / SAE reserved	Not used, reserved for future allocation.

NOTE 1. Check for other fault codes that could provide additional information.   **NOTE 2.** Communication between control units can pass via a CAN-Bus system; refer to Page 51 for CAN-Bus checks.

NOTE 3. If a fault cannot be located, it is also possible that the control unit is at fault.   **NOTE 4.** Refer to Page 53 for list of pages that show the ISO standard for component locations e.g. Sensor A - Bank 1.

Fault code	EOBD / ISO Description	Component / System Description	Meaningful Description and Quick Check
P279F		ISO / SAE reserved	Not used, reserved for future allocation.
P2800	Transmission Range Sensor "B" Circuit (PRNDL Input)	Manufacturers generally refer to the range sensor as being the sensor that indicates the selection of P - R - N - 1 -2 - 3 - 4 etc. More than one sensor can be used for back-up or for indicating additional information (depending on the system); one sensor could be located on the selector gate with the other sensor being located on the selector linkage or transmission housing.	The voltage, frequency or value of the sensor signal is incorrect (undefined fault). Sensor voltage / signal could be: out of normal operating range, constant value, non-existent, corrupt. Possible fault in the sensor or wiring e.g. short / open circuits, circuit resistance, incorrect sensor resistance or, fault with the reference / operating voltage; interference from other circuits can also affect sensor signals. Refer to list of sensor / actuator checks on page 19.
P2801	Transmission Range Sensor "B" Circuit Range / Performance	Manufacturers generally refer to the range sensor as being the sensor that indicates the selection of P - R - N - 1 -2 - 3 - 4 etc. More than one sensor can be used for back-up or for indicating additional information (depending on the system); one sensor could be located on the selector gate with the other sensor being located on the selector linkage or transmission housing.	The voltage, frequency or value of the sensor signal is within the normal operating range / tolerance but the signal is not plausible or is incorrect due to an undefined fault. The sensor signal might not match the operating conditions indicated by other sensors or, the sensor signal is not changing / responding as expected. Possible fault in the sensor or wiring e.g. short / open circuits, circuit resistance, incorrect sensor resistance or, fault with the reference / operating voltage; interference from other circuits can also affect sensor signals. Refer to list of sensor / actuator checks on page 19. It is also possible that the sensor is operating correctly but the position / gear being indicated by the sensor is unacceptable or incorrect (not as expected by the control unit with regard to information from other sensors). Check operation of range selector and mechanism, also check for any base settings that could be applicable to the sensor and selector.
P2802	Transmission Range Sensor "B" Circuit Low	Manufacturers generally refer to the range sensor as being the sensor that indicates the selection of P - R - N - 1 -2 - 3 - 4 etc. More than one sensor can be used for back-up or for indicating additional information (depending on the system); one sensor could be located on the selector gate with the other sensor being located on the selector linkage or transmission housing.	The voltage, frequency or value of the sensor signal is either "zero" (no signal) or, the signal exists but it is below the normal operating range (lower than the minimum operating limit). Possible fault in the sensor or wiring e.g. short / open circuits, circuit resistance, incorrect sensor resistance or, fault with the reference / operating voltage; interference from other circuits can also affect sensor signals. Refer to list of sensor / actuator checks on page 19. It is also possible that the control unit is detecting "no change" in the signal when the range selector is moved due to a fault in the mechanism connecting the sensor to the selector.
P2803	Transmission Range Sensor "B" Circuit High	Manufacturers generally refer to the range sensor as being the sensor that indicates the selection of P - R - N - 1 -2 - 3 - 4 etc. More than one sensor can be used for back-up or for indicating additional information (depending on the system); one sensor could be located on the selector gate with the other sensor being located on the selector linkage or transmission housing.	The voltage, frequency or value of the sensor signal is either at full value (e.g. battery voltage with no signal or frequency) or, the signal exists but it is above the normal operating range (higher than the maximum operating limit). Possible fault in the sensor or wiring e.g. short / open circuits, circuit resistance, incorrect sensor resistance or, fault with the reference / operating voltage; interference from other circuits can also affect sensor signals. Refer to list of sensor / actuator checks on page 19. It is also possible that the control unit is detecting "no change" in the signal when the range selector is moved due to a fault in the mechanism connecting the sensor to the selector.
P2804	Transmission Range Sensor "B" Intermittent	Manufacturers generally refer to the range sensor as being the sensor that indicates the selection of P - R - N - 1 -2 - 3 - 4 etc. More than one sensor can be used for back-up or for indicating additional information (depending on the system); one sensor could be located on the selector gate with the other sensor being located on the selector linkage or transmission housing.	The voltage, frequency or value of the sensor signal is intermittent (likely to be out of normal operating range / tolerance when intermittent fault is detected) or, the signal / voltage is erratic (e.g. signal changes are irregular / unpredictable). Possible fault with wiring or connections (including internal sensor connections); also interference from other circuits can affect the signal. Refer to list of sensor / actuator checks on page 19. It is also possible that the control unit is detecting "no change" in the signal when the range selector is moved due to a fault in the mechanism connecting the sensor to the selector.
P2805	Transmission Range Sensor "A" / "B" Correlation	Manufacturers generally refer to the range sensor as being the sensor that indicates the selection of P - R - N - 1 -2 - 3 - 4 etc. More than one sensor can be used for back-up or for indicating additional information (depending on the system); one sensor could be located on the selector gate with the other sensor being located on the selector linkage or transmission housing.	The signals from the two sensors are providing conflicting or implausible information e.g. one sensor indicating P position but the other sensor indicating N position. It is possible that one of the sensors is operating incorrectly, therefore check each sensor for a fault with the sensor / wiring e.g. short / open circuits, circuit resistance, incorrect sensor resistance or reference / operating voltage fault. Refer to list of sensor / actuator checks on page 19. Also check Range Selector and linkage operation, and check for a base setting of the selector, linkage and sensors.
P2806	Transmission Range Sensor Alignment	Different manufacturers use different systems to select transmission mode and range; it may therefore be necessary to refer to vehicle specific information to identify the components and gear selection system fitted to the vehicle. The term Range Sensor could refer to the main gear selection range (PRND etc) or it could refer to the High / Low range for a 4-wheel drive system.	Identify the Range Sensor. The fault code indicates that the sensor is not aligned or adjusted correctly. The fault could be detected due to information from other sensors that provide conflicting information, or it is also possible that a range has been selected and engaged but the sensor is not providing a correct or matching signal. Adjust sensor as necessary but also check that there are no short or open circuits in the wiring and that the sensor is operating correctly (a wiring or sensor fault that causes an incorrect signal value could be interpreted as an alignment problem).
P2807	Pressure Control Solenoid "G"	Pressure control solenoids are used in the transmission to regulate the pressure in the hydraulic circuits (usually for gear shift control). The pressure control solenoid can also influence / control the actual gear shift. Refer to vehicle specific information to establish which gear(s) are applicable to the identified solenoid.	The control unit has detected an undefined fault with the pressure control solenoid system; the fault could be detected due to an electrical problem with the solenoid / wiring (e.g. solenoid power supply, short / open circuit or resistance). Refer to list of sensor / actuator checks on page 19. The fault could also be detected because sensors are detecting incorrect gear shift operation; check operation of any position sensors (if applicable), check hydraulic pressures and gear selection mechanism.
P2808	Pressure Control Solenoid "G" Performance / Stuck Off	Pressure control solenoids are used in the transmission to regulate the pressure in the hydraulic circuits (usually for gear shift control). The pressure control solenoid can also influence / control the actual gear shift. Refer to vehicle specific information to establish which gear(s) are applicable to the identified solenoid.	The control unit is providing a control signal to the solenoid (signal is within the normal operating range / tolerance) but the control unit has detected that the solenoid is stuck in a fixed (off) position and is not responding to the control signal. The fault could be related to a solenoid / wiring fault. Refer to list of sensor / actuator checks on page 19. Note that the control unit could be detecting the fault because of "a lack of response / no response" from the system or mechanism being controlled by the solenoid; check system (as well as the solenoid) for correct operation, movement and response. A position / pressure sensor (if fitted) could be indicating the "stuck position"; also check sensor operation.
P2809	Pressure Control Solenoid "G" Stuck On	Pressure control solenoids are used in the transmission to regulate the pressure in the hydraulic circuits (usually for gear shift control). The pressure control solenoid can also influence / control the actual gear shift. Refer to vehicle specific information to establish which gear(s) are applicable to the identified solenoid.	The control unit is providing a control signal to the solenoid (signal is within the normal operating range / tolerance) but the control unit has detected that the solenoid is stuck in a fixed (on) position and is not responding to the control signal. The fault could be related to a solenoid / wiring fault. Refer to list of sensor / actuator checks on page 19. Note that the control unit could be detecting the fault because of "a lack of response / no response" from the system or mechanism being controlled by the solenoid; check system (as well as the solenoid) for correct operation, movement and response. A position / pressure sensor (if fitted) could be indicating the "stuck position"; also check sensor operation.

P2810	Pressure Control Solenoid "G" Electrical	Pressure control solenoids are used in the transmission to regulate the pressure in the hydraulic circuits (usually for gear shift control). The pressure control solenoid can also influence / control the actual gear shift. Refer to vehicle specific information to establish which gear(s) are applicable to the identified solenoid.	The voltage, frequency or value of the control signal in the solenoid circuit is incorrect (undefined electrical fault). Signal could be:- out of normal operating range, constant value, non-existent, corrupt. Possible fault with solenoid or wiring e.g. solenoid power supply, short / open circuit, high circuit resistance or incorrect solenoid resistance. Note: the control unit could be providing the correct signal but it is being affected by a circuit fault. Refer to list of sensor / actuator checks on page 19.
P2811	Pressure Control Solenoid "G" Intermittent	Pressure control solenoids are used in the transmission to regulate the pressure in the hydraulic circuits (usually for gear shift control). The pressure control solenoid can also influence / control the actual gear shift. Refer to vehicle specific information to establish which gear(s) are applicable to the identified solenoid.	The voltage, frequency or value of the control signal in the solenoid circuit is intermittent (likely to be out of normal operating range / tolerance when intermittent fault is detected) or, the signal is erratic e.g. signal changes are irregular / unpredictable. Possible fault with solenoid or wiring (e.g. internal / external connections, broken wiring) that is causing intermittent or erratic signal to exist. Note: the control unit could be providing the correct signal but it is being affected by a circuit fault. Refer to list of sensor / actuator checks on page 19.
P2812	Pressure Control Solenoid "G" Control Circuit / Open	Pressure control solenoids are used in the transmission to regulate the pressure in the hydraulic circuits (usually for gear shift control). The pressure control solenoid can also influence / control the actual gear shift. Refer to vehicle specific information to establish which gear(s) are applicable to the identified solenoid.	The voltage, frequency or value of the control signal in the solenoid circuit is incorrect. Signal could be:- out of normal operating range, constant value, non-existent, corrupt. The fault code indicates a possible "open circuit" but other undefined faults in the solenoid or wiring could activate the same code e.g. solenoid power supply, short circuit, high circuit resistance or incorrect solenoid resistance. Note: the control unit could be providing the correct signal but it is being affected by the circuit fault. Refer to list of sensor / actuator checks on page 19.
P2813	Pressure Control Solenoid "G" Control Circuit Range / Performance	Pressure control solenoids are used in the transmission to regulate the pressure in the hydraulic circuits (usually for gear shift control). The pressure control solenoid can also influence / control the actual gear shift. Refer to vehicle specific information to establish which gear(s) are applicable to the identified solenoid.	The voltage, frequency or value of the control signal in the solenoid circuit is within the normal operating range / tolerance, but the signal might be incorrect for the operating conditions. It is also possible that the solenoid response (or response of system controlled by the solenoid) differs from the expected / desired response (incorrect response to the control signal provided by the control unit). Possible fault in solenoid or wiring, but also possible that the solenoid (or mechanism / system controlled by the solenoid) is not operating or moving correctly. Note: the control unit could be providing the correct signal but it is being affected by a circuit fault. Refer to list of sensor / actuator checks on page 19. Note that the fault code could also be activated due to mechanical or hydraulic faults e.g. restriction in the movement of the solenoid, a mechanical fault with the gear selection / deselection mechanism, a fault with the hydraulic pressure or valves.
P2814	Pressure Control Solenoid "G" Control Circuit Low	Pressure control solenoids are used in the transmission to regulate the pressure in the hydraulic circuits (usually for gear shift control). The pressure control solenoid can also influence / control the actual gear shift. Refer to vehicle specific information to establish which gear(s) are applicable to the identified solenoid.	The voltage, frequency or value of the control signal in the solenoid circuit is either "zero" (no signal) or, the signal exists but is below the normal operating range. Possible fault with solenoid or wiring (e.g. solenoid power supply, short / open circuit or resistance) that is causing a signal value to be lower than the minimum operating limit. Note: the control unit could be providing the correct signal but it is being affected by a circuit fault. Refer to list of sensor / actuator checks on page 19.
P2815	Pressure Control Solenoid "G" Control Circuit High	Pressure control solenoids are used in the transmission to regulate the pressure in the hydraulic circuits (usually for gear shift control). The pressure control solenoid can also influence / control the actual gear shift. Refer to vehicle specific information to establish which gear(s) are applicable to the identified solenoid.	The voltage, frequency or value of the control signal in the solenoid circuit is either at full value (e.g. battery voltage with no on / off signal) or, the signal exists but it is above the normal operating range. Possible fault with solenoid or wiring (e.g. solenoid power supply, short / open circuit or resistance) that is causing a signal value to be higher than the maximum operating limit. Note: the control unit could be providing the correct signal but it is being affected by a circuit fault. Refer to list of sensor / actuator checks on page 19.
P2816	Pressure Control Solenoid "H"	Pressure control solenoids are used in the transmission to regulate the pressure in the hydraulic circuits (usually for gear shift control). The pressure control solenoid can also influence / control the actual gear shift. Refer to vehicle specific information to establish which gear(s) are applicable to the identified solenoid.	The control unit has detected an undefined fault with the pressure control solenoid system; the fault could be detected due to an electrical problem with the solenoid / wiring (e.g. solenoid power supply, short / open circuit or resistance). Refer to list of sensor / actuator checks on page 19. The fault could also be detected because sensors are detecting incorrect gear shift operation; check operation of any position sensors (if applicable), check hydraulic pressures and gear selection mechanism.
P2817	Pressure Control Solenoid "H" Performance / Stuck Off	Pressure control solenoids are used in the transmission to regulate the pressure in the hydraulic circuits (usually for gear shift control). The pressure control solenoid can also influence / control the actual gear shift. Refer to vehicle specific information to establish which gear(s) are applicable to the identified solenoid.	The control unit is providing a control signal to the solenoid (signal is within the normal operating range / tolerance) but the control unit has detected that the solenoid is stuck in a fixed (off) position and is not responding to the control signal. The fault could be related to a solenoid / wiring fault. Refer to list of sensor / actuator checks on page 19. Note that the control unit could be detecting the fault because of "a lack of response / no response" from the system or mechanism being controlled by the solenoid; check system (as well as the solenoid) for correct operation, movement and response. A position / pressure sensor (if fitted) could be indicating the "stuck position"; also check sensor operation.
P2818	Pressure Control Solenoid "H" Stuck On	Pressure control solenoids are used in the transmission to regulate the pressure in the hydraulic circuits (usually for gear shift control). The pressure control solenoid can also influence / control the actual gear shift. Refer to vehicle specific information to establish which gear(s) are applicable to the identified solenoid.	The control unit is providing a control signal to the solenoid (signal is within the normal operating range / tolerance) but the control unit has detected that the solenoid is stuck in a fixed (on) position and is not responding to the control signal. The fault could be related to a solenoid / wiring fault. Refer to list of sensor / actuator checks on page 19. Note that the control unit could be detecting the fault because of "a lack of response / no response" from the system or mechanism being controlled by the solenoid; check system (as well as the solenoid) for correct operation, movement and response. A position / pressure sensor (if fitted) could be indicating the "stuck position"; also check sensor operation.
P2819	Pressure Control Solenoid "H" Electrical	Pressure control solenoids are used in the transmission to regulate the pressure in the hydraulic circuits (usually for gear shift control). The pressure control solenoid can also influence / control the actual gear shift. Refer to vehicle specific information to establish which gear(s) are applicable to the identified solenoid.	The voltage, frequency or value of the control signal in the solenoid circuit is incorrect (undefined electrical fault). Signal could be:- out of normal operating range, constant value, non-existent, corrupt. Possible fault with solenoid or wiring e.g. solenoid power supply, short / open circuit, high circuit resistance or incorrect solenoid resistance. Note: the control unit could be providing the correct signal but it is being affected by a circuit fault. Refer to list of sensor / actuator checks on page 19.

NOTE 1. Check for other fault codes that could provide additional information. **NOTE 2.** Communication between control units can pass via a CAN-Bus system; refer to Page 51 for CAN-Bus checks.

NOTE 3. If a fault cannot be located, it is also possible that the control unit is at fault. **NOTE 4.** Refer to Page 53 for list of pages that show the ISO standard for component locations e.g. Sensor A - Bank 1.

Fault code	EOBD / ISO Description	Component / System Description	Meaningful Description and Quick Check
P281A	Pressure Control Solenoid "H" Intermittent	Pressure control solenoids are used in the transmission to regulate the pressure in the hydraulic circuits (usually for gear shift control). The pressure control solenoid can also influence / control the actual gear shift. Refer to vehicle specific information to establish which gear(s) are applicable to the identified solenoid.	The voltage, frequency or value of the control signal in the solenoid circuit is intermittent (likely to be out of normal operating range / tolerance when intermittent fault is detected) or, the signal is erratic e.g. signal changes are irregular / unpredictable. Possible fault with solenoid or wiring (e.g. internal / external connections, broken wiring) that is causing intermittent or erratic signal to exist. Note: the control unit could be providing the correct signal but it is being affected by a circuit fault. Refer to list of sensor / actuator checks on page 19.
P281B	Pressure Control Solenoid "H" Control Circuit / Open	Pressure control solenoids are used in the transmission to regulate the pressure in the hydraulic circuits (usually for gear shift control). The pressure control solenoid can also influence / control the actual gear shift. Refer to vehicle specific information to establish which gear(s) are applicable to the identified solenoid.	The voltage, frequency or value of the control signal in the solenoid circuit is incorrect. Signal could be:- out of normal operating range, constant value, non-existent, corrupt. The fault code indicates a possible "open circuit" but other undefined faults in the solenoid or wiring could activate the same code e.g. solenoid power supply, short circuit, high circuit resistance or incorrect solenoid resistance. Note: the control unit could be providing the correct signal but it is being affected by the circuit fault. Refer to list of sensor / actuator checks on page 19.
P281C	Pressure Control Solenoid "H" Control Circuit Range / Performance	Pressure control solenoids are used in the transmission to regulate the pressure in the hydraulic circuits (usually for gear shift control). The pressure control solenoid can also influence / control the actual gear shift. Refer to vehicle specific information to establish which gear(s) are applicable to the identified solenoid.	The voltage, frequency or value of the control signal in the solenoid circuit is within the normal operating range / tolerance, but the signal might be incorrect for the operating conditions. It is also possible that the solenoid response (or response of system controlled by the solenoid) differs from the expected / desired response (incorrect response to the control signal provided by the control unit). Possible fault in solenoid or wiring, but also possible that the solenoid (or mechanism / system controlled by the solenoid) is not operating or moving correctly. Note: the control unit could be providing the correct signal but it is being affected by a circuit fault. Refer to list of sensor / actuator checks on page 19. Note that the fault code could also be activated due to mechanical or hydraulic faults e.g. restriction in the movement of the solenoid, a mechanical fault with the gear selection / deselection mechanism, a fault with the hydraulic pressure or valves.
P281D	Pressure Control Solenoid "H" Control Circuit Low	Pressure control solenoids are used in the transmission to regulate the pressure in the hydraulic circuits (usually for gear shift control). The pressure control solenoid can also influence / control the actual gear shift. Refer to vehicle specific information to establish which gear(s) are applicable to the identified solenoid.	The voltage, frequency or value of the control signal in the solenoid circuit is either "zero" (no signal) or, the signal exists but is below the normal operating range. Possible fault with solenoid or wiring (e.g. solenoid power supply, short / open circuit or resistance) that is causing a signal value to be lower than the minimum operating limit. Note: the control unit could be providing the correct signal but it is being affected by a circuit fault. Refer to list of sensor / actuator checks on page 19.
P281E	Pressure Control Solenoid "H" Control Circuit High	Pressure control solenoids are used in the transmission to regulate the pressure in the hydraulic circuits (usually for gear shift control). The pressure control solenoid can also influence / control the actual gear shift. Refer to vehicle specific information to establish which gear(s) are applicable to the identified solenoid.	The voltage, frequency or value of the control signal in the solenoid circuit is either at full value (e.g. battery voltage with no on / off signal) or, the signal exists but it is above the normal operating range. Possible fault with solenoid or wiring (e.g. solenoid power supply, short / open circuit or resistance) that is causing a signal value to be higher than the maximum operating limit. Note: the control unit could be providing the correct signal but it is being affected by a circuit fault. Refer to list of sensor / actuator checks on page 19.
P281F	Pressure Control Solenoid "J"	Pressure control solenoids are used in the transmission to regulate the pressure in the hydraulic circuits (usually for gear shift control). The pressure control solenoid can also influence / control the actual gear shift. Refer to vehicle specific information to establish which gear(s) are applicable to the identified solenoid.	The control unit has detected an undefined fault with the pressure control solenoid system; the fault could be detected due to an electrical problem with the solenoid / wiring (e.g. solenoid power supply, short / open circuit or resistance). Refer to list of sensor / actuator checks on page 19. The fault could also be detected because sensors are detecting incorrect gear shift operation; check operation of any position sensors (if applicable), check hydraulic pressures and gear selection mechanism.
P2820	Pressure Control Solenoid "J" Performance / Stuck Off	Pressure control solenoids are used in the transmission to regulate the pressure in the hydraulic circuits (usually for gear shift control). The pressure control solenoid can also influence / control the actual gear shift. Refer to vehicle specific information to establish which gear(s) are applicable to the identified solenoid.	The control unit is providing a control signal to the solenoid (signal is within the normal operating range / tolerance) but the control unit has detected that the solenoid is stuck in a fixed (off) position and is not responding to the control signal. The fault could be related to a solenoid / wiring fault. Refer to list of sensor / actuator checks on page 19. Note that the control unit could be detecting the fault because of "a lack of response / no response" from the system or mechanism being controlled by the solenoid; check system (as well as the solenoid) for correct operation, movement and response. A position / pressure sensor (if fitted) could be indicating the "stuck position"; also check sensor operation.
P2821	Pressure Control Solenoid "J" Stuck On	Pressure control solenoids are used in the transmission to regulate the pressure in the hydraulic circuits (usually for gear shift control). The pressure control solenoid can also influence / control the actual gear shift. Refer to vehicle specific information to establish which gear(s) are applicable to the identified solenoid.	The control unit is providing a control signal to the solenoid (signal is within the normal operating range / tolerance) but the control unit has detected that the solenoid is stuck in a fixed (on) position and is not responding to the control signal. The fault could be related to a solenoid / wiring fault. Refer to list of sensor / actuator checks on page 19. Note that the control unit could be detecting the fault because of "a lack of response / no response" from the system or mechanism being controlled by the solenoid; check system (as well as the solenoid) for correct operation, movement and response. A position / pressure sensor (if fitted) could be indicating the "stuck position"; also check sensor operation.
P2822	Pressure Control Solenoid "J" Electrical	Pressure control solenoids are used in the transmission to regulate the pressure in the hydraulic circuits (usually for gear shift control). The pressure control solenoid can also influence / control the actual gear shift. Refer to vehicle specific information to establish which gear(s) are applicable to the identified solenoid.	The voltage, frequency or value of the control signal in the solenoid circuit is incorrect (undefined electrical fault). Signal could be:- out of normal operating range, constant value, non-existent, corrupt. Possible fault with solenoid or wiring e.g. solenoid power supply, short / open circuit, high circuit resistance or incorrect solenoid resistance. Note: the control unit could be providing the correct signal but it is being affected by a circuit fault. Refer to list of sensor / actuator checks on page 19.
P2823	Pressure Control Solenoid "J" Intermittent	Pressure control solenoids are used in the transmission to regulate the pressure in the hydraulic circuits (usually for gear shift control). The pressure control solenoid can also influence / control the actual gear shift. Refer to vehicle specific information to establish which gear(s) are applicable to the identified solenoid.	The voltage, frequency or value of the control signal in the solenoid circuit is intermittent (likely to be out of normal operating range / tolerance when intermittent fault is detected) or, the signal is erratic e.g. signal changes are irregular / unpredictable. Possible fault with solenoid or wiring (e.g. internal / external connections, broken wiring) that is causing intermittent or erratic signal to exist. Note: the control unit could be providing the correct signal but it is being affected by a circuit fault. Refer to list of sensor / actuator checks on page 19.

P2824	Pressure Control Solenoid "J" Control Circuit / Open	Pressure control solenoids are used in the transmission to regulate the pressure in the hydraulic circuits (usually for gear shift control). The pressure control solenoid can also influence / control the actual gear shift. Refer to vehicle specific information to establish which gear(s) are applicable to the identified solenoid.	The voltage, frequency or value of the control signal in the solenoid circuit is incorrect. Signal could be:- out of normal operating range, constant value, non-existent, corrupt. The fault code indicates a possible "open circuit" but other undefined faults in the solenoid or wiring could activate the same code e.g. solenoid power supply, short circuit, high circuit resistance or incorrect solenoid resistance. Note: the control unit could be providing the correct signal but it is being affected by the circuit fault. Refer to list of sensor / actuator checks on page 19.
P2825	Pressure Control Solenoid "J" Control Circuit Range / Performance	Pressure control solenoids are used in the transmission to regulate the pressure in the hydraulic circuits (usually for gear shift control). The pressure control solenoid can also influence / control the actual gear shift. Refer to vehicle specific information to establish which gear(s) are applicable to the identified solenoid.	The voltage, frequency or value of the control signal in the solenoid circuit is within the normal operating range / tolerance, but the signal might be incorrect for the operating conditions. It is also possible that the solenoid response (or response of system controlled by the solenoid) differs from the expected / desired response (incorrect response to the control signal provided by the control unit). Possible fault in solenoid or wiring, but also possible that the solenoid (or mechanism / system controlled by the solenoid) is not operating or moving correctly. Note: the control unit could be providing the correct signal but it is being affected by a circuit fault. Refer to list of sensor / actuator checks on page 19. Note that the fault code could also be activated due to mechanical or hydraulic faults e.g. restriction in the movement of the solenoid, a mechanical fault with the gear selection / deselection mechanism, a fault with the hydraulic pressure or valves.
P2826	Pressure Control Solenoid "J" Control Circuit Low	Pressure control solenoids are used in the transmission to regulate the pressure in the hydraulic circuits (usually for gear shift control). The pressure control solenoid can also influence / control the actual gear shift. Refer to vehicle specific information to establish which gear(s) are applicable to the identified solenoid.	The voltage, frequency or value of the control signal in the solenoid circuit is either "zero" (no signal) or, the signal exists but is below the normal operating range. Possible fault with solenoid or wiring (e.g. solenoid power supply, short / open circuit or resistance) that is causing a signal value to be lower than the minimum operating limit. Note: the control unit could be providing the correct signal but it is being affected by a circuit fault. Refer to list of sensor / actuator checks on page 19.
P2827	Pressure Control Solenoid "J" Control Circuit High	Pressure control solenoids are used in the transmission to regulate the pressure in the hydraulic circuits (usually for gear shift control). The pressure control solenoid can also influence / control the actual gear shift. Refer to vehicle specific information to establish which gear(s) are applicable to the identified solenoid.	The voltage, frequency or value of the control signal in the solenoid circuit is either at full value (e.g. battery voltage with no on / off signal) or, the signal exists but it is above the normal operating range. Possible fault with solenoid or wiring (e.g. solenoid power supply, short / open circuit or resistance) that is causing a signal value to be higher than the maximum operating limit. Note: the control unit could be providing the correct signal but it is being affected by a circuit fault. Refer to list of sensor / actuator checks on page 19.
P2828	Pressure Control Solenoid "K"	Pressure control solenoids are used in the transmission to regulate the pressure in the hydraulic circuits (usually for gear shift control). The pressure control solenoid can also influence / control the actual gear shift. Refer to vehicle specific information to establish which gear(s) are applicable to the identified solenoid.	The control unit has detected an undefined fault with the pressure control solenoid system; the fault could be detected due to an electrical problem with the solenoid / wiring (e.g. solenoid power supply, short / open circuit or resistance). Refer to list of sensor / actuator checks on page 19. The fault could also be detected because sensors are detecting incorrect gear shift operation; check operation of any position sensors (if applicable), check hydraulic pressures and gear selection mechanism.
P2829	Pressure Control Solenoid "K" Performance / Stuck Off	Pressure control solenoids are used in the transmission to regulate the pressure in the hydraulic circuits (usually for gear shift control). The pressure control solenoid can also influence / control the actual gear shift. Refer to vehicle specific information to establish which gear(s) are applicable to the identified solenoid.	The control unit is providing a control signal to the solenoid (signal is within the normal operating range / tolerance) but the control unit has detected that the solenoid is stuck in a fixed (off) position and is not responding to the control signal. The fault could be related to a solenoid / wiring fault. Refer to list of sensor / actuator checks on page 19. Note that the control unit could be detecting the fault because of "a lack of response / no response" from the system or mechanism being controlled by the solenoid; check system (as well as the solenoid) for correct operation, movement and response. A position / pressure sensor (if fitted) could be indicating the "stuck position"; also check sensor operation.
P282A	Pressure Control Solenoid "K" Stuck On	Pressure control solenoids are used in the transmission to regulate the pressure in the hydraulic circuits (usually for gear shift control). The pressure control solenoid can also influence / control the actual gear shift. Refer to vehicle specific information to establish which gear(s) are applicable to the identified solenoid.	The control unit is providing a control signal to the solenoid (signal is within the normal operating range / tolerance) but the control unit has detected that the solenoid is stuck in a fixed (on) position and is not responding to the control signal. The fault could be related to a solenoid / wiring fault. Refer to list of sensor / actuator checks on page 19. Note that the control unit could be detecting the fault because of "a lack of response / no response" from the system or mechanism being controlled by the solenoid; check system (as well as the solenoid) for correct operation, movement and response. A position / pressure sensor (if fitted) could be indicating the "stuck position"; also check sensor operation.
P282B	Pressure Control Solenoid "K" Electrical	Pressure control solenoids are used in the transmission to regulate the pressure in the hydraulic circuits (usually for gear shift control). The pressure control solenoid can also influence / control the actual gear shift. Refer to vehicle specific information to establish which gear(s) are applicable to the identified solenoid.	The voltage, frequency or value of the control signal in the solenoid circuit is incorrect (undefined electrical fault). Signal could be:- out of normal operating range, constant value, non-existent, corrupt. Possible fault with solenoid or wiring e.g. solenoid power supply, short / open circuit, high circuit resistance or incorrect solenoid resistance. Note: the control unit could be providing the correct signal but it is being affected by a circuit fault. Refer to list of sensor / actuator checks on page 19.
P282C	Pressure Control Solenoid "K" Intermittent	Pressure control solenoids are used in the transmission to regulate the pressure in the hydraulic circuits (usually for gear shift control). The pressure control solenoid can also influence / control the actual gear shift. Refer to vehicle specific information to establish which gear(s) are applicable to the identified solenoid.	The voltage, frequency or value of the control signal in the solenoid circuit is intermittent (likely to be out of normal operating range / tolerance when intermittent fault is detected) or, the signal is erratic e.g. signal changes are irregular / unpredictable. Possible fault with solenoid or wiring (e.g. internal / external connections, broken wiring) that is causing intermittent or erratic signal to exist. Note: the control unit could be providing the correct signal but it is being affected by a circuit fault. Refer to list of sensor / actuator checks on page 19.
P282D	Pressure Control Solenoid "K" Control Circuit / Open	Pressure control solenoids are used in the transmission to regulate the pressure in the hydraulic circuits (usually for gear shift control). The pressure control solenoid can also influence / control the actual gear shift. Refer to vehicle specific information to establish which gear(s) are applicable to the identified solenoid.	The voltage, frequency or value of the control signal in the solenoid circuit is incorrect. Signal could be:- out of normal operating range, constant value, non-existent, corrupt. The fault code indicates a possible "open circuit" but other undefined faults in the solenoid or wiring could activate the same code e.g. solenoid power supply, short circuit, high circuit resistance or incorrect solenoid resistance. Note: the control unit could be providing the correct signal but it is being affected by the circuit fault. Refer to list of sensor / actuator checks on page 19.

NOTE 1. Check for other fault codes that could provide additional information.   **NOTE 2.** Communication between control units can pass via a CAN-Bus system; refer to Page 51 for CAN-Bus checks.   321

NOTE 3. If a fault cannot be located, it is also possible that the control unit is at fault.   **NOTE 4.** Refer to Page 53 for list of pages that show the ISO standard for component locations e.g. Sensor A - Bank 1.

Fault code	EOBD / ISO Description	Component / System Description	Meaningful Description and Quick Check
P282E	Pressure Control Solenoid "K" Control Circuit Range / Performance	Pressure control solenoids are used in the transmission to regulate the pressure in the hydraulic circuits (usually for gear shift control). The pressure control solenoid can also influence / control the actual gear shift. Refer to vehicle specific information to establish which gear(s) are applicable to the identified solenoid.	The voltage, frequency or value of the control signal in the solenoid circuit is within the normal operating range / tolerance, but the signal might be incorrect for the operating conditions. It is also possible that the solenoid response (or response of system controlled by the solenoid) differs from the expected / desired response (incorrect response to the control signal provided by the control unit). Possible fault in solenoid or wiring, but also possible that the solenoid (or mechanism / system controlled by the solenoid) is not operating or moving correctly. Note: the control unit could be providing the correct signal but it is being affected by a circuit fault. Refer to list of sensor / actuator checks on page 19. Note that the fault code could also be activated due to mechanical or hydraulic faults e.g. restriction in the movement of the solenoid, a mechanical fault with the gear selection / deselection mechanism, a fault with the hydraulic pressure or valves.
P282F	Pressure Control Solenoid "K" Control Circuit Low	Pressure control solenoids are used in the transmission to regulate the pressure in the hydraulic circuits (usually for gear shift control). The pressure control solenoid can also influence / control the actual gear shift. Refer to vehicle specific information to establish which gear(s) are applicable to the identified solenoid.	The voltage, frequency or value of the control signal in the solenoid circuit is either "zero" (no signal) or, the signal exists but is below the normal operating range. Possible fault with solenoid or wiring (e.g. solenoid power supply, short / open circuit or resistance) that is causing a signal value to be lower than the minimum operating limit. Note: the control unit could be providing the correct signal but it is being affected by a circuit fault. Refer to list of sensor / actuator checks on page 19.
P2830	Pressure Control Solenoid "K" Control Circuit High	Pressure control solenoids are used in the transmission to regulate the pressure in the hydraulic circuits (usually for gear shift control). The pressure control solenoid can also influence / control the actual gear shift. Refer to vehicle specific information to establish which gear(s) are applicable to the identified solenoid.	The voltage, frequency or value of the control signal in the solenoid circuit is either at full value (e.g. battery voltage with no on / off signal) or, the signal exists but it is above the normal operating range. Possible fault with solenoid or wiring (e.g. solenoid power supply, short / open circuit or resistance) that is causing a signal value to be higher than the maximum operating limit. Note: the control unit could be providing the correct signal but it is being affected by a circuit fault. Refer to list of sensor / actuator checks on page 19.
P2A00	O2 Sensor Circuit Range / Performance Bank 1 Sensor 1	Note: Many O2 / Lambda sensor related fault codes are activated due to faults in other systems (engine, fuelling, misfires etc). O2 (Oxygen / Lambda) sensors detect the exhaust gas oxygen content, which is also an indicator of air / fuel ratio (rich / lean). Oxygen content is indicated by a "Lambda" value. The air / fuel ratio (and therefore oxygen content) is controlled by the engine control unit to maximise efficiency of catalytic converters and reduce emissions of harmful gases.	For a detailed list of checks and identification of the sensor location, Refer to list of sensor / actuator checks on page 19. Different types of O2 / Lambda sensors (e.g. Zirconia or Titania / wideband etc) provide different signals and values, it will be necessary to identify the type of sensor fitted to the vehicle. The fault code indicates that the voltage of the O2 / Lambda sensor signal is within the normal operating range / tolerance but the signal is not plausible or is incorrect (the sensor signal is not changing / responding as expected to the changes in operating conditions). The problem could be related to a sensor or wiring fault; check wiring between sensor and control unit for short / open circuits and high resistance. Note that O2 / Lambda sensor Range / Performance fault codes are frequently activated due to the exhaust gas oxygen content being incorrect for the operating conditions; this can be caused by incorrect mixtures, misfires, air leaks or other engine related problems.
P2A01	O2 Sensor Circuit Range / Performance Bank 1 Sensor 2	Note: Many O2 / Lambda sensor related fault codes are activated due to faults in other systems (engine, fuelling, misfires etc). O2 (Oxygen / Lambda) sensors detect the exhaust gas oxygen content, which is also an indicator of air / fuel ratio (rich / lean). Oxygen content is indicated by a "Lambda" value. The air / fuel ratio (and therefore oxygen content) is controlled by the engine control unit to maximise efficiency of catalytic converters and reduce emissions of harmful gases.	For a detailed list of checks and identification of the sensor location, Refer to list of sensor / actuator checks on page 19. Different types of O2 / Lambda sensors (e.g. Zirconia or Titania / wideband etc) provide different signals and values, it will be necessary to identify the type of sensor fitted to the vehicle. The fault code indicates that the voltage of the O2 / Lambda sensor signal is within the normal operating range / tolerance but the signal is not plausible or is incorrect (the sensor signal is not changing / responding as expected to the changes in operating conditions). The problem could be related to a sensor or wiring fault; check wiring between sensor and control unit for short / open circuits and high resistance. Note that O2 / Lambda sensor Range / Performance fault codes are frequently activated due to the exhaust gas oxygen content being incorrect for the operating conditions; this can be caused by incorrect mixtures, misfires, air leaks or other engine related problems. It is also possible that a failing catalytic converter can cause a Range / Performance fault code for post-cat Lambda sensors.
P2A02	O2 Sensor Circuit Range / Performance Bank 1 Sensor 3	Note: Many O2 / Lambda sensor related fault codes are activated due to faults in other systems (engine, fuelling, misfires etc). O2 (Oxygen / Lambda) sensors detect the exhaust gas oxygen content, which is also an indicator of air / fuel ratio (rich / lean). Oxygen content is indicated by a "Lambda" value. The air / fuel ratio (and therefore oxygen content) is controlled by the engine control unit to maximise efficiency of catalytic converters and reduce emissions of harmful gases.	For a detailed list of checks and identification of the sensor location, Refer to list of sensor / actuator checks on page 19. Different types of O2 / Lambda sensors (e.g. Zirconia or Titania / wideband etc) provide different signals and values, it will be necessary to identify the type of sensor fitted to the vehicle. The fault code indicates that the voltage of the O2 / Lambda sensor signal is within the normal operating range / tolerance but the signal is not plausible or is incorrect (the sensor signal is not changing / responding as expected to the changes in operating conditions). The problem could be related to a sensor or wiring fault; check wiring between sensor and control unit for short / open circuits and high resistance. Note that O2 / Lambda sensor Range / Performance fault codes are frequently activated due to the exhaust gas oxygen content being incorrect for the operating conditions; this can be caused by incorrect mixtures, misfires, air leaks or other engine related problems. It is also possible that a failing catalytic converter can cause a Range / Performance fault code for post-cat Lambda sensors.
P2A03	O2 Sensor Circuit Range / Performance Bank 2 Sensor 1	Note: Many O2 / Lambda sensor related fault codes are activated due to faults in other systems (engine, fuelling, misfires etc). O2 (Oxygen / Lambda) sensors detect the exhaust gas oxygen content, which is also an indicator of air / fuel ratio (rich / lean). Oxygen content is indicated by a "Lambda" value. The air / fuel ratio (and therefore oxygen content) is controlled by the engine control unit to maximise efficiency of catalytic converters and reduce emissions of harmful gases.	For a detailed list of checks and identification of the sensor location, Refer to list of sensor / actuator checks on page 19. Different types of O2 / Lambda sensors (e.g. Zirconia or Titania / wideband etc) provide different signals and values, it will be necessary to identify the type of sensor fitted to the vehicle. The fault code indicates that the voltage of the O2 / Lambda sensor signal is within the normal operating range / tolerance but the signal is not plausible or is incorrect (the sensor signal is not changing / responding as expected to the changes in operating conditions). The problem could be related to a sensor or wiring fault; check wiring between sensor and control unit for short / open circuits and high resistance. Note that O2 / Lambda sensor Range / Performance fault codes are frequently activated due to the exhaust gas oxygen content being incorrect for the operating conditions; this can be caused by incorrect mixtures, misfires, air leaks or other engine related problems. It is also possible that a failing catalytic converter can cause a Range / Performance fault code for post-cat Lambda sensors.

NOTE 1. Check for other fault codes that could provide additional information. **NOTE 2.** Communication between control units can pass via a CAN-Bus system; refer to Page 51 for CAN-Bus checks.

322

Code	Description		
P2A04	O2 Sensor Circuit Range / Performance Bank 2 Sensor 2	Note: Many O2 / Lambda sensor related fault codes are activated due to faults in other systems (engine, fuelling, misfires etc). O2 (Oxygen / Lambda) sensors detect the exhaust gas oxygen content, which is also an indicator of air / fuel ratio (rich / lean). Oxygen content is indicated by a "Lambda" value. The air / fuel ratio (and therefore oxygen content) is controlled by the engine control unit to maximise efficiency of catalytic converters and reduce emissions of harmful gases.	For a detailed list of checks and identification of the sensor location, Refer to list of sensor / actuator checks on page 19. Different types of O2 / Lambda sensors (e.g. Zirconia or Titania / wideband etc) provide different signals and values, it will be necessary to identify the type of sensor fitted to the vehicle. The fault code indicates that the voltage of the O2 / Lambda sensor signal is within the normal operating range / tolerance but the signal is not plausible or is incorrect (the sensor signal is not changing / responding as expected to the changes in operating conditions). The problem could be related to a sensor or wiring fault; check wiring between sensor and control unit for short / open circuits and high resistance. Note that O2 / Lambda sensor Range / Performance fault codes are frequently activated due to the exhaust gas oxygen content being incorrect for the operating conditions; this can be caused by incorrect mixtures, misfires, air leaks or other engine related problems. It is also possible that a failing catalytic converter can cause a Range / Performance fault code for post-cat Lambda sensors.
P2A05	O2 Sensor Circuit Range / Performance Bank 2 Sensor 3	Note: Many O2 / Lambda sensor related fault codes are activated due to faults in other systems (engine, fuelling, misfires etc). O2 (Oxygen / Lambda) sensors detect the exhaust gas oxygen content, which is also an indicator of air / fuel ratio (rich / lean). Oxygen content is indicated by a "Lambda" value. The air / fuel ratio (and therefore oxygen content) is controlled by the engine control unit to maximise efficiency of catalytic converters and reduce emissions of harmful gases. NOTE. Many O2 / Lambda sensor related fault codes are activated due to faults in other systems (engine fuelling, misfires etc).	For a detailed list of checks and identification of the sensor location, Refer to list of sensor / actuator checks on page 19. Different types of O2 / Lambda sensors (e.g. Zirconia or Titania / wideband etc) provide different signals and values, it will be necessary to identify the type of sensor fitted to the vehicle. The fault code indicates that the voltage of the O2 / Lambda sensor signal is within the normal operating range / tolerance but the signal is not plausible or is incorrect (the sensor signal is not changing / responding as expected to the changes in operating conditions). The problem could be related to a sensor or wiring fault; check wiring between sensor and control unit for short / open circuits and high resistance. Note that O2 / Lambda sensor Range / Performance fault codes are frequently activated due to the exhaust gas oxygen content being incorrect for the operating conditions; this can be caused by incorrect mixtures, misfires, air leaks or other engine related problems. It is also possible that a failing catalytic converter can cause a Range / Performance fault code for post-cat Lambda sensors.
P2A06	O2 Sensor Negative Voltage Bank 1 Sensor 1	Note: Many O2 / Lambda sensor related fault codes are activated due to faults in other systems (engine, fuelling, misfires etc). O2 (Oxygen / Lambda) sensors detect the exhaust gas oxygen content (an indicator of air / fuel ratio). Oxygen content is indicated by a "Lambda" value. The control unit alters the fuelling to achieve the optimum Lambda value to enable good efficiency of catalytic converters (reduce emissions of harmful gases). For the operation of "broadband" sensors, the measurement process involves applying positive or negative voltage / current (pumping current) to one part of the sensing element; the level of positive or negative voltage / current being applied is an indicator of the oxygen content.	For a detailed list of checks and identification of the sensor location, Refer to list of sensor / actuator checks on page 19. The fault code indicates that there is an undefined fault in the application of the negative voltage / current to the measuring element of the Oxygen / Lambda sensor. The problem is most likely to be related to a sensor or wiring fault; check wiring between sensor and control unit for short / open circuits and high resistance. Note that the sensor could also be contaminated or faulty. Check sensor signal and signal response under different operating conditions (load, constant rpm etc) and Refer to list of sensor / actuator checks on page 19. If the sensor does however appear to be operating correctly but the voltage value remains incorrect, this could indicate an engine / fuelling fault, or other fault causing incorrect oxygen content in the exhaust gas.
P2A07	O2 Sensor Negative Voltage Bank 1 Sensor 2	Note: Many O2 / Lambda sensor related fault codes are activated due to faults in other systems (engine, fuelling, misfires etc). O2 (Oxygen / Lambda) sensors detect the exhaust gas oxygen content (an indicator of air / fuel ratio). Oxygen content is indicated by a "Lambda" value. The control unit alters the fuelling to achieve the optimum Lambda value to enable good efficiency of catalytic converters (reduce emissions of harmful gases). For the operation of "broadband" sensors, the measurement process involves applying positive or negative voltage / current (pumping current) to one part of the sensing element; the level of positive or negative voltage / current being applied is an indicator of the oxygen content.	For a detailed list of checks and identification of the sensor location, Refer to list of sensor / actuator checks on page 19. The fault code indicates that there is an undefined fault in the application of the negative voltage / current to the measuring element of the Oxygen / Lambda sensor. The problem is most likely to be related to a sensor or wiring fault; check wiring between sensor and control unit for short / open circuits and high resistance. Note that the sensor could also be contaminated or faulty. Check sensor signal and signal response under different operating conditions (load, constant rpm etc) and Refer to list of sensor / actuator checks on page 19. If the sensor does however appear to be operating correctly but the voltage value remains incorrect, this could indicate an engine / fuelling fault, or other fault causing incorrect oxygen content in the exhaust gas. It is also possible that a failing catalytic converter can cause a Range / Performance fault code for post-cat Lambda sensors.
P2A08	O2 Sensor Negative Voltage Bank 1 Sensor 3	Note: Many O2 / Lambda sensor related fault codes are activated due to faults in other systems (engine, fuelling, misfires etc). O2 (Oxygen / Lambda) sensors detect the exhaust gas oxygen content (an indicator of air / fuel ratio). Oxygen content is indicated by a "Lambda" value. The control unit alters the fuelling to achieve the optimum Lambda value to enable good efficiency of catalytic converters (reduce emissions of harmful gases). For the operation of "broadband" sensors, the measurement process involves applying positive or negative voltage / current (pumping current) to one part of the sensing element; the level of positive or negative voltage / current being applied is an indicator of the oxygen content.	For a detailed list of checks and identification of the sensor location, Refer to list of sensor / actuator checks on page 19. The fault code indicates that there is an undefined fault in the application of the negative voltage / current to the measuring element of the Oxygen / Lambda sensor. The problem is most likely to be related to a sensor or wiring fault; check wiring between sensor and control unit for short / open circuits and high resistance. Note that the sensor could also be contaminated or faulty. Check sensor signal and signal response under different operating conditions (load, constant rpm etc) and Refer to list of sensor / actuator checks on page 19. If the sensor does however appear to be operating correctly but the voltage value remains incorrect, this could indicate an engine / fuelling fault, or other fault causing incorrect oxygen content in the exhaust gas. It is also possible that a failing catalytic converter can cause a Range / Performance fault code for post-cat Lambda sensors.
P2A09	O2 Sensor Negative Voltage Bank 2 Sensor 1	Note: Many O2 / Lambda sensor related fault codes are activated due to faults in other systems (engine, fuelling, misfires etc). O2 (Oxygen / Lambda) sensors detect the exhaust gas oxygen content (an indicator of air / fuel ratio). Oxygen content is indicated by a "Lambda" value. The control unit alters the fuelling to achieve the optimum Lambda value to enable good efficiency of catalytic converters (reduce emissions of harmful gases). For the operation of "broadband" sensors, the measurement process involves applying positive or negative voltage / current (pumping current) to one part of the sensing element; the level of positive or negative voltage / current being applied is an indicator of the oxygen content.	For a detailed list of checks and identification of the sensor location, Refer to list of sensor / actuator checks on page 19. The fault code indicates that there is an undefined fault in the application of the negative voltage / current to the measuring element of the Oxygen / Lambda sensor. The problem is most likely to be related to a sensor or wiring fault; check wiring between sensor and control unit for short / open circuits and high resistance. Note that the sensor could also be contaminated or faulty. Check sensor signal and signal response under different operating conditions (load, constant rpm etc) and Refer to list of sensor / actuator checks on page 19. If the sensor does however appear to be operating correctly but the voltage value remains incorrect, this could indicate an engine / fuelling fault, or other fault causing incorrect oxygen content in the exhaust gas.

NOTE 1. Check for other fault codes that could provide additional information.

NOTE 3. If a fault cannot be located, it is also possible that the control unit is at fault.

NOTE 2. Communication between control units can pass via a CAN-Bus system; refer to Page 51 for CAN-Bus checks.

NOTE 4. Refer to Page 53 for list of pages that show the ISO standard for component locations e.g. Sensor A - Bank 1.

Fault code	EOBD / ISO Description	Component / System Description	Meaningful Description and Quick Check
P2A10	O2 Sensor Negative Voltage Bank 2 Sensor 2	Note: Many O2 / Lambda sensor related fault codes are activated due to faults in other systems (engine, fuelling, misfires etc). O2 (Oxygen / Lambda) sensors detect the exhaust gas oxygen content (an indicator of air / fuel ratio). Oxygen content is indicated by a "Lambda" value. The control unit alters the fuelling to achieve the optimum Lambda value to enable good efficiency of catalytic converters (reduce emissions of harmful gases). For the operation of "broadband" sensors, the measurement process involves applying positive or negative voltage / current (pumping current) to one part of the sensing element; the level of positive or negative voltage / current being applied is an indicator of the oxygen content.	For a detailed list of checks and identification of the sensor location, Refer to list of sensor / actuator checks on page 19. The fault code indicates that there is an undefined fault in the application of the negative voltage / current to the measuring element of the Oxygen / Lambda sensor. The problem is most likely to be related to a sensor or wiring fault; check wiring between sensor and control unit for short / open circuits and high resistance. Note that the sensor could also be contaminated or faulty. Check sensor signal and signal response under different operating conditions (load, constant rpm etc) and Refer to list of sensor / actuator checks on page 19. If the sensor does however appear to be operating correctly but the voltage value remains incorrect, this could indicate an engine / fuelling fault, or other fault causing incorrect oxygen content in the exhaust gas. It is also possible that a failing catalytic converter can cause a Range / Performance fault code for post-cat Lambda sensors.
P2A11	O2 Sensor Negative Voltage Bank 2 Sensor 3	Note: Many O2 / Lambda sensor related fault codes are activated due to faults in other systems (engine, fuelling, misfires etc). O2 (Oxygen / Lambda) sensors detect the exhaust gas oxygen content (an indicator of air / fuel ratio). Oxygen content is indicated by a "Lambda" value. The control unit alters the fuelling to achieve the optimum Lambda value to enable good efficiency of catalytic converters (reduce emissions of harmful gases). For the operation of "broadband" sensors, the measurement process involves applying positive or negative voltage / current (pumping current) to one part of the sensing element; the level of positive or negative voltage / current being applied is an indicator of the oxygen content.	For a detailed list of checks and identification of the sensor location, Refer to list of sensor / actuator checks on page 19. The fault code indicates that there is an undefined fault in the application of the negative voltage / current to the measuring element of the Oxygen / Lambda sensor. The problem is most likely to be related to a sensor or wiring fault; check wiring between sensor and control unit for short / open circuits and high resistance. Note that the sensor could also be contaminated or faulty. Check sensor signal and signal response under different operating conditions (load, constant rpm etc) and Refer to list of sensor / actuator checks on page 19. If the sensor does however appear to be operating correctly but the voltage value remains incorrect, this could indicate an engine / fuelling fault, or other fault causing incorrect oxygen content in the exhaust gas. It is also possible that a failing catalytic converter can cause a Range / Performance fault code for post-cat Lambda sensors.
P3400	Cylinder Deactivation System Bank 1	Refer to vehicle specific information to establish the type of deactivation system and components used. The deactivation system can make use of an actuator (usually a solenoid controlling an oil pressure feed to the valve lift mechanism) to restrict the opening of the valve; this prevents cylinder operation (no compression or cylinder pumping losses when valves are closed). The system design will dictate the checks that need to be carried out.	The fault code indicates an undefined fault with the Cylinder Deactivation system. The actuator or the Cylinder Deactivation system response does not match the control signal value being provided by the control unit e.g. control signal should cause the deactivation system to operate but it does not respond; this may be due to a fault in the actuator / wiring or in other parts of the deactivation system. Note that the fault relates to one bank of cylinders and, because deactivations systems usually select individual cylinders, the fault could be related to wiring or other components affecting the bank of cylinders rather than an individual cylinder. It is also possible that the actuator (or Deactivation mechanism) is not operating or moving correctly. Check for correct movement of Deactivation mechanism and check operation of valve assembly; also check for sufficient oil pressure to the Deactivation system (if applicable). Refer to Fault code P3494 for additional information. Refer to list of sensor / actuator checks on page 19.
P3401	Cylinder 1 Deactivation / Intake Valve Control Circuit / Open	Refer to vehicle specific information to establish the type of deactivation system and components used. The deactivation system can make use of an actuator (usually a solenoid controlling an oil pressure feed to the valve lift mechanism) to restrict the opening of the valve; this prevents cylinder operation (no compression or cylinder pumping losses when valves are closed).	The voltage, frequency or value of the control signal in the actuator circuit is incorrect. Signal could be:- out of normal operating range, constant value, non-existent, corrupt. The fault code indicates a possible "open circuit" but other undefined faults in the actuator or wiring could activate the same code e.g. actuator power supply, short circuit, high circuit resistance or incorrect actuator resistance. Note: the control unit could be providing the correct signal but it is being affected by a circuit fault. Refer to list of sensor / actuator checks on page 19.
P3402	Cylinder 1 Deactivation / Intake Valve Control Circuit Performance	Refer to vehicle specific information to establish the type of deactivation system and components used. The deactivation system can make use of an actuator (usually a solenoid controlling an oil pressure feed to the valve lift mechanism) to restrict the opening of the valve; this prevents cylinder operation (no compression or cylinder pumping losses when valves are closed).	The voltage, frequency or value of the control signal in the actuator circuit is within the normal operating range / tolerance, but the signal might be incorrect for the operating conditions. It is also possible that the actuator response (or response of the Deactivation system) differs from the expected / desired response (incorrect response to the control signal provided by the control unit). Possible fault in actuator or wiring, but it is also possible that the actuator (or Deactivation mechanism) is not operating or moving correctly. Check for correct movement of Deactivation mechanism and check operation of valve assembly; also check for sufficient oil pressure to the Deactivation system (if applicable to the system). Note: the control unit could be providing the correct signal but it is being affected by a circuit fault. Refer to list of sensor / actuator checks on page 19.
P3403	Cylinder 1 Deactivation / Intake Valve Control Circuit Low	Refer to vehicle specific information to establish the type of deactivation system and components used. The deactivation system can make use of an actuator (usually a solenoid controlling an oil pressure feed to the valve lift mechanism) to restrict the opening of the valve; this prevents cylinder operation (no compression or cylinder pumping losses when valves are closed).	The voltage, frequency or value of the control signal in the actuator circuit is either "zero" (no signal) or, the signal exists but is below the normal operating range. Possible fault with actuator or wiring (e.g. actuator power supply, short / open circuit or resistance) that is causing a signal value to be lower than the minimum operating limit. Note: the control unit could be providing the correct signal but it is being affected by a circuit fault. Refer to list of sensor / actuator checks on page 19.
P3404	Cylinder 1 Deactivation / Intake Valve Control Circuit High	Refer to vehicle specific information to establish the type of deactivation system and components used. The deactivation system can make use of an actuator (usually a solenoid controlling an oil pressure feed to the valve lift mechanism) to restrict the opening of the valve; this prevents cylinder operation (no compression or cylinder pumping losses when valves are closed).	The voltage, frequency or value of the control signal in the actuator circuit is either at full value (e.g. battery voltage with no on / off signal) or, the signal exists but it is above the normal operating range. Possible fault with actuator or wiring (e.g. actuator power supply, short / open circuit or resistance) that is causing a signal value to be higher than the maximum operating limit. Note: the control unit could be providing the correct signal but it is being affected by a circuit fault. Refer to list of sensor / actuator checks on page 19.
P3405	Cylinder 1 Exhaust Valve Control Circuit / Open	Refer to vehicle specific information to establish the type of deactivation system and components used. The deactivation system can make use of an actuator (usually a solenoid controlling an oil pressure feed to the valve lift mechanism) to restrict the opening of the valve; this prevents cylinder operation (no compression or cylinder pumping losses when valves are closed).	The voltage, frequency or value of the control signal in the actuator circuit is incorrect. Signal could be:- out of normal operating range, constant value, non-existent, corrupt. The fault code indicates a possible "open circuit" but other undefined faults in the actuator or wiring could activate the same code e.g. actuator power supply, short circuit, high circuit resistance or incorrect actuator resistance. Note: the control unit could be providing the correct signal but it is being affected by a circuit fault. Refer to list of sensor / actuator checks on page 19.

NOTE 1. Check for other fault codes that could provide additional information. NOTE 2. Communication between control units can pass via a CAN-Bus system; refer to Page 51 for CAN-Bus checks.

NOTE 3. If a fault cannot be located, it is also possible that the control unit is at fault NOTE 4. Refer to Page 53 for list of pages that show the ISO standard for component locations e.g. Sensor A - Bank 1.

P3406	Cylinder 1 Exhaust Valve Control Circuit Performance	Refer to vehicle specific information to establish the type of deactivation system and components used. The deactivation system can make use of an actuator (usually a solenoid controlling an oil pressure feed to the valve lift mechanism) to restrict the opening of the valve; this prevents cylinder operation (no compression or cylinder pumping losses when valves are closed).	The voltage, frequency or value of the control signal in the actuator circuit is within the normal operating range / tolerance, but the signal might be incorrect for the operating conditions. It is also possible that the actuator response (or response of the Deactivation system) differs from the expected / desired response (incorrect response to the control signal provided by the control unit). Possible fault in actuator or wiring, but it is also possible that the actuator (or Deactivation mechanism) is not operating or moving correctly. Check for correct movement of Deactivation mechanism and check operation of valve assembly; also check for sufficient oil pressure to the Deactivation system (if applicable to the system). Note: the control unit could be providing the correct signal but it is being affected by a circuit fault. Refer to list of sensor / actuator checks on page 19.
P3407	Cylinder 1 Exhaust Valve Control Circuit Low	Refer to vehicle specific information to establish the type of deactivation system and components used. The deactivation system can make use of an actuator (usually a solenoid controlling an oil pressure feed to the valve lift mechanism) to restrict the opening of the valve; this prevents cylinder operation (no compression or cylinder pumping losses when valves are closed).	The voltage, frequency or value of the control signal in the actuator circuit is either "zero" (no signal) or, the signal exists but is below the normal operating range. Possible fault with actuator or wiring (e.g. actuator power supply, short / open circuit or resistance) that is causing a signal value to be lower than the minimum operating limit. Note: the control unit could be providing the correct signal but it is being affected by a circuit fault. Refer to list of sensor / actuator checks on page 19.
P3408	Cylinder 1 Exhaust Valve Control Circuit High	Refer to vehicle specific information to establish the type of deactivation system and components used. The deactivation system can make use of an actuator (usually a solenoid controlling an oil pressure feed to the valve lift mechanism) to restrict the opening of the valve; this prevents cylinder operation (no compression or cylinder pumping losses when valves are closed).	The voltage, frequency or value of the control signal in the actuator circuit is either at full value (e.g. battery voltage with no on / off signal) or, the signal exists but it is above the normal operating range. Possible fault with actuator or wiring (e.g. actuator power supply, short / open circuit or resistance) that is causing a signal value to be higher than the maximum operating limit. Note: the control unit could be providing the correct signal but it is being affected by a circuit fault. Refer to list of sensor / actuator checks on page 19.
P3409	Cylinder 2 Deactivation / Intake Valve Control Circuit / Open	Refer to vehicle specific information to establish the type of deactivation system and components used. The deactivation system can make use of an actuator (usually a solenoid controlling an oil pressure feed to the valve lift mechanism) to restrict the opening of the valve; this prevents cylinder operation (no compression or cylinder pumping losses when valves are closed).	The voltage, frequency or value of the control signal in the actuator circuit is incorrect. Signal could be:- out of normal operating range, constant value, non-existent, corrupt. The fault code indicates a possible "open circuit" but other undefined faults in the actuator or wiring could activate the same code e.g. actuator power supply, short circuit, high circuit resistance or incorrect actuator resistance. Note: the control unit could be providing the correct signal but it is being affected by a circuit fault. Refer to list of sensor / actuator checks on page 19.
P3410	Cylinder 2 Deactivation / Intake Valve Control Circuit Performance	Refer to vehicle specific information to establish the type of deactivation system and components used. The deactivation system can make use of an actuator (usually a solenoid controlling an oil pressure feed to the valve lift mechanism) to restrict the opening of the valve; this prevents cylinder operation (no compression or cylinder pumping losses when valves are closed).	The voltage, frequency or value of the control signal in the actuator circuit is within the normal operating range / tolerance, but the signal might be incorrect for the operating conditions. It is also possible that the actuator response (or response of the Deactivation system) differs from the expected / desired response (incorrect response to the control signal provided by the control unit). Possible fault in actuator or wiring, but it is also possible that the actuator (or Deactivation mechanism) is not operating or moving correctly. Check for correct movement of Deactivation mechanism and check operation of valve assembly; also check for sufficient oil pressure to the Deactivation system (if applicable to the system). Note: the control unit could be providing the correct signal but it is being affected by a circuit fault. Refer to list of sensor / actuator checks on page 19.
P3411	Cylinder 2 Deactivation / Intake Valve Control Circuit Low	Refer to vehicle specific information to establish the type of deactivation system and components used. The deactivation system can make use of an actuator (usually a solenoid controlling an oil pressure feed to the valve lift mechanism) to restrict the opening of the valve; this prevents cylinder operation (no compression or cylinder pumping losses when valves are closed).	The voltage, frequency or value of the control signal in the actuator circuit is either "zero" (no signal) or, the signal exists but is below the normal operating range. Possible fault with actuator or wiring (e.g. actuator power supply, short / open circuit or resistance) that is causing a signal value to be lower than the minimum operating limit. Note: the control unit could be providing the correct signal but it is being affected by a circuit fault. Refer to list of sensor / actuator checks on page 19.
P3412	Cylinder 2 Deactivation / Intake Valve Control Circuit High	Refer to vehicle specific information to establish the type of deactivation system and components used. The deactivation system can make use of an actuator (usually a solenoid controlling an oil pressure feed to the valve lift mechanism) to restrict the opening of the valve; this prevents cylinder operation (no compression or cylinder pumping losses when valves are closed).	The voltage, frequency or value of the control signal in the actuator circuit is either at full value (e.g. battery voltage with no on / off signal) or, the signal exists but it is above the normal operating range. Possible fault with actuator or wiring (e.g. actuator power supply, short / open circuit or resistance) that is causing a signal value to be higher than the maximum operating limit. Note: the control unit could be providing the correct signal but it is being affected by a circuit fault. Refer to list of sensor / actuator checks on page 19.
P3413	Cylinder 2 Exhaust Valve Control Circuit / Open	Refer to vehicle specific information to establish the type of deactivation system and components used. The deactivation system can make use of an actuator (usually a solenoid controlling an oil pressure feed to the valve lift mechanism) to restrict the opening of the valve; this prevents cylinder operation (no compression or cylinder pumping losses when valves are closed).	The voltage, frequency or value of the control signal in the actuator circuit is incorrect. Signal could be:- out of normal operating range, constant value, non-existent, corrupt. The fault code indicates a possible "open circuit" but other undefined faults in the actuator or wiring could activate the same code e.g. actuator power supply, short circuit, high circuit resistance or incorrect actuator resistance. Note: the control unit could be providing the correct signal but it is being affected by a circuit fault. Refer to list of sensor / actuator checks on page 19.
P3414	Cylinder 2 Exhaust Valve Control Circuit Performance	Refer to vehicle specific information to establish the type of deactivation system and components used. The deactivation system can make use of an actuator (usually a solenoid controlling an oil pressure feed to the valve lift mechanism) to restrict the opening of the valve; this prevents cylinder operation (no compression or cylinder pumping losses when valves are closed).	The voltage, frequency or value of the control signal in the actuator circuit is within the normal operating range / tolerance, but the signal might be incorrect for the operating conditions. It is also possible that the actuator response (or response of the Deactivation system) differs from the expected / desired response (incorrect response to the control signal provided by the control unit). Possible fault in actuator or wiring, but it is also possible that the actuator (or Deactivation mechanism) is not operating or moving correctly. Check for correct movement of Deactivation mechanism and check operation of valve assembly; also check for sufficient oil pressure to the Deactivation system (if applicable to the system). Note: the control unit could be providing the correct signal but it is being affected by a circuit fault. Refer to list of sensor / actuator checks on page 19.
P3415	Cylinder 2 Exhaust Valve Control Circuit Low	Refer to vehicle specific information to establish the type of deactivation system and components used. The deactivation system can make use of an actuator (usually a solenoid controlling an oil pressure feed to the valve lift mechanism) to restrict the opening of the valve; this prevents cylinder operation (no compression or cylinder pumping losses when valves are closed).	The voltage, frequency or value of the control signal in the actuator circuit is either "zero" (no signal) or, the signal exists but is below the normal operating range. Possible fault with actuator or wiring (e.g. actuator power supply, short / open circuit or resistance) that is causing a signal value to be lower than the minimum operating limit. Note: the control unit could be providing the correct signal but it is being affected by a circuit fault. Refer to list of sensor / actuator checks on page 19.

NOTE 1. Check for other fault codes that could provide additional information. **NOTE 2.** Communication between control units can pass via a CAN-Bus system; refer to Page 51 for CAN-Bus checks.
NOTE 3. If a fault cannot be located, it is also possible that the control unit is at fault. **NOTE 4.** Refer to Page 53 for list of pages that show the ISO standard for component locations e.g. Sensor A - Bank 1.

Fault code	EOBD / ISO Description	Component / System Description	Meaningful Description and Quick Check
P3416	Cylinder 2 Exhaust Valve Control Circuit High	Refer to vehicle specific information to establish the type of deactivation system and components used. The deactivation system can make use of an actuator (usually a solenoid controlling an oil pressure feed to the valve lift mechanism) to restrict the opening of the valve; this prevents cylinder operation (no compression or cylinder pumping losses when valves are closed).	The voltage, frequency or value of the control signal in the actuator circuit is either at full value (e.g. battery voltage with no on / off signal) or, the signal exists but it is above the normal operating range. Possible fault with actuator or wiring (e.g. actuator power supply, short / open circuit or resistance) that is causing a signal value to be higher than the maximum operating limit. Note: the control unit could be providing the correct signal but it is being affected by a circuit fault. Refer to list of sensor / actuator checks on page 19.
P3417	Cylinder 3 Deactivation / Intake Valve Control Circuit / Open	Refer to vehicle specific information to establish the type of deactivation system and components used. The deactivation system can make use of an actuator (usually a solenoid controlling an oil pressure feed to the valve lift mechanism) to restrict the opening of the valve; this prevents cylinder operation (no compression or cylinder pumping losses when valves are closed).	The voltage, frequency or value of the control signal in the actuator circuit is incorrect. Signal could be:- out of normal operating range, constant value, non-existent, corrupt. The fault code indicates a possible "open circuit" but other undefined faults in the actuator or wiring could activate the same code e.g. actuator power supply, short circuit, high circuit resistance or incorrect actuator resistance. Note: the control unit could be providing the correct signal but it is being affected by a circuit fault. Refer to list of sensor / actuator checks on page 19.
P3418	Cylinder 3 Deactivation / Intake Valve Control Circuit Performance	Refer to vehicle specific information to establish the type of deactivation system and components used. The deactivation system can make use of an actuator (usually a solenoid controlling an oil pressure feed to the valve lift mechanism) to restrict the opening of the valve; this prevents cylinder operation (no compression or cylinder pumping losses when valves are closed).	The voltage, frequency or value of the control signal in the actuator circuit is within the normal operating range / tolerance, but the signal might be incorrect for the operating conditions. It is also possible that the actuator response (or response of the Deactivation system) differs from the expected / desired response (incorrect response to the control signal provided by the control unit). Possible fault in actuator or wiring, but it is also possible that the actuator (or Deactivation mechanism) is not operating or moving correctly. Check for correct movement of Deactivation mechanism and check operation of valve assembly; also check for sufficient oil pressure to the Deactivation system (if applicable to the system). Note: the control unit could be providing the correct signal but it is being affected by a circuit fault. Refer to list of sensor / actuator checks on page 19.
P3419	Cylinder 3 Deactivation / Intake Valve Control Circuit Low	Refer to vehicle specific information to establish the type of deactivation system and components used. The deactivation system can make use of an actuator (usually a solenoid controlling an oil pressure feed to the valve lift mechanism) to restrict the opening of the valve; this prevents cylinder operation (no compression or cylinder pumping losses when valves are closed).	The voltage, frequency or value of the control signal in the actuator circuit is either "zero" (no signal) or, the signal exists but is below the normal operating range. Possible fault with actuator or wiring (e.g. actuator power supply, short / open circuit or resistance) that is causing a signal value to be lower than the minimum operating limit. Note: the control unit could be providing the correct signal but it is being affected by a circuit fault. Refer to list of sensor / actuator checks on page 19.
P3420	Cylinder 3 Deactivation / Intake Valve Control Circuit High	Refer to vehicle specific information to establish the type of deactivation system and components used. The deactivation system can make use of an actuator (usually a solenoid controlling an oil pressure feed to the valve lift mechanism) to restrict the opening of the valve; this prevents cylinder operation (no compression or cylinder pumping losses when valves are closed).	The voltage, frequency or value of the control signal in the actuator circuit is either at full value (e.g. battery voltage with no on / off signal) or, the signal exists but it is above the normal operating range. Possible fault with actuator or wiring (e.g. actuator power supply, short / open circuit or resistance) that is causing a signal value to be higher than the maximum operating limit. Note: the control unit could be providing the correct signal but it is being affected by a circuit fault. Refer to list of sensor / actuator checks on page 19.
P3421	Cylinder 3 Exhaust Valve Control Circuit / Open	Refer to vehicle specific information to establish the type of deactivation system and components used. The deactivation system can make use of an actuator (usually a solenoid controlling an oil pressure feed to the valve lift mechanism) to restrict the opening of the valve; this prevents cylinder operation (no compression or cylinder pumping losses when valves are closed).	The voltage, frequency or value of the control signal in the actuator circuit is incorrect. Signal could be:- out of normal operating range, constant value, non-existent, corrupt. The fault code indicates a possible "open circuit" but other undefined faults in the actuator or wiring could activate the same code e.g. actuator power supply, short circuit, high circuit resistance or incorrect actuator resistance. Note: the control unit could be providing the correct signal but it is being affected by a circuit fault. Refer to list of sensor / actuator checks on page 19.
P3422	Cylinder 3 Exhaust Valve Control Circuit Performance	Refer to vehicle specific information to establish the type of deactivation system and components used. The deactivation system can make use of an actuator (usually a solenoid controlling an oil pressure feed to the valve lift mechanism) to restrict the opening of the valve; this prevents cylinder operation (no compression or cylinder pumping losses when valves are closed).	The voltage, frequency or value of the control signal in the actuator circuit is within the normal operating range / tolerance, but the signal might be incorrect for the operating conditions. It is also possible that the actuator response (or response of the Deactivation system) differs from the expected / desired response (incorrect response to the control signal provided by the control unit). Possible fault in actuator or wiring, but it is also possible that the actuator (or Deactivation mechanism) is not operating or moving correctly. Check for correct movement of Deactivation mechanism and check operation of valve assembly; also check for sufficient oil pressure to the Deactivation system (if applicable to the system). Note: the control unit could be providing the correct signal but it is being affected by a circuit fault. Refer to list of sensor / actuator checks on page 19.
P3423	Cylinder 3 Exhaust Valve Control Circuit Low	Refer to vehicle specific information to establish the type of deactivation system and components used. The deactivation system can make use of an actuator (usually a solenoid controlling an oil pressure feed to the valve lift mechanism) to restrict the opening of the valve; this prevents cylinder operation (no compression or cylinder pumping losses when valves are closed).	The voltage, frequency or value of the control signal in the actuator circuit is either "zero" (no signal) or, the signal exists but is below the normal operating range. Possible fault with actuator or wiring (e.g. actuator power supply, short / open circuit or resistance) that is causing a signal value to be lower than the minimum operating limit. Note: the control unit could be providing the correct signal but it is being affected by a circuit fault. Refer to list of sensor / actuator checks on page 19.
P3424	Cylinder 3 Exhaust Valve Control Circuit High	Refer to vehicle specific information to establish the type of deactivation system and components used. The deactivation system can make use of an actuator (usually a solenoid controlling an oil pressure feed to the valve lift mechanism) to restrict the opening of the valve; this prevents cylinder operation (no compression or cylinder pumping losses when valves are closed).	The voltage, frequency or value of the control signal in the actuator circuit is either at full value (e.g. battery voltage with no on / off signal) or, the signal exists but it is above the normal operating range. Possible fault with actuator or wiring (e.g. actuator power supply, short / open circuit or resistance) that is causing a signal value to be higher than the maximum operating limit. Note: the control unit could be providing the correct signal but it is being affected by a circuit fault. Refer to list of sensor / actuator checks on page 19.
P3425	Cylinder 4 Deactivation / Intake Valve Control Circuit / Open	Refer to vehicle specific information to establish the type of deactivation system and components used. The deactivation system can make use of an actuator (usually a solenoid controlling an oil pressure feed to the valve lift mechanism) to restrict the opening of the valve; this prevents cylinder operation (no compression or cylinder pumping losses when valves are closed).	The voltage, frequency or value of the control signal in the actuator circuit is incorrect. Signal could be:- out of normal operating range, constant value, non-existent, corrupt. The fault code indicates a possible "open circuit" but other undefined faults in the actuator or wiring could activate the same code e.g. actuator power supply, short circuit, high circuit resistance or incorrect actuator resistance. Note: the control unit could be providing the correct signal but it is being affected by a circuit fault. Refer to list of sensor / actuator checks on page 19.

NOTE 1. Check for other fault codes that could provide additional information. **NOTE 2.** Communication between control units can pass via a CAN-Bus system; refer to Page 51 for CAN-Bus checks.

NOTE 3. If a fault cannot be located, it is also possible that the control unit is at fault **NOTE 4.** Refer to Page 53 for list of pages that show the ISO standard for component locations e.g. Sensor A - Bank 1.

Code	Description	Deactivation system info	Fault details
P3426	Cylinder 4 Deactivation / Intake Valve Control Circuit Performance	Refer to vehicle specific information to establish the type of deactivation system and components used. The deactivation system can make use of an actuator (usually a solenoid controlling an oil pressure feed to the valve lift mechanism) to restrict the opening of the valve; this prevents cylinder operation (no compression or cylinder pumping losses when valves are closed).	The voltage, frequency or value of the control signal in the actuator circuit is within the normal operating range / tolerance, but the signal might be incorrect for the operating conditions. It is also possible that the actuator response (or response of the Deactivation system) differs from the expected / desired response (incorrect response to the control signal provided by the control unit). Possible fault in actuator or wiring, but it is also possible that the actuator (or Deactivation mechanism) is not operating or moving correctly. Check for correct movement of Deactivation mechanism and check operation of valve assembly; also check for sufficient oil pressure to the Deactivation system (if applicable to the system). Note: the control unit could be providing the correct signal but it is being affected by a circuit fault. Refer to list of sensor / actuator checks on page 19.
P3427	Cylinder 4 Deactivation / Intake Valve Control Circuit Low	Refer to vehicle specific information to establish the type of deactivation system and components used. The deactivation system can make use of an actuator (usually a solenoid controlling an oil pressure feed to the valve lift mechanism) to restrict the opening of the valve; this prevents cylinder operation (no compression or cylinder pumping losses when valves are closed).	The voltage, frequency or value of the control signal in the actuator circuit is either "zero" (no signal) or, the signal exists but is below the normal operating range. Possible fault with actuator or wiring (e.g. actuator power supply, short / open circuit or resistance) that is causing a signal value to be lower than the minimum operating limit. Note: the control unit could be providing the correct signal but it is being affected by a circuit fault. Refer to list of sensor / actuator checks on page 19.
P3428	Cylinder 4 Deactivation / Intake Valve Control Circuit High	Refer to vehicle specific information to establish the type of deactivation system and components used. The deactivation system can make use of an actuator (usually a solenoid controlling an oil pressure feed to the valve lift mechanism) to restrict the opening of the valve; this prevents cylinder operation (no compression or cylinder pumping losses when valves are closed).	The voltage, frequency or value of the control signal in the actuator circuit is either at full value (e.g. battery voltage with no on / off signal) or, the signal exists but it is above the normal operating range. Possible fault with actuator or wiring (e.g. actuator power supply, short / open circuit or resistance) that is causing a signal value to be higher than the maximum operating limit. Note: the control unit could be providing the correct signal but it is being affected by a circuit fault. Refer to list of sensor / actuator checks on page 19.
P3429	Cylinder 4 Exhaust Valve Control Circuit / Open	Refer to vehicle specific information to establish the type of deactivation system and components used. The deactivation system can make use of an actuator (usually a solenoid controlling an oil pressure feed to the valve lift mechanism) to restrict the opening of the valve; this prevents cylinder operation (no compression or cylinder pumping losses when valves are closed).	The voltage, frequency or value of the control signal in the actuator circuit is incorrect. Signal could be:- out of normal operating range, constant value, non-existent, corrupt. The fault code indicates a possible "open circuit" but other undefined faults in the actuator or wiring could activate the same code e.g. actuator power supply, short circuit, high circuit resistance or incorrect actuator resistance. Note: the control unit could be providing the correct signal but it is being affected by a circuit fault. Refer to list of sensor / actuator checks on page 19.
P3430	Cylinder 4 Exhaust Valve Control Circuit Performance	Refer to vehicle specific information to establish the type of deactivation system and components used. The deactivation system can make use of an actuator (usually a solenoid controlling an oil pressure feed to the valve lift mechanism) to restrict the opening of the valve; this prevents cylinder operation (no compression or cylinder pumping losses when valves are closed).	The voltage, frequency or value of the control signal in the actuator circuit is within the normal operating range / tolerance, but the signal might be incorrect for the operating conditions. It is also possible that the actuator response (or response of the Deactivation system) differs from the expected / desired response (incorrect response to the control signal provided by the control unit). Possible fault in actuator or wiring, but it is also possible that the actuator (or Deactivation mechanism) is not operating or moving correctly. Check for correct movement of Deactivation mechanism and check operation of valve assembly; also check for sufficient oil pressure to the Deactivation system (if applicable to the system). Note: the control unit could be providing the correct signal but it is being affected by a circuit fault. Refer to list of sensor / actuator checks on page 19.
P3431	Cylinder 4 Exhaust Valve Control Circuit Low	Refer to vehicle specific information to establish the type of deactivation system and components used. The deactivation system can make use of an actuator (usually a solenoid controlling an oil pressure feed to the valve lift mechanism) to restrict the opening of the valve; this prevents cylinder operation (no compression or cylinder pumping losses when valves are closed).	The voltage, frequency or value of the control signal in the actuator circuit is either "zero" (no signal) or, the signal exists but is below the normal operating range. Possible fault with actuator or wiring (e.g. actuator power supply, short / open circuit or resistance) that is causing a signal value to be lower than the minimum operating limit. Note: the control unit could be providing the correct signal but it is being affected by a circuit fault. Refer to list of sensor / actuator checks on page 19.
P3432	Cylinder 4 Exhaust Valve Control Circuit High	Refer to vehicle specific information to establish the type of deactivation system and components used. The deactivation system can make use of an actuator (usually a solenoid controlling an oil pressure feed to the valve lift mechanism) to restrict the opening of the valve; this prevents cylinder operation (no compression or cylinder pumping losses when valves are closed).	The voltage, frequency or value of the control signal in the actuator circuit is either at full value (e.g. battery voltage with no on / off signal) or, the signal exists but it is above the normal operating range. Possible fault with actuator or wiring (e.g. actuator power supply, short / open circuit or resistance) that is causing a signal value to be higher than the maximum operating limit. Note: the control unit could be providing the correct signal but it is being affected by a circuit fault. Refer to list of sensor / actuator checks on page 19.
P3433	Cylinder 5 Deactivation / Intake Valve Control Circuit / Open	Refer to vehicle specific information to establish the type of deactivation system and components used. The deactivation system can make use of an actuator (usually a solenoid controlling an oil pressure feed to the valve lift mechanism) to restrict the opening of the valve; this prevents cylinder operation (no compression or cylinder pumping losses when valves are closed).	The voltage, frequency or value of the control signal in the actuator circuit is incorrect. Signal could be:- out of normal operating range, constant value, non-existent, corrupt. The fault code indicates a possible "open circuit" but other undefined faults in the actuator or wiring could activate the same code e.g. actuator power supply, short circuit, high circuit resistance or incorrect actuator resistance. Note: the control unit could be providing the correct signal but it is being affected by a circuit fault. Refer to list of sensor / actuator checks on page 19.
P3434	Cylinder 5 Deactivation / Intake Valve Control Circuit Performance	Refer to vehicle specific information to establish the type of deactivation system and components used. The deactivation system can make use of an actuator (usually a solenoid controlling an oil pressure feed to the valve lift mechanism) to restrict the opening of the valve; this prevents cylinder operation (no compression or cylinder pumping losses when valves are closed).	The voltage, frequency or value of the control signal in the actuator circuit is within the normal operating range / tolerance, but the signal might be incorrect for the operating conditions. It is also possible that the actuator response (or response of the Deactivation system) differs from the expected / desired response (incorrect response to the control signal provided by the control unit). Possible fault in actuator or wiring, but it is also possible that the actuator (or Deactivation mechanism) is not operating or moving correctly. Check for correct movement of Deactivation mechanism and check operation of valve assembly; also check for sufficient oil pressure to the Deactivation system (if applicable to the system). Note: the control unit could be providing the correct signal but it is being affected by a circuit fault. Refer to list of sensor / actuator checks on page 19.
P3435	Cylinder 5 Deactivation / Intake Valve Control Circuit Low	Refer to vehicle specific information to establish the type of deactivation system and components used. The deactivation system can make use of an actuator (usually a solenoid controlling an oil pressure feed to the valve lift mechanism) to restrict the opening of the valve; this prevents cylinder operation (no compression or cylinder pumping losses when valves are closed).	The voltage, frequency or value of the control signal in the actuator circuit is either "zero" (no signal) or, the signal exists but is below the normal operating range. Possible fault with actuator or wiring (e.g. actuator power supply, short / open circuit or resistance) that is causing a signal value to be lower than the minimum operating limit. Note: the control unit could be providing the correct signal but it is being affected by a circuit fault. Refer to list of sensor / actuator checks on page 19.

Fault code	EOBD / ISO Description	Component / System Description	Meaningful Description and Quick Check
P3436	Cylinder 5 Deactivation / Intake Valve Control Circuit High	Refer to vehicle specific information to establish the type of deactivation system and components used. The deactivation system can make use of an actuator (usually a solenoid controlling an oil pressure feed to the valve lift mechanism) to restrict the opening of the valve; this prevents cylinder operation (no compression or cylinder pumping losses when valves are closed).	The voltage, frequency or value of the control signal in the actuator circuit is either at full value (e.g. battery voltage with no on / off signal) or, the signal exists but it is above the normal operating range. Possible fault with actuator or wiring (e.g. actuator power supply, short / open circuit or resistance) that is causing a signal value to be higher than the maximum operating limit. Note: the control unit could be providing the correct signal but it is being affected by a circuit fault. Refer to list of sensor / actuator checks on page 19.
P3437	Cylinder 5 Exhaust Valve Control Circuit / Open	Refer to vehicle specific information to establish the type of deactivation system and components used. The deactivation system can make use of an actuator (usually a solenoid controlling an oil pressure feed to the valve lift mechanism) to restrict the opening of the valve; this prevents cylinder operation (no compression or cylinder pumping losses when valves are closed).	The voltage, frequency or value of the control signal in the actuator circuit is incorrect. Signal could be:- out of normal operating range, constant value, non-existent, corrupt. The fault code indicates a possible "open circuit" but other undefined faults in the actuator or wiring could activate the same code e.g. actuator power supply, short circuit, high circuit resistance or incorrect actuator resistance. Note: the control unit could be providing the correct signal but it is being affected by a circuit fault. Refer to list of sensor / actuator checks on page 19.
P3438	Cylinder 5 Exhaust Valve Control Circuit Performance	Refer to vehicle specific information to establish the type of deactivation system and components used. The deactivation system can make use of an actuator (usually a solenoid controlling an oil pressure feed to the valve lift mechanism) to restrict the opening of the valve; this prevents cylinder operation (no compression or cylinder pumping losses when valves are closed).	The voltage, frequency or value of the control signal in the actuator circuit is within the normal operating range / tolerance, but the signal might be incorrect for the operating conditions. It is also possible that the actuator response (or response of the Deactivation system) differs from the expected / desired response (incorrect response to the control signal provided by the control unit). Possible fault in actuator or wiring, but it is also possible that the actuator (or Deactivation mechanism) is not operating or moving correctly. Check for correct movement of Deactivation mechanism and check operation of valve assembly; also check for sufficient oil pressure to the Deactivation system (if applicable to the system). Note: the control unit could be providing the correct signal but it is being affected by a circuit fault. Refer to list of sensor / actuator checks on page 19.
P3439	Cylinder 5 Exhaust Valve Control Circuit Low	Refer to vehicle specific information to establish the type of deactivation system and components used. The deactivation system can make use of an actuator (usually a solenoid controlling an oil pressure feed to the valve lift mechanism) to restrict the opening of the valve; this prevents cylinder operation (no compression or cylinder pumping losses when valves are closed).	The voltage, frequency or value of the control signal in the actuator circuit is either "zero" (no signal) or, the signal exists but is below the normal operating range. Possible fault with actuator or wiring (e.g. actuator power supply, short / open circuit or resistance) that is causing a signal value to be lower than the minimum operating limit. Note: the control unit could be providing the correct signal but it is being affected by a circuit fault. Refer to list of sensor / actuator checks on page 19.
P3440	Cylinder 5 Exhaust Valve Control Circuit High	Refer to vehicle specific information to establish the type of deactivation system and components used. The deactivation system can make use of an actuator (usually a solenoid controlling an oil pressure feed to the valve lift mechanism) to restrict the opening of the valve; this prevents cylinder operation (no compression or cylinder pumping losses when valves are closed).	The voltage, frequency or value of the control signal in the actuator circuit is either at full value (e.g. battery voltage with no on / off signal) or, the signal exists but it is above the normal operating range. Possible fault with actuator or wiring (e.g. actuator power supply, short / open circuit or resistance) that is causing a signal value to be higher than the maximum operating limit. Note: the control unit could be providing the correct signal but it is being affected by a circuit fault. Refer to list of sensor / actuator checks on page 19.
P3441	Cylinder 6 Deactivation / Intake Valve Control Circuit / Open	Refer to vehicle specific information to establish the type of deactivation system and components used. The deactivation system can make use of an actuator (usually a solenoid controlling an oil pressure feed to the valve lift mechanism) to restrict the opening of the valve; this prevents cylinder operation (no compression or cylinder pumping losses when valves are closed).	The voltage, frequency or value of the control signal in the actuator circuit is incorrect. Signal could be:- out of normal operating range, constant value, non-existent, corrupt. The fault code indicates a possible "open circuit" but other undefined faults in the actuator or wiring could activate the same code e.g. actuator power supply, short circuit, high circuit resistance or incorrect actuator resistance. Note: the control unit could be providing the correct signal but it is being affected by a circuit fault. Refer to list of sensor / actuator checks on page 19.
P3442	Cylinder 6 Deactivation / Intake Valve Control Circuit Performance	Refer to vehicle specific information to establish the type of deactivation system and components used. The deactivation system can make use of an actuator (usually a solenoid controlling an oil pressure feed to the valve lift mechanism) to restrict the opening of the valve; this prevents cylinder operation (no compression or cylinder pumping losses when valves are closed).	The voltage, frequency or value of the control signal in the actuator circuit is within the normal operating range / tolerance, but the signal might be incorrect for the operating conditions. It is also possible that the actuator response (or response of the Deactivation system) differs from the expected / desired response (incorrect response to the control signal provided by the control unit). Possible fault in actuator or wiring, but it is also possible that the actuator (or Deactivation mechanism) is not operating or moving correctly. Check for correct movement of Deactivation mechanism and check operation of valve assembly; also check for sufficient oil pressure to the Deactivation system (if applicable to the system). Note: the control unit could be providing the correct signal but it is being affected by a circuit fault. Refer to list of sensor / actuator checks on page 19.
P3443	Cylinder 6 Deactivation / Intake Valve Control Circuit Low	Refer to vehicle specific information to establish the type of deactivation system and components used. The deactivation system can make use of an actuator (usually a solenoid controlling an oil pressure feed to the valve lift mechanism) to restrict the opening of the valve; this prevents cylinder operation (no compression or cylinder pumping losses when valves are closed).	The voltage, frequency or value of the control signal in the actuator circuit is either "zero" (no signal) or, the signal exists but is below the normal operating range. Possible fault with actuator or wiring (e.g. actuator power supply, short / open circuit or resistance) that is causing a signal value to be lower than the minimum operating limit. Note: the control unit could be providing the correct signal but it is being affected by a circuit fault. Refer to list of sensor / actuator checks on page 19.
P3444	Cylinder 6 Deactivation / Intake Valve Control Circuit High	Refer to vehicle specific information to establish the type of deactivation system and components used. The deactivation system can make use of an actuator (usually a solenoid controlling an oil pressure feed to the valve lift mechanism) to restrict the opening of the valve; this prevents cylinder operation (no compression or cylinder pumping losses when valves are closed).	The voltage, frequency or value of the control signal in the actuator circuit is either at full value (e.g. battery voltage with no on / off signal) or, the signal exists but it is above the normal operating range. Possible fault with actuator or wiring (e.g. actuator power supply, short / open circuit or resistance) that is causing a signal value to be higher than the maximum operating limit. Note: the control unit could be providing the correct signal but it is being affected by a circuit fault. Refer to list of sensor / actuator checks on page 19.
P3445	Cylinder 6 Exhaust Valve Control Circuit / Open	Refer to vehicle specific information to establish the type of deactivation system and components used. The deactivation system can make use of an actuator (usually a solenoid controlling an oil pressure feed to the valve lift mechanism) to restrict the opening of the valve; this prevents cylinder operation (no compression or cylinder pumping losses when valves are closed).	The voltage, frequency or value of the control signal in the actuator circuit is incorrect. Signal could be:- out of normal operating range, constant value, non-existent, corrupt. The fault code indicates a possible "open circuit" but other undefined faults in the actuator or wiring could activate the same code e.g. actuator power supply, short circuit, high circuit resistance or incorrect actuator resistance. Note: the control unit could be providing the correct signal but it is being affected by a circuit fault. Refer to list of sensor / actuator checks on page 19.

NOTE 1. Check for other fault codes that could provide additional information. **NOTE 2.** Communication between control units can pass via a CAN-Bus system; refer to Page 51 for CAN-Bus checks. 328

NOTE 3. If a fault cannot be located, it is also possible that the control unit is at fault NOTE 4. Refer to Page 53 for list of pages that show the ISO standard for component locations e.g. Sensor A - Bank 1.

P3446	Cylinder 6 Exhaust Valve Control Circuit Performance	Refer to vehicle specific information to establish the type of deactivation system and components used. The deactivation system can make use of an actuator (usually a solenoid controlling an oil pressure feed to the valve lift mechanism) to restrict the opening of the valve; this prevents cylinder operation (no compression or cylinder pumping losses when valves are closed).	The voltage, frequency or value of the control signal in the actuator circuit is within the normal operating range / tolerance, but the signal might be incorrect for the operating conditions. It is also possible that the actuator response (or response of the Deactivation system) differs from the expected / desired response (incorrect response to the control signal provided by the control unit). Possible fault in actuator or wiring, but it is also possible that the actuator (or Deactivation mechanism) is not operating or moving correctly. Check for correct movement of Deactivation mechanism and check operation of valve assembly; also check for sufficient oil pressure to the Deactivation system (if applicable to the system). Note: the control unit could be providing the correct signal but it is being affected by a circuit fault. Refer to list of sensor / actuator checks on page 19.
P3447	Cylinder 6 Exhaust Valve Control Circuit Low	Refer to vehicle specific information to establish the type of deactivation system and components used. The deactivation system can make use of an actuator (usually a solenoid controlling an oil pressure feed to the valve lift mechanism) to restrict the opening of the valve; this prevents cylinder operation (no compression or cylinder pumping losses when valves are closed).	The voltage, frequency or value of the control signal in the actuator circuit is either "zero" (no signal) or, the signal exists but is below the normal operating range. Possible fault with actuator or wiring (e.g. actuator power supply, short / open circuit or resistance) that is causing a signal value to be lower than the minimum operating limit. Note: the control unit could be providing the correct signal but it is being affected by a circuit fault. Refer to list of sensor / actuator checks on page 19.
P3448	Cylinder 6 Exhaust Valve Control Circuit High	Refer to vehicle specific information to establish the type of deactivation system and components used. The deactivation system can make use of an actuator (usually a solenoid controlling an oil pressure feed to the valve lift mechanism) to restrict the opening of the valve; this prevents cylinder operation (no compression or cylinder pumping losses when valves are closed).	The voltage, frequency or value of the control signal in the actuator circuit is either at full value (e.g. battery voltage with no on / off signal) or, the signal exists but it is above the normal operating range. Possible fault with actuator or wiring (e.g. actuator power supply, short / open circuit or resistance) that is causing a signal value to be higher than the maximum operating limit. Note: the control unit could be providing the correct signal but it is being affected by a circuit fault. Refer to list of sensor / actuator checks on page 19.
P3449	Cylinder 7 Deactivation / Intake Valve Control Circuit / Open	Refer to vehicle specific information to establish the type of deactivation system and components used. The deactivation system can make use of an actuator (usually a solenoid controlling an oil pressure feed to the valve lift mechanism) to restrict the opening of the valve; this prevents cylinder operation (no compression or cylinder pumping losses when valves are closed).	The voltage, frequency or value of the control signal in the actuator circuit is incorrect. Signal could be:- out of normal operating range, constant value, non-existent, corrupt. The fault code indicates a possible "open circuit" but other undefined faults in the actuator or wiring could activate the same code e.g. actuator power supply, short circuit, high circuit resistance or incorrect actuator resistance. Note: the control unit could be providing the correct signal but it is being affected by a circuit fault. Refer to list of sensor / actuator checks on page 19.
P3450	Cylinder 7 Deactivation / Intake Valve Control Circuit Performance	Refer to vehicle specific information to establish the type of deactivation system and components used. The deactivation system can make use of an actuator (usually a solenoid controlling an oil pressure feed to the valve lift mechanism) to restrict the opening of the valve; this prevents cylinder operation (no compression or cylinder pumping losses when valves are closed).	The voltage, frequency or value of the control signal in the actuator circuit is within the normal operating range / tolerance, but the signal might be incorrect for the operating conditions. It is also possible that the actuator response (or response of the Deactivation system) differs from the expected / desired response (incorrect response to the control signal provided by the control unit). Possible fault in actuator or wiring, but it is also possible that the actuator (or Deactivation mechanism) is not operating or moving correctly. Check for correct movement of Deactivation mechanism and check operation of valve assembly; also check for sufficient oil pressure to the Deactivation system (if applicable to the system). Note: the control unit could be providing the correct signal but it is being affected by a circuit fault. Refer to list of sensor / actuator checks on page 19.
P3451	Cylinder 7 Deactivation / Intake Valve Control Circuit Low	Refer to vehicle specific information to establish the type of deactivation system and components used. The deactivation system can make use of an actuator (usually a solenoid controlling an oil pressure feed to the valve lift mechanism) to restrict the opening of the valve; this prevents cylinder operation (no compression or cylinder pumping losses when valves are closed).	The voltage, frequency or value of the control signal in the actuator circuit is either "zero" (no signal) or, the signal exists but is below the normal operating range. Possible fault with actuator or wiring (e.g. actuator power supply, short / open circuit or resistance) that is causing a signal value to be lower than the minimum operating limit. Note: the control unit could be providing the correct signal but it is being affected by a circuit fault. Refer to list of sensor / actuator checks on page 19.
P3452	Cylinder 7 Deactivation / Intake Valve Control Circuit High	Refer to vehicle specific information to establish the type of deactivation system and components used. The deactivation system can make use of an actuator (usually a solenoid controlling an oil pressure feed to the valve lift mechanism) to restrict the opening of the valve; this prevents cylinder operation (no compression or cylinder pumping losses when valves are closed).	The voltage, frequency or value of the control signal in the actuator circuit is either at full value (e.g. battery voltage with no on / off signal) or, the signal exists but it is above the normal operating range. Possible fault with actuator or wiring (e.g. actuator power supply, short / open circuit or resistance) that is causing a signal value to be higher than the maximum operating limit. Note: the control unit could be providing the correct signal but it is being affected by a circuit fault. Refer to list of sensor / actuator checks on page 19.
P3453	Cylinder 7 Exhaust Valve Control Circuit / Open	Refer to vehicle specific information to establish the type of deactivation system and components used. The deactivation system can make use of an actuator (usually a solenoid controlling an oil pressure feed to the valve lift mechanism) to restrict the opening of the valve; this prevents cylinder operation (no compression or cylinder pumping losses when valves are closed).	The voltage, frequency or value of the control signal in the actuator circuit is incorrect. Signal could be:- out of normal operating range, constant value, non-existent, corrupt. The fault code indicates a possible "open circuit" but other undefined faults in the actuator or wiring could activate the same code e.g. actuator power supply, short circuit, high circuit resistance or incorrect actuator resistance. Note: the control unit could be providing the correct signal but it is being affected by a circuit fault. Refer to list of sensor / actuator checks on page 19.
P3454	Cylinder 7 Exhaust Valve Control Circuit Performance	Refer to vehicle specific information to establish the type of deactivation system and components used. The deactivation system can make use of an actuator (usually a solenoid controlling an oil pressure feed to the valve lift mechanism) to restrict the opening of the valve; this prevents cylinder operation (no compression or cylinder pumping losses when valves are closed).	The voltage, frequency or value of the control signal in the actuator circuit is within the normal operating range / tolerance, but the signal might be incorrect for the operating conditions. It is also possible that the actuator response (or response of the Deactivation system) differs from the expected / desired response (incorrect response to the control signal provided by the control unit). Possible fault in actuator or wiring, but it is also possible that the actuator (or Deactivation mechanism) is not operating or moving correctly. Check for correct movement of Deactivation mechanism and check operation of valve assembly; also check for sufficient oil pressure to the Deactivation system (if applicable to the system). Note: the control unit could be providing the correct signal but it is being affected by a circuit fault. Refer to list of sensor / actuator checks on page 19.
P3455	Cylinder 7 Exhaust Valve Control Circuit Low	Refer to vehicle specific information to establish the type of deactivation system and components used. The deactivation system can make use of an actuator (usually a solenoid controlling an oil pressure feed to the valve lift mechanism) to restrict the opening of the valve; this prevents cylinder operation (no compression or cylinder pumping losses when valves are closed).	The voltage, frequency or value of the control signal in the actuator circuit is either "zero" (no signal) or, the signal exists but is below the normal operating range. Possible fault with actuator or wiring (e.g. actuator power supply, short / open circuit or resistance) that is causing a signal value to be lower than the minimum operating limit. Note: the control unit could be providing the correct signal but it is being affected by a circuit fault. Refer to list of sensor / actuator checks on page 19.

Fault code	EOBD / ISO Description	Component / System Description	Meaningful Description and Quick Check
P3456	Cylinder 7 Exhaust Valve Control Circuit High	Refer to vehicle specific information to establish the type of deactivation system and components used. The deactivation system can make use of an actuator (usually a solenoid controlling an oil pressure feed to the valve lift mechanism) to restrict the opening of the valve; this prevents cylinder operation (no compression or cylinder pumping losses when valves are closed).	The voltage, frequency or value of the control signal in the actuator circuit is either at full value (e.g. battery voltage with no on / off signal) or, the signal exists but it is above the normal operating range. Possible fault with actuator or wiring (e.g. actuator power supply, short / open circuit or resistance) that is causing a signal value to be higher than the maximum operating limit. Note: the control unit could be providing the correct signal but it is being affected by a circuit fault. Refer to list of sensor / actuator checks on page 19.
P3457	Cylinder 8 Deactivation / Intake Valve Control Circuit / Open	Refer to vehicle specific information to establish the type of deactivation system and components used. The deactivation system can make use of an actuator (usually a solenoid controlling an oil pressure feed to the valve lift mechanism) to restrict the opening of the valve; this prevents cylinder operation (no compression or cylinder pumping losses when valves are closed).	The voltage, frequency or value of the control signal in the actuator circuit is incorrect. Signal could be:- out of normal operating range, constant value, non-existent, corrupt. The fault code indicates a possible "open circuit" but other undefined faults in the actuator or wiring could activate the same code e.g. actuator power supply, short circuit, high circuit resistance or incorrect actuator resistance. Note: the control unit could be providing the correct signal but it is being affected by a circuit fault. Refer to list of sensor / actuator checks on page 19.
P3458	Cylinder 8 Deactivation / Intake Valve Control Circuit Performance	Refer to vehicle specific information to establish the type of deactivation system and components used. The deactivation system can make use of an actuator (usually a solenoid controlling an oil pressure feed to the valve lift mechanism) to restrict the opening of the valve; this prevents cylinder operation (no compression or cylinder pumping losses when valves are closed).	The voltage, frequency or value of the control signal in the actuator circuit is within the normal operating range / tolerance, but the signal might be incorrect for the operating conditions. It is also possible that the actuator response (or response of the Deactivation system) differs from the expected / desired response (incorrect response to the control signal provided by the control unit). Possible fault in actuator or wiring, but it is also possible that the actuator (or Deactivation mechanism) is not operating or moving correctly. Check for correct movement of Deactivation mechanism and check operation of valve assembly; also check for sufficient oil pressure to the Deactivation system (if applicable to the system). Note: the control unit could be providing the correct signal but it is being affected by a circuit fault. Refer to list of sensor / actuator checks on page 19.
P3459	Cylinder 8 Deactivation / Intake Valve Control Circuit Low	Refer to vehicle specific information to establish the type of deactivation system and components used. The deactivation system can make use of an actuator (usually a solenoid controlling an oil pressure feed to the valve lift mechanism) to restrict the opening of the valve; this prevents cylinder operation (no compression or cylinder pumping losses when valves are closed).	The voltage, frequency or value of the control signal in the actuator circuit is either "zero" (no signal) or, the signal exists but is below the normal operating range. Possible fault with actuator or wiring (e.g. actuator power supply, short / open circuit or resistance) that is causing a signal value to be lower than the minimum operating limit. Note: the control unit could be providing the correct signal but it is being affected by a circuit fault. Refer to list of sensor / actuator checks on page 19.
P3460	Cylinder 8 Deactivation / Intake Valve Control Circuit High	Refer to vehicle specific information to establish the type of deactivation system and components used. The deactivation system can make use of an actuator (usually a solenoid controlling an oil pressure feed to the valve lift mechanism) to restrict the opening of the valve; this prevents cylinder operation (no compression or cylinder pumping losses when valves are closed).	The voltage, frequency or value of the control signal in the actuator circuit is either at full value (e.g. battery voltage with no on / off signal) or, the signal exists but it is above the normal operating range. Possible fault with actuator or wiring (e.g. actuator power supply, short / open circuit or resistance) that is causing a signal value to be higher than the maximum operating limit. Note: the control unit could be providing the correct signal but it is being affected by a circuit fault. Refer to list of sensor / actuator checks on page 19.
P3461	Cylinder 8 Exhaust Valve Control Circuit / Open	Refer to vehicle specific information to establish the type of deactivation system and components used. The deactivation system can make use of an actuator (usually a solenoid controlling an oil pressure feed to the valve lift mechanism) to restrict the opening of the valve; this prevents cylinder operation (no compression or cylinder pumping losses when valves are closed).	The voltage, frequency or value of the control signal in the actuator circuit is incorrect. Signal could be:- out of normal operating range, constant value, non-existent, corrupt. The fault code indicates a possible "open circuit" but other undefined faults in the actuator or wiring could activate the same code e.g. actuator power supply, short circuit, high circuit resistance or incorrect actuator resistance. Note: the control unit could be providing the correct signal but it is being affected by a circuit fault. Refer to list of sensor / actuator checks on page 19.
P3462	Cylinder 8 Exhaust Valve Control Circuit Performance	Refer to vehicle specific information to establish the type of deactivation system and components used. The deactivation system can make use of an actuator (usually a solenoid controlling an oil pressure feed to the valve lift mechanism) to restrict the opening of the valve; this prevents cylinder operation (no compression or cylinder pumping losses when valves are closed).	The voltage, frequency or value of the control signal in the actuator circuit is within the normal operating range / tolerance, but the signal might be incorrect for the operating conditions. It is also possible that the actuator response (or response of the Deactivation system) differs from the expected / desired response (incorrect response to the control signal provided by the control unit). Possible fault in actuator or wiring, but it is also possible that the actuator (or Deactivation mechanism) is not operating or moving correctly. Check for correct movement of Deactivation mechanism and check operation of valve assembly; also check for sufficient oil pressure to the Deactivation system (if applicable to the system). Note: the control unit could be providing the correct signal but it is being affected by a circuit fault. Refer to list of sensor / actuator checks on page 19.
P3463	Cylinder 8 Exhaust Valve Control Circuit Low	Refer to vehicle specific information to establish the type of deactivation system and components used. The deactivation system can make use of an actuator (usually a solenoid controlling an oil pressure feed to the valve lift mechanism) to restrict the opening of the valve; this prevents cylinder operation (no compression or cylinder pumping losses when valves are closed).	The voltage, frequency or value of the control signal in the actuator circuit is either "zero" (no signal) or, the signal exists but is below the normal operating range. Possible fault with actuator or wiring (e.g. actuator power supply, short / open circuit or resistance) that is causing a signal value to be lower than the minimum operating limit. Note: the control unit could be providing the correct signal but it is being affected by a circuit fault. Refer to list of sensor / actuator checks on page 19.
P3464	Cylinder 8 Exhaust Valve Control Circuit High	Refer to vehicle specific information to establish the type of deactivation system and components used. The deactivation system can make use of an actuator (usually a solenoid controlling an oil pressure feed to the valve lift mechanism) to restrict the opening of the valve; this prevents cylinder operation (no compression or cylinder pumping losses when valves are closed).	The voltage, frequency or value of the control signal in the actuator circuit is either at full value (e.g. battery voltage with no on / off signal) or, the signal exists but it is above the normal operating range. Possible fault with actuator or wiring (e.g. actuator power supply, short / open circuit or resistance) that is causing a signal value to be higher than the maximum operating limit. Note: the control unit could be providing the correct signal but it is being affected by a circuit fault. Refer to list of sensor / actuator checks on page 19.
P3465	Cylinder 9 Deactivation / Intake Valve Control Circuit / Open	Refer to vehicle specific information to establish the type of deactivation system and components used. The deactivation system can make use of an actuator (usually a solenoid controlling an oil pressure feed to the valve lift mechanism) to restrict the opening of the valve; this prevents cylinder operation (no compression or cylinder pumping losses when valves are closed).	The voltage, frequency or value of the control signal in the actuator circuit is incorrect. Signal could be:- out of normal operating range, constant value, non-existent, corrupt. The fault code indicates a possible "open circuit" but other undefined faults in the actuator or wiring could activate the same code e.g. actuator power supply, short circuit, high circuit resistance or incorrect actuator resistance. Note: the control unit could be providing the correct signal but it is being affected by a circuit fault. Refer to list of sensor / actuator checks on page 19.

NOTE 1. Check for other fault codes that could provide additional information.
NOTE 2. Communication between control units can pass via a CAN-Bus system; refer to Page 51 for CAN-Bus checks.

Code	Description		
P3466	Cylinder 9 Deactivation / Intake Valve Control Circuit Performance	Refer to vehicle specific information to establish the type of deactivation system and components used. The deactivation system can make use of an actuator (usually a solenoid controlling an oil pressure feed to the valve lift mechanism) to restrict the opening of the valve; this prevents cylinder operation (no compression or cylinder pumping losses when valves are closed).	The voltage, frequency or value of the control signal in the actuator circuit is within the normal operating range / tolerance, but the signal might be incorrect for the operating conditions. It is also possible that the actuator response (or response of the Deactivation system) differs from the expected / desired response (incorrect response to the control signal provided by the control unit). Possible fault in actuator or wiring, but it is also possible that the actuator (or Deactivation mechanism) is not operating or moving correctly. Check for correct movement of Deactivation mechanism and check operation of valve assembly; also check for sufficient oil pressure to the Deactivation system (if applicable to the system). Note: the control unit could be providing the correct signal but it is being affected by a circuit fault. Refer to list of sensor / actuator checks on page 19.
P3467	Cylinder 9 Deactivation / Intake Valve Control Circuit Low	Refer to vehicle specific information to establish the type of deactivation system and components used. The deactivation system can make use of an actuator (usually a solenoid controlling an oil pressure feed to the valve lift mechanism) to restrict the opening of the valve; this prevents cylinder operation (no compression or cylinder pumping losses when valves are closed).	The voltage, frequency or value of the control signal in the actuator circuit is either "zero" (no signal) or, the signal exists but is below the normal operating range. Possible fault with actuator or wiring (e.g. actuator power supply, short / open circuit or resistance) that is causing a signal value to be lower than the minimum operating limit. Note: the control unit could be providing the correct signal but it is being affected by a circuit fault. Refer to list of sensor / actuator checks on page 19.
P3468	Cylinder 9 Deactivation / Intake Valve Control Circuit High	Refer to vehicle specific information to establish the type of deactivation system and components used. The deactivation system can make use of an actuator (usually a solenoid controlling an oil pressure feed to the valve lift mechanism) to restrict the opening of the valve; this prevents cylinder operation (no compression or cylinder pumping losses when valves are closed).	The voltage, frequency or value of the control signal in the actuator circuit is either at full value (e.g. battery voltage with no on / off signal) or, the signal exists but it is above the normal operating range. Possible fault with actuator or wiring (e.g. actuator power supply, short / open circuit or resistance) that is causing a signal value to be higher than the maximum operating limit. Note: the control unit could be providing the correct signal but it is being affected by a circuit fault. Refer to list of sensor / actuator checks on page 19.
P3469	Cylinder 9 Exhaust Valve Control Circuit / Open	Refer to vehicle specific information to establish the type of deactivation system and components used. The deactivation system can make use of an actuator (usually a solenoid controlling an oil pressure feed to the valve lift mechanism) to restrict the opening of the valve; this prevents cylinder operation (no compression or cylinder pumping losses when valves are closed).	The voltage, frequency or value of the control signal in the actuator circuit is incorrect. Signal could be:- out of normal operating range, constant value, non-existent, corrupt. The fault code indicates a possible "open circuit" but other undefined faults in the actuator or wiring could activate the same code e.g. actuator power supply, short circuit, high circuit resistance or incorrect actuator resistance. Note: the control unit could be providing the correct signal but it is being affected by a circuit fault. Refer to list of sensor / actuator checks on page 19.
P3470	Cylinder 9 Exhaust Valve Control Circuit Performance	Refer to vehicle specific information to establish the type of deactivation system and components used. The deactivation system can make use of an actuator (usually a solenoid controlling an oil pressure feed to the valve lift mechanism) to restrict the opening of the valve; this prevents cylinder operation (no compression or cylinder pumping losses when valves are closed).	The voltage, frequency or value of the control signal in the actuator circuit is within the normal operating range / tolerance, but the signal might be incorrect for the operating conditions. It is also possible that the actuator response (or response of the Deactivation system) differs from the expected / desired response (incorrect response to the control signal provided by the control unit). Possible fault in actuator or wiring, but it is also possible that the actuator (or Deactivation mechanism) is not operating or moving correctly. Check for correct movement of Deactivation mechanism and check operation of valve assembly; also check for sufficient oil pressure to the Deactivation system (if applicable to the system). Note: the control unit could be providing the correct signal but it is being affected by a circuit fault. Refer to list of sensor / actuator checks on page 19.
P3471	Cylinder 9 Exhaust Valve Control Circuit Low	Refer to vehicle specific information to establish the type of deactivation system and components used. The deactivation system can make use of an actuator (usually a solenoid controlling an oil pressure feed to the valve lift mechanism) to restrict the opening of the valve; this prevents cylinder operation (no compression or cylinder pumping losses when valves are closed).	The voltage, frequency or value of the control signal in the actuator circuit is either "zero" (no signal) or, the signal exists but is below the normal operating range. Possible fault with actuator or wiring (e.g. actuator power supply, short / open circuit or resistance) that is causing a signal value to be lower than the minimum operating limit. Note: the control unit could be providing the correct signal but it is being affected by a circuit fault. Refer to list of sensor / actuator checks on page 19.
P3472	Cylinder 9 Exhaust Valve Control Circuit High	Refer to vehicle specific information to establish the type of deactivation system and components used. The deactivation system can make use of an actuator (usually a solenoid controlling an oil pressure feed to the valve lift mechanism) to restrict the opening of the valve; this prevents cylinder operation (no compression or cylinder pumping losses when valves are closed).	The voltage, frequency or value of the control signal in the actuator circuit is either at full value (e.g. battery voltage with no on / off signal) or, the signal exists but it is above the normal operating range. Possible fault with actuator or wiring (e.g. actuator power supply, short / open circuit or resistance) that is causing a signal value to be higher than the maximum operating limit. Note: the control unit could be providing the correct signal but it is being affected by a circuit fault. Refer to list of sensor / actuator checks on page 19.
P3473	Cylinder 10 Deactivation / Intake Valve Control Circuit / Open	Refer to vehicle specific information to establish the type of deactivation system and components used. The deactivation system can make use of an actuator (usually a solenoid controlling an oil pressure feed to the valve lift mechanism) to restrict the opening of the valve; this prevents cylinder operation (no compression or cylinder pumping losses when valves are closed).	The voltage, frequency or value of the control signal in the actuator circuit is incorrect. Signal could be:- out of normal operating range, constant value, non-existent, corrupt. The fault code indicates a possible "open circuit" but other undefined faults in the actuator or wiring could activate the same code e.g. actuator power supply, short circuit, high circuit resistance or incorrect actuator resistance. Note: the control unit could be providing the correct signal but it is being affected by a circuit fault. Refer to list of sensor / actuator checks on page 19.
P3474	Cylinder 10 Deactivation / Intake Valve Control Circuit Performance	Refer to vehicle specific information to establish the type of deactivation system and components used. The deactivation system can make use of an actuator (usually a solenoid controlling an oil pressure feed to the valve lift mechanism) to restrict the opening of the valve; this prevents cylinder operation (no compression or cylinder pumping losses when valves are closed).	The voltage, frequency or value of the control signal in the actuator circuit is within the normal operating range / tolerance, but the signal might be incorrect for the operating conditions. It is also possible that the actuator response (or response of the Deactivation system) differs from the expected / desired response (incorrect response to the control signal provided by the control unit). Possible fault in actuator or wiring, but it is also possible that the actuator (or Deactivation mechanism) is not operating or moving correctly. Check for correct movement of Deactivation mechanism and check operation of valve assembly; also check for sufficient oil pressure to the Deactivation system (if applicable to the system). Note: the control unit could be providing the correct signal but it is being affected by a circuit fault. Refer to list of sensor / actuator checks on page 19.
P3475	Cylinder 10 Deactivation / Intake Valve Control Circuit Low	Refer to vehicle specific information to establish the type of deactivation system and components used. The deactivation system can make use of an actuator (usually a solenoid controlling an oil pressure feed to the valve lift mechanism) to restrict the opening of the valve; this prevents cylinder operation (no compression or cylinder pumping losses when valves are closed).	The voltage, frequency or value of the control signal in the actuator circuit is either "zero" (no signal) or, the signal exists but is below the normal operating range. Possible fault with actuator or wiring (e.g. actuator power supply, short / open circuit or resistance) that is causing a signal value to be lower than the minimum operating limit. Note: the control unit could be providing the correct signal but it is being affected by a circuit fault. Refer to list of sensor / actuator checks on page 19.

NOTE 1. Check for other fault codes that could provide additional information.   **NOTE 2.** Communication between control units can pass via a CAN-Bus system; refer to Page 51 for CAN-Bus checks.   **331**

NOTE 3. If a fault cannot be located, it is also possible that the control unit is at fault.   **NOTE 4.** Refer to Page 53 for list of pages that show the ISO standard for component locations e.g. Sensor A - Bank 1.

Fault code	EOBD / ISO Description	Component / System Description	Meaningful Description and Quick Check
P3476	Cylinder 10 Deactivation / Intake Valve Control Circuit High	Refer to vehicle specific information to establish the type of deactivation system and components used. The deactivation system can make use of an actuator (usually a solenoid controlling an oil pressure feed to the valve lift mechanism) to restrict the opening of the valve; this prevents cylinder operation (no compression or cylinder pumping losses when valves are closed).	The voltage, frequency or value of the control signal in the actuator circuit is either at full value (e.g. battery voltage with no on / off signal) or, the signal exists but it is above the normal operating range. Possible fault with actuator or wiring (e.g. actuator power supply, short / open circuit or resistance) that is causing a signal value to be higher than the maximum operating limit. Note: the control unit could be providing the correct signal but it is being affected by a circuit fault. Refer to list of sensor / actuator checks on page 19.
P3477	Cylinder 10 Exhaust Valve Control Circuit / Open	Refer to vehicle specific information to establish the type of deactivation system and components used. The deactivation system can make use of an actuator (usually a solenoid controlling an oil pressure feed to the valve lift mechanism) to restrict the opening of the valve; this prevents cylinder operation (no compression or cylinder pumping losses when valves are closed).	The voltage, frequency or value of the control signal in the actuator circuit is incorrect. Signal could be:- out of normal operating range, constant value, non-existent, corrupt. The fault code indicates a possible "open circuit" but other undefined faults in the actuator or wiring could activate the same code e.g. actuator power supply, short circuit, high circuit resistance or incorrect actuator resistance. Note: the control unit could be providing the correct signal but it is being affected by a circuit fault. Refer to list of sensor / actuator checks on page 19.
P3478	Cylinder 10 Exhaust Valve Control Circuit Performance	Refer to vehicle specific information to establish the type of deactivation system and components used. The deactivation system can make use of an actuator (usually a solenoid controlling an oil pressure feed to the valve lift mechanism) to restrict the opening of the valve; this prevents cylinder operation (no compression or cylinder pumping losses when valves are closed).	The voltage, frequency or value of the control signal in the actuator circuit is within the normal operating range / tolerance, but the signal might be incorrect for the operating conditions. It is also possible that the actuator response (or response of the Deactivation system) differs from the expected / desired response (incorrect response to the control signal provided by the control unit). Possible fault in actuator or wiring, but it is also possible that the actuator (or Deactivation mechanism) is not operating or moving correctly. Check for correct movement of Deactivation mechanism and check operation of valve assembly; also check for sufficient oil pressure to the Deactivation system (if applicable to the system). Note: the control unit could be providing the correct signal but it is being affected by a circuit fault. Refer to list of sensor / actuator checks on page 19.
P3479	Cylinder 10 Exhaust Valve Control Circuit Low	Refer to vehicle specific information to establish the type of deactivation system and components used. The deactivation system can make use of an actuator (usually a solenoid controlling an oil pressure feed to the valve lift mechanism) to restrict the opening of the valve; this prevents cylinder operation (no compression or cylinder pumping losses when valves are closed).	The voltage, frequency or value of the control signal in the actuator circuit is either "zero" (no signal) or, the signal exists but is below the normal operating range. Possible fault with actuator or wiring (e.g. actuator power supply, short / open circuit or resistance) that is causing a signal value to be lower than the minimum operating limit. Note: the control unit could be providing the correct signal but it is being affected by a circuit fault. Refer to list of sensor / actuator checks on page 19.
P3480	Cylinder 10 Exhaust Valve Control Circuit High	Refer to vehicle specific information to establish the type of deactivation system and components used. The deactivation system can make use of an actuator (usually a solenoid controlling an oil pressure feed to the valve lift mechanism) to restrict the opening of the valve; this prevents cylinder operation (no compression or cylinder pumping losses when valves are closed).	The voltage, frequency or value of the control signal in the actuator circuit is either at full value (e.g. battery voltage with no on / off signal) or, the signal exists but it is above the normal operating range. Possible fault with actuator or wiring (e.g. actuator power supply, short / open circuit or resistance) that is causing a signal value to be higher than the maximum operating limit. Note: the control unit could be providing the correct signal but it is being affected by a circuit fault. Refer to list of sensor / actuator checks on page 19.
P3481	Cylinder 11 Deactivation / Intake Valve Control Circuit / Open	Refer to vehicle specific information to establish the type of deactivation system and components used. The deactivation system can make use of an actuator (usually a solenoid controlling an oil pressure feed to the valve lift mechanism) to restrict the opening of the valve; this prevents cylinder operation (no compression or cylinder pumping losses when valves are closed).	The voltage, frequency or value of the control signal in the actuator circuit is incorrect. Signal could be:- out of normal operating range, constant value, non-existent, corrupt. The fault code indicates a possible "open circuit" but other undefined faults in the actuator or wiring could activate the same code e.g. actuator power supply, short circuit, high circuit resistance or incorrect actuator resistance. Note: the control unit could be providing the correct signal but it is being affected by a circuit fault. Refer to list of sensor / actuator checks on page 19.
P3482	Cylinder 11 Deactivation / Intake Valve Control Circuit Performance	Refer to vehicle specific information to establish the type of deactivation system and components used. The deactivation system can make use of an actuator (usually a solenoid controlling an oil pressure feed to the valve lift mechanism) to restrict the opening of the valve; this prevents cylinder operation (no compression or cylinder pumping losses when valves are closed).	The voltage, frequency or value of the control signal in the actuator circuit is within the normal operating range / tolerance, but the signal might be incorrect for the operating conditions. It is also possible that the actuator response (or response of the Deactivation system) differs from the expected / desired response (incorrect response to the control signal provided by the control unit). Possible fault in actuator or wiring, but it is also possible that the actuator (or Deactivation mechanism) is not operating or moving correctly. Check for correct movement of Deactivation mechanism and check operation of valve assembly; also check for sufficient oil pressure to the Deactivation system (if applicable to the system). Note: the control unit could be providing the correct signal but it is being affected by a circuit fault. Refer to list of sensor / actuator checks on page 19.
P3483	Cylinder 11 Deactivation / Intake Valve Control Circuit Low	Refer to vehicle specific information to establish the type of deactivation system and components used. The deactivation system can make use of an actuator (usually a solenoid controlling an oil pressure feed to the valve lift mechanism) to restrict the opening of the valve; this prevents cylinder operation (no compression or cylinder pumping losses when valves are closed).	The voltage, frequency or value of the control signal in the actuator circuit is either "zero" (no signal) or, the signal exists but is below the normal operating range. Possible fault with actuator or wiring (e.g. actuator power supply, short / open circuit or resistance) that is causing a signal value to be lower than the minimum operating limit. Note: the control unit could be providing the correct signal but it is being affected by a circuit fault. Refer to list of sensor / actuator checks on page 19.
P3484	Cylinder 11 Deactivation / Intake Valve Control Circuit High	Refer to vehicle specific information to establish the type of deactivation system and components used. The deactivation system can make use of an actuator (usually a solenoid controlling an oil pressure feed to the valve lift mechanism) to restrict the opening of the valve; this prevents cylinder operation (no compression or cylinder pumping losses when valves are closed).	The voltage, frequency or value of the control signal in the actuator circuit is either at full value (e.g. battery voltage with no on / off signal) or, the signal exists but it is above the normal operating range. Possible fault with actuator or wiring (e.g. actuator power supply, short / open circuit or resistance) that is causing a signal value to be higher than the maximum operating limit. Note: the control unit could be providing the correct signal but it is being affected by a circuit fault. Refer to list of sensor / actuator checks on page 19.
P3485	Cylinder 11 Exhaust Valve Control Circuit / Open	Refer to vehicle specific information to establish the type of deactivation system and components used. The deactivation system can make use of an actuator (usually a solenoid controlling an oil pressure feed to the valve lift mechanism) to restrict the opening of the valve; this prevents cylinder operation (no compression or cylinder pumping losses when valves are closed).	The voltage, frequency or value of the control signal in the actuator circuit is incorrect. Signal could be:- out of normal operating range, constant value, non-existent, corrupt. The fault code indicates a possible "open circuit" but other undefined faults in the actuator or wiring could activate the same code e.g. actuator power supply, short circuit, high circuit resistance or incorrect actuator resistance. Note: the control unit could be providing the correct signal but it is being affected by a circuit fault. Refer to list of sensor / actuator checks on page 19.

Code	Fault Location		Probable Cause
P3486	Cylinder 11 Exhaust Valve Control Circuit Performance	Refer to vehicle specific information to establish the type of deactivation system and components used. The deactivation system can make use of an actuator (usually a solenoid controlling an oil pressure feed to the valve lift mechanism) to restrict the opening of the valve; this prevents cylinder operation (no compression or cylinder pumping losses when valves are closed).	The voltage, frequency or value of the control signal in the actuator circuit is within the normal operating range / tolerance, but the signal might be incorrect for the operating conditions. It is also possible that the actuator response (or response of the Deactivation system) differs from the expected / desired response (incorrect response to the control signal provided by the control unit). Possible fault in actuator or wiring, but it is also possible that the actuator (or Deactivation mechanism) is not operating or moving correctly. Check for correct movement of Deactivation mechanism and check operation of valve assembly; also check for sufficient oil pressure to the Deactivation system (if applicable to the system). Note: the control unit could be providing the correct signal but it is being affected by a circuit fault. Refer to list of sensor / actuator checks on page 19.
P3487	Cylinder 11 Exhaust Valve Control Circuit Low	Refer to vehicle specific information to establish the type of deactivation system and components used. The deactivation system can make use of an actuator (usually a solenoid controlling an oil pressure feed to the valve lift mechanism) to restrict the opening of the valve; this prevents cylinder operation (no compression or cylinder pumping losses when valves are closed).	The voltage, frequency or value of the control signal in the actuator circuit is either "zero" (no signal) or, the signal exists but is below the normal operating range. Possible fault with actuator or wiring (e.g. actuator power supply, short / open circuit or resistance) that is causing a signal value to be lower than the minimum operating limit. Note: the control unit could be providing the correct signal but it is being affected by a circuit fault. Refer to list of sensor / actuator checks on page 19.
P3488	Cylinder 11 Exhaust Valve Control Circuit High	Refer to vehicle specific information to establish the type of deactivation system and components used. The deactivation system can make use of an actuator (usually a solenoid controlling an oil pressure feed to the valve lift mechanism) to restrict the opening of the valve; this prevents cylinder operation (no compression or cylinder pumping losses when valves are closed).	The voltage, frequency or value of the control signal in the actuator circuit is either at full value (e.g. battery voltage with no on / off signal) or, the signal exists but it is above the normal operating range. Possible fault with actuator or wiring (e.g. actuator power supply, short / open circuit or resistance) that is causing a signal value to be higher than the maximum operating limit. Note: the control unit could be providing the correct signal but it is being affected by a circuit fault. Refer to list of sensor / actuator checks on page 19.
P3489	Cylinder 12 Deactivation / Intake Valve Control Circuit / Open	Refer to vehicle specific information to establish the type of deactivation system and components used. The deactivation system can make use of an actuator (usually a solenoid controlling an oil pressure feed to the valve lift mechanism) to restrict the opening of the valve; this prevents cylinder operation (no compression or cylinder pumping losses when valves are closed).	The voltage, frequency or value of the control signal in the actuator circuit is incorrect. Signal could be:- out of normal operating range, constant value, non-existent, corrupt. The fault code indicates a possible "open circuit" but other undefined faults in the actuator or wiring could activate the same code e.g. actuator power supply, short circuit, high circuit resistance or incorrect actuator resistance. Note: the control unit could be providing the correct signal but it is being affected by a circuit fault. Refer to list of sensor / actuator checks on page 19.
P3490	Cylinder 12 Deactivation / Intake Valve Control Circuit Performance	Refer to vehicle specific information to establish the type of deactivation system and components used. The deactivation system can make use of an actuator (usually a solenoid controlling an oil pressure feed to the valve lift mechanism) to restrict the opening of the valve; this prevents cylinder operation (no compression or cylinder pumping losses when valves are closed).	The voltage, frequency or value of the control signal in the actuator circuit is within the normal operating range / tolerance, but the signal might be incorrect for the operating conditions. It is also possible that the actuator response (or response of the Deactivation system) differs from the expected / desired response (incorrect response to the control signal provided by the control unit). Possible fault in actuator or wiring, but it is also possible that the actuator (or Deactivation mechanism) is not operating or moving correctly. Check for correct movement of Deactivation mechanism and check operation of valve assembly; also check for sufficient oil pressure to the Deactivation system (if applicable to the system). Note: the control unit could be providing the correct signal but it is being affected by a circuit fault. Refer to list of sensor / actuator checks on page 19.
P3491	Cylinder 12 Deactivation / Intake Valve Control Circuit Low	Refer to vehicle specific information to establish the type of deactivation system and components used. The deactivation system can make use of an actuator (usually a solenoid controlling an oil pressure feed to the valve lift mechanism) to restrict the opening of the valve; this prevents cylinder operation (no compression or cylinder pumping losses when valves are closed).	The voltage, frequency or value of the control signal in the actuator circuit is either "zero" (no signal) or, the signal exists but is below the normal operating range. Possible fault with actuator or wiring (e.g. actuator power supply, short / open circuit or resistance) that is causing a signal value to be lower than the minimum operating limit. Note: the control unit could be providing the correct signal but it is being affected by a circuit fault. Refer to list of sensor / actuator checks on page 19.
P3492	Cylinder 12 Deactivation / Intake Valve Control Circuit High	Refer to vehicle specific information to establish the type of deactivation system and components used. The deactivation system can make use of an actuator (usually a solenoid controlling an oil pressure feed to the valve lift mechanism) to restrict the opening of the valve; this prevents cylinder operation (no compression or cylinder pumping losses when valves are closed).	The voltage, frequency or value of the control signal in the actuator circuit is either at full value (e.g. battery voltage with no on / off signal) or, the signal exists but it is above the normal operating range. Possible fault with actuator or wiring (e.g. actuator power supply, short / open circuit or resistance) that is causing a signal value to be higher than the maximum operating limit. Note: the control unit could be providing the correct signal but it is being affected by a circuit fault. Refer to list of sensor / actuator checks on page 19.
P3493	Cylinder 12 Exhaust Valve Control Circuit / Open	Refer to vehicle specific information to establish the type of deactivation system and components used. The deactivation system can make use of an actuator (usually a solenoid controlling an oil pressure feed to the valve lift mechanism) to restrict the opening of the valve; this prevents cylinder operation (no compression or cylinder pumping losses when valves are closed).	The voltage, frequency or value of the control signal in the actuator circuit is incorrect. Signal could be:- out of normal operating range, constant value, non-existent, corrupt. The fault code indicates a possible "open circuit" but other undefined faults in the actuator or wiring could activate the same code e.g. actuator power supply, short circuit, high circuit resistance or incorrect actuator resistance. Note: the control unit could be providing the correct signal but it is being affected by a circuit fault. Refer to list of sensor / actuator checks on page 19.
P3494	Cylinder 12 Exhaust Valve Control Circuit Performance	Refer to vehicle specific information to establish the type of deactivation system and components used. The deactivation system can make use of an actuator (usually a solenoid controlling an oil pressure feed to the valve lift mechanism) to restrict the opening of the valve; this prevents cylinder operation (no compression or cylinder pumping losses when valves are closed).	The voltage, frequency or value of the control signal in the actuator circuit is within the normal operating range / tolerance, but the signal might be incorrect for the operating conditions. It is also possible that the actuator response (or response of the Deactivation system) differs from the expected / desired response (incorrect response to the control signal provided by the control unit). Possible fault in actuator or wiring, but it is also possible that the actuator (or Deactivation mechanism) is not operating or moving correctly. Check for correct movement of Deactivation mechanism and check operation of valve assembly; also check for sufficient oil pressure to the Deactivation system (if applicable to the system). Note: the control unit could be providing the correct signal but it is being affected by a circuit fault. Refer to list of sensor / actuator checks on page 19.
P3495	Cylinder 12 Exhaust Valve Control Circuit Low	Refer to vehicle specific information to establish the type of deactivation system and components used. The deactivation system can make use of an actuator (usually a solenoid controlling an oil pressure feed to the valve lift mechanism) to restrict the opening of the valve; this prevents cylinder operation (no compression or cylinder pumping losses when valves are closed).	The voltage, frequency or value of the control signal in the actuator circuit is either "zero" (no signal) or, the signal exists but is below the normal operating range. Possible fault with actuator or wiring (e.g. actuator power supply, short / open circuit or resistance) that is causing a signal value to be lower than the minimum operating limit. Note: the control unit could be providing the correct signal but it is being affected by a circuit fault. Refer to list of sensor / actuator checks on page 19.

Fault code	EOBD / ISO Description	Component / System Description	Meaningful Description and Quick Check
P3496	Cylinder 12 Exhaust Valve Control Circuit High	Refer to vehicle specific information to establish the type of deactivation system and components used. The deactivation system can make use of an actuator (usually a solenoid controlling an oil pressure feed to the valve lift mechanism) to restrict the opening of the valve; this prevents cylinder operation (no compression or cylinder pumping losses when valves are closed).	The voltage, frequency or value of the control signal in the actuator circuit is either at full value (e.g. battery voltage with no on / off signal) or, the signal exists but it is above the normal operating range. Possible fault with actuator or wiring (e.g. actuator power supply, short / open circuit or resistance) that is causing a signal value to be higher than the maximum operating limit. Note: the control unit could be providing the correct signal but it is being affected by a circuit fault. Refer to list of sensor / actuator checks on page 19.
P3497	Cylinder Deactivation System Bank 2	Refer to vehicle specific information to establish the type of deactivation system and components used. The deactivation system can make use of an actuator (usually a solenoid controlling an oil pressure feed to the valve lift mechanism) to restrict the opening of the valve; this prevents cylinder operation (no compression or cylinder pumping losses when valves are closed). The system design will dictate the checks that need to be carried out.	The fault code indicates an undefined fault with the Cylinder Deactivation system. The actuator or the Cylinder Deactivation system response does not match the control signal value being provided by the control unit e.g. control signal should cause the deactivation system to operate but it does not respond; this may be due to a fault in the actuator / wiring or in other parts of the deactivation system. Note that the fault relates to one bank of cylinders and, because deactivations systems usually select individual cylinders, the fault could be related to wiring or other components affecting the bank of cylinders rather than an individual cylinder. It is also possible that the actuator (or Deactivation mechanism) is not operating or moving correctly. Check for correct movement of Deactivation mechanism and check operation of valve assembly; also check for sufficient oil pressure to the Deactivation system (if applicable). Refer to Fault code P3494 for additional information. Refer to list of sensor / actuator checks on page 19.
U0000		ISO / SAE reserved	Not used, reserved for future allocation.
U0001	High Speed CAN Communication Bus	The Vehicle / CAN Communication Bus is a network for sharing signals and data / messages between the various vehicle systems, including Powertrain, Chassis and Infotainment.	Refer to vehicle specific information for identification of location and function of the specified systems / CAN-Bus components. The CAN speed relates to the transmission rate of the data messages between the various control systems. The fault code indicates that one or more control modules has detected an undefined fault on the High Speed CAN communication Bus due to the incorrect transmission of data messages. Refer to CAN-Bus information on page 51.
U0002	High Speed CAN Communication Bus Performance	The Vehicle / CAN Communication Bus is a network for sharing signals and data / messages between the various vehicle systems, including Powertrain, Chassis and Infotainment.	Refer to vehicle specific information for identification of location and function of the specified systems / CAN-Bus components. The CAN speed relates to the transmission rate of the data messages between the various control systems. The fault code indicates that one or more control modules has detected a performance fault with the data transmission on the High Speed CAN communication Bus. Refer to CAN-Bus information on page 51.
U0003	High Speed CAN Communication Bus (+) Open	The Vehicle / CAN Communication Bus is a network for sharing signals and data / messages between the various vehicle systems, including Powertrain, Chassis and Infotainment. For wire based CAN-Bus systems, signals are transmitted along 2 wires that are twisted together (twisted pair). Each wire carries a digital signal; one being a mirror image of the other. The signal does not start at a zero volts base line but, the mirrored signals can be regarded as positive or negative values relative to the base line voltage.	Refer to vehicle specific information for identification of location and function of the specified systems / CAN-Bus components. The CAN speed relates to the transmission rate of the data messages between the various control systems. The fault code indicates that there is a probable open circuit likely to be detected due to loss of signal in the specified circuit. Refer to CAN-Bus information on page 51.
U0004	High Speed CAN Communication Bus (+) Low	The Vehicle / CAN Communication Bus is a network for sharing signals and data / messages between the various vehicle systems, including Powertrain, Chassis and Infotainment. For wire based CAN-Bus systems, signals are transmitted along 2 wires that are twisted together (twisted pair). Each wire carries a digital signal; one being a mirror image of the other. The signal does not start at a zero volts base line but, the mirrored signals can be regarded as positive or negative values relative to the base line voltage.	Refer to vehicle specific information for identification of location and function of the specified systems / CAN-Bus components. The CAN speed relates to the transmission rate of the data messages between the various control systems. The fault code indicates that one or more control modules has detected that the voltage, frequency or value of the signal is permanently low, below the normal operating range. There could be other undefined faults in the CAN-Bus system that could activate the same code. Refer to CAN-Bus information on page 51.
U0005	High Speed CAN Communication Bus (+) High	The Vehicle / CAN Communication Bus is a network for sharing signals and data / messages between the various vehicle systems, including Powertrain, Chassis and Infotainment. For wire based CAN-Bus systems, signals are transmitted along 2 wires that are twisted together (twisted pair). Each wire carries a digital signal; one being a mirror image of the other. The signal does not start at a zero volts base line but, the mirrored signals can be regarded as positive or negative values relative to the base line voltage.	Refer to vehicle specific information for identification of location and function of the specified systems / CAN-Bus components. The CAN speed relates to the transmission rate of the data messages between the various control systems. The fault code indicates that one or more control modules has detected that the voltage, frequency or value of the signal is permanently high, above the normal operating range. There could be other undefined faults in the CAN-Bus system that could activate the same code. Refer to CAN-Bus information on page 51.
U0006	High Speed CAN Communication Bus (-) Open	The Vehicle / CAN Communication Bus is a network for sharing signals and data / messages between the various vehicle systems, including Powertrain, Chassis and Infotainment. For wire based CAN-Bus systems, signals are transmitted along 2 wires that are twisted together (twisted pair). Each wire carries a digital signal; one being a mirror image of the other. The signal does not start at a zero volts base line but, the mirrored signals can be regarded as positive or negative values relative to the base line voltage.	Refer to vehicle specific information for identification of location and function of the specified systems / CAN-Bus components. The CAN speed relates to the transmission rate of the data messages between the various control systems. The fault code indicates that there is a probable open circuit likely to be detected due to loss of signal in the specified circuit. Refer to CAN-Bus information on page 51.

NOTE 1. Check for other fault codes that could provide additional information.   **NOTE 2.** Communication between control units can pass via a CAN-Bus system; refer to Page 51 for CAN-Bus checks.

U0007	High Speed CAN Communication Bus (-) Low	The Vehicle / CAN Communication Bus is a network for sharing signals and data / messages between the various vehicle systems, including Powertrain, Chassis and Infotainment. For wire based CAN-Bus systems, signals are transmitted along 2 wires that are twisted together (twisted pair). Each wire carries a digital signal; one being a mirror image of the other. The signal does not start at a zero volts base line but, the mirrored signals can be regarded as positive or negative values relative to the base line voltage.	Refer to vehicle specific information for identification of location and function of the specified systems / CAN-Bus components. The CAN speed relates to the transmission rate of the data messages between the various control systems. The fault code indicates that one or more control modules has detected that the voltage, frequency or value of the signal is permanently low, below the normal operating range. There could be other undefined faults in the CAN-Bus system that could activate the same code. Refer to CAN-Bus information on page 51.
U0008	High Speed CAN Communication Bus (-) High	The Vehicle / CAN Communication Bus is a network for sharing signals and data / messages between the various vehicle systems, including Powertrain, Chassis and Infotainment. For wire based CAN-Bus systems, signals are transmitted along 2 wires that are twisted together (twisted pair). Each wire carries a digital signal; one being a mirror image of the other. The signal does not start at a zero volts base line but, the mirrored signals can be regarded as positive or negative values relative to the base line voltage.	Refer to vehicle specific information for identification of location and function of the specified systems / CAN-Bus components. The CAN speed relates to the transmission rate of the data messages between the various control systems. The fault code indicates that one or more control modules has detected that the voltage, frequency or value of the signal is permanently high, above the normal operating range. There could be other undefined faults in the CAN-Bus system that could activate the same code. Refer to CAN-Bus information on page 51.
U0009	High Speed CAN Communication Bus (-) shorted to Bus (+)	The Vehicle / CAN Communication Bus is a network for sharing signals and data / messages between the various vehicle systems, including Powertrain, Chassis and Infotainment. For wire based CAN-Bus systems, signals are transmitted along 2 wires that are twisted together (twisted pair). Each wire carries a digital signal; one being a mirror image of the other. The signal does not start at a zero volts base line but, the mirrored signals can be regarded as positive or negative values relative to the base line voltage.	Refer to vehicle specific information for identification of location and function of the specified systems / CAN-Bus components. The CAN speed relates to the transmission rate of the data messages between the various control systems. The fault code indicates that there is a probable short circuit between the two twisted wires (likely to be detected due to loss of signal in the specified circuit). Signal could be non-existent or out of normal operating range, constant value or corrupt. Refer to CAN-Bus information on page 51.
U0010	Medium Speed CAN Communication Bus	The Vehicle / CAN Communication Bus is a network for sharing signals and data / messages between the various vehicle systems, including Powertrain, Chassis and Infotainment.	Refer to vehicle specific information for identification of location and function of the specified systems / CAN-Bus components. The CAN speed relates to the transmission rate of the data messages between the various control systems. The fault code indicates that one or more control modules has detected an undefined fault on the Medium Speed CAN communication Bus due to the incorrect transmission of data messages. Refer to CAN-Bus information on page 51.
U0011	Medium Speed CAN Communication Bus Performance	The Vehicle / CAN Communication Bus is a network for sharing signals and data / messages between the various vehicle systems, including Powertrain, Chassis and Infotainment.	Refer to vehicle specific information for identification of location and function of the specified systems / CAN-Bus components. The CAN speed relates to the transmission rate of the data messages between the various control systems. The fault code indicates that one or more control modules has detected a performance fault with the data transmission on the Medium Speed CAN communication Bus. Refer to CAN-Bus information on page 51.
U0012	Medium Speed CAN Communication Bus (+) Open	The Vehicle / CAN Communication Bus is a network for sharing signals and data / messages between the various vehicle systems, including Powertrain, Chassis and Infotainment. For wire based CAN-Bus signals, signals are transmitted along 2 wires that are twisted together (twisted pair). Each wire carries a digital signal; one being a mirror image of the other. The signal does not start at a zero volts base line but, the mirrored signals can be regarded as positive or negative values relative to the base line voltage.	Refer to vehicle specific information for identification of location and function of the specified systems / CAN-Bus components. The CAN speed relates to the transmission rate of the data messages between the various control systems. The fault code indicates that there is a probable open circuit likely to be detected due to loss of signal in the specified circuit. Refer to CAN-Bus information on page 51.
U0013	Medium Speed CAN Communication Bus (+) Low	The Vehicle / CAN Communication Bus is a network for sharing signals and data / messages between the various vehicle systems, including Powertrain, Chassis and Infotainment. For wire based CAN-Bus systems, signals are transmitted along 2 wires that are twisted together (twisted pair). Each wire carries a digital signal; one being a mirror image of the other. The signal does not start at a zero volts base line but, the mirrored signals can be regarded as positive or negative values relative to the base line voltage.	Refer to vehicle specific information for identification of location and function of the specified systems / CAN-Bus components. The CAN speed relates to the transmission rate of the data messages between the various control systems. The fault code indicates that one or more control modules has detected that the voltage, frequency or value of the signal is permanently low, below the normal operating range. There could be other undefined faults in the CAN-Bus system that could activate the same code. Refer to CAN-Bus information on page 51.
U0014	Medium Speed CAN Communication Bus (+) High	The Vehicle / CAN Communication Bus is a network for sharing signals and data / messages between the various vehicle systems, including Powertrain, Chassis and Infotainment. For wire based CAN-Bus systems, signals are transmitted along 2 wires that are twisted together (twisted pair). Each wire carries a digital signal; one being a mirror image of the other. The signal does not start at a zero volts base line but, the mirrored signals can be regarded as positive or negative values relative to the base line voltage.	Refer to vehicle specific information for identification of location and function of the specified systems / CAN-Bus components. The CAN speed relates to the transmission rate of the data messages between the various control systems. The fault code indicates that one or more control modules has detected that the voltage, frequency or value of the signal is permanently high, above the normal operating range. There could be other undefined faults in the CAN-Bus system that could activate the same code. Refer to CAN-Bus information on page 51.

NOTE 1. Check for other fault codes that could provide additional information.   NOTE 2. Communication between control units can pass via a CAN-Bus system; refer to Page 51 for CAN-Bus checks.
NOTE 3. If a fault cannot be located, it is also possible that the control unit is at fault.   NOTE 4. Refer to Page 53 for list of pages that show the ISO standard for component locations e.g. Sensor A - Bank 1.

Fault code	EOBD / ISO Description	Component / System Description	Meaningful Description and Quick Check
U0015	Medium Speed CAN Communication Bus (-) Open	The Vehicle / CAN Communication Bus is a network for sharing signals and data / messages between the various vehicle systems, including Powertrain, Chassis and Infotainment. For wire based CAN-Bus systems, signals are transmitted along 2 wires that are twisted together (twisted pair). Each wire carries a digital signal; one being a mirror image of the other. The signal does not start at a zero volts base line but, the mirrored signals can be regarded as positive or negative values relative to the base line voltage.	Refer to vehicle specific information for identification of location and function of the specified systems / CAN-Bus components. The CAN speed relates to the transmission rate of the data messages between the various control systems. The fault code indicates that there is a probable open circuit likely to be detected due to loss of signal in the specified circuit. Refer to CAN-Bus information on page 51.
U0016	Medium Speed CAN Communication Bus (-) Low	The Vehicle / CAN Communication Bus is a network for sharing signals and data / messages between the various vehicle systems, including Powertrain, Chassis and Infotainment. For wire based CAN-Bus systems, signals are transmitted along 2 wires that are twisted together (twisted pair). Each wire carries a digital signal; one being a mirror image of the other. The signal does not start at a zero volts base line but, the mirrored signals can be regarded as positive or negative values relative to the base line voltage.	Refer to vehicle specific information for identification of location and function of the specified systems / CAN-Bus components. The CAN speed relates to the transmission rate of the data messages between the various control systems. The fault code indicates that one or more control modules has detected that the voltage, frequency or value of the signal is permanently low, below the normal operating range. There could be other undefined faults in the CAN-Bus system that could activate the same code. Refer to CAN-Bus information on page 51.
U0017	Medium Speed CAN Communication Bus (-) High	The Vehicle / CAN Communication Bus is a network for sharing signals and data / messages between the various vehicle systems, including Powertrain, Chassis and Infotainment. For wire based CAN-Bus systems, signals are transmitted along 2 wires that are twisted together (twisted pair). Each wire carries a digital signal; one being a mirror image of the other. The signal does not start at a zero volts base line but, the mirrored signals can be regarded as positive or negative values relative to the base line voltage.	Refer to vehicle specific information for identification of location and function of the specified systems / CAN-Bus components. The CAN speed relates to the transmission rate of the data messages between the various control systems. The fault code indicates that one or more control modules has detected that the voltage, frequency or value of the signal is permanently high, above the normal operating range. There could be other undefined faults in the CAN-Bus system that could activate the same code. Refer to CAN-Bus information on page 51.
U0018	Medium Speed CAN Communication Bus (-) shorted to Bus (+)	The Vehicle / CAN Communication Bus is a network for sharing signals and data / messages between the various vehicle systems, including Powertrain, Chassis and Infotainment. For wire based CAN-Bus systems, signals are transmitted along 2 wires that are twisted together (twisted pair). Each wire carries a digital signal; one being a mirror image of the other. The signal does not start at a zero volts base line but, the mirrored signals can be regarded as positive or negative values relative to the base line voltage.	Refer to vehicle specific information for identification of location and function of the specified systems / CAN-Bus components. The CAN speed relates to the transmission rate of the data messages between the various control systems. The fault code indicates that there is a probable short circuit between the two twisted wires (likely to be detected due to loss of signal in the specified circuit). Signal could be non-existent or out of normal operating range, constant value or corrupt. Refer to CAN-Bus information on page 51.
U0019	Low Speed CAN Communication Bus	The Vehicle / CAN Communication Bus is a network for sharing signals and data / messages between the various vehicle systems, including Powertrain, Chassis and Infotainment.	Refer to vehicle specific information for identification of location and function of the specified systems / CAN-Bus components. The CAN speed relates to the transmission rate of the data messages between the various control systems. The fault code indicates that one or more control modules has detected an undefined fault on the Low Speed CAN communication Bus due to the incorrect transmission of data messages. Refer to CAN-Bus information on page 51.
U0020	Low Speed CAN Communication Bus Performance	The Vehicle / CAN Communication Bus is a network for sharing signals and data / messages between the various vehicle systems, including Powertrain, Chassis and Infotainment.	Refer to vehicle specific information for identification of location and function of the specified systems / CAN-Bus components. The CAN speed relates to the transmission rate of the data messages between the various control systems. The fault code indicates that one or more control modules has detected a performance fault with the data transmission on the Low Speed CAN communication Bus. Refer to CAN-Bus information on page 51.
U0021	Low Speed CAN Communication Bus (+) Open	The Vehicle / CAN Communication Bus is a network for sharing signals and data / messages between the various vehicle systems, including Powertrain, Chassis and Infotainment. For wire based CAN-Bus systems, signals are transmitted along 2 wires that are twisted together (twisted pair). Each wire carries a digital signal; one being a mirror image of the other. The signal does not start at a zero volts base line but, the mirrored signals can be regarded as positive or negative values relative to the base line voltage.	Refer to vehicle specific information for identification of location and function of the specified systems / CAN-Bus components. The CAN speed relates to the transmission rate of the data messages between the various control systems. The fault code indicates that there is a probable open circuit likely to be detected due to loss of signal in the specified circuit. Refer to CAN-Bus information on page 51.
U0022	Low Speed CAN Communication Bus (+) Low	The Vehicle / CAN Communication Bus is a network for sharing signals and data / messages between the various vehicle systems, including Powertrain, Chassis and Infotainment. For wire based CAN-Bus systems, signals are transmitted along 2 wires that are twisted together (twisted pair). Each wire carries a digital signal; one being a mirror image of the other. The signal does not start at a zero volts base line but, the mirrored signals can be regarded as positive or negative values relative to the base line voltage.	Refer to vehicle specific information for identification of location and function of the specified systems / CAN-Bus components. The CAN speed relates to the transmission rate of the data messages between the various control systems. The fault code indicates that one or more control modules has detected that the voltage, frequency or value of the signal is permanently low, below the normal operating range. There could be other undefined faults in the CAN-Bus system that could activate the same code. Refer to CAN-Bus information on page 51.

NOTE 1. Check for other fault codes that could provide additional information.   **NOTE 2.** Communication between control units can pass via a CAN-Bus system; refer to Page 51 for CAN-Bus checks.

U0023	Low Speed CAN Communication Bus (+) High	The Vehicle / CAN Communication Bus is a network for sharing signals and data / messages between the various vehicle systems, including Powertrain, Chassis and Infotainment. For wire based CAN-Bus systems, signals are transmitted along 2 wires that are twisted together (twisted pair). Each wire carries a digital signal; one being a mirror image of the other. The signal does not start at a zero volts base line but, the mirrored signals can be regarded as positive or negative values relative to the base line voltage.	Refer to vehicle specific information for identification of location and function of the specified systems / CAN-Bus components. The CAN speed relates to the transmission rate of the data messages between the various control systems. The fault code indicates that one or more control modules has detected that the voltage, frequency or value of the signal is permanently high, above the normal operating range. There could be other undefined faults in the CAN-Bus system that could activate the same code. Refer to CAN-Bus information on page 51.
U0024	Low Speed CAN Communication Bus (-) Open	The Vehicle / CAN Communication Bus is a network for sharing signals and data / messages between the various vehicle systems, including Powertrain, Chassis and Infotainment. For wire based CAN-Bus systems, signals are transmitted along 2 wires that are twisted together (twisted pair). Each wire carries a digital signal; one being a mirror image of the other. The signal does not start at a zero volts base line but, the mirrored signals can be regarded as positive or negative values relative to the base line voltage.	Refer to vehicle specific information for identification of location and function of the specified systems / CAN-Bus components. The CAN speed relates to the transmission rate of the data messages between the various control systems. The fault code indicates that there is a probable open circuit likely to be detected due to loss of signal in the specified circuit. Refer to CAN-Bus information on page 51.
U0025	Low Speed CAN Communication Bus (-) Low	The Vehicle / CAN Communication Bus is a network for sharing signals and data / messages between the various vehicle systems, including Powertrain, Chassis and Infotainment. For wire based CAN-Bus systems, signals are transmitted along 2 wires that are twisted together (twisted pair). Each wire carries a digital signal; one being a mirror image of the other. The signal does not start at a zero volts base line but, the mirrored signals can be regarded as positive or negative values relative to the base line voltage.	Refer to vehicle specific information for identification of location and function of the specified systems / CAN-Bus components. The CAN speed relates to the transmission rate of the data messages between the various control systems. The fault code indicates that one or more control modules has detected that the voltage, frequency or value of the signal is permanently low, below the normal operating range. There could be other undefined faults in the CAN-Bus system that could activate the same code. Refer to CAN-Bus information on page 51.
U0026	Low Speed CAN Communication Bus (-) High	The Vehicle / CAN Communication Bus is a network for sharing signals and data / messages between the various vehicle systems, including Powertrain, Chassis and Infotainment. For wire based CAN-Bus systems, signals are transmitted along 2 wires that are twisted together (twisted pair). Each wire carries a digital signal; one being a mirror image of the other. The signal does not start at a zero volts base line but, the mirrored signals can be regarded as positive or negative values relative to the base line voltage.	Refer to vehicle specific information for identification of location and function of the specified systems / CAN-Bus components. The CAN speed relates to the transmission rate of the data messages between the various control systems. The fault code indicates that one or more control modules has detected that the voltage, frequency or value of the signal is permanently high, above the normal operating range. There could be other undefined faults in the CAN-Bus system that could activate the same code. Refer to CAN-Bus information on page 51.
U0027	Low Speed CAN Communication Bus (-) shorted to Bus (+)	The Vehicle / CAN Communication Bus is a network for sharing signals and data / messages between the various vehicle systems, including Powertrain, Chassis and Infotainment. For wire based CAN-Bus systems, signals are transmitted along 2 wires that are twisted together (twisted pair). Each wire carries a digital signal; one being a mirror image of the other. The signal does not start at a zero volts base line but, the mirrored signals can be regarded as positive or negative values relative to the base line voltage.	Refer to vehicle specific information for identification of location and function of the specified systems / CAN-Bus components. The CAN speed relates to the transmission rate of the data messages between the various control systems. The fault code indicates that there is a probable short circuit between the two twisted wires (likely to be detected due to loss of signal in the specified circuit). Signal could be non-existent or out of normal operating range, constant value or corrupt. Refer to CAN-Bus information on page 51.
U0028	Vehicle Communication Bus A	The Vehicle / CAN Communication Bus is a network for sharing signals and data / messages between the various vehicle systems, including Powertrain, Chassis and Infotainment.	Refer to vehicle specific information for identification of location and function of the specified systems / CAN-Bus components. The fault code indicates that one or more control modules has detected an undefined fault on the Vehicle Communication Bus A due to the incorrect transmission of data messages. Refer to CAN-Bus information on page 51.
U0029	Vehicle Communication Bus A Performance	The Vehicle / CAN Communication Bus is a network for sharing signals and data / messages between the various vehicle systems, including Powertrain, Chassis and Infotainment.	Refer to vehicle specific information for identification of location and function of the specified systems / CAN-Bus components. The fault code indicates that one or more control modules has detected a performance fault with the data transmission on the Vehicle Communication Bus A. Refer to CAN-Bus information on page 51.
U0030	Vehicle Communication Bus A (+) Open	The Vehicle / CAN Communication Bus is a network for sharing signals and data / messages between the various vehicle systems, including Powertrain, Chassis and Infotainment. For wire based CAN-Bus systems, signals are transmitted along 2 wires that are twisted together (twisted pair). Each wire carries a digital signal; one being a mirror image of the other. The signal does not start at a zero volts base line but, the mirrored signals can be regarded as positive or negative values relative to the base line voltage.	Refer to vehicle specific information for identification of location and function of the specified systems / CAN-Bus components. The fault code indicates that there is a probable open circuit likely to be detected due to loss of signal in the specified circuit. Refer to CAN-Bus information on page 51.
U0031	Vehicle Communication Bus A (+) Low	The Vehicle / CAN Communication Bus is a network for sharing signals and data / messages between the various vehicle systems, including Powertrain, Chassis and Infotainment. For wire based CAN-Bus systems, signals are transmitted along 2 wires that are twisted together (twisted pair). Each wire carries a digital signal; one being a mirror image of the other. The signal does not start at a zero volts base line but, the mirrored signals can be regarded as positive or negative values relative to the base line voltage.	Refer to vehicle specific information for identification of location and function of the specified systems / CAN-Bus components. The fault code indicates that one or more control modules has detected that the voltage, frequency or value of the signal is permanently low, below the normal operating range. There could be other undefined faults in the CAN-Bus system that could activate the same code. Refer to CAN-Bus information on page 51.

NOTE 1. Check for other fault codes that could provide additional information.
NOTE 3. If a fault cannot be located, it is also possible that the control unit is at fault.
NOTE 2. Communication between control units can pass via a CAN-Bus system; refer to Page 51 for CAN-Bus checks.
NOTE 4. Refer to Page 53 for list of pages that show the ISO standard for component locations e.g. Sensor A - Bank 1.

Fault code	EOBD / ISO Description	Component / System Description	Meaningful Description and Quick Check
U0032	Vehicle Communication Bus A (+) High	The Vehicle / CAN Communication Bus is a network for sharing signals and data / messages between the various vehicle systems, including Powertrain, Chassis and Infotainment. For wire based CAN-Bus systems, signals are transmitted along 2 wires that are twisted together (twisted pair). Each wire carries a digital signal; one being a mirror image of the other. The signal does not start at a zero volts base line but, the mirrored signals can be regarded as positive or negative values relative to the base line voltage.	Refer to vehicle specific information for identification of location and function of the specified systems / CAN-Bus components. The fault code indicates that one or more control modules has detected that the voltage, frequency or value of the signal is permanently high, above the normal operating range. There could be other undefined faults in the CAN-Bus system that could activate the same code. Refer to CAN-Bus information on page 51.
U0033	Vehicle Communication Bus A (-) Open	The Vehicle / CAN Communication Bus is a network for sharing signals and data / messages between the various vehicle systems, including Powertrain, Chassis and Infotainment. For wire based CAN-Bus systems, signals are transmitted along 2 wires that are twisted together (twisted pair). Each wire carries a digital signal; one being a mirror image of the other. The signal does not start at a zero volts base line but, the mirrored signals can be regarded as positive or negative values relative to the base line voltage.	Refer to vehicle specific information for identification of location and function of the specified systems / CAN-Bus components. The fault code indicates that there is a probable open circuit likely to be detected due to loss of signal in the specified circuit. Refer to CAN-Bus information on page 51.
U0034	Vehicle Communication Bus A (-) Low	The Vehicle / CAN Communication Bus is a network for sharing signals and data / messages between the various vehicle systems, including Powertrain, Chassis and Infotainment. For wire based CAN-Bus systems, signals are transmitted along 2 wires that are twisted together (twisted pair). Each wire carries a digital signal; one being a mirror image of the other. The signal does not start at a zero volts base line but, the mirrored signals can be regarded as positive or negative values relative to the base line voltage.	Refer to vehicle specific information for identification of location and function of the specified systems / CAN-Bus components. The fault code indicates that one or more control modules has detected that the voltage, frequency or value of the signal is permanently low, below the normal operating range. There could be other undefined faults in the CAN-Bus system that could activate the same code. Refer to CAN-Bus information on page 51.
U0035	Vehicle Communication Bus A (-) High	The Vehicle / CAN Communication Bus is a network for sharing signals and data / messages between the various vehicle systems, including Powertrain, Chassis and Infotainment. For wire based CAN-Bus systems, signals are transmitted along 2 wires that are twisted together (twisted pair). Each wire carries a digital signal; one being a mirror image of the other. The signal does not start at a zero volts base line but, the mirrored signals can be regarded as positive or negative values relative to the base line voltage.	Refer to vehicle specific information for identification of location and function of the specified systems / CAN-Bus components. The fault code indicates that one or more control modules has detected that the voltage, frequency or value of the signal is permanently high, above the normal operating range. There could be other undefined faults in the CAN-Bus system that could activate the same code. Refer to CAN-Bus information on page 51.
U0036	Vehicle Communication Bus A (-) shorted to Bus A (+)	The Vehicle / CAN Communication Bus is a network for sharing signals and data / messages between the various vehicle systems, including Powertrain, Chassis and Infotainment. For wire based CAN-Bus systems, signals are transmitted along 2 wires that are twisted together (twisted pair). Each wire carries a digital signal; one being a mirror image of the other. The signal does not start at a zero volts base line but, the mirrored signals can be regarded as positive or negative values relative to the base line voltage.	Refer to vehicle specific information for identification of location and function of the specified systems / CAN-Bus components. The fault code indicates that there is a probable short circuit between the two twisted wires (likely to be detected due to loss of signal in the specified circuit). Signal could be non-existent or out of normal operating range, constant value or corrupt. Refer to CAN-Bus information on page 51.
U0037	Vehicle Communication Bus B	The Vehicle / CAN Communication Bus is a network for sharing signals and data / messages between the various vehicle systems, including Powertrain, Chassis and Infotainment.	Refer to vehicle specific information for identification of location and function of the specified systems / CAN-Bus components. The fault code indicates that one or more control modules has detected an undefined fault on the Vehicle Communication Bus B due to the incorrect transmission of data messages. Refer to CAN-Bus information on page 51.
U0038	Vehicle Communication Bus B Performance	The Vehicle / CAN Communication Bus is a network for sharing signals and data / messages between the various vehicle systems, including Powertrain, Chassis and Infotainment.	Refer to vehicle specific information for identification of location and function of the specified systems / CAN-Bus components. The fault code indicates that one or more control modules has detected a performance fault with the data transmission on the Vehicle Communication Bus B. Refer to CAN-Bus information on page 51.
U0039	Vehicle Communication Bus B (+) Open	The Vehicle / CAN Communication Bus is a network for sharing signals and data / messages between the various vehicle systems, including Powertrain, Chassis and Infotainment. For wire based CAN-Bus systems, signals are transmitted along 2 wires that are twisted together (twisted pair). Each wire carries a digital signal; one being a mirror image of the other. The signal does not start at a zero volts base line but, the mirrored signals can be regarded as positive or negative values relative to the base line voltage.	Refer to vehicle specific information for identification of location and function of the specified systems / CAN-Bus components. The fault code indicates that there is a probable open circuit likely to be detected due to loss of signal in the specified circuit. Refer to CAN-Bus information on page 51.
U0040	Vehicle Communication Bus B (+) Low	The Vehicle / CAN Communication Bus is a network for sharing signals and data / messages between the various vehicle systems, including Powertrain, Chassis and Infotainment. For wire based CAN-Bus systems, signals are transmitted along 2 wires that are twisted together (twisted pair). Each wire carries a digital signal; one being a mirror image of the other. The signal does not start at a zero volts base line but, the mirrored signals can be regarded as positive or negative values relative to the base line voltage.	Refer to vehicle specific information for identification of location and function of the specified systems / CAN-Bus components. The fault code indicates that one or more control modules has detected that the voltage, frequency or value of the signal is permanently low, below the normal operating range. There could be other undefined faults in the CAN-Bus system that could activate the same code. Refer to CAN-Bus information on page 51.

NOTE 1. Check for other fault codes that could provide additional information.   **NOTE 2.** Communication between control units can pass via a CAN-Bus system; refer to Page 51 for CAN-Bus checks.

U0041	Vehicle Communication Bus B (+) High	The Vehicle / CAN Communication Bus is a network for sharing signals and data / messages between the various vehicle systems, including Powertrain, Chassis and Infotainment. For wire based CAN-Bus systems, signals are transmitted along 2 wires that are twisted together (twisted pair). Each wire carries a digital signal; one being a mirror image of the other. The signal does not start at a zero volts base line but, the mirrored signals can be regarded as positive or negative values relative to the base line voltage.	Refer to vehicle specific information for identification of location and function of the specified systems / CAN-Bus components. The fault code indicates that one or more control modules has detected that the voltage, frequency or value of the signal is permanently high, above the normal operating range. There could be other undefined faults in the CAN-Bus system that could activate the same code. Refer to CAN-Bus information on page 51.
U0042	Vehicle Communication Bus B (-) Open	The Vehicle / CAN Communication Bus is a network for sharing signals and data / messages between the various vehicle systems, including Powertrain, Chassis and Infotainment. For wire based CAN-Bus systems, signals are transmitted along 2 wires that are twisted together (twisted pair). Each wire carries a digital signal; one being a mirror image of the other. The signal does not start at a zero volts base line but, the mirrored signals can be regarded as positive or negative values relative to the base line voltage.	Refer to vehicle specific information for identification of location and function of the specified systems / CAN-Bus components. The fault code indicates that there is a probable open circuit likely to be detected due to loss of signal in the specified circuit. Refer to CAN-Bus information on page 51.
U0043	Vehicle Communication Bus B (-) Low	The Vehicle / CAN Communication Bus is a network for sharing signals and data / messages between the various vehicle systems, including Powertrain, Chassis and Infotainment. For wire based CAN-Bus systems, signals are transmitted along 2 wires that are twisted together (twisted pair). Each wire carries a digital signal; one being a mirror image of the other. The signal does not start at a zero volts base line but, the mirrored signals can be regarded as positive or negative values relative to the base line voltage.	Refer to vehicle specific information for identification of location and function of the specified systems / CAN-Bus components. The fault code indicates that one or more control modules has detected that the voltage, frequency or value of the signal is permanently low, below the normal operating range. There could be other undefined faults in the CAN-Bus system that could activate the same code. Refer to CAN-Bus information on page 51.
U0044	Vehicle Communication Bus B (-) High	The Vehicle / CAN Communication Bus is a network for sharing signals and data / messages between the various vehicle systems, including Powertrain, Chassis and Infotainment. For wire based CAN-Bus systems, signals are transmitted along 2 wires that are twisted together (twisted pair). Each wire carries a digital signal; one being a mirror image of the other. The signal does not start at a zero volts base line but, the mirrored signals can be regarded as positive or negative values relative to the base line voltage.	Refer to vehicle specific information for identification of location and function of the specified systems / CAN-Bus components. The fault code indicates that one or more control modules has detected that the voltage, frequency or value of the signal is permanently high, above the normal operating range. There could be other undefined faults in the CAN-Bus system that could activate the same code. Refer to CAN-Bus information on page 51.
U0045	Vehicle Communication Bus B (-) shorted to Bus B (+)	The Vehicle / CAN Communication Bus is a network for sharing signals and data / messages between the various vehicle systems, including Powertrain, Chassis and Infotainment. For wire based CAN-Bus systems, signals are transmitted along 2 wires that are twisted together (twisted pair). Each wire carries a digital signal; one being a mirror image of the other. The signal does not start at a zero volts base line but, the mirrored signals can be regarded as positive or negative values relative to the base line voltage.	Refer to vehicle specific information for identification of location and function of the specified systems / CAN-Bus components. The fault code indicates that there is a probable short circuit between the two twisted wires (likely to be detected due to loss of signal in the specified circuit). Signal could be non-existent or out of normal operating range, constant value or corrupt. Refer to CAN-Bus information on page 51.
U0046	Vehicle Communication Bus C	The Vehicle / CAN Communication Bus is a network for sharing signals and data / messages between the various vehicle systems, including Powertrain, Chassis and Infotainment.	Refer to vehicle specific information for identification of location and function of the specified systems / CAN-Bus components. The fault code indicates that one or more control modules has detected an undefined fault on the Vehicle Communication Bus C due to the incorrect transmission of data messages. Refer to CAN-Bus information on page 51.
U0047	Vehicle Communication Bus C Performance	The Vehicle / CAN Communication Bus is a network for sharing signals and data / messages between the various vehicle systems, including Powertrain, Chassis and Infotainment.	Refer to vehicle specific information for identification of location and function of the specified systems / CAN-Bus components. The fault code indicates that one or more control modules has detected a performance fault with the data transmission on the Vehicle Communication Bus C. Refer to CAN-Bus information on page 51.
U0048	Vehicle Communication Bus C (+) Open	The Vehicle / CAN Communication Bus is a network for sharing signals and data / messages between the various vehicle systems, including Powertrain, Chassis and Infotainment. For wire based CAN-Bus systems, signals are transmitted along 2 wires that are twisted together (twisted pair). Each wire carries a digital signal; one being a mirror image of the other. The signal does not start at a zero volts base line but, the mirrored signals can be regarded as positive or negative values relative to the base line voltage.	Refer to vehicle specific information for identification of location and function of the specified systems / CAN-Bus components. The fault code indicates that there is a probable open circuit likely to be detected due to loss of signal in the specified circuit. Refer to CAN-Bus information on page 51.
U0049	Vehicle Communication Bus C (+) Low	The Vehicle / CAN Communication Bus is a network for sharing signals and data / messages between the various vehicle systems, including Powertrain, Chassis and Infotainment. For wire based CAN-Bus systems, signals are transmitted along 2 wires that are twisted together (twisted pair). Each wire carries a digital signal; one being a mirror image of the other. The signal does not start at a zero volts base line but, the mirrored signals can be regarded as positive or negative values relative to the base line voltage.	Refer to vehicle specific information for identification of location and function of the specified systems / CAN-Bus components. The fault code indicates that one or more control modules has detected that the voltage, frequency or value of the signal is permanently low, below the normal operating range. There could be other undefined faults in the CAN-Bus system that could activate the same code. Refer to CAN-Bus information on page 51.

NOTE 1. Check for other fault codes that could provide additional information. **NOTE 2.** Communication between control units can pass via a CAN-Bus system; refer to Page 51 for CAN-Bus checks.

NOTE 3. If a fault cannot be located, it is also possible that the control unit is at fault. **NOTE 4.** Refer to Page 53 for list of pages that show the ISO standard for component locations e.g. Sensor A - Bank 1.

Fault code	EOBD / ISO Description	Component / System Description	Meaningful Description and Quick Check
U0050	Vehicle Communication Bus C (+) High	The Vehicle / CAN Communication Bus is a network for sharing signals and data / messages between the various vehicle systems, including Powertrain, Chassis and Infotainment. For wire based CAN-Bus systems, signals are transmitted along 2 wires that are twisted together (twisted pair). Each wire carries a digital signal; one being a mirror image of the other. The signal does not start at a zero volts base line but, the mirrored signals can be regarded as positive or negative values relative to the base line voltage.	Refer to vehicle specific information for identification of location and function of the specified systems / CAN-Bus components. The fault code indicates that one or more control modules has detected that the voltage, frequency or value of the signal is permanently high, above the normal operating range. There could be other undefined faults in the CAN-Bus system that could activate the same code. Refer to CAN-Bus information on page 51.
U0051	Vehicle Communication Bus C (-) Open	The Vehicle / CAN Communication Bus is a network for sharing signals and data / messages between the various vehicle systems, including Powertrain, Chassis and Infotainment. For wire based CAN-Bus systems, signals are transmitted along 2 wires that are twisted together (twisted pair). Each wire carries a digital signal; one being a mirror image of the other. The signal does not start at a zero volts base line but, the mirrored signals can be regarded as positive or negative values relative to the base line voltage.	Refer to vehicle specific information for identification of location and function of the specified systems / CAN-Bus components. The fault code indicates that there is a probable open circuit likely to be detected due to loss of signal in the specified circuit. Refer to CAN-Bus information on page 51.
U0052	Vehicle Communication Bus C (-) Low	The Vehicle / CAN Communication Bus is a network for sharing signals and data / messages between the various vehicle systems, including Powertrain, Chassis and Infotainment. For wire based CAN-Bus systems, signals are transmitted along 2 wires that are twisted together (twisted pair). Each wire carries a digital signal; one being a mirror image of the other. The signal does not start at a zero volts base line but, the mirrored signals can be regarded as positive or negative values relative to the base line voltage.	Refer to vehicle specific information for identification of location and function of the specified systems / CAN-Bus components. The fault code indicates that one or more control modules has detected that the voltage, frequency or value of the signal is permanently low, below the normal operating range. There could be other undefined faults in the CAN-Bus system that could activate the same code. Refer to CAN-Bus information on page 51.
U0053	Vehicle Communication Bus C (-) High	The Vehicle / CAN Communication Bus is a network for sharing signals and data / messages between the various vehicle systems, including Powertrain, Chassis and Infotainment. For wire based CAN-Bus systems, signals are transmitted along 2 wires that are twisted together (twisted pair). Each wire carries a digital signal; one being a mirror image of the other. The signal does not start at a zero volts base line but, the mirrored signals can be regarded as positive or negative values relative to the base line voltage.	Refer to vehicle specific information for identification of location and function of the specified systems / CAN-Bus components. The fault code indicates that one or more control modules has detected that the voltage, frequency or value of the signal is permanently high, above the normal operating range. There could be other undefined faults in the CAN-Bus system that could activate the same code. Refer to CAN-Bus information on page 51.
U0054	Vehicle Communication Bus C (-) shorted to Bus C (+)	The Vehicle / CAN Communication Bus is a network for sharing signals and data / messages between the various vehicle systems, including Powertrain, Chassis and Infotainment. For wire based CAN-Bus systems, signals are transmitted along 2 wires that are twisted together (twisted pair). Each wire carries a digital signal; one being a mirror image of the other. The signal does not start at a zero volts base line but, the mirrored signals can be regarded as positive or negative values relative to the base line voltage.	Refer to vehicle specific information for identification of location and function of the specified systems / CAN-Bus components. The fault code indicates that there is a probable short circuit between the two twisted wires (likely to be detected due to loss of signal in the specified circuit). Signal could be non-existent or out of normal operating range, constant value or corrupt. Refer to CAN-Bus information on page 51.
U0055	Vehicle Communication Bus D	The Vehicle / CAN Communication Bus is a network for sharing signals and data / messages between the various vehicle systems, including Powertrain, Chassis and Infotainment.	Refer to vehicle specific information for identification of location and function of the specified systems / CAN-Bus components. The fault code indicates that one or more control modules has detected an undefined fault on the Vehicle Communication Bus D due to the incorrect transmission of data messages. Refer to CAN-Bus information on page 51.
U0056	Vehicle Communication Bus D Performance	The Vehicle / CAN Communication Bus is a network for sharing signals and data / messages between the various vehicle systems, including Powertrain, Chassis and Infotainment.	Refer to vehicle specific information for identification of location and function of the specified systems / CAN-Bus components. The fault code indicates that one or more control modules has detected a performance fault with the data transmission on the Vehicle Communication Bus D. Refer to CAN-Bus information on page 51.
U0057	Vehicle Communication Bus D (+) Open	The Vehicle / CAN Communication Bus is a network for sharing signals and data / messages between the various vehicle systems, including Powertrain, Chassis and Infotainment. For wire based CAN-Bus systems, signals are transmitted along 2 wires that are twisted together (twisted pair). Each wire carries a digital signal; one being a mirror image of the other. The signal does not start at a zero volts base line but, the mirrored signals can be regarded as positive or negative values relative to the base line voltage.	Refer to vehicle specific information for identification of location and function of the specified systems / CAN-Bus components. The fault code indicates that there is a probable open circuit likely to be detected due to loss of signal in the specified circuit. Refer to CAN-Bus information on page 51.
U0058	Vehicle Communication Bus D (+) Low	The Vehicle / CAN Communication Bus is a network for sharing signals and data / messages between the various vehicle systems, including Powertrain, Chassis and Infotainment. For wire based CAN-Bus systems, signals are transmitted along 2 wires that are twisted together (twisted pair). Each wire carries a digital signal; one being a mirror image of the other. The signal does not start at a zero volts base line but, the mirrored signals can be regarded as positive or negative values relative to the base line voltage.	Refer to vehicle specific information for identification of location and function of the specified systems / CAN-Bus components. The fault code indicates that one or more control modules has detected that the voltage, frequency or value of the signal is permanently low, below the normal operating range. There could be other undefined faults in the CAN-Bus system that could activate the same code. Refer to CAN-Bus information on page 51.

NOTE 1. Check for other fault codes that could provide additional information.   **NOTE 2.** Communication between control units can pass via a CAN-Bus system; refer to Page 51 for CAN-Bus checks.

U0059	Vehicle Communication Bus D (+) High	The Vehicle / CAN Communication Bus is a network for sharing signals and data / messages between the various vehicle systems, including Powertrain, Chassis and Infotainment. For wire based CAN-Bus systems, signals are transmitted along 2 wires that are twisted together (twisted pair). Each wire carries a digital signal; one being a mirror image of the other. The signal baseline does not start at zero volts but, the mirrored signals can be regarded as positive or negative values relative to whatever voltage is used for the base line.	Refer to vehicle specific information for identification of location and function of the specified systems / CAN-Bus components. The fault code indicates that one or more control modules has detected that the voltage, frequency or value of the signal is permanently high, above the normal operating range. There could be other undefined faults in the CAN-Bus system that could activate the same code. Refer to CAN-Bus information on page 51.
U0060	Vehicle Communication Bus D (-) Open	The Vehicle / CAN Communication Bus is a network for sharing signals and data / messages between the various vehicle systems, including Powertrain, Chassis and Infotainment. For wire based CAN-Bus systems, signals are transmitted along 2 wires that are twisted together (twisted pair). Each wire carries a digital signal; one being a mirror image of the other. The signal does not start at a zero volts base line but, the mirrored signals can be regarded as positive or negative values relative to the base line voltage.	Refer to vehicle specific information for identification of location and function of the specified systems / CAN-Bus components. The fault code indicates that there is a probable open circuit likely to be detected due to loss of signal in the specified circuit. Refer to CAN-Bus information on page 51.
U0061	Vehicle Communication Bus D (-) Low	The Vehicle / CAN Communication Bus is a network for sharing signals and data / messages between the various vehicle systems, including Powertrain, Chassis and Infotainment. For wire based CAN-Bus systems, signals are transmitted along 2 wires that are twisted together (twisted pair). Each wire carries a digital signal; one being a mirror image of the other. The signal does not start at a zero volts base line but, the mirrored signals can be regarded as positive or negative values relative to the base line voltage.	Refer to vehicle specific information for identification of location and function of the specified systems / CAN-Bus components. The fault code indicates that one or more control modules has detected that the voltage, frequency or value of the signal is permanently low, below the normal operating range. There could be other undefined faults in the CAN-Bus system that could activate the same code. Refer to CAN-Bus information on page 51.
U0062	Vehicle Communication Bus D (-) High	The Vehicle / CAN Communication Bus is a network for sharing signals and data / messages between the various vehicle systems, including Powertrain, Chassis and Infotainment. For wire based CAN-Bus systems, signals are transmitted along 2 wires that are twisted together (twisted pair). Each wire carries a digital signal; one being a mirror image of the other. The signal does not start at a zero volts base line but, the mirrored signals can be regarded as positive or negative values relative to the base line voltage.	Refer to vehicle specific information for identification of location and function of the specified systems / CAN-Bus components. The fault code indicates that one or more control modules has detected that the voltage, frequency or value of the signal is permanently high, above the normal operating range. There could be other undefined faults in the CAN-Bus system that could activate the same code. Refer to CAN-Bus information on page 51.
U0063	Vehicle Communication Bus D (-) shorted to Bus D (+)	The Vehicle / CAN Communication Bus is a network for sharing signals and data / messages between the various vehicle systems, including Powertrain, Chassis and Infotainment. For wire based CAN-Bus systems, signals are transmitted along 2 wires that are twisted together (twisted pair). Each wire carries a digital signal; one being a mirror image of the other. The signal does not start at a zero volts base line but, the mirrored signals can be regarded as positive or negative values relative to the base line voltage.	Refer to vehicle specific information for identification of location and function of the specified systems / CAN-Bus components. The fault code indicates that there is a probable short circuit between the two twisted wires (likely to be detected due to loss of signal in the specified circuit). Signal could be non-existent or out of normal operating range, constant value or corrupt. Refer to CAN-Bus information on page 51.
U0064	Vehicle Communication Bus E	The Vehicle / CAN Communication Bus is a network for sharing signals and data / messages between the various vehicle systems, including Powertrain, Chassis and Infotainment.	Refer to vehicle specific information for identification of location and function of the specified systems / CAN-Bus components. The fault code indicates that one or more control modules has detected an undefined fault on the Vehicle Communication Bus E due to the incorrect transmission of data messages. Refer to CAN-Bus information on page 51.
U0065	Vehicle Communication Bus E Performance	The Vehicle / CAN Communication Bus is a network for sharing signals and data / messages between the various vehicle systems, including Powertrain, Chassis and Infotainment.	Refer to vehicle specific information for identification of location and function of the specified systems / CAN-Bus components. The fault code indicates that one or more control modules has detected a performance fault with the data transmission on the Vehicle Communication Bus E. Refer to CAN-Bus information on page 51.
U0066	Vehicle Communication Bus E (+) Open	The Vehicle / CAN Communication Bus is a network for sharing signals and data / messages between the various vehicle systems, including Powertrain, Chassis and Infotainment. For wire based CAN-Bus systems, signals are transmitted along 2 wires that are twisted together (twisted pair). Each wire carries a digital signal; one being a mirror image of the other. The signal does not start at a zero volts base line but, the mirrored signals can be regarded as positive or negative values relative to the base line voltage.	Refer to vehicle specific information for identification of location and function of the specified systems / CAN-Bus components. The fault code indicates that there is a probable open circuit likely to be detected due to loss of signal in the specified circuit. Refer to CAN-Bus information on page 51.
U0067	Vehicle Communication Bus E (+) Low	The Vehicle / CAN Communication Bus is a network for sharing signals and data / messages between the various vehicle systems, including Powertrain, Chassis and Infotainment. For wire based CAN-Bus systems, signals are transmitted along 2 wires that are twisted together (twisted pair). Each wire carries a digital signal; one being a mirror image of the other. The signal does not start at a zero volts base line but, the mirrored signals can be regarded as positive or negative values relative to the base line voltage.	Refer to vehicle specific information for identification of location and function of the specified systems / CAN-Bus components. The fault code indicates that one or more control modules has detected that the voltage, frequency or value of the signal is permanently low, below the normal operating range. There could be other undefined faults in the CAN-Bus system that could activate the same code. Refer to CAN-Bus information on page 51.

NOTE 1. Check for other fault codes that could provide additional information.　**NOTE 2.** Communication between control units can pass via a CAN-Bus system; refer to Page 51 for CAN-Bus checks.

NOTE 3. If a fault cannot be located, it is also possible that the control unit is at fault.　**NOTE 4.** Refer to Page 53 for list of pages that show the ISO standard for component locations e.g. Sensor A - Bank 1.

Fault code	EOBD / ISO Description	Component / System Description	Meaningful Description and Quick Check
U0068	Vehicle Communication Bus E (+) High	The Vehicle / CAN Communication Bus is a network for sharing signals and data / messages between the various vehicle systems, including Powertrain, Chassis and Infotainment. For wire based CAN-Bus systems, signals are transmitted along 2 wires that are twisted together (twisted pair). Each wire carries a digital signal; one being a mirror image of the other. The signal does not start at a zero volts base line but, the mirrored signals can be regarded as positive or negative values relative to the base line voltage.	Refer to vehicle specific information for identification of location and function of the specified systems / CAN-Bus components. The fault code indicates that one or more control modules has detected that the voltage, frequency or value of the signal is permanently high, above the normal operating range. There could be other undefined faults in the CAN-Bus system that could activate the same code. Refer to CAN-Bus information on page 51.
U0069	Vehicle Communication Bus E (-) Open	The Vehicle / CAN Communication Bus is a network for sharing signals and data / messages between the various vehicle systems, including Powertrain, Chassis and Infotainment. For wire based CAN-Bus systems, signals are transmitted along 2 wires that are twisted together (twisted pair). Each wire carries a digital signal; one being a mirror image of the other. The signal does not start at a zero volts base line but, the mirrored signals can be regarded as positive or negative values relative to the base line voltage.	Refer to vehicle specific information for identification of location and function of the specified systems / CAN-Bus components. The fault code indicates that there is a probable open circuit likely to be detected due to loss of signal in the specified circuit. Refer to CAN-Bus information on page 51.
U0070	Vehicle Communication Bus E (-) Low	The Vehicle / CAN Communication Bus is a network for sharing signals and data / messages between the various vehicle systems, including Powertrain, Chassis and Infotainment. For wire based CAN-Bus systems, signals are transmitted along 2 wires that are twisted together (twisted pair). Each wire carries a digital signal; one being a mirror image of the other. The signal does not start at a zero volts base line but, the mirrored signals can be regarded as positive or negative values relative to the base line voltage.	Refer to vehicle specific information for identification of location and function of the specified systems / CAN-Bus components. The fault code indicates that one or more control modules has detected that the voltage, frequency or value of the signal is permanently low, below the normal operating range. There could be other undefined faults in the CAN-Bus system that could activate the same code. Refer to CAN-Bus information on page 51.
U0071	Vehicle Communication Bus E (-) High	The Vehicle / CAN Communication Bus is a network for sharing signals and data / messages between the various vehicle systems, including Powertrain, Chassis and Infotainment. For wire based CAN-Bus systems, signals are transmitted along 2 wires that are twisted together (twisted pair). Each wire carries a digital signal; one being a mirror image of the other. The signal does not start at a zero volts base line but, the mirrored signals can be regarded as positive or negative values relative to the base line voltage.	Refer to vehicle specific information for identification of location and function of the specified systems / CAN-Bus components. The fault code indicates that one or more control modules has detected that the voltage, frequency or value of the signal is permanently high, above the normal operating range. There could be other undefined faults in the CAN-Bus system that could activate the same code. Refer to CAN-Bus information on page 51.
U0072	Vehicle Communication Bus E (-) shorted to Bus E (+)	The Vehicle / CAN Communication Bus is a network for sharing signals and data / messages between the various vehicle systems, including Powertrain, Chassis and Infotainment. For wire based CAN-Bus systems, signals are transmitted along 2 wires that are twisted together (twisted pair). Each wire carries a digital signal; one being a mirror image of the other. The signal does not start at a zero volts base line but, the mirrored signals can be regarded as positive or negative values relative to the base line voltage.	Refer to vehicle specific information for identification of location and function of the specified systems / CAN-Bus components. The fault code indicates that there is a probable short circuit between the two twisted wires (likely to be detected due to loss of signal in the specified circuit). Signal could be non-existent or out of normal operating range, constant value or corrupt. Refer to CAN-Bus information on page 51.
U0073	Control Module Communication Bus Off	The Vehicle / CAN Communication Bus is a network for sharing signals and data / messages between the various vehicle systems, including Powertrain, Chassis and Infotainment.	Refer to vehicle specific information for identification of location and function of the specified systems / CAN-Bus components. The fault code indicates that one or more control modules has lost communication or received no data messages due to the Vehicle / CAN communication Bus being inactive. Refer to CAN-Bus information on page 51.
U0074		ISO / SAE reserved	Not used, reserved for future allocation.
U0075		ISO / SAE reserved	Not used, reserved for future allocation.
U0076		ISO / SAE reserved	Not used, reserved for future allocation.
U0077		ISO / SAE reserved	Not used, reserved for future allocation.
U0078		ISO / SAE reserved	Not used, reserved for future allocation.
U0079		ISO / SAE reserved	Not used, reserved for future allocation.
U0080		ISO / SAE reserved	Not used, reserved for future allocation.
U0081		ISO / SAE reserved	Not used, reserved for future allocation.
U0082		ISO / SAE reserved	Not used, reserved for future allocation.
U0083		ISO / SAE reserved	Not used, reserved for future allocation.
U0084		ISO / SAE reserved	Not used, reserved for future allocation.
U0085		ISO / SAE reserved	Not used, reserved for future allocation.
U0086		ISO / SAE reserved	Not used, reserved for future allocation.
U0087		ISO / SAE reserved	Not used, reserved for future allocation.
U0088		ISO / SAE reserved	Not used, reserved for future allocation.

NOTE 1. Check for other fault codes that could provide additional information. **NOTE 2.** Communication between control units can pass via a CAN-Bus system; refer to Page 51 for CAN-Bus checks.

U0089		ISO / SAE reserved	Not used, reserved for future allocation.
U0090		ISO / SAE reserved	Not used, reserved for future allocation.
U0091		ISO / SAE reserved	Not used, reserved for future allocation.
U0092		ISO / SAE reserved	Not used, reserved for future allocation.
U0093		ISO / SAE reserved	Not used, reserved for future allocation.
U0094		ISO / SAE reserved	Not used, reserved for future allocation.
U0095		ISO / SAE reserved	Not used, reserved for future allocation.
U0096		ISO / SAE reserved	Not used, reserved for future allocation.
U0097		ISO / SAE reserved	Not used, reserved for future allocation.
U0098		ISO / SAE reserved	Not used, reserved for future allocation.
U0099		ISO / SAE reserved	Not used, reserved for future allocation.
U0100	Lost Communication With ECM / PCM "A"	The Vehicle / CAN Communication Bus is a network for sharing signals and data / messages between the various vehicle systems, including Powertrain, Chassis and Infotainment.	Refer to vehicle specific information for identification of location and function of the specified systems / CAN-Bus components. The fault code indicates that one or more control modules has lost communication / received no response / received incorrect data messages from the Engine ECM / PCM "A". Refer to CAN-Bus information on page 51.
U0101	Lost Communication with TCM	The Vehicle / CAN Communication Bus is a network for sharing signals and data / messages between the various vehicle systems, including Powertrain, Chassis and Infotainment.	Refer to vehicle specific information for identification of location and function of the specified systems / CAN-Bus components. The fault code indicates that one or more control modules has lost communication / received no response / received incorrect data messages from the Transmission Control Module. Refer to CAN-Bus information on page 51.
U0102	Lost Communication with Transfer Case Control Module	The Vehicle / CAN Communication Bus is a network for sharing signals and data / messages between the various vehicle systems, including Powertrain, Chassis and Infotainment.	Refer to vehicle specific information for identification of location and function of the specified systems / CAN-Bus components. The fault code indicates that one or more control modules has lost communication / received no response / received incorrect data messages from the Transfer Case Control Module. Refer to CAN-Bus information on page 51.
U0103	Lost Communication With Gear Shift Control Module "A"	The Vehicle / CAN Communication Bus is a network for sharing signals and data / messages between the various vehicle systems, including Powertrain, Chassis and Infotainment.	Refer to vehicle specific information for identification of location and function of the specified systems / CAN-Bus components. The fault code indicates that one or more control modules has lost communication / received no response / received incorrect data messages from the Gear Shift Control Module "A". Refer to CAN-Bus information on page 51.
U0104	Lost Communication With Cruise Control Module	The Vehicle / CAN Communication Bus is a network for sharing signals and data / messages between the various vehicle systems, including Powertrain, Chassis and Infotainment.	Refer to vehicle specific information for identification of location and function of the specified systems / CAN-Bus components. The fault code indicates that one or more control modules has lost communication / received no response / received incorrect data messages from the Cruise Control Module. Refer to CAN-Bus information on page 51.
U0105	Lost Communication With Fuel Injector Control Module	The Vehicle / CAN Communication Bus is a network for sharing signals and data / messages between the various vehicle systems, including Powertrain, Chassis and Infotainment.	Refer to vehicle specific information for identification of location and function of the specified systems / CAN-Bus components. The fault code indicates that one or more control modules has lost communication / received no response / received incorrect data messages from the Injector Control Module. Refer to CAN-Bus information on page 51.
U0106	Lost Communication With Glow Plug Control Module	The Vehicle / CAN Communication Bus is a network for sharing signals and data / messages between the various vehicle systems, including Powertrain, Chassis and Infotainment.	Refer to vehicle specific information for identification of location and function of the specified systems / CAN-Bus components. The fault code indicates that one or more control modules has lost communication / received no response / received incorrect data messages from the Glow Plug Control Module. Refer to CAN-Bus information on page 51.
U0107	Lost Communication With Throttle Actuator Control Module	The Vehicle / CAN Communication Bus is a network for sharing signals and data / messages between the various vehicle systems, including Powertrain, Chassis and Infotainment.	Refer to vehicle specific information for identification of location and function of the specified systems / CAN-Bus components. The fault code indicates that one or more control modules has lost communication / received no response / received incorrect data messages from the Throttle Actuator Control Module. Refer to CAN-Bus information on page 51.
U0108	Lost Communication With Alternative Fuel Control Module	The Vehicle / CAN Communication Bus is a network for sharing signals and data / messages between the various vehicle systems, including Powertrain, Chassis and Infotainment.	Refer to vehicle specific information for identification of location and function of the specified systems / CAN-Bus components. The fault code indicates that one or more control modules has lost communication / received no response / received incorrect data messages from the Alternative Fuel Control Module. Refer to CAN-Bus information on page 51.
U0109	Lost Communication With Fuel Pump Control Module	The Vehicle / CAN Communication Bus is a network for sharing signals and data / messages between the various vehicle systems, including Powertrain, Chassis and Infotainment.	Refer to vehicle specific information for identification of location and function of the specified systems / CAN-Bus components. The fault code indicates that one or more control modules has lost communication / received no response / received incorrect data messages from the Fuel Pump Control Module. Refer to CAN-Bus information on page 51.
U0110	Lost Communication With Drive Motor Control Module "A"	The Vehicle / CAN Communication Bus is a network for sharing signals and data / messages between the various vehicle systems, including Powertrain, Chassis and Infotainment.	Refer to vehicle specific information for identification of location and function of the specified systems / CAN-Bus components. The fault code indicates that one or more control modules has lost communication / received no response / received incorrect data messages from the Drive Motor Control Module "A". Refer to CAN-Bus information on page 51.
U0111	Lost Communication With Battery Energy Control Module "A"	The Vehicle / CAN Communication Bus is a network for sharing signals and data / messages between the various vehicle systems, including Powertrain, Chassis and Infotainment.	Refer to vehicle specific information for identification of location and function of the specified systems / CAN-Bus components. The fault code indicates that one or more control modules has lost communication / received no response / received incorrect data messages from the Battery Energy Control Module "A". Refer to CAN-Bus information on page 51.
U0112	Lost Communication With Battery Energy Control Module "B"	The Vehicle / CAN Communication Bus is a network for sharing signals and data / messages between the various vehicle systems, including Powertrain, Chassis and Infotainment.	Refer to vehicle specific information for identification of location and function of the specified systems / CAN-Bus components. The fault code indicates that one or more control modules has lost communication / received no response / received incorrect data messages from the Battery Energy Control Module "B". Refer to CAN-Bus information on page 51.

NOTE 1. Check for other fault codes that could provide additional information.   **NOTE 2.** Communication between control units can pass via a CAN-Bus system; refer to Page 51 for CAN-Bus checks.   **343**

NOTE 3. If a fault cannot be located, it is also possible that the control unit is at fault.   **NOTE 4.** Refer to Page 53 for list of pages that show the ISO standard for component locations e.g. Sensor A - Bank 1.

Fault code	EOBD / ISO Description	Component / System Description	Meaningful Description and Quick Check
U0113	Lost Communication With Emissions Critical Control Information	The Vehicle / CAN Communication Bus is a network for sharing signals and data / messages between the various vehicle systems, including Powertrain, Chassis and Infotainment.	Refer to vehicle specific information for identification of location and function of the specified systems / CAN-Bus components. The fault code indicates that one or more control modules has lost communication / received no response / received incorrect data messages from the Emissions Critical Control Information. Refer to CAN-Bus information on page 51.
U0114	Lost Communication With Four–Wheel Drive Clutch Control Module	The Vehicle / CAN Communication Bus is a network for sharing signals and data / messages between the various vehicle systems, including Powertrain, Chassis and Infotainment.	Refer to vehicle specific information for identification of location and function of the specified systems / CAN-Bus components. The fault code indicates that one or more control modules has lost communication / received no response / received incorrect data messages from the Four–Wheel Drive Clutch Control Module. Refer to CAN-Bus information on page 51.
U0115	Lost Communication With ECM / PCM "B"	The Vehicle / CAN Communication Bus is a network for sharing signals and data / messages between the various vehicle systems, including Powertrain, Chassis and Infotainment.	Refer to vehicle specific information for identification of location and function of the specified systems / CAN-Bus components. The fault code indicates that one or more control modules has lost communication / received no response / received incorrect data messages from the ECM / PCM "B". Refer to CAN-Bus information on page 51.
U0116	Lost Communication With Coolant Temperature Control Module	The Vehicle / CAN Communication Bus is a network for sharing signals and data / messages between the various vehicle systems, including Powertrain, Chassis and Infotainment.	Refer to vehicle specific information for identification of location and function of the specified systems / CAN-Bus components. The fault code indicates that one or more control modules has lost communication / received no response / received incorrect data messages from the Coolant Temperature Control Module. Refer to CAN-Bus information on page 51.
U0117	Lost Communication With Electrical PTO Control Module	The Vehicle / CAN Communication Bus is a network for sharing signals and data / messages between the various vehicle systems, including Powertrain, Chassis and Infotainment.	Refer to vehicle specific information for identification of location and function of the specified systems / CAN-Bus components. The fault code indicates that one or more control modules has lost communication / received no response / received incorrect data messages from the Electrical PTO Control Module. Refer to CAN-Bus information on page 51.
U0118	Lost Communication With Fuel Additive Control Module	The Vehicle / CAN Communication Bus is a network for sharing signals and data / messages between the various vehicle systems, including Powertrain, Chassis and Infotainment.	Refer to vehicle specific information for identification of location and function of the specified systems / CAN-Bus components. The fault code indicates that one or more control modules has lost communication / received no response / received incorrect data messages from the Fuel Additive Control Module. Refer to CAN-Bus information on page 51.
U0119	Lost Communication With Fuel Cell Control Module	The Vehicle / CAN Communication Bus is a network for sharing signals and data / messages between the various vehicle systems, including Powertrain, Chassis and Infotainment.	Refer to vehicle specific information for identification of location and function of the specified systems / CAN-Bus components. The fault code indicates that one or more control modules has lost communication / received no response / received incorrect data messages from the Fuel Cell Control Module. Refer to CAN-Bus information on page 51.
U0120	Lost Communication With Starter / Generator Control Module	The Vehicle / CAN Communication Bus is a network for sharing signals and data / messages between the various vehicle systems, including Powertrain, Chassis and Infotainment.	Refer to vehicle specific information for identification of location and function of the specified systems / CAN-Bus components. The fault code indicates that one or more control modules has lost communication / received no response / received incorrect data messages from the Starter / Generator Control Module. Refer to CAN-Bus information on page 51.
U0121	Lost Communication With Anti-Lock Brake System (ABS) Control Module	The Vehicle / CAN Communication Bus is a network for sharing signals and data / messages between the various vehicle systems, including Powertrain, Chassis and Infotainment.	Refer to vehicle specific information for identification of location and function of the specified systems / CAN-Bus components. The fault code indicates that one or more control modules has lost communication / received no response / received incorrect data messages from the Anti-Lock Brake System (ABS) Control Module. Refer to CAN-Bus information on page 51.
U0122	Lost Communication With Vehicle Dynamics Control Module	The Vehicle / CAN Communication Bus is a network for sharing signals and data / messages between the various vehicle systems, including Powertrain, Chassis and Infotainment.	Refer to vehicle specific information for identification of location and function of the specified systems / CAN-Bus components. The fault code indicates that one or more control modules has lost communication / received no response / received incorrect data messages from the Vehicle Dynamics Control Module. Refer to CAN-Bus information on page 51.
U0123	Lost Communication With Yaw Rate Sensor Module	The Vehicle / CAN Communication Bus is a network for sharing signals and data / messages between the various vehicle systems, including Powertrain, Chassis and Infotainment.	Refer to vehicle specific information for identification of location and function of the specified systems / CAN-Bus components. The fault code indicates that one or more control modules has lost communication / received no response / received incorrect data messages from the Yaw Rate Sensor Module. Refer to CAN-Bus information on page 51.
U0124	Lost Communication With Lateral Acceleration Sensor Module	The Vehicle / CAN Communication Bus is a network for sharing signals and data / messages between the various vehicle systems, including Powertrain, Chassis and Infotainment.	Refer to vehicle specific information for identification of location and function of the specified systems / CAN-Bus components. The fault code indicates that one or more control modules has lost communication / received no response / received incorrect data messages from the Lateral Acceleration Sensor Module. Refer to CAN-Bus information on page 51.
U0125	Lost Communication With Multi-axis Acceleration Sensor Module	The Vehicle / CAN Communication Bus is a network for sharing signals and data / messages between the various vehicle systems, including Powertrain, Chassis and Infotainment.	Refer to vehicle specific information for identification of location and function of the specified systems / CAN-Bus components. The fault code indicates that one or more control modules has lost communication / received no response / received incorrect data messages from the Multi-axis Acceleration Sensor Module. Refer to CAN-Bus information on page 51.
U0126	Lost Communication With Steering Angle Sensor Module	The Vehicle / CAN Communication Bus is a network for sharing signals and data / messages between the various vehicle systems, including Powertrain, Chassis and Infotainment.	Refer to vehicle specific information for identification of location and function of the specified systems / CAN-Bus components. The fault code indicates that one or more control modules has lost communication / received no response / received incorrect data messages from the Steering Angle Sensor Module. Refer to CAN-Bus information on page 51.
U0127	Lost Communication With Tire Pressure Monitor Module	The Vehicle / CAN Communication Bus is a network for sharing signals and data / messages between the various vehicle systems, including Powertrain, Chassis and Infotainment.	Refer to vehicle specific information for identification of location and function of the specified systems / CAN-Bus components. The fault code indicates that one or more control modules has lost communication / received no response / received incorrect data messages from the Tire Pressure Monitor Module. Refer to CAN-Bus information on page 51.

NOTE 1. Check for other fault codes that could provide additional information.   NOTE 2. Communication between control units can pass via a CAN-Bus system; refer to Page 51 for CAN-Bus checks.

Code	Description		
U0128	Lost Communication With Park Brake Control Module	The Vehicle / CAN Communication Bus is a network for sharing signals and data / messages between the various vehicle systems, including Powertrain, Chassis and Infotainment.	Refer to vehicle specific information for identification of location and function of the specified systems / CAN-Bus components. The fault code indicates that one or more control modules has lost communication / received no response / received incorrect data messages from the Park Brake Control Module. Refer to CAN-Bus information on page 51.
U0129	Lost Communication With Brake System Control Module	The Vehicle / CAN Communication Bus is a network for sharing signals and data / messages between the various vehicle systems, including Powertrain, Chassis and Infotainment.	Refer to vehicle specific information for identification of location and function of the specified systems / CAN-Bus components. The fault code indicates that one or more control modules has lost communication / received no response / received incorrect data messages from the Brake System Control Module. Refer to CAN-Bus information on page 51.
U0130	Lost Communication With Steering Effort Control Module	The Vehicle / CAN Communication Bus is a network for sharing signals and data / messages between the various vehicle systems, including Powertrain, Chassis and Infotainment.	Refer to vehicle specific information for identification of location and function of the specified systems / CAN-Bus components. The fault code indicates that one or more control modules has lost communication / received no response / received incorrect data messages from the Steering Effort Control Module. Refer to CAN-Bus information on page 51.
U0131	Lost Communication With Power Steering Control Module	The Vehicle / CAN Communication Bus is a network for sharing signals and data / messages between the various vehicle systems, including Powertrain, Chassis and Infotainment.	Refer to vehicle specific information for identification of location and function of the specified systems / CAN-Bus components. The fault code indicates that one or more control modules has lost communication / received no response / received incorrect data messages from the Power Steering Control Module. Refer to CAN-Bus information on page 51.
U0132	Lost Communication With Suspension Control Module	The Vehicle / CAN Communication Bus is a network for sharing signals and data / messages between the various vehicle systems, including Powertrain, Chassis and Infotainment.	Refer to vehicle specific information for identification of location and function of the specified systems / CAN-Bus components. The fault code indicates that one or more control modules has lost communication / received no response / received incorrect data messages from the Suspension Control Module. Refer to CAN-Bus information on page 51.
U0133	Lost Communication With Active Roll Control Module	The Vehicle / CAN Communication Bus is a network for sharing signals and data / messages between the various vehicle systems, including Powertrain, Chassis and Infotainment.	Refer to vehicle specific information for identification of location and function of the specified systems / CAN-Bus components. The fault code indicates that one or more control modules has lost communication / received no response / received incorrect data messages from the Active Roll Control Module. Refer to CAN-Bus information on page 51.
U0134	Lost Communication With Power Steering Control Module Rear	The Vehicle / CAN Communication Bus is a network for sharing signals and data / messages between the various vehicle systems, including Powertrain, Chassis and Infotainment.	Refer to vehicle specific information for identification of location and function of the specified systems / CAN-Bus components. The fault code indicates that one or more control modules has lost communication / received no response / received incorrect data messages from the Power Steering Control Module Rear. Refer to CAN-Bus information on page 51.
U0135	Lost Communication With Differential Control Module Front	The Vehicle / CAN Communication Bus is a network for sharing signals and data / messages between the various vehicle systems, including Powertrain, Chassis and Infotainment.	Refer to vehicle specific information for identification of location and function of the specified systems / CAN-Bus components. The fault code indicates that one or more control modules has lost communication / received no response / received incorrect data messages from the Differential Control Module Front. Refer to CAN-Bus information on page 51.
U0136	Lost Communication With Differential Control Module Rear	The Vehicle / CAN Communication Bus is a network for sharing signals and data / messages between the various vehicle systems, including Powertrain, Chassis and Infotainment.	Refer to vehicle specific information for identification of location and function of the specified systems / CAN-Bus components. The fault code indicates that one or more control modules has lost communication / received no response / received incorrect data messages from the Differential Control Module Rear. Refer to CAN-Bus information on page 51.
U0137	Lost Communication With Trailer Brake Control Module	The Vehicle / CAN Communication Bus is a network for sharing signals and data / messages between the various vehicle systems, including Powertrain, Chassis and Infotainment.	Refer to vehicle specific information for identification of location and function of the specified systems / CAN-Bus components. The fault code indicates that one or more control modules has lost communication / received no response / received incorrect data messages from the Trailer Brake Control Module. Refer to CAN-Bus information on page 51.
U0138	Lost Communication With All Terrain Control Module	The Vehicle / CAN Communication Bus is a network for sharing signals and data / messages between the various vehicle systems, including Powertrain, Chassis and Infotainment.	Refer to vehicle specific information for identification of location and function of the specified systems / CAN-Bus components. The fault code indicates that one or more control modules has lost communication / received no response / received incorrect data messages from the All Terrain Control Module. Refer to CAN-Bus information on page 51.
U0139		ISO / SAE reserved	Not used, reserved for future allocation.
U0140	Lost Communication With Body Control Module	The Vehicle / CAN Communication Bus is a network for sharing signals and data / messages between the various vehicle systems, including Powertrain, Chassis and Infotainment.	Refer to vehicle specific information for identification of location and function of the specified systems / CAN-Bus components. The fault code indicates that one or more control modules has lost communication / received no response / received incorrect data messages from the Communication With Body Control Module. Refer to CAN-Bus information on page 51.
U0141	Lost Communication With Body Control Module "A"	The Vehicle / CAN Communication Bus is a network for sharing signals and data / messages between the various vehicle systems, including Powertrain, Chassis and Infotainment.	Refer to vehicle specific information for identification of location and function of the specified systems / CAN-Bus components. The fault code indicates that one or more control modules has lost communication / received no response / received incorrect data messages from the Body Control Module "A". Refer to CAN-Bus information on page 51.
U0142	Lost Communication With Body Control Module "B"	The Vehicle / CAN Communication Bus is a network for sharing signals and data / messages between the various vehicle systems, including Powertrain, Chassis and Infotainment.	Refer to vehicle specific information for identification of location and function of the specified systems / CAN-Bus components. The fault code indicates that one or more control modules has lost communication / received no response / received incorrect data messages from the Body Control Module "B". Refer to CAN-Bus information on page 51.
U0143	Lost Communication With Body Control Module "C"	The Vehicle / CAN Communication Bus is a network for sharing signals and data / messages between the various vehicle systems, including Powertrain, Chassis and Infotainment.	Refer to vehicle specific information for identification of location and function of the specified systems / CAN-Bus components. The fault code indicates that one or more control modules has lost communication / received no response / received incorrect data messages from the Body Control Module "C". Refer to CAN-Bus information on page 51.
U0144	Lost Communication With Body Control Module "D"	The Vehicle / CAN Communication Bus is a network for sharing signals and data / messages between the various vehicle systems, including Powertrain, Chassis and Infotainment.	Refer to vehicle specific information for identification of location and function of the specified systems / CAN-Bus components. The fault code indicates that one or more control modules has lost communication / received no response / received incorrect data messages from the Body Control Module "D". Refer to CAN-Bus information on page 51.
U0145	Lost Communication With Body Control Module "E"	The Vehicle / CAN Communication Bus is a network for sharing signals and data / messages between the various vehicle systems, including Powertrain, Chassis and Infotainment.	Refer to vehicle specific information for identification of location and function of the specified systems / CAN-Bus components. The fault code indicates that one or more control modules has lost communication / received no response / received incorrect data messages from the Body Control Module "E". Refer to CAN-Bus information on page 51.

NOTE 1. Check for other fault codes that could provide additional information. **NOTE 2.** Communication between control units can pass via a CAN-Bus system; refer to Page 51 for CAN-Bus checks.

NOTE 3. If a fault cannot be located, it is also possible that the control unit is at fault. **NOTE 4.** Refer to Page 53 for list of pages that show the ISO standard for component locations e.g. Sensor A - Bank 1.

Fault code	EOBD / ISO Description	Component / System Description	Meaningful Description and Quick Check
U0146	Lost Communication With Gateway "A"	The Vehicle / CAN Communication Bus is a network for sharing signals and data / messages between the various vehicle systems, including Powertrain, Chassis and Infotainment. The "Gateway" can be the interface for the transfer of data between the high and low speed networks on the Vehicle / CAN Communication Bus. The gateway can also be used as the diagnostic interface because it receives all of the information transmitted within the network.	Refer to vehicle specific information for identification of location and function of the specified systems / CAN-Bus components. The fault code indicates that one or more control modules has lost communication / received no response / received incorrect data messages from the Gateway "A". Refer to CAN-Bus information on page 51.
U0147	Lost Communication With Gateway "B"	The Vehicle / CAN Communication Bus is a network for sharing signals and data / messages between the various vehicle systems, including Powertrain, Chassis and Infotainment. The "Gateway" can be the interface for the transfer of data between the high and low speed networks on the Vehicle / CAN Communication Bus. The gateway can also be used as the diagnostic interface because it receives all of the information transmitted within the network.	Refer to vehicle specific information for identification of location and function of the specified systems / CAN-Bus components. The fault code indicates that one or more control modules has lost communication / received no response / received incorrect data messages from the Gateway "B". Refer to CAN-Bus information on page 51.
U0148	Lost Communication With Gateway "C"	The Vehicle / CAN Communication Bus is a network for sharing signals and data / messages between the various vehicle systems, including Powertrain, Chassis and Infotainment. The "Gateway" can be the interface for the transfer of data between the high and low speed networks on the Vehicle / CAN Communication Bus. The gateway can also be used as the diagnostic interface because it receives all of the information transmitted within the network.	Refer to vehicle specific information for identification of location and function of the specified systems / CAN-Bus components. The fault code indicates that one or more control modules has lost communication / received no response / received incorrect data messages from the Gateway "C". Refer to CAN-Bus information on page 51.
U0149	Lost Communication With Gateway "D"	The Vehicle / CAN Communication Bus is a network for sharing signals and data / messages between the various vehicle systems, including Powertrain, Chassis and Infotainment. The "Gateway" can be the interface for the transfer of data between the high and low speed networks on the Vehicle / CAN Communication Bus. The gateway can also be used as the diagnostic interface because it receives all of the information transmitted within the network.	Refer to vehicle specific information for identification of location and function of the specified systems / CAN-Bus components. The fault code indicates that one or more control modules has lost communication / received no response / received incorrect data messages from the Gateway "D". Refer to CAN-Bus information on page 51.
U0150	Lost Communication With Gateway "E"	The Vehicle / CAN Communication Bus is a network for sharing signals and data / messages between the various vehicle systems, including Powertrain, Chassis and Infotainment. The "Gateway" can be the interface for the transfer of data between the high and low speed networks on the Vehicle / CAN Communication Bus. The gateway can also be used as the diagnostic interface because it receives all of the information transmitted within the network.	Refer to vehicle specific information for identification of location and function of the specified systems / CAN-Bus components. The fault code indicates that one or more control modules has lost communication / received no response / received incorrect data messages from the Gateway "E". Refer to CAN-Bus information on page 51.
U0151	Lost Communication With Restraints Control Module	The Vehicle / CAN Communication Bus is a network for sharing signals and data / messages between the various vehicle systems, including Powertrain, Chassis and Infotainment.	Refer to vehicle specific information for identification of location and function of the specified systems / CAN-Bus components. The fault code indicates that one or more control modules has lost communication / received no response / received incorrect data messages from the Restraints Control Module. Refer to CAN-Bus information on page 51.
U0152	Lost Communication With Side Restraints Control Module Left	The Vehicle / CAN Communication Bus is a network for sharing signals and data / messages between the various vehicle systems, including Powertrain, Chassis and Infotainment.	Refer to vehicle specific information for identification of location and function of the specified systems / CAN-Bus components. The fault code indicates that one or more control modules has lost communication / received no response / received incorrect data messages from the Side Restraints Control Module Left. Refer to CAN-Bus information on page 51.
U0153	Lost Communication With Side Restraints Control Module Right	The Vehicle / CAN Communication Bus is a network for sharing signals and data / messages between the various vehicle systems, including Powertrain, Chassis and Infotainment.	Refer to vehicle specific information for identification of location and function of the specified systems / CAN-Bus components. The fault code indicates that one or more control modules has lost communication / received no response / received incorrect data messages from the Side Restraints Control Module Right. Refer to CAN-Bus information on page 51.
U0154	Lost Communication With Restraints Occupant Classification System Module	The Vehicle / CAN Communication Bus is a network for sharing signals and data / messages between the various vehicle systems, including Powertrain, Chassis and Infotainment.	Refer to vehicle specific information for identification of location and function of the specified systems / CAN-Bus components. The fault code indicates that one or more control modules has lost communication / received no response / received incorrect data messages from the Restraints Occupant Classification System Module. Refer to CAN-Bus information on page 51.
U0155	Lost Communication With Instrument Panel Cluster (IPC) Control Module	The Vehicle / CAN Communication Bus is a network for sharing signals and data / messages between the various vehicle systems, including Powertrain, Chassis and Infotainment.	Refer to vehicle specific information for identification of location and function of the specified systems / CAN-Bus components. The fault code indicates that one or more control modules has lost communication / received no response / received incorrect data messages from the Instrument Panel Cluster (IPC) Control Module. Refer to CAN-Bus information on page 51.
U0156	Lost Communication With Information Centre "A"	The Vehicle / CAN Communication Bus is a network for sharing signals and data / messages between the various vehicle systems, including Powertrain, Chassis and Infotainment.	Refer to vehicle specific information for identification of location and function of the specified systems / CAN-Bus components. The fault code indicates that one or more control modules has lost communication / received no response / received incorrect data messages from the Information Centre "A". Refer to CAN-Bus information on page 51.
U0157	Lost Communication With Information Centre "B"	The Vehicle / CAN Communication Bus is a network for sharing signals and data / messages between the various vehicle systems, including Powertrain, Chassis and Infotainment.	Refer to vehicle specific information for identification of location and function of the specified systems / CAN-Bus components. The fault code indicates that one or more control modules has lost communication / received no response / received incorrect data messages from the Information Centre "B". Refer to CAN-Bus information on page 51.

NOTE 1. Check for other fault codes that could provide additional information.   **NOTE 2.** Communication between control units can pass via a CAN-Bus system; refer to Page 51 for CAN-Bus checks.

NOTE 3. If a fault cannot be located, it is also possible that the control unit is at fault.   **NOTE 4.** Refer to Page 53 for list of pages that show the ISO standard for component locations e.g. Sensor A - Bank 1.

Code	Description		
U0158	Lost Communication With Head Up Display	The Vehicle / CAN Communication Bus is a network for sharing signals and data / messages between the various vehicle systems, including Powertrain, Chassis and Infotainment.	Refer to vehicle specific information for identification of location and function of the specified systems / CAN-Bus components. The fault code indicates that one or more control modules has lost communication / received no response / received incorrect data messages from the Head Up Display. Refer to CAN-Bus information on page 51.
U0159	Lost Communication With Parking Assist Control Module "A"	The Vehicle / CAN Communication Bus is a network for sharing signals and data / messages between the various vehicle systems, including Powertrain, Chassis and Infotainment.	Refer to vehicle specific information for identification of location and function of the specified systems / CAN-Bus components. The fault code indicates that one or more control modules has lost communication / received no response / received incorrect data messages from the Parking Assist Control Module "A". Refer to CAN-Bus information on page 51.
U0160	Lost Communication With Audible Alert Control Module	The Vehicle / CAN Communication Bus is a network for sharing signals and data / messages between the various vehicle systems, including Powertrain, Chassis and Infotainment.	Refer to vehicle specific information for identification of location and function of the specified systems / CAN-Bus components. The fault code indicates that one or more control modules has lost communication / received no response / received incorrect data messages from the Audible Alert Control Module. Refer to CAN-Bus information on page 51.
U0161	Lost Communication With Compass Module	The Vehicle / CAN Communication Bus is a network for sharing signals and data / messages between the various vehicle systems, including Powertrain, Chassis and Infotainment.	Refer to vehicle specific information for identification of location and function of the specified systems / CAN-Bus components. The fault code indicates that one or more control modules has lost communication / received no response / received incorrect data messages from the Compass Module. Refer to CAN-Bus information on page 51.
U0162	Lost Communication With Navigation Display Module	The Vehicle / CAN Communication Bus is a network for sharing signals and data / messages between the various vehicle systems, including Powertrain, Chassis and Infotainment.	Refer to vehicle specific information for identification of location and function of the specified systems / CAN-Bus components. The fault code indicates that one or more control modules has lost communication / received no response / received incorrect data messages from the Navigation Display Module. Refer to CAN-Bus information on page 51.
U0163	Lost Communication With Navigation Control Module	The Vehicle / CAN Communication Bus is a network for sharing signals and data / messages between the various vehicle systems, including Powertrain, Chassis and Infotainment.	Refer to vehicle specific information for identification of location and function of the specified systems / CAN-Bus components. The fault code indicates that one or more control modules has lost communication / received no response / received incorrect data messages from the Navigation Control Module. Refer to CAN-Bus information on page 51.
U0164	Lost Communication With HVAC Control Module	The Vehicle / CAN Communication Bus is a network for sharing signals and data / messages between the various vehicle systems, including Powertrain, Chassis and Infotainment.	Refer to vehicle specific information for identification of location and function of the specified systems / CAN-Bus components. The fault code indicates that one or more control modules has lost communication / received no response / received incorrect data messages from the HVAC Control Module. Refer to CAN-Bus information on page 51.
U0165	Lost Communication With HVAC Control Module Rear	The Vehicle / CAN Communication Bus is a network for sharing signals and data / messages between the various vehicle systems, including Powertrain, Chassis and Infotainment.	Refer to vehicle specific information for identification of location and function of the specified systems / CAN-Bus components. The fault code indicates that one or more control modules has lost communication / received no response / received incorrect data messages from the HVAC Control Module Rear. Refer to CAN-Bus information on page 51.
U0166	Lost Communication With Auxiliary Heater Control Module	The Vehicle / CAN Communication Bus is a network for sharing signals and data / messages between the various vehicle systems, including Powertrain, Chassis and Infotainment.	Refer to vehicle specific information for identification of location and function of the specified systems / CAN-Bus components. The fault code indicates that one or more control modules has lost communication / received no response / received incorrect data messages from the Auxiliary Heater Control Module. Refer to CAN-Bus information on page 51.
U0167	Lost Communication With Vehicle Immobilizer Control Module	The Vehicle / CAN Communication Bus is a network for sharing signals and data / messages between the various vehicle systems, including Powertrain, Chassis and Infotainment.	Refer to vehicle specific information for identification of location and function of the specified systems / CAN-Bus components. The fault code indicates that one or more control modules has lost communication / received no response / received incorrect data messages from the Vehicle Immobilizer Control Module. Refer to CAN-Bus information on page 51.
U0168	Lost Communication With Vehicle Security Control Module	The Vehicle / CAN Communication Bus is a network for sharing signals and data / messages between the various vehicle systems, including Powertrain, Chassis and Infotainment.	Refer to vehicle specific information for identification of location and function of the specified systems / CAN-Bus components. The fault code indicates that one or more control modules has lost communication / received no response / received incorrect data messages from the Vehicle Security Control Module. Refer to CAN-Bus information on page 51.
U0169	Lost Communication With Sunroof Control Module	The Vehicle / CAN Communication Bus is a network for sharing signals and data / messages between the various vehicle systems, including Powertrain, Chassis and Infotainment.	Refer to vehicle specific information for identification of location and function of the specified systems / CAN-Bus components. The fault code indicates that one or more control modules has lost communication / received no response / received incorrect data messages from the Sunroof Control Module. Refer to CAN-Bus information on page 51.
U0170	Lost Communication With "Restraints System Sensor A"	The Vehicle / CAN Communication Bus is a network for sharing signals and data / messages between the various vehicle systems, including Powertrain, Chassis and Infotainment.	Refer to vehicle specific information for identification of location and function of the specified systems / CAN-Bus components. The fault code indicates that one or more control modules has lost communication / received no response / received incorrect data messages from the "Restraints System Sensor A". Refer to CAN-Bus information on page 51.
U0171	Lost Communication With "Restraints System Sensor B"	The Vehicle / CAN Communication Bus is a network for sharing signals and data / messages between the various vehicle systems, including Powertrain, Chassis and Infotainment.	Refer to vehicle specific information for identification of location and function of the specified systems / CAN-Bus components. The fault code indicates that one or more control modules has lost communication / received no response / received incorrect data messages from the "Restraints System Sensor B". Refer to CAN-Bus information on page 51.
U0172	Lost Communication With "Restraints System Sensor C"	The Vehicle / CAN Communication Bus is a network for sharing signals and data / messages between the various vehicle systems, including Powertrain, Chassis and Infotainment.	Refer to vehicle specific information for identification of location and function of the specified systems / CAN-Bus components. The fault code indicates that one or more control modules has lost communication / received no response / received incorrect data messages from the "Restraints System Sensor C". Refer to CAN-Bus information on page 51.
U0173	Lost Communication With "Restraints System Sensor D"	The Vehicle / CAN Communication Bus is a network for sharing signals and data / messages between the various vehicle systems, including Powertrain, Chassis and Infotainment.	Refer to vehicle specific information for identification of location and function of the specified systems / CAN-Bus components. The fault code indicates that one or more control modules has lost communication / received no response / received incorrect data messages from the "Restraints System Sensor D". Refer to CAN-Bus information on page 51.
U0174	Lost Communication With "Restraints System Sensor E"	The Vehicle / CAN Communication Bus is a network for sharing signals and data / messages between the various vehicle systems, including Powertrain, Chassis and Infotainment.	Refer to vehicle specific information for identification of location and function of the specified systems / CAN-Bus components. The fault code indicates that one or more control modules has lost communication / received no response / received incorrect data messages from the "Restraints System Sensor E". Refer to CAN-Bus information on page 51.

NOTE 1. Check for other fault codes that could provide additional information.

NOTE 3. If a fault cannot be located, it is also possible that the control unit is at fault.

NOTE 2. Communication between control units can pass via a CAN-Bus system; refer to Page 51 for CAN-Bus checks.

NOTE 4. Refer to Page 53 for list of pages that show the ISO standard for component locations e.g. Sensor A - Bank 1.

Fault code	EOBD / ISO Description	Component / System Description	Meaningful Description and Quick Check
U0175	Lost Communication With "Restraints System Sensor F"	The Vehicle / CAN Communication Bus is a network for sharing signals and data / messages between the various vehicle systems, including Powertrain, Chassis and Infotainment.	Refer to vehicle specific information for identification of location and function of the specified systems / CAN-Bus components. The fault code indicates that one or more control modules has lost communication / received no response / received incorrect data messages from the "Restraints System Sensor F". Refer to CAN-Bus information on page 51.
U0176	Lost Communication With "Restraints System Sensor G"	The Vehicle / CAN Communication Bus is a network for sharing signals and data / messages between the various vehicle systems, including Powertrain, Chassis and Infotainment.	Refer to vehicle specific information for identification of location and function of the specified systems / CAN-Bus components. The fault code indicates that one or more control modules has lost communication / received no response / received incorrect data messages from the "Restraints System Sensor G". Refer to CAN-Bus information on page 51.
U0177	Lost Communication With "Restraints System Sensor H"	The Vehicle / CAN Communication Bus is a network for sharing signals and data / messages between the various vehicle systems, including Powertrain, Chassis and Infotainment.	Refer to vehicle specific information for identification of location and function of the specified systems / CAN-Bus components. The fault code indicates that one or more control modules has lost communication / received no response / received incorrect data messages from the "Restraints System Sensor H". Refer to CAN-Bus information on page 51.
U0178	Lost Communication With "Restraints System Sensor I"	The Vehicle / CAN Communication Bus is a network for sharing signals and data / messages between the various vehicle systems, including Powertrain, Chassis and Infotainment.	Refer to vehicle specific information for identification of location and function of the specified systems / CAN-Bus components. The fault code indicates that one or more control modules has lost communication / received no response / received incorrect data messages from the "Restraints System Sensor I". Refer to CAN-Bus information on page 51.
U0179	Lost Communication With "Restraints System Sensor J"	The Vehicle / CAN Communication Bus is a network for sharing signals and data / messages between the various vehicle systems, including Powertrain, Chassis and Infotainment.	Refer to vehicle specific information for identification of location and function of the specified systems / CAN-Bus components. The fault code indicates that one or more control modules has lost communication / received no response / received incorrect data messages from the "Restraints System Sensor J". Refer to CAN-Bus information on page 51.
U017A	Lost Communication With "Restraints System Sensor K"	The Vehicle / CAN Communication Bus is a network for sharing signals and data / messages between the various vehicle systems, including Powertrain, Chassis and Infotainment.	Refer to vehicle specific information for identification of location and function of the specified systems / CAN-Bus components. The fault code indicates that one or more control modules has lost communication / received no response / received incorrect data messages from the "Restraints System Sensor K". Refer to CAN-Bus information on page 51.
U017B	Lost Communication With "Restraints System Sensor L"	The Vehicle / CAN Communication Bus is a network for sharing signals and data / messages between the various vehicle systems, including Powertrain, Chassis and Infotainment.	Refer to vehicle specific information for identification of location and function of the specified systems / CAN-Bus components. The fault code indicates that one or more control modules has lost communication / received no response / received incorrect data messages from the "Restraints System Sensor L". Refer to CAN-Bus information on page 51.
U017C	Lost Communication With "Restraints System Sensor M"	The Vehicle / CAN Communication Bus is a network for sharing signals and data / messages between the various vehicle systems, including Powertrain, Chassis and Infotainment.	Refer to vehicle specific information for identification of location and function of the specified systems / CAN-Bus components. The fault code indicates that one or more control modules has lost communication / received no response / received incorrect data messages from the "Restraints System Sensor M". Refer to CAN-Bus information on page 51.
U017D	Lost Communication With "Restraints System Sensor N"	The Vehicle / CAN Communication Bus is a network for sharing signals and data / messages between the various vehicle systems, including Powertrain, Chassis and Infotainment.	Refer to vehicle specific information for identification of location and function of the specified systems / CAN-Bus components. The fault code indicates that one or more control modules has lost communication / received no response / received incorrect data messages from the "Restraints System Sensor N". Refer to CAN-Bus information on page 51.
U0180	Lost Communication With Automatic Lighting Control Module	The Vehicle / CAN Communication Bus is a network for sharing signals and data / messages between the various vehicle systems, including Powertrain, Chassis and Infotainment.	Refer to vehicle specific information for identification of location and function of the specified systems / CAN-Bus components. The fault code indicates that one or more control modules has lost communication / received no response / received incorrect data messages from the Automatic Lighting Control Module. Refer to CAN-Bus information on page 51.
U0181	Lost Communication With Headlamp Levelling Control Module	The Vehicle / CAN Communication Bus is a network for sharing signals and data / messages between the various vehicle systems, including Powertrain, Chassis and Infotainment.	Refer to vehicle specific information for identification of location and function of the specified systems / CAN-Bus components. The fault code indicates that one or more control modules has lost communication / received no response / received incorrect data messages from the Headlamp Levelling Control Module. Refer to CAN-Bus information on page 51.
U0182	Lost Communication With Lighting Control Module Front	The Vehicle / CAN Communication Bus is a network for sharing signals and data / messages between the various vehicle systems, including Powertrain, Chassis and Infotainment.	Refer to vehicle specific information for identification of location and function of the specified systems / CAN-Bus components. The fault code indicates that one or more control modules has lost communication / received no response / received incorrect data messages from the Lighting Control Module Front. Refer to CAN-Bus information on page 51.
U0183	Lost Communication With Lighting Control Module Rear "A"	The Vehicle / CAN Communication Bus is a network for sharing signals and data / messages between the various vehicle systems, including Powertrain, Chassis and Infotainment.	Refer to vehicle specific information for identification of location and function of the specified systems / CAN-Bus components. The fault code indicates that one or more control modules has lost communication / received no response / received incorrect data messages from the Lighting Control Module Rear "A". Refer to CAN-Bus information on page 51.
U0184	Lost Communication With Radio	The Vehicle / CAN Communication Bus is a network for sharing signals and data / messages between the various vehicle systems, including Powertrain, Chassis and Infotainment.	Refer to vehicle specific information for identification of location and function of the specified systems / CAN-Bus components. The fault code indicates that one or more control modules has lost communication / received no response / received incorrect data messages from the Radio. Refer to CAN-Bus information on page 51.
U0185	Lost Communication With Antenna Control Module	The Vehicle / CAN Communication Bus is a network for sharing signals and data / messages between the various vehicle systems, including Powertrain, Chassis and Infotainment.	Refer to vehicle specific information for identification of location and function of the specified systems / CAN-Bus components. The fault code indicates that one or more control modules has lost communication / received no response / received incorrect data messages from the Antenna Control Module. Refer to CAN-Bus information on page 51.
U0186	Lost Communication With Audio Amplifier	The Vehicle / CAN Communication Bus is a network for sharing signals and data / messages between the various vehicle systems, including Powertrain, Chassis and Infotainment.	Refer to vehicle specific information for identification of location and function of the specified systems / CAN-Bus components. The fault code indicates that one or more control modules has lost communication / received no response / received incorrect data messages from the Audio Amplifier. Refer to CAN-Bus information on page 51.
U0187	Lost Communication With Digital Disc Player / Changer Module "A"	The Vehicle / CAN Communication Bus is a network for sharing signals and data / messages between the various vehicle systems, including Powertrain, Chassis and Infotainment.	Refer to vehicle specific information for identification of location and function of the specified systems / CAN-Bus components. The fault code indicates that one or more control modules has lost communication / received no response / received incorrect data messages from the Digital Disc Player / Changer Module "A". Refer to CAN-Bus information on page 51.

NOTE 1. Check for other fault codes that could provide additional information. **NOTE 2.** Communication between control units can pass via a CAN-Bus system; refer to Page 51 for CAN-Bus checks.

NOTE 3. If a fault cannot be located, it is also possible that the control unit is at fault NOTE 4. Refer to Page 53 for list of pages that show the ISO standard for component locations e.g. Sensor A - Bank 1.

		The Vehicle / CAN Communication Bus is a network for sharing signals and	Refer to vehicle specific information for identification of location and function of the specified systems / CAN-Bus components. The
U0188	Lost Communication With Digital Disc Player / Changer Module "B"	The Vehicle / CAN Communication Bus is a network for sharing signals and data / messages between the various vehicle systems, including Powertrain, Chassis and Infotainment.	Refer to vehicle specific information for identification of location and function of the specified systems / CAN-Bus components. The fault code indicates that one or more control modules has lost communication / received no response / received incorrect data messages from the Digital Disc Player / Changer Module "B". Refer to CAN-Bus information on page 51.
U0189	Lost Communication With Digital Disc Player / Changer Module "C"	The Vehicle / CAN Communication Bus is a network for sharing signals and data / messages between the various vehicle systems, including Powertrain, Chassis and Infotainment.	Refer to vehicle specific information for identification of location and function of the specified systems / CAN-Bus components. The fault code indicates that one or more control modules has lost communication / received no response / received incorrect data messages from the Digital Disc Player / Changer Module "C". Refer to CAN-Bus information on page 51.
U0190	Lost Communication With Digital Disc Player / Changer Module "D"	The Vehicle / CAN Communication Bus is a network for sharing signals and data / messages between the various vehicle systems, including Powertrain, Chassis and Infotainment.	Refer to vehicle specific information for identification of location and function of the specified systems / CAN-Bus components. The fault code indicates that one or more control modules has lost communication / received no response / received incorrect data messages from the Digital Disc Player / Changer Module "D". Refer to CAN-Bus information on page 51.
U0191	Lost Communication With Television	The Vehicle / CAN Communication Bus is a network for sharing signals and data / messages between the various vehicle systems, including Powertrain, Chassis and Infotainment.	Refer to vehicle specific information for identification of location and function of the specified systems / CAN-Bus components. The fault code indicates that one or more control modules has lost communication / received no response / received incorrect data messages from the Television. Refer to CAN-Bus information on page 51.
U0192	Lost Communication With Personal Computer	The Vehicle / CAN Communication Bus is a network for sharing signals and data / messages between the various vehicle systems, including Powertrain, Chassis and Infotainment.	Refer to vehicle specific information for identification of location and function of the specified systems / CAN-Bus components. The fault code indicates that one or more control modules has lost communication / received no response / received incorrect data messages from the Personal Computer. Refer to CAN-Bus information on page 51.
U0193	Lost Communication With "Digital Audio Control Module A"	The Vehicle / CAN Communication Bus is a network for sharing signals and data / messages between the various vehicle systems, including Powertrain, Chassis and Infotainment.	Refer to vehicle specific information for identification of location and function of the specified systems / CAN-Bus components. The fault code indicates that one or more control modules has lost communication / received no response / received incorrect data messages from the "Digital Audio Control Module A". Refer to CAN-Bus information on page 51.
U0194	Lost Communication With "Digital Audio Control Module B"	The Vehicle / CAN Communication Bus is a network for sharing signals and data / messages between the various vehicle systems, including Powertrain, Chassis and Infotainment.	Refer to vehicle specific information for identification of location and function of the specified systems / CAN-Bus components. The fault code indicates that one or more control modules has lost communication / received no response / received incorrect data messages from the "Digital Audio Control Module B". Refer to CAN-Bus information on page 51.
U0195	Lost Communication With Subscription Entertainment Receiver Module	The Vehicle / CAN Communication Bus is a network for sharing signals and data / messages between the various vehicle systems, including Powertrain, Chassis and Infotainment.	Refer to vehicle specific information for identification of location and function of the specified systems / CAN-Bus components. The fault code indicates that one or more control modules has lost communication / received no response / received incorrect data messages from the Subscription Entertainment Receiver Module. Refer to CAN-Bus information on page 51.
U0196	Lost Communication With Entertainment Control Module Rear "A"	The Vehicle / CAN Communication Bus is a network for sharing signals and data / messages between the various vehicle systems, including Powertrain, Chassis and Infotainment.	Refer to vehicle specific information for identification of location and function of the specified systems / CAN-Bus components. The fault code indicates that one or more control modules has lost communication / received no response / received incorrect data messages from the Entertainment Control Module Rear "A". Refer to CAN-Bus information on page 51.
U0197	Lost Communication With Telephone Control Module	The Vehicle / CAN Communication Bus is a network for sharing signals and data / messages between the various vehicle systems, including Powertrain, Chassis and Infotainment.	Refer to vehicle specific information for identification of location and function of the specified systems / CAN-Bus components. The fault code indicates that one or more control modules has lost communication / received no response / received incorrect data messages from the Telephone Control Module. Refer to CAN-Bus information on page 51.
U0198	Lost Communication With Telematic Control Module	The Vehicle / CAN Communication Bus is a network for sharing signals and data / messages between the various vehicle systems, including Powertrain, Chassis and Infotainment.	Refer to vehicle specific information for identification of location and function of the specified systems / CAN-Bus components. The fault code indicates that one or more control modules has lost communication / received no response / received incorrect data messages from the Telematic Control Module. Refer to CAN-Bus information on page 51.
U0199	Lost Communication With "Door Control Module A"	The Vehicle / CAN Communication Bus is a network for sharing signals and data / messages between the various vehicle systems, including Powertrain, Chassis and Infotainment.	Refer to vehicle specific information for identification of location and function of the specified systems / CAN-Bus components. The fault code indicates that one or more control modules has lost communication / received no response / received incorrect data messages from the "Door Control Module A". Refer to CAN-Bus information on page 51.
U0200	Lost Communication With "Door Control Module B"	The Vehicle / CAN Communication Bus is a network for sharing signals and data / messages between the various vehicle systems, including Powertrain, Chassis and Infotainment.	Refer to vehicle specific information for identification of location and function of the specified systems / CAN-Bus components. The fault code indicates that one or more control modules has lost communication / received no response / received incorrect data messages from the "Door Control Module B". Refer to CAN-Bus information on page 51.
U0201	Lost Communication With "Door Control Module C"	The Vehicle / CAN Communication Bus is a network for sharing signals and data / messages between the various vehicle systems, including Powertrain, Chassis and Infotainment.	Refer to vehicle specific information for identification of location and function of the specified systems / CAN-Bus components. The fault code indicates that one or more control modules has lost communication / received no response / received incorrect data messages from the "Door Control Module C". Refer to CAN-Bus information on page 51.
U0202	Lost Communication With "Door Control Module D"	The Vehicle / CAN Communication Bus is a network for sharing signals and data / messages between the various vehicle systems, including Powertrain, Chassis and Infotainment.	Refer to vehicle specific information for identification of location and function of the specified systems / CAN-Bus components. The fault code indicates that one or more control modules has lost communication / received no response / received incorrect data messages from the "Door Control Module D". Refer to CAN-Bus information on page 51.
U0203	Lost Communication With "Door Control Module E"	The Vehicle / CAN Communication Bus is a network for sharing signals and data / messages between the various vehicle systems, including Powertrain, Chassis and Infotainment.	Refer to vehicle specific information for identification of location and function of the specified systems / CAN-Bus components. The fault code indicates that one or more control modules has lost communication / received no response / received incorrect data messages from the "Door Control Module E". Refer to CAN-Bus information on page 51.

NOTE 1. Check for other fault codes that could provide additional information. **NOTE 2.** Communication between control units can pass via a CAN-Bus system; refer to Page 51 for CAN-Bus checks.

NOTE 3. If a fault cannot be located, it is also possible that the control unit is at fault. **NOTE 4.** Refer to Page 53 for list of pages that show the ISO standard for component locations e.g. Sensor A - Bank 1.

Fault code	EOBD / ISO Description	Component / System Description	Meaningful Description and Quick Check
U0204	Lost Communication With "Door Control Module F"	The Vehicle / CAN Communication Bus is a network for sharing signals and data / messages between the various vehicle systems, including Powertrain, Chassis and Infotainment.	Refer to vehicle specific information for identification of location and function of the specified systems / CAN-Bus components. The fault code indicates that one or more control modules has lost communication / received no response / received incorrect data messages from the "Door Control Module F". Refer to CAN-Bus information on page 51.
U0205	Lost Communication With "Door Control Module G"	The Vehicle / CAN Communication Bus is a network for sharing signals and data / messages between the various vehicle systems, including Powertrain, Chassis and Infotainment.	Refer to vehicle specific information for identification of location and function of the specified systems / CAN-Bus components. The fault code indicates that one or more control modules has lost communication / received no response / received incorrect data messages from the "Door Control Module G". Refer to CAN-Bus information on page 51.
U0206	Lost Communication With Folding Top Control Module	The Vehicle / CAN Communication Bus is a network for sharing signals and data / messages between the various vehicle systems, including Powertrain, Chassis and Infotainment.	Refer to vehicle specific information for identification of location and function of the specified systems / CAN-Bus components. The fault code indicates that one or more control modules has lost communication / received no response / received incorrect data messages from the Folding Top Control Module. Refer to CAN-Bus information on page 51.
U0207	Lost Communication With Moveable Roof Control Module	The Vehicle / CAN Communication Bus is a network for sharing signals and data / messages between the various vehicle systems, including Powertrain, Chassis and Infotainment.	Refer to vehicle specific information for identification of location and function of the specified systems / CAN-Bus components. The fault code indicates that one or more control modules has lost communication / received no response / received incorrect data messages from the Moveable Roof Control Module. Refer to CAN-Bus information on page 51.
U0208	Lost Communication With "Seat Control Module A"	The Vehicle / CAN Communication Bus is a network for sharing signals and data / messages between the various vehicle systems, including Powertrain, Chassis and Infotainment.	Refer to vehicle specific information for identification of location and function of the specified systems / CAN-Bus components. The fault code indicates that one or more control modules has lost communication / received no response / received incorrect data messages from the "Seat Control Module A". Refer to CAN-Bus information on page 51.
U0209	Lost Communication With "Seat Control Module B"	The Vehicle / CAN Communication Bus is a network for sharing signals and data / messages between the various vehicle systems, including Powertrain, Chassis and Infotainment.	Refer to vehicle specific information for identification of location and function of the specified systems / CAN-Bus components. The fault code indicates that one or more control modules has lost communication / received no response / received incorrect data messages from the "Seat Control Module B". Refer to CAN-Bus information on page 51.
U0210	Lost Communication With "Seat Control Module C"	The Vehicle / CAN Communication Bus is a network for sharing signals and data / messages between the various vehicle systems, including Powertrain, Chassis and Infotainment.	Refer to vehicle specific information for identification of location and function of the specified systems / CAN-Bus components. The fault code indicates that one or more control modules has lost communication / received no response / received incorrect data messages from the "Seat Control Module C". Refer to CAN-Bus information on page 51.
U0211	Lost Communication With "Seat Control Module D"	The Vehicle / CAN Communication Bus is a network for sharing signals and data / messages between the various vehicle systems, including Powertrain, Chassis and Infotainment.	Refer to vehicle specific information for identification of location and function of the specified systems / CAN-Bus components. The fault code indicates that one or more control modules has lost communication / received no response / received incorrect data messages from the "Seat Control Module D". Refer to CAN-Bus information on page 51.
U0212	Lost Communication With Steering Column Control Module	The Vehicle / CAN Communication Bus is a network for sharing signals and data / messages between the various vehicle systems, including Powertrain, Chassis and Infotainment.	Refer to vehicle specific information for identification of location and function of the specified systems / CAN-Bus components. The fault code indicates that one or more control modules has lost communication / received no response / received incorrect data messages from the Steering Column Control Module. Refer to CAN-Bus information on page 51.
U0213	Lost Communication With Mirror Control Module "A"	The Vehicle / CAN Communication Bus is a network for sharing signals and data / messages between the various vehicle systems, including Powertrain, Chassis and Infotainment.	Refer to vehicle specific information for identification of location and function of the specified systems / CAN-Bus components. The fault code indicates that one or more control modules has lost communication / received no response / received incorrect data messages from the Mirror Control Module "A". Refer to CAN-Bus information on page 51.
U0214	Lost Communication With Remote Function Actuation	The Vehicle / CAN Communication Bus is a network for sharing signals and data / messages between the various vehicle systems, including Powertrain, Chassis and Infotainment.	Refer to vehicle specific information for identification of location and function of the specified systems / CAN-Bus components. The fault code indicates that one or more control modules has lost communication / received no response / received incorrect data messages from the Remote Function Actuation. Refer to CAN-Bus information on page 51.
U0215	Lost Communication With "Door Switch A"	The Vehicle / CAN Communication Bus is a network for sharing signals and data / messages between the various vehicle systems, including Powertrain, Chassis and Infotainment.	Refer to vehicle specific information for identification of location and function of the specified systems / CAN-Bus components. The fault code indicates that one or more control modules has lost communication / received no response / received incorrect data messages from the "Door Switch A". Refer to CAN-Bus information on page 51.
U0216	Lost Communication With "Door Switch B"	The Vehicle / CAN Communication Bus is a network for sharing signals and data / messages between the various vehicle systems, including Powertrain, Chassis and Infotainment.	Refer to vehicle specific information for identification of location and function of the specified systems / CAN-Bus components. The fault code indicates that one or more control modules has lost communication / received no response / received incorrect data messages from the "Door Switch B". Refer to CAN-Bus information on page 51.
U0217	Lost Communication With "Door Switch C"	The Vehicle / CAN Communication Bus is a network for sharing signals and data / messages between the various vehicle systems, including Powertrain, Chassis and Infotainment.	Refer to vehicle specific information for identification of location and function of the specified systems / CAN-Bus components. The fault code indicates that one or more control modules has lost communication / received no response / received incorrect data messages from the "Door Switch C". Refer to CAN-Bus information on page 51.
U0218	Lost Communication With "Door Switch D"	The Vehicle / CAN Communication Bus is a network for sharing signals and data / messages between the various vehicle systems, including Powertrain, Chassis and Infotainment.	Refer to vehicle specific information for identification of location and function of the specified systems / CAN-Bus components. The fault code indicates that one or more control modules has lost communication / received no response / received incorrect data messages from the "Door Switch D". Refer to CAN-Bus information on page 51.
U0219	Lost Communication With "Door Switch E"	The Vehicle / CAN Communication Bus is a network for sharing signals and data / messages between the various vehicle systems, including Powertrain, Chassis and Infotainment.	Refer to vehicle specific information for identification of location and function of the specified systems / CAN-Bus components. The fault code indicates that one or more control modules has lost communication / received no response / received incorrect data messages from the "Door Switch E". Refer to CAN-Bus information on page 51.
U0220	Lost Communication With "Door Switch F"	The Vehicle / CAN Communication Bus is a network for sharing signals and data / messages between the various vehicle systems, including Powertrain, Chassis and Infotainment.	Refer to vehicle specific information for identification of location and function of the specified systems / CAN-Bus components. The fault code indicates that one or more control modules has lost communication / received no response / received incorrect data messages from the "Door Switch F". Refer to CAN-Bus information on page 51.
U0221	Lost Communication With "Door Switch G"	The Vehicle / CAN Communication Bus is a network for sharing signals and data / messages between the various vehicle systems, including Powertrain, Chassis and Infotainment.	Refer to vehicle specific information for identification of location and function of the specified systems / CAN-Bus components. The fault code indicates that one or more control modules has lost communication / received no response / received incorrect data messages from the "Door Switch G". Refer to CAN-Bus information on page 51.

NOTE 1. Check for other fault codes that could provide additional information. NOTE 2. Communication between control units can pass via a CAN-Bus system; refer to Page 51 for CAN-Bus checks.

Code	Description		
U0222	Lost Communication With "Door Window Motor A"	The Vehicle / CAN Communication Bus is a network for sharing signals and data / messages between the various vehicle systems, including Powertrain, Chassis and Infotainment.	Refer to vehicle specific information for identification of location and function of the specified systems / CAN-Bus components. The fault code indicates that one or more control modules has lost communication / received no response / received incorrect data messages from the "Door Window Motor A". Refer to CAN-Bus information on page 51.
U0223	Lost Communication With "Door Window Motor B"	The Vehicle / CAN Communication Bus is a network for sharing signals and data / messages between the various vehicle systems, including Powertrain, Chassis and Infotainment.	Refer to vehicle specific information for identification of location and function of the specified systems / CAN-Bus components. The fault code indicates that one or more control modules has lost communication / received no response / received incorrect data messages from the "Door Window Motor B". Refer to CAN-Bus information on page 51.
U0224	Lost Communication With "Door Window Motor C"	The Vehicle / CAN Communication Bus is a network for sharing signals and data / messages between the various vehicle systems, including Powertrain, Chassis and Infotainment.	Refer to vehicle specific information for identification of location and function of the specified systems / CAN-Bus components. The fault code indicates that one or more control modules has lost communication / received no response / received incorrect data messages from the "Door Window Motor C". Refer to CAN-Bus information on page 51.
U0225	Lost Communication With "Door Window Motor D"	The Vehicle / CAN Communication Bus is a network for sharing signals and data / messages between the various vehicle systems, including Powertrain, Chassis and Infotainment.	Refer to vehicle specific information for identification of location and function of the specified systems / CAN-Bus components. The fault code indicates that one or more control modules has lost communication / received no response / received incorrect data messages from the "Door Window Motor D". Refer to CAN-Bus information on page 51.
U0226	Lost Communication With "Door Window Motor E"	The Vehicle / CAN Communication Bus is a network for sharing signals and data / messages between the various vehicle systems, including Powertrain, Chassis and Infotainment.	Refer to vehicle specific information for identification of location and function of the specified systems / CAN-Bus components. The fault code indicates that one or more control modules has lost communication / received no response / received incorrect data messages from the "Door Window Motor E". Refer to CAN-Bus information on page 51.
U0227	Lost Communication With "Door Window Motor F"	The Vehicle / CAN Communication Bus is a network for sharing signals and data / messages between the various vehicle systems, including Powertrain, Chassis and Infotainment.	Refer to vehicle specific information for identification of location and function of the specified systems / CAN-Bus components. The fault code indicates that one or more control modules has lost communication / received no response / received incorrect data messages from the "Door Window Motor F". Refer to CAN-Bus information on page 51.
U0228	Lost Communication With "Door Window Motor G"	The Vehicle / CAN Communication Bus is a network for sharing signals and data / messages between the various vehicle systems, including Powertrain, Chassis and Infotainment.	Refer to vehicle specific information for identification of location and function of the specified systems / CAN-Bus components. The fault code indicates that one or more control modules has lost communication / received no response / received incorrect data messages from the "Door Window Motor G". Refer to CAN-Bus information on page 51.
U0229	Lost Communication With Heated Steering Wheel Module	The Vehicle / CAN Communication Bus is a network for sharing signals and data / messages between the various vehicle systems, including Powertrain, Chassis and Infotainment.	Refer to vehicle specific information for identification of location and function of the specified systems / CAN-Bus components. The fault code indicates that one or more control modules has lost communication / received no response / received incorrect data messages from the Heated Steering Wheel Module. Refer to CAN-Bus information on page 51.
U0230	Lost Communication With Rear Gate Module	The Vehicle / CAN Communication Bus is a network for sharing signals and data / messages between the various vehicle systems, including Powertrain, Chassis and Infotainment.	Refer to vehicle specific information for identification of location and function of the specified systems / CAN-Bus components. The fault code indicates that one or more control modules has lost communication / received no response / received incorrect data messages from the Rear Gate Module. Refer to CAN-Bus information on page 51.
U0231	Lost Communication With Rain Sensing Module	The Vehicle / CAN Communication Bus is a network for sharing signals and data / messages between the various vehicle systems, including Powertrain, Chassis and Infotainment.	Refer to vehicle specific information for identification of location and function of the specified systems / CAN-Bus components. The fault code indicates that one or more control modules has lost communication / received no response / received incorrect data messages from the Rain Sensing Module. Refer to CAN-Bus information on page 51.
U0232	Lost Communication With Side Obstacle Detection Control Module Left	The Vehicle / CAN Communication Bus is a network for sharing signals and data / messages between the various vehicle systems, including Powertrain, Chassis and Infotainment.	Refer to vehicle specific information for identification of location and function of the specified systems / CAN-Bus components. The fault code indicates that one or more control modules has lost communication / received no response / received incorrect data messages from the Side Obstacle Detection Control Module Left. Refer to CAN-Bus information on page 51.
U0233	Lost Communication With Side Obstacle Detection Control Module Right	The Vehicle / CAN Communication Bus is a network for sharing signals and data / messages between the various vehicle systems, including Powertrain, Chassis and Infotainment.	Refer to vehicle specific information for identification of location and function of the specified systems / CAN-Bus components. The fault code indicates that one or more control modules has lost communication / received no response / received incorrect data messages from the Side Obstacle Detection Control Module Right. Refer to CAN-Bus information on page 51.
U0234	Lost Communication With Convenience Recall Module	The Vehicle / CAN Communication Bus is a network for sharing signals and data / messages between the various vehicle systems, including Powertrain, Chassis and Infotainment.	Refer to vehicle specific information for identification of location and function of the specified systems / CAN-Bus components. The fault code indicates that one or more control modules has lost communication / received no response / received incorrect data messages from the Convenience Recall Module. Refer to CAN-Bus information on page 51.
U0235	Lost Communication With Cruise Control Front Distance Range Sensor	The Vehicle / CAN Communication Bus is a network for sharing signals and data / messages between the various vehicle systems, including Powertrain, Chassis and Infotainment.	Refer to vehicle specific information for identification of location and function of the specified systems / CAN-Bus components. The fault code indicates that one or more control modules has lost communication / received no response / received incorrect data messages from the Cruise Control Front Distance Range Sensor. Refer to CAN-Bus information on page 51.
U0236	Lost Communication With Column Lock Module	The Vehicle / CAN Communication Bus is a network for sharing signals and data / messages between the various vehicle systems, including Powertrain, Chassis and Infotainment.	Refer to vehicle specific information for identification of location and function of the specified systems / CAN-Bus components. The fault code indicates that one or more control modules has lost communication / received no response / received incorrect data messages from the Column Lock Module. Refer to CAN-Bus information on page 51.
U0237	Lost Communication With "Digital Audio Control Module C"	The Vehicle / CAN Communication Bus is a network for sharing signals and data / messages between the various vehicle systems, including Powertrain, Chassis and Infotainment.	Refer to vehicle specific information for identification of location and function of the specified systems / CAN-Bus components. The fault code indicates that one or more control modules has lost communication / received no response / received incorrect data messages from the "Digital Audio Control Module C". Refer to CAN-Bus information on page 51.
U0238	Lost Communication With "Digital Audio Control Module D"	The Vehicle / CAN Communication Bus is a network for sharing signals and data / messages between the various vehicle systems, including Powertrain, Chassis and Infotainment.	Refer to vehicle specific information for identification of location and function of the specified systems / CAN-Bus components. The fault code indicates that one or more control modules has lost communication / received no response / received incorrect data messages from the "Digital Audio Control Module D". Refer to CAN-Bus information on page 51.

NOTE 1. Check for other fault codes that could provide additional information. **NOTE 2.** Communication between control units can pass via a CAN-Bus system; refer to Page 51 for CAN-Bus checks.

NOTE 3. If a fault cannot be located, it is also possible that the control unit is at fault. **NOTE 4.** Refer to Page 53 for list of pages that show the ISO standard for component locations e.g. Sensor A - Bank 1.

Fault code	EOBD / ISO Description	Component / System Description	Meaningful Description and Quick Check
U0239	Lost Communication With Entrapment Control Module "A"	The Vehicle / CAN Communication Bus is a network for sharing signals and data / messages between the various vehicle systems, including Powertrain, Chassis and Infotainment.	Refer to vehicle specific information for identification of location and function of the specified systems / CAN-Bus components. The fault code indicates that one or more control modules has lost communication / received no response / received incorrect data messages from the Entrapment Control Module "A". Refer to CAN-Bus information on page 51.
U0240	Lost Communication With Entrapment Control Module "B"	The Vehicle / CAN Communication Bus is a network for sharing signals and data / messages between the various vehicle systems, including Powertrain, Chassis and Infotainment.	Refer to vehicle specific information for identification of location and function of the specified systems / CAN-Bus components. The fault code indicates that one or more control modules has lost communication / received no response / received incorrect data messages from the Entrapment Control Module "B". Refer to CAN-Bus information on page 51.
U0241	Lost Communication With Headlamp Control Module "A"	The Vehicle / CAN Communication Bus is a network for sharing signals and data / messages between the various vehicle systems, including Powertrain, Chassis and Infotainment.	Refer to vehicle specific information for identification of location and function of the specified systems / CAN-Bus components. The fault code indicates that one or more control modules has lost communication / received no response / received incorrect data messages from the Headlamp Control Module "A". Refer to CAN-Bus information on page 51.
U0242	Lost Communication With Headlamp Control Module "B"	The Vehicle / CAN Communication Bus is a network for sharing signals and data / messages between the various vehicle systems, including Powertrain, Chassis and Infotainment.	Refer to vehicle specific information for identification of location and function of the specified systems / CAN-Bus components. The fault code indicates that one or more control modules has lost communication / received no response / received incorrect data messages from the Headlamp Control Module "B". Refer to CAN-Bus information on page 51.
U0243	Lost Communication With Parking Assist Control Module "B"	The Vehicle / CAN Communication Bus is a network for sharing signals and data / messages between the various vehicle systems, including Powertrain, Chassis and Infotainment.	Refer to vehicle specific information for identification of location and function of the specified systems / CAN-Bus components. The fault code indicates that one or more control modules has lost communication / received no response / received incorrect data messages from the Parking Assist Control Module "B". Refer to CAN-Bus information on page 51.
U0244	Lost Communication With Running Board Control Module "A"	The Vehicle / CAN Communication Bus is a network for sharing signals and data / messages between the various vehicle systems, including Powertrain, Chassis and Infotainment.	Refer to vehicle specific information for identification of location and function of the specified systems / CAN-Bus components. The fault code indicates that one or more control modules has lost communication / received no response / received incorrect data messages from the Running Board Control Module "A". Refer to CAN-Bus information on page 51.
U0245	Lost Communication With Entertainment Control Module Front	The Vehicle / CAN Communication Bus is a network for sharing signals and data / messages between the various vehicle systems, including Powertrain, Chassis and Infotainment.	Refer to vehicle specific information for identification of location and function of the specified systems / CAN-Bus components. The fault code indicates that one or more control modules has lost communication / received no response / received incorrect data messages from the Entertainment Control Module Front. Refer to CAN-Bus information on page 51.
U0246	Lost Communication With Seat Control Module "E"	The Vehicle / CAN Communication Bus is a network for sharing signals and data / messages between the various vehicle systems, including Powertrain, Chassis and Infotainment.	Refer to vehicle specific information for identification of location and function of the specified systems / CAN-Bus components. The fault code indicates that one or more control modules has lost communication / received no response / received incorrect data messages from the Seat Control Module "E". Refer to CAN-Bus information on page 51.
U0247	Lost Communication With Seat Control Module "F"	The Vehicle / CAN Communication Bus is a network for sharing signals and data / messages between the various vehicle systems, including Powertrain, Chassis and Infotainment.	Refer to vehicle specific information for identification of location and function of the specified systems / CAN-Bus components. The fault code indicates that one or more control modules has lost communication / received no response / received incorrect data messages from the Seat Control Module "F". Refer to CAN-Bus information on page 51.
U0248	Lost Communication With Remote Accessory Module	The Vehicle / CAN Communication Bus is a network for sharing signals and data / messages between the various vehicle systems, including Powertrain, Chassis and Infotainment.	Refer to vehicle specific information for identification of location and function of the specified systems / CAN-Bus components. The fault code indicates that one or more control modules has lost communication / received no response / received incorrect data messages from the Remote Accessory Module. Refer to CAN-Bus information on page 51.
U0249	Lost Communication With Entertainment Control Module Rear "B"	The Vehicle / CAN Communication Bus is a network for sharing signals and data / messages between the various vehicle systems, including Powertrain, Chassis and Infotainment.	Refer to vehicle specific information for identification of location and function of the specified systems / CAN-Bus components. The fault code indicates that one or more control modules has lost communication / received no response / received incorrect data messages from the Entertainment Control Module Rear "B". Refer to CAN-Bus information on page 51.
U0250	Lost Communication With Impact Classification System Module	The Vehicle / CAN Communication Bus is a network for sharing signals and data / messages between the various vehicle systems, including Powertrain, Chassis and Infotainment.	Refer to vehicle specific information for identification of location and function of the specified systems / CAN-Bus components. The fault code indicates that one or more control modules has lost communication / received no response / received incorrect data messages from the Impact Classification System Module. Refer to CAN-Bus information on page 51.
U0251	Lost Communication With Running Board Control Module "B"	The Vehicle / CAN Communication Bus is a network for sharing signals and data / messages between the various vehicle systems, including Powertrain, Chassis and Infotainment.	Refer to vehicle specific information for identification of location and function of the specified systems / CAN-Bus components. The fault code indicates that one or more control modules has lost communication / received no response / received incorrect data messages from the Running Board Control Module "B". Refer to CAN-Bus information on page 51.
U0252	Lost Communication With Lighting Control Module Rear "B"	The Vehicle / CAN Communication Bus is a network for sharing signals and data / messages between the various vehicle systems, including Powertrain, Chassis and Infotainment.	Refer to vehicle specific information for identification of location and function of the specified systems / CAN-Bus components. The fault code indicates that one or more control modules has lost communication / received no response / received incorrect data messages from the Lighting Control Module Rear "B". Refer to CAN-Bus information on page 51.
U0253	ISO / SAE reserved	Not used, reserved for future allocation.	
U0254	ISO / SAE reserved	Not used, reserved for future allocation.	
U0255	ISO / SAE reserved	Not used, reserved for future allocation.	
U0256	ISO / SAE reserved	Not used, reserved for future allocation.	
U0257	ISO / SAE reserved	Not used, reserved for future allocation.	
U0258	ISO / SAE reserved	Not used, reserved for future allocation.	
U0259	ISO / SAE reserved	Not used, reserved for future allocation.	
U0260	ISO / SAE reserved	Not used, reserved for future allocation.	
U0261	ISO / SAE reserved	Not used, reserved for future allocation.	

NOTE 1. Check for other fault codes that could provide additional information.   **NOTE 2.** Communication between control units can pass via a CAN-Bus system; refer to Page 51 for CAN-Bus checks.

NOTE 3. If a fault cannot be located, it is also possible that the control unit is at fault   **NOTE 4.** Refer to Page 53 for list of pages that show the ISO standard for component locations e.g. Sensor A - Bank 1.

U0262		ISO / SAE reserved	Not used, reserved for future allocation.
U0263		ISO / SAE reserved	Not used, reserved for future allocation.
U0264		ISO / SAE reserved	Not used, reserved for future allocation.
U0265		ISO / SAE reserved	Not used, reserved for future allocation.
U0266		ISO / SAE reserved	Not used, reserved for future allocation.
U0267		ISO / SAE reserved	Not used, reserved for future allocation.
U0268		ISO / SAE reserved	Not used, reserved for future allocation.
U0269		ISO / SAE reserved	Not used, reserved for future allocation.
U0270		ISO / SAE reserved	Not used, reserved for future allocation.
U0271		ISO / SAE reserved	Not used, reserved for future allocation.
U0272		ISO / SAE reserved	Not used, reserved for future allocation.
U0273		ISO / SAE reserved	Not used, reserved for future allocation.
U0274		ISO / SAE reserved	Not used, reserved for future allocation.
U0275		ISO / SAE reserved	Not used, reserved for future allocation.
U0276		ISO / SAE reserved	Not used, reserved for future allocation.
U0277		ISO / SAE reserved	Not used, reserved for future allocation.
U0278		ISO / SAE reserved	Not used, reserved for future allocation.
U0279		ISO / SAE reserved	Not used, reserved for future allocation.
U0280		ISO / SAE reserved	Not used, reserved for future allocation.
U0281		ISO / SAE reserved	Not used, reserved for future allocation.
U0282		ISO / SAE reserved	Not used, reserved for future allocation.
U0283		ISO / SAE reserved	Not used, reserved for future allocation.
U0284		ISO / SAE reserved	Not used, reserved for future allocation.
U0285		ISO / SAE reserved	Not used, reserved for future allocation.
U0286		ISO / SAE reserved	Not used, reserved for future allocation.
U0287		ISO / SAE reserved	Not used, reserved for future allocation.
U0288		ISO / SAE reserved	Not used, reserved for future allocation.
U0289		ISO / SAE reserved	Not used, reserved for future allocation.
U0290		ISO / SAE reserved	Not used, reserved for future allocation.
U0291	Lost Communication With Gear Shift Control Module "B"	The Vehicle / CAN Communication Bus is a network for sharing signals and data / messages between the various vehicle systems, including Powertrain, Chassis and Infotainment.	Refer to vehicle specific information for identification of location and function of the specified systems / CAN-Bus components. The fault code indicates that one or more control modules has lost communication / received no response / received incorrect data messages from the Gear Shift Control Module "B". Refer to CAN-Bus information on page 51.
U0292	Lost Communication With Drive Motor Control Module "B"	The Vehicle / CAN Communication Bus is a network for sharing signals and data / messages between the various vehicle systems, including Powertrain, Chassis and Infotainment.	Refer to vehicle specific information for identification of location and function of the specified systems / CAN-Bus components. The fault code indicates that one or more control modules has lost communication / received no response / received incorrect data messages from the Drive Motor Control Module "B". Refer to CAN-Bus information on page 51.
U0293	Lost Communication With Hybrid Powertrain Control Module	The Vehicle / CAN Communication Bus is a network for sharing signals and data / messages between the various vehicle systems, including Powertrain, Chassis and Infotainment.	Refer to vehicle specific information for identification of location and function of the specified systems / CAN-Bus components. The fault code indicates that one or more control modules has lost communication / received no response / received incorrect data messages from the Hybrid Powertrain Control Module. Refer to CAN-Bus information on page 51.
U0294	Lost Communication With Powertrain Control Monitor Module	The Vehicle / CAN Communication Bus is a network for sharing signals and data / messages between the various vehicle systems, including Powertrain, Chassis and Infotainment.	Refer to vehicle specific information for identification of location and function of the specified systems / CAN-Bus components. The fault code indicates that one or more control modules has lost communication / received no response / received incorrect data messages from the Powertrain Control Monitor Module. Refer to CAN-Bus information on page 51.
U0295	Lost Communication With AC to AC Converter Control Module	The Vehicle / CAN Communication Bus is a network for sharing signals and data / messages between the various vehicle systems, including Powertrain, Chassis and Infotainment.	Refer to vehicle specific information for identification of location and function of the specified systems / CAN-Bus components. The fault code indicates that one or more control modules has lost communication / received no response / received incorrect data messages from the AC to AC Converter Control Module. Refer to CAN-Bus information on page 51.

NOTE 1. Check for other fault codes that could provide additional information. **NOTE 2.** Communication between control units can pass via a CAN-Bus system; refer to Page 51 for CAN-Bus checks.

NOTE 3. If a fault cannot be located, it is also possible that the control unit is at fault. **NOTE 4.** Refer to Page 53 for list of pages that show the ISO standard for component locations e.g. Sensor A - Bank 1.

Fault code	EOBD / ISO Description	Component / System Description	Meaningful Description and Quick Check
U0296	Lost Communication With AC to DC Converter Control Module "A"	The Vehicle / CAN Communication Bus is a network for sharing signals and data / messages between the various vehicle systems, including Powertrain, Chassis and Infotainment.	Refer to vehicle specific information for identification of location and function of the specified systems / CAN-Bus components. The fault code indicates that one or more control modules has lost communication / received no response / received incorrect data messages from the AC to DC Converter Control Module "A". Refer to CAN-Bus information on page 51.
U0297	Lost Communication With AC to DC Converter Control Module "B"	The Vehicle / CAN Communication Bus is a network for sharing signals and data / messages between the various vehicle systems, including Powertrain, Chassis and Infotainment.	Refer to vehicle specific information for identification of location and function of the specified systems / CAN-Bus components. The fault code indicates that one or more control modules has lost communication / received no response / received incorrect data messages from the AC to DC Converter Control Module "B". Refer to CAN-Bus information on page 51.
U0298	Lost Communication With DC to DC Converter Control Module "A"	The Vehicle / CAN Communication Bus is a network for sharing signals and data / messages between the various vehicle systems, including Powertrain, Chassis and Infotainment.	Refer to vehicle specific information for identification of location and function of the specified systems / CAN-Bus components. The fault code indicates that one or more control modules has lost communication / received no response / received incorrect data messages from the DC to DC Converter Control Module "A". Refer to CAN-Bus information on page 51.
U0299	Lost Communication With DC to DC Converter Control Module "B"	The Vehicle / CAN Communication Bus is a network for sharing signals and data / messages between the various vehicle systems, including Powertrain, Chassis and Infotainment.	Refer to vehicle specific information for identification of location and function of the specified systems / CAN-Bus components. The fault code indicates that one or more control modules has lost communication / received no response / received incorrect data messages from the DC to DC Converter Control Module "B". Refer to CAN-Bus information on page 51.
U0300	Internal Control Module Software Incompatibility	The Vehicle / CAN Communication Bus is a network for sharing signals and data / messages between the various vehicle systems, including Powertrain, Chassis and Infotainment. The vehicle specific software is programmed into the computer to perform various system related functions.	Refer to vehicle specific information for identification of location and function of the specified systems / CAN-Bus components. The fault code indicates that the Internal Control Module software is incompatible with one or more systems. Refer to CAN-Bus information on page 51.
U0301	Software Incompatibility With ECM / PCM	The Vehicle / CAN Communication Bus is a network for sharing signals and data / messages between the various vehicle systems, including Powertrain, Chassis and Infotainment. The vehicle specific software is programmed into the computer to perform various system related functions.	Refer to vehicle specific information for identification of location and function of the specified systems / CAN-Bus components. The fault code indicates that one or more systems have been programmed with software that is incompatible with the Engine ECM / PCM. Refer to CAN-Bus information on page 51.
U0302	Software Incompatibility With Transmission Control Module	The Vehicle / CAN Communication Bus is a network for sharing signals and data / messages between the various vehicle systems, including Powertrain, Chassis and Infotainment. The vehicle specific software is programmed into the computer to perform various system related functions.	Refer to vehicle specific information for identification of location and function of the specified systems / CAN-Bus components. The fault code indicates that one or more systems have been programmed with software that is incompatible with the Transmission Control Module. Refer to CAN-Bus information on page 51.
U0303	Software Incompatibility With Transfer Case Control Module	The Vehicle / CAN Communication Bus is a network for sharing signals and data / messages between the various vehicle systems, including Powertrain, Chassis and Infotainment. The vehicle specific software is programmed into the computer to perform various system related functions.	Refer to vehicle specific information for identification of location and function of the specified systems / CAN-Bus components. The fault code indicates that one or more systems have been programmed with software that is incompatible with the Transfer Case Control Module. Refer to CAN-Bus information on page 51.
U0304	Software Incompatibility With Gear Shift Control Module "A"	The Vehicle / CAN Communication Bus is a network for sharing signals and data / messages between the various vehicle systems, including Powertrain, Chassis and Infotainment. The vehicle specific software is programmed into the computer to perform various system related functions.	Refer to vehicle specific information for identification of location and function of the specified systems / CAN-Bus components. The fault code indicates that one or more systems have been programmed with software that is incompatible with the Gear Shift Control Module "A". Refer to CAN-Bus information on page 51.
U0305	Software Incompatibility With Cruise Control Module	The Vehicle / CAN Communication Bus is a network for sharing signals and data / messages between the various vehicle systems, including Powertrain, Chassis and Infotainment. The vehicle specific software is programmed into the computer to perform various system related functions.	Refer to vehicle specific information for identification of location and function of the specified systems / CAN-Bus components. The fault code indicates that one or more systems have been programmed with software that is incompatible with the Cruise Control Module. Refer to CAN-Bus information on page 51.
U0306	Software Incompatibility With Fuel Injector Control Module	The Vehicle / CAN Communication Bus is a network for sharing signals and data / messages between the various vehicle systems, including Powertrain, Chassis and Infotainment. The vehicle specific software is programmed into the computer to perform various system related functions.	Refer to vehicle specific information for identification of location and function of the specified systems / CAN-Bus components. The fault code indicates that one or more systems have been programmed with software that is incompatible with the Fuel Injector Control Module. Refer to CAN-Bus information on page 51.
U0307	Software Incompatibility With Glow Plug Control Module	The Vehicle / CAN Communication Bus is a network for sharing signals and data / messages between the various vehicle systems, including Powertrain, Chassis and Infotainment. The vehicle specific software is programmed into the computer to perform various system related functions.	Refer to vehicle specific information for identification of location and function of the specified systems / CAN-Bus components. The fault code indicates that one or more systems have been programmed with software that is incompatible with the Glow Plug Control Module. Refer to CAN-Bus information on page 51.
U0308	Software Incompatibility With Throttle Actuator Control Module	The Vehicle / CAN Communication Bus is a network for sharing signals and data / messages between the various vehicle systems, including Powertrain, Chassis and Infotainment. The vehicle specific software is programmed into the computer to perform various system related functions.	Refer to vehicle specific information for identification of location and function of the specified systems / CAN-Bus components. The fault code indicates that one or more systems have been programmed with software that is incompatible with the Throttle Actuator Control Module. Refer to CAN-Bus information on page 51.

NOTE 1. Check for other fault codes that could provide additional information.   **NOTE 2.** Communication between control units can pass via a CAN-Bus system; refer to Page 51 for CAN-Bus checks.

U0309	Software Incompatibility With Alternative Fuel Control Module	The Vehicle / CAN Communication Bus is a network for sharing signals and data / messages between the various vehicle systems, including Powertrain, Chassis and Infotainment. The vehicle specific software is programmed into the computer to perform various system related functions.	Refer to vehicle specific information for identification of location and function of the specified systems / CAN-Bus components. The fault code indicates that one or more systems have been programmed with software that is incompatible with the Alternative Fuel Control Module. Refer to CAN-Bus information on page 51.
U0310	Software Incompatibility With Fuel Pump Control Module	The Vehicle / CAN Communication Bus is a network for sharing signals and data / messages between the various vehicle systems, including Powertrain, Chassis and Infotainment. The vehicle specific software is programmed into the computer to perform various system related functions.	Refer to vehicle specific information for identification of location and function of the specified systems / CAN-Bus components. The fault code indicates that one or more systems have been programmed with software that is incompatible with the Fuel Pump Control Module. Refer to CAN-Bus information on page 51.
U0311	Software Incompatibility With Drive Motor Control Module	The Vehicle / CAN Communication Bus is a network for sharing signals and data / messages between the various vehicle systems, including Powertrain, Chassis and Infotainment. The vehicle specific software is programmed into the computer to perform various system related functions.	Refer to vehicle specific information for identification of location and function of the specified systems / CAN-Bus components. The fault code indicates that one or more systems have been programmed with software that is incompatible with the Drive Motor Control Module. Refer to CAN-Bus information on page 51.
U0312	Software Incompatibility With Battery Energy Control Module A	The Vehicle / CAN Communication Bus is a network for sharing signals and data / messages between the various vehicle systems, including Powertrain, Chassis and Infotainment. The vehicle specific software is programmed into the computer to perform various system related functions.	Refer to vehicle specific information for identification of location and function of the specified systems / CAN-Bus components. The fault code indicates that one or more systems have been programmed with software that is incompatible with the Battery Energy Control Module A. Refer to CAN-Bus information on page 51.
U0313	Software Incompatibility With Battery Energy Control Module B	The Vehicle / CAN Communication Bus is a network for sharing signals and data / messages between the various vehicle systems, including Powertrain, Chassis and Infotainment. The vehicle specific software is programmed into the computer to perform various system related functions.	Refer to vehicle specific information for identification of location and function of the specified systems / CAN-Bus components. The fault code indicates that one or more systems have been programmed with software that is incompatible with the Battery Energy Control Module B. Refer to CAN-Bus information on page 51.
U0314	Software Incompatibility With Four-Wheel Drive Clutch Control Module	The Vehicle / CAN Communication Bus is a network for sharing signals and data / messages between the various vehicle systems, including Powertrain, Chassis and Infotainment. The vehicle specific software is programmed into the computer to perform various system related functions.	Refer to vehicle specific information for identification of location and function of the specified systems / CAN-Bus components. The fault code indicates that one or more systems have been programmed with software that is incompatible with the Four-Wheel Drive Clutch Control Module. Refer to CAN-Bus information on page 51.
U0315	Software Incompatibility With Anti-Lock Brake System Control Module	The Vehicle / CAN Communication Bus is a network for sharing signals and data / messages between the various vehicle systems, including Powertrain, Chassis and Infotainment. The vehicle specific software is programmed into the computer to perform various system related functions.	Refer to vehicle specific information for identification of location and function of the specified systems / CAN-Bus components. The fault code indicates that one or more systems have been programmed with software that is incompatible with the Anti-Lock Brake System Control Module. Refer to CAN-Bus information on page 51.
U0316	Software Incompatibility With Vehicle Dynamics Control Module	The Vehicle / CAN Communication Bus is a network for sharing signals and data / messages between the various vehicle systems, including Powertrain, Chassis and Infotainment. The vehicle specific software is programmed into the computer to perform various system related functions.	Refer to vehicle specific information for identification of location and function of the specified systems / CAN-Bus components. The fault code indicates that one or more systems have been programmed with software that is incompatible with the Vehicle Dynamics Control Module. Refer to CAN-Bus information on page 51.
U0317	Software Incompatibility With Park Brake Control Module	The Vehicle / CAN Communication Bus is a network for sharing signals and data / messages between the various vehicle systems, including Powertrain, Chassis and Infotainment. The vehicle specific software is programmed into the computer to perform various system related functions.	Refer to vehicle specific information for identification of location and function of the specified systems / CAN-Bus components. The fault code indicates that one or more systems have been programmed with software that is incompatible with the Park Brake Control Module. Refer to CAN-Bus information on page 51.
U0318	Software Incompatibility With Brake System Control Module	The Vehicle / CAN Communication Bus is a network for sharing signals and data / messages between the various vehicle systems, including Powertrain, Chassis and Infotainment. The vehicle specific software is programmed into the computer to perform various system related functions.	Refer to vehicle specific information for identification of location and function of the specified systems / CAN-Bus components. The fault code indicates that one or more systems have been programmed with software that is incompatible with the Brake System Control Module. Refer to CAN-Bus information on page 51.
U0319	Software Incompatibility With Steering Effort Control Module	The Vehicle / CAN Communication Bus is a network for sharing signals and data / messages between the various vehicle systems, including Powertrain, Chassis and Infotainment. The vehicle specific software is programmed into the computer to perform various system related functions.	Refer to vehicle specific information for identification of location and function of the specified systems / CAN-Bus components. The fault code indicates that one or more systems have been programmed with software that is incompatible with the Steering Effort Control Module. Refer to CAN-Bus information on page 51.
U0320	Software Incompatibility With Power Steering Control Module	The Vehicle / CAN Communication Bus is a network for sharing signals and data / messages between the various vehicle systems, including Powertrain, Chassis and Infotainment. The vehicle specific software is programmed into the computer to perform various system related functions.	Refer to vehicle specific information for identification of location and function of the specified systems / CAN-Bus components. The fault code indicates that one or more systems have been programmed with software that is incompatible with the Power Steering Control Module. Refer to CAN-Bus information on page 51.
U0321	Software Incompatibility With Suspension Control Module	The Vehicle / CAN Communication Bus is a network for sharing signals and data / messages between the various vehicle systems, including Powertrain, Chassis and Infotainment. The vehicle specific software is programmed into the computer to perform various system related functions.	Refer to vehicle specific information for identification of location and function of the specified systems / CAN-Bus components. The fault code indicates that one or more systems have been programmed with software that is incompatible with the Suspension Control Module. Refer to CAN-Bus information on page 51.

NOTE 1. Check for other fault codes that could provide additional information. **NOTE 2.** Communication between control units can pass via a CAN-Bus system; refer to Page 51 for CAN-Bus checks.

NOTE 3. If a fault cannot be located, it is also possible that the control unit is at fault. **NOTE 4.** Refer to Page 53 for list of pages that show the ISO standard for component locations e.g. Sensor A - Bank 1.

Fault code	EOBD / ISO Description	Component / System Description	Meaningful Description and Quick Check
U0322	Software Incompatibility With Body Control Module	The Vehicle / CAN Communication Bus is a network for sharing signals and data / messages between the various vehicle systems, including Powertrain, Chassis and Infotainment. The vehicle specific software is programmed into the computer to perform various system related functions.	Refer to vehicle specific information for identification of location and function of the specified systems / CAN-Bus components. The fault code indicates that one or more systems have been programmed with software that is incompatible with the Body Control Module. Refer to CAN-Bus information on page 51.
U0323	Software Incompatibility With Instrument Panel Control Module	The Vehicle / CAN Communication Bus is a network for sharing signals and data / messages between the various vehicle systems, including Powertrain, Chassis and Infotainment. The vehicle specific software is programmed into the computer to perform various system related functions.	Refer to vehicle specific information for identification of location and function of the specified systems / CAN-Bus components. The fault code indicates that one or more systems have been programmed with software that is incompatible with the Instrument Panel Control Module. Refer to CAN-Bus information on page 51.
U0324	Software Incompatibility With HVAC Control Module	The Vehicle / CAN Communication Bus is a network for sharing signals and data / messages between the various vehicle systems, including Powertrain, Chassis and Infotainment. The vehicle specific software is programmed into the computer to perform various system related functions.	Refer to vehicle specific information for identification of location and function of the specified systems / CAN-Bus components. The fault code indicates that one or more systems have been programmed with software that is incompatible with the HVAC Control Module. Refer to CAN-Bus information on page 51.
U0325	Software Incompatibility With Auxiliary Heater Control Module	The Vehicle / CAN Communication Bus is a network for sharing signals and data / messages between the various vehicle systems, including Powertrain, Chassis and Infotainment. The vehicle specific software is programmed into the computer to perform various system related functions.	Refer to vehicle specific information for identification of location and function of the specified systems / CAN-Bus components. The fault code indicates that one or more systems have been programmed with software that is incompatible with the Auxiliary Heater Control Module. Refer to CAN-Bus information on page 51.
U0326	Software Incompatibility With Vehicle Immobilizer Control Module	The Vehicle / CAN Communication Bus is a network for sharing signals and data / messages between the various vehicle systems, including Powertrain, Chassis and Infotainment. The vehicle specific software is programmed into the computer to perform various system related functions.	Refer to vehicle specific information for identification of location and function of the specified systems / CAN-Bus components. The fault code indicates that one or more systems have been programmed with software that is incompatible with the Vehicle Immobilizer Control Module. Refer to CAN-Bus information on page 51.
U0327	Software Incompatibility With Vehicle Security Control Module	The Vehicle / CAN Communication Bus is a network for sharing signals and data / messages between the various vehicle systems, including Powertrain, Chassis and Infotainment. The vehicle specific software is programmed into the computer to perform various system related functions.	Refer to vehicle specific information for identification of location and function of the specified systems / CAN-Bus components. The fault code indicates that one or more systems have been programmed with software that is incompatible with the Vehicle Security Control Module. Refer to CAN-Bus information on page 51.
U0328	Software Incompatibility With Steering Angle Sensor Module	The Vehicle / CAN Communication Bus is a network for sharing signals and data / messages between the various vehicle systems, including Powertrain, Chassis and Infotainment. The vehicle specific software is programmed into the computer to perform various system related functions.	Refer to vehicle specific information for identification of location and function of the specified systems / CAN-Bus components. The fault code indicates that one or more systems have been programmed with software that is incompatible with the Steering Angle Sensor Module. Refer to CAN-Bus information on page 51.
U0329	Software Incompatibility With Steering Column Control Module	The Vehicle / CAN Communication Bus is a network for sharing signals and data / messages between the various vehicle systems, including Powertrain, Chassis and Infotainment. The vehicle specific software is programmed into the computer to perform various system related functions.	Refer to vehicle specific information for identification of location and function of the specified systems / CAN-Bus components. The fault code indicates that one or more systems have been programmed with software that is incompatible with the Steering Column Control Module. Refer to CAN-Bus information on page 51.
U0330	Software Incompatibility With Tire Pressure Monitor Module	The Vehicle / CAN Communication Bus is a network for sharing signals and data / messages between the various vehicle systems, including Powertrain, Chassis and Infotainment. The vehicle specific software is programmed into the computer to perform various system related functions.	Refer to vehicle specific information for identification of location and function of the specified systems / CAN-Bus components. The fault code indicates that one or more systems have been programmed with software that is incompatible with the Tire Pressure Monitor Module. Refer to CAN-Bus information on page 51.
U0331	Software Incompatibility With Body Control Module "A"	The Vehicle / CAN Communication Bus is a network for sharing signals and data / messages between the various vehicle systems, including Powertrain, Chassis and Infotainment. The vehicle specific software is programmed into the computer to perform various system related functions.	Refer to vehicle specific information for identification of location and function of the specified systems / CAN-Bus components. The fault code indicates that one or more systems have been programmed with software that is incompatible with the Body Control Module "A". Refer to CAN-Bus information on page 51.
U0332	Software Incompatibility With Multi-axis Acceleration Sensor Module	The Vehicle / CAN Communication Bus is a network for sharing signals and data / messages between the various vehicle systems, including Powertrain, Chassis and Infotainment. The vehicle specific software is programmed into the computer to perform various system related functions.	Refer to vehicle specific information for identification of location and function of the specified systems / CAN-Bus components. The fault code indicates that one or more systems have been programmed with software that is incompatible with the Multi-axis Acceleration Sensor Module. Refer to CAN-Bus information on page 51.
U0333	Software Incompatibility With Gear Shift Control Module "B"	The Vehicle / CAN Communication Bus is a network for sharing signals and data / messages between the various vehicle systems, including Powertrain, Chassis and Infotainment. The vehicle specific software is programmed into the computer to perform various system related functions.	Refer to vehicle specific information for identification of location and function of the specified systems / CAN-Bus components. The fault code indicates that one or more systems have been programmed with software that is incompatible with the Gear Shift Control Module "B". Refer to CAN-Bus information on page 51.
U0334	Software Incompatibility With Radio	The Vehicle / CAN Communication Bus is a network for sharing signals and data / messages between the various vehicle systems, including Powertrain, Chassis and Infotainment. The vehicle specific software is programmed into the computer to perform various system related functions.	Refer to vehicle specific information for identification of location and function of the specified systems / CAN-Bus components. The fault code indicates that one or more systems have been programmed with software that is incompatible with the Radio. Refer to CAN-Bus information on page 51.

NOTE 1. Check for other fault codes that could provide additional information.　　**NOTE 2.** Communication between control units can pass via a CAN-Bus system; refer to Page 51 for CAN-Bus checks.

U0400	Invalid Data Received	The Vehicle / CAN Communication Bus is a network for sharing signals and data / messages between the various vehicle systems, including Powertrain, Chassis and Infotainment.	Refer to vehicle specific information for identification of location and function of the specified systems / CAN-Bus components. The fault code indicates that one or more control modules has received invalid data messages. Refer to CAN-Bus information on page 51.
U0401	Invalid Data Received From ECM / PCM "A"	The Vehicle / CAN Communication Bus is a network for sharing signals and data / messages between the various vehicle systems, including Powertrain, Chassis and Infotainment.	Refer to vehicle specific information for identification of location and function of the specified systems / CAN-Bus components. The fault code indicates that one or more control modules has received invalid data messages from the Engine ECM / PCM "A". Refer to CAN-Bus information on page 51.
U0402	Invalid Data Received From TCM	The Vehicle / CAN Communication Bus is a network for sharing signals and data / messages between the various vehicle systems, including Powertrain, Chassis and Infotainment.	Refer to vehicle specific information for identification of location and function of the specified systems / CAN-Bus components. The fault code indicates that one or more control modules has received invalid data messages from the Transmission Control module. Refer to CAN-Bus information on page 51.
U0403	Invalid Data Received From Transfer Case Control Module	The Vehicle / CAN Communication Bus is a network for sharing signals and data / messages between the various vehicle systems, including Powertrain, Chassis and Infotainment.	Refer to vehicle specific information for identification of location and function of the specified systems / CAN-Bus components. The fault code indicates that one or more control modules has received invalid data messages from the Transfer Case Control module. Refer to CAN-Bus information on page 51.
U0404	Invalid Data Received From Gear Shift Control Module "A"	The Vehicle / CAN Communication Bus is a network for sharing signals and data / messages between the various vehicle systems, including Powertrain, Chassis and Infotainment.	Refer to vehicle specific information for identification of location and function of the specified systems / CAN-Bus components. The fault code indicates that one or more control modules has received invalid data messages from the Gear Shift Control module "A". Refer to CAN-Bus information on page 51.
U0405	Invalid Data Received From Cruise Control Module	The Vehicle / CAN Communication Bus is a network for sharing signals and data / messages between the various vehicle systems, including Powertrain, Chassis and Infotainment.	Refer to vehicle specific information for identification of location and function of the specified systems / CAN-Bus components. The fault code indicates that one or more control modules has received invalid data messages from the Cruise Control module. Refer to CAN-Bus information on page 51.
U0406	Invalid Data Received From Fuel Injector Control Module	The Vehicle / CAN Communication Bus is a network for sharing signals and data / messages between the various vehicle systems, including Powertrain, Chassis and Infotainment.	Refer to vehicle specific information for identification of location and function of the specified systems / CAN-Bus components. The fault code indicates that one or more control modules has received invalid data messages from the Fuel Injector Control module. Refer to CAN-Bus information on page 51.
U0407	Invalid Data Received From Glow Plug Control Module	The Vehicle / CAN Communication Bus is a network for sharing signals and data / messages between the various vehicle systems, including Powertrain, Chassis and Infotainment.	Refer to vehicle specific information for identification of location and function of the specified systems / CAN-Bus components. The fault code indicates that one or more control modules has received invalid data messages from the Glow Plug Control module. Refer to CAN-Bus information on page 51.
U0408	Invalid Data Received From Throttle Actuator Control Module	The Vehicle / CAN Communication Bus is a network for sharing signals and data / messages between the various vehicle systems, including Powertrain, Chassis and Infotainment.	Refer to vehicle specific information for identification of location and function of the specified systems / CAN-Bus components. The fault code indicates that one or more control modules has received invalid data messages from the Throttle Actuator Control module. Refer to CAN-Bus information on page 51.
U0409	Invalid Data Received From Alternative Fuel Control Module	The Vehicle / CAN Communication Bus is a network for sharing signals and data / messages between the various vehicle systems, including Powertrain, Chassis and Infotainment.	Refer to vehicle specific information for identification of location and function of the specified systems / CAN-Bus components. The fault code indicates that one or more control modules has received invalid data messages from the Alternative Fuel Control module. Refer to CAN-Bus information on page 51.
U040A	Invalid Data Received From Digital Disc Player / Changer Module "C"	The Vehicle / CAN Communication Bus is a network for sharing signals and data / messages between the various vehicle systems, including Powertrain, Chassis and Infotainment.	Refer to vehicle specific information for identification of location and function of the specified systems / CAN-Bus components. The fault code indicates that one or more control modules has received invalid data messages from the Digital Disc Player / Changer Module "C". Refer to CAN-Bus information on page 51.
U0410	Invalid Data Received From Fuel Pump Control Module	The Vehicle / CAN Communication Bus is a network for sharing signals and data / messages between the various vehicle systems, including Powertrain, Chassis and Infotainment.	Refer to vehicle specific information for identification of location and function of the specified systems / CAN-Bus components. The fault code indicates that one or more control modules has received invalid data messages from the Fuel Pump Control Module. Refer to CAN-Bus information on page 51.
U0411	Invalid Data Received From Drive Motor Control Module "A"	The Vehicle / CAN Communication Bus is a network for sharing signals and data / messages between the various vehicle systems, including Powertrain, Chassis and Infotainment.	Refer to vehicle specific information for identification of location and function of the specified systems / CAN-Bus components. The fault code indicates that one or more control modules has received invalid data messages from the Drive Motor Control Module "A". Refer to CAN-Bus information on page 51.
U0412	Invalid Data Received From Battery Energy Control Module "A"	The Vehicle / CAN Communication Bus is a network for sharing signals and data / messages between the various vehicle systems, including Powertrain, Chassis and Infotainment.	Refer to vehicle specific information for identification of location and function of the specified systems / CAN-Bus components. The fault code indicates that one or more control modules has received invalid data messages from the Battery Energy Control Module "A". Refer to CAN-Bus information on page 51.
U0413	Invalid Data Received From Battery Energy Control Module "B"	The Vehicle / CAN Communication Bus is a network for sharing signals and data / messages between the various vehicle systems, including Powertrain, Chassis and Infotainment.	Refer to vehicle specific information for identification of location and function of the specified systems / CAN-Bus components. The fault code indicates that one or more control modules has received invalid data messages from the Battery Energy Control Module "B". Refer to CAN-Bus information on page 51.
U0414	Invalid Data Received From Four-Wheel Drive Clutch Control Module	The Vehicle / CAN Communication Bus is a network for sharing signals and data / messages between the various vehicle systems, including Powertrain, Chassis and Infotainment.	Refer to vehicle specific information for identification of location and function of the specified systems / CAN-Bus components. The fault code indicates that one or more control modules has received invalid data messages from the Four-Wheel Drive Clutch Control Module. Refer to CAN-Bus information on page 51.

NOTE 1. Check for other fault codes that could provide additional information.   **NOTE 2.** Communication between control units can pass via a CAN-Bus system; refer to Page 51 for CAN-Bus checks.

NOTE 3. If a fault cannot be located, it is also possible that the control unit is at fault.   **NOTE 4.** Refer to Page 53 for list of pages that show the ISO standard for component locations e.g. Sensor A - Bank 1.

Fault code	EOBD / ISO Description	Component / System Description	Meaningful Description and Quick Check
U0415	Invalid Data Received From Anti-Lock Brake System (ABS) Control Module	The Vehicle / CAN Communication Bus is a network for sharing signals and data / messages between the various vehicle systems, including Powertrain, Chassis and Infotainment.	Refer to vehicle specific information for identification of location and function of the specified systems / CAN-Bus components. The fault code indicates that one or more control modules has received invalid data messages from the Anti-Lock Brake System Control Module. Refer to CAN-Bus information on page 51.
U0416	Invalid Data Received From Vehicle Dynamics Control Module	The Vehicle / CAN Communication Bus is a network for sharing signals and data / messages between the various vehicle systems, including Powertrain, Chassis and Infotainment.	Refer to vehicle specific information for identification of location and function of the specified systems / CAN-Bus components. The fault code indicates that one or more control modules has received invalid data messages from the Vehicle Dynamics Control Module. Refer to CAN-Bus information on page 51.
U0417	Invalid Data Received From Park Brake Control Module	The Vehicle / CAN Communication Bus is a network for sharing signals and data / messages between the various vehicle systems, including Powertrain, Chassis and Infotainment.	Refer to vehicle specific information for identification of location and function of the specified systems / CAN-Bus components. The fault code indicates that one or more control modules has received invalid data messages from the Park Brake Control Module. Refer to CAN-Bus information on page 51.
U0418	Invalid Data Received From Brake System Control Module	The Vehicle / CAN Communication Bus is a network for sharing signals and data / messages between the various vehicle systems, including Powertrain, Chassis and Infotainment.	Refer to vehicle specific information for identification of location and function of the specified systems / CAN-Bus components. The fault code indicates that one or more control modules has received invalid data messages from the Brake System Control Module. Refer to CAN-Bus information on page 51.
U0419	Invalid Data Received From Steering Effort Control Module	The Vehicle / CAN Communication Bus is a network for sharing signals and data / messages between the various vehicle systems, including Powertrain, Chassis and Infotainment.	Refer to vehicle specific information for identification of location and function of the specified systems / CAN-Bus components. The fault code indicates that one or more control modules has received invalid data messages from the Steering Effort Control Module. Refer to CAN-Bus information on page 51.
U0420	Invalid Data Received From Power Steering Control Module	The Vehicle / CAN Communication Bus is a network for sharing signals and data / messages between the various vehicle systems, including Powertrain, Chassis and Infotainment.	Refer to vehicle specific information for identification of location and function of the specified systems / CAN-Bus components. The fault code indicates that one or more control modules has received invalid data messages from the Power Steering Control Module. Refer to CAN-Bus information on page 51.
U0421	Invalid Data Received From Suspension Control Module	The Vehicle / CAN Communication Bus is a network for sharing signals and data / messages between the various vehicle systems, including Powertrain, Chassis and Infotainment.	Refer to vehicle specific information for identification of location and function of the specified systems / CAN-Bus components. The fault code indicates that one or more control modules has received invalid data messages from the Suspension Control Module. Refer to CAN-Bus information on page 51.
U0422	Invalid Data Received From Body Control Module	The Vehicle / CAN Communication Bus is a network for sharing signals and data / messages between the various vehicle systems, including Powertrain, Chassis and Infotainment.	Refer to vehicle specific information for identification of location and function of the specified systems / CAN-Bus components. The fault code indicates that one or more control modules has received invalid data messages from the Body Control Module. Refer to CAN-Bus information on page 51.
U0423	Invalid Data Received From Instrument Panel Cluster Control Module	The Vehicle / CAN Communication Bus is a network for sharing signals and data / messages between the various vehicle systems, including Powertrain, Chassis and Infotainment.	Refer to vehicle specific information for identification of location and function of the specified systems / CAN-Bus components. The fault code indicates that one or more control modules has received invalid data messages from the Instrument Panel Cluster Control Module. Refer to CAN-Bus information on page 51.
U0424	Invalid Data Received From HVAC Control Module	The Vehicle / CAN Communication Bus is a network for sharing signals and data / messages between the various vehicle systems, including Powertrain, Chassis and Infotainment.	Refer to vehicle specific information for identification of location and function of the specified systems / CAN-Bus components. The fault code indicates that one or more control modules has received invalid data messages from the HVAC Control Module. Refer to CAN-Bus information on page 51.
U0425	Invalid Data Received From Auxiliary Heater Control Module	The Vehicle / CAN Communication Bus is a network for sharing signals and data / messages between the various vehicle systems, including Powertrain, Chassis and Infotainment.	Refer to vehicle specific information for identification of location and function of the specified systems / CAN-Bus components. The fault code indicates that one or more control modules has received invalid data messages from the Auxiliary Heater Module. Refer to CAN-Bus information on page 51.
U0426	Invalid Data Received From Vehicle Immobilizer Control Module	The Vehicle / CAN Communication Bus is a network for sharing signals and data / messages between the various vehicle systems, including Powertrain, Chassis and Infotainment.	Refer to vehicle specific information for identification of location and function of the specified systems / CAN-Bus components. The fault code indicates that one or more control modules has received invalid data messages from the Vehicle Immobilzer Control Module. Refer to CAN-Bus information on page 51.
U0427	Invalid Data Received From Vehicle Security Control Module	The Vehicle / CAN Communication Bus is a network for sharing signals and data / messages between the various vehicle systems, including Powertrain, Chassis and Infotainment.	Refer to vehicle specific information for identification of location and function of the specified systems / CAN-Bus components. The fault code indicates that one or more control modules has received invalid data messages from the Vehicle Security Control Module. Refer to CAN-Bus information on page 51.
U0428	Invalid Data Received From Steering Angle Sensor Module	The Vehicle / CAN Communication Bus is a network for sharing signals and data / messages between the various vehicle systems, including Powertrain, Chassis and Infotainment.	Refer to vehicle specific information for identification of location and function of the specified systems / CAN-Bus components. The fault code indicates that one or more control modules has received invalid data messages from the Steering Angle Sensor Module. Refer to CAN-Bus information on page 51.
U0429	Invalid Data Received From Steering Column Control Module	The Vehicle / CAN Communication Bus is a network for sharing signals and data / messages between the various vehicle systems, including Powertrain, Chassis and Infotainment.	Refer to vehicle specific information for identification of location and function of the specified systems / CAN-Bus components. The fault code indicates that one or more control modules has received invalid data messages from the Steering Column Control Module. Refer to CAN-Bus information on page 51.
U0430	Invalid Data Received From Tire Pressure Monitor Module	The Vehicle / CAN Communication Bus is a network for sharing signals and data / messages between the various vehicle systems, including Powertrain, Chassis and Infotainment.	Refer to vehicle specific information for identification of location and function of the specified systems / CAN-Bus components. The fault code indicates that one or more control modules has received invalid data messages from the Tire Pressure Monitor Module. Refer to CAN-Bus information on page 51.

NOTE 1. Check for other fault codes that could provide additional information.   **NOTE 2.** Communication between control units can pass via a CAN-Bus system; refer to Page 51 for CAN-Bus checks.

Code	Description	System	Fault Details
U0431	Invalid Data Received From Body Control Module "A"	The Vehicle / CAN Communication Bus is a network for sharing signals and data / messages between the various vehicle systems, including Powertrain, Chassis and Infotainment.	Refer to vehicle specific information for identification of location and function of the specified systems / CAN-Bus components. The fault code indicates that one or more control modules has received invalid data messages from the Body Control Module A. Refer to CAN-Bus information on page 51.
U0432	Invalid Data Received From Multi-axis Acceleration Sensor Module	The Vehicle / CAN Communication Bus is a network for sharing signals and data / messages between the various vehicle systems, including Powertrain, Chassis and Infotainment.	Refer to vehicle specific information for identification of location and function of the specified systems / CAN-Bus components. The fault code indicates that one or more control modules has received invalid data messages from the Multi-axis Acceleration Sensor Module. Refer to CAN-Bus information on page 51.
U0433	Invalid Data Received From Cruise Control Front Distance Range Sensor	The Vehicle / CAN Communication Bus is a network for sharing signals and data / messages between the various vehicle systems, including Powertrain, Chassis and Infotainment.	Refer to vehicle specific information for identification of location and function of the specified systems / CAN-Bus components. The fault code indicates that one or more control modules has received invalid data messages from the Cruise Control Front Distance Range Sensor. Refer to CAN-Bus information on page 51.
U0434	Invalid Data Received From Active Roll Control Module	The Vehicle / CAN Communication Bus is a network for sharing signals and data / messages between the various vehicle systems, including Powertrain, Chassis and Infotainment.	Refer to vehicle specific information for identification of location and function of the specified systems / CAN-Bus components. The fault code indicates that one or more control modules has received invalid data messages from the Active Roll Control Module. Refer to CAN-Bus information on page 51.
U0435	Invalid Data Received From Power Steering Control Module Rear	The Vehicle / CAN Communication Bus is a network for sharing signals and data / messages between the various vehicle systems, including Powertrain, Chassis and Infotainment.	Refer to vehicle specific information for identification of location and function of the specified systems / CAN-Bus components. The fault code indicates that one or more control modules has received invalid data messages from the Power Steering Control Module Rear. Refer to CAN-Bus information on page 51.
U0436	Invalid Data Received From Differential Control Module Front	The Vehicle / CAN Communication Bus is a network for sharing signals and data / messages between the various vehicle systems, including Powertrain, Chassis and Infotainment.	Refer to vehicle specific information for identification of location and function of the specified systems / CAN-Bus components. The fault code indicates that one or more control modules has received invalid data messages from the Differential Control Module Front. Refer to CAN-Bus information on page 51.
U0437	Invalid Data Received From Differential Control Module Rear	The Vehicle / CAN Communication Bus is a network for sharing signals and data / messages between the various vehicle systems, including Powertrain, Chassis and Infotainment.	Refer to vehicle specific information for identification of location and function of the specified systems / CAN-Bus components. The fault code indicates that one or more control modules has received invalid data messages from the Differential Control Module Rear. Refer to CAN-Bus information on page 51.
U0438	Invalid Data Received From Trailer Brake Control Module	The Vehicle / CAN Communication Bus is a network for sharing signals and data / messages between the various vehicle systems, including Powertrain, Chassis and Infotainment.	Refer to vehicle specific information for identification of location and function of the specified systems / CAN-Bus components. The fault code indicates that one or more control modules has received invalid data messages from the Trailer Brake Control Module. Refer to CAN-Bus information on page 51.
U0439	Invalid Data Received From All Terrain Control Module	The Vehicle / CAN Communication Bus is a network for sharing signals and data / messages between the various vehicle systems, including Powertrain, Chassis and Infotainment.	Refer to vehicle specific information for identification of location and function of the specified systems / CAN-Bus components. The fault code indicates that one or more control modules has received invalid data messages from the All Terrain Control Module. Refer to CAN-Bus information on page 51.
U043A		ISO / SAE reserved	
U0441	Invalid Data Received From Emissions Critical Control Information	The Vehicle / CAN Communication Bus is a network for sharing signals and data / messages between the various vehicle systems, including Powertrain, Chassis and Infotainment.	Refer to vehicle specific information for identification of location and function of the specified systems / CAN-Bus components. The fault code indicates that one or more control modules has received invalid data messages from the Emissions Critical Control Information system. Refer to CAN-Bus information on page 51.
U0442	Invalid Data Received From ECM / PCM "B"	The Vehicle / CAN Communication Bus is a network for sharing signals and data / messages between the various vehicle systems, including Powertrain, Chassis and Infotainment.	Refer to vehicle specific information for identification of location and function of the specified systems / CAN-Bus components. The fault code indicates that one or more control modules has received invalid data messages from the Engine ECM / PCM "B". Refer to CAN-Bus information on page 51.
U0443	Invalid Data Received From Body Control Module "B"	The Vehicle / CAN Communication Bus is a network for sharing signals and data / messages between the various vehicle systems, including Powertrain, Chassis and Infotainment.	Refer to vehicle specific information for identification of location and function of the specified systems / CAN-Bus components. The fault code indicates that one or more control modules has received invalid data messages from the Body Control Module "B". Refer to CAN-Bus information on page 51.
U0444	Invalid Data Received From Body Control Module "C"	The Vehicle / CAN Communication Bus is a network for sharing signals and data / messages between the various vehicle systems, including Powertrain, Chassis and Infotainment.	Refer to vehicle specific information for identification of location and function of the specified systems / CAN-Bus components. The fault code indicates that one or more control modules has received invalid data messages from the Body Control Module "C". Refer to CAN-Bus information on page 51.
U0445	Invalid Data Received From Body Control Module "D"	The Vehicle / CAN Communication Bus is a network for sharing signals and data / messages between the various vehicle systems, including Powertrain, Chassis and Infotainment.	Refer to vehicle specific information for identification of location and function of the specified systems / CAN-Bus components. The fault code indicates that one or more control modules has received invalid data messages from the Body Control Module "D". Refer to CAN-Bus information on page 51.
U0446	Invalid Data Received From Body Control Module "E"	The Vehicle / CAN Communication Bus is a network for sharing signals and data / messages between the various vehicle systems, including Powertrain, Chassis and Infotainment.	Refer to vehicle specific information for identification of location and function of the specified systems / CAN-Bus components. The fault code indicates that one or more control modules has received invalid data messages from the Body Control Module "E". Refer to CAN-Bus information on page 51.
U0447	Invalid Data Received From Gateway "A"	The Vehicle / CAN Communication Bus is a network for sharing signals and data / messages between the various vehicle systems, including Powertrain, Chassis and Infotainment.	Refer to vehicle specific information for identification of location and function of the specified systems / CAN-Bus components. The fault code indicates that one or more control modules has received invalid data messages from Gateway "A". The "Gateway" is the diagnostic interface for the Vehicle / CAN Communication Bus. Refer to CAN-Bus information on page 51.

NOTE 1. Check for other fault codes that could provide additional information.   **NOTE 2.** Communication between control units can pass via a CAN-Bus system; refer to Page 51 for CAN-Bus checks.

NOTE 3. If a fault cannot be located, it is also possible that the control unit is at fault.   **NOTE 4.** Refer to Page 53 for list of pages that show the ISO standard for component locations e.g. Sensor A - Bank 1.

Fault code	EOBD / ISO Description	Component / System Description	Meaningful Description and Quick Check
U0448	Invalid Data Received From Gateway "B"	The Vehicle / CAN Communication Bus is a network for sharing signals and data / messages between the various vehicle systems, including Powertrain, Chassis and Infotainment.	Refer to vehicle specific information for identification of location and function of the specified systems / CAN-Bus components. The fault code indicates that one or more control modules has received invalid data messages from Gateway "B". The "Gateway" is the diagnostic interface for the Vehicle / CAN Communication Bus. Refer to CAN-Bus information on page 51.
U0449	Invalid Data Received From Gateway "C"	The Vehicle / CAN Communication Bus is a network for sharing signals and data / messages between the various vehicle systems, including Powertrain, Chassis and Infotainment.	Refer to vehicle specific information for identification of location and function of the specified systems / CAN-Bus components. The fault code indicates that one or more control modules has received invalid data messages from Gateway "C". The "Gateway" is the diagnostic interface for the Vehicle / CAN Communication Bus. Refer to CAN-Bus information on page 51.
U044A	Invalid Data Received From Gateway "D"	The Vehicle / CAN Communication Bus is a network for sharing signals and data / messages between the various vehicle systems, including Powertrain, Chassis and Infotainment.	Refer to vehicle specific information for identification of location and function of the specified systems / CAN-Bus components. The fault code indicates that one or more control modules has received invalid data messages from Gateway "D". The "Gateway" is the diagnostic interface for the Vehicle / CAN Communication Bus. Refer to CAN-Bus information on page 51.
U0451	Invalid Data Received From Gateway "E"	The Vehicle / CAN Communication Bus is a network for sharing signals and data / messages between the various vehicle systems, including Powertrain, Chassis and Infotainment.	Refer to vehicle specific information for identification of location and function of the specified systems / CAN-Bus components. The fault code indicates that one or more control modules has received invalid data messages from Gateway "E". The "Gateway" is the diagnostic interface for the Vehicle / CAN Communication Bus. Refer to CAN-Bus information on page 51.
U0452	Invalid Data Received From Restraints Control Module	The Vehicle / CAN Communication Bus is a network for sharing signals and data / messages between the various vehicle systems, including Powertrain, Chassis and Infotainment.	Refer to vehicle specific information for identification of location and function of the specified systems / CAN-Bus components. The fault code indicates that one or more control modules has received invalid data messages from the Restraints Control Module. Refer to CAN-Bus information on page 51.
U0453	Invalid Data Received From Side Restraints Control Module Left	The Vehicle / CAN Communication Bus is a network for sharing signals and data / messages between the various vehicle systems, including Powertrain, Chassis and Infotainment.	Refer to vehicle specific information for identification of location and function of the specified systems / CAN-Bus components. The fault code indicates that one or more control modules has received invalid data messages from the Side Restraints Control Module Left. Refer to CAN-Bus information on page 51.
U0454	Invalid Data Received From Side Restraints Control Module Right	The Vehicle / CAN Communication Bus is a network for sharing signals and data / messages between the various vehicle systems, including Powertrain, Chassis and Infotainment.	Refer to vehicle specific information for identification of location and function of the specified systems / CAN-Bus components. The fault code indicates that one or more control modules has received invalid data messages from the Side Restraints Control Module Right. Refer to CAN-Bus information on page 51.
U0455	Invalid Data Received From Restraints Occupant Classification System Module	The Vehicle / CAN Communication Bus is a network for sharing signals and data / messages between the various vehicle systems, including Powertrain, Chassis and Infotainment.	Refer to vehicle specific information for identification of location and function of the specified systems / CAN-Bus components. The fault code indicates that one or more control modules has received invalid data messages from the Restraints Occupant Classification System Module. Refer to CAN-Bus information on page 51.
U0456	Invalid Data Received From Coolant Temperature Control Module	The Vehicle / CAN Communication Bus is a network for sharing signals and data / messages between the various vehicle systems, including Powertrain, Chassis and Infotainment.	Refer to vehicle specific information for identification of location and function of the specified systems / CAN-Bus components. The fault code indicates that one or more control modules has received invalid data messages from the Coolant Temperature Control Module. Refer to CAN-Bus information on page 51.
U0457	Invalid Data Received From Information Centre "A"	The Vehicle / CAN Communication Bus is a network for sharing signals and data / messages between the various vehicle systems, including Powertrain, Chassis and Infotainment.	Refer to vehicle specific information for identification of location and function of the specified systems / CAN-Bus components. The fault code indicates that one or more control modules has received invalid data messages from the Information Centre "A". Refer to CAN-Bus information on page 51.
U0458	Invalid Data Received From Information Centre "B"	The Vehicle / CAN Communication Bus is a network for sharing signals and data / messages between the various vehicle systems, including Powertrain, Chassis and Infotainment.	Refer to vehicle specific information for identification of location and function of the specified systems / CAN-Bus components. The fault code indicates that one or more control modules has received invalid data messages from the Information Centre "B". Refer to CAN-Bus information on page 51.
U0459	Invalid Data Received From Head Up Display	The Vehicle / CAN Communication Bus is a network for sharing signals and data / messages between the various vehicle systems, including Powertrain, Chassis and Infotainment.	Refer to vehicle specific information for identification of location and function of the specified systems / CAN-Bus components. The fault code indicates that one or more control modules has received invalid data messages from the Head Up Display. Refer to CAN-Bus information on page 51.
U045A	Invalid Data Received From Parking Assist Control Module "A"	The Vehicle / CAN Communication Bus is a network for sharing signals and data / messages between the various vehicle systems, including Powertrain, Chassis and Infotainment.	Refer to vehicle specific information for identification of location and function of the specified systems / CAN-Bus components. The fault code indicates that one or more control modules has received invalid data messages from the Parking Assist Control Module "A". Refer to CAN-Bus information on page 51.
U0461	Invalid Data Received From Audible Alert Control Module	The Vehicle / CAN Communication Bus is a network for sharing signals and data / messages between the various vehicle systems, including Powertrain, Chassis and Infotainment.	Refer to vehicle specific information for identification of location and function of the specified systems / CAN-Bus components. The fault code indicates that one or more control modules has received invalid data messages from the Audible Alert Control Module. Refer to CAN-Bus information on page 51.
U0462	Invalid Data Received From Compass Module	The Vehicle / CAN Communication Bus is a network for sharing signals and data / messages between the various vehicle systems, including Powertrain, Chassis and Infotainment.	Refer to vehicle specific information for identification of location and function of the specified systems / CAN-Bus components. The fault code indicates that one or more control modules has received invalid data messages from the Compass Module. Refer to CAN-Bus information on page 51.
U0463	Invalid Data Received From Navigation Display Module	The Vehicle / CAN Communication Bus is a network for sharing signals and data / messages between the various vehicle systems, including Powertrain, Chassis and Infotainment.	Refer to vehicle specific information for identification of location and function of the specified systems / CAN-Bus components. The fault code indicates that one or more control modules has received invalid data messages from the Navigation Display Module. Refer to CAN-Bus information on page 51.
U0464	Invalid Data Received From Navigation Control Module	The Vehicle / CAN Communication Bus is a network for sharing signals and data / messages between the various vehicle systems, including Powertrain, Chassis and Infotainment.	Refer to vehicle specific information for identification of location and function of the specified systems / CAN-Bus components. The fault code indicates that one or more control modules has received invalid data messages from the Navigation Control Module. Refer to CAN-Bus information on page 51.

NOTE 1. Check for other fault codes that could provide additional information. **NOTE 2.** Communication between control units can pass via a CAN-Bus system; refer to Page 51 for CAN-Bus checks.

U0465	Invalid Data Received From Electrical PTO Control Module	The Vehicle / CAN Communication Bus is a network for sharing signals and data / messages between the various vehicle systems, including Powertrain, Chassis and Infotainment.	Refer to vehicle specific information for identification of location and function of the specified systems / CAN-Bus components. The fault code indicates that one or more control modules has received invalid data messages from the Electrical PTO Control Module. Refer to CAN-Bus information on page 51.
U0466	Invalid Data Received From HVAC Control Module Rear	The Vehicle / CAN Communication Bus is a network for sharing signals and data / messages between the various vehicle systems, including Powertrain, Chassis and Infotainment.	Refer to vehicle specific information for identification of location and function of the specified systems / CAN-Bus components. The fault code indicates that one or more control modules has received invalid data messages from the HVAC Control Module Rear. Refer to CAN-Bus information on page 51.
U0467	Invalid Data Received From Fuel Additive Control Module	The Vehicle / CAN Communication Bus is a network for sharing signals and data / messages between the various vehicle systems, including Powertrain, Chassis and Infotainment.	Refer to vehicle specific information for identification of location and function of the specified systems / CAN-Bus components. The fault code indicates that one or more control modules has received invalid data messages from the Fuel Additive Control Module. Refer to CAN-Bus information on page 51.
U0468	Invalid Data Received From Fuel Cell Control Module	The Vehicle / CAN Communication Bus is a network for sharing signals and data / messages between the various vehicle systems, including Powertrain, Chassis and Infotainment.	Refer to vehicle specific information for identification of location and function of the specified systems / CAN-Bus components. The fault code indicates that one or more control modules has received invalid data messages from the Fuel Cell Control Module. Refer to CAN-Bus information on page 51.
U0469	Invalid Data Received From Starter / Generator Control Module	The Vehicle / CAN Communication Bus is a network for sharing signals and data / messages between the various vehicle systems, including Powertrain, Chassis and Infotainment.	Refer to vehicle specific information for identification of location and function of the specified systems / CAN-Bus components. The fault code indicates that one or more control modules has received invalid data messages from the Starter / Generator Control Module. Refer to CAN-Bus information on page 51.
U046A	Invalid Data Received From Sunroof Control Module	The Vehicle / CAN Communication Bus is a network for sharing signals and data / messages between the various vehicle systems, including Powertrain, Chassis and Infotainment.	Refer to vehicle specific information for identification of location and function of the specified systems / CAN-Bus components. The fault code indicates that one or more control modules has received invalid data messages from the Sunroof Control Module. Refer to CAN-Bus information on page 51.
U0471	Invalid Data Received From "Restraints System Sensor A"	The Vehicle / CAN Communication Bus is a network for sharing signals and data / messages between the various vehicle systems, including Powertrain, Chassis and Infotainment.	Refer to vehicle specific information for identification of location and function of the specified systems / CAN-Bus components. The fault code indicates that one or more control modules has received invalid data messages from the "Restraints System Sensor A". Refer to CAN-Bus information on page 51.
U0472	Invalid Data Received From "Restraints System Sensor B"	The Vehicle / CAN Communication Bus is a network for sharing signals and data / messages between the various vehicle systems, including Powertrain, Chassis and Infotainment.	Refer to vehicle specific information for identification of location and function of the specified systems / CAN-Bus components. The fault code indicates that one or more control modules has received invalid data messages from the "Restraints System Sensor B". Refer to CAN-Bus information on page 51.
U0473	Invalid Data Received From "Restraints System Sensor C"	The Vehicle / CAN Communication Bus is a network for sharing signals and data / messages between the various vehicle systems, including Powertrain, Chassis and Infotainment.	Refer to vehicle specific information for identification of location and function of the specified systems / CAN-Bus components. The fault code indicates that one or more control modules has received invalid data messages from the "Restraints System Sensor C". Refer to CAN-Bus information on page 51.
U0474	Invalid Data Received From "Restraints System Sensor D"	The Vehicle / CAN Communication Bus is a network for sharing signals and data / messages between the various vehicle systems, including Powertrain, Chassis and Infotainment.	Refer to vehicle specific information for identification of location and function of the specified systems / CAN-Bus components. The fault code indicates that one or more control modules has received invalid data messages from the "Restraints System Sensor D". Refer to CAN-Bus information on page 51.
U0475	Invalid Data Received From "Restraints System Sensor E"	The Vehicle / CAN Communication Bus is a network for sharing signals and data / messages between the various vehicle systems, including Powertrain, Chassis and Infotainment.	Refer to vehicle specific information for identification of location and function of the specified systems / CAN-Bus components. The fault code indicates that one or more control modules has received invalid data messages from the "Restraints System Sensor E". Refer to CAN-Bus information on page 51.
U0476	Invalid Data Received From "Restraints System Sensor F"	The Vehicle / CAN Communication Bus is a network for sharing signals and data / messages between the various vehicle systems, including Powertrain, Chassis and Infotainment.	Refer to vehicle specific information for identification of location and function of the specified systems / CAN-Bus components. The fault code indicates that one or more control modules has received invalid data messages from the "Restraints System Sensor F". Refer to CAN-Bus information on page 51.
U0477	Invalid Data Received From "Restraints System Sensor G"	The Vehicle / CAN Communication Bus is a network for sharing signals and data / messages between the various vehicle systems, including Powertrain, Chassis and Infotainment.	Refer to vehicle specific information for identification of location and function of the specified systems / CAN-Bus components. The fault code indicates that one or more control modules has received invalid data messages from the "Restraints System Sensor G". Refer to CAN-Bus information on page 51.
U0478	Invalid Data Received From "Restraints System Sensor H"	The Vehicle / CAN Communication Bus is a network for sharing signals and data / messages between the various vehicle systems, including Powertrain, Chassis and Infotainment.	Refer to vehicle specific information for identification of location and function of the specified systems / CAN-Bus components. The fault code indicates that one or more control modules has received invalid data messages from the "Restraints System Sensor H". Refer to CAN-Bus information on page 51.
U0479	Invalid Data Received From "Restraints System Sensor I"	The Vehicle / CAN Communication Bus is a network for sharing signals and data / messages between the various vehicle systems, including Powertrain, Chassis and Infotainment.	Refer to vehicle specific information for identification of location and function of the specified systems / CAN-Bus components. The fault code indicates that one or more control modules has received invalid data messages from the "Restraints System Sensor I". Refer to CAN-Bus information on page 51.
U047A	Invalid Data Received From "Restraints System Sensor J"	The Vehicle / CAN Communication Bus is a network for sharing signals and data / messages between the various vehicle systems, including Powertrain, Chassis and Infotainment.	Refer to vehicle specific information for identification of location and function of the specified systems / CAN-Bus components. The fault code indicates that one or more control modules has received invalid data messages from the "Restraints System Sensor J". Refer to CAN-Bus information on page 51.
U047B	Invalid Data Received From "Restraints System Sensor K"	The Vehicle / CAN Communication Bus is a network for sharing signals and data / messages between the various vehicle systems, including Powertrain, Chassis and Infotainment.	Refer to vehicle specific information for identification of location and function of the specified systems / CAN-Bus components. The fault code indicates that one or more control modules has received invalid data messages from the "Restraints System Sensor K". Refer to CAN-Bus information on page 51.

NOTE 1. Check for other fault codes that could provide additional information. **NOTE 2.** Communication between control units can pass via a CAN-Bus system; refer to Page 51 for CAN-Bus checks.

NOTE 3. If a fault cannot be located, it is also possible that the control unit is at fault. **NOTE 4.** Refer to Page 53 for list of pages that show the ISO standard for component locations e.g. Sensor A - Bank 1.

Fault code	EOBD / ISO Description	Component / System Description	Meaningful Description and Quick Check
U047C	Invalid Data Received From "Restraints System Sensor L"	The Vehicle / CAN Communication Bus is a network for sharing signals and data / messages between the various vehicle systems, including Powertrain, Chassis and Infotainment.	Refer to vehicle specific information for identification of location and function of the specified systems / CAN-Bus components. The fault code indicates that one or more control modules has received invalid data messages from the "Restraints System Sensor L". Refer to CAN-Bus information on page 51.
U047D	Invalid Data Received From "Restraints System Sensor M"	The Vehicle / CAN Communication Bus is a network for sharing signals and data / messages between the various vehicle systems, including Powertrain, Chassis and Infotainment.	Refer to vehicle specific information for identification of location and function of the specified systems / CAN-Bus components. The fault code indicates that one or more control modules has received invalid data messages from the "Restraints System Sensor M". Refer to CAN-Bus information on page 51.
U047E	Invalid Data Received From "Restraints System Sensor N"	The Vehicle / CAN Communication Bus is a network for sharing signals and data / messages between the various vehicle systems, including Powertrain, Chassis and Infotainment.	Refer to vehicle specific information for identification of location and function of the specified systems / CAN-Bus components. The fault code indicates that one or more control modules has received invalid data messages from the "Restraints System Sensor N". Refer to CAN-Bus information on page 51.
U0481	Invalid Data Received From Automatic Lighting Control Module	The Vehicle / CAN Communication Bus is a network for sharing signals and data / messages between the various vehicle systems, including Powertrain, Chassis and Infotainment.	Refer to vehicle specific information for identification of location and function of the specified systems / CAN-Bus components. The fault code indicates that one or more control modules has received invalid data messages from the Automatic Lighting Control Module. Refer to CAN-Bus information on page 51.
U0482	Invalid Data Received From Headlamp Levelling Control Module	The Vehicle / CAN Communication Bus is a network for sharing signals and data / messages between the various vehicle systems, including Powertrain, Chassis and Infotainment.	Refer to vehicle specific information for identification of location and function of the specified systems / CAN-Bus components. The fault code indicates that one or more control modules has received invalid data messages from the Headlamp Levelling Control Module. Refer to CAN-Bus information on page 51.
U0483	Invalid Data Received From Lighting Control Module Front	The Vehicle / CAN Communication Bus is a network for sharing signals and data / messages between the various vehicle systems, including Powertrain, Chassis and Infotainment.	Refer to vehicle specific information for identification of location and function of the specified systems / CAN-Bus components. The fault code indicates that one or more control modules has received invalid data messages from the Lighting Control Module Front. Refer to CAN-Bus information on page 51.
U0484	Invalid Data Received From Lighting Control Module Rear "A"	The Vehicle / CAN Communication Bus is a network for sharing signals and data / messages between the various vehicle systems, including Powertrain, Chassis and Infotainment.	Refer to vehicle specific information for identification of location and function of the specified systems / CAN-Bus components. The fault code indicates that one or more control modules has received invalid data messages from the Lighting Control Module Rear "A". Refer to CAN-Bus information on page 51.
U0485	Invalid Data Received From Radio	The Vehicle / CAN Communication Bus is a network for sharing signals and data / messages between the various vehicle systems, including Powertrain, Chassis and Infotainment.	Refer to vehicle specific information for identification of location and function of the specified systems / CAN-Bus components. The fault code indicates that one or more control modules has received invalid data messages from the Radio. Refer to CAN-Bus information on page 51.
U0486	Invalid Data Received From Antenna Control Module	The Vehicle / CAN Communication Bus is a network for sharing signals and data / messages between the various vehicle systems, including Powertrain, Chassis and Infotainment.	Refer to vehicle specific information for identification of location and function of the specified systems / CAN-Bus components. The fault code indicates that one or more control modules has received invalid data messages from the Antenna Control Module. Refer to CAN-Bus information on page 51.
U0487	Invalid Data Received From Audio Amplifier	The Vehicle / CAN Communication Bus is a network for sharing signals and data / messages between the various vehicle systems, including Powertrain, Chassis and Infotainment.	Refer to vehicle specific information for identification of location and function of the specified systems / CAN-Bus components. The fault code indicates that one or more control modules has received invalid data messages from the Audio Amplifier. Refer to CAN-Bus information on page 51.
U0488	Invalid Data Received From Digital Disc Player / Changer Module "A"	The Vehicle / CAN Communication Bus is a network for sharing signals and data / messages between the various vehicle systems, including Powertrain, Chassis and Infotainment.	Refer to vehicle specific information for identification of location and function of the specified systems / CAN-Bus components. The fault code indicates that one or more control modules has received invalid data messages from the Digital Disc Player / Changer Module "A". Refer to CAN-Bus information on page 51.
U0489	Invalid Data Received From Digital Disc Player / Changer Module "B"	The Vehicle / CAN Communication Bus is a network for sharing signals and data / messages between the various vehicle systems, including Powertrain, Chassis and Infotainment.	Refer to vehicle specific information for identification of location and function of the specified systems / CAN-Bus components. The fault code indicates that one or more control modules has received invalid data messages from the Digital Disc Player / Changer Module "B". Refer to CAN-Bus information on page 51.
U0491	Invalid Data Received From Digital Disc Player / Changer Module "D"	The Vehicle / CAN Communication Bus is a network for sharing signals and data / messages between the various vehicle systems, including Powertrain, Chassis and Infotainment.	Refer to vehicle specific information for identification of location and function of the specified systems / CAN-Bus components. The fault code indicates that one or more control modules has received invalid data messages from the Digital Disc Player / Changer Module "D". Refer to CAN-Bus information on page 51.
U0492	Invalid Data Received From Television	The Vehicle / CAN Communication Bus is a network for sharing signals and data / messages between the various vehicle systems, including Powertrain, Chassis and Infotainment.	Refer to vehicle specific information for identification of location and function of the specified systems / CAN-Bus components. The fault code indicates that one or more control modules has received invalid data messages from the Television. Refer to CAN-Bus information on page 51.
U0493	Invalid Data Received From Personal Computer	The Vehicle / CAN Communication Bus is a network for sharing signals and data / messages between the various vehicle systems, including Powertrain, Chassis and Infotainment.	Refer to vehicle specific information for identification of location and function of the specified systems / CAN-Bus components. The fault code indicates that one or more control modules has received invalid data messages from the Personal Computer. Refer to CAN-Bus information on page 51.
U0494	Invalid Data Received From "Digital Audio Control Module A"	The Vehicle / CAN Communication Bus is a network for sharing signals and data / messages between the various vehicle systems, including Powertrain, Chassis and Infotainment.	Refer to vehicle specific information for identification of location and function of the specified systems / CAN-Bus components. The fault code indicates that one or more control modules has received invalid data messages from the "Digital Audio Control Module A". Refer to CAN-Bus information on page 51.

NOTE 1. Check for other fault codes that could provide additional information.   **NOTE 2.** Communication between control units can pass via a CAN-Bus system; refer to Page 51 for CAN-Bus checks.

NOTE 3. If a fault cannot be located, it is also possible that the control unit is at fault   NOTE 4. Refer to Page 53 for list of pages that show the ISO standard for component locations e.g. Sensor A - Bank 1.

Code	Description		
U0495	Invalid Data Received From "Digital Audio Control Module B"	The Vehicle / CAN Communication Bus is a network for sharing signals and data / messages between the various vehicle systems, including Powertrain, Chassis and Infotainment.	Refer to vehicle specific information for identification of location and function of the specified systems / CAN-Bus components. The fault code indicates that one or more control modules has received invalid data messages from the "Digital Audio Control Module B". Refer to CAN-Bus information on page 51.
U0496	Invalid Data Received From Subscription Entertainment Receiver Module	The Vehicle / CAN Communication Bus is a network for sharing signals and data / messages between the various vehicle systems, including Powertrain, Chassis and Infotainment.	Refer to vehicle specific information for identification of location and function of the specified systems / CAN-Bus components. The fault code indicates that one or more control modules has received invalid data messages from the Subscription Entertainment Receiver Module. Refer to CAN-Bus information on page 51.
U0497	Invalid Data Received From Entertainment Control Module Rear "A"	The Vehicle / CAN Communication Bus is a network for sharing signals and data / messages between the various vehicle systems, including Powertrain, Chassis and Infotainment.	Refer to vehicle specific information for identification of location and function of the specified systems / CAN-Bus components. The fault code indicates that one or more control modules has received invalid data messages from the Entertainment Control Module Rear "A". Refer to CAN-Bus information on page 51.
U0498	Invalid Data Received From Telephone Control Module	The Vehicle / CAN Communication Bus is a network for sharing signals and data / messages between the various vehicle systems, including Powertrain, Chassis and Infotainment.	Refer to vehicle specific information for identification of location and function of the specified systems / CAN-Bus components. The fault code indicates that one or more control modules has received invalid data messages from the Telephone Control Module. Refer to CAN-Bus information on page 51.
U0499	Invalid Data Received From Telematic Control Module	The Vehicle / CAN Communication Bus is a network for sharing signals and data / messages between the various vehicle systems, including Powertrain, Chassis and Infotainment.	Refer to vehicle specific information for identification of location and function of the specified systems / CAN-Bus components. The fault code indicates that one or more control modules has received invalid data messages from the Telematic Control Module. Refer to CAN-Bus information on page 51.
U049A	Invalid Data Received From "Door Control Module A"	The Vehicle / CAN Communication Bus is a network for sharing signals and data / messages between the various vehicle systems, including Powertrain, Chassis and Infotainment.	Refer to vehicle specific information for identification of location and function of the specified systems / CAN-Bus components. The fault code indicates that one or more control modules has received invalid data messages from the "Door Control Module A". Refer to CAN-Bus information on page 51.
U0501	Invalid Data Received From "Door Control Module B"	The Vehicle / CAN Communication Bus is a network for sharing signals and data / messages between the various vehicle systems, including Powertrain, Chassis and Infotainment.	Refer to vehicle specific information for identification of location and function of the specified systems / CAN-Bus components. The fault code indicates that one or more control modules has received invalid data messages from the "Door Control Module B". Refer to CAN-Bus information on page 51.
U0502	Invalid Data Received From "Door Control Module C"	The Vehicle / CAN Communication Bus is a network for sharing signals and data / messages between the various vehicle systems, including Powertrain, Chassis and Infotainment.	Refer to vehicle specific information for identification of location and function of the specified systems / CAN-Bus components. The fault code indicates that one or more control modules has received invalid data messages from the "Door Control Module C". Refer to CAN-Bus information on page 51.
U0503	Invalid Data Received From "Door Control Module D"	The Vehicle / CAN Communication Bus is a network for sharing signals and data / messages between the various vehicle systems, including Powertrain, Chassis and Infotainment.	Refer to vehicle specific information for identification of location and function of the specified systems / CAN-Bus components. The fault code indicates that one or more control modules has received invalid data messages from the "Door Control Module D". Refer to CAN-Bus information on page 51.
U0504	Invalid Data Received From "Door Control Module E"	The Vehicle / CAN Communication Bus is a network for sharing signals and data / messages between the various vehicle systems, including Powertrain, Chassis and Infotainment.	Refer to vehicle specific information for identification of location and function of the specified systems / CAN-Bus components. The fault code indicates that one or more control modules has received invalid data messages from the "Door Control Module E". Refer to CAN-Bus information on page 51.
U0505	Invalid Data Received From "Door Control Module F"	The Vehicle / CAN Communication Bus is a network for sharing signals and data / messages between the various vehicle systems, including Powertrain, Chassis and Infotainment.	Refer to vehicle specific information for identification of location and function of the specified systems / CAN-Bus components. The fault code indicates that one or more control modules has received invalid data messages from the "Door Control Module F". Refer to CAN-Bus information on page 51.
U0506	Invalid Data Received From "Door Control Module G"	The Vehicle / CAN Communication Bus is a network for sharing signals and data / messages between the various vehicle systems, including Powertrain, Chassis and Infotainment.	Refer to vehicle specific information for identification of location and function of the specified systems / CAN-Bus components. The fault code indicates that one or more control modules has received invalid data messages from the "Door Control Module G". Refer to CAN-Bus information on page 51.
U0507	Invalid Data Received From Folding Top Control Module	The Vehicle / CAN Communication Bus is a network for sharing signals and data / messages between the various vehicle systems, including Powertrain, Chassis and Infotainment.	Refer to vehicle specific information for identification of location and function of the specified systems / CAN-Bus components. The fault code indicates that one or more control modules has received invalid data messages from the Folding Top Control Module. Refer to CAN-Bus information on page 51.
U0508	Invalid Data Received From Moveable Roof Control Module	The Vehicle / CAN Communication Bus is a network for sharing signals and data / messages between the various vehicle systems, including Powertrain, Chassis and Infotainment.	Refer to vehicle specific information for identification of location and function of the specified systems / CAN-Bus components. The fault code indicates that one or more control modules has received invalid data messages from the Moveable Roof Control Module. Refer to CAN-Bus information on page 51.
U0509	Invalid Data Received From "Seat Control Module A"	The Vehicle / CAN Communication Bus is a network for sharing signals and data / messages between the various vehicle systems, including Powertrain, Chassis and Infotainment.	Refer to vehicle specific information for identification of location and function of the specified systems / CAN-Bus components. The fault code indicates that one or more control modules has received invalid data messages from the "Seat Control Module A". Refer to CAN-Bus information on page 51.
U050A	Invalid Data Received From "Seat Control Module B"	The Vehicle / CAN Communication Bus is a network for sharing signals and data / messages between the various vehicle systems, including Powertrain, Chassis and Infotainment.	Refer to vehicle specific information for identification of location and function of the specified systems / CAN-Bus components. The fault code indicates that one or more control modules has received invalid data messages from the "Seat Control Module B". Refer to CAN-Bus information on page 51.
U0511	Invalid Data Received From "Seat Control Module C"	The Vehicle / CAN Communication Bus is a network for sharing signals and data / messages between the various vehicle systems, including Powertrain, Chassis and Infotainment.	Refer to vehicle specific information for identification of location and function of the specified systems / CAN-Bus components. The fault code indicates that one or more control modules has received invalid data messages from the "Seat Control Module C". Refer to CAN-Bus information on page 51.

NOTE 1. Check for other fault codes that could provide additional information.   **NOTE 2.** Communication between control units can pass via a CAN-Bus system; refer to Page 51 for CAN-Bus checks.

NOTE 3. If a fault cannot be located, it is also possible that the control unit is at fault.   **NOTE 4.** Refer to Page 53 for list of pages that show the ISO standard for component locations e.g. Sensor A - Bank 1.

Fault code	EOBD / ISO Description	Component / System Description	Meaningful Description and Quick Check
U0512	Invalid Data Received From "Seat Control Module D"	The Vehicle / CAN Communication Bus is a network for sharing signals and data / messages between the various vehicle systems, including Powertrain, Chassis and Infotainment.	Refer to vehicle specific information for identification of location and function of the specified systems / CAN-Bus components. The fault code indicates that one or more control modules has received invalid data messages from the "Seat Control Module D". Refer to CAN-Bus information on page 51.
U0513	Invalid Data Received From Yaw Rate Sensor Module	The Vehicle / CAN Communication Bus is a network for sharing signals and data / messages between the various vehicle systems, including Powertrain, Chassis and Infotainment.	Refer to vehicle specific information for identification of location and function of the specified systems / CAN-Bus components. The fault code indicates that one or more control modules has received invalid data messages from the Yaw Rate Sensor Module. Refer to CAN-Bus information on page 51.
U0514	Invalid Data Received From Mirror Control Module "A"	The Vehicle / CAN Communication Bus is a network for sharing signals and data / messages between the various vehicle systems, including Powertrain, Chassis and Infotainment.	Refer to vehicle specific information for identification of location and function of the specified systems / CAN-Bus components. The fault code indicates that one or more control modules has received invalid data messages from the Mirror Control Module "A". Refer to CAN-Bus information on page 51.
U0515	Invalid Data Received From Remote Function Actuation	The Vehicle / CAN Communication Bus is a network for sharing signals and data / messages between the various vehicle systems, including Powertrain, Chassis and Infotainment.	Refer to vehicle specific information for identification of location and function of the specified systems / CAN-Bus components. The fault code indicates that one or more control modules has received invalid data messages from the Remote Function Actuation. Refer to CAN-Bus information on page 51.
U0516	Invalid Data Received From "Door Switch A"	The Vehicle / CAN Communication Bus is a network for sharing signals and data / messages between the various vehicle systems, including Powertrain, Chassis and Infotainment.	Refer to vehicle specific information for identification of location and function of the specified systems / CAN-Bus components. The fault code indicates that one or more control modules has received invalid data messages from "Door Switch A". Refer to CAN-Bus information on page 51.
U0517	Invalid Data Received From "Door Switch B"	The Vehicle / CAN Communication Bus is a network for sharing signals and data / messages between the various vehicle systems, including Powertrain, Chassis and Infotainment.	Refer to vehicle specific information for identification of location and function of the specified systems / CAN-Bus components. The fault code indicates that one or more control modules has received invalid data messages from "Door Switch B". Refer to CAN-Bus information on page 51.
U0518	Invalid Data Received From "Door Switch C"	The Vehicle / CAN Communication Bus is a network for sharing signals and data / messages between the various vehicle systems, including Powertrain, Chassis and Infotainment.	Refer to vehicle specific information for identification of location and function of the specified systems / CAN-Bus components. The fault code indicates that one or more control modules has received invalid data messages from "Door Switch C". Refer to CAN-Bus information on page 51.
U0519	Invalid Data Received From "Door Switch D"	The Vehicle / CAN Communication Bus is a network for sharing signals and data / messages between the various vehicle systems, including Powertrain, Chassis and Infotainment.	Refer to vehicle specific information for identification of location and function of the specified systems / CAN-Bus components. The fault code indicates that one or more control modules has received invalid data messages from "Door Switch D". Refer to CAN-Bus information on page 51.
U051A	Invalid Data Received From "Door Switch E"	The Vehicle / CAN Communication Bus is a network for sharing signals and data / messages between the various vehicle systems, including Powertrain, Chassis and Infotainment.	Refer to vehicle specific information for identification of location and function of the specified systems / CAN-Bus components. The fault code indicates that one or more control modules has received invalid data messages from "Door Switch E". Refer to CAN-Bus information on page 51.
U0521	Invalid Data Received From "Door Switch F"	The Vehicle / CAN Communication Bus is a network for sharing signals and data / messages between the various vehicle systems, including Powertrain, Chassis and Infotainment.	Refer to vehicle specific information for identification of location and function of the specified systems / CAN-Bus components. The fault code indicates that one or more control modules has received invalid data messages from "Door Switch F". Refer to CAN-Bus information on page 51.
U0522	Invalid Data Received From "Door Switch G"	The Vehicle / CAN Communication Bus is a network for sharing signals and data / messages between the various vehicle systems, including Powertrain, Chassis and Infotainment.	Refer to vehicle specific information for identification of location and function of the specified systems / CAN-Bus components. The fault code indicates that one or more control modules has received invalid data messages from "Door Switch G". Refer to CAN-Bus information on page 51.
U0523	Invalid Data Received From "Door Window Motor A"	The Vehicle / CAN Communication Bus is a network for sharing signals and data / messages between the various vehicle systems, including Powertrain, Chassis and Infotainment.	Refer to vehicle specific information for identification of location and function of the specified systems / CAN-Bus components. The fault code indicates that one or more control modules has received invalid data messages from the "Door Window Motor A". Refer to CAN-Bus information on page 51.
U0524	Invalid Data Received From "Door Window Motor B"	The Vehicle / CAN Communication Bus is a network for sharing signals and data / messages between the various vehicle systems, including Powertrain, Chassis and Infotainment.	Refer to vehicle specific information for identification of location and function of the specified systems / CAN-Bus components. The fault code indicates that one or more control modules has received invalid data messages from the "Door Window Motor B". Refer to CAN-Bus information on page 51.
U0525	Invalid Data Received From "Door Window Motor C"	The Vehicle / CAN Communication Bus is a network for sharing signals and data / messages between the various vehicle systems, including Powertrain, Chassis and Infotainment.	Refer to vehicle specific information for identification of location and function of the specified systems / CAN-Bus components. The fault code indicates that one or more control modules has received invalid data messages from the "Door Window Motor C". Refer to CAN-Bus information on page 51.
U0526	Invalid Data Received From "Door Window Motor D"	The Vehicle / CAN Communication Bus is a network for sharing signals and data / messages between the various vehicle systems, including Powertrain, Chassis and Infotainment.	Refer to vehicle specific information for identification of location and function of the specified systems / CAN-Bus components. The fault code indicates that one or more control modules has received invalid data messages from the "Door Window Motor D". Refer to CAN-Bus information on page 51.
U0527	Invalid Data Received From "Door Window Motor E"	The Vehicle / CAN Communication Bus is a network for sharing signals and data / messages between the various vehicle systems, including Powertrain, Chassis and Infotainment.	Refer to vehicle specific information for identification of location and function of the specified systems / CAN-Bus components. The fault code indicates that one or more control modules has received invalid data messages from the "Door Window Motor E". Refer to CAN-Bus information on page 51.
U0528	Invalid Data Received From "Door Window Motor F"	The Vehicle / CAN Communication Bus is a network for sharing signals and data / messages between the various vehicle systems, including Powertrain, Chassis and Infotainment.	Refer to vehicle specific information for identification of location and function of the specified systems / CAN-Bus components. The fault code indicates that one or more control modules has received invalid data messages from the "Door Window Motor F". Refer to CAN-Bus information on page 51.
U0529	Invalid Data Received From "Door Window Motor G"	The Vehicle / CAN Communication Bus is a network for sharing signals and data / messages between the various vehicle systems, including Powertrain, Chassis and Infotainment.	Refer to vehicle specific information for identification of location and function of the specified systems / CAN-Bus components. The fault code indicates that one or more control modules has received invalid data messages from the "Door Window Motor G". Refer to CAN-Bus information on page 51.

NOTE 1. Check for other fault codes that could provide additional information. **NOTE 2.** Communication between control units can pass via a CAN-Bus system; refer to Page 51 for CAN-Bus checks.

NOTE 3. If a fault cannot be located, it is also possible that the control unit is at fault. **NOTE 4.** Refer to Page 53 for list of pages that show the ISO standard for component locations e.g. Sensor A - Bank 1.

U052A	Invalid Data Received From Heated Steering Wheel Module	The Vehicle / CAN Communication Bus is a network for sharing signals and data / messages between the various vehicle systems, including Powertrain, Chassis and Infotainment.	Refer to vehicle specific information for identification of location and function of the specified systems / CAN-Bus components. The fault code indicates that one or more control modules has received invalid data messages from the Heated Steering Wheel Module. Refer to CAN-Bus information on page 51.
U0531	Invalid Data Received From Rear Gate Module	The Vehicle / CAN Communication Bus is a network for sharing signals and data / messages between the various vehicle systems, including Powertrain, Chassis and Infotainment.	Refer to vehicle specific information for identification of location and function of the specified systems / CAN-Bus components. The fault code indicates that one or more control modules has received invalid data messages from the Rear Gate Module. Refer to CAN-Bus information on page 51.
U0532	Invalid Data Received From Rain Sensing Module	The Vehicle / CAN Communication Bus is a network for sharing signals and data / messages between the various vehicle systems, including Powertrain, Chassis and Infotainment.	Refer to vehicle specific information for identification of location and function of the specified systems / CAN-Bus components. The fault code indicates that one or more control modules has received invalid data messages from the Rain Sensing Module. Refer to CAN-Bus information on page 51.
U0533	Invalid Data Received From Side Obstacle Detection Control Module Left	The Vehicle / CAN Communication Bus is a network for sharing signals and data / messages between the various vehicle systems, including Powertrain, Chassis and Infotainment.	Refer to vehicle specific information for identification of location and function of the specified systems / CAN-Bus components. The fault code indicates that one or more control modules has received invalid data messages from the Side Obstacle Detection Control Module Left. Refer to CAN-Bus information on page 51.
U0534	Invalid Data Received From Side Obstacle Detection Control Module Right	The Vehicle / CAN Communication Bus is a network for sharing signals and data / messages between the various vehicle systems, including Powertrain, Chassis and Infotainment.	Refer to vehicle specific information for identification of location and function of the specified systems / CAN-Bus components. The fault code indicates that one or more control modules has received invalid data messages from the Side Obstacle Detection Control Module Right. Refer to CAN-Bus information on page 51.
U0535	Invalid Data Received From Convenience Recall Module	The Vehicle / CAN Communication Bus is a network for sharing signals and data / messages between the various vehicle systems, including Powertrain, Chassis and Infotainment.	Refer to vehicle specific information for identification of location and function of the specified systems / CAN-Bus components. The fault code indicates that one or more control modules has received invalid data messages from the Convenience Recall Module. Refer to CAN-Bus information on page 51.
U0536	Invalid Data Received From Lateral Acceleration Sensor Module	The Vehicle / CAN Communication Bus is a network for sharing signals and data / messages between the various vehicle systems, including Powertrain, Chassis and Infotainment.	Refer to vehicle specific information for identification of location and function of the specified systems / CAN-Bus components. The fault code indicates that one or more control modules has received invalid data messages from the Lateral Acceleration Sensor Module. Refer to CAN-Bus information on page 51.
U0537	Invalid Data Received From Column Lock Module	The Vehicle / CAN Communication Bus is a network for sharing signals and data / messages between the various vehicle systems, including Powertrain, Chassis and Infotainment.	Refer to vehicle specific information for identification of location and function of the specified systems / CAN-Bus components. The fault code indicates that one or more control modules has received invalid data messages from the Column Lock Module. Refer to CAN-Bus information on page 51.
U0538	Invalid Data Received From "Digital Audio Control Module C"	The Vehicle / CAN Communication Bus is a network for sharing signals and data / messages between the various vehicle systems, including Powertrain, Chassis and Infotainment.	Refer to vehicle specific information for identification of location and function of the specified systems / CAN-Bus components. The fault code indicates that one or more control modules has received invalid data messages from the "Digital Audio Control Module C". Refer to CAN-Bus information on page 51.
U0539	Invalid Data Received From "Digital Audio Control Module D"	The Vehicle / CAN Communication Bus is a network for sharing signals and data / messages between the various vehicle systems, including Powertrain, Chassis and Infotainment.	Refer to vehicle specific information for identification of location and function of the specified systems / CAN-Bus components. The fault code indicates that one or more control modules has received invalid data messages from the "Digital Audio Control Module D". Refer to CAN-Bus information on page 51.
U053A	Invalid Data Received From Entrapment Control Module "A"	The Vehicle / CAN Communication Bus is a network for sharing signals and data / messages between the various vehicle systems, including Powertrain, Chassis and Infotainment.	Refer to vehicle specific information for identification of location and function of the specified systems / CAN-Bus components. The fault code indicates that one or more control modules has received invalid data messages from the Entrapment Control Module "A". Refer to CAN-Bus information on page 51.
U0541	Invalid Data Received From Entrapment Control Module "B"	The Vehicle / CAN Communication Bus is a network for sharing signals and data / messages between the various vehicle systems, including Powertrain, Chassis and Infotainment.	Refer to vehicle specific information for identification of location and function of the specified systems / CAN-Bus components. The fault code indicates that one or more control modules has received invalid data messages from the Entrapment Control Module "B". Refer to CAN-Bus information on page 51.
U0542	Invalid Data Received From Headlamp Control Module "A"	The Vehicle / CAN Communication Bus is a network for sharing signals and data / messages between the various vehicle systems, including Powertrain, Chassis and Infotainment.	Refer to vehicle specific information for identification of location and function of the specified systems / CAN-Bus components. The fault code indicates that one or more control modules has received invalid data messages from the Headlamp Control Module "A". Refer to CAN-Bus information on page 51.
U0543	Invalid Data Received From Headlamp Control Module "B"	The Vehicle / CAN Communication Bus is a network for sharing signals and data / messages between the various vehicle systems, including Powertrain, Chassis and Infotainment.	Refer to vehicle specific information for identification of location and function of the specified systems / CAN-Bus components. The fault code indicates that one or more control modules has received invalid data messages from the Headlamp Control Module "B". Refer to CAN-Bus information on page 51.
U0544	Invalid Data Received From Parking Assist Control Module "B"	The Vehicle / CAN Communication Bus is a network for sharing signals and data / messages between the various vehicle systems, including Powertrain, Chassis and Infotainment.	Refer to vehicle specific information for identification of location and function of the specified systems / CAN-Bus components. The fault code indicates that one or more control modules has received invalid data messages from the Parking Assist Control Module "B". Refer to CAN-Bus information on page 51.
U0545	Invalid Data Received From Running Board Control Module	The Vehicle / CAN Communication Bus is a network for sharing signals and data / messages between the various vehicle systems, including Powertrain, Chassis and Infotainment.	Refer to vehicle specific information for identification of location and function of the specified systems / CAN-Bus components. The fault code indicates that one or more control modules has received invalid data messages from the Running Board Control Module. Refer to CAN-Bus information on page 51.

NOTE 1. Check for other fault codes that could provide additional information.   **NOTE 2.** Communication between control units can pass via a CAN-Bus system; refer to Page 51 for CAN-Bus checks.

NOTE 3. If a fault cannot be located, it is also possible that the control unit is at fault.   **NOTE 4.** Refer to Page 53 for list of pages that show the ISO standard for component locations e.g. Sensor A - Bank 1.

Fault code	EOBD / ISO Description	Component / System Description	Meaningful Description and Quick Check
U0546	Invalid Data Received From Entertainment Control Module Front	The Vehicle / CAN Communication Bus is a network for sharing signals and data / messages between the various vehicle systems, including Powertrain, Chassis and Infotainment.	Refer to vehicle specific information for identification of location and function of the specified systems / CAN-Bus components. The fault code indicates that one or more control modules has received invalid data messages from the Entertainment Control Module Front. Refer to CAN-Bus information on page 51.
U0547	Invalid Data Received From Seat Control Module "E"	The Vehicle / CAN Communication Bus is a network for sharing signals and data / messages between the various vehicle systems, including Powertrain, Chassis and Infotainment.	Refer to vehicle specific information for identification of location and function of the specified systems / CAN-Bus components. The fault code indicates that one or more control modules has received invalid data messages from the Seat Control Module "E". Refer to CAN-Bus information on page 51.
U0548	Invalid Data Received From Seat Control Module "F"	The Vehicle / CAN Communication Bus is a network for sharing signals and data / messages between the various vehicle systems, including Powertrain, Chassis and Infotainment.	Refer to vehicle specific information for identification of location and function of the specified systems / CAN-Bus components. The fault code indicates that one or more control modules has received invalid data messages from the Seat Control Module "F". Refer to CAN-Bus information on page 51.
U0549	Invalid Data Received From Remote Accessory Module	The Vehicle / CAN Communication Bus is a network for sharing signals and data / messages between the various vehicle systems, including Powertrain, Chassis and Infotainment.	Refer to vehicle specific information for identification of location and function of the specified systems / CAN-Bus components. The fault code indicates that one or more control modules has received invalid data messages from the Remote Accessory Module. Refer to CAN-Bus information on page 51.
U054A	Invalid Data Received From Entertainment Control Module Rear "B"	The Vehicle / CAN Communication Bus is a network for sharing signals and data / messages between the various vehicle systems, including Powertrain, Chassis and Infotainment.	Refer to vehicle specific information for identification of location and function of the specified systems / CAN-Bus components. The fault code indicates that one or more control modules has received invalid data messages from the Entertainment Control Module Rear "B". Refer to CAN-Bus information on page 51.
U0551	Invalid Data Received From Impact Classification System Module	The Vehicle / CAN Communication Bus is a network for sharing signals and data / messages between the various vehicle systems, including Powertrain, Chassis and Infotainment.	Refer to vehicle specific information for identification of location and function of the specified systems / CAN-Bus components. The fault code indicates that one or more control modules has received invalid data messages from the Impact Classification System Module. Refer to CAN-Bus information on page 51.
U0552	Invalid Data Received From Running Board Control Module "B"	The Vehicle / CAN Communication Bus is a network for sharing signals and data / messages between the various vehicle systems, including Powertrain, Chassis and Infotainment.	Refer to vehicle specific information for identification of location and function of the specified systems / CAN-Bus components. The fault code indicates that one or more control modules has received invalid data messages from the Running Board Control Module "B". Refer to CAN-Bus information on page 51.
U0553	Invalid Data Received From Lighting Control Module Rear "B"	The Vehicle / CAN Communication Bus is a network for sharing signals and data / messages between the various vehicle systems, including Powertrain, Chassis and Infotainment.	Refer to vehicle specific information for identification of location and function of the specified systems / CAN-Bus components. The fault code indicates that one or more control modules has received invalid data messages from the Lighting Control Module Rear "B". Refer to CAN-Bus information on page 51.
U0554		ISO / SAE reserved	Not used, reserved for future allocation.
U0555		ISO / SAE reserved	Not used, reserved for future allocation.
U0556		ISO / SAE reserved	Not used, reserved for future allocation.
U0557		ISO / SAE reserved	Not used, reserved for future allocation.
U0558		ISO / SAE reserved	Not used, reserved for future allocation.
U0559		ISO / SAE reserved	Not used, reserved for future allocation.
U0560		ISO / SAE reserved	Not used, reserved for future allocation.
U0561		ISO / SAE reserved	Not used, reserved for future allocation.
U0562		ISO / SAE reserved	Not used, reserved for future allocation.
U0563		ISO / SAE reserved	Not used, reserved for future allocation.
U0564		ISO / SAE reserved	Not used, reserved for future allocation.
U0565		ISO / SAE reserved	Not used, reserved for future allocation.
U0566		ISO / SAE reserved	Not used, reserved for future allocation.
U0567		ISO / SAE reserved	Not used, reserved for future allocation.
U0568		ISO / SAE reserved	Not used, reserved for future allocation.
U0569		ISO / SAE reserved	Not used, reserved for future allocation.
U0570		ISO / SAE reserved	Not used, reserved for future allocation.
U0571		ISO / SAE reserved	Not used, reserved for future allocation.
U0572		ISO / SAE reserved	Not used, reserved for future allocation.
U0573		ISO / SAE reserved	Not used, reserved for future allocation.
U0574		ISO / SAE reserved	Not used, reserved for future allocation.
U0575		ISO / SAE reserved	Not used, reserved for future allocation.
U0576		ISO / SAE reserved	Not used, reserved for future allocation.
U0577		ISO / SAE reserved	Not used, reserved for future allocation.
U0578		ISO / SAE reserved	Not used, reserved for future allocation.

NOTE 1. Check for other fault codes that could provide additional information. **NOTE 2.** Communication between control units can pass via a CAN-Bus system; refer to Page 51 for CAN-Bus checks.

U0579		ISO / SAE reserved	Not used, reserved for future allocation.
U0580		ISO / SAE reserved	Not used, reserved for future allocation.
U0581		ISO / SAE reserved	Not used, reserved for future allocation.
U0582		ISO / SAE reserved	Not used, reserved for future allocation.
U0583		ISO / SAE reserved	Not used, reserved for future allocation.
U0584		ISO / SAE reserved	Not used, reserved for future allocation.
U0585		ISO / SAE reserved	Not used, reserved for future allocation.
U0586		ISO / SAE reserved	Not used, reserved for future allocation.
U0587		ISO / SAE reserved	Not used, reserved for future allocation.
U0588		ISO / SAE reserved	Not used, reserved for future allocation.
U0589		ISO / SAE reserved	Not used, reserved for future allocation.
U0590		ISO / SAE reserved	Not used, reserved for future allocation.
U0591		ISO / SAE reserved	Not used, reserved for future allocation.
U0592	Invalid Data Received From Gear Shift Control Module "B"	The Vehicle / CAN Communication Bus is a network for sharing signals and data / messages between the various vehicle systems, including Powertrain, Chassis and Infotainment.	Refer to vehicle specific information for identification of location and function of the specified systems / CAN-Bus components. The fault code indicates that one or more control modules has received invalid data messages from the Gear Shift Control Module "B". Refer to CAN-Bus information on page 51.
U0593	Invalid Data Received From Drive Motor Control Module "B"	The Vehicle / CAN Communication Bus is a network for sharing signals and data / messages between the various vehicle systems, including Powertrain, Chassis and Infotainment.	Refer to vehicle specific information for identification of location and function of the specified systems / CAN-Bus components. The fault code indicates that one or more control modules has received invalid data messages from the Drive Motor Control Module "B". Refer to CAN-Bus information on page 51.
U0594	Invalid Data Received From Hybrid Powertrain Control Module	The Vehicle / CAN Communication Bus is a network for sharing signals and data / messages between the various vehicle systems, including Powertrain, Chassis and Infotainment.	Refer to vehicle specific information for identification of location and function of the specified systems / CAN-Bus components. The fault code indicates that one or more control modules has received invalid data messages from the Hybrid Powertrain Control Module. Refer to CAN-Bus information on page 51.
U0595	Invalid Data Received From Powertrain Control Monitor Module	The Vehicle / CAN Communication Bus is a network for sharing signals and data / messages between the various vehicle systems, including Powertrain, Chassis and Infotainment.	Refer to vehicle specific information for identification of location and function of the specified systems / CAN-Bus components. The fault code indicates that one or more control modules has received invalid data messages from the Powertrain Control Monitor Module. Refer to CAN-Bus information on page 51.
U0596	Invalid Data Received From AC to AC Converter Control Module	The Vehicle / CAN Communication Bus is a network for sharing signals and data / messages between the various vehicle systems, including Powertrain, Chassis and Infotainment.	Refer to vehicle specific information for identification of location and function of the specified systems / CAN-Bus components. The fault code indicates that one or more control modules has received invalid data messages from the AC to AC Converter Control Module. Refer to CAN-Bus information on page 51.
U0597	Invalid Data Received From AC to DC Converter Control Module "A"	The Vehicle / CAN Communication Bus is a network for sharing signals and data / messages between the various vehicle systems, including Powertrain, Chassis and Infotainment.	Refer to vehicle specific information for identification of location and function of the specified systems / CAN-Bus components. The fault code indicates that one or more control modules has received invalid data messages from the AC to DC Converter Control Module "A". Refer to CAN-Bus information on page 51.
U0598	Invalid Data Received From AC to DC Converter Control Module "B"	The Vehicle / CAN Communication Bus is a network for sharing signals and data / messages between the various vehicle systems, including Powertrain, Chassis and Infotainment.	Refer to vehicle specific information for identification of location and function of the specified systems / CAN-Bus components. The fault code indicates that one or more control modules has received invalid data messages from the AC to DC Converter Control Module "B". Refer to CAN-Bus information on page 51.
U0599	Invalid Data Received From DC to DC Converter Control Module "A"	The Vehicle / CAN Communication Bus is a network for sharing signals and data / messages between the various vehicle systems, including Powertrain, Chassis and Infotainment.	Refer to vehicle specific information for identification of location and function of the specified systems / CAN-Bus components. The fault code indicates that one or more control modules has received invalid data messages from the DC to DC Converter Control Module "A". Refer to CAN-Bus information on page 51.
U059A	Invalid Data Received From DC to DC Converter Control Module "B"	The Vehicle / CAN Communication Bus is a network for sharing signals and data / messages between the various vehicle systems, including Powertrain, Chassis and Infotainment.	Refer to vehicle specific information for identification of location and function of the specified systems / CAN-Bus components. The fault code indicates that one or more control modules has received invalid data messages from the DC to DC Converter Control Module "B". Refer to CAN-Bus information on page 51.

NOTE 1. Check for other fault codes that could provide additional information.   **NOTE 2.** Communication between control units can pass via a CAN-Bus system; refer to Page 51 for CAN-Bus checks.

NOTE 3. If a fault cannot be located, it is also possible that the control unit is at fault.   **NOTE 4.** Refer to Page 53 for list of pages that show the ISO standard for component locations e.g. Sensor A - Bank 1.